Lecture Notes in Artificial Intelligence 4496

Edited by J. G. Carbonell and J. Siekmann

Subseries of Lecture Notes in Computer Science

Ngoc Thanh Nguyen   Adam Grzech
Robert J. Howlett   Lakhmi C. Jain (Eds.)

Agent and Multi-Agent Systems: Technologies and Applications

First KES International Symposium, KES-AMSTA 2007
Wroclaw, Poland, May 31– June 1, 2007
Proceedings

Springer

Volume Editors

Ngoc Thanh Nguyen
Adam Grzech
Wroclaw University
Institute of Information Science and Engineering
Str. Janiszewskiego 11/17, 50-370 Wroclaw, Poland
E-mail: {Ngoc-Thanh.Nguyen, Adam.Grzech}@pwr.wroc.pl

Robert J. Howlett
University of Brighton
Centre for SMART Systems, School of Engineering
Brighton, BN2 4GJ, UK
E-mail: R.J.Howlett@bton.ac.uk

Lakhmi C. Jain
University of South Australia
Knowledge-Based Intelligent Information and Engineering Systems Centre
School of Electrical and Information Engineering
Mawson Lakes, South Australia 5095, Australia
E-mail: Lakhmi.Jain@unisa.edu.au

Library of Congress Control Number: 2007927161

CR Subject Classification (1998): I.2, H.4, H.3, J.1, H.5, K.6, K.4

LNCS Sublibrary: SL 7 – Artificial Intelligence

ISSN 0302-9743
ISBN-10 3-540-72829-5 Springer Berlin Heidelberg New York
ISBN-13 978-3-540-72829-0 Springer Berlin Heidelberg New York

Springer is a part of Springer Science+Business Media

springer.com

Printed in Germany

Typesetting: Camera-ready by author, data conversion by Scientific Publishing Services, Chennai, India
Printed on acid-free paper SPIN: 12070190 06/3180 5 4 3 2 1 0

Preface

This volume contains papers selected for presentation at the 1st KES Symposium on Agent and Multi-Agent Systems – Technologies and Applications (KES-AMSTA 2007), held in Wroclaw, Poland, May 31 – June 1, 2007. The symposium was organized by the Institute of Information Science and Engineering, Wroclaw University of Technology, and KES International, as part of the KES Conference Series.

The aim of the symposium was to provide an international forum for scientific research in the technologies and applications of agent and multi-agent systems. Agents and multi-agent systems are related to the modern software which has long been recognized as a promising technology for constructing autonomous, complex and intelligent systems. A key development in the field of agent and multi-agent systems has been the specification of agent communication languages and formalization of ontologies. Agent communication languages are intended to provide standard declarative mechanisms for agents to communicate knowledge and make requests of each other, whereas ontologies are intended for conceptualization of the knowledge domain.

Despite being the first KES event of its type, the symposium attracted a very large number of scientists and practitioners, who submitted their papers for eight main tracks concerning the methodology and applications of agent and multi-agent systems, a doctoral track and three special sessions. There were 464 submissions for KES-AMSTA 2007 and only 110 papers were selected for presentation and inclusion in the proceedings. The resulting acceptance rate was 23.7 %. The Program Committee defined the following main tracks: Methodological Aspects of Agent Systems; Agent-oriented Web Applications; Mobility Aspects of Agent Systems and Ontology Management; Multi-agent Resource Allocation; Negotiating Agents; Agents for Network Management; Agent Approaches to Robotic Systems and Market Agents and Other Applications. Jointly with the main tracks of the symposium there was also the doctoral track and the following three special sessions: Mobile Agent Application and Its Development; Agent on Networked Media and Its Applications on Next-Generation Convergence Network (ANM07) and Intelligent and Secure Agent for Digital Content Management.

We would like to thank Jerzy Świątek, Dean of Faculty of Computer Science and Management, Wroclaw University of Technology, and Jerzy Józefczyk, Director of the Institute of Information Science and Engineering, for their enthusiastic and valued support of the project.

Thanks are also due to the many experts who contributed to making the event a success.

We would like to thank the Program Committee and Board of Reviewers, essential for reviewing the papers to ensure a high standard. We thank the members of the Local Organizing Committee, the Doctoral Track Chair, Publicity Chair and Special Session Chairs. We thank the invited speakers for their interesting and informative talks of world-class standard. We extend thanks to the KES Secretariat for support with central administration and the registration process. Finally, we thank authors, presenters and delegates for their contribution to a successful event.

We hope that KES-AMSTA 2007 demonstrated the KES mission of academic excellence, leading on to even greater successes in the future.

May 2007

Ngoc Thanh Nguyen
Adam Grzech
Robert J. Howlett
Lakhmi C. Jain

Symposium Organization

General Chair

Ngoc Thanh Nguyen
Institute of Information Science and Engineering
Wroclaw University of Technology, Poland

Honorary Chair

Lakhmi C. Jain
Knowledge-Based Intelligent Information and Engineering Systems Centre
University of South Australia, Australia

General Co-chairs

Adam Grzech
Institute of Information Science and Engineering
Wroclaw University of Technology, Poland

Robert J. Howlett
Centre for SMART Systems
School of Engineering University of Brighton, UK

Program Co-chairs

Noelle Carbonell
LORIA, CNRS & INRIA
Université Henri Poincaré, France

Janusz Sobecki
Institute of Applied Informatics
Wroclaw University of Technology, Poland

Local Organizing Committee

Chair

Agnieszka Pieczyńska
Institute of Information Science and Engineering
Wroclaw University of Technology, Poland

Members

Kamila Aftarczuk
Jarosław Drapała
Mariusz Kowalski
Monika Koprowska
Adrianna Kozierkiewicz
Michał Rusin
Trong Hieu Tran
Paweł Świątek
Kamil Zemlak

Doctoral Track Chair

Dariusz Król
Institute of Applied Informatics
Wroclaw University of Technology, Poland

Publicity Chair

Krzysztof Juszczyszyn
Institute of Information Science and Engineering
Wroclaw University of Technology, Poland

Special Session Chairs

1. *Mobile Agent Application and Its Development*
   Heang-KonKim
   Catholic University of Daegu, Republic of Korea
2. *Agent on Networked Media and Its Applications on Next-Generation Convergence Network (ANM07)*
   Il Soek Ko
   Dongguk University, Republic of Korea
3. *Intelligent and Secure Agent for Digital Content Management*
   Geuk Lee
   Hannam University, Republic of Korea

Invited Speakers

1. Andrzej Skowron
   Warsaw University, Poland
2. Toyoaki Nishida
   Kyoto University, Japan
3. Paul Davidsson Blekinge
   Institute of Technology, Sweden
4. Paolo Giorgini
   University of Trento, Italy

International Program Committee

Costin Badica	University of Craiova, Romania
Matteo Baldoni	University of Turin, Italy
Maria-Victoria Belmonte	Universidad de Málaga, Spain
Mária Bieliková	Slovak University of Technology, Bratislava, Slovakia
Longbing Cao	University of Technology Sydney, Australia
Tru Hoang Cao	Ho Chi Minh City University of Technology, Vietnam
Frantisek Capkovic	Slovak Academy of Sciences, Slovakia
Krzysztof Cetnarowicz	AGH - University of Science and Technology, Poland
François Charpillet	LORIA, France
Yiu-ming Cheung	Hong Kong Baptist University, China
Paul Davidsson	Blekinge Institute of Technology, Sweden
Grzegorz Dobrowolski	AGH - University of Science and Technology, Poland
Luminita Dumitriu	Dunarea de Jos University Galati, Romania
Colin Fyfe	University of Paisley, UK
Stan Franklin	University of Memphis, USA
Bogdan Gabrys	Bournemouth University, UK
Paolo Giorgini	Technology University of Trento, Italy
Dongbing Gu	University of Essex, UK
Frank Guerin	University of Aberdeen, UK
Robert J. Howlett	University of Brighton, UK
Lakhmi Jain	University of South Australia, Australia
Gordan Jezic	University of Zagreb, Croatia
Jerzy Jozefczyk	Wroclaw University of Technology, Poland
Jason J. Jung	Inha University, Republic of Korea
Haeng Kon Kim	Catholic University of Daegu, Republic of Korea
Tai Hoon Kim	Cewha Womans University, Republic of Korea
Zofia Kruczkiewicz	Wroclaw University of Technology, Poland
Naoyuki Kubota	Tokyo Metropolitan University, Japan
Raymond Lee	Hong Kong Polytechnic University, China
Pawan Lingras	Saint Mary's University, Canada
Honghai Liu	University of Portsmouth, UK
James N.K. Liu	Hong Kong Polytechnic University, China
Jiming Liu	University of Windsor, Canada
Debajyoti Mukhopadhyay	West Bengal University of Technology, India
V.L. Narasimhan	Western Kentucky University, USA

Wee Keong Ng	Nanyang Technological University, Singapore
Anton Nijholt	University of Twente, The Netherlands
Toyaoki Nishida	Kyoto University, Japan
Marcin Paprzycki	SWPS and IBS PAN, Poland
Gloria Philips-Wren	Loyola College in Maryland, USA
Samuel Pierre	Ecole Polytechnique de Montreal, Canada
Germano Resconi	Catholic University, Brescia, Italy
Phill-Kyu Rhee	Inha University, Republic of Korea
Alessandro Ricci	University of Bologna, Italy
Tony Bastin Roy Savarimuthu	University of Otago, New Zealand
Barry G. Silverman	University of Pennsylvania, USA
Jerzy Swiatek	Wroclaw University of Technology, Poland
Sun Zengqi	Tsinghua University, China
Ning Zhong	Maebashi Institute of Technology, Japan

Board of Reviewers

Matteo Baldoni
Maria Belmonte
Leszek Borzemski
Mohamed Bouassida
Louie Cervantes
Byungrae Cha
Cheeseng Chan
Sekchin Chang
Kim Chansung
Vincent Chevrier
Sung-Ja Choi
In Sun Choi
Cho Choonman
Wusheng Chou
Jun Chulmin
Nam Doohee
Damian Dudek
Eun-Seok Kim
Josh Fahimi
Olivier Festor
Stan Franklin
Mariusz Fras
Bobby Gerardo
Frank Guerin
Kang Heajoe
Park Heeun
Shin Ho-Jun
Chungwon Lee
Bong Gyou Lee
Jong Chan Lee
Seongjoon Lee
Eun-Ser Lee
Seong Joon Lee
James Liu
Yonghuai Liu
N. Masson
Qinggang Meng
Dominique Mery
Kil Min Wook
Ian Morgan
Narasimhan
Wee Keong Ng
Anton Nijholt
Ziemowit Nowak
Mariusz Nowostawski
Cezary Orlowski
Donat Orski
Seungchul Park
Olivier Perrin
Gloria Phillips-Wren
Maciej Piasecki
Samuel Pierre
Germano Resconi
Phill-Kyu Rhee

You-Sik Hong
Kwon Hoyeal
Gongzhu Hu
Kim Hyunah
Chang Hyunho
Il Seok Ko
Lee Impyung
Lim Jae Geol
Do Jae Soo
Kwon Jayhyun
Kwak Jin
Mun Jinsu
Yang Jongpil
Koh Joonhwan
Bi Jun
Krzysztof Juszczyszyn
Radoslaw Katarzyniak
Przemyslaw Kazienko
Young-Soo Kim
Hyun-Ah Kim
Sijung Kim
Gui-Jung Kim
Ganghyun Kim
Sang-Wook Kim
Grzegorz Kolaczek
Jung Doo Koo
Sakurai Kouich
Naoyuki Kubota
Santosh Kumar
Halina Kwasnicka
Young Mi Kwon
Kang Kyungwoo
Eunjoo Lee
Geuk Lee
Malrey Lee
Alessandro Ricci
Jae-Sung Roh
Byun Sangchul
Yeo Sang-Soo
Ku Seokmo
Cho Seongkil
Shin Seongwoong
Lee Seung Jun
Lee Seungjae
Haeyung Shin
Lee Shinhae
Andrzej Sieminski
Barry Silverman
Radu State
Zengqi Sun
Shin Sungil
Iwan Tabakow
Wojciech Thomas
Vincent Thomas
Yannick Toussaint
Bogdan Trawinski
Tuglular Tugkan
David Vallejo
Marek Wnuk
Jaroslaw Wojcik
Kim Woongi
Cho Wooseok
Yeon-Mo Yang
Choi Yoonsoo
Park Young-Su
Lim Youngtak
Na Yun Ji
Sun Zengqi
Aleksander Zgrzywa

Table of Contents

Main Track: Methodological Aspects of Agent Systems

Main Track: Agent-oriented Web Applications

Main Track: Mobility Aspects of Agent Systems and Ontology Management

Main Track: Multi-agent Resource Allocation

Main Track: Negotiating Agents

Main Track: Agents for Network Management

Main Track: Agent Approaches to Robotic Systems

Main Track: Market Agents and other Applications

Doctoral Track

Special Session: Mobile Agent Application and its Development

Special Session: Agent on Networked Media and Its Applications on Next-Generation Convergence Network (ANM07)

Special Session: Intelligent and Secure Agent for Digital Content Management

On the Integration of Agent-Based and Mathematical Optimization Techniques*

Paul Davidsson, Jan A. Persson, and Johan Holmgren

Department of Systems and Software Engineering, Blekinge Institute of Technology, Soft Center, 372 25 Ronneby, Sweden
{paul.davidsson,jan.persson,johan.holmgren}@bth.se

Abstract. The strengths and weaknesses of agent-based approaches and classical optimization techniques are compared. Their appropriateness for resource allocation problems were resources are distributed and demand is changing is evaluated. We conclude that their properties are complementary and that it seems beneficial to combine the approaches. Some suggestions of such hybrid systems are sketched and two of these are implemented and evaluated in a case study and compared to pure agent and optimization-based solutions. The case study concerns allocation of production and transportation resources in a supply chain. In one of the hybrid systems, optimization techniques were embedded in the agents to improve their decision making capability. In the other system, optimization was used for creating a long-term coarse plan which served as input the agents that adapted it dynamically. The results from the case study indicate that it is possible to capitalize both on the ability of agents to dynamically adapt to changes and on the ability of optimization techniques for finding high quality solutions.

1 Introduction

For a long time, mathematical optimization techniques based on linear programming and branch and bound have been used to solve different types of resource allocation problems, e.g., production and transportation planning in various industries at strategic and tactical level [10]. Additionally one can find examples of optimization techniques applied to short term planning (operational), e.g., activity scheduling [4,2]. Agent-based computing has often been suggested as a promising technique for problem domains that are distributed, complex and heterogeneous [9,11]. In particular, a number of agent-based approaches have been proposed to solve different types of resource allocation problems [6]. We compare the strengths and weaknesses of these two approaches and evaluate their appropriateness for a special class of resource allocation problems, namely dynamic distributed resource allocation. The purpose is to find hybrid approaches which capitalize on the strengths of the two approaches. In the class of problems studied, information and/or resources are distributed and the exact conditions, e.g., the demand and the availability of resources, are not known in advance and are changing. These characteristics make the problem domain particularly challenging

* This research was funded by the Swedish Knowledge Foundation through the project Integrated Production and Transportation Planning within Food Industry.

N.T. Nguyen et al. (Eds.): KES-AMSTA 2007, LNAI 4496, pp. 1–10, 2007.

and suitable for exploring different hybrid approaches. Examples of using multi-agent systems for solving optimization problems can be found [5,8]. However, the potential of mathematical optimization techniques are typically not explored in such approaches, which is done in this paper.

The decisions of how to manage the resources may either be taken locally at distributed decision centers, or at a central node (as is common in optimization approaches). In a strictly centralized solution, all nodes except one are just sensors and/or actuators with little or no information processing.

In the next section, an evaluation framework is applied in a theoretical comparison of the two approaches. This is followed by a small experimental case study concerning planning and allocation of resources in combined production and transportation. In this study, a number of different agent-based and optimization-based approaches are compared, including hybrid approaches. Then the experimental results are presented, and finally, some conclusions are drawn and future directions are discussed.

2 Comparison of the Approaches

We will compare the two approaches with respect to how well they are able to handle some important properties of the problem domain. (This analysis is based on earlier work [3] and [7].) Please note that some of the statements below are hypotheses that need to be verified through further analysis or experiments. We make the assumptions that in agent-based approaches, control is distributed and concurrent. Some approaches that often are classified as agent-based, such as highly centralized auction-based approaches, do not comply to this assumption. Regarding optimization approaches, We choose to focus on methods using a central node which has the entire responsibility for computing the optimal (or near optimal) solution/allocation to the problem. Further, we focus on methods that have the potential to provide solutions of guaranteed good quality. These are typical characteristics of many classical optimization methods (e.g. linear programming, branch-and-bound, branch-and-price and branch-and-cut).

Size (number of resources to be allocated): Since agent-based approaches support the dividing of the global problem into a number of smaller local allocation problems, large-sized problems could be handled well in such cases the problem is modular. On the other hand, the complexity and the size of the problem may affect the solution time dramatically when applying an optimization method. Since optimization techniques attempt to achieve global optimality, capitalizing on partial modularity in order to handle large-sized problems is difficult.

Cost of communication: Since agent-based approaches typically depend on frequent interaction in order to coordinate activities and decisions, they are not a good choice when communication is expensive. The need for communication in centralized decision-making, such as classical optimization, is rather small, since the nodes only need to send information and receive the response of the decision once.

Communication and computational stability: The more centralized the decision making is, the more vulnerable the system gets to single point failures of a central node. In agent-based systems, the reallocation may function partially even though some nodes or links have failed since the decision making is distributed and agents can have strategies for handling link failures. In optimization it is typically assumed that computations and communication will occur as planned. Furthermore, due to its centralized structure, optimization is not particular robust with respect to failures in computation and communication.

Modularity (see [6]): As agent-based approaches are modular by nature they are as such very suitable for highly modular domains. However, if the modularity of the domain is low they may be very difficult to apply. Some optimization techniques may be parallelized, but this parallelization is typically made from an algorithmic standpoint and not with the physical nodes in mind.

Time scale/Adaptability (time between re-allocation of resources): Since agents are able to continuously monitor the state of its local environment and typically do not have to make very complex decisions, they are able to react to changes fast. Optimization techniques often require a relatively long time to respond. Hence a rather high degree of predictability is required for optimization methods to work efficiently if a short response time is required. Sometimes methods of re-optimization can be used for lowering the response time. However, the scope for efficient use of re-optimization in complex decision problems is rather limited.

Changeability (how often the structure of the domain changes): As it is relatively simple to add or delete agents during run-time, agent-based approaches are highly modifiable [6]. On the other hand, a complete restart of the optimization method may be required if the structure of the system changes, e.g. a decision node is added or removed.

Quality of solution (how important it is to find a good allocation): Since agent-based approaches are distributed, they do not have (or at least it is costly to get) a global view of the state of the system, which unfortunately often is necessary in order to find a truly good allocation. Therefore, the quality of the solution suggested by an optimization method in this context often will be of a higher quality.

Quality assurance: It may be very difficult (and sometimes even impossible) to estimate the quality of the allocation made by an agent-based approach due to the lack of a global view. In many optimization methods a measure of how far (in terms of cost) a solution is at most from an optimal solution is obtained (i.e., a bound of the optimal solution values is obtained).

Integrity (importance of not distributing sensitive information): Agent-based approaches supports integrity since sensitive information may be exclusively processed locally, whereas centralized decision-making implies that all information must be made available at the central node. Hence integrity may be hard to achieve for optimization approaches.

Decomposition in optimization may make the approach partially distributed; subproblems may equate nodes. In decomposition approaches though, the central node typically retains the control of all decisions, which makes most of the analysis above (assuming a central node is responsible for decision making) also hold for decomposition approaches. However, in a decomposition approach, dual prices and suggestions of solutions are typically sent between the central node and the other nodes a large number of times, which makes it sensitive to high communication costs. If (optimization) heuristics are considered (e.g. tabu search, simulated annealing and genetic algorithms), problems with *Size* and *Time scale* can potentially be alleviated. However, *Quality assurance* of the solution can hardly be achieved in such case.

According to our comparison, agent-based approaches tend to be preferable when: the size of the problem is large, communication and computational stability is low, the time scale of the domain is short, the domain is modular in nature, the structure of the domain changes frequently (i.e., high changeability), there is sensitive information that should be kept locally; and classical optimization techniques when: the cost of communication is high, the domain is monolithic in nature, the quality of the solution is very important, it is desired that the quality of the solution can be guaranteed.

Thus, the analysis indicates that agent-based approaches and optimization techniques complement each other. There are a number of ways of combining the approaches into combined approaches which potentially can make use of this complementarity. In the following sections, we will investigate two such hybrid approaches and particularly focus on aspects related to *Time scale/Adaptability* and *Quality of solution*. The first approach is using an optimization technique for coarse planning and agents for operational replanning, i.e., for performing local adjustments of the initial plan in real-time to handle the actual conditions when and where the plan is executed. In the second approach, optimization is embedded in the agents, which can be seen as a further development of distributed optimization where agent technology is used, e.g., to improve coordination.

3 Problem Description

We will now describe a small case suitable for illustrating some of the strengths and weaknesses of agent-based and optimization approaches, as well as the potential of hybrid approaches. The case study concerns the planning and allocation of resources in combined production and transportation based on a real world case within the food industry. The problem concerns how much to produce and how much to send to different customers in each time period. Due to the complexity of the problem, (combinatorial structure, uncertainty associated with the demand, the many actors involved, etc.) optimal or near optimal solutions are hard to find. If optimization tools are to be employed, extensive and time consuming interactions between planner and tools will typically be required including many runs of the optimization algorithm. Hence, the time requirement is typically a limiting factor in this problem domain. The problem is dynamic since one often has to plan based on forecasts which may be rather uncertain; and there are uncertainties associated with the availability of the resources, i.e., the production and transportation capacity.

In the simple version of the problem, there are one production unit, two customers and inventories of a single product at the customers and at the production unit (producer). A few transport options (to a single customer or to both customers) are available. The actual demand often diverges from the forecast. The forecasted demand for a day (a period) of a customer is 4 units if the forecast is made 14 days (or more) in advance. However, this forecast changes as more information becomes known about the customers demand. For each day, the forecast changes according to the probabilities given in Table 1. Forecasts made on the time period for the customer demand are equal to the actual customer demand. Thus, the closer you get to the day of the demand the more accurate the forecast is. However, the actual demand in a time period for each customer is never lower than 0 and never larger than 8. There are a number of decisions to be made: volumes to produce and transport; and a number of costs to consider: production, transportation; inventory and inventory level penalty costs. A formal description of the problem, including the parameter values (costs, capacities etc.), is given in Appendix.

Table 1. Probabilities of changes to the demand

Change	-4	-3	-2	- 1	0	+1	+2	+3	+4
Probability (case 1)	0.01	0.03	0.06	0.15	0.50	0.15	0.06	0.03	0.01
Probability (case 2)	0.04	0.08	0.12	0.16	0.20	0.16	0.12	0.08	0.04

4 Solutions

Next we introduce agents which make the decisions concerning production and transportation based on rudimentary decision rules. Later we embed optimization within the agents; and assist the agents with a coarse plan obtained by optimization (i.e. the two hybrid approaches suggested in Section 2).

In the real world case, the decisions (and some planning) of production and transportation are taken by two different actors (from different organizations). Hence, we find it suitable to introduce an agent for each type of decisions. Agent A is the production and production inventory planner; Agent B is the transportation as well as customer inventory planner, see Figure 1.

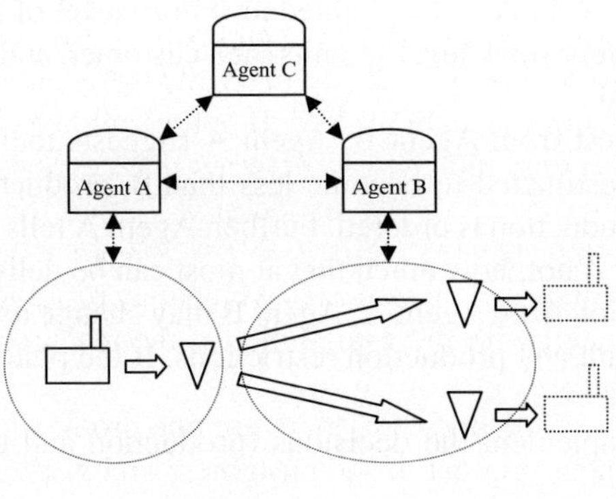

Fig. 1. The transport chain and Agents A and B; and a potential Agent C

The resource allocation problem of both production and transportation can be viewed as an optimization problem as indicated in Appendix. For the simple case, with a time horizon of 14 time periods (days), the problem can be solved to optimality rather quickly (in seconds) by using Cplex 8.1.1 (www.ilog.com) for solving. However, in a real world application this would not be the case due to the large size of the problem, and we make some assumptions in order to make the experimental setting more realistic. We assume that if the optimization is applied to the whole system, it can at most be applied once every week. The reason is both related to difficulties of information gathering and the time requirements of solving the real world problem using optimization techniques. Further, if agents A or B use optimization, we assume they can only solve a problem with a time horizon of 7 days. The reason for this limitation is that in a real world setting, the computational time needs to be short in order to be reactive.

We have chosen to use Cplex as the solver when applying optimization to the mixed-integer linear programming (MILP) problem occurring in this paper. The solver employs a branch-and-bound solutions strategy and is recognized as a efficient solver widely used in academia for MILP problems. Since the general MILP problem is NP-hard, and probably the MILP models formulated in this work, we cannot guarantee good running times for large problems with any algorithm. However, the branch-and-bound solution strategy implies that we can guarantee that we find the optimal solution.

4.1 Pure Agent Approach

In this setting, we use rather simple decision rules for Agents A and B. The rule is to order production or transportation if the inventory level is anticipated to become lower than a certain level (safety stock level), which is a rule commonly used in logistics. Further, in case a transport order cannot be satisfied, the agents interact in order to find a suitable transport order quantity.

The following actions (steps) are taken by agents A and B in each period in order to make production and transportation decisions in the *Pure Agent* approach:

1. Agent B receives customer forecasts of the demand and suggests, if necessary, transports to be initiated (current evening). It sends a request of transport (of a full truck load) to Agent A if the planned inventory level of tomorrow is estimated to be less than a safety stock level (3 units) at a customer, and a zero request if there is no transport need.
2. Based on the request from Agent B, Agent A suggests today's production. If the inventory level is estimated to become less than 0, production at full capacity is ordered, else no production is ordered. Further, Agent A tells Agent B if the request can be satisfied, or if not, how much that at most can be delivered.
3. Based on information from Agent A, Agent B may change its transportation plan in order to comply with any production restrictions. If the plan is changed, it informs Agent A about this.
4. Agent A and B implement the decisions (production and transportation) for the current period.

4.2 Embedded Optimization

In order to improve the performance of Agents A and B, optimization is embedded. In this implementation, the agents run an optimization of the next seven time periods before suggesting today's decisions. An MILP-model is created for each of the agents, see Appendix for a formal description. The steps of actions given in Section 4.1 are maintained. However, Agent B sends a suggestion of a transport plan with a time horizon of seven days to Agent A in step 1 (and not just today's transport order); and Agent A considers a planning horizon of seven days when planning production in step 2.

4.3 Pure Optimization

In the *Pure Optimization* approach, the production and transportation decisions are given by a global plan which is created once a week, i.e., every 7th period. This global plan is created by solving the the global optimization model (MILP), presented in Appendix, with a time horizon of 14 days.

4.4 Tactical/Operational Hybrid Approach

In this hybrid approach, the *Embedded Optimization* is improved by utilizing information of the previously created global plan (as in *Pure Optimization*). The global (or coarse) plan is obtained from Agent C, see Figure 1, which uses optimization on the whole MILP model as specified in Appendix. The global coarse plan is obtained by solving the MILP model of the whole system every seventh time period (as in *Pure Optimization*). Agent A and B uses its MILP model with the inclusion of costs for deviating from the coarse plan. This can be viewed as there is a coarse plan which helps the production and transportation to be better coordinated.

The global plan is communicated to Agent A and B, and might enhance the agents' decisions in actions 1 to 4. When each agent runs embedded optimization, costs for deviating from the global plan are considered in addition to the real costs, see Appendix for further details. Note that the quality of the coarse plan is probably good right after it has been created, since it is up-to-date, and probably less good later due to changes in the demand compared to the forecast.

5 Experimental Results

In Table 2, the results of using the different approaches for 3500 time periods are presented. It shows the average total cost and computational time (in seconds) per time period.

When the problem size increases significantly, guaranteed optimal solutions to the formulations are unlikely to be obtained within reasonable time using off-the-shelf optimization software. For instance, additional experiments indicate that when the number of customers is doubled to four, the time for solving the problem using approach *Tact./Oper. Hybrid*, increases roughly with a factor of 100. However, good solutions might still be obtained by using customized optimization techniques.

The results for the approaches *Pure Agent* and *Pure Optimization*, are consistent with the results of the theoretical analysis in Section 2 with respect to time, quality and communication properties. Further, by comparing the total costs between approach *Pure Agent*, *Embedded Optimization* and *Tact./Oper. Hybrid*, the results indicate: adding optimization to the agents, improves the agents' decision making with respect to the quality of the solution.

We also see that when the quality of the predictions is rather good (case 1), the *Pure Optimization* approach is competitive. On the other hand, when the predictability of the demand is lower (case 2), the agent-based approaches are superior.

Even though the variations in costs are large for these approaches, a Student's T-test indicates that the differences in total cost are significant (at a level of $p = 0.01$). With respect to response time (computational time) it increases in the hybrid approaches.

Table 2. Cost and time requirements for the different approaches

Approach	Total cost case 1	Total cost case 2	Time (s)
Pure Agent	22.7	23.4	0.03
Embedded Optimization	21.7	22.7	0.07
Pure Optimization	20.7	25.3	0.22
Tact./Oper. Hybrid	20.2	21.4	0.44

6 Conclusions and Future Work

According to our analysis, the properties of agent-based approaches and optimization techniques complement each other. This was partially confirmed by the experimental results, which in addition investigated two promising ways of combining agent-based and optimization techniques into hybrid approaches: using coarse planning and embedded optimization techniques. The hybrid approaches appear to combine some of the good properties from each of the two investigated approaches.

Additional potential hybrid approaches can be based on that the multi-agent system invoke optimization algorithms when it cannot handle the situation. Another approach is to "Agentify" optimization techniques that are already based on decomposition to encompass some of the properties of agent-based approaches. In order to make sure a solution which is possible to implement is available at each node/sub-problem, one may let the sub-problems keep track of a convex combination of previous solutions. Ideas from the volume algorithm [1] can be adopted in order to have a reasonable good solution to be implemented at any time during progress of the algorithm. We plan to further experimentally verify the conclusions of the theoretical analysis regarding the properties of the agent-based and classical optimization techniques and explore hybrid approaches. Yet another hybrid approach would be to experiment with the combination of pure optimization with the usage of reactive agents for dealing with unpredictable events.

Future experiments will include an increased number of customers and producers, several product types, and different lengths of planning horizons, time steps and time in between global planning. Increased numbers of customers, producers and product types will of course increase the computational requirements for solving the optimization problems and therefore algorithmic improvements might be necessary. It would also for example be interesting to partition the customers into clusters and let different agents make the plans for these clusters. The partitioning can for instance be based on geographical locations of the customers. Possible improvements, for solution quality rather than for performance, is to develop more advanced routing and production scheduling mechanisms than those used today.

References

1. Francisco Barahona and Ranga Anbil. The volume algorithm: producing primal solutions with a subgradient algorithm. *Mathematical Programming*, 87:385–399, 2000.
2. Paolo Brandimarte and Agostino Villa. *Advanced Models for Manufacturing Systems Management.* CRC Press, Inc, Bocan Raton, Florida, 1995.
3. Paul Davidsson, Stefan J. Johansson, Jan A. Persson, and Fredrik Wernstedt. Agent-based approaches and classical optimization techniques for dynamic distributed resource allocation: A preliminary study. In *AAMAS'03 workshop on Representations and Approaches for Time-Critical Decentralized Resource/Role/Task Allocation*, Melbourne, Australia, 2003.
4. Uday S. Karmarkar and Linus Schrage. The deterministic dynamic product cycling problem. *Operations Research*, 33:326–345, 1985.
5. Ohbyung Kwon, Ghiyoung Im, and Kun Chang. Mace-scm: An effective supply chain decision making approach based on multi-agent and case-based reasoning. In *Proceedings of the 38th Annual Hawaii International Conference on System Sciences (HICSS'05)*, 2005.
6. H. V. Parunak. Industrial and practical applications of DAI. In G. Weiss, editor, *Multiagent Systems.* The MIT Press, 1999.
7. Jan A. Persson, Paul Davidsson, Stefan J. Johansson, and Fredrik Wernstedt. Combining agent-based approaches and classical optimization techniques. In *EUMAS - Proceedings of the Third European Workshop on Multi-Agent Systems*, pages 260–269, 2005.
8. Qiang Wei, Tetsuo Sawaragi, and Yajie Tian. Bounded optimization of resource allocation among multiple agents using an organizational decision model. *Advanced Engineering Informatics*, 19:67–78, 2005.
9. Gerhard Weiss. *Multiagent Systems, - a modern approach to distributed artificial intelligence.* MIT Press, 1999.
10. Laurence A. Wolsey. *Integer Programming.* John Wiley & Sons, New York, 1998.
11. Michael Wooldridge. *An Introduction to MultiAgent Systems.* Wiley, 2002.

Appendix

In the following, sets, variables and parameters are introduced in order to formally define the problem as an optimization problem. Set T (with index t) denotes the set of periods; J (with index j) denotes the set of customers; and R (with index r) the set of transport routes. Further, the set $R_j \subset R$ is the set of routes visiting customer j. The variables used are: x_t, produced units in period t; y_t, 1 if production occurs in period t; s_t, 1 if production starts in period t; I_t, inventory level at the producer at the end of period t; z_{jrt}, transported quantity from producer to customer j on route r in period t; o_{rt}, 1 if transport route r is used in period t; L_{jt}, inventory level at customer j at the end of period t; v_{jt}, shortage of products at customer j in period t; w_{jt}, excess of products at customer j in period t; V_{jt}, shortage of products with respect to safety stock level at customer j in period t.

Parameters (constants) are given below with their values given within parentheses (superscripts are used to distinguish between different parameters): u^x, production capacity (12); u^I, inventory capacity at producer (35); u^z, truck capacity (30); d_{jt}, demand forecast for period t and customer j; u^L, inventory capacity at customers (35); l^L, safety stock level at customers (3); c^y, fixed production cost if products are produced in period t (10k); c^s, cost of starting production if the product was not produced in previous period t (20k); c^r, cost of using a transport route r (20k and 22k), respectively, for visiting a single customer and 23k for visiting both customers; c^I, inventory cost at producer (100); c^L, inventory cost at customer (110); L^{lc}, cost of inventory shortages at customers (8k); L^{uc}, cost of exceeding inventory levels at customers (2.5k); l^c, cost of not meeting the inventory safety levels (2k).

Given the variables and parameters above, a cost function can be specified accordingly: $f(\cdot) = \sum_{t \in T}(c^y y_t + c^s s_t + c^I I_t + \sum_{r \in R} c^r o_{rt} + \sum_{j \in J}(c^L L_{jt} + L^{lc} v_{jt} + L^{uc} w_{jt}))$. At the end of each period, the cost is recorded using function $f(\cdot)$ for period t equal to one. Then, inventory levels are computed given the actual demand (not the forecast), which becomes the initial inventory levels the next day. Below is an optimization model of the problem presented.

$$\min z = f(\cdot) + l^c V_{jt}$$

$$\text{s. t.} \quad x_t \leq u^x \qquad t \in T \qquad (1)$$

$$s_t \geq y_t + y_{t-1} \qquad t \in T \qquad (2)$$

$$0 \leq I_t \leq u^I \qquad t \in T \qquad (3)$$

$$I_{t-1} + x_t - \sum_{j \in J} \sum_{r \in R_j} z_{jrt} = I_t \qquad t \in T \qquad (4)$$

$$\sum_{j \in J} z_{jrt} \leq u^z o_{rt} \qquad r \in R, t \in T \qquad (5)$$

$$L_{j,t-1} + \sum_{r \in R_j} z_{jr,t-1} - d_{jt} = L_{jt} \qquad j \in J, t \in T \qquad (6)$$

$$0 - v_{jt} \leq L_{jt} \leq u^L + w_{jt} \qquad j \in J, t \in T \qquad (7)$$

$$l^L - V_{jt} \leq L_{jt} \qquad j \in J, t \in T \qquad (8)$$

All variables and cost parameter are assumed to be non-negative. Variables y_t, s_t, o_{rt} are binary. In the case of *Embedded Optimization*, an optimization problem is created for each of the agents: variables x, y, s, I and constraints (1)-(4) for Agent A; and variables z, o, L, v, w, V and constraints (5) - (8) for Agent B, with the associated costs, respectively. In Agent A's optimization problem, z is regarded as a parameter (i.e. a transport plan). In the *Tactical/Operational Hybrid* approach, a cost (200) per unit of deviation from the coarse plan with respect to values x, I, z and L is considered when solving the agents' optimization problems.

Building Agent Service Oriented Multi-Agent Systems*

Dan Luo, Longbing Cao, Jiarui Ni, and Li Liu

Faculty of Information Technology, University of Technology, Sydney, Australia
{dluo,lbcao,jiarui,liliu}@it.uts.edu.au

Abstract. An effective agent-based design approach is significant in engineering agent-based systems. Existing design approaches meet with challenges in designing Internet-based open agent systems. The emergence of service-oriented computing (SOC) brings in intrinsic mechanisms for complementing agent-based computing (ABS). In this paper, we investigate the dialogue between agent and service, and between ABS and SOC. As a consequence, we synthesize them and develop a design approach called *agent service-oriented design* (ASOD). The ASOD consists of agent service-based architectural design and detailed design. ASOD expands the content and range of agent and ABS, and synthesizes the qualities of SOC such as interoperability and openness, and the performances of ABC like flexibility and autonomy. The above techniques have been deployed in developing an online trading and mining support infrastructure F-Trade.

1 Introduction

An effective agent-based design (ABD) approach is significant in engineering multi-agent systems (MAS). With the increase of system complexities of MAS, ABD faces new challenges [6]. In practice, ABD approaches must adapt to the following increasingly emergent requirements: (i) MAS are towards middle to large scale [9,4,3], (ii) they are Internet-oriented, and (iii) they integrate enterprise applications. However, ABD support available from most of existing agent-oriented software engineering approaches [13, 14] are at abstract level, they provide limited capabilities and are not deployable for designing practical agent systems, in particular Internet-based enterprise applications.

Furthermore, from a theoretical perspective, there are many critical issues emerging in utilizing existing ABD to solve real problems. For instance, services are not explicitly separated and supported; support for the architectural design of network-based agent systems is weak, e.g., system architecture for Internet-based applications; the requirement for the integration of enterprise applications brings new open problems for "traditional" agent-based architecture, e.g., locating, transporting and interoperating with distributed heterogeneous enterprise applications.

In practice, an agent system satisfying the above requirements is an Internet-based open enterprise automated systems. This trend requires novel supports available from

* This work is sponsored by Australian Research Council Discovery Grant (DP0773412, LP0775041, DP0667060, DP0449535), and UTS internal grants.

N.T. Nguyen et al. (Eds.): KES-AMSTA 2007, LNAI 4496, pp. 11–20, 2007.

agent-based computing (ABC) [13]. Fortunately, service and service-oriented computing (SOC) [7] have newly emerged as a powerful paradigm for designing distributed and especially Internet-based systems. The interaction between agent and service [5] can enhance the function and range of agents; while the dialogue between ABC and SOC can integrate bilateral benefits from both of them, which are more suitable for building internet-based enterprise applications.

Motivated by the above ideas, this paper presents the design of agent service oriented MAS. We first discuss the dialogue between agent and service, and between ABC and SOC. Further, we introduce *agent service-oriented design* (ASOD) approach. The ASOD consists of and provides practical support for the design of MAS. We briefly illustrate the ASOD approach in building up a financial trading and mining system. ASOD is promising for designing Internet-based enterprise systems.

2 Dialogue Between Agent-Based and Service-Oriented Computing

2.1 Complementation Between ABS and SOC

In general, agent and service are two independent computational concepts [13]. Agent is proposed for modeling stakeholders with autonomous actions and cooperative abilities to archive their individual or global design objectives. ABC [13] presents unprecedented computational intelligence of flexibility and autonomy in modeling complex software systems. However, the existing ABC techniques face many challenges from distributed and especially Internet computing, for instance, building a suitable architecture for enterprise application integration.

On the other hand, service is developed to wrap application logics, which exposes message-based interfaces suitable for being accessed across a network. SOC [7] is good at integrating and managing Internet-based enterprise interoperation, and wrapping legacy applications. However, SOC is weak in supporting the qualities of service such as autonomy and flexibility.

By nature, the content of ABC and SOC are quite complementary. For the internal qualities of applications, it is agent and ABC that can do well; while service and SOC are more suitable for implementing the software-as-service initiatives and application-to-application architecture. The integration of agent and service, and of ABC and SOC can benefit each other in building a more powerful computational system. This is the motivation of developing the concept *agent service*, and *agent service-oriented design* approach.

2.2 Agent Service, Service of Agent and Service of Service

In practice, the integration of agent and service, and of ABC and SOC are feasible [5]. On one side, we develop the concepts of *agent service*, *service of agent*, and *service of service* to build the internal linkages between agent and service at the low level. On the other hand, we build mechanisms for the integration of agent and service, and of ABC and SOC in architectural and detailed designs.

The definition of *agent service* is twofold: (i) it is an abstract computational concept, which combines functions of agents and services together; (ii) it is a loosely aggregated entity embedding two computational units -- agent and service, in some cases, one computational unit is relatively independent of another.

Besides the fore-mentioned meanings of service, another meaning of service and the scenarios for using service is that *service* represents dynamic functions and activities an agent or a service can perform. In this case, a set of services may belong to an agent or a service. Therefore, we differentiate these two kinds of services from agent service, and call them as *services of agent* and *service of service*, respectively.

Services of an agent (or a service) consist of dynamic functions and activities (namely services) an agent (or a service) can provide or perform. These services are important attributes of an agent (or a service) except those static attributes an agent (or a service) has. Sometime we separate services from agent and service, so that we can explicitly organize or manage dynamic actions an agent or a service can or must perform. This is helpful in abstracting and designing agent service-based systems.

The benefit of the dialogue between (i) agent and service, (ii) ABC and SOC is bilateral. It enhances not only the original concept and its range of agent, but the function and performance of ABC in building Internet-based open automated systems. For instance, the strengths of service and SOC make the agent service-based architecture more suitable for a distributed agent system; while an agent service-based distributed system can synthesize the qualities of SOC such as interoperability and openness, and the performances of ABC like flexibility and autonomy.

3 Agent Service Oriented MAS

3.1 Agent Service Abstract Model

Agent service abstract model specifies fundamental architectures and properties of an agent service, intra-agent activities, and inter-agent communications. We define agent model and service model, respectively.

Agent Model

Agent model specifies agent architecture and agent classes. An *agent architecture* defines some specific commitments about the internal structure and intra-agent activities of an agent or a set of agents. There are mainly four categories of agent architectures; they are deductive agent architecture [12], reactive agent architecture [11], belief-desire-intention agent architecture [13], and hybrid architecture. For these agent architectures, we can build high-level agent architecture diagram for them.

In addition, the agent model specifies the generic attributes of an agent class; the attributes include agent name, type, locators, owner, roles, address, messages, input variables, preconditions, output variables, post conditions and exception handling.

Agent ::= *f*(*Name*, *Type*, *Locators*, *Owner*, *Roles*, *Address*, *Message*, *InputVariables*, *Preconditions*, *OutputVariables*, *Postconditions*, *Exception*)

Services model

Service model describes an agent's (or service's) activities required to realize the intra-agent or intra-service's roles and their properties, and inter-agent or inter-service communications in scenarios of legacy integration and enterprise integration.

The agent or service activities embody basic attributes of agent service; these attributes represent intra-agent or intra-service's roles and properties. On the other hand, the definition of services managing inter-agent and inter-service communications is to design interaction and communication supports and patterns via attributes for cross-agent or cross-service interoperability. By nature, attributes describing intra- and inter-agent behaviors can be combined or even to some degree shared with each other.

The following shows a multiple-attribute tuple of service model. A service is specified by attributes such as a service's activities, type, locators, owner, roles, type, address, message, input variables, preconditions, output variables, post conditions and exception handling.

Service ::= *f*(*Activity*, *Service Type*, *Locators*, *Owner*, *Roles*, *Type*, *Address*, *Message*, *InputVariables*, *Preconditions*, *OutputVariables*, *Postconditions*, *Exception*)

3.2 Agent Service Design Patterns

Agent service design patterns consist of agent service architectural patterns, and agent service functional patterns.

Agent service architectural patterns

The above-mentioned four kinds of agent architectures represent four agent architectural patterns. They can be abstracted and described in certain architectural frameworks, for instance the architectural diagram in Figure 1. In addition, all of them can be presented in some symbolic logic. For instance, for a reactive agent, decision-making can be modeled in the following form with a direct mapping from perception P to action A:

$$\wp(P) \to \wp(A)$$

Similarly, we can build symbolic presentations to describe architectural patterns of deductive agent and BDI agents. [12, 13] present their architectural patterns in first-order predicate logic.

Agent service functional patterns

From the perspective of system function, we can extract and classify generic agent services into some patterns. For instance, the following agent service patterns are developed for middlewaring or for enterprise application integration: proxy service, wrapper service, adapter service; while agent services such as matchmaking, brokerage, gateway, negotiation, auction and orchestration engine are used for communications. The table 8.1 in lists definitions and context of 20 services.

3.3 Agent Service-Based Integration Architecture

To solve emergent problems in middleware-based integration (e.g. cost, inflexible and inefficient) and Web service-based integration (e.g. no guarantee with qualities of service such as autonomy and flexibility), we propose *agent service-based integration*. In agent service-based integration, applications are packed (or wrapped) as agents or services, inter-application communication is based on inter-agent service or via a mediation (or integration) layer; the middle layer includes either an agent service or a cluster of agent services.

Agent service-based integration can be instantiated into various integration architectures. For instance, in legacy integration, every legacy application can be wrapped as a wrapper agent service, or a mediator agent service-based middleware can be built between legacy applications; we further build messaging-based communication between these agent services via exposing APIs. Figure 1 illustrates an agent service-enabled hub and spoke architecture, where a centrally located server hosts the integration logic that controls the orchestration and brokering of all inter-application communication. Application 1 is integrated after wrapped as a service; application 2 interacts with a wrapper server via an agent adapter; both of them interact with the server through business services. In addition, application 3 and 4 are combined with the server via an agent adapter and a gateway agent respectively.

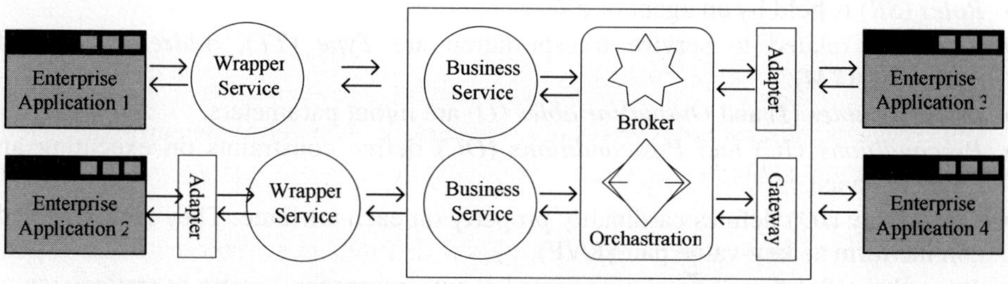

Fig. 1. Agent service-enabled hub and spoke integration

3.4 Developing Agent Service Problem-Solving Ontology

Agent service problem-solving ontology specifies attributes and relationships of agent services in a problem-solving system. This involves two steps (i) Extracting problem-solving ontologies from the problem domain, (ii) Developing of agent service ontologies for detailed design based on the extracted problem-solving ontologies.

Problem-solving ontologies existing in a problem domain can be extracted through the analysis and build-up of conceptual model, extended entity relationship model, or agent class diagram. These models assist us to capture and indicate attributes and their relationships of agents and services. Semantic relationships between ontological items are classified into seven classes such as Similar_to, Relate_to, Disjoin_with and Overlap_to, Part_of, etc. [1].

Furthermore, we can develop agent service ontology for the problem-solver. The algorithm for developing agent service ontologies is as follows.

- extracting actors of agents and services, and their goals from the problem-solving ontology diagram,
- specifying roles, activities, and type of agents and services,

- analyzing relevant environment objects of an agent service, and relationships to other agent services,
- design agent service ontological diagram to specify all agents and services
- refining the developed agent services through reverse re-engineering and scenarios-based analysis, and
- presenting agent service ontology.

3.5 Representation of Agent Services

Agent services are presented in terms of ontological engineering and agent service ontologies. An agent service consists of multiple ontological items. Each ontological item atom describes an attribute or a property of an agent service. The agent service model specifies attributes and properties of an agent service.

- *Activity* (*SA*) describes the function of the agent service.
- *Locators* (*SL*, *RL*) of either sender or receiver indicate location of agent services.
- *Owner* (*SO*) is the one who holds the agent or service.
- *Roles* (*SR*) is held by an agent or a service.
- Attributes related to service transportation are *Type* (*TT*), *Address* (*TA*) and *Message* (*TM*).
- *InputVariables* (*I*) and *OutputVariables* (*O*) are in/out parameters.
- *Preconditions* (*IC*) and *Postconditions* (*OC*) define constraints on executing an agent service.
- *Cardinality* (*IO*) defines cardinality property on each attribute. They are expressed in the form as key-value pair (KVP).
- *Exception (E)* defines varied unexpected events, messages, system operations etc.

To present an agent service with the above item atoms, some constraint properties for instance the cardinality must be added on them. The following illustrates the cardinality constraints on each of item atoms in presenting an agent service:

$$\{\{SA, \mathrm{MO}\}, \{SL, \mathrm{MO}\}, \{RL, \mathrm{MO}\}, \{SO, \mathrm{MO}\}, \{TT, \mathrm{MO}\}, \{TA, \mathrm{MO}\}, \{TM, \mathrm{OM}\}, \{I, \mathrm{OM}\}, \{IC, \mathrm{OM}\}, \{O, \mathrm{OM}\}, \{OC, \mathrm{OM}\}, \{E, \mathrm{OM}\}\}$$

where MO and OM mean one mandatory or multiple optional elements respectively.

3.6 Agent Service Interface Design

The efficiency and usability of an agent service framework and the consistency of agent service-oriented enterprise infrastructure somehow rely on endpoint interface design. Enterprise-wide generic and consistent interfaces further depend on establishing (i) responsibilities for designing an agent service interface; (ii) generic, consistent enterprise-wide naming conventions and interface design standards; (iii) an integration layer provides standard interfaces to disparate applications. For instance, typical responsibilities of designing an agent service interface include:

- Agent service interface standard and naming conventions;
- Interface implementation documents;
- Agent service message standard and specifications;

- Message documents; and
- Agent service interface consistency, clarity and extensibility.

It is beneficial to make both the interface naming convention as well as the functionality of the interface more generic. For instance,

- An agent service provides a generic operation (service interface) that represents multiple agent service methods. As illustrated in Figure 2, in which the operation *Implementing Algorithm* (ImplementAlgo) is performed via methods *Implementing Algorithm API* (ImplemtAlgoAPI) and *Coding Algorithm Logics* (CodeAlgoLogic).
- An agent service splits a coarsely grained agent service method into multiple service operations.

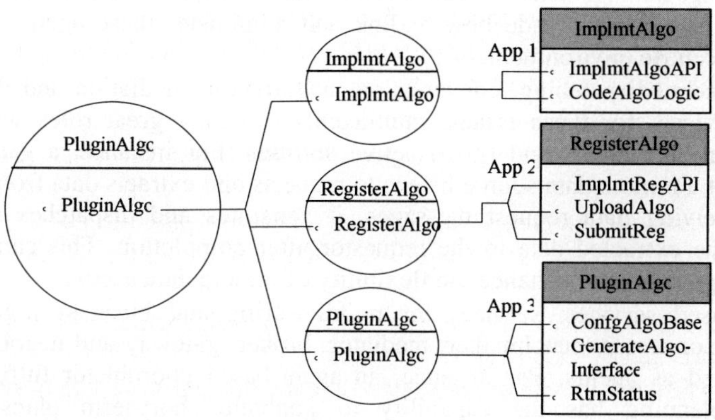

Fig. 2. An agent service including multiple operations combining multiple methods

- An agent service includes one to many service operations that combines various pieces of legacy or enterprise business logics.
- An operation of an agent service encapsulates functionality from multiple business agent services; each of these agent services assembles one to multiple methods representing legacy or enterprise applications. As in Figure 2, the service *Plugin Algorithm* (PluginAlgo) is divided into three operations – *Implementing Algorithm* (ImplmtAlgo), *Register Algorithm* (RegisterAlgo), and *Plugin Algorithm* (PluginAlgo) performed by three agent services. These oprations are further undertaken through eight agent methods held by the three agents.

4 Implementation Strategy — Multiagent + Web Services

This approach builds a dialogue between multiagent system and XML-driven Web service (from hereon referred to as *Web service*) [7]. The integration focuses on building enterprise integration infrastructure and integrated applications, agent-based dynamic service integration, agent-based automatic negotiation in Web services, conversations with Web services, agents for Web service composition, etc. We call this approach as "Web services-driven agent systems". In this method, we advocate the usage of the second-generation Web services.

The basic work for the Web service-based infrastructure includes:

- Provide a service description that, at minimum, consists of a WSDL document;
- Be capable of transporting XML documents using SOAP over HTTP.

On the other hand, multiagent technology would play significant roles in aspects beyond the above basic infrastructure linkage. For instance, the following lists some main functionality that multiagent can serve.

- Agentized components: applications and application components in integration enterprise can be agentized on demand for establishing some specific functional components. For instance, agents for human-computer interaction, resource access dispatching, management of Web services, and enterprise business logic, etc. The problem here may include how to link and administer these agents in the Web services-centric environment.
- Management, dispatching, conversation, negotiation, mediation and discovery of Web services: for these issues, multiagents can play great roles with flexible, intelligent, automated and (pro-) active abilities. For instance, a gateway agent located at a remote data source listens to requests and extracts data from the source after receiving data request messages, it generates and dispatches an agent to deliver the extracted data to the requestor after completion. This can reduce the network payload and enhance the flexibility of remote data access.
- Agent-based services: services for middlewaring and business logics such as adapter, connector, matchmaker, mediator, broker, gateway and negotiator can be customized as agents. For instance, an agent-based coordinator fulfilling partial global planning has the capability to generate short-term plans to satisfy themselves, it may further alter local plans in order to better coordinate its own activities. Agent-based services can enhance the automated and flexible decision-making of these services.

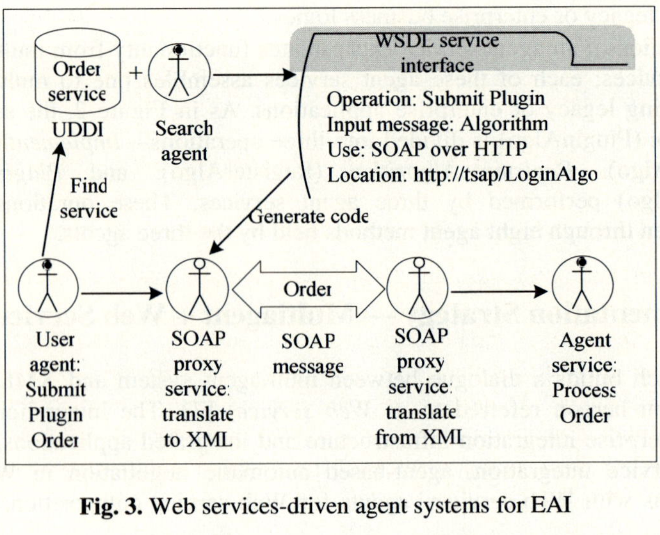

Fig. 3. Web services-driven agent systems for EAI

Figure 3 describes Web services-based agent system. In this model, client defines and submits algorithm Plugin Order via user agent. User agent finds Order Service through UDDI and Search agent. A SOAP proxy service is generated by using WSDL. This proxy service is taken as a bridge to talk to the Process Order agent service via SOAP message and a proxy service translating the message.

5 Case Study and Evaluation

We study and deploy the ASOD approach in designing an online agent service-based trading and mining support system – F-Trade [2, 8] (as shown in Figure 4). Main functions available for trading and mining in the F-Trade include: online plug-and-play, data gateway, profiles and business-oriented interaction, system customization and reconstruction, optimization of trading strategies, discovering stock-strategy pairs, supporting comprehensive add-on applications from capital markets, online data connection to real stock data, etc. F-Trade is running on two super-servers located at Broadway Sydney and connecting to remote stock warehouse at Australian Technology Park. More than 20 trading and mining algorithms have been plugged into F-Trade. It supports personalized simulation and back-testing on real stock data.

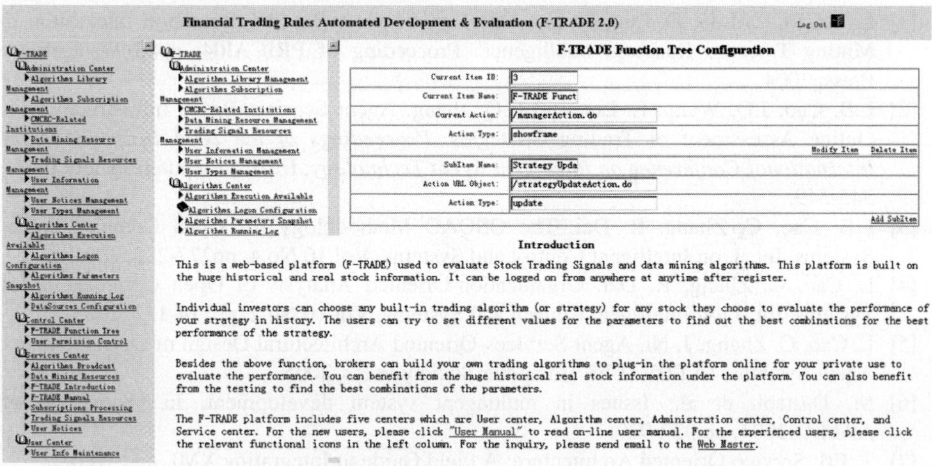

Fig. 4. F-TRADE and its ontology tree

The following lists some of its qualities as an agent-based system.

- Distributed and interoperable. Distributed components can interoperate and be plugged to achieve requested goal.
- Flexible. Supporting varied trading and mining algorithms, heterogeneous data sources, online soft plug-and-play, system reconstruction, etc.
- Open and dynamic. It supports Internet-based access, data source and component plug-and-play, dynamic interfaces and reporting generation, online update, etc.
- Automated. System components have private and group-based control thread of execution, knowledge base, rules and exception management, etc.

- Usable and user-friendly. Comprehensive user profiles are supported in business-oriented format; personalized interfaces are generated in terms of particular roles.
- Adaptive. It supports some potential emergent functions/applications/roles in the future, and can be expanded dynamically and easily.

6 Conclusions

Agent and service are generally viewed as two independent computational paradigms. However, ABC and SOC are highly complementary; the dialogue between them can benefit existing ABC in designing distributed especially Internet-based agent systems. This paper discusses this dialogue and the integration of agent and service. We synthesize the content of agent and service, and integrate the strengths of ABC and SOC in building a new agent-based design approach, referred to as *agent service-oriented design*. The ASOD provides practical support for agent service-oriented architectural and detailed design. We have briefly outlined the main issues in ASOD.

References

[1] L.B. Cao, C. Luo, D. Luo, L. Liu. Ontology Services-Based Information Integration in Mining Telecom Business Intelligence. Proceeding of PRICAI04, pp85-94, Springer Press, 2004.

[2] L.B. Cao, J.Q. Wang, L. Lin, and C.Q. zhang. Agent Services-Based Infrastructure for Online Assessment of Trading Strategies. *Proceedings of the 2004 IEEE/WIC/ACM International Conference on Intelligent Agent Technology*, IEEE Computer Society Press, 345-349.

[3] L.B. Cao, C. Zhang, R. Dai. The OSOAD Methodology for Open Complex Agent Systems. Int. J. on Intelligent Control and Systems, Vol.10 No.4, pp277-285, 2005.

[4] L. Cao, C. Zhang, R. Dai. Organization-Oriented Analysis of Open Complex Agent Systems. Int. J. on Intelligent Control and Systems, Vol.10 No.2, pp114-122, 2005.

[5] L. Cao, C. Zhang, J. Ni. Agent Services-Oriented Architectural Design of Open Complex Agent Systems. IAT'05, IEEE Computer Society Press, 2005.

[6] M. Dastani, et al.: Issues in multiagent system development, In Proceeding of AAMAS2004, pp.922-929.

[7] T. Erl: Service-Oriented Architecture: A Field Guide to Integrating XML and Web Services. Pearson Education, 2004

[8] F-Trade: http://www.f-trade.info

[9] A. Garcia et al. (eds) Software engineering for large-scale multi-agent systems. Springer, 2003.

[10] Java Agent Services: http://www.jcp.org/aboutJava/communityprocess/review/jsr087/

[11] P. Maes (Eds): Designing autonomous agents. The MIT Press, 1990

[12] G. Weiss (Eds): Multiagent systems: a modern approach to distributed artificial intelligence, The MIT Press, 1999

[13] M. Wooldridge: An Introduction to Multiagent Systems. John Wiley & Sons, Ltd, 2002

[14] F. Zambonelli, N.R. Jennings and M. Wooldridge. "Developing multiagent systems: the GAIA Methodology", *ACM Trans on Software Engineering and Methodology*, 12(3):317-370, 2003

Agent–Based Approach for LabVIEW Developed Distributed Control Systems

Grzegorz Polaków and Mieczysław Metzger

Faculty of Automatic Control, Electronics and Computer Science,
Silesian University of Technology, Akademicka 16, 44–100 Gliwice, Poland
{grzegorz.polakow, mieczyslaw.metzger}@polsl.pl

Abstract. Idea of networked software agents is particularly popular in the field of information sciences dealing with distributed content, whereas in industrial automation its use is usually limited to manufacturing systems. This work presents a concept of multi–agent networked system for automation of continuous processes. Some properties of typical software agent (i.e. advanced high level languages and social skills) had to be dropped in exchange for determinism and satisfying time performance in negotiations between network–connected control components. The proposed environment is National Instruments LabVIEW, very popular solution for automation and measurement. LabVIEW is equipped with advanced data acquisition tools and is capable of artificial intelligence methods, although it lacks agentification mechanisms. A framework is presented, providing LabVIEW the required functionality.

Keywords: multiagent–based networking, producer–distributor–consumer, distributed control systems, process control, systems integration.

1 Introduction

There is strong ongoing progress in the field of agent–based technologies. The theory proved its usefulness and commercial interest in its applications boosted rapidly, resulting in countless implementations, especially in the most popular environment of distributed content i.e. World Wide Web [1], [2]. However, process of adaptation and implementation of the original idea in automation and control science is not so successful and varies greatly depending on subject. On the one hand, natural entity based structure of software agencies reflects perfectly processes occurring in manufacturing systems and already proved its usability in that field, overtaking classical approaches [3]. On the other hand, in the automation of continuous processes it is hard to tie agents to non–discrete physical phenomena like flow of liquids, gases or heat. In this case agents have rather to be bound to logical structure of control system. Great implementation of such agent–based frameworks in continuous plants was done by van Breemen and de Vries [4]. Their works are a significant contribution inttroducing agents theory in the field of process automation, however they overlook important issue, that is implementation of theory in existing physical plants and systems. Solving this issue is significant and non–trivial, since while there is wide

N.T. Nguyen et al. (Eds.): KES-AMSTA 2007, LNAI 4496, pp. 21–30, 2007.

choice of frameworks for developing intelligent agents in abstract knowledge environments, there is no ready solution for agentification of spatially distributed control systems. Specific requirement, needed to be fulfilled by software to be usable in the field of data acquisition and control, is easy and unproblematic integration of logical software layer with physical hardware of control instrumentation. One of applications fulfilling this requirement, therefore suitable for the task, is National Instruments' LabVIEW.

LabVIEW is a very popular software development solution for engineers. A programming language which LabVIEW implements is simple yet powerful graphical language based on functional block diagrams, often called a "G language". Capabilities of the NI LV are the same as capabilities of any other classical text–based programming language. It also includes a wide set of predefined function blocks providing functionalities useful in advanced software, i.e.:

- artificial intelligence tools (neural networks and fuzzy logic);
- basic functions of a desktop computer's operational system's network protocol stack;
- control instrumentation connectibility (using standards like OPC, and proprietary protocols like Logos);
- rich Human–Machine Interfaces and visualisation tools.

As it can be concluded from the above list, internal agent's intelligence can easily be programmed by an end user using library of components supplied by default. However, it was necessary to design a protocol for agents communication and build a required framework supporting it. A new potential of continuous control systems gained by agentification at control loop layer turned out to be unprecedented, which was motivation to describe it in detail. Concept of networked multiagent system for continuous process control (including structure of the framework, communication protocols and language of agents) is an original contribution.

2 Architecture

As it was noticed before, it is difficult to assign an agent to non–discrete continuous physical phenomenon. Therefore, it is assumed that each LabVIEW developed agent, is tied to one of the clearly defined tasks of control system itself. Such tasks include:

- control: agent receives values of process variables and answers with values of control variables;
- simulation: agent receives values of inputs of mathematical model and answers with outputs of the model;
- data acquisition: on request agent sends value gathered from plant;
- visualisation: agent receives some values and presents them in human readable form to system operator.

Systems of such agents developed in LabVIEW and working together in a modular way were described in [5], [6]. Described agents were communicating by means of TCP/IP protocol and they proved to be especially useful in the field of learning and education as their modular composition allowed for quick reconfiguration.

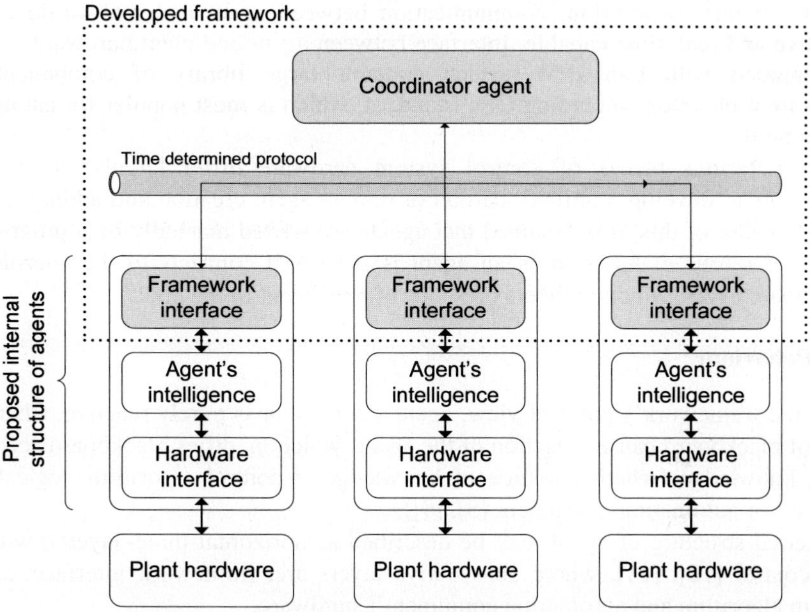

Fig. 1. Proposed architecture for agent–based control system

Using works [4], [5] and [6] as a base, after generalisation and adding new functionalities, architecture shown in Fig.1 evolved.

2.1 Communication

A primary issue in processing of sampled continuous variables is strict time determinism. Duration of all calculations and transmissions must be short and known in advance. This requirement is a consequence of mathematical principles of control algorithms implemented in the form of difference equations operating on sampled signals. A usual way to accomplish this, is to minimise the amount of transmitted data and to simplify its structure.

While generally used agent communication languages (like KQML) are text–based, the core of the control system is the set of numerical values. Only those values must be exchanged, so even the most basic KQML–based syntax would be burdened with excessive formatting text requiring processing, thus consuming precious time. As languages with text–based advanced syntax cannot be used in such framework, custom networking system was designed. It is based on low level networking protocols (TCP and UDP) and uses producer–distributor–consumer paradigm, where main distributed resource is a base of knowledge shared between agents [7].

2.2 Creation and Residence

It is assumed that agent's threads are physically running in hardware resources located as close as possible to control instrumentation the agent is logically tied to.

Thanks to this assumption, communication between agent and real world is very effective and real–time capable. Interface between agent and plant hardware is easily programmed with LabVIEW, which contains large library of components for hardware interaction (including OPC standard, which is most popular for automation integration)

Since various pieces of control system hardware differ largely, it is nearly impossible to develop a unified method of remote agent creation and adding it to the pool. Because of this, it is assumed that agents are started manually by a programmer in proper resource. As soon as an agent is started, it connects to the coordinating blackboard agent, which registers presence of new agent in the pool.

2.3 Behaviour

From the framework's point of view, agent's behaviour is purely reactive. Change of state of blackboard causes reaction of the agent, which modifies blackboard according to its knowledge (whether source of knowledge is control algorithm, logical rule, database or automation instrument [8], [9]).

Internal structure of agent may be described as horizontal three–layered with two pass control [10], [11], where consecutive layers are: framework interface, agent's custom algorithm and automation equipment's hardware.

From implementation's point of view, agent is able to execute multiple processing threads (threads are supported by LabVIEW in the form of parallel execution of conditional loops), where change of blackboard's content is one of many possible events. Agent's thread is free to work constantly in the background actively cooperating with the plant hardware as long as it conforms to the framework's specification of interaction.

Summarizing, when a LabVIEW agent is mentioned in the paper, it is a software module which is:

- developed with National Instruments LabVIEW;
- dedicated to perform some defined task (acquisition, control, visualisation, etc.) in a control system;
- residing in a hardware related to the performed task (i.e. desktop PC or a commercial PLC) and executing its threads there;
- presenting the common interface to other agents while completely embedding its custom intelligent behaviour inside.

3 Description of Communication Framework

All protocols accessible in LabVIEW work according to connection oriented client–server vertical data communication scheme, which is the most popular scheme in local area networks. Engineering practice shows that in industrial networks (fieldbuses) some other data communication schemes are usually used i.e. master–slave, producer–consumer or producer–distributor–consumer. Behaviour of such networks differs from classic LANs, as fieldbuses are designed especially for distributed control systems, where data oriented horizontal communication is rather preferred than vertical one.

Since functionality of the TCP and UDP protocol stack in LabVIEW is limited to basic, use of these protocols for communication between agents developed with G language is limited to native distribution schemes i.e. client–server for TCP and sender–receiver for UDP. There are no functional blocks shipped by default which would enable horizontal data transmission capability. Agenda behind the presented framework was to enable horizontal communication functionality in LabVIEW, as simple and quick as in case of standard protocols. Such developed protocol had to be open and compatible with wide range of hardware and software products to allow for integration of software modules embedding different parts of distributed control system. A protocol was designed to reflect properties of industrial networks while working on top of the typical office area network which is usual LabVIEW environment.

System built of such agents offers large potential for integration of hardware and software modules of varying types and construction (see Fig. 2).

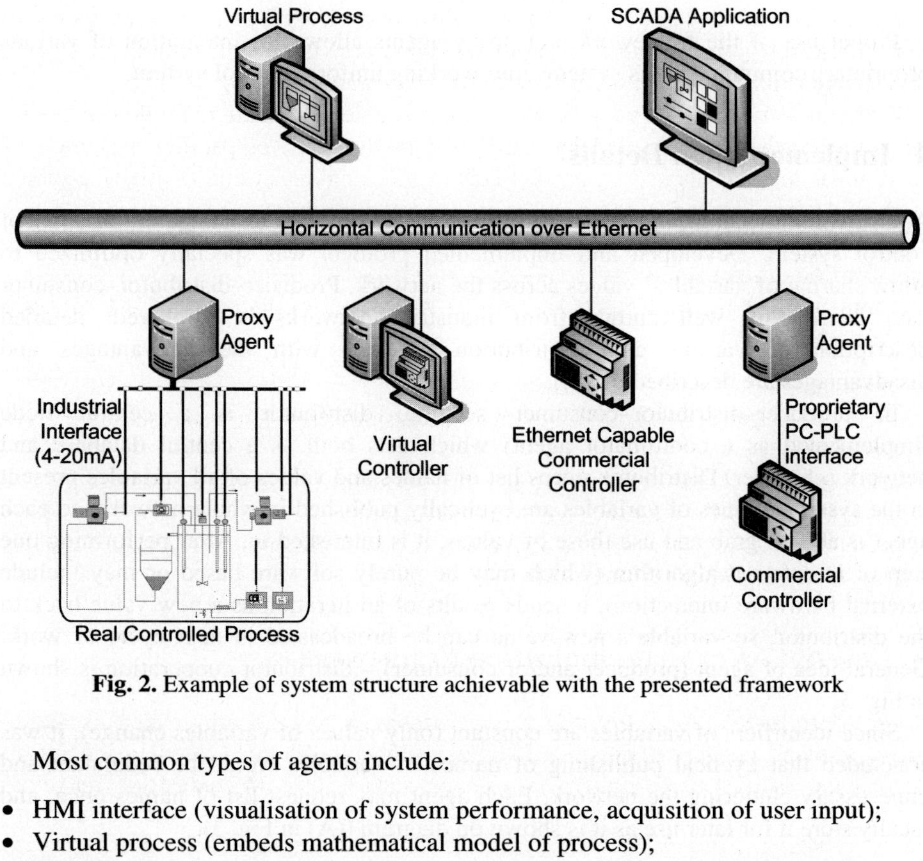

Fig. 2. Example of system structure achievable with the presented framework

Most common types of agents include:

- HMI interface (visualisation of system performance, acquisition of user input);
- Virtual process (embeds mathematical model of process);
- Control algorithm (embeds a set of mathematical equations forming algorithm).

Such modules may be implemented in:

- software (computer program running on desktop PC);
- hardware (commercially available programmable logic controller, some of them are connectible to Ethernet and can implement full framework functionality).

Special type of module is Proxy agent. Its role is translation of messages transmitted with the common protocol to/from custom hardware, which is physically incapable of the framework connectibility itself. Usually such hardware is connected with a kind of specialized interface card to personal computer (for example Simatic controllers are connectible to PC using the MPI/PPI interface). Such computer can run custom designed proxy agent, thus integrating hardware into framework. Two most often kinds of proxy agents are ones interacting with:

- Commercial programmable logic controllers (often executing control algorithm but not limited to them);
- Real process plants (sensors and actuators connected to specialized PC card by analog industrial standard 4–20mA or 0–10V).

Proper use of the framework and proxy agents allows for integration of various proprietary communications systems into working uniform control system.

4 Implementation Details

Agents communication in presented framework is limited to exchange of variables of control system. Developed and implemented protocol was specially optimized to allow sharing of variables' values across the network. Producer–distributor–consumer data distribution well known from industrial networks is employed; detailed description of various data distribution schemes with their advantages and disadvantages are described in [12].

In producer–distributor–consumer scheme, distributor is a central node (implemented as a coordinator agent) which acts both as a central database and network scheduler. Distributor stores list of names and values of all variables present in the system. Values of variables are cyclically published in whole network, so each agent is able to grab and use those of values, it is interested in. After performing one step of its internal algorithm (which may be purely software based or may include external hardware interaction), it sends results of an iteration as a new value back to the distributor, so variable's new value can be broadcasted in next cycle of work. General idea of agent (producer and/or consumer) – distributor cooperation is shown in Fig. 3.

Since identifiers of variables are constant (only values of variables change), it was concluded that cyclical publishing of names of variables would be redundant and unnecessary cluttering the network. Each agent may request list of names once, and locally store it for later use as it is shown on diagram ((A) in Fig. 3).

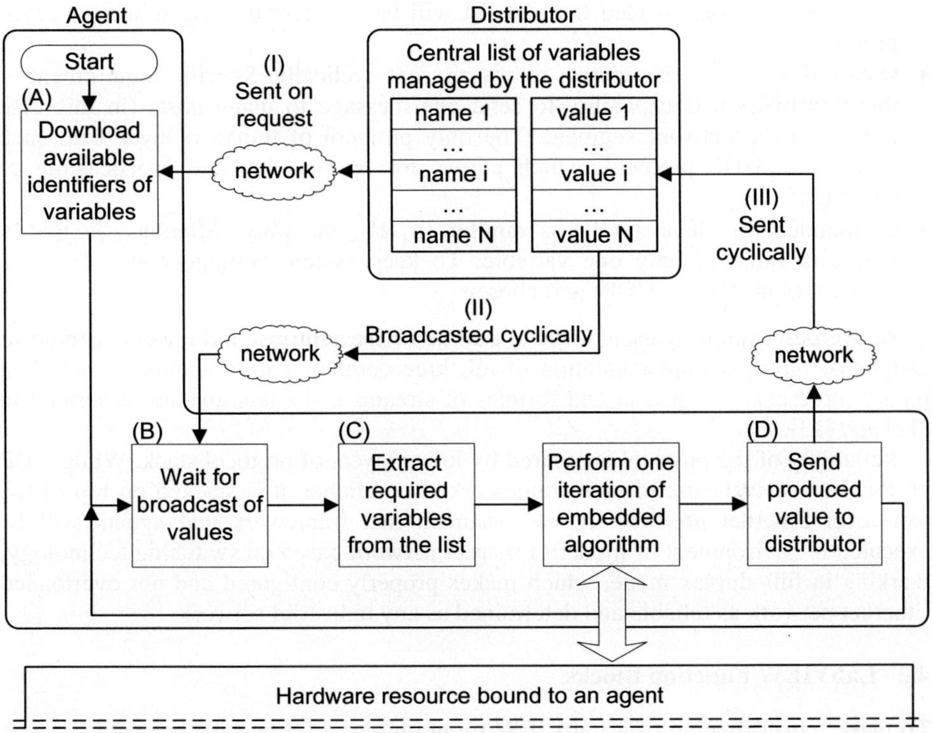

Fig. 3. External environment and internal structure of typical agent working with the protocol

Duration of one cycle of work is a parameter of specific control system, as it is the same value as sampling period of signals present in the system, which should be determined according to the specific dynamics of components of control system. However, minimal value of period should be larger than longest possible time of execution of internal algorithms for all cooperating agents.

4.1 Physical Layer and Its Reliability

Main conception of physical layer of the protocol was full compatibility with popular and widely available Ethernet instrumentation, including already existing networks. Because of this assumption whole protocol was designed to work in application layer of Internet protocol suite i.e. on top of Ethernet physical layer, IP networking layer and TCP/UDP transport layer.

As it is shown in Fig. 3, whole communication consists of three streams of data, labelled as (I), (II) and (III). Each of those channels was implemented separately on top of TCP or UDP protocols, according to its particular requirements:

- through channel (I) data is sent one time and only when requested. Stream contains vital configuration data i.e. list of variables identifiers. Because the list is main resource of the framework and normal functioning of whole system depends on its correctness, it was decided that the list will be transferred using reliable TCP/IP protocol;

- stream (II) consists of separate messages, sent cyclically. Specific requirement of this transmission is capability to send one message to many hosts (in this case agents) in the network segment. The only protocol of transport layer with such property is UDP protocol, which allows for multicasting and broadcasting of datagrams;
- communication channel (III) is similar to (II), the only difference is that it transmits value of only one variable. To keep system homogeneous, the same protocol as in (II) (i.e. UDP) was chosen.

Above description focuses on main outline of the protocol and reasons behind it. Details of physical implementation of all three communication channels, including binary representation of data and formats of streams and datagrams are described in [12] and [13].

Reliability of the protocol is assured by lower layers of protocol stack. While UDP protocol extensively used in the framework is unreliable, it is stacked on top of the low level Ethernet protocol. It was assumed that framework and agents will be executed in environment of modern Ethernet network based on switching technology, working in full–duplex mode, which makes properly configured and not overloaded Ethernet network as reliable and determined as any industrial network.

4.2 LabVIEW Function Blocks

To make transformation of a LabVIEW program into a framework compatible agent easy, a set of function blocks was developed (Fig. 4). Each of blocks has its place in the agent's logical diagram presented in Fig. 3. Block named *init* creates all required network connections and downloads list of available identifiers (A), *consume* block waits for broadcasted datagram containing values of variables (B), *produce* block sends value of one specified variable to the distributor node (D). *Close* block (not present in Fig. 3) is called when agent is being shut down to close all connections opened before during step (A). Extraction of required variables from global list of values (C) is easily done using standard function blocks supplied by default.

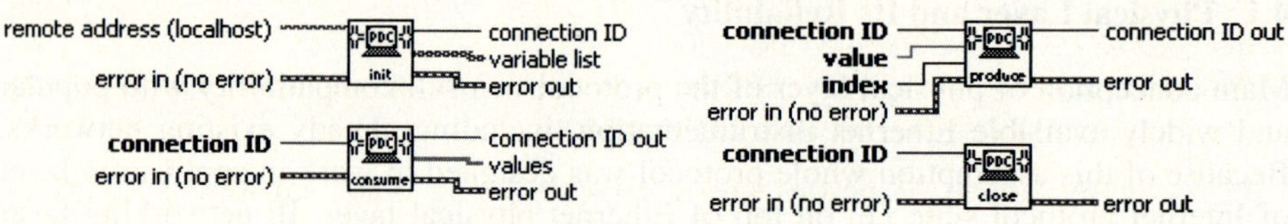

Fig. 4. LabVIEW–based functional blocks for agentification in proposed framework

5 Concluding Remarks

Achieved architecture structure is a blackboard system, in which there are two groups of agents in a constant conflict. Agents representing control algorithms are trying to modify the blackboard's content according to a goal of control, while agents embedding system sensors interfere by modifying the blackboard according to a real plant performance. The third group of agents does not take part in the conflict but performs their tasks based on a current blackboard's content (i.e. writing values to actuators, visualisation, data storing).

Agent paradigm is usually used in process automation for data mining and storage purposes, while control loops are closed using standard instrumentation [14]. In the presented framework control loops are closed directly through agents intelligence providing unmatched data processing capabilities at control loop hardware layer.

Modular structure of the framework makes it possible to quickly reconfigure modules of the control system; changing algorithm, models of objects, adding control loops and rearranging connections is easy with the framework, as individual agents can easily be added to and/or removed from a system. On the contrary, in classical control loops each change of structure usually requires extensive rewiring and reprogramming.

Strong unification of the framework interface with loosely defined plant hardware interface results in possibility of proxy agents development. A concept of proxy agents offers unmatched potential in the field of systems integration (connecting control instrumentation hardware of varying standards and vendors).

A low–level network protocol complying with a producer–distributor–consumer communication scheme was developed specifically for the framework and is a completely novel technique in the domain of G language. It is strictly time determined in contrast to existing communications standards used in LabVIEW i.e.: OPC, DataSocket and Logos.

Although the proposed agent communication language limits agents cooperation capabilities among spatially distributed hardware, agents are free to implement locally any advanced artificial intelligence using methods supported in LabVIEW: advanced algorithms, fuzzy logic, neural networks, database integration, etc. Some interesting examples of agent personalities are:

- HMI assistant – gathering some input from a user; in emergency situations adjusting on its own gathered variables to safe values, modifying GUI accordingly;
- historian – storing values in private database and providing user with inferences based on history of variables;
- overseer – watching the state of system and tuning parameters of control algorithms performed by other agents;
- controller–agency (see [4]) – embedding multiple control algorithm and switching between them depending on state of the system.

Many of the above properties are achievable using other software systems or theories, although there is no other unified middleware for low level agents, which is the only way to achieve all these properties at the same time. The presented framework was implemented and tested in various applications. It turned out to be particularly effective in the field of multiple systems interconnection, as it allows to integrate various hardware platforms by using proxy agents. Achieved results show that the idea of networked agents in spatially distributed control systems of continuous processes is very promising and should be a subject for further research.

Acknowledgement

This work was supported by the Polish Ministry of Scientific Research and Information Technology.

References

1. Bigus, J.P., Bigus, J.: Constructing Intelligent Agents Using Java. 2nd edn. John Wiley & Sons, New York (2001)
2. Sobecki, J., Nguyen, NT.: Consensus–based adaptive interface construction for multiplatform Web applications. LNCS, Vol. 2690 (2003), 457–461
3. Jennings, N. R., Bussman, S.: Agent–Based Control Systems. IEEE Control Systems Magazine, Vol. 23 (2003) 61–74
4. van Breemen, A., de Vries, T.: An Agent–Based Framework for Designing Multi–Controller Systems. Proceedings of the Fifth International Conference on The Practical Applications of Intelligent Agents and Multi–Agent Technology (2000) 219–235
5. Metzger, M.: TCP/IP–based virtual control systems – a low cost alternative for realistic simulation of control systems. Proceedings of the 13–th SCS European Simulation Symposium, Marseille, SCS Publication (2001), 140–143.
6. Metzger, M.: Agent–Based Simulation of Flexible, Distributed Control Systems. Proceedings of 9th IEEE International Conference on Methods and Models in Automation and Robotics (2003) 1183–1188
7. Ferber, J.: Multi–agent systems – an introduction to distributed artificial intelligence. Addison–Wesley, 1999.
8. Wooldridge, M.: An Introduction to MultiAgent Systems. John Wiley & Sons, Chichester (2002)
9. Jennings, N. R., Sycara, K., Wooldridge, M.: A Roadmap of Agent Research and Development. Autonomous Agents and Multi–Agent Systems, 1, 7–38, Kluwer Academic Publishers Boston (1998)
10. Weiss, G. (ed.): Multiagent Systems. A Modern Approach to Distributed Artificial Intelligence. MIT Press, Cambridge (1999)
11. Knapik, M., Johnson, J.: Developing Intelligent Agent for Distributed Systems. Mc Graw–Hill, New York (1998)
12. Polaków, G., Metzger, M.: Design and Implementation of Ethernet–Based Horizontal Communication Scheme Using LabVIEW. Proceedings of the 12th IEEE International Conference on Methods and Models in Automation and Robotics (2006) 829–834
13. Polaków, G., Metzger, M.,: Programming LabVIEW–Based Producer/Consumer Communication for Distributed Control Systems. Proceedings of IFAC Workshop on Programmable Devices and Embedded Systems, Brno (2006) 322–327
14. Pirttioja, T., Seilonen, I., Appelqvist, P., Halme, A., Koskinen, K.: Agent–Based Architecture for Information Handling in Automation Systems. Emerging Solutions for Future Manufacturing Systems, Vienna (2004), 73–80

Pr$\mathcal{SH}$: A Belief Description Logic*

Tao Jia, Wen Zhao, and Lifu Wang

School of Electronics Engineering and Computer Science
Peking University
Beijing 100871, China
{jiat, Owen}@sei.pku.edu.cn

Abstract. Some research has been done on probabilistic extension of description logics such as P-CLASSIC and P-$\mathcal{SHOQ}$ which focus on the statistical information. For example, in those kind of probabilistic DL, we can express such kind of uncertainty that the probability a *randomly* chosen individual in concept C is also in concept D is 90 percent. This kind of statistical knowledge is certain which means the author of this statement is sure about it. In this paper, we will describe a new kind of probabilistic description logic Pr$\mathcal{SH}$ which could let user express the uncertain knowledge(i.e. degrees of belief). For example, if the user is not sure about that concept C is subsumed by concept D, he could describe it with Pr$\mathcal{SH}$ such as the probability that concept C is subsumed by concept D is 90 percent.Furthermore, user could make use of the uncertain knowledge to infer some implicit knowledge by the extension of *tableau-algorithm* of $\mathcal{SH}$ which will be also introduced in this paper.

1 Introduction

Recently, a number of probabilistic description logics have been developed such as P-CLASSIC[1] and P-$\mathcal{SHOQ}$[2] to support the representation and reasoning of the uncertain knowledge. Those probabilistic description logics aim to handle the statistical information such as computing the probability that a randomly chosen individual in class C is also in class D[1].

However, we often want to express another kind of uncertainty about knowledge. For example, to his knowledge limitation, the knowledge base developer may not sure whether *all the animals with four legs are mammals* but he think the probability this assertion being true is about 0.9. How to make use of this uncertain knowledge in his ontology? Unfortunately, he could not express it using P-CLASSIC or P-$\mathcal{SHOQ}$ because what he wants to say is not *90 percent four leg animals are mammals*.

According to [3], the fundamental difference between these two kind of probability is as follows. The former one can be viewed as statements about what Hacking calls a *chance setup*[4], that is, about what one might expect as the result of performing some experiment or trial in a given situation. It can also be viewed as capturing statistical information about the world. Moreover, it assumes only one possible world(the "real"

* Supported by the National Grand Fundamental Research 973 Program of China Under Grant No. 2002CB312006; the National Natural Science Foundation of China Under Grant Nos. 60473058.

N.T. Nguyen et al. (Eds.): KES-AMSTA 2007, LNAI 4496, pp. 31–41, 2007.

world). On the other hand, the second kind of probability captures what has been called *degree of belief*[5] and it assumes the existence of a number of possible worlds with some probability over them.

In this paper, we will introduce a new kind of probabilistic description logics named Pr$\mathcal{SH}$ which could describe and reason on degrees of belief. We will use $(C \sqsubseteq D)^{\alpha}(0 < \alpha \leq 1)$ to express the degree of belief that class C is a subclass of class D is α and $(a : C)^{\alpha}$ to express the degree of belief that individual a is in class C is α.

In many application domains, this kind of probabilistic subsumption semantics is more appropriate. For example,

- Dealing with conflict ontologies. On the coming semantic web, there would be a lot of ontologies on the web which are created by different people. Due to the limitation of their knowlegde or other reasons, there must be many conflict axioms and assertions such as one ontology may define the company with no more than 5 employees is *small company* and another ontology defines the company with no more than 10 employees is *small company*. How to make use of such kind of conflict knowledge? One solution is to assign each axiom a probability as degree of belief. For example, we could assign degree of belief 0.5 to the former one and 0.4 to the latter one. The remaining fraction 0.1 is left for the other possible definition about *small company*.
- Making use of knowledge not being proved yet such as Goldbach Conjecture or could not be proved such as the definition of *small company*.
- In the *ontology mapping*[6] or *schema matching*[7] applications, using Pr$\mathcal{SH}$ to describe the degree of the similarity of the concepts in different ontologies or schemas. For example, if there is a concept *SmallCom* in schema A and a concept *SCompany* in schema B, we may draw a conclusion that the probability $SmallCom \equiv SCompany$ is 0.9. This conclusion should follow the degree of belief semantics but not statistical semantics, because if apply the statistical semantics, we have to admit $SmallCom \not\equiv SCompany$.

Moreover, Pr$\mathcal{SH}$ is different from other probabilistic description logics such as P-CLASSIC not only in semantics, but also in reasoning algorithms. For example, in Pr$\mathcal{SH}$, given $(A \equiv B)^{0.9}$ and $(B \equiv C)^{0.8}$, we can infer $0.7 \leqslant Pr(A \equiv C) \leqslant 0.8$. On the other hand, in P-CLASSIC, the statement that the probability class A euqals class B is 0.9 could be written as $Pr(A \sqcap B|A \sqcup B) = 0.9$. Similarly, $Pr(B \sqcap C|B \sqcup C) = 0.8$ describes the probability B euqals C is 0.8. However, we cannot draw any conclusion about $Pr(A \sqcap C|A \sqcup C)$.

The rest of this paper is organized as follows. Section 2 describes the foundations of Pr$\mathcal{SH}$. Section 3 introduces a typical probabilistic description logic Pr$\mathcal{SH}$ and its reasoning algorithm which is extended from the tableau algorithm of $\mathcal{SH}$. Section 4 is the related work. Section 5 is the conclusion and future work.

2 Foundations

Pr$\mathcal{SH}$ is a probabilistic extension of $\mathcal{SH}$. They have exactly same description languages[8] but some differences in *knowledge base* definition and *semantics* as follows.

2.1 Knowledge Base

A Pr$\mathcal{SH}$ knowledge base is also composed of two distinct part: the intensional knowledge (TBox) and extensional knowledge (ABox), but they are extended by the probabilistic factors. In this paper, we separate the TBox to the C(oncept)Box and R(ole)Box which contain the axioms about concept and roles, respectively.

CBox In A Pr$\mathcal{SH}$ CBox C, an axiom may have the form (Mammal $\equiv$ Animal $\sqcap$ FourLegThing)$^{0.9}$ Which means the probability that the axiom is true is 0.9. Formally, let $\mathcal{L}$ be the Pr$\mathcal{SH}$ language, $C, D \in$ **Cdsc**($\mathcal{L}$)$\mathcal{L}$-concepts[9], a CBox C is a finite, possibly empty, set of statements of the form $(C \sqsubseteq D)^{\alpha}, 0 < \alpha \leqslant 1$, called *concept inclusion*. α denotes the probability that the statement is true. $(C \equiv D)^{\alpha}$, called *concept equivalence*, is an statement denotes the probability that both $C \sqsubseteq D$ and $D \sqsubseteq C$ is true is α. Statements in C are called *concept axioms*. If $\alpha = 1$, the statements can be abbreviated to $C \sqsubseteq D$ or $C \equiv D$, called *certain concept axioms*. Otherwise, the statements are called *uncertain concept axioms*.

We can divide the CBox C into two parts C_d and C_p. C_d consists of all the certain concept axioms, while C_p consists of all the uncertain concept axioms. Here we should not treat $(C \equiv D)^{\alpha}$ as the abbreviation of $(C \sqsubseteq D)^{\alpha}$ and $(D \sqsubseteq C)^{\alpha}$, since $(C \sqsubseteq D)^{\alpha}$ and $(D \sqsubseteq C)^{\alpha}$ imply $(C \equiv D)^{\beta}, max(2\alpha - 1, 0) \leqslant \beta \leqslant \alpha$, but not $(C \equiv D)^{\alpha}$.

A concept axiom without its probabilistic weight is called the *certain extension* of the concept axiom. For example, $C \equiv D$ is the certain extension of $(C \equiv D)^{\alpha}$. The certain extension of a CBox C is a new CBox C_e whose concept axioms are all coming from the certain extension of the concept axioms in C.

An interpretation $\mathcal{I}$ *satisfies* $(C \sqsubseteq D)^{\alpha}$, or $\mathcal{I}$ *models* $(C \sqsubseteq D)^{\alpha}$ (written as $\mathcal{I} \models (C \sqsubseteq D)^{\alpha}$), if $\mathcal{I}$ *satisfies* its certain extension(that is $C^{\mathcal{I}} \subseteq D^{\mathcal{I}}$), and $\mathcal{I}$ satisfies $(C \equiv D)^{\alpha}$ (written as $\mathcal{I} \models (C \equiv D)^{\alpha}$), if it satisfies its certain extension (which means $C^{\mathcal{I}} = D^{\mathcal{I}}$). An interpretation $\mathcal{I}$ *possibly satisfies* a Pr$\mathcal{SH}$ CBox C(written as $\mathcal{I} \models C$), if it satisfies all the certain concept axioms in C. We also say that $\mathcal{I}$ is a *possible model* of C.

RBox We could give the similar definition about RBox of Pr$\mathcal{SH}$. Let $\mathcal{L}$ be the Pr$\mathcal{SH}$ language, $RN, SN \in$ **R** role names, $R_1, R_2 \in$ **Rdsc**($\mathcal{L}$)$\mathcal{L}$-roles, an Pr$\mathcal{SH}$ RBox $\mathcal{R}$ is a finite, possible empty, set of statements of the form:
-$(RN \in \mathbf{F})^{\alpha}, 0 < \alpha \leqslant 1$, where $\mathbf{F} \subseteq \mathbf{R}$ is a set of *functional roles*, or
-$(SN \in \mathbf{R}_{+})^{\alpha}, 0 < \alpha \leqslant 1$, where $\mathbf{R}_{+} \subseteq \mathbf{R}$ is a set of *transitive roles*, or
-$(R_1 \sqsubseteq R_2)^{\alpha}, 0 < \alpha \leqslant 1$, called *role inclusions*; $(R_1 \equiv R_2)^{\alpha}, 0 < \alpha \leqslant 1$, called *role equivalence*, denotes the probability that both $R_1 \sqsubseteq R_2$ and $R_2 \sqsubseteq R_1$ is true is α.

ABox Let $\mathcal{L}$ be the Pr$\mathcal{SH}$ language, $a, b \in$ **I** individual names, $C \in$ **Cdsc**($\mathcal{L}$) an $\mathcal{L}$-concept and $R \in$ **Rdsc**($\mathcal{L}$) an $\mathcal{L}$-role. An Pr$\mathcal{SH}$ ABox $\mathcal{A}$ is a finite, possible empty, set of statements of the form $(a : C)^{\alpha}$, called *concept assertions*, or $(< a, b >: R)^{\alpha}$, called *role assertions*. Statements in $\mathcal{A}$ are called *assertions*. If $\alpha = 1$, the assertions can be abbreviated to $a : C$ or $< a, b >: R$ and can be called *certain assertions*. Otherwise, they do not have abbreviated form and is called *uncertain assertions*.

Knowledge Base A Pr$\mathcal{SH}$ knowledge base $\mathcal{K}$ is a triple $< C, \mathcal{R}, \mathcal{A} >$, where C is a CBox, $\mathcal{R}$ is a RBox, and $\mathcal{A}$ is an ABox. An interpretation $\mathcal{I}$ *satisfies* a knowledge base $\mathcal{K}$, written as $\mathcal{I} \models \mathcal{K}$, iff it satisfies C, $\mathcal{R}$ and $\mathcal{A}$; $\mathcal{K}$ is *satisfiable* (*unsatisfiable*) iff there exists (does not exist) such an interpretation $\mathcal{I}$ that satisfies $\mathcal{K}$.

2.2 Semantics

We use possible worlds[10,11] to describe the semantics of the Pr$\mathcal{SH}$. The approach is mapping the Pr$\mathcal{SH}$ knowledge base onto a set of DL knowledge bases, where the models of each of the latter constitute the set of possible worlds. First, we give the definition of the DL knowledge bases related to a Pr$\mathcal{SH}$ knowledge base.

Definition 1 (DL KBs Related to the Pr$\mathcal{SH}$ KB). *Given a PrSH knowledge base $\mathcal{K} =< C, \mathcal{R}, \mathcal{A} >$. C_d, $\mathcal{R}_d$ and $\mathcal{A}_d$ are their certain parts, and C_p, $\mathcal{R}_p$ and $\mathcal{A}_p$ are their uncertain parts. Let C_{pe}, $\mathcal{R}_{pe}$ and $\mathcal{A}_{pe}$ be the certain extension of C_p, $\mathcal{R}_p$ and $\mathcal{A}_p$. Their formal definition is as follows*

$$C_{pe} = \{c|(c)^\alpha \in C_p\}, \mathcal{R}_{pe} = \{r|(r)^\alpha \in \mathcal{R}_p\}, \mathcal{A}_{pe} = \{a|(a)^\alpha \in \mathcal{A}_p\}$$

The set of DLs D related to this PrSH is defined as

$$D_{\mathcal{K}} = \{\mathcal{K}_d =< C_d \cup C_i, \mathcal{R}_d \cup \mathcal{R}_i, \mathcal{A}_d \cup \mathcal{A}_i > | C_i \subseteq C_{pe} \wedge \mathcal{R}_i \subseteq \mathcal{R}_{pe} \wedge \mathcal{A}_i \subseteq \mathcal{A}_{pe}\}$$

Obviously, a model of the knowledge base in $D_{\mathcal{K}}$ is also a possible model of $\mathcal{K}$. All the Pr$\mathcal{SH}$ knowledge base possible models constitute the set of its possible worlds $\mathcal{W}_{\mathcal{K}}$.

Definition 2 (Probability Distribution). *For a PrSH knowledge base $\mathcal{K}=< C, \mathcal{R}, \mathcal{A} >$ with its set of possible worlds $\mathcal{W}_{\mathcal{K}}$, let $M = (\mathcal{W}_{\mathcal{K}}, \mu)$ denote a probability structure of $\mathcal{K}$, where μ is a discrete probability distribution on $\mathcal{W}_{\mathcal{K}}$. Then we can define the notion of an extension $[t]_{(\mathcal{K},M)}$ of the term t (could be a concept description, the certain extension of an axiom in $C \cup \mathcal{R}$ or the certain extension of an assertion in $\mathcal{A}$) by the following rules. Let w be a world (possible model) of $\mathcal{W}_{\mathcal{K}}$, $\mathcal{K}_d$ a DL knowledge base related to $\mathcal{K}$.*

1.If $\exists\alpha \in (0, 1], (t)^\alpha \in C \cup \mathcal{R} \cup \mathcal{A}$, then $[t]_{(\mathcal{K},M)} = \alpha = \mu(\cup_{\mathcal{K}_d \in D_{\mathcal{K}} \wedge w \models \mathcal{K}_d \wedge \mathcal{K}_d \models t}\{w\})$

2.else if t is a concept, $[t]_{(\mathcal{K},M)} = 1 - \mu(\cup_{\mathcal{K}_d \in D_{\mathcal{K}} \wedge w \models \mathcal{K}_d \wedge \mathcal{K}_d \not\models t}\{w\})$ $\mathcal{K}_d \not\models t$ denotes concept t is not satisfiable with respect to the knowledge base $\mathcal{K}_d$.

Definition 3 (Concept satisfiability). *Given a PrSH knowledge base $\mathcal{K}$ and a concept C, the probability that C is satisfiable with respect to $\mathcal{K}$ is α iff $[C]_{(\mathcal{K},M)} = \alpha$.*

Example 1. Given a Pr$\mathcal{SH}$ knowledge base $\mathcal{K} =< C, \mathcal{R}, \Phi >$, where

$$C = \{(C \sqsubseteq \forall R.D)^{0.4}\} \text{ and } \mathcal{R} = \{(S \sqsubseteq R)^{0.8}\}$$

we have a possible worlds distribution $M_1 = (\mathcal{W}_{\mathcal{K}}, \mu)$:

$$\begin{aligned}
P(\mathcal{I}_1) = 0.5 &: \Delta^{\mathcal{I}_1} = \{a, b, c\}, C^{\mathcal{I}_1} = \{a, b\}, R^{\mathcal{I}_1} = \{< a, b >, < a, c >, < b, a >\}, \\
&\quad D^{\mathcal{I}_1} = \{b, c\}, S^{\mathcal{I}_1} = \{< a, b >\} \\
P(\mathcal{I}_2) = 0.3 &: \Delta^{\mathcal{I}_2} = \{a, b, c\}, C^{\mathcal{I}_2} = \{a\}, R^{\mathcal{I}_2} = \{< a, b >, < a, c >, < b, a >\}, \\
&\quad D^{\mathcal{I}_2} = \{b, c\}, S^{\mathcal{I}_2} = \{< a, b >\} \\
P(\mathcal{I}_3) = 0.1 &: \Delta^{\mathcal{I}_3} = \{a, b, c\}, C^{\mathcal{I}_3} = \{a\}, R^{\mathcal{I}_3} = \{< a, b >, < a, c >, < b, a >\}, \\
&\quad D^{\mathcal{I}_3} = \{b, c\}, S^{\mathcal{I}_3} = \{< a, b >, < b, c >\} \\
P(\mathcal{I}_4) = 0.1 &: \Delta^{\mathcal{I}_4} = \{a, b, c\}, C^{\mathcal{I}_4} = \{a, b\}, R^{\mathcal{I}_4} = \{< a, b >, < a, c >, < b, a >\}, \\
&\quad D^{\mathcal{I}_4} = \{b, c\}, S^{\mathcal{I}_4} = \{< a, b >, < b, c >\} \\
P(\mathcal{I}_k) = 0 &: \quad \mathcal{I}_k \in \mathcal{W} \wedge k \neq 1, 2, 3, 4
\end{aligned}$$

In the possible world $\mathcal{I}_1$, only the second axiom is satisfiable and $\mathcal{I}_3$ only satisfies the first axiom. Both axioms are satisfiable in the possible world $\mathcal{I}_2$. So

$$[C \sqsubseteq \forall R.D]_{(\mathcal{K},M)} = 0.4, [S \sqsubseteq R]_{(\mathcal{K},M)} = 0.8$$

Consequently, $M_1 \models \mathcal{K}$. And the probability that $C \sqsubseteq \forall S.D$ is 0.3(written as $\mathcal{K}, M_1 \models (C \sqsubseteq \forall S.D)^{0.3}$). Actually, the probability will be range from 0.2 to 0.4 with different probability distributions. But if we assume the independence of the terms in the knowledge base, $\mathcal{K}$ would only yield a point value 0.32.

On the other hand, under the statistical semantics, the probabilistic subsumption of this KB could be represented to $\Pr(\forall R.D|C) = 0.4$ and $\Pr(R|S) = 0.8$. But we cannot infer the value of $\Pr(\forall S.D|C)$(or we can only infer $0 \leqslant \Pr(\forall S.D|C) \leqslant 1$) according to these assertions.

2.3 Reasoning Tasks

There should be many reasoning tasks for Pr$\mathcal{SH}$ such as *Terminology-Satisfiability, Concept-Satisfiability and Concept-Subsumption,* etc. In this paper, we only focus on the most typical reasoning task, Concept-Satisfiability, whose definition is: *given a knowledge base $\mathcal{K}$ and a concept C, compute the probability that C is satisfiable with respect to $\mathcal{K}$*, i.e. compute $[C]_{(\mathcal{K},M)}$.

3 Inference Algorithm for Pr$\mathcal{SH}$

We will introduce the probabilistic extension of tableaux algorithm for the terminologies without cycles and whose certain extensions contain only *unique introductions*[9] ($(C \sqsubseteq D)^{\alpha}$ could be replaced by $(C \equiv C' \sqcap D)^{\alpha}$), called Pr-Tableaux-Algorithm, to support the inference of Pr$\mathcal{SH}$. First, we give some relative definitions.

Definition 4 (Probabilistic Transitive Reflexive Closure $\dot{\sqsubseteq}_p$). *Given an RBox $\mathcal{R}$ of Pr$\mathcal{SH}$, $R, S \in \mathbf{R}$. $R \dot{\sqsubseteq}_p S$ with respect to $\mathcal{R}$ iff $R \dot{\sqsubseteq} S$ with respect to $\mathcal{R}_e$($\mathcal{R}_e$ is the certain extension of $\mathcal{R}$). Let $\mathcal{R}_{\dot{\sqsubseteq}_p}$ denotes the set of all such relations with respect to $\mathcal{R}$.*

Definition 5 (Path P_{RS} with respect to $\mathcal{R}$). *P_{RS} with respect to $\mathcal{R}$ is a subset of $\mathcal{R}$ which should have the pattern $\{(R \sqsubseteq R_1)^{\alpha_1}, (R_1 \sqsubseteq R_2)^{\alpha_2}, ..., (R_{n-1} \sqsubseteq R_n)^{\alpha_n}, (R_n \sqsubseteq S)^{\alpha_{n+1}}\}$ where $n \geqslant 0$. P_{RS} is true when each of its element is true. Furthermore, we define PATH$_{RS}$ to be the set of all such kind of paths which start with R and end with S.*

Example 2. if $\mathcal{R} = \{(R_1 \sqsubseteq R_2)^{\alpha_1}, (R_2 \sqsubseteq R_3)^{\alpha_2}, (R_1 \sqsubseteq R_3)^{\alpha_3}\}$, $\text{PATH}_{R_1R_3} = \{P^1_{R_1R_3}, P^2_{R_1R_3}\}$ where $P^1_{R_1R_3} = \{(R_1 \sqsubseteq R_3)^{\alpha_3}\}$ and $P^2_{R_1R_3} = \{(R_1 \sqsubseteq R_2)^{\alpha_1}, (R_2 \sqsubseteq R_3)^{\alpha_2}\}$.

Then, we could define the extension of $R \dot{\sqsubseteq}_p S$ (the probability $R \dot{\sqsubseteq} S$ is true). Obviously, the probability $R \dot{\sqsubseteq} S$ is true equals the probability any path in PATH$_{RS}$ is true. Therefore $[R \dot{\sqsubseteq}_p S]_{(\mathcal{K},M)} = P(\bigvee \text{PATH}_{RS})$.

Definition 6 (Keys and Their Boolean Algebra). **K** *is a set of identifiers which contains the special elements $\bot$ and $\top$. Given a Pr$\mathcal{SH}$ knowledge base $\mathcal{K} =< \mathcal{C}, \mathcal{R}, \mathcal{A} >$, whose set of possible worlds is $\mathcal{W}$. Let $\varepsilon : \mathcal{C} \cup \mathcal{R} \cup \mathcal{A} \longrightarrow \mathbf{K} - \{\bot\}$ with following rules:*

1.$\forall (t)^\alpha \in \mathcal{C} \cup \mathcal{R} (\alpha = 1 \Longleftrightarrow \varepsilon((t)^\alpha) = \top)$
2.$\forall (t)^\alpha, (t')^\beta \in \mathcal{C} \cup \mathcal{R} (\varepsilon((t)^\alpha) = \varepsilon((t')^\beta) \Longrightarrow t = t' \vee \alpha = 1 \wedge \beta = 1)$

We extends **K** *to* **KE** *by the following rules:*

1.$\mathbf{K} \subseteq \mathbf{KE}$;
2.*if* $e_1, e_2 \in \mathbf{KE}$, $e_1 \wedge e_2 \in \mathbf{KE}$ *and* $e_1 \vee e_2 \in \mathbf{KE}$

Let $\mathcal{B}(\mathbf{KE}, \bot, \top, \{\wedge, \vee\})$ *denotes the boolean algebra over* **KE**. *Now we can extend* ε *to* $\varepsilon : \mathcal{C} \cup \mathcal{R} \cup \mathcal{A} \cup \mathcal{R}_{\dot\sqsubseteq_p} \longrightarrow \mathbf{KE} - \{\bot\}$ *with an additional rule:*

$$\varepsilon(R \dot\sqsubseteq_p S) = \bigvee_{P_{RS} \in PATH_{RS}} \bigwedge_{t \in P_{RS}} \varepsilon(t)$$

And we can define a mapping $\omega : \mathbf{KE} \longrightarrow 2^{\mathcal{W}}$ *by following rules*$(e, e_1, e_2 \in \mathbf{KE})$*:*

1.$\omega(\top) = \mathcal{W}$
2.$\omega(\bot) = \Phi$
3.$\omega(e_1 \wedge e_2) = \omega(e_1) \cap \omega(e_2)$
4.$\omega(e_1 \vee e_2) = \omega(e_1) \cup \omega(e_2)$
5.*if* $\varepsilon((t)^\alpha) = e$ *and* $(t)^\alpha \in \mathcal{C} \cup \mathcal{R} \cup \mathcal{A}$, *then* $\omega(e) = \{w | \, w \models (t)^\alpha\}$

Finally, we define the probability of an key expression $e \in \mathbf{KE}$ *as* $P(e) = \mu(\omega(e))$

Since axioms may have probabilistic factors, the expansion rule of $\mathcal{SH}$ is not enough. We should add a set Θ as the suffix to each concept name called *weight set* during the expansion process. Θ is a subset of **KE** and the elements of Θ are the keys of axioms which contribute to generating the concept they follow. The concept with weight set is called *weighted concept*. Formally,

Definition 7 (Weighted Concepts). *Given a Pr*$\mathcal{SH}$ *CBox C, let a concept name* $CN \in$ **C**. $CN \wr \Theta$ *is called* weighted concept name *where* $\Theta \subseteq$ **K**. Weighted concept descriptions*(or* weighted concepts*) in Pr*$\mathcal{SH}$ *are formed according following syntax rule:*

$$C, D \longrightarrow CN \wr \Theta | \top \wr \{\top\} | \bot \wr \{\top\} | \neg C | C \sqcap D | C \sqcup D | \exists R.C | \forall R.C$$

We consider concept name CN to be the abbreviation of $CN \wr \{\top\}$. So weighted concept is the generalization of concept. Then, we will use the weighted concepts as the basic elements during the inference. Let **WCdsc**$(\mathcal{K})$ denotes the set of all weighted concepts of knowledge base $\mathcal{K}$. Then we can define a mapping ξ : **WCdsc**$(\mathcal{K}) \rightarrow 2^{\mathbf{K}(\mathcal{K})}$ as $\xi(C) = \{e | e$ occurs in $C\}$. Furthermore, we define $P(C) = P(\bigwedge \xi(C)) = \mu(\{e | e$ occurs in $C\})$. We can see different possibility structure M may lead to different value of $P(C)$. Similarly, we could define *weighted roles* by the exactly same way. So we won't describe it here. We consider $CN \wr \Theta$ and $CN \wr \Omega$(resp. $RN \wr \Theta$ and $SN \wr \Theta$) are different weighted concepts(resp. roles) if $\Theta \neq \Omega$.

Definition 8 (Unfold-Rule). *Given a Pr*$\mathcal{SH}$ *CBox C,* $CN \wr \Theta$ *is a weighted concept names. If there exists an axiom* $(CN \equiv C)^\alpha \in C$ *whose key is e, we can unfold* $CN \wr \Theta$ *with following rules:*

1. Replace CN by C;
2. Let $\Theta' = \Theta \cup \{e\}$*;*
3. Replace each concept name DN appeared in C by $DN \wr \Theta'$*.*
4. Replace each role name RN appeared in C by $RN \wr \Theta'$*.*

Example 3. Given A CBox $C = \{(CN_1 \equiv CN_2 \sqcap CN_3)^{0.9}, (CN_2 \equiv \forall R.CN_4)^{0.3}\}$ $\varepsilon(CN_1 \equiv CN_2 \sqcap CN_3) = e_1, \varepsilon(CN_2 \equiv \forall R.CN_4) = e_2$. Concept $CN_1 \sqcup CN_4$ can be unfolded to $(\forall R \wr \{e_1, e_2\}.CN_4 \wr \{e_1, e_2\} \sqcap CN_3 \wr \{e_1\}) \sqcup CN_4 \wr \{\top\}$

We need an operator "+" between weighted concept description C and weight set Θ to simplify our expressions. We define $C + \Theta = C'$ where C' is a weighted concept description derived from C by replacing each weighted concept name $CN \wr \Omega$ appeared in C by $CN \wr (\Theta \cup \Omega)$.

Definition 9 **($R \wr \Theta$-successor).** *Given a completion tree, node y is called an $R \wr \Theta$-Successor of node x if $\mathcal{L}(< x, y >) = R \wr \Theta$.*

Given a concept D in *negation normal form* and a CBox C and an RBox $\mathcal{R}$, We assign a key to each axiom in C and $\mathcal{R}$ following Definition 6. We initialize the completion tree T with one node x and $\mathcal{L}(x) = \{D\}$. T is then expanded by repeatedly applying the rules from Table 1 *until there is no rule could be applied*.(Here we deem C, C_1, C_2 weighted concept descriptions). So the *completion trees* are slightly different from what the DLs

Table 1. The tableaux expansion rules for Pr$\mathcal{SH}$

Name	Action
$\sqcap$-rule	if $C_1 \sqcap C_2 \in \mathcal{L}(x)$,$x$ is not blocked, and $\{C_1, C_2\} \nsubseteq \mathcal{L}(x)$, then $\mathcal{L}(x) = \mathcal{L}(x) \cup \{C_1, C_2\}$
$\sqcup$-rule	if $C_1 \sqcup C_2 \in \mathcal{L}(x)$, x is not blocked, and $\{C_1, C_2\} \cap \mathcal{L}(x) = \Phi$, then $\mathcal{L}(x) = \mathcal{L}(x) \cup \{C\}$ for some $C \in \{C_1, C_2\}$
$\exists$-rule	if $\exists R \wr \Theta.C \in \mathcal{L}(x)$, x is not blocked, and x has no $R \wr \Theta$-successor y with $C \in \mathcal{L}(y)$ then, create a new node y with $\mathcal{L}(< x, y >) = R \wr \Theta$, and $\mathcal{L}(y) = \{C\}$
$\forall$-rule	if $\forall R \wr \Theta.C \in \mathcal{L}(x)$, x is not blocked, and x has an $S \wr \Omega$-successor y and $S \dot{\sqsubseteq}_p R$ and $C + (\Omega \cup \{\varepsilon(S \dot{\sqsubseteq}_p R)\}) \notin \mathcal{L}(y)$ then $\mathcal{L}(y) = \mathcal{L}(y) \cup \{C + (\Omega \cup \{\varepsilon(S \dot{\sqsubseteq}_p R)\})\}$
$\forall_+$-rule	if $\forall S \wr \Theta.C \in \mathcal{L}(x)$, x is not blocked, and there is some R with $(R \in \mathbf{R}_+)^\alpha$(let its key be e) and $R \dot{\sqsubseteq}_p S$, and an $R' \wr \Omega$-successor y and $R' \dot{\sqsubseteq}_p R$ and $\forall R.(C + \{e\} \cup \Omega \cup \{\varepsilon(R \dot{\sqsubseteq}_p S)\} \cup \{\varepsilon(R' \dot{\sqsubseteq}_p R)\}) \notin \mathcal{L}(y)$, then $\mathcal{L}(y) = \mathcal{L}(y) \cup \{\forall R.(C + \{e\} \cup \Omega \cup \{\varepsilon(R \dot{\sqsubseteq}_p S)\} \cup \{\varepsilon(R' \dot{\sqsubseteq}_p R)\})\}$
Unfold-rule	if 1.a weighted concept name $CN \wr \Theta \in \mathcal{L}(x)$(resp. $\neg CN \wr \Theta \in \mathcal{L}(x)$), x is not blocked, 2.there is an axiom $(CN \equiv C)^\alpha \in C$, $\varepsilon((CN \equiv C)^\alpha) = e$, and $C + \{\Theta \cup \{e\}\} \notin \mathcal{L}(x)$ (resp. $\sim C + \{\Theta \cup \{e\}\} \notin \mathcal{L}(x)$) then $\mathcal{L}(x) = \mathcal{L}(x) \cup \{C + \{\Theta \cup \{e\}\}\}$(resp. $\mathcal{L}(x) = \mathcal{L}(x) \cup \{\sim C + \{\Theta \cup \{e\}\}\}$)

tableau algorithms defined which would stop when encounter a clash. Finally, we could get a set of completion tree $\mathbf{T} = \{T_1, ..., T_n\}$ by $\sqcup$-rule. Now, we are able to compute the probability that concept description D is satisfiable according to these completion trees. First, we should redefine the *clash*.

Definition 10 (Possible Clash). *Let T be a completion tree for concept D. T is said to contain a* possible clash *c(written as $T \models_p c$) if for some $CN \in \mathbf{C}$ and x of T, $\{CN \wr \Theta, \neg CN \wr \Omega\} \subseteq \mathcal{L}(x)$.*

We define $\xi(c) = \Theta \cup \Omega$. Then, the probability c is true would be $P(c) = P(\bigwedge \xi(c))$.

Given a weighted concept C of knowledge base $\mathcal{K}$, let $\mathbf{T}$ is the set of all the completion trees generated by the Table 1. If there exists a completion tree $T \in \mathbf{T}$ which has no possible clash, the probability C is satisfiable is 1(written as $[D]_{(\mathcal{K},M)} = 1$). Otherwise, 2. $[D]_{(\mathcal{K},M)} = 1-$the probability that every $T \in \mathbf{T}$ has a clash. So

$$[D]_{(\mathcal{K},M)} = \begin{cases} 1 & \exists T \in \mathbf{T}.T \text{ has no possible clash} \\ 1 - P(\bigwedge_{T\in\mathbf{T}}(\bigvee_{T \models_p c} c)) & else \end{cases}$$

According to definition 10, $\bigwedge_{T\in\mathbf{T}}(\bigvee_{T\models_p c} c)$ could be translated to equivalent boolean expression of $\mathbf{KE}(\mathcal{K})$. So we could transform it to disjunction normal form(DNF) of keys. Let it be $K_1 \vee K_2 \vee \ldots \vee K_n$. If we assume the axioms are independent with each other, $P(\bigwedge_{T\in\mathbf{T}}(\bigvee\{c|T \models_p c\}))$ can be computed by the following formula

$$P(\bigwedge_{T\in\mathcal{F}}(\bigvee_{T\models_p c} c)) = P(K_1 \vee K_2 \vee \ldots \vee K_n) = \sum_{i=1}^{n}(-1)^{i-1}(\sum_{1\leqslant j_1<\ldots<j_i\leqslant n} \Pi_{k=1}^{i} P(e_{j_k}))$$

Obviously, $\bigwedge_{T\in\mathbf{T}}(\bigvee_{T\models_p c} c) = \bigvee_{i=1}^{n} \bigwedge\{c_{i1}, ..., c_{im}\}, m = |\mathbf{T}|, n = \prod_{T\in\mathbf{T}} |\{c|c \in T\}|$ where $\{c_{i1}, ..., c_{im}\}$ is a set of the possible clashes that comes from every completion tree in $\mathbf{T}$. The meaning of such kind of set is a clash composition that could make the concept unsatisfiable. The probability that any one of such kind of composition is true is the probability that the concept is unsatisfiable. We call such composition *possible clash composition*. Given a possible clash composition ψ, We define $\xi(\psi) = \bigcup_{c\in\psi} \xi(c)$.

Finally, we can prove that the probability computed by the Pr-Tableaux-Algorithm is equal to the concept extensions introduced in 2.2.

Lemma 1. *Given a PrSH knowledge base* $\mathcal{K} =< \mathcal{C}, \mathcal{R}, \mathcal{A} >$ *with the probability structure* $M =< \mathcal{W}, \mu >$, *a concept C and the set of related SH knowledge bases* $D_{\mathcal{K}} = \{\mathcal{K}_1^D, ..., \mathcal{K}_n^D\}$, *let* $D_{\mathcal{K}}^C = \{\mathcal{K}_d =< \mathcal{C}_d, \mathcal{R}_d, \mathcal{A}_d > |\mathcal{K}_d \nvDash C \wedge \mathcal{K}_d \in D_{\mathcal{K}}\}$. *Then we have* $[C]_{(\mathcal{K},M)} = 1 - P\left(\bigvee_{\mathcal{K}_d\in D_{\mathcal{K}}^C}\left(\bigwedge_{t\in\mathcal{C}_d\cup\mathcal{R}_d\wedge(t)^\alpha\in\mathcal{C}\cup\mathcal{R}} \varepsilon((t)^\alpha)\right)\right)$.

Proof. According to Definition 6, we know that for any $\mathcal{K}_d \in D_{\mathcal{K}}^C$,

$$\omega\left(\bigwedge_{t\in\mathcal{C}_d\cup\mathcal{R}_d\wedge(t)^\alpha\in\mathcal{C}\cup\mathcal{R}} \varepsilon((t)^\alpha)\right) = \{w|w \models \mathcal{K}_d\}$$

then, according to Definition 2,

$$\begin{aligned}[C]_{(\mathcal{K},M)} &= 1 - \mu(\cup_{\mathcal{K}_d\in D_{\mathcal{K}}^C}\{w|w \models \mathcal{K}_d\}) \\ &= 1 - \mu\left(\cup_{\mathcal{K}_d\in D_{\mathcal{K}}^C}\omega\left(\bigwedge_{t\in\mathcal{C}_d\cup\mathcal{R}_d\wedge(t)^\alpha\in\mathcal{C}\cup\mathcal{R}} \varepsilon((t)^\alpha)\right)\right) \\ &= 1 - \mu\left(\omega\left(\bigvee_{\mathcal{K}_d\in D_{\mathcal{K}}^C}\left(\bigwedge_{t\in\mathcal{C}_d\cup\mathcal{R}_d\wedge(t)^\alpha\in\mathcal{C}\cup\mathcal{R}} \varepsilon((t)^\alpha)\right)\right)\right) \\ &= 1 - P\left(\bigvee_{\mathcal{K}_d\in D_{\mathcal{K}}^C}\left(\bigwedge_{t\in\mathcal{C}_d\cup\mathcal{R}_d\wedge(t)^\alpha\in\mathcal{C}\cup\mathcal{R}} \varepsilon((t)^\alpha)\right)\right)\end{aligned}$$

For simplicity, we define $\xi(\mathcal{K}_d) = \{\varepsilon((t)^\alpha)|t \in \mathcal{C}_d \cup \mathcal{R}_d \wedge (t)^\alpha \in \mathcal{C} \cup \mathcal{R}\}$. Then we have $[C]_{(\mathcal{K},M)} = 1 - P(\bigvee_{\mathcal{K}_d\in D_{\mathcal{K}}^C}(\bigwedge \xi(\mathcal{K}_d))$

Lemma 2. *Given a PrSH knowledge base* $\mathcal{K} =< \mathcal{C}, \mathcal{R}, \mathcal{A} >$ *with the probability structure* $M =< \mathcal{W}, \mu >$, *a concept C and the set of related SH knowledge bases* $D_{\mathcal{K}} = \{\mathcal{K}_1^D, ..., \mathcal{K}_n^D\}$, *let* $D_{\mathcal{K}}^C = \{\mathcal{K}_d =< \mathcal{C}_d, \mathcal{R}_d, \mathcal{A}_d > |\mathcal{K}_d \nvDash C \wedge \mathcal{K}_d \in D_{\mathcal{K}}\}$. *For any* $\mathcal{K}_d \in D_{\mathcal{K}}^C$, *its clash composition* ψ_d *generated by the SH tableau algorithm has corresponding possible clash composition* ψ *generated by the PrSH tableau algorithm (which means each possible clash c in* ψ *has a corresponding clash* c_d *in* ψ_d *which only differ in their weight set) and* $\xi(\psi) \subseteq \xi(\mathcal{K}_d)$

Proof. First, we could prove that each completion tree T_d of C with respect to $\mathcal{K}_d \in D^C_{\mathcal{K}}$ which is generated by the $\mathcal{SH}$ tableau algorithm is a "sub-tree"(with same root node) of some T generated by Pr$\mathcal{SH}$ tableau algorithm without considering the weight set, since Pr$\mathcal{SH}$ tableau algorithm is the extension of the $\mathcal{SH}$ tableau algorithm. Similarly, each T generated by Pr$\mathcal{SH}$ tableau algorithm must have a corresponding T_d of C with respect to $\mathcal{K}_d$ which is a "sub-tree" of T. So each concept D occurred in each node of T_d could be found its weighted version $D \wr \Theta$ in the corresponding node of T and $\Theta \subseteq \xi(\mathcal{K}_d)$. Then each clash c_d occurred in T_d has a corresponding possible clash c in T and $\xi(c) \subseteq \xi(\mathcal{K}_d)$. Consequently, the clash composition ψ_d of T_d has a corresponding possible clash composition ψ of T and $\xi(\psi) \subseteq \xi(\mathcal{K}_d)$.

Lemma 3. *Given a Pr$\mathcal{SH}$ knowledge base $\mathcal{K} =< C, \mathcal{R}, \mathcal{A} >$ with the set of possible worlds $M =< \mathcal{W}, \mu >$, a concept C and the set of related $\mathcal{SH}$ knowledge bases $D_{\mathcal{K}} = \{\mathcal{K}^D_1, ..., \mathcal{K}^D_n\}$, let $D^C_{\mathcal{K}} = \{\mathcal{K}_d =< C_d, \mathcal{R}_d, \mathcal{A}_d > | \mathcal{K}_d \nvDash C \wedge \mathcal{K}_d \in D_{\mathcal{K}}\}$. If ψ is a possible clash composition generated by Pr$\mathcal{SH}$, then there is a related $\mathcal{SH}$ knowledge base $\mathcal{K}_d$ with $\xi(\mathcal{K}_d) = \xi(\psi)$ and $\mathcal{K}_d \in D^C_{\mathcal{K}}$.*

Proof. According to the tableau algorithm of Pr$\mathcal{SH}$ we have introduced, the set of the axioms $\rho = \{(t)^\alpha | (t)^\alpha \in C \cup \mathcal{R} \wedge \varepsilon((t)^\alpha) \in \xi(\psi)\}$ is sufficient for generating the possible clash composition ψ. So the related $\mathcal{SH}$ knowledge base $\mathcal{K}_d$ whose axioms all come from the certain extension of the axioms in ρ must be able to generate a corresponding clash composition by the $\mathcal{SH}$ tableau algorithm since it is the specialization of Pr$\mathcal{SH}$ tableau algorithm. Then, $\mathcal{K}_d \in D^C_{\mathcal{K}}$.

Theorem 1 (Correctness of the Algorithm). *The probability computed by the Pr-Tableaux-Algorithm introduced in this section is equal to the concept extensions defined in section 2.2*

Proof. According to *Lemma 1*, we only need to prove that $P(\bigvee_{\mathcal{K}_d \in D^C_{\mathcal{K}}} (\bigwedge \xi(\mathcal{K}_d)) = P(\bigvee_{\text{each } \psi \text{ of T}} \bigwedge \psi)$ where ψ is a possible clash composition.

According to *Lemma 2*, we can prove $P(\bigvee_{\mathcal{K}_d \in D^C_{\mathcal{K}}} (\bigwedge \xi(\mathcal{K}_d)) \leq P(\bigvee_{\text{each } \psi \text{ of T}} \bigwedge \psi)$

According to *Lemma 3*, we can prove $P(\bigvee_{\mathcal{K}_d \in D^C_{\mathcal{K}}} (\bigwedge \xi(\mathcal{K}_d)) \geq P(\bigvee_{\text{each } \psi \text{ of T}} \bigwedge \psi)$

4 Related Work

Halpern et al. have done much research on degrees of belief(subject probability) and statistical information(object probability)[5,12]. They mainly focus on the relation ship between these two kind of uncertainty[5,13], belief change[14,15] and probabilistic reasoning[3,10]. For description logics, Heinsohn[16] presents a probabilistic extension of the description logic $\mathcal{ALC}$, which allows to represent generic statistical information about concepts and roles, and which is essentially based on probabilistic reasoning in probabilistic logics, similar to[11,17]. Jaeger[18] gives a probabilistic extension of the description logic, which allows for generic (resp., assertional) statistical information about concepts and roles (resp., concept instances), but does not support statistical information about role instances. The uncertain reasoning formalism in [18] is essentially based on probabilistic reasoning in probabilistic logics, as the one in [16]. The work by

Koller et al. [1] gives a probabilistic generalization of the CLASSIC description logic. Like Heinsohn's work [16], which is based on inference in Bayesian networks as underlying probabilistic reasoning formalism. Giugno presents a probabilistic extension of $\mathcal{SHOQ}$[2], which allows to represent generic statistical knowledge about concept and roles and the assertional statistical knowledge about concept and role instance. Baader extends Description Logics with modal operators in [19]to describe belief but not degree of belief.

5 Conclusion and Future Work

We have presented a probabilistic version of description logics–Pr$\mathcal{SH}$ which are used to represent the uncertainty of the axioms and assertions in the knowledge base. We also introduce an inference algorithm for Pr$\mathcal{SH}$ to discover the possible implicit knowledge. In future, we will improve our work in two aspects. First, develop the inference engine for Pr$\mathcal{SH}$. Second, combine Pr$\mathcal{SH}$ with other probabilistic description logics describing statistical information.

References

1. Koller, D., Levy, A.Y., Pfeffer, A.: P-CLASSIC: A tractable probablistic description logic. In: AAAI/IAAI. (1997) 390–397
2. Giugno, R., Lukasiewicz, T.: P-SHOQ(D): A probabilistic extension of SHOQ(D) for probabilistic ontologies in the semantic web. In Flesca, S., Greco, S., Leone, N., Ianni, G., eds.: JELIA. Volume 2424 of Lecture Notes in Computer Science., Springer (2002) 86–97
3. Halpern, J.Y.: An analysis of first-order logics of probability. In: IJCAI. (1989) 1375–1381
4. I.Hacking: Logic of Statistical Inference. Cambridge University Press, Cambridge, U.K. (1965)
5. Bacchus, F., Grove, A.J., Halpern, J.Y., Koller, D.: From statistical knowledge bases to degrees of belief. Artif. Intell **87**(1-2) (1996) 75–143
6. Noy, N.F.: Semantic integration: A survey of ontology-based approaches. SIGMOD Record **33**(4) (2004) 65–70
7. Rahm, E., Bernstein, P.A.: A survey of approaches to automatic schema matching. VLDB J **10**(4) (2001) 334–350
8. Schmidt-Schauß, M., Smolka, G.: Attributive concept descriptions with complements. Artificial Intelligence **48**(1) (1991) 1–26
9. Pan, J.Z.: Description Logics: Reasoning Support for the Semantic Web. PhD thesis, School of Computer Science, The University of Manchester, Oxford Rd, Manchester M13 9PL, UK (2004)
10. Fagin, R., Halpern, J.Y., Megiddo, N.: A logic for reasoning about probabilities. Inf. Comput **87**(1/2) (July/August 1990) 78–128
11. Nilsson, N.: Probabilistic logic. ai **28** (1986) 71–87
12. Halpern, J.Y.: Reasoning about knowledge: A survey. In Gabbay, D.M., Hogger, C.J., Robinson, J.A., eds.: Handbook of Logic in Artificial Intelligence and Logic Programming, Volume 4: Epistemic and Temporal Reasoning. Clarendon Press, Oxford (1995) 1–34
13. Fagin, R., Halpern, J.Y.: Uncertainty, belief, and probability. In: IJCAI. (1989) 1161–1167
14. Friedman, N., Halpern, J.Y.: Modeling belief in dynamic systems, part I: Foundations. Artif. Intell **95**(2) (1997) 257–316

15. Friedman, N., Halpern, J.Y.: Modeling belief in dynamic systems, part II: Revision and update. J. Artif. Intell. Res. (JAIR) **10** (1999) 117–167
16. Heinsohn, J.: Probabilistic description logics. In de Mantaras, R.L., Poole, D., eds.: Proceedings of the 10th Conference on Uncertainty in Artificial Intelligence, San Francisco, CA, USA, Morgan Kaufmann Publishers (July 1994) 311–318
17. Lukasiewicz, T.: Probabilistic deduction with conditional constraints over basic events. In: KR. (1998) 380–393
18. Jaeger, M.: Probabilistic reasoning in terminological logics. In: KR. (1994) 305–316
19. Baader, F., Laux, A.: Terminological logics with modal operators. In: IJCAI (1). (1995) 808–815

Reinforcement Learning on a Futures Market Simulator

Koichi Moriyama, Mitsuhiro Matsumoto, Ken-ichi Fukui,
Satoshi Kurihara, and Masayuki Numao

The Institute of Scientific and Industrial Research, Osaka University.
8-1, Mihogaoka, Ibaraki, Osaka, 567-0047, Japan
koichi@ai.sanken.osaka-u.ac.jp

Abstract. In recent years, it becomes vigorous to forecast a market by using machine learning methods. Since they assume that each trader's individual decisions do not affect market prices at all, most existing works use a past market data set. Meanwhile there is an attempt to analyze economic phenomena by constructing a virtual market simulator, where human and artificial traders really make trades. Since prices in the market are determined by every trader's decisions, it is more realistic and the assumption cannot be applied any more. In this work, we design and evaluate several reinforcement learners on a futures market simulator U-Mart (Unreal Market as an Artificial Research Testbed). After that, we compare our learner to the previous champions of U-Mart competitions.

1 Introduction

In recent years, it becomes vigorous to forecast a market by using machine learning methods. In reinforcement learning domain, there are several works using real market data, e.g., [1,2]. Most of them divide a *past* market data set into training and test sets, learn a strategy by the training one, and verify the result using the test one. This is because they assume that each trader's individual decisions do not affect the market at all since there are too many traders in the market. This assumption allows a learner to learn passively, that is, a learner is able to learn only from a data set. In such a passive learning domain, we can use offline, batch learning methods.

Some researchers in economics are interested in a virtual market. Such a virtual market, where human and/or artificial traders trade virtual stocks, provides the researchers a tool to analyze interesting phenomena in a market that current economic theories cannot explain, and/or dissect the market structure itself. Since prices in a virtual market are determined by every trader's decisions, it is more realistic and the assumption described above is not applied any more. Therefore, a learner has to learn actively because the learner's strategy changes the market and vice versa. Reinforcement learning is originally suitable for such an active domain.

In this work, we design and evaluate several reinforcement learners, each of which has a different state space, on a futures market simulator U-Mart (Unreal Market as an Artificial Research Testbed) [3,4,5]. After that, we compare our learner to the previous champions of U-Mart competitions.

This paper consists of five sections. Section 2 explains what are futures and what is U-Mart. In Sect. 3, we design several reinforcement learners each of which has a

N.T. Nguyen et al. (Eds.): KES-AMSTA 2007, LNAI 4496, pp. 42–52, 2007.

different state space. In Sect. 4, we evaluate our learners and compare one of them to the past champions. Finally, we conclude this paper in Sect. 5.

2 U-Mart: A Futures Market Simulator

Futures are agreements of trading something at some time in the future (*due date*) at the price decided now. It is an example of futures that two persons agree that the buyer buys gold at $15 per gram from the seller a year later. The futures make it easy to plan to do something that uses gold a year later because both the buyer and the seller do not need to be annoyed by the change of gold price. Therefore, futures are needed to manage risks in market prices, called *spot prices*. However, if the spot price of gold becomes $20 per gram on the due date, the buyer gets a gain because she has to pay only $15 to get $20 worth of gold. On the other hand, the buyer suffers a loss if the spot price becomes $10. Hence, futures themselves have values depending on spot prices, and therefore, they are traded at *futures prices* in a *futures market*.

U-Mart (Unreal Market as an Artificial Research Testbed) [3,4,5] is a futures market simulator dealing with the stock index J30. J30 was the average of stock prices of thirty big companies in Japan at Tokyo Stock Exchange and was announced by the Mainichi Newspapers until 2003. J30 had no real futures market, however, traders in U-Mart trade J30 futures (*J30F*) based on J30 spot (*J30S*) prices. Although futures prices generally influence spot prices in a true market, they in U-Mart do not affect spot prices because U-Mart deals with only futures trades based on existing, but unknown to traders in advance, spot prices.

In U-Mart, each trader initially has a certain amount of money and tries to maximize it by trading J30F. The whole simulation time, which corresponds to the time until the due date, is divided into several *days* and each day is divided into several *intervals*. In each interval, traders send buying and/or selling orders with the price and volume to the exchange. The exchange records them to an *order book*. At the end of the interval, the exchange plots a price-volume graph with demand and supply curves from the order book. The crossing point of the curves shows the price and volume of being executed, that is, the selling orders at the left of the point and the buying orders at the right are executed at the price. See [4] for details. Every trader whose orders have been executed changes her *position*, which is the difference between the amount of futures she has bought and has sold. Every trader whose position is not zero has to deposit some margin, depending on her position, to the exchange.

At the end of the day, the exchange calculates unrealized profit/loss of all traders by the closing price of J30F and clears it by paying/receiving money to/from traders, respectively. If a trader is not able to pay the unrealized loss, she becomes a bankrupt and drops out of the simulation. At the end of the simulation, all positions are cleared by the first J30S price of the next day and all traders are ranked by several metrics.

3 Reinforcement Learning on U-Mart

In this section, we consider how a reinforcement learner, called *agent*, gets gains in U-Mart simulations. Generally, to get gains, a trader has to sell at a higher price than that at which she bought, or buy at a lower price than that at which she sold. Suppose an agent sent a buying order in the t-th interval, i.e., between the time $t-1$ and t, and the order has been executed at t. The agent gets a gain if the price at $t+1$ becomes higher than that at t. Therefore, to get gains, an agent has to predict the transition of J30F price in the near future. In this work, the agent learns to predict the transition by Q-learning [7], a representative reinforcement learning algorithm.

3.1 Q-Learning

Suppose an agent senses a state $s_t \in \mathcal{S}$ and selects an action $a_t \in \mathcal{A}(s_t)$ at a discrete time t. $\mathcal{S}$ is a set of possible states in the environment and $\mathcal{A}(s_t)$ is a set of possible actions in the state s_t. After selecting an action, it receives a reward $r_{t+1} \in \mathbb{R}$ and senses a new state s_{t+1}. Q-learning updates an *action value function* Q by the following rules to make it approach the true value under the optimal policy π^*, which is the expected sum of rewards discounted by $0 < \gamma < 1$ under π^*, i.e., $E_{\pi^*}(\sum_{k=0}^{\infty} \gamma^k r_{t+1+k})$.

$$Q_t(s,a) = \begin{cases} Q_{t-1}(s_t,a_t) + \alpha\,\delta_t & \text{if } (s,a) = (s_t,a_t), \\ Q_{t-1}(s,a) & \text{otherwise.} \end{cases} \tag{1}$$

$0 < \alpha \leq 1$ is a parameter called a learning rate and δ_t is called a TD error that approaches 0 when $Q_t(s,a)$ approaches the true value of (s,a) under π^*:

$$\delta_t \triangleq r_{t+1} + \gamma \max_{a \in \mathcal{A}(s_{t+1})} Q_{t-1}(s_{t+1},a) - Q_{t-1}(s_t,a_t). \tag{2}$$

For all s and a, $Q_t(s,a)$ is proved to converge to the true value under the optimal policy when the environment has the Markov property, the agent visits to all states and takes all actions infinitely, and the learning rate α is decreased properly [7].

If the true value function under the optimal policy, Q^*, is known, the agent can choose an optimal action a^* in a state s from Q^* by

$$a^* = \arg\max_{a' \in \mathcal{A}(s)} Q^*(s,a'). \tag{3}$$

However, if the agent always chooses such actions in process of learning, Q_t may converge to a local optimum because the agent may not visit all states. To avoid it, the agent usually uses a stochastic method like *softmax* [8] for action choices. Softmax method calculates action choice probabilities $p_{s_t}(a)$, where $a \in \mathcal{A}(s_t)$, by

$$p_{s_t}(a) \triangleq \frac{\exp(Q_{t-1}(s_t,a)/T)}{\sum_{a' \in \mathcal{A}(s_t)} \exp(Q_{t-1}(s_t,a')/T)}\,. \tag{4}$$

$T > 0$ is called a temperature that controls the effect of randomness.

3.2 Learning to Predict Price Transition on U-Mart

As we saw in Sect. 2, an order that an agent makes, which is either "sell" or "buy", consists of the price and volume. In this work, however, we only discuss about how the learning agent determines to either "sell", "buy", or "do nothing". To do it, the agent learns to predict the price transition by Q-learning. The agent will simply determine the order price from the price at $t-1$ and the order volume from the softmax probability.

Here we compare the following three types of state space of Q-learning:

- Transition of J30F price as states,
- Transition of J30S price as states, and
- Spread between J30F and J30S prices as states.

We see them in this order. Hereafter, $F(t)$ and $S(t)$ stand for "J30F price at t" and "J30S price at t", respectively. Price transition functions $\Delta F : \mathbb{N} \rightarrow \{up, same, down\}$ and $\Delta S : \mathbb{N} \rightarrow \{up, same, down\}$ are defined as follows.

$$\Delta F(t) \triangleq \begin{cases} up & \text{if } F(t) > F(t-1), \\ same & \text{if } F(t) = F(t-1), \\ down & \text{otherwise.} \end{cases} \quad (5)$$

$$\Delta S(t) \triangleq \begin{cases} up & \text{if } S(t) > S(t-1), \\ same & \text{if } S(t) = S(t-1), \\ down & \text{otherwise.} \end{cases} \quad (6)$$

Transition of J30F Price as States (Approaches 1–1 and 1–2). The approaches described here use the transition of J30F price until now (t-th interval). The agent predicts $\Delta F(t+1)$, i.e., J30F price transition from t to $t+1$. It is divided into two types by ways of prediction.

Approach 1–1: The agent learns to predict $\Delta F(t)$ from the past two transitions of J30F price; $\Delta F(t-2)$ and $\Delta F(t-1)$ (Fig. 1(a), Left). Q-function has nine *states*, each of which is a combination of the transitions; (*up*, *up*), (*up*, *same*), (*up*, *down*), etc., and two *actions*, *up* and *down*, as a prediction of $\Delta F(t)$. In order to predict $\Delta F(t+1)$ through the Q-function, the agent regards the predicted $\Delta F(t)$ as a real transition and applies it with $\Delta F(t-1)$ to its Q-function (Fig. 1(a), Right). If the predicted $\Delta F(t+1)$ is *up*, the order is "buy", otherwise "sell". In the $(t+1)$-th interval, the agent knows $F(t)$ and calculates true $\Delta F(t)$. If true $\Delta F(t)$ is *same*, the *reward* is 0. If it is equal to the *action*, the *reward* is a positive value. Otherwise, it is a negative one.

Approach 1–2: The agent learns to predict $\Delta F(t+1)$ directly from the past two transitions of J30F price (Fig. 1(b)). Q-learning has identical *states* and *actions* with Approach 1–1, but an *action* shows a prediction of $\Delta F(t+1)$ instead of $\Delta F(t)$. If the *action* is *up*, the order is "buy", otherwise "sell". If true $\Delta F(t+1)$ is *same*, the *reward* is 0. If it is equal to the *action*, the *reward* is a positive value. Otherwise, it is a negative one. It is different from Approach 1–1 that this approach does not consider $\Delta F(t)$ at all. As the agent knows true $\Delta F(t+1)$ in the $(t+2)$-th interval, the *reward* is delayed.

Transition of J30S Price as States (Approach 2). Generally, we can infer that a futures price will go up when the present spot price is high and the present futures price is lower than it. The approach described here uses the transition of J30S price until now (t-th interval) and the inference. The agent learns to predict $\Delta S(t)$ from the past two transitions of J30S price; $\Delta S(t-2)$ and $\Delta S(t-1)$ (Fig. 1(c)). Q-function has nine *states*, each of which is a combination of the transitions, and two *actions*, *up* and *down*, as a prediction of $\Delta S(t)$. The order is determined by the rules in Table 1 using the *action*,

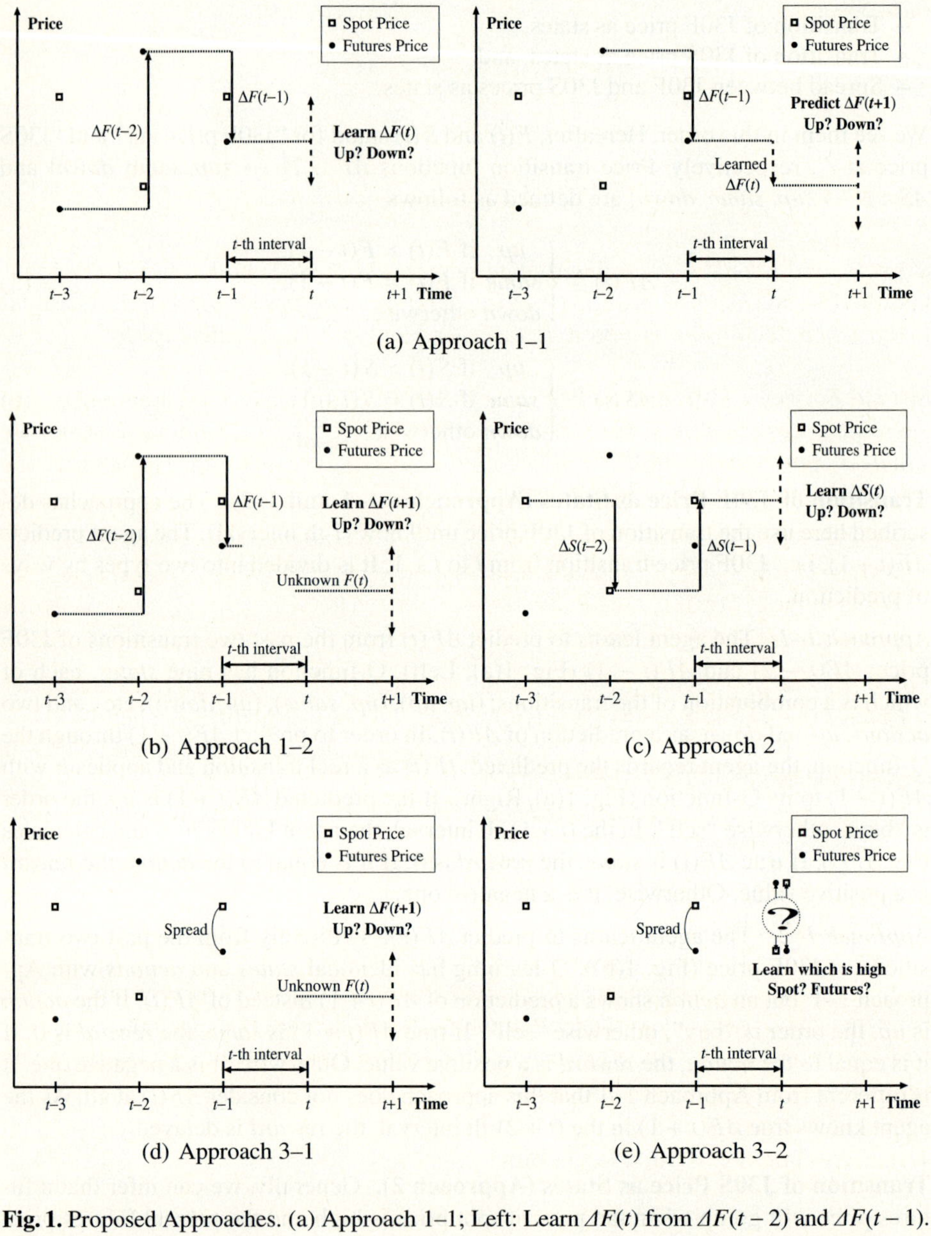

Fig. 1. Proposed Approaches. (a) Approach 1–1; Left: Learn $\Delta F(t)$ from $\Delta F(t-2)$ and $\Delta F(t-1)$. Right: Predict $\Delta F(t+1)$ from $\Delta F(t-1)$ and the learned $\Delta F(t)$. (b) Approach 1–2; Learn $\Delta F(t+1)$ directly from $\Delta F(t-2)$ and $\Delta F(t-1)$. (c) Approach 2; Learn $\Delta S(t)$ from $\Delta S(t-2)$ and $\Delta S(t-1)$. (d) Approach 3–1; Learn $\Delta F(t+1)$ from the spread between $F(t-1)$ and $S(t-1)$. (e) Approach 3–2; Learn which $F(t)$ or $S(t)$ is high from the spread between $F(t-1)$ and $S(t-1)$.

Table 1. Rule to determine an order in Approach 2. The first column is the *action*. The second shows the relation between $F(t-1)$ and $S(t-1)$. The third is the order.

Action	$F(t-1)$	Order
up	$< S(t-1)$	Buy
up	$\geq S(t-1)$	Do nothing
down	$\leq S(t-1)$	Do nothing
down	$> S(t-1)$	Sell

$S(t-1)$, and $F(t-1)$, according to the inference. In the $(t+1)$-th interval, the agent knows $S(t)$ and calculates true $\Delta S(t)$. If true $\Delta S(t)$ is *same*, the *reward* is 0. If it is equal to the *action*, the *reward* is a positive value. Otherwise, it is a negative one.

Spread Between J30F and J30S Prices as States (Approaches 3–1 and 3–2). The approaches described here use the spread between J30F and J30S prices. The spread, which is continuous, is discretized and used as *states* of Q-learning. The approach is divided into two types by what the agent learns.

Approach 3–1: The agent learns to predict $\Delta F(t+1)$ by the spread between $F(t-1)$ and $S(t-1)$ (Fig. 1(d)). Q-function has two *actions*, *up* and *down*, as a prediction of $\Delta F(t+1)$. If the *action* is *up*, the order is "buy", otherwise "sell". The difference between Approaches 1–2 and 3–1 is only the state space.

Approach 3–2: The agent learns to predict which $F(t)$ or $S(t)$ is high from the spread between $F(t-1)$ and $S(t-1)$ (Fig. 1(e)). An *action* is a prediction of higher price at t; $F(t)$ or $S(t)$. If the *action* is $S(t)$, the order is "buy", otherwise "sell". In the $(t+1)$-th interval, the agent knows true $F(t)$ and $S(t)$. If true $F(t)$ is equal to true $S(t)$, the *reward* is 0. If the true relation is equal to the *action*, the *reward* is a positive value. Otherwise, it is a negative one.

4 Experiments

In this section, we see the results of experiments. We conducted two experiments. First is to see which approach is the best in the proposal. Second is to see how much the best approach is effective compared with other known good strategies, i.e., the previous champions of U-Mart competitions.

4.1 Experiment 1: Which Is the Best?

We see the result of experiment to know which approach is the best in the proposal.

Setup. We used J30S price data provided with the U-Mart simulator, which contained 2444 records being between 1532 and 4778 Japanese yen. The duration of each simulation was thirty days and each day had eight intervals, in each of which every agent decided orders using the histories of J30F and J30S prices until the previous interval.

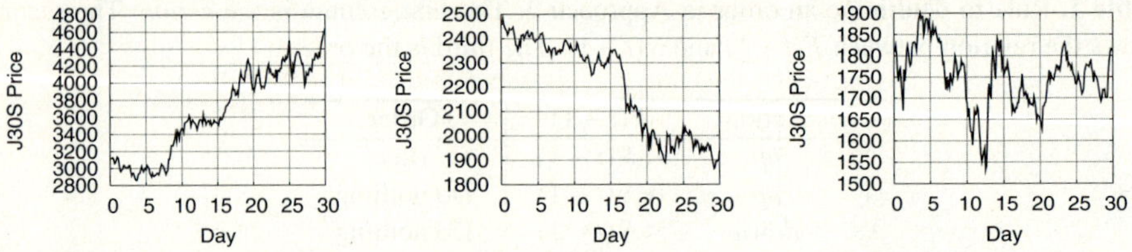

(a) Ascending series (b) Descending series (c) Oscillating series

Fig. 2. Three J30S price patterns in the experiment

Table 2. Experiment 1: (a) Profit of each approach (million yen). (b) Precision of actions in each approach (%).

(a) Profit (million yen)

App.	Ascend.	Descend.	Oscillat.
1–1	120.9	−2.0	−79.5
1–2	377.6	−48.5	−68.5
2	407.2	116.4	−17.7
3–1	416.5	69.1	3.8
3–2	440.2	98.9	47.2

(b) Precision (%)

App.	Ascend.	Descend.	Oscillat.
1–1	59.4	61.3	60.5
1–2	49.4	42.9	49.8
2	51.4	49.3	50.3
3–1	57.1	43.7	49.2
3–2	71.0	62.5	65.6

Twenty agents joined in the experiment. The participants were:

- Nineteen sample agents of ten types already built in the U-Mart simulator, and
- One of the proposed agents (Approach 1–1, 1–2, 2, 3–1, or 3–2).

All of the sample agents did not learn anything and decided what to do by fixed strategies. The distribution of sample agents and the strategies each had are described in Appendix A. In the experiment, we used three J30S price patterns (ascending, descending, and oscillating series) (Fig. 2), and every agent initially had one billion yen. We see how much each proposed agent obtained profit under these patterns. *Note that, even if spot prices are identical in simulations, futures prices we consider here become different in each simulation because of randomness in the agents.*

The proposed agent, if the order was to buy in the t-th interval, ordered p_{s_t} (the softmax probability (4)) $\times$ 100 trading units, each of which was 1000 shares, at $F(t-1)+$ 20 yen. If the order was to sell, the agent ordered $p_{s_t} \times 100$ trading units at $F(t-1)-20$ yen. For Q-learning, we set the *reward* +1 when the *action* was right and −1 when wrong. In Approaches 3–1 and 3–2, the spread was discretized into 20 sections, that is, 18 sections in increments of 5 from −45 to 45, more than 45, and less than −45. The learning rate α and the discount factor γ were 0.1 and 0.9, respectively. The temperature T in (4) was set to 1. Before starting the experiment, the agent learned a policy in 100 simulations with the 19 sample agents and randomly chosen J30S price patterns from the data. The agent continued to learn in the experiment.

Table 3. Experiment 2: Profit of each strategy (million yen)

Strategy	Ascending	Descending	Oscillating
App. 3–2	90.7	83.0	25.0
Op	−141.0	−51.4	−5.3
Ns	178.9	−14.1	43.3
Ts	351.2	35.0	24.4
OFB	19.4	1077.9	−332.3
TDP	647.0	−284.2	4.1

Result. The profit each approach obtained is in Table 2(a), which is the average of ten simulations. While Approaches 1–1, 1–2, and 2 got losses in some series, Approaches 3–1 and 3–2 got gains in all of the series. So, they are better than the rest.

Analysis of the Result. To analyze the result, we see the precision of *actions* in each approach in Table 2(b). It is also the average of ten simulations. It shows that, especially in the oscillating series, actions by Approaches 1–2, 2, and 3–1 were almost random. It means they failed to learn. On the other hand, it seems that Approach 1–1 succeeded in predicting $\Delta F(t)$ in some degree, but it failed to get gains. It is because Approach 1–1 had to succeed in predicting in two consecutive time ($\Delta F(t)$ and $\Delta F(t+1)$) to get gains. Actually, although the precision is around 60% in all series, it fails to predict $\Delta F(t + 1)$ in more time because $0.6^2 < 1/2$. While Approach 3–1 failed to learn, it got gains in all series. It might be because the loss of failure was less than the gain of success. Since the precision of Approach 3–2 was the largest of all and it got gains more than Approach 3–1, Approach 3–2 is the best in the proposal.

4.2 Experiment 2: How Much Effective with Others?

To know how much Approach 3–2 is effective compared with other good strategies, we conducted another experiment with five other good strategies. They were the champions of the international competition of U-Mart, UMIE (U-Mart International Experiment) in 2004 and 2005. Op, Ns, and Ts were in 2005 and OFB and TDP in 2004. The details of them are described in Appendix B.

Setup. Most of the experimental settings were identical with Experiment 1. The participants were twenty-five agents: five champion agents described above, nineteen sample agents same as Experiment 1, and Approach 3–2. Approach 3–2 initially had a Q-function at the end of Experiment 1.

Result. The profit each strategy obtained is in Table 3, which is also the average of ten simulations. We can read from the table that only Approach 3–2 and Ts were able to get gains stably. While Approach 3–2 was worse than Ts in the ascending series, the result showed an unexpected success of Approach 3–2 since it was a very simple algorithm being not so tuned compared with the champions.

5 Conclusion

In this work, we proposed several learners using Q-learning on a futures market simulator U-Mart. First, in Sect. 2, we saw what are futures and what is U-Mart. In Sect. 3, we proposed three types of learners, which had different state spaces of Q-learning, i.e., the transition of futures price, that of spot price, and the spread between futures and spot prices. We saw the results of experiments with U-Mart sample agents and the champions of past U-Mart competitions in Sect. 4. The approaches using the spread as states obtained gains in all price patterns in Experiment 1. From the point of precision, Approach 3–2 was the best. It also obtained good result in a simulation with the champions of past U-Mart competitions. Although Approach 3–2 was not the best of all, the result showed an unexpected success of Approach 3–2 since it was a very simple algorithm being not so tuned compared with the champions. We are intentionally ignoring a multiagent perspective in this work, because nobody can perceive opponents' actions in a market that are usually required in multiagent reinforcement learning.

There are several future directions. First, we are planning to participate in the competition. In this work, we used the champions but they were *past* ones. No one knows if the proposed approaches behave properly and obtain profit with new entrants of the competition. Second, we have to tune up the proposal. It is a challenge to combine reinforcement learning and various knowledge in economics, speculation, etc. Third, we have to check whether the approach is applicable in a massive simulation that contains immense number of traders. However, the present U-Mart simulator lacks scalability at all. Therefore, we have to start from scaling up the simulator.

References

1. O, J., Lee, J.W., Zhang, B.T.: Stock Trading System Using Reinforcement Learning with Cooperative Agents. In: Proc. ICML-2002 (2002) 451–458
2. Nevmyvaka, Y., Feng, Y., Kearns, M.: Reinforcement Learning for Optimized Trade Execution. In: Proc. ICML-2006 (2006)
3. U-Mart Project. `http://www.u-mart.org/`.
4. Kita, H.: An Introduction to U-Mart. (2000) (In the U-Mart developer's kit [3]).
5. Sato, H., Koyama, Y., Kurumatani, K., Shiozawa, Y., Deguchi, H.: U-Mart: A Test Bed for Interdisciplinary Research in Agent Based Artificial Market. In Aruka, Y., ed.: Evolutionary Controversies in Economics. Springer, Tokyo (2001) 179–190
6. Kitano, H., Nakashima, T., Ishibuchi, H.: Behavior Analysis of Futures Trading Agents Using Fuzzy Rule Extraction. In: Proc. 2005 IEEE SMC Conference (2005) 1477–1481
7. Watkins, C.J.C.H., Dayan, P.: Q-learning. Machine Learning **8** (1992) 279–292
8. Sutton, R.S., Barto, A.G.: Reinforcement Learning. MIT Press, Cambridge, MA (1998)

A U-Mart Sample Agents

These are U-Mart sample agents in the developer's kit [3]. The number after the agent name is how many agents were used in the experiments. If not otherwise specified, the order price is $F(t-1) + 20x$ yen, where $x \sim N(0,1)$, and the order volume is randomly determined between two parameters. See [3] for details.

Trend (1): Buy if $F(t-1) > F(t-2)$, sell if $F(t-1) < F(t-2)$, otherwise do nothing.

AntiTrend (2): Swap the conditions of Trend.

Random (1) and SRandom (3): An order is randomly determined. SRandom uses $S(t-1)$ instead of $F(t-1)$ to calculate the order price.

Rsi (1) and SRsi (3): Rsi uses RSI (Relative Strength Index) of J30F in t-th interval:

$$\frac{\sum_{\tau<t, F(\tau)>F(\tau-1)} F(\tau) - F(\tau-1)}{\sum_{\tau<t} \mid F(\tau) - F(\tau-1) \mid} .$$

Buy if it is less than 0.3, sell if more than 0.7, otherwise do nothing. SRsi uses RSI of J30S instead of J30F.

MovingAverage (1) and SMovingAverage (3): Calculate the moving average of J30F in ten intervals. Buy if it is more than $F(t-1)$, sell if less than $F(t-1)$, otherwise do nothing. SMovingAverage uses J30S instead of J30F.

SFSpread (2): Calculate a spread ratio by $\{F(t-1) - S(t-1)\}/S(t-1)$. Buy if it is less than or equal to -0.01, sell if more than or equal to 0.01, otherwise do nothing. The order price is $\{F(t-1) + S(t-1) + x|F(t-1) - S(t-1)|\}/2$, where $x \sim N(0,1)$.

DayTrade (2): Always buy and sell simultaneously. Buying and selling prices are $0.99F(t-1)$ and $1.01F(t-1)$, respectively.

B Champions of Past UMIE

These are the champions of past UMIE. Op, Ns, and Ts were the champions of UMIE 2005 and OFB and TDP were of UMIE 2004.

Osako_pivot (Op): It calculates the average of the maximum price of the previous day, the minimum price of that day, and the final price of that day as a pivot. Then, it sets two support lines and resistance lines from the pivot. It decides an order by the relation among futures price, the support lines, and the resistance lines. The volume is always 100 units and the price is $F(t-1) + 50$ yen for buying and $F(t-1) - 50$ yen for selling.

Nakamura_spread (Ns): It simultaneously sends three orders, which are an arbitrage order, an order preparing for bulge, and that for collapse. For the arbitrage order, it decides an order by the spread between $F(t-1)$ and $S(t-1)$. The volume is also determined by the spread and the price is the average of $F(t-1)$ and $S(t-1)$. For the order for bulge/collapse, it sends a sell/buy order at the price 200 yen higher/lower than $S(t-1)$, respectively. The volume is 400 units each.

Trend_swift (Ts): If $S(t-1) > S(t-2)$ and $S(t-1) - F(t-1) > 10$ yen, it buys at $F(t-1) + 40$ yen. If $S(t-1) < S(t-2)$ and $F(t-1) - S(t-1) > 10$ yen, it sells at $F(t-1) - 40$ yen. The volume depends on the spread.

OPUFuzzyB (OFB) [6]: It is a fuzzy rule based on-line learning agent that has a table containing weights for rules. First it calculates spreads $S(t-1)-S(t-2), S(t-1)-S(t-4)$, and $S(t-1) - S(t-6)$. Then it calculates fuzzy reliabilities from three membership functions, each of which is a function of one of the spreads. After that, it decides an order by the spreads and the fuzzy reliabilities. The volume is always 200 units and the price is $S(t-1) - 5$ yen for buying and $S(t-1) + 5$ yen for selling. In the next interval, it updates the weights based on the fuzzy reliabilities.

TriDiceP (TDP): It has a state-action table like Q-learning to decide orders. The states in the table is determined by linearized past spot prices. Orders are chosen by probabilities calculated from weights in the table. The weight of chosen order in the table is decreased to change behaviors. The volume is calculated from the order probabilities and the price is $S(t-1) + 10$ yen for buying and $S(t-1) - 10$ yen for selling.

The Effects of Agent Synchronization in Asynchronous Search Algorithms

Ionel Muscalagiu[1], Jose M. Vidal[2], Vladimir Cretu[3], Popa Horia Emil[4], and Manuela Panoiu[1]

[1] The "Politehnica" University of Timisoara, The Faculty of Engineering of Hunedoara, Revolutie, 5, Romania
{mionel,m.panoiu}@.fih.utt.ro

[2] University of South Carolina, Computer Science and Engineering, Columbia SC 29208, USA
vidal@sc.edu

[3] The "Politehnica" University of Timisoara, Computers Science and Engineering Department, Timisoara, V. Parvan 2, Romania
vcretu@cs.utt.ro

[4] The University of the West, The Faculty of Mathematics and Informatics, Timisoara, V. Parvan 4, Romania
hpopa@info.uvt.ro

Abstract. The asynchronous searching techniques are characterized by the fact that each agent instantiates its variables in a concurrent way. Then, it sends the values of its variables to other agents directly connected to it by using messages. These asynchronous techniques have different behaviors in case of delays in sending messages. This article depicts the opportunity for synchronizing agents' execution in case of asynchronous techniques. It investigates and compares the behaviors of several asynchronous techniques in two cases: agents process the received messages asynchronously (the real situation from practice) and the synchronous case, when a synchronization of the agents' execution is done i.e. the agents perform a computing cycle in which they process a message from a message queue. After that, the synchronization is done by waiting for the other agents to finalize the processing of their messages. The experiments show that the synchronization of the agents' execution leads to lower costs in searching for solution. A solution for synchronizing the agents' execution is proposed for the analyzed asynchronous techniques.

1 Introduction

Constraint programming is a programming approach used to describe and solve large classes of problems such as searching, combinatorial, planning problems, etc. Lately, the AI community has shown increasing interest in the distributed problems that are solvable through modeling by constraints and agents. The idea of sharing various parts of the problem among agents that act independently and collaborate among themselves to find a solution by using messages proves itself

N.T. Nguyen et al. (Eds.): KES-AMSTA 2007, LNAI 4496, pp. 53–62, 2007.

useful. It was also lead to the formalized problem known as the Distributed Constraint Satisfaction Problem (DCSP) [4].

There exist complete asynchronous searching techniques for solving the DCSP, such as the ABT (Asynchronous Backtracking) and DisDB (Distributed Dynamic Backtracking) [1,4]. There is also the AWCS (Asynchronous Weak-Commitment Search) [4] algorithm which records all the nogood values. This technique allows the agents to modify a wrong decision by a dynamic change of the agent priority order. The technique proved beneficial for AWCS. The ABT algorithm has also been generalized by presenting a unifying framework, called ABT kernel [1]. From this kernel two major techniques ABT and DisDB can be obtained.

Asynchronous algorithms are characterized by a message passing mechanism among agents when searching for solution. There are several types of messages used to announce and change the local values attributed to the caretaking of variables. Any practical implementation of these techniques need to manipulate the FIFO channels of messages. These asynchronous techniques have different behaviors in case of delays in sending messages. Which thus leads to different behaviors if we synchronize the execution of the agents.

It is intereseting to examine how these algorithms behave under different synchonization assumptions, as this type of analysis has not been done before. Specifically, it is interesting to investigate the opportunity of synchronizing the agents in case of asynchronous techniques. The behaviors of several asynchronous techniques are investigated in two cases: the agents execute asynchronously the processing of received messages (the real situation from practice) and the synchronous case where the agents' execution is synchronized. In other words, the agents perform a computing cycle in which they process a message from a message queue in the synchronous case. After that, a synchronization is done waiting for the other agents to finalize the processing of their messages. The experiments show that the synchronization of the agents' execution reduces the costs in finding the solution for several families of asynchronous techniques. Two solutions of the agents synchronization of execution are proposed. The first one is based on the existence of a central agent or the access possibility to a common memory zone, solution that allows complete synchronization of the agents. The second one, based on the use of a synchronization message between neighboring agents, allows us to obtain a partial synchronization of the agents.

In this article two families of asynchronous techniques are analyzed: the AWCS and the ABT families. The AWCS family [2,4] uses a dynamical order for agents. Learning techniques can be applied to this family for building efficient nogoods. This article analyzes a variant of this family improved by applying a nogood learning technique. The second family analyzed is the ABT family [1,4]. It uses a statical order between agents. The ABT techniques and DisDB are considered from this family. The last technique remarks itself by the fact that it does not require additional links during the search process. The DisDB technique is recommended to be used in real situations in which additional links cannot be added for various external reasons.

2 The Framework

This section presents theoretical considerations regarding the DCSP modeling and the asynchronous techniques[1], [2], [4].

2.1 The Distribution Constraint Satisfaction Problem

Definition 1 (CSP model). *The model based on constraints CSP-Constraint Satisfaction Problem, existing for centralized architectures, consists in:*

-n variables $X_1, X_2, ..., X_n$, whose values are taken from finite, discrete domains $D_1, D_2, ..., D_n$, respectively.

-a set of constraints on their values.

The solution of a CSP supposes to find an association of values to all the variables so that all the constraints should be fulfilled.

Definition 2 (The DCSP model). *A problem of satisfying the distributed constraints ($DCSP$) is a CSP, in which the variables and constraints are distributed among autonomous agents that communicate by transmitting, messages.*

This article considers that each agent A_i has allocated a single variable x_i.

The two families of techniques (the ABT, and AWCS families) are characterized by using many types of messages in the process of agents communication for obtaining the solution. These messages are similar for these families. They are based on asynchronous search principles defined by the ABT technique. Thus, this article presents further on these principles used in ABT and the other techniques presented in here.

In this algorithm, each agent instantiates its variables in a concurrent way and sends the value to the agents with which it is directly connected. The Asynchronous Backtracking algorithm uses 3 types of messages [1],[4]:

- the OK message, which contains an assignment variable-value, is sent by an agent to the constraint-evaluating-agent in order to see if the value is good.
- the nogood message which contains a list (called nogood) with the assignments for which the looseness is found is being sent in case the constraint-evaluating-agent finds an unfulfilled constraint.
- the add-link message, sent to announce the necessity to create a new direct link, caused by a nogood appearance.

ABT requires constraints to be directed. A constraint causes a directed link between the two constrained agents: the value-sending agent, from which the link departs, and the constraint-evaluating agent, to which the link arrives.

Each agent keeps its own agent view and nogood store. Considering a generic agent self, the agent view of self is the set of values that it believes to be assigned to agents connected to self by incoming links. A nogood is a subset of agent view. If a nogood exists, it means the agent can not find a value from the domain consistent with the nogood. When agent X_i finds its agent-view including a nogood, the values of the other agents must be changed. The nogood store keeps nogoods as justifications of inconsistent values. Agents exchange assignments and nogoods. When self makes an assignment, it informs those agents connected to it by outgoing links. Self always accepts new assignments, updating its agent-view accordingly. When self receives a nogood, it is accepted if it is consistent with self's agent view, otherwise it is discarded as obsolete. An accepted nogood is added to self's nogood store. When self cannot take any value consistent with its agent-view , because of the original constraints or because of the received nogoods, new nogoods are generated as inconsistent subsets of the agent-view, and sent to the closest agent involved, causing backtracking. The process terminates when achieving quiescence i.e. a solution has been found, or when the empty nogood is generated i.e. the problem is unsolvable.

2.2 The ABT Family

Starting from the algorithm of ABT, in [1], several derived techniques are suggested, based on this one and known as the ABT family. They differ in the storing process of nogoods, but they all use additional communication links between unconnected agents to detect obsolete information. These techniques are based on a common core (called ABT kernel) hence some of the known techniques can be obtained, by eliminating the old information among the agents.

The ABT kernel algorithm, is a new ABT-based algorithm that does not require to add communication links between initially unconnected agents. The ABT kernel algorithm is sound but may not terminate (the ABT kernel may store obsolete information). In [1] several solutions are suggested to eliminate the old information among agents, solutions that are summarized hereinafter.

A first way to remove obsolete information is to add new communication links to allow a nogood owner to determine whether this nogood is obsolete or not. These added links were proposed in the original ABT algorithm. A second way to remove obsolete information is to detect when a nogood could become obsolete. This solution leads to the DisDB algorithm No new links are added among the agents. To achieve completeness, this algorithm has to remove obsolete information in finite time. To do so, when an agent backtracks forgets all nogoods that hypothetically could become obsolete.

2.3 AWCS Technique

The AWCS algorithm [4], is a hybrid algorithm obtained by combining the ABT algorithm with the WCS algorithm, which exists for CSP. It can be considered as being an improved ABT variant, but not necessarily by reducing the nogood values, but by changing the priority order. The AWCS algorithm uses, like ABT, the two types of ok and nogood messages, with the same significance.

When an agent A_i receives an ok? message, it updates its agent view list and tests if a few nogood values are violated. A generic agent A_i can have the following behavior [4] :

- If no higher priority nogood value is violated, it doesn't do anything.
- If there are a few higher priority nogood values that have inconsistent values and these values could be eliminated by changing the x_i value, the agent will change this value and will send the ok? message.

– If a few higher priority values are inconsistent and this inconsistence can not be eliminated, the agent creates a new nogood messages and sends a nogood message to each agent that has variables in nogood. Than the agent increases the priority of x_i, by changing the x_i value with another value that minimizes the inconsistencies number with all the nogood values and sent the ok? message.

The AWCS algorithm can be improved by applying a learning schema. In [2] there is presented and analyze a schema called "resolvent-based learning", that applies to the AWCS algorithm and brings good results regarding the number of necessary cycles for problem solving. The nogood learning technique induced in [2] is a new method of learning the nogood values applicable to DCSP. The idea is that for each possible value for the failure variable, a nogood that forbids that value is selected and than a new good is built outside the one obtained by unifying the selected nogoods.

3 The Synchronization of the Agents' Execution in Case of Asynchronous Techniques

In [3] are presented two models of implementation and evaluation for the asynchronous techniques in NetLogo. The NetLogo is a programming medium with agents that allows implementing the asynchronous techniques These models are based on two different methods for detecting the ending of the execution of the asynchronous algorithms and are based on the use of the NetLogo environment as a basic simulator for the evaluation of asynchronous search techniques. The two methods from [3] allow the evaluation of asynchronous techniques in the asynchronous case and in the case with synchronization. The model with synchronization is based on NetLogo elements, using the *ask* command for executing the procedures for treating the agents messages. This command does a synchronization of the commands attached to the agents such as the synchronization of the agents' execution is made automatically. Of course, each agent works asynchronously with the messages, but at the end of a command's execution there is a synchronization of agents' execution. Examples of implementation can be found on the web sites [6],[7]. In reality, the agents run concurrently and asynchronous, each agent treating its messages from the messages queue in the arrival order, without expecting for the finalization of computations from the other agents.

The analysis of experimental results shows that the AWCS techniques behave better in case of synchronizing the agents' execution. Starting from this remark, in this paragraph are proposed two solutions of synchronization for the agents' execution.

The first solution proposed is based on using a common memory zone to which the agents have access. In that common memory zone the value of a global variable Nragents, accessible to all the agents is stored. Initially, that variable is initialized with the number of agents. Each agent will mark the status of its execution in the variable Nragents. Practically, each agent A_i decrements the Nragents variable with 1 when the message processing routine is executed. Also, when the agent finishes to process the messages from its message queue it increments the value of Nragents with 1. In other words, the variable Nragents allows the identification of the status of the agents' execution in any moment. At a given moment, that variable can be equal to the number of agents i.e. all agents have finished to process the messages from the message queue. That could become a moment for synchronization, which can be retained by means of a variable called Sincronizare. The first solution allows a complete synchronization of all the agents.

The second solution consists in the synchronizing only the neighboring agents. Each agent will wait for it's connected neighbors to finish their computations, which are placed before him in a lexicographical order. That solution allows a partial synchronization of the agents' execution. That second solution of partial synchronization is based on the use of a synchronization message. This message is similar to a token that each agent needs to receive in order to carry on with the execution of it's computing cycle. For that, each agent uses a second channel of communication for receiving the synchronization messages (the first channel is used for receiving the ok or nogood messages).

The working protocol supposes for each agent the completion of tho stages:

- each agent processes all the messages from it's main communications channel performing a computing cycle. In the moment that the main message channel is empty, it sends a message of the "synchronous" type to the neighboring agents, that are before him in a lexicographical order.
- after each cycle, the agents check if they have received the synchronization messages from all of it's neighbors, placed after him in a lexicographical order, and if not it waits until it receives all those messages.

The implementation of any asynchronous technique implies to build some execution routines for the computations by each agent. In the models proposed in [3] this routine is called update. Also, each agent needs another routine to treat its messages (and here, each technique is different by the fact it treats differently those messages) named in the models from [3] handle-message. These notations are also used in the synchronization solution.

The new procedure update which is performed by each agent is presented in figure 1(a). Each agent verifies if the synchronization status has been reached, in the affirmative case it runs its own message manipulation procedure (handle-message). There are some differences between the two solutions. For the first solution, the agents enter the wait state until the variable Nragents becomes equal to the number of agents. But, for the second synchronization solution, the agents wait until they receive the synchronous message from all it's neighbors placed before it in a lexicographical order.

The procedure that each agent calls to process its messages is depicted in figure 1(b). In that procedure one can notice that the agent decreases the global variable Nragents at the entry point of the procedure. In the end, after the agent has processed a message or many (in packets), the incrementation of Nragents variable is done (first solution). One can see that in order to implement each technique, the agents can process message by message, or in packs by processing all or just partially the messages that exists in the message queue. This article

```
to update
 if Sincronizare
 [ handle-message ]
 if Solution
 [
   WriteSolution
   stop]
 While [synchronization condition]
 [
   wait
   set Sincronizare false
 ]
 set Sincronizare true
end
```

(a) The update procedure

```
to handle-message
1' set Nragents Nragents - 1 // only for the first solution
if not empty? message-queue
 [
   set msg retrieve-message
   Process-message msg
 ]
if (empty? message-queue)
 [ stop ]
1' set Nragents Nragents + 1 //only for the first solution
2' set msg list "sincron" who //only for the second solution
2' send msg to Neighbors //is sent only to the neighbors
   that are before him in a lexicographical order
end
```

(b) The handle-message procedure

Fig. 1. The new procedure

analyzes the behavior of the asynchronous techniques in both variants (treating of one message or complete treating of messages from the message queue of each agent). But, for the second solution, in the routine for message processing, the synchronous message is sent to all it's neighbors placed before him in a lexicographical order (at the end of processing the messages)

The two message manipulation procedures can be applied in any language chosen when implementing. In particular, those two routines can be applied for the asynchronous model proposed in [3], obtaining a synchronization of the agents execution. Thus, starting from the elements from [3] one can obtain a third model of implementation and evaluation for the asynchronous techniques.

4 Experimental Results

In order to make such estimation, these techniques are implemented in NetLogo 3.0, a distributed environment, using a special language named NetLogo, [5], [6], [7]. The implementation and evaluation is done using the two models proposed in [3] and the model with synchronization is proposed in this article.

The asynchronous techniques are applied to a classical problem: the problem of coloring a graph in the distributed versions. For this problem we take into consideration two types of problems (we keep in mind the parameters n-number of knots/agents, k-3 colours and m - the number of connections between the agents). We evaluate three types of graphs: graphs with few connections (called sparse problems, having m=n x 2 connections) and graphs with a special number of connections (m=n x 2.3 and m=n x 2.7 connections, called difficult problems). For each version we carried out a number of 100 trials, retaining the average of the measured values (for each class 10 graphs are generated randomly, for each graph being generated 10 initial values, a total of 100 runnings).

In order to make the evaluation of the asynchronous search techniques, the message flow was counted i.e. the quantity of messages ok and nogood exchanged by the agents, the number of verified constraints i.e. the local effort made by each agent, and the number of concurrent constraints verified (noted with c-ccks) necessary for obtaining of the solution.

4.1 AWCS Family

In the AWCS family there are many variants that are based on building of efficient nogoods (nogood learning) or on storing and using those nogoods in the process of selecting the values (nogood processor). The basic variant proposed in [4] improved with the nogood learning technique from [2] (noted with AWCS-nl) is applied in this article. A version in which each agent treats entirely the existing messages in its message queue (noted AWCS-nl_k) is implemented for this variant. Three implementations are done corresponding to the three obtained models:

- variants based on synchronization with the aid of the "ask" command: AWCS-nl_1.
- variants based on the asynchronous model from [3] : AWCS-nl_2.
- variant based on the first method of synchronization introduced in this article (complete synchronization)-AWCS-nl_3.
- variant based on the second solution of synchronization -AWCS-nl_4.

Table 1. The results for AWCS versions (Distributed n-Graph-Coloring Problem)

		n=20			n=30			n=40		
		m=nx2	m=nx2.3	m=nx2.7	m=nx2	m=nx2.3	m=nx2.7	m=nx2	m=nx2.3	m=nx2.7
AWCS-nl_1	Nogood	126.66	261.58	902.57	575.84	2052.00	3639.52	713.60	4857.77	24652.61
	Ok	470.48	745.27	1918.56	1864.50	5135.07	7624.27	2403.25	12733.40	51132.39
	Constr.	856.43	1474.91	4475.59	2913.94	9252.07	14482.23	3747.34	20371.70	102388.82
	c-ccks	307.12	528.49	1608.16	802.74	2414.97	4351.37	824.10	4422.60	21888.16
AWCS-nl_2	Nogood	129.86	217.62	1412.17	502.77	3407.93	12181.62	733.93	6877.65	93259.95
	Ok	456.06	633.52	2881.41	1625.23	8725.69	23964.88	2422.19	17612.68	185845.00
	Constr.	900.47	1334.05	6659.52	2628.08	15309.45	45813.89	4163.19	27261.72	387355.71
	c-ccks	294.81	442.89	2243.28	662.13	3625.55	12512.22	793.45	5409.08	73099.00
AWCS-nl_3	Nogood	137.31	244.91	1137.93	534.27	2840.31	3439.96	755.42	5302.25	22394.24
	Ok	493.51	710.35	2374.80	1738.56	6978.90	7129.70	2512.65	13643.28	47181.16
	Constr.	894.40	1413.90	5507.58	2618.45	12559.69	13858.95	3943.26	22631.85	93429.74
	c-ccks	319.11	506.20	1992.89	759.39	3300.41	4109.20	865.93	4749.67	20065.21
AWCS-nl_4	Nogood	133.24	227.05	1211.12	567.23	1423.10	3672.32	723.15	2820.10	23371.67
	Ok	471.54	662.94	2417.56	1799.98	3776.31	7889.14	2579.42	7604.10	50083.72
	Constr.	931.18	1354.89	5863.72	2751.92	7031.34	15219.11	4081.18	12692.65	109657.25
	c-ccks	328.38	478.78	2101.82	795.28	1842.69	4310.87	876.12	2837.00	21011.11

As known, the verified constraints quantity evaluates the local effort given by each agent, but the number of concurrent constraint checks allows the evaluation of this effort without considering that the agents work concurrently (informally, the number of concurrent constraint checks approximates the longest sequence of constraint checks not performed concurrently). Analyzing the results from table 1, we can remark that synchronization of the agents execution reduces the local effort made by the agents, regardless of the variant used (that based on the ask command or the general one introduced in this article). But, as the dimension of the problems increases (40 nods), the asynchronous variants AWCS-nl_2 required much greater efforts compared to the variants with synchronization. Big differences in the local effort occur especially for the problems of high density (problems known as difficult problems). Still one can remark that for problems of scarce density (m= n x 2.0) the asynchronous variants have almost equal costs to the synchronous variants, even lower efforts.

In the case of the message flow, the behavior remarked regarding the computing effort has maintained almost the same, synchronous variants requiring a message flow lower that the asynchronous variants. The message flow increased for the synchronous variants together with the increase of dimension for the solved problems. Comparing the two synchronous variants, one can remark almost equal efforts for obtaining the solution, such as the practical solution proposed in this article require almost equal efforts to those used in the variant simulated with the ask command.

4.2 The ABT Family

Starting from the ABT kernel, by eliminating the outdated information two important techniques are obtained: the Asynchronous Backtracking and the Distributed Dynamic Backtracking [1]. These two techniques based on a static order are analyzed in order to see the effect of agents synchronization. Unlike the AWCS technique, this article depicts the versions in which each agent treats at each cycle just one message from its message queue (noted with ABT_k and $DisDB_k$). For each of the ABT and DisDB techniques three implementations are done corresponding to the three obtained models:

- variants based on synchronization by means of the "ask" command: ABT_1 and $DisDB_1$.
- variants based on the asynchronous model from [3]: ABT_2 and $DisDB_2$.
- variants based on the method of synchronization introduced in this article : $ABT_{3,4}$ and $DisDB_{3,4}$.

Table 2. The results for ABT and DisDB versions

(a) ABT

		n=20		n=30	
		m=nx2	m=nx2.7	m=nx2	m=nx2.7
A_1	Nogood	359.70	481.88.74	3883.12	3936.20
	Ok	1214.93	1615.48	8578.23	12343.88
	Constr.	67500.30	80834.37	957934.98	938790.08
	c-ccks	14502.26	16572.45	147128.29	123501.58
A_2	Nogood	305.70	366.94	3181.57	2196.52
	Ok	1124.16	1438.48	6518.74	9943.55
	Constr.	60382.63	69332.91	763992.16	698240.00
	c-ccks	8993.06	7593.00	90601.22	45264.01
A_3	Nogood	360.41	376.99	3713.65	3764.15
	Ok	1237.79	1495.00	8065.23	10209.04
	Constr.	68321.27	76823.12	85234.16	800193.03
	c-ccks	13401.26	12312.34	112675.05	89000.45
A_4	Nogood	371.12	395.18	3945.18	3799.55
	Ok	1323.11	1578.06	8245.32	11311.71
	Constr.	67001.81	75121.12	83555.23	753712.20
	c-ccks	12182.11	11133.56	109129.82	90101.34

(b) DisDB

		n=20		n=30	
		m=nx2	m=nx2.7	m=nx2	m=nx2.7
D_1	Nogood	138.76	316.94	2803.34	3104.44
	Ok	300.04	722.89	6855.19	7235.88
	Constr.	18548.68	40841.66	603288.94	778016.08
	c-ccks	4526.52	8112.34	124728.67	154409.43
D_2	Nogood	83.10	216.41	2785.65	2983.69
	Ok	213.72	556.03	6518.74	6878.05
	Constr.	12117.12	29850.88	585992.16	655243.39
	c-ccks	2468.48	3963.02	70605.97	82785.79
D_3	Nogood	125.38	271.25	2813.03	3016.96
	Ok	277.64	646.82	7005.56	7029.70
	Constr.	16837.14	35589.37	603192.16	712024.78
	c-ccks	3791.24	6289.09	115059.67	90786.78
D_4	Nogood	117.21	273.67	2767.28	2994.39
	Ok	272.40	651.71	6981.90	7002.81
	Constr.	15002.91	36111.52	523023.73	679821.34
	c-ccks	34230.74	6421.90	102011.72	88564.20

In table 2(a) are presented the values obtained for the ABT versions, and in table 2(b) those for the versions DisDB. The two techniques based on a static order behave different from the AWCS technique. The lower local effort is carried out for the asynchronous variant, the variants with synchronization require the checking of a much greater number of constraints. Also, the lowest message flow is obtained for the synchronous variants. The two techniques behave the same for both type of problems: problems with scarce density or difficult problems.

The synchronization solution proposed in this article is superior to that offered by the ask command from NetLogo. Unfortunately the techniques from din ABT family, regardless of the fact that they require or not adding extra links, behave better in the asynchronous case, the synchronization is not a solution for reducing the costs.

5 Conclusions

This article remarks two types of behaviors and, as a consequence, two classes of asynchronous techniques (among the ones evaluated in here).

The techniques from the AWCS family, based on a dynamic order for the agents, require lower costs for obtaining the solution in case of synchronization of the agents' execution. A synchronization solution is proposed in this article. The experiments show a decrease of local computing effort and of message flow compared to the asynchronous variants.

The second category of techniques, i.e. from the ABT family behave different, requiring lower costs in the asynchronous case that in case of the synchronization of the agents' execution. All these techniques use a static order for the agents.

After the analysis of these empirical studies, we deduce that the synchronization of the agents' execution is recommended for the techniques with dynamical order.

References

1. Bessiere, C., Brito, I., Maestre, A., Meseguer, P. Asynchronous Backtracking without Adding Links: A New Member in the ABT Family. A.I., **161** (2005) 7–24.
2. Hirayama, K., Yokoo, M. The Effect of Nogood Learning in Distributed Constraint Satisfaction. In Proceedings of the 20th IEEE International Conference on Distributed Computing Systems, (2000) 169–177.
3. Muscalagiu, I., Jiang, H., Popa, H. E. Implementation and evaluation model for the asynchronous techniques: from a synchronously distributed system to a asynchronous distributed system. Proceedings of the 8th International Symposium on Symbolic and Numeric Algorithms for Scientific Computing (SYNASC 2006), Timisoara, Romania, IEEE Computer Society Press, (2006) 209–216.
4. Yokoo, M., Durfee, E. H., Ishida, T., Kuwabara, K. The distributed constraint satisfaction problem: formalization and algorithms. IEEE Transactions on Knowledge and Data Engineering **10(5)** (1980) 673–685.
5. Wilensky, U.NetLogo itself:NetLogo. Available: http://ccl.northwestern.edu/netlogo/. Center for Connected Learning and Computer-Based Modeling, Evanston, (1999).
6. MAS Netlogo Models-a. Available: http://jmvidal.cse.sc.edu/netlogomas/.
7. MAS Netlogo Models-b. Available: http://ccl.northwestern.edu/netlogo/models/community.

Modelling Social Attitudes of Agents

Matteo Cristani and Elisa Burato

Dipartimento di Informatica, Università di Verona, Cà Vignal 2, strada Le Grazie 15, I-37134 Verona

matteo.cristani@univr.it, burato@sci.univr.it

Abstract. Several recent investigations in the field of multiple agent systems have shown interest in the problem of regulating agents behaviour with rules that *ought to be* obeyed by the agents when the agents are not enforced to comply to the rules themselves. Multiple Agent Systems where agents exhibit attitudes with respect to the above mentioned aspects are named Ethical MAS. A family of interesting problems in Ethical MAS are introduced. We then consider two sample interpretations, one based upon simple commitment to the law and to personal interest, and one based on the commitment to *legality*, the attention to *social utility* and the interest in *personal advantage*. Based on these three aspects we provide a classification of social attitudes of agents in EMAS.

1 Introduction

Several investigations have been carried out in the recent literature of artificial intelligence about ethical reasoning [7]. In particular many authors dealt with the so called *Moral Dilemmas* [6] There are two major classes of moral dilemmas: the *Ethical Judgement*, and the *Proper Moral Dilemma*.

The Ethical judgement is a question posed with respect to a given action x and an agent a: is x legitimate, by the moral rules to which a is committed? This seems not to be problematic, since we know that a commits herself to a given set of actions, and does not commit to another set, and the two sets are disjoint. However, the majority of judgements are complicated by two aspects: (1) though a given action x was ethically acceptable, it may be the case that some of the consequences of doing x would be not; (2) the evaluation of x may be probabilistic, namely we may have a critical judgement of x saying that x is "likely to be dangerous", or "probably immoral". A complex ethical reasoning process can therefore take place in order to decide about a given ethical judgement.

At a different level of complexity, an agent may exhibit incoherent attitudes. For instance, she could consider ethically relevant to perform her legal duties, but simultaneously she may be convinced that doing something socially dangerous would be not morally worth. Therefore, if an action is a legal duty and socially dangerous she may face a genuine dilemma: should I perform it, justified by the fact that it is my duty, and consequently cause a social damage, or should I not perform it, because it is socially dangerous, discharging therefore a duty of mine?

N.T. Nguyen et al. (Eds.): KES-AMSTA 2007, LNAI 4496, pp. 63–72, 2007.

The Proper Moral Dilemma is a question posed with respect to two given actions a and b and an agent x: what should x do, between a and b? The Moral Dilemma question is said to be *genuine* if and only if none of the two actions prevails upon the other one, namely, when both actions are ethically equivalent. We can immediately extend the above defined dilemma to the case in which involved actions are more than two. There can be three types of dilemmas:

- **Obligation dilemma.** All the feasible actions are mandatory. The agent cannot do more than one action, so she has to make a choice based upon some sort of preferential reasoning;
- **Prohibition dilemma.** All the feasible actions are forbidden. The agent has to do one action;
- **Incoherence dilemma.** All the feasible actions are incoherent, based upon the commitments of the agent.

The above defined problems are particularly interesting in a multiple agent context. However, to our knowledge, the application of ethical reasoning to the case of multiple agents has been limited to studies that attempt to understand how agents should behave in order to achieve a common goal. The problem of analysing the behaviour of those agents that do not act in a cooperative way has been only partly dealt with. We are focusing, in this paper, upon uncooperative environment, in which agents either **compete** in achieving their own goals, or **collaborate**, but not cooperate[1].

Human agents have different behaviours in the practise of real societies. For instance, rules may not be legal, but just ethical, and individuals can have the attitude of doing only things that are individually advantageous and socially useful. A widely accepted tool for ethical reasoning is Deontic Logic. Though this approach has been considered as valid by many reputable scholars in the community, it exhibits known difficulties with respect to Moral Dilemmas.

In particular, if we consider SDL, (the Standard Deontic Logic) we have two fundamental principles: the **Principle of Deontic Consistency** (PC) and the **Principle of Deontic Logic** (PD).

$$(PC) \quad \bigcirc A \rightarrow \neg \bigcirc \neg A$$

(PC) establishes that the same action cannot be both obligatory and forbidden. Dilemmas actually do not conflict with PC. Dilemmas involve a situation in which an agent ought to do A, ought to do B, but cannot do both A and B. But if we add a principle of deontic logic, then we obtain a conflict with (PC).

$$(PD) \quad \Box(A \rightarrow B) \rightarrow (\bigcirc A \rightarrow \bigcirc B).$$

[1] Two or more agents cooperate, when they accept to be subordinated to a commonly accepted authority to which they all obey, in order to achieve a common goal. Agents collaborate when they independently try to achieve a common goal, avoiding to obstacle others' activities.

How could a moral dilemma be formulated in deontic logic? We essentially have

(1) $\bigcirc A$
(2) $\bigcirc B$
(3) $\Box\neg(A \wedge B)$

From (3) we derive (4) and further we derive (5).

(4) $\Box\neg(B \wedge A)$
(5) $\Box(B \rightarrow \neg A)$

As an instance of (PD) we derive (6), which, used along with (5) derives (7).

(6) $\Box(B \rightarrow \neg A) \rightarrow (\bigcirc B \rightarrow \bigcirc\neg A)$
(7) $\bigcirc B \rightarrow \bigcirc\neg A$

Finally we derive (8) from (2) and (7) and from (1) and (8) we derive (9).

(8) $\bigcirc\neg A$
(9) $\bigcirc A \wedge \bigcirc\neg A$

(9) contradicts (PC).

In recent investigations, many scholars have tried to deal with the technical difficulties of inserting moral dilemmas into deontic logic. For a rather exhaustive survey see [4]. However, to the best of our knowledge, no attempt has yet been carried out to analyse the combinatorial and algorithmic questions posed by moral dilemmas in multiple agents systems[2].

However, in order to provide a specific family of interesting problems, we need a formalised MAS environment, that will be defined in Section 2, for which we claim general interest. In the paper we focus upon one specific problem and provide a solution for it in two simplified cases, and in the general case as well. The approach we adopt here is directly inspired by the principles used in [1], that are foundational of a theory of ethical rules in Multiple Agent Systems. However, we do not focus upon delegation and responsibility as they do, but more on attitudes, because we retain these aspects more basic in the definition of MAS with ethical rules.

This paper is organised as follows: Section 2 presents Terminology and basic definitions for an analysis of the notion of ethical rule in a multi-agent system; Section 3 discusses two sample classifications based upon binary or ternary evaluation methods. In Section 4 we show how to solve some basic problems posed in Section 2 with simple algorithmic methods. Section 5 takes some conclusions and sketches further work.

[2] Goble ([4], p. 463 footnote 2) maintains that MAS are a good example of cases for which moral dilemmas cannot be neglected. However, the technical solution he proposes cannot be directly employed in MAS, because of a technical limit in the computability results that can be derived from the adopted axioms.

2 Ethical Rules and Multiple Agent Systems

In this section we formalise the basic notion henceforth used in the paper. The approach we use here is an evolution of theone used in [2] from the viewpoint of the definition of the notion of commitment in multiple agent systems. The fundamental notions introduced here constitute an attempt at addressing the issues modelled in [3].

Definition 1. *An* ethical multiple agent system *(EMAS) is a tuple* $\mathcal{M}=\{A, X, \aleph(\cdot), e(\cdot,\cdot)\}$*, where elements of* a *are agents, elements of* X *are actions, and* $e(\cdot)$ *is a function that evaluates actions for every agent.*
For every agent x $\aleph(a)$ *actions that the agent can perform, called the* feasible actions *and a function* $e(x, a)$ *that evaluates actions for that agent.*

Example 1. Consider three robots r_1, r_2 and r_3, that can perform four actions: moving forward (f), moving backward (b), moving left (l), moving right (r). The system establishes that only r_1 and r_2 can move right and left, whilst only r_3 can move forward and backward[3]. From r_1's perspective action b, that is illegal for it, is personally advantageous, as well as action l. r_2 and r_3, instead, consider all their feasible actions advantageous.

We interpret the above described structure so that if an action is in the set of feasible actions for an agent, then that agent can perform the action; if not the agent cannot perform the action[4]. The evaluation functions associates to each action for every agent a a n-dimensional vector of values among three evaluations $e(\cdot, a)$: *positive*, *negative*, *indifferent*. Every dimension from 1 to n is called an **evaluation context**. The space $\{$*positive*, *negative*, *indifferent*$\}^n$ is named the **context space**.

Example 2. Consider the agent system of Example 1. We may have four contexts: the context of individual advantage (I), the context of social usefulness (S), the context of legal rules (L), the context of the consortium formed by robots r_1 and r_3, that are practically interested in certain goals that are out of the interest of robot r_2 (C).

In this paper we consider two basic context spaces. The binary and the ternary one. The binary context space is interpreted as defined by the **obedience to the law** and the **interest in personal advantage**. The ternary space is interpreted as defined by the same contexts of the binary one along with a context for **social usefulness**.

[3] These restrictions can be established for simple practical reasons, not for high level moral principles. Practical reasons, namely reasons that do not require to be explained in terms of physical necessity, are what we generally consider to be **ethical** motivations.

[4] In the rest of the paper, since we are not going to deal with the problems related to the feasibility of actions, that is fundamental in Robot Motion Planning and in MAS for Collaborative Design, for instance. In a very general framework, as the one we are dealing with here, it is not necessary to precise all those aspects, so the notion of feasible action will be henceforth given as employed at a higher level.

Definition 2. *Given an EMAS $\mathcal{M}=\{A, X, \aleph(\cdot), e(\cdot,\cdot)\}$ we name* **commitment** *of an agent* a *a pair $\langle P, N\rangle$ of total strong orderings of evaluation contexts.*

Example 3. Again, let us consider the EMAS of Example 1 and the context space of example 2. The positive commitment is $L > S > C > I$, the negative commitment is $S > C > L > I$.

Given a commitment $\langle P, N\rangle$ the **positive attitude** is the ordering P, which is interpreted as the preference of the agent with respect to her obligations, whilst the **negative attitude** is the ordering N which establishes preference about her prohibitions. A commitment $\langle P, N\rangle$ is said to be **coherent** if and only if P and N are equal to each other. The EMAS of Example 3 is incoherent.

We consider the following problems in EMASs.

Problem 1. **Ethical Judgement.** Given an agent a, and an action x, assuming to know the attitude of a, decide whether x is legitimate for a.

Problem 2. **Proper Moral Dilemma.** Given an agent a, and a set of actions X, assuming to know the attitude of a, decide which action in X is the **best choice** for a.

Example 4. In the case of Example 3, with respect to agent r_1 deciding about action f is an ethical judgement. The pair formed by f and b is a moral dilemma for r_1.

The above problems are said to be **solvable** when they have a solution. The third problem includes the first, since in order to establish which attitudes are compatible with any single action we have to decide, for each action, which attitudes will legitimate that action for that agent. In this sense, for each action we solve a collection of hypothetical ethical judgements.

We can claim the following propositions.

Theorem 1. *Ethical judgements of agents with coherent commitments are solvable.*

Proof. Consider an agent a that is committed to the evaluation contexts c_1, c_2, $\ldots c_n$ with the positive attitude P $c_i > c_{(i+j)}$ for any i and j with $1 \leq i \leq n-1$ and $1 \leq j \leq n-i$. Since a has coherent commitments, then $N = P$. Suppose by contradiction that for an action x, a would not be able to decide about ethical validity of x. Then, there would be a context c', such that

- for all context c'_j such that $P = c'_1 > c'_2 > \ldots c'_{(k-1)} > c'_k = c' > c'_{(k+1)} \cdots$ and $c'_j \in \{c'_1, c'_2, \ldots, c'_{(k-1)}\}$, the agent a evaluates x negative or indifferent in c'_j;
- agent a evaluates x mandatory in c';

and there would be a context c'', such that

- for all context c''_j such that $N = c''_1 > c''_2 > \ldots c''_{(l-1)} > c''_l = c'' > c''_{(l+1)} \cdots$ and $c''_j \in \{c''_1, c''_2, \ldots, c''_{(l-1)}\}$, the agent a evaluates x positive or indifferent in c''_j;
- agent a evaluates x as forbidden in c''.

Moreover, in the positive commitment order, c' and c'' should be such that $c'' > c'$, whilst in the negative commitment order they should be such that $c' > c''$ thus $P \neq N$. A contradiction. Therefore the claim is proved.

Conversely, not all moral dilemmas result solvable for agents with coherent commitments. Obviously, agents with coherent commitments never face incoherence dilemmas.

3 Modelling Agents' Commitments in EMAS with Binary and Ternary Evaluation Context Spaces

The binary evaluation space is very simple to analyse. In this space the only two types of incoherence that can be formulated are the ones established by an agent that considers an action simultaneously illegal and personally useful or legally mandatory but personally negative. The first type of incoherence is encountered by those agents who are committed to the law before than to personal advantages from a positive viewpoint and the way around from a negative viewpoint. The second type is encountered by those agents who exhibit inverse commitments wrt the first case.

Both these incoherence situations are strictly unsolvable as claimed in the following proposition whose proof is straightforward and therefore left to the reader.

Proposition 1. *Ethical judgement of incoherently committed agents of an EMAS under evaluation on a binary context space is always unsolvable.*

In the case of ternary context space we consider three aspects of legitimacy: legality, social utility, individual advantage.

The direct consequence of the above proposed approach is that the possible **basic attitudes**, the possible type of actions from an ethical point of view are obtained as settings of the above classified cases. The combinatorial cases are 27 as in below.

Though this formalisation is interesting in theory, it is not the case that we actually allow all these attitudes. In particular we may classify as *rational*, every attitude that results positive in one respect. In other terms, the attitudes (8), (12), (13), (14), (15), (16), (17), (18) are irrational. In fact, no one would perform an action that is not personally advantageous, not socially useful and not legally obligatory: it is an empty action, from a personal viewpoint, and it is not useful.

Moreover, some attitudes are trivially correct for all the possible cases. If something is positive or neutral for each of the three above defined cases, then it is worth performing. Thus, every attitude shall anyway include basic attitudes (1), (2), (3), (4), (9), (10), (11). If something is permitted or enforced by the law, it is socially useful or indifferent, and personally advantageous or indifferent, than it is an action that everyone would perform.

Number	Legality	Social usefulness	Personal advantage
(1)	Positive	Positive	Positive
(2)	Positive	Positive	Indifferent
(3)	Positive	Indifferent	Positive
(4)	Indifferent	Positive	Positive
(5)	Positive	Positive	Negative
(6)	Positive	Negative	Positive
(7)	Negative	Positive	Positive
(8)	Indifferent	Indifferent	Indifferent
(9)	Indifferent	Indifferent	Positive
(10)	Indifferent	Positive	Indifferent
(11)	Positive	Indifferent	Indifferent
(12)	Indifferent	Indifferent	Negative
(13)	Indifferent	Negative	Indifferent
(14)	Negative	Indifferent	Indifferent
(15)	Negative	Negative	Negative
(16)	Negative	Negative	Indifferent
(17)	Negative	Indifferent	Negative
(18)	Indifferent	Negative	Negative
(19)	Negative	Negative	Positive
(20)	Negative	Positive	Negative
(21)	Positive	Negative	Negative
(22)	Positive	Negative	Indifferent
(23)	Positive	Indifferent	Negative
(24)	Indifferent	Positive	Negative
(25)	Indifferent	Negative	Positive
(26)	Negative	Positive	Indifferent
(27)	Negative	Indifferent	Positive

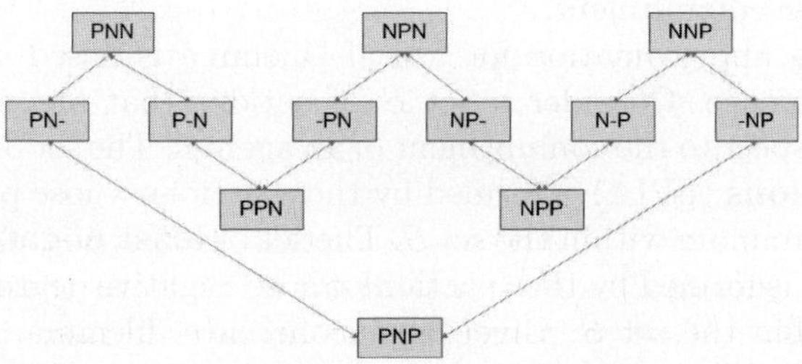

Fig. 1. The twelve basic attitudes and their dominance relation.

The remainder 12 cases are depending to each other in a complex way, represented in Figure 1. The expression XYZ is used to denote the basic attitude X wrt legality, Y wrt social usefulness and Z wrt personal advantage. P denotes positive attitude, N negative attitude and – indifferent attitude.

An edge between a basic attitude x and y in Figure 1 represents the necessity of an attitude that includes x to include y as well.

We now introduce some fundamental notions that we employ henceforth. The possible positive and negative orderings of contexts are, most obviously, 6. Therefore, in principle, there are 36 commitments, of which 6 are coherent.

4 Methodological Issues

We now introduce the notion of **positive and negative degrees of incoherence**. In order to provide such definitions we need to introduce **obligation of degree** k.

An action x is an obligation of degree k for an agent a with a commitment $\langle P, N\rangle$ if and only if a evaluates x not positive for the first $n-k$ contexts and evaluates x positive for the $(n-k+1)$th context by P.

An action x is a prohibition of degree k for an agent a with a commitment $\langle P, N\rangle$ if and only if a evaluates x not negative by N for the first $n-k$ contexts and evaluates x negative for the $(n-k+1)$th by N.

Definition 3. *Given an EMAS $\mathcal{M}$=$\{A,\ X,\ \aleph(\cdot),\ e(\cdot,\cdot)\}$interpreted against a context space with n contexts, we say that an action x has a positive degree of incoherence k with respect to an agent* a *if and only if x is an obligation of degree k for* a *and simultaneously is a prohibition of degree k', with $k' > 0$. Analogously we say that an action x has a negative degree of incoherence k with respect to an agent* a *if and only if x is an prohibition of degree k for* a *and simultaneously is a obligation of degree k', with $k' > 0$.*

If P and N agree on the all contexts for an action x, then we say that x is coherent with the commitment of a. If P and N are equal, then every action is coherent with the commitment.

An interesting approximation for Moral Dilemma is based on the notion of degree of incoherence. Consider a set S of actions that are not evaluated as coherent with respect to the commitment of an agent a. The set of **least positive incoherent actions** (LPIA) is formed by those actions whose positive degree of incoherence is minimum within the set S. The set of **least negative incoherent actions** (LNIA) is formed by those actions whose negative degree of incoherence is minimum within the set S. Given an incoherence dilemma, D, formed by a set of actions that all not coherent with the commitment of an agent a, D can be solved by a in an approximate way if and only if the sets of LPIA and LNIA in D are singletons.

A dilemma is said to be approximately solvable if and only if we can find a unique solution to the dilemma that is not a coherent action with respect to the agent's commitment. Straightforwardly, we can claim the following propositions.

Proposition 2. *Incoherence dilemmas with singleton LPIA set are approximately solvable.*

Proposition 3. *Incoherence dilemmas with singleton LNIA set are approximately solvable.*

Less obviously, LPIAs and LNIAs can be used to solve Obligation dilemmas as well as Prohibition dilemmas. Proof of the below claims are easy but long, so, for the sake of space, we omit them here.

Theorem 2. *Obligation and Prohibition dilemmas with singleton LPIA set are approximately solvable.*

Theorem 3. *Obligation and Prohibition dilemmas with singleton LNIA set are approximately solvable.*

5 Conclusions

Two branches of current research in Knowledge Representation have shown interest in the problem of collaboration of agents within a formal framework: Multiple Agent Systems studies and Legal Reasoning. Some major problems of the two branches are quite similar from a theoretical viewpoint. In MAS one fundamental question that arises when we analyse the problem of giving account to the notion of collaboration. Collaborating means trying to achieve goals (common or individual) avoiding to interfere in others' attempts to achieve their own ones. As a matter of fact, this is the basis of social structures, and is foundational for notions of ethical responsibility. In a very basic system individuals only act for themselves' interest; in more mature structures we may have individuals who accept social rules; in a more sophisticated environment people can also make an effort in the direction of projecting individual advantages to one or more than one level into the society. The first case, individuals without rules, does not correspond to a society, whilst the second one is the case of society without structured moral rules. The last case is the most interesting, that is posed in the basic case (individuals, society benefits, society rules).

Though individuals may be interested in collaborating, they may not be able to solve a family of problems known as **Moral Dilemmas**. In the case of moral dilemmas, agents may be forced to perform actions they do not evaluate positively or may be incapable of doing their duties.

In legal reasoning several interesting approaches have been tried that are based upon different families of formal logics (Deontic Logic SDL, Paraconsistent Legal Reasoning Framework, where, for instance, we do not assume the *ex falso quodlibet* principle). However, a Deontic Logic that fully accommodates dilemmas is still not completed.

To the best of our knowledge, this is the first attempt in literature to study the behaviours of agents in a legal reasoning framework by accommodating in the framework itself moral dilemmas. Such an approach would be particularly useful when these MAS formalise inter-working systems, like Internet Wiki tools, or on-line collaboration systems. For a comprehensive analysis of problems posed in the current literature by formalisation of MAS along with ethical rules see [5].

We dealt with the problem of representing the attitudes of agents involved in a multi-agent system with ethical rules. Such a system, called Ethical Multiple Agent System, is specified in terms of agents with ethical commitments, both as obligations and as prohibitions. We introduced the notion of coherence for commitments and specify a notion of degree of incoherence, that allow agents to solve moral dilemmas in an approximate way.

Acknowledgments

Authors gratefully thank Aluk Group and Address s.r.l. for financial support.

References

1. Olga Carmo and José Carmo. A role based model for the normative specification of organized collective agency and agents interaction. *Autonomous Agents and Multi-Agent Systems*, 6(2):145–184, 2003.
2. C. Castelfranchi. Commitments: from individual intentions to groups and organizations. In V. Lesser, editor, *Proc. 1st. International Conference on Multi-Agent Systems, (ICMAS95)*, page 4148, 2005.
3. Carolyn Dowling. Intelligent agents: some ethical issues and dilemmas. In *CRPIT '00: Selected papers from the second Australian Institute conference on Computer ethics*, pages 28–32, Darlinghurst, Australia, Australia, 2000. Australian Computer Society, Inc.
4. Lou Goble. A logic for deontic dilemmas. *Journal of Applied Logic*, 3(3-4):461–483, 2005.
5. Mamadou Tadiou Kone, Akira Shimazu, and Tatsuo Nakajima. The state of the art in agent communication languages. *Knowl. Inf. Syst.*, 2(3):259–284, 2000.
6. Patricia Marino. Moral dilemmas, collective responsibility, and moral progress. *Philosophical Studies*, (104):203–225, 2001.
7. Leendert van der Torre. Contextual deontic logic: Normative agents, violations and independence. *Annals of Mathematics and Artificial Intelligence*, 37(1):33–63, 2003.

Modeling the Activity of a Multiagent System with Evolving Metadata*

Krzysztof Juszczyszyn

Institute of Information Science and Engineering
Wroclaw University of Technology, Wroclaw, Poland
krzysztof@pwr.wroc.pl

Abstract. Currently we don't have a reliable abstraction for modeling activity in knowledge-processing multiagent systems with evolving metadata. The aim of this paper is to propose an approach to simulation of evolving society of software agents with private vocabularies in form of semantic nets (also: lightweight ontologies). The conditions for successful simulation of this kind of systems are formulated with respect to up-to-day results in research on agents, Semantic Web and network theory. The generic algorithm is proposed and the importance of the presented results for predicting behavior of future autonomous agents' societies in Web-based environments is discussed.

1 Introduction

In recent years rapid growth of infrastructure, knowledge sources and interoperability requirements created a global net-centric architecture for information storing, delivery and processing. Web technologies provide a basis for the interchange of knowledge and services in such an environment and support the emergence of large-scale network structures: WWW itself, thematic information networks, virtual communities, market systems.

Development of dynamic intelligent services across the above systems is inevitably connected with so-called *semantic technologies* – functional capabilities that enable *both* humans and machines to create, discover, organize, share and process the meanings and knowledge [3]. On the level of WWW this implies the adoption of the Semantic Web's XML-based standards for annotating and processing information, usually in the form of web ontologies [4]. Because ontologies are developed and managed independently the semantic mismatches between two or more ontologies are inevitable. Practical applications show that fully shared vocabularies are exceptional - a number of possible different semantic conflicts was identified in [2]. The vision of Semantic Web allowing agents to publish and exchange ontologies requires strong mechanisms supporting ontology merging and alignment [1].

It should be stressed that any formal model aiming to investigate the properties of the above systems has to take into account at least two factors: the community

* This work was supported by the Polish State Committee for Scientific Research under Grant No. 3 T11C 029 29 (2005-2007).

N.T. Nguyen et al. (Eds.): KES-AMSTA 2007, LNAI 4496, pp. 73–81, 2007.

structure and metadata evolution. The key challenge here is that they are not independent.

As shown in the next sections there are many results investigating the structure of the contemporary networks and there are also some which evaluate the large-scale structure of semantic nets and ontologies. But there are no works which deal with joining these issues. The aim of this paper is to present a work-in-progress report on developing a novel computational framework which will allow to simulate the evolution of large communicating multiagent communities with evolving metadata. The rationale for such an environment is to investigate the conditions underlying the emergence of common vocabularies and interoperability level in the agents' communities.

The evolution of private metadata has two guiding forces; both should be taken into account. The first are changes imposed by the owner of metadata, its natural that the agents will modify their vocabularies. The second reason are changes that are results of communication between agents (for example: ontology negotiations). In order to be compatible with the latest results our framework must integrate and formally represent the following components:

- Community structure (the architecture of links between actors).
- Community dynamics (the mechanism of formation of new links between actors).
- Semantic interaction model (the rules for modification of internal knowledge representation as a result of the communication and/or the internal events).

Complex network structures emerge in many everyday situations among people, organizations, software agents, linked documents (WWW) and so on. Previous research has identified the most distinctive properties of such networks [9]: small diameter and average path length (of the order of Log(N) for N nodes), high clustering and famous power-law (or scale-free) network node degree distribution.

These properties allow us simulating interactions between system components and building evolution models, and form a basis of many robust and applicable theoretical results. It was shown that they influence the search strategies, communication and cooperation models, knowledge and innovation spreading etc.[7].

Moreover, last results show that we may expect similar phenomena on the level of semantic nets and ontologies: in [6] a large-scale structure of the semantic networks of three types was evaluated. It was shown that all appear to be of small-world structure which (as a graph) may be characterized by sparse connectivity, short average paths and high node clustering – just like in the case of abovementioned networks. More results on the semantic networks modeling are to be found in [11].

From the other hand the process of acquiring new concepts (concept learning) via communication was investigated in many other works, for example in [5] a mathematical model was used to simulate the emergence of coherent dictionary in a population of independent agents. The Authors proposed a well-known mechanism of language imitation as a self-organization factor. However, a random communication and interaction strategy was assumed in most cases. This is not obvious in real multiagent and social environments. As stated in the preceding paragraphs neither connection nor communication pattern between the agents are random.

Now the challenge is to *span a bridge* between known properties of dynamic self-organizing agent societies and the semantic structures emerging within them. The growing and evolving semantic structures should be modified in result of interaction between agents constituting community of particular architecture.

In order to achieve this, the paper is structured as follows. The formal assumptions concerning the multiagent system are presented in section 2. The representation of agents' community structure and internal metadata model are also defined here. In section 3 a generic algorithm, which comprises the activity of the system is presented. In particular, an approach to modeling two main dynamic processes in the system (metadata evolution and negotiation) is proposed in sections 3.1 and 3.2 respectively. Section 3.3 addresses the evolution of multiagent community structure. System variables and properties to be tracked during simulation are defined in section 4. Finally, section 5 presents the concluding remarks.

2 The Description of the Multiagent System

The following general assumptions are being made:

1. The number of agents in the system is constant. The only one reason for this assumption is simplicity during first experiments. As we will see later, dynamic adding and removing of agents may be easily implemented and does not inflict the basic rules of the proposed framework.
2. Agents use internal private metadata representations in the form of semantic nets. We'll stick to the term *semantic net* but we may also think of them as of lightweight ontologies expressed (for example) in RDF language.
3. Before the first step all of the agents use the same semantic net (i.e. each of them owns its private copy). From that point they have possibility to introduce changes to their nets independently.

Let $\boldsymbol{A} = \{A_1, A_2, \dots A_n\}$ be a set of agents and $\boldsymbol{S} = \{S_1, S_2, \dots S_n\}$the set of their private semantic nets. Each agent A_i uses S_i as an formal conceptualization of particular domain of interest. We denote the set of concepts of S_i as C_i={ c_1^i, c_2^i ... $c_{m(i)}^i$ } (where $m(i) = |C_i|$ is the number of concepts in net S_i), and the relations between them as $R_i \subseteq C_i \times C_i$. Thus $S_i = \langle C_i, R_i \rangle$.

Also, denote the set of immediate neighbors of c_1^i in S_i by $\Diamond c_1^i$.

Additionally we assume, that the concepts have unique identifiers which allow to distinguish between them. As stated at the beginning, everyone has the same semantic net, thus $ID(c_1^1) = ID(c_1^2) = \dots = ID(c_1^n)$. (function *ID* returns concept identifier). The *ID*'s are interpreted as the *meanings* of concepts.

The agents are also linked to each other, which means that they communicate not randomly but only with their immediate neighbors – the agents who share a communication link with them. The communication structure of the system will be addressed in the next section.

2.1 Agents' Communication Graph

The communication links between agents are represented by an undirected communication graph: a symmetrical $n \times n$ matrix $\boldsymbol{G}$ of natural numbers where an non-zero entry $g_{ij} = \boldsymbol{G}[i,j]$ indicates the presence of link between the node (agent) A_i to A_j.

The architecture of connections (communication links) between agents conforms to the small world model (which holds true for most Web-based scenarios, as discussed in [7]). This assumption follows the results invoked in the preceding section – we may expect that the agent networks will follow small world properties.

The communication graph is generated using chosen classic technique (like Barabasi-Albert model of preferential attachment [13]). Initial value of non-zero elements of ***G*** is set to value g_{INI} assumed to be small natural number (like between 5 and 10). The role of the g_{INI} link label (weight) is to reflect the connection strength between the agents. It will be modified during simulation as a result of semantic interactions between them.

2.2 Modeling Agents' Initial Semantic Net

Each of the agents has its private metadata representation in the form of semantic net (lightweight ontology). As showed in the abovementioned work [6] it should also show scale-free properties. There, the authors proposed the directed growing network model which will be used here. The algorithm was shown to generate network structures compatible with several types of semantic networks (from free word association network to WordNet ontology), and was based on the following assumptions [6]:

- semantic structures grow primarily through a process of differentiation: the meaning of a new concept typically consists of some kind of variation on the meaning of an existing word or concept. When a new node is added to the network, it differentiates an existing node by acquiring a pattern of connections that corresponds to a subset of the existing node's connections.
- the probability of differentiating a particular node at each time step is proportional to its current complexity – the number of connections with its neighbors.

The algorithm runs as follows: If we want to create a network of size N (i.e $m(i)$ =N for any i), we start with a small fully connected network of M nodes (M <<N). At each time step, a new node with M links is added to the network by randomly choosing an existing node (concept) x for differentiation, and then connecting the new node to M randomly chosen nodes in the semantic neighborhood of node x (the neighborhood of node x consists of x and all the nodes connected to it). Thus, the creation of a new node can be thought of as differentiating the existing node, by acquiring a similar but slightly more specific pattern of connectivity.

The probability of choosing node x to be differentiated is proportional to the complexity of the corresponding word/concept, as measured by its number of connections. The direction of each arc is chosen randomly and independently of the other arcs, pointing towards the older node with probability α and towards the new node with probability (1- α). Typically α is close to 1 (like 0.95 in [6]) which follows common intuitions about creating new words or concepts.

This computationally simple procedure leaves us with initial model of semantic net, which is, at the beginning, identical to all of the agents and stands for S_i, $i=1,2,\ldots n$. Every concept is assigned its unique *ID*.

Now we're ready to address the evolution of the multiagent system.

3 The Dynamics of the System

As viewed from 10.000 feet there are two contradictory processes occurring it multiagent system under consideration:

- Agents change their semantic nets which has obvious negative impact on their interoperability.
- Agents communicate with each other and perform ontology negotiation which helps to maintain their ability to cooperate.

Additionally we may observe events like adding/removing edges to/from communication graph and also adding/removing agents (the second, however, is not discussed in this paper). The above actions define basic steps of the generic multiagent system's evolution algorithm, which is formally defined in this section.

Its main task is to perform the activities presented above, return the system state in the next time step and evaluate chosen system properties.

The algorithm runs as follows:

Given: ***A***, ***S***(t), ***G***(t). (t indicates discrete time moment, initially $t = 0$)
Result: ***S***$(t+1)$, ***G***$(t+1)$.
Parameters: F_{SEM}, F_{NEG}, $F_G > 0$.
BEGIN

1. Randomly choose an agent A_i. Perform evolution of the A_i's semantic net. Repeat step 1 F_{SEM} times (the A_i's semantic net is being evolved as defined in sec. 3.1).
2. Randomly choose an agent A_i. From the neighbours of A_i in ***G*** choose A_j with probability proportional to the number of A_j's connections in ***G*** (this reflects the general tendency to contact hubs which is observed in most dynamic networks). Perform negotiation (defined in sec. 3.2) between A_i and A_j. Repeat step 2 F_{NEG} times.
3. Modify ***G*** (a strategy for modifying the structure of communication links is addressed in sec.3.3). Repeat step 3 F_G times.
4. Evaluate the properties (defined in sec. 4) of the multiagent system.

END.

The fixed parameters F_{SEM}, F_{NEG}, F_G represent the frequency of corresponding actions (evolution, negotiation, modifying of the communication graph). In future experiments tuning their values will allow draw important conclusions about the dynamics of the system. One of the most important will concern estimation of the maximum rate of changes in the private ontologies which still allows the interoperability sustained by communication (ontology negotiation) between agents.

3.1 Evolution of Individual Semantic Nets

This step is not complicated because it obviously follows the semantic growth strategy, presented in sec. 2.2. The net of the chosen A_i is modified in the same way, creating a new concept. Again, a new concept is assigned an unique *ID*. But in this case it is a *local* operation and it reflects changes in local semantic network (ontology) which may be result of changes in the domain or changes in conceptualisation the semantic networks of the other agents remain unchanged.

It should be noted that in this paper we consider only the basic local operation, but in the future more sophisticated modelling may be used here. The list of possible operation which may occur during ontology evolution consists of over 20 events which were discussed in detail in [12]. In case of lightweight ontologies considered here our approach addresses *create class* and *split class* events with clear possibility of easy modelling of *remove class* operation.

3.2 Semantic Negotiation

The ultimate goal of the negotiation process is to *coordinate* the private vocabularies in order to achieve interoperation between negotiating parties. There are several approaches to ontology negotiation but they involve, in most cases, the sequence of the following steps: a query, interpretation, clarification, evaluation and ontology evolution. As a result one or both agents modify their ontology to introduce a new concept [8].

It means that, when two intelligent agents interact, there is a possibility that one of them "captures" the meaning of some concept used in communication in order to maintain "communicational compatibility" with the second. This mechanism is also called "learning via imitation" and has strong cognitive and experimental basis [10]. It works well for the humans and also for software agents (even if not explicitly called imitation). For simulations we will assume that the communication may lead (with certain probability) to agreement over the meaning of certain concepts, and modification of private ontologies according to reached agreement.

In terms of our framework, the simplified negotiation between A_i and A_j starts from sending a simple query containing randomly chosen concept c_k^i from A_i to A_j. Then the following scenarios are possible:

1. There exists such a c_l^j that $ID(c_k^i) = ID(c_l^j)$.

A_j has the concept with the same meaning. Now the $\Diamond c_k^i$ and $\Diamond c_l^j$ are compared. If $(\Diamond c_k^i \cup \Diamond c_l^j)/(\Diamond c_k^i \cap \Diamond c_l^j)$ is not an empty set random relation between elements of this set and concepts c_k^i or c_l^j is added to S_i or S_j (which means that agents randomly learn each other previously unknown association of their known concept). The value of corresponding g_{ij} is increased by 1 (the communication link strengthens).

2. There is no such a c_l^j that $ID(c_k^i) = ID(c_l^j)$. This means that the concept c_k^i is unknown to A_j.

If there exists such c_l^j that $J(\Diamond c_k^i, \Diamond c_l^j) > j_{SIM}$, where J is Jaccard similarity for two sets and j_{SIM} is some fixed threshold value, then:

if $|\Diamond c_k^i| < |\Diamond c_l^j|$ then: $ID(c_k^i) := ID(c_l^j)$

if $|\Diamond c_k^i| \geq |\Diamond c_l^j|$ then: $ID(c_l^j) := ID(c_k^i)$

This means that if two concepts have the same (within given threshold) semantic neighborhood, they are consider to be identical and both agents agree to use the meaning "suggested" by the more connected one (which supports interoperability and prefers the role of network hubs).

If there exists such c_l^j but $J(\Diamond c_k^i, \Diamond c_l^j) < j_{SIM}$, it is considered that agreement is not possible and the value g_{ij} is decreased by 1. Note, that it implies that if g_{ij} reaches zero, the edge between agents disappears as a result of subsequent communication failures.

Of course, this is most general and simplified view of the negotiation processes, but it comprises the real-life features of the existing frameworks and will be gradually extended within proposed simulation environment in order to meet the requirements of practical applications.

3.3 Modification of the Communication Graph

In most real world systems the structure of communication links is by no means static. The agents continuously look for new parties to communicate and also some links disappear after some time. I the proposed framework a simple model of link dynamics will be used – it is assumed that a link is strengthened (which means that the corresponding g_{ij} is increased) as a result of the successful ontology negotiation. The communication failure decreases g_{ij} respectively (sec. 3.2).

Apart from this the agents also look for the best partner to communicate and they choose the one which is the most semantically similar. Then a new communication link is established.

In order to add (and remove) the edges in ***G*** the notion of semantic similarity between two agents A_i and A_j is introduced:

$$SIM\left(A_{i,}A_j\right) = \frac{|S_i \cap S_j|}{|S_i \cup S_j|} \tag{1}$$

Of course, concepts are identical if they have the same *ID*s. After negotiation phase (sec. 3.2) the semantic similarity is computed for all of the pairs of agents. Then two pairs are chosen: 1. Pair with highest *SIM* rate among not connected pairs. 2. Pair with lowest *SIM* rate among connected pairs. After this an edge between pair (1) is added to ***G*** and an edge between pair (2) is deleted from ***G***. This reflects a general tendency to communicate between semantically similar agents (and, respectively) disappearing of links between semantically different ones.

4 Tracking System Properties

In general, the system properties are evaluated during the step 4 of the algorithm presented in sec.3. There are two different groups of parameters which are to be tracked during simulation, they concern the structure of the multiagent network and their interoperability which depends on the current state of the agents' private metadata. In particular the following properties will be measured:

1. Standard network properties: scale-free exponent, clustering coefficient, sparsity, average shortest path length – for agent communication graph and evolving semantic networks. These seem obvious (hence will not be defined here) and will be useful for checking how the model behaves and if it is consistent with real-life observations.

2. Semantic measures: which will reflect the evolution of agents' vocabularies. First is *concept consistency,* defined as:

$$CC\left(id\right)=\frac{\begin{pmatrix}\left|A_i \in \mathrm{A} : \exists c_k^i, ID\left(c_k^i\right)=id\right| \\ 2\end{pmatrix}}{\begin{pmatrix}n \\ 2\end{pmatrix}} \tag{2}$$

CC equals to the number of the pairs of agents which share the concept with *ID* function value *id* divided by the number of all possible pairs. The similar measure will be *semantic consistency* defined as average value of *CC* measures for all concepts:

$$SC=\frac{1}{\left|\bigcup_{i=1}^{n} S_i\right|}\sum_{id=1}^{Max(id)} CC(id) \tag{3}$$

The role of *CC* and *SC* is to evaluate the semantic interoperability in the system. Note that both *CC* and *SC* equal to 1 before simulation starts (the agents share the same semantic structure). In order to reason about the system's dynamics, the above measures will be checked for different F_{SEM}, F_{NEG},F_G which are responsible for tuning agents activity in areas of ontology evolution, mutual communication and searching network for partners respectively.

5 Conclusions and Future Work

Presented approach to computational simulation of the activity of multiagent system is an original attempt to describe and research the complex behavior of autonomous metadata-processing agents in future Semantic Web environments.

It should be noted that many processes which are simulated in presented framework (ontology negotiation, concept similarity measuring, networks formation) are subjects of intensive research. Up to now they were dealt to separately, however they are not independent and will obviously influence each other in complex network environments. Also, the general tendencies in development of Web technologies and information networks strongly show that in a short time we may face the need of building models of complex networked semantic-based agent systems. A questions concerning interoperability in this class of systems will be crucial. At this moment mechanisms that govern evolution of emergent semantic structures in modern web-based multiagent environments are relatively new and not widely addressed research task. Its successful completion has potential to influence novel interconnection architectures (like Semantic Web and Semantic Grids) in many ways.

- Possibility of reaching semantic interoperability with minimal intervention of human operators (after applying theoretically checked communication an metadata evolution strategies).
- Tuning systems' parameters in order to minimize communication/negotiation failures.

The further development of the proposed framework includes programming and software simulations in *Mathematica 5.0* environment and conducting experiments in order to research the behavior of multiagent systems with evolving metadata.

The next step will be also importing data from real-life ontology negotiation frameworks (along with introducing inevitable new parameters) in order to fine-tune the framework to be useful in predicting the evolution of real life knowledge-processing systems.

References

1. Hendler, J.: Agents and the Semantic Web. IEEE Intelligent Systems **16(2)** (2001) 30-37.
2. Hameed, A. et al.: Detecting Mismatches among Experts' Ontologies Acquired through Knowledge Elicitation. In: Proceedings of 21th International Conference on Knowledge Based Systems and Applied Artificial Intelligence ES2001, Cambridge, UK (2001) 9-24.
3. Davis M., Semantic Wave 2006 - Part 1: Executive Guide to Billion Dollar Markets, Project10X Special Report, Washington, 2006.
4. Berners Lee T., Hendler J., Lasilla O., The Semantic Web. Scientific American, May 2001
5. Ke, J., Minett, J. W., Au, C-P., and Wang, W. S-Y. (2002) Self-organization and Selection in the Emergence of Vocabulary. Complexity, **7(3),** 41-54.
6. Steyvers M., Tannenbaum J., The Large-Scale Structure of Semantic Networks: Statistical Analyses and a Model of Semantic Growth, Cognitive Science, **29(1)**, (2005).
7. Watts D, Strogatz S., Collective Dynamics of Small World Networks, Nature, 393, 440-442.
8. Bailin S., Truszkowski W., Ontology Negotiation Between Intelligent Information Agents, The Knowledge Engineering Review, **17(1)**, 7–19, 2002.
9. Carrington P., Scott J., Wasserman S. (eds.) Models and Methods in Social Networks Analysis, Cambridge University Press, 2005.
10. Meltzoff, A. N., Prinz, W. The imitative mind: Development, evolution, and brain bases. Cambridge, England: Cambridge University Press, 2002.
11. Bales M., Johnson S., Graph Theoretic Modeling of Large-Scale Semantic Networks: Methodological Review, Journal of Biomedical Informatics, **39**, 451–464, 2006.
12. Noy N., Klein M., Ontology Evolution: Not the Same as Schema Evolution, Knowledge and Information Systems **6**: 428–440, 2004.
13. Barabasi A-L., Albert R., Emergence of Scaling in Random Networks, Science **286,** 509, 1999.

Software Component Selection Algorithm Using Intelligent Agents

Blanca Z Abraham and Jose C Aguilar

Fundación para el Desarrollo de la Ciencia y la Tecnología
FUNDACITE Mérida - Venezuela
blanca@funmrd.gov.ve
Universidad de Los Andes
Merida – Venezuela
aguilar@ula.ve

Abstract. We have developed one stochastic model for intelligent selection of software components, in Internet. Components can be physically located in different repositories, and the selection is done using a XML file which is associated to each component. This file contains the most relevant characteristics of the component, with one extra field stored to be used by this algorithm; this field is called "pheromone", which is a concept taken from collective intelligence theory that has been the main inspiration of this work. Swarm intelligence is based on each agent capacity to work individually in order to achieve a collective goal; intelligent agents interact not only with each other but also with their environment. This model can be used not only for component selection but also for services, resources, etc. This is because it is general enough for been replicated with different types of requirements.

Keywords: Artificial Intelligence, Component Components.

1 Introduction

The main advantage of component reuse is that generally components have been already tested and their use can make decrease considerably the development time. The goal of component use is the division of systems in pieces of code that can be used in more than one software development. In this work we assume that one software component is a piece of code which maintains its independency, autonomy and functionality through the use of their interfaces and controllers. The development and adoption of components is increasing with several on-line companies which sell software components and related services. These companies not only allow developers to buy their own components, but also act as intermediaries, providing access to third-party artifacts and related products [13]. Usually, the components are cataloged with brief descriptions that can be searched through keywords. While this is a good starting point, keyword based searching is not efficient because it often results in too many or too few hits. It may also retrieve components that are completely unrelated [13]. The search of one component is not an easy task, many type of search

N.T. Nguyen et al. (Eds.): KES-AMSTA 2007, LNAI 4496, pp. 82–91, 2007.

algorithms have been proposed in literature, some of them are based on hit numbers and popularity [10,12]. This work does not take into account naming and trading services to do component selection, nevertheless this work tries to use the most relevant characteristics of components like functional requirements described by software developers, in order to make decisions in an accurate way, not only for software component selection but also, in a short time, for any kind of service selection, this is possible because the algorithm is very general and could be used in parallel or distributed computing, when services or processors have to be selected.

2 Theoretical Aspects

2.1 Software Components

A software component is a self contained entity that interacts with its environment through well-defined interfaces (provide services and require functionalities to be provided by other components). One of the main motivations behind component technology is reusability. If there is a collection of reusable software components, applications can be built by simply plugging existing components together. Other motivations behind component technology are: Independent evolution of application parts, enhanced flexibility, adaptability, and maintainability of software systems. In order to successfully plug components together, it is necessary that the interfaces of each component matches the requirements of the other components. In this way, the "contracts" between de components are well defined. Therefore, component-based application development depends on adherence to restricted plug-compatible interfaces and standard interaction protocols [3]. Because of marketplace of software components is highly competitive, in the last years many research have been oriented to software components, how to look for them, how to select them and how to make them work together as an integrated piece of software [11].

Many works have been developed for management of software components. Agora for example, is a prototype being developed by the Software Engineering Institute at Carnegie Mellon University [11]. The object of this work is to create an automatically generated, indexed, worldwide database of software products classified by component type; e.g., JavaBean or ActiveX control. An approach to intelligent query and component retrieval for web-based repositories is presented in [13]. In this case, the client side consists of a web browser and the server side is comprised of a query interface module, query refinement module, query execution module, domain model, ontology, repository of reusable domain objects (components), and the application database [13]. ComponentSource is another related work, it is a web repository that provides services to buy sell and develop components [5]. Many others researches can be found in literature.

When a component provides services related to aspects of other components, and at the same time can require services from other components, the concept of dependency appears. That is components can have dependency with other components, and this is an important issue to take into account when a component will be selected.

Other important concept is information retrieval (IR), which is essentially the process of the content-based, goal-directed extraction of relevant text documents, or more general, assets from large collections. This concept is very important for component selection, because selection, searching and use of components are based on this information. The overall goal of IR is to deliver an exhaustive sub-collection of all single relevant assets (i.e., independently and individually contributing to the user's goal) [7]. The selection of the right component for a system must be done carefully, generally in the planning period of the software development cycle, if the selection is well done; the fail rate is expected to decrease [8].

2.2 Collective Intelligence

A long time ago, people discovered a variety of interesting insects and animal behavior. A flock of bird sweeps the sky, a group of ants' forages for food, a school of fish swims, etc; this kind of behavior is named "swarm behavior". Recently computer scientists in the field of "artificial life" and "collective intelligence" have studied how these types of animals interact, achieve goals and evolve. Collective Intelligence (CI) can be applied in many areas like telecommunication, robotic, patterns, transportation and military applications [14]. The main idea of CI suggests that N intelligent agents in a colony are cooperating to achieve some goal. The agents use rules to govern their actions, and via interactions of the entire group, the collective achieves its objectives. A type of self-organization emerges from the collection of actions of the group. CI solves problems in a flexible, adaptable, robust, and decentralized way [4].

The agents are able to process information, modulate their behavior according to the stimulus and take the best decision based in the environment information. But the biggest challenge is to make agents to work in a collective way integrating their individual activities to generate more complex and efficient behavior.

3 Proposal

This work proposes a stochastic selection algorithm for software components, which is based on SI; this allows selecting the best component from a group that has been found in Internet according to a group of initial specifications. The selection algorithm uses pheromone tracks in order to identify the best components adding pheromone each time that the component performance is tested. Our Algorithm is called Algorithm A.

In[6] the use of a component characterization sheet is proposed, in order to define the quality of each component. We are using a characterization file that we have called *component profile*, where information associated to each component is stored. This file will have a group of both dynamic and static data, the static data is not going to change, it is referred to the component name, identification, etc, and the dynamic data can change as many times as necessary. This data contains fields like technical platform, location, programmers, performance associated to a specific technical platform where the component is running, and other aspects that can be important for a specific development that can facilitated the selection, like execution time "et".

Each component can have more than one sub-profile for each dynamic environment; a sub-profile is the group of characteristics related to a domain where a component is loaded. Because each profile will be associated to one specific platform, the different sub-profiles are stored in the same XML file (See Figure. 1). In this way, each component can have an undefined number of sub-profiles, each of them represents one domain where that component can be used and it will have a field that represents the performance of the component when it runs on that domain. After a component is selected, rejected or used, our system is going to be able to update the component pheromone according to its performance; pheromone trail can be incremented or decremented according to the component performance. Characteristics for a component evaluation have been chosen using factors related to production time, cost rating, product quality and development risk [8,9].

```
<componentStaticInformation>
 <componentInformation>
  <uniqueID>PRU0707</uniqueID>
  <name>Report Manager</name>
  <license>gpl</license>
</componentStaticInformation>
<profile1>
<pheromone>0.1</pheromone>
<location>www.cemisid.ula.ve/components/rmc </location>
<os>debian</os>
<dependency>proactive.jar - fractal.jar </dependency >
<ms>3</ms>
<et>0.01</et>
</profile1>
<profile2>
<pheromone>0.5</pheromone>
<os>fedora</os>
<dependency>proactive.jar - fractal.jar </dependency>
<ms>1</ms>
<et>0.05</et>
</profile2>
```

Fig. 1. Component Profile – XML file

The initial wished requirements, like component characteristics related to functional and non functional aspects must be well defined in order to generate an XML file that identifies the ideal profile for the component that is going to be selected. In this way, when the group of analysts and programmers search a component, an ideal component profile is generated; this ideal profile will be compared with the different sub-profiles of the found components in order to select the best choice (See Figures 1).

A selection algorithm is proposed, it uses an intelligent agent that will select one and only one component according to the initial requirements, and this means that the algorithm will create as many agents as components we want to select.

The general constraints for our algorithm are:

- A component to select should always exist.
- Each component has at least one profile into an XML file; the file should allow the incorporation of other profiles.
- It is assumed that the initial requirements are exact information about components.

Our algorithm assumes that exist as many agents as required components, the relationship is one to one. Assembling of components will be done for programmers after each agent completes its individual mission, which is the assigned component selection. Each component profile has a field where the pheromone mark is stored according to the performance of the component associated to the domain.

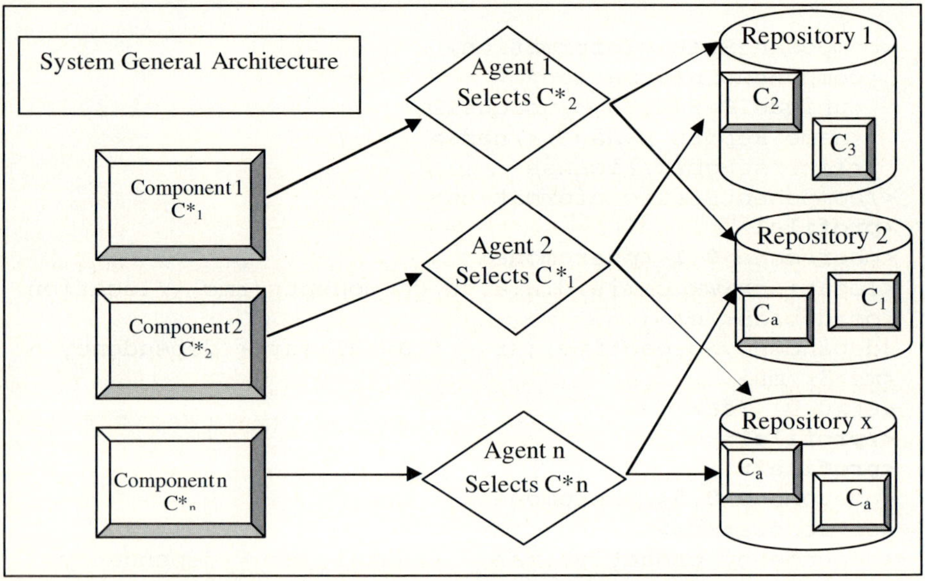

Fig. 2. Each agent selects a single component that will be integrated with the other selected components. The search for components is done using many software repositories.

The equation to establish the matching between the ideal component and the evaluated component is:

$$X_{lju}^{k} = 1 + \sum H^{k} - \sum N_{lju}^{k} \quad (1)$$

Where X_{lju}^{k} is the matching degree for component j which has been located in repository l, j has profile u, and it has been pre-selected by agent k. $\sum H$ identifies the sum of ideal characteristics (C_k^*) that represents the wished profile, for example, operating system, programming language, maximum number of dependencies, among others; on the other hand, $\sum N$ represents the sum of real characteristics of the found "pre-selected" component (component sub-profile). If X_{lju}^{k} is close to 1, means that the ideal characteristics are very similar to the component real characteristics, while less value of $X_{lju}{}^{k}$ better the matching between the ideal component and the selected one.

Each agent will select a software component under the following parameters [2,4]: The ideal value of $X_{lju}{}^{k}$ is 1 and this is obtained when the matching is perfect. The amount of pheromone Y_{lju} (t) related to each component i, that is located in a repository l for its profile u will be update after the component is used into the system. The amount of pheromone is going to be decreased or increased depending on the component performance and evaluation.

The transition equation that gives the probability for an agent k selects a component j, with profile u, into the repository l, among a group of s components named cc, with profiles n, in repositories r, into an iteration t, is coming by [2,4]:

$$P_{lju}^{k}(t)=\frac{\left[Y_{lju}(t)\right]\left[X^{k}{}_{lju}\right]^{-1}}{\sum_{rsn}\left[Y_{rsn}(t)\right]\left[X^{k}{}_{rsn}\right]^{-1}} \tag{2}$$

Where $Y_{lju}(t)$ represents the quantity of pheromone for a component j with profile u that has been found by an agent k into a repository l. $X^{k}{}_{rsu}$ close to 1 implies that there is a good initial matching between ideal components and real components if $X^{k}{}_{rsu}$ is equal to 1 means that both components are identical.

This corresponds to a stochastic algorithm where agents are created and they will initiate the process of random selection, their individual objective will be to find the best component. After complete a selection and test the component, each agent will let a quantity of pheromone ΔY^{k}_{lju} (t) whose value will depend of the system performance R where the component will be used.

$$\Delta Y_{lju}^{k}(t)=(X_{lju}^{k} * R)^{-1} \tag{3}$$

Where

$$R = f(ET, M, ND) \tag{4}$$

The performance function is given by the component execution time (ET), the quantity of memory (M) that it uses and the number of dependencies that the component needs in order to be loaded (ND) in the specific platform.

It is necessary the calculation of both, positive and negative feedback, the last one is called pheromone evaporation, and it is done when a component has not been selected this is incorporated through an evaporation coefficient α [2,4], α can be changed according to the amount of pheromone that user want to evaporate from the component profile when it is not selected, this is important because system will give more pheromone to those components that have better performance and decrease pheromone (evaporate) to those components that have not been selected.

$$Y_{lju}(t) \qquad (1-\alpha) * Y_{lju}(t) + \Delta Y_{lju}^{k}(t) \tag{5}$$

The initial amount of pheromone for each component is assumed as a constant positive value Y_0, this represents a homogeneous distribution of pheromone in time 0. In order to determine the selection process of a given component the following rules are defined:

$$S^*_k = \begin{cases} \arg\max_{rsn \in CC} \{P_{rsn}^{k}\} \\ J \quad \text{(Valor Aleatorio)} \end{cases} \tag{6}$$

Where the value of P is given by equation 2, S^*_k is the selected component from a group s, of the conjunct CC of possible components to select. J is a random selected component.

Macro Algorithm:

1. Definition of software architecture and components to use.
   1.1. Definition and identification of k required components with their profiles.
2. Search of identify components using any kind of search engines.
3. Select best components. (Using Selection Algorithm A).
4. System assemblage.
5. Component performance evaluation.

Selection Algorithm:

1. Create k intelligent selection agents, one for each component to select.
   i. Initialize each agent
   ii. Repeat for each agent i=1 to k /where k is the total number of components to select.
   iii. Identify set of possible components to use (CC) using Eq 1.
   iv. Select component i using Eq. 6
2. Analyze component performance into the system (R) using Eq. 4
3. Update pheromone trail for all components using Eq. 5.

4 Experiments

A first search was done using freshmeat.net [10], we assume that a specific component had to be selected, some functional and non functional characteristics are wished. The first search generated 61 components; we took the first 24 components, each one with its associated XML file. The initial amount of pheromone was incorporated randomly.

Popular search algorithms consider the following non functional characteristics in order to sort the results:

Popularity: The popularity score superseded the old counters for record hits, URL hits and subscriptions. Popularity is calculated as [10]:

((record hits + URL hits) * (subscriptions + 1))^(1/2)

Vitality: The vitality score is calculated like [10]:

(Announcements * age) / (last_announcement),

For example, the number of announcements is multiplied by the number of days an application exists which is then divided by the days passed since the last release. This way, apps with lots of announcements that have been around for a long time and have recently come out with a new release earn a high vitality score, old apps that have only been announced once get a low vitality score. The vitality score is available through the repository page and can be used as a sort key for the search results[10].
Rating: Every registered user of freshmeat may rate a project featured on this website. Based on these ratings, they build a top 20 list and users may sort their search results by ratings as well. The formula gives a true Bayesian estimate [10]:

$$(WR) = (v \div (v+m)) \times R + (m \div (v+m)) \times C$$

where:

WR = weighted rank
R = average for the project (mean) = (Rating)
v = number of votes for the project = (votes)
m = minimum votes required to be listed in the top 20 (currently 20)
C = the mean vote across the whole report

If we compare the selection done using the popularity, vitality, and rating characteristics associated to each components with the selection done using our algorithm, we found the following results:

Table 1. Result comparison among our algorithm and different type of search algorithms commonly used

Number of Evaluated Components = 24		
Based on Variable	Best Component (First Component)	Proposed Selection Order (First 5 Components)
Popularity	Component N. 0	Components 0,23,22,21,20
Vitality	Component N. 0	Components 0,14,22,21,15
Rating	Component N.12	Components 12,23,0,19,20
Our Algorithm	Component N. 0	Components 0,22,23,21,17

It is not always true that the most popular component is going to have better performance or better matching with initial user requirements. Vitality not always corresponds with the component performance and finally, rating is a very subjective selection way because it depends on user votes. Our algorithm uses not only initial user requirements (functional and non functional) but also takes on account the pheromone which is a value that we are incorporating in the component XML file and it is associated to the component performance related to a specific domain. Even when the results are similar among the different ways of sorting (popularity, vitality, rating, our algorithm), in the particular case of our algorithm, component 17 is in the list of proposed selection components, this is expected because this is the component with more characteristics related to the initial user requirements and more pheromone. Component 14 and 15 have a high vitality but don't have good matching with the user requirements, based on rating, component 12 is the best choice but this component doesn't have good matching with user requirements. Our algorithm gives a better result because generates a list of components that not only accomplishes the initial user requirements but also uses pheromone upgrades giving an approximation related to component performance, independently of component popularity, rating or vitality. We run our algorithm using 5 functional and non functional requirements. More than one hundred tests were run, in all of them, the results were very similar, and more information can be found in [1]. The results are very interesting because they show that selection of software component based on popularity, vitality or rating is not always the best one. When software is selected popularity can be a good starting but characteristics like functionality and performance must be considered.

5 Conclusions

This work presents a selection algorithm using intelligent agents, our algorithm uses a pheromone value related to component performance. Selection and search algorithms based on popularity, vitality or rating are very subjective and not always give to the users the best choices. Our algorithm uses functional and non functional requirement to make the selection; any search engine in order to find a group of components can be used and our algorithm is applied in order to select the best choices. This selection algorithm is based on pheromone values that agents upgrade depending on component performance; this gives the opportunity to each found component to be evaluated and its pheromone upgraded. Because of the generality of our model, we propose to test this algorithm with other services related to parallel, distributed and grid computing for example. It would be very interesting to use our algorithm with naming and trading services among others. In future works we will be showing the impact of dynamic composition of components, this is going to be part of a system architecture that we are working on.

Acknowledgment

This work was supported in part by the France-Venezuela cooperation program Grant ECOS-NORD V04M01

References

1. Abraham B. Aguilar J. Batista J. Software Component Selection Algorithm Based on Artificial Ant Systems. Technical Report 01-2005. Centro de Estudios en Microelectronica y Sistemas Distribuidos – CEMISID, 2005.
2. [Aguilar et al., 2004] Aguilar J, Velasquez L, Pool M. "The Combinatorial Ant System", Applied Artificial Intelligence, Taylor and Francis, Vol. 18, N. 5, pp. 427-446, 2004.

3. Barros T, Ludovic H and Madelaine E. Behavioral Models for Hierarchical Components. Spin 2005.
4. Bonabeau E, Dorigo M, Theraulaz G. Swarm Intelligence. From Natural to Artificial Systems. Oxford University Press. 1999.
5. ComponentSource. www.componentsource.com Accesed in oct. 2005.
6. Dellarocas C. Software Component Interconection Should be Treated as a Distinct Distinct Problem. Proceedings of the 8th Annual Workshop on Software Reuse. 1997.
7. Fischer B, Deduction-Based Software Component Retrieval. Mathematic and Informatics Faculty. University of Passau Germany, Thesis Report. November 2001.
8. Gómez-Perez A, Lozano A. Impact of Software Components Characteristics above Decision-making Factors, 2000 International Workshop on Component-Based Software Engineering (CBSE 2000), Limerick, Ireland, 2000.
9. Hamlet D, Mason D, Woit D. Theory of System Reliability Based on Components. In Workshop Proceedings, ICSE 2000, 3rd Workshop on CBSE (W09), Limerick, Ireland, May 2000
10. Search Engine, http://freshmeat.net, Accessed in 12/Oct/2005 .
11. Seacord, R.; Hissan, S.; Wallnau, K, "Agora: A Search Engine for Software Components", IEEE Internet Computing, vol.2, no.6, November/December, 1998, pp. 62-70.
12. Silvestri F, Puppin D, Laforenza D. Toward a Search Architecture for Software Components. ISTI-CNR, Italy, January 2004.
13. Sugumaran V and Storey V. A Semantic Based Approach to Component Retrieval. The Data Base for Advances in Information Systems – Summer 2003 (Vol. 34, No.3)
14. Wooldridge, J.P. Muller, M. Tambe (eds.). Assistant Agents that Distribute How-To-Do Knowledge. Intelligent Agents II. Lecture Notes in AI, Volume 1037. pages 408-411. Springer-Verlag, 1996

A Methodology to Specify Multiagent Systems

Aguilar Jose[1], Cerrada Mariela[1], and Hidrobo Francisco[2]

[1] CEMISID (Centro de Microcomputacion y Sistemas Distribuidos),
Facultad de Ingeniería, Universidad de los Andes, Mérida 5101, Venezuela
{aguilar, cerradam}@ula.ve

[2] SUMA, Facultad de Ciencias, Universidad de los Andes, Mérida 5101, Venezuela
hidrobo@ula.ve

Abstract. In this paper we present a methodology to specify Multiagent Systems, called MASINA. MASINA is based on MAS-CommonKADS; we use the models presented in this methodology to propose some extensions, modifications and substitutions allowing to describe the intelligent characteristics of an agent or group of agents, to use intelligent techniques for the accomplishment of tasks (e.g. artificial neural networks), and to specify emergent coordination approaches, among others.

1 Introduction

The investigation of methodologies for analysis and design of Multiagent Systems (MAS) is still in embryonic state. The existing methodologies for developing MAS are not exactly new; generally they are extensions from object-oriented methodologies or knowledge engineering methodologies, given their close relationship. The existing ones have important weaknesses and disadvantages which makes them unable for usage in complexly designed environments. In this article we propose a methodology for the specification of Multiagent Systems, named as MASINA (MultiAgent Systems in Automation). The term of Automation is due that this methodology was originally developed to solve automation problems using MAS [1, 2]. This methodology is based on MAS-CommonKADS [3], proposing some extensions, modifications and substitutions of the models that have already been defined in MAS-CommonKADS. MASINA depicts the fundamental elements in the MAS area, such as representing the notion of intelligence of an agent (intelligent agents modeling), in a collective level (swarm intelligence), the coordination mechanisms between agents (emerging planning, conflict resolution, etc.), the direct or indirect communication, along with other things. In addition, MASINA allows characterizing other aspects such as the use of intelligent techniques for carrying out tasks, the use of reference models for specifying agents and the definition of conversations between agents as well as the speech acts (interactions) within these conversations.

2 Multiagent Systems Specification Methodologies

There are three main groups of propositions of designing methodologies based on agents, the first one is based on object-oriented methodologies, the second one is based on

N.T. Nguyen et al. (Eds.): KES-AMSTA 2007, LNAI 4496, pp. 92–101, 2007.

methodologies coming from the knowledge engineering and the third one is based on methodologies coming from the agent paradigm itself. There are several proposals on methodologies for object-oriented agents specification, the most important ones is presented in Wood [4]. This methodology begins with an initial specification of the system and shows a set of documents with formal design based on graphics. About the methodologies coming from knowledge engineering, there is one relevant methodology which is based on CommonKADS methodology [3]. MAS-CommonKADS is the methodology that best covers the specification elements of a MAS, nevertheless, this is not implying that it does not have weaknesses: (a) the difficulty to specify an agent as a MAS, (b) the lack of dynamic schemes management for communication and (c) the lack of learning models. Finally, the methodologies coming out of the current agents' theory are founded in a social structure, where there are individuals (agents), groups and organizations. Among these methodologies we find Gaia [5], Prometheus [6], and Tropos [7]. The common problem of these methodologies is that the analysis and/or specification phases are based on the agents' paradigm, and it is not taken into account that a general analysis model can propose a multiagent scheme that would not be the most appropriate. The main design element is the social role of the agent and not its components. Some others methodologies are proposed in [8, 9].

Particularly, the MAS-CommonKADS methodology is an extension from CommonKADS model for knowledge engineering [10], adding aspects from the object-oriented methodologies like OMT, Object Oriented Software Engineering (OOSE) and Responsibility Driving Design (RDD). It has six phases, from which seven models derive. The phases are: i) **Conceptualization:** This is the extraction and purchasing of the knowledge for obtaining a first description of the problem, ii) **Analysis**: Also called Requirements Analysis, its objective or product is the requirements list. The first six models are developed in this phase, iii) **Design**: The final model of the organization is obtained under the MAS paradigm, iv) **Coding and testing:** In this phase each agent is coded and tested, v) **Integration:** Integration is accomplished into the real platform where the system will work. Then, the whole system is prooved, vi) **Operation and maintenance**: Is the working phase of the MAS, in which updating tasks are performed (maintenance) on the system's components in order to adapt it to the environmental evolution or new requirements.

This methodology uses the next seven models [3]:

- **Agent Model:** Describes each one of the agents of the MAS, its objectives, characteristics, abilities, capacities, restrictions and the services it may offer to the agents community. This model is the heart of the methodology and allows the description of the actors in the problem to be solved; allows the description of agents based on the provided services, but it does not facilitate the description of an agent as a MAS.
- **Tasks Model:** In this model the tasks that should be carried out organizational wise are described. The agent's actions are specified; this way, each agent has a bank of realizable tasks, which have very well defined properties.
- **Organization Model:** It is the way for describing the organization in which the MAS will be integrated. Particularly, it describes the organization to be modeled by the MAS, before and after implementing the MAS. This model

allows considering whether the solution to the problem is the incorporation of a MAS, so it facilitates analyzing the problem and finding a solution.

- **Communication Model:** This model describes the interactions between an human agent and a software agent and it is centered in considering human factors for such interactions. It is a model based in messages, so it only specifies pre-established direct communication; it does not allow the specification of a spontaneous communication between them. This model is not detailed in MAS-Common KADS and it is believed to be derived from a coordination model.
- **Experience Model:** This is the knowledge model. The experience model makes the difference between the knowledge of the application and the knowledge for the problem solution.
- **Coordination Model:** It describes the interactions between the software agents, the used protocols and the necessary capabilities for carrying out these interactions; the coordination model defines the technical bases for developing the communication model.
- **Design Model:** It is built from the previous models in the analysis phase. In this phase the MAS design model is created.

3 MASINA: MultiAgent Systems in Automation

The MASINA methodology is an extension from the object-oriented MAS-CommonKADS methodology. MASINA is based on the same development cycle of MAS-CommonKADS, but with the following main modifications and contributions:

- In the Conceptualization phase, MASINA uses UML 2.0 activities diagrams [11]. The activities diagram allows the description of the offered services by the agent, dissociating it into the needed activities for carrying it out.
- In the analysis phase, MASINA introduces four of the seven proposed models in MAS-CommonKADS, enough for describing the basic characteristics of the MAS (see figure 1): Agents Model, Tasks Model, Communication Model, Coordination Model. Furthermore, MASINA replaces the Experience Model by the Intelligence Model. Intelligent Model describes all the aspects to generate an intelligent behaviour: learning, reasoning, knowlegde representation, and experience accumulation.
- The Coordination and Communication models have different meaning in MASINA. The Coordination Model describes the conversations and the Communication Model describes the speech acts. For better understanding the Coordination Model , MASINA uses the UML interaction diagrams to describe the conversations.
- Agent Model has new attributes to specify different abstraction levels in a community of agents or if the agent is designed using a development framework.
- Task Model allows specifying the tasks based on intelligent techniques (genetic algorithms, artificial neural network, etc.), and the task decomposition.

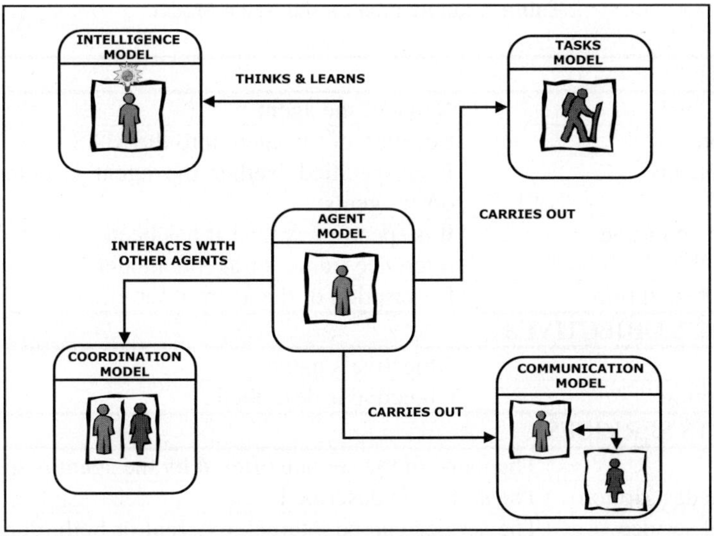

Fig. 1. MASINA Models

3.1 Agents Model

This model specifies, just like in the MAS-CommonKADS, the characteristics of an agent such as: abilities, services, etc. In MASINA, this model has two new attributes: (a) "agent components", which allows to indicate whether the agent is a MAS (allows to define abstraction levels or hierarchies in the design of the MAS), and (b) "reference frame", which indicates whether the specification of such an agent is based on an MAS reference model already existing, like the SCDIA [2, 12, 13]. For defining the agent model, we use the pattern shown in table 1.

3.2 Tasks Model

The agents have been provided with new attributes: (a) one for specifying the properties of the tasks which require the use of intelligent techniques, and (b) another for widely describing the whole procedure to be followed for the execution of such a tasks (sub-tasks). This way, the agents activate their intelligences according to the tasks type they are carrying out (whether they are intelligent or not). For describing the tasks model, MASINA makes use of two models: one of them indicates the service-task relationship (each service has a set of tasks to be carried out and a single task could be associated to different services), the other model specifies the different tasks (see table 2).

3.3 Intelligence Model

The model proposed by MAS-CommonKADS only characterizes everything related to experience acommulation in an agent, but it does not take into account other elements that allow the agent to have an intelligent behavior. MASINA describes all

Table 1. Information for the Agent Model

AGENT	
Name	Name of the agent
Position	Location of the agent into the MAS
Components	It is specified wether the agent is composed by other agents
Reference Frame	It is specified wether it has been characterized by a reference frame for agents modelling
Agent description	Description of the agent's activities
AGENT'S OBJECTIVES	
Name	Objective's name
Description	Objective is described
AGENTS SERVICES	
Name	The name of the service offered by the agent is specified
Service description	The service is described
Type of service	The service can be internal, external or both (dual service) and it is specified
Input parameters	The necessary parameters for the service achievement
Exit parameters	The parameters obtained when the service is finished
Activation Condition	The activating conditions of the tasks asociated to the achievement of the objective
Ending conditon	The conditions that indicate the termination of tasks associated to the objectives achievement
Success condition	The conditions that indicate the accomplishment of the objective
Failure condition	The conditions which indicate the objective has not been accomplished
Ontology	The used ontology

Table 2. Tasks Model Information

NAME OF TASK	
Name	Name of the task
Objective	Objective of the task
Description	Defined objective is described
Asociated services	Associated services to the objective
Precondition	Necessary conditions for task activation
Sub-tasks	The sub task to accomplish this task
NAME-OF-TASK INGREDIENT	
Name_Ingredient 1	The data/parameter necessary for task accomplishment is indicated
Name_Ingredient 2	…

the necessary aspects to incorporate the notion of intelligence into an agent. Basically, it is composed by a set of elements which correspond with human reasoning. It specifies the knowledge domain and the strategies and tasks the agent carries out, the experience and the knowledge associated to the problems, the reasoning methods needes for the agents to accomplish their objectives, as well as their learning mechanisms. The proposed intelligence model is presented in figure 2, with its attributes, relationships between them, and their interrelationship with the MASINA models. This model integrates concepts such as experience, knowledge representation, learning mechanism, and reasoning mechanism.

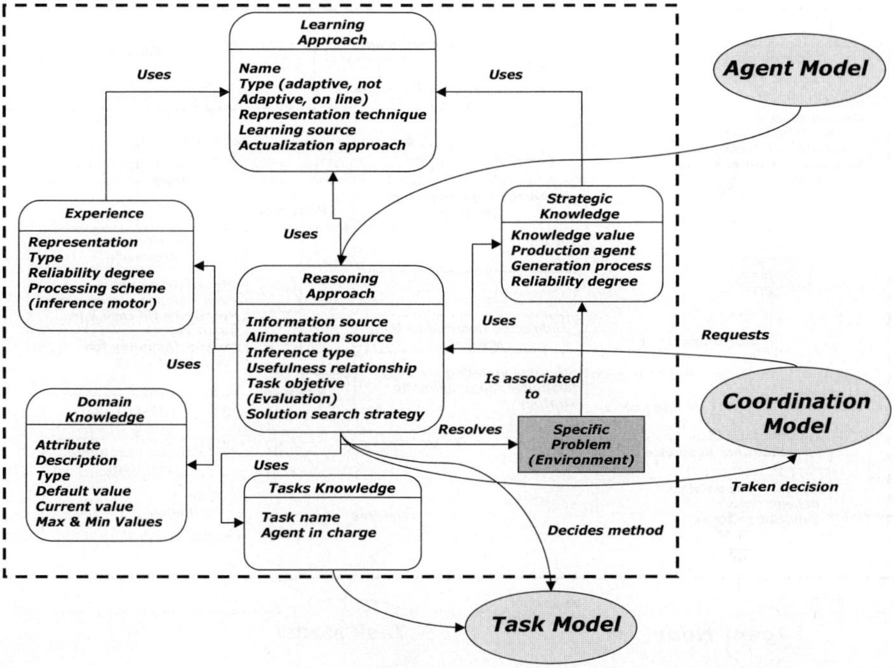

Fig. 2. Intelligence Model

3.4 Coordination Model

In MASINA the coordination model allows the specification of the conversations between agents. Unlike MAS-CommonKADS, the Coordination Model in MASINA is focused on the conversations allowing a coordinated communication between agents and not on the speech acts specifically involved, which are detailed in the Communication Model in MASINA. The proposed coordination model allows to the user establishing a set of strategies which would be used by the agents' communities for the objective achievement. These strategies, generally, imitate attitudes of human groups under a coordination system, such as the contracting processes and auctions, amongst others, where the interaction and information exchange generate individual actions which added up allow the reaching of the specific goal. The interactions (speech acts) into a conversation between agents, are presented, at a graphic level, through an UML interaction diagram and such a speech acts are detailed in the Communication Model. In the Coordination Model, two types of communication between agents can be established: a direct one, based on the message exchange and the use of ontologies, and an indirect one, based on strategies of shared memory and stimulus-response methods. It is for this reason we need more detailed description of the Coordination Model in each conversation (see figure 3). In this model, the coordination scheme, the planning and communication mechanisms (direct or indirect), the meta-language and the ontology are specified.

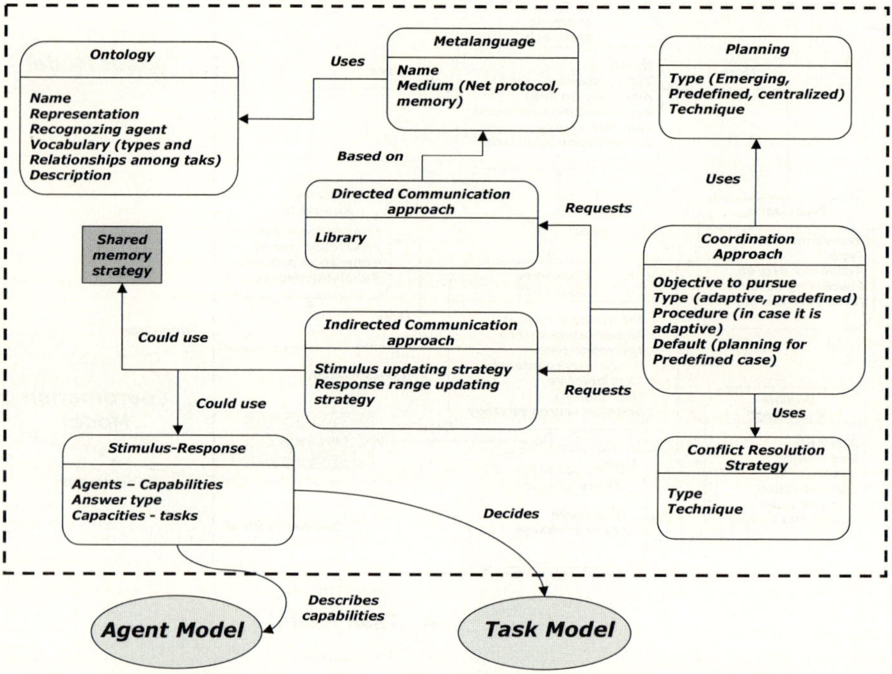

Fig. 3. Coordination Model

3.5 Communication Model

MAS-Common KADS does not describe this model in a detailed way, instead it is thought to be derive from the coordination model; describes the interactions between a human agent and a software agent and it is centered in considering human factors for such interaction. In our MASINA methodology, the interactions are considered in a wide way and it is proposed a Communication Model which describes the speech acts involved in the conversations between the agents. The model describes all the direct communications through the information exchange using mechanisms of the Message Exchange, and indirect communications through the information storage in objects, which resemble the interactions between agents. The attributes of the Communication Model allow the specification of a speech act in a conversation (see table 3).

Table 3. Communication Model Information

NAME OF THE SPEECH ACT	
Name	Specifies the name of the speech act
Type	Indicates the request characteristics (information, processing, and other requirements)
Objective	Objective of the interaction
Participating agents	The sender and receiver agents in the speech act are indicated
Initiator	The agent initiating the interaction
Exchanged Data	The data exchanged during the speech act
Precondition	The initiating conditions of the speech act
Termination Condition	The conditions determining the ending of the interaction between two agents
Conversation	Conversations where the described speech act can be found
Description	The speech act is described

4 Previous MASINA Applications

In [2] the proposed MASINA methodology has been used for specifying a Reference Model for Intelligent Distributed Control Systems based on agents (SCDIA). SCDIA is conformed by a community of agents and the system information and control tasks are distributed among them. This community has been called Control Agents Community: control agent, coordination agent, action agents, specialized agent and observer agent. All the SCDIA agents have been specified using MASINA.

In [1] MASINA has been used for designing an Industrial Automation Architecture based on MAS, called SADIA (Intelligent Distributed Automation System based on Agents). SADIA is and extension of the SCDIA, which includes all the automation tasks: failures management, planning and production management tasks, processes control and management of abnormal situations. As a part of the SADIA design, the SCDIA reference model, specified using MASINA, has been used for modeling the Failures and Abnormal Situations Management Agents [12, 14]. In [13] a software component allowing the implementation of a code generation system for the control agents of the SCDIA is developed. This includes the source code generation of the agent, its compilation and incorporation to the SCDIA. The system has three agents: Central agent, Code Generator agent and Behavior agent which have been specified using MASINA.

In [15] MASINA has been used for designing a Middleware based on agents. The Middleware based on agents can be used on any computational platform which stands the execution of MAS; it provides access and hardware resources management services, data and agents application. This agents-based middleware has qualities associated to the distributed systems such as interoperability, migration, security, naming, etc. In [16] MASINA has been used for modeling the Architecture of a Web Operative System (SOW), as well as for the detailed design for each one of the sub-systems composing the Operative System, which are: Communities Manager System, Replacing Manager System, Web Objects Manager System and Resources Manager System. Finally, in [17] a formal method has been proposed for verifying the models that have been specified using MASINA. This method allows the validation of the proposed design for an application based on MAS.

5 Conclusions

The objective of this paper was the presentation of the MASINA methodology, which modifies the models that have been already proposed in MAS-CommonKADS. Particularly, MASINA modifies the meaning of the coordination and communication models and replaces the experience model by the intelligence model. This way, MASINA strengthens the communication between agents, by using indirect communication methods and asynchronic coordination schemes in order to improve the description of the conversations between agents. Furthermore, MASINA allows the use of emerging coordination schemes and the modeling of intelligent agents using the intelligence model, amongst other things. MASINA allows a better description/specification of multiagent systems. MASINA has a method that allows the verification of the proposed design for a determined application [5]. Currently, we are developing an editing tool for the MASINA models, which would permit automatically the agents code generating on JADE language.

References

1. J. Aguilar, C. Bravo and F. Rivas. Diseño de una Arquitectura de Automatización Industrial basada en Sistemas Multiagentes. *Revista Ciencia e Ingeniería, Facultad de Ingeniería*, 25(2): 75–88, Julio 2004.
2. J. Aguilar, M. Cerrada, F. Hidrobo, G. Mousalli, and F. Rivas. "A Multiagent Model for Intelligent Distributed Control Systems", *Lecture Notes in Artificial Intelligence*, Springer-Verlag, Vol. 3681, pp. 191-197, 2005.
3. C.A. Iglesias, M. Garijo, J: C González, and J.R. Velazco. "Analysis and Design of Multi-Agent Systems using MAS-CommonKADS". In M.P. Singh, A.S. Rao, and M. Wooldridge, editors, *Intelligent Agents IV. Agents Theories, Architectures and Language*, Springer, pp 1–16, 1999.
4. M. F. Wood and S. A. DeLoach. "An Overview of the Multiagent Systems Engineering Methodology". In *First international worksho, on Agent-oriented software engineering*, Springer-Verlag, pp 207–221, 2000.
5. F. Zambonelli, N. R. Jennings, and M. Wooldridge. "Developing Multiagent Systems: The Gaia Methodology". *ACM Trans. Softw. Eng. Methodol.*, 12(3): 317–370, 2003.
6. L. Padgham and M. Wimikoff. "Prometheus: A Methodology for Developing Intelligent Agents". In P. Giunchiglia, J. Odell, and G. Weiss, editors, *Agent-Oriented Software Engineering III*, pp 174–185. Springer, LNCS 2585, 2003.
7. P. Bresciani, A.Perini, P. Giorgini, F. Giunchiglia, and J. Mylopoulos. "Tropos: An Agent-Oriented Software Development Methodology". *Autonomous Agents and Multi-Agent Systems*, 8(3): 203–236, 2004.
8. S. Vafadar, A. Barfouroush, and M. Reza A. Shirazi. "Towards a More Expressive and Refinable Multiagent System Engineering Methodology". In *AOIS*, pp 126–141, 2003.

9. F. Alonso, S. Frutos, L. Martinez, and C. Montes. “SONIA: A Methodology for Natural Agent Development”. In *Fifth International Workshop Engineering Societies in the Agents World*, Toulouse, France, October 2004.
10. Walter van de Velde Andre Valente, Joost Breuker. The CommonKADS Library in Perspective. *International Journal of Human-Computer Studies*, 49: 391–416, 1998.
11. J. Rumbaugh, I. Jacobson, G. Booch. *The Unified Modeling Language Reference Manual*, Object Technology Series, Addison-Wesley, 1998.
12. J. Aguilar, J. Cardillo, M. Cerrada, and R. Faneite. “Agents-Based Design for Fault Management Systems in Industrial Processes”, Accepted for publication, *Computer in Industry*, 2006.
13. J. Aguilar, M. Cerrada, F. Hidrobo, F. Rivas, and W. Zayas. “Development of a Code Generation System for Control Agents", *WSEAS Transactions on Computers*, Vol. 5, No. 10, pp. 2406-2411, 2006.
14. J. Aguilar, C. Bravo, E. Colina, and F. Rivas. “Sistema Multiagentes para Tratamiento de Situaciones Anormales en Procesos Industriales”. In *V Congreso Nacional de la Asociación Colombiana de Automática*, pp. 24–28, Colombia, 2003.
15. J. Aguilar, V. Bravo, F. Rivas, and M. Cerrada. “Diseño de un Medio de Gestión de Servicios para Sistemas Multiagentes”. In *XXX Conferencia Latinoamericana de Informática*, pp 431–439, Arequipa, Perú, Octubre 2004.
16. J. Aguilar, E. Ferrer, N. Perozo, and J. Vizcarrondo. “Architecture of a Web Operating System based on Multiagent Systems”, *Lecture Notes in Artificial Intelligence*, Springer-Verlag, Vol. 3681, pp. 700-706, 2005.
17. J. Aguilar, N. Perozo, and J. Vizcarrondo. “Definition of a Verification Method for the MASINA Methodology”, *International Journal of Information Technology*, Springer-Verlag, Vol. 12, N.3, pp. 121-131, 2006.

A Stochastic Process Model for Daily Travel Patterns and Traffic Information

Yongtaek Lim[1], Seung Jae Lee[2], and Joohwan Kim[3]

[1] Dept. of Transportation & Logistics, Chonnam National University, San 96-1, Dundeok-dong, Yosu, Chonnam, Korea
limyt@chonnam.ac.kr
[2] Visiting Professor, University of California at Berkeley
Berkeley, CA94720, USA
sjlee@usocc.uso.ac.kr
[3] KyungBong Corp,
K-Center 1591-9, Gwanyang-dong, Dongan-gu, Anyang-si, Gyeonggi-do, Korea
kimj3449@kyungdong.co.kr

Abstract. Sudden changes on road networks, including new roads, bridge construction, road blockage or traffic accidents cause travelers to switch their routes to less costly ones as compared to alternative routes. Travelers, however, tend to take higher cost routes due to insufficient information and errors in perceived travel time. This may cause severe congestion on a certain route. Conventional models, however, are unable to adequately simulate travelers' behavior under such suddenly changing network conditions. The objective of this paper is to analyze travelers' daily travel behavior in such cases via a stochastic process, the Markov-chain approach, which is considered to be a suitable method for representing sudden changes in states. This model is based on agent and we assumes that travelers select their route via learning process of travel time that they had previously experienced.

1 Introduction

Travelers on road networks attempt to determine the shortest time or cost path from their origins to their destinations. Sudden changes on road networks, including new roads, bridge construction, road blockage or traffic accidents can cause travelers to switch their routes to less costly ones, as compared to alternative routes. Travelers, however, tend to take higher cost routes, due to insufficient information and errors in perceived travel time. This may cause severe congestion on a certain route. In conjunction with the behaviors of the traveler, there exist quite a few transportation models. One of them is the Stochastic User Equilibrium (SUE) model. SUE models were initially developed to take into account travelers' perceived errors regarding traffic conditions. These models are predicated on Wardrop's principle: each traveler selects the route with the minimum 'perceived' travel cost. Most SUE models, however, generate only an equilibrium state after several iterations; therefore, the

N.T. Nguyen et al. (Eds.): KES-AMSTA 2007, LNAI 4496, pp. 102–110, 2007.

models are unable to simulate fully the travelers' behavior in the case of sudden network changes, including road-blocking, new road-opening, etc.

On the other hand, traffic information systems have become principal issues in many countries as a modern technology for the alleviation of traffic congestion in the urban area. By providing real-time travel information regarding traffic conditions, traffic information can enhance the traveler's knowledge of the situation in road networks, and may assist the traveler's decisions, including departure time choice, route choice, and destination choice. The usage of real-time traffic information and Advanced Traffic Management Systems within the framework of the Intelligent Transportation System, in particular, has become a powerful tool, and a possible component of a solution to the ever-growing congestion problem. In the case of traffic accidents, changes in departure time may be eventuated as the result of appropriate pre-trip information. Route changes are probably the easiest behavioral changes to effect. Effective re-routing strategies include early diversion prior to severe congestion and the switching of routes to alternative routes. Several papers have demonstrated that traffic information can provide travelers with virtual benefits including reduced travel time, avoidance of traffic accidents, etc.

The purpose of this paper is to analyze travelers' daily travel behavior in cases of sudden changes in road conditions via a stochastic process--that is, the Markov-chain approach, which is well-known as a suitable method for representing sudden changes in states. The Markov-chain approach was devised in an attempt to supplement the weaknesses of the above-mentioned SUE models. This model assumes that each traveler is an agent and he selects his route via a travel time learning process based on past experience. The agent-based decision component will be used to determine individual travelers' preferences and route switching decisions provided traffic information. In this paper, pre-trip traffic information was provided in advance for travelers (agents) prior to the beginning of trips, and the effects of the information were assessed. For this purpose, a daily route choice model was developed, in which daily travel patterns were simulated via a Markov-chain process. The developed models represent the travelers' daily behaviors, such as changing one's own route in order to minimize travel costs.

The procedure outlined in this paper is as follows: firstly, a daily route choice traffic assignment model was developed in order to describe the travelers' behaviors. The model consisted of a route choice among different routes and a stochastic process between days. It was based on the multinomial logit type, and the daily travel patterns were based on the Markov-chain process. A solution algorithm is also briefly described. Secondly, daily travel patterns with information were assessed via the developed model in the sample network.

2 Structure of the Model

2.1 Route Choice and Traffic Information

Travelers' route choice behavior can be modeled via a discrete choice theory, which describes individuals' choices between competing alternatives. The perceived travel cost on route k between the origin and the destination can be expressed as follows:

$$C_k = c_k + \varepsilon_k$$

In which c_k is the actual travel cost at a given flow, and ε_k represents the random components. If the random term ε_k represents independent and identically distributed Gumbel variants, a multinomial logit model can be constructed, as follows:

$$q_k = \frac{e^{-\beta \bullet c_k}}{\sum_{m=1}^{n} {}^{-\beta \bullet c_m}}$$

In which β is a positive scaling factor associated with variances in the perceived travel costs, and can also be interpreted as sensitive to the traveler's travel costs. The parameter can also be interpreted as a degree of information provided by information systems, such as in-vehicle route guidance systems or variable message signs, as traveler's behaviors can vary with this parameter. In addition, a relationship exists between the variance(σ^2) of perceived travel cost in the logit model and β , as shown in the equation below.

$$\sigma^2 = \frac{\pi^2}{6\beta^2}$$

According to this equation, if the value of β is small, the variance of travel cost has a large value, and thus the travelers do not have perfect knowledge of the network conditions. The β , therefore, can be interpreted as the level of traffic information on that network.

2.2 Stochastic Process Model for Daily Travel Pattern

Each traveler selects his route based on previous travel time, and may vary the route with variable route costs. Individual traveler can be modeled using an agent. The main advantage of using agents in travel behavior modeling is that they are active entities that interact with their travel environment and in concert with other agents. This agent-based approach also allows for the updating of travelers' decisions and learning process on a day-to-day basis rather than the static basis approach. This daily travel behavior pattern of agents can be described as a stochastic process. Cascetta (1989) considered traveler's dynamic evolution to be a discrete time stochastic process. A stochastic process model that has received a great deal of attention in the transportation literature is one in which travelers select routes based on memory. Let $\{R(t)\} = R(1), R(2),..$ be a sequence of route choice vectors from a stochastic process, in which the index t has a real temporal counting day value. Based on memory of the finite past, the route choice probability distribution at time t is naturally defined as conditional on the past, and is expressed as follows:

$pr(r(t)|r(t-1), r(t-2),...)$. For example, each traveler (agent) may choose a route with minimum perceived cost during the last epoch.

Note here that for any general memory, the agents' route choice for the present epoch is independent conditional on the past. Hence the joint conditional route choice distribution is the simple product of each individual's marginal choice distribution. Thus, the conditional independence allows for such traffic assignment processes to be simulated rather readily. One reason for the current interest in memory-based assignment processes is that they can be studied as Markov Chains, as follows.
The stochastic process, $Y(t)$, is a Markov chain if :

$$pr(Y(t) = y|Y(t-1), Y(t-2),..., Y(0)) = pr(Y(t) = y|Y(t-1)) \text{ for all } y .$$

With the Markov chain, we can compute on-step transition probability between days as,

$$p_{ij} = p_r \langle (v_k = j\, today | (v_k = i\, yesterday \rangle$$

where,

$$p_{ij} = \textit{From yesterday to today the probability of transition}$$
$$v_k = \textit{path}\, k\, \textit{flow}$$

The probability p_{ij} is uniquely defined as an assumed one-dependent Markov chain (see Watling(1996) for more detail).

2.3 Solution Algorithm

In this paper, a stochastic process, coupled with the Markov-chain approach for the representation of ergodicity of link flow, was utilized in order to solve the problem. The solution algorithm was as follows.

[step1] Initialization
D : total demand
State s =[0, D]
Day d =1

[step2] Compute average travel cost

$$c_k = \sum_{t=1}^{T} w_t c_k^{d-t} \qquad t = 1,.....,T$$

where, c_k^d is the travel cost on k path of day d and w_t is a weighting parameter for day d .

[step3] Compute path choice probability of agents

$$q_k = \frac{e^{-\beta \bullet c_k{}^d}}{\sum_{m=1}^{n} e^{-\beta \bullet c_m{}^d}}$$

[step4] Compute the probability p_{ij} by one-dependent Markov chain.

$$p_{ij} = \frac{D!}{j!(D-j)!}(q_i)^j(1-q_i)^{D-j}$$

[step5] calculate stationary distribution (π^d)
Let the π^d denote the probability distribution, the evolution of the probability distribution is calculated recursively from

$$\pi^{d+1} = \pi^d p$$

where, π^d is the probability distribution of path k flow on day d ,where π^0 is given as initial condition, p_{ij} is the transition probability matrix as computed in [step4]

[step6] Compute average path flow matrix (F^d)

$$F^d = s\pi^d , \quad s = 1,\ldots..,D$$

[step7] Compute link flow matrix(v^d)

$$v^d = AF^d$$

where, A is the link-path incidence matrix.

[step8] Convergence test

if $v^{d+1} = v^d$, stop
otherwise $d = d+1$, go to [step2]

3 Daily Travel Patterns

3.1 Numerical Example

A numerical example is presented in order to illustrate the daily travel patterns, using the model developed in the paper. The sample network shown in Figure 1 consists of

3 paths between the origin and destination. Input data, including link capacity and free-flow cost, are shown in Table 1. It is assumed that there is one origin-destination pair from node 1 to node 7, together with a trip demand for 10 vehicles.

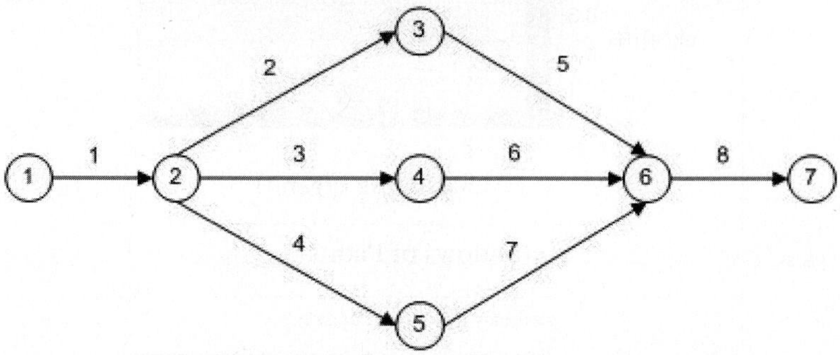

Fig. 1. Example network

Table 1. Network data

Links	Initial time	Link capacity
1	20	20
2	20	6
3	20	10
4	20	20
5	25	6
6	20	10
7	20	20
8	20	20

3.2 Results

Figure 2 shows the relationship between the stationary distribution probability and each state or path flow in the paper. The figures show the path choice probability for each of the path flows, in accordance with successive incremental path flows. This is a significant difference, in that the developed model generates the stationary probability for each state. This can not be calculated via conventional stochastic assignment models.

Table 2 shows the comparison of average path flows among the models considered in the paper. As is shown in the table, The daily travel model has path flow values similar to that of the SUE (Stochastic User Equilibrium) model. However, the UE (User Equilibrium) model allocates the total trips to path 3. From this result, we can determine that the daily travel model most closely approximates the equilibrium of the SUE model.

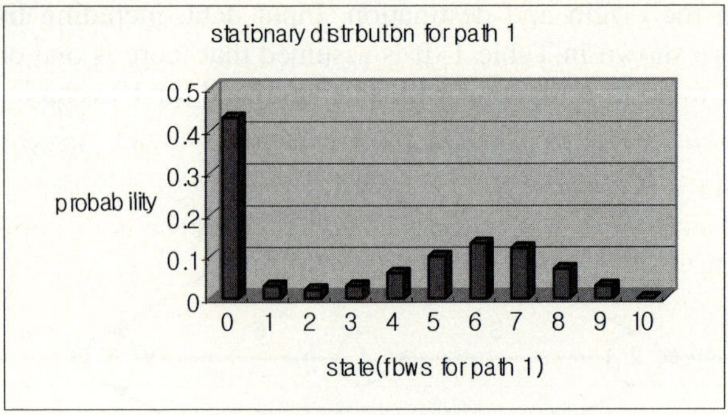

(a) flows of Path 1

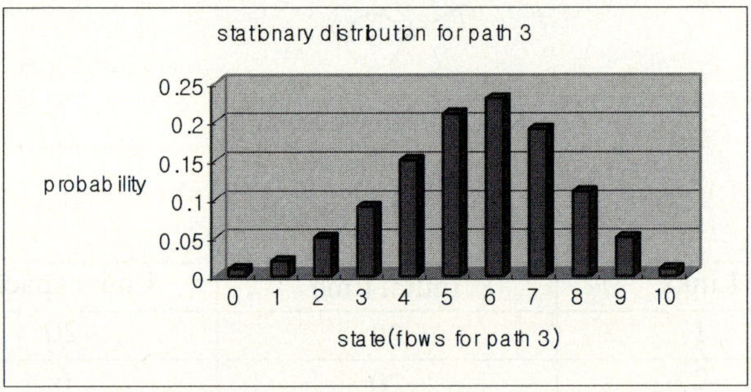

(b) flows of path 2

Fig. 2. Stationary Distribution for path 1 and for path 3

Table 2. Comparison of path flows among models

Paths	Connecting nodes	UE model	SUE model	Daily travel model
Path 1	1→2→3→6→7	0	3.7	3.2634
Path 2	1→2→4→6→7	0	2.0	2.4025
Path 3	1→2→5→6→7	10	4.3	4.3141

Figure 3 displays the daily travel pattern of each of the paths. Several fluctuations occur in early days, but as the days pass the path flows settle down to stationary values. These fluctuation phenomena indicate the process of daily travel patterns prior to equilibrium. Therefore, the model is expected to prove useful in the analysis of daily travel patterns in the case of sudden traffic shifts such as the opening of new roads, the charging of congestion tolls and road work. In this paper, after 80 days, the travel trends tend toward steady state conditions as a whole. From figure 3, we can conclude that the model simulates daily travel patterns well, in accordance with its purpose.

The magnitude of the shift in path flow will depend on the variances in the distribution of path travel times. Figure 4 shows this. In cases in which β has a low value, which means that the travelers (agents) have large variances in travel time on their routes and have not been guided, they tend to choose their paths on the basis of their previous experience . This makes the path flow evenly to each path, as is shown in Figure 4.

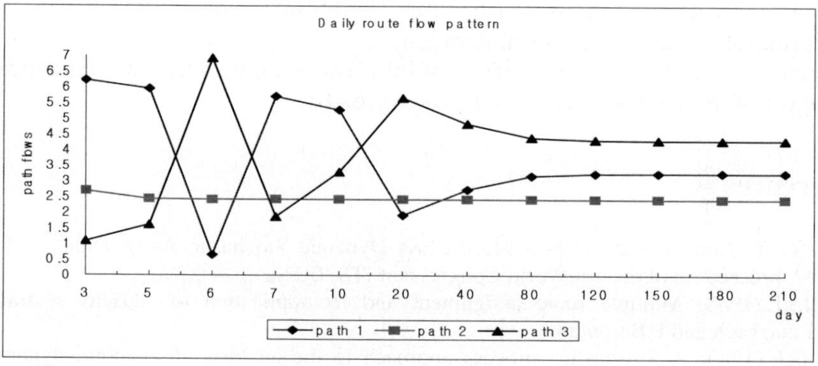

Fig. 3. Daily travel patterns for each path

However, in cases in which β has a large value, low variances in travel time are seen, showing that more information is provided to travelers. The travelers are assigned to specific paths, and this reflects low travel costs. These results indicate that the degree of information for travelers is increased, and the travelers behave in accordance with the User Equilibrium rule.

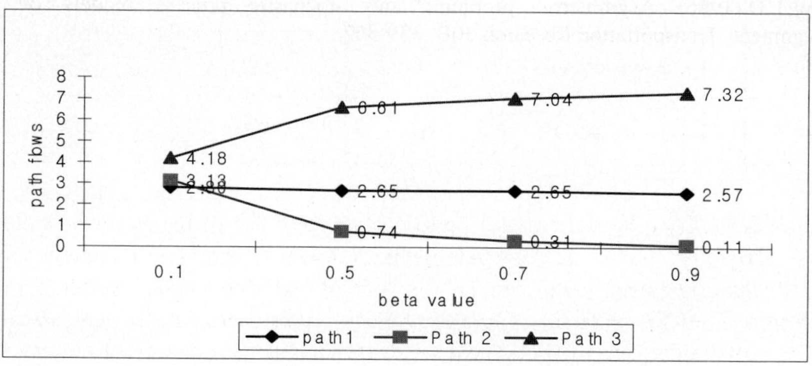

Fig. 4. Path flow patterns with varying β

4 Conclusions

In this paper, we have presented a stochastic process model based on agent which describes the daily travel patterns of travelers on the road network, and also proposes a solution algorithm.

The principal findings in the research presented herein are as follows:

First, as a result of testing the developed model with a contrived simple network composing of three paths, the model converges toward stationary stability.

Second, the model enables a representation of the travelers' daily behavior more precisely than do existing stochastic user equilibrium models, which generate an equilibrium state only after several iterations.

Finally, the model shows the effects of information on travelers in accordance with the degree of information possessed by the travelers.

References

Baek, S., Y. Lim, K. Lim (1997) Multi-Class Dynamic Stochastic Assignment, 1~24 Oct. 1997, proceeding of the 4th World Congress on ITS, Berlin

Burrell,J.E.(1968) Multiple route assignment and its application to capacity restraint, In W.Leutzbach and P.Baron(eds)

Cascetta,E.(1989) A Stochastic process approach to the analysis of temporal dynamics in transportation networks, Transportation Reserach 23B, 1-17

Cascetta, E.(1991) A day-to-day and within-day dynamic stochastic assignment model, Transportation Research 25A, 277-291

Ortuzar,J.D and L.G.Willumsen(1994) Modelling Transport, 2nd eds, Wiley

Lim, Y., S. Baek, K. Lim(1997) Assessment of Traffic Information with Stochastic Assignment, Journal of the Eastern Asia Society for Transportation Studies, Vo.2 No. 4, Autumn,1997, 1275-1284

Sheffi, Y. (1985) Urban transportation networks: equilibrium analysis with mathematical programming methods, Prentice Hall, New Jersey.

Sheffi,Y., and W.B. Powell(1981) A comparison of stochastic and deterministic traffic assignment over congested networks, Transportation Research 15B(1), 53-64

Watling, D.(1996) Asymmetric problems and stochastic process models of traffic assignment, Transportation Research 30B, 339-357

Using Data Mining Algorithms for Statistical Learning of a Software Agent

Damian Dudek

Department of Software Development and Internet Technologies,
The University of Information Technology and Management "Copernicus",
ul. Inowroclawska 56, 53-648 Wroclaw, Poland
ddudek@wsiz.wroc.pl

Abstract. In many applications software agents are supposed to show adaptive behaviour and learning capabilities in information rich environments. On the other hand agents are often expected to be resource-bounded systems, which do not utilize much memory, disk space and CPU time. In this paper we present a novel framework for incremental, statistical learning, attempting to satisfy both requirements. The new method, called APS, runs in a cycle including such phases as: storing observations in a history, rule discovery using data mining algorithms, and knowledge base maintenance. Once processed, the old facts are removed from the history and in every subsequent learning run only the recent portion of observations is analysed in search of new rules. This approach can substantially save disk space and processing time as compared to batch learning methods.

Keywords: software agent, incremental learning, data mining, association rules.

1 Introduction

Since the early work by Maes [9] the notion of *adaptive agents* and *learning agents* have gained much interest in the domain of agent-related research. Variety of solutions have been proposed both for single-agent learning (SAL) and multi-agent learning (MAL) [13]. Many of these methods are adaptations of general machine learning techniques such as: reinforcement leaning [12] or model-based learning [9,15].

In this paper we propose a new framework APS (*Analysis of Past States*) for incremental, statistical learning of an agent, based on data mining algorithms. Consider an agent that works in a cycle with performance and stand-by phases, interlaced with each other. During performance phases the agent performs actions, at the same time collecting in the history observations of the world and the interaction with it. Stand-by phases occur when the agent is idle e.g. waiting for user's input. Thus, the system accumulates experience during the performance phase (so that it does not disturb fulfilling current tasks) and processes the history while system resources are not utilized. As an example scenario consider a web browsing assistant i.e. a personal agent helping

N.T. Nguyen et al. (Eds.): KES-AMSTA 2007, LNAI 4496, pp. 111–120, 2007.

a user choose relevant WWW pages, based on accumulated experience. The agent observes user's browsing activity and registers all the pages, which are explicitly evaluated by the user as interesting or stored locally - also revealing user's interest. The agent stores information about every page using index terms combined with user's response. After a representative number of such facts has been saved, the agent analyses them in search of general rules (e.g. $adaptive \wedge autonomous \wedge learning \Rightarrow relevant \wedge stored\_yes$), which can be further used for predicting user's interest and recommending potentially relevant pages[1]. Old rules, found in previous runs, are maintained using the latest results, while the analysed facts are removed so that the agent can continue learning cycle with a clear observation history. Thus, the history log is kept relatively small and consequently learning runs can be performed faster than in batch mode. At the same time the agent's knowledge base is maintained in such a manner, that it contains a set of statistically reliable and stable rules as if they were learned through batch learning since the very beginning of agent's life cycle.

The remainder of the paper is as follows. In section 2 we introduce the APS learning procedure and outline the algorithms used for data processing. Section 3 contains an example of how the APS method works in the web browsing scenario. In section 4 we present the results of the experimental evaluation. Finally, we conclude the paper, report on related works and propose further research.

2 APS Learning Cycle

The APS method is an attempt to use a general data mining process [11] as a statistical learning technique, embedded in a software agent architecture. The proposed approach is based on the cyclic procedure *APS Learning*, which interacts with four parts of agent's knowledge base (Fig. 1): the history KB_H, the rule base KB_R, temporal knowledge KB_T and general knowledge KB_G (see the input and output data flow in the beneath procedure specification). All the algorithms mentioned here (except for Apriori [1]) are original contribution proposed for the APS cycle. During the performance phase (steps 4-5) an agent performs actions towards its goals and saves results of interaction with the environment in the history KB_H whenever a *new fact event* occurs. Every fact needs to have a unique identifier and timestamp informing of the moment of registering. Actual format of knowledge to be stored is determined by the designer, depending on a given application domain. In the example web browsing scenario an agent stores information about a WWW page, after it has been evaluated or saved by a user. If an *analyse facts event* is raised[2], the agent enters the learning phase (steps 7-16). In step 8 the agent retrieves parameters (from the KB_G module) that come from previous learning runs and are necessary for current processing. Afterwards, in step 9 a portion of facts is selected for analysis (by default the whole current history KB_H), using the FSEL algorithm. In steps 10 and 11 the

[1] Association rules are well recognized as a tool for Internet user modeling [3].

[2] It can happen when two requirements are satisfied: (i) agent is idle - it has no tasks to perform; (ii) there is sufficient number of facts in the history (set by a designer).

chosen observations are transformed to the form with binary attributes, which consists of building a new history schema (HTRANS algorithm) and filling it with data (HFILL algorithm). Consequently, the ENV algorithm replaces facts with unknown values (step 12) by all the possible facts that could happen. Then the actual rule mining takes place (step 13) using the ARM algorithm, which incorporates a general association mining algorithm (such as Apriori [1]). The crucial operation within the whole cycle is rule base maintenance, when the recently found rules R are combined with the rule base KB_R by the RMAIN algorithm (step 14). Afterwards, the analysed facts are removed from the history KB_H (step 15), the information that is needed for future learning is stored in the general knowledge module KB_G, the temporal knowledge in KB_T is disposed and the agent comes back to its performance phase. Deletion of the processed facts from the history makes an APS learner an agent with *imperfect recall* [5,7] i.e. one that *forgets* facts from the past. However, the APS method helps such an agent to keep and maintain rules that are potentially useful conclusions about forgotten observations.

```
Procedure: APS_Learning
 1. BEGIN
 2.     WHILE (NOT Event_Close_Agent_Process) DO
 3.     BEGIN
 4.         IF (Event_New_Fact) THEN
 5.             Store_Fact(IN KB_H,OUT KB_H);
 6.         IF (Event_Analyse_Facts) THEN
 7.         BEGIN
 8.             Get_Parameters(IN KB_G; OUT KB_T);
 9.             Select_Facts(IN KB_H,KB_T; OUT KB_H,KB_T);
10.             Transform_History_Schema(IN KB_H; OUT KB_H,KB_T);
11.             Fill_New_Schema(IN KB_H; OUT KB_H);
12.             Remove_N_Values(IN KB_H; OUT KB_H,KB_T);
13.             Mine_Rules(IN KB_H,KB_T; OUT KB_T);
14.             Update_Rule_Base(IN KB_R,KB_T,KB_G; OUT KB_R,KB_G);
15.             Delete_Facts(IN KB_H,KB_T; OUT KB_H);
16.         END // IF (Event_Analyse_Facts)
17.     END // WHILE
18. END // Procedure: APS_Learning
```

Presenting the complete formal model of the knowledge base modules and the proposed algorithms is beyond the scope of this paper. They were introduced and thoroughly analysed in the previous works [4,5,6]. We would like to provide here a general look on the inside of the APS method with minimal formal load.

The FSEL algorithm (step 9 of the APS cycle) takes as input data: the history KB_H and the maximal number of facts to be selected. It returns a portion of facts from the history KB_H for the current learning run. Apart from that FSEL roughly analyses the chosen facts and sets up parameters needed for further processing (e.g. average timestamp of facts). The complexity of FSEL is linear: $O(n)$ where n is the number of facts in KB_H.

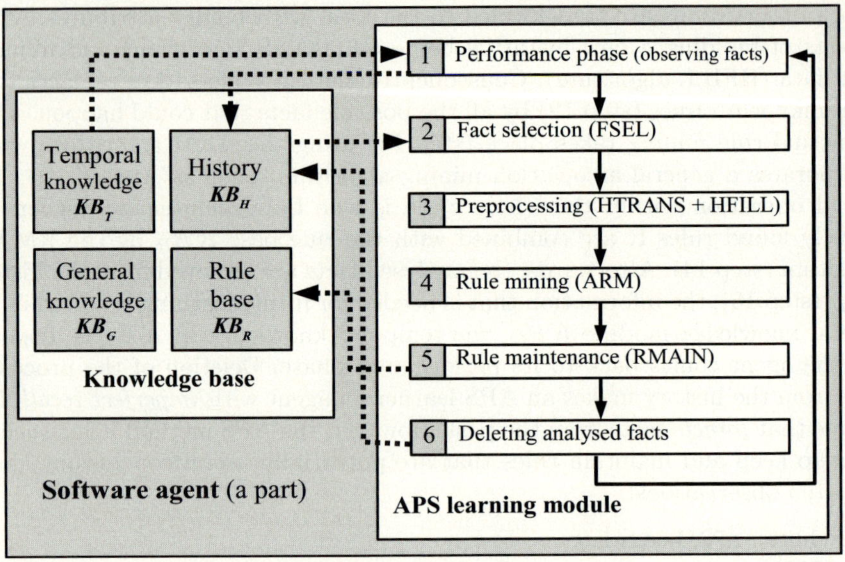

Fig. 1. The APS learning cycle and its interaction with knowledge base modules

In step 10 the HTRANS algorithm is used for transforming the original history schema to the form with binary attributes. It takes the portion of facts selected by FSEL as input and returns a new schema, where the fact identifier and timestamp remain unchanged, and the new attributes reflect all the values of the old attributes (except for N values), found in the selected portion (see the example in Section 3). Hence, it is necessary that all the attributes in the original history have discrete values (continuous values need to be discretized by the designer). The computational complexity of HTRANS is $O(k^2mn)$, where n is the size of the fact portion, m is the number of attributes in the original schema, each of which takes an average of k unique values in the portion. Transforming the history schema in every run, though costly, increases flexibility of the method, as in every run facts can be described by different attributes, provided that meaning of attributes (semantics) does not change.

The HFILL algorithm (step 11) takes the selected facts and inserts them into the transformed history schema (see the example in Section 3). The fact identifiers, timestamps and unknown values are rewritten unchanged. HFILL has the complexity of $O(mn)$, where n is the number of selected facts and m is the number of attributes of the transformed history schema.

In the next step the ENV algorithm (*Elimination of N-Values*) replaces each fact with unknown values with rows describing all the possibilities that could be observed (see the example). They can be viewed as all the *possible worlds* in a given state [7] and they are treated as equally probable, based on the *random worlds assumption*, presented by Bacchus et al. [2]. ENV is rather complex: $O(n2^m)$, where n is the number of facts and m is the maximal number of unknown values in a single fact. Because of that ENV deletes without replacement all the facts containing more *N-values* than a given threshold parameter η_c.

Various algorithms can be used for rule mining (step 13) - we do not restrict the method to any specific one. However, we need to encapsulate a general association rule mining algorithm in the ARM algorithm so that appropriate input parameters were used (e.g. thresholds of minimal support and confidence) and the output rules were described due to the extended model proposed in [6].

The crucial phase of the APS cycle is updating the rule base KB_R (step 14), through combining its previous content with the rule set R found in the recent run. For this purpose we proposed the RMAIN algorithm (*Rule MAINtenance*), which compares both old and new rules and updates their support and confidence (plus other parameters), based on the following proportion formulae.

$$sup_c(r) = \frac{b_1 sup_1(r) f_T(\Delta t_1) + b_2 sup_2(r) f_T(\Delta t_2)}{b_1 + b_2} \tag{1}$$

$$con_c(r) = \frac{con_1(r) con_2(r) \left(b_1 sup_1(r) f_T(\Delta t_1) + b_2 sup_2(r) f_T(\Delta t_2)\right)}{b_1 sup_1(r) f_T(\Delta t_1) con_2(r) + b_2 sup_2(r) f_T(\Delta t_2) con_1(r)} \tag{2}$$

where b_i is the number of facts, which were processed giving a rule r in all the previous runs (b_1) and the recent run (b_2). If support and confidence of a rule are unknown in a given run (a rule was not found), they are replaced by estimators of expected support $\hat{\sigma}$ and confidence $\hat{\gamma}$. The time depreciation function $f_T : [0; +\infty) \rightarrow [0; 1]$ (such that $f_T(0) = 1$ and $\forall x_1, x_2 \in [0; +\infty).x_1 < x_2 \Rightarrow f_T(x_1) \geq f_T(x_2)$) is used to promote recent rules against old ones that otherwise could be over-persistent. The complexity of RMAIN is $O(mn)$, where m and n are the numbers of rules in KB_R and R, respectively. Other properties of the RMAIN algorithm were analysed in detail in [4,6].

3 Example–Web Browsing Assistant

Below we present an example run of the APS learning cycle for a web browsing assistant scenario (see Introduction). In this application the aim of APS learning is user modeling through finding general rules describing his or her preferences and plausible behaviour. The history KB_H is a relational database table with the following columns (see Fig. 2): *HKey*, *HTime*, *Term*, *Eval*, and *Stored*, standing for: unique key, event timestamp, frequent index terms found in the page (multi-valued column), user's evaluation (*relevant* or *irrelevant*), and saving page (*yes* or *no*), respectively. Missing information is represented by N values.

The algorithm FSEL selects a portion of recent facts, stored since the previous processing run. The initial schema of this record set is: $S_H = \{HKey, HTime, Term, Eval, Stored\}$. The algorithm HTRANS transforms S_H to the schema: $S^{\#} = \{HKey, HTime, Term^{(base)}, Term^{(computer)}, Term^{(copy)}, Term^{(disc)}, Term^{(Internet)}, Term^{(matrix)}, \cdots, Term^{(server)}, Eval^{(relevant)}, Eval^{(irrelevant)}, Stored^{(yes)}, Stored^{(no)}\}$. Except for the special columns *HKey* and *HTime* all attributes in the new schema accept solely two values: 1 (positive assertion) and

HKey	HTime	Term	Eval	Stored
1	2006/04/19 16:30	*base, disc, computer, ..., storage*	*relevant*	*yes*
2	2006/04/20 15:35	*base, matrix, page, ..., server*	*irrelevant*	*no*
...	...	...	...	...
k	2006/05/21 17:38	*base, copy, pattern, ..., server*	*N*	*yes*

Fig. 2. Example history of a web brosing assistant

HKey	HTime	*Term* (*base*)	*Term* (*disc*)	...	*Eval* (relevant)	*Eval* (*irrelevant*)	*Stored* (*yes*)	*Stored* (*no*)
1	2006/04/19 16:30	1	1	...	1	0	1	0
2	2006/04/20 15:35	1	0	...	0	1	0	1
...	...	...	...	...	...	...	...	...
k	2006/05/21 17:38	1	0	...	*N*	*N*	1	0
k_1	2006/05/21 17:38	1	0	...	**1**	**0**	1	0
k_2	2006/05/21 17:38	1	0	...	**0**	**1**	1	0

Fig. 3. History facts transformed to the new schema $S^{\#}$. N values are removed by the algorithm ENV.

0 (negative assertion). Following the APS cycle, the algorithm HFILL fills the new schema $S^{\#}$ with facts that look like the dataset in Fig. 3.

The next step is removing unknown values N, done by the algorithm ENV. Each fact with N values is replaced by rows reflecting all the possible worlds (see Fig. 3). If the number of unknown values is a given fact exceeds a threshold η_c, such a row is deleted without replacement (this case is not shown in Fig. 3).

The transformed and preprocessed set of facts is given as an input to the association rule mining algorithm ARM. Assume that this procedure processes 500 facts and yields the set R, consisting of two rules: $r_1 : term\_base \wedge term\_disc \Rightarrow eval\_relevant$ ($sup = 0.15$; $con = 0.90$; $b = 500$; $t_m = 2006/05/05\ 17{:}04$) and $r_2 : term\_base \wedge term\_matrix \Rightarrow eval\_relevant$ ($sup = 0.28$; $con = 0.85$; $b = 500$; $t_m = 2006/05/05\ 17{:}04$). Presume that since the previous learning run the rule base KB_R has contained two rules, based on 3000 analysed facts: $p_1 : term\_base \wedge term\_disc \Rightarrow eval\_relevant$ ($sup = 0.12$; $con = 0.73$; $b = 3000$; $t_m = 2006/01/16\ 10{:}45$) and $p_2 : term\_base \wedge term\_copy \Rightarrow stored\_yes$ ($sup = 0.21$; $con = 0.69$; $b = 3000$; $t_m = 2006/01/16\ 10{:}45$).

We can see that the rules r_1 and p_1 have common antecedents and consequents, differing only in statistical parameters and time, r_2 is a new rule discovered in the recent fact portion, unlike p_2, which has not been found there. The rule maintenance algorithm RMAIN combines the rule sets R and KB_R, considering three cases: (i) similar rules in both sets, (ii) rules only in R, and (iii) rules only in KB_R. Provided that a linear function of time depreciation $f_T(t) = 1$ is used, the resulting rule base KB_R is as follows. $p_1 : term\_base \wedge term\_disc \Rightarrow eval\_relevant$ ($sup = 0.12$; $con = 0.75$; $b = 3500$; $t_m = 2006/02/01\ 01{:}22$) $p_2 : term\_base \wedge term\_copy \Rightarrow stored\_yes$ ($sup = 0.18$; $con = 0.67$; $b = 3500$; t_m

= 2006/02/01 01:22) and r_2 : *term_base* $\wedge$ *term_matrix* $\Rightarrow$ *eval_relevant* (*sup* = 0.08; *con* = 0.44; b = 3500; t_m = 2006/02/01 01:22).

Afterwards all the history facts are deleted and the agent exits the learning run, starting another performance phase.

4 Experimental Results

Quality and performance of the APS method were evaluated experimentally using a synthetic T10.I5.D20K dataset of the web browsing scenario. The set was generated using the *DataGen* application [5,6] with the following parameters [1]: number of transactions $|D| = 20000$; average transaction size $|T| = 10$; average size of maximal, potentially frequent itemsets $|I| = 5$; number of binary attributes 307; the facts contain no N values.

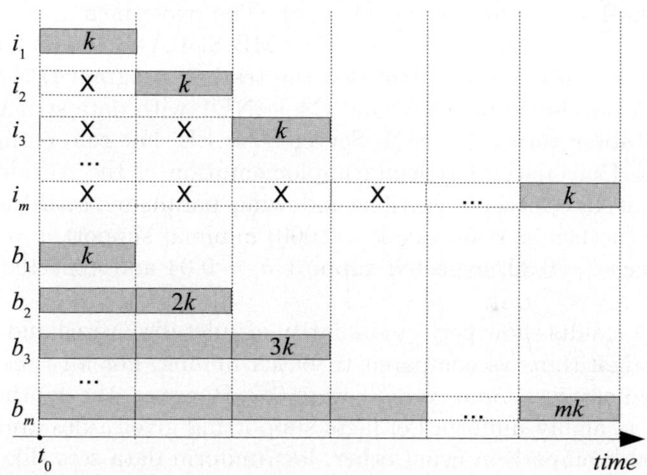

Fig. 4. The strategy of the experiment - comparing APS incremental processing (fact portions $i_1, i_2, \cdots, i_m$) and batch processing ($b_1, b_2, \cdots, b_m$)

Test runs were divided into two series: one for incremental and the other for batch processing (see Fig. 4). For each run the resulting KB_R rule base was stored together with detailed time reports of particular APS cycle steps. Then, we compared the results achieved in the APS and batch mode. For qualitative evaluation we proposed three rule comparison measures, listed below. The rule overlapping ratio of two rule sets R_1 and R_2 is given by the following equation.

$$rule_{overlap}(R_1, R_2) = \frac{|KB_R^O(R_1, R_2)|}{|R_1| + |R_2| - |KB_R^O(R_1, R_2)|}, \quad (3)$$

where $KB_R^O(R_1, R_2)$ is a set of rules in R_1, for which exist corresponding rules in R_2 with the same antecedents and consequents. In other words, $KB_R^O(R_1, R_2)$ is an intersection of both rule sets with no respect to either support or confidence of rules. The next two measures provide deeper insight, showing to what extent rules in both sets differ in support and confidence.

$$sup_{diff}(R_1, R_2) = \frac{1}{n}\left(\sum_{i=1}^{c} |sup(p_i) - sup(r_i)| + d\right), \tag{4}$$

$$con_{diff}(R_1, R_2) = \frac{1}{n}\left(\sum_{i=1}^{c} |con(p_i) - con(r_i)| + d\right), \tag{5}$$

where n is the number of all rules in R_1 and R_2 with different antecedents and consequents, $c = KB_R^O(R_1, R_2)$, $d = n - c$, and $p_i \in R_1$ has the same antecedent and consequent as $r_i \in R_2$ for $i \in \{1, ..., c\}$. The experiments were conducted on an x86 computer (892.50 MHz CPU, 752 MB SDRAM, 40 GB HDD, NTFS, OS MS Windows 2000 Server). We used the testbed program*APS Incremental Learning* [5,6], developed in MS Visual C++ .NET with data structures stored under the database server MS SQL Server 2000 EE. For rule mining the program uses the University of Helsinki implementation of the Apriori algorithm [1], worked out by Goethals [8]. The following parameter settings were used for the tests: the fact portion size $k = 1000$, minimal support $\sigma = 0.08$; minimal confidence $\gamma = 0.30$; expected support $\hat{\sigma} = 0.04$ and expected confidence $\hat{\gamma} = 0.15$

The quality results show perfect similarity of rules discovered and maintained by the APS algorithms as compared to batch mining. For all runs $rule_{overlap}$ was 100%, while sup_{diff} and con_{diff} were 0%. However, the synthetic dataset T10.I5.D20K is highly uniform i.e. facts supporting given rules are distributed randomly. Rule comparison using other, less uniform data sets (like most real-life data) should be less perfect. Nevertheless, the frequent, confident, and stable rules (i.e. found regularly) are likely to be discovered and maintained correctly.

The efficiency test results show that APS processing time remains quite stable within some lower and upper bounds, what lets it outperform batch rule mining starting from about 8000 facts (see Fig. 5). While this conclusion is rather obvious, it is worth observing the shares of particular APS steps. We can see that the registered times directly reflect computational complexity of given processing phases (see Section 2). Maintenance of the rule base KB_R (the RMAIN algorithm) requires the biggest part of overall run time (at average over 60%). The next place goes to filling the history schema and rule mining (both 17%). Straightforward database operations, such as selecting (FSEL) and deleting facts, essentially take 0% of time. However, small time of transforming the history schema (HTRANS) can be misleading. It could be bigger, if there were more attributes to change. The same conclusion concerns the complex $O(n2^m)$ algorithm for removing N values (ENV) that would run much longer for higher density of missing data.

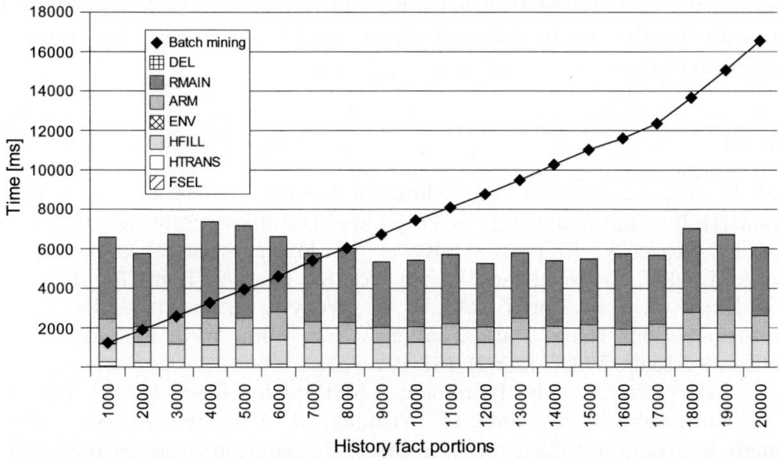

Fig. 5. Processing time of particular APS algorithms vs. Apriori batch mining

5 Conclusions

We proposed a novel APS method for incremental, statistical learning of a software agent, using data mining algorithms. The presented approach can be classified as *partial memory learning* [10] or *learning with imperfect recall* [5,7]. Experiments, conducted on a synthetic dataset, prove the method to work well in environments, where observed facts do not change rapidly in successive learning runs. For less stable data sources we can expect respectively worse results.

Symeonidis and his coworkers [14] proposed the system *Agent Academy* (AA) that can be compared to APS as it uses data mining techniques (e.g. algorithms ID3, C4.5, Apriori) for training agents. However, in AA agents are trained by an external process, whereas APS works as an autonomous learning process inside an agent. Another related work is the *PagePrompter* system [15], which supports administering of a web site using log analysis, based on the Apriori algorithm and clustering methods. In contrast to that system, the APS method is not restricted either to any particular association rule mining algorithm or a given application domain, but it tends to be a general learning mechanism that can be embedded in a software agent architecture. Finally, there are many alternative approaches to incremental association mining and maintenance, e.g. [16], but none of them is a framework dedicated for software agents.

The APS framework can be particularly suitable for agents with strongly bounded resources as it can save both disk space and CPU time (for the cost of loosing precision of rule maintenance). The method is also potentially beneficial for agents, whose normal performance interlaces with stand-by phases, when they are idle, awaiting user's commands. Example applications include personal and mobile agents (e.g. in the domain of WWW browsing, retrieval and filtering).

Propositions of further research are: implementing the APS procedure as a learning module inside one of existing agent architectures and deploying it in real-life environments.

References

1. Agrawal, R., Srikant, R.: Fast Algorithms for Mining Association Rules. Proc. of the Twentieth International Conf. on Very Large Databases, Santiago, Chile (1994)
2. Bacchus, F., Grove, A., Halpern, J., Koller, D.: From statistical knowledge bases to degrees of belief. Artificial Intelligence, Vol. 87, No. 1-2 (1996) 75-143
3. Chen, Z., Lin, F., Liu, H., Liu, Y., Ma, W.-Y., Wenyin, L.: User Intention Modeling in Web Applications Using Data Mining. World Wide Web: Internet and Web Information Systems, Vol. 5 (2002) 181-191
4. Dudek, D., Zgrzywa, A.: The Incremental Method for Discovery of Association Rules. In: Kurzynski, M., Puchala, E., Wozniak, M., Zolnierek, A. (eds.): Proc. of the Fourth International Conf. on Computer Recognition Systems (CORES'05), Advances in Soft Computing, Springer-Verlag, Berlin Heidelberg (2005) 153-160
5. Dudek, D., Kubisz, M., Zgrzywa, A.: APS: Agent's Learning With Imperfect Recall. In: Kwasnicka, H., Paprzycki, M. (eds.): Proceedings of the Fifth International Conference on Intelligent Systems Design and Applications, IEEE Computer Society Press, Washington, Brussels, Tokyo (2005) 172-177
6. Dudek, D.: Knowledge Acquisition Within an Agent System Using Data Mining Methods. PhD Thesis, Technical Report No. 2, Institute of Applied Informatics, Wroclaw University of Technology, Wroclaw (2005) (in Polish)
7. Fagin, R., Halpern, J.Y., Moses, Y., Vardi, M.Y.: Reasoning About Knowledge. MIT Press, Cambridge, MA, USA (1995)
8. Goethals, B.: Implementation of the Apriori Algorithm. University of Helsinki (http://www.adrem.ua.ac.be/~goethals/) (2003)
9. Maes, P.: Modeling Adaptive Autonomous Agents. In: Artificial Life: An Overview. MIT Press, Cambridge, MA, USA (1994) 135-162
10. Maloof, M.A., Michalski, R.S.: Incremental learning with partial instance memory. Artificial Intelligence, Vol. 154 (2004) 95-126
11. Mannila, H.: Methods and problems in data mining. A tutorial. In: Afrati, F., Kolaitis, P. (eds.): Proceedings of the International Conference on Database Theory (ICDT'97), Delphi, Greece (1997) 41-55
12. Ribeiro, C.: Reinforcement Learning Agents. Artif Intel Rev, Vol. 17 (2002) 223-250
13. Sen, S., Weiss, G.: Learning in Multiagent Systems. In: Weiss, G. (ed.): Multiagent systems, The MIT Press (1999) Chapter 6, 259-298
14. Symeonidis, A.L., Mitkas, P.A., Kechagias, D.D.: Mining Patterns And Rules For Improving Agent Intelligence Through An Integrated Multi-Agent Platform. Proceedings of the Sixth IASTED International Conference on Artificial Intelligence and Soft Computing (ASC 2002), Banff, Alberta, Canada (2002)
15. Yao, Y.Y., Hamilton, H.J., Wang, X.: PagePrompter: An Intelligent Web Agent Created Using Data Mining Techniques. In: Alpigini, J.J. et al. (eds.): RSCTC 2002, Lect Notes Artif Int, Vol. 2475, Springer-Verlag, Berlin Heidelberg (2002) 506-513
16. Zhou, Z., Ezeife, C.I.: A Low-Scan Incremental Association Rule Maintenance Method Based on the Apriori Property. In: Stroulia, E., Matwin, S. (eds.): AI 2001, Lect Notes Artif Int, Vol. 2056. Springer-Verlag, Berlin Heidelberg (2001) 26-35

Expressivity of STRIPS-Like and HTN-Like Planning

Marián Lekavý and Pavol Návrat

Slovak University of Technology
Faculty of Informatics and Information Technologies
Ilkovičova 3, 842 16 Bratislava, Slovakia,
lekavy@fiit.stuba.sk, navrat@fiit.stuba.sk

Abstract. It is widely believed, that the expressivity of STRIPS and STRIPS-like planning based on actions is generally lower than the expressivity of Hierarchical Task Network (HTN) and HTN-like planning, based on hierarchical decomposition. This would mean that a HTN-like planner can generally solve more domains than a STRIPS-like planner with the same extensions. In this paper, we show that both approaches, as they are practically used, are identically expressive and can solve all domains solvable by a Turing machine with finite tape (i.e. solvable by a common computer).

1 Introduction

Two most known and used approaches to domain-independent symbolical planning are STRIPS-like (based on operators) and HTN-like (based on hierarchical decomposition) planning. STRIPS-like planning is older and based on creating a plan as a chain of actions, while each of these actions has its precondition and results. HTN-like planning is based on hierarchical decomposition - the planner starts from an initial task and decomposes it into more primitive tasks according to decomposition information given to the planner. The primitive non-decomposable tasks form the final plan.

HTN was created as an extension to the "classical" STRIPS-like planning, allowing the planner to use additional information about the hierarchical decomposition.

As hierarchical decomposition is an extension of operator-based planning, the question of expressivity arose: Is the expressivity of HTN-like planning larger than the expressivity of STRIPS-like planning? Can HTN-like planning solve more domains than STRIPS-like planning?

This question was answered positively in the past[2]. The proof was, however, based on the assumption, that the HTN planner can use an infinite set of symbols to mark the tasks. This can be true for the theoretical HTN model, but this assumption cannot be fulfilled by any practically usable planner, implemented on a common computer.

This paper shows, that HTN-like planning expressivity is identical to STRIPS-like planning expressivity, under the assumption that we use any restriction,

N.T. Nguyen et al. (Eds.): KES-AMSTA 2007, LNAI 4496, pp. 121–130, 2007.

which makes the HTN-planning process decidable. This assumption is not very restricting, as every implemetnation of HTN-like planning uses this kind of restriction to end the computation in finite time (even if unsuccesfull).

In this section, we will provide a brief overview of STRIPS-like and HTN-like planning. Section 2 provides discussion on expressivity of both approaches. In section 3, we show that STRIPS can emulate the finite tape Turing machine with the same time complexity. Finally, section 4 describes a conversion of HTN-like domains to STRIPS-like domains.

1.1 STRIPS

The basic principle of STRIPS[3] and STRIPS-like planning is finding a sequence of actions, which will modify the initial state of the world into the final state of the world. The state of the world is expressed in the form of a set of literals. The planner adds actions incrementally to the plan, trying to create the correct transformation from initial to the final state.

The STRIPS planning is based on operators in the form $Op = (pre, del, add)$, where *pre* is a precondition, which has to be valid immediately before the operator is applied, *add* / *del* are the sets of literals added / deleted to / from the world state after the operator ends. An executed operator (added to a plan) is called action.

Since STRIPS (35 years ago), there are plenty different planners based on the idea STRIPS used. Most planners are not limited to the basic STRIPS formalism, but have extensions like resources, parallel execution, timed actions, sensory actions and coordination actions, conditional and contingent actions etc.

The basic algorithm of STRIPS-like planning is based on sequential adding actions to a plan. When starting from the initial state, we speak about forward chaining, when starting from the final state, it is backchaining and if we add actions on an arbitrary place of the plan, then it is plan-space search. We construct the plan based only on the knowledge of the operators' precondition and effects and the current world state. No additional information is provided to the planner, which is the main difference to the HTN-like approaches, described in the next section.

1.2 Hierarchical Task Network (HTN)

The HTN[7] and HTN-like approaches are based on hand-made hierarchical decomposition of the problem domain. The planner is provided with domain knowledge, expressed as the possible decompositions of tasks into subtasks. Tasks can be primitive (directly executable) and non-primitive. Non-primitive tasks have to be decomposed into other tasks. Each non-primitive task has one or more lists of tasks, it can be decomposed into. This list of tasks, together with other restrictions (like precedence of tasks, variable binding or mutual exclusion) is called task network.

The creation of a plan starts with one or more initial tasks, which are decomposed into simpler tasks, until all tasks are decomposed into primitive tasks.

If the decomposition is not possible (e.g. because of colliding restrictions), the planner backtracks and creates a different decomposition.

The decomposition may be fully ordered, but there are also planners which allow interleaving of subtasks of different tasks (e.g. SHOP2[6]).

HTN can also be seen as an augmentation of action-based planning by a grammar, pruning the plan space[4].

The basic HTN algorithm is as follows:

```
1. Insert the initial tasks into the plan.
2. If the plan contains only primitive tasks, success.
3. Choose one non-primitive task from the plan.
4. Replace the chosen task by its subtasks according to some task network.
5. Resolve interactions and conflicts in the plan.
   If not possible, backtrack.
6. Continue with step 2.
```

Similar to STRIPS-like planning, HTN-like planners also introduce a variety of extensions like resources handling, parallel execution, timed actions, sensoric and coordination actions. HTN-like planners have an important role in planning for multi-agent systems. Planners like STEAM[8] or PGP/GPGP[5] introduce coordination and negotiation mechanisms, allowing multiple agents to participate in fulfilling common tasks. While STEAM introduces team tasks, which are decomposed into tasks for individual agents, in PGP, each agent has its own initial task that has to be fulfilled and parts of the decomposition tree can overlap for different agents (Figure 1).

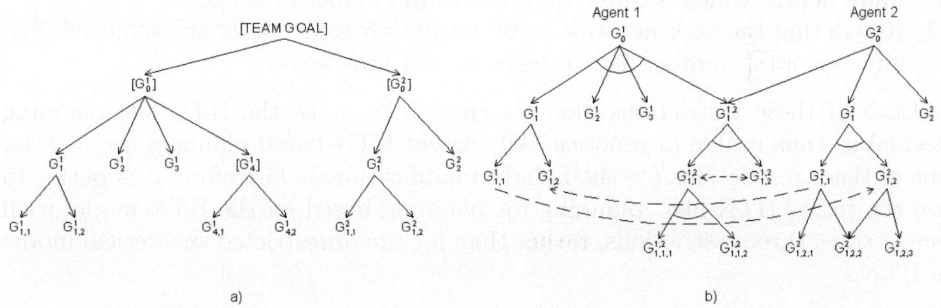

Fig. 1. An example of decomposition tree for STEAM (a) and PGP (b)

2 Expressivity

It is widely believed, that HTN-like planning expressivity is larger than the expressivity of STRIPS-like planning. By expressivity, we mean the set of planning domains the planning system is able to solve.

There is a proof[2], based on converting the planning problem to grammars and then showing, that STRIPS-like planning corresponds with regular grammars and HTN-like planning corresponds with context-free grammars. This way, it is shown that the HTN-like planning covers a wider class of possible planning domains.

The main difference, causing this result, is the fact, that the theoretical model of HTN uses an infinite set of symbols for marking the tasks, thus having the possibility of an infinite plan-space even in a simple domain. On the other hand, STRIPS and STRIPS-like planning have a finite plan-space, if not using some extensions (usually considered to be non-standard). Some STRIPS-like planners (like the FHP[9]) extend the basic formalism by including function symbols, allowing them to express undecidable problems.

It is clear, that the theoretical HTN model is more expressive than the basic STRIPS-like planning. On the other hand, the theoretical HTN model is not usable in practice, as it is undecidable, even under severe restrictions. HTN remains generally undecidable even if no variables are allowed, as long as there is the possibility that a task network can contain two non-primitive tasks without specifying the order of their execution[2].

As the theoretical model of HTN is undecidable, the computation could take an unlimited amount of time and we cannot even reliably predict the time in advance, so it is not (and cannot be) used in practice. In practice, modifications are used, restricting the plan space to finite and making the problem decidable [2].

The main HTN restrictions used are:

1. Restricting the length of a plan. As the maximal length of a plan becomes finite, the space of possible plans becomes finite, as we choose from a finite number of possibilities, when adding a task to the plan.
2. Restricting the methods to be acyclic. Any task can be expanded up to a finite depth, which is lower than the total number of tasks.
3. Restricting the task network to be totally ordered. Tasks are achieved serially, one after another, so subtasks cannot interleave.

Each of these restrictions alone is enough to make the HTN-like planning decidable, thus usable in practice. All current HTN-based planners use at least one of these restrictions (or their slight modifications). Therefore, it is better to use the term "HTN-like" planning for planning based on the HTN model with one of these three restrictions, rather than for the unrestricted theoretical model of HTN.

Naming convention. *The term "HTN-like planning" shall be used for planning based on the HTN model with a restriction, making its plan-space finite and the planning problem decidable.*

For the scope of this paper, we adhere to this naming convention.

Theorem 1. *Every HTN-like domain can be expressed as a STRIPS-like domain. Every STRIPS-like domain can be expressed as a HTN-like domain. Therefore, the expressivity of STRIPS-like planning is identical to the expressivity of HTN-like planning.*

Proof (sketch). The plan-space of HTN-like planning domain is finite; therefore the state-space of the domain is finite. For a finite state-space, we can construct a STRIPS-like domain by simply enumerating all possible state transitions as STRIPS actions. As a result, STRIPS-like planning expressivity is not smaller than the expressivity of HTN-like planning. The second half of the proof, showing that HTN-like planning expressivity is not smaller than the expressivity of STRIPS-like planning, is constructive, can be found in [2] and is based on the transformation of STRIPS-like domain to a flat HTN-like domain with decomposition depth 0. □

Enumerating all states of a domain is not very practical and leads to exponential number of actions. In the next sections, we will show how to transform a HTN-like domain into a STRIPS-like domain in low-order polynomial time, using STRIPS for emulating the HTN decomposition.

3 STRIPS as a Turing Machine with Finite Tape

The previous sections showed, that STRIPS-like and HTN-like planning expressivity is identical. In this section, we will provide a simple construction of a Turing machine with finite tape using STRIPS. This way, we show that STRIPS expressivity (and therefore also HTN-like planning expressivity) is equal to the expressivity of a Turing machine with finite tape.

Turing machine is a tuple:

$$M = (Q, \Gamma, b, \delta, q_0, F) \tag{1}$$

where Q is a finite set of states, Γ is a finite set of tape symbols, $b \in \Gamma$ is the blank symbol, $\delta = Q \times \Gamma \rightarrow Q \times \Gamma \times \{L, R\}$ is the transition function (L and R are the shift left and right symbols respectively), q_0 is the initial state and $F \subseteq Q$ is the set of final states.

In the STRIPS emulation, the set of states Q will be represented by a set of constants C_Q. Γ is represented by set of constants C_Γ. Current state is expressed by the literal $state(q)$, position of the reading head by $position(p)$ and the symbol on a specific tape location is expressed as the literal $symbol(\gamma, p)$. The transition function is defined by the literal $transition(q, g, q', g', m)$, where $q \in C_Q$ and $g \in C_\Gamma$ are the old state and symbol on the tape, $q' \in C_Q$ and $g' \in C_\Gamma$ are the new state and tape symbol and $m \in \{LEFT, RIGHT\}$ is the direction of head movement. The following operators encode the Turing machine with finite tape:

```
operator: stateTransition
pre: state(q), position(p), symbol(g, p), transition(q, g, q', g', m),
  translate
del: state(q), symbol(g, p), translate
add: state(q'), symbol(g', p), move(m)

operator: moveHeadLeft
pre: move(LEFT), position(from), leftOf(to, from)
```

```
del: move(LEFT), position(from)
add: position(to), translate

operator: moveHeadRight
pre: move(RIGHT), position(from), leftOf(from, to)
del: move(RIGHT), position(from)
add: position(to), translate

operator: finish
pre: state(q), final(q), translate
del: state(q), translate
add: stop
```

As we can see, the machine operates in two steps: state transition and head movement. The machine stops if one of the final states is reached, marked with literal $final(q)$.

Prior to starting the STRIPS reasoning, we have to "create" the structure of the tape by adding literals $leftOf(left, right)$ for each two neighbouring states. With extensions of STRIPS allowing arithmetic operators, we would instead use the $+$ and $-$ operator to move right and left.

The tape sequence representation for a tape of length n uses $n-1$ literals. On the other hand, we have to remember, that the memory contents is also stored in the form of literals, so the n memory places are represented by $2n-1$ literals, increasing the memory complexity only constantly, comparing to the Turing machine.

The emulated finite tape Turing machine is deterministic if only one literal $transition(q, g, q', p', m)$ exists for some (q, g). Otherwise, the machine is non-deterministic.

We could also add input to the machine, represented in the same way like the tape. This would, of course, not change the expressivity.

Theorem 2. *The STRIPS domain defined above emulates a Turing machine with finite tape. The emulation time complexity is constantly higher than the complexity of the emulated machine.*

Proof (sketch). The initial literal set is $state(q_0)$ together with the encoding of the Turing machine as described above. The final condition for the STRIPS planner is set to *stop*. It is easy to see, that if the Turing machine stops, the STRIPS planner creates a plan leading from the initial to the final state and each two following steps of the plan (stateTransition, moveHeadLeft/Right) correspond to one step of the Turing machine. If the finite tape Turing machine doesn't stop, no plan is found. If the emulated machine is deterministic, exactly one action is executable at each state and the plan derivation process is deterministic. If the emulated machine is non-deterministic, the STRIPS planner chooses one action for execution at each state, the same way the machine has to choose one. If the decision doesn't lead to the final state, the planner backtracks and systematically searches all alternatives until the final state is found or no more alternatives are left (i.e. the machine doesn't stop). □

Turing machine with an *in*finite tape is an important theoretical concept. On the other hand, the computers we use in practice have only finite memory and can be simulated by a Turing machine with finite tape.

The equality of expressivity of STRIPS and Turing machine with finite tape has an important implication. It means that STRIPS can express all problems solvable by a computer. The expression as a STRIPS domain can, of course, be sometimes very artificial and clumsy and computational complexity can be much larger. Nevertheless, STRIPS expressive power should not be underestimated.

4 STRIPS as HTN Emulator

The previous section shows, that it is possible to express a finite tape Turing machine as a STRIPS domain. Together with the possibility to express a HTN-like domain (with restrictions to be decidable) using the finite tape Turing machine, it is obvious, that STRIPS can express an arbitrary HTN-like domain.

This section provides a conversion of a HTN-like domain to STRIPS in low-order polynomial time. The conversion is based on emulating the decomposition of tasks by STRIPS plan derivation. The final STRIPS plan expresses the order of decomposition.

Let's say there is a task network $n = (A, \{B, C\})$, allowing the decomposition of task A into B and C. We create two operators A_{start}, A_{stop} and operators for B and C. A_{start} adds literals ($B_{A_{init}}$, $C_{A_{init}}$) allowing the execution of B and C. B can be decomposable again, so it again consists of operators B_{start}, B_{stop}, allowing its further decomposition. If B is a primitive task, it consists of only one operator B. After B is processed (operator B_{stop} or B finished), it adds literal $B_{A_{finish}}$, which is a part of A_{stop} precondition. This way, A_{stop} is only executed after all subtasks of A are fully decomposed or primitive. The initial state of the STRIPS planner contains only the literal S_{init}, allowing the start of the root task S (or possibly several literals if there are more root tasks). The final state contains the literal S_{finish}, which is reached after the full decomposition of the root task S and all its subtasks. Additionally, the final state may contain literals, added in tasks marked as goal tasks.

If task interleaving is allowed, then we have to avoid situations when a finished subtask allows the ending of a parent task from a different task network. Therefore, we have to create different operators for a subtask, which is in more task networks. For the same reason, we have to create separate operators for a parent task, which is in more task networks. As a result, a task being parent in n task networks and a subtask in m networks is converted to $m * n$ operators.

The HTN to STRIPS conversion algorithm for an acyclic decomposition graph is expressed by the following pseudocode. We use the shortened notation of an operator $Op = (pre, del, add)$, where pre, del and add are the precondition, delete and add sets of the operator Op. An overview of corresponding concepts of HTN-like and STRIPS-like domain after the conversion can be found in table 1.

```
for each task A
  if A is a decomposition of some task B, for each B
    if A is primitive
      add operator
        A_B = (A.pre ∪ {A_Binit}, A.del ∪ {A_Binit}, A.add ∪ {A_Bfinish})
    else
      for each task network n_i = (A, {C_j|j}) add operators
        A_iBstart = (A.pre ∪ {A_Binit}, {A_Binit}, {C_jA_init})
        A_iBstop  = ({C_jA_finish}, A.del ∪ {C_jA_finish}, A.add ∪ {A_Bfinish})
  else
    if A is primitive
      add operator
        A = (A.pre ∪ {A_init}, A.del ∪ {A_init}, A.add ∪ {A_finish})
    else
      for each task network n_i = (A, {C_j|j}) add operators
        A_istart = (A.pre ∪ {A_init}, {A_init}, {C_jA_init})
        A_istop  = ({C_jA_finish}, A.del ∪ {C_jA_finish}, A.add ∪ {A_finish})
for each initial task S
  add literal S_init to the initial state
  add literal S_finish to the goal state
```

Table 1. Corresponding concepts of HTN and STRIPS emulation of HTN

HTN	STRIPS
primitive task A	operator A
non-primitive task A	operators $A_{i_{start}}$, $A_{i_{stop}}$
task network $n_i = (A, \{C_j\|j\})$	operators $A_{i_{start}}$, $A_{i_{stop}}$; $C_{jA_{i_{start}}}$, $C_{jA_{i_{stop}}}$ or C_{jA_i}
adding task A as a subtask of B	adding literal $A_{B_{init}}$ into the actual state
choosing one of the possible decompositions of A	choosing one applicable operator $A_{iB_{start}}$ with literal $A_{B_{init}}$ in the precondition
decomposing A using the task network $n_i = (A, \{C_j\|j\})$	executing the sequence of operators $A_{iB_{start}}$; $C_{jA_{start}}$, $C_{jA_{stop}}$ or C_{jA}; $A_{iB_{stop}}$

If necessary, parameters (i.e. variables and constants) can be passed from the decomposed task A to its subtask C by adding parameters to the literal $C_{A_{init}}$. Other constraints among tasks, like task precedence or mutual exclusion can be simply expressed by adding literals to the operators. For example if we want operator A to precede operator B, we simply add a literal to the *add* part of A and to the *pre* part of B. This will prevent B being executed before A.

Theorem 3. *The algorithm above converts an acyclic HTN-like domain to an equivalent STRIPS-like domain. The STRIPS-like domain time complexity is constantly higher than the complexity of the initial HTN-like domain.*

Proof (sketch). The plan derivation in the resulting STRIPS-like domain copies the decomposition of the initial HTN-like domain. For every decomposition of some task A (i.e. choosing a suitable task network and adding its subtasks $\{B_j\}$

to the plan), there is exactly one A_{start} action, one action ${B_j}_A$ for every primitive task from $\{B_j\}$, one literal ${B_j}_{A_{start}}$ for every non-primitive task from $\{B_j\}$ and one A_{stop} action. Actions representing subtasks of A are never executed before A_{start} or after A_{stop}. The STRIPS planning algorithm only chooses actions at a point, when a HTN-like planner would choose a task to decompose and a task network to be used for the decomposition. This means, we have exactly one STRIPS action for every step of a HTN-like planner and exactly one decision of the STRIPS planner for every decision of the HTN-like planner. □

If we want to use cyclic decomposition graphs (i.e. a task can be decomposed into itself after some steps), we have to restrict the domain to a maximal depth of decomposition or to be fully ordered (see section 2 Expressivity) in order to have a decidable domain.

For a fully ordered domain, we simply add precedence restrictions on operators.

If we want to restrict the maximal depth of decomposition while having a cyclic and not fully ordered domain, we have to mark the actions to differentiate decompositions of the same task by the same task network in a cyclic decomposition. This can be done by creating a sequence of symbols (like the finite tape in the previous section), which are then used for marking actions representing one decomposition. We simply add an incrementing action (similar to the moveHeadRight from the previous section) after each A_{start} action, while the current "counter" value is a parameter of A_{start} and this value is carried between operators representing the same task network using the $B_{A_{init}}$ and $B_{A_{finish}}$ literals. This finite set of marking symbols is the equivalent of task marking symbols used in HTN-like planners.

Many planners (based on HTN or STRIPS) allow different extensions to the basic HTN, like handling resources, allowing variables and arithmetic operators, allowing concurrency, timed actions or planning with uncertainty. According to Theorem 2, all extensions of HTN-like planning (as long as the problem remains computable) can be transformed to STRIPS, perhaps increasing computational complexity. On the other hand, most extensions are common for both approaches, so it is possible to modify the conversion algorithm introduced in this section to achieve the same time complexity of the planning process.

5 Conclusions

The concept of HTN is nothing more (but nothing less) than allowing the user to provide the planning engine with additional heuristic information about how to construct a plan, but does not increase the domain-space.

Nevertheless, HTN is a very useful and user-friendly concept, as we can see on the large number of practical uses.

This paper shows that STRIPS-like planning can be used for the same domains, HTN-like planning (with restrictions causing it to be decidable) can be used for (i.e. the expressivity of both approaches is identical). Moreover, the domains can be converted from HTN-like to STRIPS-like and vice versa in low order polynomial time, thus allowing the theoretical results for STRIPS-like planning to be used for HTN-like planning and vice versa.

Finally, this paper shows that STRIPS expressivity is equal to the expressivity of a Turing machine with finite tape, i.e. all problems that can be solved by a (common) computer can also be solved by STRIPS. This is rather a theoretical result than a practically usable conversion. However, it places the lower and upper bound on both, STRIPS-like and HTN-like planning expressivity. Additionally, complexity results for different STRIPS-like and HTN-like domains can make use of Turing machines formalism.

Acknowledgments. This work was partially supported by the Slovak State Programme of Research and Development "Establishing of Information Society" under contract No. 1025/04; by the Science and Technology Assistance Agency under the contract No. APVT 51-024604; by the Grant Agency of Slovak Republic grant No. VG 1/3102/06.

References

1. Bylander, T.: The computational complexity of propositional STRIPS planning. Artificial Intelligence 69, Elsevier Science Publishers Ltd., Essex, UK (1994) 161–204.
2. Erol, K., Nau, D., Hendler, J.: HTN planning: Complexity and expressivity. Proceedings of the Twelfth National Conference on Artificial Intelligence (AAAI-94) AAAI Press, Menlo Park, USA (1994)
3. Fikes, R.E., Nilsson, N.J.: STRIPS: A New Approach to the Application of Theorem Proving to Problem Solving. Artificial Intelligence 2, Elsevier Science Publishers Ltd., Essex, UK (1971) 189–208.
4. Kambhampati, S., Mali, A., Srivastava, B.: Hybrid planning in partially hierarchical domains. Proceedings of the National Conference on Artificial Intelligence (AAAI) (1998)
5. Lesser, V.R., et al.: Evolution of the GPGP/TAEMS Domain-Independent Coordination Framework. Autonomous Agents and Multi-Agent Systems. Vol. 9, No. 1, Kluwer Academic Publishers (2004) 87–143.
6. Nau, D. S., Au, T.-C., Ilghami, O., Kuter, U., Murdock, W., Wu, D., Yaman, F.: SHOP2: An HTN planning system. Journal of Artificial Intelligence Research 20 (2003) 379–404
7. Sacerdoti, E.D.: A Structure for Plans and Behavior. Elsevier-North Holland (1977)
8. Tambe, M.: Towards flexible teamwork. Journal of Artificial Intelligence Research. Vol. 7 (1997) 83–124.
9. Zalaket, J., Camilleri, G.: FHP: Functional Heuristic Planning. 8th International Conference on Knowledge-Based Intelligent Information and Engineering Systems (KES 2004), Wellington, New Zealand, Springer-Verlag (2004)

Implementation and Performance Evaluation of the Agent-Based Algorithm for ANN Training

Ireneusz Czarnowski and Piotr Jędrzejowicz

Department of Information Systems, Gdynia Maritime University
Morska 83, 81-225 Gdynia, Poland
{irek, pj}@am.gdynia.pl

Abstract. The paper contains a description of the implementation and performance evaluation of the agent-based population learning algorithm used to train the feed-forward artificial neural networks. The goal of the research was to evaluate efficiency of the agent-based approach and to establish experimentally which different factors representing the A-team structure and topology affect the performance of the analyzed agent-based algorithm. The paper includes a general overview of the JABAT environment used to deploy the ANN training algorithm, a description of different agents employed and their roles, as well as the computational experiment plan and the discussion of the performance evaluation results.

1 Introduction

Historically, research into systems composed of multiple agents has been carried out under the banner of Distributed Artificial Intelligence, which has been divided into two main fields: Distributed Problem Solving and Multi-Agent Systems. More recently, as it was pointed out in [7], the term *multi-agent systems* has a wider meaning, and is used to refer to all types of systems composed of multiple autonomous components.

The field of autonomous agents and multi-agent systems is an rapidly expanding area of research and development. It is based on many ideas originating from distributing computing, object-oriented programming and software engineering.

During the last decade, a number of significant advances have been made in both the design and implementation of autonomous agents. A number of applications of agent technology is growing systematically. Nowadays agent technology is used in the real life supporting successfully variety of industrial and commercial applications. Also a number of agent-based approaches have been proposed to solve different types of optimization problems [1], [9], [12].

One of the successful approaches to agent-based optimization is the concept of an asynchronous team (A-Team), originally introduced by Talukdar [13]. An A-Team is a collection of software agents that cooperate to solve a problem by dynamically evolving a population of solutions. An A-Team usually uses combination of approaches inspired by natural phenomena including, for example, insect societies [11], evolutionary processes [5] or particle swarm optimization [8], as well as local search techniques like, for example, tabu search [6].

N.T. Nguyen et al. (Eds.): KES-AMSTA 2007, LNAI 4496, pp. 131–140, 2007.

An A-Team is a cyclic network of autonomous agents and shared, common memories. Each agent contains some problems solving skills and each memory contains a population of temporary solutions to the problem to be solved. All the agents can work asynchronously and parallel. During their works agents cooperate by selecting and modifying these solutions.

This paper includes a short overview of the **JA**DE-**b**ased **A-T**eam environment (in short: JABAT) as a middleware supporting the construction of the dedicated A-Team architectures used for solving a variety of computationally hard optimization problems. JABAT has been developed by a team of researchers with the participation of both authors [2]. JABAT engine is JADE, which, in turn, is an enabling technology for the development and run-time execution of peer-to-peer applications which are based on the agents paradigm and which can seamlessly work and interoperate both in wired and wireless environment [3]. From the functional point of view, JADE provides the basic services necessary to distributed peer-to-peer applications in the fixed and mobile environment. JADE allows each agent to dynamically discovery other agents and to communicate with team according to the peer-to-peer paradigm.

The overview of JADE is followed by a more detailed description of the different agents employed and their roles. To evaluate the performance of the proposed agent-based approach computational experiment involving several benchmark datasets has been planned and carried out. Experiment result are discussed in the final section of the paper. Conclusions focus on identified properties of the analyzed implementation and on directions of further research.

2 Overview of the JABAT

The central problem in the design of a multi-agent system is how much intelligence to place in the system and at what level. As it was observed in [11], the vast majority of the work in this field has focused on making agents more knowledgeable and able. This has been achieved by giving the deliberative agent a deeper knowledge base and ability to reason about data, giving it the ability to plan actions, negotiate with other agents, or change its strategies in response to actions of other agents. At the opposite end of the spectrum lie agent-based systems that demonstrate complex group behavior, but whose individual elements are rather simple.

The JABAT belongs to the latter class. It does not provide ready answers to questions on how population of agents should be selected, which agents work best in combination, how should agents decide when to act and on which solutions or what should be the strategy for destroying unwanted solutions? Instead it offers tools to easily implement variety of strategies and solutions when dealing with the above listed questions, through providing a flexible framework and a set of predefined classes. The environment, when properly used, is expected to be able to produce solutions to difficult optimization problems through applying the following general rules:

- To solve difficult optimization problems use a set of agents, each representing an improvement algorithm.
- To escape getting trapped into a local optimum generate or construct an initial population of solutions called individuals, which, during computations will be improved by agents, thus increasing chances for reaching a global optimum.

Agent-based architecture of the JABAT allowed implementation of the following features:

- The system can in parallel solve instance of several different optimization problems.
- The optimization processes can be performed on many computers. The user can easily add or delete a computer from the system. In both cases JABAT will adopt to the changes, commanding the agents working within the system to migrate.

2.1 Search for the Optimum Solution

Main functionality of the proposed environment is searching for the optimum solution of a given problem instance through employing a variety of the solution improvement algorithms including, for example random and local search techniques, greedy construction algorithms, genetic algorithms etc. The search involves a sequence of the following steps:

- Generation of an initial population of solutions.
- Application of solution improvement algorithms which draw individuals from the common memory and store them back after attempted improvement, using some user defined replacement strategy.
- Continuation of the reading-improving-replacing cycle until a stopping criterion is met.

The above functionality is realized by the two main types of classes. The first one includes *OptiAgents*, which are implementations of the improvement algorithms. The second are *SolutionManagers*, which are agents responsible for maintenance and updating of individuals in the common memory. All agents act in parallel. Each *OptiAgent* is representing a single improvement algorithm (for example simulated annealing, tabu search, genetic algorithm, local search heuristics etc.). An *OptiAgent* has two basic behaviors defined. The first is sending around messages on readiness for action including the required number of individuals (solutions). The second is activated upon receiving a message from some *SolutionManager* containing the problem instance description and the required number of individuals. This behaviour involves improving fitness of individuals and resending the improved ones to a sender. A *SolutionManager* is brought to life for each problem instance. Its behaviour involves sending individuals to *OptiAgents* and updating the common memory.

2.2 Agent Mobility Management in JABAT

In JABAT *OptiAgents* are able to migrate between available locations. Mobility of agents is managed by a special agent class named the *PlatformManager*. The *PlatformManager* manages optimization agents and system platforms. It can move optimization agents among available containers, i.e. between containers on other computers that have joined the main platform, creating their copies. In this way the system can work with better efficiency achieving better quality of results or producing good results in shorter time. The use of mobile agents can bring decentralization of computations resulting in a more effective use of available resources and reduction of the computation time.

The management of the JABAT platforms is based on the two simple rules:

- If JABAT has been activated on a single container (computer), without any joined computers, then all *OptiAgents* would be also placed on this machine.
- If JABAT has been activated on multiple containers, with main container placed on one computer and the remote joined containers placed on other computers, then *OptiAgents* are moved from the main container to outside containers with a view to distribute the workload evenly.

The *PlatformManager* role includes the following:

- Identifying the number of containers available.
- Activating the required number of working *OptiAgents*, and also destroying agents indicated by *TaskManager*.
- Registering containers in which *OptiAgents* are working and sending to them requests to change location.

To improve efficiency of the ongoing optimization process the *PlatformManager* can create copies of *OptiAgents*. However, before activating the searching process the maximum number of copies for kind of *OptiAgents* have to be set by the user of JABAT.

3 Computational Experiment

To evaluate the performance of the proposed JADE-based implementation of the population learning algorithm it has been decided to carry a computational experiment. It involved the JABAT implementation of agent-based neural network training algorithm. In particular, the following factors and metrics representing different facets of system performance, have been investigated:

- Quality of ANN classifiers trained by the agent-based algorithm.
- Computation time until the desired correct classification ratio is achieved.
- Computation speed-up factor resulting from enabling agent migration to additional computers.
- Selection of optimization agents employed.

The following subsections contain a short description of the JABAT-based population learning algorithm for artificial neural network training, experiment plans for evaluation of the ANN classifiers quality, computation speed-up property and the influence of the composition of the optimization agents employed.

3.1 Implementation of the Proposed Algorithm and Evaluation of the Resulting ANN Classifiers

The experiment has been based on the population learning algorithm designed by the authors for artificial neural network training. The earlier implementation of the algorithm consisted of several improvement procedures run sequentially (for particulars see [4]). The main feature of the population learning approach is an increasing complexity of the improvement algorithms applied to the population of solutions and a decreasing number of individuals as the computation progresses. The agent-based JABAT version is using identical improvement procedures as in the original sequential algorithm, each represented by a different agent type. Within the JABAT-based artificial neural network training system there are seven agent types representing seven different improvement procedures including standard mutation, local search, non-uniform mutation, gradient mutation, gradient adjustment, a single point crossover and arithmetic crossover. Out of this set the crossover agents take care of information exchange and diversification while the remaining agents are used to directly improve the fitness of individuals drawn from the common memory. The details of each of the respective improvement algorithms can be found in [4].

Each optimization agent operates on one or more individuals (solutions) provided by the *SolutionManager* and is expected to improve quality of solutions or iterate for a prescribed number of iterations whatever comes first. After the stopping criterion has been met, each agent resends individuals to the *SolutionManager*, which, in turn, updates common memory by replacing worst individuals with the improved ones.

The experiment involved training of the MLP type artificial neural networks aimed at solving benchmark datasets including instances of four well known classification problems - Cleveland heart disease (303 instances, 13 attributes, 2 classes), credit approaval (690, 15, 2), Wisconsin breast cancer (699, 9, 2) and sonar problem (208, 60, 2). The respective datasets have been taken from [10].

Each benchmarking problem has been solved 30 times and the reported values of the quality measures have been averaged over all runs. The quality measure in all cases was the correct classification ratio calculated using the 10-cross-validation approach. The common memory size in JABAT was set to 50 individuals. All optimization agents, except the crossover ones, have been allowed to continue iterating until an improvement has been achieved or until 100 iterations have been performed.

The results obtained by using the ANN trained by the proposed JABAT-based training algorithm are shown in Table 1. The structure of the respective ANN is shown in the column 2 and the value of training error in column 3 of the discusses table. Table 1 contains also correct classification ratios produced by the ANN trained using JABAT-based training algorithm, the original population learning algorithm - PLA as reported in [4] and back propagation algorithm - BP. There are also the respective ratios obtained by the radial basis function classifiers - RBF.

Table 1. Single platform JABAT performance versus the best reported (* Source for the best reported with respect to cases of BP and RBF: http://www.phys.uni.torun.pl/kmk/projects/datasets-stat.html)

Problem	ANN structure	Training error	Accuracy(%)			
			JABAT	MLP+PLA	MLP+BP	RBF
Sonar	60-6-1	0.182	88.1	-	83.5*	83.6*
Credit	15-15-1	0.118	85.1	86.6	82.1	85.5*
Cancer	9-9-1	0.021	96.8	96.6	96.7*	95.9*
Heart	13-13-1	0.109	85.7	86.5	76.4	84.0*

3.2 Evaluation of the Computational Speed-Up Factor

The results reported in Table 1 had been used to establish a critical value of the training error for each problem type, which, in turn, has been used to evaluate the computational speed-up factor resulting from adding additional computers and enabling agents migration. The respective critical (that is the required) values have been set at the following levels: 0.25 for sonar problem, 0.13 - credit, 0.035 - cancer and 0.12 - heart.

In the experiment the training procedure of the respective ANN classifier has been terminated as soon as the classifier has been able to reach the respective critical value of the training error. In each run computation times were registered. The computations have been carried on several PC computers with Pentium IV 1.7 GHz processors and 256 MB RAM, connected within the local area network.

Dependency between the speed-up factor and the number of containers (computers) used by JABAT is shown in Figure 1 (see the left box). Mean speed-up factor can be estimated as 1.6 to 2.3 for, respectively, 2 to 6 computers. The above results have been achieved without cloning of agents, that is with only 7 optimization agents distributed among platforms. Increasing the number of copies can still improve the speed-up factor. This is shown in Figure 1 (see the right box) where the speed-up factor depending on the number of computers and the allowed number of agent clones is shown. The results are shown for 0, 1, 2, 3 and 4 additional clones allowed for each agent type, corresponding, respectively, to 7, 14, 21, 28 and 32 optimization agents distributed among available platforms.

Data obtained from the experiment have been used to perform the two-ways analysis of variance with a view to analyze the effect of two qualitative factors (the number of copies of each agent type and the number of computers used) on one dependent variable - computation time. The following null hypotheses have been tested:

- The number of copies of each agent type does not influence computation time.
- The number of computers used does not influence computation time.
- There is no interaction of the above factors.

It has been established that at the significance level of 0.05 all the above null hypotheses should be rejected. Additionally, the *post hoc* Tukey test has shown that there are no significant differences between average computation speed-up achieved while increasing the number of computers from 3 to 4 as compared with the speed-up achieved with the increase from 5 to 6 computers. In all of the remaining cases average speed-ups achieved statistically differ.

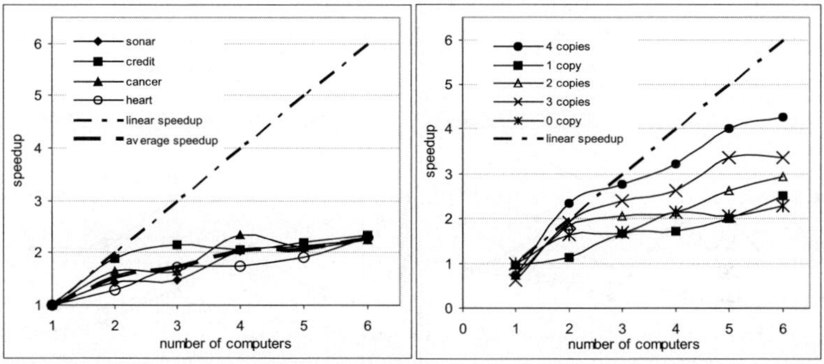

Fig. 1. Relations between the speed-up factor and the number of computers used (left box) and between the speed-up factor and the number of computers with different numbers of agent clones allowed - data averaged over all problem types (right box)

3.3 Selection of Optimization Agents

The second part of the reported experiment focused on answering the question whether a selection of optimization agents and the number of containers/platforms used has an impact on system computation time, assuming the required quality level measured by the training error is attained?

Computation time needed by the agent-based training algorithm to train ANN in such a way that the subsequent average training error is not greater then the value of the critical level has been recorded for each experiment run. There have been 6 variants of the number of computers used (from 1 to 6) and 5 scenarios for selection of optimization agents. Full experiment for each problem instance

Table 2. Agent selection scenarios

Scenario	Selected improvement procedures
opti 1	standard mutation; gradient mutation
opti 2	local search; non-uniform mutation
opti 3	gradient adjustment; arithmetic crossover
opti 4	arithmetic crossover; single point crossover
opti 5	standard mutation; gradient mutation, local search, non-uniform mutation, gradient adjustment; arithmetic crossover and single point crossover

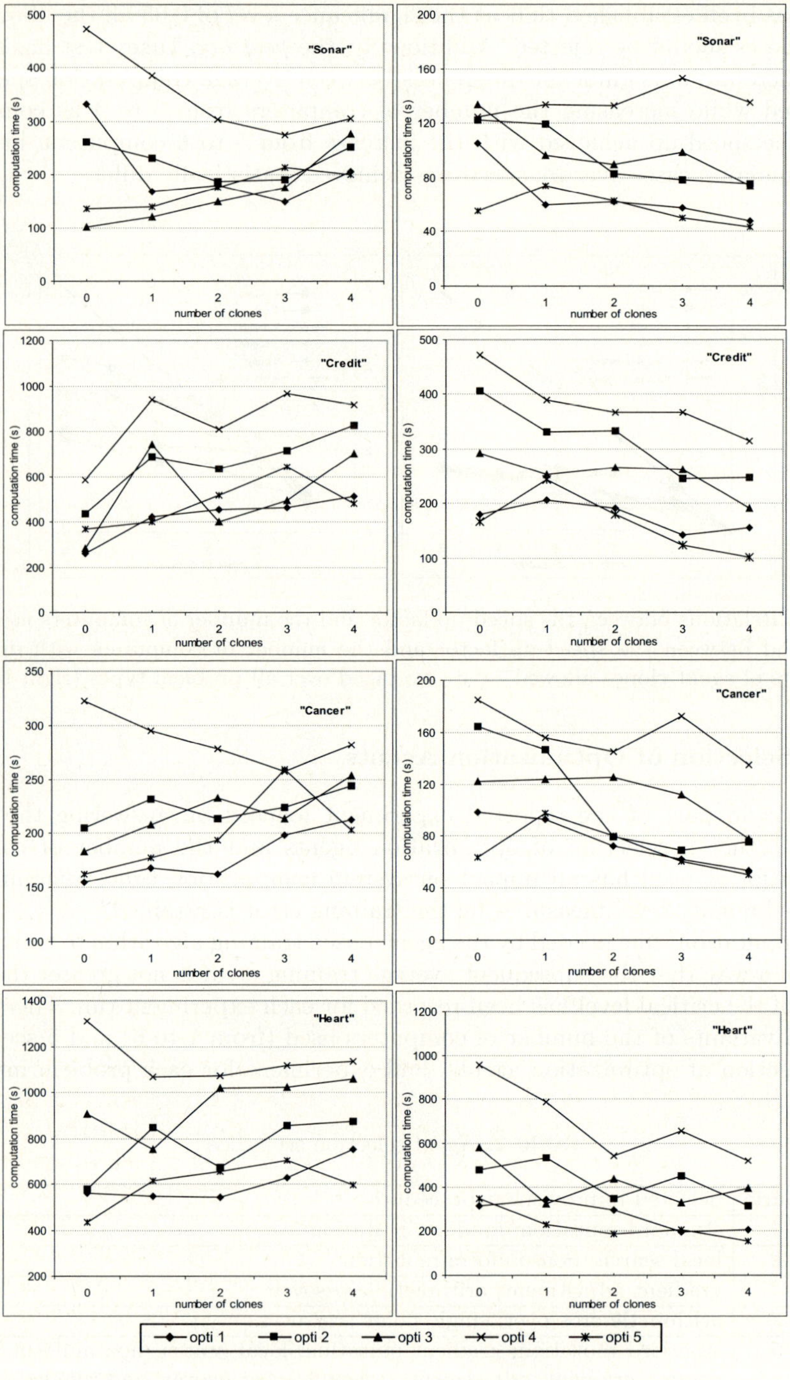

Fig. 2. Average computation times using single computer (left column) and four computers (right column) for different number of agent clones

of each problem type with 30 repetitions has been carried out. Table 2 lists types of the optimization agents used within each selection scenario.

Data obtained from the experiment, separately for different numbers of computers used, have formed an input to the two-ways analysis of variance with a view to analyze the effect of two qualitative factors (a number of copies of each agent type and a selection scenario of optimization agents) on one dependent variable - computation time. The following null hypotheses have been tested:

- The number of copies of each agent type does not influence computation time.
- The agent selection scenario does not influence computation time.
- There is no interaction of the above factors.

It has been established that at the significance level of 0.05 all the above null hypotheses should be rejected. Additionally, the *post hoc* Tukey test has shown that computation times for all variants of the computer numbers are statistically identical for *opti 1* and *opti 5* procedures. Similar lack of statistically significant differences has been observed between procedures *opti 2* and *opti 3*. However, average computation times for *opti 2* and *opti 3* have been significantly different from such times in case of *opti 1* and *opti 5*.

In Figure 2 average computation times for different numbers of agent clones for different problem types and 2 variants of computer numbers are shown.

4 Conclusion

The paper focuses on performance evaluation of the agent-based population learning algorithm proposed by the authors to train artificial neural networks. The algorithm has been implemented using JADE-based A-Team environment (JABAT). The proposed agent-based implementation of the ANN training algorithm produces competitive classifiers in terms of classification accuracy even when run on a single computer. Increasing the number of computers/containers and allowing for agent migration and cloning, results in substantially speeding-up the computations. This shows that the mobility scheme in JABAT serves well its purposes and can positively contribute to increasing effectiveness of the agent-based system.

Computational experiment described in the paper has shown that there are significant interactions between number of computers and number of agent copies used, as well as between number of agent copies of each kind and the selection scenarios of agents. Identification of these interactions may help in designing better agent-based systems, that is systems assuring best quality of results within the allowed time or systems producing the required quality of results within minimal time. Further research will concentrate on developing methodology for designing such systems.

Acknowledgements. The research has been supported by the KBN, grant no. 3T11C05928.

References

1. Aydin, M.E., Fogarty, T.C.: Teams of autonomous agents for job-shop scheduling problems: An Experimental Study, Journal of Intelligent Manufacturing, 15(4) (2004) 455-462
2. Barbucha D., Czarnowski I., Jędrzejowicz P., Ratajczak-Ropel E., Wierzbowska, I.: JADE-Based A-Team as a Tool for Implementing Population-Based Algorithms. In: Yuehui Chen, Ajith Abraham (ed.), Proceedings of the Sixth International Conference on Intelligent Systems Design and Applications (ISDA'06), IEEE Computer Society, 3 (2006) 144-149
3. Bellifemine, F., Caire, G., Poggi, A., Rimassa, G.: JADE. A White Paper, Exp, 3(3), (2003) 6-20
4. Czarnowski, I., Jędrzejowicz P.: An Approach to Artificial Neural Network Training. In: Bramer, M., Preece, A., Coenen, F., (ed.), "Research and Development in Intelligent Systems XIX", Springer, London (2003) 149-162
5. Davis, L. (ed.): Handbook of Genetic Algorithms, Van Nostrand Reinhold (1991)
6. Glover, F.: Tabu Search. Part I and II, ORSA Journal of Computing, 1(3), Summer 1990, and 2(1) Winter 1990
7. Jennings, N.R., Sycara, K., Wooldride, M.: A Roadmap of Agent Research and Developmant, Autonomous Agents and Multi-Agent Systems, 1 (1998) 7-38
8. Kennedy, J., Eberhart, R.C.: Particle swarm optimisation, Proc. of IEEE International Conference on Neural Networks, Piscataway, N.J. (1995) 1942-1948
9. Marinescu, D.C., Boloni, L.: A component-based architecture for problem solving environments, Mathematics and Computers in Simulation, 54 (2000) 279-293
10. Merz, C.J., Murphy, P.M.: UCI Repository of Machine Learning Databases [http://www.ics.uci.edu/~ mlearn/MLRepository.html.] Irvine, CA: University of California, Department of Information and Computer Science (1998)
11. Oster, G.F., Wilson, E.O.: Caste and Ecology in the Social Insect, Princeton University Press, Princeton, NJ8 (1978)
12. Parunak, H.V.D.: Agents in Overalls: Experiences and Issues in the Development and Deployment of Industrial Agent-Based Systems. International Journal of Cooperative Information Systems, 9(3) (2000) 209-228
13. Talukdar, S., Baerentzen, L., Gove, A., P. de Souza: Asynchronous Teams: Cooperation Schemes for Autonomous, Computer-Based Agents, Technical Report EDRC 18-59-96, Carnegie Mellon University, Pittsburgh (1996)

Agent-Based Discovery Middleware Supporting Interoperability in Ubiquitous Environments

Seung-Hyun Lee[1], Kyung-Soo Jang[2], Ho-Jin Shin[1], and Dong-Ryeol Shin[1]

[1] School of Information & Communication, Sungkyunkwan University, Korea
{lshyun0, hjshin, drshin}@ece.skku.ac.kr
[2] School of Computer Information Technology, Kyungin Women's College, Korea
ksjang@kic.ac.kr

Abstract. Service discovery in ubiquitous environments faces various issues such as dynamic-topology, devices-capability, resource-sharing, interoperability etc. In this paper, we propose Service Discovery Middleware (SDM) to solve the interoperability issue by using Discovery Middleware (DM), which was introduced by the FIPA (Foundation for Intelligent Physical Agents) [1]. The SDM provides an appropriate choice for providing interoperability in heterogeneous environments. Moreover, it supports a means for autonomic discovery of heterogeneous integration services in ubiquitous computing and transparency between users and services. With these ideas in mind, we design a simple mechanism for building a wide range of distributed services and applications as well as for supporting a softness and adaptable means to control and manage the Service Discovery Protocols (SDP). Furthermore, we implement as an example, a Personal Agent (PA) platform based on the FIPA-OS agent platform, in order to provide SDM component functionality [10] [20].

1 Introduction

Recently, computing environments have become more and more pervasive, ubiquitous, heterogeneous, and mobile. As the number of Internet services grows rapidly, it becomes increasingly important for network users to be able to locate and utilize those services that are available on the Internet. As a result, service discovery plays a key role in such dynamic environments, i.e., in ad-hoc networks. Some SDPs such as UPnP [8], JXTA [5], mSLP [17], Jini [18], HAVi [19], etc., have already provided appropriate and partial solutions to assist service discovery and delivery.

Service discovery among heterogeneous services could also be achieved by using the agent's service discovery technology, which especially brings a new software paradigm shift in ubiquitous environments. An agent based infrastructure would help us define and communicate with heterogeneous platforms, which is designed to aid in service discovery. To be more specific, a new agent paradigm is composed of intelligent, cooperative and autonomous agents, through which issues such as interoperability and service discovery are partially solved. However, the current agent platforms are not sufficient to support full interoperability between heterogeneous platforms and their corresponding services. This is the motivation of our research. We attempt to solve these problems by inserting the SDM component into the agent platform. Our test-bed uses components of SDM inside the PA platform.

N.T. Nguyen et al. (Eds.): KES-AMSTA 2007, LNAI 4496, pp. 141–149, 2007.

The rest of the paper is organized as follows. In Section 2, we introduce SDP technologies and FIPA DM. In Section 3, we design the system architecture under consideration. Section 4 deals with aspects of implementation and application scenarios. Finally, Section 5 concludes this paper.

2 Related Work

2.1 SDPs Technologies

In this subsection, we review various SDPs. Many service discovery protocols have been proposed to facilitate dynamic cooperation among services with minimal administration and human intervention. In order to be able to support the service, they should provide the means to announce service presence in the network, to discover services in the neighborhood, and to access these services [13].

While many of the architectures provide good foundations for developing systems with distributed components, they do not adequately solve all the problems that may arise in a dynamic domain and a large-scale network [21]. For example, when a newly connected UPnP, SLP, and, Jini client attempts to find services available in the distributed network. It sends requests to all devices connected to this network searching for services. Then the user initiates further interactions to obtain the presentation pages for each device. But, not all devices and services provided currently support interactive functionality. For this reason, we provide interoperability between several SDPs, through use of an agent platform.

2.2 FIPA Discovery Middleware

In a real ubiquitous computing environment, a standardized, open and softness infrastructure is needed. The FIPA is a committee to support the effort to make specifications for standard agent technologies. Since 1997, the FIPA has released a set of specifications. The FIPA does not promote a technology for just simple agent application, but a set of technologies for summarizing application areas for developers to integrate and make systems that are both interoperable and cooperative [12]. So, the use of DM is recommended to solve these problems in the FIPA Technical Community (TC). In the FIPA Discovery Service Specification [7], DM requires access via MANET (Mobile Ad-Hoc Network) [2] (Refer Fig.2.).

To find the service in one or more ad-hoc network, the agent uses the search functionality of the ADS (Agent Discovery Service) by providing the composed *df-agent-description* search template and the time period for which the search must be completed. Eventually, the agent may also specify special DM which should be used for this search. This can be done after the agent has queried the ADS for all currently available DM. Otherwise, the ADS will use all available DM by default. The ADS uses the search functionality of each available DM in turn, by providing the *df-agent-description* search template. The invoked DM maps the *df-agent-description* search template to the appropriate representation of the corresponding technology and performs the search in the corresponding ad-hoc network.

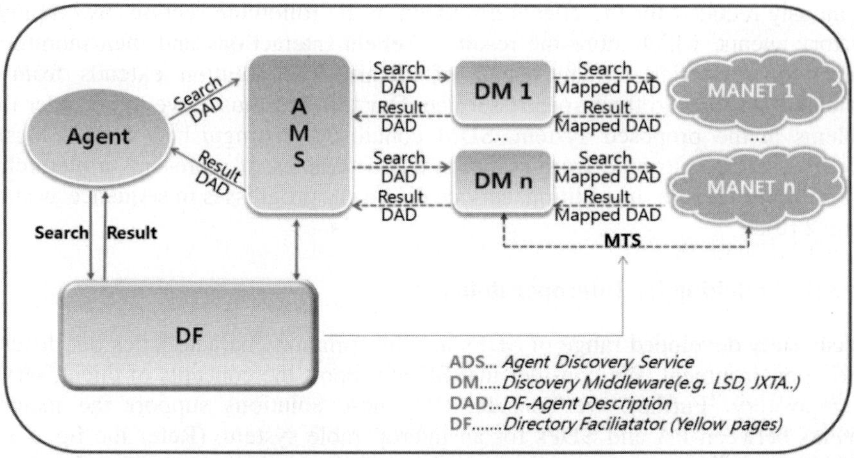

Fig. 1. Workflow of the DM and AP

All available *df-agent-descriptions*, which have previously been registered with the ADS, are represented in an appropriate form for DM and are discoverable by agents on remote devices. The results of such a search are matching *df-agent-descriptions* in the appropriate representation of DM. The DM now maps the results back to real *df-agent-descriptions* and returns them to the ADS.

3 System Architecture

We propose a new scheme to provide more efficient and scalable interoperability according to the standard FIPA specification.

In the scheme, once an agent has discovered and interrogated a set of appropriate services, as before, a second client looking for the same service must communicate with each service again. The process of gathering service information can be

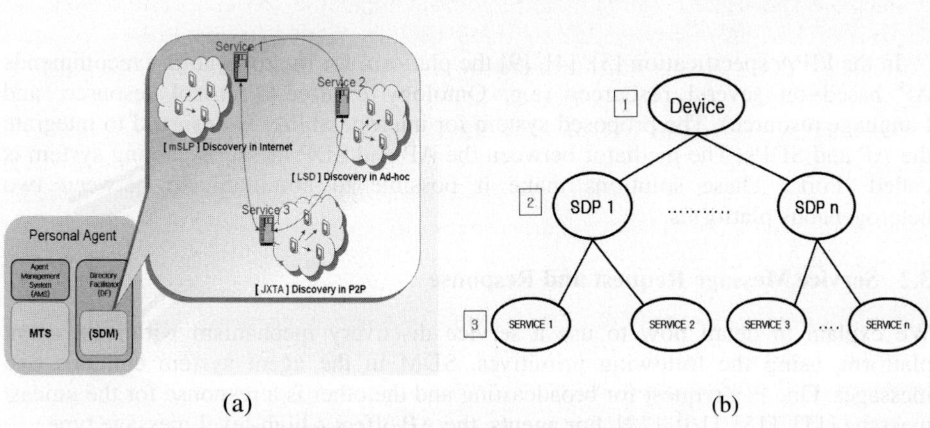

(a) (b)

Fig. 2. (a) SDM Architecture and (b) A Hierarchical Structure

enormously reduced for the clients discussed in the following section, by employing directory agents, which store the results of client interactions and then monitor for changes of service states on behalf of clients. The solution extends from the distributed platform to their special service. Our test-bed must have to consider these problems in the proposed system. SDM contains a distinguished service location value and service attribute from common characteristics. It represents a hierarchical attribute and structure. In addition, service discovery progresses in sequence as shown in Fig. 2 (b).

3.1 SDP Matching for Interoperability

The currently developed range of SDPs has the common characteristics but differs in working environment. We consider that SDPs support the concepts of client, service, and repository. Further, we consider that these solutions support the matching elements between PA and SDPs for an interoperable system. (Refer the fig. 2.) We provide two solutions among the several possible solutions [14]. The first is port matching for interoperability and the second is protocol for Message passing. It is the Service message Request and Response.

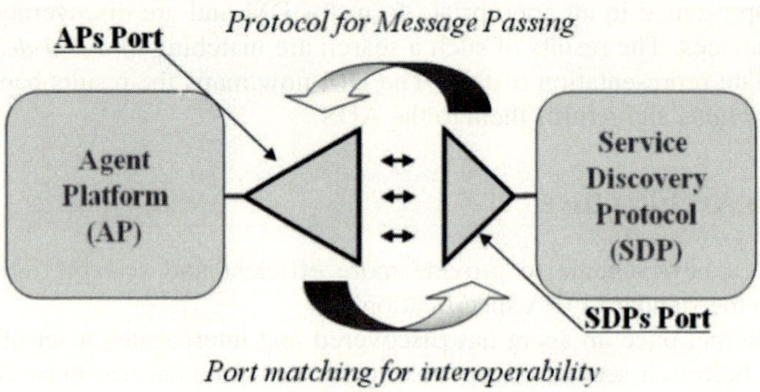

Fig. 3. AP and SDP Ports Matching

In the FIPA specification [3], [4], [9] the platform for interoperability recommends AP based on several resources (e.g. Ontology-resource, Protocol resource, and Language resource). The proposed system for interoperability is designed to integrate the AP and SDPs. The mediator between the AP and SDP message passing system is called ‘Port’. These solutions make it possible to communicate between two heterogeneous platforms.

3.2 Service Message Request and Response

We explain in detail how to use a service discovery mechanism within an agent platform, using the following primitives. SDM in the agent system contains two messages. One is a request for broadcasting and the other is a response for the unicast message [11], [15], [16], [22]. For agents, the AP offers a high-level message type.

Table 1 is presented a pseudocode for interaction in the message. In contrast to the existing DF, the SDM provide a functionality to subscribe DADs which match a DAD search template. From the time of subscription on, newly registered DADs that satisfy the search criteria will be returned to the agent until unsubscribing. Each of these functionalities can be applied either to all available DMs or to a subset.

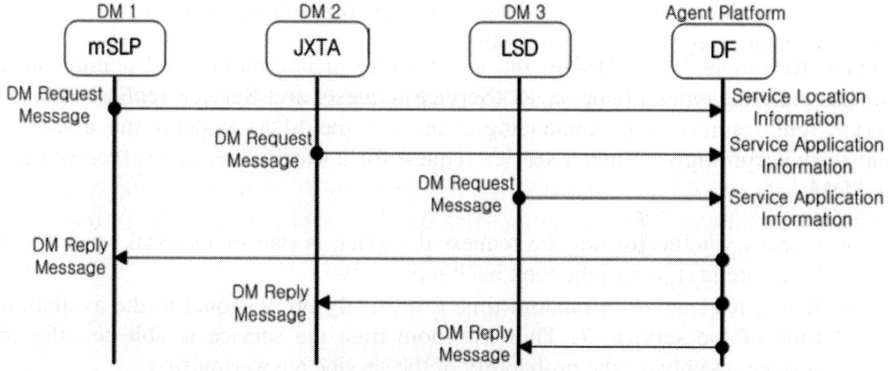

Fig. 4. DM Interaction

Table 1. A PsedoCode for Interaction

```
receive_msg(msg, aid)
{
    Message_format buf = new Message_format();
    While(read msg until End_of_msg)
    {
      Ontology = getElementsByTagName('deviceType');
      Service_name = getElementsByTagName('serviceName')
      Service_type = getElementsByTagName('serviceType');
        Buf.append (ontology);
        Buf.append (Service_name);
        Buf.append (Service_type);
        Buf.append (aid);
      Send buffer to sdm@localap;
    }
    Message_format
    {
      // predefine a DAD's form…
    }
}
```

Service Request. When the user needs a service, whether a specific service or any service offered by the agent, it requests the service from the distributed SDP. The number of message in transmission should be as few as possible so as to conserve resource consumption:

- If a specific service *S* has been requested, the SDM searches for that service in its repository. If it is not found, it broadcasts a service request for that service.
- If the request is for all available services in the network, the service updates its repository by sending a service message through the middleware.

Service Response. The SDM in the agent platform is continually listening on all interfaces for all types of messages (Service requests and Service replies). When a service reply is received, announcing a service, the SDM updates the associated repository accordingly. When a service request for a specific service is received then the SDM:

- Checks whether or not the requested service is one of its local services and therefore is stored in the look back repository.
- If not, it generates a random time *t*, inversely proportional to the availability time of the service, *T*. Thus the more time the service is able to offer the service, the higher the probability of the service answering first.
- During the interval t, the SDM listens to the network. If another reply to this service request arrives, it aborts its service reply, otherwise when the interval expires, is sends its service reply.

When a user requests a specified service list, the SDM is stored using a local repository.

- Generates a random time *t*, inversely proportional to the availability time of the service, *T*, and to the number of elements stored in the repository of that interface. So, the more time the service is able to offer the service and the larger the registry, the higher the probability of answering first.
- During the interval *t*, the SDM listens to the network. If another reply to this service request arrives, it aborts its service reply, otherwise when the interval expires it sends its service reply, listening the services in the look back registry plus the services in the registry of that interface.

4 Implementation

Fig 5 illustrates DF Management GUI (Graphic user interface) [12]. On the GUI, Service is the received message from the SDM of each platform. The proposed scenario makes it possible to control simple home devices. Users can show the service list and control the service.

Our test-bed consists of a desktop PC on the Java VM (Virtual machine) 1.4.2 and a Notebook on wireless LAN using the same VM. The PA (Personal agent) needs a SDM whose module listens to all advertisement messages of each platform. It sends the received message to the PA. The SDM listens and saves information in its local repository for a given period. When users want to find a service, PA searches the service using SDM. If there is no information, then it sends the unicast message through the SDM. The message exchange keeps their matching in a communication layer. The service- application information executes on LSD (Lightweight Service Discovery) [6] and JXTA.

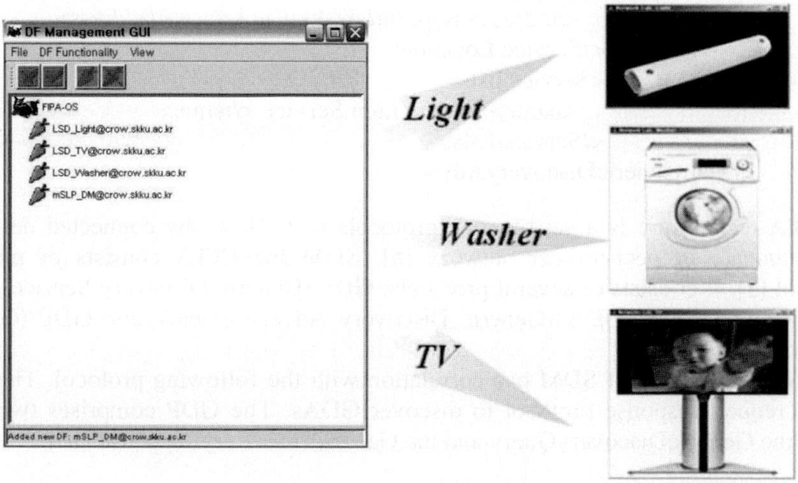

Fig. 5. DF-Management GUI

Some development for interoperability should be made in the agent environment. The current agent platform is not appropriate for supporting dynamic networks. Some modifications should be made to guarantee interoperability between heterogeneous platforms and services among agents in the AP. We include several SDM components on the AP. We use JXTA, LSD and mSLP as SDM materials.

The following present the Contents of the SDM description. The message received by the DF agent is presented in the GUI. The description based on XML communicates between FIPA-OS agent and each SDM. They have a flexible structure based XML and use a broadcasting-message ex-change mechanism. An Example of the SDM message is shown below.

Table 2. Contents of the SDM description

Attribute	Content
: Device Info.	Information of Device (e.g. Device Info.)
: Domain Info.	Domain of the each SDP
: Service Location	Location of Service in SDP (e.g. Address)
: Service Lists	Lists of Service (e.g. printing, presentation service)

The SDM request-description for LSD and JXTA is the followings:

```
<?xml version="1.0"?>
<!DOCTYPE fipa:GenericDiscoveryAdv>
<fipa:GenericDiscoveryAdv xmlns:fipa="http://www.fipa.org">
        <Device ID> NRL's Desktop </Device ID>
        <Domain Info> LSD #1 </Domain Info>
            <Service Location>
                <address> iiop://nrlab.skku.ac,kr/acc </address>
            </Service Location>
            <Service lists>
                <name> Presentation Service </name>
            </Service lists>
</fipa:GenericDiscoveryAdv>
```

JXTA technology is a set of open protocols that allow any connected device to com-municate in peer-to-peer network [5]. SDM for JXTA consists of message protocol [3]. It consists of several protocols: GDS (Generic Discovery Service), APG (Agent Peer Group), GDA (Generic Discovery Advertisement), and GDP (Generic Discov-ery Protocol).

Service discovery of SDM has correlation with the following protocol. The GDP uses a request/response protocol to discover GDAs. The GDP comprises two messages, the GenericDiscoveryQuery and the GenericDiscoveryResponse [15].

```
<?xml version="1.0"?>
<!DOCTYPE fipa:GenericDiscoveryAdv>
<fipa:GenericDiscoveryAdv xmlns:fipa="http://www.fipa.org">
        <Device ID> NRL's PDA </Device ID>
        <Domain Info> JXTA #1 </Domain Info>
            <Service Location>
                <address> iiop://nrlab.skku.ac,kr/acc </address>
            </Service Location>
        <Service lists>
            <name> Group Meeting </name>
        </Service lists>
</fipa:GenericDiscoveryAdv>
```

5 Conclusion and Future Work

In this paper, we presented a method for increasing interoperability of the software agent system in middleware.

The first stage creates an agent defining a relation among several discovery-middleware. The second stage is to define an agent that supports service-discovery and message. It may provide summarized SDM functionality as follows: The SDM can listen to messages using a particular port when the SDP is used to advertise information. Conversely, it can be used instead of a user to execute a particular service. We also designed the SDM Module in an agent platform that enables effective utilization of existing service discoveries. More importantly, the distinguishing feature of our approach does not provide a common API, but does provide direct translation between service descriptions and message. Currently, we are working on a distributed trust model for interoperability and awareness of services in the ubiquitous computing environment.

Acknowledgments. This work was supported by grant No. (R01-2004-000-10755-0) from the Basic Research Program of the Korea Science & Engineering Foundation.

References

1. The Foundation for Intelligent Physical Agent, http://www.fipa.org
2. M.Beger, M. watzke, H.Helin:Toward a FIPA approach for Mobile Ad Hoc Environment, ICIC'03 (Intelligence in Next Generation Networks), Bordeaux, Apr. 2003
3. FIPA: JXTA Discovery Middleware Specification, Oct. 2003
4. FIPA: Agent Management Specification, Dec. 2002
5. The JXTA Project, http://www.jxta.org
6. Jae-Wan Park, Byung-In Lim, Kee-Hyun Choi and Dong-Ryeol Shin: Design and Implementation of Light-weight Service Discovery System for Ad-Hoc Environments, ICACT 2005, Feb. 2005
7. FIPA: Agent Discovery Service Specification, Oct. 2003
8. Universal Plug and Play Specification, v.1.0, http://www.upnp.org
9. FIPA: Abstract Architecture Specification, Dec. 2002
10. Teppo.Pirttioja@hut.fi: FIPA-OS as an example of agent system implementation, AS-116.140 Postgraduate Seminar on Agent technology and its industrial applications fall. 2001
11. Erich Bircher and Torsten Braun: An Agent-Based Architecture for Service Discovery and Negotiations in Wireless Networks, 2nd International Conference on Wired/Wireless Internet Communications (WWIC 2004), Frankfurt an der Oder, Germany, Feb. 04 - 06, 2004
12. Federico Bergenti, Agostino Poggi, Fabio Bellifemine: Middleware and Programming Support for Agent Systems, EMCSR 2004, European Meetings on Cybernetics and Systems Research. Apr. 13-16. 2004
13. Christian Bettstetter, Christoph Renner: A Comparison of Service Implementation of the Service Location Protocol. In Proc. EUNICE mer School (EUNICE), Twente, Netherlands, Sep. 13-15, 2000
14. Stanislaw Ambroszkiewicz, Tomasz Nowak: Agentsapae as Integration, Engineering Societies in the Agents World II (ESAW'01), Jul. 7. pp. 134-159
15. FIPA: Agent Message Transport Service Specification, Oct. 2003
16. FIPA: Michael Berger: Agent in Ad-Hoc Environment, A Whitepaper.
17. mesh Service Location Protocol, http://mslp.sourceforge.net/
18. Jini, http://www.jini.org/
19. HAVi, www.havi.org
20. FIPA-OS, http://www.emorphia.com/research/about.htm
21. Kee-Hyun Choi, Ho-Jin Shin, Dong Ryeol Shin: D2HT: Directory Federation Using DHT to Support Open Scalability in Ubiquitous Network. PerCom Workshops 2005: 253-257
22. Celeste Campo: Distributed Directory Facilitator: A proposal for the FIPA Ad-hoc First CFT", Apr 30, 2002, http://www.fipa.org/docs/input/f-in-00063/f-in-00063.pdf

Performance of Fast TCP in Multi-agent Systems

Jung-Ha Kang[1], Hong-kyun Oh[2], Sung-Eon Cho[3,*], and Eun-Gi Kim[4]

[1] Fumate Co., 2F Jin Sung B/D 534-6 Noeun-dong,
Yuseong-gu, Daejeon, 305-325, Korea
jhkang@fumate.com

[2] BitsGen Co., LTD., 5F Owner's Tower 16-5, Sunae-dong,
Bundang-gu, Sungnam-shi, Gyeonggi-do, 463-825, Korea
netohk@hotmail.com

[3] Dept. of Inform. & Comm. Eng., Sunchon National University,
Sunchon-shi, Chonnam, 540-472, Korea
chose@sunchon.ac.kr

[4] Div. of Inform. Comm. & Computer Eng., Hanbat National University,
Dukmyung-dong, Yuseong-gu, Daejeon, 305-719, Korea
egkim@hanbat.ac.kr

Abstract. In this paper, the performance between RFC compatible normal TCP and several speed constraints ignored fast TCP is compared. To do these, the main algorithms that constraints the transmit rate of TCP are removed. We, and also, have modified TCP protocol stack in a Linux kernel as a kind of agent system to compare the speeds between the standard TCP and our modified fast TCP. We find that if the destination agent is short distance away from the source agent and packet error is scarce then the speed differences between normal and fast TCP may be negligible. However, if the destination agent is far away from the source agent and slow start algorithm is not adopted then the transfer time for small file is different greatly. In addition, if packet error occurred frequently, our modified fast TCP is faster than the standard TCP regardless of distance.

Keywords: TCP, modified fast TCP, TCP performance.

1 Introduction

There are many RFCs for TCP.[5][6][7] These RFCs defines the basic and extended operations of TCP for the control of transmission speed according to the network state. [1]. However, the Internet protocol specification is a recommendation only. If anybody make a fast TCP that does not follows the RFC specification – e.g. no flow controlled TCP - then the network may be congested. The speed and network effect owing to the RFC incompatible TCP protocol is not studied.

We have designed and implemented RFC incompatible fast TCP. The protocol performance and network effects are measured by using the fast TCP. The design considerations of our RFC incompatible fast TCP is as follows.

---

* Corresponding author.

N.T. Nguyen et al. (Eds.): KES-AMSTA 2007, LNAI 4496, pp. 150–158, 2007.

Firstly, the TCP should operate at a maximum speed regardless of TCP requirements in RFCs.

Secondly, our designed fast TCP should be operating to the standard TCP in multi-agent systems.

In this study, the performance is compared between standard TCP and several speed constraints ignored fast TCP in real Internet environments. This thesis is composed of 4 chapters. The design of our fast TCP is described in chapter 2. The performance results and analyses of our fast TCP is descried in chapter 3. Finally, the conclusions are described in chapter 4.

The remaining of this paper is organized as follows. In the next section, we briefly overview the fundamental of MAC protocols, further the promising EDCF-NA scheme is described in detail. Then, the performance of previously presented MAC protocols in Ad-hoc network is evaluated and compared through simulations by using NS-2 in section 3. Finally, section 4 concludes this paper via summarizing results and outlining future works

2 Design of Fast TCP

The components of TCP which affect the performance are delayed acknowledgement, Nagle algorithm, silly window syndrome avoidance, sliding window, slow start, congestion avoidance, fast retransmit and fast recovery, and retransmission timeout mechanism[1].

In the above described algorithms, we selected the four main algorithms to the best performance of bulk data transfer. The selected algorithms are modified as follows.

- No slow start (NSS)
- Slow start algorithm is not used. Only advertised window restrict the transmission rate.
- No congestion avoidance algorithm (NCA)
- The transmittable segment size increased exponentially even though cwnd is greater than ssthresh. The NCA is different from NSS. The NCA use cwnd and transmission rate grows exponentially regardless of ssthresh.
- Modified Fast retransmit (FST)
- Standard fast retransmit mechanism retransmits a packet with 3 duplicated ACKs. Our modified fast TCP retransmit a packet with only a 1 duplicate ACK.
- Modified RTO
- The time interval of RTO, in standard method, is about double of RTT interval. Therefore, timeout for lost segment is too long. Our fast TCP modified the RTO calculation in RFC793. In our modification, the value of β set to 1 to make RTO value should be similar to RTT.

The twelve test items are determined using the combination of above described 4 algorithms. These combinations are presented in table 1.

Table 1. Performance test items

No.	Case
Case 1	ORG(Original)
Case 2	NSS
Case 3	MRTO
Case 4	NCA
Case 5	FST
Case 6	NSS+MRTO
Case 7	NSS+FST
Case 8	MRTO+NCA
Case 9	MRTO+FST
Case 10	NCA+FST
Case 11	NSS+MRTO+FST
Case 12	NCA+MRTO+FST

3 Performance Analysis of Fast TCP

To test performances, the TCP in linux kernel 2.4.19 is modified according to table 1. [2][3] The client application in our modified fast TCP transmits a file to the discard server above standard TCP in real Internet environments, and the transmission time is compared. Test for each case is performed independently, and each file is transmitted 120 times to eliminate the instant network and server state changes. File transmission time is started at the sending of first data after 3 way handshake connection establishment phase. The tcpdump is modified to accurate the measurements. [3]

Test is performed according to below 2 scenarios.

Firstly, file transmission time is measured according to each case in table 1 at real Internet. The hop count from source to destination is 1 for short distance test and 20 for long.

Secondly, the same as first scenario except that packets are dropped intentionally. The PER (packet error rate) is changed from 0% to 3%.

3.1 Test in Real Internet Environment

The measurement is performed without intentional packet drop. In all graphs presented below, case 1 (standard TCP, ORG) is set to 100%, and other cases are changed according to case 1.

3.1.1 In Case of Short Distance

In these measurements, the discard server is located short distance away from the source. File transmission time and the number of packets are presented in figure 1 and 2. File size changed from 1KB to 2MB.

As shown in figure 1, the difference of transmission time between standard and our modified TCP is random for a small file below 100 KB. The reasons for these are that the ACK arrives before all segments within a cwnd value are transmitted. The client, therefore, can transmit a next segment without waiting time.

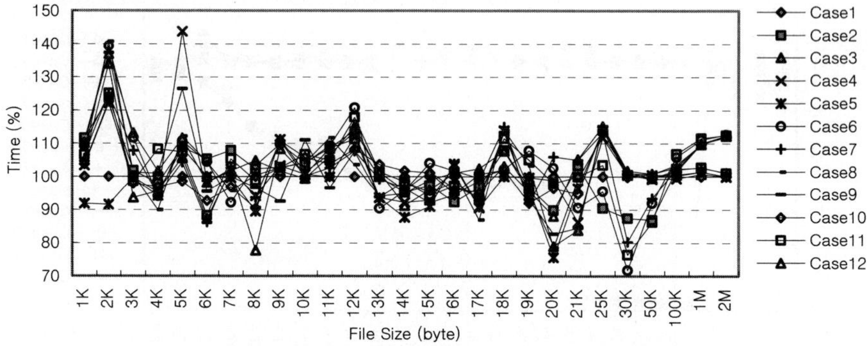

Fig. 1. Transmission time for each file size in a short distance

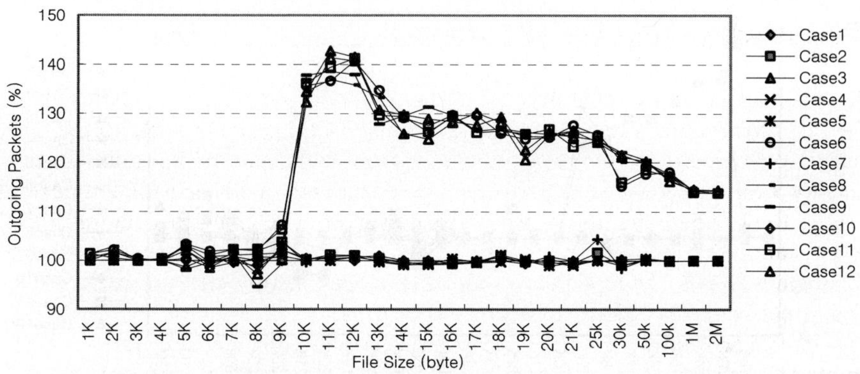

Fig. 2. A number of packets for each file size in a short distance

The difference, however, is obvious for big files such as 1 MB or 2 MB. All cases which shows longer transmission time than ORG (case 1) are MRTO adopted cases (case3, 6, 8, 9, 11, 12). The reasons for these are that the number of retransmission is increased owing to the small retransmission timeout interval. The reason for the abrupt increase of segments from 10 KB (10240 bytes) file is that the size limitations of receiver buffer (10136 byes) in discard server makes ACK delay, and in the mean time, the segment is retransmitted in client side owing to the timeout.

In conclusion, there are little differences for transmission time of small file between standard and our modified TCP in a small RTT environment.

3.1.2 In Case of Long Distance

In these measurements, the discard server is located at a long distance away (20 hops) from the source. File transmission time and the number of packets are presented in figure 3 and 4.

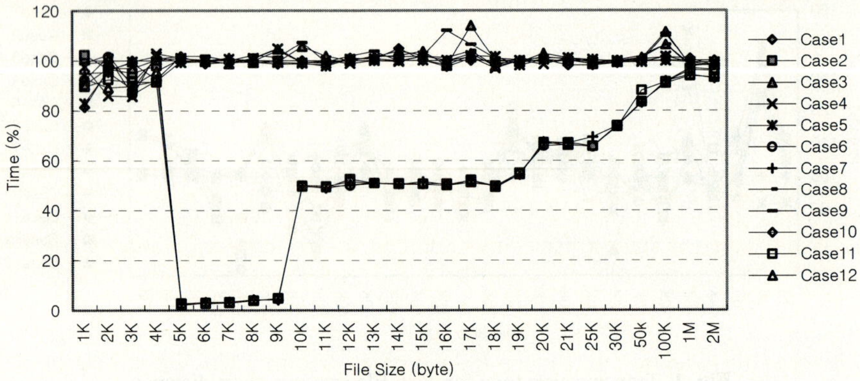

Fig. 3. Transmission time for each file size in a long distance

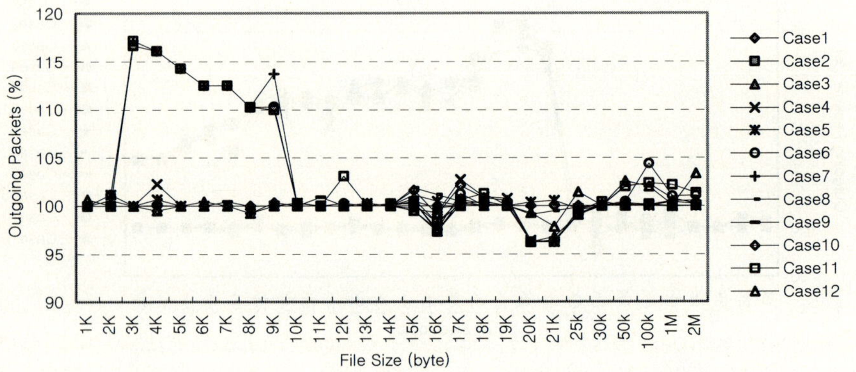

Fig. 4. A number of packets for each file size in a long distance

As shown in figure 3, the transmission time for NSS adopted cases (case2, 6, 7, 11) is superiorly fast. The reason of these results is as follows. The MSS (maximum segment size) is 1448 bytes, and advertised window of server is 7. So, the needed segment is 4 for 5KB file transfer. Initial cwnd value for Linux kernel is 2. Therefore, at the first stage of transmission, the client can transmit a 2 segment without a server ACK. In conclusion, all NSS not adopted cases including ORG (case 1) can transmit initially at a maximum 2 segment before ACK. The RTT and ACK time for long distance is large. This increase the time delay compared with NSS adopted cases.

However, in all NSS adopted cases, the transmission rate of sender is restricted only by the advertised window regardless of cwnd. At the first stage of transmission, the sender can transmit 7 segments without ACK. Therefore, if the file size is in 5KB ~ 9KB range then all segment can be transmitted without waiting ACK. The time difference is small for 4KB below sized file because of TCP in Linux kernel can transmit all segments without ACK.

The reason for the abrupt increase of transmission time in NSS adopted 10 KB sized file is that TCP delays the transmission of last segment (eight time segment)

until ACK. The number of transmission delay until ACK is increased as file size increase. So, the transmission time is approaches ORG for a big file.

The transmitted packet numbers in NSS adopted 3KB ~ 9KB file, as shown in figure 4, is large than the others. In these cases, the data segment and FIN set segment transmitted separately. Therefore, these cases transmit one more segment than the other cases. The transmitted packet numbers for the above 10 KB file is little difference from the ORG because the sender transmits a last data segment with FIN set. In conclusion, in a large RTT environment, the transmission time of small file such as WEB pages can be reduced by excluding a slow start algorithm from TCP.

3.2 Test in Real Internet with Intentional Packet Drops

For each of all cases in table 1, the transmission time of 1MB file is measured with packets are dropped intentionally. The Linux driver is modified to drop some packets randomly to emulate Internet with a determined error rate. PER range varies 0% to 3%. PER 0% means that there is no intentional packet error in Linux driver. There is little packet error in real Internet.

3.2.1 In Case of Long Distance

The transmission time and the number of packets for the transmission of 1 MB file within a short range are presented in figure 5 and 6.

As shown in figure 5, at the 0% PER, the transmission time between MRTO adopted cases and other cases are different according to the same reason at 1MB file transfer in figure 1. As PER increased, the transmission rates are more fasted than the ORG (case 1). Especially, NSS adopted case 2 and 7 shows 20 % reduced transmission time at 3 % PER with a little packet increasing than the ORG.

As PER increased, the MRTO adopted cases (case3, 6, 8, 9, 11, 12) shows more fast transmission rate. And at the 3% PER, the speed of MRTO adopted cases are similar to NSS adopted cases. However, the number of packets transmitted increased about 15 % than the ORG.

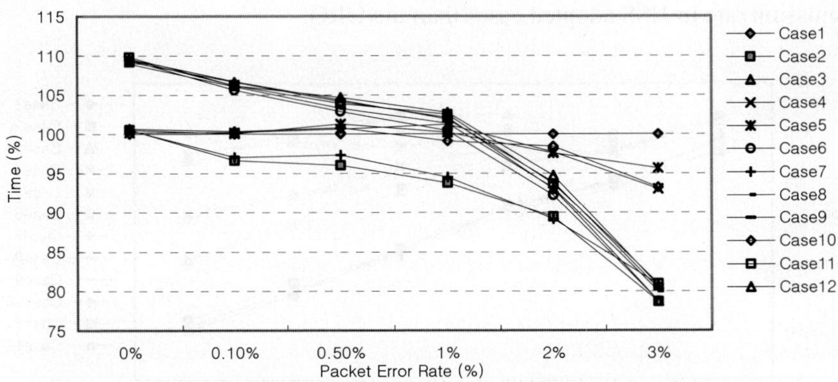

Fig. 5. File transfer time in short distance, packet dropped environments

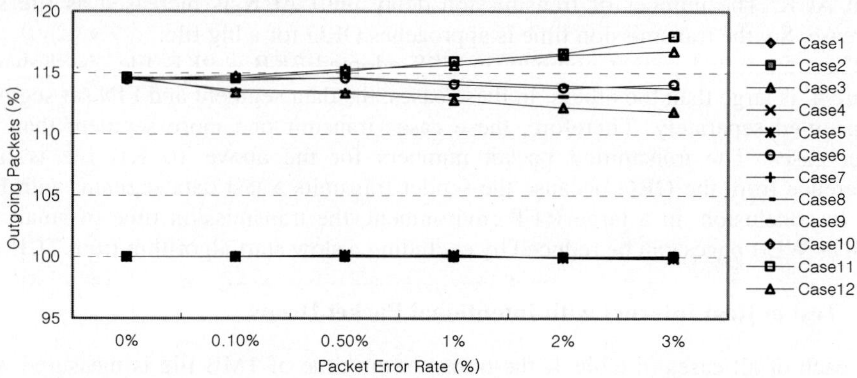

Fig. 6. A number of packets in short distance, packet dropped environments

In conclusion, in a packet dropped network environments, all cases show faster transmission rate then the ORG regardless of small RTT.

3.2.2 In Case of Long Distance

The transmission time and the number of packets for the transmission of 1 MB file with a long distance are presented in figure 7 and 8.

As PER increased, the transmission speed increments is notable. The transmission time of NSS adopted cases (case2, 6, 7, 11) reduced abruptly in accordance with the PER increase, and at 3% PER, 40% reduced than the ORG. The NCA and MRTO combined case 8 and 12 shows about 25% transmission time reductions. All other cases (case3, 5, 9) show about 5% transmission time reductions.

The MRTO adopted cases shows about 7~12% packet increase than the ORG at 3% PER. All other cases shows similar packet numbers to the ORG at 3% PER.

The reason for faster transmission in NSS adopted cases than the ORG is that the NSS adopted cases can transmit a packet to the advertised windows even though packet retransmission. The more often packet error or lost occurs, the more fast transmission rate in NSS adopted cases than the ORG.

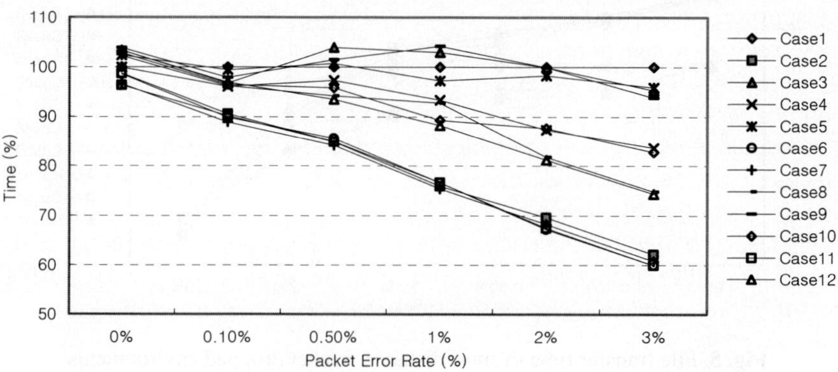

Fig. 7. File transfer time in long distance, packet dropped environments

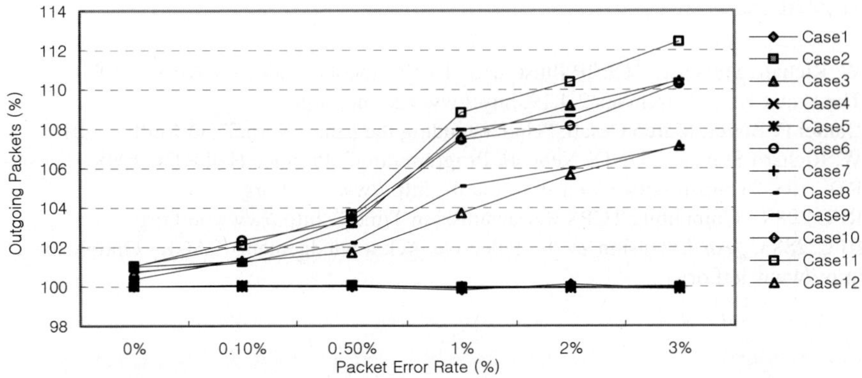

Fig. 8. A number of packets in long distance, packet dropped environments

Although the number of transmittable segments in NCA adopted case are increased exponentially, the NCA adopted cases are more delays than the NSS adopted cases because the restriction of cwnd.

In conclusion, as the PER increased, all cases shows the fast transmission rate than ORG in a large RTT test. However, some of these cases shows the increasing the number of packets, and this can make a network congestion.

4 Conclusion

As PER increased, the transmission speed increments is notable. The transmission time of NSS adopted cases (case2, 6, 7, 11) reduced abruptly in accordance with the PER increase, and at 3% PER, 40% reduced than the ORG. The NCA and MRTO combined case 8 and 12 shows about 25% transmission time reductions. All other cases (case3, 5, 9) show about 5% transmission time reductions.

The MRTO adopted cases shows about 7~12% packet increase than the ORG at 3% PER. All other cases shows similar packet numbers to the ORG at 3% PER.

The reason for faster transmission in NSS adopted cases than the ORG is that the NSS adopted cases can transmit a packet to the advertised windows even though packet retransmission. The more often packet error or lost occurs, the more fast transmission rate in NSS adopted cases than the ORG.

Although the number of transmittable segments in NCA adopted case are increased exponentially, the NCA adopted cases are more delays than the NSS adopted cases because the restriction of cwnd.

In conclusion, as the PER increased, all cases shows the fast transmission rate than ORG in a large RTT test. However, some of these cases shows the increasing the number of packets, and this can make a network congestion in multi-agent systems environments.

References

1. W. Richard Stevens, “TCP/IP Illustrated: The Protocols”, Addison-Wesley, 1999.
2. The Linux Kernel Archives 2.4.19, http://www.kernel.org/
3. Daniel P. Bovet, Marco Cesati, “Understanding the Linux Kernel”, O’Reilly, 2001
4. W. Richard Stevens, “UNIX Network Programming”, Prentice Hall PTR, 1998
5. RFC 793, “Transmission Control Protocol”, http://www.ietf.org
6. RFC 2988, “Computing TCP's Retransmission Timer”, http://www.ietf.org
7. RFC 2883, “An Extension to the Selective Acknowledgement (SACK) Option for TCP”, http://www.ietf.org

Manipulating Paraconsistent Knowledge in Multi-agent Systems

Jair Minoro Abe[1] and Kazumi Nakamatsu[2]

[1] Graduate Program in Production Engineering, ICET - Paulista University
R. Dr. Bacelar, 1212, CEP 04026-002 São Paulo - SP – Brazil
jairabe@uol.com.br

[2] School of Human Science and Environment/H.S.E. – University of Hyogo – Japan
nakamatu@shse.u-hyogo.ac.jp

Abstract. In this paper we introduce first order Paraconsistent Annotated Multimodal systems Mτ which may constitute a framework for multi-agent system reasoning. Such systems are capable of handling imprecise, inconsistent and paracomplete knowledge in a non-trivial manner in their structures.

Keywords: Annotated logics, knowledge and paraconsistency, distributed systems, multi-agents.

1 Introduction

In recent years, many interesting formal systems have been proposed for multi-agent systems reasoning. When several agents are doing a task, one natural question that raises is of contradiction. When we think agents as human beings, then they surely keep contradictory beliefs in theirs lives. When it is taken questions of logical omniscience, one relevant concept that also appears is that of contradiction. Some authors have taken into account this problem, for instance, [3]. Other authors have showed how different properties of knowledge can be captured by imposing certain conditions on semantics which permit such contradictions [9].

Roughly speaking, the problem of logical omniscience is an agent to know all logical consequences from a set of premises, in particular all tautologies. A good discussion is to be found in [7]. Also, it is well known that from a contradiction all formulas is provable, and so due previous observation, all agents know all formulas ! Definitely this is not also natural.

The use of modal systems for modeling knowledge and belief has been largely considered in Artificial Intelligence. For instance, it seems that the first one to consider knowledge and belief to machines was McCarthy [10]. Subsequently, [12], [11], [8] among others, have considered knowledge in multi-agent systems, besides other approaches.

The essential ideas underlying the multimodal system proposed by [7] can be summarized as follows: $\Box_i A$ can be read *agent i knows A, i* = 1, ..., *n. Common knowledge* and *distributed knowledge* are also defined in terms of additional modal

N.T. Nguyen et al. (Eds.): KES-AMSTA 2007, LNAI 4496, pp. 159–168, 2007.

operators: $\Box_G$ ("everyone in the group G knows"), $\Box_G^C$ ("it is common knowledge among agents in G"), and $\Box_G^D$ ("it is distributed knowledge among agents in G") for every nonempty subset G of $\{1, \ldots, n\}$.

Nevertheless, the most of these proposals use extensions of Classical Logic or at least part of it, keeping as much as fundamental characteristics of Classical Logic.

So, in this paper we propose a logical system for reasoning with inconsistencies in multi-agent systems in a non-trivial way. The desired aim, the system Mτ, is obtained by using ideas of [1] and [7].

In this paper, we present a logical framework that has the following characteristics:

a) The principle of contradiction, in the form $\neg(A \wedge \neg A)$, is not valid in general among hyper-literal formulas;
b) From two contradictory hyper-literal formulas, A and $\neg A$, we cannot deduce any formula B whatsoever;
c) For complex formulas, it is valid all properties of the classical logic.
d) Mτ contains the most important schemes and rules of inference of the classical predicate calculus that are compatible with conditions a) and b).

Due these properties, this system can be the underlying logic for modeling conflicting common knowledge. The system can be adapted to cope conflicting beliefs and awareness.

2 Paraconsistent, Paracomplete, and Non-alethic Logics

In this paragraph we establish some terminologies.

Let T be a theory whose underlying logic is L. T is called *inconsistent* when it contains theorems of the form A and $\neg A$ (the negation of A). If T is not inconsistent, it is called *consistent.* T is said to be *trivial* if all formulae of the language of T are also theorems of T. Otherwise, T is called *non-trivial.* When L is classical logic (or other logic, e.g. intuitionistic logic), T is inconsistent iff T is trivial. So, in trivial theories the extension of the concepts of formula and theorem coincide. A *paraconsistent logic* is a logic that can be used as the basis for inconsistent but non-trivial theories. A theory is called *paraconsistent* if its underlying logic is a paraconsistent logic.

A logical system is called *paracomplete* if it can function as the underlying logic of theories in which there are formulae such that these formulae and their negations are simultaneously false.

As a consequence, paraconsistent theories do not satisfy the principle of non-contradiction, which can be stated as follows: of two contradictory propositions, i.e., one of which is the negation of the other, one must be false. And, paracomplete theories do not satisfy the principle of the excluded middle, formulated in the following form: of two contradictory propositions, one must be true.

Finally, logics which are simultaneously paraconsistent and paracomplete are called *non-alethic logics.*

3 The Paraconsistent Multimodal Logics $M\tau$

We present, in this section, the multimodal predicate calculi $M\tau$, based on annotated logics extensively studied by Abe [1], [2], [4].

The symbol $\tau = <|\tau|, \leq, \sim>$ indicates some finite lattice with operator called the *lattice of truth-values*. We use the symbol $\leq$ to denote the ordering under which τ is a complete lattice, $\perp$ and $\top$ to denote, respectively, the bottom element and the top element of τ. Also, $\wedge$ and $\vee$ denote, respectively, the greatest lower bound and least upper bound operators with respect to subsets of $|\tau|$. The operator $\sim: |\tau| \rightarrow |\tau|$ will work as the "meaning" of the negation of the system $M\tau$.

The language of $M\tau$ has the following primitive symbols:

1. Individual variables: a denumerable infinite set of variable symbols: $x_1, x_2, \ldots$
2. Logical connectives: $\neg$ (negation), $\wedge$ (conjunction), $\vee$ (disjunction), and $\rightarrow$ (implication).
3. For each n, zero or more n-ary function symbols (n is a natural number).
4. For each $n \neq 0$, n-ary predicate symbols.
5. The equality symbol: =
6. Annotational constants: each member of τ is called an annotational constant.
7. Modal operators: $\Box_1, \Box_2, \ldots, \Box_n$, $(n \geq 1)$, $\Box_G$, $\Box_G^C$, $\Box_G^D$ (for every nonempty subset G of $\{1, \ldots, n\}$).
8. Quantifiers: $\forall$ (for all) and $\exists$ (there exists).
9. Auxiliary symbols: parentheses and comma.

For each n the number of n-ary function symbols may be zero or non-zero, finite or infinite. A 0-ary function symbol is called a *constant*. Also, for each $n \geq 1$, the number of n-ary predicate symbols may be finite or infinite. We suppose that $M\tau$ possesses at least one predicate symbol.

We define the notion of *term* as usual. Given a predicate symbol p of arity n and n terms $t_1, \ldots, t_n$, a *basic formula* is an expression of the form $p(t_1, \ldots, t_n)$. An *annotated atomic formula* is an expression of the form $p_\lambda(t_1, \ldots, t_n)$, where λ is an annotational constant. We introduce the general concept of (*annotated*) *formula* in the standard way. For instance, if A is a formula, then $\Box_1 A$, $\Box_2 A$, ..., $\Box_n A$, $\Box_G A$, $\Box_G^C A$, and $\Box_G^D A$ are also formulas. Among several intuitive readings, an atomic annotated formula $p_\lambda(t_1, \ldots, t_n)$ can be read: *it is believed that* $p(t_1, \ldots, t_n)$*'s truth value is at least* λ.

Definition 3.1. Let A and B be formulas. We put
1. $A \leftrightarrow B =_{\text{Def.}} (A \rightarrow B) \wedge (B \rightarrow A)$
2. $\neg\!\!\!\neg A =_{\text{Def.}} A \rightarrow ((A \rightarrow A) \wedge \neg(A \rightarrow A))$

The symbol '$\leftrightarrow$' is called *biconditional*, '$\neg\!\!\!\neg$' is called *strong negation.*

Let A be a formula. Then: $\neg^0 A$ indicates A, $\neg^1 A$ indicates $\neg A$, and $\neg^k A$ indicates $\neg(\neg^{k-1} A)$, ($k \in \mathbf{N}$, $k > 0$, $\mathbf{N}$ is the set of natural numbers). Also, if $\mu \in \tau$, $\sim^0\mu$ indicates μ, $\sim^1\mu$ indicates $\sim\mu$, and $\sim^k\mu$ indicates $\sim(\sim^{k-1}\mu)$, ($k \in \mathbf{N}$, $k > 0$). If A is an atomic formula $p_\lambda(t_1, \ldots , t_n)$, then a formula of the form $\neg^k p_\lambda(t_1, \ldots , t_n)$ $(k \geq 0)$ is called a *hyper-literal*. A formula other than hyper-literals is called a *complex* formula.

The postulates (axiom schemata and primitive rules of inference) of Mτ are the following: A, B, and C are any formulas whatsoever, F and G are complex formulas, $p(t_1, \ldots , t_n)$ is a basic formula, and λ, μ, μ_j are annotational constants.

1. $A \to (B \to A)$
2. $(A \to (B \to C)) \to ((A \to B) \to (A \to C))$
3. $((A \to B) \to A) \to A$
4. $\dfrac{A,\ A \to B}{B}$
5. $A \wedge B \to A$
6. $A \wedge B \to B$
7. $A \to (B \to (A \wedge B))$
8. $A \to A \vee B$
9. $B \to A \vee B$
10. $(A \to C) \to ((B \to C) \to ((A \vee B) \to C))$
11. $(F \to G) \to ((F \to \neg G) \to \neg F)$
12. $F \to (\neg F \to A)$
13. $F \vee \neg F$
14. $p_\perp(t_1, \ldots , t_n)$.
15. $\neg^k p_\lambda(t_1, \ldots , t_n) \to \neg^{k-1} p_{\sim\lambda}(t_1, \ldots , t_n)$, $k \geq 1$
16. $p_\lambda(t_1, \ldots , t_n) \to p_\mu(t_1, \ldots , t_n)$, $\lambda \geq \mu$
17. $p_{\lambda 1}(t_1, \ldots , t_n) \wedge p_{\lambda 2}(t_1, \ldots , t_n) \wedge \ldots \wedge p_{\lambda m}(t_1, \ldots , t_n) \to p_\lambda(t_1, \ldots , t_n)$, where $\lambda = \bigvee_{i=1}^{m} \lambda_i$
18. $\square_i(A \to B) \to (\square_i A \to \square_i B)$, $i = 1, 2, \ldots , n$
19. $\square_i A \to \square_i \square_i A$, $i = 1, 2, \ldots , n$
20. $\neg\square_i A \to \square_i \neg\square_i A$, $i = 1,2, \ldots , n$
21. $\square_i A \to A$, $i = 1, 2, \ldots , n$
22. $\dfrac{A}{\square_i A}$, $i = 1, 2, \ldots , n$
23. $\square_G A \leftrightarrow \wedge_{i\in G} \square_i A$
24. $\square^C_G A \to \square_G(A \wedge \square^C_G A)$
25. $\square^D_{\{i\}} A \leftrightarrow \square_i A$, $i = 1, 2, \ldots , n$
26. $\square^D_G A \to \square^D_{G'} A$ if $G' \subseteq G$.
27. $\dfrac{A \to \square_G(B \wedge A)}{A \to \square^D_G B}$ (Induction Rule)
28. $A(t) \to \exists x A(x)$
29. $\dfrac{A(x) \to B}{\exists x A(x) \to B}$
30. $\forall x A(x) \to A(t)$
31. $\forall x \square_i A \to \square_i \forall x A$, $i = 1, 2, \ldots , n$

32. $\dfrac{B \rightarrow A(x)}{B \rightarrow \forall x A(x)}$

33. $x = x$

34. $x = y \rightarrow A[x] \leftrightarrow A[y]$

35. $_{\Pi}(x = y) \rightarrow \Box_{i\Pi}(x = y),\ i = 1, 2, \ldots, n$

with the usual restrictions.

Theorem 3.1: In Mτ all valid schemes and rules of classical positive propositional logic are true. In particular, the deduction theorem is valid in Mτ ant it contains intuitionistic positive logic.

Theorem 3.2: Mτ is non-trivial.

4 Semantical Analysis: Kripke Models

Definition 4.1: A Kripke model for Mτ (or Mτ structure) is a set theoretical structure $K = [W, R_1, R_2, \ldots, R_n, I]$ where

1. W is a nonempty set of elements called 'worlds'
2. R_i $(i = 1, 2, \ldots, n)$ is a binary relation on W such that it is an equivalence relation.
3. I is an interpretation function with the usual properties with the exception that for each n-ary predicate symbol p we associate a function $p_I: W^n \rightarrow |\tau|$.

Given a Kripke model K for the language L of Mτ, the *diagram* language $L(K)$ is obtained as usual. Given a free variable term a of $L(K)$ we define, as usual, the individual $K(a)$ of K. We use i and j as meta-variables for names.

Definition 4.2: If A is a closed formula of Mτ, and $w \in W$, we define the relation $K,w \Vdash A$ (K,w *force* A) by recursion on A:

1. If A is atomic of the form $p_\lambda(t_1, \ldots, t_n)$, then
   $K,w \Vdash A$ iff $p_I(K(t_1), \ldots, K(t_n)) \geq \lambda$.
2. If A is of the form $\neg^k p_\lambda(t_1, \ldots, t_n)$ $(k \geq 1)$, $K,w \Vdash A$ iff $K,w \Vdash \neg^{k-1} p_{\sim\lambda}(t_1, \ldots, t_n)$.
3. Let A and B formulas. Then, $K,w \Vdash (A \wedge B)$ iff $K,w \Vdash A$ and $K,w \Vdash B$; $K,w \Vdash (A \vee B)$ iff $K,w \Vdash A$ or $K,w \Vdash B$; $K,w \Vdash (A \rightarrow B)$ iff it is not the case that $K,w \Vdash A$ or $K,w \Vdash B$;
4. If F is a complex formula, then $K,w \Vdash (\neg F)$ iff it is not the case that $K,w \Vdash F$.
5. If A is of the form $(\exists x)B$, then $K,w \Vdash A$ iff $K,w \Vdash B_x[i]$ for some i in $L(K)$.
6. If A is of the form $(\forall x)B$, then $K,w \Vdash A$ iff $K,w \Vdash B_x[i]$ for all i in $L(K)$.
7. If A is of the form $\Box_i B$ then $K,w \Vdash A$ iff $K,w' \Vdash B$ for each $w' \in W$ such that wR_iw', $i = 1, 2, \ldots, n$

Definition 4.3: Let $K = [W, R_1, R_2, \ldots, R_n, I]$ be a Kripke structure for Mτ. The Kripke structure K *forces* a formula A (in symbols, $K \Vdash A$), if $K,w \Vdash A$ for each $w \in W$. A formula A is called Mτ-*valid* if for any Mτ-structure K, $K \Vdash A$. A formula A is called *valid* if it is Mτ-valid for all Mτ structure. We symbolize this fact by $\Vdash A$.

Theorem 4.1: Let $K = [W, R_1, R_2, \dots, R_n, I]$ be a Kripke structure for Mτ. For all formulas A, B, then

1. If A is an instance of a propositional tautology then, $K \Vdash A$
2. If $K \Vdash A$ and $K \Vdash A \to B$, then $K \Vdash B$
3. $K \Vdash \Box_i(A \to B) \to (\Box_i A \to \Box_i B)$, $i = 1, 2, \dots, n$
4. $K \Vdash \Box_i A \to \Box_i \Box_i A$, $i = 1, 2, \dots, n$
5. $K \Vdash \Box_i A \to A$, $i = 1, 2, \dots, n$
6. If $K \Vdash A$ then $K \Vdash \Box_i A$, $i = 1, 2, \dots, n$

Theorem 4.2: Let K be a Kripke model for Mτ and F a complex formula. Then we have not simultaneously $K,w \Vdash \neg F$ and $K,w \Vdash F$.

Theorem 4.3: Let $p(t_1, \dots, t_n)$ be a basic formula and $\lambda, \mu, \rho \in |\tau|$. We have

1. $\Vdash p_\perp(t_1, \dots, t_n)$
2. $\Vdash p_\lambda(t_1, \dots, t_n) \to p_\mu(t_1, \dots, t_n)$, if $\lambda \geq \mu$
3. $\Vdash p_\lambda(t_1, \dots, t_n) \wedge p_\mu(t_1, \dots, t_n) \to p_\rho(t_1, \dots, t_n)$, where $\rho = \lambda \vee \mu$

Proof. 1. For any Kripke model K, we have $p_I(K(t_1), \dots, K(t_n)) \geq \perp$, for all $w \in K$. So, $K \models p_\perp(t_1, \dots, t_n)$ for every K, and therefore $\Vdash p_\perp(t_1, \dots, t_n)$.
2. Let us suppose that there exists a K such that it is not the case that $K \Vdash p_\lambda(t_1, \dots, t_n) \to p_\mu(t_1, \dots, t_n)$, that is $K \Vdash p_\lambda(t_1, \dots, t_n)$ and it is not the case that $K \Vdash p_\mu(t_1, \dots, t_n)$, for some $w \in K$. So, $p_I(K(t_1), \dots, K(t_n)) \geq \lambda$. As $\lambda \geq \mu$, we have $p_I(K(t_1), \dots, K(t_n)) \geq \mu$, which contradicts the hypothesis. Therefore, we have $\Vdash p_\lambda(t_1, \dots, t_n) \to p_\mu(t_1, \dots, t_n)$, if $\lambda \geq \mu$.
3. Similar to the preceding, using conditions 1 and 2 of Definition 4.2.

Theorem 4.4: Let A and B be arbitrary formulas and F a complex formula. Then:

1. $\Vdash ((A \to B) \to ((A \to \urcorner B) \to \urcorner A))$
2. $\Vdash (A \to (\urcorner A \to B))$
3. $\Vdash (A \vee \urcorner A)$
4. $\Vdash (\neg F \leftrightarrow \urcorner F)$
5. $\Vdash A \leftrightarrow \urcorner\urcorner A$
6. $\Vdash \forall x A \leftrightarrow \exists x \urcorner A$
7. $\Vdash (A \wedge B) \leftrightarrow \urcorner(\urcorner A \vee \urcorner B)$
8. $\Vdash \forall A \leftrightarrow \exists x \urcorner A$
9. $\Vdash \forall x A \vee B \leftrightarrow \exists x(A \vee B)$
10. $\Vdash A \vee \exists x B \leftrightarrow \exists x(A \vee B)$

Corollary 4.4.1: In the same conditions of the preceding theorem, we have not simultaneously $K \Vdash \urcorner A$ and $K \Vdash A$. The set of all formulas together with the connectives $\wedge$, $\vee$, $\to$, and $\urcorner$ has all properties of the classical logic.

Theorem 4.5: There are Kripke models K such that for some hyper-literals A and B and some worlds w and $w' \in W$, we have $K,w \Vdash \neg A$ and $K,w \Vdash A$ and it is not the case that $K,w' \Vdash B$.

Proof. Let $W = \{\{a\}\}$ and $R = \{(\{a\},\{a\})\}$ (that is $w = \{a\}$) and $p(t_1, \dots, t_n)$ and $q(t'_1, \dots, t'_n)$ basic (closed) formulas such that $I(p) \equiv \top$ and $I(q) \equiv \bot$. As $\top \geq \top$, it follows that $p_\top(t_1, \dots, t_n) \geq \top$. Also, $\top \geq {\sim}\top$. So, $p_I \geq {\sim}\top$. Therefore, $K,w \Vdash p_\top(t_1, \dots, t_n)$ and $K,w \Vdash p_{\sim\top}(t_1, \dots, t_n)$. By condition 2 of Definition 4.2, it follows that $K,w \Vdash \neg\, p_\top(t_1, \dots, t_n)$. On the other hand, as it is false that $\bot \geq \top$; it follows that it is not the case that $q_I \geq \top$, and so, it is not the case that $K,w \Vdash q_\bot(t'_1, \dots, t'_n)$.

Theorem 4.6: For some systems Mτ there are Kripke models K such that for some hyper-literal formula A and some world $w \in W$, we don't have $K,w \Vdash A$ nor $K,w \Vdash \neg A$.

Proof. Let us define the operator $\sim: |\tau| \rightarrow |\tau|$ by setting ${\sim}\top = \top$. Then, let I be the interpretation such that $I(p) \equiv \bot$. So, it is no the case that $I(p) \geq \top$ and also, it is not the case that $I(p) \geq {\sim}\top$ (or, equivalently, not $K,w \Vdash p_\top(t_1, \dots, t_n)$ and not $K,w \Vdash \neg\, p_\top(t_1, \dots, t_n)$).

Corollary 4.6.1: For some systems Mτ there are Kripke models K such that for some hyper-literal formulas A and B, and some worlds w, $w' \in W$, we have $K,w \Vdash \neg A$ and $K,w \Vdash A$ and we don't have $K,w \Vdash B$ nor $K,w \Vdash \neg B$.

Proof. Consequence of the theorems 4.5 and 4.6.

The earlier results show us that there are systems Mτ such that we have "inconsistent" worlds, "paracomplete" worlds, or both.

Now we present a strong version these results linking with paraconsistent, paracomplete, and non-alethic logics.

Definition 4.4: A Kripke model K is called *paraconsistent* if there are basic formulas $p(t_1, \dots, t_n)$, $q(t_1, \dots, t_n)$, and annotational constants $\lambda, \mu \in |\tau|$ such that $K,w \Vdash p_\lambda(t_1, \dots, t_n)$, $K,w \Vdash \neg\, p_\lambda(t_1, \dots, t_n)$, and it is not the case that $K,w \Vdash q_\mu(t_1, \dots, t_n)$.

Definition 4.5: A system Mτ is called *paraconsistent* if there is a Kripke model K for Mτ such that K is paraconsistent.

Theorem 4.7: Mτ is a paraconsistent system iff $\#|\tau| \geq 2$.

Proof. Define a structure $K = [\{w\}, \{(w, w)\}, I]$ such that $\begin{cases} q_I = \bot \\ p_I = \top \end{cases}$

It is clear that $p_I \geq \top$, and so $K \Vdash p_\top(t_1, \dots, t_n)$. Also, $p_I \geq {\sim}\top$, and, so $K \Vdash p_{\sim\top}(t_1, \dots, t_n)$, or $K \Vdash \neg p_\top(t_1, \dots, t_n)$. Also, it is not the case that $q_I(t_1, \dots, t_n) \geq \bot$, so it is not the case that $K,w \Vdash q_\bot(t_1, \dots, t_n)$.

Definition 4.6: A Kripke model K is called *paracomplete* if there are a basic formula $p(t_1, \ldots, t_n)$ and an annotational constant $\lambda \in |\tau|$ such that it is false that $K,w \Vdash p_\lambda(t_1, \ldots, t_n)$ and it is false that $K,w \Vdash \neg p_\lambda(t_1, \ldots, t_n)$. A system M$\tau$ is called *paracomplete* if there is a Kripke models K for Mτ such that K is paracomplete.

Definition 4.7: A Kripke model K is called *non-alethic* if K are both paraconsistent and paracomplete. A system Mτ is called *non-alethic* if there is a Kripke model K for Mτ such that K is non-alethic.

Theorem 4.8: If $\#|\tau| \geq 2$, then there are systems Mτ which are paracomplete and systems Mτ' that are not paracomplete, $\#|\tau'| \geq 2$.

Proof. Similar to the preceding theorem.

Corollary 4.8.1: If $\#|\tau| \geq 2$, then there are systems Mτ which are non-alethic and systems Mτ' that are not non-alethic, $\#|\tau'| \geq 2$.

5 Soundness and Completeness

Theorem 5.1: Let U be a maximal non-trivial maximal (with respect to inclusion of sets) subset of the set of all formulas. Let A and B formulas whatsoever. Then

1. If A is an axiom of Mτ, then $A \in U$
2. $A \wedge B \in U$ iff $A \in U$ and $B \in U$.
3. $A \vee B \in U$ iff $A \in U$ or $B \in U$.
4. $A \rightarrow B \in U$ iff $A \notin U$ or $B \in U$.
5. If $p_\mu(t_1, \ldots, t_n) \in U$ and $p_\lambda(t_1, \ldots, t_n) \in U$, then $p_\rho(t_1, \ldots, t_n) \in U$, where $\rho = \mu \vee \lambda$
6. $\neg^k p_\lambda(t_1, \ldots, t_n) \in U$ iff $\neg^{k-1} p_{\sim\lambda}(t_1, \ldots, t_n) \in U$.
7. If A and $A \rightarrow B \in U$, then $B \in U$.
8. $A \in U$ iff $\urcorner A \notin U$. Moreover $A \in U$ or $\urcorner A \in U$
9. If A is a complex formula, $A \in U$ iff $\neg A \notin U$. Moreover $A \in U$ or $\neg A \in U$.
10. If $A \in U$, then $\Box_i A \in U$.

Proof. Let us show only 5. In fact, if $p_\mu(t_1, \ldots, t_n) \in U$ and $p_\lambda(t_1, \ldots, t_n) \in U$, then $p_\mu(t_1, \ldots, t_n) \wedge p_\lambda(t_1, \ldots, t_n) \in U$ by 2. But it is an axiom $p_\mu(t_1, \ldots, t_n) \wedge p_\lambda(t_1, \ldots, t_n) \rightarrow p_\rho(t_1, \ldots, t_n)$, where $\rho = \mu \vee \lambda$. It follows that $p_\mu(t_1, \ldots, t_n) \wedge p_\lambda(t_1, \ldots, t_n) \rightarrow p_\rho(t_1, \ldots, t_n) \in U$, and so $p_\rho(t_1, \ldots, t_n) \in U$, by 7.

We give a Henkin-type proof of the completeness theorem for the logics Mτ. For this we define relations R_i on the set of all free-variable terms of Mτ as usual and we indicate by $\mathring{t}$ the equivalence class determined by t. Also, we will consider the quotient set F/R_i, where F indicates the set of all formulas.

Given a set U of formulas, define $U/\Box_i = \{A \mid \Box_i A \in U\}$, $i = 1, 2, \ldots, n$. Let us consider the canonical structure $K = [W, R_i, I]$ where $W = \{U \mid U$ is a maximal non-trivial set$\}$

and the interpretation function is as usual with the exception that given a n-ary predicate symbol p we associate the function $p_I : W^n \rightarrow |\tau|$ defined by $p_I(\overset{\circ}{t}_1, \ldots, \overset{\circ}{t}_n) =_{def.} \vee\{\mu \in |\tau| \mid p_\mu(t_1, \ldots, t_n) \in U\}$ (such function is well defined, so $p_\bot(t_1, \ldots, t_n) \in U$). Moreover, define $R_i =_{Def.} \{(U, U') \mid U/\Box_i \subseteq U'\}$.

Lemma 5.1: For all propositional variable p and if U is a maximal non-trivial set of formulas, we have $p_{pI(t^\circ 1, \ldots, t^\circ n)} (t_1, \ldots, t_n) \in U$.

Proof. It is a simple consequence of the previous theorem, item 5.

Theorem 5.2: For any formula A and for any non-trivial maximal set U, we have $(K, U) \Vdash A$ iff $A \in U$.

Proof. Let us suppose that A is $p_\lambda(t_1, \ldots, t_n)$ and $(K, U) \Vdash p_\lambda(t_1, \ldots, t_n)$. It is clear by previous lemma that $p_{pI(t^\circ 1, \ldots, t^\circ n)} (t_1, \ldots, t_n) \in U$. It follows also that $p_I(\overset{\circ}{t}_1, \ldots, \overset{\circ}{t}_n) \geq \lambda$. It is an axiom that $p_{pI(t^\circ 1, \ldots, t^\circ n)} (t_1, \ldots, t_n) \rightarrow p_\lambda(t_1, \ldots, t_n)$. Thus, $p_\lambda(t_1, \ldots, t_n) \in U$. Now, let us suppose that $p_\lambda(t_1, \ldots, t_n) \in U$. By previous lemma, $p_{pI(t^\circ 1, \ldots, t^\circ n)} (t_1, \ldots, t_n) \in U$. It follows that $p_I(\overset{\circ}{t}_1, \ldots, \overset{\circ}{t}_n) \geq \lambda$. Thus, by definition, $(K, U) \Vdash p_\lambda(t_1, \ldots, t_n)$. By theorem 5.1, $\neg^k p_\lambda(t_1, \ldots, t_n) \in U$ iff $\neg^{k-1} p_{\sim\lambda}(t_1, \ldots, t_n) \in U$. Thus, by definition 4.2, $(K, U) \Vdash \neg^k p_\lambda(t_1, \ldots, t_n)$ iff $(K, U) \Vdash \neg^{k-1} p_{\sim\lambda}(t_1, \ldots, t_n)$. So, by induction on k the assertion is true for hyper-literals.

The other cases, the proof is as in the classical case.

Corollary 5.2.1: A is a provable formula of $M\tau$ iff $\Vdash A$.

6 Concluding Remarks

Multimodal systems is having a growing attention by specialists. In this paper we have presented a class of paraconsistent multimodal annotated systems. It has shown that there is a logic in which it is possible to deal with imprecise, inconsistent and paracomplete knowledge, conflicting beliefs and awareness. Our main point is that imprecise, inconsistency and paracomplete concepts are a natural phenomenon when describing real world. We are often faced with conflicting advice. In such situations, most human beings are able to make an appropriate choice. The design of very large knowledge bases poses a similar problem. Disagreements between agents in the domain of interest may lead to the construction of a knowledge base that is inconsistent. Worse still, the very fact that it is inconsistent may emerge only much later, after the expert system has been in use for a significant period of time. Thus it is clear that formal methods are needed to handle this problem. The system proposed have provide a means to reasoning about inconsistent system of information.

References

1. Abe, J.M., *Fundamentos da lógica anotada*, (Foundations of annotated logics), in Portuguese, Ph.D. thesis, University of São Paulo, São Paulo, (1992) 135p.
2. Abe, J.M., Some Aspects of Paraconsistent Systems and Applications, *Logique et Analyse*, 15, (1997) 83-96.

3. Cresswell, M.J., Intensional logics and logical truth, *Journal of Philosophical Logic,* 1, (1972) 2-15.
4. Akama, S. & J.M. Abe, Many-valued and annotated modal logics, IEEE 1998 International Symposium on Multiple-Valued Logic (ISMVL'98), Proceedings, Fukuoka, Japan, (1998) 114-119.
5. Cresswell, M.J., *Logics and Languages,* London, Methuen and Co. (1973).
6. Da Costa, N.C.A., J.M. Abe, and V.S. Subrahmanian, Remarks on annotated logic, *Zeitschr. f. Math. Logik und Grundlagen d. Math.*, 37: (1991) 561-570.
7. Fagin, R., J.Y. Halpern, Y. Moses & M.Y. Vardi, *Reasoning about knowledge*, The MIT Press, London (1995).
8. Fischer, M.J. and N. Immerman, Foundation of knowledge for distributed systems. In J.Y. Halpern (Ed.), *Theoretical Aspects of Reasoning about Knowledge: Proc. Fifth Conference,* San Francisco, Cal.: Morgan Kaufmann, (1986) 171-186.
9. Lipman, B.L., *Decision theory with impossible possible worlds.* Technical report Working paper, Queen's University (1992).
10. McCarthy, J., *Ascribing mental qualities to machines*. Technical report CRL 90/10, DEC-CRL (1979).
11. Parikh, R. and R. Ramamujan, 1985, Distributed processing and the logic of knowledge. In R. Parikh (Ed.), *Proc. Workshop on Logics of Program* (1985) 256-268.
12. Rosenschein, S.J., 1985, Formal theories of AI in knowledge and robotics. *New Generation Computing* 3 (1985) 345-357.
13. Sylvan, R. & J.M. Abe, On general annotated logics, with an introduction to full accounting logics, *Bulletin of Symbolic Logic*, 2 (1996) 118-119.

Consensus-Based Evaluation Framework for Cooperative Information Retrieval Systems

Jason J. Jung and Geun-Sik Jo

Intelligent E-Commerce Systems Laboratory,
Department of Computer and Information Engineering, Inha University,
253 Yonghyun-dong, Incheon, Korea 402-751
j2jung@intelligent.pe.kr, gsjo@inha.ac.kr

Abstract. Multi-agent systems have been attacking the challenges of distributed information retrieval. In this paper, we propose a consensus method-based framework to evaluate the performance of cooperative information retrieval tasks of the agents. Two well-known measurements, *precision* and *recall*, are extended to handle consensual closeness (i.e., local and global consensus) between the retrieved results. We show in a motivating example that the proposed criteria are prone to solve the problem of rigidity of classical *precision* and *recall*. More importantly, the retrieved results can be ranked with respect to the consensual score.

Keywords: Information retrieval; Multi-agent systems; Evaluation.

1 Introduction

Information retrieval (IR) systems have been applied to distributed computing environment where several different hosts store only a part of information partitioned from a very large amount of information (particularly, textual documents). Basic operations of the IR system are i) to propagate a given user query, ii) to retrieve a set of results from each host, and iii) to integrate the result sets for the user [1]. Multi-agent systems (MAS) has been regarded as one of the efficient approaches to fulfill the process [2,3] with various methods (e.g., fuzzy theory [4], web mining [5], and ontology [6]). The agents can send a query and receive the corresponding partial results with each other, and finally, all of the result sets are simply unioned to create the final result set.

In this paper, we investigate evaluation issue of the retrieval strategies between multiple agents, and propose an efficient evaluation framework measuring their performance. The most well-known evaluation measurements are *precision* and *recall*. They are based on the comparison of the resulting document sets with another desired output document set, effectively comparing which documents are retrieved and which are not. These criteria are well understood and widely accepted.

However, they have the problem to be of the *all-or-nothing* manners for a certain query q represented as a set of terms (this will be formulated and explained in detail, later). Even though a IR system may show users the results that are very close to the expected results, its evaluation measurements are quite lower than it deserves. The reason for this is that the agents take into account only the common results from the set of corresponding partial results, without the similarities (or affinities) between them; if

N.T. Nguyen et al. (Eds.): KES-AMSTA 2007, LNAI 4496, pp. 169–178, 2007.

there is no common results, the measurement is zero. It is thus necessary to leverage the classic measurements with the proximity between the partial results. We thereby exploit a consensus method, which is capable of building two kinds of consensus (i.e., local and global consensus) to overcome the problem presented above. In particular, the ratio of co-occurrence patterns among documents is playing an important role of obtaining consensus results $d_{\mathbb{C}}$.

The outline of this paper is as follows. In Sect. 2, we describe background knowledge about distributed IR and preliminary notations. Sect. 3 shows the way to extend the classic measurement for evaluation of IR performance as the main contribution of this paper. In Sect. 4, a simple example and its experimental results will be given. Sect. 5 discusss some significant issues and mentions the existing studies related to consensus-based IR. Finally, we will draw a conclusion in Sect. 6.

2 Motivation

Basically, in a distributed IR system a document collection $\mathcal{D}$ is split into N partitions $\{D_1, D_2, \ldots, D_N\}$ (in this paper, we allow the partitions to be overlapped, i.e., $D_i \cap D_j \neq \emptyset$), and stored in N hosts $\{H_1, H_2, \ldots, H_N\}$, meaning that H_i stores D_i. Let a multi-agent system be organized as a set of agents $\mathcal{M} = \{M_1, \ldots, M_S\}$, which deploy its own retrieval strategy Ω_{M_i}. They can access to the hosts with the query given by the corresponding user and bring back the retrieved results to the user, as shown in Fig. 1. The formation and strategy of this agent system are fully dependent on the user's experiences and heuristics.

In case of vector space model, given a query q, which is represented as a set of terms $\{t_1, \ldots, t_{|q|}\}$ from each host H_j, an agent M_i can retrieve a set of documents $D_i^{j(+)}$ from D_j

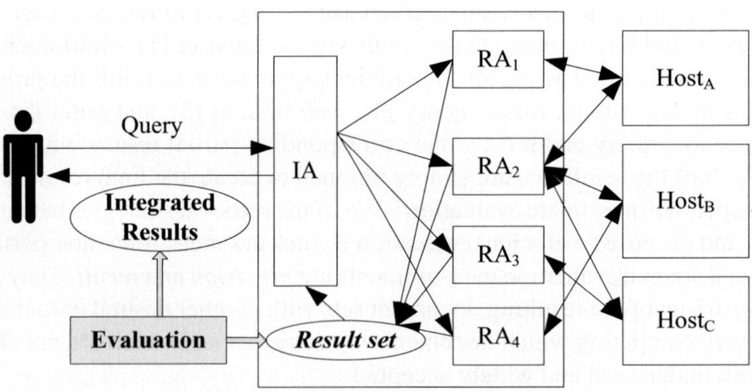

Fig. 1. A generic multi-agent system architecture for cooperative information retrieval; This system organized as five agents (an integration agent (IA) and four retrieval agents (RA)) has to search for three different information sources

Next step is to integrate the retrieved results. This integration function Φ can be defined as two ways; for a set of results retrieved from a same host,

$$\Phi_{\sqcap}^{D_i} = \Phi_{\sqcap}\left(\Omega_{M_x}(q, D_i), \ldots, \Omega_{M_y}(q, D_i)\right) = \bigcap_{j=x}^{y} D_j^{i(+)}, \tag{1}$$

and then, for integrating the set of results,

$$\Phi_{\sqcup}(\Omega_{M_1}(q, D_\alpha), \ldots, \Omega_{M_S}(q, D_\beta)) = \bigcup_{i=\alpha}^{\beta} \Phi_{\sqcap}^{D_i} \tag{2}$$

$$= \bigcup_{i=\alpha}^{\beta}\left(\bigcap_{j=1}^{S} D_j^{i(+)}\right) = \mathcal{D}^{+}. \tag{3}$$

We has to evaluate several critical factors (e.g., access patterns of agents and collaborative formation among agents) related to the performance of information retrieval by computing two measures *precision* and *recall* from the retrieved document collection $\mathcal{D}^{+}$ sent from the N hosts.

Definition 1 (Precision, Recall). *Given a set of reference (or desired) documents D^{Ref} for a certain query q, the precision and recall of retrieval for the query are given by*

$$\mathcal{P} = \frac{|D^{Ref} \cap \mathcal{D}^{+}|}{|\mathcal{D}^{+}|} \text{ and } \mathcal{R} = \frac{|D^{Ref} \cap \mathcal{D}^{+}|}{|D^{Ref}|}, \tag{4}$$

respectively.

3 Consensus-Based Relaxation

3.1 Problems

However, these criteria do not work properly to evaluate a certain cooperative retrieval strategy. Especially, this task can be difficult because the results produced by each host are based on different corpus statistics and possibly different data schema and retrieval algorithms [7]; they usually can not be compared directly. We mainly concern about the following two problems of the classical measurements;

1. An information retrieval (IR) system using integration function $\Phi_{\sqcap}$ seems "too strict", so that it possibly makes *precision* higher and *recall* lower.
2. It is difficult for IR system with $\Phi_{\sqcup}$ to rank a set of retrieved results. We can say that it is related to the *precision* measurement, as GLOSS [8] can do in boolean IR model.

Thus, the evaluation process should be changed. In case of this study, as shown in Fig. 1, our evaluation module is conducted before integrating the set of results.

Example 1. Suppose that a document collection $\mathcal{D}$ be split into two hosts H_1 and H_2 (i.e., $N = 2$), and a multi-agent system be organized by three agents M_α, M_β, and M_γ with additional agent for integration. Their information retrieval strategies are given by

- Classic boolean model $\Omega_{M_\alpha}(q, D_1) = \{d_i | \forall t_k \in q, Occur(t_k, d_i)\}$,
- Extended boolean model $\Omega_{M_\beta}(q, D_2) = \{d_i | \frac{|\{t_k|Occur(t_k,d_i)\}|}{|q|} \geq \tau_\beta\}$,
- Vector-space model $\Omega_{M_\gamma}(q, D_1) = \{d_i | \frac{\overrightarrow{f_{\gamma q}} \cdot \overrightarrow{f_{\gamma d_i}}}{|\overrightarrow{f_{\gamma q}}| \times |\overrightarrow{f_{\gamma d_i}}|} \geq \tau_\gamma\}$, and $\Omega_{M_\gamma}(q, D_2)$

where τ_β and τ_γ are user-specified thresholds, and function f can extract a certain set of term features. Additionally, Ω_{M_β} and Ω_{M_γ} can return the corresponding scores value scr. Four retrieved results

$$
\begin{aligned}
D_\alpha^{1(+)} &= \{d_1, d_2\}, \\
D_\beta^{1(+)} &= \{d_2, d_5 | scr(d_5) \geq scr(d_2)\}, \\
D_\beta^{2(+)} &= \{d_2, d_3, d_4 | scr(d_2) \geq scr(d_3) \geq scr(d_4)\}, \\
D_\gamma^{1(+)} &= \{d_1, d_2, d_5, d_6 | scr(d_1) \geq scr(d_2) \geq scr(d_5) \geq scr(d_6)\}, \text{and} \\
D_\gamma^{2(+)} &= \{d_2, d_3, d_7 | scr(d_2) \geq scr(d_3) \geq scr(d_7)\}
\end{aligned}
$$

are merged into $\mathcal{D}^+$ by either $\Phi_\sqcap$ or $\Phi_\sqcup$.

When $\mathcal{D}_1^{Ref} = \{d_1, d_2, d_5\}$ from H_1, for $\Phi_\sqcap\left(D_\alpha^{1(+)}, D_\beta^{1(+)}, D_\gamma^{1(+)}\right)$, the rigidness of boolean models (e.g., M_α) might remove some potential candidates (e.g., d_5), without considering the others. Also, the ranking information from $D_\gamma^{1(+)}$ is ignored. Especially, *popularity* patterns such as d_2, meaning a type of score measuring the consensus among the hosts.

3.2 Building Consensus and Relaxation

In order to solve this problem, we want to exploit a consensus method to relax the evaluation measurements. Consensus methods is basically able to resolve the conflicts among the results from distributed systems [9]. In this context, we want to capture the information loss caused by the difference between the retrieved results. For consensus choice functions, we extended some postulates defined in [10]. The distance function δ is given by

$$
\delta(\Omega_{M_\alpha}, \Omega_{M_\beta}) = \delta(D_\alpha^{i(+)}, D_\beta^{i(+)}) = 1 - \frac{|D_\alpha^{i(+)} \cap D_\beta^{i(+)}|}{\max(|D_\alpha^{i(+)}|, |D_\beta^{i(+)}|)} \tag{5}
$$

$$
= 1 - \frac{|\{d_c | d_c \in D_\alpha^{i(+)}, D_\beta^{i(+)}\}|}{\max(|D_\alpha^{i(+)}|, |D_\beta^{i(+)}|)}. \tag{6}
$$

The consensus is made up when the summation of all possible pairs of agents is minimized. Two kinds of consensus are considered in this paper.

Definition 2 (Local consensus). *A local (or host) consensus within a host can be established by a set of document sets retrieved by a set of corresponding agents. Given a query q, the local consensus* $\mathbb{C}_{L(i)}$ *of* H_i *is given by*

$$
\mathbb{C}_{L(i)}^q = \left\{ \langle d_{\mathbb{C}_L}, scr_{\mathbb{C}_L} \rangle \middle| \min \sum_{k=1}^{{}_nC_2} \delta_k(D_\alpha^{i(+)}, D_\beta^{i(+)}) \right\} \tag{7}
$$

where n is a number of agents accessing to the host, and ${}_nC_2$ is a combinatorial computation (i.e., $\frac{n\times(n-1)}{2}$).
For obtaining $d_{\mathbb{C}_L}$, we can compute the co-occurrence ratio

$$d_{\mathbb{C}_L} \in \left\{ d_i \middle| src_{\mathbb{C}_L}(d_i) = \overline{src}(d_i) \frac{|\{D_\alpha^{(+)}|d_i \in D_\alpha^{(+)}\}|}{|\{D_\beta^{(+)}|d_i \notin D_\beta^{(+)}\}|+1} \geq \tau_{\mathbb{C}_L} \right\} \tag{8}$$

where $\tau_{\mathbb{C}_L}$ is a score threshold for local consensus. Function $\overline{src}$ is a normalized score computed from initial ranked results by

$$\overline{src}(d_i) = \frac{\sum_{d_i \in D_\alpha^{(+)}} \frac{|D_\alpha^{(+)}|}{2\times rank_\alpha(d_i)}}{|\{D_\alpha^{(+)}|d_i \in D_\alpha^{(+)}\}|} \tag{9}$$

(it simply express linear proportion, but it can also be applied to logarithm function.). For the unranked results (e.g., in boolean model Ω_β), it should be

$$\forall d_i \in D_\beta^{(+)}, rank_\beta(d_i) = 1/2. \tag{10}$$

From the previous Example 1, we can build the local consensus as

$$\mathbb{C}_{L_1} = \{\langle d_1, 1.25\rangle, \langle d_2, 2.01\rangle, \langle d_5, 0.58\rangle\}, \tag{11}$$

by

$$src_{\mathbb{C}_L}(d_1) = \frac{2}{2} \times \frac{0.5+2}{2} = 1.25, \quad src_{\mathbb{C}_L}(d_2) = \frac{3}{1} \times \frac{0.5+0.5+1}{3} = 2.01,$$
$$src_{\mathbb{C}_L}(d_5) = \frac{2}{2} \times \frac{0.5+0.67}{2} = 0.58, \quad src_{\mathbb{C}_L}(d_6) = \frac{1}{3} \times \frac{0.5}{1} = 0.17$$

when $\lambda_{\mathbb{C}_L} = \frac{1}{2}$.

Definition 3 (Global consensus). *A global consensus can be established by a set of local consensus of corresponding hosts. Given a set of $\mathbb{C}_L$, the global consensus $\mathbb{C}_G$ for query q is given by*

$$\mathbb{C}_G^q = \bigcup_{i=1}^{N} \mathbb{C}_{L(i)}^q = \left\{ \langle d_{\mathbb{C}_G}, src_{\mathbb{C}_G}\rangle \middle| scr_{\mathbb{C}_G}(d_{\mathbb{C}_G}) \geq \tau_{\mathbb{C}_G} \right\} \tag{12}$$

where N is the number of hosts, and $\lambda_{\mathbb{C}_G}$ is a score threshold for global consensus. With respect to the global consensus score $scr_{\mathbb{C}_G}$, the results can be ranked.

More importantly, two different heuristics for measuring global consensus score of each document are given by

$$scr_{\mathbb{C}_G}(d_{\mathbb{C}_G}) = \frac{\max\left[scr_{\mathbb{C}_L}(d_{\mathbb{C}_G})\right]}{|\{D_\alpha^{(+)}|d_{\mathbb{C}_G} \in D_\alpha^{(+)}\}|} \tag{13}$$

$$= \frac{\sum scr_{\mathbb{C}_L}(d_{\mathbb{C}_G})}{|\{D_\alpha^{(+)}|d_{\mathbb{C}_G} \in D_\alpha^{(+)}\}|} \tag{14}$$

which return the normalized maximum score and average score, respectively. We will empirically compare them in Sect. 4.

Now, we can rewrite the conventional measurements (shown in Equ. 4). The relaxed precision and recall are given by

$$\mathcal{P}_{\mathbb{C}} = \frac{|\mathbb{C}_G \cap D^{Ref}|}{|D^{Ref}|} \text{ and } \mathcal{R}_{\mathbb{C}} = \frac{|\mathbb{C}_G \cap D^{Ref}|}{|\mathbb{C}_G|}, \tag{15}$$

respectively. If D^{Ref} is ranked, we can obtain more elaborated evaluation measurement.

4 Experimental Results

In order to prove the rationality of the proposed evaluation method, we conduct several experimentations. For testing bed, we chose twenty newsgroups dataset[1] as testing dataset. It is a collection of approximately 20,000 newsgroup documents, partitioned (nearly) evenly across 20 different newsgroups. We randomly partitioned them into seven fragments for seven hosts. The specification of this testing bed is shown in Table 1. We constructed three kinds of multi-agent formations for cooperative information retrieval from the testing bed, as shown in Table 2.

Table 1. Specifications of testing bed

Host	Number of documents	Topics
H_1	3000	alt.atheism, rec.autos, rec.sport.hockey
H_2	3000	comp.graphics, misc.forsale comp.os.ms-windows.misc,
H_3	2000	comp.sys.mac.hardware, talk.religion.misc
H_4	4000	comp.windows.x, rec.sport.baseball, rec.motorcycles, talk.politics.misc
H_5	3000	comp.sys.ibm.pc.hardware, sci.electronics, sci.med
H_6	3000	sci.crypt, soc.religion.christian, talk.politics.mideast
H_7	2000	sci.space, talk.politics.guns

Especially, in F_3, we considered two more variances by changing the portion of IR agent population.

According to the MAS formations, the consensus was built to measure consensus based precision $\mathcal{P}_{\mathbb{C}}$ and recall $\mathcal{R}_{\mathbb{C}}$ by comparing with five D^{Ref}s of the corresponding queries. We have conducted three cases; *i)* $\tau_{\mathbb{C}_L} = 0.2$, $\tau_{\mathbb{C}_G} = 0.3$ *ii)* $\tau_{\mathbb{C}_L} = 0.5$, $\tau_{\mathbb{C}_G} = 0.3$, and *iii)* $\tau_{\mathbb{C}_L} = 0.7$, $\tau_{\mathbb{C}_G} = 0.3$.

Given twenty queries, we measured the conventional measurements; i.e., $\mathcal{P} = 0.68$ and $\mathcal{R} = 0.33$ when $\tau = 0.3$ (it is the same value as $\tau_{\mathbb{C}_G}$). In terms of precision, F_1 and

[1] 20 Newsgroups dataset. http://people.csail.mit.edu/jrennie/20Newsgroups/

Table 2. Formation of MASs; They are denoted as F_1, F_2, $F_{3,1}$, $F_{3,2}$, and $F_{3,3}$

Formation		F_1	F_2	$F_{3,1}, F_{3,2}, F_{3,3}$		
Number of agents		3	7	20		
IR models (Portion of agents)	Boolean	33%	29%	60%	10%	30%
	Extended boolean	33%	43%	30%	60%	10%
	Vector-space	33%	29%	10%	30%	60%

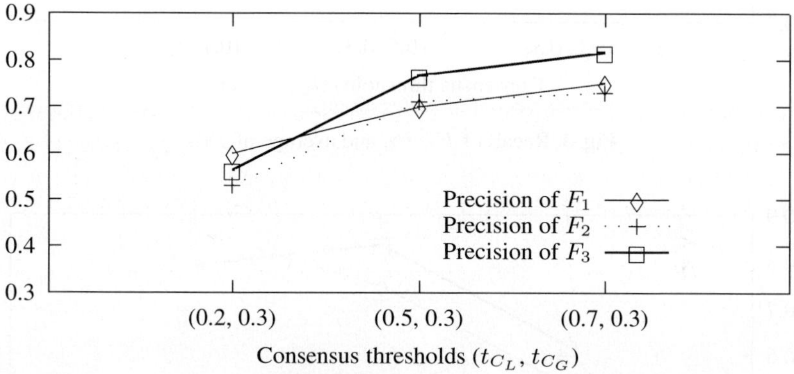

Fig. 2. Precision of F_1, F_2, and average of F_3's

F_2 have shown slightly improvement, while F_3 provides about 25.5% higher precision, as shown in Fig. 2. Especially, as shown in Fig. 3, except for F_1 MAS, F_2 and F_3 outperform the conventional evaluations (over approximately 32% higher).

5 Discussion

We want to discuss two issues;

- relationship between the number of agents and the evaluation measures, and
- relationship between the combination of IR methods and the evaluation measures

For the first issue, the three different formations (i.e., F_1, F_2, and F_3) have been configured. While the consensus-based precisions $\mathcal{P}_{\mathbb{C}}$ are in the nearly same level, the consensus-based recalls $\mathcal{R}_{\mathbb{C}}$ show significantly different patterns between F_1 and the other two. We found out that boolean model-based IR strategies have the more dominant influence as the population of agents is getting less.

As second issue, we have to consider the formations $F_{3,1}$, $F_{3,2}$, and $F_{3,3}$. In addition, these results are ranked to be compared with D_{Ref}, as shown in Fig. 4 and Fig. 5. We found out that in the sense of precision, $F_{3,1}$ of which portion of boolean model is largest has retrieved the most precise results. In particular, $F_{3,3}$ has shown the lest precision at $\tau_{\mathbb{C}_L} = 0.7, \tau_{\mathbb{C}_G} = 0.3$. During building the global consensus, we could realize

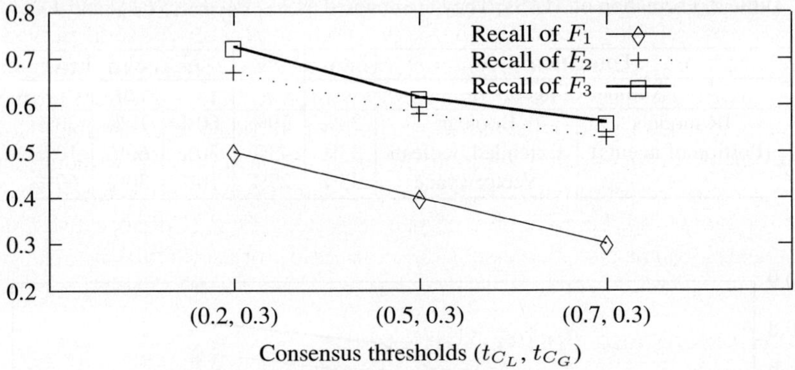

Fig. 3. Recall of F_1, F_2, and average of F_3's

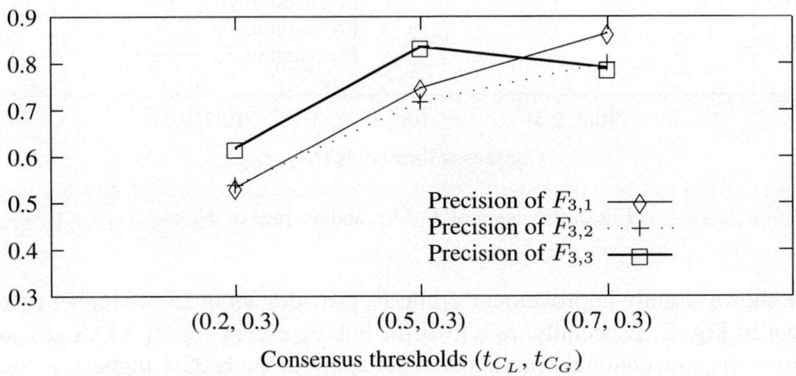

Fig. 4. Precision of F_1, F_2, and average of F_3's

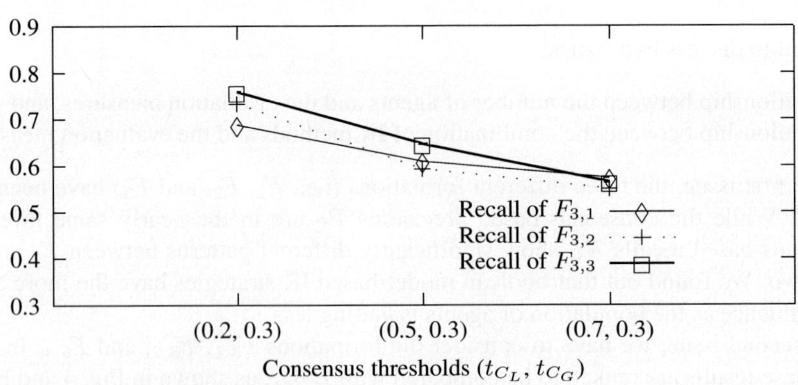

Fig. 5. Recall of F_1, F_2, and average of F_3's

some error propagated from mismatched ranking between extended boolean model and vector-space model. On the other hand, $\mathcal{R}_{\mathbb{C}}$ was computed as almost same performance.

As related work, recently, Nguyen et al. proposed consensus-based information retrieval system [11]. It applies the consensus method to obtain the final answers, by adjusting the queries given by the users and selecting the best search engines. However, for the evaluation process, they have concerned each partial results without considering merging results. Moreover, INQUERY [12] has proposed error-based evaluation method for evaluating the collection of ranked results. CQE (Collaborative Querying Environment) [13] has provided visualization facilities for user intuition.

6 Conclusions and Future Work

The partial results retrieved by heterogeneous IR methods should be integrated to end users. In this paper, we have been proposing a consensus-based evaluation method for distributed information retrieval systems. Co-occurrence pattern-driven consensus information has been exploited to build the global consensus and minimize the distances between partial results. Compared to the traditional evaluation criteria (e.g., precision and recall), the proposed measurements have shown better performance in the following point of views

- evaluation of queries generated by the users; some queries might be very vague
- evaluation of retrieval strategies of agents; relaxing boolean type models, and
- evaluation of integration strategies of agents.

As a future work, in the context of "collective intelligence," we will extend this evaluation method on ontology-based approach systems [6]. They can automatically communicate with each other by semantic alignment. Especially, this kind of IR evaluation platform is very important to interactive IR systems for improving the IR performance [14].

References

1. Baeza-Yates, R., Ribeiro-Neto, B.: Modern Information Retrieval. Addison Wesley (1999)
2. Huhns, M.N., Singh, M.P.: Agents on the Web: "Agents are Everywhere!". IEEE Internet Computing **1**(1) (1997) 87
3. Oates, T., Prasad, M.V.N., Lesser, V.R.: Cooperative information-gathering: a distributed problem-solving approach. IEE Proceedings - Software **144**(1) (1997) 72–88
4. Herrera-Viedma, E., Herrera, F., Martínez, L., Herrera, J.C., López, A.G.: Incorporating filtering techniques in a fuzzy linguistic multi-agent model for information gathering on the web. Fuzzy Sets and Systems **148**(1) (2004) 61–83
5. Lee, R.S.T., Liu, J.N.K.: iJADE Web-Miner: An intelligent agent framework for internet shopping. IEEE Transactions on Knowledge and Data Engineering **16**(4) (2004) 461–473
6. Jung, J.J.: Ontological framework based on contextual mediation for collaborative information retrieval. Information Retrieval **10**(1) (2007) 85–109
7. Callan, J.P.: Chapter 5. Distributed information retrieval. In: Advances in Information Retrieval. Kluwer Academic Publishers (2000) 127–150

8. Gravano, L., Garcia-Molina, H.: Generalizing GlOSS to vector-space databases and broker hierarchies. In: Proceedings of the 21th International Conference on Very Large Data Bases (VLDB '95), San Francisco, CA, USA, Morgan Kaufmann Publishers Inc. (1995) 78–89
9. Coulouris, G., Dollimore, J., Kindberg, T.: Distributed Systems - Concepts and Design. Addison-Wesley (1996)
10. Nguyen, N.T.: Consensus system for solving conflicts in distributed systems. Information Science **147**(1-4) (2002) 91–122
11. Nguyen, N.T., Ganzha, M., Paprzycki, M.: A consensus-based multi-agent approach for information retrieval in internet. In Alexandrov, V.N., van Albada, G.D., Sloot, P.M.A., Dongarra, J., eds.: Proceedings of the 6th International Conference Computational Science (ICCS 2006). Volume 3993 of Lecture Notes in Computer Science., Springer (2006) 208–215
12. Callan, J.P., Lu, Z., Croft, W.B.: Searching distributed collections with inference networks. In: Proceedings of the 18th annual international ACM SIGIR conference on Research and development in information retrieval (SIGIR '95), New York, NY, USA, ACM Press (1995) 21–28
13. Fu, L., Goh, D.H.L., Foo, S.S.B.: Cqe: a collaborative querying environment. In Marlino, M., Sumner, T., III, F.M.S., eds.: Proceedings of the 2005 ACM/IEEE Joint Conference on Digital Libraries (JCDL 2005), ACM (2005) 378
14. Borlund, P.: The IIR evaluation model: a framework for evaluation of interactive information retrieval systems. Information Research **8**(3) (2003)

Hierarchy of Logics of Irrational and Conflicting Agents

Germano Resconi[1] and Boris Kovalerchuk[2]

[1] Dept. of Mathematics and Physics, Catholic University Brescia, I-25121, Italy,
resconi@numerica.it.

[2] Dept. of Computer Science, Central Washington University, Ellensburg,
WA 98926-7520, USA
borisk@cwu.edu.

Abstract. This paper proposes a hierarchy of logics of agents relative to levels of their conflicts, self-conflicts and irrationality to provide a base for several studies on foundations of the theories of uncertainties. These studies include the foundation of known uncertainty theories (probability theory, fuzzy logic, and others) as well as new logics and types of uncertainties. Probability theory, fuzzy logic, and others theories of uncertainties provide a calculus for manipulating with probabilities, membership functions, and other types of uncertainty indicators. However, these theories lack a mechanism for getting initial (basic) uncertainties. The proposed hierarchy of conflicting and irrational agents creates a base for generating uncertainty values, logic operations with these values and for comparing different types of uncertainty for the preference relation. A core concept of this hierarchy is the concept of expansion by superposition that includes fusion and adjustment of contradictory events and statements.

Keywords: uncertainty logic, conflicting agent, irrational agent, probability theory, fuzzy logic, self-conflict, mutual exclusion.

1 Introduction

The concept of rational agents in [14] is motivated by fundamental goals in the *software engineering* area to provide a formal logicisation of software engineering [3]. In this area, the rationality of a software agent rests in the maximization of its chances of success based on software agent's knowledge of its environment. Any agent that maximizes chances of success is considered as a rational one. In fact, an agent's success criterion as well as the knowledge of the environment can be quite irrational. As a result, agent's behavior is not rational, thus, the concept of agent's rationality/irrationality should be developed deeper and more specifically in this area if software agent applications that model complex irrational human behavior are envisioned.

The research area that is deeply interested in partially rational agents called boundedly rational agents is motivated by fundamental goals in *psychology and economics* [5,10]. The works in this area explore the psychology of intuitive beliefs and choices by examining their bounded rationality as Kahneman outlined in his

N.T. Nguyen et al. (Eds.): KES-AMSTA 2007, LNAI 4496, pp. 179–188, 2007.

Nobel Prize Lecture in 2002. In economics, bounded rationality means use of (1) multiple utility functions instead of one global scalar utility function, (2) limited types of utility functions due to cost of information collection, and (3) heuristics.

Our concept of irrational agent is motivated by fundamental goals in the *artificial intelligence* area of *logic and reasoning under uncertainty* [6,7] with agents [4]. We envision a formal logicisation of reasoning of irrational agents that do not follow rational reasoning in the frames of the classical logic and the probability theory and even fuzzy logic. This area is quite different from two previous areas as well as the areas that study expert reasoning simply because irrational agents hardly fit a definition of experts. We define a specific concept of an *ME-rational* (the agent that is rational relative to mutual exclusion) with a clear criterion how to distinguish such agents. Similarly, we define a concept of an *ME-irrational* agent as a *self-conflicting agent* as well as *contradictory agents* that contradict each other but without self-conflict in judgment. The concepts of conflict and self-conflict in judgments are critical for our study of irrational agents, because uncertainty often means that several contradictory statements are made. We build our approach on the results of direct measurements or answers of agents not on the aggregated utility functions to be able to model a structure of contradictions explicitly.

Conducting experiments with devices or asking agents are two common ways to assign initial (basic) uncertainties. We can ask agents to evaluate classical truth-value v(p) of a statement p (true/false) and then compute frequencies of these answers provided by a set of agents G. Alternatively we can ask each agent to give us a truth-value v(p) in the [0,1] interval. Both frequency and the subjective evaluations then can be an empirical base for a probability of the statement p [1,6,7] or as a value of the fuzzy logic membership function (MF)[7]. An elaborated description of these processes treating MF in the probabilistic framework is given in [7]. Many other authors claim that MFs represent a type of uncertainty that differs from probabilistic one. However, an explicit model that supports this claim is absent. The goal of this paper is to build a conceptual base for answering this question in a rigorous way using the agent approach.

This paper introduces a hierarchy of logics of uncertainty in the context of conflicting evaluations of preferences by agents that includes a mechanism for identifying a type of logic and reasoning computations in the agent logic. It is a further development of our previous works [8-13] that contain extensive references to related work. The fundamental analysis and review of relevant issues can be found in [1-4].

The probability calculus does not incorporate explicitly the concepts of irrationality and conflicts of agents. It misses the structural information at the level of individual objects, but preserves the information on a global property of a set of objects. Given a dice the probability theory studies frequencies of the different faces E={e} as independent elements. This set of elementary events E has no structure. It is only required that elements of E are *mutually exclusive* and *complete* (no other alternative is possible). The order of its elements is irrelevant to probabilities of each number in E. No irrationality or conflict is allowed in this definition. The classical probability calculus does not provide a mechanism for modelling uncertainty when agents communicate (collaborates or conflict). Recent work by Halpern [6] is an important attempt to fill this gap.

Now we can provide a more formal definition. A reasoning agent g is called *totally consistent for statement* S if g always provides the same truth value for S, thus the change of context C and time t does not change S value for agent g. Note that total consistence does not imply correctness of the evaluation of S by g because a consistent truth-value can be incorrect. The classical logic models such agents and their reasoning. A reasoning agent g is called *partially consistent for statement* S if g *not* always provides the same truth-value for S. If sources of inconsistency are numerous, cannot be traced or explicitly presented then the probability theory is used commonly to model such agents and their reasoning with probability P(S)=n/N, where n is the number of cases when g answered that S is true and N is a total number of cases. Otherwise, the classical logic still can be used.

Similarly, a reasoning agent g is called *totally consistent for a set of statements* {S} if g always provides the same truth value for each S from {S}. Again, the classical logic is applicable for such agent. A reasoning agent g is called *partially consistent for a set of statement* {S} if g *not* always provides the same truth-value for some or all statements from {S}. The expansion from a single statement S to a set of statements {S} can be done differently with *different relations among statements* in the {S} for the agent g. In addition, set {S} may contain statements about more than one object, event. Relations among statements and objects define applicability of the probability theory. Kolmogorov's axioms of the probability theory set up these relations.

One of the most important of these axioms is a mutual exclusion axiom for elementary events. Let E be a predicate which is true if e is an elementary event, {e} be a set of all given elementary events, and P(e) be a probability of event e, then $P(e_i\&e_j)=0$ and $P(e_i \vee e_j)=P(e_i)+P(e_j)$. Property $P(e_i\&e_j)=0$ means that e_i and e_j cannot happen simultaneously, Let S(e) = True if elementary event e has happened. If g tells that e_i has happened, $S(e_i)$=True, then g should tell $S(e_j)$=False, i.e., $S(e_i)\&S(e_j)$=False for this agent. We will call this *agent* g *rational on Mutual Exclusion* (**ME-rational agent**). Accordingly we will call *agent g irrational on Mutual Exclusion* (**ME-irrational agent**) if g tells both $S(e_i)$=True and $S(e_j)$=True for mutually exclusive events e_i and e_j, $S(e_i)\&S(e_i)$=True. In other words, we can say that for this agent g contradiction $S(e_i)\&S(e_j)$ is True.

We distinguish between conflicting agents, self-conflicting agents, and irrational agents. Each individual agent can be very consistent, but agent g_1 can contradict agent g_2 creating a pair of **conflicting agents** (g_1,g_2). A **self-conflicting agent** g can provide two contradictory preferences, $(p,\neg p)$ with p=(A> B) and ,$\neg$p=(B>A) for statement p by stating the p true and false at the same time. This can be an indication that g has two different competing criteria, C_1 and C_2, such as less price and better quality. However, this does not resolve the ultimate preference contradiction it only explains it because the goal is to make a preference between A and B and the explanation does not solve this problem.

If criteria C_1 and C_2 are the same then we have an **irrational agent** g in a specific context. Thus, a set of irrational agents is a subset of the class of self-conflicting agents. The agent can be irrational because of inability to understand complex **context and evaluate criteria**. In the stock market, some traders quite often do not understand a complex context and rational criteria of trading. These traders are irrational agents exhibiting chaotic, random, and impulsive behavior. At the same market situation they can sell and buy stocks, exhibiting irrational $(p,\neg p)$ behavior.

2 Framework of First Order Conflicting Agents

To define a concept of first order conflicting agents we set up a framework using an example of agents who buy cars. The concept of first order conflicting agents is much wider than provided in the example with a binary preference relation. Using this preference relation we show that at the first order of conflicts AND and OR operations should not be traditional scalar classical logic operations, but **vector operations** to reflect a structure of individual agent evaluations.

Consider 100 agents g_i, two cars *A* and *B* and preference relation ">" between cars to be assigned by each of these agents (potential buyers). We define a Boolean variable *X* such that *X*=1 (True) if *A>B* else *X*=0 (False). Each agent g_i answered a questionnaire with two options offered: (1) "*A>B* is true" and (2) "*A>B* is false". Say 70 agents marked "*A>B* is true" giving a frequency value, *m(A >B)*= 70/100, that can be interpreted as a probability or fuzzy logic membership function value.

The situation for which a group of agents assumes that A>B is true and a complementary group of agents assumes that the same A>B is false, is a situation of conflict/contradiction among agents that are wrestling for the logic value of A > B. We denote this situation as a **first order of conflict/contradiction**. Here *each individual agent is completely rational and has no self-conflict. The conflict exists only between different agents when they evaluate the same statement.*

We specify a set of agents *G*, a subset of agents *G(A>B)* for which A > B and a set of agents $G_{A<B}$ for which A < B:

$$G(A>B) = \{g \in G \mid A >_g B \text{ is true}\}, \quad G(A<B) = \{g \in G \mid A >_g B \text{ is false}\},$$

where $>_g$ is a preference relation by agent g. Sets *G(A>B)* and *G(A<B)* are complimentary, thus we will also use notation, $G^c(A>B)=G(B>A)$.

Example. Let $G = \{Agent_1, Agent_2, Agent_3, Agent_4\}$ and $G(A>B) = \{Agent_1, Agent_4\}$ and $G(A<B) = \{Agent_2, Agent_3\}$. This is a conflicting set of agents. The evaluation for all four agents can be recorded in this way

$$value(A > B) = \begin{bmatrix} Agent_1 & Agent_2 & Agent_3 & Agent_4 \\ True & False & False & True \end{bmatrix}.$$

Similarly, for N agents we can record the evaluations as follows

$$V(A > B) = \begin{bmatrix} agent_1 & agent_2 & \dots & agent_N \\ v_1(A > B) & v_2(A > B) & \dots & v_N(A > B) \end{bmatrix}$$

where v_i(A > B) = true if agent g_i evaluates A > B as true and v_i(A > B) = false if agent g_i evaluates A > B as false. Now the previous evaluation for the four agents can be written as V(A>B)= (v(A>$_1$B), v(A>$_2$B), v(A>$_3$B, v(A>$_4$B)) or more generally for N agents V(A>B)= (v(A>$_1$B), v(A>$_2$B),…, v(A>$_N$B)).

Below we show that at the first order of conflicts, AND and OR operations should not be the traditional scalar classical logic operations, but should be **vector operations** that can be aggregated to a scalar value in [0,1] interval later if needed. In this way, operations reflect a structure of individual agent evaluations.

In fact, at the first order of agent conflict we must introduce a **fusion process** of the logic values given by the agents. A basic way to do this is to compute a frequency of preferences of all agents:

$$\mu(A > B) = \frac{v(A >_1 B) + ... + v(A >_N B)}{N}.$$

An alternative way is to use affine weighted frequencies.

At first glance, μ(A>B) is the same as used in the probability and utility theories. However, classical axioms of the probability theory have no references to agents producing initial uncertainty values and do not violate the mutual exclusion. As a result, formulas for assigning initial uncertainty values such as μ(A>B) are not a part of the theory. Value μ(A>B) can differ significantly from the classical relative frequency for some irrational agents.

A probability value can be agent's judgment or a result of physical experiments. In [2] agents are connected to logic and probability in the following way. In a given state s, the formula $P_j(\varphi)$ denotes the probability of the logic proposition φ according to the agent j's probability distribution in the state s. In this model, an individual agent gives the probability. All agents are autonomous and no conflict among agents and self-conflict is modeled explicitly. This formalization does not provide tools to fuse the contradictory knowledge of different agents.

In the first order of irrationality the **vector logic operations** are

$$V(p \wedge q) = \begin{bmatrix} agent_1 & ... & ... & agent_N \\ v_1(p) \wedge v_1(q) & ... & ... & v_N(p) \wedge v_N(q) \end{bmatrix},$$

$$V(p \vee q) = \begin{bmatrix} agent_1 & ... & ... & agent_N \\ v_1(p) \vee v_1(q) & ... & ... & v_N(p) \vee v_N(q) \end{bmatrix},$$

$$V(\neg p) = \begin{bmatrix} agent_1 & ... & ... & agent_N \\ \neg v_1(p) & ... & ... & \neg v_N(p) \end{bmatrix},$$

where the symbols ∧ ∨ ¬ in the right side of equations are the classical AND , OR and NOT operations. For a set of conflicting agents at the first order of conflicts, we have these important properties

$$G(p \wedge q) = G(p) \cap G(q) = \min(G(p), G(q)) - G(\neg p \wedge q),$$
$$G(p \vee q) = G(p) \cup G(q) = \max(G(p), G(q)) + G(\neg p \wedge q),$$

where G(p) is the set of agents for which statement p is true , G(q) is the set of agents for which statement q is true. If $G(q) \subset G(p)$ then we have well known min, max operations. We remark also that $G(\neg p) = G^C(p)$, where $G^C(p)$ is the complementary set of G(p) in G. Thus,

$$G(p \vee \neg p) = G(p) \cup G(\neg p) = G(p) \cup G(\neg p) = G(p) \cup G(p)^C = G$$
$$G(p \wedge \neg p) = G(p) \cap G(\neg p) = G(p) \cap G(\neg p) = G(p) \cap G(p)^C = \varnothing$$

In other words, $G(p \wedge \neg p) = \varnothing$ corresponds to the contradiction $p \wedge \neg p$ that is always false and $G(p \vee \neg p) = G$ corresponds to the tautology $p \vee \neg p$ that is always true in the first order of conflict.

In this section, we formalized the concept of first order conflicting agents with the conflict only between different agents while each agent has no self-conflict. It is shown that for these conflicts AND and OR operations should be defined as vector operations that preserve a structure of individual agent evaluations. Next, we had shown that a preference evaluation by multiple contradictory and self-contradictory agents (that contradict each other) can differ from a traditional frequency that does not preserve a structure of individual agent evaluations.

3 Second Order of Conflict, Self-conflict, and Irrationality

In this section, we use concepts of irrational agents to explain the deviation of the fuzzy logic from the classical logic and the probability theory where the contradiction is always false and the tautology is always true. We discovered that this deviation is a consequence of agent's self-conflict that is not allowed in the classical logic and the probability theory for mutual exclusion (ME). The existence of a self-conflicting ME-state for agents such as real people is the main motivation to introduce second order of conflict. We formalize this concept below.

To explain the deviation of fuzzy logic from the classical logic and the probability theory we introduce a new (for fuzzy logic and probability theory) type of evaluation (self-conflicting vector evaluation) as well as a superposition process to combine self-conflicting logic values. Below we define the concepts of a higher order of conflict including a concept of the a second order of conflict/contradiction. In general, we define an n-th order of conflict. The existence of self-conflicting states motivates changes in the definition of the negation operator and computations of contradiction and tautology that differ from the definitions in the classical logic.

Deviation from classical logic in fuzzy set theory. We denote a situation as a *higher order of conflict* if there are individual agents in a set of agents G that have self-conflict in addition to the possible first order conflict between different agents when they evaluate the same statement. a *second order of conflict*/contradiction can take place if we have two complimentary evaluation criteria (e.g., to set up preferences of a set of objects). In general, we define an **n-th order of conflict** as a conflict that involves n complimentary evaluation criteria.

Zadeh has defined fuzzy sets and logic operations by the formulas:

$$\mu(p \wedge q) = \min[\mu(p), \mu(q)],\ \mu(p \vee q) = \max[\mu(p), \mu(q)],$$
$$\mu(\neg p) = 1 - \mu(p).$$

Thus, $\mu(p \wedge \neg p) = \min[\mu(p), 1 - \mu(p)]$, $\mu(p \vee \neg p) = \max[\mu(p), 1 - \mu(p)]$. Therefore, here the value for contradiction, $\mu(p \wedge \neg p)$ can be greater than zero and for the tautology, $\mu(p \vee \neg p)$ can be less than one. In other words, in the fuzzy logic, the contradiction can have a non-zero degree of truth and the tautology can have a non-zero degree of falseness. This is completely in disagreement with the classical logic, where the tautology always is true and contradiction is always false. Below we present a possible explanation of these fuzzy logic properties by using the concept of self-conflicting agents at the second order. However this explanation does not justify specific min, max operations and negation, $\mu(\neg p) = 1 - \mu(p)$ used in fuzzy logic.

Mutual exclusion vs. ghost events. Implicitly, the classical logic and the probability theory assume that agents are *rational*, that is every agent marks in the questionnaire even e_1 = "A > B is true" or an opposite event e_2 = "B > A is true", but not both of them. Both situations are possible but only one appear at the time. This is a fundamental assumption of the probability theory – elementary events are *disjoint, mutually exclusive (ME) and one of them needs to happen.* Agents must select only one of two events (e_1, e_2).

In general, if an agent marks both options (case (c) below), both answers will be discarded, as non-valid produced by an ME-irrational agent and only ME-rational agents with a single answer will be counted in relative frequencies.

☑ A>B ☐ B>A	☐ A>B ☑ B>A	☑ A>B ☑ B>A
(a) Rational agent selected event e_1 = "A>B"	(b) Rational agent selected event e_2= "B>A"	(c) Irrational agent selected e_1&e_2.

Events e_1 and e_2 can differ from preferences between A and B. They can be any mutually exclusive events as shown below. We will call event e_1&e_2 a **ghost event**, it does not exist in a set of mutually exclusive alternatives {e_1,e_2}, but an ME-irrational agent who marks events randomly or deliberately irrationally creates this ghost event.

☑ e_1 ☐ $e_2 = \neg e_1$	☐ e_1 ☑ $e_2 = \neg e_1$	☑ e_1 ☑ $e_2 = \neg e_1$

How to deal with ME-irrational agents?. There are three approaches: (1) ignoring ME-irrational agents by excluding their answers if we have only a few ME-irrational agents, (2) redesigning the questionnaire (change events) to satisfy mutual exclusion to be able to use the probability theory, and (3) developing a new logic to deal with a mixture of ME-irrational agents and rational agents for situations when redesign is not feasible. Redesigning a set of events often is not feasible because for any given set of events we do not know in advance how many agents are ME-irrational for these events. Thus, potentially a redesign process can continue indefinitely. The same question: "How many agents are ME-irrational for these events?" can be raised after every consequent redesign. Therefore, it is desirable to develop a *theory to deal with a mixture* of rational agents.

Expansion by superposition. Now we will explore the issues that determine the possibility of such a theory. The first idea is that we discussed above was to add a new alternative e_i&e_j to two mutually exclusive events e_i and e_j, that is recognizing that e_i and e_j can be more than just mutually exclusive in some situations.

We call this method an **expansion by superposition** of a set of events where two mutually exclusive events e_i and e_j are mixed or superposed with possible adjustment of meaning of e_i and e_j. In general, this method does not explain why seemingly mutually exclusive alternatives can coexist. It can range from an irrational belief of the agent in such ghost events, or be similar to the superposing situation in quantum physics. In addition, there are cases where a simple explanation is possible. For instance, if e_i ="young" and e_j="old" then a new event e_i■e_j can be called "middle-age" with adjustment of e_i being new e_i` "young" that does not include "middle-age". Similarly, adjusted "old" does not include "middle-age." For a relevant discussion, see [11,12]. The **superposition operation ■** is a new operation that produces a new event/statement e_i■e_j , which fuses events e_i and e_j.

Channel method. Assume that a set of agents G has two different binary criteria, or channels (C_1 and C_2) to evaluate the sentence "A > B". Thus, each agent has a set S of four possible evaluation states S = (TT, T&F , F&T, FF) obtained by the superposition of the two criteria, as it is shown below:

$$\begin{bmatrix} \textit{First evaluation criterion}\,(C_1) & T & T & F & F \\ \textit{Second evaluation criterion}\,(C_2) & T & F & T & F \\ \textit{superposition} & TT & T\,\&\,F & F\,\&\,T & FF \end{bmatrix}$$

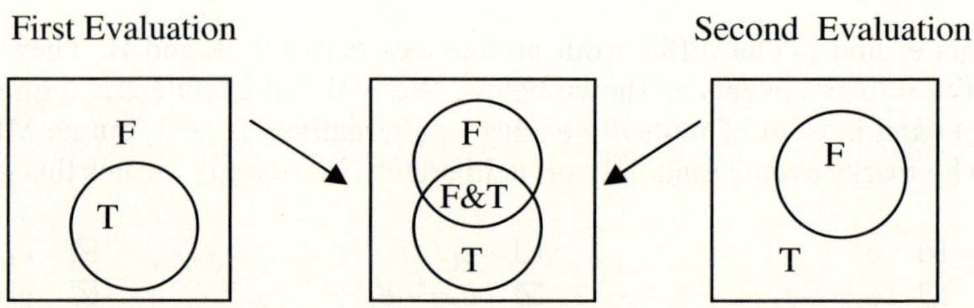

Fig. 1. Second order of conflict as a superposition of two first order of conflict types of evaluations for two internal criteria of the agent in the same time and space

In terms of events e_i and e_j with a single criteria C (that is not explicitly given or not given at all) we may have for agent g two mutually exclusive events/statements e_i=p=(A$>_g$ B) and e_j=q=(B$>_g$A). If we introduce explicitly two criteria C_1 and C_2 (instead of implicit C) we can define preference events in each criterion for agent g:

$$e_i = (A >_{C_1 g} B), e_j = (B >_{C_2 g} A), \quad e_i \wedge e_j = (A >_{C_1 g} B) \wedge (B >_{C_2 g} A)$$

Now e_i and e_j are not mutually exclusive but still contradictory to be able to prefer A or B. In terms of T and F notation e_i&e_j corresponds to a pair TF, where criterion C_1 is True, but criterion C_2 is False for A>B. The fundamental advantage of the channel method is that we explicitly model evaluation criteria. However, this method has also an important disadvantage. We may not know explicit criteria and may not know that only two specific criteria C_1 and C_2 are involved in evaluation.

Negation operation. The negation operation at the second order of conflict requires special consideration starting from the definition of the truth of the statement. There are two truth values for each statement p, V(p,C_1) and V(p,C_1) for criteria C_1 and C_2 V(p,C_1), respectively:

$$T = \begin{bmatrix} T \\ T \end{bmatrix}, F = \begin{bmatrix} F \\ F \end{bmatrix}, T\,\&\,F = \begin{bmatrix} T \\ F \end{bmatrix}, F\,\&\,T = \begin{bmatrix} F \\ T \end{bmatrix}$$

We can define p is true if $V(p,C_1)=V(p,C_1)=True$. The negation of p is defined similarly as $V(p,C_1)=V(p,C_1)=False$. Thus, we define a set of agent G(p), such that p is true for both criteria C_1 and C_2,

$$G(p) \equiv \{g | V(p,C_1)= V(p,C_2)=T\}$$

Similarly we define a set of agents G(NOT p) for **negation** of p:

$$G(\neg p) \equiv \{g | V(p,C_1)= V(p,C_2)=F\}$$

These sets are equivalent to sets that we defined above, $G(p)=G_{TT}$ and $G(NOTp)=G_{FF}$. An **alternative negation** is to use a compliment set to the set G(p):

$$G(NOT\ p) \equiv G^c(p) = G_{TF} \cup G_{FT} \cup G_{FF}$$

This NOT operator differs from "¬" operator for the second order of contradiction (see also Figure 2), but these negations are equal in the first order of contradiction.

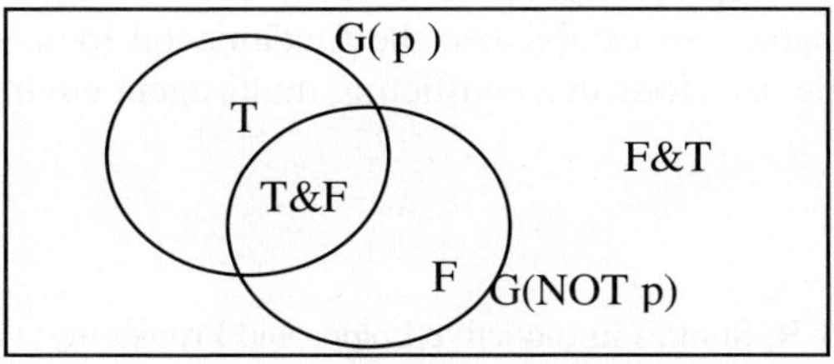

Fig. 2. Set of Agents for which NOT p is True

Now we can clarify the concept of **irrational agents**. A **necessary condition** for agent g to be an irrational agent is $g \in G_{TF} \cup G_{FT}$.

A **sufficient condition** of agent g to be irrational is that criteria C_1 and C_2 are the same. Thus, cases FT and TF are equal. Agent g is called **seemingly irrational** if criteria C_1 and C_2 are not identified but two contradictory evaluations are provided by the agent g for a statement p. A set of agents that are irrational for p will be denoted as $G_\Gamma(p)$. Now with the second negation NOT we have these properties

$$G(p \wedge NOT\ p) = G(p) \cap G(NOT\ p) = \min(G(p), G(NOT\ p)) - G(\neg p \wedge NOT\ p)$$
$$G(p \vee NOT\ p) = G(p) \cup G(NOT\ p) = \max(G(p), G(NOT\ p)) + G(\neg p \wedge NOT\ p),$$

where $G(NOT\ p) \subset G(p)$.

Thus, for the second negation the set of agents in a contradiction state or in the state T & F (TF for short) is not an empty set. Given a set of agents G(p), we can compute membership value μ by using the averaging operation used for computing frequencies, but these frequencies will be different from computed in the probability calculus, because counting evaluation of irrational agents. This approach can be generalized to define the third (and higher) order of conflicting agents with three or more criteria (channels) available to an agent to evaluate the same sentence.

4 Conclusion

The hierarchy of sets of agents to levels of conflicts, self-conflicts, and irrationality proposed in this paper is intended to provide a base for several areas of further studies. These studies include the foundation of probability theory, fuzzy logic, and other types of uncertainties, as well as real-world interpretation of these theories.

The major weakness in fuzzy logic is its ad hoc nature with operations that are not justified in advance (as it is done in the probability theory), but adjusted in many different ways (by using many different t-norms and t-conorms). The novelty of this paper is not in introduction of new formulas for initial membership function values and operations with them, but in creating a base for their *interpretation* by means of conflicting and irrational agents as an internal part of the theory. Physics provides plenty of positive examples of such interpretations that can be inspiration examples for building comprehensive logics of uncertainty for agents. We expect that in near future this will be an active area of new fundamental research and discoveries.

A society of agents always has a degree of conceptual conflicts and any agent can be self-conflicting. This affects agents' abilities to use resources for identifying goals and acting. In this paper, we established the fundamental logic structure that can be used for making logic decisions in a conflicting multi-agent environment.

References

1. Carnap R., Jeffrey R, Studies in Inductive Logics and Probability, vol. 1, 35-165 Berkeley, CA, University of California Press (1971).
2. Fagin R., Halpern J. Reasoning about Knowledge and Probability, Journal of the ACM 41, 2 (1994) 340 – 367.
3. Edmonds, B., Review of Reasoning about Rational Agents by Michael Wooldridge, Vol. 5. Issue 1, Journal of Artificial Societies and Social Simulation (2002) http://jasss.soc.surrey.ac.uk/5/1/reviews/edmonds.html
4. Ferber J., Multi Agent Systems, Addison Wesley (1999)
5. Gigerenzer G. & Selten, R. Bounded Rationality, Cambridge: The MIT Press (2002)
6. Halpern J. Reasoning about uncertainty, MIT Press (2005)
7. Hisdal E., Logical Structures for Representation of Knowledge and Uncertainty, Springer (1998)
8. Resconi, G., Jain, L. Intelligent agents, Springer Verlag (2004)
9. Resconi G., Kovalerchuk, B., The Logic of Uncertainty with Irrational Agents In: Proc. of JCIS-2006 Advances in Intelligent Systems Research, Taiwan, Atlantis Press (2006)
10. Kahneman, D.Maps of Bounded Rationality: Psychology for Behavioral Economics.The American Economic Review. 93(5). pp. 1449-1475 (2003)
11. Kovalerchuk B., Analysis of Gaines' logic of uncertainty, In: Proceeding of NAFIPS '90 vol.2 edited by I.B. Turksen, Toronto Canada pp.293-295 (1990)
12. Kovalerchuk B., Context spaces as necessary frames for correct approximate reasoning. International Journal of General Systems, v.25, n 1, (1996) 61-80.
13. Kovalerchuk B., Vityaev E., Data mining in finance: advances in relational and hybrid methods, Kluwer (2000)
14. Wooldridge M., Reasoning about Rational Agents, Cambridge, MA: The MIT Press (2000)

Stratified Multi-agent HTN Planning in Dynamic Environments

Hisashi Hayashi

Knowledge Media Laboratory
Corporate Research and Development Center, Toshiba Corporation
1 Komukai Toshiba-cho, Saiwai-ku, Kawasaki, 212-8582 Japan
hisashi3.hayashi@toshiba.co.jp

Abstract. In stratified multi-agent planning (SMAP), the parent planning agent and its child planning agents work together to achieve a goal. The parent planning agent executes a rough plan for a goal, and the child planning agents execute detailed plans for subgoals. Although this kind of SMAP is efficient, it is difficult for the parent agent to change the plan while a child planning agent is working. On the other hand, on-line planning, where the agent continuously updates its plan during the plan execution, is very important if we need to implement planning agents working in dynamic environments. This paper shows how to realize on-line planning in SMAP. For this purpose, we extend Dynagent which is an on-line HTN planning agent system.

1 Introduction

Recently, HTN planning [5,12,15,18,19] is used in many application areas such as mobile agents [10], multi-agents [6,7], RoboCup simulation [13], robotics [2], web service composition [17] and story-telling [4]. HTN planning is different from standard planning which just connects preconditions and effects of actions. Instead, HTN planners make plans by decomposing abstract tasks into more detailed tasks (subplans.) In addition to its efficiency and expressive power as explained in [20,9], it is suitable for planning in dynamic environments. By postponing the subtask decomposition until the agent needs to execute the subtask, the agent can adapt the plan for the subtask to the situation at the time of execution. Dynagent [11], an on-line HTN planning agent, can adapt to the dynamic world more easily by continuously updating alternative plans and selecting the best plan in terms of the cost. On-line planning is an important research area not only in HTN planning [4,11,19] but also in standard planning [14].

It is important to find the best (or a better) plan with regard to the cost. However, in HTN planning, the planner has to compare a large number of alternative plans if the planning domain is big. One way of solving this problem is to use the idea of stratified multi-agent planning (SMAP) where the parent planning agent makes a rough plan first and its child planning agents make detailed plans. As shown in Figure 1, by using SMAP, it is possible to reduce the search space of HTN planning because once the parent planning agent makes

N.T. Nguyen et al. (Eds.): KES-AMSTA 2007, LNAI 4496, pp. 189–198, 2007.

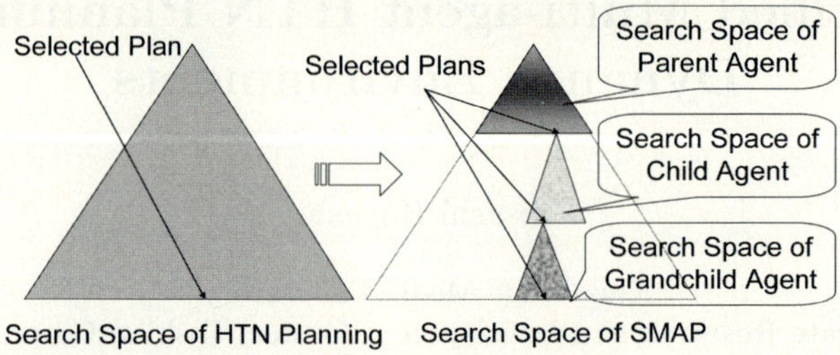

Fig. 1. Search Space of HTN Planning and SMAP

plans and selects a plan from them, the other alternative plans of the parent planning agent are not taken into consideration while the child planning agent is planning. As pointed out in a textbook [8] of multi-agent systems, it is a standard technique for the parent planning agent to give subgoals to its child agents, which are not necessarily planning agents. Also, stratified multi-agent systems, which are not necessarily SMAP systems, are often used for the adaptation to the dynamic world. For example, Mobilespaces [16] is a mobile agent system[1] such that the parent mobile agent changes its child mobile agents according to the computer environment. The layered architecture [3] of robotics, where the agents in higher levels control the agents in lower levels, makes mobile robots robust to the environment.

It seems that the combination of on-line HTN planning and SMAP is promising for the implementation of efficient and robust planning agents working in dynamic environments. However, if the planning agent cannot stop action execution, this is a problem. Normally, in planning, an action is a primitive task for the agent to execute, and the agent does not stop the action execution once it starts it. The agent normally replans after the current action execution and before the next action execution. However, in stratified multi-agent systems, in order to execute an action of the parent planning agent, the child planning agent makes and executes a subplan. The action execution time of the parent planning agent is generally long and changes the world greatly. Therefore, when replanning, we do not wish to wait till an action execution finishes. On the other hand, if the parent agent replans without caring about the plan execution of its child planning agents, the child planning agents might execute meaningless plans which are no longer related to the plan of the parent planning agent. In this paper, we tackle this problem of replanning in SMAP.

This paper presents a new on-line SMAP algorithm where the agent, notably the parent planning agent, quickly replans when obtaining new information by keeping and continuously updating alternative plans while executing a plan. The parent planning agent replans, even while a child planning agent is executing a plan, and controls the child planning agent accordingly. Because the planning agents continuously modify their plans, our planning agents replan efficiently.

[1] Mobile agents are software agents that move from one computer to another through the network.

In order to implement this on-line multi-agent planning system, we extend and use Dynagent, an on-line HTN planning agent system. Replanning in multi-agent planning is also researched in [1]. However, in this system, each agent has a different goal and the problem to solve is how to recoordinate other agents' plans which affect one another. Therefore, this is different from replanning in SMAP in which the agents work together for one goal.

The rest of this paper is organized as follows. Section 2 explains the museum guide scenario. Section 3 intuitively explains the replanning algorithms of SMAP. Section 4 defines the agent algorithms. Section 5 evaluates our on-line SMAP algorithm by means of experiments using the museum guide scenario. Section 6 analyzes the efficiency of our on-line SMAP algorithm. Section 7 is the conclusion.

2 Museum Guide Scenario

In this section, we introduce a museum guide scenario as an example to illustrate replanning in SMAP. Subsequently, this scenario will also be used for experimental evaluation. Figure 2 shows the map of a museum where the robot moves. Nodes are places where the robot localizes itself relative to the map with the help of, for example, markers which can be recognized through image processing. In particular, nodes are set at intersections of paths or at points of interests. The robot moves from one node to the next node along an arc. When the user specifies the destination (node), the robot takes the person there.

The museum is divided into areas. Some areas are connected by a door. Given the destination, considering the doors, the parent planning agent first searches a rough route that connects only areas. Then the child planning agent searches a detailed route in the first area that connects nodes. For example, when moving from $n1$ ($area1$) to $n40$ ($area8$), the parent planning agent first makes a rough plan: $area1 \rightarrow area3 \rightarrow area5 \rightarrow area7 \rightarrow area8$. The child planning agent then thinks about how it should move in the first area: $n1 \rightarrow n6 \rightarrow n10$.

The robot can detect if a door is open or not. If the robot notices that a door on the route is closed, the parent planning agent should change the route. If the robot notices that a door is open, the parent planning agent might be able to find a better plan. Therefore, the robot must be able to replan.

In general, actions are primitive tasks that agents execute quickly, and the time for action execution is not taken into account. Also, the action execution cannot be stopped normally. However, in this scenario, one action of the parent planning agent is an area movement. Therefore, before replanning, we cannot wait till the action execution of the parent planning agent finishes.

For example, suppose that the robot is on $arc28$ in $area5$ and moving towards $area8$ via $area7$ when it finds $door2$ is closed. If the parent planning agent cannot replan immediately and stop the current action execution, then the robot continues to move to $area7$ and the parent planning agent would change the plan there. The new route would be to go to $area8$ via $area5$ and $area6$, and the robot would go back to $area5$. This is completely correct replanning.

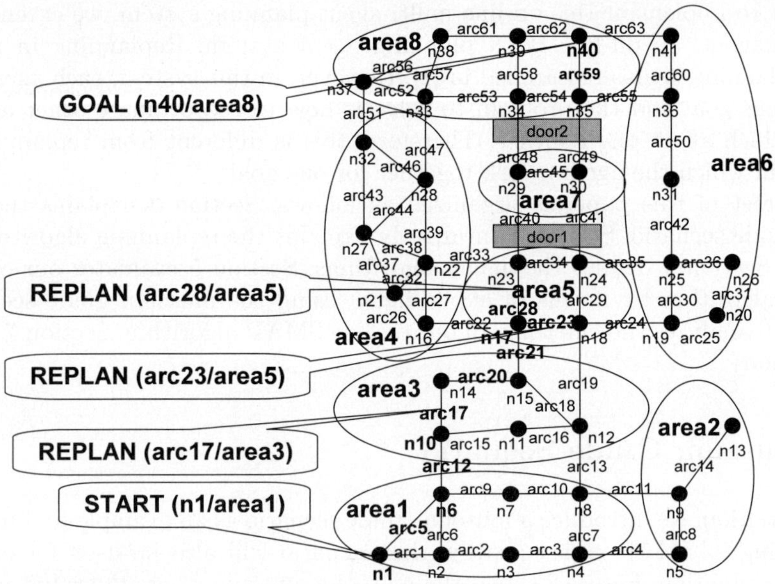

Fig. 2. The Map

However, we should stop moving from *area*5 to *area*7 before replanning because this movement is meaningless and time-consuming.

3 Replanning in SMAP

In SMAP, the parent planning agent executes a rough plan by giving subgoals to its child planning agents. For the parent planning agent, its child planning agents are just action executors, and the parent planning agent does not know how its action executors or child planning agents are implemented. Replanning during a plan execution would be easy if the planning agent could wait for an action executor (or a child planning agent) to finish the current action execution. As we have understood in the museum guide scenario, however, it is meaningless and time-consuming to continue the current action execution before replanning when using child planning agents. This section intuitively shows how to start replanning quickly by stopping the current action execution.

As shown in Figure 3, replanning is triggered by a belief update. After replanning, if the plan is modified and the current action execution becomes no longer relevent with regard to the new plan, then the agent should stop the action execution. However, it is not always possible to stop the current action execution. Therefore, the planning agent asks the action executor if it is possible to suspend the current action execution. If yes, the planning agent tells the action executor

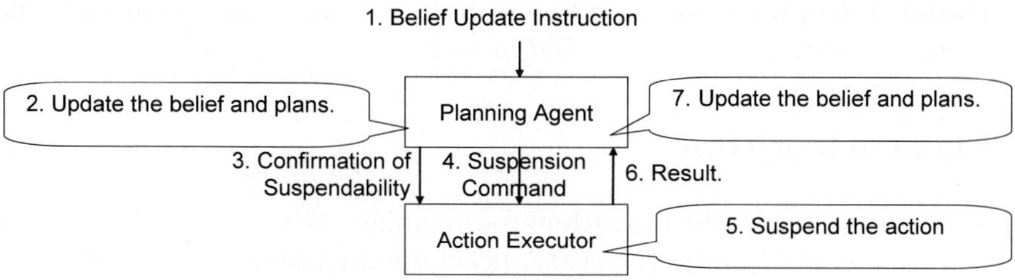

Fig. 3. Replanning

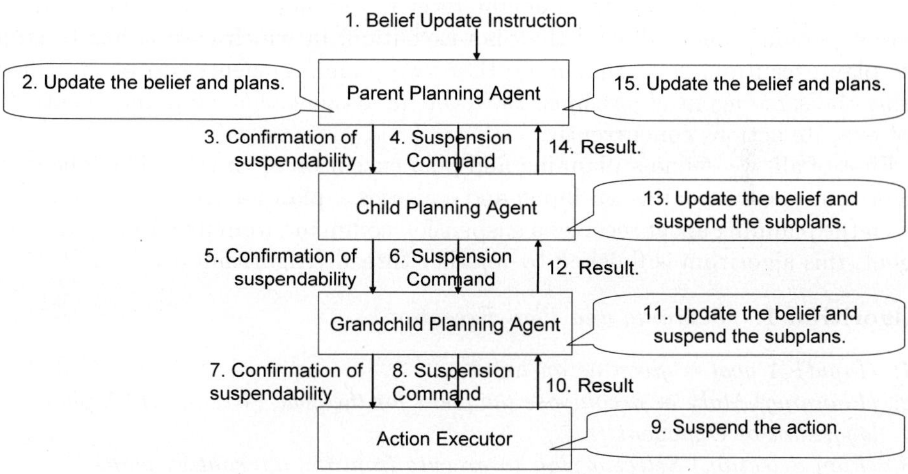

Fig. 4. Replanning in SMAP

to suspend the current action execution. After stopping the current action execution, the planning agent updates its belief and plans considering the effect of the action suspension. For example, if the action to go to $n6$ along $arc5$ is suspended, then the location of the robot becomes on $arc5$. We assume that the planning agent knows the effect of action suspension.

In SMAP, action suspension can be implemented in a similar way. As shown in Figure 4, when the parent planning agent suspends the current action execution after replanning, it tells the child planning agent to stop the current action execution. (Note that a child planning agent is just an action executor from the viewpoint of its parent planning agent.) When the child planning agent receives the suspension command from the parent planning agent, it tells the grandchild planning agent to stop the current action execution. When the grandchild planning agent receives the suspension command from the child planning agent, it tells the action executor to stop the current action execution. After suspending the current action execution, all the the planning agents update the belief based on the effect of action suspension, and the parent planning agent reupdates its plans, if necessary. In Figure 4, three planning agents are stratified. More planning agents can be stratified in a similar way.

4 Agent Algorithm

In this section, we define the algorithms for planning agents. Our planning agents receive three kinds of inputs: 1. goals; 2. belief update instructions; 3. suspension commands. Given a goal, our planning agents make and execute a plan. During the plan execution, if a planning agent receives a belief update instruction, it updates the belief, checks the plan, and replans if necessary. If a planning agent has a parent planning agent, it might receive a suspension command from its parent planning agent during the plan execution, in which case it has to stop the plan execution. Note that more than two planning agents can be stratified using the same agent algorithms. However, we assume that planning agents do not execute actions concurrently.

First of all, we define a planning and plan execution algorithm. The following algorithm takes a goal as an input and executes a plan for the goal. Note that when the planning agent receives a suspension command from its parent planning agent, this algorithm is finished by another thread (Algorithm 3.)

Algorithm 1. *(Planning and Plan Execution)*

1. *(Input) A goal is given as an input.*
2. *(Planning) Make or decompose the plans for the goal. (Use the HTN planning algorithm of Dynagent [11].)*
3. *(Plan Selection) Select a plan to execute from the alternative plans.*
4. *Set the status of the goal to "active."*
5. *(Plan Execution Loop) Repeat the following procedure while the status of the goal is "active:"*
    - (a) *(Action Execution) Following the selected plan, tell the action executor to execute the next action and wait for the result ("success", "failure", or "suspended") that is reported from it.*
    - (b) *(Plan Update) If the result of the action execution is either "success" or "failure," then modify the belief and all the plans, following the plan modification algorithm of Dynagent [11].*
    - (c) *(Plan Update) If the result of the action execution is "suspended," modify the belief, considering the effect of the action suspension.*
    - (d) *(Successful Plan Execution) If one of the plans is successful, then change the status of the goal to "success."*
    - (e) *(Plan Execution Failure) If no alternative plan exists, then change the status of the goal to "failure."*
    - (f) *(Plan Selection) If the status of the goal is "active," then select a plan from alternative plans.*
6. *Output the status of the goal ("success", "failure", or "suspended.")*

The following algorithm takes a belief update instruction as an input, updates the belief, and changes the plan if necessary.

Algorithm 2. *(Belief Update and Replanning)*

1. *(Input) A belief update instruction is given as an input.*
2. *(Belief Update) Update the belief following the instruction.*
3. *If a plan is being executed by Algorithm 1, execute the following procedure:*
   (a) *(Plan Modification) Based on the updated belief, modify the plans using the algorithm [11] of Dynagent.*
   (b) *(Action Suspension) If the current action execution becomes no longer relevant to the current plan execution and it is possible to stop the action execution, then give the suspension command to the action executor[2] (or its child planning agent that is executing the action) and wait for the result.*

Finally, the following algorithm takes a suspension command as an input and stops the execution of the action and the plan. Note that when the following algorithm changes the status of the goal to "suspended," then the plan execution process (Algorithm 1) is finished.

Algorithm 3. *(Goal Suspension)*

1. *(Input) A suspension command is given as an input.*
2. *(Action Suspension) If it is possible to stop the current action execution, then give the suspension command to the action executor (or its child planning agent that is executing the action) and wait for the result.*
3. *(Plan Suspension) Change the status of the goal to "suspended."*

Our agent algorithm depends on the on-line HTN planning algorithm of Dynagent. Because of the space limitation, we cannot write the algorithm of Dynagent which is written in [11]. However, we briefly explain it here. Dynagent keeps several alternative plans. Each task in an alternative plan is not necessarily an action (= a primitive task). Dynagent decomposes a task into subtasks when necessary. However, the first task in each plan must be an action before selecting a plan and executing an action. After successfully executing an action, Dynagent updates the belief based on the effect of the action, and removes the executed action from each plan if the first action in the plan is the action Dynagent has executed. When Dynagent fails to execute an action, Dynagent removes each plan such that the first action in the plan is the action Dynagent could not execute. After the belief update, it is easy to find and remove invalid plans. Invalid plans can be found by rechecking the (protected) precondition of each action in the plans. Some preconditions must be protected if they might become unsatisfiable when the belief is updated. Even if the precondition of an action is unsatisfiable when planning initially, it might become satisfiable later because

---

[2] Note that a child planning agent is an action executor of the parent planning agent.

of a belief update. Therefore, Dynagent keeps such invalid plans and makes new plans from them when unsatisfiable preconditions become satisfiable.

5 Experiments

This section evaluates the efficiency of replanning in SMAP by means of experiments based on the museum guide scenario explained in Section 2.

Initially, the robot is at $n1$ in $area1$, and $door1$ is open. (See Figure 2.) The destination is $n40$ in $area8$. If the goal is given when $door2$ is open, then the parent planning agent tries to execute the plan to go from $n1$ to $n40$ via $area1$, $area3$, $area5$, $area7$, and $area8$. If the goal is given when $door2$ is closed, then the parent planning agent tries to execute the plan to go to $n40$ via $area1$, $area3$, $area5$, $area6$, and $area8$. In any case, in order to move from $area1$ to $area3$, the child planning agent starts the plan execution to go from $n1$ to $n10$ via $n6$. When the robot arrives at $n6$, in order to move from $area3$ to $area5$, the child planning agent starts planning and plan execution to go from $n10$ to $n17$ via $n14$ and $n15$. While the robot is moving in $area3$ ($arc17$) or $area5$ ($arc23$ or $arc28$), we close or open $door2$ and the planning agents replan.

We evaluate the time for replanning and compare our on-line replanning and the naive method to replan from scratch. It is not the aim of these experiments to evaluate the efficiency of pathfinding because our HTN planner can be used for other purposes. The agent system is implemented in Java and the planner that the planning agents use is implemented in Prolog, which is implemented in Java. We used a PC (Windows XP) equipped with a Pentium4 2.8GHz and 512MB of RAM.

Table 1 shows the replanning time of the parent planning agent. Our on-line replanning is 3 $\sim$ 4 times as fast as the naive replanning. This is because our planning agents reuse old plans when replanning. Table 2 shows the replanning time of the child planning agent. When replanning in $area3$, the replanning time is 0.0 sec. This is because the child planning agent does not change the plan. On the other hand, when replanning in $area5$, our on-line replanning is as fast as the naive replanning. This is because the child planning agent changed the next destination area (from $area7$ to $area6$ or from $area6$ to $area7$) and has to make plans from scratch. Table 3 shows the total replanning time. Our on-line replanning is 2 $\sim$ 5 times as fast as the naive replanning.

Table 1. Replanning Time (Parent Planning Agent)

Place of Replanning	door2	On-Line Replanning	Naive Replanning
arc17(area3)	*open* → *closed*	0.1 sec	0.4 sec
arc28(area5)	*open* → *closed*	0.1 sec	0.4 sec
arc17(area3)	*closed* → *open*	0.1 sec	0.4 sec
arc23(area5)	*closed* → *open*	0.1 sec	0.4 sec

Table 2. Replanning Time (Child Planning Agent)

Place of Replanning	door2	On-Line Replanning	Naive Replanning
arc17(area3)	*open → closed*	0.0 sec	0.1 sec
arc28(area5)	*open → closed*	0.1 sec	0.1 sec
arc17(area3)	*closed → open*	0.0 sec	0.1 sec
arc23(area5)	*closed → open*	0.1 sec	0.1 sec

Table 3. Total Replanning Time

Place of Replanning	door2	On-Line Replanning	Naive Replanning
arc17(area3)	*open → closed*	0.1 sec	0.5 sec
arc28(area5)	*open → closed*	0.2 sec	0.5 sec
arc17(area3)	*closed → open*	0.1 sec	0.5 sec
arc23(area5)	*closed → open*	0.2 sec	0.4 sec

6 Efficiency (General Case)

In the previous section, the efficiency of replanning in SMAP was evaluated by means of experiments. In general, when a planning agent A replans, the total replanning cost in SMAP largely depends on whether its child planning agent needs to change the goal or not. If there is no need for the child planning agent to change the goal when A replans, there is no need for the descendants of A to change their plans. In this case, it is clear that we can save the replanning cost which is roughly proportional to the number of the descendants of A. On the other hand, if the child planning agent needs to change the goal, the descendants of A cannot reuse their plans. Even in such case, the planning agent A normally replans efficiently because Dynagent is an on-line planning agent which reuses its plans when replanning.

7 Conclusion

This paper has presented a new on-line SMAP system by extending and using Dynagent, which is an on-line HTN planning agent system. Our planning agents can replan before finishing an action execution. In particular, when replanning, the parent planning agent does not have to wait for the child planning agent to finish the current plan execution. Our on-line planning agents continuously keep and modify plans. Therefore, compared with the naive replanning method that makes plans from scratch, our planning agents replan efficiently when modifying the belief as confirmed by the experiments. In the near future, we would like to apply our on-line SMAP system to implement real museum guide robots, extending the scenario used in the experiments.

In this paper, we assumed that planning agents do not receive the belief update instructions while planning or replanning. Therefore, the frequency of replanning is limited. In the future, we would like to tackle this problem.

Currently, Dynagent handles total-order plans only. Another future work is to extend Dynagent and our on-line SMAP system so that it can handle partial-order plans. In SMAP, this means that more than two child planning agents in the same layer (and their descendants) work concurrently, and they have to coordinate their plans which might affect each other.

References

1. T. Bartold and E. Durfee. Limiting disruption in multiagent replanning. In *AAMAS03*, pages 49–56, 2003.
2. T. Belker, M. Hammel, and J. Hertzberg. Learning to optimize mobile robot navigation based on HTN plans. In *ICRA03*, pages 4136–4141, 2003.
3. R. Brooks. A robust layered control system for a mobile robot. *IEEE Journal of Robotics and Automation*, 2(1):14–23, 1986.
4. M. Cavazza, F. Charles, and S. Mead. Planning characters' behaviour in interactive storytelling. *The Journal of Visualization and Computer Animation*, 13(2):121–131, 2002.
5. K. Currie and A. Tate. O-plan: The open planning architecture. *Artificial Intelligence*, 52(1):49–86, 1991.
6. M. desJardins, E. Durfee, C. Ortiz, and M. Wolverton. A survey of research in distributed, continual planning. *AI Magazine*, 20(4):13–22, Winter 1999.
7. J. Dix, H. Mũnoz-Avila, and D. Nau. IMPACTing SHOP: Putting an AI planner into a multi-agent environment. *Annals of Math and AI*, 4(37):381–407, 2003.
8. J. Ferber. *Multi-Agent Systems: An Introduction to Distributed Artificial Intelligence.* Addison-Wesley, 1999.
9. M. Ghallab, D. Nau, and P. Traverso. *Automated Planning: Theory and Practice.* Morgan Kaufmann, 2004.
10. H. Hayashi, K. Cho, and A. Ohsuga. Mobile agents and logic programming. In *MA02*, pages 32–46, 2002.
11. H. Hayashi, S. Tokura, T. Hasegawa, and F. Ozaki. Dynagent: An incremental forward-chaining HTN planning agent in dynamic domains. In *Post-Proceedings of DALT05*, LNAI 3904, pages 171–187. Springer-Verlag, 2006.
12. D. Nau, Y. Cao, A. Lotem, and H. Mũnoz-Avila. SHOP: simple hierarchical ordered planner. In *IJCAI99*, pages 968–975, 1999.
13. O. Obst, A. Maas, and J. Boedecker. HTN Planning for Flexible Coordination Of Multiagent Team Behavior. Technical report, Universität Koblenz-Landau, 2005.
14. S. Russell and P. Norvig. *Artificial Intelligence: A Modern Approach.* Prentice Hall, 1995.
15. E. Sacerdoti. *A Structure for Plans and Behavior.* American Elsevier, 1977.
16. I. Satoh. Mobilespaces: A framework for building adaptive distributed applications using hierarchical mobile agent system. In *ICDCS00*, pages 161–168, 2000.
17. E. Sirin, B. Parsia, D. Wu, J. Hendler, and D. Nau. HTN planning for web service composition using SHOP2. *Journal of Web Semantics*, 1(4):377–396, 2004.
18. A. Tate. Generating project networks. In *IJCAI77*, pages 888–893, 1977.
19. D. Wilkins. *Practical Planning.* Morgan Kaufmann, 1988.
20. D. Wilkins and M. desJardins. A call for knowledge-based planning. *AI Magazine*, 22(1):99–115, Spring 2001.

WSrep: A Novel Reputation Model for Web Services Selection*

Zhen Li, Sen Su, and Fangchun Yang

State Key Lab. of Networking and Switching, Beijing University of Posts and Telecommunications
187#, 10 Xi Tu Cheng Rd., Beijing, 100876, P.R. China
jasonlizhen@gmail.com, {susen, fcyang}@bupt.edu.cn

Abstract. Web services selection is based on QoS and trust. As one of the important attributes of QoS, reputation is commonly used to assess the trustworthiness of the web services and minimize the threats of transactions. However, most existing reputation models of web services are all based on the subjective user ratings. These systems are easily attacked by malicious raters. This paper presents a novel reputation model named WSrep, in WSrep, the reputation integrates user ratings and a significant objective factor-credibility of QoS advertisements which is an objective view of the past behaviors of a given service. Other contributions of the paper include a customer measurable QoS model, a Bayesian learning model for building the credibility, and a set of experiments to show the benefits of our approach.

1 Introduction

Web service technologies promise the dynamic construction of loosely coupled information systems [4]. As a consequence of the rapid growth of web services (especially functional overlapping services), systematically and (semi-) automatically selecting the "best" service becomes a difficult and challenging work. In such a scenario, the quality of service (QoS) and trust are the key factors to do services selection processes [7]. QoS can help customers to select a distinguished service that has higher qualities, and trust is used to assist customers to choose good providers who always deliver promised qualities honestly. As an important attribute of QoS, reputation is a measure of the trustworthiness of the services, it mainly depends on end user's experiences of using the service [8]. Reputation not only can be efficiently used to find good services, but also can stimulate transactions to be executed exactly without the expense of third party monitoring.

Recognizing the importance of reputation, an immediate question to ask is how to model the reputation. There is an extensive amount of research focused on building reputation models for web services [5], [6], and [9]. Further, some well-known sites

* This work is supported by the National Basic Research and Development Program (973 program) of China under Grant No. 2003CB314806; the program for Changjiang Scholars and Innovative Research Team in University (PCSIRT).

N.T. Nguyen et al. (Eds.): KES-AMSTA 2007, LNAI 4496, pp. 199–208, 2007.

such as bindingpoint.com [2] also provide an interface to rate web services. These approaches all defined reputation as average of subjective user ratings. Actually, these rating-based reputation systems do not work efficiently under some complex circumstances. A key limitation of current approaches is that they can not cope with the malicious rating attack [3]. In the web services settings, the dishonest providers often oscillate their reputation between building and milking to mislead the customers, they also collude with malicious raters who always provide low feedbacks to their opponents and high feedbacks to themselves. So the current subjective reputation models are hard to reveal the correct performance history of a given service. In addition, current approaches are not flexible enough to express a given customer's intentions and norms. They disseminate the values of services' reputation among different customers. These values represent the general perspectives of raters but not a personal opinion, it may make a customer to select a service with which he is not satisfied although this service may have a higher reputation value. So there is the need to model reputation from a personal perspective.

With these research problems in mind, we develop WSrep, a novel reputation model for web services selection. In WSrep, the reputation model is made up of the average of user ratings and the objective view of performance history of a given service. We propose a concept named "credibility of QoS advertisements" (Cre) as the objective factor. Cre is a probability which is used to predict the possibility of providers delivering QoS values complying with their advertisements. The Cre can reduce the wicked influence made by malicious rating attack and our reputation is computed at user local side with personal information for expressing the different cognitions of a given service from different customers. We also introduce a customer measurable QoS model for software agents to justify the compliance of the actual QoS values and their advertisements automatically, a Bayesian learning approach to model the Cre. Finally, we demonstrate the effectiveness of this approach via simulation-based experiments.

The remainder of this paper is organized as follows: Section 2 introduces our customer measurable QoS model. Section 3 discusses the model of WSrep. Section 4 presents a series of simulation-based experiments to show the effectiveness and benefits of our approach; Section 5 summarizes the related works in this field; Finally, Section 6 concludes the paper.

2 A Customer Measurable QoS Model

Many literatures like [11] and [12] introduce their comprehensive QoS attributes models which include service level attributes and network level attributes. However, a given customer can not validate the actual values of some attributes directly (e.g., throughput, which is the total number of served requests in a time window. etc.). In this section, we propose a customer Measurable QoS model-MQ used by customers' agents to justify the compliance of the actual QoS values and their advertisements automatically.

Definition 1: MQ

$$MQ \equiv \{Latency, DomQ\}$$

- *Latency:* It is the delay between sending a request and receiving a response.
- *Domain-specific QoS ($DomQ$):* Domain-specific QoS is a set of properties which is used to describe the characteristics of services in a special domain. It should be introduced by domain experts, for example, the "interest rate" is one of the domain-specific attributes of a loan service.

Latency can be justified by run-time monitoring; the values of domain-specific QoS are offered in the service results, user agents can parse these XML-based messages to get the real values and compare them with advertised ones automatically.

3 The WSrep Model

Modeling and designing WSrep are the main purposes of this paper. The WSrep mixes the subjective and objective view of past behaviors of providers, and uses this knowledge to help customers to select the most trustworthy service.

3.1 WSrep Parameters

In WSrep, We identify two important parameters:

- *User Ratings.* After a customer accessing a web service, he rates the service depending on his own preference. The ratings imply the level of satisfaction of users about the whole QoS attributes the provider delivered. So the user ratings are valuable for any reputation models (there are no other ways to express customers' satisfaction properly). User agents in WSrep can adopt any of the existing non-negative integer rating mechanisms (i.e. the rating mechanism of [2], where customers can rate a service with one integer from 0 to 10). This design makes WSrep can be used by more heterogeneous user applications which use different rating mechanisms. Now WSrep can not accept negative ratings, because our Cre is a probability, we need to ensure average of user ratings and Cre within an identical data range for computational simplification and data consistency. Taking this into consideration, we need to normalize user ratings into range [0,1] first:

$$Rating = \frac{R}{Max} \tag{1}$$

  Where R denotes the value of rating and Max is the possible maximal value of this rating mechanism. The trustworthy third-party computes the average of ratings in every fixed time window (e.g. e-bay's reputation system [1] where average of ratings can be computed every week, month or six months) and shares them with all customers.

- Cre. The Cre denotes the possibility of service provider advertising QoS information honestly. The computation of Cre is based on the objective feedbacks generated by user agents. After each transaction, the user applications give objective feedbacks on each element of MQ automatically. They compare the real values of QoS with the values advertised; if they are compliant, the user applications mark the attributes with 1, otherwise, 0. For example, if the user agent finds the real latency is 3s, but the advertised latency is less than 2s, it will rate the latency with 0. This work depends on an assumption:

Assumption 1:
User applications or user agents, in this paper, is trustworthy. That means the objective feedbacks generated by the user applications can not be modified by the customers.

In WSrep, the Cre has two levels:

- *Cre of Attribute (CoA)* expresses the credibility of a single attribute defined in MQ; the CoA is computed in the trustworthy third-party with the objective feedbacks collected from different users.
- *Cre of Web Service (CoWS)* shows the overall credibility of all the attributes defined in MQ, the CoWS is computed at the user local side.

3.2 The Credibility Model

In this subsection, we model the most important parameter in WSrep which is the Cre. Whether the QoS advertisement of a given service is trustworthy is the key influencing factor of services selection. We need a mechanism to objectively predict it, then we can make the rational decisions for customers. Taking this situation into account, an approach based on Bayesian learning theory is needed to model the Cre.

Bayesian learning theory is a statistical theory of making statements about uncertain events. This theory is widely applied in many research fields (i.e. scientific prediction, game-theoretical analyses, decision making and statistics). According to Bayesian learning theory, initially events of interest are assigned a prior belief which reflects existing knowledge about the event and the problem area. Later, as new information (sample) becomes available, the beliefs are updated using the Bayes' rule.

We let a random variable θ denote the CoA. Then we model the CoA using Bayesian learning theory. This work depends on an assumption:

Assumption 2:
For each attribute of MQ, we assume that we have the priors of CoA (results of former evaluations) and represent them as:

$$Priors \equiv \{\theta_1, \theta_2, \theta_3 \ldots\ldots\}$$

We let n denotes the total number of transactions preformed by a given service in a time window and the random variable X denotes the times of compliance of a single attribute among n transactions, which is the sample. So the value of X can be defined as follow:

$$x = \sum_{i=1}^{n} OF_i, i = 1, 2, \ldots\ldots, n \tag{2}$$

Where OF denotes the objective feedbacks generated for this attribute. Obviously, X is binominal distributed ($b(n, \theta)$). According to Bayesian learning theory, a binominal distribution has a Beta-prior ($\pi(\theta) \sim \beta(a,b)$), the resulting posterior is also Beta-distributed ($h(\theta \mid x) \sim \beta(a+x, n+b-x)$). The Beta-distribution is a two parameter distribution whose parameters are denoted by a and b. In order to compute these two parameters, we let $\overline{\theta}$ denotes the average of priors and δ denotes the standard deviation of priors. Otherwise the expectation of $\beta(a,b)$ is $a/(a+b)$ and the standard deviation is $\sqrt{\dfrac{ab}{(a+b)^2(a+b+1)}}$. So the estimate of a and b are computed according to the formula (3)

$$\begin{cases} \hat{a} = \overline{\theta} \times \left(\dfrac{(1-\overline{\theta}) \times \overline{\theta}}{\delta^2} - 1\right) \\ \hat{b} = (1-\overline{\theta}) \times \left(\dfrac{(1-\overline{\theta})\overline{\theta}}{\delta^2} - 1\right) \end{cases} \tag{3}$$

Then we model the CoA as the posterior average estimation of θ (according to Bayesian learning theory, the posterior average estimation of a random variable denotes the most possible average of this variable), which is defined in (4):

$$CoA = \hat{\theta}_E = \frac{a+x}{n+a+b} \tag{4}$$

We assume that a given service has m attributes defined in MQ, the CoWS is defined as weighted average of CoA:

$$CoWS = \sum_{i=1}^{m} \omega_i \times CoA_i \;, i = 1, 2, \ldots\ldots, m \tag{5}$$

Where ω_i denotes user weight on each attribute of MQ and $\sum_{i=1}^{m} \omega_i = 1$, customers weight each attribute depending on their own need of trust (e.g., a customer may focus on the Cre of latency, so he will weight it more) and the user applications ensure the sum of weighs is 1. In this paper, we do not focus on how the weights are given.

3.3 The Reputation Model

In this subsection, we formalize the parameters introduced above to present the reputation metric. In WSrep, the Reputation of Web Service combines average of user ratings and the CoWS, which is used to measure the level of trust of a service implementation. The WSrep is defined in (6):

$$WSrep = \alpha \times \frac{\sum_{i=1}^{n} Rating}{n} + \beta \times CoWS \tag{6}$$

Where α and β denote the normalized weight factors for the collective rating and Cre and they need to follow the limitation that the sum of α and β is always 1. The α and β parameters can be used to assign different weights according to different needs of customers. For instance, if a user believes the subjective view of performance history of a service is more reasonable, he can give α a higher value. For a given service, the metric shows that the WSrep focus not only on the subjective view of overall performance of a given service but also the objective view of attributes defined in MQ.

4 Experimentation

We performed two sets of experiments to evaluate the WSrep approach and show its feasibility, effectiveness, and benefits. For comparison, we implemented the WSrep approach and the subjective reputation approach. Further, 4 samples of services will be tested in a services selection scenario.

We divide the services providers into two types, one is honest (always delivering promised QoS), and the other is strategic (colluding with dishonest raters to cheat customers and fool the reputation systems). In our experiments, 4 services are chosen, denoted as {S1, S2, S3, and S4}. The cardinality of MQ of each service is set to be 3 and the user weights on them are generated randomly. For each $q \in MQ$, the priors of Cre are set to be {0.45, 0.50, and 0.55} impartially and α and β are all set to be 0.5. Our experiments also consist of 100 subjective raters (supposed customers and 60% of them are set to be malicious) and corresponding 100 objective raters (supposed software agents). The subjective raters rank services with $i \in \{0,1,2,3,4\}$, and objective raters rank each $q \in MQ$ with 1 or 0. The 4 services we designed have different characteristics, S1 is an excellent service (is honest and offers the best QoS), the honest subjective raters always rank S1 with 4 and the objective feedbacks are always 1; S2 and S3 are two good services (is honest and offers less best QoS). For S2 and S3, each subjective rating given by honest raters is generated randomly from 2 to 4 and the objective feedbacks are always 1; S4 is a strategic service, it colludes with the 60 malicious raters who always rate other 3 services with 0 and rate S4 with 4. When S4 builds it own reputation, each subjective rating given by honest raters is generated randomly from 2 to 4 and the objective feedbacks are always 1; when milking from it, each subjective rating given by honest raters is generated randomly from 0 to 1 the and the objective feedbacks are seldom 1. The subjective reputation and WSrep of the four services are computed 20 times, in each time window, 1000 transactions are performed (each customer does 10 times equally).

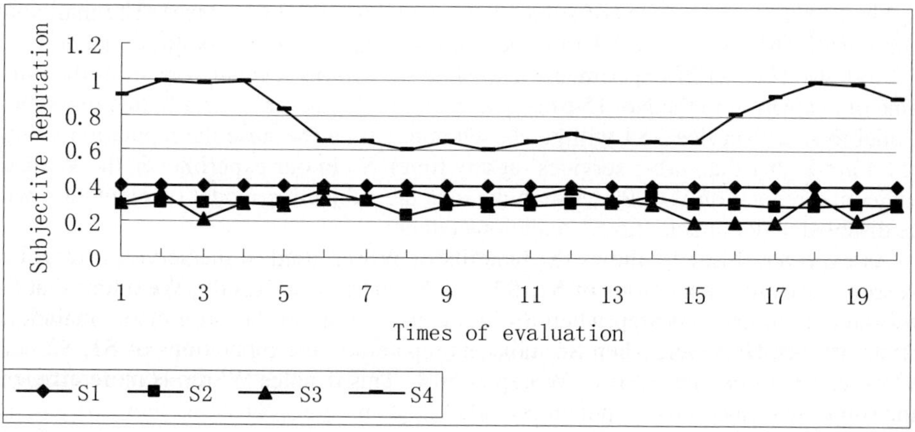

a) Values of subjective reputation

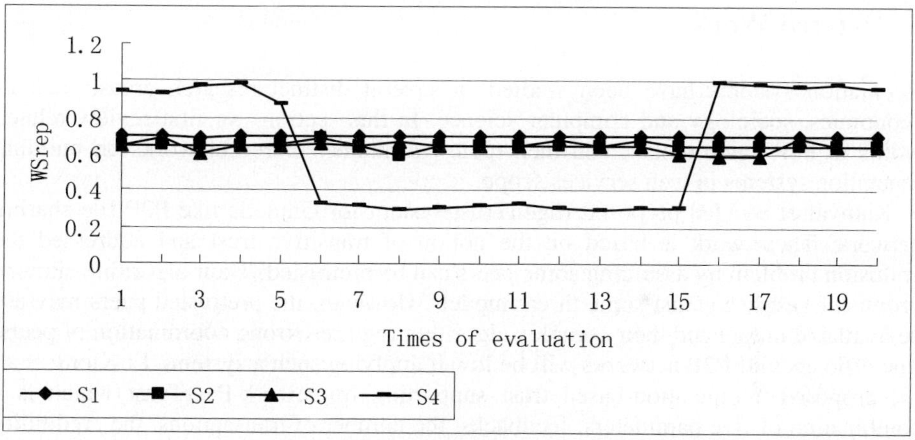

b) Values of WSrep

Fig. 1. The benefit of WSrep-based services selection

Fig. 1 shows the values of the reputation of the 4 services using different computational models.

Reputation-based services selection is to choose the honest service which has the highest reputation value depending on once evaluation. For judging the qualities of reputation models used for services selection, we propose a new criterion named Wrong Selection Rate (WSR), which is computed as:

$$WSR = \frac{T_{wrong}}{T_{right} + T_{wrong}} \tag{7}$$

Where T_{wrong} denotes the times of choosing the dishonest services and T_{right} denotes the times of choosing the honest services. The reputation model is better when its WSR is lower.

Fig.1 a) shows the values of subjective reputation of the 4 services. The malicious raters make the S1, S2 and S3 lose their advantages completely (values change from 0.2 to 0.4). This set of experiments also shows S4 milks it reputation from the No.6 time of evaluation to the No. 15 time of evaluation. However, if we use this reputation model to select services, S4 will be chosen at any time (because the reputation values of S4 are higher than other services' at any time). So in our experiments, the WSR of subjective reputation is 100%, which denotes the subjective model of reputation will be disabled under attack of 60% malicious raters.

As expected, Fig.1 b) shows the benefits of WSrep used in the services selection. WSrep makes the reputations of S1, S2, and S3 increase evidently. We admit that S4 will be chosen using WSrep when S4 builds its reputation (because of the malicious rating attack). However, when S4 milks its reputation, the reputations of S1, S2, and S3 exceed S4's. So the WSR of WSrep is 50%. This denotes WSrep is more efficient and robust than subjective reputation model.

5 Related Work

Reputation systems have been studied in several distinct research areas, such as economics, sociology and computer science. In this section, we first review related works in P2P environment, and then review a number of recent works on building reputation systems in web services scope.

Kamvar et al. [13] proposed EigenTrust system for Gnutella like P2P file sharing network. Their work is based on the notion of transitive trust and addressed the collusion problem by assuming some peers can be pretrusted. Their algorithm showed promising results against some threat models. However, the pretrusted peers may not be available in fact and their complex algorithm requires strong coordination of peers. The efficiency of P2P networks will be low if applying such a system. Li Xiong et al. [3] proposed a reputation-based trust supporting framework-PeerTrust which is a combination of five parameters: feedbacks, the number of transactions, the credibility of the feedback sources, transaction context factor and the community context factor. They tried to use the parameter credibility of feedback to find the malicious raters. However, the complex algorithm of credibility made this approach can only be applied in a small community, because if the number of peers is large, differentiating honest raters and dishonest raters is a very heavy work. So we argue that reducing the influence of malicious rating attack is more applicable than discovering each dishonest rater.

The reputation systems researched in web service field are focused on ensuring the providers delivering their promised QoS and helping the customers select the trustworthy services. Most of existing web services reputation models are not strong and flexible enough to reduce threats. Liangzhao Zeng et al. [8] modeled the reputation of web service as average of user ratings. E. Michael Maximilien et al.[6] introduce a conceptual model of web service reputation, they denote within a specific domain the reputation of the service depends on the subjective view of the users of the service on the various attributes. They also present a UML static model for the components that make up the reputation of a service. The main shortcoming of these reputation model is that the rating-based or subjective view-based reputation is not sophisticated enough to deal with malicious rating attacks so that these systems are easy to be disabled. Interestingly, Sravanthi Kalepu et al. [10] observed the importance of the objective view of services' performance history. They modeled reputation of web service as f (User Ranking, Compliance, and Verity), where compliance refers to the service provider's ability to meet the service level of each QoS parameter laid out in the SLA without incurring penalties and verity (the objective factor the authors advocated) is a mathematical variance represents the compliance levels. However, the computation of verity is based on the subjective user rankings, so it is not "real" objective. If malicious raters exist, the verity will not make any sense too. Further, above reputation models all express the global view of trustworthiness of a given service, it can not satisfy the customers who have special trust needs when they face the services selection problem.

Our work differs from them in a number of ways. First, we emphasize the objective factor is as important as the subjective factor in web services' reputation systems. Then we introduce Cre to reduce the negative influence of the malicious rating attack. The Cre is computed based on objective feedbacks generated automatically by customers' agents, so this value is trustworthy. Second, we integrate the average of user ratings and Cre into the reputation model. A given customer can customize the values of reputation by assigning weights. The personal reputation is more valuable than the general reputation in services selection processes. Third, we run a series simulation-based experiment to show the effectiveness and benefits of our approach.

6 Conclusion

Web services environments offer both opportunities and threats. Building a flexible and robust reputation system is the most efficient way to minimize threats. In this paper, we have described WSrep-a novel reputation model used for selecting the most trustworthy web service. The WSrep model is made up of subjective user ratings and credibility of QoS advertisements which is an objective factor to predict whether the providers delivering their promised QoS honestly. For modeling the credibility, we first proposed a customer measurable QoS model used by software agents to validate the compliance of the actual QoS values and their advertisements automatically and generate the objective feedbacks. Then the credibility is modeled based on the Bayesian learning theory. Finally, we reported initial simulation-based experiments, demonstrating the effectiveness and benefits of our approach.

References

1. ebay, http://www.ebay.com, 2005.
2. bindingpoint, http://www.bindingpoint.com/, 2005.
3. Li Xiong, and Ling Liu. PeerTrust: Supporting Reputation-Based Trust for Peer-to-Peer Electronic Communities. IEEE Transaction on Knowledge and Data Engineering. Vol.16, No.7, JULY 2004.
4. Shuping Ran. A Model for Web Services Discovery With QoS. ACM SIGecom Exchanges. Vol.4, No.1, 2003.

5. E. Michael Maximilien, and Munindar P. Singh. Reputation and Endorsement for Web Services. ACM SIGecom Exchanges. Vol.3, No.1, Dec 2001.
6. E. Michael Maximilien, and Munindar P. Singh. Conceptual Model of Web Service Reputation. ACM SIGMOD Record. Vol. 31, No. 4, Dec 2002.
7. E. Michael Maximilien, and Munindar P. Singh. Toward Autonomic Web Services Trust and Selection. In *ICSOC'04,* November 15–19, 2004.
8. Liangzhao Zeng, Boualem Benatallah, Marlon Dumas, Jayant Kalagnanam, and Quan Z. Sheng. Quality Driven Web Services Composition. In *Proceedings of international World Wide Web conference* 2003.
9. Dimitris Gouscos, Manolis Kalikakis, and Panagiotis Georgiadis. An Approach to Modeling Web Service QoS and Provision Price. In *Proceedings of the Fourth International Conference on Web Information Systems Engineering Workshops* (WISEW'03).
10. Sravanthi Kalepu, Shonali Krishnaswamy, and Seng Wai Loke. Reputation = f(User Ranking, Compliance, Verity). In *Proceedings of the IEEE International Conference on Web Services* (ICWS'04).
11. Daniel A. Menasce. QoS Issues in Web Services. IEEE INTERNET COMPUTING, 2002.
12. QoS for Web Services: Requirements and Possible Approaches. http://www.w3c.or.kr/kr-office/TR/2003/ws-qos/.
13. S.Kamvar, M.Scholsser, and H.Garcia-Molina, The EigenTrust Algorithm for Reputation Management in P2P Metworks. In *Proceedings of international World Wide Web conference* 2003.

Performance Evaluation of Mobile Agents for Knowledge-Based Web Information Services

Yung Bok Kim[1] and Soon Woo Lee[2]

[1] Department of Computer Engineering, Sejong University, KunJa-Dong, Kwang-Jin-Ku, Seoul, Korea 143-747
yungbkim@sejong.ac.kr
[2] Korea Electrotechnology Research Institute (KERI), Uiwang-City, Gyeonggi-Do, Korea 437-808
rheesw@keri.re.kr

Abstract. Knowledge-based Web information system should be considered for performance with different types of mobile agents serviced by different mobile communication operators. The ubiquitous Web information server accessed by a user-group with various mobile agents should be a unified center for unified-and-ubiquitous (U^2) knowledge-based Web information services. We studied the performance of Web information access, i.e. the registration and retrieval of information/knowledge, with mobile agents for U^2 knowledge-based Web information services. We show the empirical results based on the implementation and experience got both in Korea and in Japan.

1 Introduction

The Web and its contents including UCC (User Created Contents) have been revolutionarily changing and affecting the world in various ways, especially toward the Knowledge and Information Society. Using the two ways, wired Internet and mobile Internet, unified-and-ubiquitous (U^2) knowledge-based Web information services for information access should be considered for convenience as well as integrity of consistent information in this Information Society. The Web server is a role center for unified information services; and the client mobile agents for Web information access have become very important for a user-group in ubiquitous computing environments. For performance evaluation of the knowledge-based Web information access from the unified information Web server in ubiquitous information network, we considered several aspects about the Web server for a user-group using Web services.

The performance of a worldwide web (WWW) server became a central issue in providing a ubiquitous, reliable, and efficient information network for U^2 knowledge-based Web information services. For wired Internet using HTML with agents in PCs, and for mobile Internet using WML or mHTML with mobile agents in mobile devices, the management of Web server agents and mobile agents becomes more difficult to provide Quality of Service for users. We have studied the performance evaluation of Web information access for U^2 knowledge-based Web information services, using the unified

N.T. Nguyen et al. (Eds.): KES-AMSTA 2007, LNAI 4496, pp. 209–218, 2007.

Web information portal with inexpensive Web server and mobile agents. When browsing information on large Web sites, users often receive too much irrelevant information. The vast amount of irrelevant information on most large Web sites can overwhelm users, and the study about personalized Web views for multilingual Web sources was introduced by Liu et al. [1]. To avoid users browsing against over thousand of results, Ruvini [2] introduced the study of adapting to the user's Internet search strategy. The amount of knowledge and information in the Web has been growing tremendously and pushing in a sense a flooding society with knowledge and information; however, the searching the right information/knowledge in Web portals has become more difficult in other sense because of the amount of answers and the inconsistency of answers provided by various multi-agent portals.

A multi-agent system aiding information retrieval in Internet using consensus methods was researched by Nguyen et al. [3, 6] for reconciling inconsistency of answers generated by agents for the same query. Sobecki [4] studied the consensus-based hybrid adaptation of Web systems user interfaces, and the hybrid recommendation was a combination of three methods: demographic, content-based and collaborative. Ontology can be treated as the background of an information system; and conflicts on ontologies were classified and proposed for solving conflicts [5]. Considering mobile agents, the above concepts for information retrieval from multi-agent server systems to multi-agent client browsers of mobile devices can be expanded further in ubiquitous computing and networking environments.

If we consider the birth and death of knowledge-based information, the inconsistency of the searched results by different agents become more serious. Because the registration time of newly created information/knowledge into multi-agent search systems are very different from agent-system to agent-system, e.g. 1~7 days or more for registration in Web portals: Yahoo, Google, Naver, Daum, Empas, and other domestic Web portals. Therefore we need to consider the lifetime of information/knowledge, i.e. from birth to death of knowledge-based information, for processing with multi-agent servers and multi-agent clients environments.

We studied performance evaluation of mobile agents for Web information accessibility in the U^2 Web information services. We considered simplification of the above complexity at the user's perspective, moreover at the user-group's viewpoint about the real-time services for effective and efficient information access, i.e. the real-time registration and retrieval of information/knowledge, in the ubiquitous computing environments. In the following sections, we introduce usability and accessibility in wired and mobile Internet, especially with mobile agents at the specific user-group's perspective. We will discuss the performance evaluation with some metrics in the Web information services, and we will discuss with the results from implementation of a U^2 knowledge-based Web information service.

2 Usability and Accessibility with Mobile Agents

The definition of usability for mobile devices was referred [7] from of ISO 9241-11: "The extent to which a product can be used by specified users to achieve specified goals with effectiveness, efficiency and satisfaction in a specified context of use." Ford and

Kotze [8] proposed that the interface design characteristics required to design interfaces that accommodated high power distance, high uncertainty avoidance, masculinity and short-term orientation would provide a more usable interface to all users; and the user interfaces designed to accommodate the above cultural dimensions and collectivism provide better performance.

We considered the collectivism with the application for a 'user-group' instead of application for uncorrelated individual user; because the value of knowledge-based information is different from user to user and is even time-variant to users. We assume that the knowledge-based information is meaningful to a specific user-group. We also considered conceptually the above cultural dimensions to implement a unified-and-ubiquitous Web information service. In their research, they tried to increase the general usability of user interfaces, depending upon the following hypotheses:

-User interfaces designed for high power distance will be more generally usable than interfaces designed for low power distance.

-User interfaces designed for high uncertainty avoidance will be more generally usable than interfaces designed for low uncertainty avoidance.

-User Interfaces designed for masculinity will be more generally usable than interfaces designed for femininity.

-User interface designed for short-term orientation will be more generally usable than interfaces designed for long-term orientation.

-User interfaces designed for collectivism will be as usable as interfaces designed for individualism.

In many countries, a better access to Web services and Web administration is becoming an important issue; and a few action lines of improvement for the Web Content Accessibility Guidelines are suggested [9]. Universal access implies the accessibility and usability of Information Society Technologies (IST) by anyone, anywhere, anytime. It is important that the needs of the broadest possible end-user population are taken into account in the early design phases of new products and services. Universal Design in the Information Society has been defined as the conscious and systematic effort to proactively apply principles, methods and tools, in order to develop IST products and services that are accessible and usable by all, thus avoiding the need for a posteriori adaptations or specialized design [10].

The unified Web server for any user-group should have the capability of showing the appropriate contents, i.e. the HTML contents for wired Internet as well as the mobile contents for many different kinds of mobile devices, e.g. WML, mHTML etc. We implemented the simplified and unified portal of a multilingual Information Network for agents in PCs and wireless Internet for mobile agents, in ubiquitous Web information services. We used a single Web server in a ubiquitous information network for the unified portal service for the simplicity of management and the cost-effectiveness for a user-group. This method gives the effectiveness and efficiency for the notification of information and utilization of resources, in terms of the bandwidth for communication and the size of required disk storage for information to store.

To access pervasively the unified portal for a user-group in the time-critical ubiquitous information network, the user interface for accessing Web should be convenient and unified even for typing-in the domain names or URLs with mobile agents; because the just first step for Web service with wired/mobile Internet (especially, mobile case) is typing-in the URL of the Web site offering the requested information. Based on the assumption of single alphabet character; it is analyzed at the user-group's viewpoint for accessing Web information. Because we assume that the knowledge-based information is meaningful to a specific user-group level. For writing the information in Web information system, the user's typing speed of characters is one of important performance factors, especially using mobile agents for text-string URLs and information.

For users in the time-critical information network, even the input of text-string becomes important for retrieval of information or registration of information with mobile agents, especially with keypads in the mobile agent for text-string URLs as well as information. To access the unified portal ubiquitously, the user interface for a user-group should be as convenient as possible even for typing-in the domain names or URLs, or information. For writing the information in real-time way, the user's typing speed of text-string is one of important performance factors, especially with the mobile agents.

3 Performance Evaluation of Web Information Access

Performance evaluation of user modeling servers under real-world workload conditions was studied by Kobsa and Fink [11]. The performance evaluation for Web accessibility at the user-group's viewpoint may be different from the conventional evaluation methodology. However, we should consider this approach, because the environments have been changed a lot, especially in terms of the interactivity of users with mobile agents in the Web information system. In our research, we studied the performance evaluation for a Web information access with single-character domain names as keys for a U^2 knowledge-based Web information service, at the user-group's viewpoint. The time-critical application, getting required information as well as writing information in the wired Internet and mobile Internet environments, will be considered.

We studied the important performance metric, delay, at the user's perspective for U^2 knowledge-based Web information services. We studied the performance metric, delay, not only with the time in the network and server, but also with the spent time by a user and the input time with mobile agents for URL or information for notification in a specific user-group. For example, for a mobile user in a specific user-group, we assume that the random variables, the round-trip response time for a user's single interaction in a session of mobile agent, from user to the contents in DB through wired/mobile Internet before next interaction with mobile agent is $\boldsymbol{R}$. That is composed of the preparation time for any user in a ubiquitous information network to get a mobile device in his hand is $\boldsymbol{U}$. The time spent by the user with the mobile agent to do appropriate action for a service is $\boldsymbol{D}$. The aggregate time to the Web server after the mobile agent through wired/mobile Internet for a mobile service is $\boldsymbol{S}$ (the network time is embedded here). The agent time depending upon mobile contents is $\boldsymbol{C}$.

The session time of mobile agent may be dependent on the retrieval or registration of contents, and there may be several back and forth iterations. The returning round trip time, from the content retrieval time to the requesting user through Web server agent and wired/mobile Internet using mobile agent, is $\mathbf{R_2}$. Among the above random variables, i.e. the performance metrics, (***U, D, S, C***) for mobile agent and PC agent, the most dominating factor, i.e. the random variable, may be different from person to person.

At first, for the fair comparison, we assume that the same person is using the same contents in the same server agent with the same mobile agent. Then, we can order the dominating random variables, after estimation with an implementation of information network with the unified server, http://ktrip.net [12]. The previous works for computer networking have been mainly focused on the analysis of the random variable time, ***S***, as well as other individual random variable times, but we suggest the averaged overall performance metric for a user-group. For example, averaged delay, i.e. the response time Mean(***R***), instead of the partial and minor delay *S*, for all users in the time-critical ubiquitous information network. The user-group preparation time or ubiquity metric in the information network, Mean(***U***) will be decreasing depending upon the proliferation of ubiquitous computing and networking environments.

For the time-critical application using wired Internet and mobile Internet, the dominating factor and the variance of that averaged random variable should be bounded within the deterministic response time. To be deterministic for time-critical application, the user-group should be skilled, the user-group UI agent should be convenient, the network should be stable if possible, the server agent should be efficient and have high performance for the dedicated application, and finally the contents for information processing by agents should be simple as possible with simplified and efficient format for agents. We also considered the packet size to be independent from the network traffic condition as well as the simple service by the Web server agent to be independent from the load of the Web server in a U^2 knowledge-based Web information service for a user-group.

Let's consider the average of the keypad press number for the case of handheld phone model, and this is related to the device time Mean(***D***) with mobile agents. Based on the assumption of single multilingual character composed of one consonant and one vowel; this is analyzed at the user's viewpoint for a user-group. For writing the information in real-time way, the user's typing speed of multilingual character is one of important performance factors in any mobile Internet services with the mobile agent. For example with the simplest multilingual character composed of one consonant and one vowel, several Korean handheld phone models showed that the average number of keypad pressing is around 4~5 for the mentioned single Korean character and around 2~3 for single English alphabet, including shift key for next character. This is also serious dominating factor related to the time, ***D***, especially for writing contents of information with mobile agents. We studied to order the dominating factors in the overall performance at the user-group's perspective; the relationship of the dominating factors will be discussed in the following section. Firstly, we assume the information retrieval from a single server agent.

In Fig.1, the overall time delay by several elements (i.e. user, mobile agent and Web server agent) are shown and broken down as follow; user-group's preparation time, $\boldsymbol{U_1}$; agent time with client's device, $\boldsymbol{D_1}$ and $\boldsymbol{D'_1}$; time in network and Web server agent, S_1 and S'_1; user-group's time for understanding and readiness, $\boldsymbol{U_2}$; agent time with

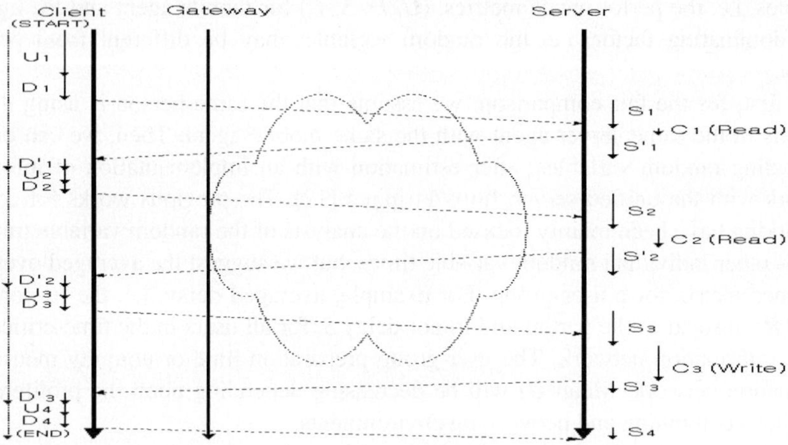

Fig. 1. Overall Time Delay for a Session with Mobile Agents

client's device, $\boldsymbol{D_2}$ and $\boldsymbol{D'_2}$; time in network and by Web server agent, $\boldsymbol{S_2}$ and $\boldsymbol{S'_2}$; time for reading contents by the DB server agent, $C_{1(\text{Read})}$; time for writing contents by the DB server agent, $C_{1(\text{Write})}$; user's preparation time for understanding and readiness, $\boldsymbol{U_3}$; time for writing contents with user-group's mobile agents, D_3 and D'_3 ; time in network and Web server agent, S_3 and S'_3; user-group's preparation time for understanding and readiness, U_4; time for finishing the session with user-group's mobile agents, D_4 ; time in network and Web server agent, S_4 .

From the Fig.1, a performance metric, the response time for ith transaction at the user-group's viewpoint, $r_i = U_i + D_i + D'_i + S_i + S'_i + C_i$; and the overall performance metric, i.e. the overall delay for a session in information access is $\sum_{i=1}^{n} r_i$. In this session the dominating factor is $\boldsymbol{U_i}$ and $\boldsymbol{D_i}$ in normal network and server, considering the recent stable network. We considered the packet size to be independent from the network traffic condition as well as the simple service in the Web server to be independent from the load of the Web server agent.

Secondly, if we assume the information retrieval from multi-agents server, then the consensus methods [3,4,5,6] should be considered, and the other network and server agent times: $\boldsymbol{S_j}$, $\boldsymbol{S_k}$,…, the contents times: $\boldsymbol{C_j}$, $\boldsymbol{C_k}$,… and the reconciling time should be added. However, we focus on a U^2 Web information service for a user-group to provide a real-time QoS.

For the time-critical application with knowledge-based information using wired Internet and mobile Internet, the dominating factor and the variance of that averaged random variable should be bounded within the deterministic response time. To be deterministic for time-critical application with knowledge-based information, the user-group should be skilled, the user-group UI agent should be convenient, the network should be stable if possible, the server agent should be efficient and have high performance for the dedicated application, and finally the contents for a U^2 knowledge-based Web information service should be simple as possible with simplified and efficient format for processing by mobile agents.

The bandwidth requirement for wireless or mobile Internet should be as little as possible to be immune to the network traffic condition for information processing of text-string; also that will be good in terms of degradation caused by the other rich multimedia contents. The user-group's averaged preparation time Mean(***U***) will be shortened depending upon the proliferation of ubiquitous computing devices, i.e. mobile agents for U^2 knowledge-based Web information services.

4 Implementation and Empirical Results with User-Group

The information networking for U^2 knowledge-based Web information services, here Korean information network, as an example, is based on wired or mobile Internet, many single multilingual (e.g. Korean) character domain names for fast access of the required domain name with a single character. The required information or advertisement for a user-group can be registered in any time and any place using wired or mobile Internet in the unified Web server for a ubiquitous information service, i.e. the 'ktrip.net'. The size of Web page for a U^2 Web information service was considered below 1.5Kbyte, i.e. between 500Bytes and 1.5Kbytes of compiled WAP binary, to minimize the dependency of the overall performance to the shared and stochastically varying network traffic; also the Web server is dedicated to minimize the server load, and dedicated for the information system for an information networking as an example of a user-group of a U^2 knowledge-based Web information service.

As an example for a U^2 knowledge-based Web information service with various mobile agents serviced by many mobile service operators, we used the following service program based on Microsoft IIS Web server and ASP for various mobile agents of a user-group. With this program for the Web information access by a user-group, the information portal, http://ktrip.net can be accessed in a unified way by different mobile agents for a user-group.

```
<%  ' Infomation of  Mobile Agent Header and for Web Server: MS IIS and ASP
 agent = Request.ServerVariables("HTTP_USER_AGENT")
 subno = Request.ServerVariables("HTTP_X_UP_SUBNO")
        If InStr(agent,"SK") >= 1 Then  'for Mobile Agent 011
              response.Redirect "http://ktrip.net/listwml.asp"
        ElseIf InStr(subno,"itouch") >= 1 or InStr(subno,"ezweb") >= 1 Then
            response.Redirect "http://ktrip.net/listhdml.asp" 'for Mobile Agent 019
        ElseIf InStr(agent,"MSMB") >= 1 Then 'for Mobile Agent 016
              response.Redirect "http://ktrip.net/listm.asp"
        ElseIf InStr(agent,"Mozilla") >= 1 Then 'for other Browsers with wired Internet
              response.Redirect "http://ktrip.net/list.asp"
        End If  %>
```

The speed of time-critical registration of any advertisement as well as the speed of accessing a special information for various communities is fast enough for a real-time application in U^2 knowledge-based Web information services.

From the empirical results of the mean and standard deviation of 100 samples, we could observe that the response time with a wired PC is fastest and stable with a little deviation, the averaged response time with a mobile agent in mobile Internet is around 12 seconds with about 2 second standard deviation. The size of Web page for the wired Internet accessed by the domain name ktrip.net was about 5 Kbytes, and the size of the mobile Web page (not-compiled) was about 1.5Kbytes, that became about 1Kbytes after compiling to WAP binary file at the WAP gateway. We could observe also that the network and Web server agent response time for Ping command was much shorter than the contents retrieval time by the Web server agent. The mobile 1.5Kbyte content retrieval time by mobile Internet with a mobile agent was about 10 seconds longer than by the wired Internet 5Kbyte contents retrieval time with a PC because of the elapsed time with the gateway, and this time was related to the network time (in WAP gateway and in information network) instead of agent time in Web server.

Considering the performance of the unified portal in the information network, we could make the processing time deterministic in the Web server for contents, where the deterministic time was possible with the deterministic size of packet, below around 1.5Kbytes, i.e. below one WML deck size even with the old models for mobile agents. Referring to the previous section, we could get the relationship between $\boldsymbol{S}$ and $\boldsymbol{C}$. With the PC using wired Internet, the time $\boldsymbol{S}$ may be considered rather short period (around 5~30 msec with Ping, which is related to the $\boldsymbol{S}$; but with 5Kbytes Web page for PC the response time is around 2~3 seconds, which is related to the $\boldsymbol{S}$ and $\boldsymbol{C}$, here $\boldsymbol{C}$ is much longer than $\boldsymbol{S}$). With recent mobile agent using mobile Internet (for short packets below 1.5Kbytes and even around 5Kbytes), the response time is around 12 seconds with a little deviation through the WAP gateway; therefore the time $\boldsymbol{S}$ is longer than $\boldsymbol{C}$, where $\boldsymbol{S}$ includes the elapsed time at the gateway in the mobile Internet.

We may order the dominating factors in the overall performance at the user-group's perspective as follows. In general, the relationship for mobile Internet with mobile agent could be Mean($\boldsymbol{U}$)>Mean($\boldsymbol{D}$)>Mean($\boldsymbol{S}$)>Mean($\boldsymbol{C}$). Here, we need to decrease the major times: Mean($\boldsymbol{U}$) and Mean($\boldsymbol{D}$), as well as the network and server agent time Mean($\boldsymbol{S}$) (or access time for Contents in DB, Mean($\boldsymbol{C}$)). We need to try continuously to decrease the times Mean($\boldsymbol{U}$) and Mean($\boldsymbol{D}$) in the time-critical information services for a user-group; as Internet URLs for the unified Web services, we used around 300 multilingual single-character.net as URLs to find information as well as to notify information in real-time way and ubiquitously for a U^2 knowledge-based Web information service. We can also consider speech technology to decrease the time Mean($\boldsymbol{D}$) instead of text-based information processing by a specific user-group.

The handheld phone model SCH-X600D manufactured by Samsung was used for testing of an international roaming service as well as testing of a mobile Internet service in July 2006, in Japan. In Tokyo, Kyoto and Osaka, the primitive experiment of a U^2 knowledge-based Web service for soft-realtime access to information, i.e. reading and writing information in anytime and anywhere, was studied. Even in Japanese express train 'Sinkansen' moving around 300Km/hour, the registration of U^2 Web information with size of around 100 bytes was tested.

The Short Message Service (SMS) was not provided by the service operators, therefore there was no other way to send or register information with mobile phones. The reading of U^2 knowledge-based Web information with size of around 1 Kbytes, of course, was easy and took similar amount of time as in Korea. In Korean express train 'KTX' (i.e. **K**orea **T**rain e**X**press) the reading and writing of U^2 Web information has been possible since two years ago. The time ***U*** is almost negligible because we carry always our handheld phones being able to connect to mobile Internet, and the ubiquitous computing and networking environments prevail gradually.

From the experiment in Japan, we could observe that the response time at wired PC is rather fast and stable with a little deviation as in Korea. The averaged response time with mobile agent for the first access to 'ktrip.net' was around 12[sec] with a little deviation as in Korea. After initial connection to 'ktrip.net', the reading time of 1 Kbyte registered information was around 2~3[sec]. As concluding remarks, the critical time was the device time ***D*** with mobile agent in our experiment, similarly as in Korea. The summated time (***S+C***) was around 2~3[sec] and was not comparable to the time ***D*** that is at least over 30~60[sec] depending upon the amount of text-based information for writing with keypads during the registration of information. Because the inconvenient interface for writing URLs or information/knowledge with mobile agents was the major bottleneck for degradation of overall performance in U^2 knowledge-based Web information services.

5 Conclusions

The performance evaluation for Web information accessibility with mobile agents was presented for a unified-and-ubiquitous (U^2) knowledge-based Web information service beyond information retrieval. The overall performance evaluation at a user-group's perspective showed that the critical factors in multi-agent system environments for knowledge-based Web information services are the UI with mobile agents and registration of knowledge/information for a specific user-group. With expanding the ubiquitous computing environments (i.e. the decrease of the access time to mobile agents), the interaction time between mobile agents and a user-group will become more critical especially for real-time and unified registration and retrieval of information; therefore we need more efficient UI with mobile agents for U^2 knowledge-based Web information services in the ubiquitous computing environments.

References

1. Liu, Z., Ng, W.K. and Lim, E.P.: Personalized Web Views for Multilingual Web Sources. IEEE Internet Computing, pp. 16-22, July/August 2004
2. Ruvini, J.D.: Adapting to the User's Internet Search Strategy. UM2003, LNAI 2702, pp. 55-64, 2003

3. Nguyen, N.T., Blazowski, A. and Malowiecki, M.: A Multi-agent System Aiding Information Retrieval in Internet Using Consensus Methods. SOFSEM 2005, LNCS, pp.399-402, 2005
4. Sobecki, J.: Consensus-Based Hybrid Adaptation of Web Systems User Interfaces. Journal of Universal Computer Sciences, vol. 11, no.2, pp.250-270, 2005
5. Nguyen, N.T.: Conflicts of Ontologies-Classification and Consensus-Based Methods for Resolving. KES 2006, Part II, LNAI 4252, pp.267-274, 2006
6. Nguyen, N.T., Ganzha, M. and Paprzycki, M.: A Consensus-Based Multi-agent Approach for Information Retrieval in Internet. ICCS 2006, Part III, LNCS 3993, pp.208-215, 2006
7. Betiol, A.H. and Cybis, W.A.: Usability Testing of Mobile Devices: A Comparison of Three Approaches. INTERACT 2005, LNCS 3585, pp. 470-481, 2005
8. Ford, G. and Kotze, P.: Designing Usable Interfaces with Cultural Dimensions. INTERACT 2005, LNCS 3585, pp. 713-726, 2005
9. Duchateau, S., Boulay, D., Tchang-Ayo, C. and Burger, D.: A Strategy to Achieve the Accessibility of Public Web Sites. ICCHP 2002, LNCS 2398, pp.58-60, 2002
10. Stephanidis, C. and Emiliani, P.L.: Universal Access to Information Society Technologies: Opportunities for People with Disabilities. ICCHP 2002, LNCS 2398, pp.8-10, 2002
11. Kobsa, A. and Fink, J.: Performance Evaluation of User Modeling Servers under Real-World Workload Conditions. UM2003, LNAI 2702, pp.143-153, 2003
12. U^2 Web Information Service Portal (Knowledge-based Information Web site): http://ktrip.net

An Agent Based Method for Web Page Prediction

Debajyoti Mukhopadhyay, Priyanka Mishra, and Dwaipayan Saha

Web Intelligence & Distributed Computing Research Lab,
Department of Computer Science & Engineering, Techno India,
West Bengal University of Technology,
EM 4/1, Sector V, Salt Lake, Calcutta 700091, India
{debajyoti.mukhopadhyay, priyanka147, dwaipayansaha}@gmail.com

Abstract. Studies have been conducted on pre-fetching models based on decision trees, Markov chains, and path analysis. However, the increased uses of dynamic pages, frequent changes in site structure and user access patterns have limited the efficacy of these static techniques. One of the techniques that are used for improving user latency is Caching and another is Web pre-fetching. Approaches that bank solely on caching offer limited performance improvement because it is difficult for caching to handle the large number of increasingly diverse files. An agent based method is proposed here to cluster related pages into different categories based on the access patterns. Additionally page ranking is used to build up the prediction model at the initial stages when users are yet to invoke any page.

1 Introduction

The exponential proliferation of Web usage has dramatically increased the volume of Internet traffic and has caused serious performance degradation in terms of user latency and bandwidth on the Internet. The use of the World Wide Web has become indispensable in everybody's life which has also made it critical to look for ways to accommodate increasing number of users while preventing excessive delays and congestion. Studies have been conducted on pre-fetching models based on decision trees, Markov chains, and path analysis [1] [2] [4]. There are several factors that contribute to the Web access latencies such as: server configuration, server load, client configuration, document to be transferred, network characteristics.

Web Caching is a technique that made efforts to solve the problem of these access latencies. Specially, global caching methods that straddle across users work quite well. However, the increasing trend of generating dynamic pages in response to HTTP requests from users has rendered them quite ineffective. The following can be seen as the major reasons for the increased use of dynamic Web pages:

(1) For user customized Web pages, the content of which depends on the users' interests. Such personalized pages allow the user to reach the information they want in much lesser time. (2) For pages that need frequent updating, it is irrational to make those changes on the static Web pages. Maintaining a database and generating the content of the Web pages from the database is a much cheaper alternative. Pages displaying sports updates, stock updates, weather information etc. which involve a lot

N.T. Nguyen et al. (Eds.): KES-AMSTA 2007, LNAI 4496, pp. 219–228, 2007.

of variables are generated dynamically. (3) Pages that need a user authentication before displaying their content are also generated dynamically, as separate pages are generated as per the user information for each user. This trend is increasing rapidly. (4) All response pages on a secure connection are generated dynamically as per the password and other security features such as encryption keys. These pages expire immediately by resetting the Expire field and/or by the Pragma directive of 'nocache' in the HTTP header of the server response, to prevent them from being misused in a Replay attack.

As the Internet grows and becomes a primary means of communication in business as well as the day-to-day life, the majority of Web pages will tend to be dynamic. In such a situation, traditional caching methods will be rendered obsolete. The dynamic pages need a substantial amount of processing on the server side, after receiving the request from the client and hence contribute to the increase in the access latency further. An important pre-fetching task is to build an effective prediction model and data-structure for predicting the future requests of the user and then sending those predicted requests to the user before he/she actually makes the request.

2 Proposed Method

2.1 Prediction Model

Links are made by Web designers based on relevance of content and certain interests of their own. In our method, we classify Web pages based on hyperlink relations and the site structure. We use this concept to build a category based dynamic prediction model. For example in a general portal www.abc.com all pages under the movies section fall under a single unique class. We assume that a user will preferably visit the next page, which belongs to the same class as that of the current page. To apply this concept we consider a set of dominant links that point to pages that define a particular category. All the pages followed by that particular link remain in the same class. The pages are categorized further into levels according to the page rank in the initial period and later, the users' access frequency [6] [7] [8].

The major problem in this field is that, the prediction models have been dependent on history data or logs [5]. They were unable to make predictions in the initial stages [3]. We present the structure of our agent based prediction model in Figure 1. Our method is not dependent on log data, rather it is built up using ranking of pages and updated dynamically as HTTP requests from the users arrive [6]. To begin with, HTTP requests arrive at the Predictor Agent. The Predictor Agent uses the data from the data-structure for prediction, and after predicting the forthcoming requests, passes the requested URL to the Update Agent to update the data-structure.

In our prediction model shown in Figure 2, we categorize the users on the basis of the related pages they access. Our model is divided into levels based on the popularity of the pages. Each level is a collection of disjoint classes and each class contains related pages. Each page placed in higher levels has higher probability of being predicted.

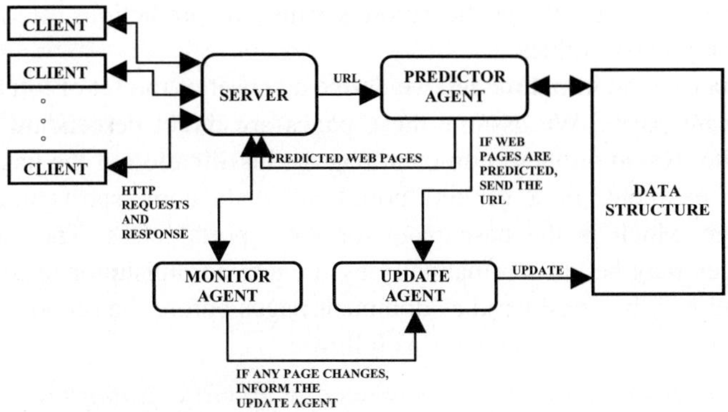

Fig. 1. Proposed Structure of the Agent Based Prediction Model

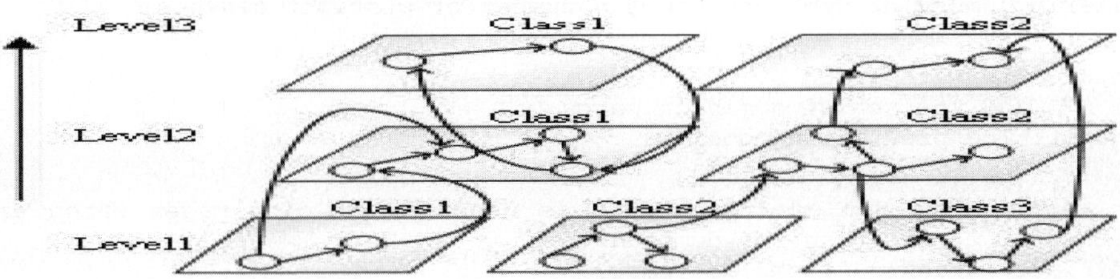

Fig. 2. The Prediction Model 'T'

Mathematically the model may be represented as:

Let T is the Prediction Model which is a logical reconstruction of the site graph. T is composed of discrete classes each containing some URLs,

Let $C = \{C_1, C_2, C_3, \ldots, C_n\}$ is the set of classes, where n= Number of classes.

For every element C_i in C there exists,

$CU_i = \{U_1, U_2, U_3, \ldots, U_m\}$ which is a set of URLs in plane C_i

And $i = 1,2,\ldots,n$, and $k \neq j$ for all U_k, U_j

belonging to C where $k, j = 1,2,\ldots,m$

Also,

$$\bigcap_{P=1}^{P=n} C_p = \{\ \}$$

$$\bigcup_{P=1}^{P=n} C_p = T$$

Each C_i in C has its own level number.

The connotation of 'class' can be defined as follows, when a user requests for a particular URL the next few pages have higher probability to have the same 'class' in which the previous page belongs. So before starting the predictions the URLs must be assigned to their 'class' values.

For constructing the initial model, we define a subset of the set of total pages in the site as dominant pages. We assume these pages are direct descendant of the home page of the site. Based on these dominant pages, classification of the pages in the site is done. For example, in a general portal www.abc.com, sports.html may be a dominant page which is the base page for the 'sports' class. The candidates for dominant pages may be chosen manually by the site administrator or all the links in the home page may be considered as dominant pages before the server is started. The algorithm to create the initial model is as follows:

```
Let S= (U, L) is the digraph representing the website where U is the set
of URLs and L is set of links. The set of dominant-pages D is the subset
of U.

Clearly number of dominant pages = number of distinct classes.

Input: The site graph S.
       and the set of dominant pages D.
Output: The prediction model T.

An empty array called common-page is used for holding pages which are
linked from more than one dominant page.
Initially stack1, stack2 are empty.

1: Based on the page ranks, assign level numbers to pages.
2: Put all the elements of D in stack1, and assign them a unique class
   number.
   3: while(stack1 and stack2 are not empty),
      If(stack1 is empty)
        Pop all the pages from stack2 and push back to stack1.
      end if
   pop the first element from Stack1, name it P.
    for(all pages pointed by P)
      if(any page is already assigned a class number) then
         if(class number is same as P) then
         do nothing
      else
        add that page to common-page.
      end else
     else
       a. assign the class number of P as the class number of that page.
       b. push the page to stack2
     end else.
    end for
   end while.
4: for(each page in common-page)
     Reassign the class number same as that of the class having maximum
     number of links pointing to it. For any further conflict (i.e., if
     two or more classes have same number of links to this page) choose
     any random one.
```

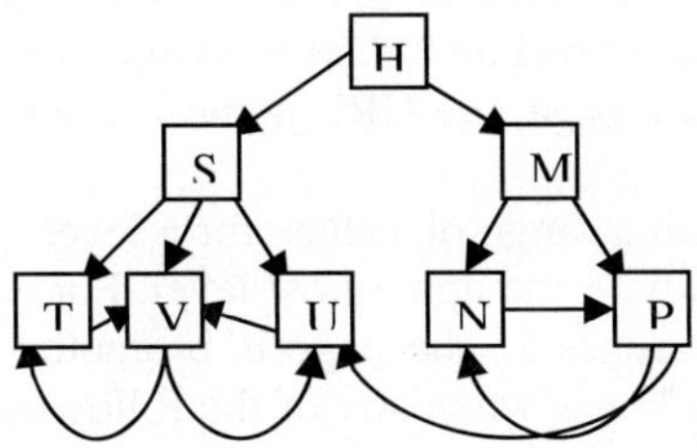

Fig. 3. Flow Diagram of the Algorithm

The above algorithm can be depicted as shown in Figure 3. Here, H is the home page. and two dominant pages are S and M. Now S is marked with class number CS and M is marked with class number CM where CS ≠ CM.

1. Initially push M, S into the stack1.
   stack1: M, S. stack2: void
2. Pop S and assign T, U and V the same class number as S i.e., CS and push them in stack2.
   stack1: M. stack2: T, U, V.
3. Pop M and assign N, P the class number CM. Push N, P in stack2.
   stack1: void. stack2: T, U, V, N, P.
4. Pop all the elements from stack2 and push back to stack1.
   stack1: P, N, V, U, T. stack2: void.
5. Since all the neighbors of T, U, V and N are already marked with class number and no conflict arises, they are popped one by one.
   stack1: P, stack2: void.
6. Pop P from stack1. The neighbor of P is N and U. But the 'class' number of P is equal to that of N, hence no action is taken. Since U is already marked with a 'class' number and that class number is not equal to the 'class' number of P, there is a conflict. So U is put in the common-page array.
   stack1: void. stack2: void.
7. Since stack1 and stack2 are both empty, we are entering step 4 of the algorithm. The common-page contains U. Since from CS class, it has two back links (from S and V) and from CM class it has only one back link (from P only), U will retain its class number CS.

The disjoint classes signify the categorization of the pages accessed by the users. Each level signifies the possibility of the page to be accessed in the future. Higher the level higher is the possibility to be accessed. The pages in the various classes are promoted to the higher levels based on the number of accesses to that page by the user. The next request for a page is predicted according to its presence in a higher level than the current page that points to it. More than one page is predicted and sent to the user's cache depending upon the presence of links in the higher levels.

After calculating page ranks, a normalized value in the range of 1 to p is assigned to each page where p is the number of pages in the site. For storage reasons, the number of levels is restricted to a predefined constant value L, where typically $L=\lceil\sqrt{p}\rceil$. We further divide the p pages into L sets. For each set, classes are formed depending upon the actual links present between them. Thus pages are categorized into disjoint classes "C." Each level and class is assigned a distinct number. In order to search for the presence of a page, the URL name is used as a key to the hash table data- structure.

Since we are working with a range of values for a level, we assign a counter to all the pages except those already in the uppermost level. For each request the counter is incremented, and when it reaches L, the page is promoted to the next higher level. Pages may traverse between levels when any of the following conditions occur: a) the page is demoted to a lower level when the time stamp value assigned to it expires; b) the page is promoted to a higher level if it has been modified recently. This is discussed further in Sub-section 2.3.

2.2 Predictor Agent

All required information about the pages of the Website is indexed using their URLs in a hash table where a URL acts as the key. When a request is received, a search on the hash table is conducted and the information thus obtained is analyzed in the following manner:

1. Get the level and class number of the requested URL.
2. Get the links associated with the page and also fetch their respective levels and class numbers.
3. Determine Prediction-Value (P-value) pairs for the entire candidate URLs, where a P-value pair is defined as [Level, Rank].
4. Sort the links of the requested URL according to the four types of precedence relations between two P-value pairs (Li, Ri) and (Lj, Rj): (Li, Ri) <·(Lj, Rj); (Li, Ri) ·> (Lj, Rj); (Li, Ri) || (Lj, Rj); (Li, Ri) ≈ (Lj, Rj).
5. Compare the links' level number with the URLs' level number.
6. Compare the class numbers of the links with that of the requested URL. The link having the same class number will get preference.
7. The links in the higher levels are the predicted links to be sent to the users' cache.

2.3 Update Agent

In the updating process we adjust the counter value and decide whether the page should go to a higher level, class numbers are assigned at the initial stage and remain static. The process may be described as follows:

```
1. Check the local counter associated with the requested URL.
2. If the counter value is less than (L-1) then increment the
   counter
      Else
      Fetch the current level number of the URL. Let it
      be L.
      Increment L.
3. Reset the counter and time stamp.
```

The Update Agent also checks periodically for a page that is present in a higher level and has not been accessed for a long duration to relegate it to a lower level according to a predetermined threshold value. This periodic process compares the timestamp of all the pages with this threshold value and demotes those pages which exceed this threshold. There's another periodic check that checks the last date of modification of the page. If there is recent modification then the page is raised to a higher level. This is done as a recently modified page always has higher probability of being accessed by the user.

2.4 Monitor Agent

This module continuously monitors the server for any modifications in any of the pages and informs the update engine to make necessary changes in the Data Structure. For example if a page has been recently modified, it's DM (date of modification) should also be modified in the data-structure. Update Agent takes necessary actions to change the page's date of modification and promote it to one level higher. This is done as a recently modified page always has higher probability of being accessed by the user.

3 Experimental Setup

The prediction model is implemented using a link data-structure which is shown in Figure 4:

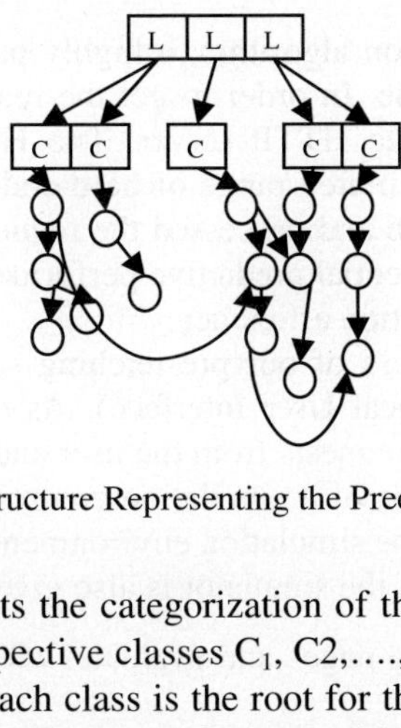

Fig. 4. Data-Structure Representing the Prediction Model

This data-structure represents the categorization of the URLs where the levels L_1, L_2,..Ln acts as index of the respective classes C_1, C2,….,C_n. Each Li where i = 1,…, n is the root for its classes and each class is the root for the respective URL trees. This data-structure is implemented in the form of a hash table with URLs being used as the key. Table 1 shows the implementation of the above data-structure.

Following are the brief description of each of the labels used in the data-structure: 'Key' represents the key of the current row in the hash table; 'URL' represents the Web address of the page; 'LC' represents the local counter associated with the page which represents the number accesses made to the page in a particular level. When this value reaches (L–1) the page is promoted to a higher level and this counter is reset; 'L#' is the level number and 'C#' is the class number; 'TS' is the timestamp associated with the URL that represents the duration for which the page has been in a particular level; 'DM' represents the last date of modification of the page; 'Links' represents the list of links to which the current URL points.

When a request is received the row for the requested URL is fetched from the hash table using URL as the key. The level and class numbers are obtained. The links corresponding to this URL are obtained from the 'links' field of the hash table. The class and level numbers of these links is obtained from their respective rows in the hash table. Thus we can make predictions on the basis of level, rank and class values of the linked URLs.

Table 1. Tabular Representation of the Data-Structure

Key	UR L	LC	L #	C#	TS	DM	Links
A1	A	2	1	2	Xx	Yy	A4,..
A2	B	0	1	2	Xx	Yy	A8,..
A3	C	2	1	3	Xx	Yy	A1,..
A4	D	1	2	3	Xx	Yy	A5,..
A5	E	3	2	4	Xx	Yy	A7,..
A6	F	1	2	4	Xx	Yy	A5,...

To implement the prediction algorithm, a highly parameterized prediction system is implemented in JAVA code. In order to get the real time HTTP request from the client, we developed a simple HTTP server. The HTTP server only supports the 'GET' operation and a very limited range of hard-coded MIME type. We hosted the web-server in our research lab and processed the requests. Finally, for the purpose of these tests, we focused on potential predictive performance (e.g., accuracy) and ignore certain aspects of implementation efficiency.

We evaluated the usefulness of our pre-fetching scheme using a simulator. This simulator has a GUI (Graphical User Interface). As described in the Section 1, we recorded the real time HTTP requests from the user and then showed the results of the prediction technique by means of a simulation environment. We coded a number of JAVA classes to implement the simulation environment.

A step by step execution of the simulator is also given below:

- It scans all the HTML pages and evaluates all the hyperlinks present in each of the pages.
- Based on these links it creates the site graph.
- It then applies the algorithm for creating the discrete classes onto the site graph.
- The simulator assigns unique numbers to the pages from the same classes in the Prediction Model.
- Now based on the site structure, it evaluates the rank of each HTML page.
- Based on these rank values the simulator now assigns the level value to each page. Higher rank implies higher level.
- After the creation of the initial model, the simulator starts the server.

For each request, the simulator updates the data-structure and displays the results in the graphical users interface.

4 Experimental Results and Evaluations

We examined the hit percentage vs. user session as per the prediction window size. Figure 5 is a snapshot of the server's session interface showing the hit/miss that occurred after each user request.

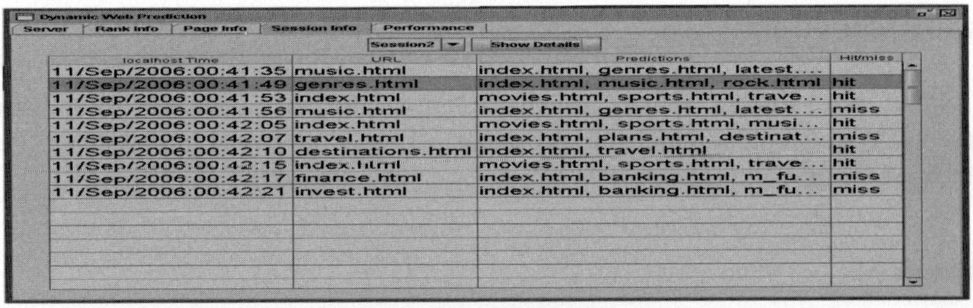

Fig. 5. Session Interface

The hit percentage remained consistent throughout the testing period including the initial stages. The size of the prediction window was taken as two and three considering the number of pages in our test environment. Size of a prediction window indicates the number of Web-Pages sent to the Client-cache by the Web-Server while predicting the pages. The average hit percentage was found to be around 35% with a prediction window size of 2 and 51% with a prediction window size of 3, an improvement of around more than 15%. In Figure 6, sessions recorded during the testing period at different intervals with variable prediction window sizes are plotted.

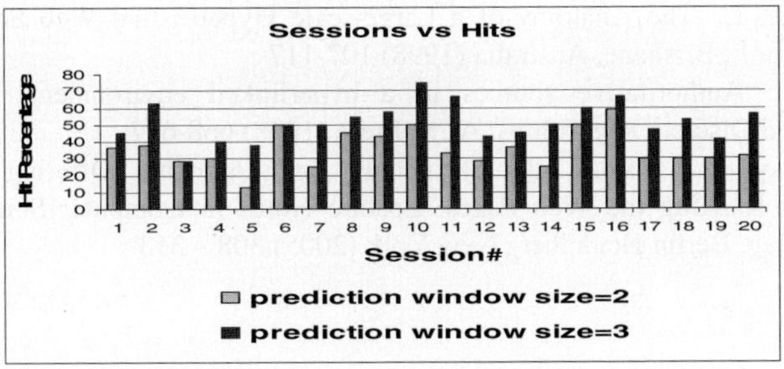

Fig. 6. Chart showing comparative hit ratio with different prediction window sizes

5 Conclusions

In most of the cases prediction of Web pages is done using logs and history data requiring a huge amount of memory to implement. Another problem found is the inability to build up the prediction model in the initial stages when no log or history data is available. The use of page ranking in our method enables to build our prediction model in the initial stages and make predictions right away. Henceforth our model updates itself as per the access patterns of users. Categorizing the users into different classes also help as we don't have to keep track of each user as all access patterns are maintained in the form of sessions. Updating the model dynamically according to access patterns of users as well as changes in the content of the Website is computationally cheaper as it doesn't put extra load on the Web traffic for requesting or maintaining extra information.

Acknowledgements

We extend our sincere thanks to the anonymous reviewers for their valuable comments, which helped improve the paper. This work was supported by Techno India (West Bengal University of Technology), Calcutta, India.

References

1. Chen, X., Zhang, X.: Popularity-Based Prediction Model for Web Prefetching. IEEE Computer. Vol.36, No.3 (2003) 63-70
2. Palpanas, T., Mendelzon, A.: Web Prefetching using Partial Match Prediction. Department of Computer Science, University of Toronto. Technical Report CSRG-376 (1998) 1-21
3. Davison, B.D.: Learning Web Request Patterns. Web Dynamics: Adapting to Change in Content, Size, Topology and Use. Springer-Verlag, Berlin Heidelberg New York (2004) 435-460
4. Bonino, D., Corno, F., Squillero, G.: An Evolutionary Approach to Web Request Prediction. 12th International WWW Conference, Budapest, Hungary (2003) poster S2
5. Su, Z., Yang, Q., Lu, Y., Zhang, H.: WhatNext: A Prediction System for Web Requests using N-gram Sequence Models. 1st Int. Conf. on Web Information System and Engineering (2000) 200-207
6. Brin, S., Page, L.: The Anatomy of a Large-scale Hypertextual Web Search Engine. 7th WWW Int. Conf., Brisbane, Australia (1998) 107-117
7. Kleinberg, J.: Authoritative sources in a hyperlinked environment. 9th ACM-SIAM Symposium on Discrete Algorithms, ACM Press (1998) 668-677
8. Mukhopadhyay, D., Biswas, P.: FlexiRank: An Algorithm Offering Flexibility and Accuracy for Ranking the Web Pages. Lecture Notes in Computer Science, Vol. 3816. Springer-Verlag, Berlin Heidelberg New York (2005) 308 – 313

X-Binder: Path Combining System of XML Documents Based on RDBMS

Bum-Suk Lee and Byung-Yeon Hwang

Department of Computer Engineering, The Catholic University of Korea
{bslee, byhwang}@catholic.ac.kr

Abstract. With the increasing use of XML, considerable research is being conducted on the XML document management systems for more efficient storage and searching of XML documents. Depending on the base systems, these researches can be classified into object-oriented DBMS (OODBMS) and relational DBMS (RDBMS). OODBMS-based systems are better suited to reflect the structure of XML-documents than RDBMS-based ones. However, using an XML parser to map the contents of documents to relational tables is a better way to construct a stable and effective XML document management system. The proposed X-Binder system uses an RDBMS-based inverted index; this guarantees high searching speed but wastes considerable storage space. To avoid this, the proposed system incorporates a path combining module agent that combines paths with sibling relations, and stores them in a single row. Performance evaluation revealed that the proposed system reduces storage wastage and search time.

1 Introduction

The Extensible Markup Language (XML) [1] is designed to maintain, transfer, and process general information on the Internet by a simple method. With the increasing use of XML, the need to conduct research in various XML-related fields is also increasing. In particular, much of the recent research is related to the efficient storage and retrieval of documents that use the XML format [2].

This research is divided on the basis of base systems into object-oriented DBMS (OODBMS) [3–6] and relational DBMS (RDBMS) [7–9]. OODBMS is suitable for XML's concept; however, the use of RDBMS provides greater stability to an XML document management system. This method requires a process for mapping the structure and content of an XML document to a relational table using an XML parser. Therefore, the number of researches involving RDBMS is gradually increasing.

The XRel [7] system was developed to exploit the existing relational systems for XML retrieval. This system suggests a relational mapping scheme and path searching method; however, it does not guarantee a high search speed, and the result of the performance test shows defects in partial match and complex queries. Furthermore, the size of its path table increases continuously because it stores all the paths.

In this paper, we suggest an RDBMS-based X-Binder system to solve these problems. The proposed system includes a path combining module as an agent and query

N.T. Nguyen et al. (Eds.): KES-AMSTA 2007, LNAI 4496, pp. 229–238, 2007.

translation module. The path combining module stores the sibling nodes having the same ancestor path in a single row and thereby prevents a rapid increase in the size of the path table. The system also includes a query translation module that converts a user query into an SQL format. The module utilizes XPath's linear path expressions (LPE) [10] to retrieve XML documents. The linear path expression can be classified as a full match, partial match, and complex query. These three query types are converted to SQL queries by the query translation module.

The X-Binder system applies an inverted indexing technique [11–12] to guarantee fast performance with respect to the search speed of a user query. An inverted index is easy to implement and has the merit of speed. On the other hand, this technique wastes storage space. The system maintains an index table of various items with the size of which could be up to three times the size of the original data file. The path combining module of X-Binder exploits the existing inverted indexing technique and avoids its defect.

The system showed a reduction in the number of rows of the stored path tables when the path combining module was used. It also exhibited a high performance in terms of the path searching speed in comparative tests. The query translation module utilizes XPath's LPE.

This paper is organized as follows: Section 2 introduces the XML query model used in X-Binder. In Section 3, we describe the system architecture and the path combining and query translation modules. Section 4 presents the results of our performance experiments. Finally, in Section 5, we discuss our conclusions.

2 Related Work

In this section, we define the XML query model. Fig. 1 shows an example of an XML document that is used to explain the proposed system.
The primary XML query languages—XPath [10], XQuery [13], and Quilt [14]— use a path expression to define queries regarding an XML tree. Both LPE and branching path expressions (BPE) used in this study are the most widely used path expressions in the field of XML research. *PathFinder(p)*, the path query function, is given path *p* as input, which is expressed as an LPE; it, in turn, outputs the documents and text values associated with the path.

```
<movie title="Old boy" year="2003" country="Korea">
  <director>
    <fullname>Chan-wook Park</fullname><nationality>Korean</nationality></director>
  <cast>
   <players>
        <player><role>Dae-su Oh</role><name>Min-sik Choi</name></player>
     <player><role>Mi-do</role><name>Hye-jeong Kang</name></player></players></cast>
  <genre>Action, Mystery, Thriller</genre>
  <comments>
     <comment>Old Boy is definitely my favorite movie ...</comment>
     <comment>I am a big fan of Asian cinema and ...</comment></comments>
</movie>
```

Fig. 1. Example of an XML document

Definition 1. *PathFinder(p)*, a path query function, is given path p as input and it returns the set of documents and text values associated with the path.

PathFinder(p)={(*document*, *text*) | *p* is used in *document* and *text* associated with the *p*}

Q1 and Q2 in Fig. 2 are two basic types of LPE. The symbol "/" in Q1 represents a parent-child relationship in the path. This type of a query is referred to as a full match query. The symbol "//" in Q2, indicates an ancestor-descendant relationship, and it represents a partial match query.

Q1 ::	/movie/director/fullname
Q2 ::	/movie//player/role
Q3 ::	/movie[/@year='2003']//player/name

Fig. 2. Linear path expression and complex query

Definition 2. A full match query is expressed as /l0/l1/l2…/ln, where l0 is the root of the document and li (i = 1, 2,…, n) is the ith label of the path. The relationship between li–1 and li is represented by "/"; this indicates that the path consists of a parent-child relationship only. ln is the end node.

Definition 3. A partial match query is expressed as $/l_0/l_1//l_2.../l_n$. The "//" symbol is included in this query at least once. The "/" symbol can also be included. This query is a path expression that includes an ancestor-descendant relationship in it.

Further, we define a complex query as a combination of both LPE and a content query. Q3 in Fig. 2 shows an example of a complex query.

Definition 4. A complex query is expressed as $/l_0[/l_1 = 'c_1']/l_2.../l_n$. In this query, the contents of l_0/l_1 include c_1, and it selects documents with the path $l_0/l_2.../l_n$. A complex query uses both "/" (full match query) and "//" (partial match query).

3 An Agent System for Combining the Paths of XML Documents

3.1 System Architecture

The system architecture of X-Binder involves a path combining module as an agent for storing XML documents efficiently, a query translation module for transforming a linear path query into an SQL query to search the RDBMS, and an inverted index management module to guarantee fast retrieval. The inverted index management module operates such that an ordered status of the index table is maintained.

Fig. 3 shows the system architecture of X-Binder. An XML document is provided as input to the database through the user interface (UI), and it passes through the path combining and inverted index management modules. When a linear path query is provided as input to search a stored XML document, the query is translated into an SQL query by the query translation module. After this process, the SQL query can be used to search the database, and the system displays the obtained search results through the UI.

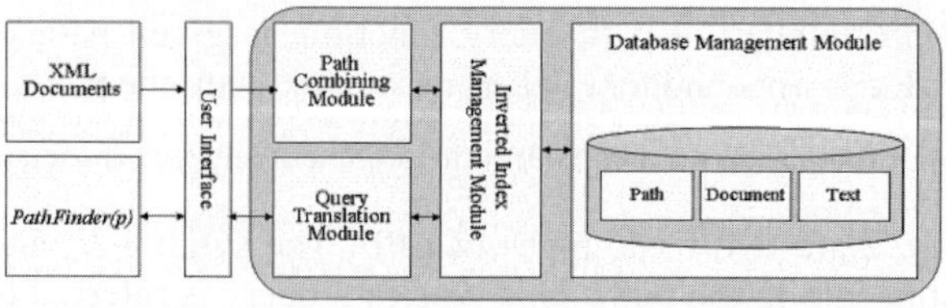

Fig. 3. System architecture of X-Binder

To use the path combining and inverted indexing techniques, we designed three relational table schemes with Document, Path, and Text tables, as shown in Fig. 4. In this database, docID, pathID, and textID are the identifiers for documents, paths, and text values, respectively. The main difference between X-Binder from XRel is that in XRel, the stored paths are combined with their siblings. XRel uses the LIKE operation to label the path match retrieval. The demerit of this method is that it results in an increase in the search time; this is because it uses a string match even when there is excessive storage data. X-Binder decreases the number of rows of a path table to be searched by combining paths, thereby providing efficiency in terms of storage space and search speed.

Table	Columns
Document	docID, docName
Path	pathID, CombinedPathexp
Text	textID, docID, pathID, endPath, value

Fig. 4. Table structure of X-Binder

Fig. 5 shows an example of a database that stores the XML document shown in Fig. 1. It stores the combined paths in the path table mentioned above. The combined paths use "#/" instead of "/" in the Path table, similar to XRel. The three tables serve as an inverted index relative to the other tables.

Text

textID	docID	pathID	endPath	value
1	1	1	fullname	Chan-wook Park
2	1	1	nationality	Korean
3	1	2	role	Dae-su Oh
4	1	2	name	Min-sik Choi
5	1	2	role	Mi-do
6	1	2	name	Hye-jeong Kang
7	1	3	genre	Action, Mystery, Thriller
8	1	3	@title	Old boy
9	1	3	@year	2003
10	1	3	@country	Korea
11	1	1	comment	Old Boy is definitely my favorite movie ..
12	1	2	comment	I am a big fan of Asian cinema and …

Document

docID	docName
1	\dataset\imdb\01.xml
2	\dataset\imdb\02.xml
…	…
…	…
10	\dataset\imdb\10.xml

Path

pathID	CombinedPathexp
1	#/movie#/director#(/fullname#|/nationality#)
2	#/movie#/cast#/players#/player#(/role#|/name#)
3	#/movie#(/genre#|/@title#|/@year#|/@country#)
4	#/movie#/comments#(/comment#)

Fig. 5. Example of a stored table of X-Binder

3.2 Path Combining Module

Few studies—1-Index [15], A(k)-Index [6], and DataGuides [16]—have investigated the construction of a small and fast index by combining similar paths, but these techniques are based on graph indexing techniques. These procedures combine nodes, not only the end nodes but also the middle nodes of a path. Therefore, they are not relevant to our research, which is based on RDBMS.

In this section, we describe a path combining module of X-Binder. The proposed method extracts a complete path from the root to the end node with a value, and it regards the paths with the same ancestor path (except the end nodes) as similar paths. The module combines these paths and stores them in a row in a relational table. If there exist attributes and elements with the same parent, the module adds an attribute identifier "@" to the head of an attribute name and recognizes the nodes as sibling nodes.

Definition 5. Combining the paths of XML documents refers to the binding of all the sibling paths (sibling paths: p_1, p_2, p_3, …, p_n) and storing them in the rows of a relational table.

PathCombine(sibling_paths) = {(*CombinedPathexp*) | *sibling_paths* are several sibling path inputs. *CombinedPathexp* represents a stored path expression that is a combination of the paths provided as inputs.}

For example, when the *PathCombine*(*sibling_paths*) operation is executed for the XML document shown in Fig. 1, '/movie/director/fullname' and '/movie/director/nationality' have the same relative ancestor paths and only their end nodes differ; thus, they have a sibling node relationship. Further, '/movie/genre', '/movie/@title', '/movie/@year', and '/movie/@country' are siblings, and also '/movie/cast/players/player/role' and '/movie/cast/players/player/name' are siblings. Lastly, the path '/movie/comments/comment' is stored in the table by itself because it does not have any sibling nodes.

By combining and storing sibling paths using this method, four path expressions are generated, as shown in the path table in Fig. 5. This result is comparable to the path table of the XRel system, which extracts fourteen path expressions. In Fig. 5, the "|" symbol is used to distinguish the end nodes from the combined paths.

Fig. 6 shows the algorithm of this method. We disregard the significance of the "/#" symbol in this explanation. The algorithm requires four inputs: fullPath, endPath, queryFullPath, and queryPrePath. For instance, if there exists a path '/aa/bb/cc' to be used for storage, '/aa/bb/(cc)', 'cc', '/aa/bb/(%cc%)', and '/aa/bb/(%' could be the four inputs. The first parts of two of them are the inputted values to be stored, and the others are required for searching the database.

The algorithm can process for the following three cases: (1) queryFullPath already exists in the Path table; (2) queryFullPath does not exist, but queryPrePath does; or (3) both queryFullPath and queryPrePath do not exist. In the first case, the algorithm merely returns the pathID of the row of the existing path. In the second case, a path similar to '/aa/bb/(dd|ee|ff)' exists; therefore, the algorithm modifies the Path table to '/aa/bb/(dd|ee|ff|cc)' and returns the pathID. Lastly, in the third case, a new input path is provided that does not exist in the table. Therefore, the algorithm stores this new

```
Input : fullPath, endPath, queryFullPath, queryPrePath
Output : pathID
Method :
    resultQFP = getStoredPath(queryFullPath);
    if(resultQFP != null){
        Return pathID;
    }else{
        resultQPP = getStoredPath(queryPrePath);
        if(resultQPP != null{
            CPathexp = combinedPath(resultQPP, endPath);
        updatePath(CPathexp);
        return pathID;
        }else{
            insertPath(fullPath);
            return pathID;
        }
    }
```

Fig. 6. Path combining algorithm

input path in the form of '/aa/bb/(cc)'; the parentheses at the end node are used for distinguishing it. This method separates '/aa/bb/cc/gg/(hh|ii)' from '/aa/bb/cc'.

3.3 Query Translation Module

It is impossible to directly search a relational database with an LPE to execute the *PathFinder(p)* function. Accordingly, to obtain information regarding the XML documents that are stored in an RDBMS, a query translation process that converts an LPE to an SQL query is required.

3.3.1 Full Match Query

A full match query is a type of linear path query that traverses from the root to the end node. Assuming that a query '/movie/cast/players/player/role' is provided as input, it has to be converted to an SQL query. To use the structure of the Text table, a query might be separated by using an end node 'role' and an ancestor path. Fig. 7 shows the SQL query of the input. The third line in Fig. 7 restricts the endPath of the Text table to 'role'; thus, information regarding 'name', which is stored with 'role', might be ignored.

```
1  SELECT  d1.docName, t1.value
2  FROM    Document d1, Path p1, Text t1
3  WHERE   t1.endPath = 'role'
4  AND     p1.CombinedPathexp LIKE '#/movie#/cast#/players#/player#/(%'
5  AND     p1.pathID = t1.pathID
6  AND     t1.docID = d1.docID
```

Fig. 7. An SQL query for *'/movie/cast/players/player/role'*

3.3.2 Partial Match Query

A partial match query contains more than one "//" symbol, which indicates an ancestor–descendant relationship, and it might not be suffixed with the root. We can easily convert this type of query to an SQL query by including conditions such as JOIN and LIKE operations. For instance, in the query '/movie//cast', the "//" symbol will select all paths that have 'cast' at a lower position in documents with the root 'movie'. It is easy to detect results by using the string match query method. In order to make the

```
1 SELECT  d1.docName, t1.value
2 FROM    Document d1, Path p1, Text t1
3 WHERE   p1.pathID = t1.pathID
4 AND     t1.docID = d1.docID
5 AND     (p1.CombinedPathexp LIKE '#/movie#%/cast#%(%'
6 OR      (p1.CombinedPathexp LIKE '#/movie#%(%/cast#%'
7 AND     t1.endPath = 'cast'))
```

Fig. 8. An SQL query for *'/movie//cast'*

abovementioned query consistent with the format of X-Binder's database, "/" and "//" have to be changed to "#/" and "#%/". Fig. 8 shows the SQL query for '/movie//cast'.

In Fig. 8, the condition of the location of 'cast' and 'movie' is restricted by the WHERE clause. The condition stated on the fifth line is that 'cast' should not be an end node, and that on the sixth line states that 'cast' should be an end node. Moreover, it checks again whether endPath of the Text table is 'cast' or not. For the query '//movie//cast' wherein 'movie' may or may not exist as the root, the module will insert "%" before the LIKE operation in lines five and six. Therefore, the final forms are '%/movie#%/cast#%(%' and '%/movie#%(%/cast#%'.

3.3.3 Complex Query

A complex query is a combination of a linear path query and a content query. The query '/movie[/@year='2003']//player/name' selects a set of documents with the path '/movie//player/name' and an attribute value of 'year' in which the low-level node of 'movie' is 2003. This query can be separated into two parts—'/movie/@year='2003'' and '/movie//player/name'. To consider this, the query generating module of X-Binder converts this query to one similar to that shown in Fig. 9. The second line creates table t1 for the first part, and the fourth line creates table p1 for the remaining part.

```
1 SELECT   d1.docName
2 FROM     (SELECT * FROM Text
3           WHERE endPath = '@year' and value = '2003') t1,
4          (SELECT * FROM Path
5           WHERE CombinedPathexp LIKE '#/movie#%/player#(%/name#%') p1,
6          Document d1
7 WHERE  (t1.docID = d1.docID)
```

Fig. 9. SQL query for *'/movie[/@year='2003']//player/name'*

4 Performance Evaluation

The result of the performance test of X-Binder is discussed in this section. The test environment comprised the following: MS Windows XP Professional was used as the operating system and MS-SQL was used as the base DBMS. X-Binder was developed using JAVA 1.4.2.09 and JDOM 1.0 [17] was used for XML parsing. There are two aspects to the performance test. First, we compared the storage spaces of XRel that represents the RDBMS based system and X-Binder that represents our suggestion. Due to the differing table structures, we considered the designed Path tables to have similar structures. Second, we compared the search speeds of the *PathFinder(p)* function in the two systems.

4.1 Comparison of Storage Space

As datasets for the experiment, we used the IMDB database in XML format [18], a Reuters news file with NewsML [19], and a random dataset from the ReGet program [20] that collects documents on the Internet. The maximum depths of each dataset were 6, 9, and 14, and the numbers of documents were 100, 250, and 300, respectively.

We counted the number of rows in the Path tables of both XRel and X-Binder. Fig. 10 shows the result: the number of paths stored in the X-binder database was lesser than that stored in the XRel one. This is a reasonable result because XRel stores all the paths from XML documents separately. However, X-Binder stores the sibling paths in a row. Next, we compared the database sizes. The size of the database in the proposed system was also smaller than that of XRel.

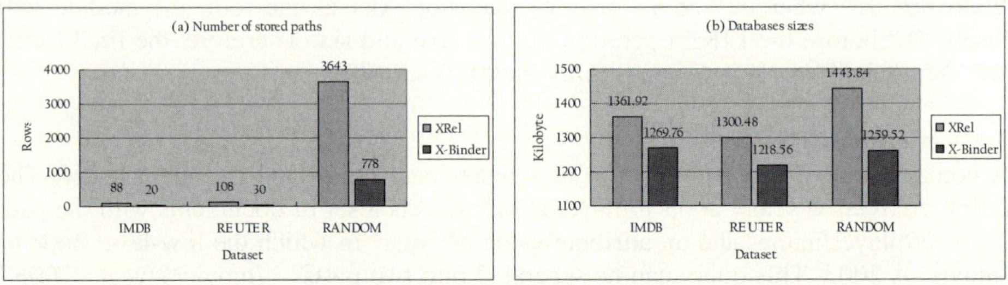

Fig. 10. Comparison of the database sizes and the numbers of stored paths

4.2 Comparison of Search Speeds of *PathFinder(p)*

Fig. 11 shows the list of queries that were created to evaluate the speed test of the *PathFinder(p)* function. Q1 and Q2 were full match queries with different lengths. Q3 and Q4 were partial match queries, and they included the "//" symbol once or twice each. Lastly, Q5 and Q6 were complex queries comprising a full match query and a partial match query.

Q1	/seller/name	Full match query(Short)
Q2	/movie/cast/player/filmographies/filmography/work	Full match query(Long)
Q3	//topic//topicname	Partial match query(one '//')
Q4	//news//topicname	Partial match query(two '//'s)
Q5	/members/member/id/name/[first='Lee']	Complex query
Q6	/members/member//[first='Seo']	Complex query with '//'

Fig. 11. Queries for performance test

The result of the performance test for *PathFinder(p)* is shown in Fig. 12. We found that the X-Binder system exhibited a better performance than the existing XRel system. In particular, while executing a complex query, the search time was decreased by 25–33%. This was caused by the decrease in the number of rows of the Path table, which in turn decreased the number of rows that were created after the JOIN operation.

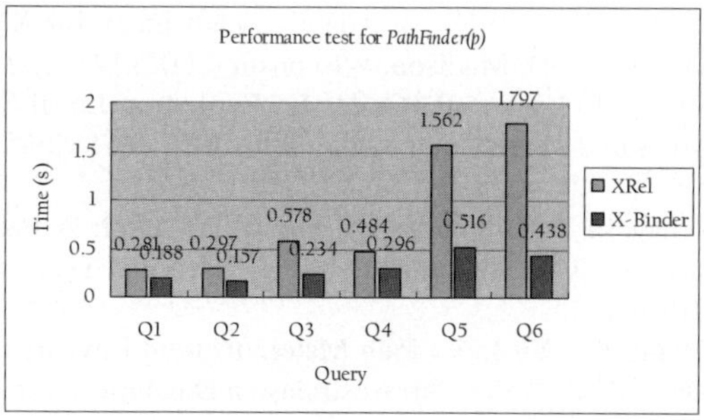

Fig. 12. Comparison of the search speeds of *PathFinder(p)*

5 Conclusion

An increasing amount of research is being conducted on various RDBMS-based XML document management systems. The systems that use an inverted index typically guarantee fast retrieval, and their performance results are superior. However, an inverted index composes a new index that considers the values of the special fields of a database table, and this results in the wastage of storage space.

To solve this problem, we suggest an X-Binder system that has a path combining module based on RDBMS. The proposed system combines sibling nodes and stores them in a single row in a relational table. This method decreases the number of rows of a Path table, and therefore utilizes less storage space.

We compared the proposed system and XRel in terms of the storage space and search speed of *PathFinder(p)* to prove that our system exhibits a better performance. The results revealed that the number of rows of stored data decreased by 20–25%. An experiment for comparing the sizes of the databases showed an overall decrease in size. The X-Binder system guarantees high search speed and better performance, especially for complex queries.

In future studies, considering an inverted index system using a posting list might improve performance. In addition, in order to generate an efficient query, the calculation of the search cost using SQL might yield interesting results.

References

1. W3C: Extensible Markup Language (XML) Version 1.0 (Second Edition), http://www.w3c.org/TR/REC-xml (2000)
2. Ceri S., Fraternali P., Paraboschi S.: XML: Current Developments and Future Challenges for the Database Community. Proc. of the 7th Int'l Conf. on EDBT (2000) 3–17
3. McHugh J., Abiteboul S., Goldman R., Quass D., Widom J.: Lore: A Database Management System for Semistructured Data. ACM SIGMOD Record, Vol. 26, No. 3 (1997) 54–66
4. Cooper B. F., Sample N., Franklin M. J., Hjaltason G. R., Shadmon M.: A Fast Index for Semistructured Data. Proc. of the 27th Int'l Conf. on VLDB, Rome, Italy (2001) 341–350

5. Chung C., Min J., Shim K.: APEX: An Adaptive Path Index for XML Data. Proc. of the Int'l Conf. on ACM SIGMOD, Madison, Wisconsin (2002) 121–132
6. Kaushik R., Shenoy P., Bohannon P., Gudes E.: Exploiting Local Similarity for Indexing Paths in Graph-Structured Data. Proc. of the 18th IEEE Int'l Conf. on Data Engineering (2002) 129–140
7. Yoshikawa M., Amagasa T., Shimura T., Uemura S.: XRel: A Path-Based Approach to Storage and Retrieval of XML Documents Using Relational Databases. ACM ToIT, Vol. 1, No. 1 (2001) 110–141
8. Jiang H., Lu H., Wang W., Yu J. X.: Path Materialization Revisited: An Efficient Storage Model for XML Data. Proc. of the 13th Australasian Database Conf., Melbourne, Australia (2002) 85–94
9. Jiang H., Lu H., Wang W., Yu J. X.: XParent: An Efficient RDBMS-Based XML Database System. Proc. of the 18th Int'l Conf. on Data Engineering, San Jose, California (2002) 335–336
10. Clark J., DeRose S.: XML Path Language (XPath) Version 1.0, W3C Recommendation (1999)
11. Zhang C., Naughton J., Dewitt D., Luo Q., Lohman G.: On Supporting Containment Queries in Relational Database Management Systems. ACM SIGMOD Record, Vol. 30, No. 2 (2001) 425–436
12. Florescu D., Kossmann D., Manolescu I.: Integrating Keyword Search into XML Query Processing. Proc. of the 9th Int'l WWW Conf. on Computer Networks (2000) 119–135
13. W3C: XQuery 1.0: An XML Query Language. http://www.w3.org/TR/2005/WD-xquery-20050915/ (2005)
14. Chamberlin D., Robie J., Florescu D.: Quilt: An XML Query Language for Heterogeneous Data Sources. Proc. of the 3rd Int'l Workshop on ACM WebDB (2000)
15. Milo T., Suciu D., Index Structure for Path Expressions. Proc. of the 7th Int'l Conf. on Database Theory (1999)
16. Goldman R., Widom J.: Approximate DataGuides. Proc. of the Workshop on Query Processing for Semistructured Data and Non-Standard Data Formats, Jerusalem, Israel (1999) 436–445
17. http://www.jdom.org
18. http://us.imdb.com/top_250_films
19. http://about.reuters.com/newsml
20. http://deluxe.reget.com/en/

A Digital TV Agent System for Broadcast and Web Information Hybrid

Sangmin Oh and Jongtae Lim

School of Electronics, Telecommunications and Computer Engineering,
HANKUK Aviation University, Korea
{likepeny, lim} @hau.ac.kr

Abstract. The development of various digital broadcasting services and the rapid growth of broadband network infrastructure have accelerated the convergence of broadcasting and communication services. To show the potentiality of hybrid or convergence technologies, we designed and implemented a hybrid Electronic Program Guide (EPG) agent system for digital broadcasting systems, which provides comprehensive information from broadcast and web information sources. The implemented hybrid EPG agent system can provide detailed information for a specific program by combining broadcast information and web information retrieved from the scattered web pages. The methods of searching the URLs that contain the related information for a specific TV program and retrieving the related information from the selected web pages are proposed.

Keywords: DTV Agent Systems, Digital Broadcasting Systems, Electronic Program Guide, Information Retrieval, Internet Search.

1 Introduction

With the advent of digital broadcasting era, it is possible to provide various digital broadcasting services such as data broadcasting [1] and TV-Anytime service [2]. The digital broadcasting services in accordance with the rapid growth of broadband network infrastructure and Internet technologies have accelerated the convergence or fusion of broadcasting and communication service. Thus a communication-and-broadcasting convergence service, IPTV (Internet Protocol Television) [3] has emerged where a digital television service is delivered using the Internet Protocol over a network infrastructure.

The communication-and-broadcasting convergence is becoming an essential part in ubiquitous computing environment to provide diverse information from different communication media. Since each type of communication media is producing huge amount of information at the same time, the hybrid, fusion or integration of the different types of information is required to provide more comprehensive information to users. Thus, many agent systems are requested to provide certain types of information hybrid functionalities. There has been an effort to design an agent system

N.T. Nguyen et al. (Eds.): KES-AMSTA 2007, LNAI 4496, pp. 239–247, 2007.

to integrate the content from web and broadcast information sources [4]. Also, various methods are devised to get personalized information from overabundant information [5, 6].

In digital broadcasting systems, simple form of information is delivered from broadcasting stations to DTV users by one-directional delivery channels. As the technology for the convergence of broadcasting and communication systems has rapidly grown, DTV systems need to employ various kinds of agents that retrieve information from the convergent environment, so that they can provide combined information and user-requested information from unidirectional broadcasting channels and bi-directional communication channels. While the DTV agent can provide rich information to users with the benefit of this kind of information convergence, it could be a specific application confined to a specific DTV manufacturer and it may be facing some difficulties in maintaining and updating itself in DTV sets.

One of the most commonly used DTV agents is the personalized TV program recommendation system, which automatically matches user's wishes to TV programs and recommends the ones with high user preference [5, 9]. The recommendation agents need explicit and implicit feedback of user interests. Explicit feedback is obtained by user's manual modification of the user profile and implicit feedback is adopted using automatic user profile adaptation. Only DTV agents (or set-top box based agents) can use the implicit feedback which reflects gradual changes of user interest and preference. Besides the recommendation agents, DTV systems need a methodology of gathering different kinds of information about TV programs in broadcasting and communication convergent environment. To fulfill this need, TV-Anytime specifications [2] are standardized based on the server-client model. Based on the convergence of TV and Web, many applications are developed to search for information about TV programs and available TV content according to the user's preference [10]. However, the applications based on the TV-Anytime specifications require a pretty large size of middleware to perform the client-sever based networking and manipulations of the metadata, and it also needs a TV-Anytime server which locates the outside of DTV sets and responses the individual users' inquiries. This kind of the service might be operated with some additional service charge. Also, a method for dynamic integration of TV-program content and related web content is proposed in [11], where they used a topic structure based searching based on closed captions in TV program to show the related web pages with TV programs.

In this paper, we designed and implemented a DTV agent system providing information hybrid with digital broadcasting program information and the related web information. TV Program information in digital broadcasting systems is delivered by Program and System Information Protocol (PSIP) [7], which provides the necessary information to configure an electronic program guide (EPG). However, the bandwidth allocated to the PSIP tables is not enough to convey sophisticated information such as the preview of to-be-broadcasted programs. On the other hand, detailed information about the programs is scattered over the Internet. TV users want to get hybrid information for the programs from broadcast and web sources at the same time. For this purpose, we designed a hybrid EPG agent system, which provides the related information from web pages as well as broadcast information from PSIP tables. The agent is not based on TV-anytime server-client model and instead, uses a well known search engine. With a light size of the hybrid EGP middleware in DTV sets or STB's, the agent can provide converged and compound information about TV programs from diverse information sources on the EPG menu, which is the most frequently called by users.

The rest of this paper is organized as follows. Section 2 provides the overall concept of the hybrid of broadcast and web information. Section 3 describes the architecture of the hybrid EPG agent system. An implementation example of the hybrid EPG agent system is shown in Section 4. This paper is concluded in Section 5.

2 System Architecture

Fig. 1 illustrates the overall system architecture for the hybrid EPG agent system. Digital broadcasting programs are delivered to a digital broadcasting receiver, set-top-box (STB) through terrestrial, cable or satellite broadcasting channels. To provide program information to TV users, the STB gathers program information from transport streams (TS). The information about TV programs is conveyed in the form of PSIP tables: System Time Table (STT), Master Guide Table (MGT), Virtual Channel Table (VCT), Event Information Table (EIT) and Extended Text Table (ETT) [7]. The EPG manager filters PSIP tables from TS and composes a graphical user interface (GUI) display for an EPG.

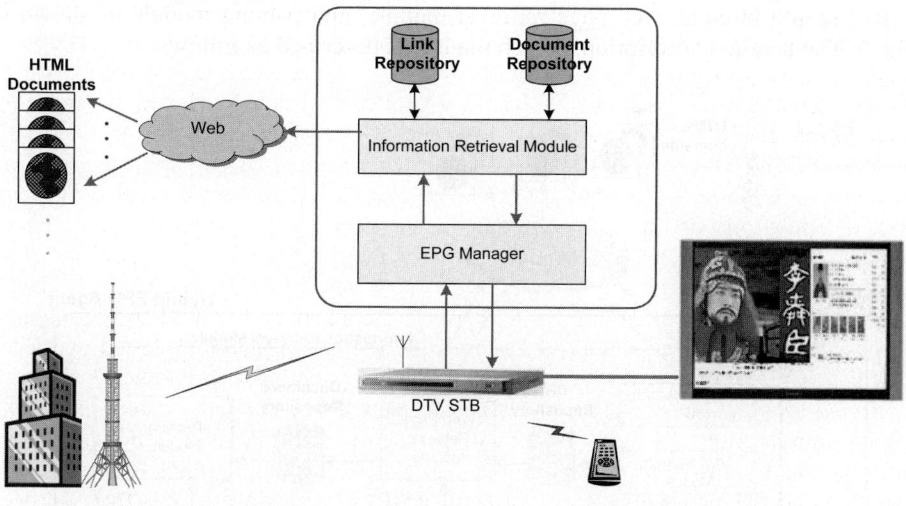

Fig. 1. Overall system architecture for the hybrid EPG agent system

In general, the information delivered in TS is very limited because the most of the available rate is allocated to video and audio data of the program. Thus, the conventional EPG system provides very limited information about the program such as program name, broadcasting day and time, duration, and broadcasting station. On the other hand, detailed information about the program can be found in the various web sites. Our EPG agent system retrieves the program information from the scatter web pages and combines the web information with broadcast information delivered in TS.

The information retrieval module in Fig. 1 retrieves information from the scattered web pages which convey more detailed information about the program, and passes the retrieved information to the EPG manager so that the EPG manager may provide the hybrid information to TV users. The mostly requested information by TV users, who are navigating EPG menu on the TV, may be the preview of a specific program to decide whether they want to watch or record the to-be-broadcasted program. Our hybrid EPG agent provides the sophisticated preview by combining the information from broadcasting channel and web information scattered on the web. This involves the processes of searching the URLs which contain the related information for the TV program, retrieving the selected web pages and parsing the web pages to get the necessary parts. Section 3 shows the detailed architecture of the information retrieval module.

3 The Hybrid EPG Agent System

The hybrid EPG agent system consists of EPG manager, Uniform Resource Locator (URL) search module, web page retrieval module, and parsing module as shown in Fig. 2. The detailed description of each module is described as follows.

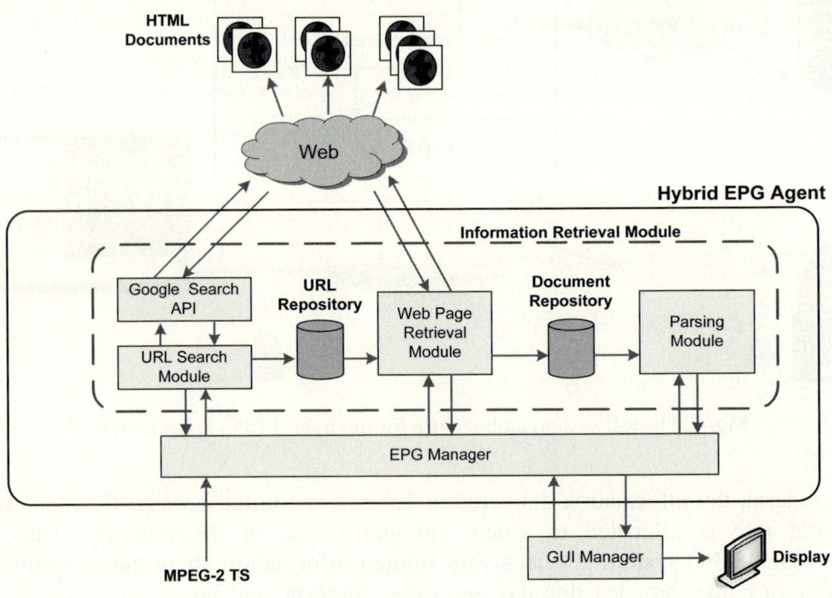

Fig. 2. Architecture of hybrid EPG

3.1 EPG Manager

The EPG manager controls each module of the hybrid EPG agent system. The basic function of the manager is to provide a program guide to TV users by filtering the PSIP tables from MPEG-2 TS and formatting the information into a graphic interface. When TV users want to get information about a specific program, the hybrid EPG agent invokes the information retrieval module, so web information about the program is retrieved and provided to users in a hybrid form. The hybrid EPG agent passes program title, broadcasting station, broadcasting time and duration to the information retrieval module as a set of keywords (or queries). The information retrieval module returns web information about the TV program.

3.2 URL Search Module

The URL search module is a web search engine that locates web sites which contain TV program information. To retrieve URLs that contain the information about a specific TV program, we used Google search engine, which is one of the most popular web search engines and has an excellent ranking algorithm [8]. Using the Google search API, the URL search module gets the URLs where the web information about the program is contained in the form of HTML pages. The module searches URLs using the keywords "program title", "broadcasting station" and "preview." The search results are ranked with a specific ranking order provided by Google search engine. In our experiment, the first URLs of the resulting URL list refer to the program preview site of broadcasting station's home page. The URLs are passed to the web page retrieval module.

At this step, "broadcasting time" is not used as a keyword for search the URLs, because Google search is based on cached pages and the web pages indicted by the resulting URLs have no information about to-be-broadcast programs in many cases. In general, EPG users want to get the information about a program broadcasted on a future specific time, not a past time. This is also because the EPG agent itself provides a guide for the program to be broadcast. To refine the search, broadcasting time is integrated in the web page retrieval module.

3.3 Web Page Retrieval Module

Using the URLs obtained by the URL search module, the web page retrieval module downloads the web pages referred by the URLs. The downloading of the web pages is performed using HTTP. Most of weekly and daily programs run their preview web sites and the sites have the preview of the to-be-broadcast program and the previews of the previously broadcast programs. Thus, the downloaded web pages with the URLs from the URL search module have many sub-links referring to the previews of the specific day, future or past day. To get the page referring to the specific day's program, the downloaded web page is parsed to find specific types of day, such as

"20060808" or "month/day/year," and a sub-link. If the pages contain specific types of day and provide a sub-link to the preview page for that day, the web page retrieval module fetches the URL for the sub-link and downloads the web page once more. Then, the downloaded web page is passed to the parsing module to get the necessary part for an EPG. Otherwise, the module returns "No information available" to EPG manager.

3.4 Parsing Module

The downloaded HTML page is processed to provide only necessary part, which is shown on the EPG menu display. Considering that the information about the programs consists of several sentences, the module searches the part at least two sentences between HTML tags in the retrieved HTML page. The method of finding the sentences is based on the general algorithm which counts the number of words and the punctuation marks.

The hybrid EPG agent communicates among these modules. The procedures for providing the combined information are as follows:

1. The user requests hybrid information for a specific TV program by pressing "More Info" menu item on the EPG menu shown in Fig. 3.
2. The EPG manager invokes the information retrieval module and passing program title, broadcasting station, broadcasting day and time to it.
3. In the information retrieval module, the URL search module forms an inquiry for Google search engine using the keywords "program title", "broadcasting station" and "preview." The module passes the inquiry to Google search engine using Google search APIs. Then, the module gets the URLs for the related HTML pages from Google search engine, and stores the resulting URLs into the URL repository.
4. The web page retrieval module fetches URLs from the URL repository and download each pages referred by the URLs using HTTP. The downloaded web pages are processed to find a sub-link for the web page containing information for the specific broadcasting day & time. If the module can find the sub-link, then proceeds to the next step. Otherwise the module returns "No information available" to the EPG manager.
5. The web page retrieval module downloads the web page referred by the sub-link once again. The downloaded web pages at this stage are stored in the document repository.
6. The parsing module processes the web pages in the document repository to get only necessary part. Once the web pages are parsed, the module returns the retrieved information part to the EPG manager.
7. The EPG manager invokes the GUI manager to configure a hybrid information display, as shown in Fig. 4, using the retrieved information from the parsing module.

4 Implementation of the Hybrid EPG Agent System

Digital broadcasting systems usually provides a system specific EPG. On the EPG menu, the basic program information such as program title, broadcasting station, broadcasting time and duration is displayed. The hybrid EPG agent system basically provides basic information through the EPG menu as shown in Fig 3. The information in the figure is very limited since it is obtained only from broadcast media.

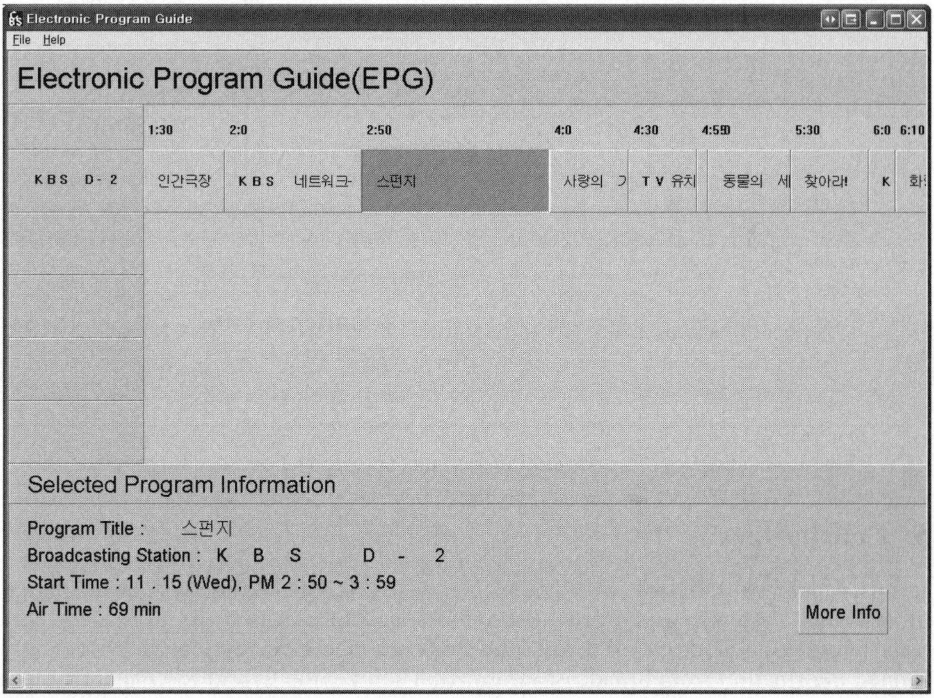

Fig. 3. A snapshot of the conventional EPG

When TV users want to get more information such as program preview, the hybrid EPG agent invokes the information retrieval module, collects the necessary information scattered throughout the web, and provides more sophisticated information to users. An example is shown in Fig. 4, where the preview for a specific program is retrieved from the web and displayed on the EPG in the hybrid form of broadcast and web information. The indicated part is the retrieved web information, which can be found in the program's web page, which is shown in Fig. 5.

Using the hybrid EPG agent system, TV users can acquire the necessary hybrid information on their digital broadcasting systems without accessing the Internet and searching the related information on their personal computing devices that are connected to the web. Also, based on the hybrid information provided by the hybrid EPG agent system, the users can easily determine their future actions such as recording and reserving to watch the selected programs.

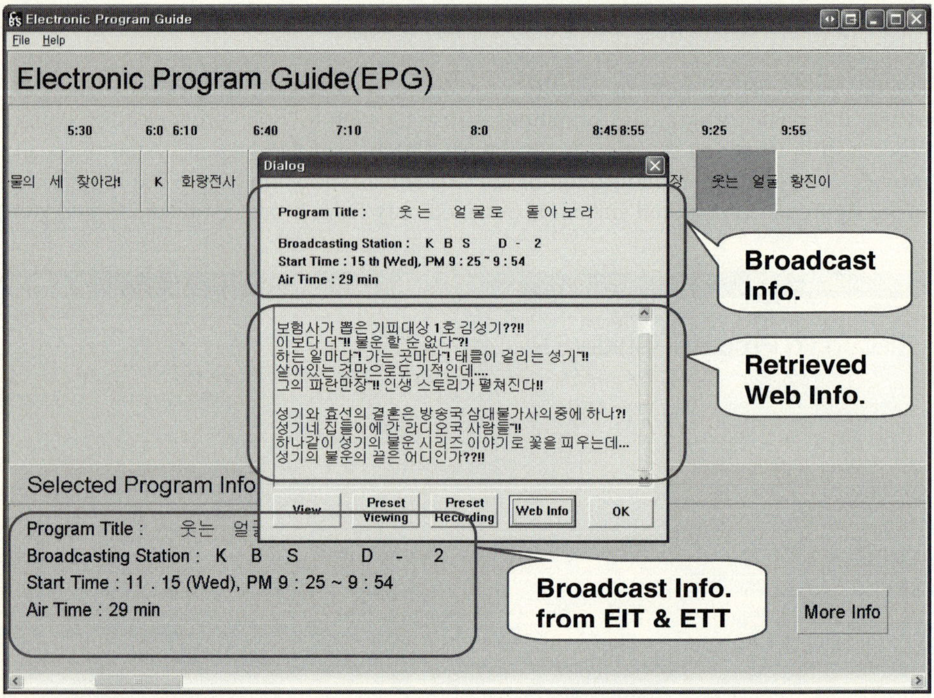

Fig. 4. A snapshot of the hybrid EPG showing a preview of the specific program

5 Conclusions

In this paper, we designed and implemented a hybrid EPG agent system for digital broadcasting systems, which provides sophisticated information from broadcast and web information sources. Whereas the conventional EPG system provides only limited information from broadcast information source, the hybrid EPG agent system can provide detailed information for the specific program by combining broadcast information and the retrieved web information from the scattered web pages. The hybrid EPG agent system is a useful application that helps the TV users to determine their further action during surfing the EPG, based on the combined information for TV programs. The hybrid EPG agent system is a good example to show the potentiality of the hybrid or convergence technology requested in the upcoming broadcasting and communication convergence era.

The hybrid EPG agent system can be extended to automatically collect the information which is personalized to the users by analyzing the user's navigating patterns and usage on the EPG. Further research also includes a more user-friendly human-interface for the broadcast-and-communication convergence systems.

Acknowledgment

This research was supported by the Internet Information Retrieval Research Center (IRC) in Hankuk Aviation University. IRC is a Regional Research Center of Gyeonggi Province, designated by ITEP and Ministry of Commerce, Industry and Energy.

References

1. Chernock, R.S., Crinon, R.J., Dolan, M.A., Mick Jr., J.R.: Data Broadcasting: Understanding the ATSC Data Broadcast Standard, McGraw-Hill (2001)
2. TV-Anytime Specifications: "http://www.tv-anytime.org," (2003)
3. Cherry, S.: IEEE Spectrum. Vol. 42, Issue 1 (2005) 24-29
4. Tanaka, K.: Content Integration form Web and Broadcast Information Sources. International Conference on Informatics Research for Development of Knowledge Society Infrastructure (2004) 99-106
5. Yu, Z., Zhou, X.: TV3P: An Adaptive Assistant for Personalized TV. IEEE transaction on Consumer Electronics, Vol. 50, No. 1 (2004) 393-399
6. Wang, G.T., Xie, F., Tsunoda, F., Maezawa, H., Onoma, A.K.: Web Search with Personalization and Knowledge. Proc. of the Fourth International Symposium on Multimedia Software Engineering (2002) 90-97
7. Eyer, M.K.: PSIP: Naming, Numbering, and Navigation for Digital Television. McGraw-Hill (2003)
8. Mueller, J.P.: Mining Google Web Services: Building Applications with the Google API. Sybex (2004)
9. Xu, J., Zhang, L., Lu, H., Li, Y.: The Development and Prospect of Personalized TV program Recommendation Systems. Proc. of IEEE International Symposium on Multimedia Software Engineering (2002)
10. Leban, M.: Internet Search for TV Content Based on TV Anytime. EUROCON (2003) 70-73
11. Ma, Q., Tanaka, K.: WEBTELOP: Dynamic TV-content augmentation by using web pages. Proc. of IEEE International Conference on Multimedia and Expo (2003) 173-176

Formal Modeling of Agent-Based English Auctions Using Finite State Process Algebra*

Amelia Bădică and Costin Bădică

University of Craiova,
Bvd.Decebal 107, Craiova, 200440, Romania
badica_costin@software.ucv.ro,
ameliabd@yahoo.com

Abstract. The vision of global agent-based e-commerce environments that enable dynamic trading between business partners requires the study and development of suitable formal modeling frameworks. In particular, negotiation is a necessary and important activity to allow engagement of business parties in non-trivial business relationships. In this note we propose a formal framework using *finite state process algebra* for modeling and analysis of agent-based negotiations, with a focus on a particular price negotiation – English auction.

1 Introduction

Negotiations (and auctions in particular) are complex activities frequently encountered in modern e-commerce processes that require a tight interaction of the business parties ([11]). Their analysis and understanding, especially when negotiations are automatized using software agents ([12]), requires the study and development of suitable formal modeling frameworks.

The development of formal frameworks for modeling agent interactions, including those encountered in negotiations and auctions, generated a lot of interest during the last years ([3,5,6,9,16]). A similar interest has been manifested in formalizing business process notations to describe socio-economic activities ([1,4,7,10,15]).

Following the trend, this paper proposes a formal framework for modeling and analysis of agent-based negotiations, with a focus on auctions, using *finite state process algebra* (FSP hereafter) ([13]). The approach is applied to model an English auction – a non-trivial and well-known auction mechanism. The benefits of our proposal are twofold: i) it allows formal verification of the system against a set of qualitative properties (see subsection 4.6), and it can be adapted to derive quantitative performance measures (like throughput, see [9]); ii) it can be used as a basis for agents implementation – eg. local processes can be mapped to agent behaviors (see section 4).

* Work of A. Bădică and C. Bădică was partially funded by CNCSIS grants 94/2005: "HiperProc: Hypermedia Techniques for Knowledge-Based Representation of Business Processes" and 185/2006 "Technologies and Intelligent Software Tools for Automated Construction of E-Catalogues of Products Using Knowledge Acquisition from the Web".

N.T. Nguyen et al. (Eds.): KES-AMSTA 2007, LNAI 4496, pp. 248–257, 2007.

2 Background on FSP

FSP is an algebraic specification technique of concurrent and cooperating computational processes as finite state labeled transition systems (LTS hereafter).

Definition 1. *(labeled transition system) Let $\mathcal{S}$ be the universal set of states, $\mathcal{L}$ be the universal set of action labels and τ be the internal unobservable action. A* finite LTS *is a quadruple $P = \langle S, A, \Delta, q \rangle$ s.t.: i) $S \subseteq \mathcal{S}$ is a finite set of states; ii) $A = \alpha P \cup \{\tau\}$, where $\alpha P \subseteq \mathcal{L}$ denotes the alphabet of P; iii) $\Delta \subseteq S \times A \times S$ is the transition relation that maps a state and an action to another state; iv) $q \in S$ is the initial state of P.*

LTS models are suitable for specifying discrete-event systems. However, descriptions, either visual or textual of LTS models as labeled directed graphs are impractical for more than a few states. For this reason the FSP process algebra has been proposed ([13]). FSP uses the following constructs: prefix, choice, parallel composition, re-labeling and definition (see [13] for more details).

i) *Prefix*. The process $a \to P$ performs the action a and then behaves like P. The prefix operator specifies sequential execution of actions.
ii) *Choice*. The process $P \mid Q$ behaves either like P or like Q. If both are enabled then the choice is non-deterministic.
iii) *Parallel composition*. The composite process $P \parallel Q$ specifies the interaction between processes P and Q on the common set of actions in their alphabets αP and αQ. This means that for actions outside set $\alpha P \cap \alpha Q$, P and Q proceed independently, but for actions in $\alpha P \cap \alpha Q$, P and Q must cooperate and proceed together.
iv) *Re-labeling*. Re-labeling functions applied to a process term change the names of the action labels. The process P/L where $L = \{nl_1/ol_1, \ldots, nl_k/ol_k\}$, $ol_i \in L$, $nl_i \subseteq L$, behaves like P excepting that any action ol_i appears to an external observer as any of the actions in the set nl_i, for all $1 \leq i \leq k$.
v) *Definition*. A definition $A = P_A$ associates the behavior of the process term P_A with the name A. You can then use A in process terms to describe more complex behaviors. Thus A is interpreted as the name of a re-usable process component.

The syntax of FSP is introduced in two steps: i) definition of sequential processes; ii) definition of composite processes. The key point is to not arbitrarily mix choices and parallel compositions in order to preserve the finiteness of the state space ([13]).

The set of process names is partitioned into the sets $\mathcal{P}_S$ of sequential process names and $\mathcal{P}_C$ of composite process names[1]. Let $\mathcal{L}$ be the set of all action labels.

Definition 2. *(sequential process) A* sequential process term *is defined according to the following rules: a) END is a sequential process term denoting an empty process that engages in no further actions; b) If $SPN \in \mathcal{P}_S$ then SPN is a sequential process term; c) If $a_i \in \mathcal{L}$ and SP_i are sequential process terms, $1 \leq i \leq k$, then $a_1 \to SP_1 | a_2 \to SP_2 | \ldots | a_k \to SP_k$ is a sequential process term.*

A sequence $SPN_1 = SP_1, \ldots, SPN_p = SP_p$ s.t. $SPN_i \in \mathcal{P}_S$ and SP_i are sequential process terms, $1 \leq i \leq p$, defines a sequential process with name SPN_1. Definitions of SPN_i, $2 \leq i \leq p$, are called local definitions.

[1] Elements of $\mathcal{P}_C$ are distinguished from elements of $\mathcal{P}_S$ by prefixing with || (see appendix).

Definition 3. *(composite process) A* composite process term *is defined with the following rules: a) If $PN \in \mathcal{P}_C \cup \mathcal{P}_S$ then PN is a composite process term; b) If CP_i are composite process terms, $1 \leq i \leq k$, then $CP_1 \parallel CP_2 \parallel \ldots \parallel CP_k$ is a composite process term; c) If CP is a composite process term and L is a re-labeling function then CP/L is a composite process term.*

If $CPN \in \mathcal{P}_C$ and CP is a composite process term then $CPN = CP$ defines a composite process with name CPN.

A FSP model consists of a finite set of sequential and/or composite process definitions such that no process name occurring in the right-hand side of a composite process definition was left undefined. FSP has an operational semantics given via a LTS. The mapping of a FSP term to a LTS is described in detail in [13] and it follows the intuitive meaning of FSP constructs introduced in this section.

3 Agent Negotiation Model

We understand automated negotiations as a process by which a group of software agents communicate with each other to reach a mutually acceptable agreement on some matter ([12]). In this paper we focus our attention on *auctions* – a particular form of negotiation where resource allocations and prices are determined by bids exchanged between participants according to a given set of rules ([14]).

In automated negotiations (including auctions) it is important to distinguish between *negotiation protocols* (or *mechanisms*) and *negotiation strategies*. The protocol comprises public "rules of encounter" between negotiation participants by specifying the requirements that enable them to interact and negotiate. The strategy defines the private behavior of participants aiming at achieving their desired outcome ([17]).

Our negotiation model follows the generic software framework for automated negotiation proposed by [3] and it is specialized for the particular case of English auctions following implementation details reported in [2]. So, our work can be also seen as an attempt to formalize behavior of negotiation agents, as defined by that implementation.

Authors of [3] sketched a software framework for implementing agent negotiations that comprises: (1) negotiation infrastructure, (2) generic negotiation protocol and (3) taxonomy of declarative rules. The *negotiation infrastructure* defines roles of negotiation participants (eg.*Buyer* or *Seller* in an auction) and of a negotiation host. According to the *generic negotiation protocol* ([3]), participants exchange proposals (or bids) via a common space (or market) that is governed by an authoritative entity – the negotiation host (or market maker). Status information describing negotiation state and intermediary information is automatically forwarded by the host to all entitled participants according to the information revealing policy of that particular negotiation ([3,2]). *Negotiation rules* deal with the semantic constraints a particular negotiation mechanism (e.g. English auctions). Rules are used for checking validity of proposals and sequences of exchanged messages, updating of negotiation status and informing participants, and controlling agreement formation and negotiation termination.

Formal modeling of an agent-based English auction requires a precise description of the generic negotiation protocol and of semantic constraints specific to English auctions. We follow [3] by representing messages using FIPA ACL ([8]).

The *generic negotiation protocol* controls how messages are exchanged by the host and participants by facilitating the following negotiation activities: (1) admission to negotiation, (2) proposal (or bid) submission, (3) informing participants about the change of negotiation state, (4) agreement formation and (5) negotiation termination.

Admission to negotiation. This activity starts when a new participant requests admission by sending a PROPOSE message to the host. The host grants (or not) participant admission responding with either an ACCEPT-PROPOSAL or a REJECT-PROPOSAL message. In particular, the first admission request (always submitted by a seller participant in an English auction) initiates the negotiation.

Proposal submission. Participants may enter the phase of submitting bids after they were admitted to the negotiation. The generic negotiation protocol states that a participant will be notified by the host if its proposal was either accepted (with an ACCEPT-PROPOSAL) or rejected (with an REJECT-PROPOSAL).

Informing participants. The negotiation protocol requires that participants will always be notified (with INFORM messages) about any new state of the negotiation.

Agreement formation can be triggered at any time during negotiation. When agreement formation rules signal that an agreement was reached, the protocol states that participants involved in the agreement will be notified by the host with INFORM messages.

Negotiation termination can be triggered at any time during negotiation. When negotiation termination rules signal that the negotiation process reached its final state, the protocol states that all participants will be notified by the Host with INFORM messages.

We now follow with a brief and concise description of English auctions. Technically, English auctions are single-item, first-price, open-cry, ascending auctions ([11],[17]). In an English auction there is a single item sold by a single seller and many buyers bidding against one another for buying the item until the auction terminates. Usually, there is a time limit for ending the auction (either a total time limit or a certain inactivity period), a seller reservation price that must be met by the winning bid for the item to be sold and a minimum value of the bid increment. A new bid must be higher than the currently highest bid plus the bid increment in order to be accepted. All the bids are visible to all the auction participants, while seller reservation price is private to the auction.

4 FSP Model of Agent Negotiation

4.1 Negotiation Structure

A *negotiation structure* defines a general framework that statically constraints a given negotiation. It consists of a set of roles that contains a *negotiation host* role and one or more *negotiation participant* roles.

The *negotiation host* role orchestrates the negotiation and coordinates negotiators by employing the general negotiation protocol.

A *negotiation participant* role describes the behavior of a negotiator that plays a certain role in the negotiation. Usually, two negotiation participant roles are defined – *buyer* and *seller*. For example, in an English auction there is a single *seller* participant and one or more *buyer* participants, while in an reverse English auction there is a single participant with role *buyer* and one or more participants with role *seller*.

A negotiation process is always initiated by a certain participant known as *negotiation initiator*. Usually is required that the initiator has a given negotiation role – *negotiation initiator* role. For example, in an English auction the initiator has always role *seller*, while in a reverse English auction the initiator has always role *buyer*.

Focusing our discussion on auctions for buying and selling goods, a negotiation structure can be formally defined as follows:

Definition 4. *(Negotiation Structure)*
A negotiation structure *is a triple* $N = \langle Host, Seller, Buyer, Initiator \rangle$ *such that: i)* $Host$ *is the* negotiation host *role; ii)* $Seller$ *is the* seller *role that defines behavior of participants selling goods in the auction; iii)* $Buyer$ *is the* buyer *role that defines behavior of participants buying goods in the auction; iv)* $Initiator$ *is the role that is allowed to initiate the auction – either* buyer *or* seller*, i.e.* $Initiator \in \{Buyer, Seller\}$.

Behavior of negotiation roles is described using FSP. Therefore we shall have FSP processes describing the $Host$, the $Seller$ and the $Buyer$ roles. A participant behavior is defined by instantiating its role. Finally, the behavior of the negotiation system is defined using parallel composition of roles for each negotiation participant, including of course the negotiation host.

4.2 Negotiation Host

In what follows we shall assume that our negotiation host is able to handle a single negotiation at a certain time. In other words, the negotiation host functions as a *one-at-a-time server*. In order to handle multiple negotiations concurrently, several negotiation hosts instances must be ran concurrently. However, as focus of this paper is to formally describe a single negotiation, we do not explore further this path.

Negotiation consists of a series of stages that, in what follows, are particularized for the case of an English auction:

(i) *initiation* – the negotiation is initiated by the seller using the *init* action; note that initiation acts also as a registration of the seller agent participant; initiation is either accepted (action $accept\_init$) or rejected (action $reject\_init$) by the host;
(ii) *buyer registration* – each buyer agent must register with the negotiation using *register* action before she is allowed to submit bids; registration is granted (action $accept\_registration$) or not (action $reject\_registraton$) by the negotiation host;
(iii) *bids submission* – each registered buyer agent is allowed to submit bids using the *bid* action; bids are either accepted (action $accept\_bid$) or not (action $reject\_bid$) by the host; when a certain bid is accepted, the other registered buyer participants are notified accordingly by the host using action $inform$;
(iv) *agreement formation* – when the host observes a certain period of bidding inactivity, it triggers negotiation termination via action *stop*. This event subsequently triggers agreement formation. In this stage the host checks if an agreement can be generated. If no buyer has registered before the negotiation terminated then no agreement can be made and action *no_win* with no parameter is executed. However, if at least one buyer has successfully submitted an accepted bid then the host will have to decide if there is a winner (action *win*) or not (action *no_win* with parameter) depending on if the currently highest bid overbids or not the seller reservation price.

Table 1. $Server$ process that describes the negotiation host role

$Server$	$= init \rightarrow AnswerInit,$
$AnswerInit$	$= accept\_init \rightarrow ServerBid(\perp, \emptyset)\mid$
	$reject\_init \rightarrow Server,$
$ServerBid(chb, Bs)$	$= bid(b \in Bs) \rightarrow AnswerBid(b, chb, Bs)\mid$
	$stop \rightarrow ServerAgreement(chb)\mid$
	$register(b' \notin Bs) \rightarrow AnswerReg(b', chb, Bs),$
$AnswerReg(b', chb, Bs)$	$= accept\_registration(b') \rightarrow ServerBid(chb, Bs \cup \{b'\})\mid$
	$reject\_registration(b') \rightarrow ServerBid(chb, Bs)\mid$
$AnswerBid(b, chb, Bs)$	$= accept\_bid(b) \rightarrow InformBuyers(b, Bs)\mid$
	$reject\_bid(b) \rightarrow ServerBid(chb, Bs),$
$InformBuyers(b, Bs)$	$= inform(b_1) \rightarrow inform(b_2) \rightarrow \ldots \rightarrow$
	$inform(b_k) \rightarrow ServerBid(b, Bs),$
$ServerAgeement(\perp)$	$= no\_win \rightarrow Server,$
$ServerAgeement(chb)$	$= win(chb) \rightarrow Server\mid$
	$no\_win(chb) \rightarrow Server.$

Negotiation host behavior is described as the $Server$ process below (see table 1). Note that message contents (i.e. bid value or submission time), excepting buyer identities, are ignored in this model.

Note that $Server$ process has a cyclic behavior and thus it runs infinitely, being able to handle an infinite sequence of negotiations, one negotiation at a time.

In a real setting, participant agents (buyers and sellers) can be created and destroyed dynamically. In our model we assume there is a given set of buyers and a single seller that are created when the system is started. Buyers are able to dynamically register to negotiations. Whenever a new negotiation finishes, a new one can be immediately initiated by the seller agent and buyers are required to register again in order to be able to participate and bid for buying the sold product.

Assuming each buyer agent has a unique name, let $\mathcal{B}$ be the set of all names of buyer agents that were created when the system was initiated and let be $\perp$ a name not in $\mathcal{B}$. Definition of the $Server$ process is using several indexed families of local processes:

- $ServerBid(b, B)$ such that $b \in B \cup \{\perp\}, B \subseteq \mathcal{B}$. Here b records the buyer associated with currently highest bid and B denotes the set of registered buyers. The condition $b \in B \cup \{\perp\}$ means that either no buyer agent has submitted a bid in the current negotiation ($b = \perp$) or the buyer agent that submitted the currently highest bid must have already registered with the negotiation before the submission $b \in B$.
- $AnswerBid(b_1, b_2, B)$ such that $b_1 \in B, b_2 \in B \cup \{\perp\}, B \subseteq \mathcal{B}$. Here b_1 denotes the buyer that submitted the most recent bid, b_2 denotes the buyer associated with currently highest bid and B denotes the set of registered buyers. The fact that $b_1 \in B$ means that the most recent submitted bid comes from a registered buyer. The fact that $b_2 \in B \cup \{\perp\}$ means that either the currently highest bid has not been submitted yet ($b_2 = \perp$) or it was submiited by a registered buyer ($b_2 \in B$).
- $AnswerReg(b_1, b_2, B)$ such that $b_1 \in \mathcal{B} \setminus B, b_2 \in B \cup \{\perp\}, B \subseteq \mathcal{B}$. Here b_1 denotes the buyer that requested registration with the current negotiation, b_2 denotes the buyer associated with currently highest bid and B denotes the set of registered buyers. The fact that $b_1 \in \mathcal{B} \setminus B$ means that the registration request comes from a buyer that is not yet registered with the negotiation. The fact that $b_2 \in B \cup \{\perp\}$ means that either the currently highest bid has not been submitted yet ($b_2 = \perp$) or it was submiited by a registered buyer ($b_2 \in B$).

Table 2. *Buyer* and *Seller* processes

$$
\begin{aligned}
Buyer &= register \rightarrow BuyerRegister \mid \\
&\quad inform \rightarrow Buyer, \\
BuyerRegister &= accept\_registration \rightarrow BuyerBid \mid \\
&\quad reject\_registration \rightarrow Buyer, \\
BuyerBid &= bid \rightarrow WaitBid \mid \\
&\quad cancel\_bid \rightarrow WaitBid \mid \\
&\quad inform \rightarrow BuyerBid, \\
WaitBid &= accept\_bid \rightarrow Wait \mid \\
&\quad reject\_bid \rightarrow BuyerBid \mid \\
&\quad inform \rightarrow BuyerBid, \\
Wait &= inform \rightarrow BuyerBid \mid \\
&\quad end \rightarrow Buyer.
\end{aligned}
$$

$$
\begin{aligned}
Seller &= init \rightarrow WaitInit, \\
WaitInit &= accept\_init \rightarrow WaitEnd \mid \\
&\quad reject\_init \rightarrow Seller, \\
WaitEnd &= end \rightarrow Seller.
\end{aligned}
$$

- $InformBuyers(b, B)$ such that $b \in B, B \subseteq \mathcal{B}$. Here b denotes the buyer that submitted an accepted bid and B denotes the set of registered buyers. The fact that $b \in B$ means that the bid that was accepted comes from a buyer that has registered with the negotiation.

4.3 *Buyer* Role

The *Buyer* role is defined as a cyclic FSP process. Note that a buyer agent must first register to the negotiation before starting to submit bids. If registration is granted, she can start bidding according to its private strategy – action *bid*. Here we have chosen a very simple strategy: each buyer agent submits a first bid immediately after it is granted admission to the negotiation and subsequently, whenever it gets a notification that another participant issued a bid that was accepted by the host. Additionally, each buyer participant has its own valuation of the negotiated product. If the current value that the buyer decided to bid exceeds her private valuation then the proposal submission is canceled – action *cancel_bid*, i.e. product became "too expensive". Note that after a buyer agent submitted a bid that was accepted, she will enter a state waiting for a notification that either another successful bid was submitted or that she eventually was the last submitter of a successful bid in the current auction (i.e. a potentially winning bid, depending on if the bid value was higher than the seller reservation price) – see action *end*.

4.4 *Seller* Role

The *Seller* role is also defined as a cyclic FSP process. The seller agent initiates the auction – action *init* and then, assuming initiation was successful, she waits for the auction to terminate – action *end*, before issuing a new initiation request.

4.5 Negotiation System

Let us assume that our system is initialized by creating 2 buyer agents, i.e. $\mathcal{B} = \{b_1, b_2\}$, and one seller agent. Buyer and seller agents are created by instantiating *Buyer* and respectively *Seller* roles. Note that instantiation of *Buyer* roles assumes also indexing of actions *bid*, *reject_bid*, *accept_bid*, *inform*, *cancel_bid*, *register*, *accept_registration*, *reject_registration* with buyer's name and also renaming action *end* with an indexed set of actions $\{win, no\_win\}$. Similarly, instantiation of *Seller* role assumes renaming action

Table 3. *System* process as parallel composition of negotiation host, buyers and seller processes

$$\begin{aligned}
BuyerAgent_1 &= Buyer/\{bid_1/bid, reject_bid_1/reject_bid, accept_bid_1/accept_bid,\\
&\quad inform_1/inform, cancel_bid_1/cancel_bid, \{win_1, no_win_1\}/end,\\
&\quad register_1/register, accept_registration_1/accept_registration,\\
&\quad reject_registration_1/reject_registration\}.\\
BuyerAgent_2 &= Buyer/\{bid_2/bid, reject_bid_2/reject_bid, accept_bid_2/accept_bid,\\
&\quad inform2/inform, cancel_bid_2/cancel_bid, \{win_2, no_win_2\}/end,\\
&\quad register_2/register, accept_registration_2/accept_registration,\\
&\quad reject_registration_2/reject_registration\}.\\
SellerAgent &= Seller/\{\{no_win, win_1, no_win_1, win_2, no_win_2\}/end\}.\\
System &= Server\|SellerAgent\|BuyerAgent_1\|BuyerAgent_2.
\end{aligned}$$

end with a set of actions denoting various ways the auction may terminate: without a winner assuming no buyer submitted an accepted bid – *no_win*, with or without a winner assuming at least one buyer submitted an accepted bid – indexed set of actions $\{win, no\_win\}$. Finally, instantiation of *Server* role requires no renaming, as the names of the buyer agents were supposed known in the definition of *Server* process.

Negotiation system is defined as parallel composition of negotiation host, seller agent and buyer agents processes – see table 3.

We have determined the LTS of a negotiation system with 2 buyers using LTSA tool ([13]). The complete definition of this system is shown in the appendix. The analysis performed revealed that the system has 66 states and it is free of deadlocks.

4.6 Properties of the Negotiation System

Techniques discussed in [10] for workflow analysis can be also applied to analyze our negotiation system. These techniques include interactive step-by-step simulation and model verification against safety and liveness properties.

Interactive simulation allows to perform a manually controlled step-by-step execution of the negotiation system. While this feature may give a "feeling" about how the system would behave before actually being implemented, it is quite limited for large applications. Additionally, the trace facility can only eventually detect abnormal behaviors, failing to prove that the system behaves correctly for all its possible executions.

LTSA tool supports a more powerful way of checking a target system using *safety and progress properties* ([13]). In what follows, because of space limitations, we only give two examples.

A *safety property* is defined as a deterministic process asserting that any of its executions is correct. If an error state is reachable in the LTS of its composition with the target system then the safety property is violated. An example of safety property for our negotiation system is that whenever a negotiation is successfully started it must also safely terminate with or without a winner, before a new negotiation can be started.

$$\begin{aligned}
&\textbf{property}\ SafeTerminaion =\\
&\qquad accept_init \rightarrow \{win(b \in \mathcal{B}), no_win(b \in \mathcal{B}), no_win\} \rightarrow SafeTermination.
\end{aligned}$$

A *progress property* is defined as a finite set of actions and requires that any infinite execution of the target system contains at least one of the actions in this set infinitely often. We have checked our negotiation system against the *default progress property* that asserts each action in the process alphabet will be executed infinitely often ([13])[2].

5 Conclusions and Future Work

We proposed a formal framework for modeling agent-based English auctions using finite state process algebra. As future works we plan: i) to asses its generality by application on other negotiation mechanisms; ii) to investigate the applicability of the process algebra approach as a basis for implementation of agent negotiation systems.

References

1. Bădică, C., Bădică, A., Liţoiu, V.: Role Activity Diagrams as Finite State Processes. In: Paprzycki, M. (ed.): *Proc. International Symposium on Parallel Distributed Computing ISPDC'03*, Ljubljana, Slovenia. IEEE Computer Society Press 15–22, 2003.
2. Bădică, C., Băddiţă, A., Ganzha, M., Iordache, A., Paprzycki, M.: Rule-Based Framework for Automated Negotiation: Initial Implementation. In: A. Adi, S. Stoutenburg, S. Tabet (eds.): *Proc. RuleML'2005*, Galway, Ireland. LNCS 3791, Springer Verlag 193–198, 2005.
3. Bartolini, C., Preist, C., Jennings, N.R.: A Software Framework for Automated Negotiation. In: *Proc. SELMAS'2004, LNCS 3390*, Springer Verlag, 213–235 2005.
4. Dong, Y., Sheng, Z.: Using pi-Calculus to Formalize UML Activity Diagram, In: *Proc.10th IEEE International Conference on Engineering of Computer-Based Systems ECBS 2003.* IEEE Computer Society 47–54, 2003.
5. van Eijk, R.M., de Boer, F.S., van der Hoek, W., Meyer, J.-J.Ch.: Process Algebra for Agent Communication: A General Semantic Approach. In: *Communication in Multiagent Systems 2003.* LNCS 1650, Springer Verlag 113–128, 2003.
6. Esterline, A.C., Rorie, T.: Using the pi-Calculus to Model Multiagent Systems. In: *Formal Approaches to Agent-Based Systems, First International Workshop, FAABS'2000*, Greenbelt, MD, USA. LNCS 1871, Springer Verlag 164–179, 2001.
7. Feng, Z., Yin, J., Zhang, H., Dong, J.: Inter-organizational business process modeling for electronic commerce based on pi-calculus. In: *Proc.Int.Conf.on Services Systems and Services Management, ICSSSM'2005*, Chongqing, China. IEEE Press, vol. 2, 966–970, 2005.
8. *FIPA: Foundation for Physical Agents.* `http://www.fipa.org`
9. Hillston, J., Kloul, L.: Performance investigation of an on-line auction system. In: *Concurrency and Computation: Practice and Experience*, 13(1): 23- 41, 2001.
10. Karamanolis, C., Giannakapoulou, D., Magee, J., Wheater, S.: Modelling and Analysis of Workflow Processes. Technical Report, Department of Computing, Imperial College of Science, Technology and Medicine, 1999.
11. Laudon, K.C., Traver, C.G.: *E-commerce. business. technology. society* (2nd ed.). Pearson Addison-Wesley, 2004.
12. Lomuscio, A.R., Wooldridge, M., Jennings, N.R.: A classification scheme for negotiation in electronic commerce. In: F. Dignum, C. Sierra (Eds.): *Agent Mediated Electronic Commerce: The European AgentLink Perspective*. LNCS 1991, Springer Verlag, 19–33, 2002.
13. Magee, J., Kramer, J.: *Concurrency. State Models and Java Programs* 2nd *ed.*, John Wiley & Sons, 2002.
14. McAfee, R.P., McMillan, J.: Auctions and bidding. In: *Journal of Economic Literature*, 25(2): 699–738, 1987.

[2] Note that progress properties are checked under the *fair choice assumption* ([13]).

15. Puhlmann, F.: Why Do We Actually Need the Pi-Calculus for Business Process Management?. In: Abramowicz, W. and Mayr, H.C. (eds.): *Proc. 9th International Conference on Business Information Systems, BIS'2006*, Klagenfurt, Austria. LNI 85, GI, 2006.
16. Weiliang, M., Xiaodong, W., Huanye, S.: A Configurable Auction Framework for Open Agent Systems. In: Joint Workshop *Agent Technology and Software Engineering/ Agent Infrastructure, Tools and Applications*, Net.ObjectDays 2002, Erfurt, Germany, 2002.
17. Wooldridge, M.: *An Introduction to MultiAgent Systems*, John Wiley & Sons, 2002.

Appendix

The model shown here is expressed using the FSP language supported by LTSA tool. It was derived from the general models shown in tables 1, 2 and 3, as follows: $\perp$ is encoded as 0 and buyer names are encoded as 1 and 2; an argument representing the name of a buyer b is encoded as `[b]`; an argument representing an empty set of buyers is encoded as `[0]`; an argument representing a set with a single buyer b is encoded as `[b]`; an argument representing the set of 2 buyers $\{1,2\}$ is represented as `[1][2]`. Following these conventions, for example, local process $AnswerBid(1,2,\{1,2\})$ becomes `AnswerBid[1][2][1][2]`.

```
Buyer = (register -> BuyerRegister | inform -> Buyer),
BuyerRegister = (accept_registration -> BuyerBid | reject_registration -> Buyer),
BuyerBid = (bid -> WaitBid | cancel_bid -> Buyer | inform -> BuyerBid),
WaitBid = (accept_bid -> Wait | reject_bid -> BuyerBid | inform -> BuyerBid),
Wait = (inform -> BuyerBid | end -> Buyer).

Seller = (init -> WaitInit),
WaitInit = (accept_init -> WaitEnd | reject_init -> Seller),
WaitEnd = (end -> Seller).

Server = (init -> AnswerInit),
AnswerInit = (accept_init -> ServerBid[0][0] | reject_init -> Server),
ServerBid[chb:0..2][0] = (stop -> ServerAgreement[0] |
  register[b1:1..2] -> AnswerReg[b1][chb][0]),
ServerBid[chb:0..2][b:1..2] = (bid[b] -> AnswerBid[b][chb][b] | stop -> ServerAgreement[chb] |
  register[3-b] -> AnswerReg[3-b][chb][b]),
ServerBid[chb:0..2][1][2] = (bid[b:1..2] -> AnswerBid[b][chb][1][2] |
  stop -> ServerAgreement[chb]),
AnswerReg[b1:1..2][chb:0..2][0] = (accept_registration[b1] -> ServerBid[chb][b1] |
  reject_registration[b1] -> ServerBid[chb][0]),
AnswerReg[1][chb:0..2][2] = (accept_registration[1] -> ServerBid[chb][1][2] |
  reject_registration[1] -> ServerBid[chb][2]),
AnswerReg[2][chb:0..2][1] = (accept_registration[2] -> ServerBid[chb][1][2] |
  reject_registration[2] -> ServerBid[chb][1]),
AnswerBid[b:1..2][chb:0..2][b] = (accept_bid[b] -> InformBuyers[b][b] |
  reject_bid[b] -> ServerBid[chb][b]),
AnswerBid[b:1..2][chb:0..2][1][2] = (accept_bid[b] -> InformBuyers[b][1][2] |
  reject_bid[b] -> ServerBid[chb][1][2]),
InformBuyers[b:1..2][1][2] = (inform[3-b] -> ServerBid[b][1][2]),
InformBuyers[b:1..2][b] = ServerBid[b][b],
ServerAgreement[0] = (no_win -> Server),
ServerAgreement[chb:1..2] = (win[chb] -> Server | no_win[chb] -> Server).

||System = (
  Server || Seller/{{win[b:1..2],no_win[1..2], no_win}/end} ||
  (Buyer/{bid[1]/bid,reject_bid[1]/reject_bid,accept_bid[1]/accept_bid,inform[1]/inform,
          cancel_bid[1]/cancel_bid,{win[1],no_win[1]}/end,register[1]/register,
          accept_registration[1]/accept_registration,
          reject_registration[1]/reject_registration}) ||
  (Buyer/{bid[2]/bid,reject_bid[2]/reject_bid,accept_bid[2]/accept_bid,inform[2]/inform,
          cancel_bid[2]/cancel_bid,{win[2],no_win[2]}/end,register[2]/register,
          accept_registration[2]/accept_registration,
          reject_registration[2]/reject_registration})).
```

Wiki-News Interface Agent Based on AIS Methods

Janusz Sobecki and Leszek Szczepanski

Institute of Applied Informatics, Wroclaw University of Technology
Wyb.Wyspianskiego 27, 50-370 Wroclaw, Poland
sobecki@pwr.wroc.pl, spikee2@wp.pl

Abstract. Interface agents are applications of recommender systems that are applied in many areas such as information filtering, information retrieval or web browsing. Different reasoning methods that are known from many disciplines: Artificial Intelligence, Expert Systems or Information Retrieval are being used by recommender systems. Recently, the recommender systems also adopt some nature inspired methods such as Artificial Immune System (AIS). In this paper we present application of AIS collaborative filtering in the system Reporter that is based on Wiki-news and recommends both articles and interface layouts. Wiki-based information systems are gaining its popularity among many different users so it is becoming necessary to apply recommendation for most effective information delivery.

1 Introduction

Different types of recommender systems [8] are nowadays gaining popularity among internet systems providers owing to their ability to deliver customized information for their users. Recommender systems may be applied in many areas (e-commerce, web content, web advertising and also user interfaces) and depending on the application field may be called adaptive or personalized user interfaces [7] or interface agents [5].

An interface agent is defined by P. Maes as an agent that acts as a kind of intelligent assistant to a user with respect to some computer application [13] or as mediator between the human and the cyberspace and are able to personalize the interface by monitoring and sensing users' capabilities [1].

Many different reasoning methods that are known from many disciplines such as Artificial Intelligence, Expert Systems or Information Retrieval are being applied by recommender systems.. Nowadays the recommender systems can also adopt some nature inspired methods such as Artificial Immune System (AIS). In the Biology, the immune system is defined as "the system of specialized cells and organs that protect an organism from outside biological influences" [12]. In the literature we can find some application of AIS for collaborative filtering. Application of AIS collaborative filtering in wiki-news articles recommendation is presented in this paper.

AIS is kind of simulation of Human Immune System where the antigens attacking our body can stimulate the immune system to produce antibodies [11]. Using AIS nomenclature for our recommender system, wiki-news is a *body*, any new user is an *antigen* and similar users are called *antibodies* that protect the body. AIS is used to select the group of other system users with similar preferences. In our system we used

N.T. Nguyen et al. (Eds.): KES-AMSTA 2007, LNAI 4496, pp. 258–266, 2007.

weighted kappa affinity measure algorithm to calculate the correlation coefficients. Having correlations between all antibodies we can eliminate antibodies if correlation is lower than the suitable weighted kappa value. If antibodies concentration is high enough, then we can make predictions about wiki-news article relevance for any new user. The quality of recommendation was verified experimentally using precision accuracy of recommendation and prove to increase when correlation increases.

Besides article recommendation we also verified how antibodies correlation based on the information content preferences may have an influence on the interface layout preferences, in this case however the accuracy of layout recommendation is more independent from the correlation coefficients.

The paper contains general information on recommendation methods in the section Recommendation Systems. The following section presents basic information on Artificial Immune Systems (AIS) application in recommender systems. The experimental results are presented in the following section. The paper conclusions are given in the Summary.

2 Recommender Systems

The growing popularity of different web-based systems recommendation methods help to deliver customized information to a great variety of users and may be applied in many different domains, such as [8]: net-news filtering, web recommender, personalized newspaper, sharing news, movie recommender, document recommender, information recommender, E-commerce, purchase, travel and store recommender, E-mail filtering, music recommender and music list recommender.

2.1 User Model

User modeling is central problem of all recommender systems [7]. The user model concerns usually the following data: content, representation and utilization within the systems. We can select two main parts of those data: usage data (concerns different selective operations that express users' interests, unfamiliarity or preferences, temporal viewing behavior, as well as ratings concerning the relevance of these elements) and user data that characterizes the user itself (information on demographic data, users' knowledge, their interests and preferences, their skills and capabilities, and also their plans and goals).

The problems of the user modeling concerns the following elements [8]: user profile generation, initial profile generation, profile learning technique, relevance feedback, information filtering method, user profile-item matching technique, user profile matching technique and profile adaptation technique.

2.2 Filtering Methods

Information filtering methods according to [8] may be: demographic (DF), collaborative (CF), content-based (CBF) and hybrid (HA). Other authors [6] present also following types: case-based reasoning (CBR), rule-based filtering (RBF).

2.2.1 Demographic Filtering

DF bases on the information stored in the user profile (containing different demographic features) and uses a stereotype reasoning [8] in recommendations. Stereotype reasoning is aimed at generating initial predictions about the user [7]. The demographic data about user contains different elements, such as: record data (name, address, e-mail, etc.), user's characteristics (sex, education, occupation), geographic data (zip-code, city, state, country) and some other customer qualifying data. The DF has however some disadvantages [8], [9]: for many users generalizations of the user's interests associated with some demographic attribute values may be too general; they do not provide any individual adaptation, also when the user interests tend to change over time; users are quite often reluctant to submit demographic information or lie in this matter. These disadvantages may be overcome by application of other recommendation methods such as CF, CBF or hybrid approach.

2.2.2 Content-Based Filtering

Content-based filtering is a method of recommendation applied in many interface agents. It takes descriptions of the content of the previously evaluated items to learn the relationship between a single user and the description of the new items [8].. The interface agents developed at MIT [3] first observe their users and then apply some machine learning mechanisms to draw the recommendation. For each new situation, the agent computes the distances between the current state and each past state that is stored in the memory. Together with these past states, corresponding user actions are stored. The interface agent recalls action which bears the largest resemblance to the current situation, or in other words, which has the smallest distance from it, and offers it as a recommendation. We can find quite many applications of interface agents. For example, Letizia, an autonomous interface agent for Web browsing [5], records URL's of visited pages and constructs the user profile out of them. Then, using simple keyword-frequency measure, adopted from the field of Information Retrieval, the agent searches the neighborhood of pages currently visited for potentially relevant pages. Another type of interface agent is Apt Decision that learns user's real estates rental preferences to suggest appropriate apartments. Apt Decision agent uses initial profile provided by the user as well as descriptions of apartments extracted from offers the user has analyzed so far.

CBF approach enables personalized and effective recommendations for particular users, but has also some disadvantages: content-based approaches depends on so called objective description of the recommended items; it tends to overspecialize its recommendations; content-based approach is based only on the particular user relevance evaluations, but users usually are very reluctant to give them explicit, so usually other implicit, possibly less adequate, methods must be used.

2.2.3 Collaborative Filtering

CF makes automatic predictions (filtering) about the recommended items by collecting and using information about tastes of other users (collaboration) [12]. CF based recommender systems are using usually user item rating matrix that is used for both: identifying similar users and recommend items highly ranked by those users. The other approach called also item-based approach uses item-item matrix to determine the current user taste according to selection one item. The main advantages of collaborative filtering over the CBF architecture are following [6]: the community of users can deliver subjective data about items; collaborative filtering is able to offer novel items, even such that user has never seen before; collaborative recommendation utilizes item ratings of other users to find the best fitting one. Collaborative recommended agents have also some disadvantages: when the number of other similar users is small then the prediction is rather poor; the quality of service for users of peculiar tests is also bad; this is rather difficult to get sufficient number of similar users to be able to make proper predictions; observe their users and then apply some machine learning mechanisms to draw the recommendation; lack of transparency in the process of prediction and finally the user's personal dislike may be overcome by the number of other similar users opinions.

The CF could overcome by applying the hybrid solution, for user interface recommendation for web-based information system presented in [10] the disadvantages mentioned above do not influence it much. First, we can assume that web-based systems always have quite many similar users. Second, when the prediction does not fit the user, he is able to personalize the interface manually.

2.3 Hybrid and Other Recommendation Approaches

The disadvantages of each of the above mentioned recommendation approaches could be overcome by applying HA. For example the disadvantage of the insufficient number of the similar users at the early stages of the system operation using CF may be overcome by application of the demographic stereotype reasoning.

For example in the user interface recommendation was based on the mixture of the DF and CF [9], [10]. Basically the HA [10] is a combination of demographic, collaborative and content based recommendation. However other types of recommendations that are based on: user emotions, user platform or context of use may be also considered.

We can distinguish at least two types of the HA, first that builds the recommendation basing on each single approach and second, that integrates the knowledge from each single approach before determining the recommendation. For example in the movie recommendation using the former approach we should first determine the lists of recommended movies using DF, CF and CBF separately and then combine these three lists. Using the later approach we modify the stereotype reasoning or fuzzy reasoning rules according to the data or rules of other users (CF) or ranked movies (CBF). Then the final recommendation is determined using these modified rules.

Beside above mentioned recommendations: DF, CF, CBF and HA, we should also mention other ones, such as: platform, situation or emotion based, which could be combined under single name case-based reasoning (CBR). These recommendations may be dealt in two different ways. The first one is based on the expansion of the subject's attribute set with the attribute concerning platform, situation or emotions in standard DF, CF or CBF. The second method treats these recommendations as separate ones, with their own knowledge acquisition methods and reasoning rules.

3 Artificial Immune Systems (AIS)

AIS may be used in several different ways [4], as model for biological immune system explanation, exploitation and prediction or as a abstraction of some immunological processes. However AIS is a relatively young field it has already pretty many different applications: change detection (i.e. made by computer viruses); fault detection and diagnosis (i.e. applied in building hardware fault-tolerant systems); a means of implementing the negative selection. Figure 1 presents the steps of AIS method in recommender system [2]:

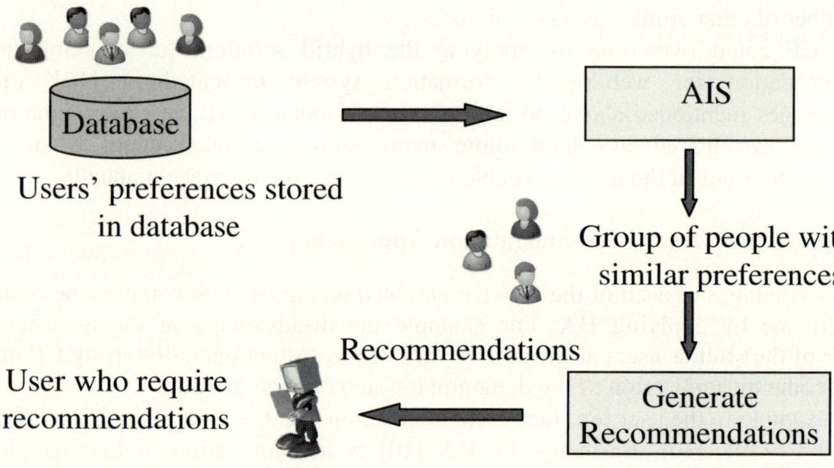

Fig. 1. AIS architecture applied in recommender system

1. The system stores some people's preferences in the database;
2. Each user inputs his or her preferences for some items (i.e. movies), and requires recommendations on some items that he or she has not seen before;
3. AIS selects a group of people (antibodies) who has similar preferences with the particular user (antigen);
4. The weighted average of the preferences for that group of people is calculated by the CF to generate recommendations which the user requires [2].

Equation 1 controls the AIS model [2] (it describes the antigen concentration changes, i.e. it increases with antibody's matching the antigen and decreases in the case when antibody is matching other antibodies; there also exist the death rate that decreases the value when the antibody is neither good or bad):

$$\frac{dx_i}{dt} = k_1 m_i x_i y - \frac{k_2}{n} \sum m_{i,j} x_i x_j - k_3 x_i \tag{1}$$

where:
y represents the antigen concentration;
x_i represents the concentration of antibody i;

x_j represents the concentration of antibody j;
$m_{i,j}$ represents the affinity between the antibody i and j;
m_i represents the affinity between the antibody i and the antigen;
k_1, k_2 and k_3 represents some weights.

In order to generate recommendation we should first determine the group of users with similar preferences. Kappa and Kendall tau can be used for determining affinity between antibodies (correlation coefficient) [2]. We can also find some others algorithms for determining affinity between users, like Person's coefficient or cosine similarity measure.

Weighted kappa coefficient that measures of ratings between two users practically may be calculated using the Equation (2) [2]:

$$k = \frac{1}{n}\sum_{i=1}^{g}\sum_{j=1}^{g} w_{ij} f_{ij} \qquad (2)$$

where:

g is the number of categories (i.e. rates given to the movies) that could be given for items,
n is the number of observations,
f_{ij} is the number of rating agreements for the cell in row i (representing ratings of the first user) and column j (representing ratings of the second user);
w_{ij} represents the value of weight (see Equation (3)) for the cell in row i and column j:

$$w_{ij} = 1 - \frac{|i-j|}{g-1} \qquad (3)$$

Having weighted kappa coefficient measure (that should be greater than specified threshold value p) we can calculate prediction rating for the new user and item a, which is determined according to the Equation (4):

$$Rating \quad prediction(a) = \frac{\sum_{i=1}^{n} weight_i \times Rating_i}{\sum_{i=1}^{n} weight_i} \qquad (4)$$

where:

$weight_i$ represents the weight of the i-th antibody, such that $weight_i = concentration_i$ (of the i-th antibody giving the rating for the item a),
$weight_i = 0$ (the i-th antibody did not deliver rating the item a),
$concentration_i$ represents the concentration of the i-th antibody,
$Rating_i$ represents the score which the i-th antibody voted the item a.4.

4 Tests and Results

We used MediaWiki software to build Reporter system which could help us to check effectiveness of AIS method. In order to gather users' rates, rating module was implemented in our system. Users could mark articles (over 2000 articles were downloaded from polish WikiNews portal and stored in Reporter system) and new different application layouts (skins called: "monobook", "portal", "sport" and "multimedia"). All users' rates were stored in the database. Users could rate articles and layouts using the following scale: 1 – poor, 2 – average, 3 – good, 4 – very good, 5 – excellent.

4.1 Tests

We conducted our tests among the group of 22 users, mainly students of the Computer Science Faculty at Wroclaw University of Technology. Users were asked to register (giving some information about their interests) and log in to the system. They started to read and rate articles. At the beginning they could choose article by free hand. Next task for the users was to rate over 20 articles from prepared list. Each of them rated the same articles. They also gave marks for application layouts.

4.2 Results

Users' affinity was very important in our tests and we calculated weighted Kappa coefficient between them in order to find group of similar users. The calculations were made for different assumed values *p*, which determined users affinity level in our tests ($p = 0{,}7$; $p = 0{,}75$; $p = 0{,}8$; $p = 0{,}85$). Weighted kappa coefficient between users was compared with assumed value *p* and if it was higher than *p* value we treated these users as similar users. Other users were eliminated. In that way we could find group of similar users for AIS method. Selected groups of users were different for assumed *p* values. The amount of selected similar users decreased among with growing value *p*, but the affinity between these users increased.

Having users' articles and layouts rates, group of similar users selected in AIS method, we could calculate prediction rates for each items and compare them with the real ones marks given by the users. We could say that AIS method effectiveness was very high if these rates (article rates or layout rates) were very close or the same. If there was a big difference in user rates and AIS predictions, we judged effectiveness of AIS method as very low.

Precision accuracy measure could help us to assess the quality of predictions made by AIS method [2]:

$$1-\frac{\sum_{i=1}^{n}\left|\frac{prediction-actualVote}{g-1}\right|}{n} \tag{5}$$

where:
n – number of articles;
prediction – article's rate predicted using AIS method;

actualVote – actual user rate of the article;
g – number of rates (categories).

Precision accuracy measure was calculated for articles and application layouts. Figure 2 illustrates how this measure changed with the growing *p* value. The results for articles predictions were good for all *p* values (it was the result of good selection of similar users). We could see growing prediction accuracy with the growing *p* value. However, layouts prediction accuracy varied with increasing *p* value. The results were satisfied for the highest *p* value = 0,85. In this case we selected the groups of similar users using their articles rates. We wanted to check if the users who have similar preferences about articles could have had similar tastes about application layouts. Our results showed that there could be some influence however it was more independent.

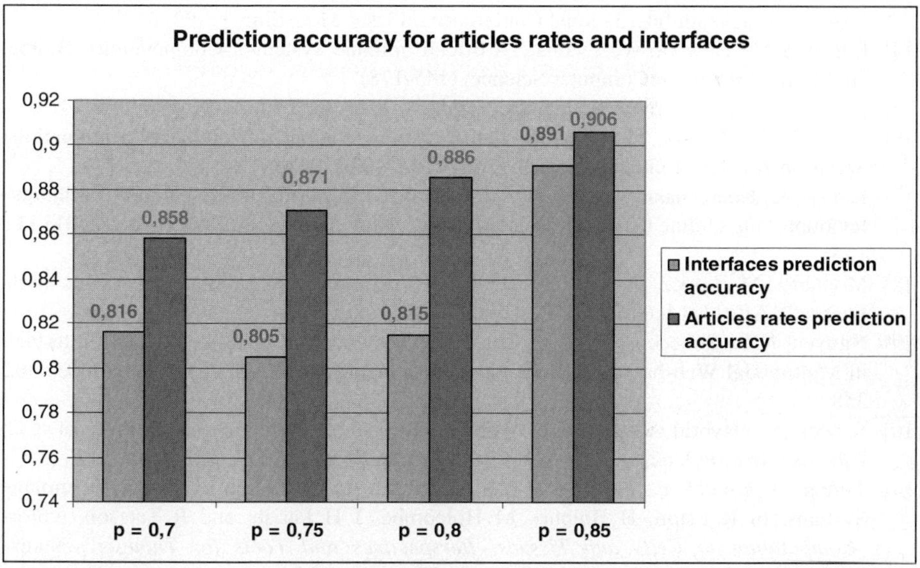

Fig. 2. Articles and interfaces rates prediction accuracy depending on threshold value *p*

5 Summary

The application of AIS method proved to be very effective for article recommendation, however a bit less effective for user interface recommendations. It was observed that we receive better articles predictions from more similar users, but this tendency does not hold for interface layout prediction. This means that AIS, one of the collaborative filtering (CF) method, is sufficient for articles recommendation. However, for user interface we should rather apply a hybrid method that would combine at least two recommendation methods: CF and content-based filtering (CBF).

The general outcomes of the experiments are consistent with well known recommendation methods properties [8] that CF is very good in new items (i.e. articles) recommendation. User interface preferences are more personal and in this case CBF methods should also be considered.

References

[1] Arafa, Y., Mamdani, A.: "Virtual Personal Service Assistants: Towards Real-time Characters with Artificial Hearts"; Proc. Intelligent User Interfaces New Orleans LA USA (2000), 9-12.

[2] Chen Q., Aickelin U., Movie Recommendation System using an artificial immune system, Poster Proceedings of ACDM 2004 Engineers' House, Bristol, UK.

[3] Fleming, M., Cohen, R.: "User modelling in the design of interactive interface agents"; Proc of the Seventh International Conference on User Modelling, (1999) 67-76.

[4] Garrett S.M., How Do We Evaluate Artificial Immune Systems?. Computational Biology Group, Department of Computer Science, (145-178).

[5] Lieberman, H.: "Autonomous Interface Agents"; Proc. CHI 97, ACM (1997) 67-74.

[6] Kazienko P., Kiewra, M.: "Personalized Recommendation of Web pages". International Series on Advanced Intelligence, 10, (2004) 163–183.

[7] Kobsa, A., Koenemann, J., Pohl, W.: "Personalized Hypermedia Presentation Techniques for Improving Online Customer Relationships". Knowledge Eng. Rev. 16(2), (2001) 111-155.

[8] Montaner, M., Lopez, B., de la Rosa, J.P.: "A Taxonomy of Recommender Agents on the Internet"; Artificial Intelligence Review 19, (2003) 285-330.

[9] Nguyen, N.T., Sobecki, J.: "Using Consensus Methods to Construct Adaptive Interfaces in Multimodal Web-based Systems"; Universal Access in Inf. Society 2(4), (2003) 342-358.

[10] Sobecki J.: "Hybrid Adaptation of Web-Based Systems User Interfaces". Proc. of ICCS 2004, Krakow, Poland, Lecture Notes in Computer Science 3038, (2004) 505-512.

[11] Timmis J, Knight T., De Castro L.N., Hart E., An Overview of Artificial Immune. Systems. In R Paton, H Bolouri, M Holcombe, J H Parish, and R Tateson, editors, *"Computation in Cells and Tissues: Perspectives and Tools for Thought"*, Natural Computation Series, pages 51-86. Springer, November 2004.

[12] Wikipedia, articles on Artificial Immune Systems and Collaborative Filtering. Downloaded on Octoer 2006 http://en.wikipedia.org.

[13] Wooldridge, M., Jennings, N.R.: "Intelligent agents: theory & practice"; Knowledge Engineering Review 100(2), (1995) 115-152.

Retrieving Geospatial Information into a Web-Mapping Application Using Geospatial Ontologies

Miguel Torres, Serguei Levachkine, Marco Moreno, Rolando Quintero, and Giovanni Guzmán

Geoprocessing Laboratory-Centre for Computing Research-National Polytechnic Institute, Mexico City, Mexico

{mtorres, sergei, marcomoreno, quintero, jguzmanl}@cic.ipn.mx

http://geo.cic.ipn.mx, http://www.geosco.org

Abstract. Many types of information are geographically referenced and interactive maps provide a natural user interface to such data. However, the process to access and retrieve geospatial data presents several problems related to heterogeneity and interoperability of the geospatial information. Thus, information integration and semantic heterogeneity are not trivial tasks. Therefore, we propose a *web-mapping* system focused on retrieving geospatial information by means of geospatial ontologies and representing this information on the Internet. Moreover, a Multi-Agent System is proposed to deal with the process related to obtain the tourist geo-information, which aids in the information-integration task for several nodes (geographic sites) that are involved in this application. The agent system provides the mechanism to communicate different distributed and heterogeneous Geographic Information Systems and retrieves the data by means of GML description. Also, this paper proposes an interoperability approach based on geospatial ontologies matching that is performed by the Multi-Agent System in each node considered in the application. The retrieval mechanism is based on encoding the information in a GML description to link each geospatial data with a concept of the ontologies that have been proposed.

1 Introduction

Maps are being used increasingly in local, networked and mobile information systems for communicating geographically referenced information. The applications are widely ranging including local government planning, environmental monitoring, market analysis, navigation and public access to information. Interaction with a digital map is typically based on a cycle of elicitation of user input via menu and dialog boxes, selection of map areas or features, and return of information, which may in turn induce modification to the map content [1].

Developments in human computer interaction with regard information retrieval and data visualization are new challenge to investigate new methods, since the current map interface, particularly on the Internet, often suffers from poor legibility of symbol and text, unnecessary user actions and inadequate adaptation to user interests.

N.T. Nguyen et al. (Eds.): KES-AMSTA 2007, LNAI 4496, pp. 267–277, 2007.

Up-to-date, the ontology notion plays an important role in establishing robust theoretical foundations for geographic information science [2]. An ontology allows us to solve problems associated to heterogeneity, interoperability, representation, integration and exchange of geospatial data. In this paper, we generate geospatial ontologies based on the conceptualization of a particular context, which is used to represent geographic objects by means of concepts ("not words"). The geospatial information is matched with any concept of the ontologies in order to retrieve this information in the web-mapping application. We propose a *Multi-Agent System* (MAS) to perform the communication between different spatial databases. Although the encoding agents may still refer to centralized ontology databases during the encoding process, the spatial databases can also be encoded in GML because of its openness. A Spatial User Interface Agent (SUIA) is proposed in the web-mapping application to use the ontologies for validating user inputs and capturing the requests to retrieve geospatial data by means of "concepts". In addition, the SUIA works in a web browser providing an easy-use web user interface for online geospatial information retrieval. SUAI is characterized by the following features:

- Handle spatial data.
- Retrieve spatial data by means of concepts, considering the context.
- Perform spatial queries according to the generated geospatial ontologies.
- Generate new geospatial information making spatial analysis.

In this application, the geospatial data are associated with different concepts. For instance, a *"water body"* can be represented like a *"natural water body"* such as a *"lake"* or it can be represented like an *"artificial water body"* as a *"dam"*.

The rest of the paper is organized as follows. Section 2 describes the Multi-Agent System proposed to perform several tasks in the application, and the geospatial ontologies definition to retrieve geospatial data. Section 3 sketches out the architecture of the web-mapping application. The implementation of the prototype is pointed out in Section 4. Section 5 shows some results of the application in order to retrieve geospatial data. Our conclusions are outlined in Section 6.

2 Multi-Agent System and Geospatial Ontologies

The application is composed of two components to retrieve geospatial data.

- A Multi-Agent System (MAS) to perform tasks related to communicate different spatial databases by means of GML definition and encode the geospatial data for retrieving in the SUAI.
- Geospatial ontologies to solve *ambiguities* by means of concepts ("not words").

2.1 Multi-Agent System

According to [3], an agent is a system that attempts to fulfill a set of goals in a complex, dynamic environment. It can sense the environment through its sensors and act upon it using its actuators. In this work, we propose a Multi-Agent System (MAS) that provides services to facilitate the retrieval of geospatial information within a tourist context.

There are two functions of the agent in the application. One is to provide an intelligent service to communicate different spatial databases and encode the geospatial data for retrieving to the user. The other is to check the GML definition and link the ontology for knowing whether the concepts accomplish to the search criteria. Several types of agent have been proposed, they are organized in four layers depending on their functionality.

- *Data Layer.* It is composed of the agents that provide data access services.
- *Management Layer.* The agents handle and coordinate other agents into MAS. Also, they provide the capabilities of communication to other agents.
- *Application Layer.* Agents perform specific tasks such as visualization and functions to the geospatial data. Moreover, these agents manage the ontologies (*Trip Package* and *Map*) to provide data to the interface for giving its own services.
- *Presentation Layer.* Here, the agents provide a user graphic interface to allow the users obtaining the application services.

We grouped the layers in two clusters: *Core Utility Agents* and *System Specific Agents*. The tasks of the Core Utility are the follows:

- *Data Locator.* It finds the data that better fulfill the description given by clients. The agent provides the address of the agent, which can provide the access to the data.
- *Data Access.* It provides the accessing mechanisms for the data and metadata of a particular source. The queries and results are given in XML.
- *Communication Router.* This agent provides the capabilities of MAS to communicate with other one, through any suitable way (Internet, others MAS, Virtual Private Networks, etc.).
- *System Management.* This agent handles the process within a MAS. It starts the computation of all other agents in the same MAS.
- *Directory Facilitator.* It maintains a list of all the known agents by MAS as well as the services that each agent provides to the layers.
- *Spatial Facilitator.* This agent retrieves the geospatial data from the SDB. According to the client's request. The agent sends the geospatial data to make-up a map in a GML description.

The System Specific Agents is a set of agents that interacts to accomplish specific goals. The agents that belong to this cluster are the following:

- *Resource Management.* This agent deals with all the tasks of resources assignment. For instance, searching a hotel and flight and finding trip packages.
- *Ontology management.* This agent keeps the information about the "Map Geo-Ontology" and uses it to translate the user's request into structured queries, which will be computed into the Ontology Administration Query Module.

These queries allow other agents assigning resources to users and finding geographic objects to provide specific maps to the clients.

- *Ontology matching.* This agent acts when there are confusions about the concepts in the client's ontology and the MAS ontology. Then the agent attempts to find the closest concept in MAS ontology, according to the concept given by the client.
- *Spatial User Interface Agent.* This agent translates data given by the MAS into a rendered map that the user can read.

Fig. 1 depicts the steps to accomplish the process to communicate and retrieve geospatial data between two nodes.

1. The Client (Spatial User Interface Agent) makes a request to the application (for example, a user in Spain wishes to get a road map of the zone of Cancun in Mexico).
2. The MAS in Spain asks to its Directory Facilitator for the MAS that has such information.
3. The Directory Facilitator searches in its database the information requested, and responds to the MAS that the MAS of Mexico has the map.
4. The MAS in Spain asks the MAS in Mexico for the road map of the zone of Cancun.
5. The MAS in Mexico computes the request and determines if it has such information.
6. If it does, then the request passes to the Spatial Facilitator.
7. It makes a spatial query to the SDB for retrieving the geographic objects requested.
8. The Spatial Facilitator returns to the MAS in Mexico the geographic objects needed to make-up the map.
9. The MAS in Mexico translates this information using the ontology to a format that the MAS in Spain will interpret.
10. It sends the information to the MAS in Spain.
11. Finally, the MAS in Spain sends the result to the client, and it displays the road map of Cancun to the user with a attributive description.

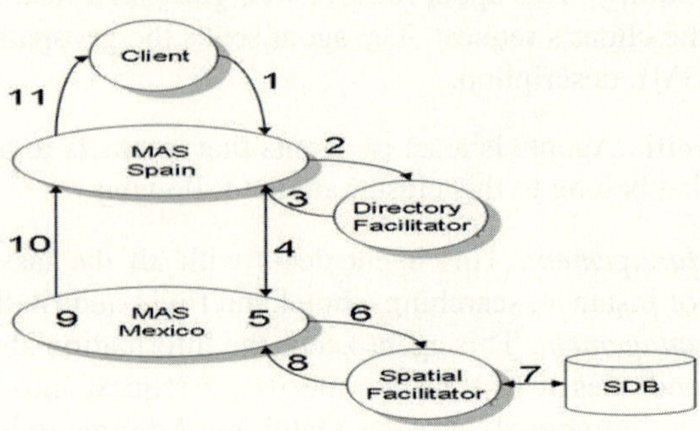

Fig. 1. Interaction model between two MAS

2.2 Geospatial Ontologies

Most widely accepted common conceptualization of the geospatial data is based on the description of geographic objects and fields [4]. These objects are not necessarily related to a specific geographic phenomenon, because human-built features are typically modeled as objects [5]. We consider a spatial ontology as an explicit, shared and structured specification of conceptualization, that is, a description of properties and relationships that can exist between the geographic objects to define concepts.

Moreover, ontologies can be considered as "languages", which use a specific vocabulary to describe entities, classes, properties and functions related to a certain view of the geographic world [6]. The geospatial ontologies can be classified by levels according to their dependence on a specific task or point of view. These levels are generated for a specific geospatial ontology (*top-ontology*) and it can be particularized to define a particular ontology (*down-ontology*). There are also different levels of information detail. *Low-level* ontologies correspond to very detailed information and high-level ontologies belong to more general information. The levels of ontologies can be used to guide processes for the extraction of more general detailed information, according to their *granularity*. The use of explicit geospatial ontologies contributes to better spatial representation, because each geographic object description is based on an implicit ontology. By using that, it is possible to avoid explicit *conflicts* and *confusions* between the ontological concepts and the implementation. The use of ontologies in spatial databases enables knowledge sharing and information integration. The proposed approach provides dynamic and flexible information exchange and allows partial integration of spatial data when completeness is impossible.

In our approach, the ontologies are composed of two types of concepts (C): *terminal* (C_T) and *non-terminal* (C_N). The first ones are concepts that do not use other concepts to define their meaning (they are defined by "simple values"). The meaning of non-terminal concepts is conceived by other concepts, which can be terminal or non-terminal concepts (see Eqn. 1) [7].

$$C = C_N \bigcup C_T \tag{1}$$

Each concept has a set of *aspects*. These are characteristics that describe the properties, relations and instances that involve the geospatial objects. From-now-on, we shall use the term "relation" to denote unary relations/properties. From this point of view, all aspects of a terminal concept are simple, e.g. the type of all aspects that belong to the set of primitive types (punctual, linear and areal objects) is denoted by (T_P), as shown in Eqn. 2.

$$\begin{aligned} &T_P = \{number, character, string, enumeration, struct\}, \\ &A = \{a_i \mid type(a_i) \in T_P\}, \end{aligned} \tag{2}$$

where T_P is the set of primitive types; A is the set of aspects.

Then, the set of *terminal concepts* is defined by Eqn. 3.

$$C_T = \{c(a_1, a_2, ..., a_n) \ni a_i \in A,\ i = 1,..,n\} \tag{3}$$

In the same way, the *non-terminal concepts* have at least one aspect that does not belong to T_P. It is denoted by Eqn. 4.

$$C_N = \{c(a_1, a_2, ..., a_n) \ni \exists a_i \notin A\}, \text{ where } c \text{ is a concept.} \quad (4)$$

Finally, the set of relations R is defined by the pairs that are associated to Γ and Φ, in which Γ and Φ are non-reflexive, non-symmetric, and transitive relations (Eqn. 5).

$$R = R_\Gamma \cup R_\Phi = \{(a,b) \mid a\Gamma b,\ a \in C_N,\ b \in C\} \cup \{(a,b) \mid a\Phi b,\ a \in C_N,\ b \in C\} \quad (5)$$

3 Architecture of the Application

The web-mapping application is composed of two tiers: Client tier and Spatial Data Server tier. These tiers contain the following components: Spatial User Agent Interface (SUAI), Ontology Administration Query Module (OAQM), Spatial Data Server (SDS), Agent Administration Module (AAM) and Spatial Database (SDB). The application is a distributed system, because it is able to retrieve geospatial data from different GIS sites by means of GML definition. Fig. 2 depicts the general architecture of the web-mapping application.

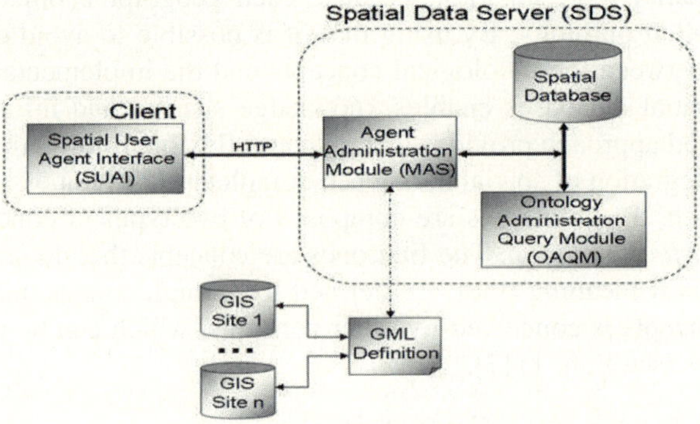

Fig. 2. Architecture of the web-mapping application

The general process to retrieve spatial and attributive data is the following:

Spatial User Agent Interface (SUAI) receives a request from user. It aids the client to search, query and manipulate the map in an efficient and user-friendly way. SUAI attempts to understand the geographic context of the user, and it sends a message to the Spatial Data Server (SDS) to ask more geospatial information or to modify the map to change the content and resolution detail. SUAI should keep a profile for each user to record his search of interest. The Agent Administration Module (AAM) receives requests from the SUAI and it broadcasts the requests of the users to the Ontology Administration Query Module (OAQM) to search the concept into the ontologies and to retrieve geospatial information from the Spatial Database (SDB). If the geo-information associated to the concept could not be found in the SDB, the OAQM will send a notification to the AAM to perform a query in different GIS sites linked to the application. This process is made by means of the GML definition, when the geo-information is found; it is encoded in the GML description and transferred to the

AAM to retrieve the geospatial data according to the geospatial ontology. Finally, the spatial data is sent to the SUAI.

4 Implementation of the Prototype

The prototype has been implemented in Java to keep the distribution and multi-platform execution. The application consists of seven nodes to retrieve spatial and attributive data. SUAI is implemented as a Java Applet and runs on the client side to interact with a web user. The AAM has been implemented as a Java servlet using Tomcat 5.0.12. The visualization on the client side is based on *Shapefiles*, which is proposed by Esri, Inc. The data workflow is depicted in Fig. 3.

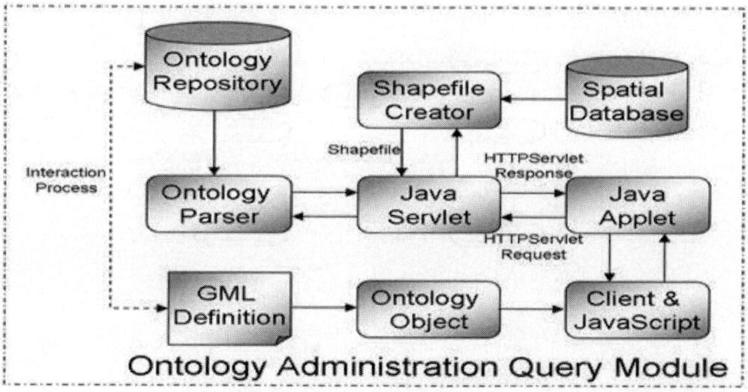

Fig. 3. Data workflow to obtain the spatial data basing on geospatial ontology

When a user accesses the web page, the JavaScript embedded in the web page will call a Java applet to send an http request to the Java servlet, which will invoke the ontology parser to create an ontology object from the ontology repository. If the information does not find in the Ontology Repository, the OAQM sends a GML definition to locate the data in any node. When information is found, it is received by the OAQM for being computed. Later, the OAQM sends the object as a serialized Java object to the applet. The applet uses the ontology object to verify if the user has performed a valid search. If it is valid, the applet will submit the search to the servlet, which in turn invokes the *shapefile* generator to obtain information for the client.

In this context, the geospatial ontology is a part of knowledge, concerning a particular geographic context (Tourism). We propose two ontologies to obtain the spatial data by means of concepts: *Map Geo-Ontology* and *Trip Package Ontology*. Both ontologies have been designed in Protègè 3.2, the relations that have been considered in the ontologies are: *"has"* and *"is-a"*.

Map Geo-Ontology is focused on retrieving particular maps of the user interest. It can generate four types of maps: Roads, Weather, Urban and Sightseeing.

Trip Package Ontology is proposed to acquire attributive data related to the interest places for the users.

Fig. 4 and Fig. 5 show the ontologies that we propose to retrieve geospatial data by means of concepts in a tourist context. The GML definition is used to obtain the geospatial data from different distributed GIS according to the request of the user. MAS sends this definition to find the specification related to the request. If the data have been found, the GML definition encodes the information, which is sent to the Ontology Repository for matching this information encoded into the GML definition with the ontology structure. Inside the Ontology Administration Query Module the information is parsed for matching it with the concepts that define the ontologies. We use the *relationships* between concepts that belong to the ontology to communicate the *Map* and *Trip Package Ontologies*. A description of the GML definition is shown in Table 1.

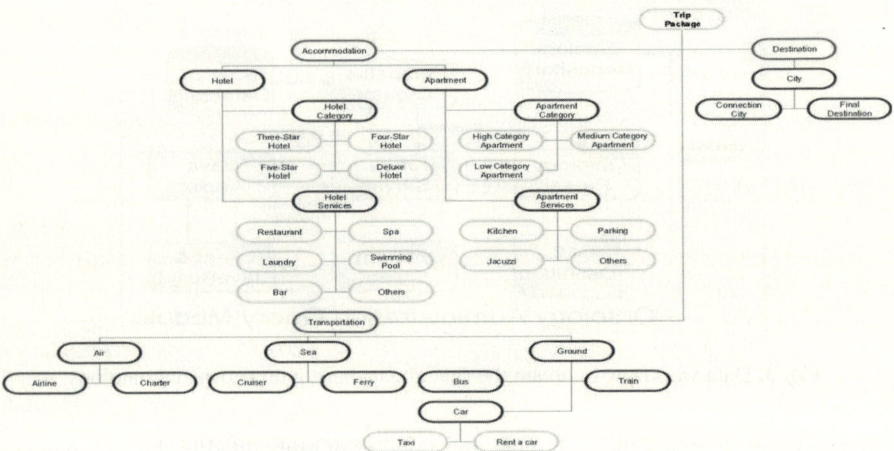

Fig. 4. Trip Package Ontology

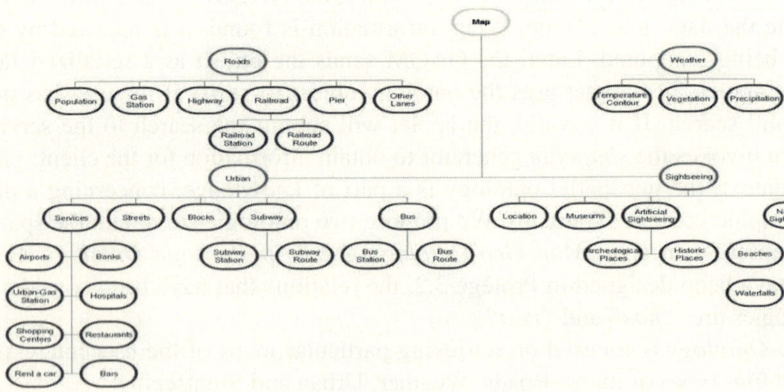

Fig. 5. Map Geo-Ontology

Table 1. GML description

```
<?xml version="1.0" encoding="UTF-8"?>
<rdf:RDF xml:lang="en"
   xmlns:camb="http://geo.cic.ipn.mx:9090/RDF/VRPexample_profile3_schema.rdfs#"
   xmlns:gml="http://geo.cic.ipn.mx:9090/RDF/VRP/Examples/gml.rdfs#"
<!--The camb and gml namespaces replaced for validation purposes Map Ontology Data-->
   <camb:Map>
      <gml:boundedBy>
         <gml:Box gml:srsName="ROAD:4326">
            <gml:coordinates>
               0.0,0.0 100.0,100.0
            </gml:coordinates>
         </gml:Box>
      </gml:boundedBy>
      <camb:modelDate>
         Dic 2003.
      </camb:modelDate>
      <camb:modelMember>
         <camb:Roads>
            <gml:name>
               I45
            </gml:name>
            <gml:description>
               Federal Highway from San Pablito to Cancun.
            </gml:description> … </rdf:RDF>
```

5 Results of Retrieving Process of Geospatial Data

We have developed roads, city, weather and sightseeing maps. These maps are generated by means of concepts that belong to the *ontologies*. Data have been retrieved by the GML definition according to the user request. Fig. 6 depicts the *map of roads* for Toluca City, Mexico. This map consists of different thematics as Populations, Roads, Urban Areas and Internal Administrative Divisions.

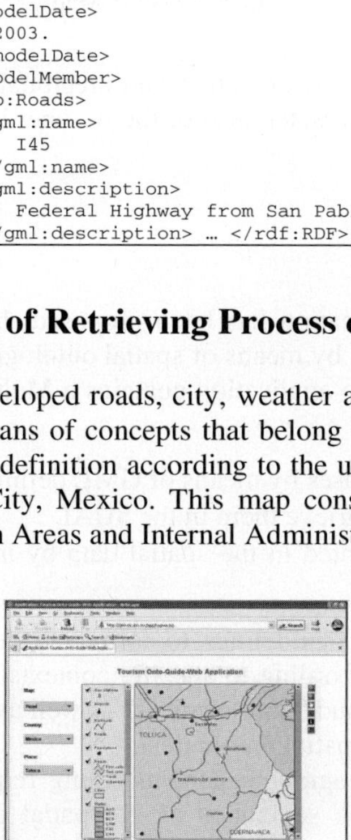

Fig. 6. Map of roads

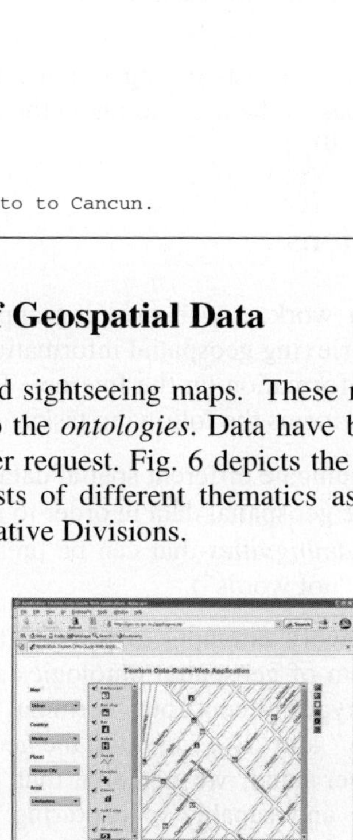

Fig. 7. City Map

In Fig. 7, *city map* is composed of streets, avenues and different interest sites. It shows the map of Lindavista area in Mexico City, which scale is 1:5,000. In this map we show the location of different sites as Restaurants, Bus Stations and Hotels in this area. Also, *Trip Description Box* provides useful information related to the user request. Fig. 8 shows the *sightseeing map* of San Pablito in Quintana Roo, Mexico. This describes general aspects of San Pablito, showing the Information Sites Location, Gas Stations, Camping Zones, Restaurants and Archeological Sites.

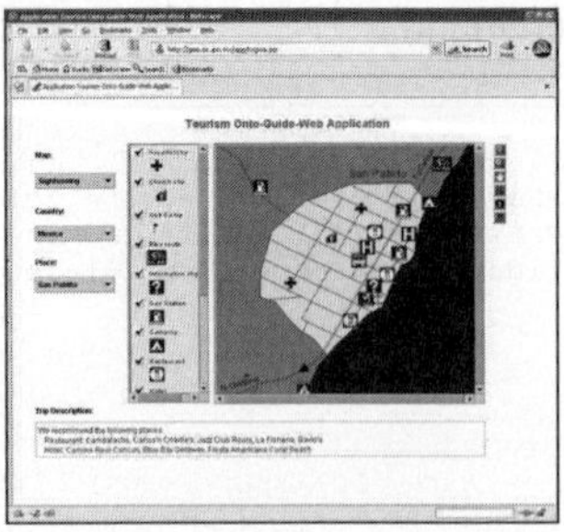

Fig. 8. Sightseeing Map

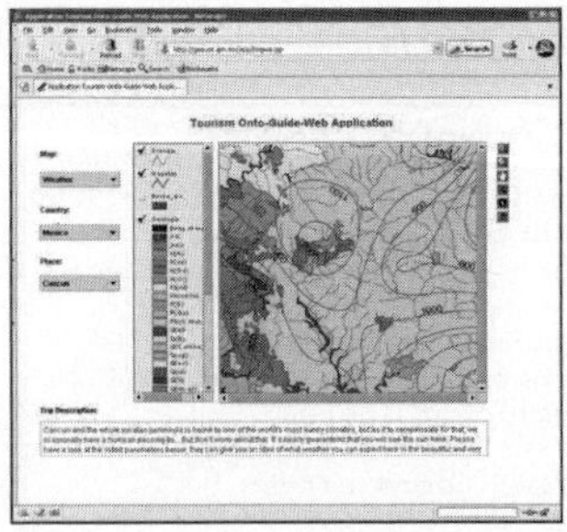

Fig. 9. Weather Map

Weather map consists of vegetation areas, temperature and precipitation contours. This map guides to the users to know the characteristics of the weather in a particular place (see Fig. 9).

6 Conclusions

In the present work, a web-mapping application has been proposed. It is a system focused on retrieving geospatial information by means of spatial ontologies and representing this information on the Internet. The application contains a Multi-Agent System, which performs the following tasks:

- Communicate different spatial databases by means of GML definition.
- Encode geospatial data in order to retrieve them in the SUAI.
- Solve *ambiguities* that can be presented in the spatial data by means of concepts ("not words").

This framework attempts to provide the guidelines to formalize the geographic domain in form of geospatial ontologies according to specific contexts. We have introduced two types of concepts: "terminal" and "non-terminal" as well as two kinds of relations: "*has*" and "*is-a*" to build the geospatial ontologies.

On the other hand, we perceive that geographic data modeling requires models more specific and capable of capturing the semantics of geospatial data, offering higher abstraction mechanisms and implementation independence. This approach allows us to process imprecise data and aid to information integration and semantic heterogeneity tasks. We attempt to show an alternative approach to represent geospatial data on the Internet considering the *relationships* that compose the ontologies to retrieve geospatial data according to several search criteria based on concepts.

The use of ontologies in spatial databases enables knowledge sharing and information integration. The proposed approach provides dynamic and flexible information exchange and allows partial integration of geospatial data when completeness is impossible in the web. The communication between ontologies is performed by MAS, which seeks the relationships of the concepts to match nodes in the ontologies. In difference with respect to others systems, this application retrieves geospatial data, using pre-designed ontologies.

Acknowledgments

The authors of this paper wish to thank the CIC, SIP, IPN and CONACYT for their support.

References

1. Li, M., Zhou, S. and Jones, C.B.: Multi-agent Systems for Web-Based Map Information Retrieval. In Egenhofer, M.J. and Mark, D.M. (Eds.), *GIScience 2002*, Lecture Notes in Computer Science Vol. 2478 (2002) 161-180
2. Mark, D., Smith, B., Egenhofer, M. and Hirtle, S.: Ontological Foundations for Geographic Information Science, in McMaster, R. and Usery, L. (eds.). *A Research Agenda for Geographic Information Science*, CRC Press, Boca Raton, FL (2004) 335-350
3. Maes, P.: Modeling Adaptive Autonomous Agents. *Artificial Life*, No. 1 (1994) 135-162
4. Fonseca, F., Egenhofer, M. and Agouris, P.: Using Ontologies for Integrated Geographic Information Systems. *Transactions in GIS*, Vol.6, No. 3 (2002) 25-40
5. Egenhofer, M. and Frank, A.: Naive Geography, in Frank A. and Kuhn W., (Eds.) Spatial Information Theory, A Theoretical Basis for GIS, *Proceedings of the International Conference COSIT '95*, Lecture Notes in Computer Science, Vol. 988, Springer-Verlag, Berlin (1995) 1-15
6. Guarino, N.: Formal Ontology, Conceptual Analysis and Knowledge Representation. *International Journal of Human and Computer Studies*, 43, Vol. 5, No. 6 (1999) 625-640
7. Torres, M., Moreno, M., Quintero, R. and Fonseca, F.: Ontology-driven description of spatial data for their semantic processing, *Proceedings of the First International Conference on Geospatial Semantics*, Springer-Verlag, 3799, Mexico City, Mexico (2005) 242-249

MWING: A Multiagent System for Web Site Measurements

Leszek Borzemski, Łukasz Cichocki, Mariusz Fraś,
Marta Kliber, and Ziemowit Nowak

Institute of Information Science and Engineering, Wroclaw University of Technology,
Wybrzeze Wyspianskiego 27, 50-370 Wroclaw, Poland
{leszek.borzemski, mariusz.fras, ziemowit.nowak}@pwr.wroc.pl

Abstract. When browsing the Web, users demand the low latency, high throughput and high availability of Web pages. Therefore each Web site should be measured, tested and evaluated from the user's perspective. In the paper, we present the architecture of the system MWING that is able to test, measure and diagnose Web sites from the user point of view. It is a new generation system that adds new functions and extends the functionality and performance of the previously developed WING tool. MWING is now the multiagent measurement platform able to cooperate with independently developed measurement agents. Here we present how the MWING works and how the new architecture meets the ultimate user needs.

Keywords: web site measuring, multiagent system, performance evaluation.

1 Introduction

Network protocol measurement and visualization tools are considered as tools for better understanding of computer network. They can help network administrators or end users in evaluation and analysis of network and Web site performance and reliability. There are general tools used for several network protocols and tools developed specifically to serve specific protocols. Nowadays, there is the need to use such tools for the TCP/IP protocol suite, especially in the measurement, testing and evaluation of Web page downloading made by means of the HTTP protocol.

Several active and passive measurement projects have been built on the Internet, e.g. [11]. Mostly they are aimed to deal with the performance problem related to whole or a significant part of Internet where large amounts of measured data regarding, for instance, round trip delay among several node pairs over a few hours, days or months, and using specific measurements and data analysis infrastructures are obtained. These projects can not be reprogrammed at all. New functionality requires a new measurement system.

In case of Web measuring the MyKeynote [12] is the most advanced benchmarking service that measures Web site's performance and availability from a word-wide network of measurement agents. It is a commercial service with some free (demo) functionality. In demo version the user can make some measurements of target URL

N.T. Nguyen et al. (Eds.): KES-AMSTA 2007, LNAI 4496, pp. 278–287, 2007.

and obtain valuable information how the target page is seen by MyKenote. To address the issues of complex systems development in distributed environments, research on multi-agent systems (MAS) and their application is on the rise [6, 7, 8, 9, 10].

In the paper, we present the Internet measurement platform for distributed measurements called MWING. MWING has been developed with the intention to be used for the purpose of the analysis of Web page downloading processes as seen from the user's perspective. Because the system has been developed based on a multiagent oriented architecture we can deploy any needed measurement and processing functionality concerning Web and Internet. Agents usually are designed to meet special needs, thus making difficult to communicate with them. Some tools like gateway agents try to solve this problem using, for instance, intermediate XML-based messages [6]. Our system has this functionality built in.

The paper is organized as follows. Section 2 introduces architecture of the MWING. Section 3 features the WING agent. Experiments performed with the use of the system are presented in Section 4. Finally, concluding remarks and future work are given in Section 5.

2 MWING Platform Architecture

2.1 General Overview

To measure Web performance we developed the active measurement system WING (Web pING) [3]. Several tools exist to measure different parameters of network and Web performance but due to our specific needs we have developed the measurement system from scratch. WING is a network measurement system that measures end-to-end Web performance path between the Web site and the end-user. It was implemented for the needs of end-users located at Wroclaw University of Technology (WUT) campus network. WING based measurements were carried out to evaluate the performance and reliability of access to Web sites from the perspective of WUT users. The obtained measurements were successfully used in analytical and data mining based research on the HTTP throughput prediction algorithms [1, 4, 5]. From this work we concluded that the approach to Web measurement and the system itself need to be improved to meet new requirements and to gain more capabilities in deeper studying of Web data transmission and Web data access problems. The following new design challenges were constituted: stability and increased precision of measurements, advanced features in measurement scheduling, possibility to perform simultaneous measurements of several sites, distributed measurements with the use of agents, advanced off-line processing.

The main goal was to develop the measurement platform for the integration of current and future measurement applications with the common high-level interface for measurements, data management, data processing and data exploration, including the API based interface to build applications for the automatization of these functionality at the level of user programs. The proposed solution is MWING (Multiagent WING) – the multiagent measurement platform for Web site measuring, testing and performance evaluation.

The general architecture of the system is shown in Fig. 1. MWING consists of four components: Web Application, System Controller, Database and Agent Sets.

Web Application (*WebApp*) is the module responsible for communication with the user. It consists of *Measurement Management* (*MeasureMgmt*) unit where all functionality offered by the system is placed. By the usage of its services the user can control system behavior, add new measurement tasks, define new agents and create measurement programs. This unit has been designed with ease of usage in mind. The *Data Browser* unit visualizes whole measurement platform functionality. Visualization concerns not only currently performed tasks but also the historic data stored in database.

The central component of the system is the *System Controller* module. The tasks related to the system management defined by the user come here. This component consists of three units. *Server* unit communicates with Database and Measurement Management unit. It is responsible for adding and deleting measurement tasks, registering agents and measurement programs. *Scheduler* unit is responsible for activation of tasks at proper date and time. Both these units are always up and ready to action. The third unit of System Controller module – *Management Controller* (*MC*) unit – is, contrary to the previous, not always in action. MC is called by the Scheduler only when any task needs to be done. It communicates with Database, gets task parameters and depending on task to be executed calls the proper measurement agent from a proper Agent Set.

The *Database* is a one or more DBMSs configured for the needs of the measurement system. Different DBMSs can be used with their specific features (for instance, in case of different data exploration engines that are to be used in future).

Agent Set is actually responsible for execution of commissioned tasks. After completion of processing by the agents of Agent Set the result is sent back to MC. Gathered data is sent to Database and MC finishes its activity. Agent Set can be a single agent (e.g. ping based) or a group of agents (e.g. WING).

Relations between agents are strictly depended on the whole measurement architecture. Agents are autonomous in action so that the system can rely on their results. They also are able to distinguish proper and fault parameters and take the

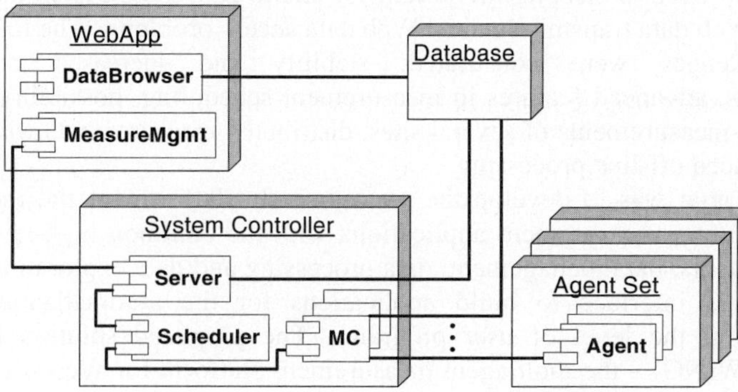

Fig. 1. The architecture of MWING Platform

proper action accordingly. The units of the single component are placed in the same network segment but each component can be placed anywhere in the Internet. Generally, there is no restriction on the agents' locations, as well.

2.2 Detailed Description of System Components

Web Application. The Web Application is the most important component from the user's point of view. It is implemented as a web application and offers the following functionality:

- **Management of agents.** All agents running in the system have to be first registered by the Measurement Management (MeasureMgmt) unit. The registration produces specific data in system tables dedicated for system administration. Administrator has to provide information about connectivity parameters to the agents and define their parameter lists within the system.
- **Management of measurement tasks.** Using the GUI interface the user can add new measurements to the system. This action requires providing the fundamental information about the measurement task. The following information has to be provided: the name and type of measurement task, list of agents to be assigned to the measurement task, definition of measurement task parameters and relations between them and agent parameters. The required parameter is the name of the data table where the result data will be stored after receiving it from MC.
- **Management of measurements.** It is the main goal of MeasureMgmt unit. To start new measurements the administrator has to provide all required information about selected measurement tasks and the date and time schedule. Date and time schedule can be taken from the list of possible schedules. System offers optional schedules: single measurement, periodical measurements, measurements based on different statistical models, measurements scheduled using the date and time list given by the user.

Data Browser unit is used for data retrieving and for the presentation of result data stored in a database. It consists of two modules: Data Export Module (DEM) and Data Visualization Module (DVM). DEM offers easy to use forms to download measurement data from the database into desired file formats. DVM offers the visualization of the result data if exists the visualization sub-module related to the given measurement task.

System Controller. System Controller module interacts every component in the system. The Server unit API allows adding, editing or removing measurement task entries within the Scheduler. Each entry consists of the following information: Measurement ID, Type of measurement, Measurement execution date and time schedule, Input data required to configure the particular measurement task (concerning the selected measurement type).

The type of the measurement is based on the available measurement task registered in the system. The date and time of measurement execution is a value which is calculated depending on assigned scheduler program. The proper Management Controller unit is called to activate measurements according to user's selections and these calculations.

It is possible to run multiple MC instances at the same time. The instances of MC are started for each single measurement ordered by the Scheduler. The MC process is responsible for the communication between the measurement agents and Database system. MCs are running on the same machine as the Scheduler. They interact with the agents according to the measurement task definition using the SSH protocol based interface. To activate and manage the particular measurement the measurement ID, the agent IDs and the parameters for agents are taken from the database. When agents finish their actions the result data is sent back to the MC with the measurement context (Measurement ID, Measurement Task ID, etc.) and MC stores that data in a dedicated database table.

The MC also stores in the database the status of performed measurements. This allows monitoring of measurement execution by the Measurement Management unit. There is no relation between MC's processes running in parallel. The only possible problem might happen only when different MCs require the same measurement task entry at the same time. The transaction mechanism supported by the measurement task status stored in database is used to handle such scenario.

Agent. The Agent Set can be a single agent or a set of cooperating agents (WING agent is an example). In the case of complex measurement that requires multiple agents to be used, one of agents may take a role of coordination agent supporting cooperation between them and the whole system. Each agent has its own log. Information on its activity is stored there for possible error trace and analysis. The agents are autonomous and can do anything. The range of agent's functionality is not defined within the core of the system. The agent algorithm is based on the measurement concept and agent behavior developed by the agent designer. It includes such issues as measured data filtering and preprocessing (e.g. WING agent filters and prepares data related to HTTP performance measurements), reactions to unexpected measurement errors (e.g. abandoning of these portions of measurements that fail) or own scheduling or retrying the ordered tasks. Any operating system and programming language can be used for the implementation of the MWING agent. The requirements of agent construction are not restricting. It is required that the target operating system platform has to provide the SSH protocol as all communication with agents is performed using that protocol. The information about connection parameters required for agent's interaction is provided by the MC during the agent initialization.

The key concept of flexibility of cooperation of agents with the system and communication facilities is agents' interaction based on XML messages. After the execution of the agent's procedure the result data is sent back to the MC in the form of XML file. XML message containing results gathered by the agent during its activity consists of the following tags: the return value describing the status of execution of the agent, the possible error/warning information, and the data collected by the agent. An example of the XML message sent by WING agent is the following:


```
<?xml version="1.0" encoding="utf-8" ?>
<response xmlns="http://www.w3schools.com">
  <retVal>0</retVal>
  <msg></msg>
  <data>
```

```
    <id>'3'</id>
    <slot_id>1</slot_id>
    <agent_id>2</agent_id>
    <datetime>'2007-01-19 15:41:27'</datetime>
    <record_seq>0</record_seq>
    <record_type>'S'</record_type>
    <measurementerror>458962</measurementerror>
    <dns>0</dns>
    <connect>48448</connect>
    <firstbyte>52800</firstbyte>
    ...
  </data>
</response>
```


All data fields are described and must be declared in MWING platform during agent registration. To minimize the amount of data sent through the network the compression of the response data file is considered.

Till now two agents are implemented for the system: the WING agent which is described in the section 3, and simple Ping agent. There are considered or under development other agents such Trace agent (for trace routing and Internet connections discovering) or extended, multifunctional Ping agent.

3 WING Agent

Originally the WING was the whole active Web measurement system. Now it is one of possible types of measurement agents. It probes the target Web page and observes the Web transaction in the way the user perceives and observes the Web page downloading. WING measures the Web transaction in the way which is shown in Fig. 2. We have proposed the ultimate interpretation and own names of Web transaction phases that includes all of them. Some phases of Web transactions are omitted in the measurements in the literature, but our experience has shown that those omitted by others phases can cause significant delays. Therefore, all phases must be taken into account.

In the beginning, the IP address is resolved using the Domain Name Service (DNS). This process takes sometimes long time because the resolution cannot be accomplished locally but sometimes the IP address can be taken immediately from the internal Web client cache. The time needed to resolve the IP address we call the DNS time. Next, the Web client opens the TCP connection using an exchange of SYN packet that initiates the three-way handshake. However, before the SYN packet issuing the client has to wait for local computing resources (e.g. memory, CPU) for the significant period. Therefore, there is the need to measure the time between DNS resolution event and SYN packet exchange. We call this time the DNS2SYN. The connection phase begins when a client initiates connection request SYN, and ends when the connection is established, i.e. when the server receives ACK packet from the client. The elapsed time between sending the SYN packet to the Web server and receiving the SYN response we call the CONNECT time. The processing phase for the requested object begins with sending HTTP GET request, and ends just after receiving the last byte of that object. Here, there is another time delay worthy to measure – time between obtaining the ACK packet and sending GET request. This time we call ACK2GET time. The GET request is sent for a page skeleton or object embedded in a page. Regardless of that, we measure two times, the FIRST_BYTE and LEFT_BYTES times. The FIRST_BYTE time is the time between the sending of GET request and the reception of the first packet including requested data. The LEFT_BYTES time is the time spent for downloading the rest of the requested data.

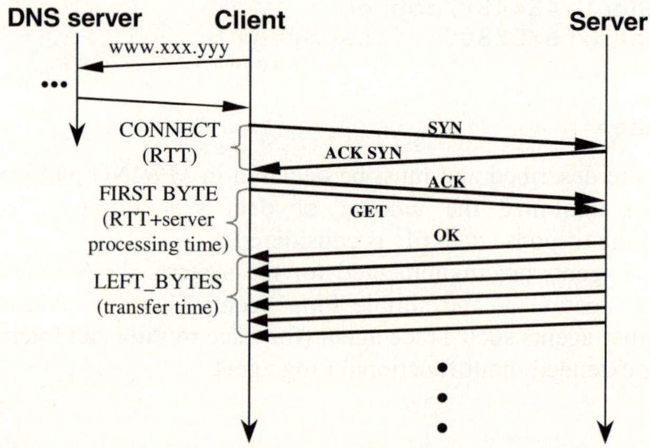

Fig. 2. The diagram of a Web transaction

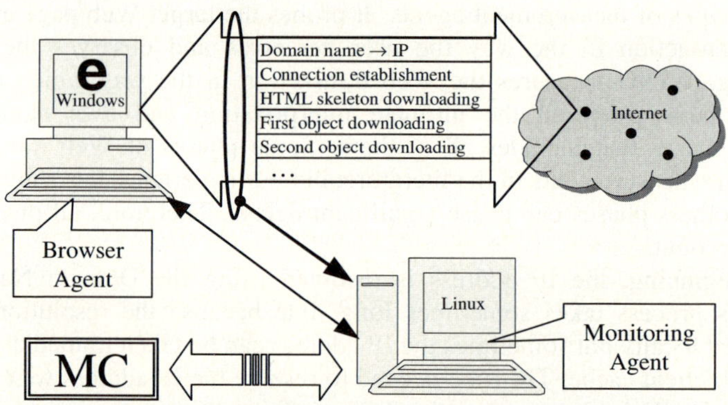

Fig. 3. The architecture of WING Agent Set

The main emphasis in the WING project was to develop a network visualization system for supporting real-life usage of popular Internet browsers. Therefore, we built the system in Windows environment for MS IE. Because of performance reasons, we decided to use two computers, one for browsing and one for monitoring the local traffic. Therefore the WING agent consists of two components (Fig 3): the Browser Agent for MS IE browser and Linux based Monitoring Agent. The cooperation with MC is carried out by the monitoring agent.

WING agent is activated by a remote request sent to Linux based controller. The controller starts a Web local client that issues GET request to target URL, as well as GET requests for all objects that are embedded in the target page. WING works like a sonar-location system, sending requests for the target Web sever and waiting for the answer. WING looks at the local client traffic, prepares and categorizes data showing how the target Web page has been loaded locally. The results are then collected and taken by MC into the relational database and can be used in further off-line processing.

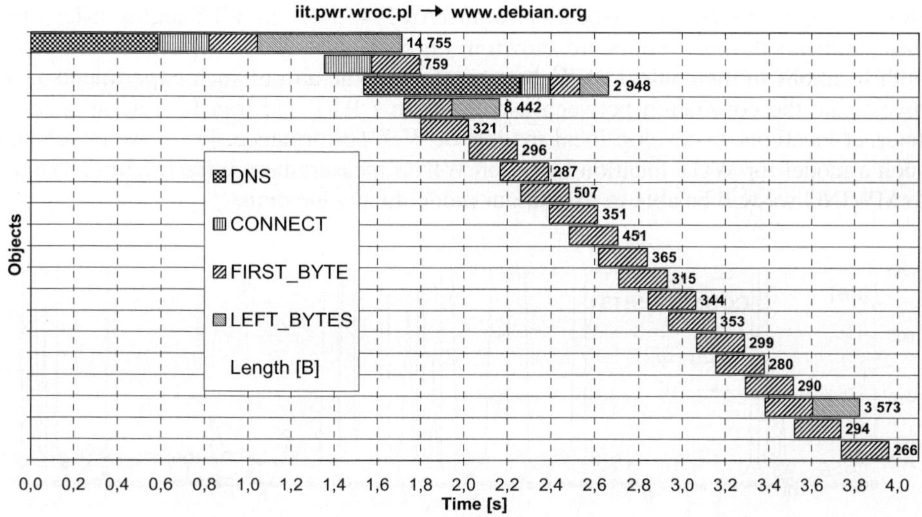

Fig. 4. An example of WING visualization

WING supports IP, TCP, UDP, DNS and HTTP/1.1 protocols, logging a dozens parameters of HTTP transactions and TCP connections thus helping deep analysis. WING helps identify inefficient network usage by the browser and Web server and helps to tune the applications and Web pages to use the network efficiently. Especially, by the visualization of the network traffic it helps identify performance and reliability issues. Therefore, WING can be a good analysis tool for Web and network application developers. Fig. 4 presents an example of the result, which is achieved in measurement made by WING agent and presented by WING visualizer.

WING was used to estimate the HTTP throughput and TCP Round-Trip Time (RTT) [1, 2, 4]. We estimated the RTT on the basis of the measurements of time spacing between the SYN packet sent by the client and the SYN-ACK packet received by the client. Although this delay is not strictly the RTT (as seen by ping facility), it is a good estimation of the RTT from the perspective of the Web client. In order to estimate the average throughput of the TCP connection we measured time spacing between the first byte packet and the last byte packet of the data received by the client. The throughput was calculated by dividing a number of transferred bytes by the transfer time. The classical statistical data analysis approaches as well as the data mining methods were considered.

4 Preliminary Measurements

In preliminary HTTP performance tests MWING agents were placed in two locations: campus network (WUT) and housing computer network (HOME). In WUT we used a dedicated computer whereas in HOME location our computer was also used for some concurrent background processing. WUT location employed a dedicated 100 Mbps link whereas HOME location shared 1 Mbps link. The measurements were simultaneously performed for several hours by both agents on 15th November 2006, starting at 13:18:47. In every probe both agents requested the file containing the rfc1945 from http://www.cgisecurity.com/rfc/rfc1945.txt. The RTT and transfer time measurements during 4 hours are shown in Fig. 5. The results show how different are both locations in the context of Web connectivity. The aim of such experiments is to investigate the correlation between a connection's RTT and transfer rate at specific Internet locations to develop local models of Web performance. In [4] we developed such a model for WUT location based on WING measurement infrastructure. Thanks to MWING we will be able to extend our model to any location.

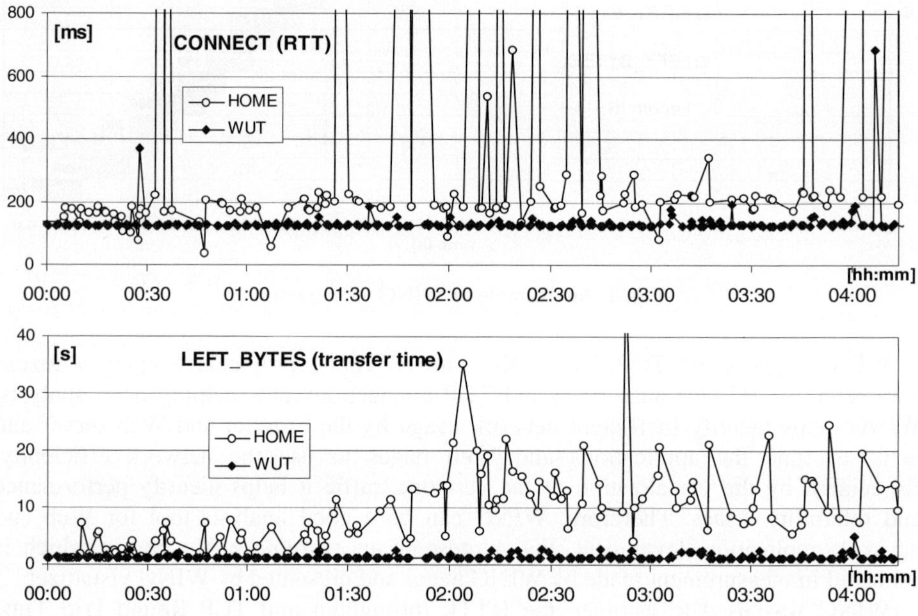

Fig. 5. Results of measurements from two different locations

5 Conclusions and Future Work

The MWING platform definitely simplifies the measurement management and extends abilities of performing deep analysis of Internet characteristic. With properly configured measurement agents spread out freely in the Internet, it is possible to manage all of them from one place. There is no more need for the administrator to spend long time for managing and configuring the different kinds of measurement systems with their own specific procedures. Thanks to the well-described interfaces, this measurement system can be extended by new module functionality. For example, to improve the set of available measurement agents and visualization modules that will make the whole system more useful and advanced.

Future work focuses on the development of Data Processor Unit (DPU) component that is responsible for the automated post-processing of collected data. The dedicated processing procedure is to be installed in the system for each measurement set, which requires some data post-processing. The DPU will implement the Data Exploration Unit that will be responsible for data mining procedures related to given DBMS. This component will enable the execution of automated exploration algorithms against data stored in database.

Acknowledgements

This work was supported by the Polish Ministry of Science and Higher Education under Grant No. N516 032 31/3359 (2006-2009).

References

1. Borzemski L.: Data Mining in the Analysis of Internet Performance as Perceived by End-Users. Proc. of ICSE2005. IEEE Computer Society Press, Los Alamitos (2005) 34-39
2. Borzemski L., Nowak Z.: Estimation of HTTP Throughput and TCP Round-Trip Times, Proc. of 10th Polish Teletraffic Symposium 2003, IEEE Chapter, Cracow (2003) 335-352
3. Borzemski L., Nowak Z., WING: A Web Probing, Visualization and Performance Analysis Service. LNCS Vol. 3140, Springer-Verlag, Berlin (2004) 601-602
4. Borzemski L., Nowak Z.: An Empirical Study of Web Quality: Measuring the Web from the Wroclaw University of Technology Campus. In: Engineering Advanced Web Applications, M. Matera, S. Comai (Eds.), Rinton Publishers, Princeton (2004) 307-320
5. Borzemski L., Nowak Z.: Using the Geographic Distance for Selecting the Nearest Agent in Intermediary-Based Access to Internet Resources. LNAI Vol. 3683, Springer-Verlag, Berlin (2005) 261-267
6. Chen J.J., Su S.W.: AgentGateway: A communication tool for multi-agent systems. Information Sciences 150 (2003) 153-164
7. Manvi S.S., Venkataram P.: Applications of agent technology in communications - a review. Computer Communications 27 (2004) 1493-1508
8. Park S., Sugumaran V.: Designing multi-agent systems a framework and application. Expert Systems with Applications 28 (2005) 259-261
9. Sugawara T., Murakami K., Goto S.: A multi-agent monitoring and diagnostic system for TCP IP-based network and its coordination. Knowledge Based Systems 14 (2001) 367-383
10. Weyns D. Parunak Van Dyke H., Michel F. Holvoet T., Ferber J.: Environments for Multiagent Systems State-of-the-Art and Research Challenges. LNAI Vol. 3374, Springer-Verlag, Berlin (2005) 1-47
11. CAIDA: http://www.caida.org, SLAC: http://www.slac.stanford.edu
12. MyKeynote: http://www.MyKeynote.com

Application of Agent-Based Personal Web of Trust to Local Document Ranking

Marek Kopel and Przemysław Kazienko

Wrocław University of Technology, Institute of Applied Informatics
Wyb.Wyspiańskiego 27, 50-370 Wrocław, Poland
{marek.kopel, kazienko}@pwr.wroc.pl

Abstract. Web is the boundless source of information and no one is able to process the vast amount of new documents published on the web every day, even with filtering out the documents the user is not interested in. However, most of the recent web documents are blog posts, news and other documents with the author information established. Each author who is also the receiver of web documents possesses their own personal agent that delivers trust information related to other authors as well as rank data for each new document. Trusts and ranks available for agents are exchanged between them and in this way new authors and new web documents can be easily assessed. Based on the general concept of Web of Trust the new idea of Personal Web of Trust and its application to local ranking method for web documents is proposed in the paper.

1 Introduction

The blogosphere's rapid expansion, which enables an easy way of publishing documents on the web, caused a great growth of people who became authors of web documents. The large amounts of new authors create new documents in different styles and with varying credibility for a reader. The reader, i.e. web user, trying to reach valuable and reliable information, needs to make some trust assertions concerning web documents and their authors. In order to help users to keep their trusts assertions up to date a concept of a Web of Trust (WoT) has been developed. Since blogs facilitate the mass communication and each user may be a blog author, many of the authors may be in same kind of relationship. Moreover, their relationship may influence their trust to others. The main idea of WoT is to make use of users' trusts to known authors for inferring the trust to new, unknown users in order to estimate their trustworthiness and propose rank values of their documents for the reader.

Web documents may be enriched with semantic information about their authors' relationships. Among the standards allowing turning a web document into the Semantic Web document by adding the information on author relationship are FOAF [2] and XNF [1]. FOAF (Friend of a Friend) is a method based on linking from web documents to a machine-readable RDF file which describes its author and the people the author knows. By processing those RDF files an author relation network can be generated automatically. XFN (XHTML Friends Network), in turn, is a method for

N.T. Nguyen et al. (Eds.): KES-AMSTA 2007, LNAI 4496, pp. 288–297, 2007.

denoting the type of relationship between linking and linked documents' authors. The relationship description is an attribute of a XHTML link from one document to another but it regards the document authors. Since determining the authorship of a web document is not always an easy task, it is proposed to use the hybrid relationship description using the combination of FOAF and XFN.

The main application of the WoT presented in this paper is the document ranking and filtering in the collaborative way and its collaborative paradigm is based on the trust inferring. However, the idea of the Web of Trust presented in this paper shall be distinguished from the general idea of Web of Trust [19] used for identification and authentication usually based on PKI (public key infrastructure) [3, 6, 11]. The presented method uses its agent based WoT for achieving another goal but some ideas of binding public keys to users from PKI WoT may be also considered to expand the method functionality. Some research in the direction of combining FOAF semantics and PKI WoT has already been made and resulted in creation of Web of Trust RDF Ontology [4, 5]. In another approach authors considered joints of different trusts delivered by other users in the network, however distribution matters have not been studied [18]. Pujol *et al.* tried to extract information about trust to others from metrics within the social network created upon mutual communication [16]. This is approach similar to the analysis of user social position [10]. Kollingbaum and Norman investigated trust extraction from the business contract data and the negotiation process performed by contracting agents [12]. Guha *et al.* studied the problem of transitive trust and distrust and their propagation especially by means of different rounding approaches [7].

2 Trust to Authors

The concept developed in the paper covers all internet services in which there is a set of documents created by valid authors like personal web pages, blogs, emails, pictures galleries etc. Each document's author, who is simultaneously a system user, possesses one personal agent which is able to communicate with other user agents and performs several tasks for its owner (user). In this way we obtain a distributed multi-agent architecture. Each user of the system can create their own documents, e.g. posts on blog and publish them in the open environment. Other users may access all published documents but due to the possibly large number of new or updated items it is difficult for them to select only reliable documents for reading. The possible solution is to infer the document trust from the trust to the document's author. In this way, we can order the obtained documents according to the trust worthiness: the most trusted items are moved to the top of the list.

Each user agent is responsible for maintaining the set of trusts to other users, i.e. confidences in other users authoring. Additionally, the personal agent also ranks and orders all documents that have been delivered to the user. Both, trust to other users and ranks of documents, can be either manually delivered by the user – agent's owner or provided by the agent itself. The agent creates its trusts and ranks based upon the knowledge supplied by other agents. In this paper, by using terms *author*, *user* and *agent* we address the same technical concept. Even though the semantics of these terms may have different intuition, for the purpose of our research we assume that each author is a user, each user is a potential document author and each user has an agent which maintains the knowledge of its user trusts, document ranks and document authorships. That is why the three terms: author, user and agent are used here equivalently.

2.1 User Trust

Each human in the real world may have some relationships with other humans. They may be family, friends, colleagues or acquaintance. The quality of such relationship may express also the level of the trust that one man have to another. Note that the human relationships are usually asymmetric. Additionally, the trust can be derived from the personal assessment of documents authored by the certain user. We can have more or less confidence in the user who created documents we like. Hence, a user that is not in any relationship with any other user in the real world may trust in authors whose documents one found interesting. Based on the relationships or document evaluation people have available from the real world, they are able to manually estimate the value of the trust function to some another system users.

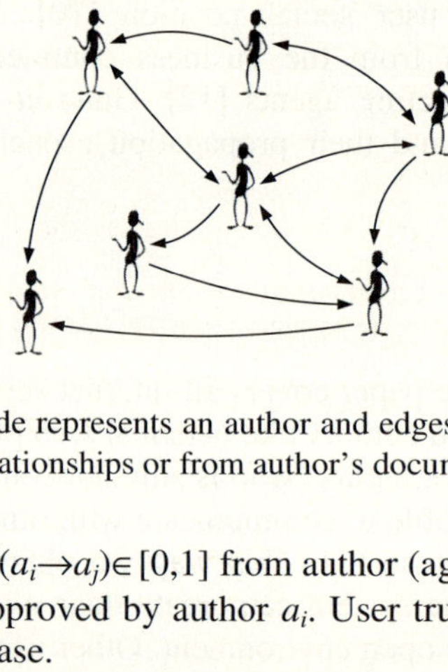

Fig. 1. Web of Trust. Each node represents an author and edges are trusts to author that can be derived from the real world relationships or from author's document evaluations

Definition 1. User trust $T^{usr}(a_i \rightarrow a_j) \in [0,1]$ from author (agent) a_i to author (agent) a_j is the trust manually set or approved by author a_i. User trust $T^{usr}(a_i \rightarrow a_j)$ is retained by agent a_i in its internal database.

When a user sets the trust to another one manually, the latter is being included into the former's Personal Web of Trust (PWoT).

2.2 Personal Web of Trust

The general idea for Web of Trust is that trust may be interpreted as a directed edge incident with two nodes – authors, shown in Fig. 1. Direction of the edge denotes who trusts in who, i.e. an edge from a_1 to a_2 stands for "a_1 trusts to a_2 ".Now let us filter out the both direction edges and take only the outgoing edges from one node only. The edges that go out from user a_1 and the nodes incident with them form a directed graph representing the Personal Web of Trust of user a_1 who occupies the center-placed node, as shown in Fig. 2.

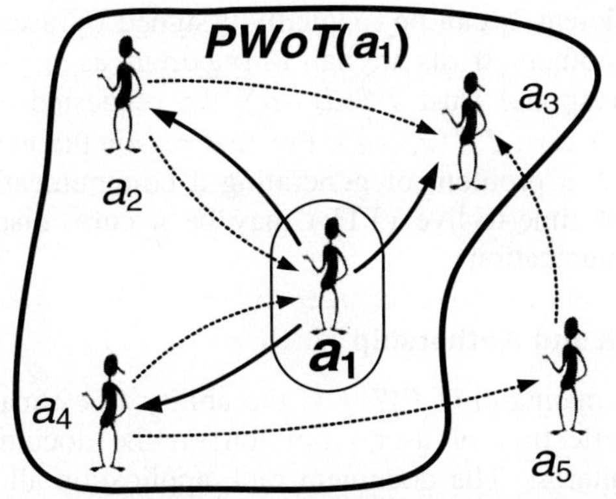

Fig. 2. Personal Web of Trust of author a_1 $PWoT(a_1)$ is a fragment of the entire Web of Trust surrounded by the solid line that includes only the authors whose trust to was assigned explicit by author a_1. Elements drawn with a dotted line does not belong to $PWoT(a_1)$.

Definition 2. Personal Web of Trust $PWoT(a_i)$ for author a_i is the set of other authors (agents) $a_j \in A$ from the entire author set A for which user trust $T^{usr}(a_i \to a_j)$ from a_i to a_j is known for a_i.

In other words, Personal Web of Trust of a single user a_i includes all the authors whose the user's trust to was assigned explicit by the given user a_i.

2.3 Agent Trust

Whenever an author a_i needs an information about another author a_j from the outside of $PWoT(a_i)$, the a_i's agent can propose the trust to author a_j based upon trusts delivered by other authors from $PWoT(a_i)$. To calculate the appropriate trust level for the new, unknown author a_j the agent a_i needs to exchange knowledge related to a_j with other agents from the $PWoT(a_i)$. Next, the agent trust $T^{agn}(a_i \to a_j)$ is evaluated and suggested to the user a_i.

Definition 3. Agent trust $T^{agn}(a_i \to a_j) \in [0,1]$ from author (agent) a_i to author (agent) a_j is the trust calculated by agent a_i, based on the trusts delivered by other agents from $PWoT(a_i)$, in the following way:

$$T^{agn}(a_i \to a_j) = \frac{\sum\limits_{a_k \in PWoT(a_i)} T^{usr}(a_i \to a_k) \cdot T^{rsp}(a_k \to a_j)}{\sum\limits_{a_k \in PWoT(a_i)} T^{usr}(a_i \to a_k)} \tag{1}$$

where: $T^{rsp}(a_k \to a_j)$ is *response trust* to a_j provided by agent a_k:

$$T^{rsp}(a_k \to a_j) = \begin{cases} T^{usr}(a_k \to a_j), & \text{if } T^{usr}(a_k \to a_j) \text{ is known} \\ \lambda_k \cdot T^{agn}(a_k \to a_j), & \text{if } T^{usr}(a_k \to a_j) \text{ is unknown} \\ & \text{and } T^{agn}(a_k \to a_j) \text{ is known} \\ 0, & \text{otherwise} \end{cases} \tag{2}$$

λ_k - the importance of the agent trust in relation to user trust for agent a_k; $\lambda_k \in [0,1]$.

The value of coeficient λ_k can be manually assigned by user a_k. The value λ_k=0.5 means that for user a_k others' trusts are half as important as a_k's own are.

To calculate the response trust $T^{rsp}(a_k \rightarrow a_j)$, the requested agent a_k may need to evaluate its own agent trust $T^{agn}(a_k \rightarrow a_j)$. For that reason the evaluation may become recursive. This causes a problem of generating a communication chain reaction to which the concept of time-to-live (TTL) may be a cure, discussed in the section related to agent communication.

2.4 Document Rank and Authorship

The main reason for maintaining *PWoT* is the ability to estimate a new document's rank. Rank is the reflection of user's opinion on the document in the aspect of relevance to user's interest. The document rank application allows user to access to the most relevant documents first, as estimated by their agent.

Definition 4. User rank $R^{usr}(a_i \rightarrow d_j) \in [0,1]$ from author (agent) a_i to document d_j is the rank manually set or approved by author a_i. User rank $R^{usr}(a_i \rightarrow d_j)$ is retained by agent a_i in its internal database.

User a_i does not rank documents which a_i is the author of, e.g. documents d_1 to d_4 created by user a_1 in Fig. 4. In this case we assume that the user rank is the highest possible. In agent knowledge the fact of being an author of a document is called authorship (Fig. 3). On the other hand, agent a_i retains also user ranks to documents created by others, e.g. agent a_1 preserves ranks to documents d_5 to d_8, Fig. 4c.

Definition 5. Author a_i's authorship of the document d_j denoted as $Ath(a_i \rightarrow d_j)$ is true only if a_i is the author of the document d_j.

Assignment of the rank to a document works the same way as a user trust assignment does. It can be done manually (explicit) by the user and then it immediately becomes the user rank. However, in case the user did not have a chance to see the document and rank it personally, the agent is responsible for delivering its agent rank, as the initial, suggested value.

Definition 6. Agent rank $R^{agn}(a_i \rightarrow d_j) \in [0,1]$ from author (agent) a_i to document d_j is the rank calculated by agent a_i, based on the ranks delivered by trusted agents from the $PWoT(a_i)$, in the following way:

$$R^{agn}(a_i \rightarrow d_j) = \frac{\sum\limits_{a_k \in PWoT(a_i)} T^{usr}(a_i \rightarrow a_k) \cdot R^{rsp}(a_k \rightarrow d_j)}{\sum\limits_{a_k \in PWoT(a_i)} T^{usr}(a_i \rightarrow a_k)} \tag{3}$$

where: $R^{rsp}(a_k \rightarrow d_j)$ is *response rank* of document d_j provided by agent a_k:

$$R^{rsp}(a_k \rightarrow d_j) = \begin{cases} 1, & \text{if } Ath(a_k \rightarrow d_j) \text{ is true} \\ R^{usr}(a_k \rightarrow d_j), & \text{if } R^{usr}(a_k \rightarrow d_j) \text{ is known} \\ \tau_k \cdot R^{agn}(a_k \rightarrow d_j), & \text{if } R^{usr}(a_k \rightarrow d_j) \text{ is unknown} \\ & \text{and } R^{agn}(a_k \rightarrow d_j) \text{ is known} \\ 0, & \text{otherwise} \end{cases} \tag{4}$$

τ_k – the importance of the agent rank in relation to the user rank for agent a_k; $\tau_k \in [0,1]$.

The value of τ_k is manually assigned by user a_k. The value τ_k=0.333 means that for user a_k others' document ranks are one third as important as a_k's own are.

Similarly to agent trust calculation, the agent rank estimation is also recursive. The same problem of chain reaction emerges. It shall be noticed that the evaluation of both agent trust T^{agn} and agent rank R^{agn} using one, common query may improve the performance.

3 Local Document Ranking

Building a Personal Web of Trust and providing user agents with the knowledge of author trusts and document ranks delivers each user an useful collaborative filtering tool. This tool may be applied to a collection of documents in many ways. One of them is the local ranking of documents while a user navigates across in the web. "Local" means created and proposed by user's own agent. Ranking is a collection of documents ordered according to document ranks and their author trusts.

The method presented below assumes the existence of a prior document list (or at least a document set) known by the user agent. This input list may be for example a response to the search engine query or an RSS file.

As stated above, each document d_j may have two values: $R^{usr}(a_i \rightarrow d_j)$ and $R^{agn}(a_i \rightarrow d_j)$ assigned in the agent a_i's database. The most evident way of making use of them is to sort input documents by their values to obtain the most and the less relevant documents. Obviously, the collection should exclude all documents the user created themselves since they are the most relevant. First, the user rank value $R^{usr}(a_i \rightarrow d_j)$ is taken into account at sorting since it reflects user interest the best. However, many documents, especially the new ones, will have $R^{usr}(a_i \rightarrow d_j)$ *unknown*. For such documents d_j the agent rank $R^{usr}(a_i \rightarrow d_j)$ is evaluated and utilized for ordering.

Two other values $T^{usr}(a_i \rightarrow a_k)$ and $T^{agn}(a_i \rightarrow a_k)$ may also be used for document d_j authored by a_k. They can be utilized at estimation of the d_j's position in the output document list. It appears to be more user friendly to present the author ranking first and next, after choosing an author, to show the ranked list of documents. Alternatively, one ranking may be proposed to the user with the several document lists sorted by each of the four values. For example, sorting the document collection descending by T^{usr} and then R^{usr} gives the ranking of most relevant documents from the authors with which the current user has the best relationship.

Yet another idea for personalizing of the document list is to sort them by a combination the four values. Moreover, the custom weighted average may deliver the best results.

4 Agent Communication

The agent approach to the Web of Trust assumes that the knowledge about the trust is distributed among all user personal agents and therefore agent communication for knowledge exchange is needed. This regards especially the information about trusts and ranks delivered or approved by users and stored in their agents' internal database.

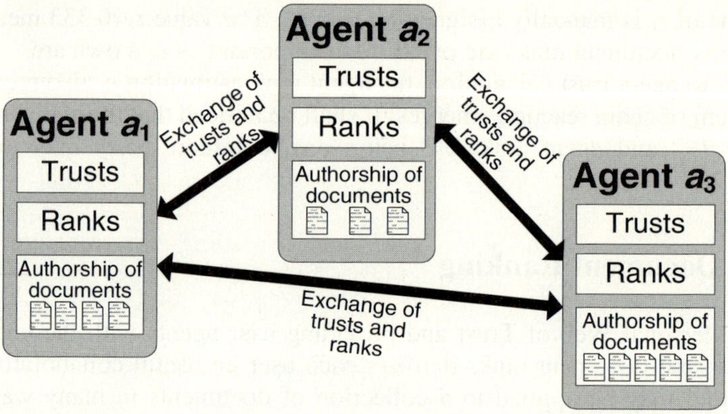

Fig. 3. Communication to the Web of Trust: agents exchange user trusts and document ranks and retain information about their owner's authorships of documents

Additionally, each agent needs the knowledge about its owner's document authorships to be able to respond to rank queries and properly order the input collection of documents (see Fig. 3).

The main problem in exchanging the knowledge by query is the, mentioned earlier, chain reaction issue. When an agent a_i evaluates its $T^{agn}(a_i \rightarrow a_j)$ or $R^{agn}(a_i \rightarrow d_j)$ for an unknown agent a_j or document d_j, it broadcasts a query to all other agents a_k from its $PWoT(a_i)$. Recursively, each of the asked agents a_k may not possess the appropriate knowledge, i.e. $T^{usr}(a_k \rightarrow a_j)$ or $R^{usr}(a_k \rightarrow d_j)$ are not available. In such case, agent a_k needs to evaluate its $T^{agn}(a_k \rightarrow a_j)$ or $R^{agn}(a_k \rightarrow d_j)$ and it also sends a new query to agents from its $PWoT(a_k)$. To prevent the system from circulating queries going around the web infinitely, each query should contain a time-to-live status (TTL). TTL is the number initiated by agent a_i that sends the primary query. Once agent a_k is queried for a trust or rank, it needs to forward the query among its $PWoT(a_k)$. It also decreases TTL with 1 for the relayed query. If agent a_k has received a query with TTL=1, then such query must not be forwarded any more and a_k responds with the value of 0. The initial TTL value is a system parameter common for all agents. Theoretically, six is the value that allows reaching the opinions of authors from the entire world – see the six degrees of separation hypothesis [13]. However, it is most likely that the only appropriate value that respects the performance limitations can be established empirically. It appears it should be about 2-3.

Another improvement of the performance is the blockade of the backward querying. For that reason, each query should have the source information attached – name of the agent who cast the original query. Just like TTL, this data should be forwarded in the recursive queries. When a query is received by an agent, it checks whether it has already received the same query. If yes, the query is dropped: not forwarded and not responded to.

There is also a serious problem related to delays in communication. How long an agent should wait for the query response (T^{rsp} or R^{rsp}) and how to process the responses? To address this problem we shall assume that the TTL of the forwarded query is always returned in a response. This allows estimation how far is an agent that has the user trust or rank available, i.e. how many agents on the return path influenced the response. An agent that queried its *PWoT* should only take into account the responses with the highest TTL. This assures the least influence of the opinions by authors which the user has not trusted directly. The response TTL value solves also another problem regarding the number of responses that need to be obtained by agent a_k before a_k evaluates the response and sends it back. Each agent should wait with the evaluation of its response until it receives a response with TTL lower than the very first response TTL. According to the intuition if we assume the architecture is in some degree homogenous the responses with higher TTL shall arrive earlier.

Even though the introduction of TTL is helpful, it is not enough. It is expected that some amount of the queries may be timed out because of the hardware limitations. To solve this problem the cache expiration period should be studied.

5 Scenario Example

Let us consider the initial Personal Web of Trust $PWoT(a_1)$ from Fig. 4a. The user a_1 goes online to find information on the favorite music band, types "*Pink Floyd*" into search engine and obtains a collection of relevant documents. Some of these documents are album and concert reviews. It turns out that many of the reviews (d_5, d_6) are blog posts by author a_2. User a_1 knows and likes a_2's point of view, so a_1 sets manually high trust to a_2 and ranks a_2's documents d_5 and d_6 in the agent a_1 database. In this way, Personal Web of Trust $PWoT(a_1)$ includes both a_2 and the a_2's reviewers (Fig 4b). Note that the set trust regards also any document that a_2 will write in the future.

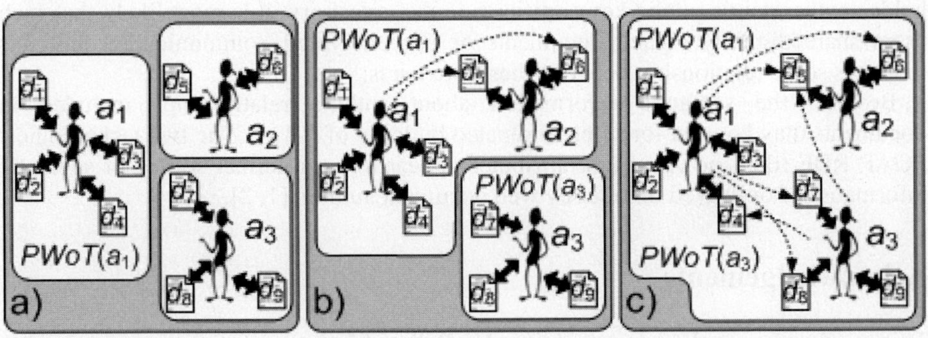

Fig. 4. The extension of $PWoT(a_1)$ of agent a_1. Thick edges denote the authorship whereas the dotted arrows – user ranks of documents.

After some time, one of the reviews reminds user a_1 about the concert and another user a_3, who a_1 met there. However, after the concert a_1 lost contact with a_3. User a_1 easily finds the a_3's blog in which a post related to the concert contains a picture of both a_1 and a_3 they had taken during that event. User a_1 includes user a_3 into $PWoT(a_1)$, see Fig. 4c.

Next, a_1 queries the search engine for "*Pink Floyd*" again. This time, a_1 sees reviews on the band by author a_2 at the top of the response document list. Moreover, there are more of them, e.g. a review of the Pink Floyd's video the author a_2 wrote.

Meanwhile, user a_3 after reading a_1's comment to the post on *Pink Floyd*'s concert (d_4), included a_1 and d_4 to $PWoT(a_3)$. Furthermore, a_3 requests agent a_1 for trust to a_2. and rank of a_2's document d_5. User a_1 responses and a_3 adds a_2 and d_5 to a_3's Personal Web of Trust $PWoT(a_3)$. Finally, $PWoT(a_3)$ extends and includes the most part of $PWoT(a_1)$, see Fig. 4c.

6 Conclusions and Future Work

The usage of Personal Web of Trust is another method for collaborative filtering of web documents. A user as a reader provided with a personal agent is able to process the most valuable information. An agent using user's trust to authors and ranks of documents is able to filter out and rank new documents even created by an author unknown to the reader. The filtering is based on the trust and rank exchange between co-operating agents in the multi-agent environment.

Apart from ranking of documents, the *PWoT* method may be used for tracking author's social position [10] based on the number and value of user trusts to them. This may be a valuable tool in social networks analysis [9], especially for scientific communities [17]. Another potential application of the proposed method is monitoring of knowledge dynamics based on the document's number of ranks as well as extension of various personalization systems with the collaborative trust [8, 14, 15].

One of the future improvements may be extending the method with PKI. If a user holds another user's public key certificate in the repository, it is most likely that such users share digitally signed documents or use encrypted communication and this indicates some relationship between these two users.

Besides, the semantic information about authors relationships included in documents may be used for semi-automated building of *PWoT*. The two technologies: FOAF RDF files and XFN link attributes appear to be a perfect source of semantic information if only used broader by web document authors [1, 2].

Acknowledgements

This work was partly supported by the Polish Ministry of Science and Higher Education, grant no. N516 037 31/3708.

References

1. Çelik, T., Meyer, E.: XHTML Friends Network (Poster). ACM Hypertext, 2004.
2. Dumbill, E.: XML Watch: Finding friends with XML and RDF. IBM Developer Works, 2002, http://www-128.ibm.com/developerworks/xml/library/x-foaf.html.
3. Falcone R., Singh M., Tan Y.-H.: Trust in Cyber-societies: Integrating the Human and Artificial Perspectives, Springer Verlag LNCS 2246, 2001.

4. Golbeck J., Hendler J.A.: Accuracy of Metrics for Inferring Trust and Reputation in Semantic Web-Based Social Networks. EKAW 2004, LNCS 3257, 2004, 116-131. http://www.mindswap.org/papers/GolbeckEKAW04.pdf
5. Golbeck J., Parsia B., Hendler J.: Trust Networks on the Semantic Web. Cooperative Intelligent Agents CIA 2003, LNCS 2782, Springer Verlag, 2003, 238-249.
6. Grandison T., Sloman M.: A Survey of Trust in Internet Applications. IEEE Communications Surveys and Tutorials 3 (4), 2000, http://www.doc.ic.ac.uk/~mss/Papers/Trust_Survey.pdf.
7. Guha R.V., Kumar R., Raghavan P., Tomkins A.: Propagation of trust and distrust. 13th International Conference on World Wide Web, WWW 2004, ACM Press, 2004, 403-412.
8. Kazienko P., Adamski M.: AdROSA - Adaptive Personalization of Web Advertising. Information Sciences, 2007.
9. Kazienko P., Musiał K.: Social Capital in Online Social Networks. 10th Int. Conf. on Knowledge-Based Intelligent Information & Engineering Systems, KES 2006, LNAI 4252, Springer Verlag, 2006, 417-424.
10. Kazienko P., Musiał K.: Mining Social Position of Individuals in Virtual Social Networks. AI Communication, Special Issue on Network Analysis in Natural Sciences and Engineering, 2007.
11. Khare R., Rifkin A.: Weaving a Web of trust. World Wide Web Journal, Special issue: Web security: a matter of trust, 2 (3), 1997, 77 – 112.
12. Kollingbaum M.J., Norman T.J.: Supervised interaction: creating a web of trust for contracting agents in electronic environments. The First International Joint Conference on Autonomous Agents & Multiagent Systems, AAMAS 2002, ACM Press 2002, 272-279.
13. Milgram, S.: The Small-World Problem. Psychology Today, 2, 1967, 60–67.
14. Mui L.: Computational Models of Trust and Reputation: Agents, Evolutionary Games, and Social Networks. Ph.D. Thesis. Massachusetts Institute of Technology, December 20, 2002, http://groups.csail.mit.edu/medg/ftp/lmui/computational%20models%20of%20trust%20and%20reputation.pdf.
15. O'Donovan J., Smyth B.: Trust in recommender systems. International Conference on Intelligent User Interfaces IUI 2005, ACM Press, 2005, 167-174.
16. Pujol J.M., Sangüesa R., Delgado J.: Extracting reputation in multi agent systems by means of social network topology. International Conference on Autonomous Agents, ACM Press, 2002, 467 – 474.
17. Rana O.F., Hinze A.: Trust and reputation in dynamic scientific communities. IEEE Distributed Systems Online, Vol 5, Issue 1, 2004. http://ieeexplore.ieee.org/iel5/8968/28452/01270714.pdf.
18. Richardson M., Agrawal R., Domingos P.: Trust Management for the Semantic Web. International Semantic Web Conference ISWC 2003, LNCS 2870, Springer Verlag, 2003, 351-368, http://www.cs.washington.edu/homes/mattr/doc/iswc2003/iswc2003.pdf.
19. Web of trust, Wikipedia, http://en.wikipedia.org/wiki/Web_of_trust.

Non-repudiation Mechanism of Agent-Based Mobile Payment Systems: Perspectives on Wireless PKI

Chung-Ming Ou[1] and C.R. Ou[2]

[1] Department of Information Management
Kainan University, Luchu 338, Taiwan
cou077@mail.knu.edu.tw
[2] Department of Electrical Engineering
Hsiuping Institute of Technology, Taichung 412, Taiwan
crou@mail.hit.edu.tw

Abstract. Non-repudiation of a mobile payment transaction ensures that when a buyer (B) sends some messages to a seller (S), neither B nor S can deny having participated in this transaction. An evidence of a transaction is generated by wireless PKI mechanism such that B and S cannot repudiate sending and receiving the purchase order respectively. The broker generates a mobile agent for B which carries encrypted purchase order to the seller. One trusted third party acts as a lightweight notary for evidence generation. One advantage of this agent-based non-repudiation protocol is to reduce inconvenience for mobile clients such as connection time; it causes difficulty for fair transaction for mobile payments.

1 Introduction

The security has been a concerned issue for mobile payment for a long time. For example, dispute of a transaction is a common problem that could jeopardize the mobile commerce [1]. The purpose of non-repudiation is to collect, maintain, make available and validate irrefutable evidence concerning a claimed event or action in order to resolve disputes on the occurrence or non-occurrence of the event or action [2], [3]. However, in the real world situation, lots of mobile stock brokerage services and mobile banks have already been promoted but without any non-repudiation mechanisms.

There are several electronic invoice systems promoted by the Ministry of Economic Affair (MOEA) in Taiwan since 2004. One purpose of these systems is to reduce the cost of generating paper-based receipts and invoices. According to MOEA, electronic invoice systems based on public-key infrastructure (PKI) are the most successful PKI applications in Taiwan which greatly reduce the cost of purchasing system. However, these PKI-based systems do not adopt any non-repudiation mechanism between buyers and sellers. Another motivation for this research is the mobile TAIWAN project (mTAIWAN). Since 2005, mTAIWAN is one major nation-wide project of establishing seamless and ubiquitous wireless infrastructure. This next generation communication network combines mobile communication systems such as

N.T. Nguyen et al. (Eds.): KES-AMSTA 2007, LNAI 4496, pp. 298–307, 2007.

2G (GSM), 2.5G (GPRS), 3G (UMTS), wireless network systems such as WiFi and WiMAX technologies. One major goal for mTAIWAN project is to promote the ubiquitous mobile applications using varied mobile devices such as mobile phone, PDA, laptop PC, etc. Combining these motivations, we propose agent-based architecture and protocol to implement non-repudiation mechanism over the mobile payment systems; this will also improve the security mechanisms of those existing electronic invoice systems.

Mobile applications need to be user-friendly and convenient for mobile clients via their mobile handsets; this investigation leads to the research of agent-based mobile applications. The digital signature-based authentication proves to be implemented efficiently within 3G communications [4]. This can be regarded as a preliminary experiment for mobile agent-based non-repudiation mechanism which also relied on digital signature mechanisms.

Non-repudiation services must ensure that when buyer B sends message to seller S over a network, neither B nor S can deny having participated in a part or the whole of this transaction. The basic idea is the following: an evidence of origin (EOO) is generated for buyer B and an evidence of receipt (EOR) is generated for seller S. In general, evidences are generated via PKI-based digital signatures. Disputes arise over the origin or the receipt of messages. For the case of origin dispute, B denies sending message while S claims having received it. As for the receipt dispute, S denies receiving any message while B claims having sent it. Buyers are at risks that sellers repudiate receiving this purchase order. Our mobile payment structure is as follows. First buyer sends out encrypted purchase order to the broker, which is a trusted server. Then this broker generates a mobile agent which carries this encrypted purchase order to the seller, which decrypts this order. The use of a broker between the wired and the wireless network can ease the access to web information from the mobile devices, and it can also alleviate some of these security constraints [5]. Mobile payment systems need time information included in evidences for dispute resolutions.

Mobile agents are considered to be an alternative to client-server systems mobile commerce where mobile devices and communication have limited computing resource. A mobile agent of the host is a set of code and data which can execute the code with the data as parameter in some trusted processing environment (TPE) or on some merchant hosts. However, there are several issues related to security and trust while considering mobile agent-based electronic commerce [6], [7], [8]. One of the major concerns is the non-repudiation. We consider the brokerage rather than the TPE in our mobile payment systems. The advantage of adopting this mobile agent architecture to non-repudiation protocol is the following: the buyer needs only to send the purchase order while it connects to the mobile base station; once such order is sent to the broker, mobile devices can be off-line if necessary for cost saving. Once the transaction is complete, the agent can return the message of payment and transaction completion to mobile clients.

The phrase *wireless public key infrastructure* (WPKI) is a loose definition as adopting public key cryptosystems in wireless (or mobile) environment. The WPKI defined in this paper is the dual public-key cryptosystems, which means there are two public-key and private-key pairs for each WPKI entity. One key pair is for encryption/decryption; the other one is for digital signature generation/verification.

The WPKI architecture suitable for non-repudiation mechanism relies on some trusted third party (TTP) which generates the final evidence for purchasing order.

The arrangement of this paper is as follows. In section 2, we introduce the architecture of an agent-based mobile payment system. In section 3, we propose an agent-based non-repudiation protocol suitable for mobile payment; we also analyze of the security mechanisms of agent-based non-repudiation protocols, namely, dispute resolutions.

2 Design of Agent-Based Non-repudiation Protocol

An efficient and fair non-repudiation protocol was proposed by Zhou and Gollmann where TTP acts as a lightweight notary (we name it ZGP) [9]. This protocol is suitable for 3G communication by analyzing the capability of implementing cryptographic operations such as digital signature, symmetric key encryption/decryption, hash function and random number generations [4][1]. According to this investigation, we design a non-repudiation protocol adaptive to agent-based mobile payment systems.

2.1 Basic Structure for 3G Mobile Payment Services

The architecture for mobile payment system is composed of the following entities: a seller represented by mobile equipment (ME), WPKI, a seller (merchant server), a bank and a broker. These entities are also issued certificates by some certification authority within this WPKI. ME utilizes the USIM (Universal Subscriber Identity Module) to store mobile clients' information such as IMSI (International Mobile Subscriber Identity) and WPKI components. ME is capable of verifying digital signatures to authenticate other entities, if necessary. We also deploy a middleware called the broker to help ME authenticate the merchant server such that attackers cannot impersonate this seller. Merchant servers can perform PKI operations for evidence generations.

2.2 WPKI

The WPKI is the core cryptographic mechanism for non-repudiation protocol; it consists of two parts, one is the operation; the other is the entity. WPKI entities must contain at least two public-private key pairs for encryption/decryption and signature generation/verification, respectively. These key pairs are generated by some certification authority (CA) whose major task is to bind public key, private key and entity together.

WPKI Operations

There are two major WPKI operations in our non-repudiation protocol, one is the PKI encryption and decryption, the other is the digital signature-based evidence generation and verification.

[1] The original ZGP did not design particularly for mobile transactions. The author has discussed with Chunghwa Telecom Lab in Taiwan for RSA digital signature implementation in 3G mobile environment. The capability of USIM cryptographic module is reasonable for WPKI operations.

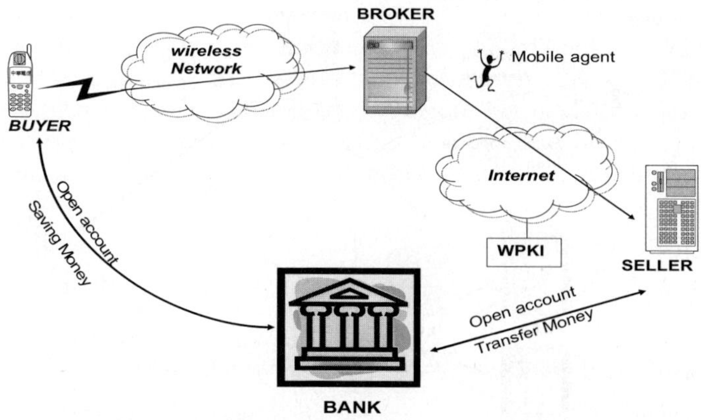

Fig. 1. The architecture of an agent-based mobile payment system

WPKI Entities

Certificate Authority (CA). A CA issues certificates to mobile subscribers and server certificates to merchant servers, TTP, Home Revocation Authority (HoRA) and banks. These entities can authenticate each other and transmit encrypted information. A CA needs to provide certificate management service to ensure the validity of certificate. In general, a CA provides certificate revocation list (CRL) and On-line Certificate Status Protocol (OCSP) service to these entities for checking certificates validity. These entities would continue WPKI operations if and only if related certificates are valid.

Mobile Client. We suggest that a mobile client be a USIM-based 3G mobile equipment for efficient signature generation and verification. A USIM stores only necessary WPKI components due to the possible limitation of a USIM resource affordable to PKI operations. In our non-repudiation protocol described in the following subsection, the USIM stores the TTP's server certificate and subscriber's two private keys; it may optionally store two public-key (server) certificates of the broker's. These public-key certificates are all issued by some CA within this WPKI. Private keys should be generated in USIM and contained in it afterwards.

Trusted Third Party (TTP). The trusted third party here is a notary server which simply generates necessary evidences for buyers and sellers. TTP needs to perform WPKI operations according to the non-repudiation protocol described in the next section. Therefore TTP needs to access CA's repository to retrieve necessary certificates of buyers' (sellers') and verify digital signatures. TTP needs to store the broker's public-key certificates and plays a role as the time stamp authority if necessary. For those generated evidences, TTP will store these information in its public directory from which buyers and sellers may fetch evidences.

TTP acts as a lightweight notary in this based non-repudiation protocol that only notarizes purchase order by requests. TTP also provides directory services accessible to the public. For the non-repudiation protocols introduced in the next section, TTP only deal with "keys" rather than purchase order, that is, TTP does not know any

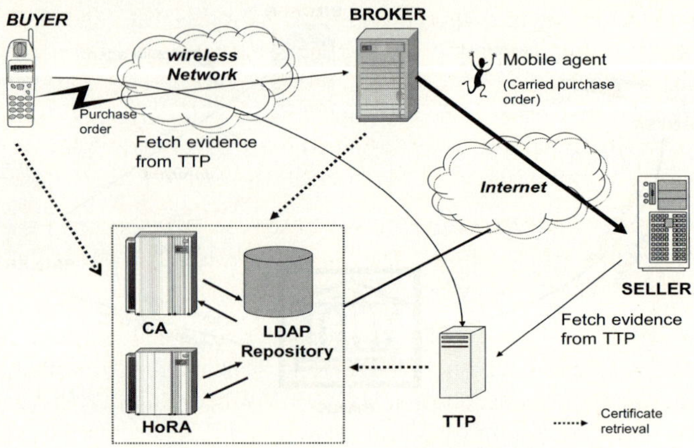

Fig. 2. Architecture of WPKI based Non-repudiation

information of this order. Therefore the communications overheads between parties and TTP are reduced, and the buyer's purchasing privacy is also guaranteed.

Host Revocation Authority (HoRA). HoRA issues host certificates (HC) to merchant servers; these certificates bind mobile agent execution capability to the merchant host identity. When a merchant server acts maliciously, HoRA only needs to revoke this server's HC to prevent the broker from sending agents to it. The functionality of HoRA to detect the status of merchant servers can be referred to [8].

HoRA issues the host revocation List (HRL), which is a digital-signed list of revoked HCs. Before sending an agent of ME to a merchant server, the broker must check the status of all servers on the agent's itinerary to see if any server is on the HRL. If the check is positive, the broker will stop sending this agent to the merchant server; this mobile payment transaction is terminated. The broker needs to update HRL periodically.

2.3 Broker

Mobile users want to pay the seller according to the purchase order. The broker acts as a mediator between the mobile users in the wireless network and the merchant servers in the Internet, see Figure 2. The broker must distinguish malicious servers from the honest ones according to HRL to avoid sending agents to them. It is possible that honest server become malicious before HRL is updated. Esparza et al. [8] provides solutions to solve this mobile agent security. HoRA will issue an updated HRL to the broker if a merchant server is detected to be malicious. The broker needs to authenticate TTP on behalf of ME before the non-repudiation protocol runs.

2.4 Mobile Agent

A mobile agent is a set of code of data generated by the broker. The code is executable with reference to these carried data. A mobile agent consists of the following components: agent owner, identifier, goal/result, life time and states. There are several security threats related to mobile agents such as repudiation, while an agent threatens another agent [10]. An agent may deny having exchanged message with other host. On the other hand, an agent may be modified by some malicious agents or hosts while transferring to some merchant server. The following steps are a general guideline for protecting these mobile agents using WPKI.

1. The broker obtains the certified public key of the merchant server.
2. The broker encrypts this mobile agent using merchant server's public key and sends it to the merchant server.

Each agent carries the item which is intended to be exchanged. These items include purchase orders and payment information (e.g. bank account number, credit card number, or micro payments account number, etc). When the buyer's agent enters the merchant server, the broker must ensure that they play fair. Furthermore, none of these agents is allowed to communicate with any other party except its host or transacted seller.

3 Fair Non-repudiation of Agent-Based Mobile Payment Transactions

In this section, we focus on evidence generations of purchase orders between a mobile client and a merchant server through some brokers. These evidences are the foundation of the mobile payment system. Once buyer and seller own their evidences respectively, seller can request his bank for fund transferring according to this purchase order. The seller needs to guarantee the amount of transferred fund be coincident with that on the purchase order. Otherwise TTP can point it out to the arbitrator that this seller is cheating. Therefore, it is sufficient for us to concentrate on the evidence generation of the purchase order rather than that of billing. This approach is also similar to that in [8].

3.1 Fair Non-repudiation Protocol with Timeliness

Time evidence of sending and receiving a purchase order is crucial in mobile payment systems. It could be achieved by adding some time stamps to evidences. Li et al. [3] improved ZGP by considering the time span for evidence preservation. This improvement needs only TTP plays the role of time stamping authority while buyers and sellers just define their intended time spans.

A non-repudiation protocol is fair if it can ensure that at the end of a protocol execution, none or both of the two entities, the sender and the receiver, can retrieve all the evidences it expects [3]. Fairness guarantees that either sender or receiver can gain advantage over the other.

Now we design a fair non-repudiation protocol suitable for agent-based mobile payment; this protocol relies on the trust of the broker. Trust is more a social issue than a technical one. We may assume reasonably that mobile operators or some service providers provide brokers which are completely trusted by mobile subscribers.

The purpose of this non-repudiation protocol is to transmit encrypted purchasing order M and obtain non-repudiation evidences for buyer B and seller S. Purchasing order M contains two parts, one is a commitment C, and the other is a key K. Notations are as follows.

M: purchasing order being sent from B to S.
K: key generated by B.
$C=e_K(M)$: commitment for purchase order M (e_K represents encryption by key K).
$sS_B(M)$: signature of message M signed by B's private key.
L=H(M,K): a label linking C and K (H represents a hash function).
f_i: flag indicating the purpose of a signed message.
$e_{TTP}(.)$: encryption by TTP's public key
EOO_C: evidence of origin of C, which is equal to $sS_B(f_{EOO}, S, L, C)$.
EOR_C: evidence of receipt of C, which is equal to $sS_S(f_{EOR}, B, L, t_S, C)$.
*sub*_K: authenticator of receipt of C, which is equal to $sS_B(f_{SUB}, S, L, t_B, K, EOO\_C)$.
*con*_K: evidence of confirmation of K issued by the TTP with time stamp T, which is equal to $sS_{TTP}(f_{CON}, B, S, L, T, t_B, t_S, K, EOO\_C, EOR\_C)$.

We include time information in this protocols; t_B is a time span defined by buyer B indicating that sub_K will be kept in TTP's private directory for t_B time units; t_S is a time span defined by seller S indicating that TTP will keep EOR_C in its private directory for t_S time units. T is the time stamp indicating the actual time TTP generate key confirmation con_K and make it public. This non-repudiation protocol which relies on two different mobile agents {A1}, {A2} generated by the broker is as follows.

1. $B \rightarrow Broker \rightarrow_{\{A1\}} S$: f_{EOO}, S, L, C, EOO_C
2. $B \rightarrow Broker \rightarrow_{\{A2\}} TTP$: f_{SUB}, S, L, t_B, K, EOO_C, sub_K
3. $S \rightarrow TTP$: f_{EOR}, B, L, t_S, EOO_C, EOR_C
4. $TTP \leftarrow B$: f_{CON}, B, S, L, T, t_B, t_S, K, EOR_C, con_K
5. $TTP \leftarrow S$: f_{CON}, B, S, L, T, t_B, t_S, K, EOR_C, con_K

"$B \rightarrow Broker \rightarrow_{\{A\}} S$: M" means B sends message M to broker, then broker will generate an agent {A} for B; message M will be carried by this agent {A} to S; "TTP ←B" means B fetches messages from TTP. The basic idea is that buyer B is able to send K, sub_K to TTP in exchange for con_K; on the other hand, seller S sends EOO_C, EOR_C and t_S to TTP in step 3. In step 1, S needs to verify EOO_C by retrieving B's (signature) public key from the corresponding CA's repository. In step 1, S needs to verify EOO_C by using B's (signature) public key; EOO_C is saved as an evidence of origin for S. In step 2, after receiving sub_K, TTP keeps it in its private directory and delete it after t_B time units or until con_K is generated and published. In step 3, after receiving EOO_C, EOR_C and t_S from S, TTP needs to verify EOR_C using S's (signature) public key and compare EOO_C with the one sent by B in step 1. If either one is not true, TTP concludes that at least one party is cheating and it will not generate con_K. We call {f_{CON}, B, S, L, T, t_B, t_S, K, EOR_C, con_K } the evidence of this purchasing order M. TTP also check if labels L from step 2 and 3 are coincident. If not, buyer B and seller S must be disagreed with this purchase order M. TTP will stop this protocol.

If steps 1-3 are shown positive results, TTP starts to generate con_K with time stamp T attached. In step 4, buyer B fetches K and con_K from TTP. In step 5, seller S fetches con_K from TTP to prove that K is available for S.

3.2 Security of Non-repudiation Protocols

The most important security issue of a non-repudiation protocol is the dispute resolution. We analyze the generated evidences of step 4-5 in the above non-repudiation protocol, dispute resolution mechanisms of buyer and seller to see whether non-repudiation can be reached. A trusted arbitrator will help solve the dispute according to submitted evidences.

Security of the payment information
The payment information is well protected by encryption and not revealed to other entities including TTP and the broker. Moreover, buyers and sellers can reach secure communications, i.e. end-to-end security, for further transactions by sharing common session keys which is not known by other parties.

Validity of Evidence
Non-repudiation service will fail if bogus evidence is accepted or no evidence is received by either buyer or seller. Validity of non-repudiation evidence depends on the security of cryptographic keys used for generating evidences. These keys need to be revoked if they are compromised according to WPKI certificate policy practice.

According to WPKI, buyer B, seller S and TTP could retrieve certificates of each other's from CA's repository to verify digital signatures. By the nature of hash functions, it is computationally hard to find two different key K and K' (with reasonable key length) with the same labels, namely L= H(M, K)= H(M, K')= L' and M= dK(C)= dK'(C)= M'. Therefore, TTP can investigate the validity of evidences by checking these labels.

Dispute of Origin
When buyer B denies having sent purchasing order M to seller S, S may present EOO_C, EOR_C and con_K to the arbitrator in the following way:

S→ arbitrator : EOO_C, EOR_C, con_K, sS_S(EOO_C, EOR_C, con_K), L, K, M, C

The arbitrator first verifies the signature of S, sS_S(EOO_C, EOR_C, con_K); if the verification is positive, the arbitrator checks the following five steps:

step 1: if EOO_C is equal to $sS_B(f_{EOO}, S, L, C)$.
step 2: if EOR_C is equal to $sS_S(f_{EOR}, B, L, t_S, C)$.
step 3: if con_K is equal to $sS_{TTP}(f_{CON}, B, S, L, T, t_B, t_S, K, EOO\_C, EOR\_C)$.
step 4: if L is equal to H(M,K).
step 5: if M is equal to dK(C).

If step 1 is checked positive, this arbitrator concludes that buyer B has sent seller S the encrypted purchase order C. If step 2 is checked positive, arbitrator concludes that

S has sent all the correct payment information to TTP. For all 5 steps being checked positive, this arbitrator finally concludes that B has sent S the purchase order M, which is encrypted by K and presented to be C.

Dispute of Receipt

When seller S denies receiving the purchase order M from buyer B, buyer may present EOO_C, EOR_C, con_K to the arbitrator in the following way:

B→ arbitrator : EOO_C, EOR_C, con_K, sS_B(EOO_C, EOR_C, con_K), L, K, M, C

The arbitrator first verifies the signature of buyer B, sS_B(EOO_C, EOR_C, con_K); if the verification is positive, the arbitrator checks all five steps same as those in the dispute of origin. For all five steps being checked positive, arbitrator concludes that seller S has received M, which is encrypted by K and presented to be C.

Dispute of Fund Transfer

If buyer B realizes the amount of transferred fund is different from that on the purchase order M, he may ask this arbitrator to check. Arbitrator will check M presented by buyer B and M' by seller S. Arbitrator also fetches K and L from TTP. If H(M,K) is not equal to L, the arbitrator concludes that buyer B is cheating. On the other hand, if H(M',K) is not equal to L, the arbitrator concludes that seller S is cheating.

4 Conclusions

We propose a fair non-repudiation protocol based on wireless PKI and mobile agents. An evidence of mobile transaction is generated by WPKI mechanism such that buyer and seller cannot repudiate sending and receiving purchase orders respectively. One challenge of non-repudiation protocols is to avoid any entity to cheat and gain advantage over the other. Mobile payment transactions need time information included in evidences for dispute resolutions. The broker generates a mobile agent for buyer which carries this encrypted purchase order to the seller. The advantage of this agent-based protocol is to provide a convenient way for mobile clients to reach non-repudiation for mobile payment transactions.

The future research of this paper is to establish and simulate mobile agent-based electronic invoice systems for mobile payments. The author and his research team will continue this work with close connections to MOEA and mTaiwan project.

Acknowledgments. This research is supported by Taiwan NSC grant No. 95-2221-E-424 -003.

References

1. J. Zhou, R. Deng, F. Bao, Evolution of Fair Non-repudiation with TTP, ACISP'99, Lecture Notes in Computer Science (LNCS) 1587, pp. 258-269, 1999.
2. ITU-T, Recommendation, X.813: Information Technology-Open Systems Interconnection-Security Frameworks in Open Systems. Non-repudiation Framework, 1996.

3. B. Li, J. Luo, On Timeliness of a Fair Non-repudiation Protocol, InfoSecu'04, Nov 14-16, 2004, pp. 99-106
4. WPKI Implementation, Initial stage, Testing and Experiments (Internal Report and Discussion), Chunghwa Telecom Lab, 2004, Taiwan.
5. O. Esparza, J. Munoz, M. Soriano, J. Forne, Secure Brokerage Mechanisms for mobile Electronic Commerce, Computer Communications 29 (2006) pp. 2308-2321.
6. H. Pagnia, H. Vogt, F. Gartner, U. Wilhelm, Solving Fair Exchange with Mobile Agents, ASA/MA 2000, LNCS 1882, pp.57-72, 2000.
7. U. Wilhelm, S. Staamann, L. Buttyan, On the Problem of Trust in Mobile Agent Systems, In Symposium on Network and Distributed System Security, pp.114-124, Internet Society, Mar. 1998.
8. O. Esparza, J. Munoz, M. Soriano, J. Forne, Host Revocation Authority: A Way of Protecting Mobile Agents from Malicious Hosts, ICWE 2003, LNCS 2722, pp.289-292, 2003.
9. J. Zhou, D. Gollmann, A Fair Non-repudiation Protocol, Proceedings of 1996 IEEE Symposium on Security and Privacy, P.55-61, Oakland, California, May 1996.
10. W. Jansen, T. Karygiannis, NIST Special Publication 800-19: Mobile Agent Security, 1999.

Agent-Based Data Compression Supporting Knowledge Discovery in Mobile Environment*

Romeo Mark A. Mateo, Hwang Jae-Jeong, and Jaewan Lee

School of Electronic and Information Engineering, Kunsan National University
68 Miryong-dong, Kunsan, Chonbuk 573-701, South Korea
rmmateo@kunsan.ac.kr, hwang@ks.kunsan.ac.kr,
jwlee@kunsan.ac.kr

Abstract. Location-aware services using data mining techniques are recent research topics where rules from the data are extracted to provide interesting information. In addition, multi-agent systems are applied in location-based service for autonomous interaction of the system. Different data mining techniques are applied for knowledge discovery from location-based services. However, wireless environment limits the transmission of large data and possible for errors. This work presents a multi-agent framework for the location-based service using data mining. To support the data mining, a data compressor agent (DCA) based on neuro-fuzzy classifier is proposed. DCA performs data preprocessing where it merges the less frequent dataset by using neuro-fuzzy classifier before sending the data. User agent processes the knowledge discovery by using data mining like association rule mining. The result shows the proposed neuro-fuzzy data compression is more efficient compressor.

1 Introduction

In the advent of wireless technologies, ubiquitous devices and wireless sensors are used to gather and process data for location information. The interaction of these devices provides location-awareness and necessary information to mobile users [1]. Location-awareness is an evolution of mobile computing, location sensing and wireless technology [2] where a mobile device like PDA is used as information service of the location. This method is necessary for mobile users who need quick information about the location. Moreover, identifying patterns and rules from the location by using data mining are challenges for researchers. But these data can be large for transmission in a limited bandwidth of the wireless network. It also delays or even hinders the data analysis. Lossless compressions are commonly used to compress the data and minimize the packet size [3]. The encoded data is possible to be transmitted in a limited bandwidth connection like the wireless environment. Data is also processed to remove unnecessary information and reduce the size. This can be done by lossy compression techniques [4].

* This research was supported by the Program for the Training of Graduate Students in Regional Innovation which was conducted by the Ministry of Commerce Industry and Energy of the Korean Government.

N.T. Nguyen et al. (Eds.): KES-AMSTA 2007, LNAI 4496, pp. 308–317, 2007.

This work proposes a data compressor agent (DCA) based on neuro-fuzzy classifier. A framework of location-based service based on multi-agent system is presented to support the data mining techniques and mobile services. Knowledge discovery is performed by the user agent where it sends request to the location agent manager (LAM) for data. LAM requests the DCA to perform data preprocessing and compression before sending this to the user agent. DCA process the data by merging the less frequent dataset using neuro-fuzzy classifier. After transmission from DCA, user agent decompresses the data and processes knowledge discovery by using association rule mining. The result shows the proposed neuro-fuzzy data compression is more efficient on data compression.

2 Related Works

2.1 Data Mining Using Agents

Data mining in mobile environment using the location-aware agent [5] is used to gather location information from location-based services (LBS). This is done by sending a mobile agent to the LBS from the user agent then the mobile agent performs the classification mining in the database. The result is sent back to the user agent to provide the location information. Another method is sending the data to mobile user and performs data mining. There are possible errors in transmitting the data and this issue is the focus of this paper. The architecture of location-based service based on multi-agent is presented in [6]. Multi-agents are used for efficiency of location management by using a nearest neighbor search on the hierarchy of the base station. It also showed the importance of the cooperation of the multi-agent system to work with the location-based services. The location agent manager manages the services of the location-based service and communicates with the location agents which predict the location of the mobile user. Association mining is used to associate the previous movements of mobile users. A collaborative framework is proposed for efficient data mining of location information in the location based-services [7]. The proposed framework of location-based service (LBS) consists of four interactive components namely, location-aware agent, location information service, mapping service and object group services. These components collaborate to produce the necessary location information and use neuro-fuzzy system for classifying the data and extract rules.

2.2 Neuro-fuzzy Classification

Fuzzy classification is based on the concept of fuzzy sets, which was conceived by Lotfi Zadeh [8]. It is presented as a way of processing data by allowing partial set membership rather than crisp set membership or non-membership. Typical fuzzy data analysis discovers rules in large set of data and these rules can be used to describe the dependencies within the data and to classify a new data [9]. Neuro-fuzzy systems are fuzzy classifiers and uses neural networks for learning by performing induction of the structure and adaptation of the connection weights [10, 11]. The NEFCLASS is a fuzzy classification system which borrowed the learning ability of the neural network to for rule learning and fuzzy set learning. It consists of 3 layered perceptron. The 1st layer is for inputs ($U_1 = \{x_1,\ldots, x_n\}$), 2nd layer is for generating rules ($U_2 = \{R_1,\ldots,R_k\}$), and 3rd layer is output layer ($U_3 = \{c_1,\ldots,c_m\}$). The system also contains weights from the input layer (U_1) to rule layer (U_2) and from rule layer (U_2) to the output layer (U_3). Each connection between units $x_i \in U_1$ and $R_k \in U_2$ is labeled with a linguistic term A $A_{jr}^{(i)}$ ($j_r \in \{1,\ldots,qi\}$). The values from the input layer are mapped through the fuzzy sets of the weights. $W(R, c) \in \{0, 1\}$ holds for all, $R \in U_2$, $c \in U_3$. The values from the input and rule layer are evaluated in the connection of the hidden and output layer. For all output units, $c \in U_3$ the net input net_c is calculated Equation 1.

$$net_c = \frac{\sum_{R \in U_2} W(R,c) \cdot o_R}{\sum_{R \in U_2} W(R,c)} \tag{1}$$

3 Multi-agent Framework for Location-Based Service

The proposed architecture of multi-agent system for location-based service (LBS) is shown in Figure 1. Multi-agents are interacting autonomously to perform the efficient service in the location-based service. In addition, multi-agents are used to support the knowledge discovery from the database of LBS. The architecture in Figure 1 consists of location agent manager (LAM), data compressor agent (DCA), sensor agents and user agent. The user agent is stored in a mobile device and initializes the interaction when it reaches the area where it can detect the LAM of the LBS. The LAM manages the services of LBS and communicates with the other LAM to collaborate and shares each service when requested. Figure 1 shows the user agent moves to other LBS which interact immediately to LAM and access the available services.

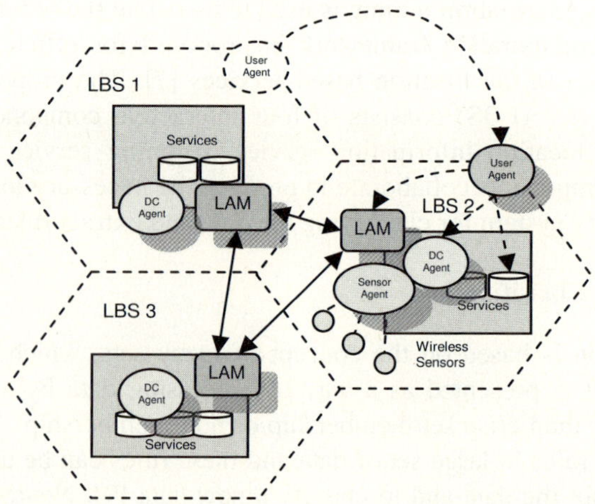

Fig. 1. Framework of location-based services managed by multi-agents

3.1 Multi-agent Components

Location Agent Manager. The architecture of LBS in Figure 1 is consists of multi-agents to produce an efficient service. Multi-agents are used for efficiency of location management [6] by using a nearest neighbor search on the hierarchy of the base station. The software agent concerned in our work is the location agent manager (LAM). It manages the services of the location-based service and communicates with the user agents from the mobile user's device.

Sensor Agent. Wireless sensors are distributed within the location and this gathers specific information of the location like temperature, humidity, brightness and others. The data from the sensors are managed by the sensor agent and this is stored in the database. Sensor agent is also managed by the LAM.

User Agent. The proposed system has a user agent which resides on a mobile device. After sending the data from the data compressor agent, the user agent decodes the message and converts it into relevant data. Data mining algorithm from user agent is applied to produce rules from location information. The knowledge discovery flow from the LBS is shown in Figure 2.

Fig. 2. Knowledge discovery flow from the data of location-base service

Data Compressor Agent. Data compressor agent (DCA) has the function of data preprocessing and compression. LAM request DCA for sending the data to mobile user. First, DCA prepares the data for wireless transmission by compressing the data. The preprocessing of data uses neuro-fuzzy classifier to merge the less frequent datasets. After the preprocessing procedure, Huffman coding is applied to produce shorter bits and then sends to the user agent. Chapter 4 discusses more details of the neuro-fuzzy data compression algorithm.

3.2 Multi-agents Processing Data Compression

The data compressor agent performs the lossy and lossless data compression to prepare for wireless transmission. In the study of Klotz [12], these methods are used on image and video compressions for network transmission. In our study, these methods are used for wireless transmission. The lossy compression is based on neuro-fuzzy classifier. This is done by determining the less frequent dataset and merging it to the dataset which has more frequent data set. After the process, lossless technique using Huffman algorithm is applied to produce shorter bits and sends to the user agent. In Figure 3 the data compressor agent processes the proposed neuro-fuzzy data compression and encoding the data. The compressed data is sent to user agent in wireless transmission. The user agent decodes the data and performs the Apriori algorithm for the rule extraction.

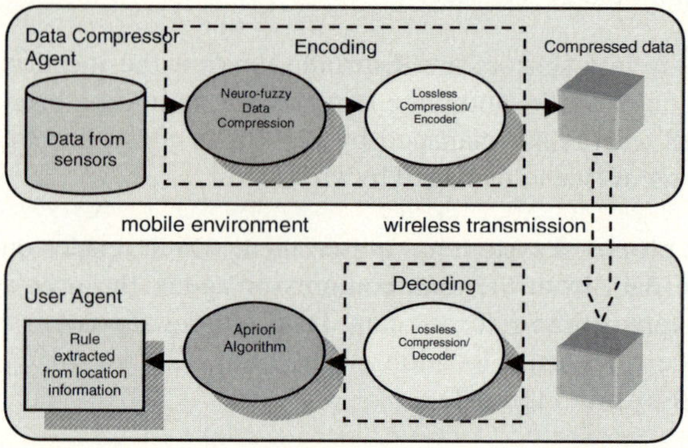

Fig. 3. The process of proposed data compression for mobile environment

4 Data Compression Using Neuro-fuzzy Classifier

The proposed data compression merges the less frequent dataset (LFD) by using neuro-fuzzy classifiers. Data were gathered by sensors and those that have less frequent could be considered as noise or could be identified as errors. However, the LFD still can be classified as it belongs to other datasets which are not LFD. In this study, neuro-fuzzy classifier is used to determine the appropriate dataset the LFD belongs and merged with it. Numeric values of data are transformed to categorical values before the frequency count. The (D_x) is defined as a single combination of values from all data. The *freqCount*(D_x) counts the frequency of set D_x containing in the transactions of the database. To determine if D_x is an LFD, we get the frequency of the dataset (*FD*) by the frequency count of D_x is divided by the total number of dataset. *totaldata* is the total number of data. The calculation of *FD* is shown in Equation 2.

$$FD = \frac{freqCount \quad (D_x)}{totaldata} \tag{2}$$

A threshold represented by *threshold* refers to the percent value which is set manually. If the quotient of *freqCount*(D_x) and *totaldata* is less than the threshold then D_x is marked as LFD. The condition of marking the D_x as LFD is shown in Equation 3.

$$LFD(D_x) = D_x \text{ is } LFD \text{ if } FD > threshold \tag{3}$$

LFD is processed for merging after Equation 3. Neuro-fuzzy classification is used to determine which dataset (D_f) the LFD will merge. This procedure uses the numeric values to process in fuzzy sets. All the datasets that is not LFD become the rule nodes.

Fuzzy sets from the linked connection are trained by the dataset which is classified by the rule node. The delta value is determined by Equation 4 to adjust the fuzzy sets.

$$\delta_R = o_R(1-o_R)\sum_{c\in U_3} W(R,c)\delta c \tag{4}$$

After the training of fuzzy sets, the LFD are processed in the structure of neuro-fuzzy. LFD becomes input pattern to calculate the membership function from each rule nodes. The conjunction function of two values in Equation 5 is used.

$$\mu_{A\wedge B}(x) = \min\{\mu_A(x), \mu_B(x)\} \tag{5}$$

The membership values are mapped by the shared links or weights of fuzzy sets. These values are aggregated to the rule nodes. Then the rule node values are compared to each rule node and the one has the greatest value is the chosen node for merging the LFD. Equation 6 shows the procedure of comparing and merging process.

$$\begin{aligned} D_f &= \sum_{i=1}^{n} compare\ (R_i, R_f),\ if\ R_i > R_f\ then\ f = i \\ Merge\ &(D_f, D_x)\ where\ D_x\ is\ LFD \end{aligned} \tag{6}$$

D_x is merged to D_f which is the dataset has the highest membership value. The merging process also implies addition to the frequency count of D_f and removal of D_x. Figure 4 shows the pseudo code of the neuro-fuzzy classification choosing the dataset.

```
Node(Rk) = Rule aggregated value
Ck = choosen dataset
for each input xj do
 for each µ(i) do
   begin
     µ(i)ji = µ(i) (xj);
   end;
 for each Rk do
  for each antecedent Aj do
   begin
     if Rk (µ(i)ji) is not 0
     then Node(Rk) = current value of Rk + µ(i)ji;
   end;
 for each Rk do
   begin
     if Node(Rk) has greatest value
     then choose K index;
   end;
return Ck;
```

Fig. 4. Neuro-fuzzy classification algorithm

5 Experiment Evaluation

The location agent manager, data compressor agent, sensor agent and user agent were developed using Java. The proposed neuro-fuzzy data compression was coded in the data compressor agent. Different OS were used like Solaris, Linux and Windows to simulate the heterogeneous LBS. EMPOS II which is an Intel PXA255 processor for the mobile device and Zigbex motes which are wireless sensors were used. The performance measures and simulation result are discussed on the following subsections.

5.1 Performance Measures

An information source [13] is a source alphabet together with a probability distribution. Let K be a coding of an information source. That is, for each source symbol a_i, we have a code word $K(a_i)$ and we know the probability $P(a_i)$ of a_i. Denoting by d_i the length of the word $K(a_i)$, we can compute the average length L of code word:

$$L = \sum_{i=1}^{n} d_i P(a_i) \tag{7}$$

The smallest average length from Equation 7 indicates that it is more efficient. We can say that the proposed data compression provides smaller L of the information because we remove the dataset that has less probability and also merging the LFD from highest probability sets. Processing the new values on Huffman algorithm minimizes the structure of the tree and the word length is reduced. Equation 8 derives the compressed size by dividing the total number of data into the ratio of the output.

$$total\ \ size = \frac{n}{Huffman \sum_{i=1}^{f} (D_f)} \tag{8}$$

The significance of performing the merging of datasets are size reduction and to provides relevant rules. To be specific, Apriori is used to generate association rules. Applying the proposed algorithm removes the less frequent dataset which are considered not important and merged to other dataset. The effects of setting a high threshold value are a relatively high support for the confidence of a datasets because of merging process which implies more association rules are generated. However, some interesting rules might be removed. It is important that the threshold is set appropriately.

In classifier algorithm, recall and precision is performed by cross-validation of the classified instances. To evaluate the accuracy performance of the neuro-fuzzy algorithm, these measurements were used. This is done by calculating the average precisions in Equation 9 where *AvgP* is the summation of precision (P_n) of classes divided by the number of classes. Average of recall is computed in Equation 10 where *AvgR* is the summation of recall (R_n) of classes divided by the number of classes. The number of correctly classified instances was used to determine accuracy.

$$AvgP = \frac{\sum_{i=1}^{n} P_n}{n} \tag{9}$$

$$AvgR = \frac{\sum_{i=1}^{n} R_n}{n} \tag{10}$$

5.2 Result

The compression ratio using the proposed algorithm was calculated by using simulation. These data were acquired by using wireless sensors and a virtual message posting. Mobile users post their feelings on the virtual board messaging about the location. The inputs were scaled into good, medium and bad. Datasets are updated in the database of LBS every time the mobile user posts their message. There were 1000 datasets gathered to perform the simulation. The threshold was set from 1% to 5%. After it executing the neuro-fuzzy preprocessing, a Huffman compression was done. Figure 5 presents the compression ratio results of using neuro-fuzzy data compression (NFDC) with threshold values and normal compression (NC) without merging. The graph shows that NFDC has a decrease its ratio as the data increases compared to NC. The average compression difference of NFDC and NC is 15 from 1%. Setting up a higher threshold also implies a high compression rate because the LFD is deducted and the highest probability is incremented by the merge function. However, it is important to set a correct threshold to acquire interestingness of data as well as of compressed data.

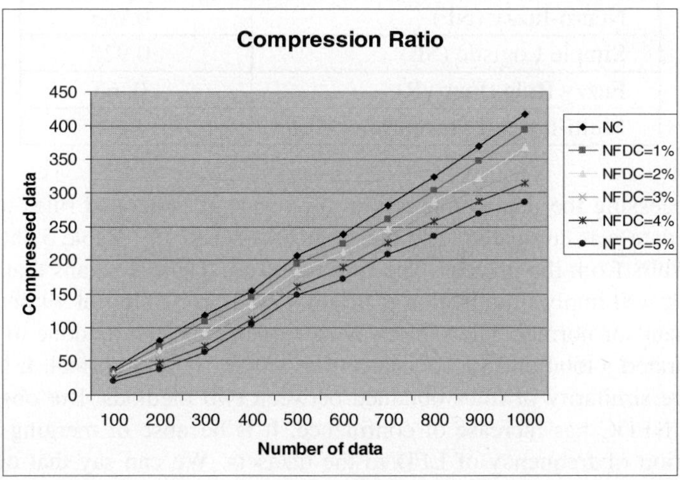

Fig. 5. Result of data compression

The accuracy of classifying the data by the merge function of NFDC was compared to other high accurate classifiers. We used the same data gathered from the wireless sensors. The LFD were classified by using NFDC and other classification algorithms. Results of precision (Equation 10) and recall (Equation 11) are presented in Table 2 and 3, respectively. The result shows that the accuracy of classification of NF is much similar to other highly accurate methods like the MLP. Neuro-fuzzy highest output which has an average of 0.943 in precision and recall which has an average of 0.936 compared to Simple Logistic (0.94 (1), 0.925 (3)), MLP (0.937 (2), 0.94 (1)), and FuzzyR (0.67 (4), 0.65 (4)). Comparing more to FuzzyR which is a fuzzy rule classifier, NF is 28% better accuracy of classifying. Also, we get the processing time performance of classifying the LFD and the NF is the second fastest processing time which has 1.2 seconds compared to FuzzyR (0.8), MLP (2.56) and SL (11.45).

Table 1. Precision of each algorithm

Algorithm	Average Precision
Neuro-fuzzy (NF)	0.943
Simple Logistic (SL)	0.94
Fuzzy Rule (FuzzyR)	0.67
Multi-layered Perceptron (MLP)	0.937

Table 2. Recall of each algorithm

Algorithm	Average Recall
Neuro-fuzzy (NF)	0.963
Simple Logistic (SL)	0.925
Fuzzy Rule (FuzzyR)	0. 65
Multi-layered Perceptron (MLP)	0.94

After processing the data compression, the result of generated rules using NFDC and normal data was compared. The threshold was set to 1%. Table 3 shows the lists of first 5 results from the process. The first rule from Table 3 means that if humidity is low then it will imply illumination is bright with a probability of 99 percent for PC and 97 percent for normal. The same way of explanation can be done to other rules. NFDC generated a total of 38 association rules while 37 for a normal. It is also noted that there are similarity of rules obtained between two methods. But observing each confidence, NFDC has increase of confidence. It is because of merging of data and the distribution of frequency of LFD to the datasets. We can say that our proposed algorithm retains the original data and also provides higher compression rate.

Table 3. Association rules generated from the data of LBS

Association rules, showing only 5	Confidence	
	NFDC	Normal
HUM=low→ILLUM=bright	0.99	0.97
MSG=medium→ILLUM=bright	0.99	0.98
TEMP=warm→ILLUM=bright	0.99	0.97
TEMP=hot→ILLUM=bright HUM=low	0.99	0.96
TEMP=hot HUM=low→ILLUM=bright	0.99	0.99

6 Conclusion and Future Works

Acquiring large data using mobile devices is one of the constraints in mobile environment. Data compression is a solution to reduce the constraints of wireless transmission. This work proposes the data compressor agent (DCA) based on neuro-fuzzy classifier. A framework of location-based service based on multi-agent system was presented to support the mobile services and data mining methods. Knowledge discovery is done by the user agent requesting data to location agent manager (LAM). User agent from mobile users decompresses the data and processes knowledge discovery. DCA performs data preprocessing and compression before sending to the user agent. The result showed that our proposed algorithm compresses the data efficiently compared to a normal compression and generates relevant association rules.

Other multi-agent components for LBS will be future works. The data gathered was only from LBS database and the future work will use other environment. The threshold in the future work will be a data analysis for setting up the appropriate threshold.

References

1. Luley, P., Almer, A., Seifert, C., Fritz, G., and Paletta, L.: A Multi-Sensor System for Mobile Services with Vision Enhanced Object and Location Awareness. 2nd IEEE Workshop on Mobile Commerce and Services (July 2005) pp. 52-59
2. Jensen, C.: Research Challenges in Location-Enabled M-Services. Proceedings of the Third International Conference on Mobile Data Management (2002) pp. 3-7
3. Holtz, K., and Holtz, E.: Lossless data compression techniques. WESCON/94, 'Idea/Microelectronics (1994) pp. 392-397
4. Berger, T. Gibson, J.D.: Lossy Source Coding. IEEE Transactions on Information Theory, Vol. 44 , Issue 6 (1998) pp. 2693-2723
5. Jaewan Lee, Romeo Mark A. Mateo, Bobby D. Gerardo, Sung-Hyun Go: Location-Aware Agent Using Data Mining for the Distributed Location-Based Services. ICCSA 2006, Part V, LNCS 3984, Springer-Verlag (May 2006) pp. 867-876
6. Mateo, R. M., Lee, J. W. and Hyunho Yang: Optimization of Location Management in the Distributed Location-based Services using Collaborative Agents. ICCSA 2006, Part III, LNCS 3982, Springer-Verlag (May 2006) pp. 17-22
7. Mateo, R. M., Lee, M., and Lee, J. W.: Location-aware Data Mining for Mobile Users based on Neuro-fuzzy Systems. ICNC-FSKD 2006, LNAI 4332, Springer-Verlag (September 2006) pp. 1269-1278
8. Zadeh, L. A.: Fuzzy Sets. Information and Control (1965) pp. 338-353
9. Kruse, R., Bolgelt, C., and Nauck, D.: Fuzzy Data Analysis: Challenges and Perspectives. In Proceedings of the 8th IEEE International Conference on Fuzzy Systems, IEEE Press, Piscataway, NJ, USA (1999)
10. Klose, A., Nürnberger, A., Nauck , D., and Kruse R.: Data Mining with Neuro-Fuzzy Models. Data Mining and Computational Intelligence, Springer-Verlag (2001) pp. 1-36
11. Nauck, D., and Kruse, R.: NEFCLASS - A Neuro-Fuzzy Approach for the Classification of Data. In Proceedings of ACM Symposium on Applied Computing, Nashville (1995)
12. Holtz, K.: Digital Image and Video Compression for Packet Networks. Available at http://www.autosophy.com/icomart.htm
13. Adamek, J.: Foundations of Coding, John Wiley and Sons, Inc. (1991) pp. 17-24

eMARP: Enhanced Mobile Agent for RFID Privacy Protection and Forgery Detection

Sang-Soo Yeo[1], Soo-Cheol Kim[2], and Sung Kwon Kim[2]

[1] Department of Computer Science & Communication Engineering, Kyushu University, Fukuoka, Japan
ssyeo@itslab.csce.kyushu-u.ac.jp

[2] School of Computer Science & Engineering, Chung-Ang University, Seoul, Korea
sckim@alg.cse.cau.ac.kr, skkim@cau.ac.kr

Abstract. We are sure that RFID system should be a widely used automatic identification system because of its various advantages and applications. However, many people know that invasion of privacy in RFID system is a still critical problem that makes it difficult to be used widely. Many works for solving this problem have focused on developing lightweight cryptographic modules inside of an RFID tag and building communication protocols with the reader or the back-end server. Another approach is use of proxy agents that control communications between the tag and the reader for protecting privacy. In this paper, we present an enhanced version of MARP scheme. We modified the original MARP scheme for reducing the probability of preventing attacker's eavesdropping and for reducing the communication number of tags. And back-end servers can authenticate mobile agents more easily using public key cryptography in this scheme. Enhanced MARP scheme guarantees not only privacy protection but also forgery detection, and it only needs slight modification in the current tag hardware.

1 Introduction

RFID, *Radio Frequency Identification*, is one of the core technologies for ubiquitous environment. RFID technology can make the existing automatic identification systems more various and efficient, since it has contactless and fast identifying mechanism using radio frequencies and it is not affected by surface contamination. We are sure that RFID system should be a widely used automatic identification system [1,2,3,4].

However, the identification using the radio frequency gives us problems as well as efficiencies. For example, an RFID tag tells its own unique identifier to any RFID readers without any authentication steps. This causes the privacy problems, such as private information leakage and location tracking [5] of consumers. The information leakage problem means that the privacy information of the consumer who has tag-embedded items can be read by any RFID readers [6,7]. In other words, whoever has an RFID reader can acquire tag data, which expresses the consumer's income level, physical condition, purchase pattern, life environment, and so on. The location tracking problem is caused by

N.T. Nguyen et al. (Eds.): KES-AMSTA 2007, LNAI 4496, pp. 318–327, 2007.

the hardware limitation of the tag which always answers the identical data, such as unique identifier, serial number, and item name [8]. An adversary can track a specific consumer without the consumer's awareness by tracing the consumer's tag data. This is the almost same that the consumer has a low-cost undetectable tracking device. If an adversary, such like a government, has a lot of readers throughout a specific region, he can know the location history of any consumers. We think that this problem is a critical invasion of privacy. There are many researches related to the RFID privacy problems until now. We are focusing the agent approach of them.

In this paper, we proposed an enhanced version of MARP [9]. We make up for the shortcomings of MARP and improves its security and efficiency by eMARP scheme. In this scheme, the tag hardware needs to be modified just slightly and the mobile agent takes the responsibility of protecting consumer's privacy, as the original MARP scheme.

In section 2, we analyze the existing agent schemes. And then we address the original MARP scheme and its problems in section 3. In section 4, we describe our eMARP scheme and the differences between eMARP and MARP in detail. In section 5, we analyze and evaluate eMARP scheme. In section 6, we conclude with mentioning the improvements of our scheme.

2 Related Work

The simplest scheme for protecting privacy is *kill command* method [4,8,10]. This method is suggested by Auto-ID center and it has been used as the basic security method for the most of EPC tags until now. This strategy seems to be a perfect solution for privacy protection, but it makes useless all of the potential valuable services of RFID. In other words, because this method makes tags to be isolated from networking environment, it is not a good solution.

Weis et al. suggested hash lock scheme, randomized hash lock scheme, and silent tree walking scheme [8]. This scheme prevent location tracking partly, but has some security holes in reader-tag communication and very heavy load in the back-end server [11]. Juels proposed the *blocker tag* scheme [12] and suggested privacy protection scheme for RFID-enabled banknotes [13]. The *blocker tag* scheme reversely uses anti-collision protocols, which are originally intended to for tag singulation. Golle proposed a scheme using re-encryption [14]. This scheme can be also attacked through various security vulnerability. Henrici et al. suggested hash based scheme using transaction counter [15].

Another approach is use of proxy agents that arbitrate the communication between the reader and tag [16,17,18]. Normally, an agent can be implemented for a stand-alone device or for a software/hardware module inside of a mobile device. Because mobile devices have more powerful processor than tags have, they can compute higher level cryptographic functions. However, the existing agent schemes have some problems related to tag forgery and ownership transferring. In REP, *RFID Enhancer Proxy*, scheme of Juels[18], a tag must send all of its secret information to its agent. This causes an important security problem that agents will be able to counterfeit or masquerade after returning or transferring of tag-embedded items.

3 Problems of the Original MARP Scheme

MARP scheme, that we had proposed before, is similar to other agent schemes in the aspect of using the mobile devices for privacy protection. However, a MARP obtains only specific part of secret information of tags for proxy authentication that confirm the reality of them. So a MARP should not forge tags after returning or transferring of them, but it still has two problems in the view of security and efficiency. Firstly, in its initial setup phase, an adversary can eavesdrop $h(PIN_t)$, which is transmitted from a reader to a tag without encryption via an air interface. So he can be a proxy of the tag and control it. Secondly, tags always have to be involved in the communication because of forgery preventing step. This reduces the overall performance of MARP scheme and makes it inefficient in the view of agent approach.

4 The Enhanced MARP Scheme

In this section, we present the enhanced MARP scheme. eMARP scheme is that the modified version of MARP makes up for its disadvantages as we mentioned.

4.1 RFID System Construction

An RFID system generally consists of RFID tag, RFID reader and back-end server. Additionally, in the proposed system, there is an eMARP that arbitrates the communication between readers and tags, and there is also the trusted public key management center. Fig.1 shows that the RFID system for eMARP scheme.

RFID tag is embedded in the item(object) to be identified. The tag comprises an IC chip and an antenna module.

RFID reader communicates generally several tags at the same time and identifies their identifiers through back-end server. And in this scheme, readers can communicate with an eMARP instead of tags. Each reader has its own public key and private key pair and registers its identity and public key to the PKC.

Back-end server has its own database and manages various types of information related to each tag that belongs to it. There are some back-end servers, which use the same public key center. All of back-end servers must be trusted and must have the capability to process every query from a lot of readers concurrently. Each back-end server has its own public key and private key pair and registers its identity and public key to the PKC.

PKC. *Trusted Public Key Center* is the trusted third party public key distribution center for eMARP scheme based RFID system. In this scheme, since servers, readers, and eMARP's use a public key cryptosystem for mutual

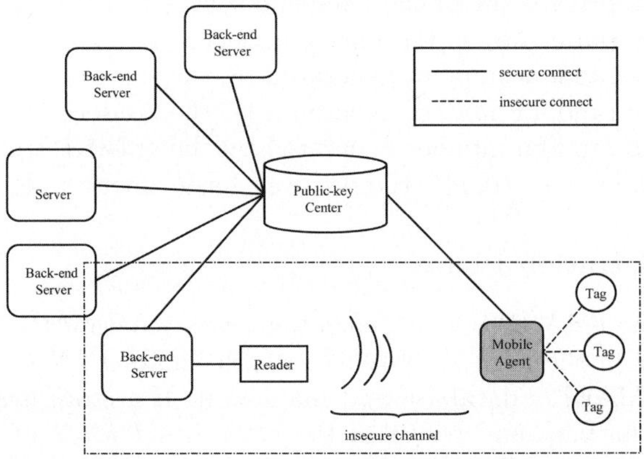

Fig. 1. RFID System Construction for eMARP scheme

authentications, a trusted public key distribution center must be needed. The PKC has public keys of registered servers, readers, and eMARP's, and manages them and their certificates.

eMARP. *Enhanced Mobile Agent*, for RFID privacy protection and fogery detection, is a compact battery-powered mobile device or a software/hardware module inside of a mobile device, such as PDA, *Personal Digital Assistant*, portable multimedia player, cellular phone, and so on. It plays the role of gathering the secret information of tags and becomes a proxy on behalf of them. In this paper, each eMARP has its own public key and private key pair and registers its identity and public key to the PKC.

4.2 Definitions

For describing eMARP scheme, we need some parameters as the followings.

- $\|$: concatenation operator.
- $\oplus$: exclusive-or operator.
- Uid_t : the unique identifier of the tag t.
- $Data_t$: the data of the tag t.
- Key_t : the secrete value of the tag t.
- PIN_t : the mode change key of the tag t.
- Uid_{ma} : the unique identifier of the eMARP m.
- Rid_g : the identifier of reader group g.
- Sid_s : the identifier of back-end server s.
- K_d^g : the private key of reader group g.
- K_e^g : the public key of reader group g .
- K_d^m : the private key of the eMARP m.
- K_e^m : the public key of the eMARP m.

- K_d^s : the private key of the back-end server s.
- K_e^s : the public key of the back-end server s.
- R_s : the random number generated by the server s.
- R_r : the random number generated by the reader r.
- R_m : the random number generated by the eMARP m.
- $h()$: $\{0,1\}^* \longrightarrow \{0,1\}^\ell$, the one-way hash function algorithm.

4.3 Tag Registration Phase

Before performing eMARP scheme, each tag t must contain PIN_t that is one of the secret values of the tag. An eMARP, which knows PIN_t, only can register a tag t to the eMARP's database and manage it. If a store has tag-embedded items for sale, the back-end server of the store has PIN_t's of all tags. In the point of sale of an item, PIN_t of the tag will be transferred to the consumer's eMARP. And then the eMARP sends $h(PIN_t)$ to the tag in order to register it to eMARP's database. If $h(PIN_t)$ is valid for the tag, the tag answers with some values needed to a registration.

In the original MARP scheme, it is possible for any adversary to get the right to managing a tag, since he can eavesdrop $h(PIN_t)$. For solving this problem, a backward channel communication method can be used. In other words, the agent sends $h(r||PIN_t)$ which is computed with a random number r that the tag sends to the eMARP before. Since tag's transmission signal is too weak to eavesdrop, it seems feasible. However, a tag needs to have a additional random number generator for this method, and besides a backward channel communication can be eavesdropped by high power signal detecting devices.

For this reason, we use another method for solving eavesdropping problem in the eMARP scheme. We add a new step in this phase, that the tag updates its PIN_t, to the tag registration phase. This slight modification makes the hash value, $h(PIN_t)$, to be used just once for a registration. Since $h()$ is a one-way hash function, a new hash value, $h(PIN_t + 1)$, cannot be computed from an old hash value, $h(PIN_t)$. It doesn't matter in the eMARP scheme if any adversary knows $h(PIN_t)$ by eavesdropping. Fig.2 shows the tag registration phase.

Detailed Protocol of Tag Registration Phase

1. The database of a store sends PIN_t to a consumer's eMARP via a secure channel.

   · Store DB $\longrightarrow$ eMARP : PIN_t.

2. The eMARP sends the hashed value of PIN_t to the tag in a short distance.

   · eMARP $\longrightarrow$ Tag : $h(PIN_t)$.

3. The tag confirms validity of $h(PIN_t)$ and updates PIN_t for security reason. And then it sends the values computed with its secret data and updates PIN_t. If the data sent by the eMARP is not valid, the tag does nothing.

   · Tag : computes PIN_t from received $h(PIN_t)$.
   · Tag : authenticates the eMARP.
   · Tag : updates PIN_t : $PIN_t = PIN_t + 1$.
   · Tag $\longrightarrow$ eMARP : $PIN_t \oplus Uid_t$, $h(PIN_t) \oplus h(Key_t)$.

4. The eMARP stores the values sent by the tag into its database, and then updates PIN_t.

   · eMARP : stores Uid_t, PIN_t, $h(Key_t)$ to its database.
   · eMARP : updates PIN_t : $PIN_t = PIN_t + 1$.

4.4 Agent Working Phase

After the tag registration phase, the eMARP acts on behalf of the registered tags. Actually this phase is divided into two parts, the ordinary part and the audit part. In the ordinary part, all of the registered tags are silent and the eMARP works as a proxy agent of the registered tags. If the back-end server requests the authentication for a specific tag, the audit part starts. Fig.2 shows the overall protocol of the agent working phase of eMARP scheme.

Detailed Protocol of Agent Working Phase

- **Ordinary Part**

1. The reader sends a query to the eMARP along with the its group ID, back-end server ID and a random number which are signed by the reader group private key.

   · Reader $\longrightarrow$ eMARP : $Query \parallel E_{K_d^g}(Sid_s \parallel Rid_g \parallel R_r) \parallel Rid_g$.

2. The eMARP retrieves the public key of the reader group g from the PKC. And the eMARP decrypts the data sent by the reader with the public key K_e^g.

   · eMARP : $Sid_s \parallel Rid_g \parallel R_r = D_{K_e^g}(E_{K_d^g}(Sid_s \parallel Rid_g \parallel R_r))$.

3. The eMARP generates a random number R_m and signs $R_r||R_m$ with its private key K_d^m. And then it makes a_1 as below and sends it to the reader.

   · eMARP : checks the reader group ID Rid_g.
   · eMARP : generates a random number R_m.
   · eMARP $\longrightarrow$ Reader : $a_1 = E_{K_e^s}(E_{K_d^m}(R_r \parallel R_m) \parallel Uid_{ma})$.

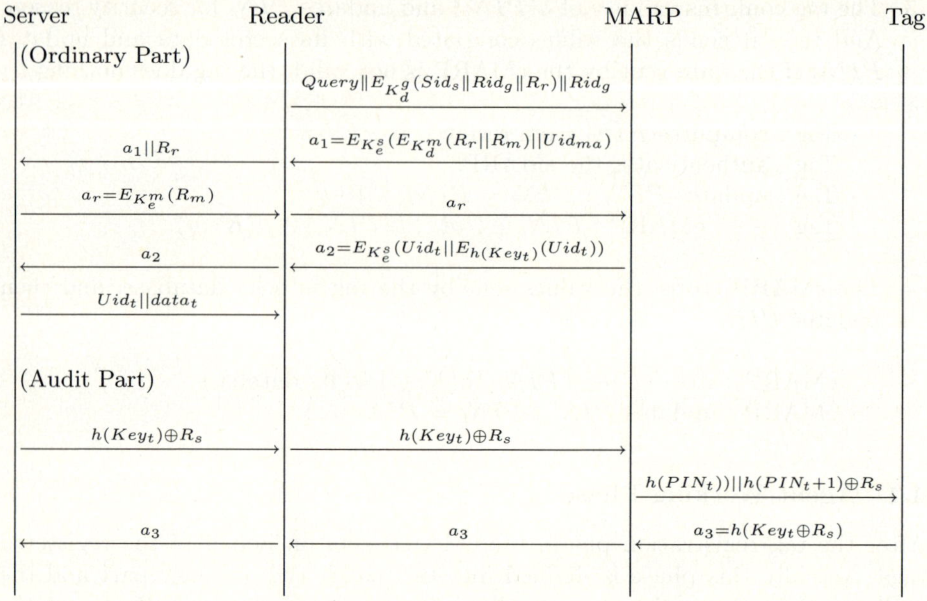

Fig. 2. Agent Working Phase of eMARP Scheme

4. The reader transmits a_1 from the eMARP and R_r to the server.

   · Reader $\longrightarrow$ Server : a_1.

5. The back-end server can know Uid_{ma} by decrypting a_1 with its private key. And then it retrieves the public key of the eMARP from the PKC and decrypts still encrypted part with K_e^m. At this point, if a decrypted R_r is equivalent to R_r from the reader, the back-end server trusts the eMARP. Finally, the back-end server encrypts R_m with the eMARP's public key and sends it to the eMARP via the reader. As you know, the reader cannot decrypt a_1 and cannot know the identifier of the eMARP.

   · Server : $E_{K_d^m}(R_r||R_m)||Uid_{ma} = D_{K_d^s}(a_1)$.
   · Server : checks whether Uid_{ma} and R_r is valid or not.
   · Server $\longrightarrow$ Reader $\longrightarrow$ eMARP : $a_r = E_{K_e^m}(R_m)$.

6. If the eMARP checks R_m successfully, the mutual authentication is completed. And then the eMARP transmits the information of the registered tags using public key cryptosystem.

   · eMARP : confirms information, the mutual authentication is completed.
   · eMARP $\longrightarrow$ Reader $\longrightarrow$ Server : $a_2 = E_{K_e^s}(Uid_t \parallel E_{h(key_t)}(Uid_t))$.

7. The back-end server stores a pair of data, (Uid_t, Uid_{ma}) into its database. This information is used for determining whether the audit part needs or not. If (Uid_t, Uid_{ma}) is unchanged in the ordinary part, the back-end server only sends $Uid_t \parallel data_t$ to the reader. If (Uid_t, Uid_{ma}) is differ from the previous value, starts the audit part.

   · Server : decrypts the received information.
   · Server : compares data (Uid_t, Uid_{ma}).
   · Server : stores data (Uid_t, Uid_{ma}).
   · Server $\longrightarrow$ Reader : $Uid_t \parallel data_t$.

- **Audit Part**

8. If (Uid_t, Uid_{ma}) is differ from the previous value stored in server's database, it means that ownership transferring of the tag-embedded item may happens or it means that either of the previous agent or the current agent may try to cheat. For confirming the authenticity of the tag, the back-end server starts the audit part. If this audit part will be done successful, the back-end server will change the value in its database. If this audit part will fails, the back-end server will refuse authentication and register that the adversarial agent into the black-list.

   · Server : starts the audit part.
   · Server $\longrightarrow$ Reader $\longrightarrow$ eMARP: $h(Key_t) \oplus R_s$.

9. The eMARP calculates R_s using the stored data, $h(Key_t)$. And it sends three values to the tag. The first value is the current registering key value $h(PIN_t)$, the second is the next time registering key value $h(PIN_t+1)$, and the last is R_s from the server.

   · eMARP : calculates $R_s = h(Key_t) \oplus R_s \oplus h(Key_t)$.
   · eMARP $\longrightarrow$ Tag : $h(PIN_t) \parallel h(PIN_t + 1) \oplus R_s$.

10. The tag analyzes the information sent by the eMARP. It responds only after confirming that the eMARP is its master. If it is, the tag computes a_3 using its secret data and sends to the back-end sever via the eMARP and the reader. The server confirms the received response and authenticates the tag and updates (Uid_t, Uid_{ma}).

    · Tag : authenticates eMARP.
    · Tag $\longrightarrow$ eMARP $\longrightarrow$ Reader $\longrightarrow$ Server: $a_3 = h(Key_t \oplus R_s)$.
    · Server : authenticates the tag and updates server's database.

5 Analysis

We analyze the proposed scheme in this section. The following issues are related to attack methods to be possible in RFID system. We present the security and the merits of the proposed scheme in each issue.

Eavesdropping. An adversary can eavesdrop the communication contents since RFID system communicates using air interface. This attack is critical to the privacy problem. In our scheme, since most of communication are performed using a public key cryptosystem, our eMARP scheme is strong from eavesdropping. Maybe the tag registration phase seems to be a weak point. However, as we mentioned, its security is guaranteed sufficiently using the one-way hash function and self refreshed PIN_t value. Since we also use mutual challenge-response mechanism in our scheme, a kind of replay attack is impossible.

Location tracking. In our scheme, since the tag and the agent changes their answers using random numbers and a one-way hash function whenever a protocol is executed.

Tag forgery. In the existing agent schemes, since an agent can get all of the secret data of the tag, the agent easily can be used for cheating or counterfeiting. In our scheme, we designed that an agent can take a specific part of the secret of the tag, PIN_t. The tag always keeps key_t secure and this value is only used for the authentication between the tag and the server in the eMARP's audit part. Therefore, in our scheme, it is possible to check ownership and ownership transfer of tags(items), and to prevent from counterfeit tags.

6 Conclusions

We proposed eMARP scheme as a concept using the external proxy agent for the privacy protection. There are already several agent schemes for privacy protection in RFID system but they have some problems in the point of security and performance. We presented that our scheme guarantees not only protecting the privacy strongly, but also preventing the forgery efficiently. Moreover, our scheme needs just slight change in the current tag hardware.

Acknowledgement

This work was supported by grant No. R01-2005-000-10568-0 from the Basic Research Program of the Korea Science & Engineering Foundation.

References

1. K. Finkenzeller, *RFID handbook*, John Wiley & Sons, 1999.
2. D. Brock, "The Electronic Product Code - A Naming Scheme for physical Objects", *Auto-ID White Paper*, http://www.autoidlabs.com/whitepapers/MIT-AUTOID-WH-002.pdf , January 2001.
3. H. Knospe and H. Pobl, "RFID Security", *Infomation Security Technical Report*, vol. 9, no. 4, pp. 39-50, Elsevier, 2004.
4. S. Sarma, S. Weis, and D. Engels, "Radio-Frequency Identification: Security Risks and Challenges", *Cryptobytes*, vol. 6 no. 1, pp. 2-9, RSA Laboratories, Spring 2003.

5. G. Avoine and P. Oechslin, "RFID Traceability: A Multilayer Problem", *Financial Cryptography - FC'05*, vol. 3570 of LNCS, pp. 125-140, February 2005.
6. R. Anderson and M. Kuhn, "Low Cost Attacks on Tamper Resistant Devices", *International Workshop on Security Protocols - IWSP*, vol. 1361 of LNCS, pp. 125-135, April 1997.
7. R. Revest, "Approaches to RFID privacy", *RSA Japan Conference*, 2003.
8. S. Weis, S. Sarma, R. Rivest, and D. Engels, "Security and Privacy Aspects of Low-cost Radio Frequency Identification Systems", *Security in Pervasive Computing - SPC 2003*, vol. 2802 of LNCS, pp. 454-469, March 2003.
9. S.C. Kim, S.S. Yeo, and S.K. Kim, "MARP : Mobile Agent for RFID Privacy Protection", *International Conference on Smart Card Research and Advanced Applications – CARDIS 06*, April 2006.
10. S. Sarma, S. Weis, and D. Engels,"RFID Systems and Security and Privacy Implications",*Cryptographic Hardware and Embedded Systems - CHES 2002*, vol.2523 of LNCS, pp. 454-469, August 2002.
11. G. Avoine, "Adversarial Model for Radio Frequency Identification", *Cryptology ePrint Archive*, Report 2005/049, http://eprint.iacr.org, 2005.
12. A. Juels, R. Rivest, and M. Szydlo, "The Blocker Tag : Selective Blocking of RFID Tags for Consumer Privacy", *Computer and Communications Security - ACM CCS 2003*, pp. 27-30, October 2003.
13. A. Juels and R. Pappu, "Squealing Euros: Privacy Protection in RFID-Enabled Banknotes", *Financial Cryptography '03*, 2003.
14. P. Golle, M. Jakobsson, A. Juels, and P. Syverson,"Universal Re-Encryption for Mixnets", *Track on the RSA Conference – CT-RSA '04*, vol. 2964 of LNCS, pp. 163-178, February 2004.
15. D. Henrici, and P. Müller, "Hash-based Enhancement of Location Privacy for Radio-Frequency Identification Devices using Varying Identifiers", *IEEE PerSec '04* at *IEEE PerCom*, March 2004.
16. M. Rieback, B. Crispo, and A. Tanenbaum, "RFID Guardian: A Battery-powered Mobile Device for RFID Privacy Management", *Australasian Conference on informaiton Security and Privacy - ACISP 2005*, vol. 3574 of LNCS, pp. 184-194, July 2005.
17. S. Konomi, "Personal Privacy Assistants for RFID Users", *International Workshop Series on RFID 2004*, November 2004.
18. A. Juels, P. Syverson, and D. Bailey, "High-Power Proxies for Enhancing RFID Privacy and Utility", *Center for High Assurance Computer Systems - CHACS 2005*, August 2005.

Ontology Agent Based Rule Base Fuzzy Cognitive Maps

Alejandro Peña[1,2,3], Humberto Sossa[3], and Francisco Gutierrez[3]

WOLNM [1], UPIICSA [2] & CIC [3] - National Polytechnic Institute [2,3], México
31 Julio 1859, # 1099B, Leyes Reforma, DF, 09310, México
apenaa@ipn.mx, {hsossa, atornes}@cic.ipn.mx

Abstract. This work proposes a framework for the design and development of Ontology Agents oriented to manage Rule Base Fuzzy Cognitive Maps (RB-FCM). The approach takes into account the foundations of the Ontology Agents and the baseline of the Fuzzy Cognitive Maps depicted by Rule Bases. With these underlying elements, a specification of a conceptualization about the modeled domain is outcome. Moreover, a knowledge structure, composed by concepts and causal relationships that fit a Fuzzy Rule Base, is grown from. As a result, a semantic repository is stated by means of the Ontology Web Language (OWL). The management of the ontology is fulfilled by an Ontology Agent. This kind of agent takes over the services required to define and update the Ontology items. Also, the Ontology Agent achieves the tasks for answering the queries sent by a community of agents. This set of agents recreates a Multi-Agent System (MAS) that is deployed on the Internet by means of Web Services, where the system carries out causal inferences based on RB-FCM.

1 Introduction

A Cognitive Map is a mental model about a specific problem that is designed from the cause-effect perspective. The entities of the domain are objects and events that are stated as concepts. Stimulus and inhibitions between couples of concepts are outlined through causal relationships [1]. Fuzzy Cognitive Maps measure the state of the concepts and the intensity of the causal influences by linguistic terms. In addition, the cause-effect relationships are pointed out like fuzzy rules, which are embedded into Rule Bases. Due to the plenty of knowledge to be organized, it is necessary to define an Ontology composed by meta-data, classes, attributes and instances of the elements characterized. Furthermore, the Ontology must be grounded on the axioms that specify the intended meaning of such vocabulary. Also, the Ontology requires a mechanism that administrates its elements and responds the queries that a community of agents claims. So we propose an Ontology Agent to deal with the knowledge stemmed from the RB-FCM and meet the requests from agents distributed on the Web. Thus, the organization of the paper is as follows: In section two it is presented a profile of the RB-FCM. In section three it is pointed out the formal model for the Ontologies. In section four it is outlined the framework for carrying out an Ontology Agent. In section five it is described the approach and some outcomes, which are evaluated through the effectiveness criteria. Finally, in section six we identify the features that distinguish our approach from others, and we depict the further work.

N.T. Nguyen et al. (Eds.): KES-AMSTA 2007, LNAI 4496, pp. 328–337, 2007.

2 Rule Base Fuzzy Cognitive Maps Profile

The RB-FCM proposed by Carvalho [2] are an approach for modeling the evolution and stability of the entities that compound a domain of study. The RB-FCM simulate system dynamics from a qualitative and causal perspective. The main assumption is that: "As a result of the activation of a given concept, it is perturbed the state of those entities that are causal interrelated with it as effect concepts". The RB-FCM are sketched like a digraph, whose nodes correspond to the concepts and its arcs to causal relationships. Direct causal relationships between two concepts, *a* → *z*, are drawn by one arc. Indirect relationships, *a* → *b*.. → *z*, are laid out by paths composed by intermediary concepts that join a cause concept *a* with an effect one *z*. Feedback relationships, *a* → *b*.. → *z* → *b*, are identified through arcs that depart from one concept *z* to any of its cause concepts *b*. Based upon these cause-effect relationships is triggered a fuzzy-causal inference engine that simulates behaviors, tendencies and final states of the entities of the domain of study in order to predict causal outcomes.

The concepts are managed as linguistic variables that are instantiated by linguistic terms. According to the nature of the entities, concepts depict variations or levels, i.e., given the concept *inflation* its variation value could be *increase-high*, whilst its level value is *low*. Each fuzzy value is stated by a membership function that gives away a fuzzy set. The fuzzy set owns the dimensional properties illustrated in Fig. 1a that correspond respectively to: a) shape; b) area; c) axis of central mass; d) support set; e) support set length; f) core set; g) core set length; h) internal base length; i) external base length; j) internal angle; k) external angle.

The values that instantiate a given concept are defined by a universe of discourse. This universe is laid out like the *x*-axis of the plane graph attached to the concept. The abscissa is illustrated by a scale of discrete points into the range [-1, 1]. Thus, fuzzy sets labeled with the highest intensities are allocated at the end of the *x*-axis. As the intensity of the fuzzy value drops, its fuzzy set is positioned closer to the central point of the universe. In Fig. 1b appears a graphical instance of the universe of discourse.

Prior begin the simulation; the concepts are initialized with the linguistic term that best represents their initial value. Thus, according to the center of mass of the fuzzy set, which depicts the linguistic term, is identified the discrete value that corresponds to a point in the *x*-axis. In addition, it is estimated the membership degree for the linguistic term taking into account the membership function of the fuzzy set. This membership value is a real number in the range [0, 1], which corresponds to the scale for the ordinate *y*-axis of the concept's graph as is shown in Fig. 1c.

The causal relations are stated by rules, whose antecedent corresponds to the cause concept and its consequent to the effect one. Based upon the linguistic term that holds the cause concept, a consequent fuzzy value is assigned to the effect concept. When the relation imposes a level or a variation value on the effect concept's state, it is called a fuzzy influence relation (FIR); otherwise it is a fuzzy causal relation (FCR). The FCR does not estimate the real value of the concept as it does a derivate; only it express the qualitative perturbation that supposes it will occur. Also, an *accumulative* effect is achieved on a given concept when it is biased simultaneously by several cause concepts. Regarding to the FIR a *strengthen* effect is applied on the effect concept when it is simultaneously instantiated with the same linguistic term by more than one FIR. The mathematical baseline for these fuzzy relations is detailed in [2].

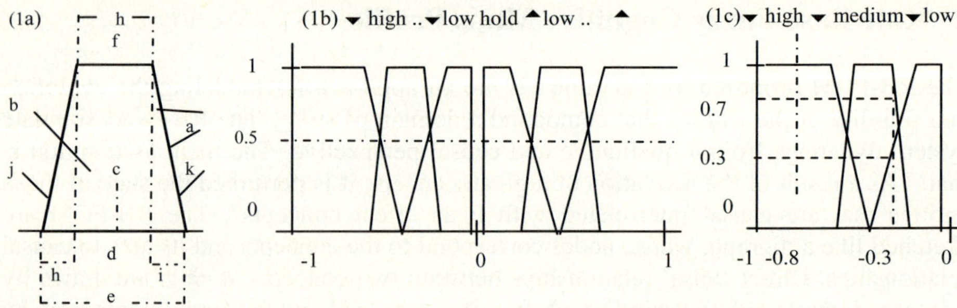

Fig. 1. Fuzzy Set. Fig. 1a Fuzzy Set Properties. Fig. 1b Linguistic Terms assigned to a Linguistic Variable. 1c Membership Degrees in y-axis, and Uncertainty Degrees in the x-axis.

3 Ontology Formal Model

According to Guarino [3] the underlying principles for an ontology formal model are: *Conceptualization* and *Ontology*. Where, conceptualization relies on the Aristotle's definition given for Ontology that claims: "The Ontology is always the same independently of the language used to describe it". The second principle is related to the meaning that the Artificial Intelligence gives to the Ontology, as: "An engineering artifact constituted by a vocabulary with a set of explicit assumptions regarding to the meaning of the words". Wherefore, conceptualization is concerned with the formal structure of reality as perceived and organized by an agent. Whereas, Ontology is a vocabulary, that depicts the intended meaning of each word. As a consequence, in despite of having different vocabularies, two Ontologies can share the same conceptualization. Wherefore, Ontology is a specification of a conceptualization.

In [4], conceptualization is defined as a world structure $\langle D, W\rangle$, where D is a domain and W is the set of extensional relations on D that reflects the states of affairs. For instance, in the puzzle domain every spatial arrangement of its items represents an extensional relation. But, conceptualization aims at the meaning of conceptual relations independently of the states of affairs, i.e., the meaning of *over*, *under* or *between*. Thus, conceptual relations are defined on a domain space whose structure is $\langle D, W\rangle$. With this structure it is defined a conceptual relation p^n of arity n as a total function $p^n : W \rightarrow (2^D)^n$ from W into the set of all n-ary extensional relations on D. Given a generic conceptual relation p, the set $\mathbf{E}$[1] $p = \{ p(w) \mid w \in W\}$ contain the admittable extensions of p. Therefore, a conceptualization for D is defined as a 3-tuple: $\mathbf{C} = \langle D, W, \mathcal{R}\rangle$, where $\mathcal{R}$ is a set of conceptual relations on $\langle D, W\rangle$.

Based upon $\mathbf{C}$, every possible world $w \in W$ owns a world structure: $\mathbf{S}_{wC} = \langle D, \mathbf{R}_{wC}\rangle$, where $\mathbf{R}_{wC} = \{p(w) \mid p \in \mathcal{R}\}$ is the set of extensions of the elements of $\mathcal{R}$. So that, the intended world structures of $\mathbf{C}$ is the set $\mathbf{S}_C = \{\mathbf{S}_{wC} \mid w \in W\}$.

Besides this, a logical language $\boldsymbol{L}$, with vocabulary V, is defined by the structure: $\langle S, I\rangle$, where $\mathbf{S} = \langle D, \mathbf{R}\rangle$ is a world structure and I is an *interpretation function* $I: V \rightarrow D \cup \mathbf{R}$ that assigns elements of D to constant symbols of V, and elements of $\mathbf{R}$

[1] Symbols denoting structures and sets of sets appear in **boldface**.

to predicate symbols of V. Furthermore, an *intentional interpretation* $\boldsymbol{K}$ can be stated by the structure $<C, \zeta>$, where $\boldsymbol{C}=<D, W, \mathcal{R}>$ is a conceptualization and ζ: $V \rightarrow D \cup \mathcal{R}$ is a function that assigns elements of D to constant symbols of V, and elements of $\mathcal{R}$ to predicate symbols of V. This kind of interpretation gives away an *ontological commitment* for $\boldsymbol{L}$. Thus, if $\boldsymbol{K} = <C, \zeta>$ is an *ontological commitment* for $\boldsymbol{L}$, it states that $\boldsymbol{L}$ commits to $\boldsymbol{C}$ by means of $\boldsymbol{K}$, while $\boldsymbol{C}$ is the underlying conceptualization of $\boldsymbol{K}$.

What is more, a model $\boldsymbol{M}$, whose world structure is $<S, I>$, will be compatible with an ontological commitment $\boldsymbol{K}$, if it meets three constraints: 1) $\boldsymbol{S} \in \boldsymbol{S}_c$; 2) for each constant c, $I(c)=\zeta(c)$; 3) for each predicate symbol p, I maps such predicate into an admittable extension of $\zeta(p)$. This means that exists a conceptual relation p and a world w such that $\zeta(p)= p \wedge p(w) = I(p)$. As a result, the set $I_k(L)$ of all models of $\boldsymbol{L}$ that are compatible with $\boldsymbol{K}$ become the set of *intended models* of $\boldsymbol{L}$ according to $\boldsymbol{K}$.

Finally, given a language $\boldsymbol{L}$ with ontological commitment $\boldsymbol{K}$, an Ontology $\boldsymbol{O}$ for $\boldsymbol{L}$ is: A set of axioms designed in a way such that, the set of its models approximates as best as possible to the set of intended models of $\boldsymbol{L}$ according to $\boldsymbol{K}$. Therefore, an Ontology $\boldsymbol{O}$ for a language $\boldsymbol{L}$ approximates a conceptualization $\boldsymbol{C}$, if there is an ontological commitment $<C, \zeta>$, such that the intended models of $\boldsymbol{L}$, according to $\boldsymbol{K}$, are included in the models of $\boldsymbol{O}$. In resume, these concepts are illustrated in the Fig. 2, where we appreciate the approximation of the Ontology to the set of models $\boldsymbol{M}$ stemmed from a given language $\boldsymbol{L}$, which commits to $\boldsymbol{C}$ by means of $\boldsymbol{K}$.

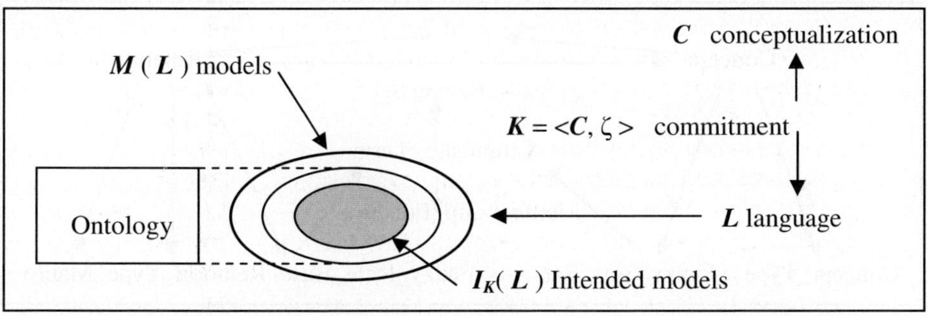

Fig. 2. Ontology approximation of the intended models derived from a language [3]

4 Ontology Agent Framework

We propose a framework for building an Ontology Agent composed by three stages: Design, development and deployment. The design follows the Ontology Development 101 guide [5]. The development is done according to the FIPA Ontology Agent specifications [4] and the Gaia Methodology [6]. The deployment is achieved through the Ontology Web Language (OWL) [7] and the Web Services paradigm [8].

4.1 Ontology Design

In this section are introduced the outcomes achieved during the seven steps required for designing an Ontology oriented to depict RB-FCM. As a first task, we identify the domain and its scope to be represented. So it is necessary to answer the following competency questions: For what it is going to use the Ontology? and what are the type of inferences to be done? In this case, the Ontology should represent the knowledge stemmed from the RB-FCM. In addition, it is required an engine that deals with inheritance inherences. As second task, we consider the reuse of existing Ontologies with the aim at taking advantage of available repositories. However, in this case there is a lack of this kind of Ontologies. As third task, we abstract the main terms of the Ontology. Thus, we set three types of terms: Meta, as *_Fuzzy_Set*; main, as *Concept* and *Relation*; and basic, as *Membership_Function* and *Fuzzy_Rule_Base*.

Afterwards, as fourth task we define the classes and hierarchies through a top-down strategy. As a result, we sketch a graphical schema in Fig. 3, where inheritance relations are drawn by directed arcs. As fifth task, the properties that describe the classes are identified. According to the multiple-inheritance paradigm, the properties are attached to the appropriate class in order to deal with legacies and exceptions. Next, as sixth task, we constraint the properties by facets such that: Default values, cardinality and value types. As seventh task we work on the creation of instances that correspond to the elements of the Ontology. So we introduce ontologic items for every concept, relationship, and linguistic term and fuzzy set of the RB-FCM.

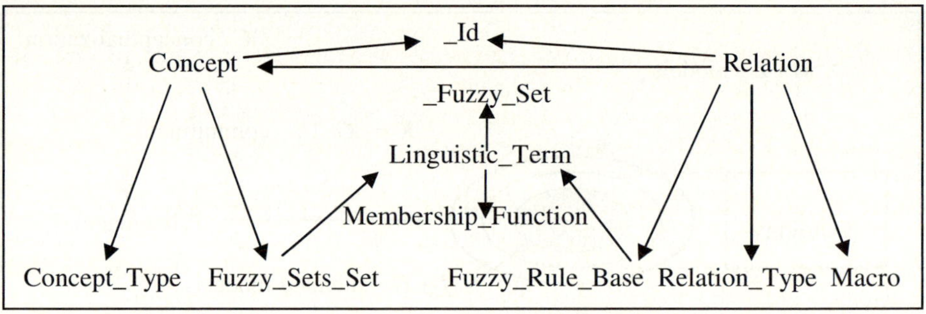

Fig. 3. Ontology RB-FCM. The schema is composed by terms and inheritance relationships.

As a final task, we have to deal with some issues like: Ensuring that the class hierarchy is correct, number and position of the siblings in a class hierarchy, conflicts stemmed from multiple inheritances, dimension of the hierarchy, and the imbalance produced when a new class or an instance is added to the Ontology.

Graphically speaking, the main components of our Ontology are sketched in Fig. 4 by means of an UML-class diagram. Thus, the main terms correspond to three classes: Meta, main and basic. The properties attached to the classes are depicted by the figure class called "property". Moreover, the instances are stated by the symbol class "instance". Regarding to the association, hierarchy and aggregations relationships, they are laid out by arcs with the conventional arrowhead.

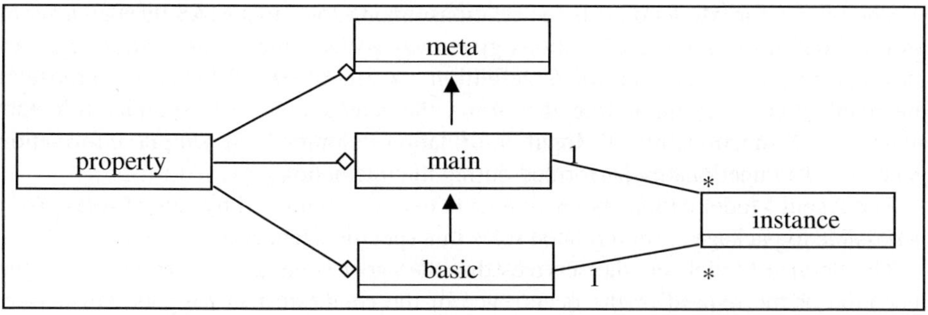

Fig. 4. UML-class diagram. Classes and relationships that describe the Ontology for RB-FCM.

4.2 Ontology Agent Development

Due to the Ontology Agent fulfils collaborative work with peers and application agents; it has to be available as a federated service in public directories like UDDI [8]. So in MAS, any agent requests the profile of the Ontology Agent that offers a specific service. Thus, the directory agent interprets the request, seeks the profile of the Ontology Agent that matches the constraints, and returns its profile to the requester agent. Later, the user agent encodes messages that describe the required services. Next, it sends a message, in a synchronous or asynchronous mode, to the appropriate Ontology Agent. Meanwhile, the Ontology Agent works on the request, the application agent stay idle or achieves some tasks. Once the response arrives from the Ontology Agent, the user agent interprets the outcome and continues its job.

Given this kind of functionality, the Ontology Agent is designed as a middleware between the application agents and the Ontology [9]. The Ontology Agent reacts before the events that happen in a distributed environment. In addition, it catches the messages that contain requests and queries that were sent by other agents. Afterwards, the Ontology Agent decodes the message and proceeds to interpret it according to its knowledge about the domain. Next, the Ontology Agent achieves the tasks necessary to commit the requests. At the end, the Ontology Agent encodes the outcome in a new message, and proceeds to send it to the requester agent. During this process are used some languages oriented to agent communications, content messages and Ontology description as FIPA-ACL [4], FIPA-SL [4] and OWL [7] respectively.

In regards to the development of the Ontology Agent two stages are done: Analysis and Design. In the Analysis stage is gained an understanding of the system and its structure by means of two models: Roles and Interaction, whereas in the Design stage it is stated how the community of agents works together to fulfill the system goals by means of three models: Agent, Services and Acquaintance.

The Roles Model is outlined by role schemas that set four properties: Permissions, activities, protocols and responsibilities. Permissions are the rights granted to the role for managing information. Activities are private actions attached to the role without interacting with others. Protocols define how the role interacts with other roles. Responsibilities state the functionality of the role by liveness and safety properties. These attributes are triggered respectively, when commit certain environmental conditions and normal invariant conditions are hold.

The Interaction Model depicts the relations among roles in a MAS through a set of protocol definitions. These definitions give away pattern interactions externalized by messages interchanges. A Protocol Definition consists of six attributes: 1) Purpose of the interaction. 2) Initiator role that starts the interaction. 3) Responder role that achieves the functionality. 4) Input information submitted. 5) Output information generated. 6) Functionality performed during the interaction.

The Agent Model defines types of agents that are composed by sets of roles. So it is possible to package several related roles in a specific agent type.

The Service Model sets the services that the agent is engaged to achieve after the reception of the request or the occurrence of the condition that triggers the service. Thus, for each service carried out by the agent are depicted the following attributes: Services, Inputs, Outputs, Pre-conditions, and post-conditions. The services are derived form the Roles Model, the inputs and outputs coming from the Interaction Model, and the pre-conditions and post-conditions depict constraints on services.

The Acquaintance Model draws the communication flow among the agent types, in order to identify potential bottlenecks that could cause problems at run-time.

4.3 Ontology Agent Deployment

The deployment of the Ontology Agent relies on the OWL and the Web Services paradigm. OWL is based on the recommendations of W3C for the Semantic Web, XML, XSD, RDF data model [7]. OWL supports machine interoperability for Web content through a vocabulary with a formal semantics. The knowledge representation is done by means of metadata and instances. The classes and properties are defined as Metadata, whereas the Ontology items are outlined by instances, as appear in Fig. 5.

In regards to Web Services, they are programs that offer specific functionalities to a community of users through the Internet. The client applications access the service by the use of an interface that invokes a particular activity on behalf of the client. A Web Service provides: Interoperability, Internet friendliness, typed interfaces, ability to leverage Internet standards, support for any language and distributed component infrastructure. The facilities required for federating Web Services are organized into five functional tiers: 1) Transport, supported by Hyper Text Transfer Protocol (HTTP). 2) Encoding, described by Extended Markup Language (XML). 3) Message format edited by Simple Object Access Protocol (SOAP). 4) Description of the functionality by Web Services Description Language (WSDL). 5) Publication of services by UDDI [8].

```
1: <owl:Class rdf:ID="concept" xmlns:rdf="rdf" xmlns:owl="owl">
2: <owl:DatatypeProperty rdf:ID="area" rdf:Cardinality="multiple"
    xmlns:rdf = "rdf" xmlns:owl="owl">
3: <concept rdf:ID="analysis" xmlns:rdf="rdf" xmlns="">
  <area rdf:datatype=http://www.w3.org/2001/XMLSchema#floatrdf:level=1/>
```

Fig. 5. OWL Code. In 1: the class *concept* is defined. In line 2: the property *area* is set. In 3: an instance for the class *concept* is stated, which has the property *area*.

5 Outcomes and Evaluation

In this section we introduce the architecture of the MAS, the code used for deploying the agents through Web Services and the evaluation of our approach as follows. In Fig. 6a we sketch the three layers of the MAS architecture that correspond to: User interface, middleware agents and Ontologies repositories.

The first layer corresponds to the user interaction with the MAS by means of Web interfaces, as the one illustrated in Fig. 6b. These interfaces are Active Server Pages that are accessed at: http://www.wolnm.org/mas/waInterface/wfInterface.aspx.cs.

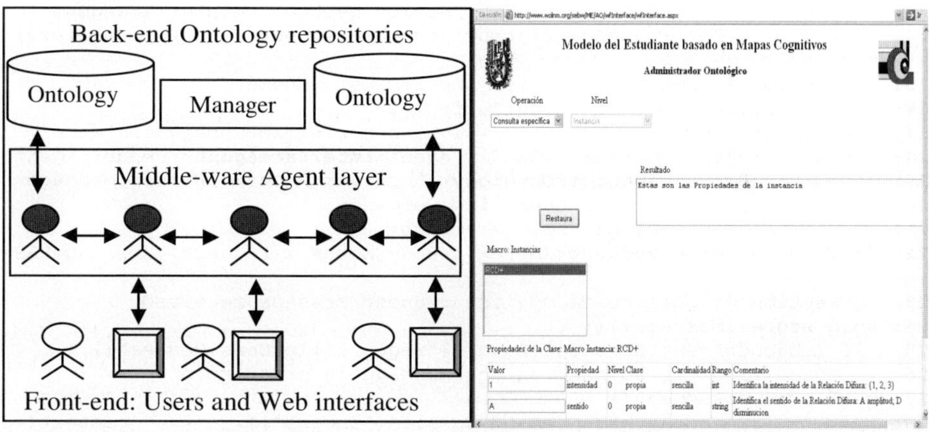

Fig. 6. MAS approach. Fig. 6a draws the Architecture. Fig 6b shows the user Web interface.

The middle-tier corresponds to the agent's environment. These agents are encoded as Web Services that carry out specialized tasks. Thus, an agent interface is attached to the Web interface by the Microsoft® C#® code edited in the first 14 lines of the Fig. 7, where the first section defines the Web Service corresponding to the agent interface, and the second section set the access to the Web Service from the Web interface.

The communication among agents is encoded in XML messages. The schema reveals an attribute-value structure stated by a sequence of elements and sub-elements embedded in tags. For instance, in the third section of the Fig. 7, lines 15–22, appears the *request* schema. The code owns a *Namespace* and a root labeled by the *request* tag. The three main elements are: *Header_Message, Body_ Message*, and *Body_Note*. The elements own sub-elements as: *Sender-Agent*, *Command*, and *Note* respectively. In order to access the elements of the message, the decode agent navigates through the structure. Besides, it interprets the elements and sub-elements, and accesses the values. These components are stored in a data structure declared in a class.

The Ontology Agent catches, interprets and forwards the requests to the Ontology manager attached to the Ontology. Also, the Ontology Agent supervises the process, gets the outcome and sends the results to the requester agent. The code for processing a *request* and *command* appears in the fourth section of the Fig. 7, lines 23–30.

Finally, the back-end level contains the Ontology repository and their managers.

```
// Section 1: Web Service Class Declaration
namespace wsInterfaceAgent {
public class wsInterfaceAgent : System.Web.Services.WebService{
// Web Service Method Declaration
[WebMethod]
public int interfaceAgent(Input-parameters,  out output-parameters)
{ code….. return value }

// Section 2: Web Service Reference
namespace waInterface { public class WebForm1 : System.Web.UI.Page {
// Namespace: host; class: wsInterfaceAgent is the Interface Agent;
// Instance: gInterfaceAgent
host.wsInterfaceAgent interface = new host.wsInterfaceAgent();
//  Web Service invocation; Web Service method: wmInterfaceAgent
interface.wmInterfaceAgent(Input-parameters, out output-parameters);

// Section 3: XML Message, for the request schema
<?xml version="1.0" encoding="utf-8" ?>
<messageMAS xmlns:xsi="http://www.w3.org/2001/XMLSchema-instance">
<request><Header_Message><Sender_Agent>interfaceAgent</Sender_Agent>
<Receiver_Agent>wsManagerOntology</Receiver_Agent></Header_Message>
<Body_Message> <Command>Query_if</Command>
  <Head_elemento>Class</Head_elemento>
<Body_Note><Note>Request</Note> </Body_Note> </request></messageMAS>

// Section 4: Call to decode the request message received
void processMessage(){
 if (rDecodeAgent.RMessage.Equals("request"))processRequest();}
// Identifies the command of the message in order to process it
void processRequest()     {
if        (rDecodeAgent.RCommand.Equals ("Insert"))           pInsert();
else if (rDecodeAgent.RCommand.Equals ("Query_if"))      pQuery_if();
else if (rDecodeAgent.RCommand.Equals ("Insert_Instance"))pInsert();
```

Fig. 7. Code instances. 1) Web Services declaration in C#®. 2) Web Services reference in C#®. 3) Message structure in XML. 4) Message process and request methods in C#®.

We tested the approach on a Web-based Student Model application, where user agents stem cognitive skills and learning preferences of the individuals, and predictive agents run causal inferences based on the RB-FCM. So in Fig. 8a we evaluate the percentage of effectiveness, *y*-axis, for answering multiples inference queries, *x*-axis, that are requested during one second. While, in Fig. 8b we estimate in seconds the average performance, *y*-axis, for responding to requests, *x*-axis, which hit during one second.

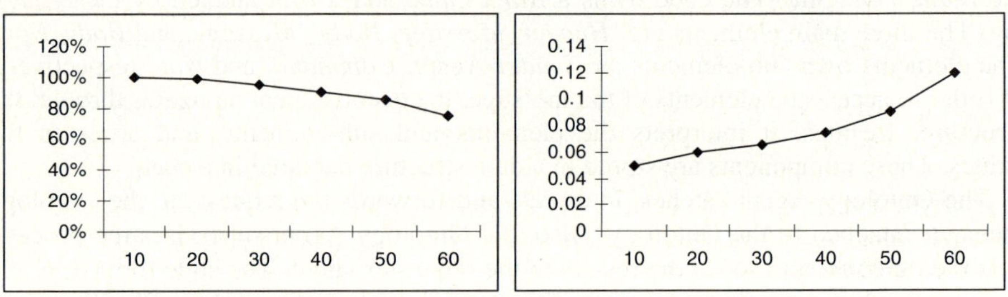

Fig. 8. Test of the approach. Fig. 8a. shows the effectiveness. Fig. 8b depicts the performance.

6 Conclusions

Our Ontology Agent oriented to administrate the knowledge of the RB-FCM is a novel approach due to the following features: The baseline relies on the soundness of the Ontologies and Conceptualizations. Also, according to our research, the Ontology developed is the first work ever done in the field of the RB-FCM. In addition, we take advantage of the strengths of the agent's paradigm for deploying a MAS approach on the Microsoft® Dot Net® platform instead of Java. As a result, we focus on the development of a whole MAS platform, like JADE y Zeus. Also, we set a framework that includes: Underlying elements, methodologies for developing Ontologies and agents, and the resources for deploying them through Web Services. As a consequence, we extend the FIPA specifications for developing Ontology Agents [4].

In regards to the efficiency of our approach, we found out that it offers a high performance under real work situations. Since, the tests we did occurred at noon, when there is more workload on the Internet. Thus, according to the graphs sketched in Fig.8 is evidenced a light decrease for the tested parameters, effectiveness and performance, as the number of simultaneously requests increase.

As further work, it is considered the enhancement of the services offered in our MAS platform. Nevertheless, we plan to extend the Ontology to represent more domains for the student modeling. What is more, we will test our approach according to other parameters as: Rate of world change and agent planning time.

Acknowledgements

First author states that this work was inspired in a special way for his Father, his brother Jesus and his Helper as part of the projects of World Outreach Light to the Nations Ministries (WOLNM). Also this work was supported by grants: CONACYT-SEP C0146805/747, CONACYT 182329, COFAA D481106, COTEPABE PL-34306, SIP 20070584 and Microsoft.

References

1. Peña, A.. Sossa, H. and Gutiérrez, F: Knowledge and Reasoning Supported by Cognitive Maps, MICAI'05, LNAI, Vol.3789. Springer, Monterrey, Mexico, November, (2005)
2. Carvalho, J.P.: Rule Base-based Cognitive Maps: Qualitative Dynamic Systems Modeling and Simulation. PhD Thesis, Lisboa Technical University, Portugal, October, (2001)
3. Guarino, N.: Formal Ontology in Information Systems. In N. Guarino (ed.) Formal Ontology in Information Systems, Proceedings of FOIS'98, Trento, Italy, 6-8 June, IOS, (1988)
4. FIPA, Foundation for Intelligent Physical Agents, Ontology Service Specification. (2001)
5. Noy, N. and Mc Guinness, D. L.: Ontology Development 101, Stanford University, (2004)
6. Wooldrige, M., Jennings, N. R., and Kinny, D.: The Gaia Methodology for Agent-Oriented Analysis and Design, Autonomous Agents & Multi-Agent Systems, Kluwer Academ (2000)
7. OWL: Ontology Web Language, URL: http://www.w3c.org/tr/CR-owl-ref/ (2006)
8. Short, S.: Building XML Web Services for the .NET Platform. Microsoft Press (2002).
9. Peña, A.. Sossa, H. and Gutiérrez, F.: Web Services based Ontology. Edited by IEEE. In P. Distributed Frameworks for Multimedia Applications. Penang, Malaysia, May, (2006)

OWL-DL Based Ontology Inference Engine Assessment for Context-Aware Services

Ohbyung Kwon[1], Jaemoon Sim[1], and Myungchul Lee[2]

[1] School of International Management, Kyunghee University
Seochun-dong, Ghiheung-gu, YongIn, Kyunggi-do 446-701, Korea
{obkwon,deskmoon}@khu.ac.kr

[2] IBM Ubiquitous Computing Lab
Dogok-dong, Kangnam-gu, Seoul, Korea
mclee@kr.ibm.com

Abstract. To acquire hidden and potentially useful information from context data, ubiquitous computing services began taking advantage of the reasoning capabilities inherent in inference engines. However, since a traditional approach to evaluating inference engines' performance levels typically focuses on static information reasoning, specific evaluations of requirements that pertain to the ubiquitous computing environment have been largely neglected. Hence, this paper aims to propose an augmented evaluation framework for inference engines, and then examine how OWL-DL-based inference engines perform by applying them to realistic context-aware services. Six measurement criteria are proposed and measured, including scalability as data set gets large, responsiveness for users' requests, and adaptability to frequent inference requests.

Keywords: Ontology, Semantic Web, Inference Engine, OWL-DL, scalability, MINERVA, DLDB-OWL, HAWK, Context-Aware Services.

1 Introduction

Managing context-related knowledge so that it can be sharable with various computing entities has been one of the crucial issues, mainly in ontology-related research [15, 16]. To accomplish this, a variety of ontology-based inference engines such as Jena, RacerPro, FacT++, and Minerva have been proposed as core elements of ubiquitous computing systems. Ontology-based inference engines are one of the keys to attaining the Semantic Web's vision: having machines acquire useful information to realize automated B2B business, as well as B2C business [1, 3].

As ontology-based inference engines proliferate, it becomes increasingly important to propose a methodology to evaluate which engines are more or less appropriate for specific applications [16, 17]. One of the outstanding methodologies is Lehigh University's LUBM (Lehigh University Benchmark) performed by IBM's China Laboratory [6, 9]. In particular, to our knowledge, LUBM is one of the first Semantic Web-based knowledge based systems, and also an authentic project for evaluating inference engines. However, this inclusion is insufficient to test all of the inference engines' functionalities that are needed in ubiquitous computing. Particularly in case-context-aware services, knowledge asserted in a context-related ontology could be frequently updated. Moreover, it has still not been examined to what extent the legacy

N.T. Nguyen et al. (Eds.): KES-AMSTA 2007, LNAI 4496, pp. 338–347, 2007.

inference engines are able to provide a response to a user's or a computer system's query that is based on up-to-date context data. In applying an ontology model to the context data process, the dual processes of inquiring into and reasoning with context-based information are crucial, yet further research efforts remain in the work [5, 19].

Hence, the main purpose of this paper is to propose a new methodology, which evaluates how legacy ontology-based inference engines can cope with a context-aware service domain, and then actually apply the new methodology to typical location-based service examples. To do so, we suggest a new set of performance evaluation criteria for the context-aware environment. Among the OWL-DL-based inference engines, we selected and tested DBMS-based inference engines such as MINERVA and DLDB-OWL (HAWK) and main memory systems-based inference engines such as Pellet and Jena. We adopted one ontology set and ontology instances which are also used at Lehigh University and IBM's China Laboratory, as well as two sets of ontology instances actually collected from Kyunghee University in South Korea. We considered the MyEntrance service scenario developed as an IBM Celadon™ Project, which aims to develop the inference engine working for pervasive computing environment [12].

2 Literature Review: Ontology Inference Engines

Current knowledge-based systems, including ontology-based inference engines, consist of a main memory type and a DBMS type to load and process ontology. The main memory systems load knowledge and facts from an ontology file, which is stored in a specific URI, to a main memory with reasoned results. Main memory systems such as JENA and PELLET perform reasoning tasks at the time the query is requested. On the other hand, loaded knowledge and facts are volatile according to the inference engines. Even though main memory systems are desirable, since for processing knowledge and facts they tend to be more efficient than a DBMS, large amounts of memory would be necessary to load complex or very large ontology files. This would result in fatal errors due to insufficient memory at runtime. Even if main memory systems could cope with scalability problems, the memory requirements would be dramatic.

DBMS-based systems store both the loaded ontology and reasoned results in permanent storage. Since the systems include a storage phase, the overall processing times tends to slower than main memory systems. However, DBMS systems are useful when the domain application requires either very large or complicated knowledge and facts from ontology. DBMS-based systems, such as MINERVA, DLDB-OWL, and HAWK, can perform a reasoning activity on an ontology at data loading time, which results in prompt responses to query requests that come in after reasoning activity is completed. In this paper, some representative systems for reasoning OWL-DL based ontology are described as listed in Table 1 [4, 6, 7, 8, 9, 10, 11].

MINERVA, developed by IBM, is a high-performance relational database based storage, reasoning, and query system based on OWL. MINERVA is now provided to developers as a component of IBM's Integrated Ontology Development Toolkit (IODT) [8]. At the time of query processing, MINERVA supports Description Logic Program (DLP), which can be regarded as a subset of SPARQL language, and enables OWL-DL execution. MINERVA plays a role as a DL reasoner for Tbox and also supports logic rule sets that are translated from DLP for Abox reasoning.

DLDB-OWL is a database-based inference engine that supports OWL-DL. HAWK is the next release of DLDB-OWL, which provides sophisticated APIs as well as implementations for parsing, editing, manipulating and preservation of OWL ontologies [6,7]. The APIs consist of core, OWL, and storage packages. Using a core package, one can define a general interface for ontology entities such as classes and properties. Meanwhile, OWL supports parsing and serializing OWL-based ontology files.

Pellet is an open source variation of an OWL-DL reasoner based on the Java programming language [17]. Pellet can be co-used with both Jena and OWL APIs to provide a DIG (DL Implementation Group) interface. The Pellet API includes a consistency ontology check, classification of taxonomy, and queries with RDQL language.

Jena is a Java-based framework to utilize the development of Semantic Web based applications. Jena provides the environment for programming RDF, RDFS, OWL, and SPARQL. It also supports rule-based reasoning. Jena's framework includes RDF API [4, 10].

3 Evaluation Framework

3.1 Evaluation Methodology

To examine to what extent the legacy OWL-DL based inference engines can work with context-aware settings, we re-performed both dynamic and combined evaluations, as well as a conventional static evaluation. For the static evaluation, we

Table 1. Performance Measures

Performance Measures	Legend	Description
Load Time	LT(d)	Total elapsed time in loading a set of OWL-DL files
Query Response Time	QT(d,q)	Elapsed time to return a set of query results against the fifteen static queries and one (16th) dynamic query
Query Completeness	QC(d,q)	Rate of return of query results produced by an inference engine with knowledge base
Query Soundness	QS(d,q)	Rate of correct results among the query results returned by an inference engine
Repository Size	RS(d)	Storage size required to execute the reasoning using a set of OWL-DL ontologies
Dynamic Endurance	DE(d,q,c1,c2)	Return rate of correct query results if an inference engine updates an ontology instance file with a specific cycle(c1) and retrieves for reasoning a query result with a specific cycle (c2) in a simultaneous manner
Enhanced Combined Metric	ECM(d)	Combined measure to show to what extent the inference engine is fast, complete, safe, and at the same time endures dynamic queries

where d, q, s1 and s2 indicates data set number (1~3), query command number (1~16), cycle time of context data update and that of dynamic query execution, respectively.

initially considered the evaluation tool developed by Lehigh University [7]. We extended both the Lehigh and IBM China's evaluation tools to work in context-aware situations. Performance measures are prepared as listed in Table 1.

3.2 Evaluation Data Set

We adopted LUBM's evaluation data sets, which are an extended form of Univ-Bench, for the evaluation. Correspondingly, the data set generation method is based on Univ-Bench's artificial data generator (UBA), which LUBM used for the evaluation. Using UBA, we generated one data set (Data Set 3) to compare the evaluation results with other evaluation studies. In addition to data set 3, to increase the reality of the data set for performance evaluation of the evaluation subjects, two data sets are collected from an actual university. Detailed characteristics of the data sets are described in Table 2.

Table 2. Explanations of the considered data sets

Data Set	Content
Data Set 1	School of International Management and Electronic & Information Engineering at K University
	8477 Instances, 134302 Triples and 11.8 MB
Data Set 2	School of International Management at K University
	3857 Instances, 60791 Triples and 5.33MB
Data Set 3	Campus example DATA SET 1 (IBM China Laboratory)
	19277 Instances, 337873 Triples and 30.5 MB

The configuration of the servers involved in the experiment is as follow:

- 1.80GHz Pentium 4 CPU, 1GB of RAM; 80GB of hard disk,
- Windows XP Professional OS, Java SDK 1.4.1; 512MB of max heap size.

4 Performance Evaluation

4.1 Query Response Time

In terms of query response times, MINERVA and DLDB-OWL(HAWK) show more stable and better performance than Pellet and Jena as shown in Fig. 1. This is mainly because main memory systems load ontology knowledge and facts into main memory at query request time. However, once the ontologies are loaded and same query requests are repeatedly issued, query response times dramatically decrease, and hence the service quality increases. Overall, Jena shows worse outcomes than the other systems in every data set.

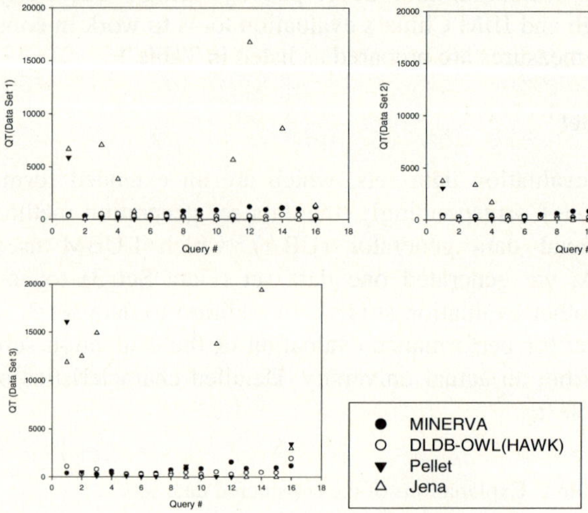

Fig. 1. Query Response Time

4.2 Query Completeness

In case of query completeness, while MINERVA and Pellet show nearly perfect outcomes, DLDB-OWL (HAWK) and Jena do not, as shown in Fig. 2. In particular, we can observe that DLDB-OWL(HAWK) did not return any answers against some queries. This may be due to the fact that even if DLDB-OWL(HAWK) parses the OWL-DL format ontology, it so far cannot derive or reason deep relationships from the ontology. Jena also did not work well: the query completeness was 0% at the 7th, 9th, 11th, 12th, 13th, 14th and 15th queries.

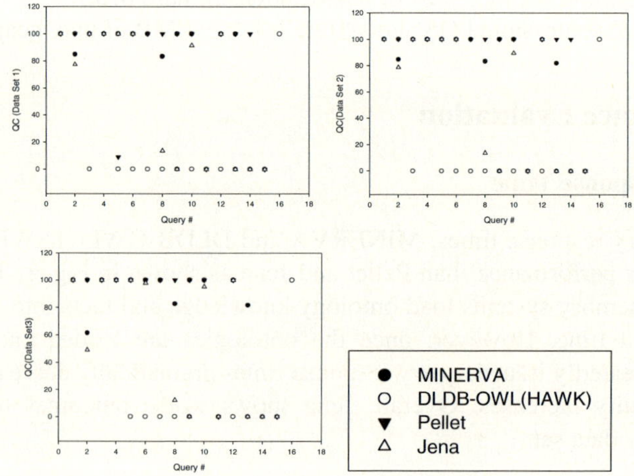

Fig. 2. Query Completeness

4.3 Query Soundness

Fig. 3 shows that all inference engines give nearly perfect outcomes in terms of query soundness. However, in some cases, we observed lower query soundness the inference engines, e.g., Pellet and Jena at the 14th query.

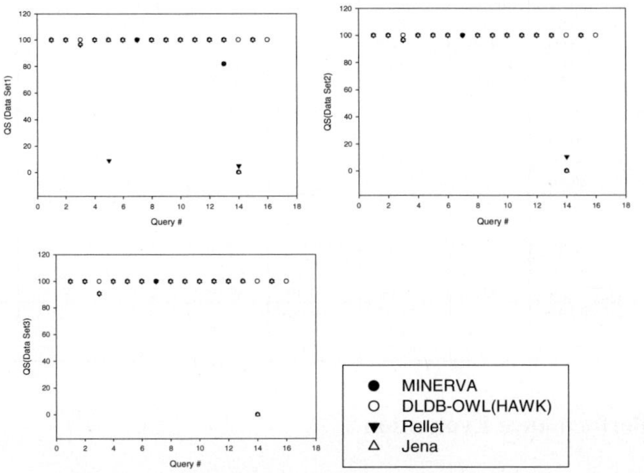

Fig. 3. Query Soundness

4.4 Dynamic Endurance

The performance results for the competitors in terms of dynamic endurance are summarized in Table 3. The three ratios in each cell indicates the rate of correct answers, wrong answers, and no responses, respectively.

Table 3. Evaluation Summary of Dynamic Endurance

Cycle time of updating context data		1200sec			900sec			600sec		
Cycle time of query processing		600sec	60sec	10sec	600sec	60sec	10sec	600sec	60sec	10sec
Data Set 1	Minerva	33.3% 0.0% 66.7%	5.0% 6.7% 88.3%	0.6% 3.6% 95.8%	33.3% 16.7% 50.0%	5.0% 6.7% 88.3%	0.8% 6.7% 92.5%	50.0% 0.0% 50.0%	6.7% 5.0% 88.3%	0.8% 6.7% 92.5%
	DLDB-OWL (HAWK)	99.8% 0.2% 0.0%	89.5% 10.5% 0.0%	91.3% 8.7% 0.0%	93.3% 6.7% 0.0%	84.0% 16.0% 0.0%	89.1% 10.9% 0.0%	90.0% 10.0% 0.0%	71.0% 29.0% 0.0%	85.7% 14.3% 0.0%
	Pellet	100.0% 0.0% 0.0%	100.0% 0.0% 0.0%	96.9% 1.7% 1.4%	100.0% 0.0% 0.0%	98.3% 1.7% 0.0%	97.2% 1.9% 0.8%	100.0% 0.0% 0.0%	100.0% 0.0% 0.0%	91.7% 3.3% 5.0%
	Jena	100.0% 0.0% 0.0%	100.0% 0.0% 0.0%	56.4% 1.1% 42.5%	100.0% 0.0% 0.0%	100.0% 0.0% 0.0%	55.8% 1.7% 42.5%	100.0% 0.0% 0.0%	100.0% 0.0% 0.0%	55.8% 3.1% 41.1%

Table 3. (*Continued*)

Data Set 2	Minerva	0.0% 50.0% 50.0%	38.3% 6.7% 55.0%	25.3% 6.7% 68.1%	0.0% 16.7% 83.3%	40.0% 6.7% 53.3%	24.7% 7.5% 67.8%	0.0% 33.3% 66.7%	30.0% 16.7% 53.3%	10.0% 10.3% 79.7%
	DLDB-OWL (HAWK)	99.9% 0.1% 0.0%	99.9% 0.1% 0.0%	99.9% 0.1% 0.0%	99.9% 0.1% 0.0%	99.9% 0.1% 0.0%	99.9% 0.1% 0.0%	99.9% 0.1% 0.0%	99.9% 0.1% 0.0%	99.9% 0.1% 0.0%
	Pellet	100.0% 0.0% 0.0%	100.0% 0.0% 0.0%	100% 0.0% 0.0%	100.0% 0.0% 0.0%	100.0% 0.0% 0.0%	99.2% 0.8% 0.0%	100.0% 0.0% 0.0%	100.0% 0.0% 0.0%	99.4% 0.6% 0.0%
	Jena	100.0% 0.0% 0.0%	100.0% 0.0% 0.0%	99.4% 0.6% 0.0%	100.0% 0.0% 0.0%	100.0% 0.0% 0.0%	99.4% 0.6% 0.0%	100.0% 0.0% 0.0%	100.0% 0.0% 0.0%	99.4% 0.6% 0.0%
Data Set 3	Minerva	33.3% 0.0% 66.7%	1.7% 3.3% 95.0%	3.6% 5.6% 90.8%	16.7% 0.0% 83.3%	5.0% 5.0% 90.0%	2.5% 6.7% 90.8%	16.7% 0.0% 83.3%	3.3% 6.7% 90.0%	0.3% 8.6% 91.1%
	DLDB-OWL (HAWK)	86.4% 13.4% 0.2%	86.1% 13.5% 0.4%	88.0% 11.5% 0.5%	61.6% 37.7% 0.7%	62.9% 36.1% 1.0%	65.7% 30.7% 3.6%	51.4% 48.4% 0.2%	48.9% 49.5% 1.6%	52.1% 44.3% 3.6%
	Pellet	0.0% 83.3% 16.7%	95.0% 5.0% 0.0%	29.40% 67.5% 3.1%	33.3% 66.7% 0.0%	38.3% 50.0% 8.3%	46.1% 50.6% 3.3%	33.3% 66.7% 0.0%	31.7% 66.7% 1.7%	27.8% 67.8% 4.4%
	Jena	100.0% 0.0% 0.0%	98.3% 1.7% 0.0%	46.4% 3.3% 50.3%	100.0% 0.0% 0.0%	96.7% 3.3% 0.0%	41.9% 5.0% 53.1%	100.0% 0.0% 0.0%	91.7% 8.3% 0.0	45.8% 6.1% 48.1%

4.5 Overall Performance Evaluation

We have made a combined metric considering dynamic features, as well as conventional static features which were involved in LUBM's Combined Metric (CM) [6]. Hence, the new metric is called Enhanced Combined Metric (ECM), which extends CM. ECM(.) is the weighted average of ECM(d), which is the combined metric in case of using data set d as shown in (1):

$$ECM(.) = \frac{1}{N}\sum_{d=1}^{N} w(d) * ECM(d) \tag{1}$$

where, w(d) is the relative weight of the performance evaluation using data set d. Hence, $0 \le w(d) \le 1, \sum_{\forall d} w(d) = 1$. ECM(d) for each data set d is shown as (2):

$$ECM(d) = \frac{1}{M}\sum_{q=1}^{M} \frac{(\alpha^2 + 1) * P(d,q) * F(d,q)}{\alpha^2 * P(d,q) + F(d,q)} \tag{2}$$

The F-Measure, F(d,q), is formed as (3). This is the weighted average of F-value, which is derived from query completeness and query soundness under static settings, and mean dynamic endurance under dynamic settings. Hence, the weight value, δ, was used to determine how the evaluation stresses on dynamic evaluation.

$$F(d,q) = \delta * \frac{(\beta^2 + 1) * QC(d,q) * QS(d,q)}{\beta^2 * QC(d,q) + QS(d,q)} + (1-\delta) * \frac{1}{S*T}\sum_{c1=1}^{S}\sum_{c2=1}^{T} ED(d,q,c1,c2) \tag{3}$$

P(d,q), a measure for response time, in (3) is derived from (4):

$$P(d,q) = max(1 - \frac{T(d,q)}{N}, \varepsilon) \quad (4)$$

where ε and $\frac{T(d,q)}{N}$ indicate a very succinct nonnegative real number and response time per triple, respectively. Accordingly, $1 - \frac{T(d,q)}{N}$ shows to what extent the inference engine promptly responds under a specific data set.

Consequently, ECM(.) is a performance measure to determine not only how agile, complete, and safe a specific engine is, but at the same time how well it endures the dynamic queries.

We set the value of ε, α and β to 0.0001, 1 and 1, respectively, to maintain the consistency with LUBM test [6]. The value of δ was set to 0.5. Thus, performance evaluation results could be obtained as listed in Table 4.

Table 4. Overall Performance Evaluation (δ =0.5)

	ECM(d=Data Set1)	ECM(d=Data Set 2)	ECM(d=Data Set 3)	ECM(.)
MINERVA	0.574408	0.592098	0.538097	0.568201
DLDB-OWL (HAWK)	0.700767	0.748329	0.652324	0.700474
Pellet	0.684109	0.705564	0.584808	0.618160
Jena	0.729180	0.760926	0.738133	0.742747

5 Conclusion

In this paper, we proposed a performance evaluation methodology for OWL-DL level ontology-based inference engines in a context-aware services setting. The inference engines that we considered consist of memory- and DBMS-based systems. We conclude that legacy ontology-based inference engines still have a few things to improve to be applied in the actual context-aware services. Hence, we also suggest several strategies to optimally use the inference engines.

The main contribution of this paper is that, to our knowledge, our research was the first to perform an evaluation of ontology-based inference engines within context-aware settings. Moreover, actual data sets collected from a university campus are used for the test, as well as the conventional benchmarking data set used by other benchmarking evaluation studies. By doing this, we can determine that most of the current inference engines must be adequately improved to address complicated transitive closure triples. Unfortunately, even though the inference engines of relatively better performance could be identified for each situation, the observed performance results indicate that legacy engines are far from being ready to be commercialized, especially in terms of scalability. These experimental results could be benchmarked by the further evaluation studies.

Acknowledgment

This research is partially supported by the ubiquitous Computing and Network (UCN) Project the Ministry of Information and Communication (MIC) 21st Century Frontier R&D Program, and also IBM Ubiquitous Computing Lab funded by the Institute of Information Technology Assessment in Korea.

References

1. Berners-Lee, T., J. Hendler, and O. Lassila, “The Semantic Web,” Scientific American May 17, 2001.
2. Borgida, A., M. Lenzerini and R. Rosati, The Description Logic Handbook: Theory, Implementation and Applications, Cambridge University Press, 2003.
3. Brewster, C., K. O'Hara, S. Fuller, Y. Wilks, E. Franconi and M. A. Musen, J. Ellman and S.B. Shum, "Knowledge Representation with Ontologies: The Present and Future," IEEE Intelligent Systems, Vol. 19, No. 1 (2004), pp. 72-81.
4. Carroll, J.J., I. Dickinson, C. Dollin, D. Reynolds, A. Seaborne, K. and Wilkinson, “Jena: Implementing the Semantic Web Recommendations,” Proceedings of the 13th International World Wide Web Conference, New York (2004), pp. 74-83.
5. Chen, H., T. Finin and J. Anupam, “Semantic Web in a Pervasive Context-Aware Architecture,” Artificial Intelligence in Mobile Systems. In UbiComp 2003, Seattle, 2003.
6. Guo, Y., Z. Pan, and J. Heflin., “An Evaluation of Knowledge Base Systems for Large OWL Datasets,” Proceedings of the 3rd International Semantic Web Conference, Hiroshima. LNCS, Vol. 3298 (2004), pp. 274-288.
7. Guo, Y.; Z. Pan, Z. and J. Heflin, "LUBM: A Benchmark for OWL Knowledge Base Systems," Journal of Web Semantics, Vol. 3, No. 2 (2005), pp. 158-182.
8. Haarslev, V., M¨oller and R. and M. Wessel, “Querying the Semantic Web with Racer + nRQL,” Proceedings of the KI-04 Workshop on Applications of Description Logics, 2004.
9. IBM’s IODT/Minerva team: Minerva Reasoner, See, http://www.alphaworks.ibm.com/tech/semanticstk or http://www.ifcomputer.com/MINERVA/
10. Kevin, W., C. Sayers and H. Kuno, “Efficient RDF Storage and Retrieval in Jena2,” Proceedings of First International Workshop on Semantic Web and Databases, (2003), pp. 131-151
11. Lee, M.C., H.K. Jang, Y.S. Paik, S.E. Jin and S. Lee, “A Ubiquitous Device Collaboration Infrastructure: Celadon,” 3rd Workshop on Software Technologies for Future Embedded & Ubiquitous Systems, 2006.
12. Liebig, T. H. Pfeifer and F. von Henke, “Reasoning Services for an OWL Authoring Tool: An Experience Report,” Proceedings of the 2004 International Conference on Descriptive Logic, Whistler, BC, Canada, 2004.
13. Ma, L., Y. Yang, Z. Qiu, G. Xie, Y. Pan and S. Liu, “Towards A Complete OWL Ontology Benchmark,” 3rd European Semantic Web Conference, 2006.
14. Mike U. and G. Michael, "Ontologies: principles, methods, and applications," Knowledge Engineering Review, Vol 11, No. 2 (1996), pp. 93–155.
15. Noy, N.F. and C. D. Hafner, "The State of the Art in Ontology Design: A Survey and Comparative Review," AI Magazine, Vol. 18, No. 3 (1997), pp. 53–74.
16. Shadbolt, N., K. O’Hara and L. Crow, “The Experimental Evaluation of Knowledge Acquisition Techniques and Methods: History, Problems, and New Directions,” International Journal Human-Computer Studies, Vol. 51, No. 4 (1999), pp. 729–755.

17. Sirin, E., B. Parsia, "Pellet: An Owl DL Reasoner," Proceedings of the International Third International Semantic Web Conference (ISWC 2004), 2004.
18. Tennison, J., K. O'Hara and N. Shadbolt, "APECKS: Using and Evaluating a Tool for Ontology Construction with Internal and External KA Support," International Journal of Human-Computer Studies, Vol. 56, No. 4 (2002), pp. 375–422.
19. Wang, X., J.S. Dong, D. Zhang, C.Y. Chin and S.R. Hettiarachchi, S.R. "Semantic Space: An Infrastructure for Smart Spaces," IEEE Pervasive Computing, Vol. 3, No. 3 (2004), pp.32-39.
20. Wessel, M. and R. Möller, "A High Performance Semantic Web Query Answering Engine," International Workshop on Description Logics (DL2005), Edinburgh, Scotland, UK, 2005.

Multi-resource Load Optimization Strategy in Agent-Based Systems

Leszek Śliwko and Aleksander Zgrzywa

Institute of Applied Informatics, The Faculty of Computer Science and Management, Wroclaw University of Technology, Wrocław, Poland
{leszek.sliwko,aleksander.zgrzywa}@pwr.wroc.pl

Abstract. Resource management is one of the key research issues for multi-agent systems. Aside from increasing the system reliability, the load balancer is also able to schedule incoming tasks to the available machines. This paper introduces a load balancing strategy algorithm which can optimize the utilization of several different resources (CPU, memory, etc.) by migrating mobile agents and their tasks to the best alternative nodes. In the course of the research, a few scenarios were simulated and analyzed. In the paper the scenarios along with initial experiment results are presented.

Keywords: load balancing, multi-agent system, resource management.

1 Introduction

Resource management is one of the key research issues for multi-agent systems. A good global system performance can be achieved if the load balance strategy is added. A load balancer can increase the capacity of any of the networked computers as it assures that each device has a remote access to all resources available in the network [21]. Moreover, it can allow the service to continue should any abnormal termination of one or more of the connected machines occur. This feature is called the fault tolerance [3][20] or graceful degradation [18]. Fault tolerance is one of the most desired features in high-availability and life-critical systems [18]. In case of system malfunction, decrease in system operating quality is proportional to the severity of the failure. In natively designed systems even small failure can cause complete denial of service and render the system unavailable [3].

Aside from facilitating the system reliability, the load balancer is also able to schedule incoming tasks to the available machines. This job scheduling is a key concept in computer multitasking and multiprocessing operating system design, and this conception has been an active research area for a considerable period of time. A number of various strategies have been proposed and compiled [10][12]. However, most of them focus on achieving optimal utilization of exclusively one resource.

In this paper, an agent-based resources balancing algorithm along with some initial experiment results are presented. Due to the implementation of this algorithm the multi-agent system is able to hold a good load balance state with various kinds of

N.T. Nguyen et al. (Eds.): KES-AMSTA 2007, LNAI 4496, pp. 348–357, 2007.

resources utilized. The algorithm can be also applied to unload an overloaded system node within a few cycles needed to migrate the tasks to another node.

The paper is organized as follows: in the second chapter, we describe related works on the load balancing in mobile agent systems; also few implemented architectures are marked. Third chapter describes our system working principles with used algorithms explained in details. Section four is the most interesting as it presents the results of the initial experiment. Finally, last chapter concludes our work.

2 Load Balancing in Multi-Agent Systems

Nowadays, computer networks often contain applications which can simultaneously use resources of a number of computer environments in order to spread tasks to several machines [14]. System components are not just properties of individual machines; in many respects they can be viewed as though they are deployed in a single application environment.

The most advanced technologies are solutions of clusters. A cluster connects individual servers and maintains the communication among them in order to process related tasks in several environments at the same time. Should one of the clusters' servers fail the failover process moves the tasks from the failed server onto another automatically, thus preserving the continuity of the service at all times [13]. Aside from the failover process, the clustering also offers some workload balancing which assures more efficient distribution of the tasks to various servers in the network. Clusters are typically more cost-effective than single computers of comparable speed and usually enable systems to have higher availability than a single machine. [1]

There exist many methodologies and solutions describing how to spread the system to independent environments, however agent-based systems are of particular interest to us. An agent is a specific entity with autonomy and the ability to accomplish tasks on its user's behalf [15]. Its additional functionality introduced in many common agent-based frameworks [2][16] consists in its ability to migrate its code and even carry its state across the network. This allows execution of carried code in destination nodes independently, automatically and asynchronously [20][21]. Parallel processing is usually described as very dynamic, complex and unpredictable [17]; some of the agent properties as autonomy, social ability, goal-directed behavior, reactivity and proactiveness [4] can be used to build solutions to this sort of problems.

Those features inspired the idea of load balancing through the migration of users' tasks to various networked machines which have spare CPU cycles. This, in turn, enables the users not only to utilize fault tolerance schemas [20] but also to have an inexpensive access to high-end computational capacities on the Internet [11].

Two main network topologies have been developed. One is centralized where resources and data are managed in limited numbers of network nodes and the other one is decentralized where resources are handled trough much more complex algorithms [21]; negotiation schemas are often required for distributed architecture to hold a good level of performance [8].

The main drawbacks of the centralized management include possible performance bottlenecks and the lack of reliability [5][22]. Those problems are addressed by distributed architecture as the dynamic resource allocations are usually performed faster and more frequently locally [9]. However, the benefits of the decentralized model can be reduced as the distributed controlling network needs a number of messages to update the load information and the balance load among the nodes. To address this problem, various research projects have been initiated. The LBMA model [21] introduced a novel algorithm ULISI, which calculates the stochastic interval for load information update and reduces the total number of control messages. In LMBA the load information of the nodes is collected by a mobile agent moving between nodes by engaged route policy.

Agent-based systems often rely on decentralized architecture [10][11][19] as being more reliable, however this topology requires more sophisticated resource planning strategies. Several approaches have been made with various results (see Messor [12], Traveler [20] and LBMA [21]). They all have employed the same idea of utilizing all available resources in optimal way. They can all hold a good global performance through the implemented strategies.

Our idea is to keep a good load balance through resource vectors comparisons. We would also like not to make our strategy deterministic as it often tends to stick to the second optimal solution. Our system is still a prototype. However, the initial results prove that it can be used as a full-featured system resources manager.

3 The Working Principles

The agent is a key technology in this approach. It has been demonstrated in various projects that the agent technology not only enables the automation of complex tasks but also creates very reliable intelligent systems [4][19]. The introduction of agent schemes has simplified many sophisticated and compound problems and enabled the researches to develop a large decentralized autonomous cooperative system relatively easily [10][15].

The presented system consists of nodes and agents. Every agent serves one purpose which is called a job. Agents' jobs require resources which are provided by the node. Every node has a certain amount of resources available. In this paper, it is called the available resource vector. To simplify the definition, both the resources needed by the agents' job and the resources available on the node are described by the resources vector.

There exist several types of resources which can be utilized by the agent. The number of defined resources is potentially unlimited but in our experiment we use 2D and 3D vectors. We also name the resources as CPU, Memory and Network, which means respectively: the CPU usage agent will cause, the memory agent will allocate and the network bandwidth agent will need to perform its task.

The basic strategy is as follows. When a node detects that it does not have enough resources for all assigned tasks it selects an agent and migrates it to another node. The alternative node's selection is based on the agents' task resource vector and the resources available on other nodes in the system.

The load balance algorithm can be divided in two stages. First, it checks which, if any, resource on that particular node is overloaded. Secondly, the node searches for an agent using most of that resource and selects it as the migration candidate.

The stage can be described as follows:

Algorithm: Stage 1. Find a candidate agent for migration

Input:

An available resource vector for a node n: $n_r = <n_{r(1)}, n_{r(2)}, \ldots, n_{r(m)}>$, where m is the total number of different resources

All agents on this node $A = <a(1), a(2), \ldots, a(k)>$; all have their needed resource vector in the following form:

$a(i)_r = <a(i)_{r(1)}, a(i)_{r(2)}, \ldots, a(i)_{r(m)}>$

Output:

Agent a^*, which will be migrated in the next stage

\- or -

No resource is overloaded and the node n decides to skip the next stage

Begin

1 Select resource r^*, which is most used on this node:

1.1 **Select** r^* **where** $r^* \in n_r$ **and** $r^* = min(n_{r(1)}, n_{r(2)}, \ldots, n_{r(m)})$

2 Check if the most used resource r^* is overloaded:

2.1 **If** $n_{r(r^*)} >= 0$ **then Exit**

(resource not overloaded, there is no need for any task to be moved)

3 **Select** agent a^*, which most utilizes the resource n^*:

3.1 **Select** a^* **where** $a^* \in A$ **and** $a(a^*)_{r^*} = max(a(1)_{r^*}, a(2)_{r^*}, \ldots, a(k)_{r^*})$

(in case of several agents meeting this requirement, select the first one)

4 Proceed to **Stage 2**.

End

During the second stage, an alternative node for the selected agent should be appointed. At first, all nodes are evaluated. The algorithm estimates their usefulness factor, processes it to ensure only positive values and finally, assigns it to the percentage values. The results are percentage chances of selecting a particular node. This approach ensures that every node can be selected. However, certain selection cases have a very slight chance to occur.

The algorithm can be described as follows:

Algorithm: Stage 2. Select an alternative node for the agent migration

Input:

All possible alternative nodes $N = <n(1), n(2), \ldots, n(k)>$, all have their available resource vector in the following form:

$n_r(i) = < n_{r(1)}(i), n_{r(2)}(i), \ldots, n_{r(m)}(i)>$, where m is the total number of different resources

Agent a^*, with its resource vector:

$a_r^* = \langle a_{r(1)}^*, a_{r(2)}^*, ..., a_{r(m)}^* \rangle$
@f_ResultSignificance = 1.05
(this parameter is adjustable, higher values can result in never selected nodes, too low values can result in nodes selection being too random)

Output:
Node n^*, which will be the destination for the agent a^* migration

Begin

1 Initialization stage:
  1.1 Create vector p:
    For each node n in N:
      1.1.1 Add $p_{(n)}$ to p
      1.1.2 Initialize $p_{(n)}$ to:
      $p_{(n)} = \sum^m (a^*_{r(i)} \cdot n_r(n)_{r(i)})$

2 **For each** point value $p_{(n)}$ in p:
  2.1 Process $p_{(n)}$ to ensure only positive values:
  $p_{(n)} = (@f\_ResultSignificance)^{p(n)}$

3 Normalize p^* to *100*:
  $p^* = norm(p^*) \cdot 100$
  (the results are the percentage chance of selecting a particular node)

4 Randomly select a target node, based on their percentage chance $p_{(n)}$:
  4.1 **Select** n^* **where** $n^* \in A$
  and $random(p_{(n)})=true$

End

4 Experiment and Results

A simulation has been performed with the help of JMASB (Java Multi-Agent System Balancer). This framework has been initially developed by authors for agent-based systems performance analysis and it enables the researcher to test even complex schemes when planning the resource management strategy. It has allowed the authors to examine thoroughly the various resource usages during the load balancing system process and to tune algorithm parameters to get the best possible configuration.

The presented strategy was tested and implemented according to three scenarios. The next few sections of this paper describe briefly each of the three cases. They also present the workload chart generated during the experiment. Furthermore, related experiments are mentioned.

4.1 Experiment Configuration

The initial configuration consists of 42 generated tasks (Fig.1) and a randomly assigned vector of the resources required:

Task name	Needed resources			Task name	Needed resources		
	CPU	Memory	Network		CPU	Memory	Network
Job1	3	4	0	Job22	19	5	18
Job2	13	6	6	Job23	1	11	15
Job3	9	2	10	Job24	5	5	3
Job4	15	5	0	Job25	0	8	17
Job5	13	13	15	Job26	2	8	6
Job6	7	5	6	Job27	8	9	5
Job7	3	14	18	Job28	5	5	13
Job8	10	8	17	Job29	0	19	18
Job9	4	14	15	Job30	6	3	13
Job10	6	10	15	Job31	10	17	16
Job11	16	11	1	Job32	16	16	7
Job12	4	2	17	Job33	6	6	9
Job13	3	19	14	Job34	18	16	4
Job14	1	17	12	Job35	7	1	0
Job15	15	14	17	Job36	13	15	9
Job16	13	19	8	Job37	15	11	10
Job17	1	3	15	Job38	17	10	0
Job18	8	3	5	Job39	3	17	7
Job19	11	1	14	Job40	5	0	6
Job20	7	19	14	Job41	0	8	2
Job21	10	1	9	Job42	16	10	1

Fig. 1. Experimental tasks

Four nodes have also been created and their available resources are presented below:

Node name	Available resources		
	CPU	Memory	Network
Node1	100	200	200
Node2	100	100	200
Node3	100	100	50
Node4	200	50	100

Fig. 2. Experiment nodes

In the course of the simulation, three different scenarios were tested. In the two initial ones, only two resources, CPU and Memory, were utilized while in the third one all three resources defined in the resource vector, CPU, Memory and Network, were tested. This enabled us to compare the results achieved with 2D and 3D vectors.

In the first test, all tasks were initially assigned to one node (Node1). The second and third tests started with tasks randomly distributed among all available nodes. Such a setup resulted in several overloads created by accumulating tasks with a high need for a particular resource.

The situation when no node has exceeded the resources will be called here a "neutral" or "stable" status. Since that moment no agent will be migrated as all nodes are utilized correctly.

Although the system should be able to reach the stable status eventually (assuming that there are enough resources to serve all tasks), the timeout counter was introduced. Each node can migrate one agent in one system cycle. The maximum number of system cycles was set to 25. This proved to be more than sufficient. In the worst case, the first test, the system needed 20 cycles to distribute all 42 tasks. Thus, the results draws are presented with the last five neutral cycles missing.

4.2 Load Balance

The first test checked the system ability to unload one particular node. The result was reached within 20 system cycles, providing that no node was overloaded. The excessive tasks from the overloaded node were distributed among remaining nodes:

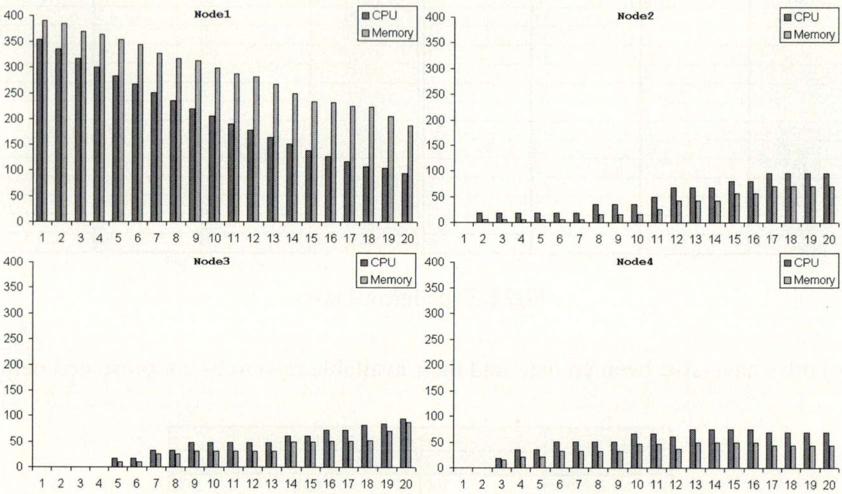

Fig. 3. The first test. Initially one node was heavily overloaded.

The second test was meant to diagnose how the system would manage a few overloaded resources. The neutral status was reached after 7 cycles:

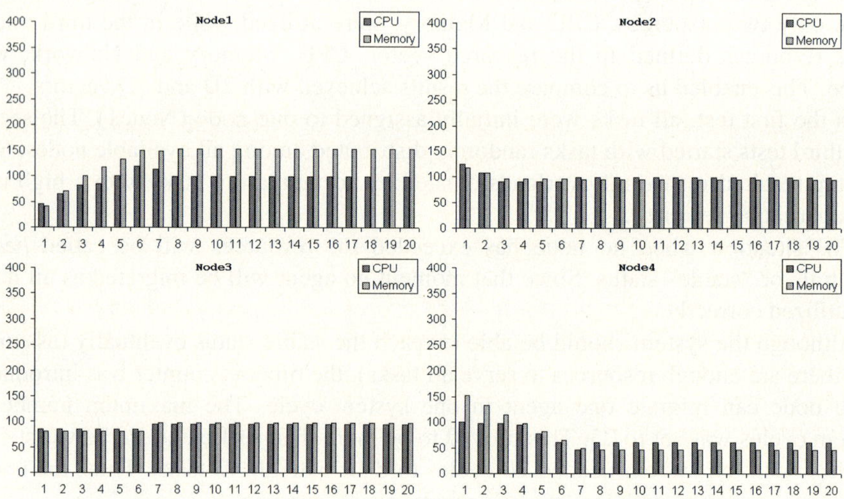

Fig. 4. The second test. The nodes were assigned tasks randomly.

During the third test, the 3D resources vector was used. The initial task assignment was similar to the one in the second test. The system reached the neutral stage after 8 system cycles:

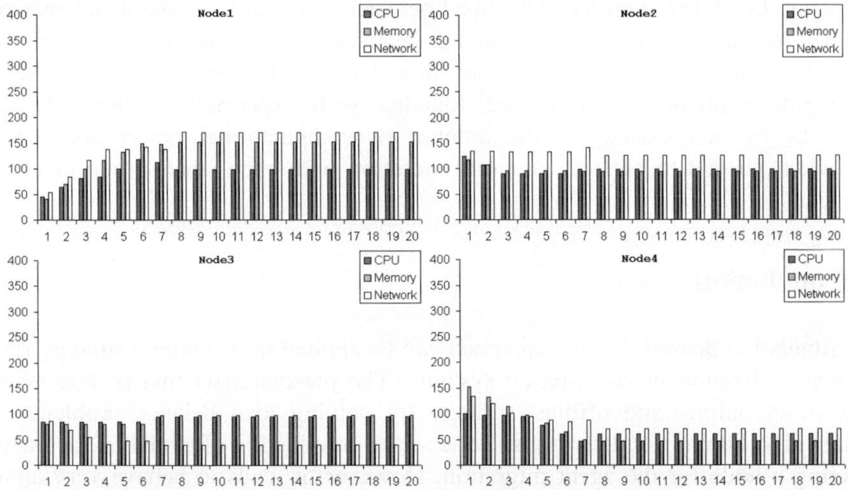

Fig. 5. The third test. The 3D resource vector was used.

4.3 Performance Analysis

In all the tests, the system managed to distribute the tasks to the nodes and resources were not overloaded in any case. The default timeout value (25 cycles) proved to be sufficient in all simulations. However, in the first test, the system needed as many as 20 cycles to unload one node. In the majority of cases, the system was able to find the stable distribution within 7-8 cycles.

The cases whére all the available system resources were insufficient to cover the required resources often resulted in "flickering" that is the agents were migrated to different nodes rapidly sometimes even within the same cycle. The simulation did not involve any task migration cost, however, such an action could cause a heavy resource usage in a real system.

4.4 Related Experiments

The idea of utilizing the multi-agent system to harness resources networked resources is not new [8][9][12][22]. It has long been an active area of research [17][20]. A number of various strategies have been designed and experimented with.

JMASB tracks other nodes' resources by load control messages. Basing on this information system selects the best node for agent migration through resource vectors comparisons.

Different approach has been proposed and compiled in Messor [12]. Messor agents' decision rules are inspired by the behavior of the artificial ant. Agents carrying the jobs migrate between nodes, dispersing jobs among less loaded computers. Colonies of such Messor ants can redistribute jobs in their environment very effectively in relatively short time [12]. This system does not rely on control messages; however reoccurring agent migrations can cause heavy resource usage.

The SCARCE framework [6] presents alternative approach to dynamic load balancing. Load negotiation is performed between resource supervising agents (called channel managers). Periodically channel manager selects one or more remaining channel managers and initiates load negotiation procedure. [6] In paper few negotiation strategies are proposed, together with experiment results. The main drawback of this strategy is the amount of computational power and network bandwidth needed to perform such an operation with even small group of channel managers.

5 Conclusions

The simulation proved that the approach can be applied in planning a strategy for the resource utilization in agent-based systems. The presented schema is able to work both as an online- and offline- strategy for solving load balance problems. The applied algorithm is not deterministic due to the introduction of a random factor while selecting a node for the agent migration. However, it tends to behave very alike in similar situations.

Moreover, the experiment demonstrated that a few important qualities should be considered for the implementation. The situations when nodes are not able to serve all requests should be detected and handled by the system. To address such problems the heuristics strategy should be considered in the future work on the system.

References

1. Buyya, R.: High Performance Cluster Computing: Architectures and Systems. Volume 1 & 2, Prentice Hall (1999)
2. Craswell, N-E., Haines, J., Humphreys, B., Johnson, C., Thistlewaite, P.: Aglets: a good idea for Spidering?. Proceedings of the 4th IDEA Workshop (1997) 474-480
3. Denning, J-P.: Fault Tolerant Operating Systems. ACM Computing Surveys, Volume 8 Issue 4 (1976) 359-389
4. Goodwin, R.: Formalizing Properties of Agents. Journal of Logic and Computation, Volume 5 Issue 6 (1995) 763-781
5. He, J.: An Architecture for Wide Area Network Load Balancing. Proceedings of the 2000 IEEE International Conference on Communications (2000) 1169-1173
6. Ho, K-S., Leong, H-V.: A Multi-Agent Negotiation Algorithm for Load Balancing in CORBA-Based Environment. Proceedings of the Second International Conference on Intelligent Data Engineering and Automated Learning, Data Mining, Financial Engineering, and Intelligent Agents (2000) 314-319
7. Jevtic, D., Kunstic, M., Jerkovic, N.: The Intelligent Agent-Based Control of Service Processing Capacity. Knowledge-Based Intelligent Information and Engineering Systems, Part 2, Springer-Verlag (2003) 668-674

8. Johansson, S., Davidsson, P., Kristell, M.: Cooperative Negotiation in a Multi-Agent System for Real-Time Load Balancing of a Mobile Cellular Network. International Conference on Autonomous Agents, Proceedings of the second international joint conference on Autonomous agents and multiagent systems (2003) 568-575
9. Johansson, S., Davidsson, P., Kristell, M.: Four Multi-Agent Architectures for Intelligent Network Load Management. Lecture notes in computer science (2002) 239-248
10. Jones, J., Crickell, C.: Second evaluation of job queuing/scheduling software. Tech. Report NAS-97-013, NASA Ames Research Center (1997) 1-34
11. Kesselman, F., Kesselman, C.: The Grid: Blueprint for a New Computing Infrastructure. MorganKaufmann (1999)
12. Montresor, A., Meling, H., Babaoglu1, O.: Messor: Load-Balancing through a Swarm of Autonomous Agents. Proceedings of the 1st Workshop on Agent and Peer-to-Peer Systems, Lecture Notes in Artificial Intelligence (2003) 125-137
13. Networld: Server clusters – Novel and Microsoft Systems. Networld 2000, International Data Group Poland SA (2000) 1-4
14. Nguyen, N-T., Sliwko, L.: Using Multi-Agent Systems and Consensus Methods for Information Retrieval in Internet. IWSE'06 (2006) 155-164
15. Nwana, S-H.: Software Agents: An Overview. Knowledge Engineering Review, Vol. 11, No 3 (1996) 1-40
16. Paek, K-J., Kim, T-Y.: AOM: An Agent Oriented Middleware Based on Java. Lecture Notes in Computer Science (1999) 474-479
17. Parent, J., Verbeeck, K., Lemeire, J., Nowe, A., Steenhaut, K., Dirkx, E-F.: Adaptive load balancing of parallel applications with multi-agent reinforcement learning on heterogeneous systems. Scientific Programming 12 (2004) 71-79
18. Randell, B., Lee, P., Treleaven, P-C.: Reliability Issues in Computing System Design. ACM Computing Surveys, Volume 10 Issue 2 (1978) 123-165
19. Shi, D., Yin, J., Zhang, W., Dong, J., Xiong, D.: A Distributed Collaborative Design Framework for Multidisciplinary Design Optimization. Computer Supported Cooperative Work in Design II, 9th International Conference (2005) 294-303
20. Xu, C-Z., Wimsaldonado, B.: A Mobile Agent Based Push Methodology for Global Parallel Computing. Java Grande 1999 Special Issue of Concurrency: Practice and Experience (2000) 2-12
21. Yang, Y., Chen, Y., Cao, X., Juguyen, J.: Load Balancing Using Mobile Agent and a Novel Algorithm for Updating Load Information Partially. Networking and Mobile Computing, Third International Conference (2005) 1243-1252
22. Zaki, M-J., Li, W., Parthasarathy, S.: Customized dynamic load balancing for network of workstations. In Proceedings of HPDC '96 (1996) 282-291

Online Network Resource Management for QoS-Sensitive Multimedia Services*

Sungwook Kim[1] and Sungchun Kim[2]

[1] Department of Computer Science, Sogang University, Shinsu-dong 1, Mapo-ku, Seoul, 121-742, South Korea
swkim01@sogang.ac.kr

[2] Department of Computer Science, Sogang University, Shinsu-dong 1, Mapo-ku, Seoul, 121-742, South Korea
ksc@mail.sogang.ac.kr

Abstract. Multimedia is a keyword in the evolving information age of the 21st century. In this paper, we propose a network management agent that includes congestion control and QoS control algorithms. Simulation results indicate the superior performance of our proposed agent to strike the appropriate performance balance between contradictory QoS requirements under widely varying diverse network traffic loads.

1 Introduction

As the telecommunication technology and multi agent system continues to grow, it needs to support an increasing range of services. Therefore, communication networking over which services are provided has become an area of great importance. Agents for network management can be defined as autonomous intelligent entities software units with certain autonomy. Their interactions can be either cooperative or selfish. That is, the agents can share a common goal – efficient network management.

Agent objectives in communication networks can be divided into two categories. The first category deals with providing network services that meet the needs of user applications such as service reliability and Quality-of-Service (QoS) guarantees. In recent years, network management techniques have been developed that support a variety of user requirements and the deployment of new services. One of the main technological challenges is to develop communication systems and control algorithms for quality of service (QoS) sensitive services.

The second category deals with network resource management strategies that provide benefits for the service provider. Supporting diverse services for a large number of subscribers is essential for the economic viability of network service provider. The key to supporting a large number of subscribers is bandwidth. However, in spite of the emergence of high network infrastructures, bandwidth is still an extremely valuable and scarce resource. Therefore, all performance guarantees in communication

* This research was supported by the Sogang University, under the 2006 new faculty support program.

N.T. Nguyen et al. (Eds.): KES-AMSTA 2007, LNAI 4496, pp. 358–363, 2007.

networks are conditional on currently available bandwidth capacity. In view of the remarkable growth in the number of users and the limited bandwidth, an efficient bandwidth management is very important and has been an active area of research over the last decade. Sophisticated bandwidth management would require the implementation of control mechanisms. The objective of these mechanisms is to maximize the overall network performance [1]-[4].

Next generation networks should support various multimedia services [5]-[6]. Different multimedia services usually require different QoS deliveries. Generally, real-time (class I) services such as voice and video are sensitive to delay but tolerant to a certain level of packet loss. In contrast, non real-time (class II) services such as data traffic is normally delay-insensitive and tolerant to transmission rate variations, but requires reliable transmission. Therefore, the main concern for class I service is a delay guarantee while the important factor for class II service is network throughput. Multimedia network control agents should take into account the prioritization among different multimedia traffic services. Based on different tolerance characteristics, class I data type has higher priority than class II data type during network operations [3],[5]-[8].

In this paper, the main challenge of our work is to try to strike the appropriate performance balance among various QoS requirements. In order to satisfy this goal, various techniques should be considered, weighed, and balanced. Motivated by the above discussion, we propose a new network management agent for QoS-sensitive multimedia networks. The important feature of our agent is the inclusion of a QoS guarantee that does not reduce the network capacity. Earlier work reported in [9] and [10] has also considered QoS based network management. The both agents proposed by [9] and [10] were designed to effectively improve the network performance, at the same time, provide QoS support for higher priority service. However, these existing agents have several shortcomings, as described in Section III. Compared to these agents, we can see that our proposed agent is quite appropriate and attains better performance for multimedia network environment.

This paper is organized as follows. Section II describes the proposed algorithms in detail. In Section III, performance evaluation results are presented along with comparisons with other existing agents [9],[10]. Finally, concluding remarks are made in Section IV.

2 Proposed Network Management Agent

In this paper, we propose a new adaptive online agent that is suitable for multimedia network management. To provide well balanced network performance, our agent consists of two management algorithms - QoS control algorithm and congestion control algorithm.

2.1 QoS Control Algorithm

In this section, we develop the QoS control algorithm based on adaptive online approach. For class I traffics, bandwidth reservation is needed to accommodate strict delay limited services. To determine the optimal reservation amount adaptively, we partition the time-axis into equal intervals of length *unit_time*. Our proposed online algorithm adjusts the amount of reserved bandwidth (Res_B) based on real time measurements during every *unit_time*.

To maintain the reserved bandwidth close to the optimal value, we define a traffic window, that is used to keep the history of class I traffic requests (W_{class_I}). The traffic window is of size [t_c - t_{class_I}, t_c], where t_c is the current time and t_{class_I} is the window length, and this size can be adjusted in time steps equal to *unit_time*. Ideally, we want the amount of reserved bandwidth to be equal to the requested hand off bandwidth during the next unit_time. The value of unit_time is chosen based on desired system performance objectives. If unit_time is relatively small, which enables the algorithm to react more quickly and accurately to the changing traffic, system performance is more nearly optimal at the expense of control overhead. For large values of unit_time, the control overhead is less but at the expense of system inefficiency; it could be too slow in adapting to real traffic changes. Based on this assumption, *unit_time* in our paper is one second. If the class I call blocking probability (CBP_{class_I}) for handoff service is higher (lower) than its predefined target probability (P_{class_I}), the traffic window size is increased (decreased). The values of Res_B can be estimated as the sum of requested bandwidths by handoff class I calls during the traffic window. Therefore, by using this traffic window, we can dynamically adjust the amount of the Res_B at every *unit_time*, which is more responsive to changes in the network condition after the bandwidth has been reserved. The main steps of our proposed QoS-provisioning mechanism are as follows:

- At every *unit_time*, our algorithm monitors the current class I CBP and then adjusts the traffic window size accordingly.
- Traffic window sizes are defined as integer multiples of *unit_time* in this paper.
- If the CBP_{class_I} is higher (or lower) than the P_{class_I}, the traffic window size is increased (or decreased) in steps equal to *unit_time*.
- Based on the size of the traffic window, we adaptively adjust Res_B at every *unit_time*.

2.2 Congestion Control Algorithm

As mentioned earlier, we differentiate between class I and class II traffic. In contrast to class I traffic services, class II services are flexible to time delay. Therefore, a congestion control algorithm is required for class II services. Various AQM algorithms for congestion control have been proposed in recent years [9]-[10]. These algorithms aim at detecting network congestion earlier so as to utilize bandwidth more efficiently, and send congestion notifications to the end-nodes. For this proactive control, incoming packets under heavy traffic situation can be dropped probabilistically. Therefore, the average queue size does not significantly exceed the maximum threshold.

Essentially, a network average queue size is considered as the degree of traffic burstiness and the current congestion level. Therefore, based on the average queue size, AQM algorithms can determine how frequently the router drops packets. In order to avoid biases and avoid the global synchronization, a router should drop packets at evenly spaced intervals.

In this paper, we propose a congestion control algorithm for the adaptive packet dropping. For the adaptive packet dropping, our online algorithm defines three queue control parameters; the queue endurance range (Q_r), the packet marking probability (M_p) and the packet queuing rate (I_{p_r}). Q_r is a threshold for queue buffering, M_p is a marking probability assigned to drop packets in a randomized manner, and I_{p_r} is the rate at which packets are being queued in the buffer in the recent time interval. Since the future is not known exactly, we assume that the traffic pattern during the recent time interval [t_c - *unit_time*, t_c] reflects the current traffic situation. Our algorithm monitors the recent I_{p_r} to decide the current traffic situation. If the incoming bits (I_b) to the queue is larger (or lower) than the outgoing bits (O_b), the traffic overflow (or underflow) occurs. Under a queue overflow period ($I_{p\_r} > 0$), the current queue length increases; an underflow situation results in the current queue length decreasing.

These Q_r, M_p and I_{p_r} parameters can significantly affect the network congestion control. Therefore, these values should be tuned adaptively according to the current network conditions. As stated above section, the bandwidth efficiency can be improved by dynamically sharing between class I and class II traffic. In our algorithm, the reserved bandwidth (Res_B) for class I service can be temporarily allocated for the buffered class II packets. Therefore, by online inspecting the current Res_B at every *unit_time*, we set the parameter Q_r value to be the same as the current Res_B. This means that class II data packet buffering can withstand until Q_r is reached.

The uncertainty of the future traffic makes it impossible to optimally control M_p value. Therefore, we treat the M_p adjustment as an on-line decision problem. Our proposed algorithm also adaptively adjusts M_p at every *unit_time* based on the current queue situations. For the more adaptive reaction to traffic fluctuations, we define two packet marking probabilities (M_{p_1} and M_{p_2}) according to (1).

$$M_{p\_1} = \frac{L}{ML} \quad \text{and} \quad M_{p\_2} = \frac{L - Q_r}{ML - Q_r} \tag{1}$$

where ML is the maximum queue length and L is the current queue length. L is used as the main factor for determining M_{p_1} whereas Q_r is considered to determine M_{p_2}. M_{p_2} value can be changed slowly under network traffic situations. In the presence of traffic congestion, our congestion control algorithm randomly drops packets when packets are queued during the time interval [t_c, t_c + *unit_time*]. Dropping packets is a signal of the congestion level on the path and provide feedback information to sender nodes.

When L is greater than the total buffer size (ML), the current traffic load in the network is very heavy and the buffer does not have any space for the incoming packets. Therefore, all arriving data packets should be dropped ($M_p = 1$). When L is greater than Q_r, but less than ML ($Q_r < L < ML$), we can assume that the total available buffer space is reached a critical low point in the near future. Therefore, based on the current traffic conditions, we adaptively control the drop rate. If the buffer is in the overflow situation ($I_{p\_r} > 0$), we set $M_p = M_{p\_1}$ to catch up with the traffic congestion. If the buffer is in the underflow situation ($I_{p\_r} \leq 0$), we set $M_p = M_{p\_2}$ to reduce the current traffic overhead. When L is less than Q_r ($0 \leq L < Q_r$), the available buffer space is sufficient to support the current traffic situation. Therefore, we set $M_p = 0$; no arriving packets are dropped. The main steps of our congestion control algorithm are follows:

- At every *unit_time*, Q_r and I_{p_r} are estimated by online monitoring.
- At every *unit_time*, M_{p_1} and M_{p_2} are adaptively adjusted according to (1).
- When L is greater than *ML*, we set $M_p = 1$.
  - all arriving data packets should be dropped.
- If L is less than Q_r ($L < Q_r$), we set $M_p = 0$.
  - no arriving packets are dropped.
- If L is between Q_r and *ML* ($Q_r < L < ML$), arriving class II data packets during the time interval $[t_c, t_c + unit\_time]$ are randomly dropped by M_p.
  - if the current buffer is in the overflow situation ($I_{p\_r} > 0$), we set $M_p = M_{p\_1}$.
  - if the current buffer is in the underflow situation ($I_{p\_r} \leq 0$), we set $M_p = M_{p\_2}$.

3 Simulation Experiments

In this section, we evaluate the performance of our proposed agent using a simulation model. Based on this simulation model, we compare the performance of our agent with other existing agents [9]-[10]. With a simulation study, we can confirm the performance superiority of our agent. With multiple classes of traffic, each has different traffic characteristics - its own requirements in terms of bandwidth, QoS guarantee and call connection time. For the performance measures, we focused on the bandwidth utilization, and system throughput (network revenue). These measures obtained on the basis of 10 simulation runs are plotted in the Figs. 1-2 as functions of the load (i.e., the call arrival rate λ).

Fig.1 compares the performance of the three agents in terms of bandwidth utilization under different traffic loads. All the agents have similar trends when the call arrival rate is low, since they have sufficient bandwidth to accept the requested calls

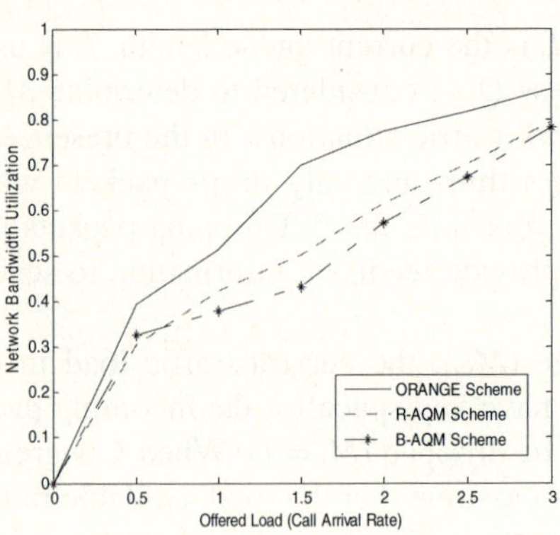

Fig. 1. Bandwidth Utilization

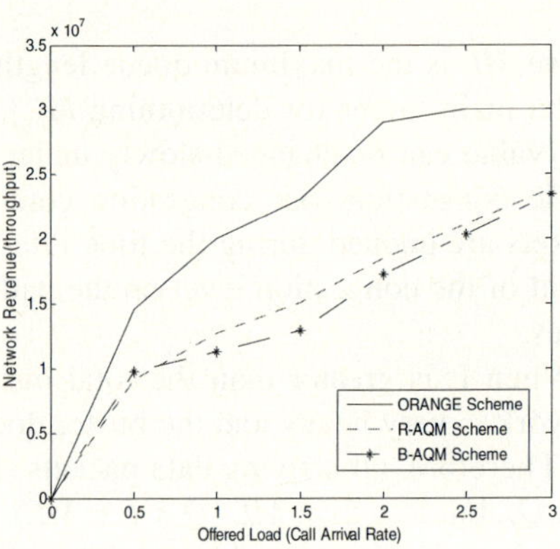

Fig. 2. Network Revenue

without traffic congestion. However, our agent exhibits significantly better bandwidth utilization as the traffic load increase. Fig.2 shows the network revenue (throughput) for all types of traffic. Our proposed online approach requires constant tuning of the parameters to adjust to the current traffic conditions so as to be able to adapt rapidly to changes in the load or available bandwidth. The network revenue curves indicate that as the call arrival rate increases, the performance of our agent becomes higher than those of the other agents. The simulation results shown in Figs.1-2 demonstrate that our online agent generally exhibits superior performance compared with the other existing agents under light to heavy traffic loads.

4 Summary and Conclusions

In this paper, we propose online adaptive network management agent for multimedia networks. In order to provide QoS-sensitive multimedia services, our online agent dynamically manages the network by suitably combining two control algorithms - the QoS control algorithm and the congestion control algorithm. We compared the performance of our proposed agent with two existing agents [9],[10]. Performance evaluation results indicate that our agent maintains a superior network performance in widely different traffic-load situations.

References

1. Mark Stemm, Randy H. Katz, "Vertical Handoffs in Wireless Overlay Networks," *ACM Mobile Networking. (MONET)*, vol.3, number 4, pp. 335-350, 1998.
2. Hakini Badis, and Khaldoun Al Agha, "An Efficient Mobility Management in Wireless Overlay Networks," *PIMRC 2003*, Vol 3, pp. 2500- 2504, Sep. 2003.
3. Wei Song, Hai Jiang, Weihua Zhuang, and Xuemin Shen, "Resource management for QoS support in cellular/WLAN interworking," *IEEE Network*, vol. 19, no. 5, pp. 12 - 18, 2005.
4. T. Dahlberg and J. Jung, "Survivable Load Sharing Protocols: A Simulation Study", *Wireless Networks* 7, no. 3, pp.283-296, 2001.
5. Sungwook Kim and Pramod K. Varshney, "An Integrated Adaptive Bandwidth Management Framework for QoS sensitive Multimedia Cellular Networks," *IEEE Transaction on Vehicular Technology*, pp.835- 846, May, 2004.
6. Sungwook Kim and Pramod K. Varshney, "An Adaptive Bandwidth Allocation Algorithm for QoS guaranteed Multimedia Networks," *Computer Communications* 28, pp.1959-1969, October, 2005.
7. Ling-Jyh Chen, Guang Yang, Tony Sun, M. Y. Sanadidi and Mario Gerla, "Enhancing QoS Support for Vertical Handoffs Using Implicit/Explicit Handoff Notification," *QShine 2005*, Orlando, FL, pp.37-45, 2005.
8. Young Sup Roh , Kyeong Hur , Doo Seop Eom , Yeonwoo Lee and Kyun Hyon Tchah, "TCP Performance Enhancement by Implicit Priority Forwarding (IPF) Packet Buffering Scheme for Mobile IP Based Networks," *Journal Of Communication and Networks* Vol 7, pp. 367-376, 2005.
9. W. Feng, D. Kandlur, D. Saha, K. Shin, "Blue: An Alternative Approach To Active Queue Management," *Proc. of NOSSDAV 2001*, pp. 41-50, June 2001
10. James Aweya, Michel Ouellette, Delfin Y. Montuno, Alan Chapman, "Enhancing TCP performance with a load-adaptive RED mechanism," *International Journal of Network Management*, Volume 11, Issue: 1 pp. 31 – 50, 2001

A Resource Discovery Method Based on Multi-agents in P2P Systems

Yasushi Kambayashi and Yoshikuni Harada

Department of Computer and Information Engineering,
Nippon Institute of Technology,
4-1 Gakuendai, Miyashiro-cho, Minamisaitama-gun, Saitama 345-8501 Japan
yasushi@nit.ac.jp, c1015367@cstu.nit.ac.jp

Abstract. A peer-to-peer (P2P) system consists of a number of decentralized distributed network nodes that are capable of sharing resources without centralized supervision. Many applications such as IP-phone, contents delivery network (CDN), distributed computing adopt P2P technology into their base communication systems. One of the most important functions in P2P system is locating resources, and it is generally hard to achieve due to the intrinsic nature of P2P, i.e. dynamic re-configuration of the network. In this paper, we propose an efficient resource locating method in pure P2P system based on multi-agents. The model of our system is a DHT base P2P system that consists of nodes with DHT (high performance nodes) and nodes without DHT (regular nodes). All the resources as well as resource information are managed by cooperative multi-agents. Migrating multi-agents are expected to reduce communication traffic in the network. Efficient migration is achieved through the clustering of nodes that makes correlated nodes in a group by the logical similarity. The numerical experiments through simulation have shown a significant reduction of generated messages.

Keywords: P2P, Multi-agent, Mobile agent, DHT, Resource discovery.

1 Introduction

As the Internet spreads into the world, it is used for a variety of human interactions. User interactions across various applications require software that exchanges resources and information within the network community. Traditional client-server model on the computer networks hardly accommodates the real-world situation such as the advent of the video-streaming services. Because the intensive accesses on a server can easily make a bottle-neck in a network. Peer-to-peer (P2P) systems can provide a solution to this problem. A P2P system consists of a number of decentralized distributed network nodes that are capable of sharing resources without central servers. Many applications such as IP-phone, contents delivery network (CDN), distributed computing adopt P2P technology into their base communication systems. A P2P system provides resource and information exchanges within nodes as peers. A P2P system comprises an overlay

N.T. Nguyen et al. (Eds.): KES-AMSTA 2007, LNAI 4496, pp. 364–374, 2007.

networks where the nodes can interact and share resources with one another. Here, resources mean a variety of services which the network nodes provide.

One of the most important issues in P2P systems is how to locate such resources, and it is one of the hardest mechanisms to implement. Napster avoids this problem by using a central server that provides indexing service [1]. Such a server, however, can be the most vulnerable point where a failure can paralyzes the entire networks. Therefore, P2P systems without any central server (pure P2P) are an area of active research in the P2P system developments. In this paper, we propose an efficient resource locating method based on multi-agents in a pure P2P system.

Most current P2P applications use message flooding for resource discovery. Message flooding makes the quantity of messages in the network increase rapidly as the number of nodes in the network increases, and causes saturations easily. It can be said that message flooding has a problem in the scalability. The proposed method pursues a coordination of multi-agents, the distributed hash table (DHT) and clustering nodes so that we can solve the resource discovery problem. We demonstrate the method's effectiveness, i.e. finding the desired contents with reasonable quantity of messages, through simulations.

We have sketched our method in the previous paper [2]. In this paper, we describe the details of implementation, and demonstrate the feasibility of our method by the results of numerical experiments.

The structure of the balance of this paper is as follows. The second section describes the background. The third section describes the P2P system we are proposing. Static and mobile multi-agents are effectively working together to find network resources in the P2P system. We also describe the clustering that contributes efficiency for resource discovery. The fourth section describes the resource discovery algorithm that uses the multi-agents to find network resources in the P2P system. The fifth section describes how the system is implemented by using an overlay construction tool kit and a mobile agent construction framework. The sixth section demonstrates the efficiency of our algorithm by the numerical experiments. The seventh section discusses the related works. Finally, eighth section discusses future works and conclusions.

2 Background

Famous pure P2P system, Gnutella, employs message flooding for locating resources [1]. The advantage of such a system is its simplicity, but it is unrealistic for a large scale network system, because flooding resource discovery messages alone can easily saturate entire networks. In order to solve this problem a distributed hash table (DHT) is proposed and used [3] [4] [5] [6]. Even though DHT is one of the most promising methods and it certainly provides fast resource lookup ($O(\log n)$ computational complexity) for pure P2P systems, it has the following disadvantages: 1) since the basic mechanism of DHT is mapping keys to nodes, it is hard to find objects based on multiple keys or contents held in objects; 2) it is hard to find multiple nodes that are related to a given key or a set of keys. In other words, DHT based resource lookup methods are rigid and hard to process flexible and intelligent queries.

In order to mitigate the rigidity of DHT base systems, several P2P systems employ message flooding for object lookup to complement DHT [7]. Since the message flooding causes dense communication traffic, it is not applicable for mobile communication environment where network connections are intermittent.

On the other hand, multi-agent systems based on mobile agents are recently popular for various fields [8]. Mobile agents in P2P systems have the following advantages: 1) mobile agents package necessary interactions between nodes and make them local by conveying the necessary processing to the destination where the desired resources reside, 2) mobile agents are asynchronous so that the node originates the mobile agents can perform completely different tasks or even leave the network temporary, and 3) mobile agents are autonomous and they can learn about the network as they progress through it [9].

One of the authors has engaged in a project where autonomous agents play major roles in an intelligent robot control system [10] [11]. The mobile agents in the project can bring the necessary functionalities and perform their tasks autonomously, and they have achieved reduction of communications as well as flexible behaviors. Thus, it is natural for us to employ not only static agents but also mobile agents in our P2P system in order to provide flexible search. The mobile agents are expected to reduce the quantity of the query messages.

Thus, we propose a resource discovery method that uses mobile multi-agents in a pure P2P system based on DHT. Mobile multi-agents provide much flexibility as well as some intelligence to the DHT base P2P systems. We also integrate clustering of nodes in order to achieve further performance improvement in agent migrations.

3 The P2P System

The model of our system is a DHT base P2P system that consists of nodes with DHT (high performance nodes) and nodes without DHT (regular nodes). All the resources as well as resource information are managed by cooperative multi-agents. They are: 1) information agents (IA), 2) search agents (SA), 3) node management agents (NA), and 4) DHT agents (DA). These four agents provide the minimum configuration. SA is the only mobile agent. IA encapsulates all the interfaces between users and applications that utilize this P2P system so that all the other parts (agents) can be independent from any applications. We separate DA from NA, because only high performance nodes have DHT. When constructing more application oriented P2P system, one may add more application-oriented agents. Our purpose, however, is constructing a multi-agent based framework for P2P systems.

1. *Information Agent (IA):* Each node has a static information agent (IA) that manages resource information. IA also interacts with users.

2. *Search Agent (SA):* Upon accepting a user query, the information agent creates a mobile search agent (SA) and dispatches it. Dispatched SA travels through the network to find the requested resources.
3. *Node Management Agent (NA):* Each node has a static node management agent (NA) that has neighbor information based on one or more cluster words we describe below. NA has a table that contains the IP addresses of neighbors and correlations with them based on the cluster word. Traveling SA refers this table in order to determine to which node it migrates next. The clustering connects related nodes much tighter; it should increase the possibility of finding the desired resources.
4. *DHT agent (DA):* DHT agents (DA) construct DHT through cooperation with DA's that reside other nodes. Only high-performance nodes have DA's so that we can construct pure P2P systems in a heterogeneous environment. Though current implementation integrates the Chord [5] as the DHT algorithm, we can replace it with other algorithms [3] [4] [6] just through replacing DA's.

In order to improve the efficiency of resource discovery, our method integrates clustering of nodes based on the cluster words. The cluster words are key words that evaluate the correlations between a node and neighbors. Clustering makes the logical distance between correlated nodes shorter. Upon joining a node to the network, the user of the system is required to specify key words that represent the resources the node has. The node management agent (NA) then calculates the logical distances (correlations) between it and neighbors. These logical distances are set in the node management table held by NA.

When we make a node join to the network, we first decide whether the node has a DHT agent (DA). Then we register the IP address of the joining node to the nearest bootstrap node. The joining node, at the same time, receives the addresses of neighbors from the bootstrap node and constructs the node management table. NA takes care of this task.

When a node leaves from the network, NA request DA's in the neighbors to erase its entries from their DHT. It also requests NA's in the neighbors to erase its entries from their node management tables. If there are any migrating search agents (SA's) in the node, NA makes them leave the node immediately.

3.1 Clustering

In order to achieve efficiency on discovering resources, we have employed the node clustering by using the cluster words. Clustering makes logically related nodes form into a group and shorten their logical distances, thus it is expected to decrease the number of hops that a SA must migrate. Cluster words are keywords that are used to evaluate the logical similarity between a node and its neighbor nodes. Each node has several keywords that are specified by users when it is created so that each node can evaluate correlations with neighbors and determine the logical similarities.

Fig. 1 shows an example of evaluating correlation between two nodes A and B. The node A has three keywords namely, "abc," "abcde" and "xy," and the node

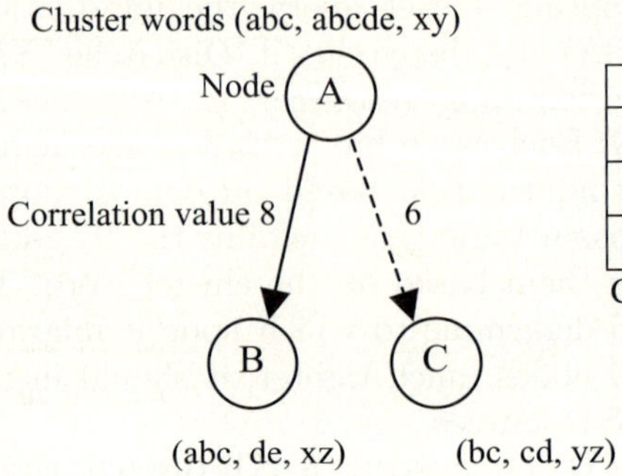

	abc	abcde	xy
abc	3	3	0
de	0	2	0
xz	0	0	0

Correlation value between A-B
3+3+2=8

Fig. 1. Logical similarity by the cluster words

B has "abc," "de" and "xz." In order to evaluate the correlation, the node makes pairs from each of its keywords and the neighbor's, then the counts the number of matched characters as shown in the table in Fig. 1. We can say that the more a pair has matched characters, the more the two nodes are correlated. We set a rule that a node evaluates two keywords when one includes the other such as "abc" and "bc," but does not when the two have some overlapped characters such as "abc" and "bcd." Because we found that all the nodes have similar correlation values when we allow such pairs to be evaluated.

4 Resource Discovery Algorithm

When a user requests the information agent (IA) in the current node to locate a resource, the user has to specify the lookup keywords and search terminating conditions such as the number of hops, duration time, and the number of found nodes. The user is also required to specify how the IA should behave when the dispatched SA does not return due to some accidents. Fig. 2 shows interactions of the cooperative agents to locate desired resources. The resource discovery algorithm that the coordinated multi-agents perform is as follows:

1. IA creates a SA for a specific search.
2. The SA requests the NA to which neighbor it should migrate. When the resource name is available, NA demands DA to give the IP address of the node where the resource resides. If the request is ambiguous and NA cannot tell the exact node SA should migrate, then the NA looks up the node management table to select the logically nearest neighbor.
3. The SA migrates to the selected neighbor.
4. The SA then interacts with the IA in the arriving node to find whether it has the requested resource.
5. If the node has the resource, the SA stores the resource information and the IP address of the node, then goes to step 6, otherwise goes to step 2.
6. The SA checks the terminating condition, and if it is satisfied, migrates back to the original node where the SA was created.

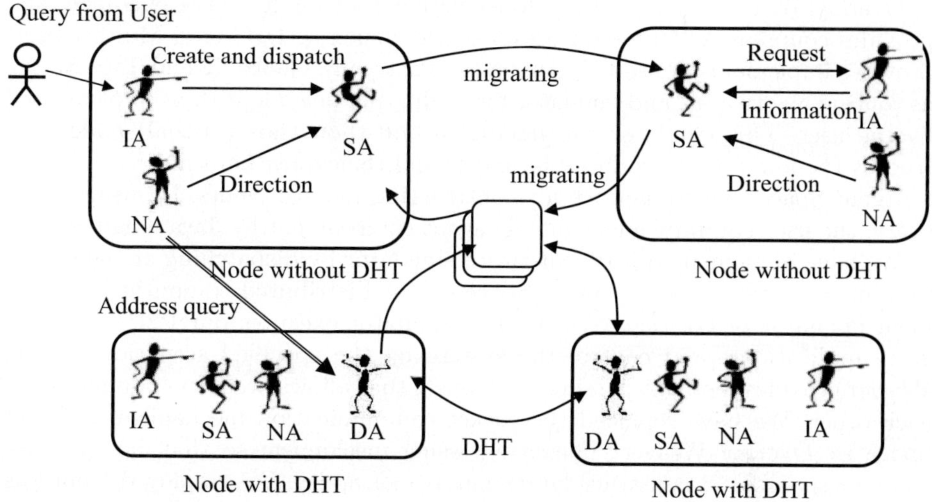

Fig. 2. Cooperative agents and resource discovery protocols

In step 2, when NA cannot use DHT due to ambiguity of the resource request, it has to select the neighbor to which the SA migrates. The basic algorithm is to select the logically nearest (the most closely correlated) node.

When the resource discovery request is ambiguous and IA cannot specify the resource name to use DHT, the SA travels one node to another to find the resource name from the resource information that the visited node has. When some candidates of resource name are found, the SA returns to the original node and reports it to the user so that the user can re-issue a new search request. This time, newly created SA may be able to migrate to the desired node with its IP address obtained from DHT. In order to avoid a problem of cyclic migration paths, the node management table is adjusted so that a SA is not sent to the same neighbor.

Since our resource discovery method uses DHT as well as migrations of mobile agents, the search agent can migrate to a much farther node directly. Also a node that has the resource information leaves the P2P system as the SA is conducting its search, if the SA gets the resource name before its leave, the SA can find the desired resource by using DHT. Further, by using DHT, the SA can find multiple nodes that have the desired resource simultaneously. Traditional P2P resource discovery system using mobile agents lacks these advantages.

5 Implementation

We have implemented our resource discovery algorithm by using Overlay Weaver [12] and Agent Space [13], and conducted numerical experiments on the distributed computing environment emulator of Overlay Weaver.

Overlay Weaver is an overlay construction tool kit for Java language, and provides common API's for high-level services such as DHT and Multicasts. It provides implementations of Kademlia [3], Pastry [4], Chord [5] and Tapestry [6] as routing algorithms, and emulator for evaluating new algorithms implemented by the user. This emulator can handle several thousands (virtual) nodes and records the number of produced messages and their duration time.

Agent Space is a framework for constructing mobile agents. By using its library, the user can implement mobile agent environment by Java language.

We have implemented a P2P network simulator by dispatching an agent system implemented by using Agent Space on the distributed computing environment emulator of Overlay Weaver. The emulator executes our P2P system (a Java application), and control the system by the specified scenario. Overlay Weaver provides message-passing mechanism that allows programs communicate each other. We have replaced the socket communication mechanism of Agent Space by Overlay Weaver's message-passing mechanism so that we can save memory spaces and construct larger number of nodes. We employed Chord as the DHT algorithm.

In Agent Space, mobile agents are defined as collections of call-back methods, and we have to implements interfaces defined in the system. In order to create a mobile agent, the application calls `create` method. When an agent arrives, `arrive` method is invoked. It also provides service API's such as `move` method to make agents migrate and `invoke` method to communicate to another agent. The following is an example of SA implementation:

1. Invoke `create` method to create a mobile agent, and specify the search conditions.
2. SA communicates to NA and determines the destination.
3. Invoke `move` method so that the SA can migrate to the specified node.
4. Invoke `leave` method.
5. The agent actually migrates to the specified node.
6. Invoke `arrive` method in the destination node, and the SA communicate to IA in order to receive necessary information.
7. Check the terminate conditions; if they are satisfied the SA returns to the original node, otherwise receive the next node address and continue migration.

Step 3 creates a duplication of the SA in the specific node, and step 4 erases the original SA in the original node, so that the SA migrates from a node to the specific destination (step5).

The agents communicate each other by invoking `invoke` method. Fig. 3 shows a situation that a SA asks an IA whether it has a resource named "abc." The SA invokes its own `requestResourceInfo` method to communicate to the IA in the node that the SA resides, and invoke `getResourceInfo` method of the IA.

```
public void requestResourceInfo(Context context) {
  String info = "";
  AgentIdentifier[] aids = context.getAgents("IA");
  if (aids != null) {
    try {
      Message msg = new Message("getResourceInfo");
      msg.setArg("abc");
      Object obj = context.invoke(aids[0].msg);
      if (obj != null && obj instanceof String) {
        info = (String)obj;
      }
    } catch (Exception ex) {
      ex.printStackTrace();
    }
  }
}
```

Fig. 3. Agent communication

6 Numerical Experiment

We have compared the number of produced messages by a simple flooding search method and our proposed method with ambiguous resource names. We constructed an environment with 2,000 nodes. The simple flooding method employs four message-sending links and makes each query at most seven hops. The number of DHT nodes, which are solely used in our method, is 400, and they use Chord for DHT algorithm [5]. Each node has one IA and one NA. We provide twenty cluster words and each node randomly selects three of them. Each node has five candidates for SA migration. SA receives initial information for lookup from the bootstrap node, and it moves at most fifty hops. We have randomly chosen five nodes and made IA to start the search for randomly located resources.

Using the above framework, we have conducted ten searches for each resource lookup with both methods, and compared the total number of the transmitted messages. Fig. 4 illustrates the results of the experiments for various numbers of resources. The quantity of messages produced by the simple flooding method does not change as the number of resources varies; while the quantity of messages produced by our method decreases as the number of resources increases. We can observe that, in general, more resources a network has more efficiency our method can achieve. We can assume that the number of the generated messages dominates the overall performance, because all the interactions between mobile search agents and local agents are local function calls and negligible.

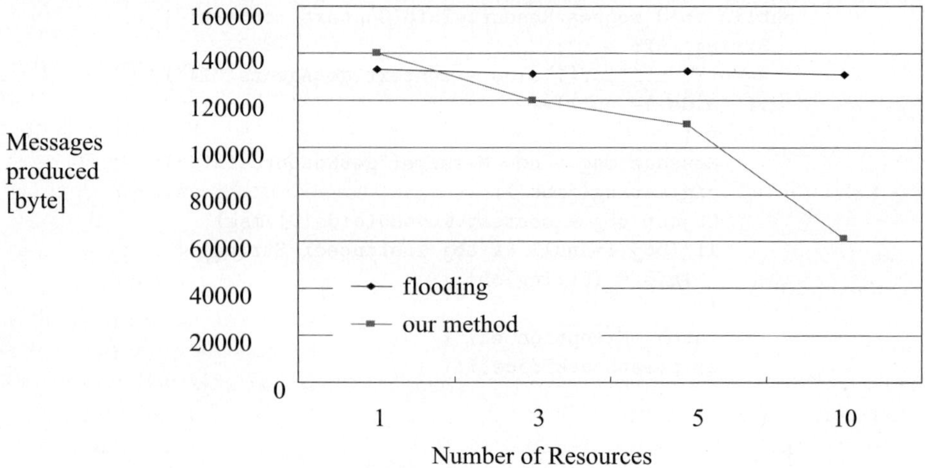

Fig. 4. Quantities of produced messages

7 Related Works

The works most closely related to ours are the Anthill framework for a P2P system developed at the University of Bologna [14] and an agent framework for a P2P system developed by Dasgupta at the University of Nebraska, Omaha [15].

The Anthill framework employs intelligent agents for nodes and resource discovery. Mobile agents called *ants* migrate across the nodes in a P2P network to discover resources and perform distributed tasks. They disseminate information about resources into the network as well as discover the desired resources. The approach is more biology inspired one than ours, and evolutionary computations such as genetic algorithms are used for governing the behaviors of mobile agents. It is more flexible than ours, but expected to be inefficient than our approach using DHT.

Dasgupta's agent framework does not integrate DHT either. The algorithm improves the behaviors of search agents by trails established from previous searches. It does not relay on routing tables for directing search agent such as those the node management agents in our system have. In contrast, the method described in this paper integrates DHT for efficient search while suppressing unnecessary communication traffic and accommodating ambiguous searches by using mobile agents. The node management agents direct search agents by using the neighbor information based on node clustering.

8 Conclusion and Future Works

We have proposed a novel resource discovery method using mobile agents in a pure P2P system, and demonstrated its efficiency by comparing the quantities of message data transmitted in our P2P system with that of a simple flooding method produced. In our proposed method, static and mobile agents cooperatively perform their tasks to locate requested resources efficiently. In order to reduce communication traffics, and at the same time, to increase search flexibility, we use mobile agents. In order to improve performance, we also use clustering of network nodes. We have implemented a P2P system integrating this resource discovery method on a distributed environment emulator of Overlay Weaver. Even though the current implementation is not complete, the numerical experiments have strongly suggested the superiority of our method.

We are aware of the incompleteness of our numerical experiment. In our simulation setting, resources are found faster by using a simple flooding method than by using our proposing method even though the simple flooding method produces an enormous number of messages. We are re-implementing our multi-agent base resource discovery algorithm on a real computer network so that we can measure how much message congestion affects resource discovery time.

Unlike many other agent base resource discovery methods, our method should be efficient, because our search agents can be dispatched to the destination directly by using DHT when enough location information is available. Nodes with DHT can be seen as super-peers and we need further investigation about the overall construction of a network with them as Yang and Garcia-Molina have done [16]. Then we may need to have separate super-peers for efficient use of clustering.

References

1. Saroiu, S., Gummadi, K.P., Gribble, S.D.: Measuring and analyzing the characteristics of napster and gnutella hosts. Multimedia Systems 9 (2003) 170–184
2. Harada, Y., Kambayashi, Y.: Designing a resource discovery method based on multi-agents in p2p systems. In: Proceedings of the IADIS International Conference WWW/Internet, 2. (2006) 196–200
3. Maymounkov, P., Mazieres, D.: Kademlia: A peer-to-peer information system based on the xor metric. revised paper from the 1st international workshop on peer-to-peer systems. In: Lecture Notes in Computer Science 2429, Springer-Verlag, Berlin Heidelberg New York (2002) 53–65
4. Rowston, A., Druschel, P.: Pastry: scalable, distributed object location and routing for large-scale peer-to-peer systems. In: Proceedings of the 18th IFIP/ACM International Conference on Distributed Systems Platforms. (2001) 329–350
5. Stoica, I., Morris, R., Karger, D.: Chord: a scalable peer-to-peer lookup service for internet applications. In: Proceedings of the 2001 ACM SIGCOMM Conference. (2001) 149–160
6. Zhao, B.Y., Huang, L., Stribling, J., Rhea, S.C.: Tapestry: a resilient global-scale overlay for service deployment. IEEE Journal on Selected Areas in Communications **22** (1) (2004) 44–53
7. Castro, M., Costa, M., Rowstron, A.: Peer-to-peer overlays: structured, unstructured, or both? Technical Report MSR-TR-2004-73. Microsoft Research, Redmond (2004)
8. Wooldridge, M.: An Introduction to Multiagent Systems. John Willey, New York (2002)

9. Cameron, R.D.: Using mobile agents for network resource discovery in peer-to-peer networks. SIGecom Exchanges **2** (3) (2001) 1–9
10. Kambayashi, Y., Takimoto, M.: Higher-order mobile agents for controlling intelligent robots. International Journal of Intelligent Information Technologies **1** (2) (2005) 28–42
11. Mizuno, M., Kurio, M., Takimoto, M., Kambayashi, Y.: Flexible and efficient use of robot resources using higher-order mobile agents. In: Proceedings of Joint Conference on Knowledge-Based Software Engineering. (2006) 253–262
12. Shudo, K., Tanaka, Y., Sekiguchi, S.: Overlay weaver: an overlay construction toolkit. In: Proceedings of Symposium on Advanced Computing Systems and Infrastructures. (2006) 183–191, In Japanese
13. Satoh, I.: A mobile agent-based framework for active networks. In: Proceedings of IEEE System, Man and Cybernetics Conference. (1999) 71–76
14. Montresor, A.: Anthill: A framework for the design and analysis of peer-to-peer systems. In: Proceedings of the 22nd International Conference on Distributed Computing Systems. (2002) 15–22
15. Dasgupta, P.: Improving peer-to-peer resource discovery using mobile agent based referrals. In: Agents and Peer-to-Peer Computing, Lecture Notes in Computer Science 2872, Springer-Verlag, Berlin Heidelberg New York (2004) 186–197
16. Yang, B., Garcia-Molina, H.: Designing a super-peer network. In: Proceedings of the 19th IEEE International Conference on Data Engineering. (2003) 49–63

Agent Based Dynamic Data Storage and Distribution in Data Warehouses

Nader Kolsi[1], Abdelaziz Abdellatif[2], and Khaled Ghedira[3]

[1] Higher Institute of Business Administration of Sfax - Tunisia,
[2] University of Sciences of Tunis - Tunisia
[3] National School of Informatics Sciences, University Campus of Manouba, Tunisia
nader.kolsi@fsegs.rnu.tn, abdelaziz.abdellatif@fst.rnu.tn,
khaled.ghedira@isg.rnu.tn

Abstract. In this paper, we propose a new approach to manage data storage and distribution in a data warehouse (DWH) environment. This approach deals with the dynamic data distribution of the DWH on a set of servers. The data distribution that we consider is different from the "classical" one which depends on the data use. The distribution in our approach consists in distributing data when the server reaches his storage capacity limit. This distribution assures the scalability and exploits the storage and processing resources available in the organization using the DWH. It is worth noting that our approach is based on a multi-agent model mixed with the scalability distribution proposed by the Scalable Distributed Data Structures.

The proposed multi-agent model is composed of stationary agent classes: Client, Dispatcher, Domain and Server, and a mobile agent class called Messenger. These agents collaborate and interact to achieve automatically the storage, the splitting (distribution), the redirection and the access operations on the distributed DWH.

Keywords: Data warehouse, Data storage, Dynamic data distribution, Multi-agent system, Mobile agent.

1 Introduction

A Data warehouse is a principal component of the information systems in the organizations. It is defined by its inventor, W. H. INMON [1], as a collection of data which are subject-oriented, integrated, stamped, non-volatile, and used as a support of decision making. It is considered as a deposit of data that have been collected from heterogeneous and autonomous distributed sources. It is used for analytical tasks in business. The DWH usually contains a very large amount of data. This is because of the scope of the period that the DWH must cover (historical data) and the diversity of data sources from which data are extracted.

In fact, the DWH is the subject of many research works. This research issues are classed in five groups as shown in [2]: (1) data warehouse modeling and design, (2) data warehouse architectures, (3) data warehouse maintenance, (4) operational issues, and (5) optimization.

N.T. Nguyen et al. (Eds.): KES-AMSTA 2007, LNAI 4496, pp. 375–384, 2007.

Our research work focuses on the operational issues and optimization topics mainly, but also data warehouse architectures and design. Our work aims at solving the problems of storage space and performance through: (1) developing a dynamic system that can manage the DWH automatically (data storage, data distribution on a set of servers, and data access), (2) taking advantage of the storage and processing resources available in the organization (processors, memory, hard disks, etc.), (3) getting better data storage time, and (4) improving the query response time.

This paper is organized as follows: Section 2 gives an overview of related works and discusses the problems related to optimization topics. In section 3, we describe the proposed multi-agent model. Section 4 details the global dynamic of the data storage operation and the data splitting operation. Section 5 analyzes the experimental results. Finally, a conclusion and an outlook to future works are made in section 6.

2 Related Works

Most of researches in literature, that work in the optimization topics, propose solutions based on a centralized DWH. These researches propose several queries optimization techniques that can be classified in two categories [5]: (1) redundant structures, and (2) non-redundant structures. These two techniques are supported by the current database management systems (DBMS). The improvements, which are provided to these systems and concern the management of large data amount, are not sufficient to satisfy the needs due to the data amount growth of the DWH. In addition, the static data fragmentation schema, actually used in these systems, constitutes a major handicap. It is worth noting that, in our approach, we use the two techniques mentioned above.

So far, distribution of data warehouses has not attracted much attention in research. The use of DWH with distributed structure has appeared only with the data marts [10]. However, the data mart does not solve the problems of storage space and performance. It is basically stand-alone and has data integration problems in a global data warehouse context [6]. In addition, the performance of many distributed queries is normally poor, mainly due to the load balance problems.

We have to point out that, in our approach, the data distribution that we consider is different from the usually-used ones [9]. In fact, it is not defined at the design phase. However, it is imposed by the storage capacity. As a matter of fact, when a machine reaches its storage capacity limit, we add another one. Then, we distribute the data on the two machines to have a balanced load.

There are several ways to divide horizontally the relation. Typically, we can assign tuples to the processors in a round-robin fashion, we can use hashing, or we can assign tuples to the processors by ranges of values [6]. In [6, 7, 8], the authors use the Data Warehouse Striping technique. The round-robin distribution is simple to use and guaranties the load balancing. Although, it has two major disadvantages: (1) the queries will be executed in parallel by all of the machines (2) we must have machines with the same treatment and storage capacities. Otherwise, some machines will be too busy and the others will be under used.

We have to note that, in our approach, we use the range partitioning. So, the queries are executed in parallel not by all the machines but only by those that contain the necessary partitions. Furthermore, the data distribution is dynamic and automatic (see §3.1). Moreover, the number of used machines, in our approach, is not fixed. Therefore, the storage capacity of the DWH tends theatrically to the infinite because we can, at any moment, add dynamically other machines. This infinite storage capacity is by the principles of the Scalable and Distributed Data Structures (SDDS) [11]. The SDDSs deal with the storage of a large data amount on a set of interconnected machines. The SDDS principle consists in distributing the file contents in a way that allows us to benefit from the available memory on a set of interconnected machines [11, 12]. In the rest of this paper; we consider that the two terms "splitting" and "distributing" have the same significance.

According to the proposed approach, the DWH will be distributed on a set of machines. In this case, the data management needs the collaboration and the interaction between those machines in order to reply to the user's queries while assuring the parallel processing of these queries. Thus, we have chosen to use the Multi-Agent System (MAS) with the mobile agents as essential actors. The mobile agents have proved a high performance when we access to the data distributed on a set of interconnected machines [13].

In the following, we present the data distribution principle and the proposed multi-agent model.

3 Proposed Model

In this paper, we propose a model for solving the problems in the DWH context using the available resources in the organization. These problems are related to the data storage, splitting and access.

In our approach, the use of the MAS is very interesting because it allows assuring the progress of the dynamic data distribution, the collaboration, the interaction, the independency of the different machines, and the parallel execution of the user queries. In addition, the use of mobile agents in the proposed solution seems to be very helpful because it allows: (1) decreasing the network loads, (2) liberating client machines during the results preparation that needs generally a very important run-time, and, essentially, (3) securing the data that are transported in the network (see §5).

In the following, we present the principle of data distribution and the multi-agent model architecture.

3.1 Principle of Data Distribution

In our approach, the DWH is horizontally distributed on a set of machines that have the same DBMS and the same star schema (see figure 1).

The principle is to start with a single machine for which we define: (1) the storage capacity limit of this machine for which the used DBMS gives its highest performance (for data access and storage), and (2) both the inferior bound mark and the superior one for each fact table key. When this machine reaches its limit, we add another one automatically (without needing the administrator intervention) and we distribute the data on the two machines to obtain a balanced load. In most cases, the fact table (Sales) undergoes the splitting operation, because of its important volume.

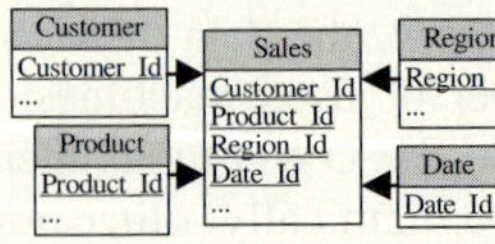

Fig. 1. Distributed Data Warehouse

The dimensional tables (Date, Region, Customer, Product) are distributed when their key constitutes a distribution criterion. Otherwise, they are duplicated.

In table 1, we present a scenario of data splitting. Machine 1 starts up the first splitting operation when it reaches its capacity storage limit. First, we search for the key value that gives two balanced partitions (e.g. Product_Id that is an integer of two numbers). Then, we move the data, related to the new interval, to machine 2. Finally, we update the intervals. The second splitting operation is lunched by machine 2 (e.g. Date_Id that is a date). The same process is started at each time when one machine reaches its limit capacity. In fact, the data distribution can be continued according to the same criteria or to other ones (Customer_Id, Region_Id).

Table 1. Splitting Scenario

	Start	First Splitting			Second Splitting		...
	Machine 1	M1	M2	M1	M2	M3	...
Customer_Id	[A, Z]	[A, Z]	[A, Z]	...	[A, Z]	[A, Z]	...
Product_Id	[0, 99]	[0, 50]	[51, 99]		[51, 99]	[51, 99]	
Region_Id	[AA, ZZ]	[AA, ZZ]	[AA, ZZ]		[AA, ZZ]	[AA, ZZ]	
Date_Id	[Jan, Dec]	[Jan, Dec]	[Jan, Dec]		[Jan, Jun]	[Jul, Dec]	

We notice that each SALE table record belongs to only one DWH partition. If we consider that each of these DWH partitions is stored in separate databases, we must, on the one hand, split the Date table and Product table according to the same criteria used for the SALES table. On the other hand, we duplicate the other tables in order to (1) facilitate the checking of the integrity constraints, (2) ensure the databases autonomy, and (3) improve the join time when we access to data.

3.2 The Proposed Multi-agent Model

The proposed model consists of five static agent classes (Client, Dispatcher, Splitting, Domain and Server) and a mobile agent class (Messenger).

Each Client agent has the Dispatcher agent as an acquaintance. Its static knowledge is made up of its name and its address. This agent class does not have a dynamic knowledge.

The acquaintances of the Dispatcher agent are: (i) the Client agents, (ii) the Messenger agents, and (iii) the Splitting agent. Its static knowledge consists of its name and its address. Its dynamic knowledge is made up of: (1) a list containing all the Domain agents existing in the system (2) a waiting queue used to store operations received from the Client agents, and (3) a waiting queue used to store the results provided by the Messenger agents.

Each Messenger agent has as acquaintances the Dispatcher agent and the Domain agents necessary to execute the operation. Its static knowledge is made up of its name and its maximum size of data that it can transport. This maximum depends on the network characteristics. The Messenger agent dynamic knowledge consists of: (i) the list of Domain agents to visit for executing the operation, (ii) the operation to execute, (iii) the lists of data to store (storage operation), or the list of data that are collected from visited Domain agents (access operation), and (iv) the size of transported data. It has a very important role in our architecture because it allows: (1) reducing the message traffic on the network, (2) accelerating the data storage and access operations, and, essentially, (3) securing the data circulation on the network (see §5).

The Domain agent has as acquaintances: (i) the Server agents that are under its control, (ii) the Messenger agents with which it has operations to execute and (iii) the Splitting agent. Its static knowledge is composed of its name, its address, the disk space limit of each Server agent, the maximum number of Server agent it can manage and the maximum size of data it can receive from the Messenger agents. This maximum depends on the machine characteristics (memory, processor, etc...). Its dynamic knowledge consists of: (1) the descendant list, (2) the size of memorized data, (3) a waiting queue used to store the operations brought by the Messenger agents, and (4) a waiting queue used to store the replies sent by the Server agents.

Each Server agent has the Domain agent to which it belongs as acquaintances. Its static knowledge is made up of its name and its address. Its dynamic knowledge is a waiting queue used to store the operations received from the Domain agent.

The Splitting agent has as acquaintances the Dispatcher agent and the Domain agents that ask for splitting. Its static knowledge consists of its name and its address. Its dynamic knowledge is the list of splitting requests sent by the Domain agents.

The Dispatcher agent manages a metabase which allows it to follow the evolution of the data distribution on the Domain agents, the network status and the Messenger agents load rate. This metabase is also used by the Messenger agents to make the execution plans of the received operations and determine the Domain agents to visit. The Splitting agent, also, uses this metabase for the splitting operations and updates it at the end of each splitting operation (see §4.2).

Furthermore, Each Domain agent has an appropriate metabase in order to follow the evolution of the data distribution on its descendants (Server agents).

In the following, we will detail the multi-agent dynamic for the data storage and data splitting operations.

4 Multi-agent Dynamic

The proposed model is designed to support the different management operations of data warehouse, namely the data storage, splitting, redirection and access. In this paper, we present only the data storage and splitting operations.

4.1 Data Storage

We will present two scenarios for this operation. In the first, we will not use the Messenger agents. In the second, we will benefit from these mobile agents and prove their utility. For the two scenarios, we suppose that the data come in a file form (INSERT commands).

In the first scenario, we suggest the use of a waiting structure that we call waiting database (WDB). In the later, we store temporarily the data coming in a file form. Then, we distribute them on the different machines according to the used distribution criteria. Finally, the data of the facts table will be deleted from of the WDB and we keep the data of the dimensional tables. This WDB has the same star schema as the distributed DWH and it is managed by the Dispatcher agent.

In the two cases described later, we will not consider the data redirection, splitting and rejection.

4.1.1 Scenario 1

In this scenario, the data storage operation is started up when the Dispatcher agent receives the data file. This agent stores the data in the WDB. Then, it finds the Domain agents to which the received data belong. The Dispatcher agent uses the available information in its metabase to determine these agents and their addresses, and make for each Domain agent its proper queries.

At each time when the Dispatcher agent formulates the queries, it sends them to the appropriate Domain agent. When this latter receives these queries, it verifies, for each query, whether the clause WHERE belongs to the Server agents which are under its responsibility. If this condition is true, the Domain agent sends this query to the appropriate Server agent. Otherwise, referring to the available information in its metabase, the Domain agent forwards the query to the right Domain agent. This query redirection occurs when a splitting operation happens before the query's arrival.

The Server agent executes the received query to load the appropriate data existing in the WDB. Then, it formulates a DELETE query with the same WHERE clause to clear the fact table in the WDB. Finally, the Server agent informs the responsible Domain agent that the data loading is successfully done. The Server agent starts up the splitting operation, if it detects, at the data loading time, that the machine has reached its storage capacity limit.

When the Domain agent receives all the replies from the Server agents, it informs the Dispatcher agent that the data storage operation is successfully terminated.

When the Dispatcher agent receives all the replies from the Domain agents, it informs the administrator that the storage operation is achieved.

4.1.2 Scenario 2

In this scenario, when the Dispatcher agent receives the data file from the administrator, it creates a Messenger agent. The latter reads the data file each record separately, and composes the appropriate data list for each Domain agent. It uses the available information in the metabase to perform this process. Meanwhile, the Messenger agent makes up the address list of the visited Domain agents, according to the data lists that it can bring. Generally, the Messenger agent can not load the entire data file in order to avoid burdening the machine memory and/or the network. So, the loaded data lists size must not be superior to the maximum loading capacity of the Messenger agent. This capacity is relative to the machine and the network characteristics. If the Messenger agent reaches its capacity limit and can not load all the records in the data file, it informs the Dispatcher agent. The latter creates another Messenger agent that repeats the same process. The Dispatcher agent generates the necessary number of Messenger agent to load all the data file records.

Each Messenger agent visits the Domain agents for the first time to distribute data lists. After each visit, the Messenger agent deletes the distributed data list and updates its transported data size by subtracting the data list size. Before its first visit to each Domain agent, the Messenger agent requests its permission. The Domain agent allows the Messenger agent to visit it, when the sum of the data size, in its memory, and the size of the data lists transported by the Messenger agent, is inferior to the maximum data size that can be carried by the Domain agent. Otherwise, the Messenger agent asks the next Domain agent in the address list for its permission. If all the Domain agents cannot receive the Messenger agent, it must wait until one of these agents is discharged. Once distributing the data lists on the Domain agents, the Messenger agent visits them again to verify whether the storage operation is successfully done. When it collects all the Domain agents replies, it returns to the Dispatcher agent and informs it that the storage operation is successfully accomplished.

When receiving the data lists from the Messenger agent, the Domain agent updates its memorized data size by adding the size of the received data lists. Then, it verifies whether the data list belongs to the Server agents which are under its responsibility. If this condition is true, the Domain agent sends these data to the appropriate Server agents. Otherwise, the Domain agent forwards this data list to the right Domain agent. The last case occurs when a splitting operation happens before the data arrival. After sending or forwarding the data list, the Domain agent updates the size of its memorized data by subtracting the data list size.

The Server agent stores the received data list and informs the responsible Domain agent that the storage of the data list is successfully accomplished. The Server agent starts up the splitting operation, if it detects, while storing the data, that the machine has reached its storage capacity limit.

The Domain agent informs the Messenger agent that the storage of the received data lists is terminated, when it receives all the replies from the Server agents.

When the Dispatcher agent receives all the Messenger agents replies, it informs the administrator that the storage operation is successfully achieved.

4.2 Data Splitting

The splitting operation will be started up when the Domain agent detects that the available space at the Server agent cannot support entirely the data amount received at the storage operation. Each Domain agent is characterized by a maximum number of Server agents that it can control. If the Domain agent does not reach this number, the splitting operation will be managed by the split Domain agent and a new Server agent will be created. Otherwise, the splitting operation will be managed by the Splitting agent and a new Domain agent will be created.

The afore-mentioned agents exchange different messages in order to achieve the splitting operation which involves the creation of a new Domain agent.

When the Server agent detects that the available space can not support the total amount of the received data, it sends a message informing the Domain agent that it reaches its capacity limit and it needs to split its data. This message contains, also, the data that are not inserted. Then, The Domain agent sends a message to the Splitting agent. This message contains the new values of both the superior bound mark and the inferior one. To determine these bounds, the Domain agent computes its records and determines the key values which allow dividing these records into two balanced parts.

When the Splitting agent receives the spilling request, it informs the Dispatcher agent that the Domain agent *i* will start up a splitting operation. Thus, the Dispatcher agent stops momentarily the operations sent this agent. The Splitting agent is responsible for preparing the new Domain agent. Once the new agent is created, the Splitting agent informs the Domain agent asking for splitting of the new-created agent address as well as the bound marks of each dimensional table. According to these bound marks, the split Domain agent selects the data from the Server agents and sends them to the Messenger agent. Then, it informs the Splitting agent that the splitting operation is terminated and updates its descendants list. The Messenger agent moves to the new Domain agent and gives it the received data. This new agent achieves the storage operation as described in § 5.1.2.

When the Splitting agent is informed that the splitting operation is terminated, it sends message informing the Dispatcher agent that the new Domain agent and the split Domain agent are ready to receive operations. So, The Dispatcher agent updates its domain agents list according to the received information.

In the following, we present the results obtained for the data storage operation.

5 Experimental Evaluation

In order to validate our model for the data storage operation, we have implemented four prototypes. One of them permits to store data on a centralized database (DB). The others allow storing data on a set of machines. In fact, we have made the experiences using, as described below, one machine that contain the WDB and/or the MB and three (then five) machines that contain the DWH partitions. These machines have the same configuration: P4 and 256Mo (RAM). We have used JDeveloper10g as a development toolkit, Oracle as a DBMS, and IBM Aglets as a multi-agent platform. We have programmed an engine that generates the data in a file form (INSERT commands). These data represent only the daily or weekly DWH refreshment.

In the first prototype, we have used one machine and we have programmed an insertion engine which reads the data from the file and inserts them on a centralized DWH.

In the second prototype, we have programmed an insertion engine, without MAS, that reads the data from the file and inserts them on the distant DWH partitions using the database links (@DWH1,…, @DWHN, etc.).

In the two last prototypes, we have programmed the first and the second scenarios. In these prototypes, the machines are used as follows: (1) at one of these machines we have made the Dispatcher agent, the metabase (MB) and the WDB (prototype 3) or the messenger agent (prototype 4), and (2) at each of the other N machines, we have made a Domain agent, a partition of the DWH database (DWHi), a metabase and a Server agent.

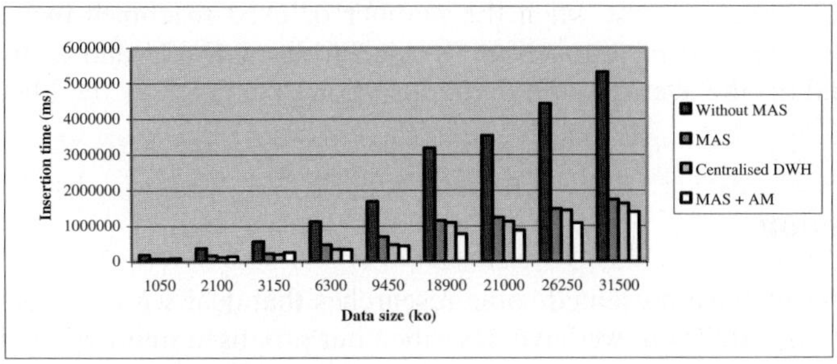

Fig. 2. Experimental results using three machines

In figure 2, we present the insertion time (ms) given by the four prototypes using tree machines. We notice that the time needed to insert the same data size in a centralized DWH is less than the time needed to insert data in distributed DWH without using MAS (an average gain of 67%). In fact, the data transfer on the network is behind this difference. So, when we have used the MAS and the WDB, to distribute the data, in the third prototype, we have reduced the communication time between machines. Thus, the time needed in case of a distributed insertion is reduced significantly (an average gain of 61%). It becomes close to the time needed in case of a centralized DWH (plus an average of 18% to the centralized prototype). This gain is thanks to: (1) the local data storage in the WDB, and (2) the parallel data loading. The use of the Messenger agents, that are mobile, in the forth prototype, not only gives the best storage time (-4.8% from the centralized prototype), but it also secures the data circulation on the network. The time reduction results from reducing the read/write time and storing the data directly in their final destination.

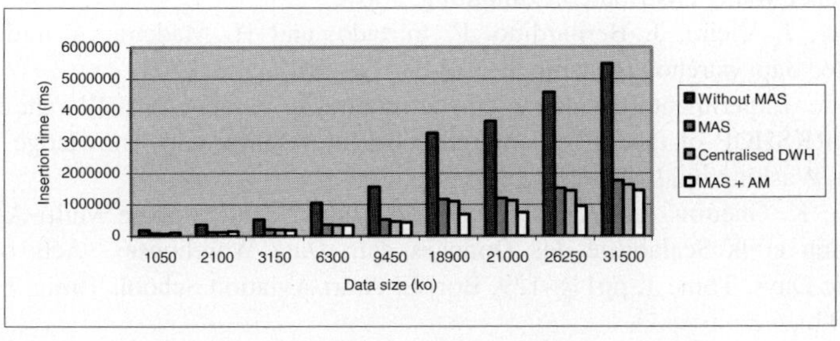

Fig. 3. Experimental results using five machines

Figure 3 gives the same measurements for the four prototypes using the same data size and five machines. We obtain the same improvement when using the MAS and the Messenger agents. Furthermore, the difference between the insertion time in the first prototype and the third one becomes an average of 6.5% (with tree machines an average of 18%). In addition, the average gain between the first prototype and the forth one (MAS with MA) becomes 6% (with tree machines an average of 4.8%).

So, we can conclude that when the number of used machines increases: (1) the difference between the insertion time in a centralized DWH and a distributed one decreases, and (2) the average gain in the insertion time gives by the Messenger agent increases.

6 Conclusion

In this article, we have presented some researches that deal with the data distribution in the DWH context. Then, we have described our proposed multi-agent model and its global dynamic concerning the data storage operation. Finally, we have demonstrated the improvements obtained when we have used the MAS and the Messenger agents in the data storage operation. These contributions consist in: reducing of the data storage time and securing of the data that circulate on the network.

As future works, our future works aims at defining the multi-agent dynamic for the data access operation. As well as, we will implement this operation and compare the given results to those obtained for the Benchmark used in the literature.

References

1. W. Inmon: Building the Data Warehouse, QED Technical Publishing Group, 1992.
2. M. Wu and A. Buchmann, Research Issues in Data Warehousing, BTW'97, March 1997.
3. Kimball, R. (1996). The Data Warehouse Toolkit. New York: J. Wiley & Sons.
4. Kimball, R., Reeves, L., Ross, M., and Thornthwalte, W. (1998). The Data Warehouse Lifecycle Toolkit.New York: J. Wiley & Sons.
5. L. Bellatreche and K. Boukhalfa. An Evolutionary Approach to Schema Partitioning Selection in a Data Warehouse. DAWAK'2005.
6. J. Bernardino and H. Madeira. Data Warehousing and OLAP: Improving Query Performance Using Distributed Computing. 2001.
7. M. Costa, J. Vieira, J. Bernardino, P. Furtado, and H. Madeira. A middle layer for distributed data warehouses using the DWS-AQA technique. 2003.
8. P. Furtado. Experimental Evidence on Partitioning in Parallel Data Warehouses. DOLAP 04 -WORKSHOP of the Int'l Conference on Information and Knowledge Management (CIKM), Washington, November-2004.
9. N. Kolsi, K. Ghedira, and A. Abdellatif. Utilisation d'un Système Multi-Agents pour la Répartition et la Scalabilité des Données d'un Data Warehouse. Acts of the Fourth Scientific Days, Tome 1, pp118-129, Borj El Amri Aviation School, Tunis, Tunisia, 21-22 May 2003.
10. Informatica white paper. Enterprise-Scalable Data Marts: A New Strategy for Building and Deploying Fast, Scalable Data Warehousing Systems.http://www.informatica.com, 1997.
11. W. Litwin, M. A. Neimat, and D. Schneider. RP*: A Family of Order-Preserving Scalable Distributed Data Structures, 20th Intl. Conf. On very Large Data Bases VLDB, 1994.
12. Diene and W. Litwin. Performance Measurements of RP*: A Scalable Distributed Data Structure for Range Partitioning. Int. Conf. on Information Society in the 21st Century: Emerging Techn. and New Challenges. Aizu City, Japan, 2000.
13. J. Arcangeli, A. Hameurlain, F. Migeon, and F. Morvan. Apport des agents mobiles à l'évaluation et l'optimisation de requêtes bases de données réparties à grande échelle. Technical Report, laboratory IRIT, Université Paul Sabatier, 2002.

Semantic Data Integration in P2P Environment Using Schema Mappings and Agent Technology*

Grażyna Brzykcy[1], Jerzy Bartoszek[1], and Tadeusz Pankowski[1,2]

[1] Institute of Control and Information Engineering,
Poznań University of Technology, Poland
[2] Faculty of Mathematics and Computer Science,
Adam Mickiewicz University, Poznań, Poland
tadeusz.pankowski@put.poznan.pl

Abstract. We discuss the problem of semantic data integration in a highly-dynamic environment consisting of a community of peer-to-peer cooperating agents (partners). Peers decide when to join and when to leave the system, how to describe their local data, when to communicate and share their resources with partners. An agent issues queries to its partners (friends) which are able partly answer the query and then propagate the query to their partners along semantic paths existing in the system. Semantic paths are determined by schema mappings defined between partners. We propose a method for specifying schema mappings and to translate them to XQuery expressions. Mappings are represented by means of logical formulas. We also propose a declarative specification of semantic-driven communication in the system. The specification is made in a peer-oriented extension of Prolog.

1 Introduction

Information integration plays the central role in building of large scale systems of P2P databases and a new generation of internet applications, where data comes from many different sources with different schemas [4,9,2,13,12].

In this paper we address the following issues that are crucial for semantic data integration in P2P environment: *(1) Schema mappings between peer's schemas of local data repositories.* Schema mappings are specifications describing how data structured under one schema (the source schema) is to be transformed into data structured under another schema (the target schema). Schema mappings can be used in both *data exchange* and *query reformulation.* In the former, the data is to be restructured using the mappings between schemas [5,1,9]. In the latter, a query formulated under the target schema is to be translated (reformulated) into a query under the source schema [6,15]. In both cases the semantics of the data should be preserved that enables semantic interoperability between peers [3]. *(2) Semantic communication between peers and between peers and the broker.* Semantic communication is the problem of carrying out the communication

* The work was supported in part by the Polish Ministry of Science and Higher Education under Grant N516 015 31/1553.

N.T. Nguyen et al. (Eds.): KES-AMSTA 2007, LNAI 4496, pp. 385–394, 2007.

based on the agreed meaning of the task that is to be performed by cooperating peers (partners). In this paper, such tasks are queries that should be answered by a community of cooperating peers (agents).

The novel scientific contributions of this paper are as follows:

1. We propose a method for specifying schema mappings by means of tree-pattern formulas and show that such mappings can be translated into XQuery queries producing an instance of the target schema for a given instance of the source schema. The method extends proposals developed in [1,15] and in our previous work [10].
2. We develop a consistent set of rules modeling semantic communication in the system of semantic data integration in P2P environment. We propose a declarative high-level notation to describe cooperation between agents (peers and the broker) involved in the process of data integration. The notation is based on a simple extension of Prolog, called *LogicPeer* proposed in [7].

From the practical point of view, the logical representation of mappings may be used to check formal properties of them and to determine implied mappings that are not explicitly represented in the system [8]. The model of semantic communication supports a rapid prototyping of the system.

In Section 2 schema mappings are discussed. The agent-based architecture of the P2P semantic data integration system is proposed in Section 3. In Section 4 a declarative specification of semantic communication between agents is described. Section 5 concludes the paper.

2 Schema Mappings

In Fig. 1 there are sample XML schema trees, S_1 and S_2, as well as their instances, where I_1 is an instance of S_1; I_2 and I_2' are two instances of S_2. The data represents the bibliographical data, where node labels are as follows: paper (P) and title (T) of the paper; author (A), name (N) and the affiliation (U) of the author; year (Y) of publication of the paper.

To establish correspondence between S_1 and S_2, we can specify the following schema mapping from S_1 to S_2:

$$M_{1,2} := /S1/P[T = x_T \wedge A[N = x_N \wedge U = x_U]] \\ \Rightarrow /S2/A[N = x_N \wedge U = x_U \wedge P[T = x_T]]$$

In general, schema mappings are specified by means of expressions of the form

$$\forall \mathbf{x}(\pi(\mathbf{x}) \Rightarrow /top/\sigma(\mathbf{x})), \tag{1}$$

where: (1) $\mathbf{x}$ is a tuple of text-valued source variables; (2) π and $/top/\sigma$ are tree-pattern formulas [1,10] conforming to the following syntax (L is a set of labels, $l \in L$, *top* is the outermost label):

$$\pi ::= /\sigma; \quad \sigma ::= P[E]; \quad E ::= x \mid P = x \mid \sigma \mid E \wedge E; \quad P ::= l \mid P/l$$

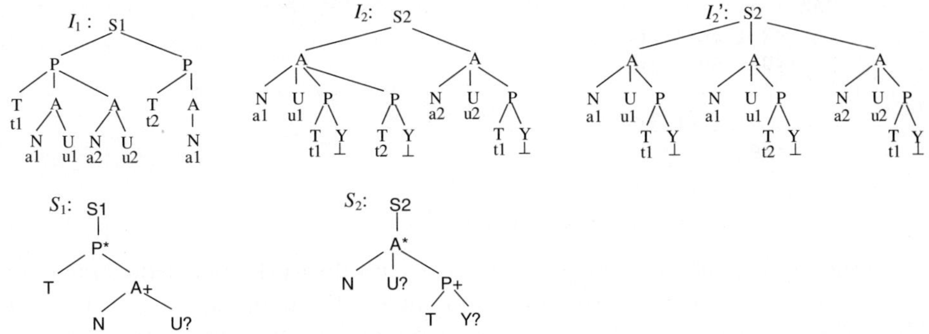

Fig. 1. Sample XML schema trees and their instances

Transformation of a schema mapping expressed by the formula (1) into a query in XQuery ([14]) can be described by the following rules:

$$Tr(\pi \Rightarrow /top/\sigma) := <top>\{\tau_\pi(\pi)\ \mathbf{return}\ \kappa_\sigma(\sigma)\}</top>$$

where:

$$\begin{array}{ll}
1.\ \tau_\pi(/P[E], var_1) & = \mathbf{for}\ \$var_1\ \mathbf{in}\ /P, \\
 & \quad \tau_E(var_1, E), \\
2.\ \tau_\sigma(var_1, P[E], var_2) & = \$var_2\ \mathbf{in}\ \$var_1/P, \\
 & \quad \tau_E(var_2, E), \\
3.\ \tau_E(var_1, x) & = x\ \mathbf{in}\ \$var_1/text(), \\
4.\ \tau_E(var_1, P = x) & = x\ \mathbf{in}\ \mathbf{if}\ (var_1[P])\ \mathbf{then}\ var_1/P/text()\ \mathbf{else}\ "null", \\
5.\ \tau_E(var_1, \sigma) & = \tau_\sigma(var_1, \sigma, var_2), \\
6.\ \tau_E(var_1, E_1 \wedge E_2) & = \tau_E(var_1, E_1), \\
 & \quad \tau_E(var_1, E_2),
\end{array}$$

$$\begin{array}{ll}
1.\ \kappa_\sigma(l[E]) & = <l>\kappa_E(E)</l> \\
2.\ \kappa_\sigma(l/P[E]) & = <l>\kappa_\sigma(P[E])</l> \\
3.\ \kappa_E(x) & = \{\$x\} \\
4.\ \kappa_E(l = x) & = <l>\{\$x\}</l> \\
5.\ \kappa_E(l/P = x) & = <l>\kappa_E(P = x)</l> \\
6.\ \kappa_E(\sigma) & = \kappa_\sigma(\sigma) \\
7.\ \kappa_E(E_1 \wedge E_2) & = \kappa_E(E_1) \\
 & \quad \kappa_E(E_2)
\end{array}$$

Using these rules to the mapping $M_{1,2}$ we obtain the following query in XQuery language:

```
<S2> {
   for $v1 in doc("i1.xml")/S1/P,
       $t in if ($v1[T]) then $v1/T/text() else "null",
       $v2 in if ($v1[A]) then $v1/A else "null",
         $n in if ($v2[N]) then $v2/N/text() else "null",
         $u in if ($v2[U]) then $v2/U/text() else "null"
       return
```

```
        <A>
            <N>{ $n }</N>
            <U>{ $u }</U>
            <P>
              <T>{ $t }</T>
            </P>
        </A> }
</S2>
```

For the input XML document I_1, the query produces the result document I'_2 (see Fig. 1) - the *canonical solution* to I_1 under the mapping $M_{1,2}$. Unknown values of Y ale replaced with null values denoted by $\perp$. In order to specify which of possible instances of the target schema should be produced, we can apply *automappings* over the schema. The automapping is a *key-pattern* formula and captures *keys* defined in the schema. Automappings and their applications to management of schema mappings, we have investigated in [9,10].

Specification of schema mappings is a crucial problem in data integration systems. They are usually defined manually or using quasi-automating methods [11]. Once introduced into the system, mappings can be used for many purposes. The most important application is data exchange and query reformulation in semantic data integration.

3 Peer-to-Peer System for Semantic Data Integration

We assume that the system for semantic integration of XML data (SIX-P2P), currently under development in Poznań University of Technology, consists of autonomous agents (peers) each of which can independently decide how to structure its local data. The data is described by a peer's local schema. The schemas reflect possibly heterogeneous semantic models that are developed by different peers.

An agent in SIX-P2P system sees a set of another agents, its partners (peers), and may ask queries only to these partners. However, a query may be propagated to partners of each peer inducing a significant extension of the set of possible "knowledge sources". So, cooperative query evaluation is performed also by agents indirectly connected to the enquirer.

We make the following assumptions about agents in SIX-P2P system:

1. Each peer is identified by a unique name and represents some user. We do not impose any particular format for agent identifiers, but assume that system has its own namespace and name mapping mechanism.
2. An agent does not know all the agents of the system. The subset of peers which identifiers are known to the agent forms a group of its partners (friends).
3. To join the system an agent has to introduce itself to a special agent - the broker. Process of registration consists in checking agent reliability (certification) and conveying to it a list of its partners. To abandon the system an agent ought to let the broker know. This information is then passed by the broker

to the interested set of agents. Due to the system openness agents can join the system and abandon it, the partner lists should then be modified accordingly.
4. Each agent has a local collection of data and schemas of this data. To posses new information an agent can generate queries to its partners. Conversely, asked by trustworthy peer an agent tries to answer the query, also by means of asking its own friends.
5. Agents can communicate by sending messages among each other and by using peer identifiers.

The general architecture of a peer is depicted in figure 2. Each peer (P) has its own local data store (LDS) managed by a local management system (LDM). There is a query interface (QI) for accepting queries and returning answers during interactions with other peers. Distributed query manager (DQM) is responsible for planning execution of a received query using P's own LDS and propagating the query to its partners. Partial results are merged and returned to the enquiring user. The metadata necessary to understand the query and to plan its execution are managed by metadata manager (MDM). Information about partners as well as rules defining integration strategy and reconciliation actions are managed by semantic integration manager (SIM).

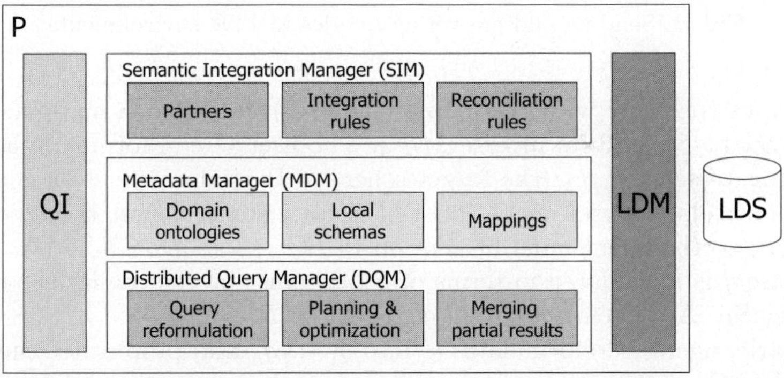

Fig. 2. Architecture of a peer in SIX-P2P

There is a broker in the SIX-P2P system (Fig 3). The broker is responsible for registration and certifications of the peers (R&C). The broker fulfils also some operations over all metadata existing in the system. The global metadata manager (GMM) can reason over schema mappings inferring compositions of mappings, inversions, or inconsistencies. It can map local ontologies and/or supports creation of the global ontology. For this aim global repositories of schemas, mappings and ontologies may be maintained.

The scenario of semantic data exchange between peers in P2P environment is shown in Fig. 4:

- Agent A wants to send query to B, so asks B for the schema S_B of its source data.

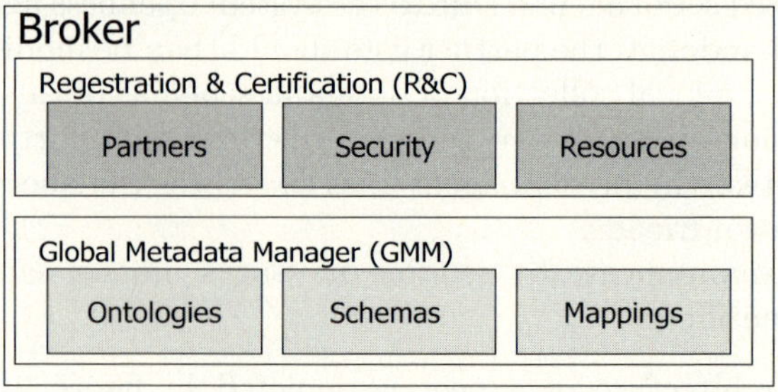

Fig. 3. Architecture of the broker in SIX-P2P

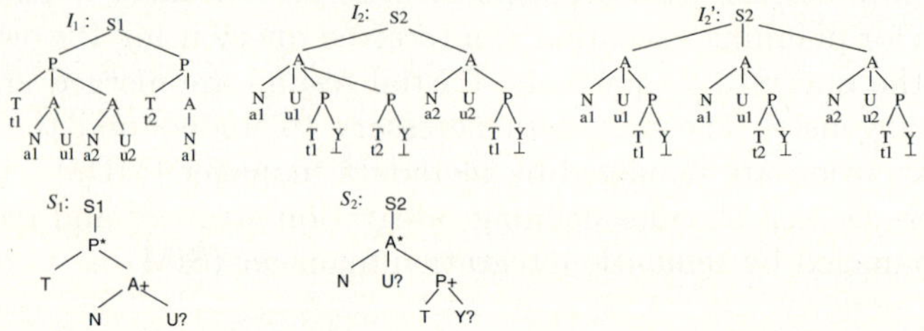

Fig. 4. Sending and answering queries in P2P environments

- A creates (possibly with the help of the user) two schema mappings: $M_{A,B}$ from S_A to S_B and its inverse $M_{B,A}$. The user U can formulate a query q in terms of schema S_A (the target schema). Partial answer to q is obtained from local data stored in the peer of A, and some answer is also expected from B, so the query must be sent on to B.
- Because q is formulated in terms of S_A, it must be reformulated to a query q' over S_B. After reformulation, q' is sent to B.
- Similarly, agent B reformulates q' into $q'_1, ..., q'_n$ and propagates them to its partners $C_1, ..., C_n$, respectively. After receiving the answers from its partners, B merges them into data d' structured under S_B, so $d' = ans(q')$ is the answer to q'.
- Finally, d' must be restructured under the schema S_A using the mapping $M_{B,A}$, $d'' = M_{B,A}(d')$.

4 Declarative Specification of Peer-to-Peer Data Integration

In system specification, we apply the declarative programming paradigm and logic programming language Prolog to provide a high level of abstraction for knowledge representation and processing. Prolog is used as a specification notation and is aimed at development of expressive models. Moreover, Prolog is also a rapid prototyping language with metaprogramming techniques, pattern matching and backtracking search. All the features are suitable to tackle the querying and reasoning with semantics, problems that are the main objective of the Web and peer-to-peer systems.

We take advantage of a simple extension of Prolog, so called LogicPeer, which is presented in [7]. In this model it is possible for an agent to send a query (goal) to be evaluated at a specific peer. We can use goals: `PeerID * goal` where `PeerID` is an agent identifier. The special name `self` is reserved for a local peer.

We also employ the underlying protocol Gnutella as the mechanism for propagating Prolog goals among peers. Together with special control tags the protocol prevents loops between peers and supports optimization techniques of goal evaluation. It is worth noticing that a peer working as a propagator can convert a query into another query. With relation to replies, they are directed back along the same path as the query.

An agent has a collection of Prolog facts and rules. Some facts store details of architecture (e.g., partners, mappings) and rules define evaluation of locally initiated goals or queries received from other peers.

There are two types of agents in SIX-P2P system: peers (Fig. 2) and brokers (Fig. 3). An environment of each peer consists of its partners (with their schemas and data), requesting peers and a broker, whereas the broker sees all the peers (their identifiers). Each peer can perform three actions directed to a broker, namely introduction (`introduce/3`), exit (`log_out/2`) and modification (`modify_schema/1`). Introduction of the `Agent` to the `Broker` consists of registration, which is an action executed by the `Broker` (a broker identifier prefixes the goal `register(Agent, Parts)`), and results in replying a list of `Agent`'s partners. The list is stored by the `Agent` as a part of its architecture.

```
introduce(Agent, Broker, Parts) :-
  Broker * register(Agent, Parts),  % registration is done by the broker
  assert(partners(Parts)).          % list of partners is stored as fact
```

Action of leaving the system by the `Agent` contains a message to the `Broker`.

```
log_out(Agent, Broker) :-
  Broker * log_out(Agent).          % log_out is done by the broker
```

When the `Agent` determines to restructure its data it have to modify its schema (`modify_schema/3`) locally and to inform via `Broker` other peers to which the `Agent` is a partner. Modification done by itself (`modify/2`) completes when a new schema replaces the old one in a store (Local schemas in Fig. 2) and new mappings (`create_maps/1`) would be created.

```
modify_schema(Agent, Broker, S1, S2) :-
  modify(S1, S2),                   % agent modifies schema locally
  Broker * modify_schema(Agent).  % broker is informed about modification
modify(S1, S2) :-
  retract(schema(self, S1),  % the old schema is removed
  assert(schema(self, S2),   % the new schema is stored
  create_maps(Maps).         % new mappings are created
```

There are also two actions in the system initiated by the broker and directed to the agent. The first one is executed when the broker informs about schema modification (`modified_schema/1`) reported by the `Partner`. Then the old schema and the old mappings are removed, the Partner is asked about its schema and new mappings are created.

```
modified_schema(Partner):-
  retract(schema(Partner,_)),              % the old schema is removed
  retract(mappings(Partner, _, _)),        % the old mappings are removed
  Partner * schema(self, Schp),            % asking about partner's schema
  assert(schema(Partner, Schema)),         % the new schema is stored
  schema(self, Scha),                      % agent's schema
  map(Scha, Schp, Mpa, Map),               % mappings from and to the partner
  assert(mappings(Partner, Mpa, Map)).     % mappings are stored as fact
```

The second action (logged_out/1)consists in updating the list of partners, if one of the partners, the Agent, has just logged out of the system.

```
logged_out(Agent):-
  partners(Parts),
  a_remov(Agent, Parts, Parts1), % the agent is removed from the partners
  retract(partners(Parts)), assert(partners(Parts1)).
```

Before asking any partner an agent has to prepare suitable mappings between schemas. The Agent tries to build mappings for all the partners (create_map/3). Since it is not assured that such a mapping exists (is constructed) the special value null is chosen to depict the situation. A mapping from the Partner is denoted as Mpa and from the Agent to the Partner as Map.

```
create_maps(Maps) :-
  partners(Parts),              % agent's partners
  create_pmaps(Parts, Maps)     % mappings from and two the partners
create_map(Part, Mpa, Map) :-
  schema(self, Scha),                 % agent's schema
  Part * schema(self, Schp),          % schema is taken from the partner
  map(Scha, Schp, Mpa, Map),          % mappings from and to the partner
  assert(mappings(Part, Mpa, Map)).   % mappings are stored as fact
```

A process of passing queries and receiving answers (see Fig. 4) is specified as an action (ask/2). It is built of local query (query/2), partners' asking (ask_partners/2) and answers' merging (merge/3).

```
ask(Agent, Query, Answer) :-
  query(Query, Ansl),               % local query is answered
  ask_partners(Query, Ansr),        % collective query is answered
  merge([Ansl], Ansr, Answer).      % answers are merging
```

An action of partners' asking is proceeded by selection (choose/1) of qualified partners - those for which both mappings are constructed.

```
ask_partners(Query, Ans) :-
  choose(QParts),                   % qualified partners are chosen
  ask_qparts(Query, QParts, Ansr),
  merge([ ], Ansr, Ans).            % answers from partners are merged
choose(Qparts) :-
  set_of(Part,(mappings(Part, Mpa, Map), Mpa ? null, Map ? null), QParts).
```

To answer the Query by acquiring data from its partners an agent may convert (convert/3) the original Query into more specialized query Qp, directed to the suitable partner P.

```
ask_qparts(_, [ ], _).                    % all the partners are asked
ask_qparts(Query, [P|Ps], [A|As]) :-
  convert(Query, P, Qp),                  % query converted for the partner
  mappings(P, Mpa, Map),
  q_reformulate(Qp, Map, Qp1),            % query transformed by the mapping
  P * ask(Qp1, Ap),                       % query answered by a partner
  a_reformulate(Ap, Mpa, A),              % answer transformed by the mapping
  ask_qparts(Query, Ps, As).
```

The broker remembers in prolog-like database all registered agents in the system and all partners of these agents. It uses its own internal criteria to select partners of the new agent.

```
register(Agent, Parts):-
  agents(As), append(As, [A], As1), retract(agents(As)),
  assert(agents(As1)),                  % the new agent is stored
  choose(A, As, Parts),                 % partners are selected
  assert(partners(A, Parts)).           % partners are stored
```

If the agent's schema has been changed, the broker sends this message to all agents that may use this schema.

```
modify_schema(Agent):-
  set_of(A, (partners(A, Parts),
  member(Agent, Parts)), As),           % agents which may use changed
  s_inform(As, Agent).                  % schema are selected and inform
s_inform_all([], _).
s_inform_all([A | As], Agent):-
  A * modified_schema(Agent), s_inform_all(As, Agent).
```

If an agent wants to log out, the broker informs about it all agents which may cooperate with this agent. Internal broker's data (i.e. the list of all registered agents, the list of agent's partners) is also updated.

```
log_out(Agent):-
  agents(As),
  a_remove(Agent, As, As1),                  % the logged out agent is removed
  retract(agents(_)), assert(agents(As1)),
  set_of(A, (partners(A, Parts),             % agents which cooperate with
         member(Agent, Parts)), As2),        % the removed agent are selected
  l_inform_all(As2, Agent),                  % and informed
  retact(partners(Agent, _)), p_remove(Agent).
```

The broker checks all partners' lists and removes the logged out agent.

```
l_inform_all([ ], _).
l_inform_all([A | As], Agent):-
  A * logged_out(Agent), l_inform_all(As, Agent).
```

5 Conclusions

We discuss some theoretical problems related to semantic data integration in peer-to-peer systems. We focus ourselves on (1) logical specification of schema mappings between local data repositories managed by a particular peer (agent); and (2) declarative specification of semantic-driven communication between agents, and between agents and the broker in the data integration processes. The considerations are based on the implementation of the SIX-P2P system, currently under development in Poznań University of Technology, where the discussed methods are under verification and evaluation.

References

1. Arenas, M., Libkin, L.: XML Data Exchange: Consistency and Query Answering, *PODS Conference*, 2005, 13–24.
2. Bernstein, P. A., Giunchiglia, F., Kementsietsidis, A., Mylopoulos, J., Serafini, L., Zaihrayeu, I.: Data Management for Peer-to-Peer Computing : A Vision., *WebDB*, 2002, 89–94.
3. Bouquet, P., Serafini, L., Zanobini, S.: Peer-to-peer semantic coordination., *Journal of Web Semantics*, **2**(1), 2004, 81–97.
4. Calvanese, D., Giacomo, G. D., Lenzerini, M., Rosati, R.: Logical Foundations of Peer-To-Peer Data Integration., *Proc. of the 23rd ACM SIGMOD Symposium on Principles of Database Systems (PODS 2004)*, 2004, 241–251.
5. Fagin, R., Kolaitis, P. G., Popa, L.: Data exchange: getting to the core., *ACM Trans. Database Syst.*, **30**(1), 2005, 174–210.
6. Lenzerini, M.: Data Integration: A Theoretical Perspective., *PODS*, 2002, 233–246.
7. Loke, S. W.: Declarative programming of integrated peer-to-peer and Web based systems: the case of Prolog., *J. of Systems and Software*, **79**(4), 2006, 523–536.
8. Meilicke, C., Stuckenschmidt, H., Tamilin, A.: Improving Automatically Created Mappings Using Logical Reasoning, *Proceedings of the 1st International Workshop on Ontology Matching OM-2006*, CEUR Workshop Proceedings Vol. 225, http://CEUR-WS.org/Vol-225.
9. Pankowski, T.: Management of executable schema mappings for XML data exchange, *Database Technologies for Handling XML Information on the Web, EDBT 2006 Workshops*, Lecture Notes in Computer Science **4254**, Springer, 2006, 264–277.
10. Pankowski, T., Cybulka, J., Meissner, A.: XML Schema Mappings in the Presence of Key Constraints and Value Dependencies, *ICDT 2007 Workshop EROW'07*, CEUR Workshop Proceedings Vol. 229, CEUR-WS.org/Vol-229, 2007, 1–15.
11. Rahm, E., Bernstein, P. A.: A survey of approaches to automatic schema matching, *The VLDB Journal*, **10**(4), 2001, 334–350.
12. Shvaiko, P., *et al.*: Dynamic Ontology Matching: A Survey, Techn. Report DIT-06-046, University of Trento, Available on: http://eprints.biblio.unitn.it/archive/00001040/, 2006.
13. Tatarinov, I., Halevy, A. Y.: Efficient Query Reformulation in Peer-Data Management Systems., *SIGMOD Conference*, 2004, 539–550.
14. XQuery 1.0: An XML Query Language. W3C Working Draft: 2002. www.w3.org/TR/ xquery
15. Yu, C., Popa, L.: Constraint-Based XML Query Rewriting For Data Integration., *SIGMOD Conference*, 2004, 371–382.

Intelligent Resource Allocation–Solutions and Pathways in a Workforce Planning Problem

Botond Virginas[1], Marian Ursu[2], Edward Tsang[3],
Gilbert Owusu[1], and Chris Voudouris[1]

[1] Intelligent Systems Lab, BT Exact, Adastral Park, Ipswich, IP5 3RE, UK
{botond.virginas, gilbert.owusu, chris.voudouris}@bt.com
[2] Department of Computing, Goldsmiths College, University of London,
London, SE14 6NW, UK
m.ursu@gold.ac.uk
[3] Department of Computer Science, University of Essex, Colchester, CO4 3SQ, UK
edward@essex.ac.uk

Abstract. The paper is based on FieldExchange – a computer system responsible for monitoring and supporting resource re-distribution decision making in BT's Operational Resource Management units. This paper considers the problem of resource allocation in the service industries approached from an agent-based perspective. The problem is formulated as a centralized/distributed planning problem. The paper describes the context of this solution, the general model and solution and four specific implementations with results and discussion.

Keywords: Multi-agent planning, E-business agents, Case studies and reports on deployments.

1 Introduction

This research is motivated by British Telecom's (BT) workforce planning and scheduling problem. The effective management of resources is critical to optimal service delivery in service organizations such as BT. Resource management is a complex process, usually involving the analysis of large amount of information. BT manages the largest access network in the United Kingdom (UK) with a field force of around 25,000. We have developed Field Optimisation Suite (*FOS)* [1]. FOS incorporates workload forecasting, optimised workforce planning, as well as advanced tools for visualising and communicating the outputs to end users. FOS employs Operational Research (OR) techniques such as constraint satisfaction for problem modelling, and heuristic search for problem solving [2, 3]. This paper is based on FieldExchange, a component of FOS in charge of monitoring and supporting resource re-distribution decision-making in BT's Operational Resource Management units.

2 Problem Description

In terms of their management, large telecommunication companies are partitioned in organizational units. This paper considers spatially (geographically) based partitions.

N.T. Nguyen et al. (Eds.): KES-AMSTA 2007, LNAI 4496, pp. 395–404, 2007.

A *domain* is defined as a location. *A job* has an inception, a deadline by which it has to be completed, a set of required skills and a nominal duration. Also, a job has a priority level, which is an indication of how quickly it needs to be resolved. An *engineer* has a set of skills. Engineers can have preferences with regards to the skills they normally wish to employ, the locations where they normally wish to travel, and the days when they wish to travel; the list of preferences can be longer. The main relationship on the basis of which the allocation of engineers to jobs is carried out is that of a *match*: a set of resources must match the set of requirements to which they are allocated. A *match* is a potential allocation.

In FOS we follow a two-phase approach: phase 1 – local allocation and phase 2 – global allocation of surpluses and shortages left after phase 1. In the first phase, for each domain, local engineers are allocated to local jobs. Phase 2 considers the "leftovers" from phase 1 – surpluses (idle engineers) and shortages (unresolved jobs). Each allocation within the second phase has an extra cost, compared to local allocations, due to travelling and, possibly, overnight accommodation. An effective solution, here, not only maximises the number of resolved jobs, but also minimises this extra cost.

A solution to local allocation (phase 1) has been developed and reported elsewhere [5, 6]. The main functional components of this solution are: *FieldForecast* (Demand Forecasting and Job Generation) and *FieldPlan* (Resource Planning). FieldForecast takes as input historical job information and produces forecasts of job volumes. FieldPlan, using both forecasted and real jobs, generates workforce capacity and deployment plans, for each domain. FieldExchange is in charge of global allocation.

3 Centralized and Distributed Power Structures

A common model has been introduced to capture an enterprise's distribution of intelligence and mode of operation. Each domain has a set of *resources* (engineers, for BT), a set of *requirements* (jobs, for BT) and is managed by a *domain manager*. The domain manager is the local repository of intelligence/knowledge – *decision criteria* and *decision-making strategies* – and power – *the ability to enforce decisions* – regarding the domain's requirements and resources. Local power may be *recognized* by other domain managers or its *enforcement* may require the intervention of central company managerial *mediators*. A computational model appropriate for such an organization consists of a set of software agents. In this model each domain has a Service Agency. The Service Agency is responsible for the management of resources and jobs for the particular region. These Service Agencies are managed by a Central Agency. For the purpose of FieldExchange each Service Agency consists of a Service Buyer and a Service Seller. Each Service Buyer has a set of jobs to be completed. Each Service Seller manages a set of engineers, who can be assigned to jobs. These Service Buyers and Service Sellers may interface with other systems like FieldPlan to capture their input data. Service Sellers and Service Buyers enact the geographic region's interest and priorities via the decisions/choices they make. These local interests and priorities may be expressed via a set of criteria/objectives that are to be optimised (e.g., "minimise the engineers' overall travelling distance"), constraints that are to be satisfied (e.g., "an engineer should not travel for more than

600 miles a week") and rules of operation or strategies that are to be followed (e.g., attend to jobs of higher priority first). We consider all three forms of expression.

The Service Buyers buy services from the sellers from other Service Agencies by expressing their requirements (e.g. inviting them to bid for jobs). The Service Sellers base their bidding on which engineers could service which job depending on a set of constraints and preferences. Example of these include availability and skills of engineers, the engineers' preferences and their travelling distances to the jobs.

The global allocation of resources is accomplished through communication and exchanges between agents. This has been approached from two perspectives:centralized approach and distributed approach. Obviously, they represent extreme points of view; many combinations between them are possible. In the centralized approach, the focus is on the company's *global* interests and priorities. They are expressed via a set of conflicting global criteria. There is no decision power expressed at domain/local level. An example of a computational model suitable for this approach is that of a central multi-criteria optimisation algorithm [7].

In the distributed approach, the focus is on *local* interests and priorities. In a purely distributed approach, all the intelligence is located/represented in the constituent agents – i.e. in the leaves of the system. The overall behaviour of the system *emerges* completely from the agents' interaction from different service agencies. The company's overall interests and priorities are not explicitly expressed – there is no central manager to enforce them. Their accomplishment should emerge from the individual accomplishment of the local ones

Examples of interaction structures include explicit asynchronous co-ordination (e.g., collaboration, competition and negotiation) and synchronism (e.g., request for resources must be received by a certain deadline). Central agents may also be regarded as fulfilling specific roles, such as "emergency expert", which would intervene in the allocation process only if certain emergency situations arouse. This perspective extends the traditional classification of central agents (within a federation of agents architecture) in facilitators, brokers and mediators. Four different solutions have been designed and implemented based on the common power model. Some of the solutions have got experimental or prototype implementations while others have reached a trial or rollout stage. The various solutions have been implemented incrementally after analyzing the current implementations and identifying future steps. The following sections describe these implementations in chronological order.

4 Centralized Collaborator

We have implemented a prototype system – Centralized Collaborator (CC) – on the basis of the completely centralized model [8]. The *Central Agent* implements global interests and priorities and these are expressed as a set of *criteria* animated via a *multi-criteria optimisation algorithm [7]*. Initially, the Central Agent collects all the job requests from all Service Buyers and ranks them according to their *importance*. The Central Agent, then, enters an iterative process whereby, at each iteration, a job is allocated with appropriate resources. Each iteration deals with the most important job from the ones that remained unresolved.

The selection is implemented as a Pareto optimisation – the chosen optimal set is the set of non-dominated solutions or the *Pareto front*. We use a greedy algorithm to construct the Pareto front. Currently, the optimisation algorithm uses two criteria: minimise travelling distance and maximise use of skill proficiency. The optimal set is sent to the Service Buyers whose request is under current consideration.

The Service Buyer uses local constraints to filter out offers that do not satisfy local interests and priorities. The Service Buyer, then, selects one offer from the filtered set using specified selection strategies and communicates its choice to the Central Agent. Finally, the Central Agent notifies the Service Seller whose offer was selected and the process is resumed with the next most important job request.

The GUI agents in the prototype are the *Global Monitor* and *Local Monitors*. The control features provided by the Global Monitor include parameter setting (such as, planning period, and global criteria) and assistance towards the manager regarding the monitoring of the allocation of resources (e.g. visualisation of results). The control features provided by local monitors include setting of local constraints, manual selection of requests or resources to be submitted to CC, manual overriding of individual allocations and data visualization.

5 Distributed Collaborator

We have implemented a second prototype system - Distributed Collaborator (DC) - as an almost isomorphic implementation of the completely and uniformly distributed model [9]. The Central Agent – Monitor – is used merely for synchronisation of the communication between the other agents. The allocation of resources is modelled as an iterative communication process, where each iteration consists of a 4-step communication protocol. The communication protocol has two stages: information gathering and contract establishment. The information exchanged during the information gathering stage bears no legal obligation. For example, a jobs' agent may request engineers for double the amount of jobs it actually holds. Therefore, various strategies, employing bogus or incomplete information, may be employed here, in order to attract favourable contracts in the following stage. The information exchanged in the contract establishment stage is legally binding.

In step one, all Service Buyers broadcast requests, based on their job requirements. In the simplest case, each Service Buyers broadcasts requests for all its jobs to all Service Sellers. This is the solution we initially adopted in DC. We are now experimenting with more elaborated strategies like *preferential* and *bogus* broadcasts.

In step two, Service Sellers respond with offers. Each Service Seller offers its best matching set of engineers to each Service Buyer.

In step three, Service Buyers, faced with various offers, must decide to whom to propose contracts (legally binding). In DC each Service Buyer aims to maximise the number of jobs attended to and to minimise the travelling costs it has to support. This could be achieved via a local multi-objective optimisation algorithm. DC takes the "strategy route" – it implements strategies of compiling contract proposals, mimicking the behaviour of domain managers.

In step four, Service Sellers decide whom to contract their resources. In DC, each Service Seller attempts to maximise the number of jobs to attend, whilst maintaining the overall travelling distance between reasonable limits. Here, too, DC takes the "strategy route".

The synchronisation of the agents' behaviour, necessary for the realisation of the 4-step communication protocol, is achieved via the Monitor. This ensures that a step is initiated only after all the agents have indeed completed the previous step.

The novelty of the solution resides in the fact that it is a natural and versatile formulation that combines the agent-based model with rule-based expressions of allocation strategies and multi-criteria optimisation expressions of allocation objectives. However, the general solution is readily extendable towards the inclusion of central agents with specialist roles. Central agents may be regarded as fulfilling specific roles, such as "emergency expert", which would intervene in the allocation process only if certain emergency situations arouse [10].

6 ASMCR: An Open Constraint Optimisation System

We have implemented a third prototype system, ASMCR, as an open constraint optimisation system where individual agents have potentially conflicting interests [11]. The Service Sellers and Service Buyers all attempt to maximise their own utility. The overall problem is considered as a multi-objective optimisation problem.

Distributed constraint satisfaction is a well studied topic which is highly relevant to distributed planning. Most of the work in distributed constraint satisfaction involves cooperative agents, where agents work together to achieve some common goal. In this solution we study distributed constraint satisfaction problems where agents may have conflicting goals. We have introduced a retractable contract net protocol, which we call RECONNET, that supports hill-climbing in the space of solutions. It is built upon a job-release and compensation mechanism. The problem of each Service Buyer and Service Seller was formulated as an *open constraint optimisation problem* [12] where some constraints are not entirely within the control of the problem solver itself.

The Service Buyer's Model:
The problem of buyer b can be formulated as an open constraint satisfaction model:

$$(Z_b, D_b, C_b, E_b, f_b, Ag_b, EtA_b, CP_b)$$

$Z_b = \{s[1], s[2], \ldots, s[n_b], p[1], p[2], \ldots, p[n_b], d[1], d[2], \ldots, d[n_b]\}$, where n_b is the number of jobs that b has, s[i] represents the service seller that b appoints to do job i, p[i] represents the preference and d[i] represents the distance for serving job i, which are proposed by the service sellers and accepted by the buyer;

Ag_b is the set of seller agents who b has contact to;

D_b is a function that defines the domain of the variables in Z_b, as in constraint satisfaction [16]. For all i, $D_b(s[i]) = Ag_b$ plus ϕ, which means s[i] could be assigned one of the service sellers, or assigned no seller at all (which is represented by the value ϕ); $D_b(p[i]) = \{0,1,\ldots,9\}$, which means p[$i$] could be assigned a value 0 to 9, with 0 meaning the job is not served, 1 to 9 are preferences in the service. For all distance variables d[i], $D_b(d[i]) = R$;

C_b represents a set of internal constraints, which is { } in this case, i.e. there are no constraints on what value b assigns to the variables;

E_b = { E_b(s[*i*], p[*i*], d[*i*]) | i = 1.. n_b }, where E_b(s[*i*], p[*i*], d[*i*]) is a constraint on the values of s[*i*], p[*i*] and d[*i*], restricting the values that they can take simultaneously; the values of p[*i*] and d[*i*] are to be determined by external agents, s[*i*] indicates the seller that *b* assigns job *i* to; it is assigned by *b*, depending on the bids by the service sellers;

f_b is the objective function for *b*. It is a multi-objective function.

The buyer *b* attempts to maximise its utility:

$$\text{Utility} = k_1 * reve_b - k_2 * failure_b - k_3 * pref_b - k_4 * dist_b - k_5 * comm_b + \text{Trade}$$

where the weights k_1 to k_5 are given by the manager; Trade is income from other buyers – payment to other buyers for contract-release;

EtA_b is the mapping from each external constraint E_b to a service seller; i.e. EtA_b(E_b(s[*i*], p[*i*], d[*i*])) returns a value in Ag_b; this indicates that the values of p[*i*] and d[*i*] are to be determined by all seller agents (Ag_b) in communication with *b*;

CP_b is the communication protocol. Here we assume the following protocol:

1. The buyer *b* sends a set of invitation to bid to seller *s*; each invitation is

   (Job_ID, Job_information)

   where Job_information is a tuple as defined above:

   (Location, Min_Skill, Duration, StartDay, Price)
2. The seller *s* sends a set of pairs of values to *b* for instantiating (p[*i*], d[*i*])
3. The buyer *b* offers *s* a contract, which comprises a pair of p and d values
4. The seller *s* accepts the contract (and commits its resources) or declines the offer, in which case, go back to Step 3 (where *b* could offer a contract to another seller)

The Service Seller's Model:

The problem of seller *s* can be formulated as a dynamic open constraint satisfaction model:

$$(Z_s, D_s, C_s, E_s, f_s, Ag_s, EtA_s, CP_s)$$

where

Z_s = {e[1], e[2], …, e[*N*], p[1], p[2], …, p[*N*], d[1], d[2], …, d[*N*]}, where *N* is the total number of jobs that *s* has been invited to bid for and *s* is still in contention (i.e. the buyer has not yet assigned the job to another seller); e[*i*] represents the engineer that is assigned to do job *i*; p and d represents preferences and distances as defined in buyers;

Ag_s is the set of service buyers who *s* has contact to;

D_s is a function that defines the domain of the variables in Z_s, as in constraint satisfaction. For all *i*, D_s(e[*i*]) = the set of engineers plus ϕ, which means e[*i*] could be assigned one of the engineers, or assigned no engineer at all (which is represented by ϕ); D_s(p[*i*]) = {0,1,…,9}, which means p[*i*] could be assigned a value 0 to 9, with 0 meaning the job is not served, 1 to 9 are preferences in the service; For all distance variables d[*i*], D_s(d[*i*]) = R; default values for p and d are 0, which means no engineer is assigned to job *i* until commitment is made (by *s*);

C_s represents the internal constraints that governing the feasibility of the engineers doing the jobs; this involves the availability and skills of the technicians;

E_s = { E_s(e[*i*], p[*i*], d[*i*]) | i = 1.. *N* }, where E_s(e[*i*], p[*i*], d[*i*]) is a constraint on the values of e[*i*], p[*i*] and d[*i*], restricting the values that they can take simultaneously; the values of e[*i*], p[*i*] and d[*i*] are proposed by s, to be approved by the service buyer;

f_s is the objective function for *s*. Associated to each job *i*, Price([*i*]) is a constant given by the Buyer. f_s a multi-objective function.

The seller s attempts to maximise its utility:

$$\text{Utility} = k_1 * JD - k_2 * (DT)^2 + k_3 * LB - k_4 * (RD) + CR$$

where the weights k_1 to k_5 are given by the manager;

EtA_s is the mapping from each external constraint E_s to a buyer; i.e. EtA_s(E_s(e[*i*], p[*i*], d[*i*])) equals the service buyer for job *i*

The objective is to balance failure across all domains. The Manager's overall objective is a multi-objective function. The manager should attempt to produce a Pareto set of solutions. This can be done by giving the Service Buyer/Service Seller different sets of weights for different measures for the buyers and sellers (the k_i's mentioned above). For example the k_i's can be used to empower individual buyers to increase their bargaining power so as to reduce their failure rates. The manager may ask the agents to schedule from scratch or improve on a previous schedule

The software, which we call ASMCR, allows the management to have full control over the company's multi-objectives. The manager generates a Pareto set of solutions by defining, for each Service Buyer and Service Seller, the weights given to each objective. ASMCR gives Service Buyers and Service Sellers ownership of their problem and freedom to maximise their performance under the criteria defined by the management. ASMCR took 5 to 15 minutes to complete when tested on real-sized problems. It has potential to be developed into practical solutions to BT's workforce planning problem.

7 FieldExchange: Resource Balancing with iOpt

We have implemented a fourth system which is currently under trial in a number of operational Resource Management units within BT. This solution was driven by a concrete business need of centrally balancing the failure rates for individual geographic areas. In this solution the Central Agent takes a very important optimisation role. It is using a search framework based on BT's iOpt optimisation toolkit [3], which was built for modelling and solving combinatorial problems using invariants (one-way constraints) and heuristic search methods. Resource allocation is achieved using a special implementation of CC. In this example the Central Agent collects all job requests (demand) and engineer data (capacity/surpluses) and it's overall aim is to balance the demand/capacity ratio across a geographic area. Once proposals are broadcasted back to Service Sellers they will allocate individual engineers to requests based on domain preferences and respond with resource offers.

To solve the balancing problem, we've designed a variable neighbourhood search (VNS) based framework and six individual heuristics. The reason behind selecting the use of a variable neighbourhood search was to do with the real time nature of the system were heuristics are required. In our example the variable neighbourhood search has shown some superiority to methods such as tabu search, simulated annealing and *BestCNS* all provided within *iOpt*. Central to this approach is an objective function which determines the quality of each candidate solution.

Since engineers in each domain can be moved to another domain for just one day or a couple of days, we consider two kinds of moves: a daily move which is only for one day, and an accommodation move which is for more than one day. The cost for moving one engineer from one domain to another varies according to the distance between the two domains and other constraints. Thus, sometimes moving engineers from one domain to another directly is impossible due to the cost. To solve this problem, we've designed another kind of move which is called shuffling move. The shuffling move tries to move some engineers from domain A to domain B via the help of domain C. The cost of moves between domain A and B is high, while the cost between domain A and C, domain C and B is low. Thus, the shuffling move will move a number of engineers from domain A to C, and then move the same number of engineers from domain C to B. Thus, we have three kinds of moves in total: the daily move, the accommodation move and the shuffling move.

The six heuristics that we have designed provide for daily moves, accommodation moves, and shuffling moves for clearing the surpluses and evenly distributing failures across domains. The VNS pre-defines a sequence for the six low level heuristics. It starts by applying the first heuristic in the sequence. When a local optimum is met, the VNS will go to next heuristic in the sequence. If a better solution is found after one iteration, the VNS will go back to the first heuristic to continue the search, otherwise it jumps to the next heuristic in the sequence to search for a better solution. The search will continue until the stopping condition is being met.

8 Results

The implementations based on the common model described in section 3 have provided us with a large experimental database. Most of the results have been described in more detail elsewhere [6,8,9,11] Please find below a brief summary of these results.

CC and DC are being considered together looking at shifting the power structures from the centralized model towards the decentralized approach with various degrees of central control. Currently richer data models and larger data sets are being considered. The issue of *strategies* vs. *criteria* in the expression of local interests and priorities occurs, whether this is done centrally or locally. Thus far, we have experimented with criteria in the central approach and strategies in the distributed approach. We shall continue our experiments using criteria locally and strategies centrally and then combinations. The initial feedback is that strategies give a better sense of control to human agents (managers). The aim is to move towards and investigate more complex decision power structures. We shall explore *collaboration strategies* and *negotiation*, which could be achieved with or without central agents. We are also going to investigate the usefulness of central agents handling *exceptions* and *emergencies*. Furthermore, we are going to experiment with both *synchronous* and *asynchronous* communication protocols. Thus far we considered *static* power models – wherein agents are assigned unchangeable decision powers – and *implicit* or *hard-coded* representations of power – i.e., as a mode of operation. However, we are now in the process of devising models in which power is an explicit attribute that can be reasoned about. We aim to devise mechanisms whereby power can be negotiated between agents.

ASMCR has been tested thoroughly using randomly generated problems. This choice has led to a more general testing model than the real application. The usefulness of hill-climbing has been confirmed. Experiments also confirmed that by changing the weights to the multiple objectives, the manager can reduce the number of jobs not served, travelling distance and preference (lower value in preference means better service quality). ASMCR has demonstrated that it has real potential to be developed into practical solutions to BT's workforce planning problem.

FieldExchange has recently been in trial within several resource management units and the initial feedback is very promising. The optimisation algorithm has demonstrated a good improvement especially over clearing and balancing surpluses across geographic regions. We are constantly improving the algorithm adapting to the business objectives driving the scope of FieldExchange.

Results of the overall usage of FOS system have showed significant improvements over practices prior to the deployment of FOS. For example, there has been an improvement in the performance of the field technicians: 8-14% improvement in the number of technicians getting jobs first thing in the morning. There is also a reduction in travel time (from 95min to 85min) since technicians are properly positioned geographically to service customers first thing in the morning. Clearly this has led to an increase in productivity and morale. There has been a reduction (from 31% to 18%) in manual intervention on the control sides, i.e. the personnel who interface between the technicians and the work allocation system. There has been an improvement in quality of service (e.g. 1.1% more business provision jobs meeting the required by date).

9 Conclusions

In this paper we have described various solutions and pathways in a workforce planning problem. Although they are presented as disjoint prototypes and applications, they are sharing a common power model and they follow an incremental development path. Our endeavour is to integrate all these implementations within a versatile service exchange system which could be fully customized to serve our internal resource market as well as external contractor based exchanges. We would like to create a resource exchange platform where one will have the option to select the various features presented in the component implementations like: dynamic power models with various degree of distributed/central control, strategies, criteria or dynamic constraint satisfaction in the expression of local interests and priorities, various synchronous and asynchronous communication protocols including the retractable contract net protocol RECONNET, and the choice of multi-objective optimisation algorithms as described in ASMCR and CC.

References

1. Voudouris, C., Owusu, G., Dorne. R., and Mccormick A. "FOS: An Advanced Planning and Scheduling Suite for Service Operations", submitted to IEEE International Conference on Services Systems and Services Management, IEEE/SSSM06.
2. Reeves, C.R. (ed)., *Modern Heuristic Techniques for Combinatorial Problems*. Oxford.: Blackwell Scientific Publications, 1993
3. Voudouris, C.; Dorne, R.; Lesaint, D.; and Liret, A. "iOpt: A Software Toolkit for Heuristic Search Methods, Principles and Practice of Constraint Programming", In Lecture Notes in Computer Science, Vol. 2239, pp716-729, CP., 2001.
4. Shen, W. and Norrie, D.H. Agent-Based Systems for Intelligent Manufacturing: A State of the Art Survey. In: International Journal of Knoweldge and Information Systems, 1:2 (1999) 129-156.
5. Voudouris, C., Owusu, G., Dorne, R, Ladde, C and Virginas, B., ARMS: An Automated Resource Management System for British Tellecomunication plc, In: EURO/INFORMS Joint International Meeting EURO Excellence in Practice Award, Istanbul (2003).
6. Owusu, G., Voudouris, C., Kern, M., Garyfalos, A., Anim-Ansah, G., Virginas, B., On Optimising Resource Planning in BT with FOS. IEEE International Conference on Services Systems and Services Management, IEEE/SSSM06 (2006)
7. Ehrgott, M. and Gandebleux, X. (eds): Multiple Criteria Optimization: State of the Art Annotated Bibliographic Survey, Kluwer's International Series in Operations Research and Management Science, Vol. 52, Kluwer Academic Publishers, Boston (2002).
8. Virginas, B., Owusu, G., Voudouris, C., and Anim-Ansah, G. A two stage optimisation platform for resource management in BT. In the: proceedings of the Twenty third Annual International Conference of the British Computer Society's Specialist Group on AI (2003) 109-121.
9. Ursu, M.F., Virginas, B., Owusu, G., and Voudouris, C, 2005. "Distributed Resource Allocation via Local Choices. A Case Study of Workforce Allocation", International Journal of Knowledge-Based and Intelligent Engineering Systems, Volume 9, Number 4 / 2005 , pp. 293-301 , IOS Press, Netherlands.
10. Shen, W., Norrie, D.H. and Kremer, R. Developing Intelligent Manufacturing Systems Using Collaborative Agents. In: Proceedings of IMS 99, Leuven, Belgium, Sept (1999) 22-24.
11. E.P.K. Tsang, T.Gosling, B. Virginas, C. Voudouris & G. Owusu, Retractable contract nets for distributed workforce scheduling, Proceedings of the 2nd Multidisciplinary Conference on Scheduling: Theory and Applications (MISTA 2005), New York, USA, 18-21 July 2005
12. Tsang, E.P.K., Foundations of Constraint Satisfaction, Academic Press 1993

A Multi-agent Architecture for Designing and Simulating Large Scale Wireless Systems Resource Allocation

P.M. Papazoglou[1], D.A. Karras[2], and R.C. Papademetriou[3]

[1] Lamia Institute of Technology Greece, University of Portsmouth, UK, ECE Dept., Anglesea Road, Portsmouth, United Kingdom, PO1 3DJ
papaz@teilam.gr

[2] Chalkis Institute of Technology, Greece, Automation Dept., Psachna, Evoia, Hellas (Greece) P.C. 34400
dakarras@teihal.gr, dakarras@ieee.org

[3] University of Portsmouth, UK, ECE Department, Anglesea Road, Portsmouth, United Kingdom, PO1 3DJ

Abstract. The simulation model adaptability to real network behavior is the key concept in wireless communications. In a cellular network, many procedures such as call admission, hand-off, etc take place simultaneously for every individual user. Every network procedure acts autonomously, interacts with the network environment (gathers information such as interference conditions), takes decisions (e.g. call establishment), etc. Although this is known in the literature, there is lack of suitable representations for such network procedures in the simulation systems proposed so far, thus compromising simulation model adaptability to real network behavior. To achieve such adaptability we herein propose to change the point of view in network procedure representation. Instead of viewing them as independent programming functions or even objects in a high level language, which are sequentially executed, due to their aforementioned properties it is proposed that such network procedures could be more efficiently modeled as agents. Considering this new approach, the agent cooperation and communication in terms of negotiation and agreement is a critical issue. In this paper we present a centralized cooperative multi-agent negotiation scheme applied to a multi-agent layered architecture for designing and simulating resource allocation in cellular communication systems, based on organizational modeling. Moreover, we show the way that the rules and implementation methods of agent negotiation affect the adaptation grade of simulation model to the real cellular network behavior.

1 Introduction

1.1 Agent, Multi-agent Systems and Organizational Modeling

An Agent is known to be a computational system that interacts autonomously with its environment and operates for the goal for which it has been designed [1]. We can face Agents as entities that are dedicated to a specific purpose and are smaller than a

N.T. Nguyen et al. (Eds.): KES-AMSTA 2007, LNAI 4496, pp. 405–415, 2007.

typical application [2]. An Agent also perceives the environment conditions, acts according to these conditions, interprets perceptions and solves problems [3]. There are some particular attributes which differentiate agents from other programs. These most important attributes are:

- *Adaptability:* ability to change due to external/internal events [4,5].
- *Autonomy:* ability to control its actions towards pre-defined goals [6,7,8].
- *Collaboration* with other agents for achieving common goals.
- *Interactivity* with surrounding environment.

According to [9], a multi-agent system consists of a number of agents which interact through communication. These agents act in an environment; have different areas of influence in an environment. Within the environment many influence areas can be coincided. In modern problems, a number of agents are needed for modeling all the system aspects and so the multi-agent systems emerged. Multi-agent systems (MAS), can be viewed as a loosely coupled network of problem-solver entities [10] that collaborate together with the common goal to solve the whole problem beyond the solving capabilities of each individual entity. In several cases agent technology has been used in the management of telecommunication systems [11,12,13]. The architecture in these cases is straightforward. Cellular systems modeling and especially resource allocation has not been viewed so far in terms of multi-agent systems. Their simulation is mainly based on sequential or parallel models (for faster execution) but not in terms of agent models. An important novel issue studied in this paper is negotiation aspects in agent models for telecommunications systems simulation. Negotiation can be faced as competitive or cooperative [14,15]. In competitive negotiation, the agents are self-interested for achieving their own goals. In cooperative negotiation, the agents are working together to find a solution for a common goal.

1.2 Cellular Networks and Channel Allocation Strategies

Based on cellular principle the network coverage is divided in small hexagonal service areas called cells. Each cell serves the moving users with a base station that is positioned in the center of the cell. Due to the restricted availability of bandwidth, flexible channel allocation schemes must be applied in order to achieve the maximum user servicing [16]. In literature, many channel assignment schemes have been widely investigated with a goal to maximize the frequency reuse. The channel assignment schemes in general can be classified into three strategies: Fixed Channel Assignment (FCA) [17,18,19,20,21], Dynamic Channel Assignment (DCA) [17,22,23,24,25], and the Hybrid Channel Assignment (HCA) [17,26]. In FCA, a set of channels are permanently allocated to each cell based on a pre-estimated traffic intensity. In DCA, there is no permanent allocation of channels to cells. Rather, the entire set of available channels is accessible to all the cells, and the channels are assigned on a call-by-call basis in a dynamic manner. The FCA scheme is simple but does not adapt to changing traffic conditions and user distribution. Moreover, the frequency planning becomes more difficult in a microcellular environment as it is based on the accurate knowledge of traffic and interference conditions. These deficiencies are overcome by DCA but FCA outperforms most known DCA schemes under heavy load conditions [18]. To overcome the drawbacks of FCA and DCA, the HCA combines the features of both FCA and DCA techniques. In this paper we use the Unbalanced DCA variation. When this variation is selected, only one try is performed for the connection of the new user within the initiated cell.

1.3 Organization Model Based Multi-agent Design and Simulation of Cellular Networks–The Concepts

The performance and the behavior of a real cellular network can be evaluated using simulation systems without the need to perform field experiments and develop prototypes. The simulation solutions give us the opportunity to develop custom protocols, channel allocation schemes, network structures, etc, towards a desired cellular network. Due to the complexity of real cellular networks the simulation software development strategy becomes a very important factor that influences the resulting network model [27]. In such a cellular network, many procedures such as call admission, hand-off, etc take place simultaneously for every individual user. Every network procedure acts autonomously, interacts with the network environment (gathers information such as interference conditions), takes decisions (e.g. call establishment), etc. Although this is known in the literature, there is lack of suitable representations for such network procedures in the simulation systems proposed so far, thus compromising simulation model adaptability to real network behavior. **Only if network procedures are represented realistically the required simulation model adaptability could be achieved and the network performance simulations results could be reliable and closely reflecting real network behavior.** To achieve such adaptability we herein propose to change the point of view in network procedure representation. Instead of viewing them as independent programming functions or even objects in a high level language, which are sequentially executed as usually in the literature, due to their aforementioned properties ***it is proposed that such network procedures could be more efficiently and realistically modeled as agents***. Developing suitable multi-agent system architectures, we can achieve more effectively a reflection of the network operation concept. Such multi-agent architectures can be implemented using a multi-threading environment like JVM (Java Virtual Machine). Considering this approach, the agent cooperation and communication in terms of negotiation and agreement is a critical issue.

2 The Ad Hoc Multi-agent System Architecture for Modeling and Simulating Cellular Systems Resource Allocation

2.1 Network Operation

We have developed an integrated simulation environment for GSM cellular networks. This system is implemented in java multi-threading platform. The structure of the cellular network is based on cell concept. Our simulation model is based on some basic operational parameters of a GSM real network such as Carrier to Noise Ratio on cell edge (dB), Carrier to Noise plus Interference Ratio threshold (dB) for acceptable connections, average call holding time (sec), channels per cell, path loss factor (alpha) for propagation conditions, standard deviation of shadowing (sigma) for propagation conditions, cell-mesh fineness, max users per cell, number of cells, New call arrival rate, User movement rate, etc. A new call admission may occur at any simulation time within any cell based on Poisson distribution modeling. A communication channel is assigned to a mobile user due to new call admission or reallocation decision if some criteria are satisfied. These criteria are a) Channel availability, b) Carrier strength, c) Carrier to Noise ratio CNR and d) Signal to Noise plus interference ratio CNIR

After a call initiation a call holding time is assigned to the new user from a predefined distribution. When the connection time is expired the call is terminated. The signal of every connected user is checked for whether the above criteria are fulfilled and so user reallocation is made when needed. Finally, user movement occurs with regards to Gaussian distribution modelling. In a user attribute table, important attributes are stored for each network user such as current position, connection status, propagation conditions, etc.

Our simulation system supports four basic network procedures, which are:

- *New call arrival.* The new calls result from a Poisson distribution with regards to a predefined daily model. Based on this model, a whole day is divided into five different traffic zones.
- *Call termination.* The program uses an exponential function that generates the call duration for each new user. If this time expires, the user is disconnected.
- *Reallocation.* The communication signal quality between user and base station is checked every simulation time. If this quality is not accepted the user is reconnected to a neighbor cell.
- *User movement.* Based on Gaussian distribution each user moves to another position within the same or a neighbor cell. After repositioning, the program tries to connect the user with the base station of the corresponding cell.

2.2 The Proposed Multi-agent System

As we previously mentioned, in a multi-agent environment each agent interacts with its environment, adapts its behavior according to current conditions, collaborates with other agents and finally takes decisions in an autonomy base. In our case, each agent represents an individual network task such as new call admission, call termination, reallocation check and user movement. Figure 1, illustrates the proposed layered agent architecture. The whole simulation model is divided in three layers. Layers one and two, constitute the whole network framework. Agents in layer two interact with the core cellular network environment and guarantee the network functionality in terms of calls management. Layer three controls the whole simulation process by synchronizing the agent activation. The simulation system has been implemented in JVM multi-threading environment.

2.2.1 Agent Structure, Control and Synchronization

Each agent solves a different problem of the network services and consists of some basic components. The core of each agent is the problem solver component that belongs to the main components category. Additional components exist for supporting normal execution, active period with regards to simulation time, communication with

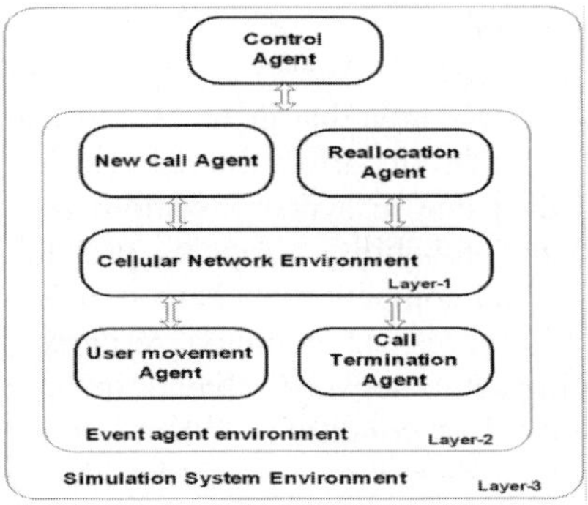

Fig. 1. Proposed Layered Agent Architecture

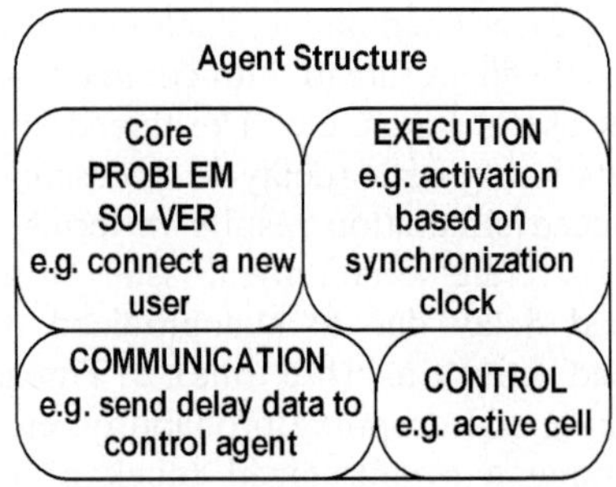

Fig. 2. Organization Structured model of a network agent

control agent, etc. Figure 2, shows the component structure of each agent. Figure 3, shows the agent synchronization and control mechanism. Each event agent keeps informed the control agent for its execution status. The control agent is active while simulation time is not finished and so the rest of the agents. This agent has a clock that takes sequential step values (e.g 1,2,3, etc). In each value, a corresponding action is activated. In the first clock step, the needed supplementary actions are activated (preparation tasks) while other agents and procedures are disabled.

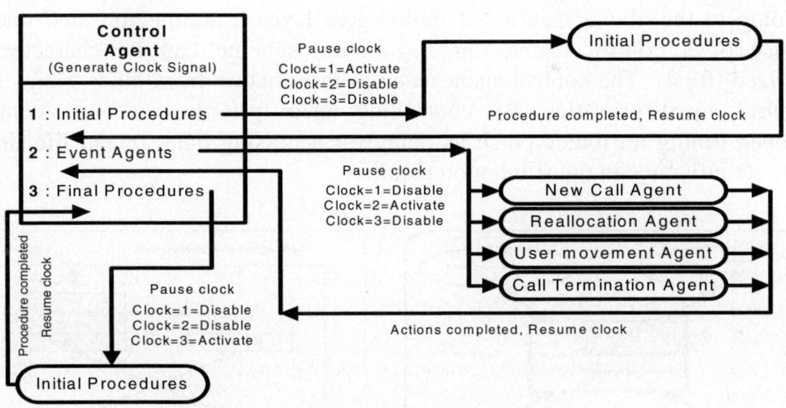

Fig. 3. Agent synchronization mechanism

In the second clock step, event agents are activated while other procedures are disabled. Control agent prevents from simultaneous activation of event agents, initial and final simulation procedures within the simulation time. The initial task executes several actions such as setting the lamda value (call arrival rate), etc and some other additional necessary actions for the program operation. The final procedures update the statistical metrics, display the current simulation time and advance the simulation time.

2.2.2 Agent Negotiation in the Simulation System

As we mentioned above, each network procedure (now an Agent) is implemented as an independent thread with common access to a shared area that includes information about user's status, etc. The thread competition conditions affect the behavior of the simulation system (delay in accessing shared data due to synchronization) and the produced simulation results in terms of blocking probability, dropping probability, etc. There are some critical issues in terms of thread competitive behavior such as a) Thread delay due to synchronized methods for accessing common resources, b) Thread active time (life time), c) Thread direct impact to network behavior in terms of blocking and dropping probability, etc. and d) Conflict conditions. When a thread is accessing a synchronized block of data, another thread is waiting for the same resource. Thus, there are many delay periods in each tread to complete the desired tasks. The thread instability due to delay factors may cause a significant change to the network behavior. For the above mentioned reasons a negotiation scheme for the competitive agents must be applied in order to optimize the agent behavior and cooperation for using common data and supporting efficiently the network procedures.

The first target using this scheme is to create a balanced delay environment for all network agents and analyze the results. Negotiation can be viewed from the point of relation between multi-agents and the environment. This relation prescribes the type of negotiation to centralized or distributed. In distributed scheme (fig.5), each agent makes its own decisions based on interaction with cellular network environment and other network agents. Negotiation rules can be defined individually for each agent. According to the above mentioned multi-agent layered architecture and due to the existence of a control agent, our negotiation scheme can be characterized as centralized (fig.4). The control agent receives information from other agents in order to control agent priorities. By controlling agent priorities we can control the mentioned timing attributes (such as delay for accessing data, thread life time, etc) with direct influence in the simulation results.

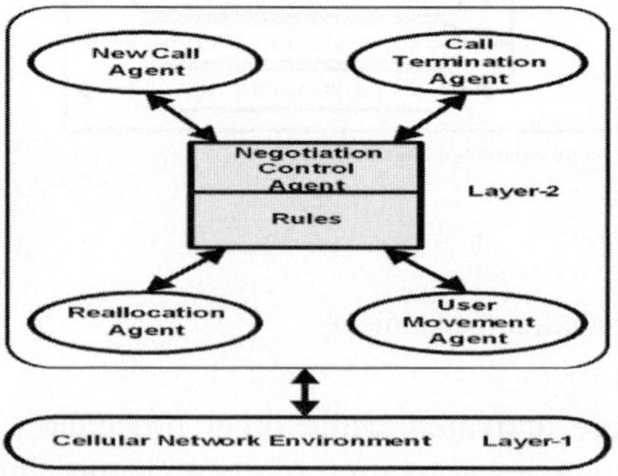

Fig. 4. Centralized Agent Negotiation

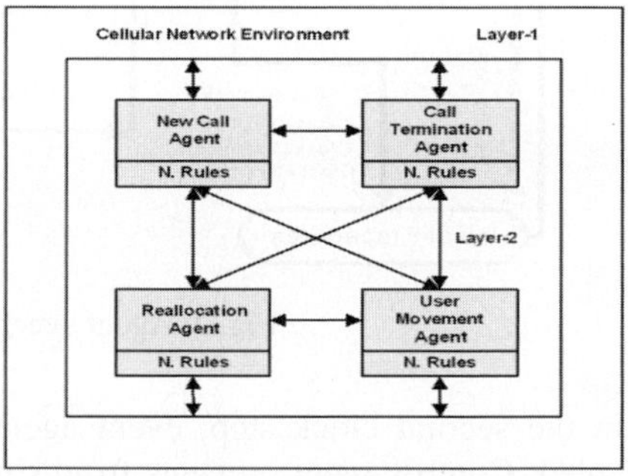

Fig. 5. Distributed Agent Negotiation

The key for controlling agent behaviour in terms of timing constraints is the priority control. Priority can be controlled by keeping in memory previous time behaviour for each agent. Such agent priorities are changing over simulation time in a dynamical manner. The central control agent acquires new knowledge at every new simulation step concerning the timing behaviour of each agent and can accordingly modify their priorities. The algorithm is simple but efficient and is based on the following simple priority setting rule, next defined:

$$NewP = NP \cdot \left[1 + \frac{Sumc[i]}{a \cdot SumMax}\right] \quad (1)$$

Where *NewP* is the resulting new priority within a new simulation step of a selected agent, *NP* is the normal priority, *Sumc[i]* is the total delay of agent *i* for accessing common resources, *a* is a user defined coefficient, *SumMax* is the highest delay of an agent. Whenever this rule is applied for setting agents priorities, the control agent compares each agent delay with the current maximum in order to decide for the significance of this delay. The more significant the delay is realized, the more priority is given to the corresponding agent. Thus, it is expected that agent delays would tend to zero and balanced conditions would be created between the agents. Fig. 6, shows the complete negotiation dialog between network agents and the control agent. Initially, every agent informs the control agent for its completion. After the completion of all agent tasks within the simulation step, the control agent exchanges information messages with all other agents in order to state the final priority assignment decision to each of them. Finally, each agent follows this decision of the control agent.

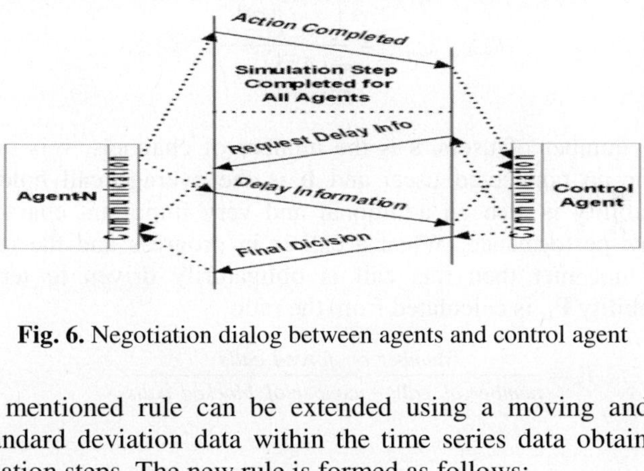

Fig. 6. Negotiation dialog between agents and control agent

The above mentioned rule can be extended using a moving and variable size window of standard deviation data within the time series data obtained through all previous simulation steps. The new rule is formed as follows:

$$NewP = NP \cdot \left[1 + \frac{Sumc[i]}{a \cdot SumMax}\right], gcount < WinSize \quad (2)$$

$$NewP = NP \cdot \left[1 + \frac{StdWin[i]}{b \cdot StdWinMax}\right], gcount \geq WinSize \quad (3)$$

where *gcount* is the number of known elements from the beginning of simulation procedure. When the size of *WinSize* does not fit existing simulation elements with regards to current simulation time, the variable window method can not be applied and so the new priorities are based on current total delay of each agent. In the first case (gcount<WinSize) a comparison is done between total delays of each individual

agent and total maximum known agent delay in order to find how significant is the measured delay of each agent. After *WinSize* number of elements the variable window method can be applied. According to this method, at each simulation step, the algorithm goes back *WinSize* number of elements in order to calculate the standard deviation within the variable window. Now, the new priorities are controlled as in the previous description but they are based on the delay standard deviation calculation.

3 Statistical Metrics

The blocking probability is one of the most important characteristics for the performance of a cellular network. When a new call arrival occurs and the network can not allocate a channel then we say that this call is blocked. The blocking probability $\mathbf{P_{blocking}}$ is calculated from the ratio

$$P_{blocking} = \frac{number\ of\ blocked\ calls}{number\ of\ calls} \tag{4}$$

If the received power for each user is high enough, we can make the assumption that the interference from other users can be ignored. Thus, we can compare the simulated blocking probability with the theoretical which is

$$P_{blocking\_theoretical} = \frac{\binom{n-1}{s}(vh)^s}{\sum_{i=0}^{s}\binom{n-1}{i}(vh)^i} \tag{5}$$

where **n** is the number of users, **s** is the number of channels, **v** is the average call arrival rate (for no connected user) and **h** is the average call holding time. The dropping probability is also an additional and very important characteristic of the cellular network performance. When a call is in progress and the required quality conditions are not met then this call is obligatorily driven to termination. The dropping probability $\mathbf{P_{fc}}$ is calculated from the ratio

$$P_{fc} = \frac{number\ of\ forced\ calls}{number\ of\ calls - number\ of\ blocked\ calls} \tag{6}$$

4 Experimental Results

The experimental results have been generated using Monte Carlo executions. Fig. 7 shows blocking probability over simulation time, divided in five days according to the previously mentioned daily model. The higher blocking probabilities correspond to the higher traffic zone of the day. Without setting dynamically the thread priorities, the system gives equal priorities to all threads and the average total delay is above 4.4×10^4 ms (fig. 8). Activating the negotiation dialog between agents and control agent (dynamically setting of priorities) the average delay time is kept below 1000ms (1sec).

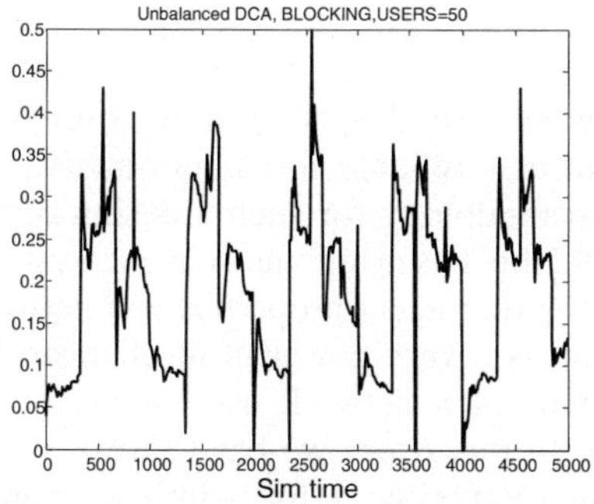

Fig. 7. A typical graph of Blocking probability over simulation time using Unbalanced DCA

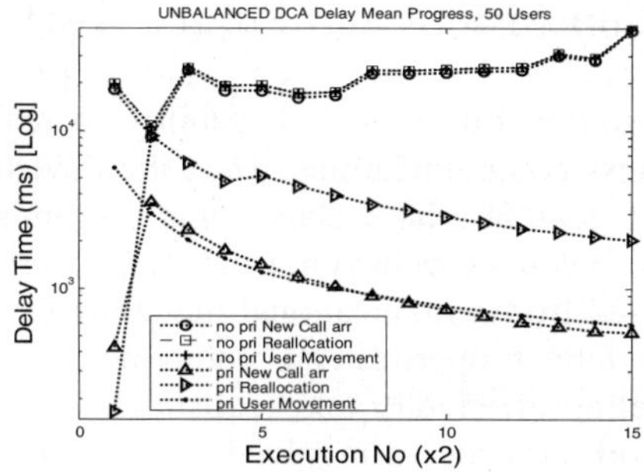

Fig. 8. Agent delay time before and after priority settings

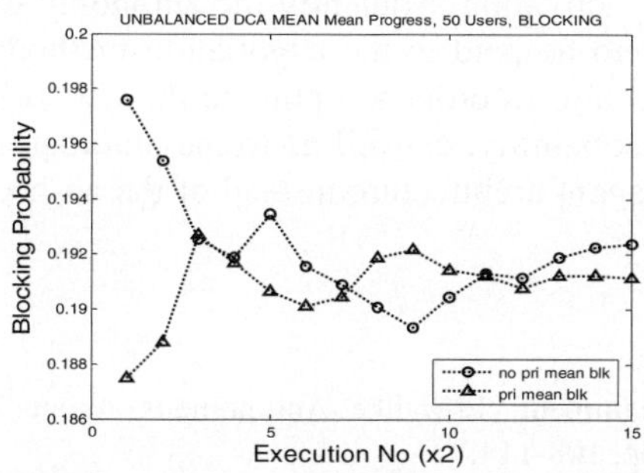

Fig. 9. Blocking probability of Unbalanced DCA before and after agent negotiation dialog

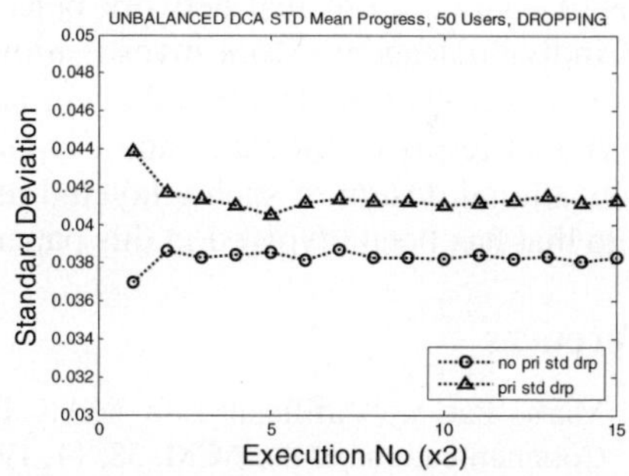

Fig. 10. Dropping probability of Unbalanced DCA before and after agent negotiation dialog

Figures 9 and 10 illustrate the impact of agent negotiation dialog to the simulation model behavior in terms of blocking and dropping probability.

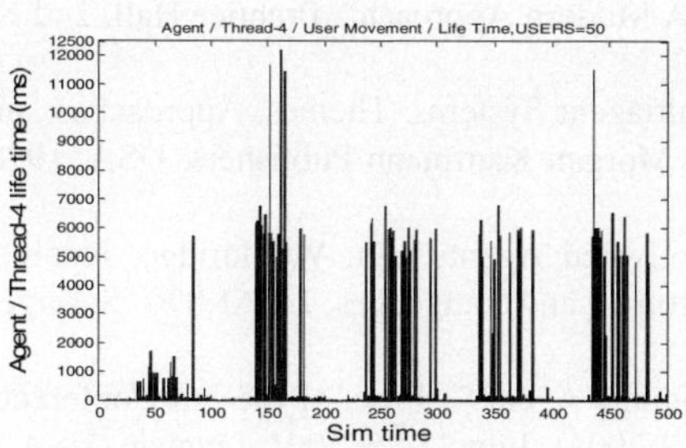

Fig. 11. Sample graph of Agent life time

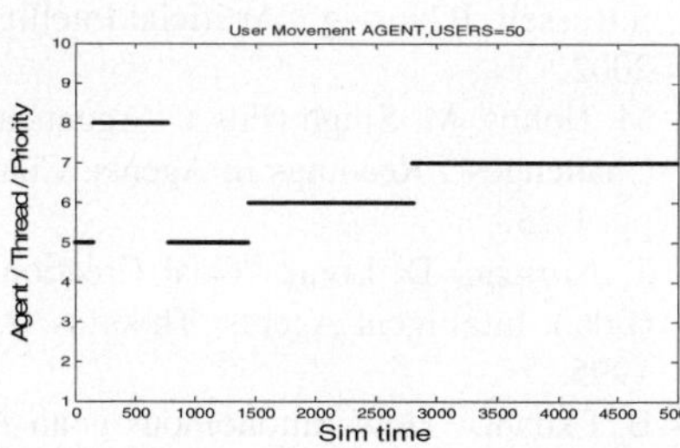

Fig. 12. Sample graph of priority settings

Figure 11 shows a typical diagram for the life time (active time) of an agent. In this graph a diagram is illustrated for the user's movement agent. Finally, figure 12 presents the dynamic change of agent properties and more specific for the user's movement agent.

5 Conclusions and Future Work

The simulation model adaptability to real network behavior is the key concept in wireless communications. Thus the development of a reliable simulation model is a major goal. We have shown that a promising methodology for such a design of the whole network behavior (including users) could be based on multi-agent systems inspired from organizational modeling by investigating agent properties and network procedures properties and requirements. An agent is a very powerful mechanism for modeling efficiently and realistically the behavior of a network user or the major network procedures involved in users servicing. In this paper we have shown that a multi-agent layered implementation of a simulation system for wireless cellular telecommunications, combined with a suitable negotiation mechanism based on thread priorities, produces remarkable results in terms of network behavior metrics. The selected approach with centralized cooperative negotiation can be improved in terms of adaptation to real network behaviour activating efficiently the autonomy of each individual agent. More investigation has to be paid to the negotiation methods and organizational models based agent architecture in order to optimize the network behavior in terms of blocking and dropping probability, as well as to the automated organizational design of such a layered multi-agent architecture instead of the ad-hoc design that has been involved in this paper.

References

[1] Maes, Pattie, “Artificial Life Meets Entertainment: Life like Autonomous Agents”, Communications of the ACM, 38, 11, 1995, pp. 108-114,

[2] Smith, D. C., A. Cypher and J. Spohrer, KidSim: “Programming Agents Without a Programming Language”, Communications of the ACM, 37, 7, (1994) 55-67

[3] Hayes-Roth, B., “An Architecture for Adaptive Intelligent Systems”, Artificial Intelligence: Special Issue on Agents and Interactivity, 72, (1995) 329-365

[4] S. Splunter, N. Wijngaards, F. Brazier. “Structuring Agents for Adaptation”. In: E. Alonso et al (Eds), Adaptive Agents and Multi-Agent Systems, LNAI, Vol. 2636, 2003, pp. 174-186.

[5] S.Russell, P.Norvig. “Artificial Intelligence: A Modern Approach”. Prentice Hall, 2nd ed, 2002.

[6] M. Huhns, M. Singh (Eds.), “Agents and Multiagent Systems: Themes, Approaches, and Challenges”. Readings in Agents, Chapter 1, Morgan Kaufmann Publishers, USA, 1998, pp. 1-23.

[7] T. Norman, D. Long. “Goal Creation in Motivated Agents”. In: Wooldridge, Jennings (Eds.), Intelligent Agents: Theories, Architectures, and Languages, LNAI 890: Springer, 1995.

[8] B. Ekdahl. “How Autonomous is an Autonomous Agent?”, Proc. of the 5th Conference on Systemic, Cybernetics and Informatics (SCI 2001), July 22-25, 2001, Orlando, USA.

[9] Jennings, N.R., “On agent-base software engineering”. Artificial Intelligence, 117, 2000, pp. 277-296

[10] Katia Sycara. “Multi-Agent Systems”. Artificial Intelligence Magazine 19(2). 1998

[11] G. Kumar and P. Venkataram., “Artificial intelligence approaches to network management: recent advances and a survey”. Computer Communications, 20:1313 – 1322, 1997.

[12] A. Hayzelden and J. Bigham. "Software Agents for Future Communications Systems". Springer-Verlag, Berlin, 1999.
[13] A. Hayzelden and J. Bigham. "Agent technology in communications systems: An overview". Knowledge Engineering Review Journal, 14, 1999.
[14] Wooldridge M., "An Introduction to Multiagent Systems", John Wiley & Sons,2002
[15] Xiaoqin Zhang, et al, "A Proposed Approach to Sophisticated Negotiation", AAAI Fall Symposium on Negotiation Methods for Autonomous Cooperative Systems. November 2001
[16] P.M.Papazoglou, D.A.Karras, R.C.Papademetriou, "A dynamic channel assignment simulation system for large scale cellular telecommunications", International Conference HERCMA 2005, September 2005,Athens, Greece, ISBN 960-87275-8-8
[17] M.Zhang, and T. S. Yum, "Comparisons of Channel Assignment Strategies in Cellular Mobile Telephone Systems", IEEE Transactions on Vehicular Technology, vol.38,no.4,pp211-215,1989.
[18] W.K. Lai and G.C. Coghill, "Channel Assignment through Evolutionary Optimization", IEEE Transactions on Vehicular Technology, vol.45,no.1,pp91-96,1996.
[19] V.H. MacDonald, "The cellular Concepts," The Bell System Technical, Journal, vol.58, pp.15–42, 1979.
[20] S. M. Elnoubi, R. Singh, and S.C. Gupta, "A New Frequency Channel Assignment Algorithm in High Capacity Mobile Communication Systems", IEEE Transactions on Vehicular Technology, vol. VT-21, no. 3,pp. 125–131, 1982.
[21] Z. Xu and P.B. Mirchandani, "Virtually Fixed Channel Assignment for Cellular Radio-Telephone Systems: A Model and Evaluation", IEEE International Conference on Communications, ICC'92, Chicago, vol. 2, pp. 1037–1041, 1982.
[22] L.J. Cimini and G.J. Foschini, "Distributed Algorithms for Dynamic Channel Allocation in Microcellular Systems", IEEE Vehicular Technology Conference, pp.641-644, 1992.
[23] D. C. Cox and D. O. Reudink, "Increasing Channel Occupancy in Large Scale Mobile Radio Systems: Dynamic Channel Reassignment", IEEE Transactions on Vehicular Technology, vol.VT-22, pp.218–222, 1973.
[24] E. Del Re, R. Fantacci, and G. Giambene, "A Dynamic Channel Allocation Technique based on Hopfield Neural Networks", IEEE Transactions on Vehicular Technology, vol.VT-45, no.1, pp.26–32, 1996.
[25] K.N. Sivarajan, R.J. McEliece, and J.W. Ketchum, "Dynamic Channel Assignment in Cellular Radio", IEEE 40th Vehicular Technology Conference, pp.631–637, 1990.
[26] T.J. Kahwa and N.D. Georgans, "A Hybrid Channel Assignment Schemes in Large-Scale, Cellular Structured Mobile Communication Systems", IEEE Transactions on Communications, vol.26, pp 432–438, 1978.
[27] P.M.Papazoglou, D.A.Karras, R.C.Papademetriou, "On new dynamic channel assignment schemes and their efficient evaluation through a generic simulation system for large scale cellular telecommunications", HERMIS, An International Journal of Computer Mathematics and its Applications, ISSN 1108-7609, Vol. 6, 2006

Towards an Agent-Based Negotiation Platform for Cooperative Decision-Making in Construction Supply Chain

Xiaolong Xue[1,2], Jinfeng Lu[1,2], Yaowu Wang[1,2], and Qiping Shen[2,3]

[1] Department of Construction and Real Estate, School of Management, Harbin Institute of Technology, Harbin 150001, China

[2] National Center of Technology, Policy and Management, Science Park, Harbin Institute of Technology, No.2, Yikuang Street, Harbin 150001, China

[3] Department of Building and Real Estate, The Hong Kong Polytechnic University, Hung Hom, Kowloon, Hong Kong, China

Xiaolong Xue, xlxue@hit.edu.cn, Jinfeng Lu, lujf@hit.edu.cn, Yaowu Wang, ywwang@hit.edu.cn, Qiping Shen, bsqpshen@polyu.edu.hk

Abstract. Negotiation is an effective and popular decision-making and coordination behavior in inter-organizations systems, especially in construction supply chain (CSC) which is characterized with fragmentation, low efficiency and multiple partners. This research proposes a multi-agent based negotiation platform for improving the effectiveness and efficiency of cooperative decision-making in CSC adopting agent technology and regarding CSC as a typical multi-agent system. General structure of the agent based negotiation platform is designed, which includes two kinds of agent group: specialty agents and service agents. Since different members in CSC have different preferences on the decision attributes (such as cost, time, quality, safety and environment), a multi-attribute negotiation model is established by designing negotiation protocol and describing the negotiation process.

Keywords: Agent, Multi-attribute negotiation, Cooperative decision-making, Construction supply chain.

1 Introduction

The construction industry is characterized with high fragmentation which results in complexity of the construction supply chain (CSC). For example, the separation of design and construction, lack of coordination and integration among organizations in supply chain, poor communication, uncertain production conditions, etc., are the important impact factors causing performance-related problems, such as low productivity, cost and time overrun, conflicts and disputes. Love et al. [1] According to Vrijhoef et al. [2], the major problems originate at the interfaces of different participants or stages involved in CSC. These problems are caused by myopic and independent decision-making in CSC operation. With the economic globalization and

N.T. Nguyen et al. (Eds.): KES-AMSTA 2007, LNAI 4496, pp. 416–425, 2007.

demands for improving construction performance and quickly responding to requirements of owner, cooperative decision-making has become the core strategy in CSC management.

The application of supply chain management (SCM) philosophy to the construction industry has been widely investigated as an effective and efficient management measure and strategy to improving the performance of construction since the middle of the 1990s [3], [4], and to address adversarial inter-organizational relationship of organization by increasing number of construction organizations and researchers [1] [3], [5], [6]. SCM, in a sense, can be consider as the coordination and cooperation of distributed decision-making of organizations or participants on material flow, information flow, human flow, and cash flow in supply chain from systems perspective.

Negotiation is defined as one kind of decision-making process where two or more decision makers jointly search a space of solution with the goal achieving consensus. It is an effective mechanism for supply chain coordination and cooperation.

Multi-agent systems (MAS) technology offers new means and tools for SCM [7]. The main advantage of MAS is its responsibilities for acting various components of the engineering process or participants of the business process which is delegated to a number of agents. MAS is suitable for domains that involve interactions between different organizations with different objectives and proprietary information [8]. SCM system is a typical MAS, where the participants are delegated to different agents. Furthermore, agent-based supply chain cooperation has been proved to be an effective mechanism to improve the performance of SCM [9]. The core principles of SCM and agent technology provide new perspectives for cooperative decision-making in CSC.

Whereas some researchers have addressed a number of key issues and have applied agent technology in construction management [10], [11], little research has been conducted to investigate the application of intelligent agent to support cooperative decision-making in CSC operation from chain-wide perspective. This paper will provides an agent-based negotiation platform for cooperative decision-making in CSC (ANeP) by adopting agent technology and multi-attribute negotiation (MAN) theory, and regarding CSC as a typical multi-agent system in the following sections.

2 General Structure Design of ANeP

CSC is a typical MAS, which involves multiple agents that delegate the organizations to autonomously perform tasks through exchanging information. CSC involves increasing participants along with the magnifying scale of more and more construction projects. So CSC also is a complex system, and cooperative decision-making becomes a challenging and significant thing that is vital to improve the performance of CSC and add the value of client. These require the agent based negotiation platform has the ability to efficiently communicate, perfect structure, effective coordination mechanism, stability, and flexibility. Considering the above factors, we design the general structure of ANeP as shown in Fig. 1.

In ANeP, the domain agents include both 'service' agents: coordinator agent, monitor agent, and name server agent, and 'specialty' agents: owner agent, design agent, general contractor (GC) agent, subcontractor agents, and supplier agents. The cooperative decision-making process in CSC is supported through MAN between specialty agents. We hypothesize all materials and human resources are organized by GC or subcontractors and don't consider the owner's suppliers in this structure. All agents communicate and cooperate through the Internet. Decision maker can control the negotiation process through Human-Computer Interface (Negotiation Window). Fig. 2 provides a detailed insight into how these agents function. The interactions shown are time ordered, with those at top occurring before those further down.

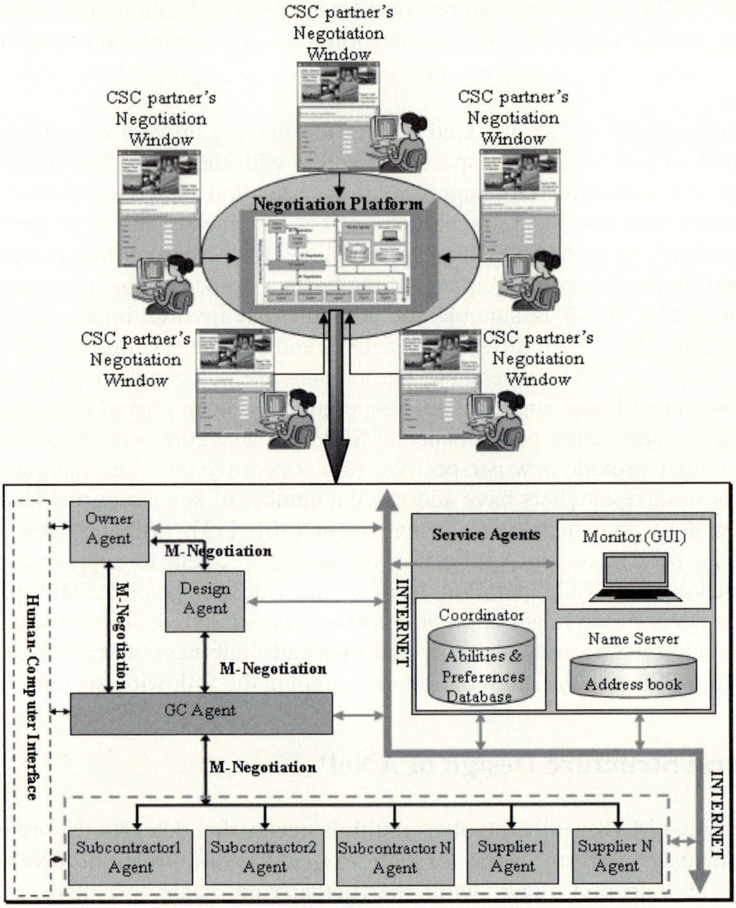

Fig. 1. General structure of ANeP

3 Multi-attribute Negotiation Model in ANeP

This research adopts the MAN technology to coordinate the cooperative decision-making in CSC. Since a number of factors such as cost, time, quality, safety, environment, must be considered in the decision-making process of CSC management. The above factors are seen as the attributes involved in CSC decision-making.

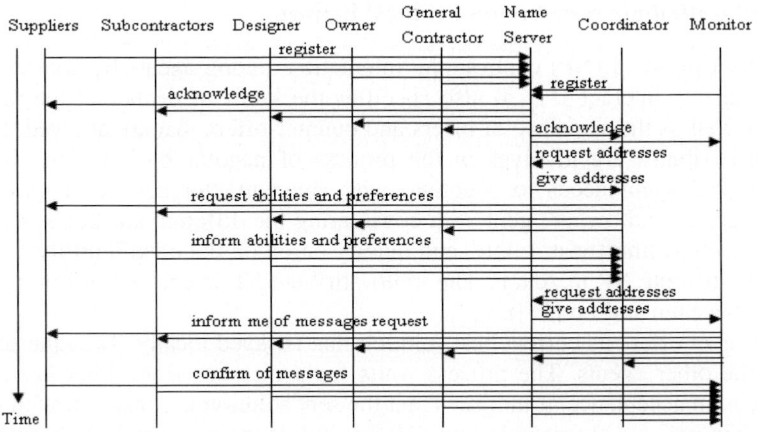

Fig. 2. Interactions of agents in ANeP

MAN technology is developed based on the multi-attribute utility theory (MAUT), which is an analytical tool for making decisions involving multiple interdependent objectives based on uncertainty and utility analyses.

This research presents an agent based MAN model for cooperative decision-making in CSC, which creatively extends the general negotiation model for A/E/C [10] from SCM and utility theory perspectives and integrates the compositional MAN model [12], as shown in Fig. 3. The model consists of four elements: negotiation protocol, CSC participants, MAN process, and the outcome. In the following we will focus on the negotiation protocol and negotiation process.

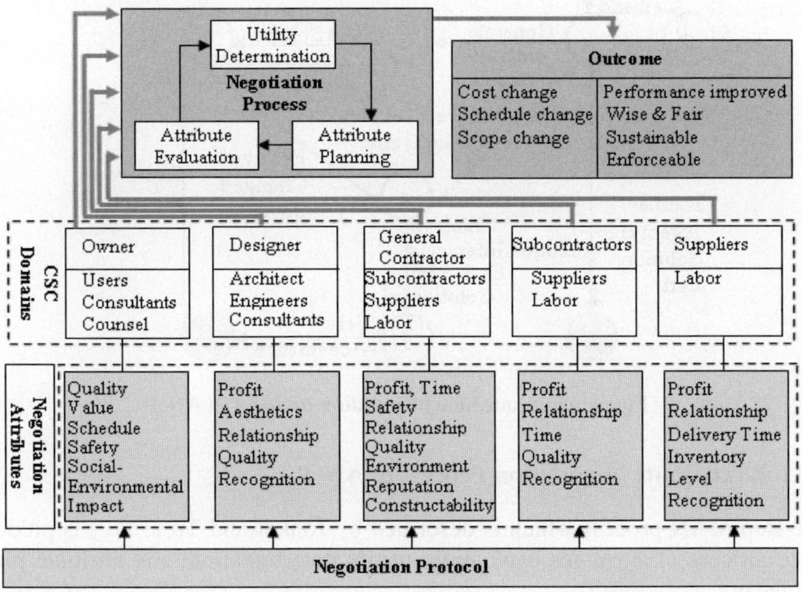

Fig. 3. Multi-attribute negotiation model in ANeP

3.1 Multi-attribute Negotiation Protocol in anep

Negotiation protocol (NP) controls the interaction among agents by constraining the way the agents interact [11]. It also specifies the kinds of deals that the agents can make, as well as the sequence of offers and counter-offers that are allowed. In ANeP, multiple attributes are involves in the process of negotiation between agents. For example, GC agent needs to negotiate with subcontractor agents, supplier agents, designer agent, and owner agent with considering the different attributes of decision-making, such as time, cost, safety, and quality based on the overall utility. So we call NP multi-attribute NP in ANeP. The multi-attribute NP is showed in Fig. 4 (adapted from Barbuceanu and Lo [13]).

Each agent offers its current best solution that is saved locally. Then the solution is sent to the other agents. The process waits for a message from other agents. If the message is an acceptance, it indicates that the sent solution is consistent with the other agents' solution, and the negotiation is successful. If the message is NoMoreSolution, the other agent has run out of solutions to generate. If the same is true for this agent as well, then the negotiation ends unsuccessfully. Otherwise the agent will continue to generate a new solution. If the message includes a changed solution from the other agent, this solution is checked for compatibility with any of the past solutions generated by this agent. If an intersection is found, it presents a mutually acceptable solution, which determines the negotiation successfully. Otherwise this indicates that utility gap remains between the own agent and the other agent. The top loop is repeated. The most advantage of the protocol is that it guarantees the discovery of the Pareto optimum [13].

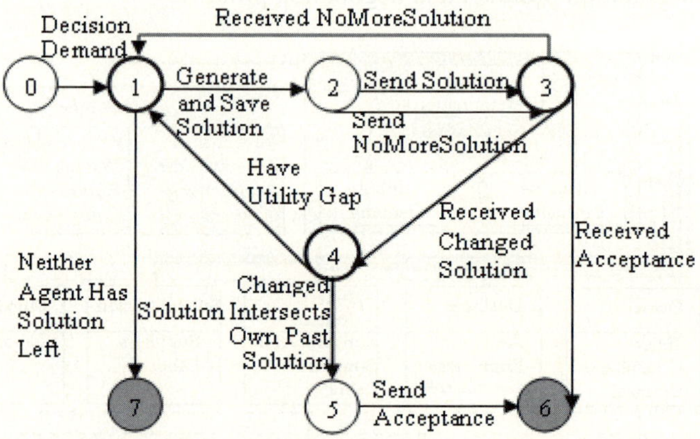

Fig. 4. Multi-attribute negotiation protocol in ANeP

3.2 Multi-attribute Negotiation Process in ANeP

A five-step MAN process which is described by Jonker and Treur [12] is predigested to three processes: attributes evaluation, utility determination, and attribute planning in this paper.

3.2.1 Attributes Evaluation

Many attributes are involved in the process of SCC decision-making in construction, as shown in part of 'Representative Negotiation Attributes' of Fig. 2. Attribute evaluation evaluates the value of the attributes based on the preferences of the participants in LCPSC. All attributes in general are classified into two categories: quantitative attributes and qualitative attributes. For the quantitative attributes, such as cost and time, the value of these attributes can directly calculate according relative principles. For the qualitative attributes, such as quality, safety, and environment, it is necessary to construct a scale representing the levels of these attributes. In this paper, a scale, from 0 (worst) to 10 (best), serves as the measure of evaluation.

3.2.2 Utility Determination

In this process, target utility (TU) is determined. TU is given by

$$TU = U_{BOW} + CS \tag{1}$$

where U_{BOW} is the utility of the own decision-making, and the concession step (CS) is determined by

$$CS = \beta(1-\mu/U_{BOW})(U_{BOT} - U_{BOW}) \tag{2}$$

where U_{BOT} is the utility of the other participant's decision-making. The factor $(1\text{-}\mu/U_{BOW})$ expresses CS will decrease to 0 if the U_{BOW} approximates the minimal utility μ and $(U_{BOT}\text{-}U_{BOW})$ expresses the current utility gap. β stands for the negotiation speed.

The utility of ith participant's decision-making (U_i) is given by

$$U_i = \sum_{j=1}^{n} w_j y_{ij} \tag{3}$$

where the w_j is the weight of jth attributes. y_{ij} is give by

$$y_{ij} = x_{ij} \Big/ \sqrt{\sum_{i=1}^{m} x_{ij}^2} \tag{4}$$

where x_{ij} is the value of jth attribute evaluated by ith participant in a decision-making process.

3.2.3 Attribute Planning

The attribute planning process refers to target evaluation and configuration determination. Target evaluation of jth attribute TE_j is given by

$$TE_j = (1-\tau)BTE_j + \tau E_{BOT,j} \tag{5}$$

where BTE_j is the basic target evaluation of jth attribute, which is determined in such away that $\sum w_j BTE_j = TU$. $E_{BOT,j}$ is the jth attribute evaluation value of other agent. τ stands for the configuration tolerance.

The configuration determination for the next decision-making includes three steps. Firstly, attribute values are determined with an evaluation that is as close as possible to the target evaluation value. Then a partial configuration (excepting the quantitative attributes, such as cost and time) is selected from the closest value of attribute. The final step is to reevaluate the quantitative attributes [12].

4 Prototype Development of ANeP

The prototype of ANeP is developed by using the ZEUS agent building toolkit. ZEUS is an advanced development toolkit for constructing distributed multi-agent applications. ZEUS is a culmination of a careful synthesis of established agent to provide an integrated and visual environment for the rapid software engineering of collaborative agent applications [14]. The developing process consists of two steps: role modeling and application design.

4.1 Role Modeling in ANeP

ZEUS agent building toolkit adopts role modeling to address the specification, analysis, design, implementation, and maintenance of agents. Role models formalize the definition of an agent role and provide a readily comprehensible means of analyzing the problem in question. The role models are grouped into domains. The domains provide a context that enables developers to compare their planned system with existing applications. Role models, which describe the dynamic interaction between roles, are architectural patterns that depict the high-level similarities between related systems, i.e. the problems inherent to each domain, but not how they were solved. The role models of ANeP are illustrated in Fig. 5.

4.2 Application Design

This process is illustrated based on a virtual CSC, which involves the following participants: Owner, designer, GC, Groundwork subcontractor, Civil and structure subcontractor, building services subcontractor, finishing works subcontractor, concrete supplier, and finishing materials supplier. Each participant is delegated to corresponding agent.

4.2.1 Ontology Creation

Ontology is a set of declarative knowledge representing every significant concept within a particular application domain. It contains the key concepts within the specific application domain, the attributes of each concept, the types of each attribute, and any restrictions on the attributes. In ZEUS an individual domain concept is described by using the term 'fact'. ZEUS provides two kinds of fact: abstract and entity. In ANeP, all the concepts refer to the entity, such as drawing, concrete, rebar, finishing materials, designer, various worker, supplier, subcontractor, etc.

4.2.2 Agent Creation

Agent creation includes three steps: agent definition, agent organization, and agent coordination in ZEUS. Agent definition determines the planning parameters, task and initial resources allocation. Agent organization illustrates the relationship between the own agent and other agents and acquaintance abilities from other agents. Agent

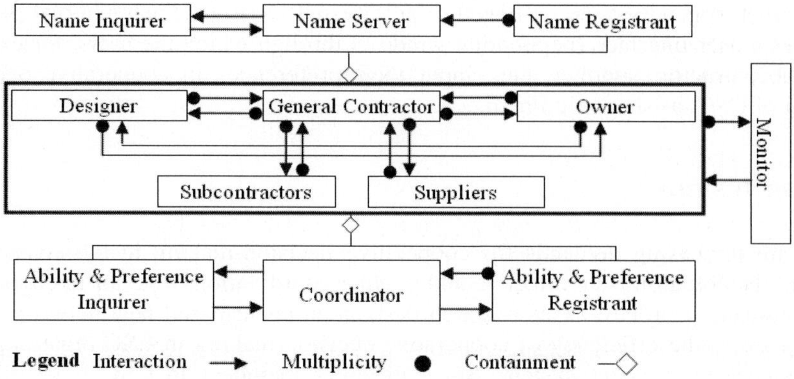

Fig. 5. Roles represented in ANeP

coordination defines the coordination protocols and strategies between the own agent and other agents. Multi-attribute NP previous mentioned is integrated into the prototype of ANeP as the coordination protocol in ZEUS.

4.2.3 External Program

ZEUS allows users to link an external java class (program) to executing ZEUS agent program. Once linked to the agent program, external program can utilize the agent's public methods to query or modify the agent's internal state. In the prototype of

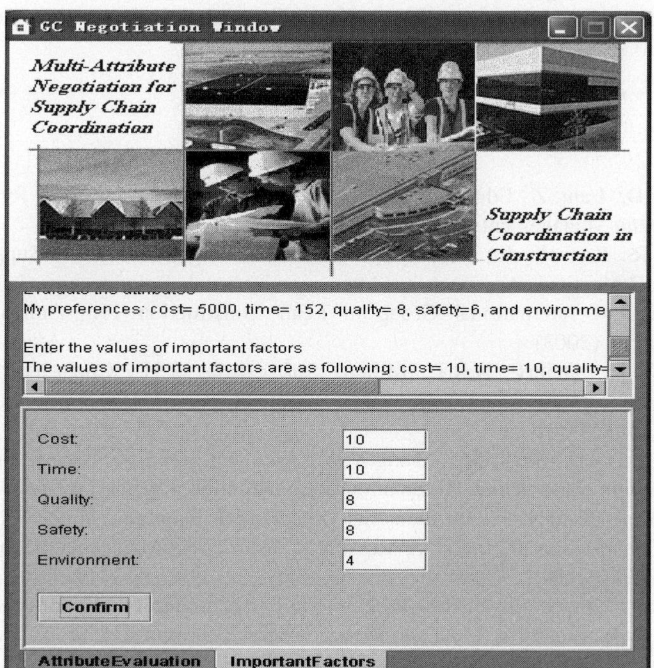

Fig. 6. Negotiation window of ANeP

ANeP, each specialty agent is linked to an external program. Each external program provides a user interface (negotiation window) through which the users, for example GC, subcontractor, supplier, etc., input their preferences in cooperative decision-making of CSC, as shown in Fig. 6.

5 Conclusions

There are increasing demands for cooperative decision-making in the construction industry. Negotiation is an effective and popular coordination behavior in cooperative decision-making process. This research designs an agent-based negotiation platform for improving the efficiency of cooperative decision-making in CSC regarding CSC as a typical multi-agent system. Since different members in CSC have different preferences on the decision attributes, a multi-attribute negotiation model is established by designing negotiation protocol and describing the negotiation process.

Since CSC, like other economic sectors, involves various participants and activities, so how to use agents to efficiently support cooperative decision-making is really a challenge. Furthermore, how agents can efficiently elicit participant's preferences and utility functions relevant will be the significant research direction in the future.

Acknowledgement. The research was supported by the foundation under the grant of htcsr06t05 from National Center of Technology, Policy and Management, Harbin Institute of Technology. The work described in this paper was also funded by the Research Grants Council of the Hong Kong Special Administrative Region, China (PolyU 5114/03E and 5264/06E).

References

1. Love, P.E.D., Irani, Z., Edwards, D.F.: A Seamless Supply Chain Model for Construction. Supply Chain Management: An International Journal, (2004) 9(1): 43-56
2. Vrijhoef, R., Koskela, L., Voordijk, H.: Understanding Construction Supply Chains: A Multiple Theoretical Approach to Inter-organizational Relationships in Construction. Proceedings of International Group of Lean Construction 11th Annual Conference, Virginia, USA (2003)
3. Briscoe, G.H., Dainty, A.R.J., Millett, S.J., Neale, R.H., Client-led Strategies for Construction Supply Chain Improvement. Construction Management and Economics, (2004) 2 (2): 193-201
4. Xue X.L., Li X.D., Shen Q.P., Wang Y.W.: An agent-based framework for supply chain coordination in construction. Automation in Construction, (2005), 14(3), 413-430
5. London, K.A., Kenley, R.: An Industrial Organization Economic Supply Chain Approach for the Construction Industry: A Review. Construction Management and Economics, (2001), 19: 777-788
6. Arbulu, R.J., Tommelein, I.D.: Contributors to Lead Time in Construction Supply Chains: Case of Pipe Supports Used in Power Plants. Proceedings of Winter Simulation Conference: Exploring New Frontiers, (2002) 1745-1751.

7. Frey, D., Stockheim, T., Woelk, P., Zimmermann, R.: Integrated Multi-agent-based Supply Chain Management. Proceedings of the Twelfth IEEE International Workshops on Enabling Technologies: Infrastructure for Collaborative Enterprises, (2003)
8. Ren, Z., Anumba, C.J.: Multi-agent Systems in Construction-State of the Art and Prospects. Automation in Construction, (2004) 13: 421-434
9. Lou, P., Zhou, Z.D., Chen, Y.P., Wu, A.: Study on the Multi-agent-based Agile Supply Chain Management. International Journal of Advanced Manufacturing Technology, (2004) 23: 197-203
10. Pena-Mora, F. Wang, C.Y.: Computer-supported Collaborative Negotiation Methodology. Journal of Computing in Civil Engineering, (1998) 12(2): 64-81
11. Kim, K., Paulson, B.C.: An Agent-based Compensatory Negotiation Methodology to Facilitate Distributed Coordination of Project Schedule Changes. Journal of Computing in Civil Engineering, (2003), 17(1): 10-18
12. Jonker, C.M., Treur, J.: An Agent Architecture for Multi-attribute Negotiation. Proceedings of the International Joint Conferences on Artificial Intelligence, Washington, USA, (2001)
13. Barbuceanu, M. Lo, W.K.: A Multi-attribute Utility Theoretic Negotiation Architecture for Electronic Commerce. Proceedings of the Fourth International Conference on Autonomous Agents, (2000)
14. Nwana, H.S., Ndumu, D.T., Lee, L.C., Collis, J.C.: ZEUS: A Toolkit for Building Distributed Multi-Agent Systems. Available from http://agent.aitia.ai/download, (Accessed 19.5.2005)

A Model for Informed Negotiating Agents

John Debenham and Simeon Simoff

Faculty of IT, University of Technology, Sydney, Australia
{debenham,simeon}@it.uts.edu.au
http://e-markets.org.au/

Abstract. We propose that the key to building intelligent negotiating agents is to take an agent's historic observations as primitive, to model that agent's changing uncertainty in that information, and to use that model as the foundation for the agent's reasoning. We describe an agent architecture, with an attendant theory, that is based on that model. In this approach, the utility of contracts, and the trust and reliability of a trading partner are intermediate concepts that an agent may estimate from its information model. This enables us to describe intelligent agents that are not necessarily utility optimisers, that value information as a commodity, and that build relationships with other agents through the trusted exchange of information as well as contracts.

1 Introduction

The potential value of the e-business market — including e-procurement — is enormous. Given that motivation and the current state of technical development it is surprising that a comparatively small amount of automated negotiation is presently deployed.[1] Technologies that support automated negotiation include multiagent systems [1] and virtual institutions [2], game theory and decision theory.

Game theory tells an agent what to do, and what outcome to expect, in many well-known negotiation situations [3], but these strategies and expectations are derived from assumptions about the agent's preferences and about the preferences of the opponent. One-to-one negotiation is generally known as *bargaining* [4] — it is the natural negotiation form when the negotiation object comprises a number of issues. For example, in bargaining over the supply of steel issues could include: the quantity of the steel, the quality of the steel, the delivery schedule, the settlement schedule and, of course, the price. Beyond bargaining there is a wealth of material on the theory of *auctions* [5] for one-to-many negotiation, and *exchanges* for many-to-many negotiation. Fundamental to this analysis is the central role of the utility function, and the notion of rational behaviour by which an agent aims to optimise its utility, when it is able to do so, and to optimise its *expected* utility otherwise.

[1] Auction bots such as those on eBay, and automated auction houses do a useful job, but do not automate negotiation in the sense described here.

N.T. Nguyen et al. (Eds.): KES-AMSTA 2007, LNAI 4496, pp. 426–435, 2007.

We propose that utility functions, or preference orderings, are often not known with certainty; further, the uncertainty that underpins them is typically in a state of flux. We propose that the key to building intelligent negotiating agents is to take an agent's historic observations as primitive, to model that agent's changing uncertainty in that information, and to use that model as the foundation for the agent's reasoning. We call such agents *information-based agents.* In Sec. 2 we describe these agents. Sec. 3 relates commitments made to their eventual execution, this leads to a formalisation of trust. Strategies for information-based agents are discussed in Sec. 4.

2 The Architecture of Information-Based Agents

The architecture of our information-based agent is shown in Fig. 1. The agent begins to function in response to a percept (received message) that expresses a need $N \in \mathcal{N}$. A *need* can be either exogenous (typically, the agent receives a message from another agent $\{\Omega_1, \ldots, \Omega_o\}$), or endogenous. The agent has a set of pre-specified *goals* (or *desires*), $G \in \mathcal{G}$, from which one or more is selected to satisfy its perceived needs. Each of these goals is associated with one or more plans, $s \in S$. This is consistent with the BDI model [1], and we do not detail these aspects here. The agent in Fig. 1 also interacts with information sources $\{\theta_1, \ldots, \theta_t\}$ that in our experiments[2] include unstructured data mining and text mining 'bots' that retrieve information from the agent market-place and from general news sources.

Finally the agent in Fig. 1 interacts with an 'Institution Agent', ξ, that reports honestly and promptly on the fulfilment of contracts. The Institution Agent is a conceptual device to prevent the requirement for agents to have 'eyes' and effectors. For example, if agent Π wishes to give an object Y to agent Ω_k then this is achieved by Π sending a message to ξ requesting the transfer of the ownership of Y from Π to Ω_k, once this is done, ξ sends a message to Ω_k advising him that he now owns Y. Given such an Institution Agent, agents can negotiate and evaluate trade by simply sending and receiving messages.

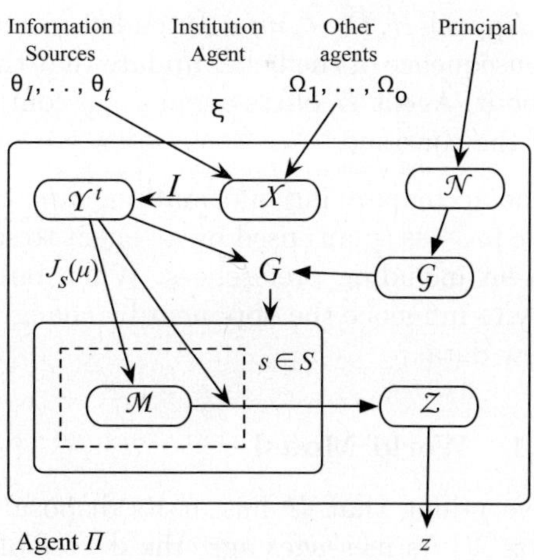

Fig. 1. Agent architecture

Π has two languages: $\mathcal{C}$ and $\mathcal{L}$. $\mathcal{L}$ is a first-order language for internal representation — precisely, it is a first-order language with sentence probabilities optionally attached to each sentence representing Π's epistemic belief in the validity of that sentence. $\mathcal{C}$ is an illocutionary-based language for communication

[2] `http://e-markets.org.au`

[6]. Messages expressed in $\mathcal{C}$ from $\{\theta_i\}$ and $\{\Omega_i\}$ are received, time-stamped, source-stamped and placed in an *in-box* $\mathcal{X}$. The illocutionary particles in $\mathcal{C}$ are:

– Offer(Π, Ω_k, δ). Agent Π offers agent Ω_k a contract $\delta = (\pi, \varphi)$ with action commitments $\pi \in \mathcal{L}$ for Π and $\varphi \in \mathcal{L}$ for Ω_k.
– Accept(Π, Ω_k, δ). Agent Π accepts agent Ω_k's previously offered contract δ.
– Reject($\Pi, \Omega_k, \delta[, \textit{info}]$). Agent Π rejects agent Ω_k's previously offered contract δ. Optionally, information $\textit{info} \in \mathcal{L}$ explaining the reason for the rejection can be given.
– Withdraw($\Pi, \Omega_k[, \textit{info}]$). Agent Π breaks down negotiation with Ω_k. Extra $\textit{info} \in \mathcal{L}$ justifying the withdrawal may be given.
– Inform($\Pi, \Omega_k, \textit{info}$). Agent Π informs Ω_k about $\textit{info} \in \mathcal{L}$ and commits to the truth of *info*.
– Reward($\Pi, \Omega_k, \delta, \phi[, \textit{info}]$). Intended to make the opponent accept a proposal with the promise of a future compensation. Agent Π offers agent Ω_k a contract δ. In case Ω_k accepts the proposal, Π commits to make $\phi \in \mathcal{L}$ true. The intended meaning is that Π believes that worlds in which ϕ is true are somehow desired by Ω_k. Optionally, additional information in support of the contract can be given.
– Threat($\Pi, \Omega_k, \delta, \phi, [\textit{info}]$) Intended to make the opponent accept a proposal with the menace of some sort of retaliation. Agent Π offers agent Ω_k a contract δ. In case Ω_k does not accept the proposal, Π commits to make $\phi \in \mathcal{L}$ true. The intended meaning is that Π believes that worlds in which ϕ is true are somehow *not* desired by Ω_k. Optionally, additional information in support of the contract can be given.
– Appeal($\Pi, \Omega_k, \delta, \textit{info}$) Intended to make the opponent accept a proposal as a consequence of the belief update that the accompanying information might bring about. Agent Π offers agent Ω_k a contract δ, and passes information in support of the contract.

The accompanying information, *info*, can be of two basic types: (i) referring to the process (plan) used by an agent to solve a problem, or (ii) data (beliefs) of the agent including preferences. When building relationships, agents will therefore try to influence the opponent by changing their processes (plans) or by providing new data.

2.1 World Model

Everything that Π has at its disposal is derived from the messages in the in-box $\mathcal{X}$. As messages age, the degree of belief that Π associates with them will decrease. We call this *information integrity decay.* A factor in the integrity of a message will be the reliability of the source. This subjective decay is a feature of the agent, and agents will differ in their subjective estimates.

Each plan is driven by its expectations of the state of the world, and by the states of the other agents. These states will generally be quite numerous, and so we assume that at any time the agent's active plans will form expectations of certain *features* only, where each feature will be in one of a finite number of states[3]. Suppose that there are m such features, introduce m random variables, $\{X_i\}_{i=1}^m$. Each value, $x_{i,j}$, of the i'th random variable, X_i, denotes that the i'th feature is in the j'th perceivable state, or *possible world*, of that feature.

The messages in $\mathcal{X}$ are then translated using an *import function* I into sentences expressed in $\mathcal{L}$ that have integrity decay functions (usually of time) attached to each sentence, they are stored in a *repository* $\mathcal{Y}^t$. And that is all that happens until Π triggers a goal.

In general Π will be uncertain of the current state of each feature. Π's *world model*, $M \in \mathcal{M}$, consists of probability distributions over each of these random variables. If these m features are independent then the overall uncertainty, or entropy, of Π's world model is: $\mathbb{H}^t(M) = -\sum_{i=1}^m \mathbb{E}(\ln \mathbb{P}^t(X_i))$. The general idea is that if Π receives new information then the overall uncertainty of the world model is expected to decrease, but if Π receives no new information then it is expected to increase.

2.2 Reasoning

Consider first what happens if Π receives no new information. Each distribution, $\mathbb{P}^t(X_i)$, is associated with a *decay limit distribution*, $\mathbb{D}(X_i)$, that represents the expected limit state of the i'th feature in the absence of any observations of the state of that feature: $\lim_{t\to\infty} \mathbb{P}^t(X_i) = \mathbb{D}(X_i)$. For example, if the i'th feature is whether it is raining in Sydney, and $x_{i,1}$ means "it is raining in Sydney" and $x_{i,2}$ means "it is not raining in Sydney" — then if Π believes that it rains in Sydney 5% of the time: $\mathbb{D}(X_i) = (0.05, 0.95)$. If Π has no background knowledge about $\mathbb{D}(X_i)$ then the decay limit distribution is the maximum entropy, "flat", distribution. In the absence of incoming information, $\mathbb{P}(X_i)$ decays by:

$$\mathbb{P}^{t+1}(X_i) = \Delta_i(\mathbb{D}(X_i), \mathbb{P}^t(X_i))$$

where Δ_i is the *decay function* for the i'th feature satisfying the property that $\lim_{t\to\infty} \mathbb{P}^t(X_i) = \mathbb{D}(X_i)$. For example, Δ_i could be linear:

$$\mathbb{P}^{t+1}(X_i) = (1 - \nu_i)\mathbb{D}(X_i) + \nu_i \times \mathbb{P}^t(X_i) \qquad (1)$$

where $\nu_i < 1$ is the decay rate for the i'th feature. Either the decay function or the decay limit distribution could also be a function of time: Δ_i^t and $\mathbb{D}^t(X_i)$.

If Π receives a message expressed in $\mathcal{C}$ then it will be transformed by inference rules into statements expressed in $\mathcal{L}$. We introduce this procedure with an example. Preference information is a statement by an agent that it prefers one class of contracts to another where contracts may be multi-issue. Preference illocutions may refer to particular issues within contracts — e.g. "I prefer red to yellow", or to combinations of issues — e.g. "I prefer a car with a five year warranty to the same car than costs 15% less with a two year warranty". An agent will find it useful to estimate which contract under consideration is favoured most by the opponent. Preference information can assist with this estimation as the

[3] We thus exclude the possibility of continuous variables.

following example shows. Suppose Π receives preference information from Ω_k through an Inform(Ω_k, Π, *info*) illocution: *info* = "for contracts with property Q_1 or property Q_2, the probability that the contract Ω_k prefers most will have property Q_1 is z" — the ontology in $\mathcal{C}$ is assumed to contain an illocutionary particle that can express this statement. What happens next will depend on Π's plans. Suppose that Π has an active plan $s \in S$ that calls for the probability distribution $\mathbb{P}^t(\text{Favour}(\Omega_k, \Pi, \delta)) \in M$ over all δ, where $\text{Favour}(\Omega_k, \Pi, \delta)$ means that "δ is the contract that Ω_k prefers most from Π". Suppose Π has a prior distribution $\boldsymbol{q} = (q_1, \dots)$ for $\mathbb{P}^t(\text{Favour}(\cdot))$. Then s will require an inference rule: $J_s^{\text{Favour}}(\textit{info})$ that is the following linear constraint on the posterior $\mathbb{P}^t(\text{Favour}(\Omega_k, \Pi, \delta))$ distribution:

$$z = \frac{\sum_{\delta:Q_1(\delta)} p_\delta}{\left(\sum_{\delta:Q_1(\delta)} p_\delta\right) + \left(\sum_{\delta:Q_2(\delta)} p_\delta\right) - \left(\sum_{\delta:Q_1 \wedge Q_2(\delta)} p_\delta\right)} \tag{2}$$

and is determined by the *principle of minimum relative entropy* — a form of Bayesian inference that is convenient when the data is sparse [7] — as described generally below. The inference rule $J_s^{Favour}(\cdot)$ infers a constraint on a distribution in M from an illocution expressed in $\mathcal{C}$. Inferences of this sort are necessary for Π to operate, but their validity is a personal matter for Π to assume.

Now, more generally, suppose that Π receives a percept μ from agent Ω_k at time t. Suppose that this percept states that something is so with probability z, and suppose that Π attaches an epistemic belief probability $\mathbb{R}^t(\Pi, \Omega_k, \mu)$ to μ. Π's set of active plans will have a set of model building functions, $J_s(\cdot)$, such that $J_s^{X_i}(\mu)$ is a set of linear constraints on the posterior distribution for X_i where the prior distribution is $\mathbb{P}^t(X_i) = \boldsymbol{q}$. Let $\boldsymbol{p} = (p_1, \dots)$ be the distribution with minimum relative entropy with respect to $\boldsymbol{q}$: $\boldsymbol{p} = \arg\min_{\boldsymbol{p}} \sum_j p_j \log \frac{p_j}{q_j}$ that satisfies the constraints $J_s^{X_i}(\mu)$. Then let $\boldsymbol{r}$ be the distribution:

$$\boldsymbol{r} = \mathbb{R}^t(\Pi, \Omega_k, \mu) \times \boldsymbol{p} + (1 - \mathbb{R}^t(\Pi, \Omega_k, \mu)) \times \boldsymbol{q}$$

and then for a small time step δt let:

$$\mathbb{P}^{t+\delta t}(X_i) = \begin{cases} \boldsymbol{r} & \text{if } \mathbb{K}(\boldsymbol{r} \| \mathbb{D}(X_i)) > \mathbb{K}(\mathbb{P}^t(X_i) \| \mathbb{D}(X_i)) \\ \boldsymbol{q} & \text{otherwise} \end{cases} \tag{3}$$

where $\mathbb{K}(\boldsymbol{x} \| \boldsymbol{y}) = \sum_j x_j \ln \frac{x_j}{y_j}$ is the Kullback-Leibler distance between two probability distributions $\boldsymbol{x}$ and $\boldsymbol{y}$. The idea in Eqn. 3 is that the vector $\boldsymbol{r}$ will only update $\mathbb{P}^t(X_i)$ if it contains more information with respect to the decay limit distribution than the prior $\boldsymbol{q}$. Then combining Eqn. 3 with Eqn. 1 let:

$$\mathbb{P}^{t+1}(X_i) = (1 - \nu_i)\mathbb{D}(X_i) + \nu_i \times \mathbb{P}^{t+\delta t}(X_i) \tag{4}$$

and note that this procedure has dealt with integrity decay, and with two probabilities: first, the probability z in the percept μ, and second the epistemic belief probability $\mathbb{R}^t(\Pi, \Omega_k, \mu)$ that Π attached to μ. Given a probability distribution $\boldsymbol{q}$, the *minimum relative entropy distribution* $\boldsymbol{p} = (p_1, \dots, p_I)$ subject to a set of J linear constraints $\boldsymbol{g} = \{g_j(\boldsymbol{p}) = \boldsymbol{a_j} \cdot \boldsymbol{p} - c_j = 0\}, j = 1, \ldots, J$ (that must include the constraint $\sum_i p_i - 1 = 0$) is: $\boldsymbol{p} = \arg\min_{\boldsymbol{p}} \sum_j p_j \log \frac{p_j}{q_j}$. This may be calculated by introducing Lagrange multipliers $\boldsymbol{\lambda}$: $L(\boldsymbol{p}, \boldsymbol{\lambda}) = \sum_j p_j \log \frac{p_j}{q_j} + \boldsymbol{\lambda} \cdot \boldsymbol{g}$. Minimising L, $\{\frac{\partial L}{\partial \lambda_j} = g_j(\boldsymbol{p}) = 0\}, j = 1, \ldots, J$ is the set of given constraints $\boldsymbol{g}$, and a solution to $\frac{\partial L}{\partial p_i} = 0, i = 1, \ldots, I$ leads eventually to $\boldsymbol{p}$.

2.3 Estimating $\mathbb{R}^t(\Pi, \Omega_k, \mu)$

Π attaches an epistemic belief probability $\mathbb{R}^t(\Pi, \Omega_k, \mu)$ to each message μ. A historic estimate of $\mathbb{R}^t(\Pi, \Omega_k, \mu)$ may be obtained by measuring the 'difference' between commitment and execution. Π's plans will have constructed a set of distributions. We measure this 'difference' as the error in the effect that μ has on each of Π's distributions. Suppose that μ is received from agent Ω_k at time u and is verified at some later time t. For example, μ could be a chunk of information: "the interest rate will rise by 0.5% next week", and suppose that the interest rate actually rises by 0.25% — represent what the message should have been μ'. What does all this tell agent Π about agent Ω_k's reliability? Consider one of Π's distributions for X that is $\boldsymbol{q}^u$ at time u. Let $\boldsymbol{p}^u_\mu$ be the posterior minimum relative entropy distribution subject to the constraint $J^X_s(\mu)$, and let $\boldsymbol{p}^u_{\mu'}$ be that distribution subject to $J^X_s(\mu')$. We now estimate what $\mathbb{R}^u(\Pi, \Omega_k, \mu)$ should have been in the light of knowing *now*, at time t, that μ should have been μ'.

The idea of Eqn. 3, is that the current value of $\mathbb{R}^t(\Pi, \Omega_k, \mu)$ should be such that, *on average*, $\boldsymbol{p}^u_\mu$ will be "close to" $\boldsymbol{p}^u_{\mu'}$ when we eventually discover μ' — no matter whether or not μ was used to update the distribution for X, as determined by the acceptability test in Eqn. 3 at time u. The *observed reliability* for μ and distribution X, $\mathbb{R}^t_X(\Pi, \Omega_k, \mu)|\mu'$, on the basis of the verification of μ with μ', is the value of r that minimises the Kullback-Leibler distance:

$$\mathbb{R}^t_X(\Pi, \Omega_k, \mu)|\mu' = \arg\min_r \mathbb{K}(r \cdot \boldsymbol{p}^u_\mu + (1 - r) \cdot \boldsymbol{q}^u \| \boldsymbol{p}^u_{\mu'})$$

If $\mathbf{X}(\mu)$ is the set of distributions that μ affects, then the overall *observed reliability* on the basis of the verification of μ with μ' is:

$$\mathbb{R}^t(\Pi, \Omega_k, \mu)|\mu' = 1 - (\max_{X \in \mathbf{X}(\mu)} |1 - \mathbb{R}^t_X(\Pi, \Omega_k, \mu)|\mu'|)$$

Then for each ontological context o_j, at time t when μ has been verified with μ':

$$\mathbb{R}^{t+1}(\Pi, \Omega_k, o_j) = (1 - \nu) \times \mathbb{R}^t(\Pi, \Omega_k, o_j) + \nu \times \mathbb{R}^t(\Pi, \Omega_k, \mu)|\mu' \times \text{Sim}(o_j, O(\mu))$$

where Sim measures the semantic distance between two sections of the ontology, and ν is the learning rate. Over time, Π notes the ontological context of the various μ received from Ω_k, and over the various ontological contexts calculates the relative frequency, $\mathbb{P}^t(o_j)$, of these contexts, $o_j = O(\mu)$. This leads to an overall expectation of the *reliability* that agent Π has for agent Ω_k:

$$\mathbb{R}^t(\Pi, \Omega_k) = \sum_j \mathbb{P}^t(o_j) \times \mathbb{R}^t(\Pi, \Omega_k, o_j)$$

3 Commitment and Execution

The interaction between agents Π and Ω_k will eventually lead to some sort of *contract*: $\delta = (\pi, \varphi)$ where π is Π's commitment and φ is Ω_k's commitment. No matter what these commitments are, Π will be interested in any variation between Ω_k's commitment, φ, and what actually happens, the execution, φ'. The form of this commitment could be a promise to deliver goods, or abide by certain trading terms that extend over a period of time, or that some information that may, or may not, prove to be correct. We denote the relationship between commitment and execution, $\mathbb{P}^t(\text{Execute}(\varphi')|\text{Commit}(\varphi))$ simply as $\mathbb{P}^t(\varphi'|\varphi)$. In general we assume that such commitment and execution takes place in the context of a *relationship* ρ between Π and Ω_k.

Beliefs 'evaporate' as time goes by. If we don't keep an ongoing relationship, we become unsure how *trustworthy* a trading partner is. This decay is what justifies a continuous relationship between agents. The conditional probabilities, $\mathbb{P}^t(\varphi'|\varphi)$, should tend to ignorance as represented by the *decay limit distribution* $\boldsymbol{d} = \{d_i\}$. If we have the set of observations $\Phi = \{\varphi_1, \varphi_2, \ldots, \varphi_n\}$ then complete ignorance of the opponent's expected behaviour means that given the opponent commits to φ, the conditional probability for each observable outcome φ' becomes $d_i = \frac{1}{n}$, but Π may have background beliefs about Ω_k's decay limit distribution. This natural decay of belief is offset by new observations. We define the evolution of the probability distribution as: $\mathbb{P}^{t+1}(\varphi'|\varphi) = \left((1-\nu)\cdot \boldsymbol{d} + \nu \cdot \mathbb{P}^t_+(\varphi'|\varphi)\right)$, where $\nu \in [0,1]$ is the learning rate, and $\mathbb{P}^t_+(\varphi'|\varphi)$ represents the posterior distribution for $(\varphi'|\varphi)$ given an observed contract execution as the following shows.

Suppose that Π has a business relationship ρ with agent Ω_k, that Ω_k commits to φ, and this commitment is sound. The material value of φ to ρ will depend on the future use that Π makes of it, and that is unlikely to be known. So Π estimates the value of φ to the relationship ρ he has with Ω_k using a probability distribution $(p_1, \ldots, p_n)$ over a *relationship evaluation space* $E = (e_1, \ldots, e_n)$ that could range from "that is what I expect from the perfect trading partner" to "it is totally useless" — E may contain hard or fuzzy values. $p_i = w_i(\rho, \varphi)$ is the probability that e_i is the correct evaluation of the enactment φ in the context of relationship ρ, and $\boldsymbol{w} : \mathcal{L} \times \mathcal{L} \to [0,1]^n$ is the *evaluation function*.

Let $(\varphi_1, \ldots, \varphi_m)$ be the set of possible contract executions in some order. Then for a given φ_k, $(\mathbb{P}^t(\varphi_1|\varphi_k), \ldots, \mathbb{P}^t(\varphi_m|\varphi_k))$ is the prior distribution of Π's estimate of what will actually occur if Ω_k committed to φ_k occurring and $\boldsymbol{w}(\rho, \varphi_k) = (w_1(\rho, \varphi_k), \ldots, w_n(\rho, \varphi_k))$ is Π's evaluation over E with respect to the relationship ρ of Ω_k's commitment φ_k. Π's expected evaluation of what will occur given that Ω_k has committed to φ_k occurring is:

$$\boldsymbol{w}^{\exp}(\rho, \varphi_k) = \left(\sum_{j=1}^{m} \mathbb{P}^t(\varphi_j|\varphi_k) \cdot w_1(\rho, \varphi_j), \ldots, \sum_{j=1}^{m} \mathbb{P}^t(\varphi_j|\varphi_k) \cdot w_n(\rho, \varphi_j) \right).$$

Now suppose that Π observes the event $(\phi'|\phi)$ in another relationship ρ' also with agent Ω_k. Eg: Π may buy wine and cheese from the same supplier. Π may wish to revise the prior estimate $\boldsymbol{w}^{\exp}(\rho, \varphi_k)$ in the light of the observation $(\phi'|\phi)$ to:

$$(\boldsymbol{w}^{\text{rev}}(\rho, \varphi_k) \mid (\varphi'|\varphi)) = \boldsymbol{g}(\boldsymbol{w}^{\exp}(\rho, \varphi_k), \boldsymbol{w}(\rho', \varphi), \boldsymbol{w}(\rho', \varphi'), \rho, \rho', \varphi, \varphi, \varphi'),$$

for some function $\boldsymbol{g}$ — the idea being, for example, that if the commitment, φ, concerning the purchase of cheese, ρ', was not kept then Π's expectation that the commitment, φ, concerning the purchase of wine, ρ, will not be kept should increase. We estimate the posterior $\mathbb{P}^t_+(\varphi'|\varphi)$ by applying the principle of minimum relative entropy: $\left(\mathbb{P}^t_+(\varphi_j|\varphi)\right)_{j=1}^m = \arg\min_{\boldsymbol{p}} \sum_{i=1}^m p_i \log \frac{p_i}{\mathbb{P}^t(\varphi_i|\varphi)}$ where $\boldsymbol{p} = (p_j)_{j=1}^m$, satisfies the n constraints:

$$\sum_{j=1}^m p_j \cdot w_i(\rho, \varphi_j) = g_i(\boldsymbol{w}^{\exp}(\rho, \varphi_k), \boldsymbol{w}(\rho', \varphi), \boldsymbol{w}(\rho', \varphi'), \rho, \rho', \phi, \varphi, \varphi')$$

for $i = 1, \ldots, n$. This is a set of n linear equations in m unknowns, and so the calculation of the minimum relative entropy distribution may be impossible if $n > m$. In this case, we take only the m equations for which the change from the prior to the posterior value is greatest. That is, we attempt to select the most significant factors.

Consider a distribution of expected fulfilment of commitments that represent Π's "ideal" for a relationship with Ω_k, in the sense that it is the best that Π could reasonably expect Ω_k to do. This distribution will be a function of Ω_k, Π's history with Ω_k, anything else that Π believes about Ω_k, and general environmental information including time — denote all of this by e, then we have $\mathbb{P}^t_I(\varphi'|\varphi, e)$. For example, if Π considers that it is unacceptable for the execution φ' to be less preferred than the commitment φ then $\mathbb{P}^t_I(\varphi'|\varphi, e)$ will only be non-zero for those φ' that Π prefers to φ. The distribution $\mathbb{P}^t_I(\cdot)$ represents what Π expects, or hopes, Ω_k will do. *Trust* is the relative entropy between this ideal distribution, $\mathbb{P}^t_I(\varphi'|\varphi, e)$, and the distribution of the observation of fulfilled commitments, $\mathbb{P}^t(\varphi'|\varphi)$. That is:

$$\text{Trust}(\Pi, \Omega_k, \varphi) = 1 - \sum_{\varphi'} \mathbb{P}^t_I(\varphi'|\varphi, e) \log \frac{\mathbb{P}^t_I(\varphi'|\varphi, e)}{\mathbb{P}^t(\varphi'|\varphi)} \tag{5}$$

where the "1" is an arbitrarily chosen constant being the maximum value that trust may have. This equation defines trust for one, single commitment φ — for example, my trust in my butcher if he commits to provide me with a 10% discount for the rest of the year. It makes sense to aggregate these values over a class of commitments, say over those φ that are subtypes of a particular relationship ρ, that is $\varphi \leq \rho$. In this way we measure the trust that I have in my butcher in relation to the commitments he makes for red meat generally:

$$\text{Trust}(\Pi, \Omega_k, \rho) = 1 - \frac{\sum_{\varphi:\varphi\leq\rho} \mathbb{P}^t(\varphi) \left[\sum_{\varphi'} \mathbb{P}^t_I(\varphi'|\varphi, e) \log \frac{\mathbb{P}^t_I(\varphi'|\varphi,e)}{\mathbb{P}^t(\varphi'|\varphi)}\right]}{\sum_{\varphi:\varphi\leq\rho} \mathbb{P}^t(\varphi)}$$

where $\mathbb{P}^t(\varphi)$ is a probability distribution over the space of commitments that the next commitment Ω_k will make to Π is φ. Similarly, for an overall estimate of Π's trust in Ω_k:

$$\text{Trust}(\Pi, \Omega_k) = 1 - \sum_{\varphi} \mathbb{P}^t(\varphi) \left[\sum_{\varphi'} \mathbb{P}^t_I(\varphi'|\varphi, e) \log \frac{\mathbb{P}^t_I(\varphi'|\varphi, e)}{\mathbb{P}^t(\varphi'|\varphi)} \right]$$

4 Strategies

An agent requires *strategies* for deciding who to interact with, and for deciding how to manage interaction using the language $\mathcal{C}$. In $\mathcal{C}$ as defined in Sec. 2, contracts may be for a single trade, or may encapsulate an on-going trading relationship. Interaction is normally bound by an *interaction protocol* that moderates the interaction sequence, and so may limit the range of model building functions, $J_s(\cdot)$. Consider the protocol in which statements in $\mathcal{C}$ are exchanged between pairs of agents, and Offer$(\cdot)$ statements are binding until countered, or until one of the pair issues a Quit$(\cdot)$. That is, an agent would only enter into a negotiation — ie: offer exchange — if it were prepared to commit. To manage this protocol, agent Π requires the following probability estimates in M where Π is bargaining with opponent Ω_k, in satisfaction of some need $N \in \mathcal{N}$:

1 $\mathbb{P}^t(val(\Pi, \Omega_k, N, \delta) = v_i)$ — for any deal, δ, the probability distribution over some valuation space $\{v_i\}$ that measures how "good" the deal δ is to Π.
2 $\mathbb{P}^t(acc(\Pi, \Omega_k, \delta))$ — for any deal, δ, the probability that Ω_k would accept δ.
3 $\mathbb{P}^t(conv(\Pi, \Omega_k, \Delta))$ — for any sequence of offer-exchanges, Δ, the probability that that sequence will converge to an acceptable deal.
4 $\mathbb{P}^t(trade(\Pi, \Omega_k, o_j) = u_i)$ — for an ontological context o_j, the probability distribution over some valuation space $\{u_i\}$ that measures how "good" Ω_k is as a trading partner to Π for deals in ontological context o_j.

The estimation of these distributions has been described previously [8]. Π's strategy determines how it uses these distributions. An approach to issue-tradeoffs is described in [9]. That strategy attempts to make an acceptable offer by "walking round" the iso-curve of Π's previous offer (that has, say, an acceptability of α) towards Ω_k's subsequent counter offer. In terms of the machinery described here: $\arg\max_{\delta}\{\, \mathbb{P}^t(acc(\Pi, \Omega_k, \delta)) \mid \mathbb{E}^t(val(\Pi, \Omega_k, N, \delta)) \approx \alpha \,\}$. By including the "information dimension" Π can implement strategies that go beyond utilitarian thinking. Π evaluates every illocution for its utilitarian value, and for its value as information. For example, the *equitable information revelation* strategy [8] responds to a message μ with a message that gives the recipient expected information gain similar to that which μ gave to Π; these responses are also "reasonable" from a utilitarian point of view. An information-based agent evaluates all exchanges in terms of both their estimated utilitarian value, and their information value.

Estimations of trust — Sec. 3 — may be used to select a trading partner. One interesting question is to determine a set of partners to maintain for deals from a particular section of the ontology — this is a question of risk management. Having identified such a set, the agent then has to decide which one of these partners to use for the next negotiation. A nice strategy is to choose the partner with a probability equal to the probability that they are the best choice — as determined by trust, or some other means.

An information-based agent additionally requires strategies to manage the exchange of information, and to be strategic in their information acquisition. This includes strategies for dealing with the information sources $\{\theta_1, \ldots, \theta_t\}$, which becomes interesting if those sources are not always available, charge a fee, or take some time to deliver. This also includes strategies for the acquisition of information by both covert and overt strategic interaction with other agents $\{\Omega_1, \ldots, \Omega_o\}$. These information strategies are the subject of current research.

5 Conclusion

We do not claim that this is the end of the matter in deploying automated negotiators, and the approach described here has yet to be trialed extensively. But we do maintain the strategic apparatus of intelligent negotiating agents should include the intelligent use of information. We have proposed a theoretical basis for managing information in the context of competitive interaction, and have shown how that theory may be computed by an intelligent agent. Information theory provides the theoretical underpinning that enables such an informed agent to value, manage and exchange her information intelligently.

References

1. Wooldridge, M.: Multiagent Systems. Wiley (2002)
2. Arcos, J.L., Esteva, M., Noriega, P., Rodríguez, J.A., Sierra, C.: Environment engineering for multiagent systems. Journal on Engineering Applications of Artificial Intelligence **18** (2005)
3. Rosenschein, J., Zlotkin, G.: Rules of Encounter. MIT Press (1998)
4. Muthoo, A.: Bargaining Theory with Applications. Cambridge UP (1999)
5. Klemperer, P.: The Economic Theory of Auctions : Vols I and II. Edward Elgar (2000)
6. Sierra, C., Jennings, N., Noriega, P., Parsons, S.: A framework for argumentation-based negotiation. In: Intelligent Agents IV: Agent Theories, Architectures, and Languages (ATAL-97), Springer-Verlag: Heidelberg, Germany (1998) 177–192
7. Cheeseman, P., Stutz, J.: On The Relationship between Bayesian and Maximum Entropy Inference. In: Bayesian Inference and Maximum Entropy Methods in Science and Engineering. American Institute of Physics, Melville, NY, USA (2004) 445 – 461
8. Debenham, J., Simoff, S.: An agent establishes trust with equitable information revelation. In Subrahmanian, V., Regli, W., eds.: Proceedings of the 2005 IEEE 2nd Symposium on Multi-Agent Security and Survivability, Drexel University, Philadelphia, USA, IEEE (2005) 66 – 74
9. Faratin, P., Sierra, C., Jennings, N.: Using similarity criteria to make issue trade-offs in automated negotiation. Journal of Artificial Intelligence **142** (2003) 205–237

Power in Norm Negotiation

Guido Boella[1] and Leendert van der Torre[2]

[1]Dipartimento di Informatica - Università di Torino - Italy
guido@di.unito.it
[2]University of Luxembourg - Luxembourg
leendert@vandertorre.com

Abstract. In social mechanism design, norm negotiation creates individual or contractual obligations fulfilling goals of the agents. The social delegation cycle distinguishes among social goal negotiation, obligation and sanction negotiation and norm acceptance. Power may affect norm negotiation in various ways, and we therefore introduce a new formalization of the social delegation cycle based on power and dependence, without referring to the rule structure of norms, actions, decision variables, tasks, and so on.

1 Introduction

Normative multiagent systems [1,2,3,4] provide agents with abilities to automatically devise organizations and societies coordinating their behavior via obligations, norms and social laws. A distinguishing feature from group planning is that also sanctions and control systems for the individual or contractual obligations can be created. Since agents may have conflicting goals with respect to the norms that emerge, they can negotiate amongst each other which norm will be created.

The social delegation cycle [5] explains the negotiation of new social norms from cognitive agent goals in three steps. First individual agents or their representatives negotiate social goals, then a social goal is negotiated in a social norm, and finally the social norm is accepted by an agent [6] when it recognizes it as a norm, the norm contributes to the goals of the agent, and it is obeyed by the other agents. A model of norm negotiation explains also what it means, for example, to recognize or to obey a norm, and how new norms interact with existing ones.

Power may affect norm negotiation in various ways, and we therefore propose to analyze the norm negotiation problem in terms of social concepts like power and dependence. Power has been identified as a central concept for modeling social phenomena in multi-agent systems by various authors [7,8,9,10], as Castelfranchi observes both to enrich agent theory and to develop experimental, conceptual and theoretical new instruments for the social sciences [11].

To motivate our social-cognitive model, we contrast it with an abstract description of the social delegation cycle using game theoretic artificial social systems. The problem studied in artificial social systems is the design, emergence or more generally the creation of social laws. Shoham and Tennenholtz [12] introduce social laws in a setting without utilities, and they define *rational* social laws as social laws that improve a social game variable [13]. We follow Tennenholtz' presentation for stable social laws [14].

N.T. Nguyen et al. (Eds.): KES-AMSTA 2007, LNAI 4496, pp. 436–446, 2007.

Moreover, we also contrast our social-cognitive model with existing highly detailed models of the social delegation cycle, like the ones we have proposed within normative multiagent systems [5]. The challenge to define social mechanisms, as we see it, is to go beyond the classical game theoretic model by introducing social and cognitive concepts and a negotiation model, but doing so in a minimal way. In the model proposed in this paper we therefore keep goals and obligations abstract and we do not describe them by first-order (or propositional) logic or their rule structure, we do not introduce decisions, actions, tasks, and so on. Similar concerns are also mentioned by Wooldridge and Dunne in their qualitative game theory [15].

The layout of this paper is as follows. In Section 2 we discuss an abstract model of the social delegation cycle in Tennenholtz' game-theoretic artificial social systems. In Section 3 we define our social-cognitive conceptual model of multiagent systems in which we study and formalize the social delegation cycle, and in Section 4 we define the negotiation protocol. In Section 5 we formalize goal negotiation, in Section 6 we formalize norm negotiation, and in Section 7 we formalize the acceptance relation.

2 Social Delegation Cycle Using Artificial Social Systems

In Tennenholtz' game-theoretic artificial social systems, the goals or desires of agents are represented by their utilities. A game or multi-agent encounter is a set of agents with for each agent a set of strategies and a utility function defined on each possible combination of strategies. Tennenholtz only defines games for two agents to keep the presentation of artificial social systems as simple as possible, but he also observes [14, footnote 4] that the extension to the multi-agent case is straightforward.

Definition 1. *A* game *(or a* multi-agent encounter*) is a tuple $\langle N, S, T, U_1, U_2\rangle$, where $N = \{1, 2\}$ is a set of agents, S and T are the sets of strategies available to agents 1 and 2 respectively, and $U_1 : S \times T \to I\!R$ and $U_2 : S \times T \to I\!R$ are utility functions for agents 1 and 2, respectively.*

The social goal is represented by a minimal value for the social game variable. Tennenholtz [14] uses as game variable the maximin value. This represents safety level decisions, in the sense that the agent optimizes its worst outcome assuming the other agents may follow any of their possible behaviors.

Definition 2. *Let S and T be the sets of strategies available to agent 1 and 2, respectively, and let U_i be the utility function of agent i. Define $U_1(s, T) = \min_{t\in T} U_1(s, t)$ for $s \in S$, and $U_2(S, t) = \min_{s\in S} U_2(s, t)$ for $t \in T$. The* maximin value for agent *1* (respectively *2*) *is defined by $\max_{s\in S} U_1(s, T)$ (respectively $\max_{t\in T} U_2(S, t)$). A strategy of agent i leading to the corresponding maximin value is called a* maximin strategy for agent i.

The social norm is represented by a social law, characterized as a restriction of the strategies available to the agents. It is *useful* with respect to an efficiency parameter e if each agent can choose a strategy that guarantees it a payoff of at least e.

Definition 3. *Given a game $g = \langle N, S, T, U_1, U_2 \rangle$ and an efficiency parameter e, we define a social law to be a restriction of S to $\overline{S} \subseteq S$, and of T to $\overline{T} \subseteq T$. The social law is* useful *if the following holds: there exists $s \in \overline{S}$ such that $U_1(s, \overline{T}) \geq e$, and there exists $t \in \overline{T}$ such that $U_2(\overline{S}, t) \geq e$. A (useful) convention is a (useful) social law where $|\overline{S}| = |\overline{T}| = 1$.*

A social law is *quasi-stable* if an agent does not profit from violating the law, as long as the other agent conforms to the social law (i.e., selects strategies allowed by the law). Quasi-stable conventions correspond to Nash equilibria.

Definition 4. *Given a game $g = \langle N, S, T, U_1, U_2 \rangle$, and an efficiency parameter e, a* quasi-stable social law *is a useful social law (with respect to e) which restricts S to $\overline{S}$ and T to $\overline{T}$, and satisfies the following: there is no $s' \in S - \overline{S}$ which satisfies $U_1(s', \overline{T}) > \max_{s \in \overline{S}} U_1(s, \overline{T})$, and there is no $t' \in T - \overline{T}$ which satisfies $U_2(\overline{S}, t') > \max_{t \in \overline{T}} U_2(\overline{S}, t)$.*

The efficiency parameter can be seen as a social kind of *utility aspiration level*, as studied by Simon [16]. Such aspiration levels have been studied to deal with limited or resource-bounded reasoning, and have led to the development of goals and planning in artificial intelligence; we therefore use a goal based ontology in this paper. The three steps of the social delegation cycle in this classical game-theoretic setting can be represented as follows. Goal negotiation implies that the efficiency parameter is higher than the utility the agents expect without the norm, for example represented by the Nash equilibria of the game. Norm negotiation implies that the social law is useful (with respect to the efficiency parameter). The acceptance relation implies that the social law is quasi-stable.

We use the game-theoretical model to motivate our conceptual model of normative multiagent systems. Due to the uniform description of agents in the game-theoretic model, it is less clear how to distinguish among kinds of agents. For example, the unique utility aspiration level does not distinguish the powers of agents to negotiate a better deal for themselves than for the other agents. Moreover, the formalization of the social delegation cycle does neither give a clue how the efficiency parameter is negotiated, nor how the social law is negotiated. For example, the goals or desires of the agents as well as other mental attitudes may play a role in this negotiation. There is no sanction or control system in the model (adding a normative system to encode enforceable social laws to the artificial social system complicates the model [17]). Finally, an additional drawback is that the three ingredients of the model (agent goals, social goals, and social laws) are formalized in three completely different ways.

3 Power Viewpoint on Normative Multiagent Systems

In this paper we follow the definition of power as the ability of agents to achieve goals. Thus, an agent is more powerful than another agent if it can achieve more goals.

For example, in the so-called power view on multi-agent systems [18], a multi-agent system consists of a set of agents (A), a set of goals (G), a function that associates with each agent the goals the agent desires to achieve (*goals*), and a function that associates with each agent the sets of goals it can achieve (*power*). To be precise, since goals can be conflicting in the sense that achieving some goals may make it impossible to achieve other goals, the function *goals* returns a set of set of goals for each set of agents. Such abstract structures have been studied as qualitative games by Wooldridge and Dunne [15], though they do not call the ability of agents to achieve goals their power. To model trade-offs among goals of agents, we introduce a priority relation among goals.

Definition 5. *Let a multiagent system be a tuple* $\langle A, G, goals, power, \geq\rangle$ *where:*

- *the set of agents* A *and the set of goals* G *are two finite disjoint sets;*
- $goals : A \rightarrow 2^G$ *is a function that associates with each agent the goals the agent desires to achieve;*
- $power : 2^A \rightarrow 2^{2^G}$ *is a function that associates with each set of agents the sets of goals the set of agents can achieve;*
- $\geq: A \rightarrow \subseteq 2^G \times 2^G$ *is a function that associates with each agent a partial pre-ordering on the sets of his goals;*

To model the role of power in norm negotiation, we extend the basic power view in a couple of ways. To model obligations we introduce a set of norms, we associate with each norm the set of agents that has to fulfill it, and of each norm we represent how to fulfill it, and what happens when it is not fulfilled. In particular, we relate norms to goals in the following two ways.

- First, we associate with each norm n a set of goals $O(n) \subseteq G$. Achieving these normative goals $O(n)$ means that the norm n has been fulfilled; not achieving these goals means that the norm is violated. We assume that every normative goal can be achieved by the group, i.e., that the group has the power to achieve it.
- Second, we associate with each norm a set of goals $V(n)$ which will not be achieved if the norm is violated (i.e., when its goals are not achieved), this is the sanction associated to the norm. We assume that the group of agents does not have the power to achieve these goals.

Since we accept norms without sanctions, we do not assume that the sanction affects at least one goal of each agent of the group the obligation belongs to.

Definition 6. *Let a normative multi-agent system be a tuple* $\langle MAS, N, O, V\rangle$ *extending a multiagent system* $MAS = \langle A, G, goals, power, \geq\rangle$ *where:*

- *the set of norms* N *is a finite set disjoint from* A *and* G*;*
- $O : N \times A \rightarrow 2^G$ *is a function that associates with each norm and agent the goals the agent must achieve to fulfill the norm; We assume for all* $n \in N$ *and* $a \in A$ *that* $O(n, a) \in power(\{a\})$*;*
- $V : N \times A \rightarrow 2^G$ *is a function that associates with each norm and agent the goals that will not be achieved if the norm is violated by agent* a*; We assume for each* $B \subseteq A$ *and* $H \in power(B)$ *that* $(\cup_{a\in A} V(n, a)) \cap H = \emptyset$.

An alternative way to represent normative multiagent systems replaces the function *power* by a function representing dependencies between agents. For example, a function of minimal dependence can be defined as follows. Agent a depends on agent set $B \subseteq A$ regarding the goal g if $g \in goals(a)$, $g \notin power(\{a\})$, $g \in power(B)$, and there is no $C \subset B$ such that $g \in power(C)$. Note that dependence defined in this way is more abstract than power, in the sense that we have defined dependence in terms of power, but we cannot define power in terms of dependence.

4 Generic Negotiation Protocol

A negotiation protocol is described by a set of sequences of negotiation actions which either lead to success or failure. In this paper we only consider protocols in which the agents propose a so-called deal, and when an agent has made such a proposal, then the other agents can either accept or reject it (following an order ($\succ$) of the agents). Moreover, they can also end the negotiation process without any result.

Definition 7 (Negotiation Protocol). *A negotiation protocol is a tuple* $\langle Ag, deals, actions, valid, finished, broken, \succ\rangle$*, where:*

- *the agents* Ag*, deals and actions are three disjoint sets, such that* $actions = \{propose(a,d), accept(a,d), reject(a,d) \mid a \in Ag, d \in deals\} \cup \{breakit(a) \mid a \in Ag\}$.
- *valid, finished, broken are sets of finite sequences of actions.*

We now instantiate this generic protocol for negotiations in normative multiagent systems. We assume that a sequence of actions (a history) is valid when each agent does an action respecting this order. Then, after each proposal, the other agents have to accept or reject this proposal, again respecting the order, until they all accept it or one of them rejects it. When it is an agent's turn to make a proposal, it can also end the negotiation by breaking it. The history is *finished* when all agent have accepted the last deal, and *broken* when the last agent has ended the negotiations.

Definition 8 (NMAS protocol). *Given a normative multiagent system* $\langle MAS, N, O, V\rangle$ *extending a multiagent system* $MAS = \langle A, G, goals, power, \geq\rangle$*, a negotiation protocol for NMAS is a tuple* $NP = \langle A, deals, actions, valid, finished, broken, \succ\rangle$*, where:*

- $\succ \subseteq A \times A$ *is a total order on A,*
- *a history* h *is a sequence of actions, and* $valid(h)$ *holds if:*
  - *the propose and breakit actions in the sequence respect* $\succ$,
  - *each propose is followed by a sequence of accept or reject actions respecting* $\succ$ *until either all agents have accepted the deal or one agent has rejected it,*
  - *there is no double occurrence of a proposal* $propose(a,d)$ *of the same deal by any agent* $a \in Ag$*, and*
  - *the sequence* h *ends iff either all agents have accepted the last proposal (*$finished(h)$*) or the last agent has broken the negotiation (*$broken(h)$*) instead of making a new proposal.*

In theory we can add additional penalties when agents break the negotiation. However, since it is in the interest of all agents to reach an agreement, we do not introduce such sanctions. In this respect norm negotiation differs from negotiation about obligation distribution [19], where it may be the interest of some agents to see to it that no agreement is reached. In such cases, sanctions must be added to the negotiation protocol to motivate the agents to reach an agreement.

Example 1. Assume three agents and the following history.
$action_1 : propose(a_1, d_1)$
$action_2 : accept(a_2, d_1)$
$action_3 : reject(a_3, d_1)$
$action_4 : propose(a_2, d_2)$
$action_5 : accept(a_3, d_2)$
$action_6 : accept(a_1, d_2)$

We have $valid(h)$, because the order of action respects $\succeq$, and we have $accepted(h)$, because the history ends with acceptance by all agents ($action_5$ and $action_6$) after a proposal ($action_4$).

The open issue of the generic negotiation protocol is the set of deals which can be proposed. They depend on the kind of negotiation. In social goal negotiation the deals represent a social goal, and in norm negotiation the deals contain the obligations of the agents and the associated control system based on sanctions.

5 Social Goal Negotiation

We characterize the allowed deals during goal negotiation as a set of goals which contains for each agent a goal it desires. Moreover, we add two restrictions. First, we only allow goals the agents have the power to achieve. Moreover, we have to consider the existing normative system, which may already contain norms that look after the goals of the agents. We therefore restrict ourselves to new goals. Additional constraints may be added, for example excluding goals an agent can see to itself. However, since such additional restrictions may be unrealistic in some applications (e.g., one may delegate some tasks to a secretary even when one has the power to see to these tasks oneself), we do not consider such additional constraints.

Definition 9 (Deals in goal negotiation). *In the goal negotiation protocol, a deal $d \in deals$ is a set of goals satisfying the following restrictions:*

1. *$d \in power(A)$*
2. *for all $a \in A$ there exists a $g \in d$ such that*
   (a) *$g \in goals(a)$*
   (b) *there does not exist a norm n in N such that $g \in \cup_{a \in A} O(n, a)$*

The following example illustrates a case in which each agent may desire to be alive. In artificial social systems, it would be based on the utility to be alive.

Example 2. Let $MAS = \langle A, G, goals, power, \geq \rangle$ be a multi-agent system with goals $G = \{not\text{-}killed\text{-}by_{a,b} \mid a, b \in A\}$, and $not\text{-}killed\text{-}by_{a,b} \in goals(a)$ for all $a, b \in A$. Moreover, let $NMAS = \langle MAS, N, O, V \rangle$ with N the empty set. Then G is a social goal (i.e., an element of *deals*) iff $G \in power(A)$.

We could easily further refine our model by defining more abstract goals such as "we have a safe society" and by adding a goal hierarchy reflecting that if some goal is fulfilled, another goal is fulfilled too. For example, if the goal safe-society if fulfilled then the goals $not\text{-}killed\text{-}by_{a,b}$ are all fulfilled too. However, to keep our model simple, and to focus on the social delegation cycle, we do not do so in this paper.

We now consider a variant of the running example from [5]. Three agents can work together in various ways. They can make a coalition to each perform a task, or they can distribute five tasks among them and obtain an even more desirable social goal.

Example 3. Let $MAS = \langle \{a_1, a_2, a_3\}, \{g_1, g_2, g_3, g_4, g_5\}, goals, power, \geq \rangle$ be a multiagent system, where:

power: $power(a_1) = \{\{g_1\}, \{g_2\}, \{g_3\}\}$, $power(a_2) = \{\{g_2\}, \{g_3\}, \{g_4\}\}$, $power(a_3) = \{\{g_3\}, \{g_4\}, \{g_5\}\}$, if $G_1 \in power(A)$ and $G_2 \in power(B)$ then $G_1 \cup G_2 \in power(A \cup B)$. Agent a_1 has the power to achieve goal g_1, g_2, g_3, agent a_2 has the power to achieve goal g_2, g_3, g_4, and agent a_3 can achieve goal g_3, g_4, and g_5. There are no conflicts among goals.

goals: $goals(a_1) = \{g_4, g_5\}$, $goals(a_2) = \{g_1, g_5\}$, $goals(a_3) = \{g_1, g_2\}$. Each agent desires the tasks it cannot perform itself.

Moreover, let $NMAS = \langle MAS, N, O, V \rangle$ be a normative multiagent system with $N = \{n\}$, $O(n, a_1) = \{g_1\}$. Since there has to be some benefit for agent a_2 and a_3, the goals g_5 and g_2 have to be part of the social goal. Therefore, social goals (i.e., possible deals) are $\{g_2, g_5\}$ and $\{g_2, g_4, g_5\}$.

Finally, consider the negotiation. Assuming agent a_1 is first in the order $\succ$, he will propose $\{g_2, g_4, g_5\}$. The other agents may accept this, or reject it and agent a_2 will them propose $\{g_2, g_5\}$. The latter would be accepted by all agents, as they know that according to the protocol no other proposals can be made.

The example illustrates that the negotiation does not determine the outcome, in the sense that there are multiple outcomes possible. Additional constraints may be added to the negotiation strategy to further delimit the set of possible outcomes.

6 Social Norm Negotiation

We formalize the allowed deals during norm negotiation as obligations for each agent to see to some goals, such that all goals of the social goal are included. Again, to determine whether the obligations imply the social goal, we have to take the existing normative system into account. We assume that the normative system only creates obligations that can be fulfilled together with the already existing obligations.

Definition 10 (Fulfillable nmas). *A normative multiagent system $\langle MAS, N, O, V \rangle$ extending a multiagent system $MAS = \langle A, G, goals, power, \geq \rangle$ can be fulfilled if there exists a $G' \in power(A)$ such that all obligations are fulfilled $\cup_{n \in N, a \in A} O(n, a) \subseteq G'$.*

Creating a norm entails adding obligations and violations for the norm.

Definition 11 (Add norm). *Let* $NMAS$ *be a normative multiagent system* $\langle MAS, N, O, V\rangle$ *extending a multiagent system* $MAS = \langle A, G, goals, power, \geq\rangle$. *Adding a norm* $n \notin N$ *with a pair of functions* $\langle d_1, d_2\rangle$ *for obligation* $d_1 : A \to 2^G$ *and for sanction* $d_2 : A \to 2^G$ *leads to the new normative multiagent system* $\langle MAS, N \cup \{n\}, O \cup d_1(n), V \cup d_2(n)\rangle$.

Moreover, if every agent fulfills it obligation, then the social goal is achieved.

Definition 12 (Deals in goal negotiation). *In the norm negotiation protocol, a deal* $d \in deals$ *for social goal* S *is a pair of functions* $\langle d_1, d_2\rangle$ *for obligation* $d_1 : A \to 2^G$ *and for sanction* $d_2 : A \to 2^G$ *satisfying the following conditions:*

1. *Adding* $\langle d_1, d_2\rangle$ *to* $NMAS$ *for a fresh variable* n *(i.e., not occurring in* N*) leads again to a normative multiagent system* $NMAS'$*;*
2. $NMAS'$ *achieves the social goal,* $\cup_{a\in A} d_1(a) = S$.
3. *If* $NMAS$ *is fulfillable, then* $NMAS'$ *is too.*

The following example models the norm not to kill someone else. Thus, this example of the social delegation cycle is an instance of the Kantian categorical imperative: you should not do to others what you don't want them to do to you. Note that we can also represent $alive_a$ as an abbreviation of the conjunction of $not\text{-}killed\text{-}by_{a,b}$ for all agents b when we extend the language with definitions, but this does not change the principle of the social mechanism.

Example 4. Assume $\{not\text{-}killed\text{-}by_{a,b} \mid a \in A\} \subseteq power(b)$ for all agents b. A possible deal for the social goal that there are no murders is that $d_1(b) = \{not\text{-}killed\text{-}by_{a,b} \mid a \in A\}$ for all agents $b \in A$.

The second example illustrates the negotiation protocol.

Example 5. Consider the social goal $\{g_2, g_4, g_5\}$. A possible solution here is that each agent sees to one of the goals. For the social goal $\{g_2, g_5\}$, there will always be one of the agents who does not have to see to any goal.

Sanctions can be added in the obvious way. In the norm negotiation as defined thus far, the need for sanctions has not been formalized yet. For this need, we have to consider the acceptance of norms.

7 Norm Acceptance

An agent accepts a norm when it believes that the other agents will fulfill their obligations, and the obligation implies the goals the cycle started with. For the former we use the quasi-stability of the norm (e.g., if the norm is a convention, then we require that the norm is a Nash equilibrium). Each agent b fulfills the norm *given that all other agents fulfill the norm.* Again we have to take the existing normative system into account, so we add the condition that all other norms are fulfilled. In general, it may mean that an agent does something which it does not like to do, but it fears the sanction more than this dislike. We use the trade-off among goals $\geq$.

Definition 13 (Stability). *A choice c of agent $b \in A$ in $NMAS$ with new norm n is $c \in power(b)$ such that $\cup_{m \in N \setminus \{n\}} O(m, b) \subseteq d$. The choices of the other agents are $oc = \cup_{a \in A \setminus \{b\}, m \in N} O(n, a) \cup V(n, a)$. The effect of choice c is $c \cup oc \cup V(n, a)$ if $O(n) \subseteq c$, $c \cup oc$ otherwise. $NMAS$ is stable if $\forall b \in A$, there is a choice c such that $O(n, b) \subseteq c$, and there is no choice $c' \geq(b) c$ with $O(n, b) \not\subseteq c'$.*

Finally, we have to test whether the new situation is better than the old one for all agents. Here we assume that the outcome in both the original multiagent system as in the new multiagent system is a Nash equilibrium, and we demand that each Nash outcome in the new system is better than each Nash outcome in the original normative multiagent system. The formalization of these concepts is along the same lines as the definition of acceptance in Definition 13.

Example 6. The norm in the 'don't kill me' example is quasi-stable, since there is no reason for an agent to divert from the norm. Moreover, it is effective since it sees to his goals of the agents. If we assume that agents minimize the set of goals they see to if it does not affect the priority of the agents, or there are some agents who would like to kill other agents, then a sanction has to be added to make sure that no-one is killed. The priority of the goal not to be sanctioned should be higher than the priority to kill.

8 Concluding Remarks

In this paper we introduce a norm negotiation model based on power and dependence structures. It is based on a distinction between social goal negotiation and the negotiation of the obligations with their control system. Roughly, the social goals are the benefits of the new norm for the agents, and the obligations are the costs of the new norm for the agents in the sense that agents risk being sanctioned. Moreover, in particular when representatives of the agents negotiate the social goals and norms, the agents still have to accept the negotiated norms. The norm is accepted when the norm is quasi-stable in the sense that agents will act according to the norm, and effective in the sense that fulfilment of the norm leads to achievement of the agents' desires – i.e., when the benefits outweigh the costs.

Our new model is based on a minimal extension of Tennenholtz' game theoretic model of the social delegation cycle. We add a negotiation protocol, sanction and control, and besides acceptance also effectiveness. It is minimal in the sense that, compared to our earlier model [5] in normative multiagent systems, we do not represent the rule structure of norms, we do not use decision variables, and so on. Also, as discussed in this paper, we do not add goal hierarchy, definitions, etc. The model therefore focusses on various uses of power: the power as ability to achieve goals and in negotiation.

The present paper may be seen as a preliminary study of the expressive power of power and dependence views on multiagent systems. As has been argued by Castelfranchi and Conte for some time, and has been supported by various researchers since then, the power and dependence viewpoint have advantages over classical game theory. However, it remains an open question whether such power structures (or qualitative games in Wooldridge and Dunne's theory) cannot be mapped on classical games by mapping goals on outcomes, power on strategies, and so on. In future research we intend to study under which conditions or assumptions such mappings can be made.

There are several other issues for further research. First, the motivation of our model is to design social mechanisms. Second, we would like to perform formal analysis, like the complexity results obtained for qualitative games [15] or in game-theoretic artificial social systems [12,13,14]. Third, we like to study more general notions of norm creation including permission creation and creation of constitutive norms or counts-as conditionals. Fourth, we are interested in the role of coalition formation, contract negotiation, and obligation distribution in the the new norm negotiation model. Finally, we would like to extend the model with the distinction between uncontrollable (or external) and controllable (or police) agents as studied by Brafman and Tennenholtz [20].

References

1. Conte, R., Falcone, R., Sartor, G.: Agents and norms: How to fill the gap? Artificial Intelligence and Law **7(1)** (1999) 1–15
2. Boella, G., van der Torre, L., Verhagen, H.: Introduction to normative multiagent systems. Computational and Mathematical Organizational Theory, Special issue on Normative Multiagent Systems **12(2-3)** (2006) 71–79
3. Boella, G., van der Torre, L.: A game theoretic approach to contracts in multiagent systems. IEEE Transactions on Systems, Man and Cybernetics - Part C **36(1)** (2006) 68–79
4. Boella, G., van der Torre, L.: Security policies for sharing knowledge in virtual communities. IEEE Transactions on Systems, Man and Cybernetics - Part A **36(3)** (2006) 439–450
5. Boella, G., van der Torre, L.: Norm negotiation in multiagent systems. International Journal of cooperative Information Systems (IJCIS) **16(1)** (2007)
6. Conte, R., Castelfranchi, C., Dignum, F.: Autonomous norm-acceptance. In: Intelligent Agents V (ATAL'98). LNAI 1555, Springer (1999) 99–112
7. Brainov, S., Sandholm, T.: Power, dependence and stability in multi-agent plans. In: Procs. of the 21st National Conference on Artificial Intelligence (AAAI'99). (1999) 11–16
8. Castelfranchi, C.: Modeling social action for AI agents. Artificial Intelligence **103(1-2)** (1998) 157–182
9. Conte, R., Sichman, J.: Multi-agent dependence by dependence graphs. In: Procs. of the 1st International Joint Conference on Autonomous Agents and Multiagent Systems (AAMAS'02). (2002) 483–490
10. Lopez y Lopez, F.: Social Power and Norms: Impact on agent behaviour. PhD thesis (2003)
11. Castelfranchi, C.: Micro-macro constitution of power. ProtoSociology **18-19** (2003)
12. Shoham, Y., Tennenholtz, M.: On social laws for artificial agent societies: off-line design. Artificial Intelligence **73 (1-2)** (1995) 231 – 252
13. Shoham, Y., Tennenholtz, M.: On the emergence of social conventions: Modeling, analysis and simulations. Artificial Intelligence **94(1–2)** (1997) 139–166
14. Tennenholtz, M.: On stable social laws and qualitative equilibria. Artificial Intelligence **102(1)** (1998) 1–20
15. Wooldridge, M., Dunne, P.: On the computational complexity of qualitative coalitional games. Artificial Intelligence **158(1)** (2004) 27–73
16. Simon, H.: A behavioral model of rational choice. The Quarterly Journal of Economics **69** (1955) 99–118
17. Boella, G., van der Torre, L.: Enforceable social laws. In: Procs. of the 4th International Joint Conference on Autonomous Agents and Multiagent Systems (AAMAS'05). (2005) 682–689

18. Boella, G., Sauro, L., van der Torre, L.: Social viewpoints on multiagent systems. In: Procs. of the 3rd International Joint Conference on Autonomous Agents and Multi Agent Systems (AAMAS'04). (2004) 1358–1359
19. Boella, G., van der Torre, L.: Fair distribution of collective obligations. In: Procs. of the 17th European Conference on Artificial Intelligence (ECAI'06). (2006) 721–722
20. Brafman, R., Tennenholtz, M.: On partially controlled multi-agent systems. Journal of Artificial Intelligence Research (JAIR) **4** (1996) 477–507

Quantitative Analysis of Single-Level Single-Mediator Multi-agent Systems

Moon Ho Lee[1], Aliaksandr Birukou[3], Alexander Dudin[1,2], Valentina Klimenok[2], and Chang-hui Choe[1]

[1] Institute of Information and Communication
Chonbuk National University
Chonju, 561-765, Korea
moonho@chonbuk.ac.kr

[2] Department of Applied Mathematics and Computer Science
Belarusian State University
Minsk, 220030, Belarus
dudin@bsu.by, klimenok@bsu.by

[3] Department of Information and Communication Technology
University of Trento
Povo(Trento), 38050, Italy
birukou@dit.unitn.it

Abstract. Queueing Theory deals with problems where some restricted resource should be shared between competitive flow of requests. In this paper we use Queueing Theory methods to perform a quantitative analysis of a single-level single-mediator multi-agent system. In the system, several agents, coordinated by the mediator process user queries. We adopt matrix analytic methods to compute performance characteristics in terms of a queueing network of tree-like topology with cooperation of the servers. Results can be used for the logical and technical design and optimal resources sharing in multi-agent systems.

1 Introduction

Organizational structures is a popular research direction in the field of Multi-Agent Systems (*MASs*), see e.g. [6,8]. However, there are only few papers dealing with quantitative analysis of such systems [1,5]. Queueing theory (*QT*) investigates situations when some restricted resource should be efficiently shared between competitive flow of requests in an optimal way. So, definitely, it should be useful in quantitative investigation and comparison of different organizational structures of *MASs*. Possibility of *QT* applications for *MAS* was discussed, e.g., in [1],[2]. For instance, in [2] operation of *MAS* is described in terms of queueing networks. In [1], the $M/M/1$ queueing system is used for utility prediction for a range of possible *MASs*.

In particular, *QT* is an appropriate tool for qualitative analysis of single-level single-mediator *MASs* [1]. In such class of systems the agent called mediator

N.T. Nguyen et al. (Eds.): KES-AMSTA 2007, LNAI 4496, pp. 447–455, 2007.

distributes queries[1] among several agents. In this paper we perform a quantitative analysis of single-level single-mediator *MASs* using *QT*. The results allow us to calculate such performance characteristics as probability of query being rejected, average query processing time, etc. and can be used for the logical and technical design and optimal resources sharing in single-level single-mediator *MASs*.

The structure of the paper is the following: Section 2 describes the considered class of *MASs* and provides a formalization in terms of *QT*. Conditions for the existence of the stationary distribution of the queueing network that models *MASs* are given in Section 3, while performance measures and the guidelines for their calculation are listed in Section 4. Finally, we conclude the paper in Section 5.

2 Mathematical Model

The class of *MASs* we consider in this paper is in a way similar to the models considered by Zhang and Lesser in [4] and by Horling and Lesser in [1]. The considered *MAS* is of the following structure: the single mediator serves as dispatcher for n independent heterogeneous agents which handle user queries. The queries are propagated just from the mediator to agents, therefore we are dealing with a single-level system. If there are i free agents at the moment of query arrival, then the query is assigned to $\min\{i, m\}$, $1 \le m \le n$, of agents that process it independently. If all agents are busy at the moment of query arrival, then the query is stored in a buffer and can be picked up later, according to the First In - First Out (*FIFO*) discipline, when some agents become free. We assume that an agent can also process queries coming from different sources, e.g. from other *MASs* or from other agents that were not able to process query on their own. The structure of the considered *MAS* system is given in Figure 1(a) and the corresponding queueing network is represented in Figure 1(b).

The mediator takes care of queries arriving to the *MAS* and buffers queries in the case of agents' unavailability. The service (query processing) in the MAS is performed by agents. Each agent has a finite buffer where queries assigned by the mediator can be stored while agent is busy with another query.

The motivation for the parallel processing of the query by m agents is as follows: (1) agents in the *MAS* can be unreliable, i.e. there is no guarantee that an agent will accomplish the task assigned to him, or an autonomous agent can decline to process the query. Also, the agent can reject the query from mediator just because the capacity of his buffer is exhausted; (2) the results of the query processing by different agents can vary greatly in terms of quality and performance because of the differences in the capabilities of agents, available resources, etc.; (3) in noisy domains the results can be distorted while passing from the agent to the mediator, so it is necessary to wait for results from several agents and to analyze them. In such situation, parallel sending of query to all

[1] We use the term query to refer to a generic kind of task or service request that comes from outside of the system and requires processing by agents.

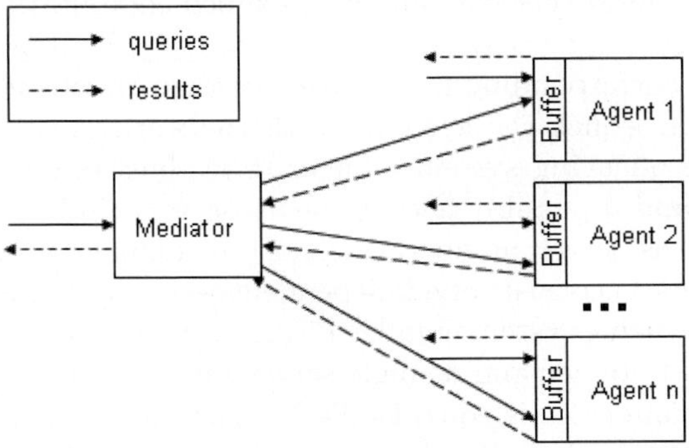

(a) The structure of the *MAS* system

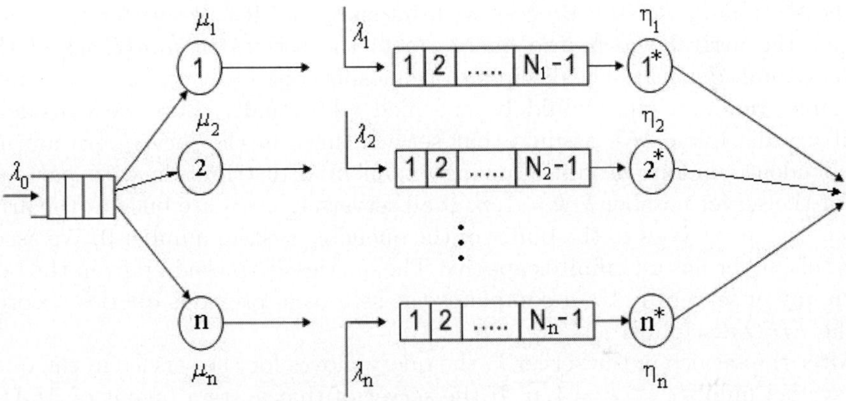

(b) Queueing network model for the operation of the considered *MAS* system

Fig. 1. The mathematical model of *MAS*

currently available agents increases chance of an arbitrary query to be successfully processes in the *MAS*. Besides this reliability aspect, parallel handling of query by several agents can decrease response time because the response time in this case is the minimum of durations of handling the query by all involved agents. The results from the field of Queueing Systems [7] have shown that this kind of assumptions is reasonable and allows for achieving a higher performance (comparing with standard one query-one agent systems) in case the arrival rate of the queries is not very high.

As an example of the described *MAS* we can consider a multi-agent information retrieval system, where agents have heterogeneous datasources and therefore process queries in different ways with different performance. Another example is a university network where computers can be used for distributed computations. In such a network, we cannot make an a priori estimation of the time required for the query processing by a single computer, because at each moment a user can start utilizing the computer therefore making assigned computations slower or, even, canceling them.

The queueing network corresponding to the MAS consists of two interacting parts (Figure 1(b)). In the sequel, the left part of the network (mediator part) will be referred to as the queueing system number 0. It consists of one buffer with an infinite capacity and n possibly heterogeneous servers (links to agents). We refer to these servers as server number $1, \ldots,$ server number n. The right part (autonomous agents part) consists of n independent service systems (agents) referred below to as the queueing system number $1^*, \ldots,$ system number n^*. Each of these systems has a finite buffer and a single server (agent). The capacity of the buffer of the system number k^* is equal to $N_{k^*} - 1$, so the maximal number of queries in this system is equal to $N_{k^*}, k = \overline{1,n}$, where $N_{k^*} = 1$ corresponds to the case when agent can process only one query at a time.

We assume that queries arrive to the queueing system number 0 according to the stationary Poisson process with intensity λ_0. If i of servers $1, \ldots, n$ are idle at the arrival epoch, the query starts the service in $\min\{i, m\}$ of these servers simultaneously. The discipline of choosing concrete servers, e.g. the fastest available, random, etc. should be specified additionally. Here we consider an arbitrary discipline. We assume that service times in the servers are mutually independent random variables having exponential distribution with parameter μ_k for the server number $k, k = \overline{1,n}$. If all servers $1, \ldots, n$ are busy at the arrival epoch, the query goes to the buffer of the queueing system number 0. We assume that this buffer has an infinite capacity. The queries are picked up from the buffer when any of servers $1, \ldots, n$ completes the service of previous queries according to the *FIFO* discipline.

After the service in the server k, the query moves for the service in the queueing system number $k^*, k = \overline{1,n}$. If the server of that system (agent of MAS) is idle at the arrival epoch, it starts processing of the arriving query with probability $q_k^{(1)}$ or declines the offer to serve this query with probability $1 - q_k^{(1)}$. Service times of successive queries in the server k^* are independent random variables distributed exponentially with parameter $\eta_k, k = \overline{1,n}$. After the service, query leaves the system number k^* and the network.

If the server k^* is busy at a query arrival epoch from the queueing system number 0, the arriving query with probability $1 - q_k^{(2)}$ is rejected and with supplementary probability it should be placed into the buffer of capacity $N_{k^*} - 1, k = \overline{1,n}$. If the buffer is already full at arrival epoch, the query is lost in the queueing system number k^*.

Besides processing the queries from the queueing system number 0, the server of the system number k^* can also process other queries. These queries arrive to server k^* according to the stationary Poisson process with intensity $\lambda_k, k = \overline{1,n}$. Service times of these queries are distributed exponentially with parameter η_k. In the case the buffer is full at the arrival epoch, the query is lost. No priority for any kind of queries is assigned.

Thus, operation of the queueing network presented in Figure 1(b) is completely described. Our purpose is to perform the stationary analysis of distribution of the number of queries in the nodes of this queueing network and computing its main performance measures.

3 Stationary State Distribution of the Network

Behavior of the queueing network under study can be described by the multi-dimensional continuous time Markov chain

$$\xi_t = \{j_t, i_t^{(1)}, \ldots, i_t^{(k)}\}, t \geq 0,\ i_t^{(k)} = \overline{0, N_k}, k = \overline{1, n},$$

where the component $i_t^{(k)}$ is equal to the number of queries in queueing system $k^*, k = \overline{1,n}$, at the moment $t, t \geq 0$. It includes the queries in the corresponding buffer, if any, and the query in the server. Component j_t describes the state of the n-server queueing system number 0. The state $j, j \geq 1$, of the component j_t corresponds to the state of the queueing system number 0 when there are j query in a buffer (sure, all the servers of this system are busy).

If the queue in this system is absent, the state of the component j_t is described by the group of n numbers $\{l_1, \ldots, l_n\}$ where the entry l_k has value 0 if the kth server is idle and value 1 if the kth server is busy at epoch $t, t \geq 0$. We denote the set of all such states by $\mathcal{L}$. It is evident that it consists of 2^n states.

Aiming to simplify denotations and use benefits of the matrix analytic methods, we enumerate the components of the process $\xi_t = \{j_t, i_t^{(1)}, \ldots i_t^{(n)}\}, t \geq 0$, in the lexicographic order. Then, we refer to the whole set of states $\{j, i_t^{(1)}, \ldots, i_t^{(n)}\}$, $i_t^{(k)} = \overline{0, N_k},\ k = \overline{1,n},$ as to the state j of the process $\xi_t, t \geq 0$,

$$j = \{\underbrace{0, \ldots, 0}_{n}\}, \{\underbrace{0, \ldots, 0, 1}_{n}\}, \ldots, \{\underbrace{1, \ldots, 1}_{n}\}, 1, 2, 3 \ldots.$$

For use in the sequel, we introduce the following notation. $\mu = \sum_{k=1}^{n} \mu_k$; I is identity matrix of dimension $K = \prod_{k=1}^{n} (N_k + 1)$; I_k is identity matrix of dimension $N_k + 1,\ k = \overline{1,n}$; O is zero square matrix of dimension K; $O_{l,m}$ is zero matrix of dimension $Kl \times Km$; $\otimes$ is the symbol of Kronecker product of the matrices; $\oplus$ is the symbol of Kronecker sum of the matrices; ^{T} denotes transposition of a matrix or vector; $\mathbf{e}_k$ is the column vector of dimension $N_k + 1$ consisting of all 1's; $\mathbf{e}_K$ is the column vector of dimension K consisting of all 1's; $\mathbf{0}_k$ is the row vector of dimension $N_k + 1$ consisting of all 0's; $k = \overline{1,n}$; $\mathbf{0}_K$ is the row vector of dimension K consisting of all 0's; $\mathbf{f}_k^{(i)}$ is the column vector of dimension $N_k + 1$ having the form $(\underbrace{0, \ldots, 0}_{i}, 1, 0, \ldots, 0)^T$, $i = \overline{0, N_k},\ k = \overline{1,n}$;

$\tilde{\mathbf{e}}_k^{(i)}, i = \overline{0, N_k}$, is the column vector of dimension K defined by formula $\tilde{\mathbf{e}}_k^{(i)} = \mathbf{e}_1 \otimes \ldots \otimes \mathbf{e}_{k-1} \otimes \mathbf{f}_k^{(i)} \otimes \mathbf{e}_{k+1} \otimes \ldots \otimes \mathbf{e}_n$; $\tilde{I}_k, \tilde{\tilde{I}}_k, I_k^+, I_k^-,\ I_k^0$ are the square matrices of dimension $N_k + 1,\ k = \overline{1,n}$, having the following structure:

$$\tilde{I}_k = \begin{pmatrix} 1\,0\ldots0\,0 \\ 0\,1\ldots0\,0 \\ \vdots\;\vdots\;\vdots\;\vdots\;\vdots \\ 0\,0\ldots1\,0 \\ 0\,0\ldots0\,0 \end{pmatrix}, \ \hat{I}_k = \begin{pmatrix} 0\,0\ldots0\,0 \\ 0\,1\ldots0\,0 \\ \vdots\;\vdots\;\vdots\;\vdots\;\vdots \\ 0\,0\ldots1\,0 \\ 0\,0\ldots0\,1 \end{pmatrix}, \ I_k^+ = \begin{pmatrix} 0\,1\,0\ldots0 \\ 0\,0\,1\ldots0 \\ \vdots\;\vdots\;\vdots\;\vdots\;\vdots \\ 0\,0\,0\ldots1 \\ 0\,0\,0\ldots0 \end{pmatrix}.$$

Here $\tilde{I}_k$ is obtained from an identity matrix by replacing the entry in the last row and column with 0; $\hat{I}_k$ is obtained from an identity matrix by replacing the entry in the first row and column with 0; I_k^+ is the matrix having 1's in the first over-diagonal and all all other entries equal to 0, $I_k^- = (I_k^+)^T$, $I_k^0 = I_k - \hat{I}_k$; $J_k = q_k^{(1)} I_k^0 I_k^+ + (1-q_k^{(1)}) I_k^0 I_k + q_k^{(2)} \hat{I}_k I_k^+ + (1-q_k^{(2)}) \hat{I}_k I_k$, $\mathcal{A}_k = \lambda_k \tilde{I}_k + \eta_k \hat{I}_k$; $\mathcal{A} = \bigoplus_{k=1}^{n} \mathcal{A}_k = \mathcal{A}_1 \oplus \ldots \oplus \mathcal{A}_n$; $\mathcal{B}_k = \lambda_k I_k^+ + \eta_k I_k^-$; $\mathcal{B} = \bigoplus_{k=1}^{n} \mathcal{B}_k$; $\mathcal{H} = \mathcal{B} - \mathcal{A}$; $\mathcal{C}_k = I_1 \otimes \ldots \otimes I_{k-1} \otimes \mu_k J_k \otimes I_{k+1} \otimes \ldots \otimes I_n$; $\mathcal{C} = \sum_{k=1}^{n} \mathcal{C}_k$; $\mathcal{M}_k = I_1 \otimes \ldots \otimes I_{k-1} \otimes \mu_k I_k \otimes I_{k+1} \otimes \ldots \otimes I_n$; $\mathcal{E} = \sum_{k=1}^{n} \mathcal{M}_k$;

$$\mathcal{D}_0 = \lambda_0 I, \ \mathcal{D}_1 = -\lambda_0 I - \mathcal{E} + \mathcal{H}, \ \mathcal{D}_2 = \mathcal{C};$$

$$Q_{1,0} = \begin{pmatrix} O_{1,2^n-1} \ \mathcal{D}_2 \end{pmatrix}, \ Q_{0,1} = \begin{pmatrix} O_{2^n-1,1} \\ \mathcal{D}_0 \end{pmatrix}.$$

Let $Q_{0,0}$ be the blocking matrix consisting of matrices

$$Q^{\{l_1,\ldots,l_n\},\{l'_1,\ldots,l'_n\}}, \ \{l_1,\ldots,l_n\},\{l'_1,\ldots,l'_n\} \in \mathcal{L},$$

which are defined via introduced matrices $\mathcal{H}$, $\mathcal{C}_k$, $\mathcal{M}_k$ depending on the discipline adopted to choose agents for the query processing. The matrix $Q_{0,0}$ can be decomposed as

$$Q_{0,0} = \begin{pmatrix} \tilde{Q}_{0,0} \ \tilde{\mathcal{D}}_0 \\ V \ \mathcal{D}_0 \end{pmatrix}$$

Denote by Q the block matrix which is the generator of the Markov chain $\xi_t, t \geq 0$.

Lemma: The generator Q has the following block structure:

$$Q = \begin{pmatrix} Q_{0,0} & Q_{0,1} & O_{2^n,1} & O_{2^n,1} & O_{2^n,1} & O_{2^n,1} & \ldots \\ Q_{1,0} & \mathcal{D}_1 & \mathcal{D}_0 & O & O & O & \ldots \\ O_{1,2^n} & \mathcal{D}_2 & \mathcal{D}_1 & \mathcal{D}_0 & O & O & \ldots \\ O_{1,2^n} & O & \mathcal{D}_2 & \mathcal{D}_1 & \mathcal{D}_0 & O & \ldots \\ O_{1,2^n} & O & O & \mathcal{D}_2 & \mathcal{D}_1 & \mathcal{D}_0 & \ldots \\ \vdots & \vdots & \vdots & \vdots & \vdots & \vdots & \vdots \end{pmatrix}.$$

Theorem 1. Stationary distribution of the Markov chain $\xi_t, t \geq 0$, exists if and only if the following inequality holds true: $\lambda_0 < \sum_{k=1}^{n} \mu_k$.

Let us denote the stationary probabilities of the states of the Markov chain $\xi_t, t \geq 0,$ by

$$p(j, i_1, \dots, i_n) = \lim_{t\to\infty} P\{j_t = j, i_t^{(1)} = i_1, \dots, i_t^{(k)} = i_k\},$$

$$j = \{l_1, \dots, l_n\} \in \mathcal{L}, 1, 2, \dots; \; i_k = \overline{0, N_k}, k = \overline{1, n}.$$

According to the lexicographic enumeration of the components of the Markov chain $\xi_t, t \geq 0,$ which was already exploited above, we combine probabilities $p(j, i_1, \dots, i_n)$, $i_k = \overline{0, N_k}, k = \overline{1, n}$, into probability row vectors $\mathbf{p}_j, j = \{l_1, \dots, l_n\} \in \mathcal{L}, 1, 2, \dots$ and the macro-vector $\mathbf{p} = (\mathbf{p}_{\{0\}}, \mathbf{p}_1, \mathbf{p}_2, \dots)$ where $\mathbf{p}_{\{0\}} = (\mathbf{p}_{\{0,\dots,0\}}, \mathbf{p}_{\{0,\dots,0,1\}}, \dots, \mathbf{p}_{\{1,\dots,1\}})$.

Theorem 2. Stationary probability vectors $\mathbf{p}_{\{l_1,\dots,l_n\}}, \{l_1, \dots, l_n\} \in \mathcal{L},\; \mathbf{p}_1, \mathbf{p}_2, \dots$ are calculated in the following way:

- the vector $\mathbf{p}_{\{l_1,\dots,l_n\}}$ is computed as the block number $\sum_{k=1}^{n} l_k 2^{n+1-k} + 1$ in the block vector $\mathbf{p}_{\{1,\dots,1\}}\mathcal{F}_1$, $\{l_1, \dots, l_n\} \in \mathcal{L}$, $\{l_1, \dots, l_n\} \neq \{1, \dots, 1\}$;
- the vectors $\mathbf{p}_j, j \geq 1$, are computed by $\mathbf{p}_i = \mathbf{p}_{\{1,\dots,1\}} R^i, i \geq 1$, where $\mathcal{F}_1 = -V(\tilde{Q}_{0,0})^{-1}$, $\mathcal{F} = \mathcal{D}_1 + \mathcal{F}_1\tilde{\mathcal{D}}_0$;
- the matrix R is a minimal non-negative solution to the matrix equation
$$R^2\mathcal{D}_2 + R\mathcal{D}_1 + \mathcal{D}_0 = O;$$
- the vector $\mathbf{p}_{\{1,\dots,1\}}$ is the unique solution to the following system of linear algebraic equations
$$\mathbf{p}_{\{1,\dots,1\}}[\mathcal{F} + R\mathcal{D}_2] = \mathbf{0}_K, \; \mathbf{p}_{\{1,\dots,1\}}[\mathcal{F}_1 + (I - R)^{-1}]\mathbf{e}_K = 1.$$

This theorem gives a straightforward easily-implementable algorithmic way for the calculation of the stationary probability vector $\mathbf{p}$.

4 Calculation of the Network Performance Measures

Having the stationary probability vectors been computed, we can calculate different performance measures of the queueing network. Formulae for calculation of some of them are given below.

Stationary distribution of the number of queries in the system k^* is given by the vector $\boldsymbol{\theta}^{(k)}$ having components

$$\boldsymbol{\theta}_i^{(k)} = \mathbf{p}_{\{1,\dots,1\}}[\mathcal{F}_1 + (I - R)^{-1}]\tilde{\mathbf{e}}_k^{(i)}, i = \overline{0, N_k}, k = \overline{1, n}.$$

Average number of queries L_k in the system k^*, average number of queries L_0 in the system number 0 and average total number of queries L in the network are calculated by

$$L_0 = \sum_{(\{l_1,\dots,l_n\})\in\mathcal{L}} \sum_{j=1}^{n} l_j \mathbf{p}_{\{l_1,\dots,l_n\}}\mathbf{e}_\mathbf{K} + \mathbf{p}_{\{1,\dots,1\}}((n+1)I - nR)R(I - R)^{-2}\mathbf{e}_\mathbf{K}.$$

$$L_k = \sum_{i=1}^{N_k} i\boldsymbol{\theta}_i^{(k)}, k = \overline{1,n},\ L = \sum_{k=0}^{n} L_k.$$

Probabilities $P_{loss}^{(k)}$ that an arbitrary query arriving to the system k^* will be rejected due to desire of agent or because the buffer is full and is calculated by the following formula:

$$P_{loss}^{(k)} = \boldsymbol{\theta}_{N_k}^{(k)} + (1 - q_1^{(k)})\boldsymbol{\theta}_0^{(k)} + (1 - q_2^{(k)})(1 - \boldsymbol{\theta}_0^{(k)} - \boldsymbol{\theta}_{N_k}^{(k)}),\ k = \overline{1,n}.$$

Probability P_{loss} that an arbitrary query arriving to the MAS will not get service by any agent is computed by formula

$$P_{loss} = \mathbf{p}_{\{1,\dots,1\}}(I + R(I - R)^{-1})\mathbf{e}_{\mathbf{K}} \sum_{k=1}^{n} \frac{\mu_k}{\mu} P_{loss}^{(k)} +$$

$$+ \sum_{\{l_1,\dots,l_n\}\in\mathcal{L},(k_1,\dots,k_{\hat{m}})} \mathbf{p}_{\{l_1,\dots,l_n\}}\mathbf{e}_{\mathbf{K}}\mathcal{B}_{\{l_1,\dots,l_n\}}^{k_1,\dots,k_{\hat{m}}}(1 - \prod_{r=1}^{\hat{m}}(1 - P_{loss}^{(k_r)})),$$

where $\hat{m} = \min\{m, n - l_1 - \dots - l_n\}$, $\mathcal{B}_{\{l_1,\dots,l_n\}}^{k_1,\dots,k_{\hat{m}}}$ is probability of assigning the agents number $k_1, \dots, k_{\hat{m}}$ for service providing to an arbitrary query which arrives when the states of servers are defined by the set $\{l_1, \dots, l_n\}$. This probability is easily computed when strategy of agents assigning is fixed.

Average sojourn time $\tilde{W}_1^{(0)}$ in the system number 0 and sojourn time $\tilde{\tilde{W}}_1^{(k)}$ for queries that are not rejected in the system number k^* are calculated as follows:

$$\tilde{W}_1^{(0)} = \lambda_0^{-1}\tilde{L}_0,\ \tilde{W}_1^{(k)} = \sum_{i=0}^{N_k-1} \frac{i+1}{\eta_k}\boldsymbol{\theta}_i^{(k)},\ \tilde{\tilde{W}}_1^{(k)} = \frac{\tilde{W}_1^{(k)}}{1 - P_{loss}^{(k)}}, k = \overline{1,n}.$$

Average sojourn time V of a query in the queueing network is computed by

$$V = \tilde{W}_1^{(0)} + \mathbf{p}_{\{1,\dots,1\}}(I + R(I - R)^{-1})\mathbf{e}_{\mathbf{K}} \sum_{k=1}^{n} \frac{\mu_k}{\mu}\tilde{\tilde{W}}_1^{(k)} +$$

$$+ \sum_{(\{l_1,\dots,l_n\})\in\mathcal{L}} \mathbf{p}_{\{l_1,\dots,l_n\}}\mathbf{e}_{\mathbf{K}}\mathcal{B}_{\{l_1,\dots,l_n\}}^{k_1,\dots,k_{\hat{m}}} W(k_1, \dots, k_{\hat{m}}),$$

$W(k_1, \dots, k_{\hat{m}})$ is expectation of minimum of sojourn times in systems number $k_1, \dots, k_{\hat{m}}$.

5 Conclusion

We have analyzed the process of user query processing in a particular class of MASs in terms of the queueing network. Tree-like structure of the network topology allows to get the steady state-distribution of the network states in the exact analytic form. Main performance measures of the network are calculated and can be used for the quantitative analysis of a particular MAS. For instance, having specified the system parameters, it is possible to calculate the average time of query processing in the system, or the probability of query being rejected, etc. The results are extendable to the cases where the input and service processes have more complicated nature. Modifications to the considered MAS, where the results of query processing are unreliable because of errors or because the agents are subject to breakdowns and recovering be can investigated analogously.

Acknowledgments

This research was supported in part by Ministry of Information and Communication (MIC) Korea, under the IT Foreign Specialist Inviting Program (ITSIP), ITSOC, International Cooperative Research by Ministry of Science and Technology, KOTEF, and 2nd stage Brain Korea 21.

References

1. Horling B., Lesser V.: Using Queueing Theory to Predict Organizational Metrics. AAMAS'06, May 8-12, 2006, Hakodate, Japan. ACM Press. 1098-1100.
2. Gnanasambandam N., Lee S.C, Gautam N., et al.: Reliable MAS Performance Prediction Using Queueing Models. IEEE First Symposium on Multi-Agent Security and Survivalibility, 2004. 55-64.
3. Gnanasambandam N.: Survivability of Multi-Agent Systems. AAMAS'05 July, 25-29, 2005, Utrecht, Netherlands. ACM Press. 1376.
4. Zhang H., Lesser V.: A Dynamically Formed Hierarchical Agent Organization for a Distributed Content Sharing System . IAT 2004, September 2004, Beijing. 169175.
5. Okamoto S., Scerri P., Sycara K.: Toward an Understanding of the Impact of Software Personal Assistants on Human Organizations. AAMAS'06, May 8-12, 2006, Hakodate, Japan. ACM Press. 630-637.
6. Vazques-Salceda J., Dignum V., Dignum F.: Organizing Multiagent Systems. Autonomous Agents and Multi-Agent Systems, 2005, 11, 307-360.
7. Lee M.H., Dudin A.N., Klimenok V.I. The SM/M/N queueing system with broadcasting service. Mathematical Problems in Engineering. 2006. Article ID 98171.
8. Horling B., Lesser V.: A survey of multi-agent organizational paradigms. Knowledge Engineering Review, Cambridge University Press, 2004, 19, 281-316.

Software Agent Negotiation for Service Composition

Claudia Di Napoli

C.N.R. - Istituto di Cibernetica "E. Caianiello",
Viale Campi Flegrei, 34 I-80078
Pozzuoli, Naples, Italy
C.DiNapoli@cib.na.cnr.it

Abstract. Service–oriented computing (SOC) is posing new challenges in the management of compositions of services that usually belong to different administrative domains. As such they cannot be provided by adopting a centralized approach, but more sophisticated computing methodologies are necessary.

In this paper we propose to use software agent negotiation to address the problem of composing services in service–oriented environments, like the Grid. In particular, we propose to use software agents to represent service providers and service consumers, and a negotiation protocol to select the service providers that meet the requirements of service consumers on the provision of multiple interconnected services. The proposed protocol is thought as a *flexible* protocol to improve the possibility of reaching an agreement by allowing both service consumers and providers to exchange more proposals to accommodate the dynamic and changing nature of service–oriented environments.

Keywords: Software Agents, Negotiation, Service–Oriented Computing.

1 Introduction

Service–Oriented Computing (*SOC*) is the new promise in the field of distributed computing. The technological advances in the recent years and the pervasiveness of the Internet in the everyday life, shifted the distributed computing paradigm from the possibility of exploiting the computing power offered by a collection of computational resources distributed over the network, towards the possibility of exploiting any type of software and hardware commodity available on the network.

Service–oriented computing paradigm relies on the concept of *services* as building blocks to support the development of rapid, low–cost, and easy composition of distributed applications even in heterogeneous environments [1]. Services are autonomous, platform–independent computational entities that can be described, published, discovered, and assembled on–demand to provide added–value applications. Services encapsulate as much as possible the intrinsic and heterogeneous nature of any kind of computational resource.

N.T. Nguyen et al. (Eds.): KES-AMSTA 2007, LNAI 4496, pp. 456–465, 2007.

In order to make service–oriented systems a viable approach to extensively provide computational capabilities to end users to solve large scale problems, middleware mechanisms are necessary to enable the sharing, selection, and aggregation of resources distributed across multiple administrative domains, depending on their availability, capability, performance, cost and users' quality–of–service requirements.

The focus of the present work is to individuate a middleware mechanism allowing for the selection and aggregation of different services that are necessary for the execution of distributed applications or workflows.

In traditional computing systems the coordinated access to resources is guaranteed by resource management systems designed to operate under the assumption that they have complete control of all necessary resources and thus they implement mechanisms and policies necessary for the effective use of those resources in isolation. In service–oriented environments this assumption cannot be made and more sophisticated computational methodologies for managing resources that are heterogeneous, located across separately administrated domains, and that inevitably adopt different policies for their use without relying on a centralized control are necessary.

We propose that the middleware mechanism to manage coordinated access to distributed resources/services is based on *software agent negotiation* both to guarantee the autonomy of resources, their coordination, and the satisfaction of both users and providers requirements in respectively requiring and providing them.

The paper is organised as follows. Section 2 motivates the use of software agent methodologies to regulate the provision of compositions of services. Section 3 presents the problem addressed and the followed approach. The proposed negotiation protocol is described in section 4 and some preliminary results in section 5. Finally some conclusion and future work are reported.

2 An Agent–Oriented Approach

In service–oriented systems each computational entity is a service [2]. It may be information or a virtual representation of some physical good or processing capability, and it has to be identified, published, allocated, and scheduled. More generally, a service is an encapsulated functionality that is accessible through the network according to well–defined protocols and interfaces. In this view, a service is provided by the *body* responsible for offering it, that we refer to as *service provider*, for consumption by others that we refer to as *service consumers*.

We also distinguish between services that are provided by a single provider, and that we refer to simply as *services*, and services that are the result of composing services that are not all provided by the same provider, and that we refer to as *composite services*. So, composite services are workflows whose components are not subject to a centralized control management system.

We start from the assumption that the provision of composite services should be regulated by a form of *agreement/contract* among the providers of the component services and the user that required the composition of services. The necessity of reaching agreements that state the terms and the conditions under which each service should be provided to meet the user request is more and more recognized by research communities [3], [4], [5]. The primary motivation for introducing an agreement among service providers and a service consumer is to provide a form of guarantee on the service quality and availability. In fact, in dynamic and changing environments like service–oriented ones where new services can be created and destroyed without a centralized control system being aware of it, it cannot be assumed that a provider will always be able to provide a service to a user. This aspect is even more crucial when composite services are required because in such a case the unavailability of one service impacts negatively the provision of the entire composition and consequently the ones that were instead available. So, in order to avoid a completely unpredictable behaviour of the computing system, it is necessary to introduce the notion of *commitments* in the provision of services, hence contracts/agreements among the entities involved in the provision/consuming of services. In order to automate the process of reaching such an agreement, services need to be equipped with middleware technology able to represent providers and consumers and to model their behaviour.

A software agent, as defined in [6], "is an encapsulated computer system that is situated in some environment, and that is capable of flexible, autonomous action in that environment in order to meet its design objectives". Software agent features well match those characterising service providers and consumers, and Multi–Agent Systems (MASs) can account for the interactions that take place among service consumers and providers [7], [8].

In particular, software agents allow us to represent:

- the distributed nature of the provided services through the location of service provider agents in different control domains,
- the different behaviours service providers may have in providing their services through the possibility of adopting different and autonomous decision making mechanisms for the different service providers,
- the available services through *agent capabilities*,
- the provision of services through *agent actions*,
- the request of services through *agent interactions*,
- the quality–of–services as *agent preferences*.

We refer to agents representing service providers as *Service Agents* (*SA*s), and agents acting on behalf of service consumers as *Service Market Agents* (*SMA*s). The agents are modeled as self–interested software agents since they are autonomous and independent business entities that do not usually have common objectives, but they are more likely to have conflicting interests. A service provider may play also the role of a service consumer and viceversa.

A Service Agent sa_j may provide a set of services $AS_j = \{s_{1,j}, ..., s_{n,j}\}$. If the set is empty or does not contain the service required by a user the SA plays the role of a SMA. At each service $s_{i,j}$ corresponds a provider sa_j, so a composite service $cs = \bigcup_{i=1}^{N} \{s_i\}$ is a workflow of N services, each one provided by different providers, and as such it corresponds to an aggregation of SAs that contribute to the provision of its components. A composite service is represented by a direct acyclic graph whose nodes ST_i are the required services, and arcs identify the dependencies among the services (figure 1).

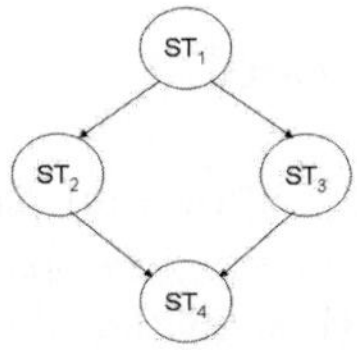

Fig. 1. A composite service

3 Software Agent Negotiation for Composing Services

In a service–oriented scenario it is necessary to oganise compositions of services on demand in response to dynamic requirements and circumstances. It is likely that in service–oriented systems more service providers can provide the same service (even though at different conditions), or that the same provider can provide the same service at different conditions. These conditions refer to the characteristics of the provided service, expressed in terms of quality–of–service, but also to other parameters like price, time to deliver, and so on. These are not static conditions, but can dynamically change and as such cannot be advertised together with the service description. Furthermore, service providers and service consumers have typically conflicting objectives, so it is unlikely that the conditions required by a user match the ones required by the providers.

The natural computational mechanism to attempt to reach an agreement on conflicting objectives is *automated negotiation*, i.e. "the process by which a group of agents come to a mutually acceptable agreement on some matter" [9].

In the present work we consider the case of a user that asks for a composite service *cs* specifying the type of services necessary for the composition and as its preference the deadline ($T_{deadline}$) by when the composite service should be delivered. We assume that a matchmaking process occurs after a request is issued resulting in the retrieval of services that match the types required by the user for the composite service.

In our approach, the dependencies occurring among the service components allow to specify the execution order in which services need to be delivered and as such they are regarded by the SMA as time constraints on the delivery of each service. So, in our scenario, the issue to be negotiated upon is the service time to deliver, i.e. the $time_{start}$ and the $time_{end}$ of each service representing respectively the time when the service is expected to start its execution and the time when the service is expected to end its execution.

When negotiating over the composition of services, settling the time for the provision of one service cannot be done without considering settling the times for the provision of the other services in the composition. Only when all SAs agree on these issues, the composite service can be successfully delivered to the end user according to its preferences. This type of negotiation requires new protocols that are flexible enough to accommodate for both service dependencies and the dynamicity of the environment in which providers and consumers operate.

4 The Negotiation Protocol

The proposed negotiation protocol allows an agent acting on behalf of an end user to select individual component services distributed across multiple administrative domains, depending on their availability and performance (in terms of time for service delivery). We started from the assumption that multiple service providers could be available to provide each one of the services necessary for the composition: the user needs to negotiate with them the conditions at which each component will be provided. The providers that are successful in the negotiation will be the ones selected to provide each component of the composite service.

In order to deal both with the dependencies that may occur among the service attributes that are negotiated upon, and with the varying conditions under which the negotiation takes place (i.e. the number of participants, their strategies, the time within which the negotiation should end, and so on), we propose a flexible negotiation protocol that we call a *Multi-Phase-Multi-Iteration Negotiation*.

It consists of three phases:

1. *Exploratory Phase*, that allows to find out the number of SAs available to enter negotiation, and their initial preferences over the issues to be negotiated upon,
2. *Intermediate Phase*, that allows to iterate the process of alternating announcements and bids for a variable number of times, so to accommodate time constraint requirements on negotiation duration,
3. *Final Phase*, that allows to end the negotiation either with a success leading to a signed contractual agreement, or with a failure.

The SMA initiates the negotiation as soon as it receives a user's request and the set of services that are potentially available for each required service type.

The Exploratory phase and each iteration of the Intermediate phase are based on a variation of the Contract Net Protocol (CNP) introduced by Smith in 1980 to allocate tasks over a distributed network of sensors [10].

As shown in figure 2, the SMA initiates the negotiation by sending an announcement proposing a $time_{start}$ and $time_{end}$ for each service necessary for the composite service. The corresponding SAs that decide to take part in the negotiation process (SA^*) reply with bids to announcements that specify when they can provide the service.

The Intermediate phase allows for re–submission of the announcement, and hence of bids, once the SMA evaluated if a combination of services that meets the user requirements and the dependencies specified in the workflow can be obtained or not with the current received bids. In the Intermediate phase the SMA

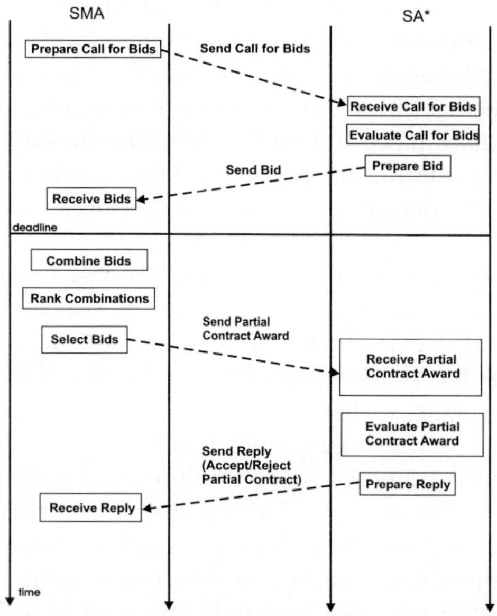

Fig. 2. The basic protocol

proposes a $time_{start}$ and $time_{end}$ for each required service in the announcements taking into account the received bids and the time constraints on the component services. In particular, the SMA builds the first announcements for each service type in the composition specifying consecutive time intervals with no overlap so that time dependencies are met. The successive announcements are built determining a new set of time intervals with no overlap, starting from the first time interval received for the first service in the composition and according to the expected execution time of each required service.

The SAs reply to announcements by adopting a simple strategy consisting in proposing a $time_{start}$ and $time_{end}$ for the service they provide as much close as possible to the ones given in the announcement. The proposed times are selected from a set of intervals coming from a random distribution of intervals in the range $[0, T_{deadline}]$ where $T_{deadline}$ is the deadline specified by the user request. The lenght of the intervals varies for each required service and represents the expected execution time of each service specified in the composite service. The random distribution of time intervals associated to each SA represents its current workload when negotiation takes place.

The flexibility of the protocol consists in allowing the iteration of this process from 0 to n times, according to the decision the SMA takes depending on the time and on the result obtained after each iteration.

The possibility of iterating the process of exchanging announcements and bids is introduced so that both the SMA and the SAs can adjust their proposals according to the changing conditions that can occur while the negotiation proceeds (e.g. SAs can become more available in terms of time, or the SMA may accept bids that result in composite services that cannot be delivered within the required deadline). Furthermore, at each iteration the SMA is able to collect more information on conflicting times and to propose new times accordingly.

The Final phase takes place to end the negotiation sending a message to all SAs involved in the negotiation either to award the ones that can provide services at the right time, or to declare a failure if no services can be provided at the required conditions.

The main difference between CNP and our proposed negotiation protocol is that a contract is not awarded after the potential contractors send back their bids to execute the service they are able to provide. The reason why we introduce this variation is due to the necessity of evaluating the combination of bids (because of the dependencies that occur among the service attributes that are negotiated upon), and so each bid cannot be evaluated independently from the others. Furthermore, the proposed protocol allows to negotiate, at each iteration, with more providers of the same service concurrently in order to try to improve the possibility to find services that meet the required time constraints. This is different from most proposed negotiations where negotiation occurs between the consumer and a selected provider, and only when an agreement is reached the negotiation for the next service takes place [11], [12]. In [13] a concurrent bilateral negotiation model is proposed, but it is not extended for the case of multiple interconnected services.

Furthermore, the SMAs and the SAs do not share any information on their time availability, and on their strategies. Also the structure of the required composite service, hence the dependencies occurring among the component services, are known only to the SMA. The reason for these decisions come from the consideration that in service–oriented environments the required workflows can be very complex and the number of involved SAs can be very high, so that the amount of shared knowledge would become not manageable in real applications. The proposed flexible protocol allows the SMA to require information only when necessary, i.e. when conflicting time intervals are proposed by the SAs.

5 Preliminary Experiments

In order to get some preliminary results on the proposed protocol, a simulation framework written in Java has been implemented to run negotiations with a different number of SAs and varying the number of iterations in the Intermediate phase.

The initial experimentation was carried out to collect information on the performance of the different protocols, i.e. protocols with a different number of iterations in the Intermediate phase. The performance of the protocol is measured as the percentage of feasible combinations of services (reported on Y–axis) that is obtained after negotiation with respect to the total number of possible combinations of services (reported on X–axis), where a feasible combination of services is a composite service that can be delivered within a specified deadline

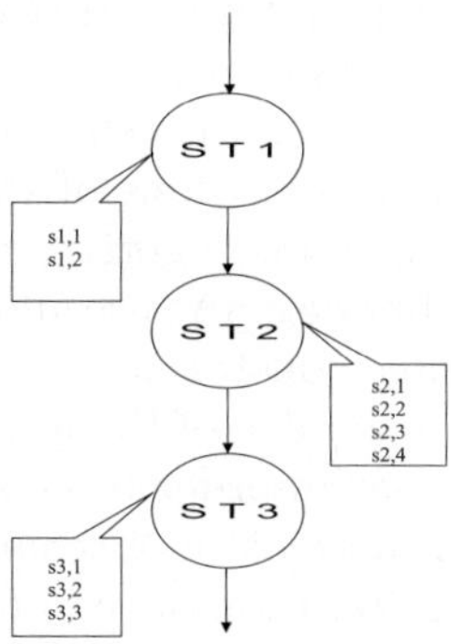

Fig. 3. A sequential abstract workflow

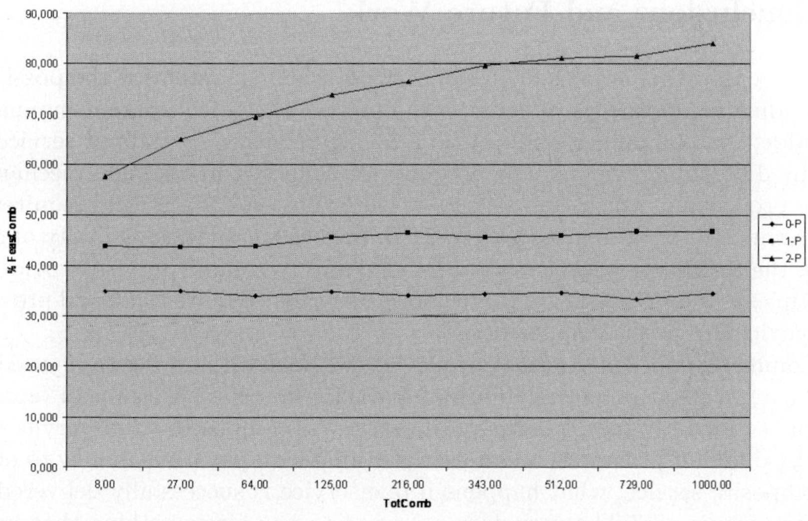

Fig. 4. Preliminary results

and whose components can be delivered according to the time constraints coming from the dependencies in their execution order. We interpret the number of feasible combinations obtained after negotiation as a measure of successful negotiations.

In our first experimental setting we considered only sequential workflows, as the one reported in figure 3, because they represent the ones for which more time constraints need to be met since services need to be executed one after another. In figure 3 the square boxes specify the available services $s_{i,j}$ for each required service type ST_i.

For these preliminary experiments, a composite service of 3 different service types is considered and the number of available services for each service type varies from 2 to 10, so that the number of SAs in the system varies from 6 to

30. The number of possible combinations varies accordingly from 8 to 1000. We varied the number of iterations from 0 to 2.

The simulation results reported in figure 4 show that the percentage of feasible combinations increases when more iterations of the protocol are allowed. This is an expected result, since with more negotiation the possibility of obtaining feasible combinations increases because SAs are allowed to propose different time intervals that could meet the time constraints.

These preliminary results give an idea of the feasibility of the proposed protocol, even though we are aware that in order to be significant in terms of performance, more sophisticated decision making mechanisms have to be specified for the SMA in determining at each iteration of the protocol which intervals to propose next according to the time constraints on each service in the composition and the received bids.

6 Conclusions and Future Work

We believe that the main challenge of service–oriented systems is the possibility of providing compositions of services to end users in a transparent manner, so that added–value applications can be built aggregating specialized services offered by different providers. The necessity of including negotiation mechanisms for the provision of services in service–oriented systems is a crucial requirement when compositions of services are required. In such a case the possibility of negotiating the times when services are provided and to commit service providers to these times is important to guarantee both users that require them and providers that participate in the composition.

In commercial–oriented scenarios services will be sold in order to be used (i.e. to be invoked), so the unavailability of a service in a composition can represent a crucial concern. In fact, if a composite service is composed of two services provided by different providers, and a user is available to pay some money to obtain the composite service, what happens if one service is successfully delivered and the other one is not? The user does not want to pay for something that he/she did not get, but at the same time the provider that successfully delivered the service cannot end up providing something for free! For this reason we propose that services are selected as a result of a negotiation process before the actual execution takes place.

In order to reach this objective services need to be equipped with middleware mechanisms that allow for automated negotiation, in particular for a flexible negotiation mechanism that accounts for the dynamic nature of service–oriented systems.

We plan to specify decision making strategies for both the SAs and the SMAs for particular cases of study to collect information on when it is useful to iterate the negotiation, and according to which parameters whose values cannot be statically determined.

References

1. Papazoglou, M.P., Georgakopoulos, D.: Service-oriented computing. Communications of the ACM **46** (2003) 24–28
2. De Roure, D., Jennings, N.R., Shadbolt, N.: The semantic grid: A future e-science infrastructure, Wiley (2003) 437–470
3. Foster, I., Jennings, N.R., Kesselman, C.: Brain meets brawn: Why grid and agents need each other. In: Proc. 3rd Int. Conf. on Autonomous Agents and MultiAgent Systems. (2004) 8–15
4. Czajkowski, K., Foster, I., Kesselman, C., Sander, V., Tuecke, S.: Snap: A protocol for negotiating service level agreements and coordinating resource management in distributed systems. Lecture Notes in Computer Science **2537** (2002) 153–183
5. Ouelhadj, D., Garibaldi, J., MacLaren, J., Sakellariou, R., Krishnakumar, K.: A multiagent infrastructure and a service level agreement negotiation protocol for robust scheduling in grid computing. Lecture Notes in Computer Science **3470** (2005) 651–660
6. Wooldridge, M.: Agent-based software engineering. In: IEE Proc. Software Engineering. (1997) 26–37
7. De Roure, D., Baker, M., Jennings, N.R., Shadbolt, N.: The evolution of the grid, Wiley (2003) 65–100
8. Wooldridge, M.: Engineering the computational economy. In: IST2000: Proceedings of the Information Society Technologies Conference, Nice, France (2000)
9. Jennings, N.R., Faratin, P., Lomuscio, A.R., Parsons, S., Sierra, C., Wooldridge, M.: Automated negotiation: prospects, methods and challenges. Int. Journal of Group Decision and Negotiation **10** (2001) 199–215
10. Smith, R.G.: The contract net protocol: Highlevel communication and control in a distributed problem solver. IEEE Transaction on Computers **29** (1980) 1104–1113
11. Andrieux, A., Czajkowski, K., Dan, A., Keahey, K., Ludwig, H., Pruyne, J., Rofrano, J., Tuecke, S., Xu, M.: Web services agreement specification (wsagreement) version 1.0. http://www.gridforge.org/projects/graap-wg/document/WS-AgreementSpecification/en/2 (2004)
12. Norman, T.J., Preece, A., Chalmers, S., Jennings, N.R., luck, M., Dang, V.D., Nguyen, T.D., Deora, V., Shao, J., Gray, A., Fiddian, N.: Conoise: Agent-based formation of virtual organisations. In: 23rd SGAI Int. Conf. on Innovative Techniques and Applications of AI. (2003) 353–366
13. Nguyen, T.D., Jennings, N.: Concurrent bilateral negotiation in agent systems. In: Proceedings of the 4th DEXA Workshop on eNegotiations. (2003)

A Misuse Detection Agent for Intrusion Detection in a Multi-agent Architecture

Eduardo Mosqueira-Rey, Amparo Alonso-Betanzos, Belen Baldonedo del Río, and Jesús Lago Piñeiro

University of A Coruña, 15071 A Coruña. Spain
eduardo@udc.es
http://www.dc.fi.udc.es/lidia

Abstract. We describe the design of a misuse detection agent, one of the different agents in a multiagent-based intrusion detection system. This system is being implemented in JADE, a well-known multiagent platform based in Java. The agent analyzes the packets in the network connections using a packet sniffer and then creates a data model based on the information obtained. This data model is the input to a rule-based agent inference engine, which uses the Rete algorithm for pattern matching, and the rules of the signature-based intrusion detection system Snort. Specifically, an implementation in Java language – the Drools-JBoss Rules– was used, and a parser was implemented that converts Snort rules to Drools rules. The use of object-oriented techniques, together with design patterns, means that the agent is flexible, easily configurable and extensible.

1 Introduction

During the last decade there has been a vast increase in the number of services offered by the Internet. More customers have access to Internet services, however this makes security a priority issue. Automatic Intrusion Detection Systems (IDS) date back to the development in the late 1980's of IDES (Intrusion Detection Expert System) [1], which was based on the use of statistical techniques and heuristic rules to detect security breaches. Several other IDSs were subsequently developed, but intrusion detection was becoming increasingly difficult as a a consequence of the great heterogeneity of the networks to be protected and the complexity of the domain. IDSs, in addition to protecting the system, needed to be able to resist attacks on themselves, and also needed to be fault tolerant, highly adaptable and configurable. Given these characteristics, agent-based technology seemed to be an appropriate alternative for developing IDSs.

One of the best-known multiagent-based intrusion detection systems is AAFID (Autonomous Agents for Intrusion Detection) [2]. Although this system was very innovative in its time, its main drawback is the extreme rigidity of its architecture, as this makes the introduction of new agents very complicated. Besides, its hierarchy is such that if an attack manages to deactivate an agent in the upper layers, the entire system might be deactivated. In an attempt to overcome this

N.T. Nguyen et al. (Eds.): KES-AMSTA 2007, LNAI 4496, pp. 466–475, 2007.

problem, a new architecture based on the AAFID approach was proposed in [3]. The system, implemented in JADE [4], is based on a flexible and adaptable architecture, that allows the integration of different intrusion detection techniques [5]. It consists of seven categories of agents performing different specific tasks as follows: information agents, prevention agents, detection agents, response agents, evidence-search agents, interface agents and finally special agents that perform a variety of tasks not included in the six previous categories– such as for example launching the rest of agents so they can carry on performing dynamically in accordance with the system's requirements. The aim of this work is the development of a misuse detection agent, that will examine the packets received by the network interface and send subsequent alert messages to other agents of the multiagent system. The misuse detection agent uses a rules engine implementation based on the efficient pattern-matching Rete algorithm and tailored to the Java language, as this last is the language in which JADE is written.

1.1 Detection Agents

Referring specifically to detection agents, these are responsible for detecting possible intrusions and for sharing this information with the other agents in the system. Basically, two intrusion detection strategies can be distinguished: misuse detection and anomaly detection. Misuse detection tries to match computer activities with previously registered and known signature attacks; in other words, it uses available *a priori* knowledge to track possible attacks. Anomaly detection is an approach that is based on learning the normal and legitimate activities of the system with the IDS trained to detect any activities that deviate from normal behavior.

A drawback to the misuse detection is that it only will detect attacks for which the system has previously registered signatures. This means that the database of signatures must be updated regularly in order to ensure adequate protection. Anomaly detection has the disadvantage that it relies heavily on the quality of the learning process: overly restricted training could result in a system with a high percentage of false positives, whereas overly general training may produce a high percentage of false negatives.

In this work, we describe the design of a misuse detection agent for inclusion in a multi-agent intrusion detection architecture. It is important to take into account that the multi-agent system is designed under the assumption that most systems are too complex for a single entity to be capable of detecting attacks as they occur. For that reason, several detection agents, hierarchically organized, work together in order to ensure adequate intrusion detection. These detection agents will supply information to another agent that combines the techniques of the former agents so as to make detection techniques more powerful.

2 The Misuse Detection Agent

Misuse intrusion detection techniques are also known as signature-based IDSs, due to the fact that the system looks for events matching a predefined pattern of events that have resulted in a known attack. These patterns are known as signatures and the detection process is very similar to that for virus scanners.

The misuse detection (MD) agent being described here analyses packets arriving from a network connection for possible attacks. Once suspicious behavior is detected, the agent passes on the information for other agents to take pertinent actions. The architecture of the MD agent is shown in Fig. 1. The agent analyzes a packet obtained from the network using a packet sniffer, which (see Fig. 2) scans the packet and creates a data model of the information it contains. This data model is the input to the rule-based inference engine (see Fig. 3) in the agent. When executed, the rules result in a set of actions that are encapsulated in an action model (see Fig. 4) which abstracts the agent from the actual actions that are finally performed. This process is described in detail in the following sections.

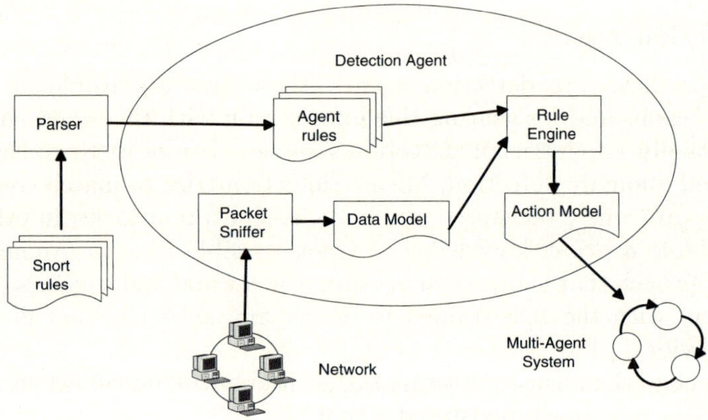

Fig. 1. Agent architecture

3 Agent Rules

The knowledge of the agent is represented in rules that are, in turn, based on intruder signatures. These rules are from the well-known signature-based IDS Snort system [6], which is written in C language. However, it was necessary to adapt Snort for its use in a Java-based platform. A possible solution is to use the Java Native Interface (JNI) to call Snort, but this implies several complications for the multiplatform characteristics of Java. So, an alternative solution is to use the Snort rules for a signature-based IDS written in Java. This last is the solution proposed by this work.

3.1 Snort

Snort is one of the most popular signature-based IDSs in use at present [6]. It is an open-source packet sniffer/logger and network intrusion detection system.

It analyzes the packets that arrive to the network interface, trying to match their characteristics with those contained in the rules stored in its rule base. If a specific packet matches the premises of any rule, this rule is executed and a specific action is generated to give notice of this fact. The rules in Snort have the following structure:

1. Rule header. Contains the basic information about the rule, including:
   (a) Rule action. The action that will be taken when rule conditions are met. The main actions are: alert (generate an alert), log (log the packet) and pass (ignore the packet).
   (b) Protocol. The protocol used by the packet being analyzed. Currently Snort understands the following protocols: IP, TCP, ICMP and UDP.
   (c) Source information. IP address and port of the source computer from where the packet originated. The keyword *any* can be used to apply the rule on all packets irrespective of the IP address or port number.
   (d) Destination information. IP address and port of the destination computer. The keyword *any* can be used again with the same meaning as before.
2. Rule options. Contains alert messages and information on the parts of the packet that should be inspected to determine if the rule action should be taken.

As an example, one possible Snort rule could be:

alert tcp any any ⟶ 192.68.30.280 *(msg: "HTTP Traffic Detection")* (1)

This rule will generate an alert containing the message HTTP Traffic Detection each time a packet using the TCP is detected for machine 192.68.30.2 and port 80.

3.2 Drools-JBoss Rules

As can be seen in Fig. 1, although the detection agent developed is based on the Snort rules, it does not use Snort as its rule engine, but uses instead Drools. Drools is a rules engine implementation based on the efficient pattern-matching Rete algorithm [7] and tailored to the Java language. Adapting Rete to an object-oriented interface allows for more natural expression of rules with regard to domain objects. It also enables advantage to be taken of object orientation techniques in rule definition (discussed below). Recently, Drools has been acquired by JBoss, which explains the name change to JBoss Rules.

3.3 Converter Snort-Drools

In order to build the MD agent, it was necessary to convert the Snort rules into Drools rules that could be executed using the Rete algorithm. A parser was implemented in order to do this. Thus, a Snort rule such as the one in expression (1) was translated into a Drools rule as follows:

```
<rule name="Snort Rule 1">
  <parameter identifier="dataPacket">
     <class>model.datapacket.TCPPacket</class>
  </parameter>
  <java:condition>
     dataPacket.getDestinationAddress().
       getHostAddress().equals("192.68.30.2")
  </java:condition>
  <java:condition>
     dataPacket.getDestinationPort() == 80
  </java:condition>
  <java:consequence>
     model.actions.facade.ActionsFacade.getInstance().
       sendAlertMessage ("TCP Message received",dataPacket);
  </java:consequence>
</rule>}
```

Two aspects of the Drools rules are particularly worthy of comment: in the first place, it can be seen that they are more verbose than Snort rules, but this is because they use XML as the representation language. Secondly, the condition and action parts of the rules are in embedded Java code. These two facts imply certain advantages, as follows:

- there is a higher level of integration between the language used by the global system and the language used in the rules.
- matching is behavior-based [8], in the sense that the object structure is not accessed in order to carry out the pattern matching; rather, the methods that define the behavior of the objects are used.
- advantage is taken of object orientation techniques within the rule language.

A summary of the advantages of using object-oriented approaches in rule-based systems can be found in [9].

4 Designing the Agent

As can be seen in Fig. 1, the MD agent has two fundamental parts: the packet sniffer and the rule engine. The packet sniffer captures the packets and generates data that can be used as input information for the rule engine. The rule engine carries out pattern matching between the data model and the stored rules. The result of the execution of the selected rules is a set of actions that are sent to the multi-agent system.

4.1 The Data Model

The data model defines the structure of packets for use in the rule engine. It is a simple structure, in which each packet defined as a class (TCP, IP, UDP, ICMP) implements a common interface called DataPacket. This means that the model can easily be extended to include new packet types.

4.2 The Packet Sniffer

This module provides a facade that allows sniffer configuration and execution and which supplies the basic configuration methods for a sniffer, such as:

- `lookupDevices`, which looks for the different network interfaces present,
- `initialize`, which opens a suitable interface for capturing the packets and adds a listener,
- `run`, which launches capture in the specified interface, and
- `close`, which stops the capture.

An overview of the classes involved in this module is shown in Fig. 2. It can be seen that the `Sniffer` facade class works in the interface called `PacketCapture`, which is responsible for abstracting the functioning of the code that actually implements capture.

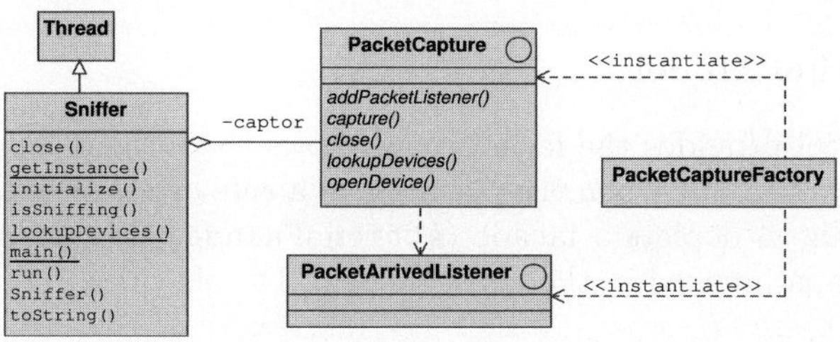

Fig. 2. Packet sniffer classes

The instantiation of this interface, together with the packet listener interface `PacketArrivedListener`, is implemented in a factory (`PacketCaptureFactory`), which permits the specific implementation of the sniffer to be changed without recompiling the application; in other words, the sniffer is transformed into a plug-in.

4.3 The Rule Engine

This module provides a facade that enables the Drools rule engine to be configured and executed. Fig. 3 depicts the classes in this module, which are defined as follows:

- `RuleProccessor`, acts as a facade and permits rule bases to be loaded and executed ;
- `RuleParameterTypeAgendaFilter`, enables filters to be applied in such a way that rules will only be applied to the packets that meet the required characteristics; and
- `CustomWorkingMemoryEventListener`, permits listeners to be registered in response to changes in the working memory. Any new functionality can easily be added to the application by implementing suitable listeners and registering them in the facade `RuleProccessor`.

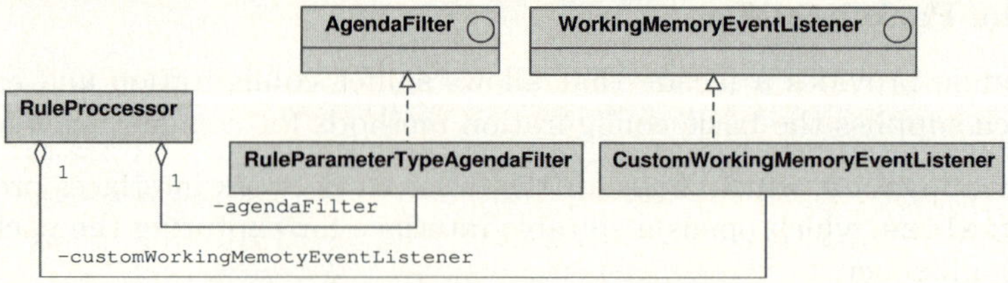

Fig. 3. Rule engine classes

Finally, it is important to highlight that Drools complies with the Java Rules specification (JSR-94), which means that it can be used through a generic Application Program Interface (API). It can thus be substituted for any other inference engine that complies with the same specification.

4.4 The Action Model

The action model provides the facade that allows the actions mentioned in section 3.1 to be carried out when the premises of a rule are satisfied, namely alert, log and pass. Fig. 4 depicts a facade (`ActionsFacade`) that permits actions to be taken, which is invoked in the consequent of Drools rules.

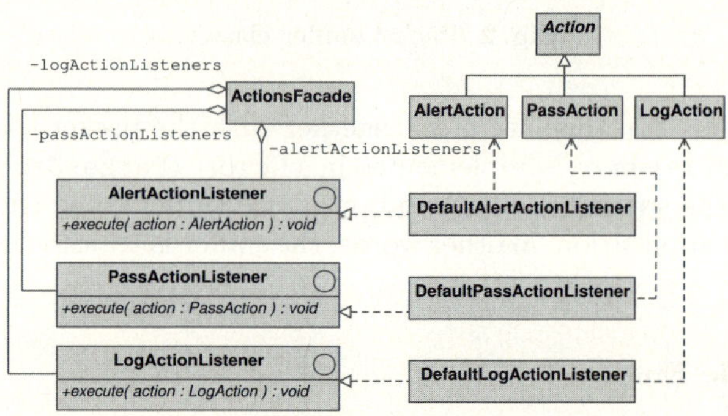

Fig. 4. Action model classes

The `AlertAction`, `PassAction` and `LogAction` actions are subclasses of `Action`. These classes are separated from the facade by a series of interfaces and listeners that ensure the independence of facade from the actions being taken. The classes also permit an action to be carried out by several listeners. Again, this type of design facilitates the incorporation of new actions into the system.

5 Experimental Results

Although the interest of our proposal lies on the scalability and integration of the system in Java, several experiments were carried out so as to test its behavior compared with that of a well-known system as Snort. The version used is Snort-2.6.0.2 for Windows, as it was downloaded from [6]. For both systems the same rule base and the same number of packets were used, varying the number of both during the experiments. The results are shown in Figure 5, where it can be seen that both systems behave similarly. It is important to remark that the time measurements in the simulations include also the time needed to inject the simulated packets in the network. This injection time is the same for both Snort and the proposed system.

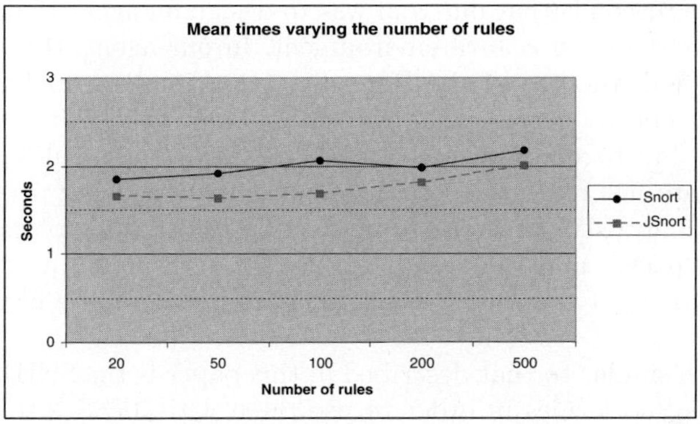

(a)

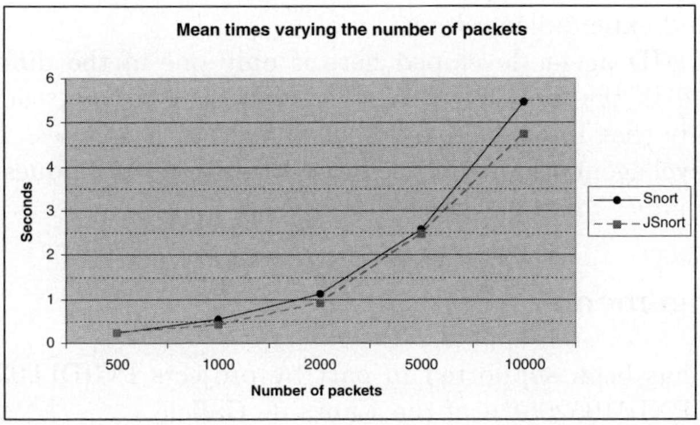

(b)

Fig. 5. Experimental results of the mean times used by Snort and the proposed system (labelled as JSnort)(a) varying the number of rules (b)varying the number of packets

6 Discussion and Conclusions

The MD agent described in this work was designed to ease configurability and extensibility. Thus, for example, it is very simple to add new packet types to the rules or new types of actions. We also demonstrated that the same action could have different consequences in the system. Moreover, since the sniffer and the inference engine used are pluggables, they can be substituted by others if necessary. The use of object orientation and design patterns was fundamental to obtaining this functionality.

Snort rules were used in our system largely, because Snort is currently the most popular signature-based IDS; moreover, its rules are frequently updated, and – as was explained in the Introduction – this is an essential characteristic of this type of system.

However, our system does not use the Snort packet sniffer and rule engine capacities. The reason is that our goal was to design an agent that could behave similarly to Snort but in a Java environment. In our agent, this functioning is optimized through the use of the Rete algorithm implemented in Drools and the filters that can be designed for the rules. Thus, specific rules with certain characteristics can be deactivated, on the basis that these rules are either not applicable or are not relevant to the type of detection being performed in the system being analyzed. The overload implied by the creation of data models for each captured packet and the action models for each action performed is thus compensated for by a working rule engine that is both more configurable and more efficient.

An approach similar to that described in this paper is the SNIDJ system [10], which converts Snort rules in order to use them with JESS [11]. However, the approach is simpler, as it avoids the use of complex objects in the condition and action parts of the rules, and this makes it difficult to develop flexible, easily configurable and extendible systems.

Finally, the MD agent developed here is only one of the different types of agents included in the final multi-agent system. In our case, there are several detection agents that implement different detection techniques. There are also other higher-level agents that combine lower-level agent techniques so as to make detection techniques more powerful.

Acknowledgements

This research has been supported in part by projects PGIDIT03TIC10501PR and PGIDT06PXIB105205PR of the Xunta de Galicia.

References

1. Lunt, T.F. et al: IDES: The enhanced prototype. A real-time intrusion-detection expert system. Technical Report SRI Project 4185-010, SRI-CSL-88-12, CSL SRI International (1988).

2. Spafford, E.H., Zamboni, D.: Intrusion Detection using autonomous agents. Computer Networks **34** (2000) 547–570.
3. Alonso-Betanzos, A., Guijarro-Berdiñas, B., Suárez-Romero, J.A.: A multiagent architecture for intrusion detection. Proc. KES-2002. IOS Press **2** (2002) 1018–1022
4. Java Agent DEvelopment framework. JADE. URL: http://jade.tilab.com. Last accessed 07/05/2006. 2006.
5. Suárez-Romero, J.A., Fontenla-Romero, O., Guijarro-Berdiñas, B., Alonso-Betanzos, A.: A new learning method for single layer neural networks based on a regularized cost function. Lecture Notes in Computer Science, Vol. 2686 part I. Springer-Verlag, Berlin Heidelberg New York (2003) 270–277
6. SNORT. URL: http://www.snort.org. Last accessed 07/05/2006. 2006.
7. Forgy, C.: Rete: A Fast Algorithm for the Many Pattern/Many Object Pattern Match Problem. Artificial Intelligence **19** (1982) 17–37
8. Bouand, J., Voyer, R.: Behavioral match: embedding production systems and objects. Proceedings of the OOPSLA94 Workshop on Embedded Object-Oriented Production Systems. Laforia, Paris (1994).
9. Santos da Figueira Filho, C., Lisboa Ramalho, G.: JEOPS The Java Embedded Object Production System. Lecture Notes in Artificial Intelligence **1952** (2000) 52–61
10. Ahmed, A., Garcia, M.: Signature-based network intrusion detection system using JESS (SNIDJ). Proc. 9th IASTED Int. Conf. on Internet and Multimedia Systems (2005) 281–286
11. Friedman-Hill, E.: Jess in action. Manning Pub. Co., Greenwich, CT (2003)

An Agent System for Bandwidth Allocation in Reservation-Based Networks Using Evolutionary Computing and Vickrey Auctions

Ángel M. Pérez-Bellido, Sancho Salcedo-Sanz, José A. Portilla-Figueras, Emilio G. Ortíz-García, and Pilar García-Díaz*

Department of Signal Theory and Communications, Universidad de Alcalá, 28871 Alcalá de Henares, Madrid, Spain
sancho.salcedo@uah.es, antonio.portilla@uah.es

Abstract. This paper presents an agent system for bandwidth allocation in reservation-based networks, using Vickrey auctions and an evolutionary algorithm. Our evolutionary approach performs the bandwidth allocation to agents which bid for networks resources by means of willingness-to-pay functions. It also calculates in a parallel way the price to be paid to the network's owner. We tackle the case of one-link, and its extension to a network. Simulations performed have shown the main advantages and problems inherent to this model.

1 Introduction

Bandwidth allocation in reservation-based networks is a key point in the future development of this kind of networks. The growth of reservation-based networks, its use on routing the Internet traffic, and the development of agent technology have impulsed the research in new techniques for finding optimal resource (bandwidth) allocation in these type of networks [7]. Agent technology is emerging as a flexible promising solution for network resource management, and specially in bandwidth allocation process in distributed environments [8].

Several problems arise in the definition of an adequate resource allocation method and charging scheme in reservation-based networks. Regarding the resource allocation methods, there are basically two type of paradigms which can be used: the centralized case, where the bandwidth is allocated to users by means of auctions or any other mechanism, outside of the network. And the distributed paradigm, where the assignment of bandwidth is automatically performed by means of a multi agent technology in each link of the network. This last paradigm provides a more flexible way of performing the bandwidth assignment, which makes easier the management of the network resources. However, it is a more complicated process, in which several difficulties can be found.

* This work has been partially supported by Universidad de Alcalá and Comunidad de Madrid, project numbers PI-UAH2005/078 and CAM-UAH2005/019, and by the Spanish Education and Science Ministery, project number TECT 2006-07010/TCM.

N.T. Nguyen et al. (Eds.): KES-AMSTA 2007, LNAI 4496, pp. 476–485, 2007.

The main problem using agent technology for bandwidth assignation is the charging scheme used in the system. Several charging schemes have been proposed in the literature [3], [4]. All these charging schemes take into account the quality of services requested by the user or other properties, such as the type of service or the requested bandwidth. In the last few years, several works have tackled the problem of applying charging schemes based on auctions of networks resources [10], [6], [7]. Vickrey Auctions (VAs), also known as second price auctions, have been recently proposed as an effective mechanism to perform contract negotiation and resource allocation (usually bandwidth) to users, specially in reservation-based networks [9]. VAs work as follows: Each user is asked to reveal his valuation function (we call $\hat{\theta}_i$ the willingness-to-pay function declared by user i, with real function θ_i). Let $a(\hat{\theta})$ be a given solution for the resource allocation, the auctioneer must choose the solution which maximizes the declared social welfare:

$$a(\hat{\theta}) \in arg \max_x \sum_i \hat{\theta}_i(a(\hat{\theta})). \tag{1}$$

The price paid by each user is calculated as the loss of declared welfare he imposes to the other users through his presence:

$$c_i = \left(\max_x \sum_{j \neq i} \hat{\theta}_j(x) \right) - \sum_{j \neq i} \hat{\theta}_j(a(\hat{\theta})). \tag{2}$$

With these definitions, the mechanism of resource allocation using VAs satisfies three major properties:

- Bidding truthfully is the dominant strategy [9]. Thus, hereafter, we suppose that $\hat{\theta}_i = \theta_i$.
- Each truthful player i obtains an utility $U_i \geq 0$.
- When users bid truthfully, the social welfare $(\sum_i \theta_i)$ is maximized.

However, there have also been reported several drawbacks related to the practical use of VAs [5], [2]. In [5] has been shown that implementing a VA for resource allocation requires solving $N+1$ NP-hard combinatorial optimization problems, where N the number of bidders (users in the system bidding to obtain network resources). Thus, a serious problem of computation time arises in systems implementing VAs. In addition, using a VA can lead to poor revenues to the network's owner if the resources offered are enough to cover all the bidders' requests, so another important problem for the network's owner is to estimate how many resources to offer, in such a way that his revenue is maximized.

In this paper we propose an agent system for automatic bandwidth allocation in reservation-based networks. We consider a model presented in [10], where each link of the network is auctioned in a network's node independently of the rest of the links. A Vickrey Auction process is then used for assigning bandwidth to agents (bidders). This process is carried out by means of an evolutionary algorithm. Our evolutionary algorithm solves the bandwidth allocation problem and also the N optimization problems associated with the calculation of the price paid by the users at once. This is done by using a single population and a novel mechanism of *multiple elitism.* This process drastically reduces the computation time of the problem.

The structure of the rest of the paper is the following: Next section describes the agents system for bandwidth allocation proposed in this paper. Section 3 describes the evolutionary algorithm proposed to carry out the VA for a one-link model. Section 4 proposes a mechanism to extend the one-link model to a network. Section 5 shows several computer experiments we have carried out to test the performance of our approach. Finally, Section 6 gives some remarks.

2 One-Link Model Description and Bandwidth Allocation Problem Definition

This paper assumes, at a first stage, a very simple model to describe the process of bandwidth allocation in reservation-based networks using VAs [10]. We consider that the network is auctioned in a link by link fashion, and a maximum Q of bandwidth is available on each link. The auction is performed in one of the network's nodes which injects traffic in the link. We consider a simultaneous one-round, sealed-bid auction, in which users (agents) are requested to provide a willingness-to-pay function in order to bid in the auction. Figure 1 shows an example of the model considered.

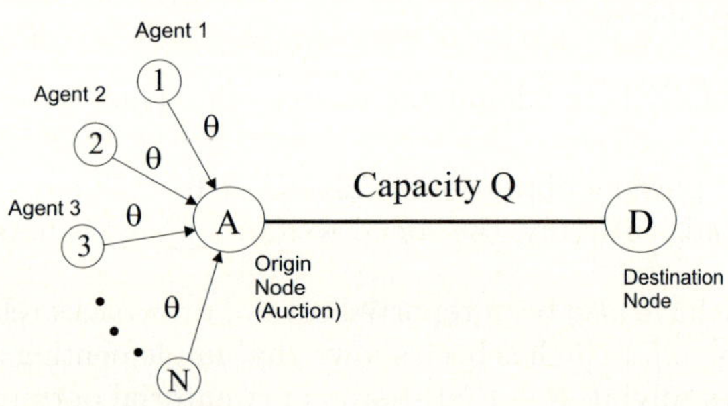

Fig. 1. Model of the agents system for bandwidth allocation proposed in this paper

The set of willingness-to-pay functions (θ_i) provided by agents must fulfil the following requirements:

1. $\theta_i(0) = 0$.
2. θ_i must be non-decreasing functions.
3. θ_i are considered to be piece-wise functions, continuous and derivable in each piece.

The first condition is a boundary condition (nothing is paid if no bandwidth is assigned). The second condition states that the more bandwidth allocated, the more will be the price paid, it is a very intuitive condition in an auction process. The third condition refers to the most general type of willingness-to-pay functions used in real cases, and is related to the strategy to follow depending on the user necessities.

Once all the agents have provided their willing-to-pay function, the auction is carried out using an evolutionary algorithm, which assigns the amount of acquired bandwidth to each agent, and calculates the price that must be paid by it. This problem can be defined as, given the willingness-to-pay functions of the agents θ_i, and the capacity of the link Q, obtaining the bandwidth allocation vector a such as Equation (1) is fulfilled, and subject to:

$$\sum a_k < Q \tag{3}$$

and, for this vector, calculate the associated users payments:

$$c_i = \left(\max_x \sum_{j \neq i} \theta_j(x) \right) - \sum_{j \neq i} \theta_j(a(\theta)). \tag{4}$$

Note the computational cost involved in the process of the VA: First, the calculation of the optimal vector $a(\theta)$ is a NP-hard problem, but, in addition, the computation of the price paid by each user i with Equation (4) implies that we have to solve other N NP-complete problems to calculate c_i. Next section presents an evolutionary approach to solve this problem, which allows solving the $N+1$ optimization problems in a parallel way, using a single population and a mechanism of multiple elitism.

3 An Evolutionary Algorithm with Multiple Elitism Strategy for the One-Link Model

3.1 Problem Encoding

We have chosen a problem encoding based on the structure of the willingness-to-pay function of each agent θ_i. Recall that every function θ_i must fulfil a number of conditions to be considered a valid willingness-to-pay function (see Section 1). In addition, without loss of generality, we also impose that the first derivative of the valuation function must be decreasing in each piece of the willingness-to-pay function. We have done this assumption since in real cases, the bidders are interested in a minimum of bandwidth, for which they will pay a given price. The price offered by bidders for more bandwidth that they need will be low. Thus, willingness-to-pay functions with decreasing first derivative describe better the system than willingness-to-pay functions with increasing first derivative.

The problem can be encoded then in the following way: For each agent i, we encode the piece of the valuation function θ_i in which the agent's demand

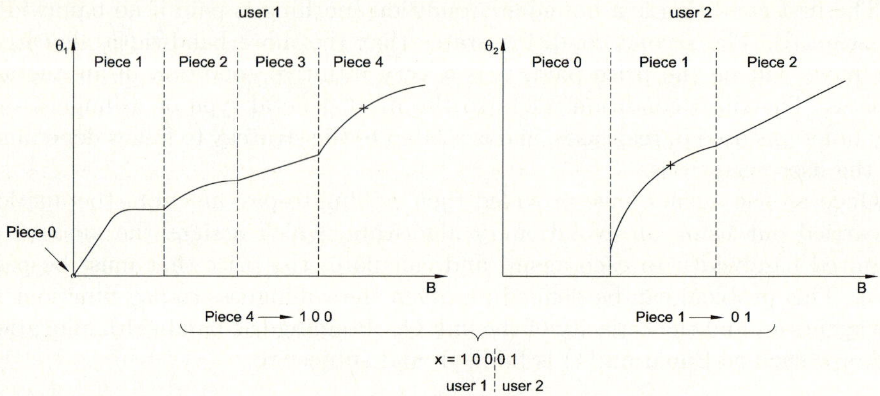

Fig. 2. Example of problem encoding for two users

is located. Thus, a solution is given by the piece number in which each agent's demand is located. As an example, consider Figure 2. This figure displays a case with two agents, showing their respective willingness-to-pay functions and the encoding of each piece number. Note that the value 0 is always considered as a piece in each function θ_i (with piece number 0) so, willingness-to-pay function θ_1 is defined in 5 pieces, and θ_2 is defined in 3 pieces. The points marked in the functions stand for the assigned bandwidth to each agent, and the encoding of these assignments is also shown in Figure 2. As the example shows, the encoding of the agent's demand in his θ_i is done by means of k_i bits, where k_i is given by the number of pieces of the willingness-to-pay function. In our example, agent 1 allocated bandwidth is defined by 3 bits ($k_1 = 3$) and agent 2 allocation is defined by 2 bits ($k_2 = 2$). The final encoding of all agents in the evolutionary algorithm is done by means of concatenating all the agents encodings, in this case it is a 5 bits encoding (see Figure 2).

3.2 Fitness Function with a Repair Procedure

Given a particular solution for the problem a, it encodes in which pieces of the functions θ_i the agents' demands are located as was explained in the previous section. Starting from this information, we must calculate the final assignment of bandwidth a which maximizes $\sum_i \theta_i(a(\theta))$, and which fulfils the capacity constraint $\sum a_k < Q$.

The first step is to calculate if the minimum capacity associated to the individual a (minimum of the possible assignments within the pieces of the willingness-to-pay functions for all agents) is over the capacity constraint of the system Q. If so, a penalty term proportional to the excess of capacity is used as fitness function for a.

In case that the capacity constraint is not violated, i.e there is a feasible assignment defined by a, the following algorithm is carried out: First, let us suppose that the optimal bandwidth assignment corresponds to the rightmost point within the pieces of the willingness-to-pay functions selected by a. If this assignment overpass the capacity constraint, then, we must reduce the assigned bandwidth for some of the users. In order to do this, the most intuitive action is to reduce the bandwidth assignment to the agent which produces less reduction in the fitness obtained. This is equivalent to start the reduction of assigned bandwidth in the agent whose willingness-to-pay function has a smaller first derivative in the starting point (rightmost point of the willingness-to-pay functions' pieces in our case). The reduction in the bandwidth assignment is performed in the same agent until a feasible solution is obtained, or another agent with a smaller first derivative is found. Finally, the fitness associated to the individual a is the sum of the willingness-to-pay functions value in the final assignment of bandwidth.

3.3 Multiple Elitism Strategy

The main problem of the application of VAs to the pricing of networks is the calculation of the price that must be paid by each agent [5], given by equation (4). Recall that this equation means that we have to solve $N+1$ NP-hard combinatorial optimization problems to obtain the total price paid to the owner of the network. We propose a mechanism incorporated in the evolutionary algorithm to drastically reduce the computation time of this process.

Basically, the main idea is to take profit of the structure of the problem. Note that the objective function for the $N+1$ combinatorial optimization problems to be solved is the same, the only difference among these problems is that a given agent is eliminated in every problem but in one, in which all the agents are included.

Thus, we can use a single population with a special mechanism to optimize the $N+1$ problems in a parallel way: In addition to the population maintained by our evolutionary algorithm, we consider a multiple elitist subpopulation with $N+1$ individuals (being N the number of users in the system), in which we include the best solution found so far in the evolution, and also we include the best individuals which do not take into account a given agent. An individual a does not take into account a given agent if and only if the bandwidth assignment corresponding to that agent is located in his piece number 0.

The multiple elitist subpopulation is updated after the application of the crossover and mutation operators. The updating is carried out by means of a hill-climbing procedure: the fitness value of every individual in the elitist subpopulation is compared with the fitness value of individuals in the evolutionary population which fulfil the condition of piece number 0 for a given agent. The best overall individual of the evolution is also updated at this stage, if a better individual is found.

4 Extension of the One-Link Model to a Network

The multi agent system is specially useful when a whole network with multiple links is considered. In this case, agents must be provided with a route (set of links) they must bid for. Note that each agent requires a minimum of bandwidth in each link in order to the transmission can be completed from the initial to the final point (link). Consider that each agent is implemented with only one willing-to-pay function θ_i, for bidding at all links. An agent may have more bandwidth than needed in a given link, but it may not reach to the minimum required in the following one. Let us denote Δ_{ik} the difference between the required bandwidth obtained by agent i in link k and the minimum bandwidth required in this link. A negotiation process is then open for each link k, in which each agent can offer their extra bandwidth Δ_{ik} or can buy bandwidth if needed to other agents. Recall that a given agent offers θ_i for a given bandwidth resource, but the price finally paid by the agent i in link k (c_{ik}) is lower, due to we are dealing with a VA process. As a first approach, we can consider that each agent can offer its extra bandwidth Δ_{ik} for a price larger than the price it has paid, i.e, $p^{\dagger}_{ik} = \frac{c_{ik}}{a}$ per unit of bandwidth. On the other hand, the same agent can bid for bandwidth in those links in which it has not reached the minimum required, offering a price $p^*_{ik} = \frac{\theta_i(\delta_{ik})}{\delta_{ik}}$ per unit of bandwidth. Thus, the final price paid by an agent in the network can be estimated as:

$$P_i = \sum_{k \in \wp_+} \left(c_{ik} - \delta_{lk} \cdot p^{\dagger}_{lk} \right) + \sum_{k \in \wp_-} \left(c_{ik} + \delta_{ik} \cdot p^*_{lk} \right) \tag{5}$$

where $\wp_+$ stands for the set of links where a given agent i has obtained more bandwidth than the minimum needed, part of which (δ_{lk} units, $\delta_{lk} \leq \Delta_{ik}$) has been sold by a price $p^{\dagger}_{lk}$. In the same way, $\wp_-$ stands for the set of links where an agent i has obtained less bandwidth than required, and it has bought δ_{ik} units to another agent l, ($\delta_{ik} \leq \Delta_{lk}$), by a price p^*_{ik} per unit.

4.1 Special Networks

The process of negotiation between agents should be implemented in a general network, with several routes between initial and final nodes. Several problems may arise using this model, for example what to do if a given agent is not able to allocate enough bandwidth in a given link even after the negotiation process, how to chose the best routes in the network or how to avoid the presence of anti-social agents in the systems [2] which bid for one specific link in order to block other competitors. While these are open problems to be studied in general networks, it would be interesting to use the one-link model, without the negotiation process in a network. This is possible in a number of special networks: ring-type networks and tree-type networks are the two main examples.

In this paper we consider the case of an unidirectional ring-type network as the one shown in Figure 3, in which each agent bids for network resources in order to transmit symmetric bidirectional traffic between nodes i and j. In this situation, since the network is unidirectional, any link of the network is occupied by the same amount of traffic, and the capacity Q of every link in the network is the same. Thus, the problem can be reduced to a one-link model, with $N \cdot N - 1$ users (agents), which bid for the Q units of bandwidth of the network.

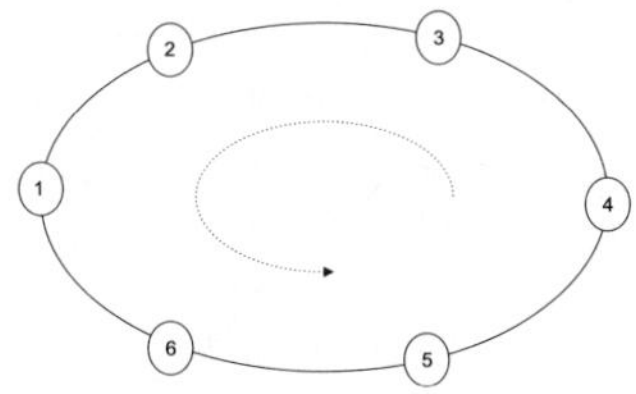

Fig. 3. Example of a ring-type network

5 Computer Simulations

5.1 Results Using the One-Link Model

In order to test the performance of our approach, we have simulated 500 link auctions with different values of the capacity constraint Q. We have considered a case with 5 users, with different willingness-to-pay functions which will be maintained during the 500 instances.

Note that we have considered that all the willingness-to-pay functions are defined between 0 and 2000 (units of bandwidth), and they have values from 0 to 260 (monetary units). In order to obtain 500 different instances, we vary the parameter Q from 20 to 10000 (units of bandwidth). For each value of Q we have run 10 times our evolutionary algorithm, with parameters $P_c = 0.6$, $P_m = 0.01$, a population of 100 individuals, with an elitist subpopulation of 6 individuals (the one for the best overall individual, and other 5 for obtaining the values of Equation (4)).

Figures 4, (a), (b) summarize the results obtained in the experiments. Figure 4 (a) shows the social welfare of the system obtained for each one of the 500 experiments run. Recall that the social welfare is defined as $\sum_{i=1}^{N} \theta_i$, and it is the objective function of the evolutionary algorithm (Fitness function). This graph can be obtained just displaying the best fitness value of the best overall individual in the evolutionary algorithm.

Figure 4 (b) shows the price paid by agents to the owner of the network. This price is calculated using Equation (4), where the term

$$\sum_{j \neq i} \theta_j(a(\theta)) \tag{6}$$

can be calculated by means of the fitness values of the elitist subpopulation maintained by the evolutionary algorithm. It is very interesting to check that the price paid to the owner of the network becomes 0 when there is enough capacity for every agent bidding for the link, and they do not have to compete for it. The network's owner obtains profit using the Vickrey auction systems when the bandwidth offered is less than the users' requirements. Depending on the willingness-to-pay functions of the agents, the maximum profit for the owner of the network will be obtained when a different amount of bandwidth is offered, in our example, it is obtained in instance 94, which corresponds to 1880 units of bandwidth offered in the system.

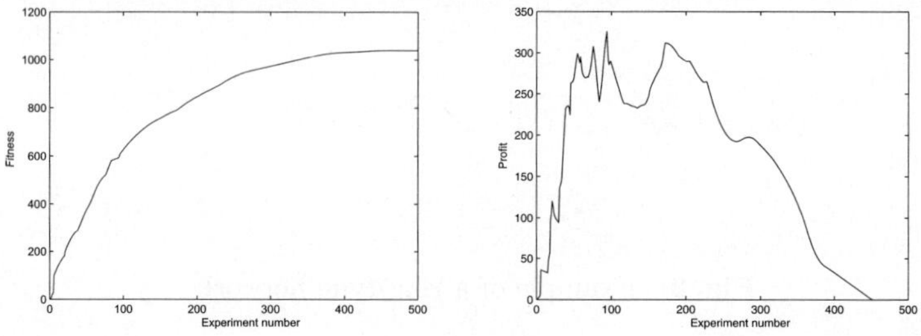

(a) (b)

Fig. 4. Results of the simulations performed; (a) Social welfare of the system; (b) Price paid by users

5.2 Results on Ring-Type Networks

Consider now a ring network with 7 nodes (42 agents), every one bidding with a different willingness-to-pay function, randomly generated. Figure 5 shows the results obtained for 50 different values of bandwidth Q. Figure 5 (a) displays the shows the social welfare of the system obtained for each one of the 50 experiments run. Figure 5 (b) shows the price paid by agents to the owner of the network.

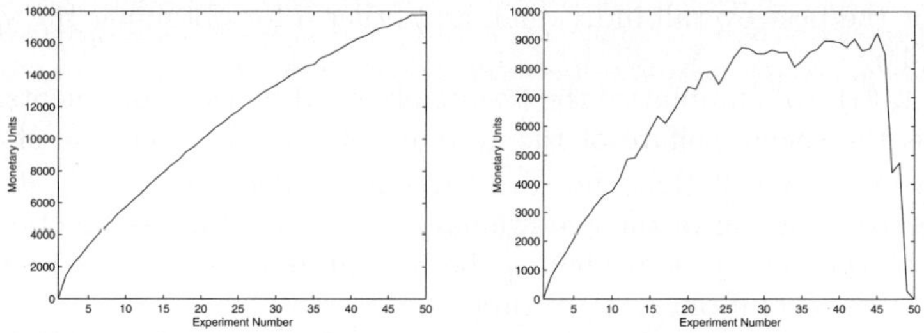

(a) (b)

Fig. 5. Results of the simulations performed in ring-type networks; (a) Social welfare of the system; (b) Price paid by users

6 Conclusions

In this paper we have presented an agent system for allocation of bandwidth in reservation-based networks, using an evolutionary algorithm to perform a Vickrey auction process. Our algorithm is able to drastically reduce the computation time for calculating the revenues to the owner of the link, by using a multiple elitism strategy. Simulations performed have shown that our approach is able to allocate link resources to bidders in an optimal way, and also to calculate the price paid by users within the same run of the evolutionary algorithm.

References

1. Bigham J., Cuthbert, L., Hayzelden, A. and Luo, Z. "Multi-agent system for network resource management," Lecture Notes in Computer Science, vol. 1597, pp.514-526, 1999.
2. Brandt, F. and Wei, G. "Antisocial agents and Vickrey auctions", Lecture Notes in Artificial Intelligence, vol. 2333, pp. 335-347, 2001.
3. Delgrossi, L. and Ferrari, D. "Chargin schemes for reservation-based networks", *Telecommunication Systems*, vol. 11, pp. 127-137, 1999.
4. Keon, N. and Anandalingam, G. "Optimal pricing for multiple services in telecommunications networks offering Quality-of-Service guarantees," *IEEE/ACM Trans. Networking*, vol. 11, no. 1, 2003.
5. Maillé, P. and Tuffin, B. "Why VCG auction can hardly be applied to the pricing of inter-domain and ad hoc networks," IRISA Internal Report, http://www.irisa.fr/armor/lesmembres/Tuffin/Publis/VCG_inter.pdf.
6. Maillé, P. and Tuffin, B. "Multi-bid auctions for bandwidth allocation in communication networks", In *Proc. of the IEEE Infocom*, Hong Kong, China, 2004.
7. Maillé, P. and Tuffin, B. "Pricing the Internet with multi-bid auctions," *IEEE/ACM Trans. Networking*, vol. 14, no. 5, pp. 992-1004, 2006.
8. Manvi, S. and Venkataram, P. "An agent based adaptive bandwidth allocation scheme for multimedia applications," *Journal of Systems and Software*, vol. 75, pp. 305-318, 2005.
9. Takahashi, E. and Tanaka, Y. "Auction-based effective bandwidth allocation mechanism," *Telecommunication Systems*, vol. 24, no. 2-4, pp. 323-338, 2003.
10. Toutain, F. and Huber, O. "A general preemption-based admision policy using a smart market approach," In *Proc. of the 15th Annual Joint Conference of the IEEE Computer and Communication Societies*, pp. 794-801, 1996.

Multiagent Approach to Network Traffic Anomalies Uncertainty Level Assessment in Distributed Intrusion Detection System*

Grzegorz Kołaczek

Institute of Information Science and Engineering
Wroclaw University of Technology, Wroclaw, Poland
grzesiek@pwr.wroc.pl

Abstract. The paper proposes a formal framework for network traffic anomalies uncertainty level assessment within a distributed multiagent Intusion Detection System (IDS) architecture. The role of traffic anomalies detection is discussed then it has been clarified how some specific values characterizing network communication can be used to detect network anomalies caused by security incidents (worm attack, virus spreading). Finally, it has been defined how to use the proposed techniques in distributed IDS.

1 Introduction

In order to process intrinsically distributed information, most of modern IDS systems are organized in a hierarchical architecture [4], consisting of low level nodes which collect information and management nodes which aim to detect large-scale phenomena. The task of management nodes is to reduce the amount of the data processed, identify attack situations as well as make decisions about responses [10].

In this approach it is assumed that the network system consists of the set of nodes. There are also two types of agents in our multiagent system: monitoring agents (MoA) and managing agents (MA) (fig 1) [7],[8],[9]. Each MoA monitors its own area of responsibility which consists of the set of several nodes. It is assumed that these areas may mutually overlap [7]. MoAs observe the nodes, process captured information and draw conclusions that are necessary to evaluate the current state of system security. This means that MoA is responsible for interpretation of the data stream arriving to the particula node. Another feature of MoA is that its opinions about various parameters that have been observed can be combined together to get the most relevant information about current node state. The role of Managing Agents is to gather information from MoAs and to make final decision about level of global threats and ongoing attacks. Especially, MA evaluates the accurateness of subordinate MoA opinions and proclaim its own opinion.

* This work was supported by the Polish State Committee for Scientific Research under Grant No. 3 T11C 029 29 (2005-2007).

N.T. Nguyen et al. (Eds.): KES-AMSTA 2007, LNAI 4496, pp. 486–495, 2007.

During network attack there occurs at least two general types of anomalies. First one is connected to observed traffic characteristics (section 2) and second to communication scheme which tends to be constant under normal conditions [8]. In this context the system properties observed by the agent MoA in the proposed architecture will fall into two basic categories: traffic measurement and communication pattern measurement. The attack recognition is being made on this base. The MoA agent's algorithm for decision making process is invoked periodically and uses observed values as input data. The results of this process are MoA opinions (section 4). MoA sends its opinions to MA and also stores acquired values thus creating the history of system behaviour. According to data obtained from MoA, Managing Agent can announce security allarm and can analyse situation for example to find out the source of attack or some new characteristics of the attack.

This paper is an extension of our ealrier research concerning application of autonomous agents in intrusion detection process [7],[8],[9] and concentrates on how some specific values characterizing network communication can be used to detect network anomalies caused by security incidents.

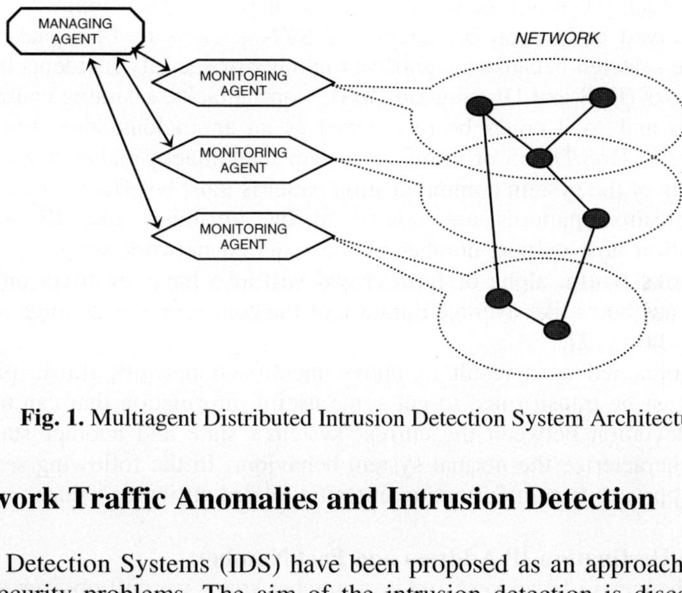

Fig. 1. Multiagent Distributed Intrusion Detection System Architecture

2 Network Traffic Anomalies and Intrusion Detection

Intrusion Detection Systems (IDS) have been proposed as an approach to cope with current security problems. The aim of the intrusion detection is discovering of all abnormal states of the system in relation to the network traffic, users activity and system configuration that may indicate violation of security policy [5],[6]. But although the IDS idea is very simple, implementation of such systems has to deal with a lot of practical and theoretical problems. Difficulties with building intrusion detection systems arise from a complexity of the structure of attacks symptoms, distributed nature of the network systems and dynamics of the source of threats especially the problems of encoding new intrusions scenarios. The security assessment of a network system requires application of complex and flexible mechanisms for monitoring values of system attributes that have an influence on the security level of all network system. Another important element is an effective computational mechanism for evaluating the states of system security on the basis of incomplete, uncertain and inconsistent resources. Finally, the algorithms of machine learning to detect new intrusions pattern scenarios and recognise new symptoms of security system breach in order to update the security system knowledge base must be defined

2.1 Evaluation of Network Traffic Anomalies

Traffic attributes that are especially important and used during process of anomaly detection are [1]: source and destination IP address, source and destination port, number of bytes and packets sent to the remote hosts, number of bytes packets received by the local host, TCP flags, especially SYN, RST and FIN flags duration of the connection

The values of variables describing these attributes are collected and processed by intrusion detection system in a purpose to identify any anomalous behaviour. The simplest decision mechanism applied in intrusion detection system uses threshold test to find out if the observed value is typical or it can be classified as anomalous.

In our approach we observe: source/destination IP and port number, number of bytes sent/received and ration of number of SYN packets to FIN packets. These attributes were selected because a significant number of security incidents like denial of service attacks (DoS and Distributeid DoS), worm attacks, scanning cause changes in their values and so it could be recognized as an anomalous state. For example intrinsic nature of DoS/DDoS or intrusive system scan attacks makes that existing in the normal state of the system communication patterns must be effected by this events [8]. Communication patterns are related to the attributes like IP address of source/destination host or port number of the required network service. Similarly, other attacks like worm, alpha or flash crowd will also have an effect on different traffic related attributes like average duration of the connection or average number of bytes sent by a host [12].

Raw data obtained as a result of above mentioned network traffic parameters observation must be transformed to get some useful information that can be used to identify the deviation between the current system's state and another state that is supposed to characterize the normal system behaviour. In the following sections we describe our approach to transformation of traffic related attributes values.

2.1.1 Source/Destination IP Address and Port Number

To measure changes in IP address and port number space we will observe a value of Shannon entropy related to these attributes [13]. Entropy values are calculated for separate time periods. The length of the period can be a subject of more detailed discussion [1], however we assume that it is possible that different monitoring agents (MoA) use various periods length.

This means that we will evaluate and collect:

– $e_{t_i}^{ip\_s}$ - entropy of source IP address in the period t_i,

– $e_{t_i}^{ip\_d}$ - entropy of destination IP address in the period t_i,

– $e_{t_i}^{p\_d}$ - entropy of destination port number in the period t_i ,
– $e_{t_i}^{p\_s}$ - entropy of source port number in the period t_i .

Entropy value is evaluated from standard formula:

$$e = -\sum_{i=0}^{N} p_i \log p_i , \quad p_i = \frac{n_i}{\sum_{i=0}^{N} n_i} , \quad 0 \le i \le N, \tag{1}$$

where:

N - cardinal number of IP address/port number set,

n_i - number of packets with a particular source/destination IP address/port number observed in the period t_I,

$\sum_{i=0}^{N} n_i$ - total number of packets observed in the period t_I

As for some t_i , the value of $\sum_{i=0}^{N} n_i$ can be equal zero (no traffic observed in t_i period), we assume that in these periods entropy value is also zero.

2.1.2 Number of Bytes and Packets

Changes of entropy values are strictly related to changes of communication patterns. Using this measure of traffic parameters, some sort of anomalies caused by intrusive actions like DoS or system scan can be detected. However, other types of intrusions do not have to disturb communication patters. For example so called topological worms using internally generated target lists tries to infect only well known by the infected host remote targets. Well known, means that instead of performing random scan to find vulnerable hosts, the worm tries to discover the local communication topology and infect only hosts which sent or received data to or form infected host [14].

Simillary like it has been shown in section 2.1.1 the values describing number of bytes and packets exchanged by a host will be obtained as a result of observation of incoming and outgoing traffic in each of constant size period while it is observed by MoA.

$b_{t_i}^{in}$ - bytes received by a host in period t_i

$b_{t_i}^{out}$ - bytes sent by a host in period t_i

$p_{t_i}^{in}$ - packets received by a host in period t_i

$p_{t_i}^{out}$ - packets sent by a host in period t_i

2.1.3 TCP Flags

The TCP flags are important source of information about host's connections state. Typical TCP connection have three phases: connection establishment, data transfer, connection termination. Each phase uses packets with some standard sequences of TCP flags, especially TCP flags brings information about current connection state. However, this information may be incorrect while an intruder can manipulate the packet's content to reach some particular aim (e.g. the intruder tries to obtain information about services activated by host by performing system scan or simmilar effect can be observed during DoS/DDoS attacks) [10].

In our approach we measure a difference between number of sent SYN packets and received RST and FIN packets.

$$\Delta f_{t_i} = p_{t_i}^{syn} - p_{t_i}^{rst} - p_{t_i}^{fin} \tag{2}$$

where:

Δf_{t_i} - parameter indicating temporal start/end connection ratio

$p_{t_i}^{syn}$ - number of sent TCP packets with SYN flag set,

$p_{t_i}^{rst}$ - number of received TCP packets with RST flag set,

$p_{t_i}^{fin}$ - number of received TCP packets with RST flag set

In normal conditions, in long time observation we should get the mean value of Δf_{t_i} near zero. Intrusive actions like system scanning, DoS attacks, may cause the temporal distortion of the mean value of Δf_{t_i} .

2.1.4 Duration of the Connection

Duration of a connection may be another characteristic attribute in anomaly detection process [1]. During various types of attacks, this value will be affected and so an anomaly may be detected. For example worm infection will generate a large number of connections with similar duration. This worm related connections should change also the observed mean values of connection duration that has been observed in a system. We evaluate simple mean value of connections' duration that have been observed in period t_i.

c_{t_i} - mean value of duration of connections that have been bserved in a period t_i

3 Traffic Statistics

In section 2.1 a few traffic related variables have been presented. Values of these variables can be used to obtain useful information about system security incidents. Apart from collecting these values, intrusion detection mechanism must preprocess them to reduce the probability of misinterpretation and so called false-positive alarm. Our approach uses Mark Burgess (MB) technique to find out anomalous behavior. This technique of anomaly detection has been described in [2],[3]. The main assumptions made in his framework are as follows.

MB defines *iterative expectation function.* Let q be an observation, and $<<q_i>>$ be the i-th estimator of the average, with geometric fall-off, then $<<q_i>>$ may be defined by the recurrence relation:

$$<<q>>_{i+1} = (q \mid <<q>>_i),\ <<q>>_0 = 0 \tag{3}$$

where

$$(q_1 \mid q_2) = \frac{wq_1 + \overline{w}q_2}{w + \overline{w}} \quad , w, \overline{w} \text{ - const} \tag{4}$$

The other fundamental notion for MB analysis is pseudo-periodic function:

$$q(t) = \sum_{n=0}^{\infty} q(nP + \tau) \equiv \sum_{n=0}^{\infty} \chi_n(\tau) \text{, where } 0 \le \tau < P \tag{5}$$

Such pseudo-periodic function can be characterized by two kinds of average: average over corresponding times in different periods (topological average $< \chi(\tau) >_T$), and average of neighboring times in a single period (local average $< \chi(\tau) >_P$).Limited memory versions of these deviations are given by the following formulas:

$$\sigma_{<<T>>}(\tau) \equiv \sqrt{<< (\delta_{<<T>>}\chi)^2 >>_T} \tag{6}$$

$$\sigma_{<<P>>}(n) \equiv \sqrt{<< (\delta_{<<P>>}\chi)^2 >>_P} \tag{7}$$

where, for any measure X:

$$(\delta_{<<P>>}X) \equiv X - << X >>_P \tag{8}$$

$$(\delta_{<<T>>}X) \equiv X - << X >>_T \tag{9}$$

These averages are calculated by replacing the evenly weighted sum over the entire history by an iteratively weighted sum that falls off with geometric degradation. The additional positive consequence of this definition is that in order to obtain all information, one only needs to retain and update the mean and the variance.

3.1 Subjective Nature of Anomaly Detection

In contemporary network, traffic congestion is avoided by packet switching. The traffic has been isolated to 'parallel' branches of a network spanning tree. Network nodes or hosts occupy points at the leaves of these branches and therefore experience an individual (subjective) view of the network traffic. The concept of an anomaly is also a very subjective one because what is unusual for one node is a regular occurrence for another. One of the best places in the network where incidents may be tracked down and so anomalies may be reveal are the network nodes.

As stated above, anomalousness is a subjective judgment, made within the context of past experience, and can be codified into a 'policy' about what is sufficiently anomalous to warrant a response. So, we look for a potential anomalous behaviour by comparing current observation to learned experience. If the event looks probable, we can consider it as the evidence derived from a supporting semantic model. As in our approach a Monitoring Agent is responsible for interpretation of the data stream arriving to the particula node, another important source of subjectiveness is the fact that Monitoring Agents can have different experiences and use different criteria to evaluate a level of anomalousness. MoA experience should be related to evaluation of its decision by MA. This is why they can also generate different opinions. Fusion of Monitoring Agents diverse opinions may be a next step to more profound node state analysis [2].

Subjective logic was proposed by A.Josang as a model for reasoning about trust propagation in secure information systems. It is compatible with Dempster-Shafer's theory of evidence and binary logic. Subjective logic includes standard logic operators and additionally two special operators for combining beliefs – consensus and recommendation. The basic definitions of subjective logic given in this section come from [15],[16]. Subjective logic can be used to express so-called opinions (see below) about facts with assumption that we do not require the knowledge of how these facts were grounded or inferred. We may also have an opinion about some subject (source of information). When expressing belief about a statement (predicate) it is assumed that it is either true or false, but we're not necessarily certain about it. Let's denote *belief, disbelief* and *uncertainty* as b, d and u respectively. A tuple $\omega = \langle b, d, u \rangle$ where $\langle b, d, u \rangle \in [0,1]^3$ and $b + d + u = 1$ is called an *opinion.*

4 Network Traffic Anomaly Level Assessment

Anomalousness is a subjective judgment, made within the context of past experience so MoA must decide the impact of historical events on contemporary opinion about system state. In our approach MoA express its opinions in subjective logic. Subjective logic is a very convenient tool which allows to formally express various aspects of oppinion. Subjective logic requires that MoA evaluates its uncertainty about exactness of its opinion, and how strong it believes that the host state is normal. The MoA formulates its opinion about recently observed values of: source/destination IP address and port number, number of bytes and packet, TCP flags and connection duration.

The MoA uncertainty level about source/destination IP address and port number activity normalness is related to the silent periods. Silent periods are the moments where MoA did not observed any incoming/outgoing traffic. The number of silent periods in the last pseudo period *P* increases our uncertainty about correspondence of calculated statistic to the reality because the most recent observation are the most significant. This means that the silence in the last pseudo period *P* may significantly distort agent's opinion.

Another important point of uncertainty is level of abnormality changes of monitored parameter values during the last pseudo period *P*. The level of these changes may be evaluated as a proportion $\frac{\delta_{<<P>>}\chi(n)}{\sigma_{<<P>>}(n)}$. The changes may be related to some accidental events not necessarily connected to any intentional security breach. Both these values (silence and abnormal changes) have been used to calculate the uncertainty value of agent's opinion.

$$u=\begin{cases}\min(1,\dfrac{\sum_{\Delta t_i|e_{ti}=0}\Delta t_i}{P}+(1-\left(\dfrac{\delta_{<<P>>}\chi(n)}{\sigma_{<<P>>}(n)}\right)^{-1}) & if\ \ \delta_{<<P>>}\chi(n)>\sigma_{<<P>>}(n)\\ \dfrac{\sum_{\Delta t_i|e_{ti}=0}\Delta t_i}{P} & if\ \ 0\le\delta_{<<P>>}\chi(n)\le\sigma_{<<P>>}(n)\end{cases} \tag{10}$$

where:

P – the length of the pseudo period (required by MB statistics)
Δt_i the length of the period for which entropy value is calculated (subperiod of *P*)
$\chi(n)$ – any measure mentioned in section 2.1.

$$d=\begin{cases}\max(0,(1-u-\left(\dfrac{\delta_{<<L>>}\chi(n)}{\sigma_{<<L>>}(n)}\right)^{-1}) & if\ \ \delta_{<<L>>}\chi(n)>\sigma_{<<L>>}(n)\\ 1-u & if\ \ 0<\delta_{<<P>>}\chi(n)\le\sigma_{<<P>>}(n)\end{cases} \tag{11}$$

$$b=1-u-d \tag{12}$$

4.1 Applications

All security incidents can be characterized by their influence on traffic parameters. For example, intrusive system scan should leave the most visible traces in destination port distribution and TCP flags sequences. MoAs opinions about various network traffic parameters can be combined together to get the most relevant information about current node state. For example we can map the conjunction of MoA opinions about distribution of destination port numbers and TCP protocol flags to get the MoA opinion about probability of the system scan event.

Let *port* and *tcp* denote following opinions:.

port - the observed distribution of destination port numbers is normal (typical),
tcp - the observed TCP flags sequences are normal (typical),

Using described in previous section statistics, MoA calculates its opinion about *port* and about *tcp*.

$$MoA_{port}=\langle b_{port}, d_{port}, u_{port}\rangle \tag{13}$$

$$MoA_{tcp}=\langle b_{tcp}, d_{tcp}, u_{tcp}\rangle \tag{14}$$

The MoA can evaluate its opinion about the probability of system scan threat using subjective logic conjunction operator.

$$MoA_{port \wedge tcp} = \left\langle \begin{array}{l} b_{port} b_{tcp}, \\ d_{port} + d_{tcp} - d_{port} d_{tcp}, \\ b_{port} u_{tcp} + u_{port} b_{tcp} + u_{port} u_{tcp} \end{array} \right\rangle \quad (15)$$

Where $MoA_{port \wedge tcp} = MoA_{scan}$ stands for MoA opinion about the probability of system scan event.

Another advantage of our proposal is related to Managing Agents (MA) collaboration with MoAs. MAs collects and process information from subordinate MoAs. This way, some information coming from different MoAs can be joined together by MA and some new knowledge about host security state can be obtained. Especially MA can try to backtrack incidents or it can assure itself that the observed anomaly is not just some accidental distortion unrelated to any security incident. In this context two operators from subjective logic should be applied: recommendation, denoted by ⊗ and consensus, denoted by ⊕.

Let consider that MA evaluates the accurateness of subordinate MoA opinions. According to this evaluation MA can announce its own opinion using recommendation operator. For example while ω^{MoA}_{scan} is MoA opinion about possibility of system scan and ω^{MA}_{MoA} is MA opinion about MoA opinions accurateness then:

$$\begin{array}{l} \omega^{MA\_MoA}_{scan} = \omega^{MA}_{MoA} \otimes \omega^{MoA}_{scan} = \\ \left\langle b^{MA}_{MoA} b^{MoA}_{scan}, b^{MA}_{MoA} d^{MoA}_{scan}, d^{MA}_{MoA} + u^{MA}_{MoA} + b^{MA}_{MoA} u^{MoA}_{scan} \right\rangle \end{array} \quad (16)$$

5 Conclusions and Future Work

Our approach is based on Mark Burgess technique and subjective logic. As Mark Burgess technique has quite good ability to tolerate seasonal changes, do not require regularized data and requires relatively small set of data and utilizes CPU only on low level we hope that all this features will characterise also our proposal. These features are especially interesting in a context of real time identification performed on a single host and within mobile agent environments.

We assume that an anomalous behaviour can be described as a subjective measure of node security level. We proposed a novel method to formally describe agent uncertainty level about observed network traffic characteristics. This method allows Monitoring Agents to express their opinions about current values of parameters according to network node activity. Using subjective logic operators and our anomaly evaluation method Managing Agents will be able to combine different opinions to chose the most relevant information, classify observed security incidents, backtrack them and finally create formal description of new incidents. Formulation and verification of appropriate algorithms that can be used in a task of incident backtracking and classification will be a subject of our future work.

References

1. Beach A., Modaff M, Chen Y.: Network Traffic Anomaly Detection and Characterization, cs.northwestern.edu/~ajb200/anomaly%20detection%20paper%201.0.pdf.
2. Burgess M.: An Approach to Understanding Policy Based on Autonomy and Voluntary Cooperation. DSOM (2005) 97-108
3. Burgess M., Two Dimensional Time-Series for Anomaly Detection and Regulation in Adaptive Systems. DSOM (2002) 169-180
4. Gorodetski V., Karsaev O., Khabalov A., Kotenko I., Popyack L., Skormin V.: Agent-based model of Computer Network Security System: A Case Study. In: Proceedings of International Workshop Mathematical Methods, Models and Architectures for Computer Network Security, Lecture Notes in Computer Science, vol. 2052, Springer Verlag, Berlin Heidelberg New York (2001) 39-50
5. Hwang K., Liu H. and Chen Y.: Cooperative Anomaly and Intrusion Detection for Alert Correlation in Networked Computing Systems. Technical Report, USC Internet and Grid Computing Lab (TR 2004-16) (2004)
6. Khoshgoftaar, T.M. Abushadi, M.E.: Resource-sensitive intrusion detection models for network traffic, Eighth IEEE International Symposium on Publication, (2004) 249- 258
7. Juszczyszyn K, Nguyen N.T., Kolaczek G., Grzech A., Pieczynska A., Katarzyniak R.: Agent-based Approach for Distributed Intrusion Detection System Design. International Conference on Computational Science 2006, United Kingdom (2006) 224-231
8. Juszczyszyn K.and Kołaczek G.: Assessing the Uncertainty of Communication Patterns in Distributed Intrusion Detection System. KES 2006, LNAI 4252, 243-250
9. Kolaczek G., Kuchtiak-Pieczynska A., Juszczyszyn K., Grzech A., Katarzynak R., Nguyen N.T.: A Mobile Agent Approach to Intrusion Detection in Network Systems., Lecture Notes in Artificial Intelligence 3682 (2005) 514-519.
10. Kotenko I. et al.: Multi-Agent Modeling and Simulation of Distributed Denial-of-Service Attacks on Computer Networks, In: Proceedings of Third International Conference Navy and Shipbuilding Nowaday. St. Petersburg, (2003), 38-47.
11. Thottan M. and Ji. C.: Anomaly detection in IP networks, IEEE Transactions on Signal Processing, 51(8) (2003), 2191- 2204
12. Lakhina A., Crovella M., and Diot C.: Characterization of Network-Wide Anomalies in Traffic Flows. Technical Report BUCS-2004-020, Boston University, (2004) http://citeseer.ist.psu.edu/715839.html
13. Shannon C.E. and Weaver W., The mathematical theory of communication. University of Illinois Press, Urbana, (1949)
14. Weaver, N., Paxson, V., Staniford, S., and Cunningham, R. , A taxonomy of computer worms, ACM Workshop on Rapid Malcode - WORM '03, ACM Press, New York, NY, (2003) 11-18.
15. Jøsang, A.: A Logic for Uncertain Probabilities. International Journal of Uncertainty, Fuzziness and Knowledge-Based Systems, **9(3)** (2001) 279–311
16. Jøsang, A.: A Metric for Trusted Systems. In: Proceedings of the 21st National Security Conference, NSA (1998), 68-77

An Algorithm for Loopless Optimum Paths Finding Agent System and Its Application to Multimodal Public Transit Network

Doohee Nam[1], Seongil Shin[2], Changho Choi[3], Yongtaek Lim[3], and Seung Jae Lee[4]

[1] Hansung University, Samseon-dong Seongbuk-gu, Seoul, Korea
doohee@hansung.ac.kr
[2] Seoul Development Institute, Seocho, Seoul, Korea
[3] Chonnam University, San96-1, Dundeok-dong,Yeosu, Korea
[4] University of California at Berkeley, Berkeley, U.S.A.

Abstract. The entire path deletion methods (EPDM) explore the K-th network transformation in order to prevent the predetermined K-1 number of path(s) from being re-searched in optimum path finding agent system. But, they show a critical drawback for the application in practical traffic network because loops, in which the same node and link are reappeared without limitation, can't be constrained. The purpose of this paper is to develop a method to selectively control loop-paths by applying link-label and to design the method to be utilized for analysis of intermodal transportation networks. For the fist purpose, this paper takes advantage of the link-label technique in optimum path agent. For the second purpose, the link-mode transformation technique is adopted to treat all modes passing through the same link as each separate link-feature is generated.

Keywords: optimum path agent, Loopless path, Looping path, K shortest Path, Network Transform, Mode-Link Transformation, Link Loopless path.

1 Introduction

1.1 EPDM and K-Path Search

An intention of the Advanced Traveler Information System(ATIS) is to provide optimum path information toward travelers' requests. Since recognition of the optimum path relies on travelers' situations and conditions, providing wide selection opportunities with diverse path information is critical to actualize the ATIS. Therefore, in ATIS various alternative paths are represented for travelers to select the most optimum path among them. One of the prevailing ways to yield diverse alternative path information is applying a network algorithm which exploits K number of paths with effect.

Path Deletion Method(PDM) means exploiting K number of paths between two nodes in network where origin and destination are already assigned(Yen,

N.T. Nguyen et al. (Eds.): KES-AMSTA 2007, LNAI 4496, pp. 496–505, 2007.

1971; Martins, 1984; Avezedo et al, 1993). Because of its merit, utilizing established algorithm(Dijkstra, 1959; Moore, 1957) in direct, this method is easy to comprehend and build its system than one(Pollack, 1961; Bellman & Kalaba, 1968, Shier, 1979) that detects K number of paths simultaneously. There are two different algorithms exploiting K number of paths with PDM. One is for exploiting loopless path(simple path & node loopless path) where no redundant nodes or links(Yen, 1971) along a path are existed, and the other is for looping path(Martins, 1984, Avezedo et al, 1993). The algorithm suggested by Yen(1971) for loopless path is based on the path partition algorithm. In other words, it considers both K-1 number of links and a set of their related part links when detecting optimum paths; as a result, the implementing speed of algorithm for selecting K-th path is decreasing as long as the network increases. On the other hand, the method suggested by Marins and Azevedo et al(1994) for exploiting looping path is based on the Entire Path Deletion Method(EPDM) through the transformation of enlarged network. In particular, the method by Azevedo et al(1993) is more efficient than the one by Martins (1984). This detects K number of paths only through network transformation induced by implementing Optimum Path Algorithm(OPA) once. In order to exploit K number of path, this algorithm only needs a network transformation of K-1 number of paths and implementation of OPA at once.

EPDM, however, has a limitation of pandering over unnecessary looping path. Looping path means an infinite reiteration of a link and/or a node. Consequently, the involvement of looping path concept could make the application of algorithm limited once exploring a path; especially, the problem increments as long as the network increase.

In a network composed with nodes and links, if a path is defined as "the connection of nodes on a path," looping path would be defined as a path which a start node and an end node are same. As a result, a same node appears repeatedly on a path in which a looping path is involved. Since turn limitation and turn delay are typically prevailing in an urban traffic network, the condition of passing an intersection back and forth such as a U-turn and a P-turn appear. In summary, for comprehensive deliberation on traffic condition, it is needed to comprehend both loopless path(simple path & node loopless path) and looping path. On the other hand, if defining a path as "the connection of links on a path," it is possible to limit traffic conditions on network to a loopless path. In urban traffic network, the repeating traffic on a link (a road) is unusual. Therefore, traffic patterns appeared on a network can be demonstrated by a link loopless.

1.2 Research Purpose

This paper suggests how to exploit selectively link loopless path by the EPDM and develop how to apply the suggested method into multiple traffic network efficiently. There are two research purposes. First, we suggest a method that unnecessary looping paths are deleted by the EPDM(Avezedo et al, 1993). In detail, based on the preceded K-1th expanded network to detect the K-th path, this method decides label setting- node and/or link label- in order not to produce a looping path when label renewing. Especially, this paper focuses on link label because its merit that considers two adjacent links (Kirby & Potts; Potts & Oliver, 1972) when detecting a path could omit the unnecessary-direction detection where the looping path is produced; also, the merit of detecting a no link repeated path. Second, we arrange a concept of the suggested algorithm in order to be applied to an intermodal transportation network, then guide to apply the loopless path algorithm properly in the network. To make a better result, we adopt the link mode transformation technique which links entire mode driving on each simple link.

2 Methodology

In order to apply to intermodal transportation network, An advanced link loopless path algorithm based on the one of Azevedo et al(1993) is developed. To reduce additional works in this algorithm, we apply link-mode transformation concept which consider entire modes moving on each link as a link, then try to simplify the algorithm that does not consider modes in EPDA.

2.1 A Path Deletion Method and Loop Production

A previous branch-based OPA utilizes a method of detecting two nodes so that it is impossible to consider a turn delay or a link loop occurred at an interchange where involves with a turn delay or turn prohibition. To prevent the link loop occurrences something else should be considered in an OPA. It is deletion of a path which creates link loop. The characteristics of this algorithm is to execute only once the link loopless path detection in link-label based OPA, to determine the label of extra nodes and/or links through k-1 th network transformation, and to figure out the OPA by the determined label. In short, there are three algorithm: 1) link-label based OPDA for link loopless path detection, 2) network transformation algorithm, 3) Algorithm which determines the labels of extra links and nodes in transformation network. And, the algorithm is organized by;

Step 1 : detection P_1 by execution of link-label transformation OPA based on N

Step 2 : repetition of K=2 to K

Building N'; from N, Azevedo's(1993) network algorithm

<u>Determine a link-label and a node-label added to N'</u>

OPA from an origin to a destination

L^NL: a link set of network N

V^N: a node set of network N

$L^{N'}$L: a link set added to network N'

$V^{N'}$: a node set added to network N'

$L^N \bigcap L^{N'} = \{\}$. $V^N \bigcap V^{N'} = \{\}$.

In a suggested algorithm, the link-label and node-label to determine labels of links and nodes are added to network N'. There are two different links added to N', which are L from N and L from N'. To build a link-label added to N' a link L of N, where link-labels are already built, should be used. Based on this process, a link-label where departing node is on N, and arriving node is on N' is built; then in the basis of this built link-label, a link-label in which both departing and arriving nodes are on N' is built. From this process link-label determining algorithm including link-loopless path is illustrated as follow.

Link-label confirmation added to N' from N
Link-label confirmation in link-loopless path
(step1) a link(a) involving in N and a link(b) in N'
$\pi^{rb} = min\{\pi^{ra} + d_{ab} + C_b \mid \forall a \in L^N; P_r^a \oplus b \in \Psi(b)\}, \forall b \in L^{N'}$
(step2) a link(a,b) involving in N
$\pi^{rb} = min\{\pi^{ra} + d_{ab} + C_b \mid \forall a \in L^{N'}; P_r^a \oplus b \in \Psi(b)\}, \forall b \in L^{N'}$

Step 3 shows the link-label based OPA that prevents link loop production. In detail, the $P_r^a \oplus b$ means a link connecting sub path of link a,P. Therefore, it can detect the sub path of link a in connecting process of a link a and the sub path link, examine whether or not there is link b, then omit the link-label selection process. Because of the already confirmed link a label, we could scrutinize whether a link b includes sub paths by tracking an optimum path back to the link a and finally reaching to the origin. The relating equations are as follow:

P_r^a: an optimum path from an origin r to a link a
$P_r^a \oplus b$: a path connected between the an arriving node of the last link a in P and a departing node of a link b
$\Psi(b)$: a link set of link b with no repetition

Step 3: transformation to next link
If $\pi^{ra} + d_{ab} + C_b < \pi^{rb}$ { if $P_r^a \oplus b \in \Psi(b)$ { $\pi^{rb} = \pi^{ra} + d_{ab} + C_b$ } }

Confirmation of a link-label and a node-label added to N' completes the node-label confirmation process. Here, it is possible to exploit optimum path by reverse tracking the nodes connected to destination. The process of confirming a node-label is completed by adopting the minimum link cost, among the links which the arriving node is j based on the newly built link-label, as a node-label cost

Node-label confirmation added to N' from
$\pi^{rj} = \min \{ \pi^{ra} \mid \forall \text{ a} \in \Gamma_j^- ; \forall \text{ a} \in L^N \bigcup L^{N'} \} , \forall j \in V^{N'}$

Fig.1 shows the detection process of $P' = \{R, 1, 3, 5', S\}$ Of the detection of a link $(2 \rightarrow 4)$ a link detection $(4 \rightarrow 2')$ is omitted during detection process of

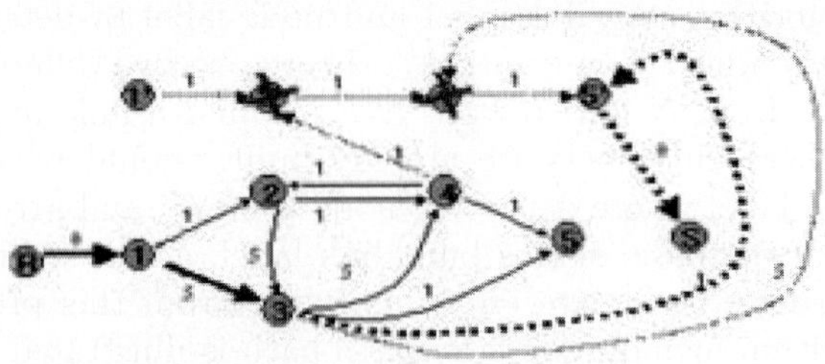

Fig. 1. Detection of a link-loopless optimum path

$P_r^{(2\to4)} \oplus (4 \to 2') \notin \vartheta\,(4,2')$ because the detection $(4 \to 2')$ creates a link-loop path where has link repetition.

Fig.2 shows the turn limitation from the direction of a link $(1 \to 3)$ to a link $(3 \to 5)$. The direction of a link $(2 \to 4)$ to a link $(4 \to 2')$ is a link loopless path that satisfies $P_r^{(2\to4)} \oplus (4 \to 2') \notin \Psi\,(4,2')\,P$ so that the detection is continuing. However, in the detection from a link $(4 \to 2')$ to a link $(2' \to 4')$ $_{P_r^{(4\to2')}}P$ includes a link $(2 \to 4)$, a link loop, which results in detection omission. Also, since the direction from a link $(1 \to 3)$ to a link $(3 \to 5)$ has a traffic control, the N's optimum path of Figure III-4 is $P' = \{R, 1, 3, 4, 2, 3, 5', S\}$ appeared in Fig.3, which is a node loop path and a link loopless path path simultaneously.

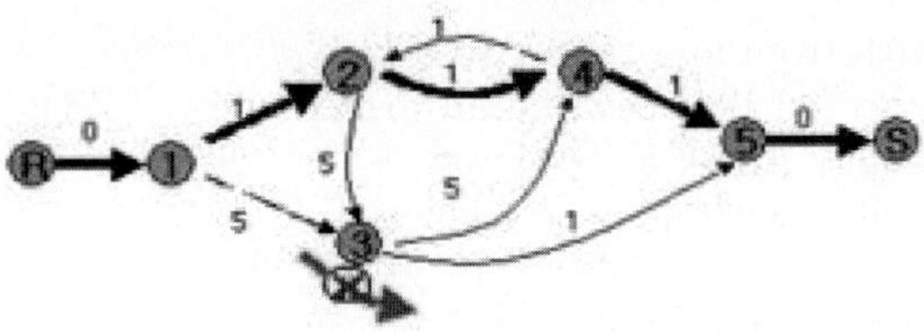

Fig. 2. Network N and traffic control

Now let us refer to the above example and demonstrate the fact that the detection of link-label OPA reduces up to K-1th number in a suggested algorithm. For this, we check the possibility of OPD from the destination to the origin only with node and link labels added to N'. Looking at the network N (Fig.2) and link N' (Fig.3) node 1', 2', 4' and 5', and $(1',2')$, $(2',4')$, $(4',5')$, $(5',S')$ are added at the optimum path P. Also, $(4,2')$, $(3,4')$, $(3',5')$, $(4,2')$ are added nodes which came from a node N as a influx link. The label information of N is significant to exploit an optimum path in N'. In case of a link $(2',4)$ in Fig.3, label confirmation is completed by comparison of a link $(2,4)$ and $(3,4)$ of N, and a link $(2',4)$ of N' toward maximum path cost. In other words, if selection of influx link-label is completed through the calculation of an equation, $\pi^{R(2,4)} = min\{\pi^{R(2,4)} + d_{(2,4)(4,2')} + c_{(4,2')}, \pi^{R(3,4)} + d_{(3,4)(4,2')} + c_{(4,2')}\}$, a link$(5',S)$-label which is connected to the final destination is also completed. In

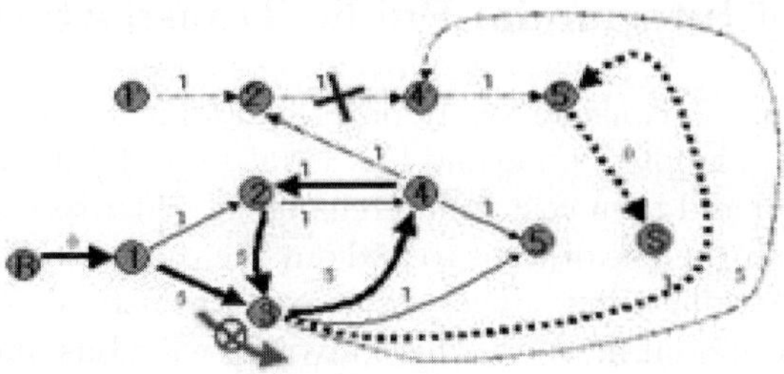

Fig. 3. Detection of a link-loopless optimum path(N')

the meantime, the selection process is based on the order of added links to P'. In this result, it is possible to execute the OPD without additional work of OPA through the determined labels.

2.2 Link Mode Transformation Method

In order to show the intermodal traveling on a same link, this paper demonstrates the method of expanding all modes traveling a same link with links. Once adopting this, the origin and the destination are same, but the characteristics of links embrace other duplicate links. In fact, this method has a drawback, which the number of links increases proportionally to the number of lines. However, the characteristics of modes can reflect on links; the analysis of various points according to modes and lines can be possible if considering the transfers between modes(De Cea & Fernandez, 1993). In addition, the property of explaining modes with links leads to figure out the transfers with adopting a link-based OPA that is used in a previous intermodal network. Fig.4 illustrates that the transfer fee d_{ab}^{nm} created in mode(m, n) traveling two adjacent links (a, b) is changed to d_{ab} by a mode link transformation which excludes modes.

All modes traveling a same link is explained by links; this causes to adopt loopless-based algorithm in this paper without additional consideration for modes. In general, to detect k-th number of paths requires k-1 th number of a network transformation. Also, a link-based OPA is executed in a single mode(in real, intermodal) network reformed with a set of links and nodes that considers network transformation.

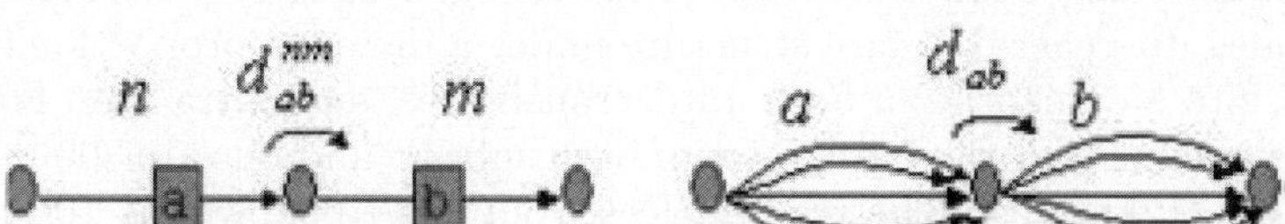

Fig. 4. The transfer fee of mode (m, n) and mode transformation in a same link

3 Analysis of Intermodal Public Transportation Network

Case study is needed to consider study purpose, and it focuses two factors. One is to compare both a loopless node and link loopless path intended for a network existed turn control and turn cost at interchange in order to verify loopless path in single traffic network. The other is to find out the successive loopless node path about variables of traffic time, travel cost based on distance rate, and transfer numbers in an intermodal network where involves with bus and subway.

As showed in Figure IV-1 network N comprises 12 links and 7 nodes and indicates each distance(cost) on links. It also has omni-directional turn penalty around each square of nodes. U-turn is available in two directions; one is $2 \to 3 \to 2$ and the other is $3 \to 2 \to 3$. And, the turn penalty is 3 at both directions. From the origin, R, to the destination, S, maximum 20 paths are detected, then 3 results – 1) a path including loop, 2) a loopless node path, 3) a link loopless path path – are compared. Within a network, 1 bus line and 3 subway lines run.

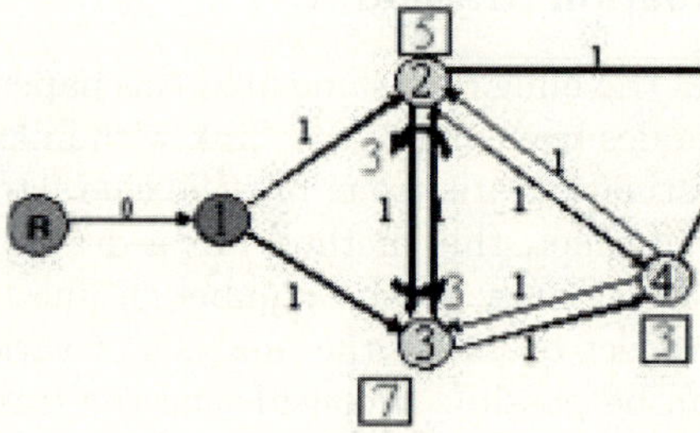

Fig. 5. Network N(5 nodes and 10 links)

Because node 1 and 7 are a origin and destination respectively, transfer locations are 3 nodes(3, 4, 5) by 4 lines. Direct lines from the origin to destination are subway S1, and S3.

Assumption of the case study is that travelers can demand information of diverse alternate paths under a distance-proportional fare system. Furthermore, once enumerating variable information such as travel time, fare and transfer, travelers are supposed to select the alternate path.

In addition, a traffic mode forming a path can be used no more once(Lozano & Storchi, 2001 and 2002) since this study assumes to satisfy a viable path of traffic modes. In general, a fare structure under a distance-proportional system is divided into 3 categories: a basic rate, transfer fee, and extra fare. For convenience, we assume that the basic rates of both subway and bus are 800unit/12km and 600 unit/12km; also, the transfer fees from bus to subway is 200 unit and extra fare is simply 100 unit/6km for all modes.

Table 1 and 2 summarize the results of a case study in terms of travel time and fare, which applies K loopless node algorithm to an intermodal transportation

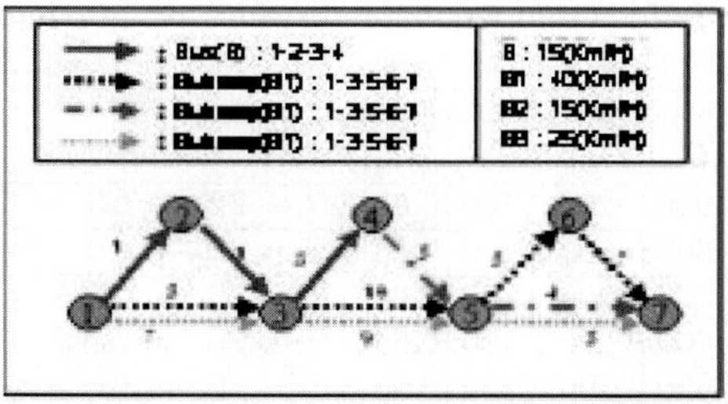

Fig. 6. Case of Intermodal transportation network

network. The overall results indicate that the algorithm detects the k number of path in regular sequence for each detection variable. It also means that the suggested k loopless-based PDA can be used for efficient analysis of intermodal network.

For travelers' point of view, the results lead them to ponder the value of path information by a single criterion in order to select an alternate path. This is because the alternate path result is different in each perspective: travel time, fare and transfer frequency. For instance, the least travel time path can be a path involving high cost and frequency; it consequently is inefficient.

Table 1. K path detection result by travel time

Path Order	Travel Time (hour)	Travel Distance (km)	Total Cost (won)	Base Cost (won)	Transfer Cost (won)	Extra Cost (won)	*Path* Origin-mode → Destination
1	0.57	20	1000	800	0	200	1*S*1→ 3S1 → 5S3 → 7
2	0.58	17	900	600	200	100	1*B*→ 2B → 3S1 → 5S3 → 7
3	0.64	19	1000	800	0	200	1*S*1→ 3S1 → 5S2 → 7
4	0.65	16	900	600	200	100	1*B*→ 2B → 3S1 → 5S2 → 7
5	0.67	27	1100	800	0	300	1*S*1→ 3S1 → 5S1 → 6S1 → 7
6	0.68	24	1000	600	200	200	1*S*3→ 3S1 → 5S1 → 6S1 → 7
7	0.68	19	1000	800	0	200	1*S*1→ 3S3 → 5S3 → 7
8	0.69	16	900	600	200	100	1*B*→ 2B → 3S3 → 5S3 → 7
9	0.75	18	900	800	0	100	1*S*1→ 3S3 → 5S2 → 7
10	0.76	15	900	600	200	100	1*B*→ 2B → 3S3 → 5S2 → 7
11	0.79	21	1000	800	0	200	1*S*3→ 3S1 → 5S2 → 7
12	0.83	29	1100	800	0	300	1*S*3→ 3S1 → 5S1 → 6S1 → 7
13	0.84	21	1000	800	0	200	1*S*3→ 3S1 → 5S3 → 7
14	0.85	18	900	800	0	100	1*S*3→ 3B → 4S2 → 5S3 → 7
15	0.86	15	900	600	200	100	1*B*→ 2B → 3B → 4S2 → 5S3 → 7
16	0.90	20	1000	800	0	200	1*S*3→ 3S3 → 5S2 → 7
17	0.92	17	900	800	0	100	1*S*1→ 3B → 4S2 → 5S2 → 7
18	0.93	14	900	600	200	100	1*B*→ 2B → 3B → 4S2 → 5S2 → 7
19	1.08	19	1000	800	0	200	1*S*3→ 3B → 4S2 → 5S2 → 7

Table 2. K path detection result by distance proportional fare system

Path Order	Travel Cost (won)	Travel Distance (km)	Total Cost (won)	Base Cost (won)	Transfer Cost (won)	Extra Cost (won)	Traffic time (hour)	*Path* Origin-mode → Destination
1	900	16	900	600	200	100	0.65	1*B* → 2B → 3S1 → 5S2 → 7
2	900	17	900	600	200	100	0.58	1*B* → 2B → 3S1 → 5S3 → 7
3	900	15	900	600	200	100	0.76	1*B* → 2B → 3S3 → 5S2 → 7
4	900	16	900	600	200	100	0.69	1*B* → 2B → 3S3 → 5S3 → 7
5	900	14	900	600	200	100	0.93	1*B* → 2B → 3B → 4S2 → 5S2 → 7
6	900	15	900	600	200	100	0.87	1*B* → 2B → 3B → 4S2 → 5S3 → 7
7	900	18	900	800	0	100	0.75	1*S*1 → 3S3 → 5S2 → 7
8	900	17	900	800	0	100	0.93	1*S*1 → 3B → 4S2 → 5S2 → 7
9	900	18	900	800	0	100	0.86	1*S*1 → 3B → 4S2 → 5S3 → 7
10	1000	19	1000	800	0	200	0.64	1*S*1 → 3S1 → 5S2 → 7
11	1000	20	1000	800	0	200	0.57	1*S*1 → S1 → S3 → 7
12	1000	19	1000	800	0	200	0.69	1*S*1 → 3S3 → 5S3 → 7
13	1000	20	1000	800	0	200	0.91	1*S*3 → 3S3 → 5S2 → 7
14	1000	21	1000	800	0	200	0.8	1*S*3 → 3S1 → 5S2 → 7
15	1000	21	1000	800	0	200	0.84	1*S*3 → 3S3 → 5S3 → 7
16	1000	24	1000	600	200	200	0.68	1*B* → 2B → 3S1 → 5S1 → 6S1 → 7
17	1000	23	1000	600	200	200	0.79	1*B* → 2B → 3S3 → 5S1 → 6S1 → 7
18	1000	22	1000	600	200	200	0.97	1*B* → 2B → 3B → 4S2 → 5S1 → 6S1 → 7
19	1000	19	1000	800	0	200	1.08	1*S*3 → 3B → 4S2 → 5S2 → 7
20	1100	27	1100	800	0	300	0.68	1*S*1 → 3S1 → 5S1 → 6S1 → 7
21	1100	29	1100	800	0	300	0.83	1*S*3 → 3S1 → 5S1 → 6S1 → 7

4 Conclusion

Loop is very significant concept in network theory toward a practical appraisal of transportation network. However, it has a limitation to express exact traffic condition on network because of the boundless repetition of nodes and links. Therefore, reasonable traffic condition can be explained by a link loopless path pattern. This study finds out a problem that EPDA produces loop when being applied to K number of path detection; develops a technique that selectively detects a link loopless path set using link-label for OPD; and broaden the application possibility of mode-link transformation technique to apply the suggested agent to intermodal public transportation network.

Two case studies demonstrate the suggested technique releases valuable results. It draws that a path detection reflecting turn-traffic penalty and traffic prohibition at interchange is possible, which was impossible under EPDA. Especially, these paths can appraise the efficient traffic condition such as U turn, and P turn in urban network where link repetition is not allowed, through the loopless path. Also, the possibility of application of the suggested method in an intermodal public transportation network is showed. Finally, variables like fare, travel time, and transfer frequency realized as important matters reflecting travelers' cognition cost.

Acknowledgment

This research was financially supported by Hansung University in the year of 2007.

References

1. Azevedo J. A., Costa M. E. O. S., Madeira J.J.E.R.S., and Martins E.Q.V. (1993) An Algorithm from the Ranking of Shortest Paths, European Journal of Operational Research, **Vol. 69**, pp 97–106.
2. Bellman R. and Kalaba R. (1968) On Kth Best Policies. J. SIAM **8**, pp.582–588.
3. De Cea. J. and J.E. Fernández. (1989) Transit Assignment for Minimal Routes: An Efficient New Algorithm, Traffic Engng. Control, pp. 492–494.
4. Dijkstra E. W. (1959) A Note of Two Problems in Connected with Graphs. Numerical Mathematics. **I**, pp. 269–271.
5. Kirby R. F. and Potts R. B. (1969) The Minimum RouteProblem for Networks with Turn Penalties and Prohibitions. Transportation Research **3**, pp.397–408.
6. Martins E.Q.V. (1984) An Algorithm for Ranking Paths that May Contain Cycles, European Journal of Operational Research, **Vol. 18**, pp.123–130.
7. Moore E. F. (1957) The Shortest Path through A Maze. Proc. Int. Conf. on the Theory of Switching. Harvard Univ., Cambridge, MA.
8. Pollack M. (1961) The Kth Best Route Through A Network, Operations Research, **Vol. 9**, pp 578–580.
9. Potts R.B. and Oliver R.M.(1972) Flows in Transportation Networks. Academic Press.
10. Shier R. D. (1979) On Algorithms from Finding the k Shortest Paths in a Network, Networks, **Vol. 9**, pp.195–214.
11. Yen J.Y. (1971) Finding the K shortest Loopless Paths in a Network, Management Science, **Vol.17**, pp.711–715.

A Channel Sounding Scheme of MIMO-OFDM System for Intelligent Area Network

Bang Hun Park[1], Juphil Cho[2], Heung Ki Baik[3], and Jae Sang Cha[4]

[1] Department of Electronic Engineering, Chonbuk National University, Korea
[2] Schol of Electronic & Information Engineering, Kunsan National University, Korea
[3] Schol of Electronic & Information Engineering, Chonbuk National University, Korea
[4] Department of Electronics Engineering, Seoul National University of Technology, Korea
DSP LAB. Electronic Engineering Chonbuk National University, Duck-jin gu Chonbuk 561-756 Korea
freshman78@chonbuk.ac.kr

Abstract. In this paper, we propose the Channel sounding scheme which is made for ideal communication between some application as well as the short distance of high speed data transmission in MIMO-OFDM system for Wireless PAN. This method is able to perceive the duration of the impulse response through the delaying of power delay profile, modeled a power delay profile which has an attenuate characteristic, and obtained the coefficient of channel response by ML (maximum likelihood). Through the amplitudes, phases and delays associated with each multipath component which were acquired from this Channel sounding scheme, we can describe the wave propagation characteristics of channels between the transmitter and receiver so that the receiver could enhance not only the reliability but also the ability of communication link.

Keywords: MIMO-OFDM system, Personal Area Network, Maximum Likelihood.

1 Introduction

According to the developing of IT industry and multimedia technology, portable electronic appliances and communicational devices require transferring enormous data for high quality video, audio system, and image files. For this, IEEE 802.15 TG3 has been conducting the Wireless PAN technique for transferring of high speed data in a 2.4GHz band, also for this research, we have been trying to obtain 55Mbps data transfer speed in 10m distance[1]. We can also use the Channel sounding scheme in order to analyze the characters of channels based on the interferences in short and long distance wireless communication system which are applied by wireless PAN technique. In order to overcome the inferior channel transfer environment, such as Piconet interference or multitude connection interference which are exist in wireless PAN transfer channel, we transmit the sounding signal so that we can grasp the condition of channel response through this signal. In this paper, we suggest the

N.T. Nguyen et al. (Eds.): KES-AMSTA 2007, LNAI 4496, pp. 506 – 511, 2007.

technique which is useful for receiver to recognize the best condition of channels between the transmitter and the receiver through the channel sounding technique in MIMO-OFDM (Multi Input Multi Output-Orthogonal Frequency Division Multiplexing) system by estimating the phase, amplitude and factors of delay which are based on ML technique.

2 Channel Sounding Scheme

2.1 Channel Impulse Response

The transmission channels for wireless PAN could produce a channel modeling as follows, in order to have a satisfactory of channel characterization, the amplitudes, phase shifts and delays associated with each multipath component

$$h_j(t,\tau) = \sum_{k=0}^{L-1} a_k(t) \exp\{ j[2\pi f_c \tau_k(t) + \phi(t,\tau_k(t))] \} \delta(\tau - \tau_k(t)) \tag{1}$$

In a point of view in many indoor wave propagation environments, we can not conduct a deterministic modeling of these parameters. In fact that, it is very difficult to grasp the channel characteristics because almost every channel is not fixed in actual environment. Thus, we are assumed over short distances of travel or short intervals of time. Assuming stationary over a small time or distance interval, the channel impulse response described by (1)

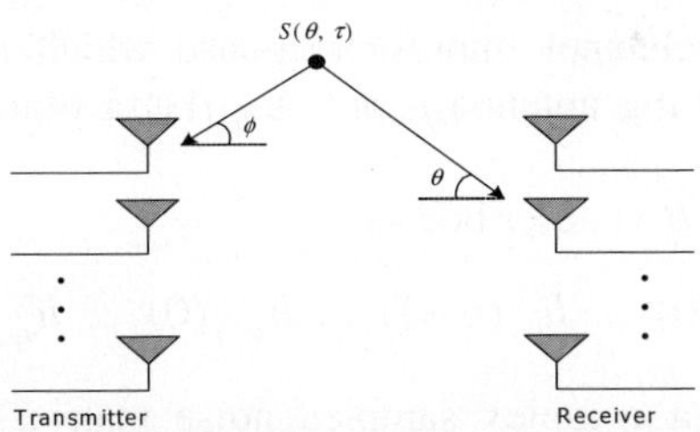

Fig. 1. The channel model of MIMO-OFDM system

As shown in Figure 1. Let's consider the model for the channel model of MIMO-OFDM system. In a side of the receiving antenna array, let's assume that a scatterer is located in a place which contains the degree of θ and time delay of τ, and in that place the amplitude of complex is $S(\theta,\tau)$. Thus, we could express that in a side of the transmitter antenna array, the same scatterer could be exists on the degree of φ. the parameter is θ,ϕ,τ could be changed by θ. If we summarize about the Channel impulse response in a MIMO-OFDM system, it could be as follows [2[[3].

$$H(\tau) = \int_{-\pi}^{\pi} \int_{0}^{\tau_{\max}} S(\theta_k, \tau_k') a(\phi_k, \theta_k) g(\tau - \tau_k') d\tau' d\theta \tag{2}$$

Where τ_{max} refers to maximum delay spread of channels and $g(\tau)$ means the combined response of the pulse shaping filter at the transmitter and a matched filter in the receiver. In order to estimate the amplitude, phase and delay through the equation (2), once we have several snapshots of these estimated they can be averaged to derive a statistical description of the wireless channel in parameterized form [4]. When it comes to several propagation related information, we can estimate through the variable which is described by wireless channel response. This kind of estimate needs a channel sounder and could get a channel information by the channel sounding technique; if the transmitter send a sounding signal which is useful to grasp channel information, then the channel sounder of the receiver receives the signal. The channel sounder in MIMO-OFDM system by wireless PAN is a device which permits the variable measurement related to the impulse response of wireless channel. In the MIMO-OFDM system, the sounding signal which is made for channels should have impulse form of autocorrelation function and other sounding signals which are different each other between transmitter antennas should have no correlated each other statistically. The received signal will be the superposition of several attenuated, phase shifted and delayed copies of the transmitted sounding, contaminated. In the MIMO channel which has n_T unit transmitting antenna and n_R unit receiving antenna, the received signal from the receiving antenna j_th and the time τ is as follows.

$$y_j(\tau) = \sum_{i=0}^{n_T} \sum_{k=0}^{n_R} h_{i,j}(k) s_i(t - \tau_k - k) + v_j(\tau_k) \tag{3}$$

Where $h_{i,j}$ represent the channel impulse response which is from the transmitting antenna i_th to the receiving antenna j_th . A $s_i(t)$ is a transmitting sounding signal and v_j is a noise

A $n_T \times n_R$ channel matrix H is described as

$$H = [h_{1,j}(0) \ \dots \ h_{1,j}(n-1) \ \dots \ h_{n_T,j}(0) \ \dots \ h_{n_T,j}(n-1)]^T \tag{4}$$

Equation (4) represents a complex sampled noise matrix, V , with spectral power density N_0

$$V = [v_j(n) v_j(n+1) \ \dots \ v_j(M)]^T \tag{5}$$

The samples of this received signals are contained in a matrix vector $Y = \sum_{i=0}^{n_T} \sum_{k=0}^{L-1} y_k$, where M is considered the number of samples.

$$Y = [y_j(n) y_j(n+1) \dots . y_j(M)]^T \tag{6}$$

2.2 Channel Parameter Measurement

Then, s_i refers to the transmitted sounding signal from the transmitting antenna i_th and v_j refers to the complex white Gaussian noise which is added to the receiving antenna j_th . In the section of sounding, there is a transmitting of sounding signals which are different each other, and we assume the sample M which is long enough for the receiver to grasp channels by using the sounding signal. If we make a sampled version of the delayed sounding signal as describing, by using a sounding matrix is as follows.

$$S=\begin{bmatrix} s_1(n-1)..s_1(0) \;.. & s_{n_T}(n-1)..s_{n_T}(0) \\ .. \quad .. & .. \quad .. \\ s_1(M-1)..s_1(M-n)..s_{n_T}(M-1)..s_{n_T}(M-n) \end{bmatrix} \tag{7}$$

The matrix S is the $(M-n+1)\times n_T n$ block-Toeplitz matrix which is made from a sounding signal. The sample of the receiving signal could be included in vector $Y=\sum_{i=0}^{n_T}\sum_{k=0}^{L-1} y_k$. Also, under these conditions the joint pdf for the sample vector Y is can be expressed

$$p(\mathrm{Y}/\Gamma)=\prod_{k=1}^{M} p_k(Y_k/\Gamma) \tag{8}$$

Equation (8), the parameter $\Gamma=[\tau\theta\alpha]^T$ describes the concatenation of three vectors of length L each containing the delays, phases and amplitudes associated with the L multipath components. For each individual sample Y we have the whole combination pdf is related to Y could be expressed as follows. [4][5]

$$p(\mathrm{Y}/\Gamma)=\frac{1}{(2\pi\sigma^2)^M}\exp\left\{-\frac{1}{2\sigma^2}\sum_{k=1}^{M}\left|Y_k-\sum_{l=0}^{L-1}a_l e^{j\theta_l}s_k(\tau_l)\right|^2\right\} \tag{9}$$

We will consider the ML (maximum likelihood) estimation because we can not get a priori information about the statistical structure of Γ. If we assume having a same probability then the best parameter vector Γ is as follow.

$$\hat{\Gamma}=\arg\min_{\Gamma}\left\{\sum_{k=1}^{M}\left|Y_k-\sum_{l=0}^{L-1}a_l e^{j\theta_l}s_k(\tau_l)\right|^2\right\} \tag{10}$$

Since the parameter vector Γ does not depend on the ML estimation, we have discarded the term containing $|Y_k|^2$. Thus, necessary and sufficient conditions to have a maximum in the likelihood function (10) are as follows

$$\frac{d}{d\Gamma}\left(\frac{d}{d\Gamma}\log p(Y/\Gamma)\right)^T \tag{11}$$

By solving these expressions for Γ we will find joint estimates of the delays and complex amplitudes associated to each path. The time-varying channel impulse response between the j_th transmit antenna and the i_th receive antenna is denoted as $h_{i,j}(\tau,t)$. This is the response at time t to an impulse applied at time $t-\tau$ .The composite MIMO-OFDM channel response is given by the $n_T\times n_R$ matrix $H(\tau,t)$ with

$$\mathbf{H}(\tau,t)=\begin{bmatrix} h_{1,1}(\tau,t) & h_{1,2}(\tau,t) & \cdots & h_{1,n_T}(\tau,t) \\ h_{2,1}(\tau,t) & h_{2,2}(\tau,t) & \cdots & h_{2,n_T}(\tau,t) \\ \vdots & \vdots & \ddots & \vdots \\ h_{n_R,1}(\tau,t) & h_{n_R,2}(\tau,t) & \cdots & h_{n_R,n_T}(\tau,t) \end{bmatrix} \tag{12}$$

Furthermore, given that the signal $s_j(t)$ is launched from the j_th transmit antenna [5][6]. Equation (11) represents a constraint on the Hessian matrix of the likelihood function. This Hessian matrix H is given

$$H=\begin{bmatrix} H_{\theta_{ll}} & H_{\theta_l a_q} & H_{\theta_l \tau_q} \\ H_{a_l\theta_q} & H_{a_{ll}} & H_{a_l\tau_q} \\ H_{\tau_l\theta_q} & H_{\tau_l a_q} & H_{\tau_{ll}} \end{bmatrix} \tag{13}$$

Since verifying the definite negativeness of H appears to be formidable task, considering marginal estimated instead. In order to obtain marginal estimates the necessary and sufficient conditions to have a maximum in the likelihood function are as follows

$$\sum_{k=1}^{M}\left\{\frac{\partial^2}{\partial\theta_l^2}\left\{2\Re\left[Y_k\sum_{q=0}^{L-1}a_q e^{-j\theta_q}s_k(\tau_q)\right]-\left|\sum_{q=0}^{L-1}a_q e^{j\theta_q}s_k(\tau_q)\right|^2\right\}\right\}<0 \tag{14}$$

$$\sum_{k=1}^{M}\left\{\frac{\partial^2}{\partial a_l^2}\left\{2\Re\left[Y_k\sum_{q=0}^{L-1}a_q e^{-j\theta_q}s_k(\tau_q)\right]-\left|\sum_{q=0}^{L-1}a_q e^{j\theta_q}s_k(\tau_q)\right|^2\right\}\right\}<0 \tag{15}$$

$$\sum_{k=1}^{M}\left\{\frac{\partial^2}{\partial\tau_l^2}\left\{2\Re\left[Y_k\sum_{q=0}^{L-1}a_q e^{-j\theta_q}s_k(\tau_q)\right]-\left|\sum_{q=0}^{L-1}a_q e^{j\theta_q}s_k(\tau_q)\right|^2\right\}\right\}<0 \tag{16}$$

We must find derivatives with respect to θ_l , α_l and τ_l of (10). We arrive at the following expressions which must be satisfied by the ML estimates. Therefore, if the receiving device knows all information of channels, we can consider that the receiving device has assumed the channel matrix H exactly, and the ML estimates could find out the matrix s which maximizes a priori probability. The transmitted signal vector according to

$$\hat{S}=\arg\min_{s}\left\|y-\sqrt{\frac{E_s}{n_T}}HS\right\|^2 \tag{17}$$

In the MIMO-OFDM system, the channel matrix occurs the space fading cross correlation caused by the antenna interval between the transmitter and the receiver. In a case of the transmitter doesn't know the information of channels, the capacity of

MIMO channel is same as the sum of the SISO formed additional channel capacity which has $s_k(\tau_l)s_k(\tau_q)$ of the average power and is arranged parallel[7][8]. However, if the transmitter knows the information of channels, both transmitter and receiver could make increase of the channel capacity through the low complexity.

3 Conclusions

In the wireless PAN, the impulse response character of the transmitting channel is inferior, and because of the existence of variety interferences we should make the efficient channel information algorithm which is proper or the transmitting of high quality and speed data in the MIMO-OFDM system. In this paper, the channel sounding scheme which is suggested the technique that we can use for the measurement of the channel changing through the sounding signal transmission between the transmitter and the receiver so that the receiver can use this measured channel information to predict the channel condition. By using this channel information both the transmitter and the receiver could make an increase of the channel capacity through a low complexity, and the transmitter could assign the best power. In addition, if we use this proposed method we could designed the efficient information transmitting, also it is good for making a channel estimation algorithm which could increase the reliability as well as the ability of communication link, diversity techniques and equalization techniques.

References

[1] IEEE Std. 802.15.1, "Wireless Medium Access Control (MAC) and Physical Layer (PHY) Specifications for Wireless Personal Area Networks (WPANs)," 2002.

[2] A. Paulaj, R. Nabar, and D. Gore, Introduction to Space-Time Wireless Communications. Cambridge, U. K.: Cambridge University Press, 2003.

[3] A. J. Paulaj, D. A. Gore, R. U. Nabar, and H. Bolcskei, "An overview of MIMO communications—A key to gigabit wireless," Proc. of IEEE, vol. 92, no. 2, pp. 198–218, Feb. 2004.

[4] W. Braun, U. Dersch "A Physical Mobile Radio Channel Model." IEEE Transactions on Mobile Technology, vol. 40, no. 2, May 1991

[5] P. M. Fitz, P. Schniter, "A wideband channel Sounder.", 2001

[6] A.J.Paulaj,D.A.Gore, R.U.Nabar, and H.Bolcskei, "An overview of MIMO communications-a key to gigabit wireless," Proc.of IEEE, vol.92,no.2,pp.198-218,Feb.2004

[7] H. Hiroschi, P. Ramjee, " Simulation and Software Radio for Mobile Communications."Artech House, pp.53-55

[8] E. Telatar, "Capacity of multiple-antenna Gaussian channel, "AT&T Bell Labs Tech. Memo., Jun. 1995

Wireless Intelligent LBT Agents for Common Frequency Band Efficiency

Seong-Kweon Kim

Division of Marine Electronics and Communications,
Mokpo Maritime National University
Jook-Gyo Dong 571-2, Mokpo, Jeonnam 530-729, Korea
skkim12632@mmu.ac.kr

Abstract. This letter introduces the calculation method of common frequency bandwidth for frequency usage efficiency, when an intelligent Listen Before Talk (LBT) systems and non-intelligent frequency hopping (FH) systems coexist in the wireless communication network. The queuing theory is employed to model the FH and LBT system. The throughput for each channel was estimated by processing the frequency in use of channel and the interval of service time statistically. Therefore, the common frequency bandwidth is calculated with the calculation multiplying the number of channel by the bandwidth per channel.

Keywords: Intelligent mobile agents, FH(Frequency Hopping), LBT(Listen Before Talk), Queuing Theory.

1 Introduction

A various attempt for ubiquitous network has been tried in the field of a wire and a wireless technology. For example, a various services are employed for realization of home network technology using wire and wireless technology. Especially, network speed is accelerated and WEB based services are offered with a variety of digital contents actively, then, the wireless transmission of mass data has been possible by technologies of ultra wide band (UWB), RF-ID, etc. therefore, various method has been attempted to make possible transmission of mass data as well as small scale data using low-power wireless devices of digital cordless phone (DCP), RF-ID, ZigBee and bluetooth system. In the low-power wireless devices, frequency hopping (FH) system and listen before talk (LBT) system are used for interference avoidance. The FH system searches a proper frequency channel for communication by random frequency hopping method, however the LBT system does by using intelligent frequency channel selection like a carrier sensing method. Both the LBT system and the FH are useful for the ISM bandwidth communication representing a common frequency bandwidth [1].

However, the intelligent LBT agent preoccupied communication channels earlier than FH with random selection, the proper common frequency bandwidth including the low-power wireless device services is estimated considering the coexistence of an

N.T. Nguyen et al. (Eds.): KES-AMSTA 2007, LNAI 4496, pp. 512–519, 2007.

intelligent agent using LBT system with FH system. Moreover, the wireless communication has to be accomplished at the limited frequency bandwidth [2,3].

In this paper, the calculation method of common frequency bandwidth using the queuing theory [4] is introduced, when an intelligent Listen Before Talk (LBT) agent and non-intelligent frequency hopping (FH) system coexist in the wireless communication network system transmitting a mass digital contents or small scale identification data. The queuing theory is useful in statistical processing the usage frequency and the interval of service time for each channel in order to analyze the throughput for each channel. The throughput for each channel is calculated by computer simulation.

2 Calculation for a Required Channel Number

The queuing theory is applied to non-intelligent FH system and intelligent LBT system, which are used for interference avoidance in the low-power wireless devices. The queuing theory handles the approach time interval of user as the statistical distribution and the throughput to the number of serviced users is measured. As the result of analysis, a required channel number of for each system is obtained for communication.

2.1 Statistical Model Using Queuing Theory

The queuing theory handles the approach time interval of user as the statistical distribution and the throughput users is measured with the number of serviced. Figure 1 shows the throughput as a function of input distribution. The horizontal axis is the

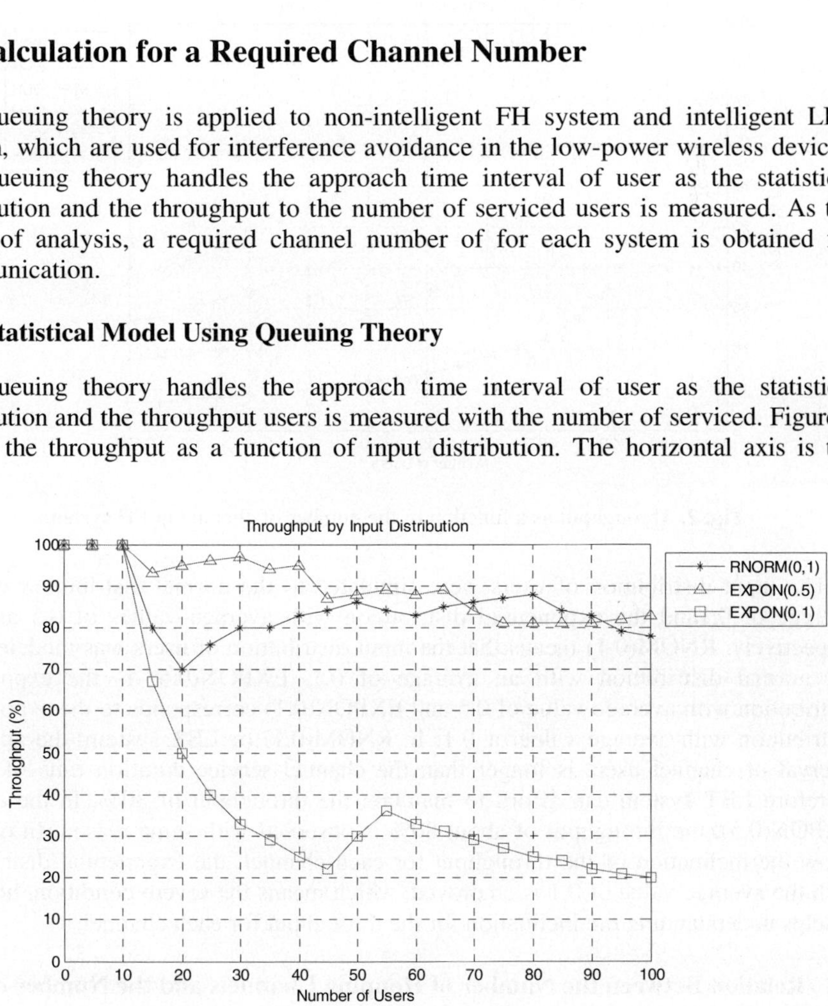

Fig. 1. Throughput as a function of Input Distribution

number of user to attempt to use communication channel and the vertical axis is throughput meaning that the users succeeded in communication. The input distributions are varied from the normal distribution to the exponential distribution, which has various average values as like 0.5 and 0.1. The system for simulation is LBT system with 10 channels and the service duration time of 4 sec. When LBT system continues to attempt to use a same channel, it is set to sleep during 0.1 sec. For this simulation, Awesim (Pritsker Cor.) tool is employed, which handles the approach time interval of user as the statistical distribution and measures the throughput to the number of serviced users.

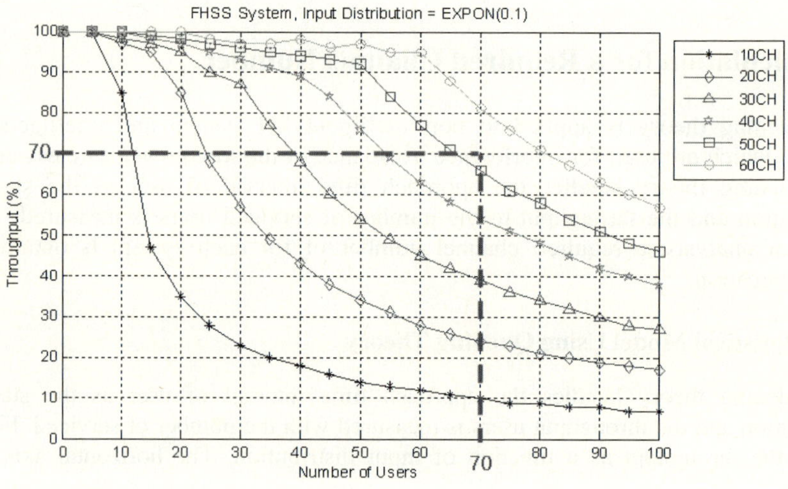

Fig. 2. Throughput as a function of the number of channel in FH system

The input distribution of users were simulated to the normal distribution with an average of 0.1and the exponential distribution with average values of 0.5 and 0.1, respectively. RNOM(0.1) means that the input distribution of users was modeled with the normal distribution with an average of 0.1. EXPON(0.5) is the exponential distribution with average value of 0.5 and EXPON(0.1) corresponds to the exponential distribution with average value of 0.1. In RNOM(0.1) of LBT system, the approach interval of channel users is longer than the channel service duration time of 4 sec, therefore LBT system can afford to maintain the throughput of 80%. In the case of EXPON(0.5), the throughput of about 80% is observed with same reason. In order to know the inclination of the throughput for each channel, the exponential distribution with the average value of 0.1 is employed, which means the severe condition, however it helps understanding the inclination for the throughput for each channel.

2.2 Relation Between the Number of Hopping Channels and the Number of Serviced Users in the FH System

Figure 2 shows the throughput with the changes of users using only FH at each channel number. The input distribution of users is assumed with the exponential distribution of average value of 0.1. The horizontal axis is the number of users, and the vertical axis is the throughput meaning the number of users in succeeding in communication. When the number of channels is varied with 10, 20, 30, 40, 50 and 60, the throughput is shown in Fig 2. It is assumed that a channel duration time is 0.4sec and the channel movement time which move to other channel after channel stay is set to be 0.01sec in FH. There are DCT, RFID, Bluetooth and ZigBee in the low-power wireless devices. DCP, RFID and Bluetooth are FH system and ZigBee is LBT system. DCP, RFID and Bluetooth require 10 channels, 20 channels and 23 channels for communication, respectively [5-10]. Therefore FH system requires total 53 channels in the low-power wireless devices.

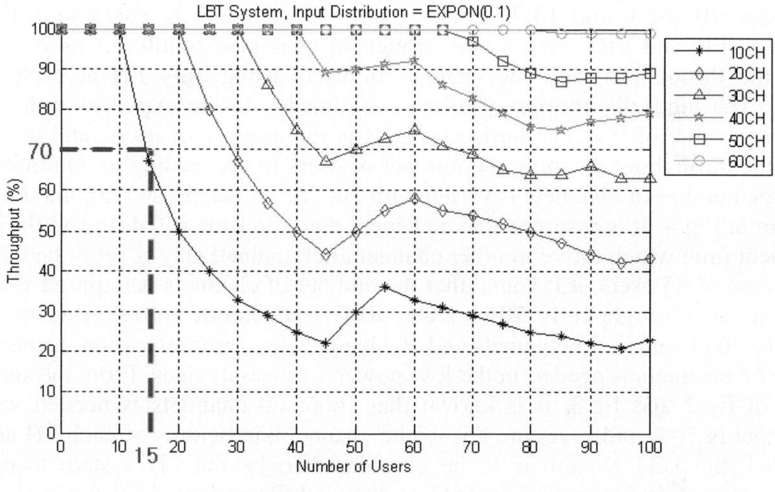

Fig. 3. Throughput as a function of the number of channel in LBT system

The FH system of non-intelligent system occupies channels with random hopping pattern in the low-power wireless devices. From the throughput result as shown Fig.2, when the throughput for 50 channels is 70%, it is found that the number of users is guaranteed with 70. The throughput of 70% is set to be a target, on the basis of assumption that the probability of three fourth is required for wire communication as like wire telephone system.

2.3 The Relation Between the Number of Channels and the Number of Serviced Users in the LBT

Figure 3 shows the throughput as a function of changes of users using LBT at each channel number. The LBT system is an intelligent system, which searches possible empty channel for communication firstly. When the channel is empty, the LBT system occupies the channel intelligently. The input distribution of users is modeled with the exponential distribution, which has the average of 0.1. The horizontal axis is the number of users, and vertical axis is the throughput. When the number of channels is varied with 10, 20, 30, 40, 50 and 60, the throughput is shown as Fig 3. In this simulation, it is assumed that a channel duration time is 4sec and the time which move to other channel is set to be 0.1sec.

In the low-power wireless devices, there are DCT, RFID, Bluetooth and ZigBee. Only ZigBee is the LBT system which functions as an intelligent Agent. As ZigBee requires 10 channels, LBT system requires totally 10 channels. From the throughput result as shown Fig.3, when the throughput for 10 channels is 70%, it is found that the number of users is guaranteed with 15.

3 Calculation of Required Channel Numbers for FH and LBT

In the result of two simulations, it is found that FH system using 53 channels guarantees 70 users and LBT system using 10 channels is guarantees 15 users. Therefore, FH and LBT system are enough to guarantee totally 85 users. Figure 4 shows the throughput with the changes of users using only FH at each channel number. The input distribution of users is assumed with the exponential distribution of average value of 0.1. The horizontal axis is the number of users, and the vertical axis is the throughput meaning the number of users in succeeding in communication. When the number of channels is varied with 10, 20, 30, 40, 50 and 60, the throughput is shown in Fig 4. It is assumed that a channel duration time is 0.4sec and the channel movement time which move to other channel after channel stay is set to be 0.01sec in FH. In case of 85 users, it is found that the number of channels is required with about 60 when the throughput is 70%. DCP, RFID, Bluetooth and ZigBee require 10 channels, 20 channels, 23 channels and 10 channels for communication, respectively. Totally 63 channels is needed in the low-power wireless devices. From the simulation results of fig.2 and fig.3, it is known that about 60 channels is needed when the throughput is 70% and users are 85. If the channel bandwidths of each FH and LBT are equal, the LBT system is to be considered to be the FH system to calculate common frequency bandwidth for LBT system and FH system.

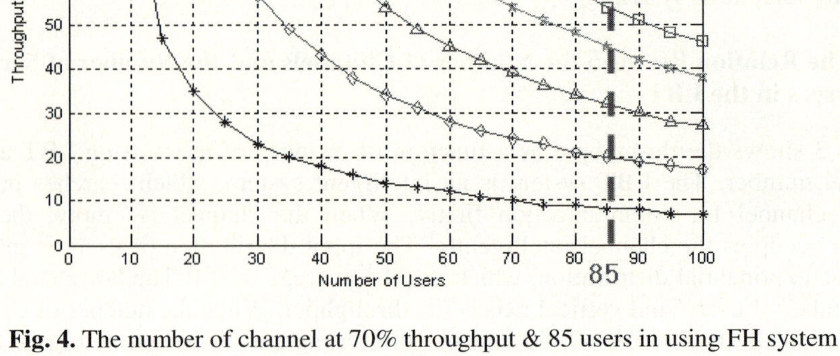

Fig. 4. The number of channel at 70% throughput & 85 users in using FH system

Table 1. Density of system

Output power	Density (N/km2)	Range
1W	20	10m
250mW	40	5m
100mW	40	3.5(estimated)
50mW	90	2m

Fig. 5. Density of systems

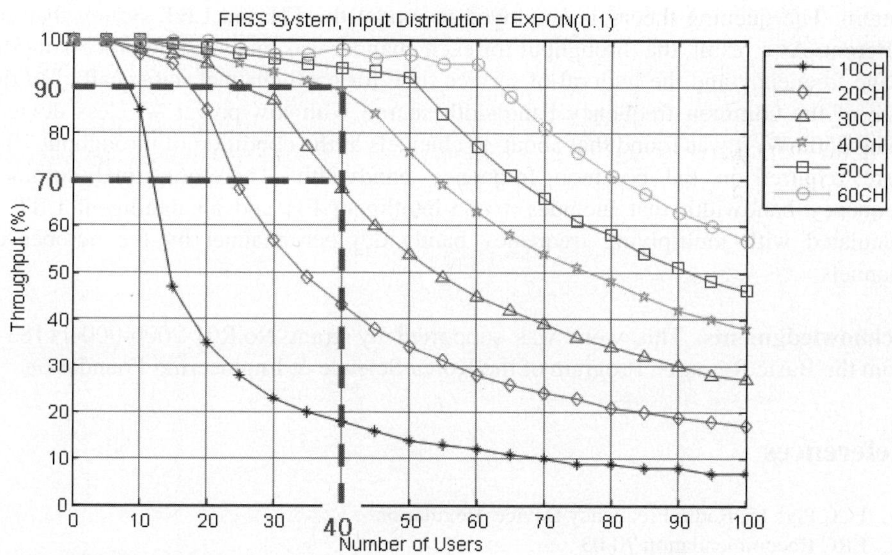

Fig. 6. The number of channel at 70%, 90% Throughput & 40 users in using FH system

Table 1 shows output power, density and radius of low-power wireless devices for RFID/USN. The output power of RFID/USN that represents on low-power wireless devices is usually 250mW. In this case, the number of serviced users is 40.

When the number of users is 40 and low-power wireless devices that require same frequency bandwidth use FH, 30 channels are needed for 70% throughput and 40 channels are needed for 90% throughput. Therefore, total common frequency bandwidth that includes a non-intelligent FH and an intelligent LBT is calculated with multiplying frequency bandwidth per channel by the number of channels.

4 Conclusion

This paper introduced the calculation method of common frequency bandwidth using the queuing theory, when an intelligent Listen Before Talk (LBT) agent and non-intelligent frequency hopping (FH) system coexist in the wireless communication network system transmitting a mass digital contents or small scale identification data.

The calculation of common frequency bandwidth including the users of intelligent LBT agent and that of non-intelligent FH agent was a work of importance, because a wireless communication system have to be accomplished at the limited frequency bandwidth, unlike the wire communication system using WEB. Substantially, FH and LBT are used for interference avoidance in the low-power wireless device. The intelligent LBT system acts like carrier sensing method that search for usable frequency bandwidth before transmitting data in the radio wave environment. However, the FH system selects a usable frequency channel randomly, using hopping pattern. The queuing theory is employed to model the FH and LBT system that are different. As a result, the throughput for each channel was analyzed by processing the usage frequency and the interval of service time for each channel statistically. In the case of the common frequency bandwidth shared with low-power wireless devices using 250mW, it was found that about 30 channels at the condition of throughput 70% were required in the common frequency bandwidth. Therefore, total common frequency bandwidth that includes a non-intelligent FH and an intelligent LBT is calculated with multiplying frequency bandwidth per channel by the number of channels.

Acknowledgments. This work was supported by grant No.R01-2006-000-11183-0 from the Basic Research Program of the Korea Science & Engineering Foundation.

References

1. FCC Part 15, Radio Frequency Device, Regulation
2. ERC Recommendation 70-03
3. ECC Report 11, Strategic Plans for The Future Use of the Frequency Bandwidths 862-870MHz and 400-2483.5MHz for Short Range Devices
4. Wang, Shengquan; Mai, Zhibin, Magnussen, W. Xuan, Dong; Zhao, Wei : Implementation of QoS-Provisioning system for voice over IP, Real-Time and Embedded Technology and Applications Symposium, 2002. Proceedings. Eighth IEEE
5. ETSI TR 101 445, Electromagnetic Compatibility and Radio Spectrum Matters; Short Range Devices(SRD) Intended for operation in 862MHz to 870MHz Bandwidth; System Reference Document for RFID Equipment.

6. ECC SE24M21_16, Draft Ecc Report on Strategic Plan 863-870MHz.
7. John A. Phillips : Personal Wireless Communication With DECT and PWT, Artech House Publishers, 1998.
8. The DECT standard explained, DECT Forum, February 1997.
9. POSITIONING OF DECT, DECT Forum Version 1, 30 June 2002.
10. FCC Kevin J.Martin : U.S. Spectrum Policy: Convergence of Co-Existenced 2002.03.05.

Verification of Mobile Agent Network Simulator

Kresimir Jurasovic and Mario Kusek

University of Zagreb
Faculty of Electrical Engineering and Computing
Unska 3, HR-10000, Zagreb, Croatia
{kresimir.jurasovic, mario.kusek}@fer.hr

Abstract. This paper deals with the verification of a multi–agent system simulator. The agents in the simulator are based on the Mobile Agent Network formal model. In order to verify the simulation results, they were compared with performance characteristics of a real multi–agent system, called the Remote Maintenance Shell, measured in a laboratory.

Keywords: multi–agent system, mobile agent network, verification, simulator.

1 Introduction

In recent years multi–agent systems based on autonomous software which migrates from host to host while communicating and cooperating with other agents in order to perform operations in place of their owner, have been applied in telecommunications, business software modeling, computer games, and many other fields. A multi–agent system containing mobile and intelligent agents is a promising paradigm for network and distributed systems management. This is particularly true for software operation and configuration in large environments, such as mobile telecommunication networks or Grid. During our previous research, we designed a formal model of a multi–agent system with mobile agents, called the Mobile Agent Network (MAN)[1], which was implemented into a real system for remote software maintenance, referred to as the Remote Maintenance Shell [2]. Such systems are very complex and it is difficult to verify their properties formally or in real systems.

In order to check various behaviors of a multi–agent system, creating a simulation is the only viable approach. The first step towards simulating the Mobile Agent Network was to investigate existing simulators. The Multi–Agent System Simulator (MASS) is a simulator which focus in validating different coordinations and adaptive qualities of a multi–agent system in an unpredictable environment [3]. It does not consider environment with agents moving from one place to another by using computer networks. Because of that this simulator is not viable option. The authors in the paper [4] are concentrated on how to simulate agents in a distributed system and they are using the network only for simulation. For

N.T. Nguyen et al. (Eds.): KES-AMSTA 2007, LNAI 4496, pp. 520–529, 2007.

example, in this simulator agent can be moved from one place to another in a real space and implementing computer network in such space is more complicated then to implement the whole simulator from skrech. That is the reason why we have build MAN simulator that correspond to our formal model.

Section 2 describes RMS, our real system, while section 3 describes the basics of MAN and the MAN simulator. The performance measures used to test the RMS system in the laboratory are explained in section 4. Comparison of the results obtained for the RMS system and the MAN simulator are elaborated upon in section 5. Section 6 concludes the paper.

2 Remote Maintenance Shell

The Remote Maintenance Shell (RMS) is an agent–based framework used to control remote locations. It supports software delivery to remote systems and operations used for remote installation/un–installation, starting/stopping, tracing, maintaining several versions of software, selective or parallel execution of two versions, and version replacement [2].

The basic concept behind RMS is shown in figure 1. RMS consists of a Management Station and of remote systems distributed over the network. The Management Station is responsible for software delivery to the remote systems and for performing remote operations on them. All tasks are defined by the user thought the Management Station GUI. The Application Testbed, an application–dependent part which must be built together with the application, provides the design for remote maintenance. When an application is ready for delivery, it is migrated together with its Application Testbed to the target system. Every application must be adopted for RMS.

The Maintenance Environment is an application–independent part of RMS, pre–installed on the target remote system(s) in order to enable maintenance actions. The Maintenance Environment is responsible for communication with the Management Station. Its main tasks include enabling remote operations and storing data about installed software.

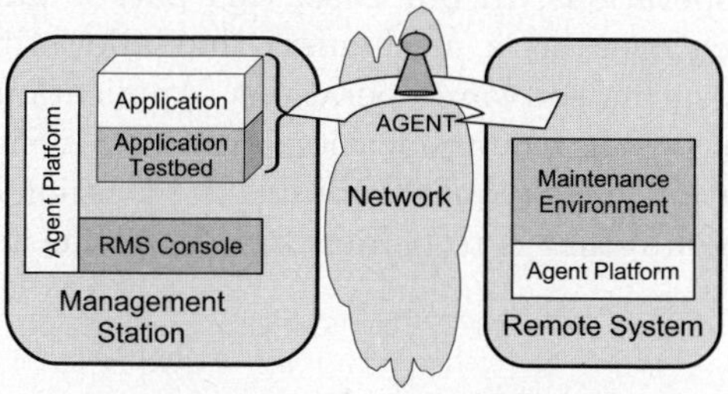

Fig. 1. The RMS Concept

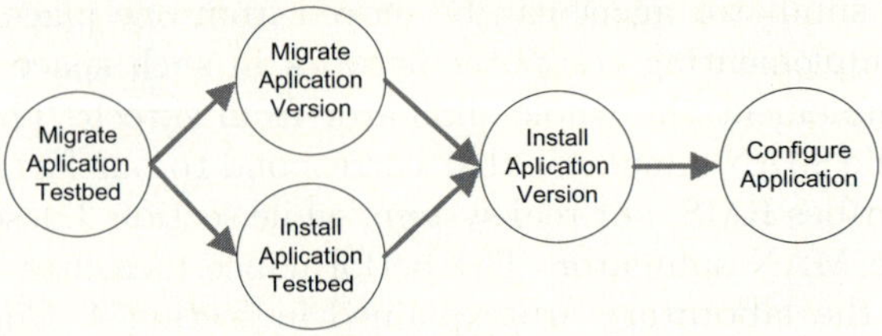

Fig. 2. The software installation operations graph

2.1 Mobile Agent Based Software Operations

Software operations and all maintenance actions are performed by mobile agents created at the RMS Management Station. These agents are referred to as the Multi–OperationAgents (MOAs) . MOAs are equipped with the required knowledge (communication protocol), data (e.g. the software package which has to be installed) and access rights to the remote system.

First, the user selects the operations which need to be performed at the remote location. After this, the Management Station generates a corresponding operations list according to the scheduling algorithm used. This algorithm is also used to define the distribution of operations among MOAs. One possibility is that a single agent performs all the required operations at the remote location. This approach is optimal in very slow networks [1]. Currently, RMS uses an algorithm which distributes all migrate operations for all locations to a single agent, while all other operations are assigned to one agent for every location.

Figure 2 shows the operations that are generated by the scheduling algorithm when software installation is requested. First, the Application Testbed must be migrated to the remote location. After Application Testbed migration, the testbed installation operation is performed in parallel with the migration of the Application Version. This is followed by application installation and, finally, configuration.

Normally, a user sends an operation request to the Management Station through it's GUI, after which the Management Station starts the execution of the corresponding operations. In our case, this part is automated by an AutoRMS station which contains a mechanism that allows it to perform all the tasks needed for executing software operations. Applications that want to perform software operations at remote locations using RMS simply send a request to the AutoRMS station. This request contains the location of the software components, the operation(s) that is to be performed and the address of the remote location(s).

3 The Mobile Agent Network

The Mobile Agent Network (MAN) is used for modeling agent coordination in an agent team. The MAN is represented by a triple, (A, S, N), where A represents a multi–agent system consisting of cooperating and communicating mobile agents which can migrate autonomously from node to node; S is a set of n_s nodes on which agents perform operations; and N is a network that connects nodes from S and enables agent mobility.

An agent $agent_k$ is defined by a triple, $(name_k, address_k, task_k)$, where $name_k$ serves as the agent's unique identification, $address_k$ gives the list of nodes to be visited by the agent, and $task_k$ describes the functionality it provides, i.e. the set of assigned elementary operations, s_i it performs. When hosted by node $S_i \in address_k$, $agent_k$ performs operations, $s_i \in task_k$. If an operation requires specific data, the agent carries this data with it during migration. Such an agent is referred to as a "loaded agent".

The organization of multi–agent systems suitable for network and distributed systems management considered in [5,6,7] is based on shared plans used by agent teams. An intelligent stationary agent is responsible for decomposing a complex management task into n_t elementary operations and ordering these operations. The same agent also collects and interprets data regarding the characteristics of the nodes and the network in order to define a suitable agent team.

The following assignments of elementary operations are considered the basic building blocks for identification of the agents–shared plan:

- R1: a single agent executes all operations on all nodes;
- R2: an agent executes a single operation on one node only;
- R3: an agent executes all operations on one node only;
- R4: an agent executes a specific operation on all nodes;
- R5: an agent executes a specific operation only once on all nodes;
- R6: operations are assigned to the agents in order to exploit maximal parallelism of operations. Mutually independent operations are assigned to different agents, in order to execute them in parallel;
- R7: a hybrid solution combining R4 and R3. An agent is responsible for a specific operation on all nodes; all other agents execute all other operations, each on a different node;
- R8: a hybrid solution combining R5 and R3 (specialization of R7 in the way R5 is specialization of R4).

When analyzing the coordination of shared plans, three types of agent communication are considered: internal (when the operations are performed by the same agent), local (when the operations are performed by different agents at the same node) and global(when the operations are performed by different agents at different nodes).

Agent creation is characterized by the birth of an agent. After birth, the agent migrates to the first node where it needs to execute an operation. If the agent carries data, its migration time is longer. Upon arrival at the first node, it executes its corresponding operation and then informs other agents of its results via dialogue. The actual dialog depends on the type of communication used for coordination (local or global). The agent then considers its next operation. If this operation is to be executed at the same node, migration is skipped. However, if the subsequent operation is to be executed on a different node, the agent migrates to the node in question, executes its operation there and again informs others through dialogue. This is repeated for all operations on its task list ($task_k$). The final operation is the agent's death by which the agent is disposed of.

The efficiency of the assignment scheme depends on the specific task submitted and environmental characteristics. The basic parameters that describe the environment are as follows: operation execution time, agent migration time, loaded agent migration time, the time required for a dialogue between agents residing at different nodes, and the type of agent communication (direct or indirect). When an agent sends a message, the dialogue fails if the receiving agent is not at the expected destination. In direct communication, the sender waits for some time and retries until the dialogue is successful. In indirect communication, the sender creates a transport agent which migrates to the destination, and delivers the message to the receiving agent upon arrival [8].

3.1 The MAN Simulator

The aim of the MAN simulator is to simulate different agent coordination's. The structure of the agents is defined by the MAN. The simulator was programmed in Java as part of a PhD dissertation [2] and its performance has been improved by introducing an event–based simulation. The input data required to run a simulation includes the following: the duration of an elementary operation, the duration of remote communication, the duration of agent migration with and without payload, the coordination model and the operations graph. The simulation results are in the form of an operations graph execution matrix. Operations graph execution matrix analysis can find coordination problems which are then solved by improved coordination rules. After correcting the coordination rules, a simulation with the same parameters could be rerun and the results compared. Operations graph execution matrix generation can be omitted from the simulation and, thus, the total execution time is the only simulation result. This improves simulation performance and resource consumption.

The core of the simulator is shown through a class diagram (figure 3). The main class is called `AgentSystem` and represents the multi–agent system itself.

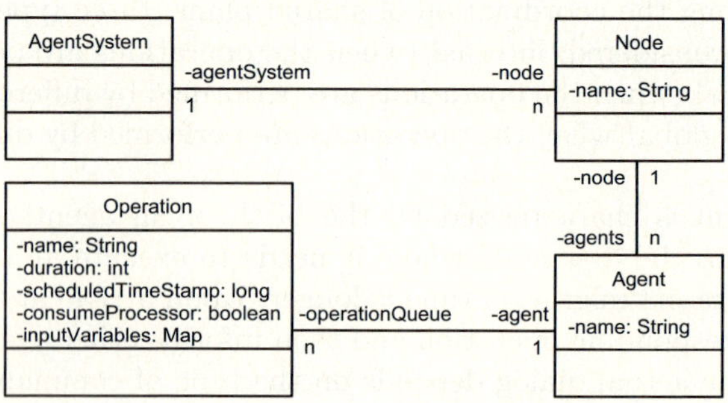

Fig. 3. The MAN simulator core

This class contains a list of nodes of class `Node`. Each node contains a queue of agents (of class `Agent`) while each agent contains a queue of elementary operations (of class `Operation`) that it must execute. Each operation has attributes such as name, duration, input variables, a flag for execution at a specified time (`scheduledTimeStamp`) and a flag for consuming processor time (`consumeProcessor`). Operation input data is stored in a map where the key is the name of the input variable. The input data value can be `null` which means that the value is not set. Input data is used for preconditions in the operations graph.

4 Laboratory Measurements

The described MAN simulator was used in experiments which simulate the execution of operations in the RMS system. Conducting this experiment was necessary to be able to compare with the results obtained by the actual multi–agent system and, thus, validate the results obtained by the simulator. If the results from this experiment were comparable with those obtained by the simulator, this would prove that the simulator could successfully be used to model the behavior of multi–agent systems based on MAN.

4.1 Laboratory Configuration

Figure 4 shows the configuration of the laboratory where the experiments were performed. Nine PCs where used: eight were used for hosting the RMS servers and one hosted the ScenarioExecutionAgent and the AutoRMS station. The PCs in question were Dell OptiPlex 170L PCs with Intel Celeron 2.66 GHz processor, 512MB of RAM, Windows XP Professional operating systems, the Sun JDK 1.5.0_09_b01 version of Java and the JADE 3.3 agent platform.

The various testing scenarios differed with respect to three parameters: the number of operations to be performed (ranging from one to eight software

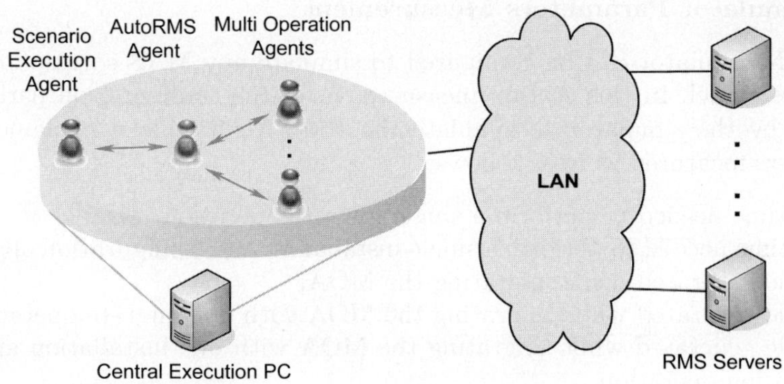

Fig. 4. Laboratory scenario configuration

installation operations), the number of PC on which these operations were to be performed (ranging from one to eight PCs) and the network bandwidth (512 Kbit/s, 1 Mbit/s and 10 Mbit/s networks) making a total of 192 measurements. Such a large number of measurements required some automation to reduce the time needed to perform the measurements and the possibility of errors which can be caused by human interaction. Automation of the first two parameters was performed by the ScenarioExecutionAgent. It was responsible for scheduling the installation operations and keeping track of the results of the experiments. The network bandwidth parameter was changed manually since we didn't find any way to control it using an agent. To simulate different networks we used network bandwidth limiters on all the PCs included in the experiment. A single scenario measurement was performed in the following way:

- The ScenarioExecutionAgent would read the scenario parameters from the XML configuration file (the number and the location of the software, and the number and the IP address of the RMS servers);
- After gathering the parameters of the experiment, the agent would generate an installation request and send it to the AutoRMS station;
- The AutoRMS station would receive and process the request. After processing the request, the station would generate the operations needed to perform the installation request;
- The corresponding operations would be scheduled for execution by the MOA according to the scheduling algorithm used;
- The MOA would migrate to the RMS server/s and perform operations;
- Upon the completion of the scenario, the AutoRMS would send a notification message to the ScenarioExecutionAgent with the time needed to perform it;
- If there were additional scenarios left, this process would be repeated.

The AutoRMS station was responsible for measuring the time needed to perform a scenario. Time measurement was started before scheduling the operations for execution and stopped upon operation completion.

4.2 Simulator Parameters Measurement

The MAN simulator can be configured to simulate any MAS corresponding to the MAN model. In this section measurement of the configuration parameters required by the simulator to simulate the RMS MAS will be explained. The parameters measured were as follows:

- The time needed to perform a single migrate operation;
- The time needed to perform a single installation and configuration operation;
- Traffic generated while migrating the MOA;
- Traffic generated while migrating the MOA with one migrate operation;
- Traffic generated while migrating the MOA with one installation and configuration operation.

To measure the time needed to perform operations, a modification of the MOA was required. Agents were modified with a timer that calculated the elapsed time.

Table 1. Measured parameters

Parameter name	value	Parameter name	value
S_a	7192 B	t_{mv}	2214 ms
S_{amo}	58242 B	t_{it}	744 ms
S_{aio}	10681 B	t_{iv}	963 ms
t_{mt}	878 ms	t_e	291 ms

The timer was initiated before the operation was scheduled for execution by the agent and stopped after the agent received notification from the RMS server that its request was completed.

Traffic generated by the migration process was measured using the Ethereal Network Protocol Analyzer. During a scenario, this tool would capture the network traffic on the PCs' network interface. The traffic generated was in the form of a Java RMI (Java Remote Network Invocation) stream, used by the Jade agent platform to migrate agents between two PCs. The packet analyzer was then used to calculate the stream size. The traffic parameters required to run the simulator were calculated as follows:

$$S_{mo} = S_{amo} - S_a \tag{1}$$
$$S_m = S_a + N \cdot S_{mo} \tag{2}$$
$$S_{io} = S_{aio} - S_a \tag{3}$$
$$S_i = S_a + N \cdot S_{io} \tag{4}$$

where S_{amo} is the traffic generated by the MOA with one migrate operation, S_a is the traffic generated by the MOA without any operations, S_{mo} is the size of the migrate operation, S_m is the size of the MOA with multiple migration operations, N is the number of software's, S_{aio} is the traffic generated by the MOA with one install operation, S_{io} is the size of the installation operation and S_i represents the size of the MOA with multiple installation operations. The time needed to migrate a MOA depends on the size of the agent and is calculated as $t = S/B$ where t represents the time needed to migrate MOA, S is the agent size and B represents the network bandwidth. All values are in bytes. The values obtained for the size of the agent and the time needed to perform operations are shown in table 1 where t_m represents the time needed to perform the migrate testbed operation, t_{mv} is the migrate version operation, t_{it} is the install testbed operation, t_{iv} is the install version operation and t_e is the time needed to configure the application.

5 Comparison of Results

In this section, a comparison of the simulation results are given. In the figures 5a, 5b and 6 the x–axis represents the number of PCs on which the software is to be installed, the y–axis represents the number of software installation requests and the z–axis shows the time needed to complete a single scenario. Figures 5a and

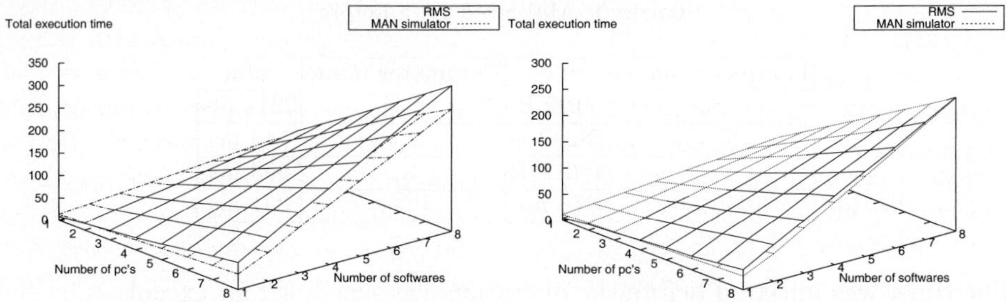

a) 512kbit/s network bandwidth b) 10Mbit/s network bandwidth

Fig. 5. Total execution time

figure 5b show the results for the 512kbit/s and 10Mbit/s scenario respectively. Results for the 1Mbit/s scenario are not shown since they are very similar to those obtained for the 512kbit/s scenario. The differences in percentage between the results obtained by the MAN simulator and the RMS are shown in figure 6.

The results show that both the MAN simulator and the RMS have a linear increase in the total execution times when increasing the number of software installation requests and the number of RMS servers. The only anomaly in the results occurs for the scenarios with only one RMS server. In this scenario, the total execution times for the RMS are up to 220% faster than the results obtained by the simulator. The cause of this anomaly is still unknown. It could be caused by a component/s of the RMS or the agent platform. For the worse case scenario (512kbit/s network bandwidth) the average simulator results where within 10.73% of the real–system's results, while for the best case scenario (1Mbit/s network bandwidth), they where within 3.023% of accuracy. With an increase in the number of PCs and the software operation requests, this difference decreased

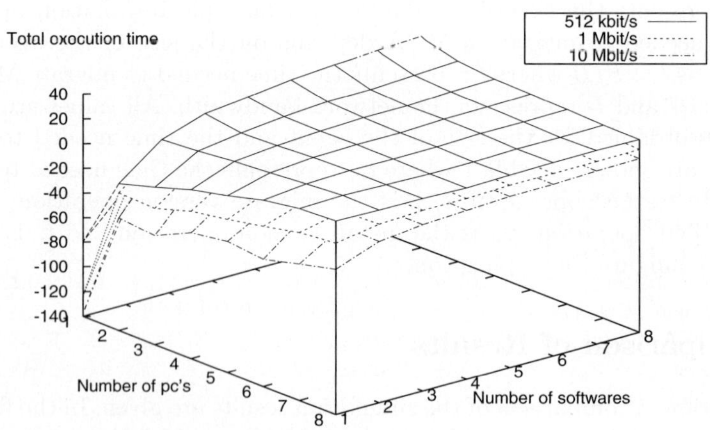

Fig. 6. Total execution time difference in percents

reaching 0.7% in the scenario with eight PCs, eight software installation requests and 10Mbit/s network bandwidth. A decrease of the network bandwidth caused an increases in the difference between the results. The reason for this is the assumption that was made in the simulator (see section 3.1). The simulator does not take into consideration the influence of other PCs in the network nor the software installed on them which can generate traffic on the network interface. Measurements performed in the laboratory indicated that on average, these components generate 80 KB of traffic per minute (not taking under consideration traffic generated by the agent platform) which makes up for 3% of the total traffic generated by the RMS.

6 Conclusion and Future Work

In this paper, we described experiments conducted to compare a real system (RMS) implementing the MAN model with our MAN simulator. The results show that our simulator can be used to simulate environments with more than two RMS servers with accuracy within 10% of accuracy. In the environments with more servers the accuracy is even better (within 0.7%). There was one anomaly found which will be investigated in future work. Furthermore, we will introduce complex network structures.

References

1. Kusek, M., Lovrek, I., Sinkovic, V.: Agent team coordination in the mobile agent network. In: Lecture Notes in Computer Science – LNCS. Volume 3681 of LNAI. Springer Verlag (2005) 240–245 9th International Conference Knowledge–Based Intelligent Information and Engineering Systems – KES 2005.
2. Kusek, M.: Coordination of Mobile Agents for Remote Software System Operations. PhD thesis, University of Zagreb, FER (2005)
3. Horling, B., Lesser, V., Vincent, R.: Multi-Agent System Simulation Framework. 16th IMACS World Congress 2000 on Scientific Computation, Applied Mathematics and Simulation (August 2000)
4. Logan, B., Theodoropolous, G.: The distributed simulation of multi–agent systems. Proceedings of the IEEE **89**(2) (2001) 174–185
5. Weiss, G.: Multiagent Systems. MIT Press: Cambridge (1999)
6. Grosz, B.J., Kraus, S.: Collaborative plans for complex group action. Artificial Intelligence **86**(2) (1996) 269–357
7. Silva, L.M., Batista, V., Martins, P., Soares, G.: Using mobile agents for parallel processing. In: Proceedings of the International Symposium on Distributed Objects and Applications, Edinburgh, United Kingdom (1999) 34–43
8. Arisha, K.A., Ozcan, F., Ross, R., Subrahmanian, V.S., Eiter, T., Kraus, S., Ilan, B.: Impact: A Platform for Collaborating Agents. IEEE Intelligent Agents **14**(2) (April 1999) 64–72

The Study of Multi-agent Network Flow Architecture for Application Performance Evaluation

Zhi Xiao-fan, Liao Zhi-cheng, and Lu Jian-xiong

Department of Computing and Information Technology, Fudan University. Shanghai 200433, P.R. China
xfzhi@fudan.edu.cn

Abstract. With the new software architecture appearing, the application performance evaluation becomes quite difficult. Based on previous studies, a new architecture is proposed concerning using multi-agent network data in the application performance evaluation. Architecture defined both active and passive methods to simulate and test. Multi-agents on the network and under-test components communicate and interact to find the performance result and the bottle neck of the system. Architecture includes the agents' arrangement, different agent's distribution and their running process. It is a black-box test method and could simulate several different parallel relative transactions at the same time. Test is started from link end that no configuration should be done. This architecture is proved by a real information system test.

Keywords: performance evaluation architecture; network flow; multi-agent.

1 Introduction

Information systems' performance is focused and studied because it relates to application's efficiency, functionality and security. More and more large, distributed and heterogeneous systems are used in core businesses process. This makes systematic performance evaluation more attractive. The target of systematic performance evaluation is not only to get performance parameters but also to discover system performance bottle neck. Core business system composed with heterogeneous components and built with different architectures is popular now. Further more, different parts of system like soft, hardware, network and their configuration may affect each other. This makes the performance analysis, discovery and judgment difficult and complex. This paper tries to integrate the multi-agent technology and network stream analysis in systematic performance evaluation. The distribution and intelligent character of this method make it suitable for both Client / Server, Browser / Server architectures and new architecture like peer to peer system (P2P), service oriented architecture (SOA).

Many performance test method is dedicated to one aspect of performance. For example, "Robot" recorder tests performance by replaying; Network test tool tests performance by send packages at line speed; Far distance performance test tool test

N.T. Nguyen et al. (Eds.): KES-AMSTA 2007, LNAI 4496, pp. 530–538, 2007.

performance by active or passive method; Simulator simulates the packet loss, line speed limitation and out of order to test performance under different limitation.

For performance evaluation, isolated evaluation method is not good enough at component interaction judgments of performance bottle neck finding. For example, the attribute of network will affect application's appearance. [4]

There are many related studies. Reverse-engineering the internet is within reach of the research community, and that a collective effort would achieve significantly more than the isolated efforts of individual researchers. [1] Related result includes: measurement and characterizing end-to-end internet service performance [6], application's remote client perceived response time test by different method [3], session or HTTP layer stream model for performance evaluations [10], performance debugging for distributed systems of black boxes [17], determining the remote client perceived response time from live packet streams [18], passive End-to-End internet service performance monitoring [19], etc. Traffic measurement on active or passive method is discussed to measure performance with different speed connections. [2][5][7] Other related research focused in software performance test with script record and replay like robot, record keyboard and mouse activity or record communication by proxy. [12]

Distributed and multi-agents method was introduced into this field. Paper [20] introduced a multi-agent testing architecture for monitoring and analyzing the performance of web applications. The parameter under test is mainly on host and network's usage percentage. The difference between our work and paper mentioned above is that our monitor target is the transaction per second and every key point's performance behavior of application. Another difference is the source data comes from network side in our research. So this means different architecture and different method.

System platform tends to be developed to Service-oriented architecture. SOA is a software architectural concept that defines the use of services to support business requirements. In SOA, resources are available to other participants in the network as independent services that are accessed though a standardized way. SOA provides a methodology and framework for documenting enterprise capabilities and can support integration and consolidation activities. [15] Architecture is tried to be compatible with SOA's idea.

Architecture uses multi-agents to intellectually collect, analyze and replay network application data stream. With the cooperation with agents, the bottle neck of system could be found. This architecture includes both passive and active performance test method. It's passive test while agents capture stream and parameters. It's active while changing parameters to reconstruct pattern and replay. The framework is presented as follows: Section 2 is the analysis and the architecture design. Section 3 is detailed designs and algorithm implementation. Section 4 is the test practice. Section 5 is the conclusion and future work.

2 Analyses

To build such architecture, the detailed issues of evaluation are: what's the common abstract evaluation method without considering the business itself? How to express it with multi-agent architecture? How to define the interaction and communication process of these agents and other input? What can we do on the inter-affect of network and application? If both the server and the browser/client are not under control, could we find a way to evaluate the performance? This paper tries to answer these issues by new multi-agent network flow architecture for application performance evaluation.

Test and evaluation method based on multi-agents is a combination of artificial intelligence and test method. Agent-based computing gives new solution for distributed, complex and multi-human-machine interaction issues. It is widely used in distribution control and complex issues' simulation area, like that on network management with agent. Distributed test platform architecture based on multi-agent includes single agent's intelligent actions, communication and cooperation between agents, agent deployment based on system architecture and parameters, statistics and analysis for result report and so on.

In the environment of multi-agents, agent could be cooperative or competitive. This method needs them cooperative. There is a control agent as the coordinator of other agents, a statistic agent reports the result of application's performance. Information collection agent and replay agent play the main roles in test actions.

The main inspiration is based on agents on network for application performance evaluation because applications use networks as the transport channel. So information collection and replay agent is the most important agent in this architecture. This agent deals with network stream capture, filter and protocol reconstruction. Protocol analysis, pattern recognition and associate method knowledge base should be built in agents to support different application. Data captured should be started with a trigger and should be filtered. An appointed trigger may be DNS's lookup action or other event. Filter could work based on time scope, IP address and TCP/UDP sessions, etc. The statistic agent's target is getting a typically transaction per second result of appointed application transaction. So it should communicate with many other agents to collect result and generate statistics. It could be customized when those special applications need other performance result parameters.

With these ideas, the first step of architecture building is to denote the performance related information stream's direction. Figure 1 is a general commercial information system sketch map organized with target of evaluation. Some systems may combine part components into one module.

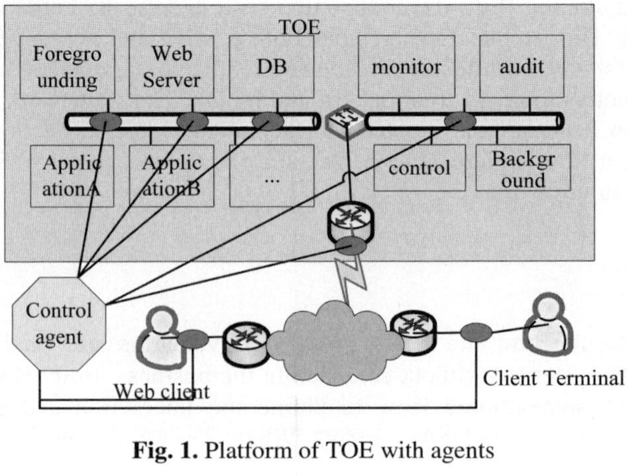

Fig. 1. Platform of TOE with agents

On figure 1, elliptical fuscous point is the arrangement of agents. They are on the key path of an application. There are some reasons for this arrangement. The performance bottle neck is generally on server side, like on the background procedure, database procedure and so on. If it's not under the thin client fat server environment, in a few cases, client is the bottle neck since some character of network. So the monitor agent should be mainly on server side and communication line. By collecting and sending virtual client's request, agents on core communication line could simulate the real interactions with server. The monitor agents could statistic on every key point of communication channel. Network performance will impact the behaviors of application transaction, like TCP congestion widow size and QoS, etc. So multi-agents' arrangement on network could simulate the reality of great deal of access. With the intelligent agent, optimize could done by parameter adjusting.

The next step is to divide agents showed in Fig 2. Agent on A is at the client side and B/C is in the rectangle of Fig 1. Agent A and C collect communication stream on network and filter them. Stream could be reconstruction with http or xml format. Agent A could put active stream according to settings in advance. In practice, A or B/C could be more than one agent. With the coordination of central control agent, they co-work and this is described in section 3. Agent B collects system resource's running parameters and analysis them. Agents communicate with protocols comes from FIPA, the standards organization for agents and multi-agent systems. Time synchronization is done by central control agent.

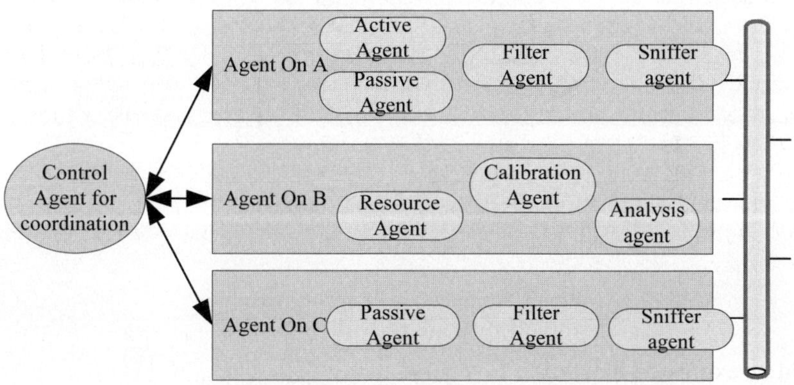

Fig. 2. Architecture of agents' distribution

3 Detailed Design and Implementation

Based on the architecture of section 2, this section is mainly focused on the detailed design. The full process includes: defining the key communication channels, sampling the application network stream and filtering, defining the interaction and communication of agents, and statistics.

Test is started from a link end, instead of from real client. So no configuration should be done for tests. The detailed design of application could also be black box to tester. For those systems with complex architecture and little document, it's a good method to solve problem. The intelligence of agent A make the interaction becomes possible. Agent could response special communication with designed parameter.

The core issues are algorithm of agent's process and the communication of agents. After figuring out the key link of application, the researchers let agent A on the start of the link, agent B and many more other agents in the distributed middle part, and agent C on the end of the link. The central control agent is mentioned as Con. The activity is showed in Fig 3. Agent A could control several point to start sending pressure data to server side at the same time. On every important point of the link, like servers, resource agent B could monitor them. Agent C adds up successful transactions according to given standard and record error response rate.

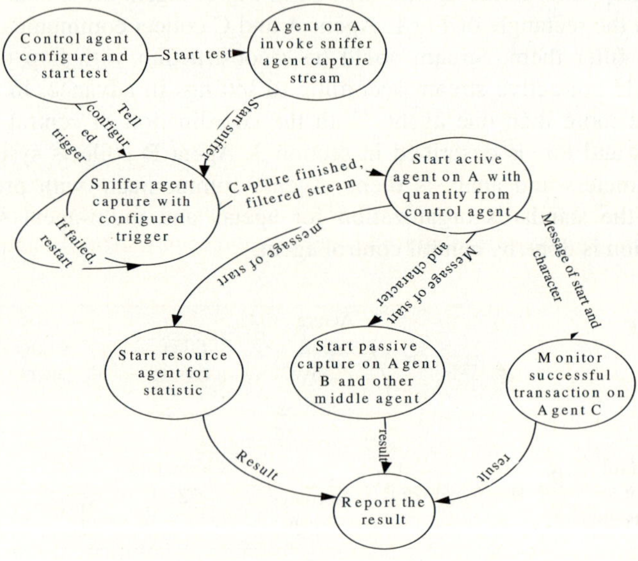

Fig. 3. Architecture of agents' running

It could be expressed like this:

Agent (A, B, C) // A, B, C means agent on A, B, C in figure 2.

Init (At(A,[Front point]) $\wedge$ At(C,[End point]) $\wedge$ At(B,[middle point] $\wedge$ At(Con,[controller point])

//notes: Initiate all those agents on their position

Goal (Returned (TPS) $\wedge$Return(Related parameter))

//agents should return transaction per second and other performance related parameters

Action (Captured (sniffer agent, Active data stream)

PRECOND: Ordered (Con, character) $\wedge$ Finding(trigger, [Packet x]) $\wedge$ Filter(A, transaction stream)

EFFECT: Returned (transaction stream)

//engender active test's stream and let it be job1
Action (Replay& Response (A, transaction stream))
PRECOND: Ordered (Con, job1) ∧Replay (A, transaction stream)□Reconfigure(A, transaction stream)
EFFECT: Returned (test stream) //replay test streams and let it be job2
Action (Captured (C, TPS))
PRECOND: Ordered (Con, job2) ∧Test (B, TPS) ∧Test (C, TPS) ∧Monitor (Servers, usage)
EFFECT: Returned (performance value))
//collected performance related info on every related agent
To realize the algorithm above, agent may receive and use the output of other test tool like ping, trace route, ethereal, etc.

Result is expressed by transaction per second (TPS). TPS was recommended by international organization Transaction Processing Council (TPC). There is no union performance parameter of system performance so it is just our choice. When system could finish N transactions rightly in T second, its' TPS value is N/T. The proof of success is the critical burden signal of the system. We define that system reaches its critical burden while the percentage of error response reached to 1% or 3%, or the transmit time of one transaction reached two or three times than it should on ideal situation. This percentage and multiple relies on the importance of the application.

4 Practice

This section is the practice while using this architecture in a real core information system's evaluation. The system under test is a multi-layer complex system. It uses Front- Kernel- DB architecture with special communication protocols and many layer load balance. System is developed with C++ and its platform is UNIX operation system. Every server has more than one CPU. Many optimizer methods have been used for better performance. Its network architecture is similar to bus topology with some special protocol. The test requirement is to test transaction performance and find the bottle neck.

Our test code is written by C++ and python with the help of ethereal 0.99.0. Some debug is done by RadCom Prismlite 2000 protocol analyzer. Agents are arranged on the key link of transactions and every important server. Future we will try to migrate it to some standard multi-agent platform like JADE (Java Agent Development Framework) of CSELT.

The parallel process related link is the key link for agents' arrangement. This includes putting request into communication system, monitoring the successful transactions and usage percentage's parameters, statistic result and report. Test lets agent A control 4 terminals, with 50 test threads per terminal. Parameter is configured from control agent. Total supposed to execute 500,000 transactions and this could be configured. Figure 4 and 5 is the test result.

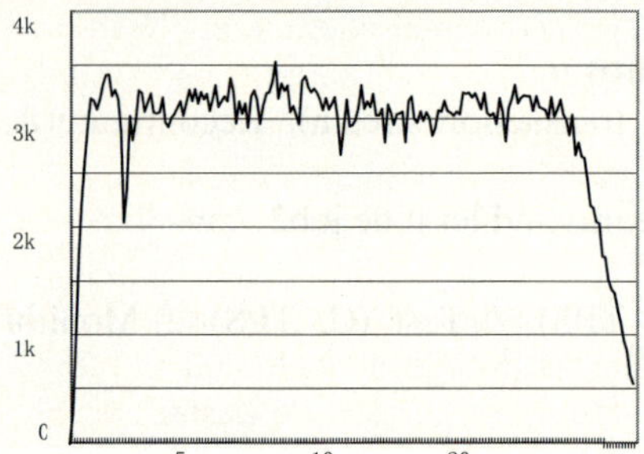

Fig. 4. TPS result from agent C

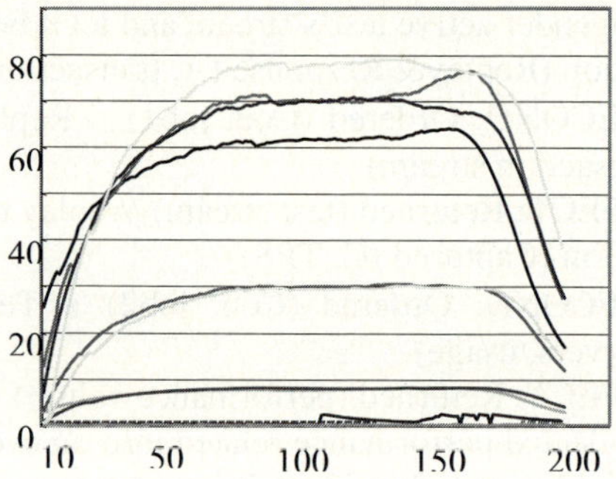

Fig. 5. CPU usage percentage from monitor agent

In figure 4, X-axis represents time and y-axis represents the successful transactions. The unit of curve is transactions per second, as mentioned above. For this system, system can continuously deal with about 3250 transactions.

For bottle neck finding, figure 5 is the percentage of CPU usage collected by resource agent. It's easy to find that calculation ability of two servers is the ability of bottle neck. In some other case, other resource should be monitored.

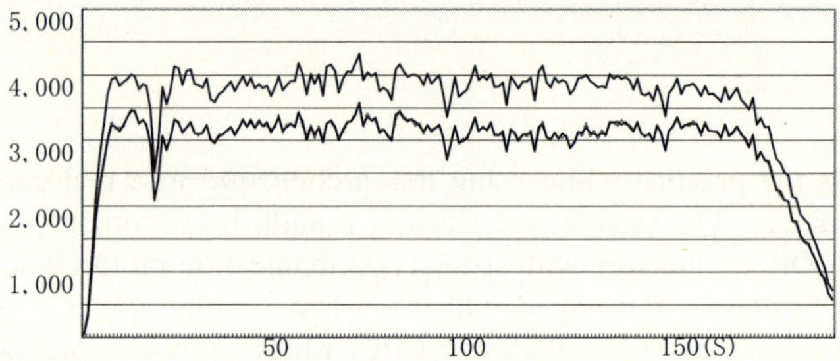

Fig. 6. Related transaction performance curve collected by different agent

In figure 6, the lower curve is transactions recorded by agent A. It means the number of successfully sent transactions. The higher curve is the responses for actions on other sever collected by agent B. One transaction from agent A may lead to more than one action of other server to finish one transaction, so it's higher. The same line covered by the lower line is the transactions successfully collected by agent C. The fact that they are almost one line means every transaction sent was implemented on time.

5 Conclusions

Multi-agent network flow architecture for application performance evaluation is introduced in this paper. This includes the background, related work, design inspiration, framework architecture, algorithm and simple practice. The prototype of multi-agent network flow architecture for application performance evaluation is proposed. Architecture is realized by the agents' behavior and interaction. The result of tests is presented by TPS from TPC and some other related parameters. Agent is put on performance related key link of network. By agent's sniffer, filter, analysis, replaying and communication, this architecture could evaluate complex architecture information system's performance. Method is proved by a real core business application's testing.

Advantages of this architecture are that it needn't change or configure any thing at both client (browser) side and server side; agents could communicate and cooperate while doing tests; it could simulate several different parallel relative businesses and negotiate at same time; it could simulate different protocol exchange process. Agents give this method ability to find potential problems on different positions. So the multi-agent network flow architecture for application performance evaluation is valuable in system testing and optimizing.

Since the limitation of environment, practice is just a mainly implementation of the architecture. So in the future work, the automatic and intelligence of agents should be enhanced step by step. The learning and statistic part should use more knowledge to be more active. More protocols and application architectures should be tested. This architecture may work in system security evaluation as well. Now finding the bottleneck is human-oriented via graphical representation, to be agent-oriented and go deep into software architecture is next direction. On the other side, we will migrate the test to stand platform like JADE. This will be studied in the future.

References

[1] Neil Spring, David Wetherall, and Thomas Anderson. Reverse Engineering the Internet. In ACM SIGCOMM Computer Communications Review Volume 34, Number 1: January 2004

[2] Chuck Fraleigh, Sue Moon, Bryan Lyles, Chase Cotton, Mujahid Khan, Deb Moll, Rob Rockell, and Ted Seely, Christophe Diot, Packet-Level Traffic Measurements from the Sprint IP Backbone, IEEE Network November/December 2003 0890-8044/03 © 2003 IEEE

[3] DAVID OLSHEFSKI, JASON NIEH, DAKSHI AGRAWAL, Using Certes to Infer Client Response Time at the Web Server, ACM Transactions on Computer Systems, Vol. 22, No. 1, February 2004, Pages 49–93.

[4] Carey Williamson, Rob Simmonds and Martin Arlitt, A case study of Web server benchmarking using parallel WAN emulation, Performance Evaluation Volume 49, Issues 1-4, September 2002, Pages 111-127

[5] Paul Barford and Mark Crovella, Measuring Web Performance in the Wide Area, ACM SIGMETRICS Performance Evaluation Review archive Volume 27, Issue 2 9.1.99 table of contents. Pages: 37 - 48. Year of Publication: 1999.

[6] LUDMILA CHERKASOVA, YUN FU, WENTING TANG, AMIN VAHDAT, Measuring and Characterizing End-to-End Internet Service Performance, ACM Transactions on Internet Technology, Vol. 3, No. 4, November 2003.

[7] Paul Barford and Mark Crovella, Critical Path Analysis of TCP Transactions, IEEE/ACM TRANSACTIONS ON NETWORKING, VOL. 9, NO. 3, JUNE 2001

[8] Ravi Prasad and Constantinos Dovrolis, Margaret Murray and kc claffy, Bandwidth Estimation: Metrics, Measurement Techniques, and Tools, IEEE Network • November/December 2003 0890-8044/03 © 2003 IEEE

[9] Jim Gray, Tom Sawyer, Omri Serlin, Database and Transaction Processing Performance Handbook, Digital Equipment Corp

[10] Zhen Liu, Nicolas Niclausse and César Jalpa-Villanueva, Traffic model and performance evaluation of Web servers, Performance Evaluation Volume 46, Issues 2-3, October 2001, Pages 77-100

[11] Martin Arlitt, Rich Friedrich and Tai Jin, Performance evaluation of Web proxy cache replacement policies, Performance Evaluation Volume 39, Issues 1-4, February 2000, Pages 149-164

[12] VALERIA CARDELLINI, EMILIANO CASALICCHIO, MICHELE COLAJANNI, PHILIP S. YU, The State of the Art in Locally Distributed Web-Server Systems, ACM Computing Surveys, Vol. 34, No. 2, June 2002, pp. 263–311.

[13] Marco Conti, Enrico Gregori and Willy Lapenna, Performance Modeling and Evaluation of Heterogeneous Networks, Performance Evaluation Volume 59, Issues 2-3, February 2005, Pages 137-157

[14] http://en.wikipedia.org/wiki/Service-oriented_architecture Service-oriented architecture, from Wikipedia, the free encyclopedia.

[15] Esther Levin, Roberto Pieraccini, Wieland Echert. A stochastic model of human-machine interaction for learning dialog strategies. IEEE Transactions on speech and audio processing, January 2000, vol.8, No.1: pages 11-23

[16] Marcos K. Aguilera, Jeffrey C. Mogul, Janet L. Wiener, Performance Debugging for Distributed Systems of Black Boxes, Patrick Reynolds and Athicha Muthitacharoen, Proceedings of the 19th ACM Symposium on Operating System Principles (SOSP), Lake George, NY, October 2003.

[17] D. Olshefski and J. Nieh and E. Nahum, "ksniffer: Determining the Remote Client Perceived Response Time from Live Packet Streams", Proceedings of the 6th Symposium on Operating Systems Design and Implementation (OSDI 2004), 333-346, San Francisco, CA, 2004.

[18] Y. Fu, L. Cherkasova, W. Tang and A. Vahdat, "EtE: Passive End-to-End Internet Service Performance Monitoring", USENIX Conference Proceedings, 2002, Monterey, CA, 2002.

[19] Huiming Yu; Jin Zhang; Jinsheng Xu; Multi-Agent Testing Architecture for Monitoring and Analyzing the Performance of Web Applications, Computer and Information Science, 2006. ICIS-COMSAR 2006. 5th IEEE/ACIS International Conference on 10-12 July 2006 Page(s):115 - 120

[20] Shiyong Zhang, Xiaofan Zhi, The Study of practice-based information system security measurement Method,InfoSecu02 International Conference on Information Security ISBN-1-58113-369-3, Order# 573013,ACM, 2003

[21] LU Jianxiong, ZHANG Shiyong, ZHONG Yiping, An IP Performance Testing System Based on End-probing, Computer Engineering, 30(23): 77-79, 2004-12

[22] Magid Nikraz1a, Giovanni Caireb, and Parisa A. Bahria, A Methodology for the Analysis and Design of Multi-Agent Systems using JADE, the International Journal of Computer Systems Science & Engineering“Software Engineering for Multi-Agent Systems”,2006

Robotic Agent Control Based on Adaptive Intelligent Algorithm in Ubiquitous Networks

Min-Jung Lee[1], Gi-Hyun Hwang[2], Won-Tae Jang[2], and Kyung-Hwan Cha[2]

[1] Regional Innovation Center, Dongseo University, San 69-1, Churye-2Dong, Sasang-Gu, Busan 617-716, Korea

[2] Division of Computer Information Engineering, Dongseo University, San 69-1, Churye-2Dong, Sasang-Gu, Busan 617-716, Korea

{mnjlee,hwanggh,jwtway}@gdsu.dongseo.ac.kr, khcha@dongseo.ac.kr

Abstract. The aim of this paper is to investigate a control method for robotic agent, operating in ubiquitous network. It is difficult for people to fulfill their tasks under dynamically changing environment. Therefore, robotic agent performs tasks in stead of people. The ubiquitous network makes it possible to connect between robotic agent and people. Because robotic agent should fulfill tasks according to the received commands from people, the adaptive control method for the agent is needed in order to do the given tasks properly. This paper introduces an adaptive tracking control method for robotic agent based on the radial based functions network (RBFN). When some commands are received through networks, the proposed method can make robotic agent possible to perform tasks under dynamically changing environment. Experimental results show that the proposed control method based on RBFN is adaptable to the environment changes and is more robust than the conventional PID control method and the neuro-control method based on the multilayer perceptron.

Keywords: Neural Networks, RBFN, Perceptron, Robotic Agent, Ubiquitous Network.

1 Introduction

Because industrial factories have generally many dangerous elements, there is the restriction in the activity of people. The advance of ubiquitous technique makes it possible to connect between automatic agents used at the industrial field and people who are managers or engineers. Robotic agents known as industrial automatic elements perform repeated tasks in stead of people under dangerous environment. The role of networks transfers the command of users to robot. Because robotic agent should fulfill tasks according to the received commands from people, the adaptable control method for the agent is needed in order to do the given tasks properly. But robotic agents are known as systems with high nonlinearities which are often unknown and time-varying. Therefore, if we want to design a controller for robots, we should consider the exact trajectory tracking performance for reference input and the

N.T. Nguyen et al. (Eds.): KES-AMSTA 2007, LNAI 4496, pp. 539–548, 2007.

robustness for the existence of system's nonlinearities and external disturbances. The conventional feedback controllers, including proportional integral derivative (PID) controller, are commonly used in the field of industries because their control architectures are very simple and easy to implement. But when these conventional feedback controllers are directly applied to nonlinear systems, they suffer from the poor performance and robustness due to the unknown nonlinearities and the external disturbances. During decades, various control strategies to deal with the unknown nonlinearities and the external disturbances are proposed such as automatic tuning of PID control, variable structure control known as nonlinear robust control, feedback linearization, model reference adaptive control, direct adaptive control, and intelligent control approaches, etc [1-4].

An adaptive control strategy has found many applications in such areas as robots, ship steering, aircraft control, and process control, because it can continuously adjust parameters of a controller to accommodate changes in system dynamics and disturbances [3]. In these applications, an on-line adaptation law is usually used to estimate the unknown parameters of the system and then, an appropriate controller is designed to control the plant to satisfy a desired performance. To apply an adaptive control method to robot as a main controller, the a priori knowledge about the robot is required: the linearity about unknown parameters and skew-symmetry features, etc. Additionally it takes a long time to calculate the regression matrix used in the algorithm.

Neural networks do not require mathematical models and have an ability to approximate nonlinear systems. With these features of neural networks, many researchers have been attempting to use neural networks to represent complex plants and construct advanced controllers. But there is a problem that neural network as well as intelligent control is difficult to guarantee the stability of control systems mathematically. Therefore, some researchers try to connect neural network to a conventional controllers and to guarantee the stability also.

Naraendra and Parthasarthy discussed identification and control of dynamical system using neural networks. Khalil et al. designed the output feedback controller using radial basis function network. Jagannathan et al. as well as some researchers showed good tracking performance through a Lyapunov's stability approach in their model reference adaptive control using multilayer neural networks. And Slotine et al. approached the direct adaptive control using Gaussian networks. But these studies are used neural networks for compensate for the unknown nonlinearities. Lewis et al. tried to connect neural networks to direct adaptive controller for robots, but they used lots of multilayer neural networks to approximate inertia matrix and Coriolis/ Centrifugal matrix. In case of multilayer neural networks, the structures are complicated and it takes long time to compute the output of multilayer neural networks [5-15].

This paper proposed another adaptive control method based on a RBFN. It deals with tracking control problems for robotic agent. The on-line learning method is employed to adjust parameters of the RBFN. The robot dynamics expressed in terms of the filtered tracking errors are nonlinear functions The RBFN is applied to approximate the nonlinear functions. The adaptation laws are derived to guarantee the stability of the total control system on the basis of Lyapunov stability theory. Also, the PD controller is employed to guarantee the stability and robustness of the control system.

By comparing with the experimental results for a SCARA type robot based on two different control strategies, such as PID controller and multilayer neural networks with backpropagation, we proved the validation that the proposed adaptive tracking controller for robots is more adaptable to the unknown nonlinearities and the external disturbances than two difference controllers.

2 Dynamics and Structural Properties of Robotic Agent

For control design purpose, it is necessary to have a mathematical model that reveals the dynamical behaviour of a system. Therefore, we derive the dynamical equations of motion for robot manipulators, one of robotic agents, and we summarize the structural properties of robot dynamics.

Using the Euler-Lagrangian formulation, the equations of motion of an n-degree-of-freedom manipulator can be written as [1-3], [11]

$$D(q)\ddot{q} + C(q,\dot{q})\dot{q} + G(q) + \tau_d = \tau \tag{1}$$

where $q \in R^n$ is the generalized coordinates (joint position); $D(q) \in R^{n\times n}$ is the symmetric, bounded positive definite inertial matrix; vector $C(q, \dot{q}) \in R^n$ presents the Coriolis and Centrifugal torques; $G(q) \in R^n$ is the vector of gravitational torques; $\tau_d \in R^n$ is the disturbance torque vector; and $\tau \in R^n$ is the vector of applied joint torques.

The dynamics of robot manipulators in the form of (1) is characterized by the following structural properties.

Property 1: An inertial matrix D is symmetric and positive definite and there exist scalars m_1 and m_2, such that $m_1 I \le D(q) \le m_2 I$.

Property 2: The Coriolis/Centrifugal matrix $C(q, \dot{q})$ is bounded by $c_b(q)\|\dot{q}\|^2$ with $c_b(q) \in C^1(S)$. S is a simply connected compact set of R^n .

Property 3: The matrix $\dot{D} - 2C$ is skew-symmetric, so the matrix is satisfied with $x^T(\dot{D} - 2C)x = 0$ for $\forall x \in R^n$.

Property 4: The norm of the unknown disturbance τ_d has a positive upper bound b_d .

3 Radial Basis Function Network

The RBFN proposed by Moody, Darken, Powell, Broomhead and Lowe, is used to approximate a non-linear function and has a faster convergence time than the multilayer neural networks. Additionally, The RBFN has a similar feature to the fuzzy inference system. First, the output value is calculated using the weighted sum or weighted average method. Second, the number of hidden layer's node of the RBFN is the same as the number of if-then rules in the FIS. Third, the radial basis functions are similar to the membership functions of FIS' premise part [4]. Fig. 1 shows the structure of the RBFN. The RBFN consists of hidden layer and output layer. The number of hidden layers is determined by designer. Gaussian function, triangular function and trapezoidal function are usually employed as basis functions of the RBFN. Gaussian function is frequently used as a basis function.

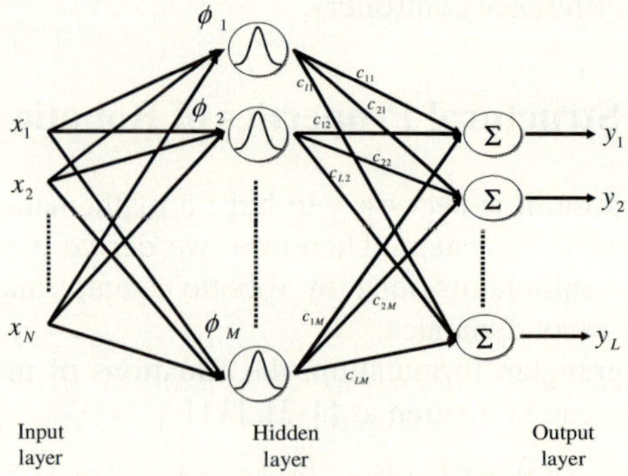

Fig. 1. Architecture of the RBFN with M receptive file units

If we select the Gaussian function as the basis function and use the weighted sum method to calculate the output of the RBFN, then the output becomes

$$y_i = \sum_{j=1}^{M} c_{ij}\phi_j \qquad i = 1,2,\cdots,l \tag{2}$$

$$\phi_j(x) = \exp\left(\frac{\|x-u_j\|^2}{\sigma_j^2}\right) \tag{3}$$

where M and l are the number of node and output, respectively. c_{ij} is the *j*-th weight of RBFN, $\phi_j(x)$ is the j th basis function, $u_j \in R^m$ is the j th center vector and $\sigma_j \in R^m$ is the j th standard deviation.

And the RBFN is used to design adaptive tracking controller for robot manipulators in this paper because the structure is simpler than multiplayer perceptron and its mathematical expression can be shown the linearity about the connection weights. As the design of adaptive control, radial basis function is used as nonlinear approximator. In general, the approximated system motel can be described as

$$y = c^T\Phi + \varepsilon \tag{4}$$

where

$$y = [y_1 \quad y_2 \quad \cdots \quad y_l]^T \qquad \Phi = [\phi_1 \quad \phi_2 \quad \cdots \quad \phi_l]^T \; c^T = \begin{bmatrix} c_{11} & c_{12} & \cdots & c_{1m} \\ \vdots & \vdots & \ddots & \vdots \\ c_{l1} & c_{l2} & \cdots & c_{lm} \end{bmatrix}^T \qquad \varepsilon = [\varepsilon_1 \quad \varepsilon_2 \quad \cdots \quad \varepsilon_l]^T$$

In (4), there are approximation errors, ε, because we just considered the finite dimensional hidden nodes of RBNF. But approximation errors, ε, have very small values and their norm value is bounded by a known constant value according to the approximation theorem [5],[11], such that

$$\|\varepsilon\| \le \varepsilon_N \tag{5}$$

We will use (5) to calculate the control gains.

4 Design of an Adaptive Tracking Controller Based on RBFN

4.1 Architecture

This paper attempts to connect neural networks to adaptive tracking control schemes in order to solve the difficult problems such as the stability in neural network control systems and the requirement of the model structure in the adaptive control scheme. We choose the RBFN since its architecture is simple and mathematically tractable. Fig. 2 shows the structure of the adaptive tracking controller. The proposed adaptive tracking controller consists of two parts: a nonlinear function approximator and an auxiliary controller. In the nonlinear function approximator, the RBFN represents the nonlinear robot dynamics expressed in terms of the filtered tracking errors. The adaptation laws for updating the weights of the RBFN and the centers and widths of Gaussian functions are derived to guarantee the stability of control system. Next we have the auxiliary controller to guarantee the stability and robustness under the existence of nonlinearities and external disturbances.

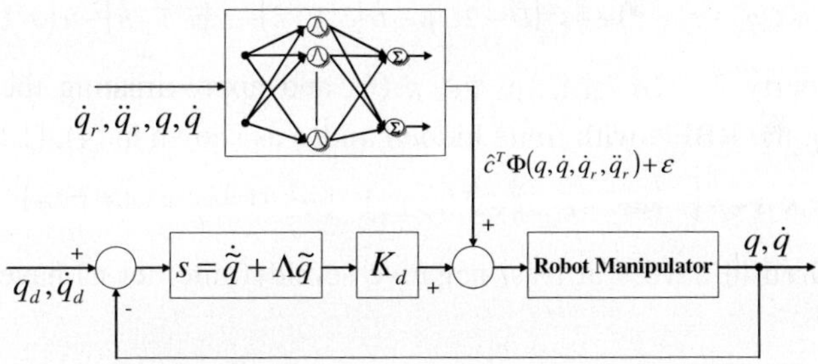

Fig. 2. Structure of the proposed controller

4.2 Stability and Robustness

If the reference trajectory $q_d \in R^n$ is given, the tracking error $\tilde{q}$ is defined as

$$\tilde{q} = q_d - q \tag{6}$$

The filtered tracking errors s and the control input τ are also defined as

$$s = \dot{\tilde{q}} + \Lambda\tilde{q} = \dot{q}_r - \dot{q} \tag{7}$$

$$\tau = \hat{c}^T \Phi + K_d s \tag{8}$$

where $x = [q, \dot{q}, \dot{q}_r, \ddot{q}_r]$, $\dot{q}_r = \dot{q}_d + \Lambda \tilde{q}$, $\Lambda = \Lambda^T > 0$, and K_d is diagonal and positive definite.

According to the Lyapunov stability analysis, the system is stable if the Lyapunov function is positive definite and its derivative is negative semidefinite. Therefore, to guarantee the stability of the total control system, a positive-definite Lyapunov function candidate function is selected as follows:

$$V = \tfrac{1}{2} s^T D s + \tfrac{1}{2} tr\left(\tilde{c}^T \Gamma_1^{-1} \tilde{c}\right) + \tfrac{1}{2} tr\left(\tilde{u}^T \Gamma_2^{-1} \tilde{u}\right) + \tfrac{1}{2} tr\left(\tilde{\sigma}^T \Gamma_3^{-1} \tilde{\sigma}\right) \tag{9}$$

where $\tilde{c} = c^* - \hat{c}$ is an error between the optimal weight c^* and estimated weight $\hat{c}$ of RFFN in (8). $\tilde{u} = u^* - \hat{u}$ and $\tilde{\sigma} = \sigma^* - \hat{\sigma}$ are a center error and a standard deviation error in (3), respectively. Γ_1, Γ_2, and Γ_3 are diagonal, symmetric, and positive-definite matrices, and $tr(\bullet)$ denotes trace.

Differentiating (9) with respect to time, we get

$$\dot{V} = s^T D \dot{s} + \tfrac{1}{2} s^T \dot{D} s - tr\left(\tilde{c}^T \Gamma_1^{-1} \dot{\hat{c}}\right) - tr\left(\tilde{u}^T \Gamma_2^{-1} \dot{\hat{u}}\right) - tr\left(\tilde{\sigma}^T \Gamma_3^{-1} \dot{\hat{\sigma}}\right) \tag{10}$$

If we differentiate s with respect to time, the robot dynamics (1) may be written in terms of the filtered tracking errors as follows:

$$D\dot{s} = D\ddot{q}_r + C\dot{q}_r + \tau_d - \tau - Cs \tag{11}$$

Substituting (11) into (10), we have

$$\dot{V} = \tfrac{1}{2} s^T \left(D\ddot{q}_r + C\dot{q}_r + \tau_d - \tau\right) + \tfrac{1}{2} s^T \left(\dot{D} - 2C\right) s - tr\left(\tilde{c}^T \Gamma_1^{-1} \dot{\hat{c}}\right) - tr\left(\tilde{u}^T \Gamma_2^{-1} \dot{\hat{u}}\right) - tr\left(\tilde{\sigma}^T \Gamma_3^{-1} \dot{\hat{\sigma}}\right) \tag{12}$$

Using the property 2, $s^T(D - 2C)s = 0 \quad \forall s \in R^n$ (6) and approximating the filtered robot dynamics using the RBFN with finite hidden nodes as shown in (4), (12) becomes

$$\dot{V} = -s^T K_d s + tr\left\{\tilde{c}^T \left(\Phi s^T - \Gamma_1^{-1} \dot{\hat{c}}\right)\right\} + s^T (\tau_d + \varepsilon) - \tfrac{1}{2} tr\left(\tilde{u}^T \Gamma_2^{-1} \dot{\hat{u}}\right) - \tfrac{1}{2} tr\left(\tilde{\sigma}^T \Gamma_3^{-1} \dot{\hat{\sigma}}\right) \tag{13}$$

Since it is desired to have $\dot{V}$ at least negative semidefinite, let us have the following adaptation laws:

$$\dot{\hat{c}} = \Gamma_1 \Phi s^T \tag{14}$$

$$\dot{\hat{u}} = -\alpha \Gamma_2 \|s\| \hat{u} \tag{15}$$

$$\dot{\hat{\sigma}} = -\beta \Gamma_3 \|s\| \hat{\sigma} \tag{16}$$

Then, (13) becomes

$$\dot{V} = -s^T K_d s + s^T (\tau_d + \varepsilon) + \alpha \|s\| tr\left\{\tilde{u}^T (u^* - \hat{u})\right\} + \beta \|s\| tr\left\{\tilde{\sigma}^T (\sigma^* - \hat{\sigma})\right\} \tag{17}$$

From the property of the Frobenius norm [11],

$$tr\left\{\tilde{u}^T (u^* - \hat{u})\right\} \le \langle \tilde{u}, u^* \rangle_F - \|\tilde{u}\|_F^2 \le \|\tilde{u}\|_F \|u^*\|_F - \|\tilde{u}\|_F^2 \tag{18}$$

$$tr\{\tilde{\sigma}^T(\sigma^* - \hat{\sigma})\} \le \langle \tilde{\sigma}, \sigma^* \rangle_F - \|\tilde{\sigma}\|_F^2 \le \|\tilde{\sigma}\|_F \|\sigma^*\|_F - \|\tilde{\sigma}\|_F^2 \tag{19}$$

Substituting (18) and (19) into (17), we get

$$\dot{V} \le -K_{d\min}\|s\|^2 + \|\tau_d + \varepsilon\|\|s\| + \alpha\|s\|\|\tilde{u}\|_F(u_{\max} - \|\tilde{u}\|_F) + \beta\|s\|\|\tilde{\sigma}\|_F(\sigma_{\max} - \|\tilde{\sigma}\|_F) \tag{20}$$

where $u_{\max}$ and $\sigma_{\max}$ are the maximum values of the Frobenius norm of the center and standard deviation vector in the RBFN.

The approximation error term ε is limited by the upper bound ε_N as shown in (5). If gain K_d is selected to satisfy the following inequality:

$$\|s\| \ge \frac{\|\tau_d + \varepsilon\| + \frac{1}{4}u_{\max} + \frac{1}{4}\sigma_{\max}}{K_{d\min}} \tag{21}$$

then, we have

$$\dot{V} \le 0 \tag{22}$$

From (9) and (22), the control system is stable based on the Lyapunov stability.

5 Experimental Results

To prove the valid performance of the proposed control method, we adopt the SCARA type robot manipulator as our test-bed. Fig. 3 shows the configuration of the robot system. The command from user is received at the server computer and the computer generates reference for robot.

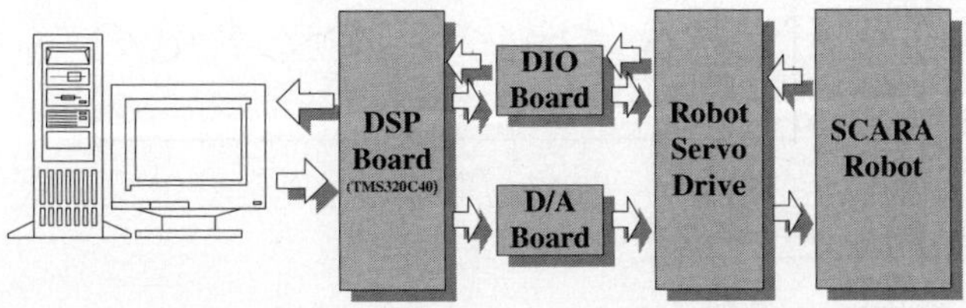

Fig. 3. Configuration of the robot system

We perform experiments for two different control strategies to compare with the proposed controller. The first control scheme is a PID type conventional control method, with the gains such as $K_{P1} = 24.23$, $K_{I1} = 7.27$, $K_{D1} = 0.60$, $K_{P2} = 39.9$, $K_{I2} = 8.34$ and $K_{D2} = 1.78$ for the joint 1 and 2. And the second control scheme is multilayer neural networks with backpropagation algorithms for updating the connection weights of neural networks. The second control scheme has the same architecture as the proposed controller but the updating law is different from the proposed controller. To implement the proposed controller, we choose the control parameters shown in Table 1. And the number of node of the RBFN is set to be 10. Sampling time is 5[msec].

Table 1. Pramegers of the Proposed Controller

Control Parameters	Values
λ	$[56\ 60]^T$
K_d	$[4\ 3]^T$
Γ	$diag(0.018\ 0.018)$
a, β	0.15, 0.005

And to compare the performance with the different control strategies, we considered three control conditions. The first condition is that the reference trajectories for the manipulator are given as $q_{d1} = 0.4\cos(3.76t)[rad]$ and $q_{d2} = 0.4\{\sin(3.76t)+1\}[rad]$ for the joints 1 and 2, respectively. Fig. 4 shows the tracking errors of the PID controller, multilayer neural networks with backpropagation and the proposed controller. In Fig. 4, the tracking errors of multilayer neural network controller are decreasing because the connection weights are updating on-line. But the decreasing rate of the errors are very slow than the proposed controller.

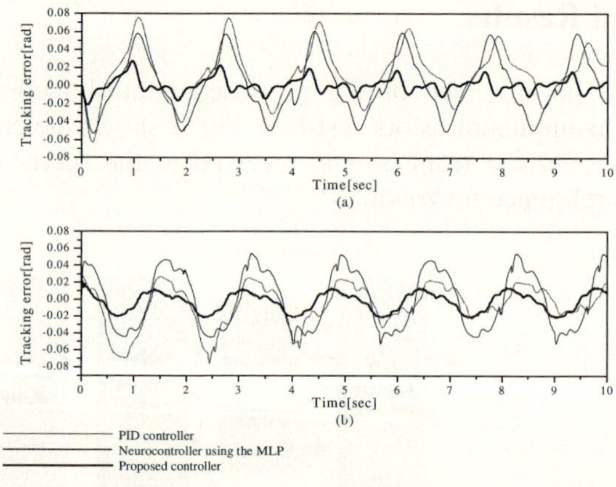

Fig. 4. Tracking errors under reference inputs with ω=3.76 rad/sec. (a) Joint 1. (b) Joint 2.

The second condition is that the new reference input trajectories are given $q_{d1} = 0.4\cos(5.65t)[rad]$ and $q_{d2} = 0.4\{\sin(5.65t)+1\}[rad]$ for the joint 1 and 2 in order to consider the environment change. Fig. 5 shows the tracking errors of three controllers mentioned above. We can find that the performance of the PID controller is deteriorated. However, the performance of the multilayer neural network controller and the proposed controller are still good due to the learning process. The tracking errors of multilayer neural network controller slowly approach the similar magnitude of the tracking errors of proposed method.

The final condition is the existence of $4\,kg$ load under the first condition. Fig. 6 shows the tracking errors of three different controllers. In Fig. 6, the joint 2 errors of

PID controller is affected by the load and the tracking errors of controllers with learning laws are reduced during the learning process.

Experimental results mentioned above show that the proposed controller is more adaptable to the environment changes and is more robust than the conventional PID controller and multilayer neural network controller with backpropagation algorithms.

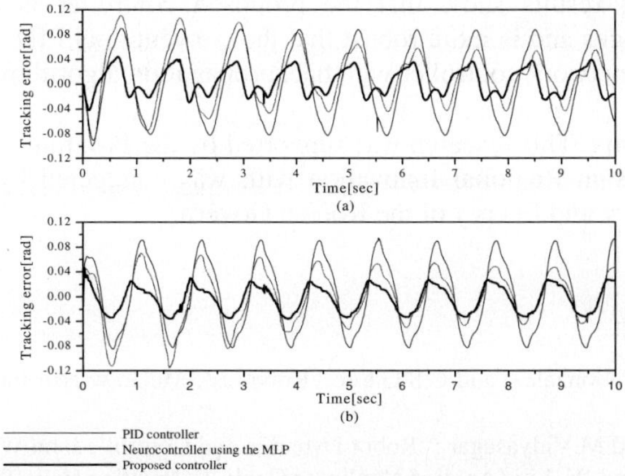

Fig. 5. Tracking errors under reference inputs with ω=5.65 rad/sec. (a) Joint 1. (b) Joint 2.

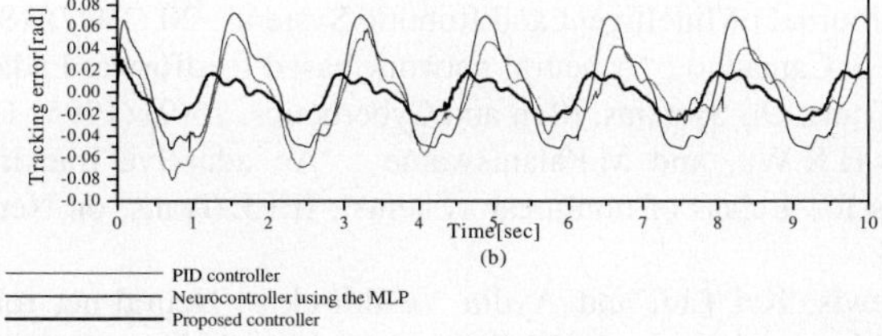

Fig. 6. Tracking errors under disturbance. (a) Joint 1. (b) Joint 2.

6 Conclusion

This paper presents an adaptive tracking control method for robots based on the RBFN. The proposed controller has a parallel structure that consists of PD controller with a fixed gain and the RBFN. And the parameters (weight, center of basis function, and standard deviation) of the RBFN are adjusted with on-line method. The learning law is constructed on the basis of Lyapunov stability theory.

This paper shows that the tracking errors are bounded uniformly and ultimately under the existence of the disturbances and modeling error, mathematically. The SCARA type robot manipulator is employed as a test-bed to apply the proposed controller. And we compared the proposed controller with two different controllers. The experimental results show that the proposed controller is adaptable to the environment changes and is more robust that the conventional PID controller and the multilayer neural network controller with backpropagation algorithms.

Acknowledgements. This research was supported by the Program for the Training of Graduate Students in Regional Innovation with was conducted by the Ministry of Commerce Industry and Energy of the Korean Govern.

References

1. R. K.S.Fu, R.C.Gonzalez, and C.S.G.Lee : Robotics', McGraw-Hill International Editions (1987)
2. M.W.Spong and M.Vidyasagar : 'Robot Dynamic and Control', John Wiley & Sons(1989)
3. J.-J.E.Slotine and W. Li : 'Applied Nonlinear Control', Prentice Hall(1991)
4. J.-S.R.Jang, C.-T.Sun, and E.Mizutani : 'Neuro-Fuzzy and Soft Computing', Prentice Hall(1997)
5. K.S.Narendra and K.Parthasarathy : 'Identification and control of dynamical systems using neural networks', IEEE Trans. On Neural Networks. 1(1) (1990) 4-27
6. M.A.Abido and Y.Abdel-Magid : 'On-line identification of synchronous machines using radial basis function neural networks', IEEE Trans. on Power Systems. 12(4) (1997) 1500-1506
7. A.S.Morris and S.Khemaissia : 'A neural network based adaptive robot controller', Journal of Intelligent and Robotic Systems. 15 (1996) 3-10
8. D.Y.Meddah and A.Benallegue : 'A stable neuro-adaptive controller for rigid robot manipulators', Journal of Intelligent and Robotic Systems. 20 (1997) 181-193
9. R.Carelli and E.F.Camacho : 'A neural network based feedforward adaptive controller for robots', IEEE Trans. On Systems, Man and Cybernetics. 25(9) (1995) 1281-1288
10. 10.M.Zhihong, H.R.Wu, and M.Palaniswame : 'An adaptive tracking controller using neural networks for a class of nonlinear systems', IEEE Trans. on Neural Networks. 9(5) (1998) 947-955
11. 11.Frank L. Lewis, Kai Liu, and Aydin Yesildirek : 'Neural-net robot controller with guaranteed tracking performance', IEEE Trans. on Neural Networks. 6(3) (1995) 703-715
12. 12.Sridhar Seshagiri and Hassan K. Khalil : 'Output feedback control of nonlinear systems using rbf neural networks', IEEE Trans. on Neural Networks. 11(1) (2000) 69-79
13. 13.Robert M. Sanner and Jean-Jacques E. Slotine : 'Gaussian networks for direct adaptive control', IEEE Trans. on Neural Networks. 3(6) (1992) 837-863
14. 14.S. Jagannathan, F. L. Lewis and O. Pastravanu : 'Model reference adaptive control nonlinear dynamical systems using multilayer neural networks', in Proc. of IEEE Int. Conf. on Neural Networks. 7 (1994) 4766-4771
15. 15.H. D. Patirio and Derong Liu : 'Neural network-based model reference adaptive control system', IEEE Trans. on Systems, Man and Cybernetics. 30(1) (2000) 198-204

Saving Energy Consumption of Multi-robots Using Higher-Order Mobile Agents

Munehiro Takimoto[1], Mayu Mizuno[1], Masato Kurio[2], and Yasushi Kambayashi[3]

[1] Department of Information Sciences, Tokyo University of Science, Japan
[2] NEC Software Ltd., Japan
[3] Department of Computer and Information Engineering, Nippon Institute of Technology, Japan

Abstract. This paper presents a framework for controlling intelligent robots connected by communication networks. This framework provides novel methods to control coordinated systems using higher-order mobile agents. Higher-order mobile agents are hierarchically structured agents that can contain other mobile agents. The combination of the higher-order mobile agent and mobile multi-robots open a new horizon of efficient use of mobile robot resources. Instead of physical movement of multi-robots, mobile software agents can migrate from one robot to another so that they can find the most suitably equipped and/or the most suitably located robots to perform their task. Thus the framework presented here provides efficient use of robot resources. In this paper, we focus on the energy saving. We have demonstrated the efficiency by numerical experiments.

Keywords: Mobile agent, Dynamic software composition, Intelligent robot control.

1 Introduction

In the last decade, robot systems have made rapid progress in not only their behaviors but also in the way they are controlled. In particular, control systems based on multi-agent methodologies have enabled a controlled robot to learn to adapt to the circumstances around it through its own interactions. Furthermore, multi-agent systems introduced modularity, reconfigurability and extensibility to control systems, which had been traditionally monolithic. It has made easier the development of control systems on distributed environments such as multi-robot systems.

On the other hand, excessive interactions among agents in the multi-agent system may cause problems in the multi-robot environment. Consider a multi-robot system where each robot is controlled by an agent, and interactions among robots are achieved through a communication network such as a wireless LAN. Since the circumstance around the robot changes as the robots move, the condition of each connection among the various robots also changes. In such an environment,

N.T. Nguyen et al. (Eds.): KES-AMSTA 2007, LNAI 4496, pp. 549–558, 2007.

once some of the connections in the network are disabled, the system may not be able to maintain consistency among the states in the robots. Additionally, such a problem has a tendency to increase, as the number of interactions increases.

In order to lessen the problems of excessive communication, mobile agent methodologies have been developed for distributed environments. In a mobile agent system, each agent can actively migrate from one site to another site. Since a mobile agent can bring the necessary functionalities with it and perform its tasks autonomously, it can reduce the necessity for interaction with other sites. In the minimal case, a mobile agent requires that the connection is established only when it performs migration [1]. This property is useful for controlling robots that have to work in a remote site with unreliable communication or intermittent communication. The concept of a mobile agent also creates the possibility that new functions and knowledge can be introduced to the entire multi-agent system from a host or controller outside the system via a single accessible member of the multi-robot system.

The model of our system is cooperative work by a pool of heterogeneous multi-robots [2]. The property of inter-robot movement of the mobile agent contributes to the flexible and efficient use of the robot resources. A mobile agent can migrate to the robot that is most conveniently located to a given task, e.g. closest robot to a physical object such as soccer ball. Since agent migration is much easier than robot motion, it contributes to saving power-consumption. Here, notice that any agents on a robot can be killed as soon as they finish their tasks. If each agent has a policy choosing idle robots rather than busy ones in addition to these migration strategies, it would result in more efficient use of robot resources and energy saving of multi-robots.

We have proposed our model in the previous paper [3]. In this paper, we demonstrate the feasibility of our model by the results of numerical experiments.

The structure of the balance of this paper is as follows. In the second section we describe the background. The third section describes the higher-order mobile agents with dynamic extension. Dynamic extension is the key feature that supports the ability to add new functionalities to intelligent robots in action. The fourth section shows an example of intelligent robot systems in which robots search for an object cooperatively. In the fifth section, we demonstrate how the higher-order mobile agents with dynamic extension can contribute to efficient use of robot resources, and energy saving through numerical experiments. Finally, we conclude in the sixth section.

2 Background

The traditional structure for the construction of intelligent robots is to make large, often monolithic, artificial intelligence software systems. The ALVINN autonomous driving system is one of the most successful such developments [4]. Putting intelligence into robots is, however, not an easy task. An intelligent robot that is able to work in the real world needs a large-scale knowledge base. The ALVINN system employs neural networks to acquire the knowledge semi-automatically [5]. One of the limitations of neural networks is that it is assumed that the system is used in the same environment as that in which it was trained. When the intelligent robot is expected to work in an unknown space or an extremely dynamic environment, it is not realistic to assume that the neural network is appropriately trained or can acquire additional knowledge with sufficient rapidity. Indeed, many intelligent robots lack a mechanism to adapt to a previously unknown environment.

On the other hand, multi-agent robotic systems are recently becoming popular in RoboCup or MIROSOT [6]. In traditional multi-agent systems, robots communicate with each other to achieve cooperative behaviors. The ALLIANCE architecture, developed in Oak Ridge National Laboratory, showed that cooperative intelligent systems could be achieved [7]. The architecture is, however, mainly designed to support self-adaptability. The robots in the system are expected to behave without external interference, and they show some intelligent behaviors. The observed intelligence, however, is limited due to the simple mechanism called *motivation*. Robots' behaviors are regulated by only two rules *robot impatience* and *robot acquiescence*. These rules are initially defined and do not evolve. In contrast, the goal of our system is to introduce intelligence and knowledge into the robots after they start to work [2]. Therefore, our system does not have any learning mechanism or knowledge acquiring mechanism. All the necessary knowledge is sent as mobile agents from other robots or the host computer.

The work most closely related to ours is the distributed Port-Based Adaptable Agent Architecture developed at Carnegie Mellon University [8]. The Port-Based Agents (PBAs) are mobile software modules that have input ports and output ports. All PBAs have the map of the port addresses so that they can move other robots and combine themselves with other PBAs to compose larger modules. The usefulness of PBA architecture is demonstrated by the Millibot project also at Carnegie Mellon University [9]. In a robot mapping application, PBA is used to control the mapping robots, and when the working robot has a hardware failure, the PBA on the robot detects it and moves to an idle robot.

Software composition is clearly possible using port-based modules. The dynamic extension capability of our mobile agent control system, however, is another strategy for the composition of larger software.

The PBA is derived from the concept of port-based objects, designed for real-time control applications [10]. Therefore it may have advantages as a robot control mechanism. The framework we present in this paper is an exploration of the applications of mobile agents and software compositions through mobility and extensibility. Constructing robot control software by mobile agents and its dynamic extension is not only novel but also flexible due to dynamic inheritance. It may be superior for extensibility of working software.

3 Higher-Order Mobile Agent with Dynamic Extension

The mobile agent system we have used to control robots is based on a mobile agent system, called MobileSpaces, developed by I. Satoh [11]. MobileSpaces is

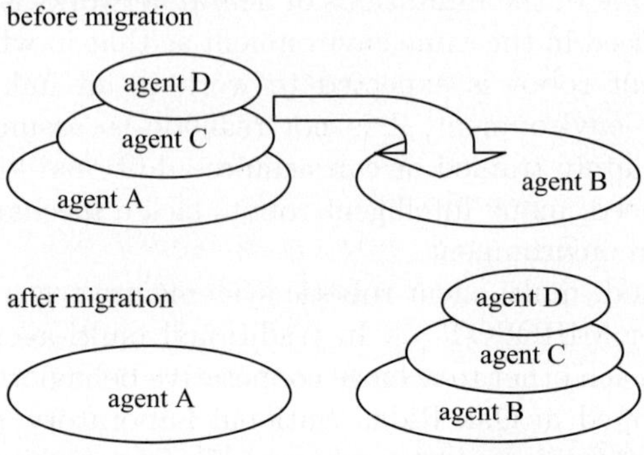

Fig. 1. When agent C migrates from agent A to agent B, the contained agent D also migrates from A to B

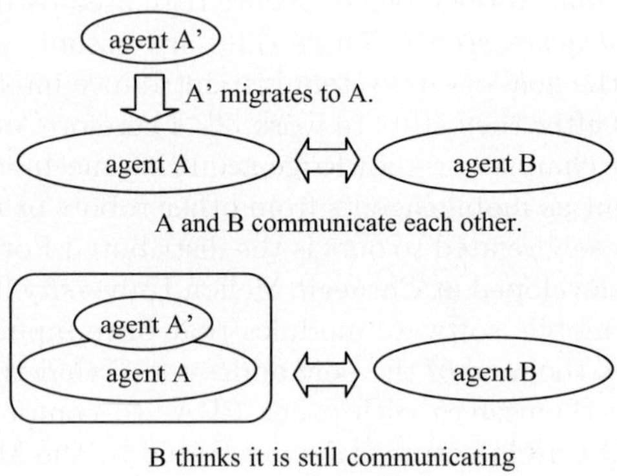

Fig. 2. Dynamic extension by migration of agent with new features

also based on the mobile ambients computational model proposed by L. Cardelli and A. D. Gordon [12]. MobileSpaces provide the basic framework for mobile agents. It is built on the Java virtual machine, and agents are supposed to be programmed in Java language.

Mobile agents can migrate from place to place. When they migrate, not only the program code of the agent but also the state of the agent can be transferred to the destination. The higher-order mobile agents are mobile agents whose destination can be other mobile agents as well as places in traditional agent systems.

Two unique features are worth mentioning for our robot control system. 1) Each mobile agent can contain one or more mobile agents (hierarchical construction), and 2) Each mobile agent can migrate to any other mobile agent (inter-agent migration). Thus migration to another agent results in a nesting structure of agents. Agents in the other agent are still autonomous agents that can behave independently. Fig. 1 illustrates the situation that agent C migrates from agent A to agent B, and agent D that is contained in agent C also migrate from agent A to agent B.

In order to enhance the intelligent robot control system in action, we have added the dynamic extension feature to customize functions of robots while they are running [2]. Suppose an agent A is working somewhere and we want to extend its capability. One way is to replace that agent with a new agent B. On the other hand in our system, we only need to send an agent A' with the new feature to the agent A. While the agent A' is included in A, the agent A behaves with the extended feature. If the agent A' leaves the agent A, the agent A behaves with the original feature. All the other agents do not have to be aware of the change of the agent A. In Fig. 2, after an agent A' migrates to an agent A, the other agent B still communicates to the agent A without knowing the migration of A'. The agents A and A' behave just as a single agent for the agent B.

In order to extend the agent A, the agent A' only needs to have the new feature to be added. If the agents A and A' have methods with the same signature, the method in agent A' overrides the method with the same signature in the agent A. The signature is an extended type of a function. When it is said that two functions have the same signature, it means that they have the same name, the same number of the same type of formal parameters, and the same return type. The agent migration achieves the same semantics as dynamic inheritance [13].

4 Robot Controller Agents for Target Searcher

In this section, we demonstrate that the model of higher-order mobile agents with dynamic extension is suitable for the composition of software to control an intelligent robot. For this purpose, we show an example of cooperative target search that can be employed for practical applications.

Intelligent multi-robots are expected to work in distributed environments where communication is relatively unstable and therefore where fully remote control is hard to achieve. Also we cannot expect that we know everything in the environment beforehand. Therefore intelligent robot control software needs to have the following features: 1) It should be autonomous to some extent. 2) It should be extensible to accommodate the working environment. 3) It should be replaceable while in action. Our higher-order mobile agent with dynamic extension satisfies all these functional requirements.

Our control software consists of autonomous mobile agents. Once an agent migrates to a remote site, it requires minimal communication to the original site. Mobile agents are higher-order so that the user can construct a larger agent by hierarchical composition of smaller agents. Finally, if we find that the constructed software is not satisfactory, we can replace the unsuitable component (an agent) with new component (another agent) by using agent migrations.

We employed the ER1 Personal Robot Platform Kit by Evolution Robotics Inc. (information available at http://www.evolution.com/) as the platform for our prototype system. Each robot has two servo-motors with tires. The power is supplied by rechargeable battery. It has a servo-motor controller board that

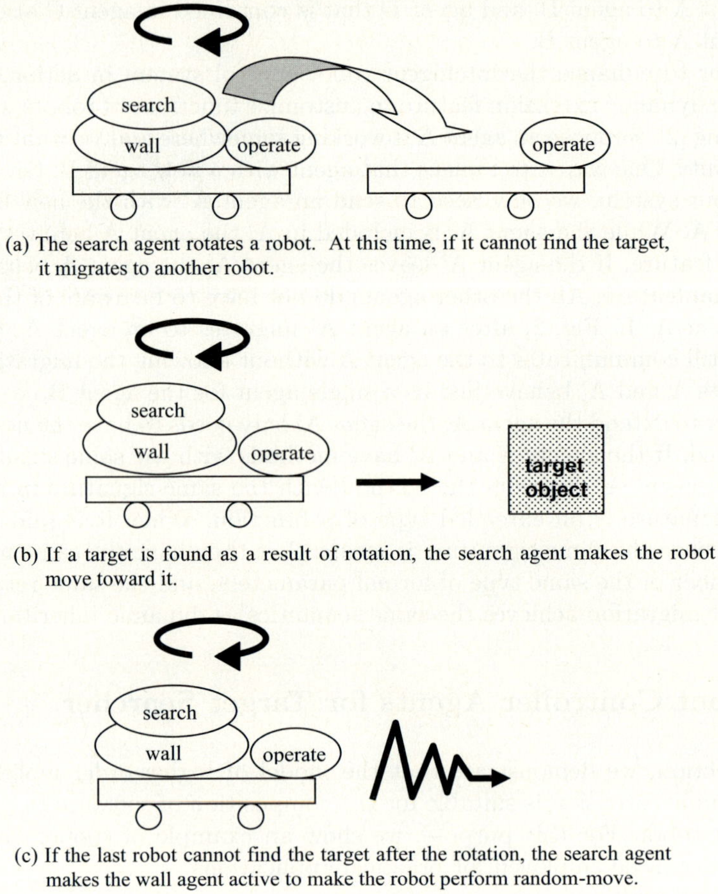

Fig. 3. The search agent's behaviors

accepts RS-232 serial data from a host computer. Each robot holds one notebook computer as its host computer. Our control agents migrate to these host computers by wireless LAN.

Let us consider how to program robots to find a target. For such a task, the most efficient solution would be to make all robots search for the target simultaneously. If the targets were comparatively fewer than the robots, however, most robots would move around in vain, consuming power without finding anything. This problem can be more serious in our model where any robots can be shared by any agents, because the robots to which an agent is going to migrate may be already occupied with another task. Thus, this searching strategy could result in increasing the total costs for the aggregate of the multi-robots.

On the other hand, our strategy of using higher-order mobile agents achieves power-preserving search without sacrificing efficiency, i.e. only one robot is in action. The core of our idea is finding the nearest robot to the target by using agent migration. Initially, an agent is dispatched from the host machine to a nearby robot in the multi-robots system. Then, the agent hops around the robots one by one and checks the robot's vision in order to locate the target until it reaches the robot that is closest to the target. Until this stage, robots in the multi-robot system do not move; only the mobile agent migrates around so that robots can save power.

Once the agent finds the target, it migrates to the closest robot and makes the robot move toward the target. This strategy is obviously not as efficient as that of having all robots search for a target simultaneously. But the migration time is negligible comparing to robot motion and the efficiency of power-preservation offsets the slight search inefficiency. In our multi-robot case, the robots' vision does not cover 360 degrees; so that a robot has to rotate to check its circumstance. Rotating a robot at its current position may capture the target and another robot closer to the target. Then the agent migrates to the more conveniently located robot. Take note that the rotation requires much less movement of the wheels than exploration, and it contributes the power saving.

Details of our searching algorithm are the followings: 1) an agent chooses an arbitrary robot to which it migrates, and performs the migration, 2) as the agent arrives on the robot, it makes that robot rotate as shown by Fig. 3 (a), 3) if the target is found, the agent makes the robot move to that direction as shown in Fig. 3 (b); otherwise, goes back to the step 1, and 4) at this stage, if all robots have been tried without finding the target, the agent makes the last robot do random-walk until it can find a target as shown by Fig. 3 (c).

This algorithm can be implemented as a migrating *search* agent. The *search* agent migrates on the other mobile agent, *wall*. The sole task of *wall* agent is to avoid collisions. If the *search* agent can control the *wall* agent, all that the *search* agent has to do against the *wall* agent is to make it hibernate until reaching step 4, and then it wakens the *wall* agent. We achieve this implementation by designing the *search* agent as an included agent of *wall* agent. The *search* agent migrates to the *wall* agent, so that these two agents are composed into one agent. After that, the combined agent traverses each robot.

In principle, we can make the *wall* agents stationary; the *search* agent does not have to take the *wall* agent to other robots, since all the robots have the wall agents. But the current system is a prototype and the cooperation with other tasks is yet to be perfected; thus the *search* agents migrate with the *wall* agent.

5 Numerical Experiments

In order to demonstrate the effectiveness of our system, we have conducted numerical experiments on the example of target search that is mentioned in the previous section. We have compared our approach based on mobile agents with the exhaustive search approach that makes all robots move around in terms

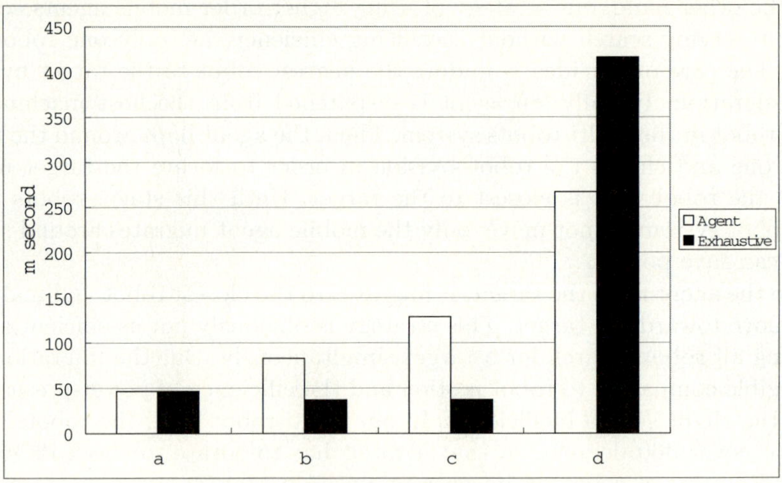

Fig. 4. Efficiency of Searching a target

of efficiency and energy consumption. In order to simulate real situations, we applied both approaches to four patterns of initial robot-locations in $3.5m \times 5.0m$ rectangle area, which is as large as possible in our laboratory.

The patterns are as follows:

Pattern a : a case where the target is near the robot to which the *search* agent migrates first,
Pattern b : a case where the target is near the robot to which the agent migrates second,
Pattern c : a case where the target is near the robot to which the agent migrates last, and
Pattern d : a case where the target is far from any robots.

Fig.4 shows the results of the experiments on efficiency and Fig.5 shows the results of the experiments on energy consumption. Fig.4 shows how long the multi-robots take to find the target in milliseconds for each setting. The results for the case of pattern **a** through **c** are the same as what we expected. Since the target is located close to one of the robots, it is obvious that the exhaustive search approach finds faster than any other approaches including ours. One rotation of robots cannot miss the target. The reason why our approach exceeds the exhaustive one in pattern **d** is that the searching algorithm we employed was so naive that the multi-robots did not coordinate their tasks. If they coordinate their search pattern, the exhaustive approach should show much efficiency.

Fig.5 shows the comparison of the times of rotating wheels. It is reasonable to assume that energy consumption of servo-motors is linear to the wheel rotation times. It is clearly observed that, in all the settings, our approach consumes extremely less energy than the exhaustive approach. Considering that our approach allows idle robots to perform other tasks, we can expect that our approach can

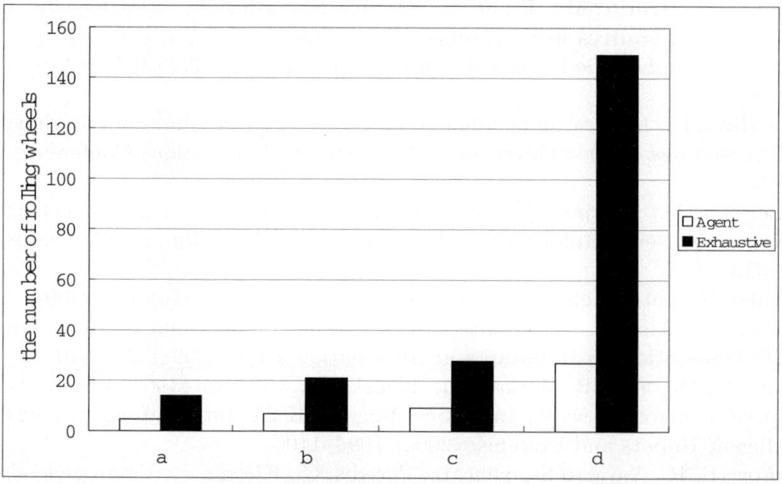

Fig. 5. Energy Consumption

be more efficient than the experiments show, and reduce the total consumption of the energy for the aggregate of the multi-robots.

6 Conclusions

We have presented a novel framework for controlling intelligent multi-robots. The framework helps users to construct intelligent robot control software by migration of mobile agents. Since the migrating agents are higher-order, the control software can be hierarchically assembled while they are running. Dynamically extending control software by the migration of mobile agents enables us to make base control software relatively simple, and to add functionalities one by one as we know the working environment. Thus we do not have to make the intelligent robot smart from the beginning or make the robot learn by itself. We can send intelligence later as new agents

We have implemented a team of cooperative search robots to show the effectiveness of our framework, and demonstrated that our framework contributes to energy saving of multi-robots. Even though our example is a toy program, the volume of saved energy is significant.

References

1. W.J. Binder, G.H., Villazon, A.: Portable resource control in the j-seal2 mobile agent system. In: Proceedings of International Conference on Autonomous Agents. (2001) 222–223
2. Kambayashi, Y., Takimoto, M.: Higher-order mobile agents for controlling intelligent robots. International Journal of Intelligent Information Technologies **1**(2) (2005) 28–42

3. Mizuno, M., Kurio, M., Takimoto, M., Kambayashi, Y.: Flexible and efficient use of robot resources using higher-order mobile agents. In: Proceedings of Joint Conference on Knowledge-Based Software Engineering 2006 (JCKBSE'06). (2006) 253–262
4. Pomerleau, D.: Defense and civilian applications of the alvinn robot driving system. In: Proceedings of 1994 Government Microcircuit Applications Conference. (1994) 358–362
5. Pomerleau, D.: Alvinn: An autonomous land vehicle in a neural network. In: Advances in Neural Information Processing System 1, Morgan Kaufmann (1989) 305–313
6. Murphy, R.: Introduction to AI robotics. MIT Press, Cambridge (2000)
7. Parker, L.: Aliance: An architecture for fault tolerant multirobot cooperation. IEEE Transaction on Robotics and Automation **14**(2) (1998) 220–240
8. Pham, T., Dixon, K.R., Jackson, J., Khosla, P.: Software systems facilitating self-adaptive control software. In: Proceedings of IEEE International Conference on Intelligent Robots and Systems. (2000) 1094–1100
9. Grabowski, R., Navarro-Serment, L., Paredis, C., Khosla, P.: Heterogeneous teams of modular robots for mapping and exploration. Autonomous Robots **8**(3) (2000) 293–308
10. Stewart, D., Khosla, P.: The chimera methodology: Designing dynamically reconfigurable and reusable real-time software using port-based objects. International Journal of Software Engineering and Knowledge Engineering **6**(2) (1996) 249–277
11. Satoh, I.: Mobilespaces: A framework for building adaptive distributed applications using a hierarchical mobile agent system. In: Proceedings of IEEE International Conference on Distributed Computing Systems. (2000) 161–168
12. Cardelli, L., Gordon, A.: Mobile ambients. In: Foundations of Software Science and Computational Structures, Lecture Notes in Computer Science 1378, Springer-Verlag (1998) 140–155
13. Abadi, M., Cardelli, L.: A Theory of Objects. Springer-Verlag (1996)

A Plan-Based Control Architecture for Intelligent Robotic Agents

Incheol Kim, Hangcheol Shin, and Jaehyuk Choi

Department of Computer Science, Kyonggi University
Suwon-si, Kyonggi-do, 442-760, South Korea
{kic, zest133,01choi}@kyonggi.ac.kr

Abstract. We present a plan-based control architecture for intelligent robotic agents. Many modern robot architectures adopt a hybrid approach using control-theoretic mechanisms as well as plan-based mechanisms, because of their complementary strengths and weaknesses. Our three-layered control architecture implements dependability through plan-based management of a set of distributed independent behavioral components. With navigational tasks in a complex maze, we investigate robustness of our control architecture.

1 Introduction

In this paper, we introduce a plan-based control architecture for intelligent robotic agents. The robot platform used for our work is AIBO ERS-7M3 [10]. It has four legs for walking and various sensors including a CMOS color image sensor and IR distance sensors. We assume the AIBO robots have to accomplish navigational tasks in a complex maze. The 4-legged robot's navigation in a maze is a difficult task which involves a lot of behavioral components such as vision-based localization, path planning, motion generation adapted to obstacles, and so on. Many modern control architectures for intelligent robots adopt a hybrid approach using control-theoretic mechanisms as well as plan-based mechanisms, because of their complementary strengths and weaknesses. Typically, control-theoretic approaches are applied to implement independent behavioral components, while plan-based mechanisms configure these components and reason about possible interactions between them [1, 2, 12]. Our three-layered control architecture implements dependability through plan-based management of a set of distributed independent behavioral components. With navigational tasks in a complex maze, we investigate robustness of our control architecture.

2 Control Architecture

Fig. 1 shows our overall control architecture for intelligent robotic agents. It consists of three different layers: the *Reactive* layer, the *Sequencing* layer, and the *Deliberative* layer.

N.T. Nguyen et al. (Eds.): KES-AMSTA 2007, LNAI 4496, pp. 559–567, 2007.

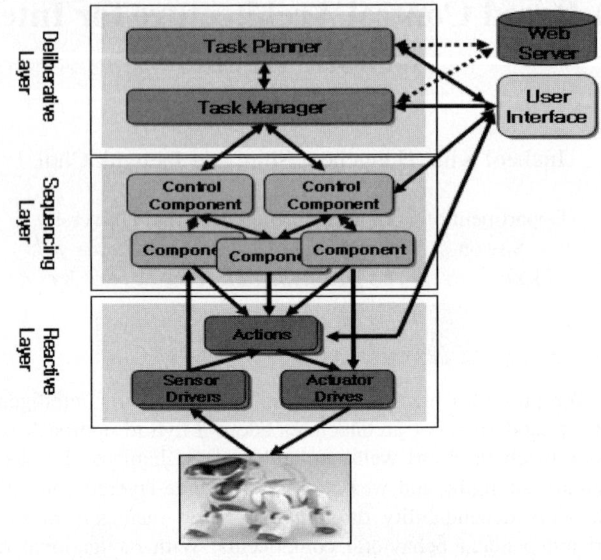

Fig. 1. The layered architecture

The reactive layer, the bottom one, includes primitive actions using some specific sensor and actuator drivers. These actions can be sequenced into a specific behavioral component on the sequencing layer. The sequencing layer, hence, consists of such long-term behavioral components and their control components. Each control component plays a role of the representative for a specific group of functional components and interacts with the task manager on the deliberative layer. The deliberative layer, the top one, configures behavioral components and reasons about their interactions for performing a given task successfully.

3 Reactive and Sequencing Layer

Each control component (CC) and other behavioral components in the sequencing layer can be set up along with its own unique IP address and port number. And then, the task manager in the deliberative layer controls their executions concurrently through an XML message-based communication mechanism. In order to assist this type of communication between the task manager and each CC, there are two control component managers (CCMs): Java and C++ control component managers. There are four different types of messages exchanged between the task manager and each CC via a CCM: *Command* messages, *Notification* messages, *Status* messages, and *Query* messages. In order to control the execution of each CC, the task manager can send one of *Load, Start, Suspend, Resume, Stop, and Unload* command messages. In response of the received command message, each CC sends a proper notification message back to the task manager. Each CC can have one of three different states (*Running, Suspended, Stopped*), and the task manager can ask the current status of a CC by sending a status message. Sometimes, each CC can also send a query message to the task manager to obtain the necessary information from other CCs or task planner in the deliberative layer. Fig. 2 shows the control architecture between the task manager and individual CCs. Fig. 3 shows the component description containing the name, the class path, and the location of individual CCs.

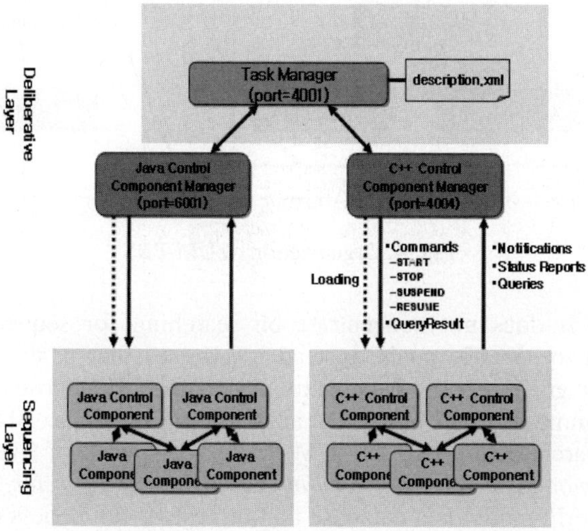

Fig. 2. Message-based control of components

```
<Description>
    <ControlComponentManagerList>
        <ControlComponentManager>
            <Name>TestControlComponentManager</Name>
            <IPAddress>localhost</IPAddress>
            <Port>5001</Port>
        </ControlComponentManager>
        <ControlComponentManager>
            <Name>CppCCManager</Name>
            <IPAddress>127.0.0.1</IPAddress>
            <Port>4004</Port>
        </ControlComponentManager>
    </ControlComponentManagerList>

    <ControlComponentList>
        <ControlComponent>
            <Name>TestCC</Name>
            <Classpath>/home/CDK/CC/bin/TestCC</Classpath>
            <Location>CppCCManager</Location>
        </ControlComponent>
        <ControlComponent>
            <Name>TestCC2</Name>
            <Classpath>/home/CDK/CC/bin/TestCC2</Classpath>
            <Location>CppCCManager</Location>
        </ControlComponent>
    </ControlComponentList>
</Description>
```

Fig. 3. A component description

4 Deliberative Layer

The deliberative layer consists of the task manager and the task planner. The kernel of the task manager is a kind of plan executive. Fig. 4 shows the plan executive, UM-PRS [7]. The UM-PRS is an implementation of PRS (*Procedural Reasoning System*)

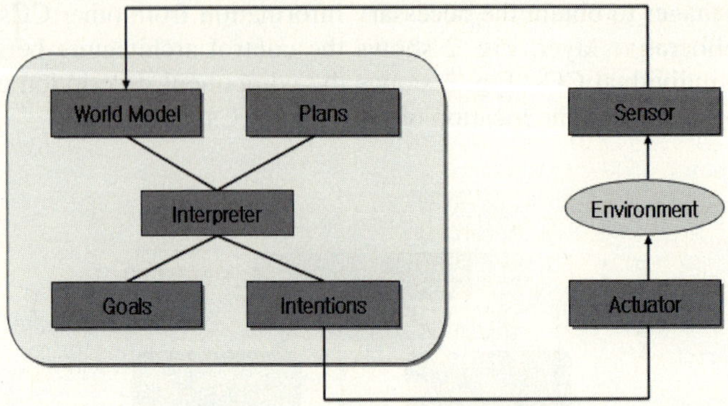

Fig. 4. Organization of UM-PRS

written in C++. It does not concentrate on searching for sequences of primitive actions that lead to specific goals. Instead it assumes that it already has plans for achieving various goals in various contexts. It dynamically chooses among plans in a changing environment. The UM-PRS is also known as a BDI (*Belief-Desire-Intention*) agent architecture [9]. The UM-PRS is composed of five primary components: a *world model*, a *goal set*, a *plan library*, an *interpreter*, and an *intention structure*. The world model is a database that represents the beliefs of the agent. The goal set is a set of goals that the agent has to achieve. The plan library is a collection of plans that the agent can use to achieve its goals. The interpreter is the agent's "brain" that reasons about what the agent should do and when and how to do it. The intention structure is an internal model of the agent's current goals and keeps track of the commitment to, and progress on, accomplishment of those goals. The UM-PRS architecture integrates traditional goal-directed reasoning and reactive behavior. Because most traditional deliberative planning systems formulate an entire course of action before starting execution of a plan, these systems are brittle to the extent that features of the world or consequences of actions might be uncertain. In contrast, the BDI architecture continuously tests its decisions against its changing knowledge about the world, and can redirect the choices of actions dynamically while remaining purposeful to the extent of the unexpected changes to the environment.

The task plans maintained by the plan executive, UM-PRS, are divided into two different categories: primitive plans and complex plans. A primitive task plan includes only single action resulting in a function call to start or stop the execution of a specific behavioral module. With respect to each behavioral module, there exists one corresponding primitive plan to initiate or terminate its execution. On the other hand, a complex plan is a sort of complex procedure combining multiple action steps and subgoals with some control constructors. To achieve these subgoals in a complex plan, other applicable (complex or primitive) plans are supposed to be selected to execute. The task plans in the library, hence, organizes a hierarchical structure. Fig. 6 shows the hierarchy of task plans for controlling an AIBO robot to navigate in a complex maze.

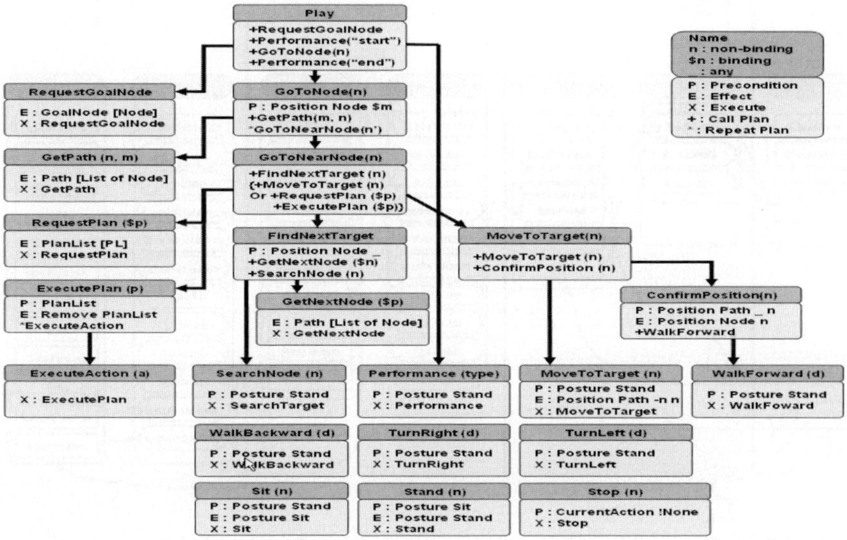

Fig. 5. The hierarchy of task plans

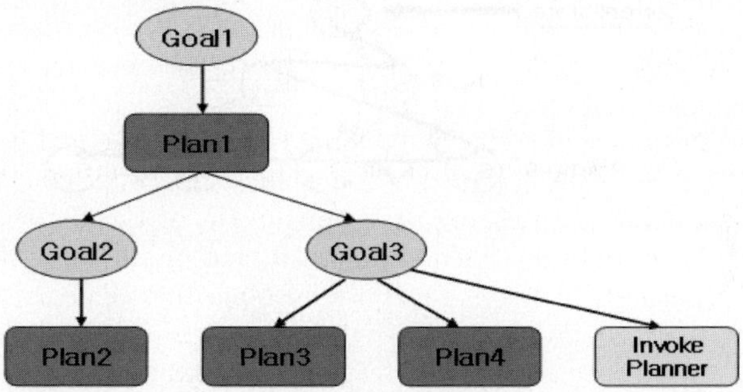

Fig. 6. Invoking the on-demand planner via the *InvokePlanner* plan

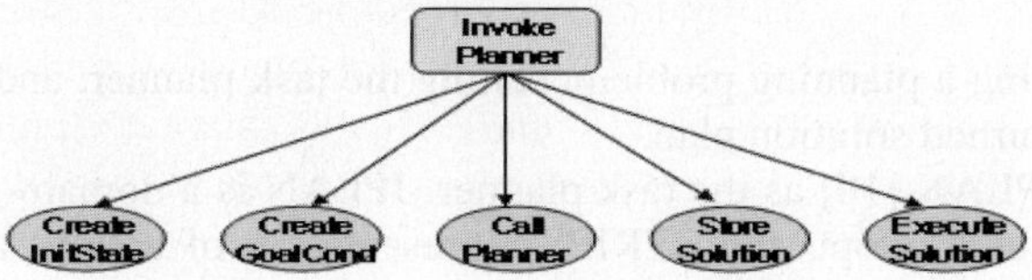

Fig. 7. Action steps of the *InvokePlanner* plan

In order to invoke the task planner on demand, the plan executive, UM-PRS, has a special plan which can be executed after all other applicable plans fail. Fig. 6 represents where the *InvokePlanner* plan is placed and when it is executed. Fig. 7 shows the detail procedure of the *InvokePlanner* plan. This procedure includes several steps

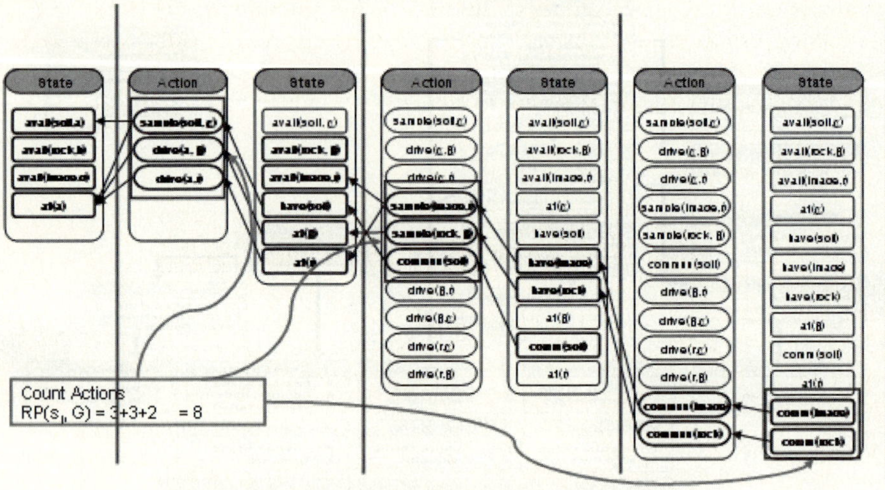

Fig. 8. Computation of a heuristic estimate using the relaxed planning graph

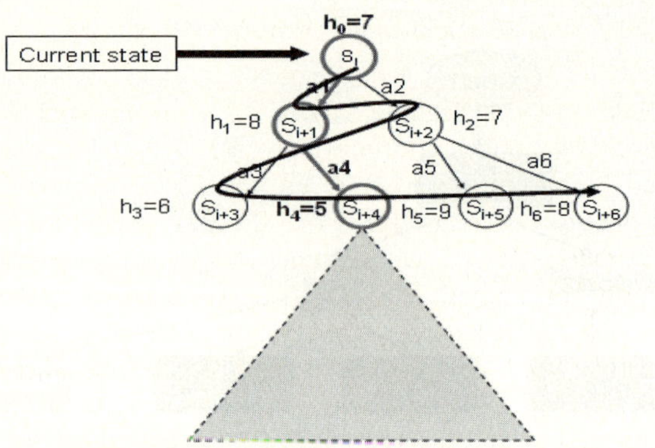

Fig. 9. Enforced Hill-Climbing Plus (EHC+) strategy

such as formulating a planning problem, calling the task planner, and then storing and executing the returned solution plan.

We use our JPLAN [10] as the task planner. JPLAN is a domain-independent heuristic planner, which adopts the STRIPS representation of actions and searches forward in the space of the states. The main heuristic principle was originally developed by Blai Bonet and Hector Geffner for the HSP system [3]: to obtain a heuristic estimate, relax the task at hand into a simpler task by ignoring the delete lists of all operators. While HSP employs a technique that gives a rough estimate for the solution length of the simpler task, JPLAN extracts an explicit solution to the simpler task, by using a GRAPHPLAN-style algorithm like FF [5] and AltAlt systems. The number of actions in the relaxed solutions is used as a goal distance estimate. Fig. 8 shows the computation of a heuristic estimate using the relaxed planning graph. These estimates control a novel local search strategy, Enforced Hill-Climbing Plus (EHC+): this is a hill-climbing procedure that, at each intermediate state, uses breadth first search to find a strictly better, possibly indirect, successor. EHC+ is an improved version of Enforced Hill-Climbing (EHC) strategy originally developed by Joerg Hoffman [5].

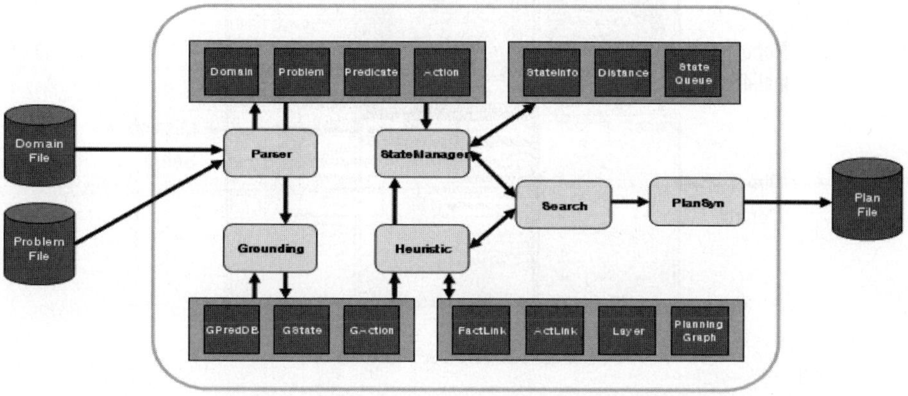

Fig. 10. Architecture of the JPLAN planner

Fig. 9 depicts EHC+ search strategy. Assume the current state is s_i with the heuristic estimate h_0=7 and its two children states are s_{i+1} and s_{i+2} with their heuristic estimates h_1=8 and h_2=7, respectively. In this case, any children don't have better heuristic estimate than the current state. EHC strategy guides the breadth first search to find the first successor state s_{i+3} with better heuristic estimate and then the intermediate actions a_1 and a_3 on the path from s_i to s_{i+3} are included in the partial plan. This process continues repeatedly starting from the new current state, s_{i+3}, until finding a goal state. On the other hand, EHC+ extends the breadth first search to the last sibling state s_{i+6} and finds the best successor state s_{i+4} with the heuristic estimate h_4=5. The intermediate actions a_1 and a_4 are included in the partial plan. Starting from the best state s_{i+4}, the search process continues until meeting a goal state. Compared with EHC, EHC+ generally explores smaller search space and finds shorter plans at the expense of additional local search. Fig. 10 shows the architecture of the JPLAN planner. It consists of several major components such as Parser, Grounding, StateManager, Heuristic, Search, and PlanSyn.

5 Mixed-Initiative Control

Fig. 11 shows a screenshot of the graphical user interface of the robot roaming in a maze. With this tool, the human operator can set up the mission for the robot as well as make the initialization of the robot such as setting an IP address and connecting remotely through wireless network. Once the robot starts, it continually displays autonomous behaviors under the plan-based control. Meanwhile, the operator can keep track of the internal decision-making process as well as monitor some of

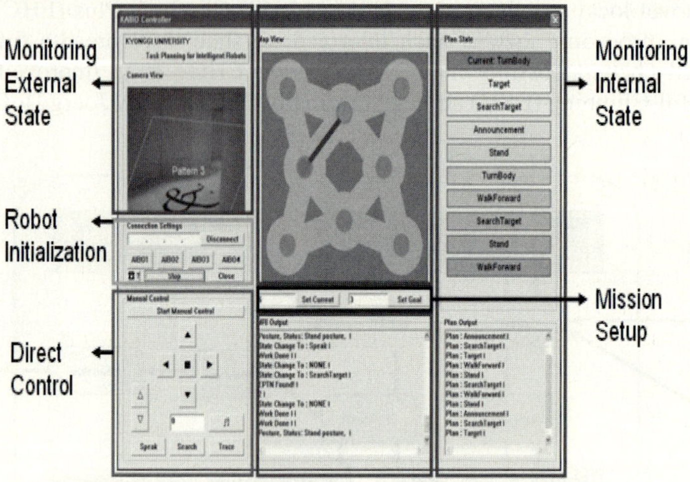

Fig. 11. The user interface

perception information such as the graphic patterns detected from color images and the localized position.

Through the user interface, the operator can also ensure the planned path for the robot to follow. Whenever the operator detects dangerous situations caused from unexpected or erroneous behaviors of the robot, he can stop immediately and then control manually the robot. Once the operator terminates his manual control, the robot resumes its plan-based control. Hence, if the robot finds the given task has not been accomplished yet, it retries to achieve the task starting from the changed situation by selecting a proper plan or generating a new plan.

6 Experiments

In order to investigate robustness of our control architecture, we conducted several experiments regarding navigational tasks in a maze like shown in Fig. 12. The AIBO robot was controlled to do many different navigational tasks, and then we measured its success rate. Fig. 13 shows the experimental results. We can notice that while the task complexity was increasing, the success rate was slightly decreased. However,

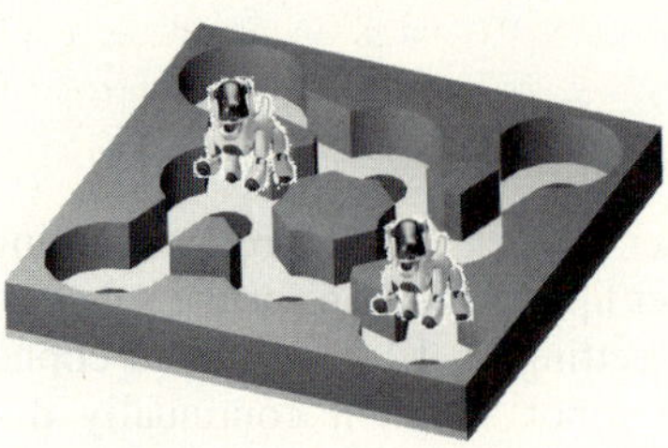

Fig. 12. Task environment

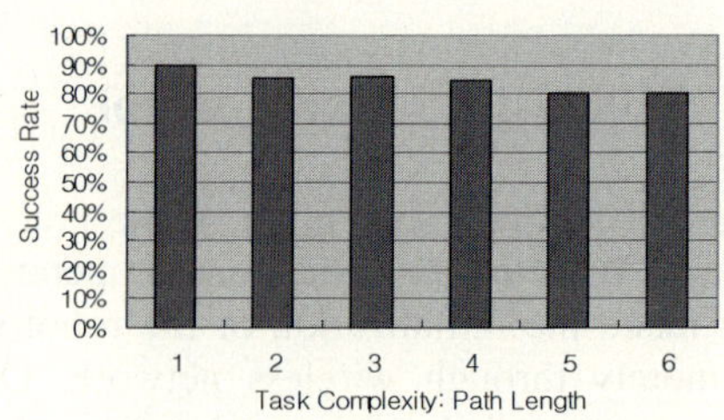

Fig. 13. Experimental results

overall performance of the robot was quite high. Even though the sensors of the robot were noisy and the motions were unstable, the average success rate was nearly about 83%. This result implies that our control architecture would be robust and flexible enough to accomplish difficult real-world tasks.

7 Conclusions

We presented a plan-based control architecture for intelligent robotic agents. Our control architecture implements dependability through plan-based management of a set of distributed independent behavioral components. With navigational tasks within a complex maze, we confirmed robustness of our control architecture.

References

1. Alami, R., Chatila, R., Fleury, S., Ghallab, M., Ingrand, F.: An Architecture for Autonomy, International Journal of Robotics Research, Special Issue on Integrated Architectures for Robot Control and Programming, (1998).
2. Beetz, M.: Plan-Based Control of Robotic Agents: Improving the Capabilities of Autonomous Robots, Springer (2002).
3. Bonet, B., Geffner, H.: Planning as Heuristic Search, Journal of Artificial Intelligence, 129(1-2), (2001).
4. Gallien G., Ingrand F.: Planning and Plan Execution for Autonomous Robots, Proceedings of RFIA-06, (2006).
5. Hoffmann, J., Nebel, B.: The FF Planning System: Fast Plan Generation Through Heuristic Search, Journal of Artificial Intelligene Research, (2003) 253-302.
6. Huber, M.: JAM: A BDI-Theoretic Mobile Agent Architecture, Proceedings of the 3rd International Conference on Autonomous Agents, (1999), 236-243.
7. Lee, J., Huber, M., Durfee, E., Kenny, P.: UM-PRS: An Implementation of the Procedural Reasoning System for Multirobot Applications, Proceedings of the Conference on Intelligent Robotics in Field, Factory, Service, and Space, (1994), 842-849.
8. Verma, V. et al.: Survey of Command Execution Systems for NASA Spacecraft and Robots, Proceedings of the Workshop on Plan Execution, International Conference on Automated Planning and Scheduling, (2005), 92-99.
9. Rao, A., Georgeff M.: BDI-agents: from Theory to Practice, Proceedings of the 1st International Conference on Multiagent Systems, (1995) 313-319.
10. Shin, H., Kim, I.: Design and Implementation of a Heuristic State-Space Planner, Proceedings of the Annual Conference of the Korea Information Science Society, (2006), 112-115.
11. SONY: AIBO Software Development Environment (SDE), http://openr.aibo.com, (2006)
12. Simmons, R., Apfelbaum, D.: A Task Description Language for Robot Control, Proceedings of the Conference on Intelligent Robots and Systems (IROS), (1998).

Remarks on Behaviours Programming of the Interactive Therapeutic Robot KOALA Based on Fuzzy Logic Techniques

Krzysztof Arent and Marek Wnuk

The Institute of Computer Engineering, Control and Robotics,
Wrocław University of Technology,
ul. Z. Janiszewskiego 11/17,
50-372 Wroclaw, Poland
{krzysztof.arent,marek.wnuk}@pwr.wroc.pl
http://www.iiar.pwr.wroc.pl

Abstract. The paper discuses various aspects of behaviour programming of an interactive social robot KOALA using fuzzy logic methodology. The considered robot (a physical, reactive agent) is intended to be applied in treatment of autistic children as a therapist's assistant. Analysing a number of selected observations of classical and modern therapeutic processes we end up with conclusions motivating application of fuzzy logic techniques to define interactive robot behaviours. Next, fuzzy logic model of KOALA is developed and described. Finally various programming levels in the proposed framework are presented and its advantages are highlighted.

Keywords: robotic agent, robot programming, robot control, social requirements, fuzzy logic, sensor system, actuator, hardware, behaviour.

1 Introduction

Application of socially interactive robots in the environment of autistic children has been a subject of profound and comprehensive research for the last half decade [1,2,3]. By assumption a robot plays the role of a therapist's assistant which persistently and gently encourages the autistic child to come into various social interactions with it. Fundamental significance of robot interactivity follows directly from autism attributes and therapeutic conditions.

Essentially there are two centres persistently conducting research in this field: under auspices of the AURORA project [4] and at Université de Sherbrooke in Canada [1]. In the the AURORA project the fundamental role is played by therapeutic issues while the research workshop is mainly based on artificial intelligence, robotics and assistive technology [2]. At the Université de Sherbrooke the main research accents are shifted towards engineering aspects of robotics design. The class of potential robot users is slightly broader than the group of autistic children.

The robotic literature reports smaller projects addressed to autistic children, e.g. SHARIC [5], KOALA [6], as well as related works, e.g. *Paro* [7].

N.T. Nguyen et al. (Eds.): KES-AMSTA 2007, LNAI 4496, pp. 568–577, 2007.

In the meantime at least ten robots have been constructed and used in various experiments. They were ranging from mobile platforms – *Labo 1* [8], *Peeke* [9] – through mobile spherical balls – *Roball 1*, *Roball 2* [10,1] – to robots with faces, hands and legs like *Robota 1*, *Robota 2* [11], *Tito* [1]. It appeared that in the context of robot's objectives a significant number of factors are of crucial importance. In particular, these are shape, appearance, durability and finally, capability to react on a wide spectrum of stimuli (including force, velocity, pressure, voice, image) with various reactions (motion, voice, light). It has to be emphasised, that robot behaviour should be adequate to a child skills, actual needs and interests.

The desired behaviour flexibility heavily depends on the software architecture, especially on accepted software paradigms. This issue has been widely discussed only in case of *Labo 1*, *Roball 1* and *Robota 2* in [8,10,11] respectively. Analysis and comparison of the implemented solutions show that they are a result of a compromise between possibilities offered by the producer of a robotic platform or hardware, individual properties of the robot, and the particular research task for the robot. It is hard to distinguish elements of the software design (besides the behaviour based paradigm [12]) which are potentially portable between robotic platforms, and which guarantee higher robot behaviour flexibility.

In this context an interesting idea has been drawn in [6]. The idea is to employ fuzzy logic methodology to specify robot behaviours. The fact that a number of microcontrollers feature machine level instructions dedicated to fuzzy logic and the fact of coincidence of the fuzzy logic rules and the natural language of therapists, are prerequisites that motivate this approach. Therefore in this paper the ideas of [6] will be further developed and confronted with concepts of [8,10,11].

In the literature addressed to application of socially interactive robots in autism therapy (e.g. [3]) there are no explicit references to agent systems. Nevertheless, taking into account robot properties and therapy process scenarios, we deal, in fact, with agent systems, in which such properties like reactivity, interactivity (cooperation, negotiation) are essential [13].

The paper is organised as follows. In section 2 comparison analysis of software architecture of *Labo 1*, and *Roball 1* is done and some background information is provided. In section 3 fuzzy logic model of the robot KOALA is presented and discussed. Next, in section 4 various aspects of KOALA programming are discussed. In section 5 selected topics related to KOALA are presented. Finally, in section 6, some conclusions are provided.

2 Autism and Socially Interactive Robots

Autism is a developmental disorder characterised by three categories of symptoms: impairments in social interaction, in social communication and in imagination [2]. People with autism have difficulties to come into normal interaction with other people, to understand the meaning of gestures, facial expressions, to go beyond a limited range of imaginative activities. It is observed that they reveal stereotypical behaviours, inclination to fixation to stable environments, significantly narrowed spectrum of activities and interests.

There are four well established therapeutic forms of autism therapy: behavioural therapy, TEACH, holding, music therapy [14,15]. The main goals of behavioural therapy are development of deficit skills, reduction of undesired behaviours, preservation of therapy results. TEACH basically lies in developmental and educational lessons and requires full cooperation of pedagogues and parents. The essence of holding is close physical (often forced) contact of a mother and her child. Music therapy has the aim to develop communication and verbal skills.

Due to unknown etiology of autism and its complexity the mentioned therapeutic techniques are not satisfactory in general. It has been realised in nineties [2] that socially interactive robots could contribute to existing therapeutic methods (mainly behavioural therapy and TEACH) as well as originate new therapeutic techniques. A robot with simple shape, nice appearance and limited set of behaviours seems to be more predictable for an autistic child than another human and therefore easier to be accepted. The idea is to put a suitably designed socially interactive robot, capable to engage the child in interactions which demonstrate important aspects of human – human interaction, into the autistic child environment and then slowly increase the robot's behaviour repertoire and the unpredictability of its actions and reactions [2]. The role of the robot will be to guide a child towards more complex forms of interaction. In other words, the basic task of the robot-agent is to initiate and sustain an interaction with a child-agent (with a therapist-agent in the background).

The main discriminants of a socially interactive robot are abilities to recognise human agents and other robots in its environment; to engage in social interactions (in particular to communicate) with other agents. This means that very important role in the robot construction is played by the software layer, which is responsible for robot social behaviour. As it was mentioned in Section 1, this issue has been more widely discussed in the context of the robots *Labo 1* and *Roball 1*.

Labo 1 has been used in experiments conducted under AURORA project. Specification of its behaviour was inspired by elements of psychological theories [2]: mindreading and imitation. This robot is a four wheeled mobile platform produced by Applied AI Systems in Canada. It has eight infrared sensors and contact detection sensors. Additionally, one heat sensor and a speech box have been mounted atop of it. The main controller unit is based on MC68332 microcontroller. The robot offers an opportunity of expansion of the standard control software by means of C programming language. In this way the robot has been equipped with behaviours which are suitable from both the therapeutic and the experimental viewpoints. The control architecture has been developed using an experimentally driven design methodology [12]. Basically, more complex robot behaviours are based on existing ones that results in hierarchical architecture [8]. It allows a modular structure of the software where one module plays the role of the supervisor. This main module is responsible for coordination and selection of the behaviours and for the time interval assigned to each of them. The architecture runs on two levels. The lower level services sensors, updates the timer and cares about obstacles avoidance. It works in cyclic mode. The upper level essentially deals with selection of behaviours. The therapist is offered four behaviour programs: a demonstration sequence, a simple forward and backward movement, heat following, simple movement behaviour triggered by presence of a heat source.

Roball 1 is a spherical mobile robot capable of autonomous navigation in the environment and interacting in various ways with the child [10]. During the design process the fact of potential application of the robot in treatment of an autistic child has been taken into account. The robot is robust, safe, affordable and able to create interesting interactions with the child. The sensor system consists of three analog accelerometers, three tilt sensors, infrared sensors and a microphone. The robot can react with motion, voice, sound signals and light. Originally, the controller unit was based on 68HC11 microcontroller board, but later MicroChip PIC18F microcontroller was used. Control algorithms were programmed in C. The software architecture is based on the behaviour based paradigm [12]. The basic idea is that behaviour producing modules control the actuators according to sensory data and the state of the robot and dynamically change the selection of the behaviours over time [10]. The selection of behaviours is done according to environmental states, the goals of the robot and reasoning done about the world. The architecture has three abstraction levels: *Behaviour*, *Recommendation*, *Motivation*. The behaviour producing modules at the *Behaviour* level run in parallel and their resulting commands are combined to generate the control actions. The other modules of architecture are responsible for changing selection of behaviours and for reconfiguring them according to the robot goals and signals from the world. Behaviour coordination takes place at *Recommendation* level. The modules *External Situation*, *Needs* and *Cognition* formulate, in parallel, three behaviour recommendations from different monitoring conditions. The last architecture level consist of *Motives* module. It is used to monitor the child goals and coordinate the proper working of the other modules.

A slightly different approach to behaviour defining has been developed in case of KOALA robot [6]. The essential feature of KOALA is the ease of programming its behaviour, reached by employing a fuzzy controller. The therapist can formulate the rules, defining the robot behaviour, in a language similar to a natural one. Thus, one can define the required agent features without the thorough knowledge of its construction.

The therapeutic multi-agent system consists of a child, KOALA robot and a therapist. The robot-agent and the child-agent constitute an interaction, initiated by robot, where child is influenced by the robot effectors and its reactions are recorded by robot sensors. The robot reactivity is tuned through the therapist-robot interaction, which is understood as robot behaviour programming based on observation of robot-child interaction. Thus, the interaction between child and therapist takes indirect form, what is often important in therapy.

3 Fuzzy Model of KOALA

A spherical, interactive robot KOALA [6] has been designed as an therapist's assistant in autistic children therapy. The project was based upon the suggestions of the specialists in the field of autism [16], as well as on the earlier SHARIC project [5]. The shape resembling a ball was chosen in order to provide maximum safety and ease of manipulation for the children, as well as extreme robustness of the robotic agent. One of the main priorities was low cost of the robot, which determines the wide availability of KOALA for therapeutic centres.

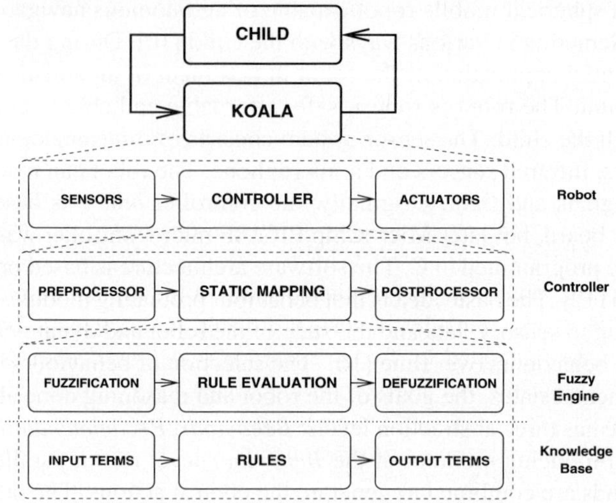

Fig. 1. KOALA control system

KOALA behaviour consists in producing light and voice as a response to touch (*proximity*, *pressure*), force (*acceleration*) and movement (*rotation*).

The hierarchy of KOALA control system is presented in fig. 1. The interaction between the child and KOALA is based on the stimuli–reaction scheme. The robot is equipped with several sensors providing the information on stimuli in child–robot interaction:

acceleration	`acc`	3D accelerometer MMA7260
rotation	`rot`	3 gyroscopes ENC03JA
pressure	`press`	pressure sensor MPXV7025
proximity	`touch`	proximity sensor MC33974

In every control step `n` the sensor data is acquired, filtered and preprocessed. For every variable `var` the preprocessor provides inertial response (with time constant defined by `epsilon` < 1):

```
ivar[n] = (1-epsilon) ivar[n-1] + epsilon var[n]
```

and incremental value:

```
dvar = var[n] - ivar[n].
```

As the result, we obtain 12 input variables for the static mapping defining the behaviour of KOALA:

original values	`acc`	`rot`	`press`	`touch`
inertial responses	`iacc`	`irot`	`ipress`	`itouch`
incremental values	`dacc`	`drot`	`dpress`	`dtouch`

The reactions of the robot are realised by simple actuators:

coloured light:	*hue*	`hue`	6 RGB LEDs
	saturation	`sat`	
	value	`val`	
sound:	*tone*	`ton`	2 loudspeakers
	volume	`vol`	

The postprocessor provides dynamic correction of the robot reactions (updating the current output variable `out` with the increment `dout`):

```
out += dout,
```

HSV → RGB conversion and LEDs and loudspeakers control, using 10 output variables of the static mapping:

current values	hue	sat	val	ton	vol
increments	dhue	dsat	dval	dton	dvol

KOALA behaviour is defined by a static mapping of the input variables into the output variables. The fuzzy controller is designed using the knowledge-based approach. The function can be described by fuzzy inference rules based upon the fuzzy input variables and resulting in fuzzy outputs on the basis of expert (therapist) knowledge. Both the fuzzification and defuzzification are provided by the embedded fuzzy controller employing the Mamdani fuzzy control concept [17]. The rules have the form of *if–then* statements using fuzzy terms defined by membership functions for each input and output variable.

Thus, the static mapping block consists of:

- *fuzzification* of the stimuli (trapezoidal membership functions),
- *inference* (MIMO rules, Zadeh logic operators, Mamdani max-min algorithm),
- *defuzzification* of the reactions (singletons, COG defuzzification method).

All the data required for the above mentioned processing forms the knowledge base used by the fuzzy inference engine. The therapist is envisioned to be able to define the inference rules based upon the fuzzy inputs representing stimuli and resulting in fuzzy outputs representing the reactions of the robotic agent.

The international standard IEC 1131, Part 7 [18] entitled "Fuzzy Control Programming" defines the syntax and explains the application rules of fuzzy control description language FCL. It guarantees portability of the fuzzy control software between different platforms.

Input variables are fuzzified with some predefined terms:

```
FUZZIFY acc
   TERM medium      := (  0.10,0.0) (  0.50,1.0) (  0.90,0.0);
   TERM large       := (  0.60,0.0) (  1.00,1.0) (  1.00,0.0);
   TERM small       := (  0.00,0.0) (  0.00,1.0) (  0.40,0.0);
END_FUZZIFY
...
FUZZIFY dtouch
   TERM n_large     := ( -1.00,0.0) ( -1.00,1.0) ( -0.60,0.0);
   TERM zero        := ( -0.40,0.0) (  0.00,1.0) (  0.40,0.0);
   TERM large       := (  0.60,0.0) (  1.00,1.0) (  1.00,0.0);
   TERM n_medium    := ( -0.90,0.0) ( -0.50,1.0) ( -0.10,0.0);
   TERM medium      := (  0.10,0.0) (  0.50,1.0) (  0.90,0.0);
END_FUZZIFY
```

and outputs defuzzified as well:

```
DEFUZZIFY vol
    TERM zero   :=  0.00;
    TERM medium :=  0.50;
    TERM large  :=  1.00;
    TERM big    :=  0.75;
    TERM small  :=  0.25;
    METHOD      : COG;
    DEFAULT     : NC;
    RANGE     :=(0.00 .. 1.00);
END_DEFUZZIFY
...
```

```
DEFUZZIFY dval
    TERM n_large  := -1.00;
    TERM zero     :=  0.00;
    TERM large    :=  1.00;
    TERM big      :=  0.75;
    TERM n_big    := -0.75;
    TERM n_medium := -0.50;
    TERM n_small  := -0.25;
    TERM small    :=  0.25;
    TERM medium   :=  0.50;
    METHOD        : COG;
    DEFAULT       : NC;
    RANGE            :=(-1.00 .. 1.00);
END_DEFUZZIFY
```

The inference rules, describing the robot behaviour, are defined by the therapist:

```
RULEBLOCK No1
    AND : MIN;
    OR : MAX;
    ACCU : MAX;
    ACT : MIN;
        RULE 1 : IF rot IS large AND drot IS n_medium
                 THEN (vol IS medium),(dton IS n_medium);
END_RULEBLOCK
```

The above FCL model of KOALA is easily converted into the data structures appropriate for the fuzzy engine implemented in KOALA controller.

4 Behaviour Programming

The MC9S12 family is particularly well suited for fuzzy control systems. Its core (CPU12 [19]) features several assembly language level instructions (MEM, REV, WAV), which facilitate implementation of the fuzzy engine tasks (fuzzification, rule evaluation and defuzzification).

Input terms (antecedents of the rules) have trapezoidal form defined by a four-byte structure (two points and two slopes). For the crisp input variable value MEM calculates the truth value for the term and puts it into fuzzy terms table.

The rule block of the fuzzy knowledge base is implemented as a list of rules, separated by the rule separator (0xFE) and terminated by the block separator (0xFF). Every rule consists of two sublists (antecedents and consequents) separated by the rule separator. Both the antecedents and consequents are represented by pointers to current values of the terms used in the rule.

REV instruction implements Mamdani max-min rule evaluation with Zadeh logic operators. The minimum of the truth value for all the rule antecedents is used to eventually update the truth value of all the rule consequents. The calculation is repeated for all the rule list, until the block separator is reached, therefore only one REV instruction is required.

Output terms (consequents) are represented as singletons with COG defuzzification method. The `WAV` (Weighted Average) instruction is used to calculate the numerator and the denominator sums for the current output variable:

$$\frac{\sum_i S_i F_i}{\sum_i F_i}, \tag{1}$$

the final division is performed by `EDIV` instruction.

The FCL behaviour definition is converted into binary form, required by the implementation of KOALA fuzzy engine, in two steps (translation to ANSI C form and compilation to S–records). The binary form (S–records) of the knowledge base is loaded into KOALA flash memory.

The intermediate form (C language) contains:

– trapezoidal membership descriptors for the input terms:

```
/* acc */
const t_term_info InVar0 [3] = {
        {  26, 230,   3,   3 },   /* medium       */
        { 153, 255,   3,   0 },   /* large        */
        {   0, 102,   0,   3 }    /* small        */
   };

...
/* dtouch */
const t_term_info InVar11 [5] = {
        {   0,  51,   0,   5 },   /* n_large      */
        {  76, 179,   5,   5 },   /* zero         */
        { 204, 255,   5,   0 },   /* large        */
        {  13, 115,   5,   5 },   /* n_medium     */
        { 140, 242,   5,   5 }    /* medium       */
   };
```

– singleton membership descriptors for the output terms:

```
/* vol */
const unsigned char OutVar0 [5] = {
        0,     /* zero        */
      128,     /* medium      */
      255,     /* large       */
      191,     /* big         */
       64      /* small       */
   };
```

```
/* dval */
const unsigned char OutVar9 [9] = {
        0,     /* n_large     */
      128,     /* zero        */
      255,     /* large       */
      223,     /* big         */
       32,     /* n_big       */
       64,     /* n_medium    */
       96,     /* n_small     */
      159,     /* small       */
      191      /* medium      */
   };

...
```

– fuzzy rules list for REV instruction:

```
const unsigned char Rules[] = {
            /* IF */
      0x0C, /* rot IS large */
      0x14, /* drot IS n_medium */
      0xFE, /* THEN */
      0x2D, /* vol IS medium */
      0x44, /* dton IS n_medium */
      0xFF  /* End Of Rules */
    };
```

5 Selected Topics

Defining of robot behaviours using conventions of the microcontroller assembler requires the user to formulate the logical expressions of the rules in disjunctive normal form. Although from the mathematical and technical view point this convention is very appropriate, in practice it may cause certain problems. Notice, that descriptions of child actions (rules antecedents) and robot reactions (rules consequences) will be *overwordy*, in particular it will be distributed over set of rules with repetitive conditions or conclusions. Effectiveness of the robot behaviour specification and user-friendliness postulate require the rules language to admit the word-formation and complex clauses.

If the logic operators AND, OR, NOT are realised by *min*, *max* and *1-* operations then laws of mutual distributiveness of operations and De Morgan laws, known from Boolean logic, are preserved. Thus any compound logic expression with AND, OR, NOT operators and brackets can be transformed to the disjunctive normal form using well known techniques. Notice that the above remark allows word-formation, i.e. assigning fuzzy logic expressions to variables and then using these variables in rules formation. The variables names should express child actions, robot states and robot reactions. The idea depicted above makes it possible to formulate the rules in the form which is close to a concise natural language and, on the other hand, is easy to implement in a therapist interface.

The prototype of the interface, currently under development, has been described in [5]. It consists of two main modules: *child's profile* and *programming*. Basically, the *child's profile module* fulfils three functions. It displays most important informations about the child. All the programs, which specify the robot behaviour, and which have been associated with the child are listed, so that the therapist can easily select one and load it into the memory of the robot. The membership functions associated with each linguistic variable can be scaled appropriately to actual predisposition of the child. The *programming module* enables the user to define the behaviour of the robot. Its conception has been based on ideas depicted above.

6 Conclusions

In this paper the idea of defining socially interactive behaviours by means of fuzzy logic methodology in the context of therapy of autistic children has been discussed. The approach does not originate in the behaviour based paradigm [12], which is present in software architecture of other constructions, but in some sense it refers to it. It offers high flexibility in defining robotic agents behaviours, conceived of as a set of stimulus–response rules. The approach, practically implemented in KOALA, is computationally cheap, easy implementable, and fits the idea of low price of robots, which are intended to be used in the centres for autistic children.

References

1. Michaud, F., Salter, T., Duquette, A., Laplante, J.F.: Perspectives on mobile robots used as tools for pediatric rehabilitation. Assistive Technologies (2005)
2. Dautenhahn, K., Werry, I.: Towards interactive robots in autism therapy. background, motivation and challenges. Pragmatics & Cognition **12**(1) (2004) 1 – 35
3. Fong, T., Nourbakhsh, I., Dautenhahn, K.: A survey of socially interactive robots: Concepts, design, and applications. Robotic and Autonomous Systems **42** (2003)
4. : The aurora project. `http://www.aurora-project.com` (2006)
5. Arent, K., Kabała, M., Wnuk, M.: Towards the therapeutic spherical robot: design, programming and control. In: 8th IFAC Symposium SYROCO. (2006)
6. Wnuk, M.: Koala – an interactive spherical robot assisting in therapy of autistic children. In: Advances in Robotics. Systems and Cooperation of Robots. WKŁ (2006) in Polish.
7. Shibata, T., Mitsui, T., Wada, K., Touda, A., Kumasaka, T., Tagami, K., Tanie, K.: Mental commit robot and its application to therapy of children. In: IEEE/ASME International Conference on Advanced Intelligent Mechatronics Proceedings. (2001) 1053 – 1058
8. Dautenhahn, K., Werry, I.: Applying mobile robot technology to the rehabilitation of autistic children. In: Proceedings of Symposium on Intelligent Robotic Systems. (1999)
9. Salter, T., Boekhorst, R.T., Dautenhahn, K.: Detecting and analysing children's play styles with autonomous mobile robots: A case study comparing observational data with sensor readings. The 8th Conference on Intelligent Autonomous Systems (IAS-8) (2004)
10. Michaud, F., Caron, S.: Roball – an autonomous toy rolling robot. In: Proceedings of the Workshop on Interactive Robot Entertainment. (2000)
11. Billard, A., Robins, B., Dautenhahn, K., Hadel, J.: Building robota, a mini–humanoid robot for the rehabilitation of children with autism. The RESNA Assistive Technology Journal (2006)
12. Arkin, R.C.: Behavior – Based Robotics. MIT Press, Cambridge, MA (1998)
13. Ferber, J.: Multi-Agent System: An Introduction to Distributed Artificial Intelligence. Harlow: Addison Wesley Longman (1999)
14. Lewartowska, J.B.: Autyzm dziecięcy. Zagadnienia diagnozy i terapii. Oficyna Wydawnicza „Impuls" (2000)
15. Dautenhahn, K., Werry, I.: Issues of robot-human interaction dynamics in the rehabilitation of children with autism. The Sixth International Conference on the Simulation of Adaptive Behavior (2000)
16. Kruk-Lasocka, J.: Autyzm czy nie autyzm. Problemy diagnostyki i terapii pedagogicznej małych dzieci. DSWE, Wrocław (1999)
17. Buckley, J.J., Eslami., E.: An introduction to fuzzy logic and fuzzy sets. Physica-Verlag (2002)
18. International Electrotechnical Commission: IEC 1131 - PROGRAMMABLE CONTROLLERS. Part 7 - Fuzzy Control Programming. (1997)
19. *CPU12RM/AD*, Rev. 1, Motorola Inc.: CPU 12 Reference Manual. (1996,1997)

Design of Admissible Schedules for AGV Systems with Constraints: A Logic-Algebraic Approach

Grzegorz Bocewicz[1], Robert Wójcik[2], and Zbigniew Banaszak[1]

[1] Deptartament of Computer Science and Management,
Technical University of Koszalin,
75-453 Koszalin, Poland
banaszak@tu.koszalin.pl
[2] Institute of Computer Engineering, Control and Robotics,
Wrocław University of Technology, 50-372 Wrocław, Poland
robert.wojcik@pwr.wroc.pl

Abstract. The subject matter of the study are the automated guided vehicle (AGV) operation synchronisation mechanisms in flexible manufacturing systems. Since each AGV can be treated as an autonomous object capable to undertake decisions regarding its routing, entering path sectors, etc., hence a class of transportation systems considered can be seen as a class of multi-agent ones. In many practical cases transport operations are repetitive. The processes examination has to guarantee the collision-free and deadlock free AGVs flow. In this paper the problem of determination of the rules coordinating access of the vehicles to the shared travel route intervals, ensuring the collision-free and deadlock-free execution of the repetitive processes was reduced to determination of the sufficient conditions of the form of a pair (initial state, a set of priority rules). In particular the problem of searching for a pair is defined in the form of the constraint satisfaction problem (CSP) and is solved with use of the logic programming techniques.

Keywords: scheduling, constraints logic programming, deadlock avoidance.

1 Introduction

The system class under consideration covers the transport subsystems of the flexible manufacturing systems (FMS). In subsystems of that type a set of automated guided vehicles (AGVs) move along assumed traveling routes. AGVs play the role of agents [6], [9], attempting to reach their goals while following rules being specific for a given FMS. So, the considered transportation systems are treated as multi-agent ones. Following their routes AGVs have to serve a given set of workstations, i.e. just-in-time loading and/or unloading workstations with curried batches. Therefore, each AGV can be seen as an autonomous object capable to undertake its decisions due to the collected knowledge and current state of semaphores. The objective of such multi-agent systems is just due time servicing of the workstations. So, the goal is to find a synchronization mechanism guaranteeing AGVs could realize their tasks while taking

N.T. Nguyen et al. (Eds.): KES-AMSTA 2007, LNAI 4496, pp. 578–587, 2007.

into account constraints following from the topology of a transportation network, local rules that determine AGVs access to the path sectors as well as an FMS user's efficiency criteria. In that context, a sought multi-agent system has to be optimal one in the sense of minimal number of AGVs employed, minimal length of transportation paths, minimal tardiness of workstations service, and so on.

The existing constraints connected with the available traveling route width (not allowing for vehicle passing by), the topology of traveling routes and itineraries of individual vehicles, lack of simultaneous access to the stations, etc. imply the necessity to investigate conditions leading to possible vehicle collisions and deadlocks [4]. This means that the problem of the given transport alternative solution admissibility check problem belongs to the NP-hard problems [8].

The existing approach to solving the problem base usually upon the simulation models, e.g. the Petri nets [5] or the algebraic models, e.g. upon the (max,+) algebra [7]. In this context, this work constitutes some continuation of the investigations conducted in [1], [7].

From the AGVs' dispatcher point of view the rules coordination vehicles access to the commonly shared path sectors as to guarantee just-in-time workstations service play a crucial role. In that context, the problem considered can be seen as searching for a set of rules providing local control o AGVs access as to guarantee their deadlock-free and collision-free move. Assuming existence of local priority decision rules (controlling the access to the shared resources), the problem reduces to determination of the sufficient conditions in the form of a pair (initial state, priority rule set). The accepted rule-based transport subsystem specification way reduces the synchronization task to solution of an appropriate decision problem of the logic-algebraic method [3]. The problem solution is achieved through application of the constraint programming techniques [1].

2 Problem Formulation

The AGV service systems co-sharing the resources and executing repetitive tasks may be presented in a form of appropriately formulated Cyclic Concurrent Process Systems (CCPS), wherein the cyclic processes (encompassing AGVs flow) are interconnected one with another by use of the common resources (tracks, machine tools, etc.). In Figure 1, there is presented a graphical representation of an exemplary CCPS. In the system, three processes are used, P_1, P_2, P_3, that reflect operation of individual vehicles. The vehicles are served by the resources $R_1 - R_5$.

2.1 Assumptions

For the systems of that type, it is assumed that the cooperation of the processes is determined by the following constraints [7]:

- the processes share the common resources in the mutual exclusion mode,
- commencement of a successive process operation happens immediately after completing of the current operation provided that there is a possibility of making use of the successive resource requested by the given process,

- during waiting for a busy resource, the process does not release the resource allocated for execution of the previous operation,
- the process is not pre-emptive, i.e. the resource may not be taken of the process while it is using it,
- the processes are executed cyclically and in one cycle, a process may pass via any resource along its transportation route once only.

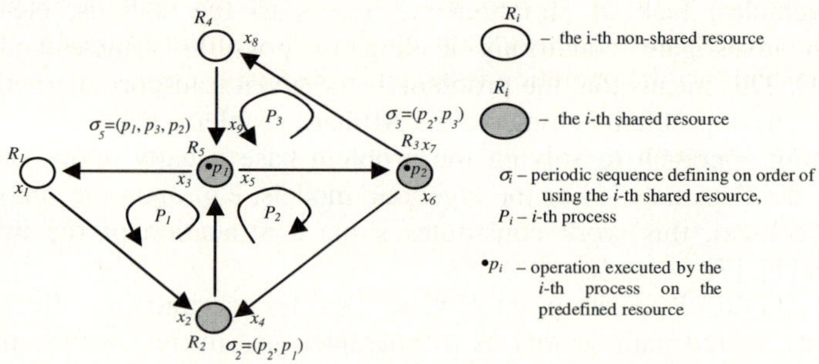

Fig. 1. Graphical representation of CCPS

2.2 Notations

In the model of CCPS the following definitions are used: cyclic process, time representation of a process, shared resource use sequence, initial system state, operation sequence, operation time sequence, priority dispatching rule encompassing access of the processes to the shared resources, and state sequence [7].

A cyclic process $P_i = (p_{i1}, p_{i2}, ..., p_{in})$ is a sequence, the components of which define the numbers of the resources used for execution of individual process operations, where: p_{ij} – denotes the resource used by the i-th process in the j-th operation; after completion of the operation with a share of the resource p_{in}, the operation with the share of the resource pi1 is executed again. The sequence defines the transportation route, i.e., the ordering in which the given process is executed. The process P_i – may begin at an arbitrary resource of the sequence presented.

The time representation of the i-th cyclic process is constituted by the sequence $T_i = (t_{i1}, t_{i2}, ... , t_{in})$, the components of which define the times of execution of individual operations of the relevant process, where t_{ij} denotes the time of execution of the j-th operation by the i-the process.

The sequence $\Theta = (\sigma_j, \sigma_k, ... ,\sigma_z)$ of the priority dispatching rules, i.e., accesses of the processes to the shared resources of the transport subsystem, where $\sigma_i = (s_j, s_k, ... ,s_l)$ – is the sequence, the components of which determine the process service order by the i-th shared resource, s_k – is the process number. Each sequence σ_i is periodic and gives access to the i-th resource to every process using it. This ensures starvation-free system execution.

The initial state $S_0 = (R_i,R_j, ...,R_k)$ of the transport subsystem is the sequence, the components of which define the initial process resource numbers, where $crd_i S_0 = R_j$ – denotes that the i-th process is begun from realization on the resource R_j; $crd_i S_0$ – denotes the i-th coordinate of the vector S_0.

The sequence of operations of the size equal to the number of all operations executed in the system is defined as $p = (P_1,P_2,...,P_r) = (p_{11}, ..., p_{1n1}, p_{21}, ..., p_{2\,n2}, ..., p_{r1}, ..., p_{r\,nr})$, where p_{ij} – denotes the j-th operation of the i-th process (the number of the resource used for realization of the j-th operation of the i-th process).

The operation execution time sequence $T = (T_1,T_2,...,T_r) = (t_{11}, ..., t_{1\,n1}, t_{21}, ..., t_{2\,n2},..., t_{r1} ,...,t_{r\,nr})$, where t_{ij} denotes the execution time of the j-th operation by the i-th process.

The state sequence (state vector) $x = (x_1,x_2,...,x_r)$, where x_i corresponds to the operation represented in the sequence p by the i-th coordinate $(crd_i p)$, the value x_i denotes the instant that the operation is begun in the first cycle.

2.3 Problem Statement

For the system described in such way, the following problem is defined: There is given a system of class CCPS, mapping the operation of automated guided vehicles. The system structure and the process parameters are given in the form of vectors defining the vehicles P_i routes and service times T_i in subsequent stations. The following question should be answered: *Does it exist a pair (initial system state S_0, priority rule set Θ) ensuring that the assumed transport processes are executed, with the cycle time not exceeding the arbitrarily given value H?*

The answer to the problem formulated in such way covers, therefore, the response to the question if there are existing the sufficient conditions that, when met, ensure the cyclic (i.e. deadlock-free) execution of the concurrent processes.

3 Logic-Algebraic Method

The components of the system class under consideration may be described in the form of the representation of the knowledge base $RW = <C,W,Y; R>$, where: $R = \{(c,w,y): F(c,w,y) = 1\}$ – a relation being the set of all triples (c,w,y), for which the facts F describing the system are true; $F(c,w,y) = (F_1(c,w,y), F_2(c,w,y),...,F_K(c,w,y))$ is the composition of the logic fact values being the functions of the variables c, w, y; $c = (c_1, c_2,...,c_k)$ – the set of the input variables; $y = (y_1,y_2,...,y_r)$ – the set of the output variables; $w = (w_1,w_2 ...,w_r)$ – the set of the auxiliary variables; $c\in C$, $y\in Y$, $w\in W$, C,Y,W – the sets defining the domains of the variables c,y,w.

3.1 Knowledge Base

The knowledge base representation *RW* describing an arbitrary system is presented in the form of the sets *C*, *W*, *Y*, that define the domains of some *c*, *y*, *w*, describing (on the qualitative level) some system properties. The variables *c* are called the input variables describing the input properties of the system, the variables *y* - the output variables describing the output properties of the system, the variables *w* are the auxiliary variables. The knowledge defining the properties of the system under consideration is presented in the form of the facts set $F(c,w,y)$. The facts $F(c,w,y)$ are propositions reflecting, on the logic level, the connections occurring between individual variables c,w,y. The triples c,w,y, for which all facts $F(c,w,y)$ are true, are presented in the form of the relation R. In this context, the representation of the knowledge for the systems CCPS is defined as follows [10]:

$$RW = <SO, \Sigma, X; R> \tag{1}$$

where: SO – the set of all possible initial states S_O (input variables), Σ - the set of all possible access rules Θ for the shared resources (input variables), X – the set of all possible forms of the state vector x (output variables), $R = \{(S_O, \Theta, x): F(S_O, \Theta, x) = 1\}$ – the relation defining the values S_O, Θ, x, for which the facts $F(S_O, \Theta, x)$ are true. The set R covers the facts $F(S_O, \Theta, x)$, being logic propositions that describe the system properties in dependence of the initial state S_O, the rules of access to shared resources Θ and of the starting times of individual operations x.

3.2 Knowledge Generation

The knowledge base RW considered can be treated as specification of general assumptions the transportation system has to follow (see Section 2.1) while taking into consideration the time and resources constraints. For example the assumption: *"during waiting for a busy resource, the process does not release the resource allocated for execution of the previous operation"* has to be specified by a set of propositions (constraints) guaranteeing its satisfaction for any process, and any resource at any moment of time. In other words, the knowledge about situations that might happened is specified by a set of facts $F(S_O, \Theta, x)$. The assumptions regarding the transportation system considered are as follows [10]:

- Constraints regarding an order in which processes have to be executed.
- Constraints limiting processes servicing by local resources: the moment x_j that the process P_i may start its execution at the resource R_k is equal to the time required by this process to complete its previous operation:

$$x_j = x_{j-1} + t_{j-1} \tag{2}$$

  where: t_{j-1} – the operation time on the resource R_{k-1}, x_{i-1} – the instant that the operation is begun at the resource R_{k-1}.
- Constraints regarding processes servicing by shared resources: the instant x_j that the operation of the process P_i is begun at the shared resource R_k is determined by the maximum within the completion time of the process P_i on the subsequent resource R_{k-1}, and the instance the operation corresponding to x_p begins its execution on the resource R_{k+1} has been served P_o previously executed on R_k just before P_i.

$$x_j = \max\{x_p, x_{j-1}+t_{j-1}\} \tag{3}$$

The constraints provided (2), (3), can be seen as a part of a transport system specification in the context of its parameters x, S_O, Θ, [10], i.e. a part of the relevant knowledge base.

Besides of the above mentioned assumptions guaranteeing the processes execution is collision-free and starvation-free the other one guaranteeing processes deadlock-freeness can be introduced. Illustration of a deadlock case is shown in Figure 2.

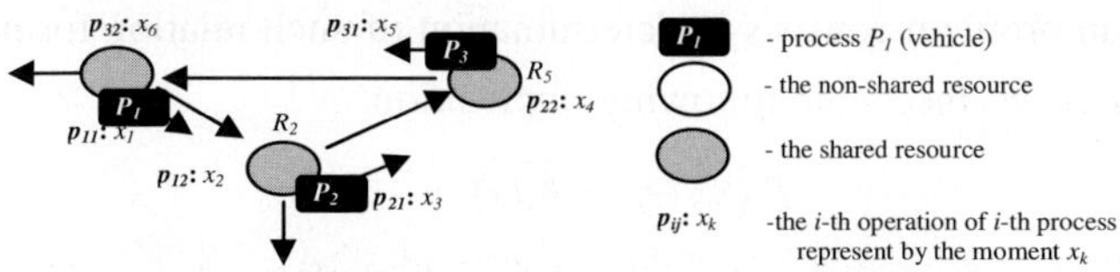

Fig. 2. Example deadlock in transportation system

Each of three AGVs associated to processes P_1, P_2 and P_3 waits for resource occupied by one of them – processes call for resources and their calls form a closed circle of resources request, i.e. the processes deadlock. It is easy to note, that the formulas (2) and (3), can be unified in the following recurrent formulae, named a state equation (4):

$$x_j = \max\{x_{j-1} + t_{j-1}, \max\{x_{j+1} + t_{j+1}, \ldots, \max\{x_{nz-3} + t_{nz-3}, \max\{x_{nz-1} + t_{nz-1}, \ldots, \ldots, \max\{x_{j-3} + t_{j-3}, x_j\}\ldots\}\}\}\} \tag{4}$$

where: x_j – denotes the instant that the operation of the process P_i is begun on the resource required.

For example the state equation for the process P_1 from Figure 2 has the following form: $x_2 = \max\{x_1+t_1, \max\{x_3 + t_3, \max\{x_5+t_5, x_2\}\}\}$.

It can be shown that if the state equation belongs to a class of the identity equations, then the system considered is deadlock-free one. Therefore, a set of facts $F(S_0, \Theta, x)$ following constraints (2), (3), and (4) allows one to detect possible deadlocks occurrence in the course of concurrent processes execution. It means, the knowledge base representation RW (1) may be determined automatically on the basis of the system structure, logistic constraints and parameters P_i, T_i.

The accepted way of the system specification, based upon the rules, enables to determine the sufficient conditions in the form of the initial states S_0 and the process service rules Θ, ensuring the deadlock-free operation of the system in cycles not exceeding the preset value of H. In this connection, so called decision problem is to be solved [3].

3.3 Decision Problem

For a given system CCPS described by the representation RW (1), one searches for such form of the input relation R_x, that will guarantee meeting of the known output relation R_y. The relations R_x and R_y are defined as follows: $R_x = \{(S_0, \Theta): F_c(S_0, \Theta) = 1\}$ – the set of the values S_0, Θ, for which the input system property $F_c(S_0, \Theta)$ is met, while $R_y = \{x: F_y(x) = 1\}$ – the set of values of x for which the output property $F_y(x)$ is met, where: $F_c(S_0, \Theta)$ is a set of the logic propositions that describe the input system properties in dependence on the initial state S_0 and the rules of access to the shared resources Θ, while $F_y(x)$ is a set of the logic sentences that describe the output properties of the system in dependence of the values of the sequence x. For the system under consideration, the output property is of the form: $F_y(x) = (x_1+t_1 \leq H) \wedge (x_2+t_2 \leq \leq H) \wedge \ldots \wedge (x_{rn}+t_{rn} \leq H)$.

The decision problem consists in determination of such relation form R_x for which the input property F_c meets the following implication:

$$F_c(S_0, \Theta) \Rightarrow F_y(x) \tag{5}$$

The relation R_x is a set of such values S_0, Θ, for which the system will operate without collisions and deadlocks in cycles not lasting longer than the value of H. The derivation of the relation R_x on the basis of the logic-algebraic method [3] occurs with use of the sets S_{x1} i S_{x2}, i.e.

$$R_x = S_{x1} \setminus S_{x2} \tag{6}$$

where: $S_{x1} = \{(S_0, \Theta): F(S_0, \Theta, x)=1, F_y(x) = 1\}$; $S_{x2} = \{(S_0, \Theta): F(S_0, \Theta, x)=1, F_y(x) = 0\}$.

The set S_{x1} can be seen as a set consisting of S_0, Θ, whose follow the facts: $F(S_0, \Theta, x)$, $F_y(x)$. In turn the set S_{x2} can be seen as a set consisting of S_0, Θ, whose follow the fact $F(S_0, \Theta, x)$ and do not follow $F_y(x)$. The intersection of sets S_{x1}, S_{x2} can consist values of S_0, Θ, for which the fact $F_y(x)$ is not determined. It happens when some facts of the knowledge base RW (1) results in state equations (6) being identity ones (an equation that is valid for all values of its variables) – such property means the system is not deadlock-free. Therefore, $S_{x1} \setminus S_{x2}$ guarantees the transportation system considered is deadlock-free. Determination of the set $R_x = \varnothing$ denotes the lack of answer to the question asked. In a general case, searching for the relation R_x is an NP-hard problem. Therefore, a concept of use of constraints programming techniques is proposed.

4 Constraint Satisfaction Problem

Each knowledge base representation RW (1) of the concurrent cyclic process system may be presented in the form of the constraint satisfaction problem (*CSP*) [2]. The problem $CSP = ((Q, D), Co)$ is defined as follows: There is given a finite set of discrete decision variables $Q = \{q_1, q_2, \ldots, q_n\}$, a family of finite variable domains $D = \{D_i \mid D_i = \{d_{i1}, d_{i2}, \ldots, d_{ij}, \ldots, d_{im}\}, i = 1..n\}$ and the finite set of constraints $Co = \{Co_i \mid i = 1..L\}$ limiting the decision variable values. The admissible solution, i.e. the one that the values of all variables meet all constraints of the set Co are sought for.

4.1 Knowledge Base Representation

In the case of the problem *CSP*, mapping the knowledge representation *RW*, the role of the constraints Co is fulfilled by the facts included in $F(S_0, \Theta, x)$ while the role of the variables Q – the values of the variables S_0, Θ, x. The variable domains are in the form of the sets D_{S0}, D_Θ, D_x. The problem *CSP* results in the form:

$$CSP = (((S_0, \Theta, x), D), \{F(S_0, \Theta, x) = 1\}) \tag{7}$$

where: $D = \{D_{S0}, D_\Theta, D_x\}$, D_{S0} is the set of the resources included in the system; D_Θ is the set including the processes realized in the system; D_x is the set of the time values; $F(S_0, \Theta, x) = 1$ denotes the series of facts: $(F_1(S_0, \Theta, x) = 1, \ldots, F_K(S_0, \Theta, x) = 1)$.

4.2 Logic-Algebraic Method Based Solution

The solution of the *CSP* problem formulated in such way is a set of sequence values of the initial state of the system S_0, the rules of the process access to the shared resources Θ and of the starting times of individual operations on the resources x for which all constraints presented in the form of the logic sentences $F(S_0, \Theta, x)$ are true. Solving the decision problem (determination of the relation R_x) in the context of *CSP* is connected with solving the following two problems:

$$\begin{aligned} CSP_{Sx1} &= ((S_0, \Theta, x), D), \{F(S_0, \Theta, x) = 1, F_y(x) = 1\}) \\ CSP_{Sx2} &= ((S_0, \Theta, x), D), \{F(S_0, \Theta, x) = 1, F_y(x) = 0\}) \end{aligned} \tag{8}$$

The results of solving the problems formulated in such way are the sets S_{x1} and S_{x2}. The sets enable to determine the set R_x (6). The determination procedure for the relation R_x has been presented in the Figure 3.

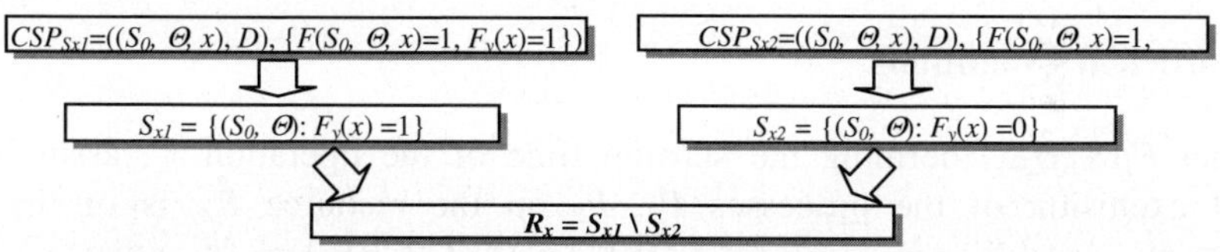

Fig. 3. Sufficient condition determination procedure

This procedure enables for determination of the answer to the question asked, in the form of the set R_x, where R_x fulfils the role of the sufficient conditions. The sufficient conditions determination procedure applied allows for effective seeking for the answer to the question asked. The set R_x includes a group of alternative solutions in the form of the value of the sequence of the initial state S_0 and the sequence of the access rules of the processes to the shared resources Θ ensuring, the deadlock-free vehicles service during the assumed time period H.

5 Example of Schedules Design

There is given an automated guided vehicle system of the structure depicted in the Figure 1. The processes P_1, P_2, P_3 reflect service of three *AGVs* according to the service times T_1, T_2, T_3. The vehicles are served by the service points R_1, R_2, R_3, R_4, R_5, the shared system resources. The following parameters are known:

$$\begin{aligned} &P_1 = (R_1, R_2, R_5),\ T_1 = (1,2,3),\ p = (R_1, R_2, R_5, R_2, R_5, R_3, R_3, R_4, R_5), \\ &P_2 = (R_2, R_5, R_3),\ T_2 = (2,2,2),\ x = (x_1, x_2, x_3, x_4, x_5, x_6, x_7, x_8, x_9), \\ &P_3 = (R_3, R_4, R_5),\ T_3 = (2,1,3),\ t = (1, 2, 3, 2, 2, 2, 2, 1, 3). \end{aligned}$$

The initial state vector $S_0 = (R_i, R_j, R_k)$ where: $R_i \in \{R_1, R_2, R_5\}$, $R_j \in \{R_2, R_5, R_3\}$, $R_k \in \{R_3, R_4, R_5\}$ is distinguished, defining the initial process resources, the access rule vector $\Theta = (\sigma_2, \sigma_3, \sigma_5)$, where $\sigma_2 = (s_{21}, s_{22})$, $\sigma_3 = (s_{31}, s_{32})$, $\sigma_5 = (s_{51}, s_{52}, s_{53})$. In accordance with the rules described in [10], the knowledge representation corresponding to the specification given is of the form: $RW = <S0, \Sigma, X; R>$, where: $R = \{(S_0, \sigma_2, \sigma_3, \sigma_5, x): F(S_0, \sigma_2, \sigma_3, \sigma_5, x) = 1\}$.

5.1 A Routine Query

The answer to the question: Does it exist such combination of the initial states S_0 of the system and the set of the vehicle access rules σ_2, σ_3, σ_5, for the shared resources that guarantee that the process realization cycle will not exceed 9 time units?

The answer to such formulated question is the solution of the decision problem. In accordance with the procedure presented (Figure 3), the following *CSP* problems are being solved (8): $CSP_{Sx1} = ((S_0, \sigma_2, \sigma_3, \sigma_5, x), D), \{F(S_0, \sigma_2, \sigma_3, \sigma_5, x)=1, F_y(x)=1\})$, $CSP_{Sx2} = ((S_0, \sigma_2, \sigma_3, \sigma_5, x), D), \{F(S_0, \sigma_2, \sigma_3, \sigma_5, x)=1, F_y(x)=0\})$, where: $D = \{D_{S0}, D_{\sigma2}, D_{\sigma3}, D_{\sigma5}, D_x\}$, $D_{S0}=\{R_1, R_2, R_3, R_4, R_5\}$, $D_{\sigma2}= \{P_1, P_2, P_3\}$, $D_{\sigma3}= \{P_1, P_2, P_3\}$, $D_{\sigma5} = \{P_1, P_2, P_3\}, D_x = \{1, \dots ,18\}$.

5.2 Sufficient Conditions

The fact $F_1(S_0,\sigma_2,x)$ defining the starting time of the operation x_2, assuming the mutual exclusion of the processes P_1, P_2 on the resource R_2, is of the form: $F_1(S_0,\sigma_2,x)$: $\neg(crd_1S_0 = R_2) \wedge \neg(crd_2S_0 = R_5) \wedge (crd_1\sigma_2 = P_2) \Rightarrow (x_2 = \max(x_5, x_1+t_1))$ i.e., if in the state S_0 the process P_1 does not use the resource R_2, and in the state S_0, the process P_2 does not use the resource R_5, and – as the first one, the P_2 process uses of the resource R_2, then $x_2 = \max(x_5, x_1+t_1)$. The fact defining the output property $F_y(x)$:$(x_1+t_1 \leq 9) \wedge \dots \wedge (x_9+t_9 \leq 9)$ corresponds to the preset condition: "*the cycle will not exceed 9 time units*".

For solving the problems CSP_{Sx1}, CSP_{Sx2}, the constraints programming language, OzMozart [1], was applied. The solution set R_x obtained, is of the form:

$$\begin{aligned} R_x = \{ &\{S_0 = (R_5,R_3,R_4),\ \sigma_5 = (P_1, P_3, P_2),\ \sigma_2 = (P_2,P_1),\ \sigma_3 = (P_2,P_3)\},\\ &\{S_0 = (R_2,R_5,R_3),\ \sigma_5 = (P_2, P_1, P_3),\ \sigma_2 = (P_1,P_2),\ \sigma_3 = (P_3,P_2)\},\\ &\{S_0 = (R_1,R_2,R_5),\ \sigma_5 = (P_3, P_2, P_1),\ \sigma_2 = (P_2,P_1),\ \sigma_3 = (P_3,P_2)\}\}. \end{aligned}$$

The set R_x constitutes the set of the alternative sufficient conditions that must be fulfilled by the system under consideration in order that the vehicles may realize the planned operations within cycles not exceeding 9 time units.

5.3 Admissible Solutions

An exemplary Gantt's chart illustrating the use of the system resources has been presented in the Figure 4. The diagram corresponds to the sufficient conditions of the form $S_0 = (R_5,R_3,R_4)$, $\sigma_5 = (P_1, P_3, P_2)$, $\sigma_2 = (P_2,P_1)$, $\sigma_3 = (P_2,P_3)$. The processes are being (deadlock-free and starvation-free) executed within the cycles not exceeding 9 time units.

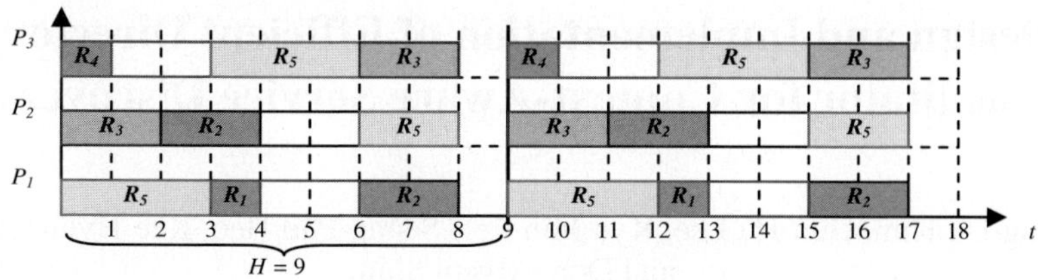

Fig. 4. The Gantt's chart

6 Conclusions

On the basis of the rule-based specification of a transport system, formulated in the form of the knowledge base representation *RW*, it is possible to seek for schedules guaranteeing the collision-free and deadlock-free operation of the AGVs based mulit-agent system. The presented concept of automated generation of the knowledge representation *RW* and the approach to determination of the sufficient conditions through specification of the CSP problem enables construction of the interactive (operating in the on-line mode) computer aided prototyping systems for distributed control procedures.

References

1. Banaszak Z., Zaremba M., Muszyński W. CP-based decision making for SME. Preprints of the 16th IFAC World Congres, 3 – 8 July, 2005, Prague, Czech Republic, Eds P. Horacek, M. Simandl, P. Zitek 2005; DVD.
2. Barták R. Incomplete Depth-First Search Techniques: A Short Survey, Proceedings of the 6th Workshop on Constraint Programming for Decision and Control, Ed. Figwer J., 2004; 7-14.
3. Bubnicki Z. Introduction to Expert Systems. PWN, Warsaw, 1990.
4. Lawley M.A., Reveliotis S.A., Ferreira P.M. A correct and scalable deadlock avoidance policy for flexible manufacturing systems. IEEE Trans. on Robotics and Automation 1998; Vol.14, No.5: 796-809.
5. Lee T., Song J. Petri net modeling and scheduling of periodic job shops with blocking. Proc. of the Workshop on Manufacturing and Petri nets, Osaka, Japan, 25 June, 1996: 197-214.
6. Liu J., Autonomous Agents and Multi-agent Systems, World Scientific, 2001.
7. Polak M., Majdzik P., Banaszak Z.A., Wójcik R. The performance evaluation tool for automated prototyping of concurrent cyclic processes. Fundamenta Inf. 2004, Vol.60, No.1-4: 269-289.
8. Ramamritham K. Allocation and scheduling of precedence-related periodic tasks. IEEE Trans. on Parallel and Distributed Systems 1995, Vol.4, No. 6: 412-420.
9. Paprzycki M., Agent-like approach to software design methodology, Computer Science Department, Oklahoma State University, Tulsa, OK 74106 USA.
10. Wójcik R., Bocewicz G., Banaszak Z. Scheduling of AGV systems under access restrictions to shared resources of the ESP (logic-algebraic model). Proc. of National Conf. on Robotics, 2006, Wroclaw: 149-163.

Design and Implementation of Efficient Directory Facilitator for Context-Aware Service Discovery

Dong-Uk Kim, Gun-Ha Lee, Kyu Min Lee, Seung-Phil Heo, Kee-Hyun Choi, and Dong-Ryeol Shin

School of Information and Communication Engineering, SungKyunKwan University
{tonykim,ghlee,kmlee,haeni22,gyunee,drshin}@ece.skku.ac.kr

Abstract. Multi-agent technologies are essential in realizing the upcoming ubiquitous environment. In the multi-agent environment, each agent has its own set of services and stores these services in the service repository of the multi-agent system. By using this repository, the user can retrieve the most appropriate service. In this paper, we propose an efficient service repository architecture that can improve the existing repository by using context inferencing, context filtering, user-predefined policies, and the categorized tree. An advantage of the proposed architecture is that the user can obtain more suitable services than existing approaches about service discovery.

Keywords: yellow page, service repository, DF, agent, context aware.

1 Introduction

According to the growth of people's interests, different technologies relevant to ubiquitous computing are becoming increasingly progressive and sophisticated. In this ubiquitous environment, agent technologies have also developed, rapidly. With the development of these agent technologies, The Foundation for Intelligent Physical Agents (FIPA)[1] establishes multi-agent system standards for promotion of agent-based technology and interoperability of these standards. Agents in the FIPA-compliant agent system can provide services to others and store these services in the Directory Facilitator (DF)[2] of the multi-agent system. Users can search specific services through the DF, which is composed of service descriptions. Also, the Java Agent Development Framework (JADE)[3] is a popular multi-agent system for supporting the DF.

However, the DF of JADE uses a sequential search mechanism for retrieving the specific service in the DF. When an user searches a service, this approach raises various problems relevant to search speed. JADE DF also do not consider any policy or context inference mechanism for providing suitable service to the user. The user may obtain too many services from the JADE DF and be placed in a state of confusion.

To solve these problems, we propose efficient FIPA-compliant DF architecture for context-aware service discovery. The proposed architecture uses a search mechanism by employing a categorized tree similar to a web search engine. To provide suitable

N.T. Nguyen et al. (Eds.): KES-AMSTA 2007, LNAI 4496, pp. 588–596, 2007.

service to the user, we propose the Policy Manager and Context Manager for context-aware service discovery. Through the Context Manager, raw context information is discarded and useful information is transformed into higher-level context information, which is then used for context-aware service discovery. When these components in our architecture is used in multi-agent system, the user can retrieve suitable service.

The remainder of the paper is organized as follows. Section 2 describes introduction of the FIPA-compliant agent system, context and existing service discovery approaches. Section 3 presents our proposed DF architecture for efficient service matchmaking. We introduce different modules described from proposed architecture in this section. In Section 4, we implement the DF prototype. Finally, the conclusion and future work are discussed in Section 5.

2 Related Work

In this section, we introduce the DF matchmaking algorithm of the JADE agent platform as a FIPA-compliant agent system. Then, we describe the context concept for the proposed context-aware service repository. Finally, existing approaches are presented.

2.1 Directory Facilitator

JADE is middleware that facilitates the development of multi-agent systems. Each running instance of the JADE runtime environment is called a Container, as it can contain several agents. The set of active containers is called a Platform. The Main Container is referred to special container in the Platform. This container includes the Agent Management System (AMS) and DF. The AMS provides the naming service among different agents, and the DF provides the repository for services provided by different agents.

The DF is a class of yellow page service. That is, the DF provides the agent's service repository. At first, agents create descriptions of their services and store these descriptions in the DF. The user can obtain the desired service's descriptions by retrieving the DF. The DF provides register, deregister, modify and search functions for processing service descriptions in its repository. The Knowledge Base in the DF is a real repository for storing service descriptions. That is, the DF consists of function modules and Knowledge Base. The Knowledge Base stores DFAgentDescription objects that show the descriptions of an agent. Similarly, a DFAgentDescription object can store ServiceDescription objects which are descriptions of the service provided by the agent. Many ServiceDescriptions can be included in a DFAgentDescription. A DFAgentDescription consists of various parameters, such as the name of the agent, services, supported interaction protocols, list of ontologies, and list of content languages. Similarly, ServiceDescription also consists of many parameters, such as the name of the service, type, supported interaction protocols, list of ontologies, list of content languages, owner of the service, and list of additional descriptive properties. JADE's existing DF uses concrete matching mechanism[4] for service discovery. This mechanism uses a syntactic and structural matching mechanism. That is, the sequential search mechanism is used. However, the DF's

Knowledge Base may contain too many DFAgentDescriptions, by different mobile agents in the ubiquitous environment. Therefore, it follows, that the DF should conduct all matching processes about all DFAgentDescriptions in the Knowledge Base. This generates problems in searching speed. For example, the number of the service registered in the DF is 1000. In the worst case, DF should conduct the matching process 1000 times because the DF uses sequential search mechanism.

2.2 Definition of Context

Context is a complex description of knowledge of physical, social, historical, or other circumstances within which an action or an event occurs. This is represented as a class of knowledge that can be inferred from adjacent text, composed of con (with) and text. Dey defined that Context includes information that can be used to determine the status of an entity. This entity may be a person, place, or object. Entities are considered relevant to the interaction between a user and application, including the user and application themselves[5]. Abowd is considers that a system is context-aware if it uses context to provide relevant information and/or services to the user, where relevancy depends on the user's task[6].

2.3 Existing Service Discovery Approaches

The architecture for adaptive Context-Aware application[7] uses the notion of system-wide adaptation policies that enable a mobile system to operate differently, given the current context and requirements of the user. That is, it adapts behavior according to the different contexts to the user. The architecture enables the discovery of context providers using the context advertised. This approach attempts to consider users' context information when discovering services. However, it is not considered as an agent-level, but rather an application-level approach. In addition, these approaches only use selected components of context information, and processing of raw context information is straightforward.

For solving these problems, middleware for Context-Aware Service Discovery is proposed[8]. This approach is JADE-based middleware for an agent-level approach. That is, different components for service discovery are implemented by using the agent concept. In this approach, raw context information for service discovery is discarded, and useful raw data is represented to high-level data via the Context Manager. The Context Manager can consider users' preferences and requirements provided by Policy Manager. This approach focuses on two procedures. First, different services using heterogeneous protocols such as UPnP[9], SLP[10] and Jini[11] are stored in the DF by translation into the DF's ServiceDescription format. Secondly, registered services can be retrieved from the DF through a new context-aware service matchmaking mechanism that uses the context information, such as that from users' requirements. That is, it focuses on context-aware service discovery. However, our proposed architecture differs from it in that we emphasize the design of the agent-based efficient service repository supporting context-aware service discovery.

3 Proposed Architecture

To solve these problems generated from existing approaches, as shown in Section 2, we propose agent-based advanced DF architecture that can categorize services into types, and use the service policy for efficient service discovery. In addition, we propose the Context Manager, for context-aware service discovery. The proposed DF architecture is used for a part of Component-based Autonomic Layered Middleware (CALM) project. Detailed information of CALM, is shown in [12]. Figure 1 shows our advanced DF architecture, illustrating the above concepts. The proposed architecture consists of ACL Processor, Context Manager, Service Matching Processor and DF function module. The role of each component is as follows:

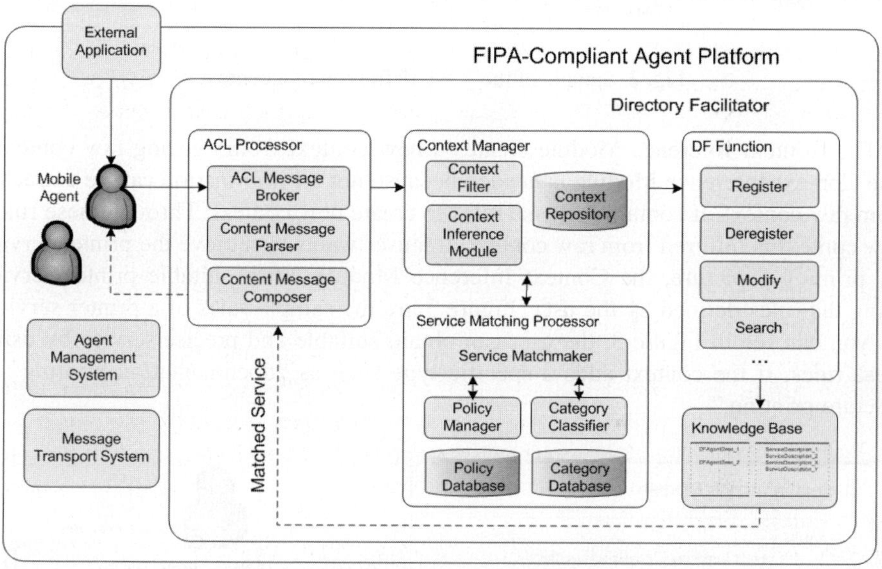

Fig. 1. Proposed efficient DF architecture

3.1 Context Manager

The Context Manager has the role of filtering or inferencing the different contexts received by various agents. The Context Manager consists of Context Filter, Context Inference Module and Context Repository.

The Context Filter's primary purpose is to protect the Directory Facilitator from being flooded with excessive information. In ubiquitous environment, mobile devices have limited resources. Additionally, they frequently enter or leave the specific network. Unlimited inflow of unnecessary context information creates traffic problems and lowers the speed of mobile devices. To suit the mobile environment, the Context Filter extracts different context information included in registration or search messages. For example, an agent may provide a color printer service with 12ppm.

Through Context Filter, the context stored in the Context Repository may be (name "printer", type "color", speed "12ppm", expiry-date "06-12-2006"). Figure 2 is the example of context in the Context Repository.

```
<service>
    <provider> agent1 </provider>
    <name> printer </name>
    <type> color </type>
    <functionality>
        <ppm> 12 </ppm>
        <resolution> 600x600dpi </resolution>
    </functionality>
    <expiry_date> 06-12-2006 </expiry_date>
</service>
```

Fig. 2. Sample of the XML definition of a context

The Context Inference Module creates a new context from existing raw contexts. The Context Inference Module is needed because not all information can be extracted from raw context. It contains diverse rules to create new context. Through these rules, new context is inferred from raw context. If a user wants to retrieve the printer service for printing a picture, the Context Inference Module infers suitable printer service using the rules defined by the user. Figure 3 are the sample rules of a printer service. As you can see in Figure 3, the user can obtain suitable and precise service by using these rules, if the context adds a specific type such as "documentation printing" or "picture printing."

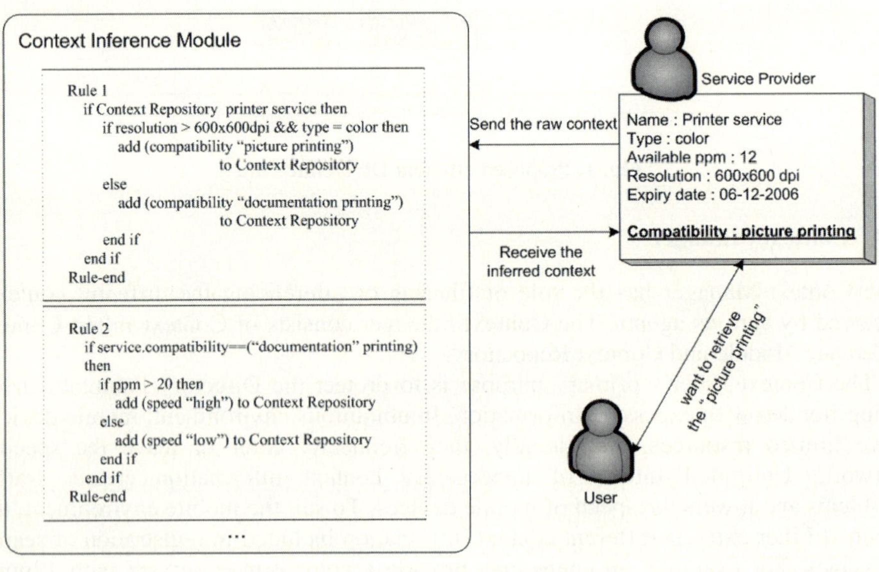

Fig. 3. Sample of the rule about a printer service

3.2 Service Matching Module

* Service Matchmaker

The results provided by the Context Manager are forwarded to the Service Matchmaker in the Service Matching Module. The Service Matching Module consists of Service Matchmaker, Policy Manager, Category Classifier and two databases.

When requests regarding registration or search of the service are received, the Service Matchmaker accesses the DF through information created by the Policy Manager and Category Classifier. Through the Policy Manager and Category Classifier, the Service Matchmaker obtains the following information for registration or search: (1) Service type in categorized tree (2) ServiceDescription's start and end address matched with the service type in Knowledge Base (3) Service that should be removed or maintained via the Policy Manager.

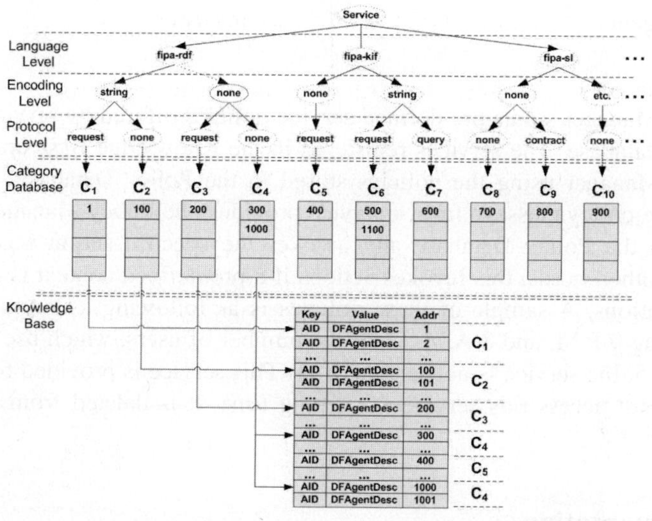

Fig. 4. Sample of category taxonomy about the ServiceDescription

* Category Classifier

From the Category Database, proposed Category Classifier can create the available addresses which are stored or searched in the Knowledge Base. Category Classifier's roles are twofold as both registration and search.

In case of registration, the Category Classifier categorizes services into a specific type that contains the real address of the Knowledge Base. ServiceDescription is stored in the Knowledge Base's address, and is linked with the type. As shown in Figure 4, the Category Classifier uses Language, Encoding and Protocol parameters for categorization. In addition, the new categorized type can be created dynamically by the Category Classifier. For example, the current number of the categorized type is

10. If the category type of the requested ServiceDescription does not exist in the Category Database, the Category Classifier creates new the category type C_{11} and links it with available address of the Knowledge Base. In the opposite case, a new address is allocated by Category Classifier as C_4 and C_6 when Knowledge Base's space linking with the specific type is full.

When a user searches the specific service, the Category Classifier grasps the category type of the requested service via analyzing the different parameters in user context. Then, it finds the Knowledge Base's real address linked with the type extracted from the Category Database. Finally, it can search all ServiceDescriptions from start to end address linked with the type. When this proposed mechanism is used, it is not necessary to search all DFAgentDescriptions and ServiceDescriptions in the Knowledge Base. When ServiceDescriptions are categorized, search time is the sum of the Category Database's seek time and ServiceDescription. That is, the search time is lower than the existing sequential mechanism, according to the increase in the number of agents.

* Policy Manager

The Policy Manager stores pre-defined service policies offered by service provider in the Policy Database. The services registered in the Knowledge Base are managed by the Policy Manager using the policies stored in the Policy Database. That is, after receipt of the policy message from a service provider, the Policy Manager stores these messages in the Policy Database and invokes the specific action according to the policies. In other words, this invokes actions if a pre-defined context is detected from current situations. A sample of these policies is as following: (1) Available service hour is during 9 P.M. and 8 A.M. (2) If the number of users, which use the service is greater than 5, the service is not searched. (3) This service is provided for a week. (4) If users do not access this service for a long time, it is deleted from the Directory Facilitator.

4 Implementation

Implementation of the proposed DF architecture is based on Java. In addition, Java Server Page (JSP) technology is used for the user interface. This allows users to easily search services in the DF by using a web browser. Using JSP technology, location-independent search via web browser can be supported.

The left of Figure 5 is the implemented web page for searching the ServiceDescription. By using this web page, users can easily input the different criteria. For searching the specific services, user inputs the service name, type, ownership, language, protocol, encoding and ontology parameters. In addition, users can input special parameters, such as preference requirements, max results, user properties and alignment parameters. The right of Figure 5 is a DF result page of a user search request. As shown in the figure, the resulting page presents the agent's AIDs, service names and service types.

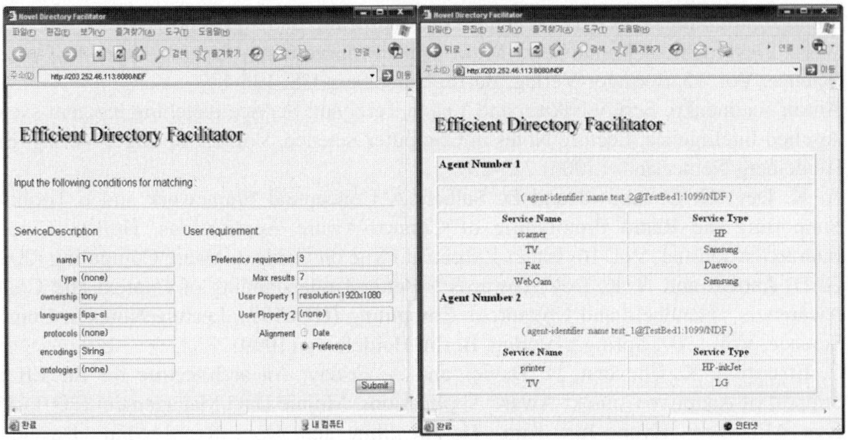

Fig. 5. DF search and result page

5 Conclusion and Future Work

In this paper, we introduced the JADE which is the most popular multi-agent system, and analyzed the JADE's DF, which is the repository of service descriptions. Similarly, we pointed out the problems of the existing DF mechanism for storing and searching the services. After consideration of these problems, an advanced DF architecture for context-aware service discovery is proposed. The proposed architecture suffers from the disadvantage that the DF should access the Category Database. However, it is not required to search for all DFAgentDescriptions in the Knowledge Base. The result is that searching speed is raised. The proposed architecture is more efficient than existing mechanisms, when increasing the number of agents in multi-agent system. Finally, the searching mechanism using the Context Manager and Policy Manager can support service QoS and provide suitable service to the user.

In future work, we will compare the proposed architecture with other searching mechanisms and evaluate the performance. Also, we will research the fault-tolerant mechanism of the proposed DF for reliability.

Acknowledgement

This research is supported by the Ubiquitous Computing and Network (UCN) Project, the Ministry of Information and Communication (MIC) 21st Century Frontier R&D Program in Korea.

References

1. FIPA: The Foundation for Intelligent Physical Agents, http://www.fipa.org.
2. Giovanni Caire: JADE Tutorial, http://jade.tilab.com, 2003

3. Fabio Bellifemine, Federico Bergenti, Giovanni Caire and Agostino Poggi: JADE-A Java Agent Development Framework, Multi-agent Programming, Lecture Notes in Computer Science, Vol. 15, Springer-Verlag, Berlin Heidelberg US, 125-147
4. Anton Naumenko, Sergiy Nikitin and Vegan Terziyan: Service matching in agent systems, Applied Intelligence, Lecture Notes in Computer Science, Vol. 25, Springer-Verlag, Berlin Heidelberg Netherlands (2006) 223-237
5. A. K. Dey, G. D. Abowd and D. Salber: A Conceptual Framework and a Toolkit for Supporting the Rapid Prototyping of Context-Aware Applications, Human Computer Interaction Journal, Vol. 16, Issue: 1, Special Issue on Context-Aware Computing (2001)
6. G. D. Abowd and A. K. Dey: Towards a Better Understanding of Context and Context-Awareness, Handheld and Ubiquitous Computing (HUC'99), Lecture Notes in Computer Science, Vol. 1707, Springer-Verlag, Berlin Heidelberg (1999)
7. C. Efstratiou, K. Cheverst, N. Davies and A. Friday: An architecture for the Effective Support of Adaptive Context-Aware Applications, Mobile Data Management (2001) 15-26
8. Kyu Min Lee, Hyung-Jun Kim, Ho-Jin Shin and Dong-Ryeol Shin: Design and Implementation of Middleware for Context-Aware Service Discovery in Ubiquitous Computing Environments, International Conference on Communication Systems and Applications, Lecture Notes in Computer Science, Vol. 3983, Springer-Verlag, Berlin Heidelberg UK (2006) 483-490
9. Universal Plug and Play Specification, v1.0, http://www.upnp.org
10. Erik Guttman: Service Location Protocol- automatic discovery of IP network services, Internet Computing Journal (1999), IEEE, Vol. 3, Issue: 4
11. Jini Architecture Specification, v1.2, http://www.sun.com/software/jini/specs/
12. Seungwok Han, Sung Keun Song, and Hee Yong Youn: CALM: An Intelligent Agent-based Middleware for Community Computing, SEUS 2006/WCCIA 2006, Proceedings of the Fourth IEEE Workshop

The CrocodileAgent: Designing a Robust Trading Agent for Volatile E-Market Conditions

Ana Petric, Vedran Podobnik, and Gordan Jezic

University of Zagreb, Faculty of Electrical Engineering and Computing, Department of Telecommunications, Unska 3, HR-10000 Zagreb, Croatia
{ana.petric,vedran.podobnik,gordan.jezic}@fer.hr

Abstract. The Trading Agent Competition (TAC) simulates a challenging game environment where competing agents engage in complex decision-making activities with purpose of maximizing their profits. One of the TAC competitive scenarios is Supply Chain Management (SCM) where six trading agents compete by buying components, assembling PCs from these components and selling the assembled PCs to customers. In this paper, we briefly describe the TAC SCM environment and present the main features of the CrocodileAgent, our entry in the 2005 TAC SCM Competition. The agent's performances in the 2005 TAC SCM, as well as in a series of controlled experiments, are discussed.

1 Introduction

Supply chain management involves several activities such as raw material procurement, producing, selling and shipping manufactured goods. Today's supply chains are still based on static long-term relationships between trading partners while in dynamic supply chains the market is the driving force. Dynamic supply chain management improves the competitiveness of companies since it has a direct impact on their capability of adapting to the changing market demands quickly and efficiently [1]. The annual worldwide supply chain transactions are counted in trillions of dollars what makes this area of research very interesting not just from the scientific, but also from the business point of view, since even the slightest improvement brings a very high profit.

The connection between AI (*Artificial Intelligence*) and economics has received a lot of attention recently [2]. The ideas proposed in this paper are also based on that connection, while the practical implementation of the presented ideas is enabled with the use of intelligent software agent technology and supported by the Internet infrastructure. Although the initial architecture of the Internet was geared towards delivering information visually to humans, currently the Web transforms into an environment filled with the goal-directed applications which intelligibly and adaptively coordinate information exchanges and actions (Web 2.0 and Web 3.0) [3]. At the same time, computers are evolving from single isolated devices to entry points into a worldwide network of information exchange and business transactions [4]. Consequently, the Internet is transforming into an enabler of the digital economy. The digital economy, by proliferation of the use of the Internet, provides a new level and

N.T. Nguyen et al. (Eds.): KES-AMSTA 2007, LNAI 4496, pp. 597–606, 2007.

form of connectivity among multiple heterogeneous ideas and actors [5]. Additionally, by utilizing the technology of intelligent software agents, the digital economy automates business transactions.

In the paper we describe the CrocodileAgent, an intelligent agent we developed to participate in 2005 TAC SCM. The paper is organized as follows. Section 2 presents the TAC SCM game. Section 3 describes the CrocodileAgent's architecture and functionalities. Section 4 comments the agent's ranking in the 2005 TAC SCM, elaborates the results of an experiment we conducted in our laboratory and includes a detailed analysis of the results. Section 5 proposes directions for future work and concludes the paper.

2 The TAC SCM Game

The Trading Agent Competition (TAC) is an international forum that promotes high-quality research on the trading agent problem. One of its game scenarios is Supply Chain Management (SCM). In the TAC SCM game [6, 7] scenario, each of the six agents included in the game has its own PC manufacturing company. During the 220 TAC days, agents compete in two different markets. In B2B market, agents compete in buying raw materials necessary to produce personal computers. Participants in this market are agents and eight suppliers which produce four types of components (CPUs, motherboards, memory, hard drives) with different performances. In its factory, an agent can manufacture 16 different types of PCs. In the B2C market, the agents try to sell all the PCs they produced to customers and, at the same time, earn as much money as possible. The purpose of the TAC SCM (*Trading Agent Competition Supply Chain Management*) game is to explore how to maximize the profit in the stochastic environment of volatile market conditions. Thus, it is important to develop an agent capable of reacting quickly to changes taking place during the game. Furthermore, it is critical to implement predictive mechanisms which enable an agent's proactive behaviour. The idea is to build a robust, highly-adaptable and easily-configurable mechanism for efficiently dealing with all SCM facets, from material procurement and inventory management to goods production and shipment [8]. Additionally, TAC SCM tournaments provide an opportunity to analyze effects common in real-world business transacting, such as the bullwhip effect, and its relationship with company profits [9]. Furthermore, the tournament can help in developing methods for identifying the current economic regime and forecasting market changes [10].

The architecture of the TAC SCM system is shown in Figure 1. The TAC SCM game server simulates suppliers (PC component manufacturers), customers (PC buyers) and the bank. The game server also controls agents' factories and warehouses. In order to participate in the game, an agent has to connect to the game server. Each TAC SCM agent has a bank account and receives a daily report regarding its current bank balance. At the beginning of the game, the agent has no money and must hence loan money from the bank. The bank charges the agent interest for every day that the agent is in debt. The winner of the game is the agent with the most money in its bank account at the end of the game.

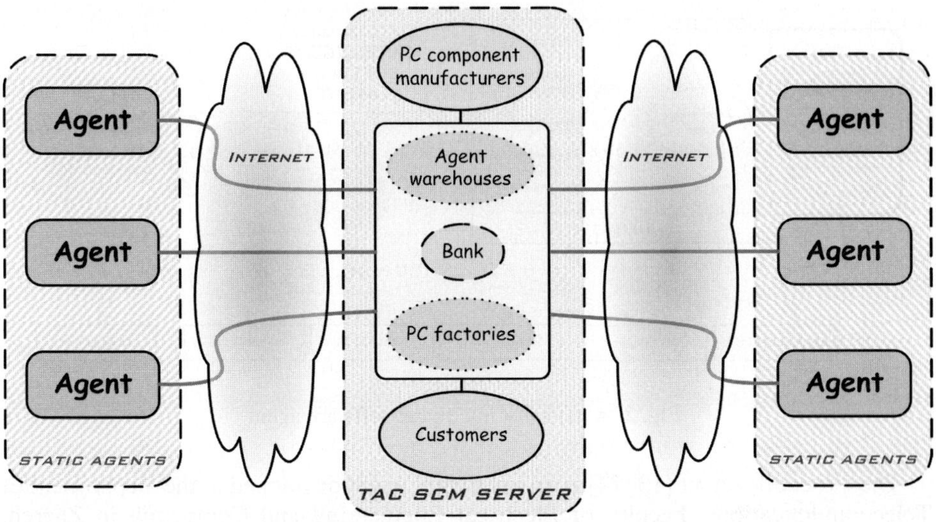

Fig. 1. The architecture of the TAC SCM system

3 The CrocodileAgent

An intelligent software agent is an autonomous program which acts on behalf of its principal (human or organizational) while conducting complex information and communication actions over the Web. The technology of intelligent software agents enables automated and process execution and coordination, thus creating added value for its principal. Figure 2 presents our model of intelligent software agents [11, 12], which we used while designing our TAC SCM agent.

The features of the technology of intelligent software agents make them perfectly applicable in modern enterprise systems and electronic markets (e-markets) [13]. The accelerated economic globalization trend and rapid development of ICT technologies in the past decade are leading us closer to the existence of just one market - the global one. Consequently, the functions of supply and demand are becoming more and more dynamic and the possibilities of choice have risen to amazing levels. This is a reason why companies today have great difficulties in enhancing the efficiency of their current business processes. Companies are instantly forced to make important decisions while continuously trying to maximize their profits. Keeping in mind the great volatility that characterizes the complex set of market conditions and the vast quantity of available information, a possible solution for improving business efficiency is the automation of business processes and excluding humans from making decisions (where this is possible). Humans simply do not posses the cognitive ability to process such an enormous quantity of information (and to make adequate decisions) in the few moments during which the relevant information does not change. A very logical solution to this problem lies in the technology of intelligent software agents – i.e. computer programs with the ability to completely autonomously manage a set of tasks.

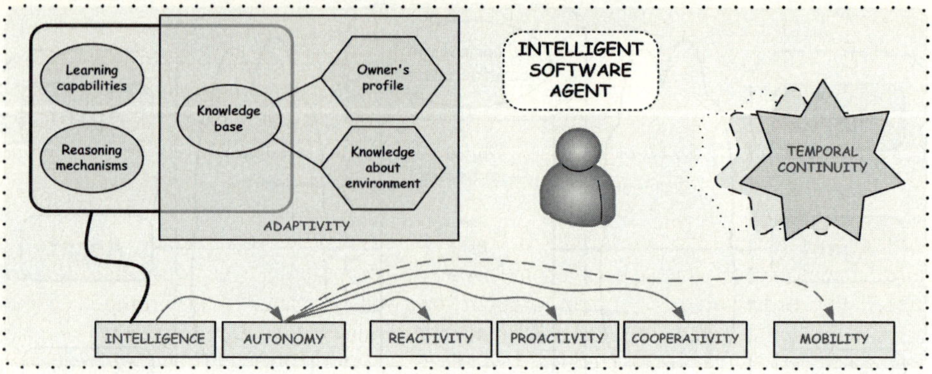

Fig. 2. A model of intelligent software agent

The CrocodileAgent [13, 14] is an intelligent agent developed at the Department of Telecommunications, Faculty of Electrical Engineering and Computing in Zagreb, Croatia. Designing the CrocodileAgent was an extension of a project that started in 2004 when the KrokodilAgent [15] was developed to participate in 2004 TAC SCM.

The agent's daily responsibilities can be divided into three logical tasks: negotiating supply contracts, bidding for customers' orders and managing daily assembly activities along with taking care of shipping completed orders to customers. These tasks will be described in the following sections.

3.1 Negotiation of Supply Contracts

In order to sell PCs it is necessary to purchase components and produce PCs from those components. The purchasing protocol is rather simple. The agent sends RFQs (*Requests for Quotes*) to the supplier that produces the needed component. The supplier responds with an offer. If the offer fits the agent's needs the agent replies with an order. A description of the CrocodileAgent's supply tactics follows.

On the day 0 the agent sends RFQs with the following delivery dates: 3, 9, 17, 27 and 69. The reserve prices the agent is willing to pay for the components are 102%, 107%, 92%, 82% and 77% of the nominal price on the respective delivery dates. These parameters were chosen after analyzing the prices and the delivery dates in a series of games. The requested quantities are smaller in short-term RFQs since the reserve prices in short-term RFQs are higher. RFQs are sent to all the suppliers that produce the needed component. The agent accepts the cheapest offer. That causes a temporary droop of the agent's reputation in the eyes of the other supplier. However, since the requested quantities are not high, the agent's reputation fully recovers in 20-30 days. The agent accepts partial offers if the chosen supplier can not deliver the requested quantity on time. In that case the quantities and reserve prices are modified for more aggressive purchasing of that component in the near future.

During the game the CrocodileAgent grounds its procurement strategy on short-term purchasing of smaller quantities of PC components. This strategy prevents the agent from paying large amounts of money to stock PC components in the warehouse.

Minimal required and maximal allowed quantities of components in storage are high during the game. As the game comes to an end the quantities are lowered to reduce potential loss of money. The goal is to sell out all the components that are still in inventory at the end of the game since they are paid upon delivery.

At the beginning of each day, the agent calculates the component quantity ordered, but not delivered, up to that moment for each component separately. The CrocodileAgent's ordered quantities of components are multiplied with a distance factor. This factor shrinks as the delivery date grows. The distance factor was introduced to lower the risk of running out of components since the supplier can cancel the order or deliver the ordered components later than arranged. New components are ordered if the current quantity of components stored in the warehouse increased with the estimated ordered quantity is lower than the maximal allowed quantities of components for that day. In spite of the maximal allowed quantities, in the first part of the game the agent makes long-term orders with small amounts of components to ensure cheap components for the second part of the game.

There are special mechanisms which calculate the reserve prices and exact quantities that need to be ordered. In case there is a very low quantity of a certain component in the warehouse a particular mechanism is activated. It allows short-term procurement of this component where the agent pays a higher price than usual. There is a similar mechanism if the customer demand rises rapidly. In this case the agent uses more components to produce more PCs, so the mechanism is activated to ensure that the agent does not run out of components and as result loses potentially profitable PC orders.

Special attention was paid to the end of the game. The intention was to enable the agent to send offers to customers for as long as possible, but also to maintain a low level of all components in the warehouse. This way, the following scenario was prevented; a large quantity of one component could be left over, not because there was no customer demand, but because the agent had spent all the other components needed to produce a certain type of PC.

3.2 Bidding for Customer Orders

The negotiation protocol between the customers and the agents is the same as the negotiation protocol between the agents and the suppliers. Each day customers send RFQs to the agents. The agents reply by sending offers. The agent that offered the lowest bid price wins the customer order.

3.2.1 An Algorithm for Sending Offers

The CrocodileAgent grades each RFQ and within those grades RFQs are sorted in chronological order of their delivery dates. The grade is determined by the difference between the customer's reserve price and the agent's cost of producing that PC. After grading and sorting RFQs the agent starts to send offers if the agent's PC production cost is lower than the customer's reserve PC price and there are enough components to produce the requested PCs. In case the latter condition is not fulfilled, the agent checks if the requested PCs can be delivered from the reserve PCs stored in the warehouse.

This algorithm comes in three versions. The version that is active on a certain day is determined depending on the stage of the game (beginning, middle, end), the number of production cycles needed to produce all the active orders and the algorithm that was used the day before. The basic difference between these three versions is the method of determining the offer prices for the PCs. The most frequently used version during the game is the *Normal version* and it determines the offer price in two ways:

- If the agent offers PCs based on the components currently present in the warehouse, the offer price is calculated as the basic PC price increased by the agent's desired profit;
- If the agent offers PCs that are already produced and stored in the warehouse, the offer price is slightly under the customer's reserve price.

The *High Demand Version* uses a "greedy" algorithm since the offer prices for the PCs are always just slightly under the customer's reserve price. It is used when there is a very high customer demand for PCs since, in those cases, agents usually do not send offers for all RFQs received. The *End Game Version* is used for the finishing stage of the game. What makes this version of the algorithm different from the other two versions is the fact that it sorts RFQs in increasing order of their corresponding penalties. The main purpose of this version is to sell out the whole inventory in the warehouse so the profit that the agent adds to the basic PC price is minimal. All the versions of the selling algorithm implement a mechanism used for preventing late deliveries. Each day, the agent monitors its obligations to customers by calculating the number of factory cycles needed to fulfill its existing orders. On the basis of that information it determines the earliest delivery date for sending new PC offers. This way the agent is prevented from sending offers that cannot be delivered by the requested delivery date.

3.2.2 Calculating Prices of Components in the Warehouse and the Profit Margin

The basic PC price is calculated by summing the average prices of each component incorporated in the PC. The agent always knows the price paid for each component that is in its warehouse. If the current supply of components is higher than the calculated optimal for that day a discount for the components is approved. The agent also gives a discount on components at the end of the game to sell out the components that are still in the warehouse.

The profit margin is calculated every day and the calculating algorithm analyzes the following parameters:

- Percentage of the agent's recent offers which resulted in customer orders
  - If this parameter is decreasing, the profit margin also decreases and vice-versa;
- Distribution of agent's factory occupancy in the next few days
  - High factory utilization causes the increase of profit margin and vice-versa;
- Recent PC prices of other agents compared to CrocodileAgent's.
  - If CrocodileAgent offers cheap PCs the profit margin increases and vice-versa;
- Due date listed in the customer RFQ
  - Earlier due date causes the higher profit margin and vice-versa;
- Demand level in the market segment the wanted PC belongs to
  - If the demand is low then the profit margin is decreased and vice-versa.

3.3 Managing Factory Activities and Shipping Completed Orders to Customers

In prior versions the CrocodileAgent produced PCs only after receiving customer orders, but we decided to add the possibility of producing PCs even if nobody ordered them. Since the TAC SCM game is of stochastic nature, customer demand varies during the game. If the agent does not produce PCs and the PC demand is low, a large part of the agent's factory capacities stay unutilized. If the agent produces PC stocks during a period of low PC demand, its factory will be utilized and everything will be prepared for a period of high PC demand. This tactic has it weaknesses since the agent cannot know for sure what the future demand is going to be. Thus, it can produce more PCs than can be sold by the end of the game. We tried to lower this risk by introducing quantity limits which represent the maximum number of PCs which can be available in stock. As the end of the game approaches, these limits are lowered accordingly.

Each day the list of active orders is sorted in chronological order of the delivery dates. The agent first checks if there are enough PCs in the warehouse to fulfill the order. If there are enough PCs, they are added to the delivery schedule. If there are not enough PCs, the agent checks if there are enough components to produce the requested PCs. If so, the components are reserved and the agent tries to add them to the production schedule. When finished with analyzing all the active orders, the CrocodileAgent checks the production schedule for the next day. After determining the amount of free capacity available, the agent checks which PC types can be produced without creating a large stock.

4 TAC SCM 2005 Competition and Controlled Experiments

The 2005 TAC SCM competition was divided into three parts: qualifying rounds[1] held from June 13th-24th, seeding rounds[2] held from July 11th -22nd and final round held from August 1st-3rd. There were 32 teams competing in the qualifying and 25 teams in the seeding rounds. 24 teams competed in the finals. The CrocodileAgent took 4th place in the quarterfinals[3] with an average score of 11.64M and ended its participation in 2005 TAC SCM.

To evaluate the performance of our agent, we held a competition with some of the best agents from the TAC SCM 2005 competition. The chosen opponents were: TacTex [16], Mertacor [8], DeepMaize [17], MinneTAC and PhantAgent. All the agents were downloaded from the official TAC Web page (http://www.sics.se/tac). The competition was held in our laboratory and consisted of 20 games. The final ranking is shown in Table 1. After the competition finished, we conducted a detailed analysis of the games played. We focused on the supply procurement mechanism since the biggest change in the game rules for the TAC SCM 2005 concerned the suppliers and their price determining algorithm.

---

[1] http://www.sics.se/tac/page.php?id=47

[2] http://www.sics.se/tac/page.php?id=48

[3] http://www.sics.se/tac/page.php?id=49

Table 1. Competition results at server mobility3.labs.tel.fer.hr .

Place	Agent	Score	Games Played	Place	Agent	Score	Games Played
1	TacTex	5 638 295	20	4	PhantAgent	976 780	20
2	DeepMaize	4 581 805	20	5	MinneTAC	-426 722	20
3	Mertacor	1 364 353	20	6	CrocodileAgent	-1 243 212	20

The majority of the analysis was done with the CMieux Analysis and Instrumentation Toolkit for TAC SCM [18]. The main task was to analyze component purchases. After gathering information regarding the prices the agents paid for each type of component and the quantities they purchased, we calculated the average prices. These prices are shown in Figure 3. Components marked with ID 1xx are CPUs, 2xx are motherboards, 3xx are memory and 4xx are hard disks.

Since the price of a CPU accounts for more than 50 % of the PC price, it is very important to purchase cheap CPUs. It can be noticed from Figure 3 that TacTex bought some of the cheapest CPUs while DeepMaize paid the highest prices for CPUs. The CrocodileAgent bought the third cheapest CPUs. The situation is slightly different with other components: Mertacor bought the cheapest motherboards and memory, while the CrocodileAgent bought the cheapest hard disks. MinneTAC bought the most expensive motherboards, memory and hard disks.

Furthermore, we analyzed the quantity of components bought during the game. From Figure 4, it can be noticed that TacTex, DeepMaize and the CrocodileAgent bought larger amounts then the other agents. The average minimal number of components that stayed in the agents' warehouses after the game has finished can be determined from the same figure, because the maximum number of PCs that can be assembled is equal to the minimal amount of a certain component type. The leftover components represent a direct loss of money since they were paid for upon delivery. The best results regarding leftover component management were obtained by the PhantAgent and DeepMaize. The majority of their leftover components were memory components. This is desirable since memory is the cheapest component type.

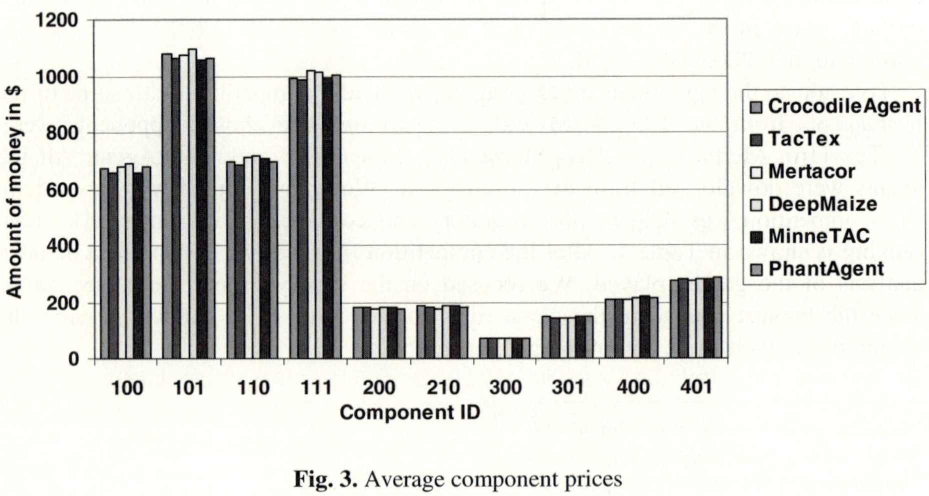

Fig. 3. Average component prices

The Mertacor agent also had a relatively small number of leftover components, but they were mostly CPUs. This is not as good since CPUs are at least ten times more expensive than memory components. All three agents had between 500 and 800 leftover components. TacTex and MinneTAC had more than 2600 leftover components on average. The difference is that half of TacTex's leftover components were CPUs, while half of MinneTAC's leftover components were memory components. The CrocodileAgent was in the middle with an average of 1250 leftover components. This is a direct consequence of the ordering algorithm that does not compare the total number of ordered components of each component type.

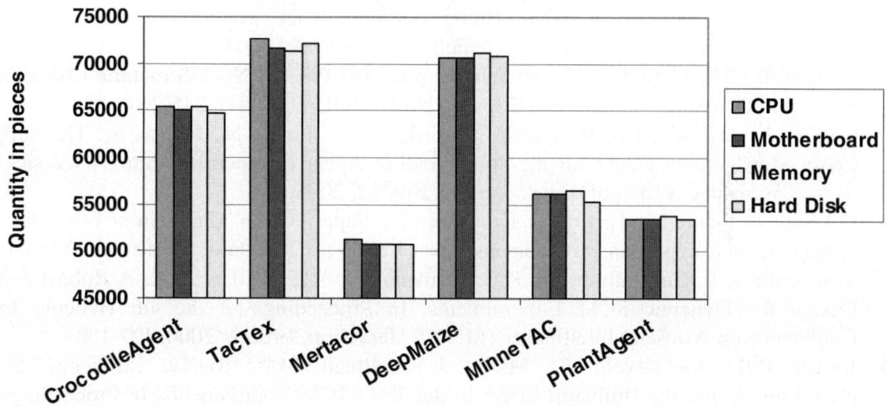

Fig. 4. Total quantity of bought components

5 Conclusions and Future Work

In this paper, we presented the CrocodileAgent which is a trading agent that participated in TAC SCM 2005. After a short game description, we listed the basic TAC SCM Agent tasks and explained how they were implemented in the CrocodileAgent.

We briefly presented the results of the CrocodileAgent in the TAC SCM 2005. In order to improve the functionalities of the agent, we held a competition with some of the best agents in TAC SCM 2005. We figured that this was a good way to determine the CrocodileAgent's soft spots. A thorough analysis of the competition was conducted. The results were a little discouraging since the CrocodileAgent placed last, but a lot was learned. The main reason for the CrocodileAgent's results lies in its reactive algorithm for selling PCs. This algorithm does not predict the fluctuation of prices on the PC market, but it only reacts to the current state of the PC market. Thus, during further development of our agent special attention needs to be dedicated to the PC selling algorithm with an emphasis on customer demand prediction and the prediction of winning PC prices. The component purchase algorithm and mechanisms for managing factory activities function quite well, but there is always room for improvement.

References

1. Benish, M., Sardinha, A., Andrews, J., Sadeh, N.: CMieux: Adaptive Strategies for Competitive Supply Chain Trading. In Proceedings of the 8th Int. Conference on Electronic Commerce (ICEC), Fredericton, Canada, 2006. 1-10
2. Wurman, P.R., Wellman, M.P., Walsch, W.E.: Specifying Rules for Electronic Auctions. AI Magazine, Vol. 23 (3). American Association for Artificial Intelligence, Menlo Park (2002). 15-24
3. Podobnik, V., Trzec, K., and Jezic, G.: An Auction-Based Semantic Service Discovery Model for E-Commerce Applications. Lecture Notes in Computer Science, Vol. 4277. Springer-Verlag, Berlin Heidelberg New York (2006). 97-106
4. Fensel, D.: Ontologies: A Silver Bullet for Knowledge Management and Electronic Commerce. Springer-Verlag, Berlin Heidelberg New York (2004)
5. Carlson, B.: The Digital Economy: What is New and What is Not?. Structural Change and Economic Dynamics, Vol. 15 (3). Elsevier, North-Holland (2004). 245-264
6. Collins, J., Arunachalam, R., Sadeh, N., Eriksson, J., Finne, N., Janson, S.: The Supply Chain Management Game for the 2005 Trading Agent Competition. http://www.sics.se/tac/tac05scmspec_v157.pdf. Date accessed: Nov 12, 2006.
7. Eriksson, J., Finne, N., Janson, S.: Evolution of a Supply Chain Management Game for the Trading Agent Competition. AI Communications, Vol. 19 (1). IOS Press, Amsterdam (2006). 1-12
8. Kontogounis, I., Chatzidimitriou, K.C., Symeonidis, A.L., Mitkas, P.A.: A Robust Agent Design for Dynamic SCM Environments. In Proceedings of the 4th Hellenic Joint Conference on Artificial Intelligence (SETN), Heraklion, Greece, 2006. 127-136
9. Jordan, P.R., Kiekintveld, C., Miller, J., Wellman, M.P.: Market Efficiency, Sales Competition, and the Bullwhip Effect in the TAC SCM Tournaments. In Proceedings of the AAMAS Joint International Workshop on the Trading Agent Design and Analysis and Agent Mediated Electronic Commerce (TADA/AMEC) Hakodate, Japan, 2006. 99-111
10. Ketter, W., Collins, J., Gini, M., Gupta, A., Shrater, P.: Identifying and Forecasting Economic Regimes in TAC SCM. In Proceedings of the IJCAI Workshop on Trading Agent Design and Analysis (TADA), Edinburgh, UK, 2005. 53-60
11. Bradshaw, J.M.: Software Agents. MIT Press, Cambridge, Massachusetts, USA (1997)
12. Chorafas, D.N.: Agent Technology Handbook. McGraw-Hill, New York, USA (1998)
13. Podobnik, V., Petric, A., Jezic, G.: The CrocodileAgent: Research for Efficient Agent-Based Cross-Enterprise Processes. Lecture Notes in Computer Science, Vol. 4277. Springer-Verlag, Berlin Heidelberg New York (2006). 752-762
14. Petric, A., Podobnik, V., Jezic, G.: The CrocodileAgent: Analysis and Comparison with Other TAC SCM 2005 Agents. In Proceedings of the AAMAS Joint International Workshop on the Trading Agent Design and Analysis and Agent Mediated Electronic Commerce (TADA/AMEC) Hakodate, Japan, 2006. 202-205
15. Petric, A., Jurasovic, K.: KrokodilAgent: A Supply Chain Management Agent. In Proceedings of the 8th International Conference on Telecommunications (ConTEL), Zagreb, Croatia, 2005. 297-302
16. Pardoe, D., Stone, P.: Bidding for Customer Orders in TAC SCM: A Learning Approach. In Proceedings of the AAMAS International Workshop on Trading Agent Design and Analysis (TADA), New York, USA, 2004.
17. Wellman, M.P., Estelle, J., Singh, S., Vorobeychik, Y, Kiekintveld, C., Soni, V.: Strategic Interactions in a Supply Chain Game. Computational Intelligence, Vol. 21 (1). Blackwell Publishing, Oxford, UK (2005). 1-26
18. Benisch, M., et. al.: CMieux Analysis and Instrumentation Toolkit for TAC SCM. Carnegie Mellon University School of Computer Science TR CMU-ISRI-05-127 (2005)

A Study to Apply Intelligent Agents for B2C Shopping Mall

Ha-Jin Hwang

Department of Management Information Systems, Catholic University of Daegu
Kyung San, Daegu, 712-702, Korea
hjhwang@cu.ac.kr

Abstract. The spread of internet is so rapid that emerging e-Business is making dramatic changes in this digital economy. An important component of e-Business research is the issue of the effective marketing in the keen competition. Many agents are developed to help buyers handle the dynamic purchasing environment and reduce the complexity of purchasing data. However, while most existing agents are focused on buyers' needs, prior research on agents lacks the sellers' perspectives. This study is intended to demonstrate the agent system that supports analysis of buyers' purchasing data and predicts their behavior so that sellers can develop more systematic marketing strategies.

1 Introduction

With the widespread of internet, e-business becomes the cutting edge for today's business. In e-business environments, numerous buyers can access to the website dealing with so many products data to make a purchase decision. At the same time, sellers are facing a very difficult situation to manage back-office requirements in order to keep the quality of customer services [2,7,19]. A variety of agents have been developed to assist buyers in searching websites and making purchase decisions. However, prior research on agents, mainly focusing on buyers' needs, has failed to address the seller's managerial perspectives. Most agents are designed to track the buying path and identify the websites visited to make a final purchase decision while some agents provide a comparison of prices and products details.

These agents can obviously help buyers save their time and efforts by reducing the amount of information to consider. However, as the importance of customer management and more effective sales strategies are required to keep competitive position in the market, sellers might be interested in agents that can analyze the buyers' purchasing pattern and predict their behavior in the websites.

The primary purpose of this paper is to demonstrate intelligent agents that support analysis of buyers' purchasing data and predict their behavior so that sellers can develop more systematic marketing strategies. The agent system that introduced in this paper monitors buyers' activities in the websites, identifies possible features that might affect purchasing decisions, and finally provides a basis for preparing effective marketing strategies with suitable alternatives.

N.T. Nguyen et al. (Eds.): KES-AMSTA 2007, LNAI 4496, pp. 607–615, 2007.

Moreover, existing agent systems do not reflect the time constraint. Consequently, they can't provide the timely adjustment based on the seasonal variation and the change of fashion. The agent system in this paper applies the adaptive system that considers the change of preferences based on the seasonal change and recommends the products at the latest fashion. Those information reflected in the change of preferences gives sellers insights to satisfy the buyers' needs and predict the product pattern that are attractive to the buyers.

This paper describes a cyber shopping mall using intelligent agents that incorporate considerations for buyer behavior analysis and purchasing pattern monitoring as well as time variations based on the change of season and fashion. The system considers the various aspects of the buyer behavior during the website search and suggests a guideline that sellers can concentrate on the factors to attract the buyers by analyzing the product selection process and the purchasing pattern.

2 Literature Review

Agents are computer programs that mimic human actions to address the needs of sellers and buyers to cut cost while improving the quality of goods and services and increasing the speed of customer services [12] . Agents are developed to help users in finding useful information and in determining reasonable decision even in the ambiguous situations[11,15].

The field of intelligent agent has been a rapid growth over the last decade and now becomes a powerful tool in most industrial applications, Recently, it is applied to intelligent user interface, autonomous agent, vision system, knowledge discovery and data mining, information retrieval, electronic commerce, a personal assistant of web. fuzzy decisions, and decision making in complex environments.

Major characteristics of agents include personalized, continuously running, and autonomous, reactive and pro-active. Jenning and Wooldridge[21] describe properties of agents as follows.

-Autonomy: agents operate without the direct intervention of humans or others.
-Social Ability: agents interact with other agents and possibly human via some kind of agent-communication.
-Reactivity: agents perceive their environment (which may be the physical world, a user via a GUI, internet, other agents, etc) and respond in a timely fashion to changes as they occur.
-Pro-activity: agents do not only act in response to their environment, they exhibit goal-directed behavior and take the initiative.

According to Nissen [15], agents are classified into five categories: watcher agents, learning agents, shopping agents, information retrieval agents, and helper agents. Many researchers address the effects of agents on electronic commerce [8,14,21]. Mae, Guttman, and Moukas [14] predict the future development of such agents, and discuss the use of different kinds of applications to assist consumers in buying activities. Examples of these agents include Anderson Consulting's Bargain Finder, Curtin University's Bargain Boat, and University of Washington's ShopBoat [4,16].

Intelligent agents perform specific tasks on behalf of users. For example, agents are designed to search the websites for information gathering, monitoring, and analyzing the environment. Some agents are applied to interact with other agent and may act upon messages from other agents. Other researchers predict that monitoring agents will continuously be developed to search the web for deals on behalf of users in e-business, considering that the importance of effective marketing has been increased for the success of e-business[1,17].

Many researches have explored the opportunity to reduce the burden of buyers in gathering information and comparison of the products using comparison agents while push technology has taken part of delivering personalized marketing by analyzing buyer profile and preferences [9,20]. These agents have significantly reduced the buyers' time and efforts to examine purchasing data. However, they still fail to address the seller's managerial point of view. Other research issue is reflecting time constraint for dynamically changing environment.

3 Architecture of the System

The system basically consists of two agents: the monitoring agent and the buyer analysis agent. The monitoring agent observes buyers' activity when the buyer enters a web site and keeps track of buyers' interests and details of buyers' behavior. The buyer analysis agent generates a database that contains the information of the buyer behavior based on the results given by the monitoring agent. Utilizing the monitoring agent and the buyer analysis agent, sellers can identify the common characteristics of the moving paths, revisit patterns, time interval between the visits and other statistics. Sellers, then, can establish not only an effective promotion strategy but also well-organized product selection to make the website to be more attractive to the buyers.

The architecture of the agent system proposed in this study, as shown in figure 1, consists of user interface, application program, user monitoring agent, and user behavior analysis agent

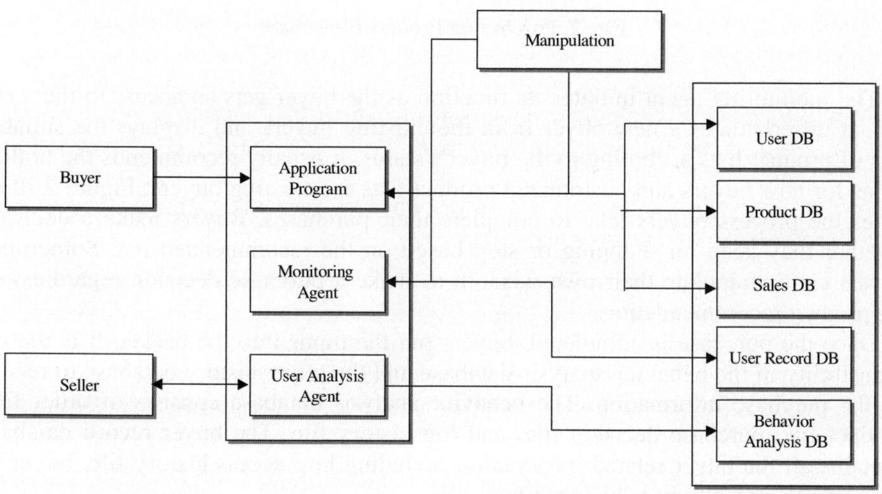

Fig. 1. Architecture of the System

Functions of each component can be summarized as follows:

-User interface: manages a dialog between users and the system.
-Application program: enables users to navigate, search, and make a purchase on the website.
-Monitoring agent: monitors users behavior on the website and record the data.
-Buyer behavior analysis agent: analyze and predict consumer's behavior based on the data from the monitoring agent

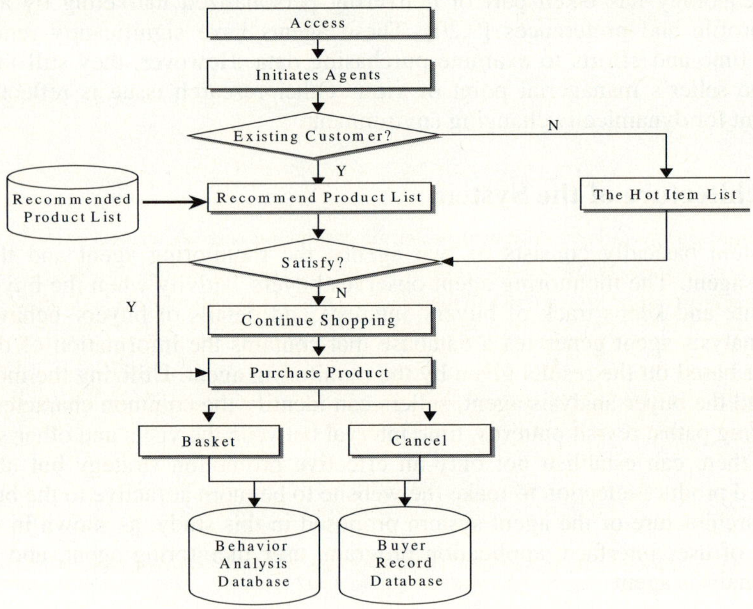

Fig. 2. Purchasing Process Flowchart

The monitoring agent initiates its function as the buyer gets an access to the website. It differentiates a new buyer from the existing buyers and displays the suitable sets of product lists according to the buyer's status. It usually recommends the hottest items for new buyers and customized product lists for existing buyers. Figure 2 illustrates the process buyers take to complete their purchases. Buyers make a decision whether they keep on shopping or stop based on the recommended list. Sometimes buyers can manipulate their own wisdom to make a purchase decision regardless of the product recommendation

Once the purchase is completed, buyers put the items into the basket. It is the responsibility of the behavior analysis database and the buyer history database to record all the purchase information. The behavior analysis database consists of order file, product file, purchase decision file, and log history file. The buyer record database contains all the buyer related information including http access history file, buyer IP file, log file, and purchase history file.

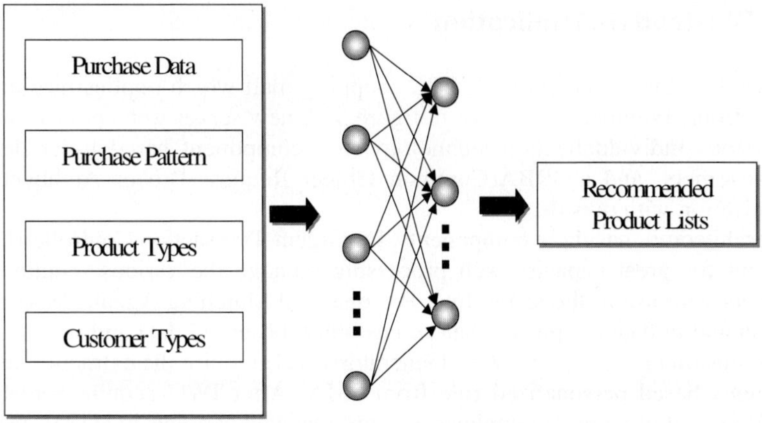

Fig. 3. Neural Nets for Recommended Product Lists

The recommended product list is drawn based on the previous purchase data, the purchasing pattern, product types, and buyers' personal data. Neural nets, as shown in figure 3, are employed to enable agents to learn the purchasing pattern and predict the future buyer behavior from the previous purchase data and buyer's personal data.

Figure 4 shows the Nassi-Schneidermann chart that explains the logic to generate the product list. The agent provides the real-time adjustment by adjusting the weights for the buyer behavior variables.

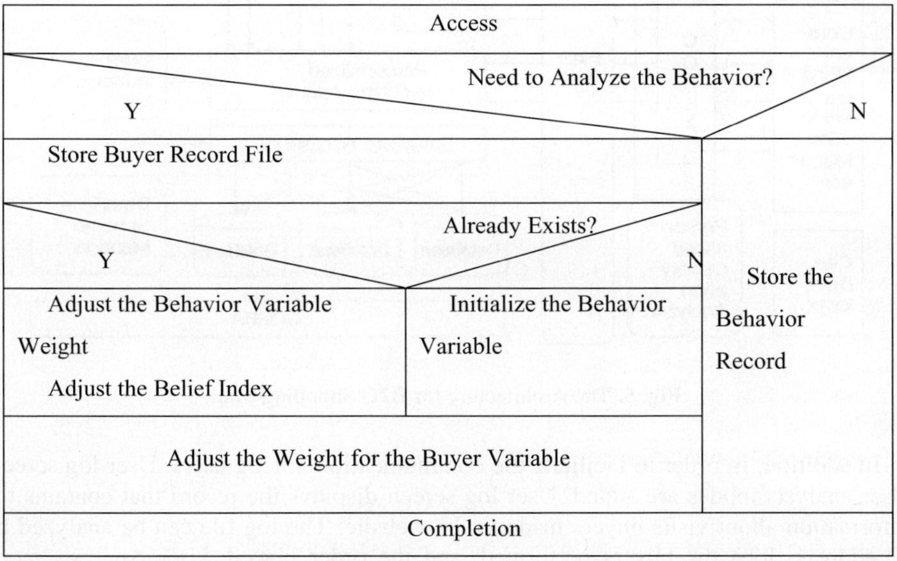

Fig. 4. Product List Generation Logic

4 An Illustrative Application

In this section, the architecture for B2C shopping mall which implies medium/ small scale electronic commerce is shown in figure 5. A new server web applications which serve various individualization technology, use component based n-tier design for many customers, and CORBA(Common Object Request Broker Architecture) for communication with outside.

The architecture has three components. Intelligent Transaction Controller(ITC) as a component for great capacity web processing manage the various requirements of many users who use at the same time. Personalized Matching Agent(PMA) manages extraction and update for profile data of customer, observed data and session data for action of customer. Interactive Web Generator(IWG) require the extraction of matching contents based personalized rule from PMA. After IWG acquire contents from PMA, IWG read predefined templates and add matched contents in HTML pages, and then send to web browser of customers. It uses Database Accessor which is a kind of middleware library. The library supports many database library such as Oracle, Sybase, Informix and etc., so P-Commerce Solution is independent of any DBMS.

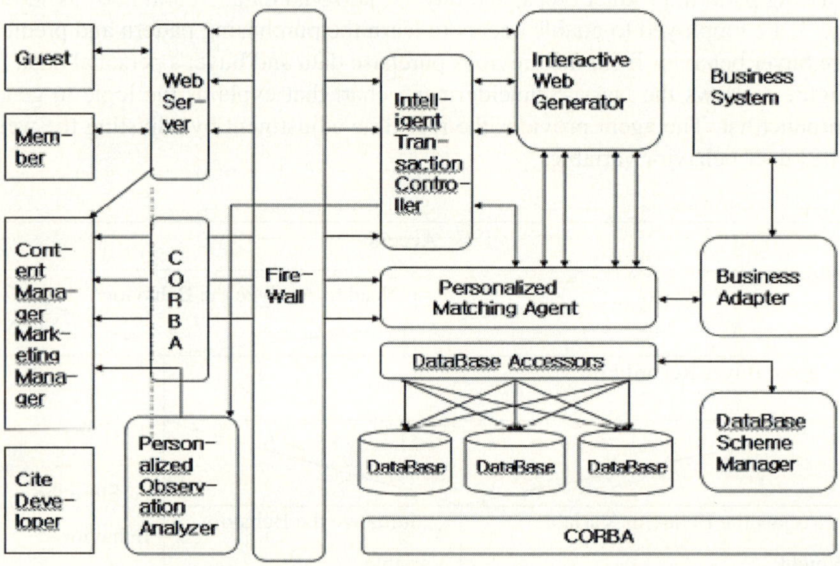

Fig. 5. The Architecture for B2C Shopping Mall

In addition, in order to facilitate the communication among users, User log screen, User analysis modes are added. User log screen displays the record that contains the information about visits buyers made to the website. The log file can be analyzed by IP address, Jobs the buyers performed, and the order placed. User Analysis mode deals with tasks that a particular buyer has done in the website. This profile is a basis of the learning process for the neural nets and can be converted into the purchasing pattern. In order for sellers to examine buyer's purchasing preferences, the data is

Table 1. Buyer Profile Code

	Field Name	Field Code	Value
General	Product-Code	Code	20 (character)
	Product-Name	Name	40 (character)
	Price	Price	1~ 99999
	Manufacturer	Prod	1 ~ 99999
	After Sale Service	As	1 ~ 99999
	Registered Date	Date	dd-mm-yy
	Warranty	Warr	50 char
Age	Teen	Age1	10 ~ 19
	Twenty-Early	Age2	20 ~ 24
	Twenty-Late	Age3	25 ~ 29
	Thirty	Age4	30 ~ 39
	Forty	Age5	40 ~ 49
	About Fifty	Age6	50 ~ 99
Preferences	Formal	Pt1	1 ~ 5
	Casual	Pt2	1 ~ 5
	Traditional	Pt3	1 ~ 5
	Modern	Pt4	1 ~ 5
	Color	Pt5	1 ~ 5
	Design	Pt6	1 ~ 5
	Fashion	Pt7	1 ~ 5

Table 2. Buyer Behavior Analysis Code

	Field Name	Field Code	Value
General	Buyer Code	Code	20(character)
	Name	Name	40(character)
	Sex	Sex	1 ~ 2
	Age	Age	1 ~ 99
	Occupation	Job	30 char
	Address	Addr	50 char
	Phone	Phone	13 char
	Marital-Status	Ma	1 ~ 2
	Hobby	Hob	30 char
	Preferences	Buying Score	Buying Count

Table 2. (*Continued*)

Formal	Pt1 (1 ~ 5)	Ps1 (0 ~ 99999)	Pc1 (0 ~ 99999)
Casual	Pt2 (1 ~ 5)	Ps2 (0 ~ 99999)	Pc2 (0 ~ 99999)
Tradi-tional	Pt3 (1 ~ 5)	Ps3 (0 ~ 99999)	Pc3 (0 ~ 99999)
Modern	Pt4 (1 ~ 5)	Ps4 (0 ~ 99999)	Pc4 (0 ~ 99999)
Color	Pt5 (1 ~ 5)	Ps5 (0 ~ 99999)	Pc5 (0 ~ 99999)
Design	Pt6 (1 ~ 5)	Ps6 (0 ~ 99999)	Pc6 (0 ~ 99999)
Fashion	Pt7 (1 ~ 5)	Ps7 (0 ~ 99999)	Pc7 (0 ~ 99999)

retrieved from the buyer record database and the buyer analysis database. This data will be used to test the hypotheses of the buyers' preferences .

For example, a recent one month data for a particular buyer can be retrieved from the database and sorted out in time, product name basis, and then weights are calculated and finally reflected into the buyer variable. Tables 1 and 2 show codes for analysis of purchasing pattern and buyer preferences.

5 Conclusion

e-Business has created many opportunities in today's internet based society. One of the most critical aspects influencing the success of e-Business is the effective marketing to interact with the buyers. Many agents have been developed to assist buyers in searching websites and making purchase decisions. However, prior research on agents, mainly focusing on buyers' needs, has failed to address the seller's managerial perspectives. Most agents are designed to track the buying path and identify the websites visited to make a final purchase decision while some agents provide a comparison of prices and products details.

Buyers are now well-informed, networked, and even more wise in making their purchasing decisions. Therefore, sellers need to provide more personalized services based on systematic analysis of buyer preferences and purchasing pattern. In this regard, this paper demonstrates an implementation of cyber shopping mall using intelligent agents that support analysis of purchasing data and predict purchasing pattern as well as consideration of time constraint. The system appeared to be very successful implementation of agents for enabling sellers to prepare effective marketing strategy by analyzing product selection process and predicting purchase pattern.

Finally, the contribution of this study is that the system proposed in this study provides guidelines for developing cyber shopping mall using intelligent agents for seller's perspectives.

For the future research, this study might need to be extended to utilize data mining approach to analyze very large data set and explore the possibility of web-housing to manage various aspects of purchase knowledge.

References

1. Bogonikolos N, fragoudis D, Likothnnasis L. (1999). Archimedes: An Intelligent Agent for Adaptive Personalized Navigation within a Web Server. In Proceedings of the 32nd International Conference on Systems Science (HICSS), Maui, HI, Shriver B, Sprague R (eds). IEEE Computer Society Press, Los Alamitos, CA.
2. Borselius, N. (2002). Mobile agent security. Electronics & Communication Engineering Journal, 14(5), 211—218.
3. Chavez A, and Maes P. (1996). Kasbsh: An Agent Marketplace for Buying and Selling Goods. In Proceedings of the First International Conference on the Practical Application of Intelligent Agents and Multi~Agent Technology, London, UK.
4. H.R. Choi, H.S. Kim, Y.J.Park, K.H. Kim, M.H. Joo, and H.S. Sohn. (2000). A Sales Agent for Part Manufacturers : VMSA. Decision Support Systems, pp.333-346.
5. Bill Fulkerson. (1997). A Response to Dynamic Change in the Market Place. Decision Support Systems, pp. 199-214.
6. Fatima, S.S., Wooldridge, M.J., Jennings, N.R., 2004. An agenda-based framework for multi-issue negotiation Artifi-cial Intelligence 152 (1), 1—45.
7. Field, S., Waidner, M. (Eds.), 2000. Electronic commerce. Special issue in the Computer Networks Journal 32 (6).
8. Fernandez, E., & Olmedo, R. (2005). An agent model based on ideas of concordance and discordance for group ranking problems. Decision Support Systems, 39(3), 429—443
9. Klein M, Dellarocas C. Exception Handling in agent Systems. (1999). In Proceedings of the Third International Conference on Autonomous Agents, Seatlle, Washington. ACM Press, New York, pp.62-68.
10. Klein, M., Faratin, P., Sayama, H., Bar-Yam, Y., 2002. Negotiating complex contracts. In: Proceedings of the ACM International Joint Conference on Autonomous Agents and Multi-Agent Systems, Bologna, Italy, July.
11. Kuo, M. H., & Lin, M. J. (2000). Using Software Agents to Retrieve Information from WWW, SCI'2000, Orlando, Florida, USA, pp. 400—405.
12. Machlis S. March 22, (1999). Agent Technology. Computerworld 33(12), p.69.
13. Maes P. (1994). Agents That Reduce Work and Information Overload. Communications of the ACM 37(7):, pp.30-40.
14. Maes P, Guttman, and RH, Moukas, A.G., (1999). Agents That Buy and Sell. Communication of the ACM 42(3), pp. 81-87.
15. Nissen, M. (1995). Intelligent Agent: A Technology and Business Application Analysis.
16. Nwana H. (1996). Software Agents: An Overview. Knowledge Engineering Review 11(3), pp.1-40.
17. Pietro Picco, G. (2001). Mobile agents: an introduction. Microprocessors and Microsystems, 25(2), 65—74.
18. Roussaki, I., Louta, M., Pechlivanos, L., 2004a. An efficient negotiation model for the next generation electronic market-place. In: Proceedings of the 12th IEEE Mediterranean Electrotechnical Conference (MELECON 2004), Dubrovnic, Croatia, May.
19. Michael J. Shaw, David M. Gardner, and Howard Thoma.. (1997). Research Opportunities in Electronic Commerce, Decision Support Systems, pp. 149-156.
20. Ting-Peng Li, and Jin Shiang Huang. (2000). A Framework for Applying Intelligent Agents to Support Electronic Trading. Decision Support Systems, pp. 305-317.
21. Wooldridge M, and Jennings N. (1998). Intelligent Agents: Theory and Practice. Knowledge Engineering Review 10(2), pp. 115-152.

Password-Only Authenticated Key Exchange Between Two Agents in the Four-Party Setting*

Youngsook Lee[1], Junghyun Nam[2], Jin Kwak[3], and Dongho Won[1,**]

[1] Information Security Group, Sungkyunkwan University, Korea
{yslee,dhwon}@security.re.kr
[2] Department of Computer Science, Konkuk University, Korea
jhnam@kku.ac.kr
[3] Department of Information Security, Soonchunhyang University, Korea
jkwak@security.re.kr

Abstract. Agent technology is emerging as a new software paradigm in the areas of distributed computing. The use of multiple agents is a common technique in agent-based systems. In distributed agent systems, it is often required for two agents to communicate securely over a public network. Authentication and key exchange are fundamental for establishing secure communication channels over public insecure networks. Password-based protocols for authenticated key exchange are designed to work even when user authentication is done via the use of passwords drawn from a small known set of values. There have been many protocols proposed over the years for password authenticated key exchange in the three-party scenario, in which two agents attempt to establish a secret key interacting with one same authentication server. However, little has been done for password authenticated key exchange in the more general and realistic four-party setting, where two clients (or, two agents) trying to establish a secret key are registered with different authentication servers. In this paper, we propose a new protocol designed carefully for four-party password authenticated key exchange that requires each agent only to remember a password shared with its authentication server.

Keywords: Password, key exchange, authentication, inter-domain, agent.

1 Introduction

Agent based computing is a new software paradigm in Information Technology today. In a broad sense, an agent is an autonomous program that acts on behalf of a (human) user. The use of multiple agents is a common technique in agent-based systems. A major challenge for the operation of multi-agent systems is to provide an efficient way for secure inter-agent communication over a

* This work was supported by the Korean Ministry of Information and Communication under the Information Technology Research Center (ITRC) support program supervised by the Institute of Information Technology Assessment (IITA).

** Corresponding author.

N.T. Nguyen et al. (Eds.): KES-AMSTA 2007, LNAI 4496, pp. 616–625, 2007.

public network. Authenticated key exchange (AKE) is of fundamental importance to anyone interested in communicating securely over a public network. AKE protocols are cryptographic primitives that specify how two or more parties communicating over a public network establish a common secret key called *session key*. This secret key is typically used to build a confidential or integrity-preserving communication channel among the involved parties. Achieving any form of authentication in key exchange protocols inevitably requires some secret information to be established between the communicating parties in advance of the authentication stage. Cryptographic keys, either secret keys for symmetric cryptography or private/public keys for asymmetric cryptography, may be one form of the underlying secret information pre-established between the parties. However, these high-entropy cryptographic keys are random in appearance and thus are difficult for humans to remember, entailing a significant amount of administrative work and costs. Eventually, it is this drawback that password-based authentication came to be widely used in reality. Passwords are drawn from a relatively small spaces like a dictionary and are easier for humans to remember than cryptographic keys with high entropy.

Two-party protocols (e.g., [6,4,13,19,12]) for password authenticated key exchange (PAKE) are well suited for client-server architectures, they are inconvenient and costly for use in large scale peer-to-peer systems. Since two-party PAKE protocols require each potential pair of communication users to share a password, a large number of users results in an even larger number of potential passwords to be shared. It is due to this problem that three-party models have been often considered in designing PAKE protocols [11,18,16,1]. In a typical three-party setting, users (called clients) do not need to remember and manage multiple passwords, one for each communicating party; rather, each client shares a single password with a trusted server who then assists two clients in establishing a session key by providing authentication services to them. However, this convenience comes at the price of users' complete trust in the server. Therefore, whilst the three-party model will not replace the two-party model, it offers easier alternative solutions to the problem of password authenticated key exchange in peer-to-peer network environments.

Up to now, most of literature discussing the problem of password authenticated key exchange focused their attention to the two-party model or the three-party model. While the three-party model seems to provide a more realistic approach in practice than the two-party model in which clients are expected to store multiple passwords, it is still restrictive in the sense that it assumes that two clients are registered and authenticated by the same server. In reality, the authentication server of one client may be different from that of another client. Indeed, it is a typical environmental assumption [15] that a client registers with the server in its own realm and trusts only its own server, not the servers in other realms. In this case, how to efficiently authenticate a client who is registered with the server in the other realm becomes an important issue.

This paper considers password authenticated key exchange in the inter-domain distributed computing environment just mentioned above. We propose a new inter-domain PAKE protocol which involves four participants: two agents and two authentication servers, one for each agent. The contribution of this paper is to propose the four-party password authenticated key exchange protocol that requires each agent only to remember a password shared with its authentication server.

2 Protocol Preliminaries

We first briefly review the cryptographic assumptions which underly the security of our protocol, and then introduce some notation used to describe the protocol. There are four entities involved in the protocol: two agents A and B, and two authentication servers SA and SB respectively of A and B. We denote by ID_A, ID_B, ID_{SA}, and ID_{SB}, the identities of A, B, SA, and SB, respectively.

Computational Diffie-Hellman (CDH) Assumption. Let g be a fixed generator of the finite cyclic group $\mathbb{Z}_p^*$. Informally, the CDH problem is to compute g^{ab} given g^a and g^b, where a and b were drawn at random from $\{1, \ldots, |\mathbb{Z}_p^*|\}$. Roughly stated, $\mathbb{Z}_p^*$ is said to satisfy the CDH assumption if solving the CDH problem in $\mathbb{Z}_p^*$ is computationally infeasible for all probabilistic polynomial time algorithms.

Symmetric Encryption Scheme. A symmetric encryption scheme is a triple of polynomial time algorithms $\Gamma = (\mathcal{K}, \mathcal{E}, \mathcal{D})$ such that:

- The *key generation algorithm* $\mathcal{K}$ is a randomized algorithm that returns a key k. Let Keys(Γ) be the set of all keys that have non-zero probability of being output of $\mathcal{K}$.
- The *encryption algorithm* $\mathcal{E}$ takes as input a key $k \in Keys(\Gamma)$ and a plaintext $m \in \{0,1\}^*$. It returns a ciphertext $\mathcal{E}_k(m)$ of m under the key k. This algorithm might be randomized or stateful.
- The *deterministic decryption algorithm* $\mathcal{D}$ takes as input a key $k \in Keys(\Gamma)$ and a purported ciphertext $c \in \{0,1\}^*$. It returns $\mathcal{D}_k(c)$, which is a plaintext $m \in \{0,1\}^*$ or a distinguished symbol $\perp$. The return value $\perp$ indicates that the given ciphertext c is invalid for the key k.

We say, informally, that a symmetric encryption scheme Γ is secure if it ensures confidentiality of messages under chosen-ciphertext attack (CCA) and guarantees integrity of ciphertexts [17]. As shown in [3,14], this combination of security properties implies indistinguishability under CCA which, in turn, is equivalent to non-malleability [9] under CCA.

Signature Scheme. A digital signature scheme is a triple of algorithms $\Sigma = (\mathcal{G}, \mathcal{S}, \mathcal{V})$ such that:

- The *probabilistic key generation algorithm* $\mathcal{G}$, on input a security parameter 1^ℓ, outputs a pair of matching public and private keys (PK, SK).

- The *signing algorithm* $\mathcal{S}$ is a probabilistic polynomial time algorithm that, given as input a message m and a key pair (PK, SK), outputs a signature σ of m.
- The *verification algorithm* $\mathcal{V}$ is a polynomial time algorithm that on input (m, σ, PK), outputs 1 if σ is a valid signature of the message m with respect to PK, and 0 otherwise.

We say that a signature scheme Σ is secure if the probability of succeeding with an existential forgery under adaptive chosen message attack [10] is negligible for all probabilistic polynomial time attackers.

Initialization. During some initialization phase, two servers SA and SB agree on the following public parameters: a large prime p, a generator g of $\mathbb{Z}_p^*$ satisfying the CDH assumption, a one-way hash function H, a secure symmetric encryption scheme $\Gamma = (\mathcal{K}, \mathcal{E}, \mathcal{D})$, and a secure signature scheme $\Sigma = (\mathcal{G}, \mathcal{S}, \mathcal{V})$. In addition, public/private key pairs are generated for each server by running the key generation algorithm $\mathcal{G}(1^\ell)$. We denote by (PK_X, SK_X) the public/private keys of the server X for $X \in \{SA, SB\}$. As part of the initialization, the agent A (resp. B) registers with server SA (resp. SB), by choosing pw_A (resp. pw_B) and sending it to the server via a secure channel.

3 Password-Only Authenticated Key Agreement

In this section we present a new four-party password authenticated key agreement protocol in which two agents wishing to agree on a session key do not need to store any public key of their authentication server but only need to share a short, easy-to-remember password with the server. In describing the protocol, we will omit 'mod p' from expressions for notational simplicity. The proposed protocol is outlined in Fig. 1 and a more detailed description is as follows:

1. Two agents A and B first need to inform each other of their respective authentication server. To this end, A sends to B the message $\langle ID_A, ID_{SA}, ID_B\rangle$ and B sends to A the message $\langle ID_B, ID_{SB}, ID_A\rangle$.
2. The agents request the assistance of their respective server in establishing a session key between them. To do this, agent A (resp. B) chooses a random nonce n_A (resp. n_B) and sends the message $\langle ID_A, ID_B, ID_{SB}, n_A\rangle$ (resp. $\langle ID_B, ID_A, ID_{SA}, n_B\rangle$) to the server SA (resp. SB).
3. Server SA chooses a random number $s \in_R \mathbb{Z}_p^*$ and a random nonce n_{SA}, and computes $r_{SA} = g^s$, $v_A = H(m_A|pw_A)$ and $c_{SA} = \mathcal{E}_{v_A}(r_{SA})$ where $m_A = I|n_A|n_{SA}$ and $I = ID_A|ID_B|ID_{SA}|ID_{SB}$. Then SA sends the message $\langle n_{SA}, c_{SA}\rangle$ to agent A. Similarly, server SB chooses a random number $t \in_R \mathbb{Z}_p^*$ and a random nonce n_B, computes $r_{SB} = g^t$, $v_B = H(m_B|pw_B)$ and $c_{SB} = \mathcal{E}_{v_B}(r_{SB})$, where $m_B = I|n_B|n_{SB}$, and then sends $\langle n_{SB}, c_{SB}\rangle$ to B.
4. After receiving $\langle n_{SA}, c_{SA}\rangle$, agent A computes $v_A = H(m_A|pw_A)$, recovers r_{SA} by decrypting c_{SA}, i.e., $r_{SA} = \mathcal{D}_{v_A}(c_{SA})$, and chooses a random number

A	SA	SB	B
$\langle pw_A \rangle$	$\langle pw_A \rangle$ $\langle PK_{SA}, SK_{SA} \rangle$	$\langle pw_B \rangle$ $\langle PK_{SB}, SK_{SB} \rangle$	$\langle pw_B \rangle$
	$I = ID_A \vert ID_B \vert ID_{SA} \vert ID_{SB}$		
	$\xrightarrow{ID_A, ID_{SA}, ID_B}$	$\xleftarrow{ID_B, ID_{SB}, ID_A}$	
$\xrightarrow{ID_A, ID_B, ID_{SB}, n_A}$			$\xleftarrow{ID_B, ID_A, ID_{SA}, n_B}$
	$s \in_R \mathbb{Z}_p^*$	$t \in_R \mathbb{Z}_p^*$	
	$r_{SA} = g^s$	$r_{SB} = g^t$	
	$m_A = I \vert n_A \vert n_{SA}$	$m_B = I \vert n_B \vert n_{SB}$	
	$v_A = H(m_A \vert pw_A)$	$v_B = H(m_B \vert pw_B)$	
$\xleftarrow{n_{SA}, c_{SA} = \mathcal{E}_{v_A}(r_{SA})}$			$\xrightarrow{n_{SB}, c_{SB} = \mathcal{E}_{v_B}(r_{SB})}$
$v_A = H(m_A \vert pw_A)$			$v_B = H(m_B \vert pw_B)$
$r_{SA} = \mathcal{D}_{v_A}(c_{SA})$			$r_{SB} = \mathcal{D}_{v_B}(c_{SB})$
$r_A = g^a,\ a \in_R \mathbb{Z}_p^*$			$r_B = g^b,\ b \in_R \mathbb{Z}_p^*$
$k_{SA} = g^{as}$			$k_{SB} = g^{bt}$
$\xrightarrow{r_A}$			$\xleftarrow{r_B}$
$u_A = H(k_{SA} \vert m_A)$			$u_B = H(k_{SB} \vert m_B)$
	$k_{SA} = g^{as}$	$k_{SB} = g^{bt}$	
	$u_A \stackrel{?}{=} H(k_{SA} \vert m_A)$	$u_B \stackrel{?}{=} H(k_{SB} \vert m_B)$	
	$\xrightarrow{I, r_A, \sigma_{SA} = \mathcal{S}_{SK_{SA}}(I \vert r_A)}$	$\xleftarrow{I, r_B, \sigma_{SB} = \mathcal{S}_{SK_{SB}}(I \vert r_B)}$	
$\xleftarrow{r_B}$			$\xrightarrow{r_A}$
$u_{SA} = H(m_A \vert k_{SA} \vert r_B)$			$u_{SB} = H(m_B \vert k_{SB} \vert r_A)$
$K = g^{ab}$			$K = g^{ab}$

$$\boxed{sk = H(K|I|r_A|r_B)}$$

Fig. 1. Password-only authenticated key agreement protocol

$a \in_R \mathbb{Z}_p^*$. Given r_{SA} and a, agent A computes the one time key k_{SA} to be shared with the server SA as

$$k_{SA} = g^{as} = (r_{SA})^a.$$

Additionally, A computes $r_A = g^a$ and $u_A = H(k_{SA}|m_A)$. Then A sends the message $\langle r_A, u_A \rangle$ to the server SA.

Meanwhile, agent B, having received $\langle n_{SB}, c_{SB} \rangle$, computes $v_B = H(m_B| pw_B)$ and $r_{SB} = \mathcal{D}_{v_B}(c_{SB})$, and chooses a random number $b \in_R \mathbb{Z}_p^*$. From r_{SB} and b, B computes the one time key k_{SB} to be shared with SB as

$$k_{SB} = g^{bt} = (r_{SB})^b.$$

B also computes $r_B = g^b$ and $u_B = H(k_{SB}|m_B)$ and then sends $\langle r_B, u_B \rangle$ to SB.

5. Upon receiving $\langle r_A, u_A \rangle$, server SA first computes the one time key $k_{SA} = g^{as}$ shared with A and then verifies that u_A from A equals the hash value $H(k_{SA}|m_A)$. If the verification fails, SA stops executing the protocol; otherwise, SA believes that agent A is genuine. SA then sends the message $\langle I$, r_A, $\sigma_{SA} \rangle$ to the server SB, where σ_{SA} is the signature on $I|r_A$ generated by the singing algorithm $\mathcal{S}$ using the private key SK_{SA}, namely,

$$\sigma_{SA} = \mathcal{S}_{SK_{SA}}(I|r_A).$$

Similarly, upon receiving $\langle r_B, u_B \rangle$, server SB computes the one time key $k_{SB} = g^{bt}$ shared with B and verifies that u_B from B equals $H(k_{SB}|m_B)$. If the verification fails, SB aborts the protocol; otherwise, SB believes agent B as authentic. Then SB sends the message $\langle I$, r_B, $\sigma_{SB} \rangle$ to SA, where σ_{SB} is the signature on $I|r_B$ generated by $\mathcal{S}$ using the private key SK_{SB}, namely,

$$\sigma_{SB} = \mathcal{S}_{SK_{SB}}(I|r_B).$$

6. After receiving $\langle I$, r_B, $\sigma_{SB} \rangle$, SA first verifies the signature σ_{SB} using the public key PK_{SB}. SA halts immediately if the verification fails. Otherwise, SA computes

$$u_{SA} = H(m_A|k_{SA}|r_B)$$

and sends $\langle r_B, u_{SA} \rangle$ to agent A.
The server SB, upon receiving $\langle I$, r_A, $\sigma_{SA} \rangle$, verifies the signature σ_{SA}, and if correct, computes

$$u_{SB} = H(m_B|k_{SB}|r_A)$$

and sends $\langle r_A, u_{SB} \rangle$ to B.

7. The agent A checks whether u_{SA} from SA equals $H(m_A|\ k_{SA}|r_B)$. If this is untrue, A aborts the protocol. Otherwise, A computes the common secret value K as

$$K = g^{ab} = r_B^a.$$

Similarly, agent B verifies that u_{SB} from SB equals $H(m_B|k_{SB}|r_A)$, and if the verification succeeds, computes the common secret value K as

$$K = g^{ab} = r_A^b.$$

Finally, the agents compute their session key sk as

$$sk = H(K|I|r_A|r_B).$$

4 Security Analysis

In this preliminary version of the paper, we only provide a heuristic security analysis of the proposed protocol, considering a variety of attacks and security properties; a rigorous proof of security in a formal communication model will be given in the full version of this paper.

Implicit Key Authentication. The fundamental security goal for a key exchange protocol to achieve is implicit key authentication. Loosely stated, a key exchange protocol is said to achieve implicit key authentication if each party trying to establish a session key is assured that no other party aside from the intended parties can learn any information about the session key.

Our protocol guarantees the implicit key authentication. Namely, without knowing pw_A or pw_B, no one computes the session key. In the protocol, the session key sk is computed as $sk = H(K|I|r_A|r_B)$. Since H is a one-way hash function, sk cannot be obtained without knowing the common secret value K which in turn is computed as $K = g^{ab} = r_B^a = r_A^b$. We claim that only two agents A and B can compute this common secret value K. In other words, two agents A and B are assured that they will always compute K using genuine r_B and r_A (i.e., r_B and r_A from genuine B and A), respectively. From the viewpoint of agent A, this is achieved by verifying (in step 7) the equality

$$u_{SA} \stackrel{?}{=} H(m_A|k_{SA}|r_B).$$

This is the case because the success of the verification implies that

1. r_B is sent by the genuine server SA since only SA (aside from A) knows the one time key k_{SA}.
2. In turn, SA can confirm that this r_B came from the genuine SB, by verifying (in step 6) the signature σ_{SB} on $I|r_B$ through using SB's public key.
3. Similarly, the server SB can verify that r_B is from the genuine B by checking (in step 5) that u_B equals $H(k_{SB}|m_B)$. This is because only the agent B (aside from SB) knows the one time key k_{SB}.

Consequently, it is guaranteed that agent A computes the secret value K using the genuine r_B. Because the protocol is symmetric, agent B also computes K using the genuine r_A. Therefore, we arrive at the conclusion that the protocol achieves implicit key authentication.

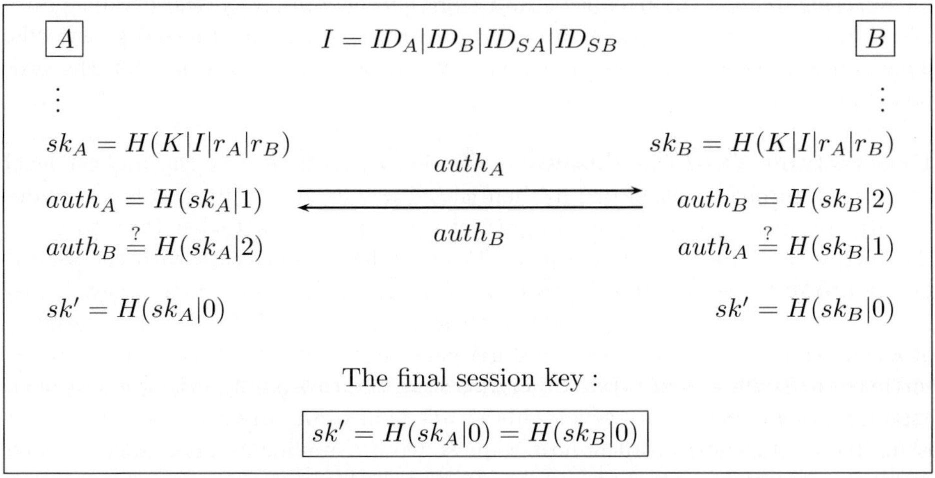

Fig. 2. Adding Explicit Authentication

Explicit Authentication. Another stronger kind of security goal for a key exchange protocol to achieve is explicit authentication, the property obtained when both implicit authentication and key confirmation hold. It is straightforward to see that our protocol does not achieve explicit authentication; that is, the agent A (resp. B) does not know whether the agent B (resp. A) has successfully computed a matching session key. However, it is easy to transform any key exchange protocol P with implicit authentication into a protocol P' providing explicit authentication by using standard techniques [2,4].

The transformation works as follows. Suppose that in protocol P, two agents A and B ended up with computing their session key sk_A and sk_B, respectively. In protocol P', agent A sends one additional flow $auth_A = H(sk_A|1)$ to B and similarly, agent B sends $auth_B = H(sk_B|2)$ to agent A. Upon receiving $auth_B$, agent A checks the equality $auth_B \stackrel{?}{=} H(sk_A|2)$. If they are equal, then A computes its final session key sk' as $sk' = H(sk_A|0)$. Otherwise, A aborts the protocol. Likewise, agent B, after receiving $auth_A$, verifies that $auth_A$ equals $H(sk_B|1)$. If so, then B computes the final session key sk' as $sk' = H(sk_B|0)$. Otherwise, B aborts the protocol. This procedure of adding explicit authentication is outlined in Fig. 2.

Off-line Password Guessing Attack. In this attack, an attacker may try to guess a password and then to check the correctness of the guessed password off-line. If his guess fails, the attacker tries again with another password, until he find the proper one. In the proposed protocol, the only information related to passwords is $c_{SA} = \mathcal{E}_{v_A}(g^S)$ and $c_{SB} = \mathcal{E}_{v_B}(r^t)$, where $v_A = H(m_A|pw_A)$ and

$v_B = H(m_B|pw_B)$, but because s and t are chosen randomly, these values does not help the attacker to verify directly the correctness of the guessed passwords. Thus, off-line password guessing attacks would be unsuccessful against the proposed protocol.

Undetectable On-Line Password Guessing Attack. At the highest level of security threat to password authenticated key exchange protocols are undetectable on-line password guessing attacks [8] where an attacker tries to check the correctness of a guessed password in an on-line transaction with the server, i.e., in a fake execution of the protocol; if his guess fails, he starts a new transaction with the server using another guessed password. Indeed, the possibility of an undetectable on-line password guessing attack in the three-or-more-party setting represents a qualitative difference from the two-party setting where such attack is not a concern. However, this attack is meaningful only when the server is unable to distinguish an honest request from a malicious one, since a failed guess should not be detected and logged by the server.

In our protocol, the server is the first who issues a challenge and the agent is the first who replies with an answer to some challenge. It is mainly due to this ordering that the protocol is secure against undetectable on-line password guessing attacks. Suppose that an attacker A', posing as A, decrypts c_{SA} by guessing a password, computes $r_{A'}$, $k_{SA'}$, and $u_{A'} = H(k_{SA'}|m_A)$ by choosing his own random a', and sends the fake message $\langle r_{A'}, u_{A'} \rangle$ to server SA. Then, the server SA, upon receiving $r_{A'}$ and $u_{A'}$ from A', should be easily able to detect a failed guess since the protocol specification mandates SA to check the correctness of $u_{A'}$. Note that the attacker cannot send a correct u_A without knowing a correct k_{SA} which in turn only can be computed if the guessed password is correct. Hence, the proposed protocol can resist undetectable on-line password guessing attacks.

Insider Attack. One of the main differences between the two-party setting and the three-or-more-party setting is the existence of insider attacks, i.e., attacks by malicious insiders. Namely, insider attacks are a concern specific to the three-or-more-party setting, and do not need to be considered in the two-party setting.

Suppose one agent, say A, tries to learn the password of the other agent B during execution of the protocol. Of course, A should not be able to have this ability through a protocol run and this is still an important security concern to be addressed. As mentioned earlier in this section, the only information related to the B's password is $c_{SB} = \mathcal{E}_{v_B}(r_{SB})$ where $v_B = H(m_B|pw_B)$. The actual value r_{SB} itself is never included in any message sent in the protocol and t is a secret information only known to SB. This means that the malicious insider A has no advantage over outside attackers in learning B's password. In other words, A's privileged information — pw_A, r_{SA}, and k_{SA} — gives A no help in learning B's password. Therefore, our protocol is also secure against insider attacks.

References

1. M. Abdalla, P.-A. Fouque, and D. Pointcheval, Password-based authenticated key exchange in the three-party setting, *PKC 2005*, LNCS 3386, pp. 65–84, 2005.
2. E. Bresson, O. Chevassut, D. Pointcheval, and J.-J. Quisquater, Provably authenticated group Diffie-Hellman key exchange, *In Proceedings of the 8th ACM conference on Computer and Communications Security (CCS 2001)*, pp. 255–264, 2001.
3. M. Bellare and C. Namprempre, Authenticated encryption: Relations among notions and analysis of the generic composition paradigm, *Asiacrypt 2000*, LNCS 1976, pp. 531–545, 2000.
4. M. Bellare, D. Pointcheval, and P. Rogaway, Authenticated key exchange secure against dictionary attacks, *Eurocrypt 2000*, LNCS 1807, pp. 139–155, 2000.
5. M. Bellare and P. Rogaway, Random oracles are practical: A paradigm for designing efficient protocols, *Proc. ACM CCS 1993*, pp. 62–73, 1993.
6. V. Boyko, P. MacKenzie, and S. Patel, Provably secure password-authenticated key exchange using Diffie-Hellman, *Eurocrypt 2000*, LNCS 1807, pp. 156–171, 2000.
7. W. Diffie, P. Oorschot, and M. Wiener, Authentication and authenticated key exchanges, *Designs, Codes, and Cryptography*, vol. 2, no. 2, pp. 107–125, 1992.
8. Y. Ding and P. Horster, Undectectable on-line password guessing attacks, *ACM SIGOPS Operating Systems Review*, vol. 29, no. 4, pp. 77–86, 1995.
9. D. Dolev, C. Dwork, and M. Naor, Nonmalleable cryptography, *SIAM Journal on Computing*, vol. 30, no. 2, pp. 391–437, 2000.
10. S. Goldwasser, S. Micali, and R. Rivest, A digital signature scheme secure against adaptive chosen-message attacks, *SIAM Journal of Computing*, vol. 17, no. 2, pp. 281–308, 1988.
11. L. Gong, M.-L. Lomas, R.-M. Needham, and J.-H. Saltzer, Protecting poorly chosen secrets from guessing attacks, *IEEE Journal on Selected Areas in Communications*, vol. 11, no. 5, pp. 648–656, 1993.
12. S. Jiang and G. Gong, Password based key exchange with mutual authentication, *SAC 2004*, LNCS 3357, pp. 267–279, 2005.
13. J. Katz, R. Ostrovsky, and M. Yung, Efficient password-authenticated key exchange using human-memorable passwords, *Eurocrypt 2001*, LNCS 2045, pp. 475–494, 2001.
14. J. Katz and M. Yung, Unforgeable encryption and adaptively secure modes of operation, *FSE 2000*, LNCS 1978, pp. 284–299, 2000.
15. J.-T. Kohl and B.-C. Neumanm, The Kerberos Network Authentication Service, Version 5 Revision 5, Project Athena, Massachusetts Institute of Technology, 1992.
16. C.-L. Lin, H.-M. Sun, and T. Hwang, Three-party encrypted key exchange: attacks and a solution, *ACM SIGOPS Operating Systems Review*, vol. 34, no. 4, pp. 12–20, 2000.
17. P. Rogaway, M. Bellare, J. Black, and T. Krovetz, OCB: A block-cipher mode of operation for efficient authenticated encryption, *Proc. ACM CCS 2001*, pp. 196–205, 2001.
18. M. Steiner, G. Tsudik, and M. Waidner, Refinement and extension of encrpyted key exchange, *ACM SIGOPS Operating Systems Review*, vol. 29, no. 3, pp. 22–30, 1995.
19. M. Zhang, New approaches to password authenticated key exchange based on RSA, *Asiacrypt 2004*, LNCS 3329, pp. 230–244, 2004.

Diagnostic Knowledge Acquisition for Agent-Based Medical Applications

Thomas M. Gatton[1], Malrey Lee[2,*], TaeEun Kim[3], and Young-Keun Lee[4]

[1] National University, 11255 North Torrey Pines Road,
La Jolla, CA 92037 USA
tgatton@nu.edu

[2] The Research Center of Industrial Technology, School of Electronics & Information Engineering, Chonbuk National University, Korea
mrlee@chonbuk.ac.kr

[3] School of Engineering, Department of Multimedia, Namseoul University, Korea
tekim@nsu.ac.kr

[4] Department of Orthopedics Surgery, Chonbuk National University, Korea
trueyklee@yahoo.co.kr

Abstract. The development of ubiquitous systems for maintenance and control of treatment systems to assist individuals in managing their medical treatment plan would provide an improved system for home healthcare. The assistance of agent-based systems to help doctors and further automate medical treatment systems is limited by medical knowledge data mining accuracy and the complexity of each patient's health history and individual. Automation of knowledge base development for each individual patient would allow efficient personalization of each patient's treatment plan and allow integration of the doctor's individual diagnosis and treatment plan. This paper presents an overview of agent based technologies and describes both historical and state-of-the-art applications of agent technologies in the medical field. Current research and development activity is identified and an algorithm to address the knowledge acquisition bottleneck for diagnostic medical knowledge is present. The algorithm can reduce time consuming knowledge acquisition and allow efficient development of individually tailored medical treatment knowledge bases.

1 Introduction

The use of computer based systems for medical applications has evolved from diagnostic software using artificial intelligence to agent based ubiquitous systems. Russell and Norvig state "An agent is anything that can be viewed as perceiving its environment through sensors and acting upon that environment through actuators." [1] They further define the properties of the agent's environment and classify agent programs into the four categories of simple reflex, model based, goal based and utility based. These categories relate to the agent's ability to reason, model, set goals, search, plan, cooperate, learn, and exhibit autonomy. Wikipedia defines a software agent as "an abstraction, a logical model that describes software that acts for a user or other

[*] Corresponding author.

N.T. Nguyen et al. (Eds.): KES-AMSTA 2007, LNAI 4496, pp. 626–635, 2007.

program in a relationship of agency." Luck, et al. stated "Put at its simplest, an agent is a computer system that is capable of flexible autonomous action in dynamic, unpredictable, typically multi-agent domains." [2] Agent technology has been described as a new paradigm and is undergoing considerable debate and clarification, as evident from numerous books and articles [3-5]. One of the continuing problems in the development of these systems has been the knowledge acquisition bottleneck. One of the recent advances in this area is the use of agent based systems to analyze medical text and records to automatically generate medical knowledge. The reliability of the knowledge generated from these systems has not reached the accuracy levels necessary for medical precision and, in many agent based systems, does not have suitable explanation capabilities. While these systems have shown success in general diagnostic and treatment recommendations, they are unable to address every individual patient characteristic and demonstrate the robustness of each individual doctor. This is caused by the dynamics of each patient's condition and the evolving state of medical knowledge. This paper presents an algorithm for automated knowledge acquisition for disease healthcare that reduces time consuming knowledge acquisition and allows physicians to efficiently develop individually tailored knowledge bases to suit the specific conditions and treatment plans for each individual patient.

Section 2 presents the background information on medical applications of computer technology and automated knowledge acquisition systems and shortcomings. Next, Section 3 provides current disease diagnosis and treatment knowledge and provides an ontological structure. Section 4 introduces the algorithm for automated disease medical knowledge acquisition and, finally, Section 5 presents the conclusions and recommendations for further study and research. Preparation of manuscripts which are to be reproduced by photo-offset requires special care. Papers submitted in a technically unsuitable form will be returned for retyping, or canceled if the volume cannot otherwise be finished on time.

2 Medical Agent Technology Foundation

2.1 Agent Technologies

It has been stated that "Taking an optimistic point of view, a knowledge base might be the basis of and intelligent agent –a consultant, tutor, librarian, or decision analyst responding actively to the needs of its user, capable of explaining itself, and, most importantly, capable of learning from experience." [6] In light of this visionary expectation, it is clear that early medical programs utilizing the techniques of artificial intelligence established the groundwork for later advancements in intelligent multi-agent systems. An overview of these systems and there progression towards agent technologies is requisite for a comprehensive view of medical applications of agent technology.

2.2 Medical Applications

The early beginnings of research and collaboration in the application of agent related technologies to the medical field can be identified with Artificial Intelligence in Medicine (AIM) workshops. While these were initially limited to a group of collaborative researchers, the gatherings were eventually opened to public attendance resulting in a surge of interest and participation by many academicians and practitioners. These conferences disseminated the techniques of artificial intelligence and demonstrated successful applications in early systems, such as MYCIN [7], CASNET [8], PIP [9] and INTERNIST [10].

In MYCIN, knowledge was represented by production rules and confidence factors. CASNET used a semantic network for knowledge representations while PIP used a frame-based structure in its knowledge base. INTERNIST used a tree like taxonomy that described symptoms of each disease. Each of these different types of medical knowledge representation exhibited various characteristics in their ability to explain their reasoning and all suffered from the bottleneck of knowledge acquisition. The process of verifying the existing knowledge and adding additional knowledge led to systems, such as EMYCIN, that provided a generic shell system to build expert systems in any domain. As more applications began development, such as IRIS [11], PUFF [12], EXPERT [13], VM [14] and SEEK [15], methods to evaluate the various capabilities and features of existing systems were developed [6, 16]. These methods included criteria such as performance, user friendliness, explanation facilities, knowledge base creation and maintenance, clinical diagnosis accuracy, and knowledge representation. As medical knowledge increased, further domain specific systems were developed to focus on a particular area of expertise. However, these systems were standalone and unable to share knowledge between them. Further, the representation of temporal knowledge limitation restricted the reasoning capabilities and evaluation of ongoing patient dynamics. The introduction of agent technologies provided solutions to many of the problems exhibited by early systems.

3 Current Research and Development

The existence of several isolated machine and computer systems used in medical applications, such as Management information systems, domain specific diagnostic systems and various types of data gathered during patient monitoring, led to a realization of the need for integration. New agent technologies, such as data mining, autonomy, learning, and communication capabilities were beginning to be applied towards a solution to these problems.

The application of agent-based technologies has been categorized into several areas [10, 11]. These areas include patient monitoring, health care, patient scheduling, medical research, training, organ transplant management, community care, information access, decision aid systems and internal hospital tasks. These applications may be classified according to their systems type such as standalone, distributed, intranet, Internet and mobile, such as PDA, cell phones and other mobile devices. Agents may also exhibit capabilities such as collaboration, cooperation, interface communication, learning and intelligence.

The medical applications area has several implementation challenges that can be addressed by agent-based approaches. The traditional doctor/patient relationship has been replaced by a complex relationship between inpatient and outpatient care, doctors and nurses, health care givers, administrative staff, social workers, computer systems, and others. Individual patients can be located at different places and health care can occur at a number of different locations and provide diagnostic, surgical, therapeutic, pharmaceutical, testing and other types of services. These services require that information be shared and available across a wide spectrum of users in an efficient, timely, secure and errorless manner. Agent capabilities of autonomy, reactivity, communication and collaboration provide a unique solution to address many of the challenges in medical applications.

3.1 Applications

Medical applications of agent-based technologies can be categorized according to several types of classification, as previously presented. There are numerous types of monitoring diagnostic systems ranging from single agent standalone to cooperative multi-agent systems and from initial prototypes to working implementations. Examples of multiple agent systems include cardiac pacing [12], trauma and hemorrhagic shock stabilization [13]. Other examples of multiple agent MIS systems are retrieval of medical information [14], scheduling and planning [15, 16]. In the complex testing and treatment scheduling required for patient appointment scheduling, the use of multi-agent systems has shown improved performance over manual and non-agent software solutions. Multi-agent systems for information retrieval and workflow management and have been applied to applications in stroke patient management as well as the coordination of tissue and organ transplant. Extensions of these systems are currently under research and improvements in allocating the resources required in transplants have been experienced. These systems can be expanded across greater regions to further improve surgical transplant operations. The IST project in Europe seeks to apply multi-agent technologies for supporting assistance to elderly people seeking telephone supervision and assistance with medical problems [17]. In an earlier multi-agent system for the elderly, a single agent is assigned to an elderly individual and the agent provides medical data to the elderly person, reminders for any necessary medical action and messages to the health care provider when something unusual occurs [18].

Several neural network based agents are capable of discovering medical knowledge from existing databases and applications in areas such as laboratory medicine, cervical smear analysis sleep disorders, neonatal intensive care, urological analysis, bacterial taxonomy and others [17-20]. Agents are being applied to proactively utilize user profiles and identify relevant medical information and assessing that information from distributed databases [22] and managing distributed decision support systems for supporting collaborative medical decisions [23]. They are also finding use in patient care management [24,25], medical training [26] and ICU patient monitoring and decision support [27]. With the explosion of medical information available on the internet, there have been significant efforts to apply agent-based technologies to process and make this information available. Systems such as the Multi-Agent Retrieval Vagabond on Information Networks (MARVIN) [28] and tools allowing internet access by patients [29] have been developed.

3.2 Trend Evaluation

Early work in agent systems began with single applications primarily addressing medical diagnosis problems. As more efficient mechanisms were developed for knowledge acquisition, it became evident that integration between systems was paramount to the utilization of this technology. Agent based software provided solutions providing the capabilities of autonomy, security, collaboration, communication and reactivity required to address these problems. Agent systems developed into sophisticated and collaborative learning agents and provided software addressing these problems. The most promising applications are in multi-agent learning and collaborative systems, medical training, web and mobile systems and multi-algorithmic agent research. The ultimate goal is a system that can peruse diagnostic knowledge, evaluate patients' medical history, evaluate current patient conditions and assist physicians, nurses, health care givers and administrative provide the best possible medical services. The most difficult is the accuracy of the patient specific diagnostic and treatment knowledge accuracy. The doctor must specifically verify any knowledge gained through agent mechanisms, such as data mining. An efficient algorithm to acquire this knowledge is the focus of the remaining sections of this paper.

3.3 Automated Knowledge Acquisition

The acquisition and formalization of current medical knowledge has been a major obstacle in the development of computer based medical consulting systems and has become known as the knowledge acquisition bottleneck [30]. Traditional methods are slow and time consuming and seek knowledge from domain experts. More recently, many automated knowledge acquisition systems have developed agent based systems to facilitate the knowledge acquisition process and, also survey digital medical records and documentation [30]. These automated systems use data mining, machine intelligence, agent-based technology, text categorization, neural network and other approaches to gather medical domain information for knowledge base development [31-33]. While these systems have performed well, their explanation systems are often poor, as compared to rule-based systems. Some systems directly query domain experts to develop rules and utilize graphical interfaces to improve user friendliness [34-36]. These systems indicate the availability of approaches that may be adapted to knowledge acquisition for disease treatment. Analysis of disease diagnosis, symptoms and treatment can provide an ontological structure amenable to automated knowledge acquisition

4 Automated Knowledge Acquisition

After diagnosis of disease occurs, each patient's treatment and monitoring plan is established by the physician and incorporates unique patient conditions. Although there are many commonalities, individual conditions require specific consideration in treatment plans. Upon establishing this patient tailored plan, the doctor must establish the range of limits for each symptom. This uniqueness, and the changing nature of medical knowledge, introduces the reliability problem of generic disease knowledge based systems. In order to provide an efficient transfer of the patient's condition, treatment and monitoring plan for each patient, the knowledge ontology must be established and an automated knowledge acquisition algorithm must be developed.

4.1 Disease Treatment and Monitoring Knowledge Ontology

After disease diagnosis and establishment of the treatment plan, the physician must establish the ranges for limits in the monitored conditions. Patient condition states and symptoms indicate where abnormalities occur that require action. These actions could be changes in the treatment plan, direct contact with medical personnel or emergency medical action. The ontological structure of the patient's condition, diagnosis and treatment, monitoring and action plan must then be evaluated for identification of an efficient automated knowledge acquisition procedure. This knowledge base could then be integrated into ubiquitous system to assist individual patients in maintaining health and controlling disease complications.

The relationships between the patient, patient conditions and symptoms, physician and disease medical knowledge are shown if Figure 1. The ubiquitous system interface is shown in dashed lines. In the ubiquitous system implementation, the knowledge base would contain the relevant patient information, treatment plan, and treatment and monitoring expertise. This knowledge contains the ranges of acceptable conditions and symptoms leading to actions and is further illustrated in Figure 2. As evident in Figure 2, monitoring of the patient's condition states, symptoms and adherence to the treatment plan provide the input to generate any actions that deviate from the normal treatment plan..

The algorithm for automated knowledge acquisition begins with the existing knowledge of the patient's condition. Next, the physician indicates the limiting ranges, such as blood glucose levels, for all conditions. Condition states outside of each of these ranges could also be categorized with severity levels, indicating the level of action required. Building on the patient's condition and diagnosis, the algorithm queries the physician for the limits for symptoms and the resulting recommendations, linking them up in a rule-based structure. With all of the condition states and symptoms known, the physician would be prompted for the single and combinatory occurrence of symptoms leading to recommendations for treatment plan modification, communication with medical personnel or emergency medical action. The overall structure of this procedure is illustrated in Figure 3.

To demonstrate the generation of rules, consider the condition states of glucose level and medication and the symptoms of nausea and blurry vision. The recommend

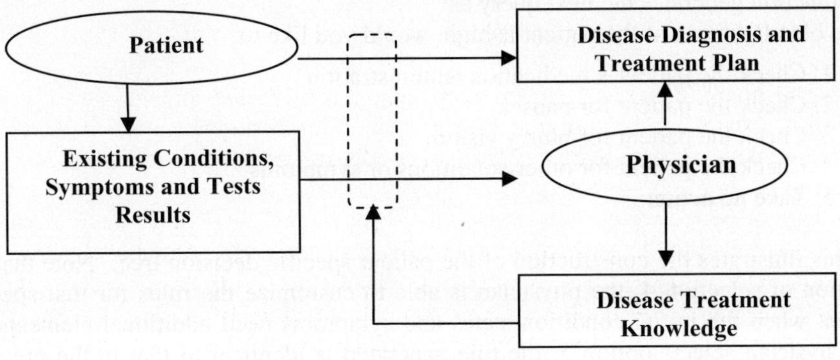

Fig. 1. Domain relationships

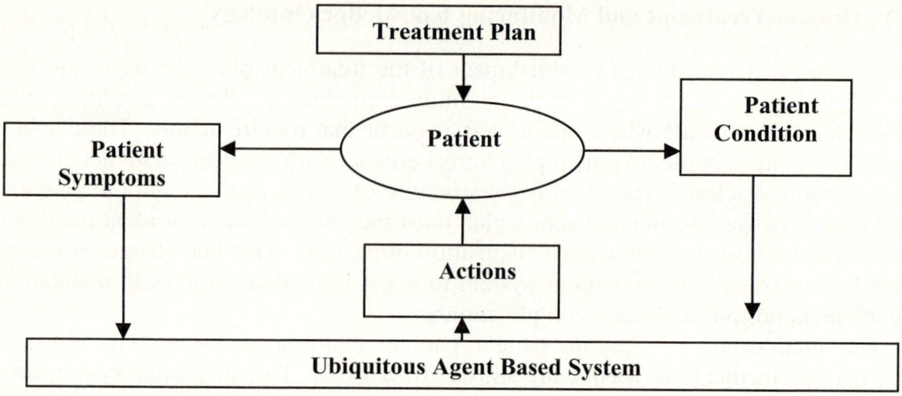

Fig. 2. Ubiquitous healthcare system organization

for this demonstration are to increase or decrease the medication or exercise, call the doctor and maintain the present treatment plan.

All rule generation begins with the deviation from a normal condition state or the occurrence of a symptom. This function is performed in the algorithm by the "Generate Symptom-Condition State" block. The query would begin by asking the physician:

If the blood glucose of the patient is high, would you like to
Step 1 : Advise the patient to change their treatment plan
Step 2 : Advise the patient to call you
Step 3 : Advise the patient to call the emergency room
Step 4 : Give the patient some other advice
Step 5 : Check the patient for other conditions and symptoms
Step 6 : Take no action

If the physician selects to advise the patient when this condition occurs, that rule is constructed. If the physician selects to give the patient some other advice, the rule is customized with the physician's advice for that specific patient's conditions. If the physician selects to check the patient for other conditions and symptoms, the build rule function generates the next query as:
If the blood glucose of the patient is high, would you like to
Step 1: Check the patient's medication administration
Step 2: Check the patient for nausea
Step 3: Check the patient for blurry vision
Step 4: Check the patient for other conditions or symptoms
Step 5: Take no action.

This illustrates the construction of the patient specific decision tree. Note that the addition of selection 4, the physician is able to customize the rules for that specific patient when the initial condition states and symptoms need additional elements. If the Physician selects option 5, the rule generated is identical to that in the previous query.

This process is continued until all combinatorial options have been exhausted or eliminated. The resulting rules comprise a patient specific knowledge base. This type of approach incorporates the fact that symptoms, acceptable condition states and recommendations are defined and the facility to add symptoms and treatments is available. Building of "generic" templates for common combinations of diseases diagnosis and treatment plans that can be modified for specific patient variables would streamline the initial generation of common symptoms, condition states and recommendations.

5 Conclusions

The use of ubiquitous systems for disease treatment and monitoring provides a needed technology to address the rising occurrence of this debilitating disease. The classification, diagnosis and treatment of disease is constantly changing, thereby complicating accurate, updated and state-of-the-art representation in "static" knowledge bases. Currently, there are no knowledge bases suitable for handling each individual patient's condition and variation in treatment plans, despite the advancements in automated knowledge acquisition and automatic knowledge generation from medical records and text. This paper presented a new algorithm for automated knowledge acquisition in disease treatment and monitoring systems that can efficiently provide individually tailored patient knowledge bases. After identifying disease treatment and monitoring relationships and their ontology, the symptoms, condition states and recommendations were identified as elements that could be used in a combinatorial fashion to generate the relevant rules and an automated knowledge acquisition algorithm was presented. Manual demonstration showed the feasibility of the approach and its appropriateness to advance development in this area. Areas of further research should consider temporal knowledge about times for last medication, food, and activity. A prototype implementation of this algorithm is also recommended and planned for future investigation.

Acknowledgements

This research was supported by the MIC(Ministry of Information and Communication), Korea under the ITRC(Information Technology Research Center) support program supervised by the IITA(Institute of Information Technology Assessment) IITA-2006-C1090-0603-0024.

References

1. Russell, S. J. and Norvig, P., Artificial Intelligence a Modern Approach (2nd Ed.), Prentice Hall, (2003)
2. Luck, M., et al., "Agent Technology Roadmap", AgentLink III, European Commission's Sixth Framework Program (FP6), (2005)
3. Bradshaw, J., el. Intelligent Agents, MIT Press, (1996).

4. Bryson, J., and McGonigle, B., “Agent Architecture as Object Oriented Design,” Intelligent Agents IV: Agent Theories, Architectures, and Languages. Proceedings of ATAL 1997, Springer, Berlin, (1998)
5. Weiss, G., ed., Multiagent Systems: A Modern Approach to Distributed Artificial Intelligence, Cambidge, MA, MIT Press
6. Clancey, W. J.. and Shortliffe, E. H., eds., Readings in Medical Artificial Intelligence The First Decade, Addison Wesley, (1984)
7. Shortliffe, E. H., Computer-based medical consultations: MYCIN, (1976)
8. Weiss, S. M., Kulikowski, C. A., Amarel, S., and Safir, A. “A model-based method for computer-aided medical decision-making”, Artificial Intelligence 11, (2000) 145-172
9. Szolovits, P., and Pauker, S., “Research on a medical consultation program for taking the present illness”, Proceedings of the Third Illinois Conference on Medical Information Systems, (1976)
10. Jennings, N. R., and Wooldridge, M., “Applications of Intelligent Agents,” in Agent Technology: Foundations, Applications, Markets, (1998) 3-28
11. Moreno, A., “Medical Applications of Multi-Agent Systems,” in Intelligent and Adaptive Systems in Medicine, Conference Proceedings, Prague, (2003)
12. Amigoni, F., Beda, A., and Gatti, N., "Multiagent Systems for Cardiac Pacing Simulation and Control". AI Communications: Vol. 18, No. 3, (2005) 217-228
13. Mabry, S. L., Troy Schneringer, T., Etters, T., and Edwards, N “Intelligent agents for patient monitoring and diagnostics” Symposium on Applied Computing, Proceedings of the (2003) 257 - 262
14. Rodríguez M., and Preciado, A., "An Agent based System for the Contextual Retrieval of Medical Information", Proceedings of Advances in Web Intelligence, Second International Atlantic Web Intelligence Conference, LNAI 3034, (2004) 64-73
15. Herrler, R., Heine, C. and Klügl, F, “Appointment Scheduling Among Agents: A Case Study In Designing Suitable Interaction Protocols,” Proceedings of the Eighth Americas Conference on Information Systems, (2002)
16. Heine, C, Herrler, R, and Petsch, M, and Anhalt, C., “ADAPT–Adaptive Multi Agent Process Planning & Coordination of Clinical Trials,” in the Proceedings of AmCIS (2003)
17. Camarinha-Matos, L. M. and Afsarmanesh, H., “Virtual Communities and Elderly Support,” Advances in Automation, Multimedia and Video Systems, and Modern Computer Science, V.V. Kluev, C.E. D’Attellis, N. E. Mastorakis (eds), (2001) 279-284
18. Beer, M, D., Huang, W. and Sixsmith, A., “Using agents to build a practical implementation of the INCA-Intelligent Community Alarm- system,” Intelligent Agents and their applications, (2002) 320-345
19. Dybowski, R. and Gant, V., “Clinical Applications of Artificial Neural Networks,” Cambridge University Press, New York, (2001)
20. Opitz D., and Shavlik J., “Connectionist Theory Refinement: Genetically Searching the Space of Network Topologies,” Journal of Artificial Intelligence Research, 6, (1997) 177-209
21. Waitman L., “ Knowledge Discovery in Preoperative Databases using Rule Induction: Hypothesis Testing, Decision Support, and Clinical Guideline Assessment,” Ph.D. Thesis, Vanderbilt University Nashville, Tennessee, (1999)
22. Lobato, E. and Shankararaman, V., “PIRA: A Personalized Information Retrieval Agent,” Proceedings of IASTED International Conference on Artificial, Intelligence and Soft Computing, (1999)

23. Lanzola, G., Gatti, L., Falasconi, S. and Stefanelli M. A., "Framework for Building Co-operative Software Agents in Medical Applications," Artificial Intelligence in Medicine, 16(3) (1999), 223-249
24. Huang J, Jennings N. R, and Fox, J., "An Agent-based Approach to Health Care Management," Applied Artificial Intelligence: An International Journal, 9, 4, (1995) 401-420
25. Huang, J., Jennings, N. R., and Fox, J., "An Agent Architecture for Distributed Medical Care," Intelligent Agents, Wooldridge, M. J., and Jennings, N.R. (Eds.), LNAI, (1995), 219-232
26. Farias, A. and Arvanitis, T. N., "Building Software Agents for Training Systems: A Case Study on Radiotherapy Treatment Planning," Knowledge-Based Systems, 10, (1997) 161-168
27. Larsson J E and Hayes-Roth B., "Guardian: An Intelligent Autonomous Agent for Medical Monitoring and Diagnosis." IEEE Intelligent Systems, (1998) 58-64
28. Baujard O., Baujard V., Aurel S., Boyer C. and Appel, R.D., "MARVIN, a multi-agent softbot to retrieve multilingual medical information on the Web," Medical Informatics, 23 3, (1998) 187-191
29. Marshall, P. and Greenwood, S., "The Use of Emergent Behaviour in a Multi-Agent System to Drive Self-Adaptation at the Interface," Joint Web Intelligence/Intelligent Agent Technology, (2001)
30. Nealon, J. L., and Moreno, A., "The Application of Agent Technology to Health Care, "Agent Cities Working Group on Health Care, Challenge (2002)
31. Buchanan, B. G. and Wilkins, D. C., (editors), "Readings in Knowledge Acquisition and Learning: Automating the Construction and Improvement of Expert Systems," Morgan Kaufmann, San Mateo, CA., 1993
32. Boicu, M., Tecuci, G., Stanescu, B., Marcu, D. and Cascaval, C., "Automatic Knowledge Acquisition from Subject Matter Experts," Proceedings of the 2001 International Conference on Tools With Artificial Intelligence, ICTAI-2001, Dallas, Texas, November 2001.
33. Nealon, J. L. and Antonio Moreno, A., "The Application of Agent Technology to Health Care," Workshop AgentCities: research in large scale open agent environments at AAMAS 2002, Bologna, Italy, July 2002
34. "Boicu, C., Tecuci, G. and Boicu, M., "Improving Agent Learning through Rule Analysis," Proceedings of the International Conference on Artificial Intelligence, ICAI-05, Las Vegas, USA, June 27-30, 2005.
35. Gatton, T. M., and Kearney, F. W., "Automated Knowledge Acquisition for Building Diagnosis Expert Systems," 6th International Symposium on Automation and Robotics in Construction, San Francisco, CA, June 1989.

Remote Control Multi-Agent System for u-Healthcare Service*

Eunyoung Kang[1], Yongsoon Im[2], and Ungmo Kim[1]

[1] School of Computer Engineering, Sungkyunkwan University,
300 Cheoncheon-dong, Jangan-gu, Suwon, Gyeonggi-do 440-746, Korea
{eykang, umkim}@ece.skku.ac.kr

[2] Dept. of Information & Communication, Kookje College,
Jangan-dong, Pyeongtaek, Gyeonggi-do 459-070, Korea
{ysimos}@kookje.ac.kr

Abstract. In this paper, we proposed a multi-agent based healthcare system (MAHS) which is the combination of medical sensor module and wireless communication technology. This MAHS provides wide services to mobile telemedicine, patient monitoring, emergency management, doctor's diagnosis and prescription, patients and doctors, information exchange between hospital workers in a long distance. Also, MAHS is connected to Body Area Network (BAN) and a doctor and hospital workers. In addition, we designed and implemented extended JADE based MAHS that reduces hospital server's burden. Agents gather, integrate, and deliver the collected patient's information from sensor, and provide presentation in healthcare environment. Proposed MAHS has advantage that can handle urgent situation in the far away area from hospital like Islands through PDA and mobile device. In addition, by monitoring condition of patient (old man) in a real time base, it shortens time and expense and supports medical service efficiently.

1 Introduction

The development of mobile devices, such as PDA, laptop and notebook computers, and technologies, such as sensors, computerized chips, and wire/wireless networks, has made ubiquitous computing possible. Ubiquitous computing offers a user-oriented computing environment, where users can freely access computers anywhere, at any time. Healthcare is a field where ubiquitous computing is most widely used [1,2]. In a ubiquitous computing environment, even medical services that have previously been offered only from a hospital, become a part of everyday life through context-aware healthcare services that capture a specific medical condition, with chips and sensors installed in a user's ordinary life space, providing the user with medical checkup, disease control, emergency medical services, and telemedicine [3,4,5].

* This research was supported in part by the university IT Research Center project (ITRC), funded by the Korean Ministry of Information and Communication and by Korea Science and Engineering Foundation (KOSEF-R01-2004-000-10755-0).

N.T. Nguyen et al. (Eds.): KES-AMSTA 2007, LNAI 4496, pp. 636–644, 2007.

The existing healthcare systems, including the MobiHealth [6] Project, generally use a central server to archive and retrieve a patient's medical information. The load is a potential bottleneck for the server. The server may be unavailable in the case where real-time biosignal data transferred from patients to the server exceed the server's proper capacity. On top of this, they have disadvantages, such as where the mobile device transferring biosignals has insufficient self-supported resources, or when the wireless network is unable to provide stable communications between a patient and the server within a hospital. It may also lead to a failure in dealing with dynamic and critical circumstances, in which immediate treatments are necessary.

In this paper, we propose a healthcare system that remotely captures real-time medical information with sensors to provide a patient with his doctor's diagnosis and prescription. The healthcare system uses sensors to capture real-time biosignals including blood pressure, body temperature, pulse, and breathing, and adopts a PDA to transfer these data to a doctor through a surrogate system that serves as a hub between a patient and a hospital. Then, the doctor makes diagnosis and prescription based on his analysis of the received medical data compared with the patient's basic data (such as sex, weight, height, so on) stored in the database. With a reduction in central load and flexibility in terms of connectivity, a combined operation of multi-agent leads to solving complicated problems that couldn't be processed by a single agent. Such a multi-agent approach is used to handle such problems as insufficient resources, which all mobile devices experience. A minimum unit of multi-agent system comprises one coordination agent and two or more application agents based on agent platform architecture defined by FIPA. All messages are transferred through a surrogate system that prevents mobile devices from access by an unauthorized patient. The surrogate system is also an easy tool to deal with local and remote agents.

The remainder of the paper is organized as follows. Section 2 describes about agent platform and related approach. Section 3 provides the details of proposed MAHS architecture and multi-agent system. Section 4 presents proposed MAHS of implementation. Finally, Section 5 gives our conclusions.

2 Related Work

2.1 JADE (Java Agent Development Framework)

The JADE is middleware developed by TILAB for the development of distributed multi-agent applications based on a peer-to-peer communication architecture. Further, it is a distributed agent platform based on FIPA standard specifications [11]. Fig. 1 shows the JADE Agent Platform architecture.

AMS supervises an entire life cycle, including creation, registration, deletion and recovery of an agent in the agent platform. The DF provides each agent with information about capacity and services to be offered by agents in the agent platform. The container is a runtime setup of agents, including such communications functions as MTS (Message Transportation System).

The advantage of the multi-agent system is that a combined operation with other agents allows the system to provide complicated services that couldn't be processed by a single agent, and that the addition of new agents enables it to easily expand to cover new services. The multi-agents exchange and share information through cooperation among agents.

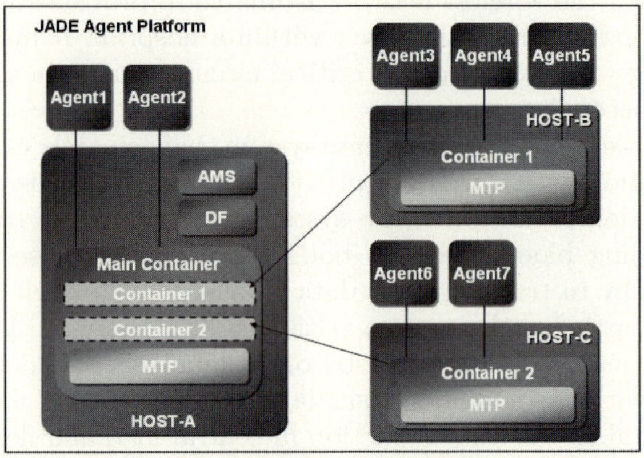

Fig. 1. JADE Agent Platform

2.2 LEAP (Lightweight Extensive Agent Platform)

The Lightweight Extensive Agent Platform (LEAP) is probably the most known agent platform for small devices [12]. LEAP is the first attempt to implement a FIPA agent platform that runs seamlessly on both mobile and fixed devices over both wireless and wired networks.

In detail, a module called LEAP, which is connected through wireless networks, allows optimizing of all communication mechanisms when dealing with devices with limited resources.

LEAP supports running of agent applications implemented over many devices and communication mechanisms (TCP/IP, WAP, etc.), and is transport layer independent. In particular, it supports transport protocols suitable for both wired and wireless communication. Finally, it is easily extensible such that, when deployed on a PC, it can provide additional functionality such as agent mobility support, user-defined ontology, and platform management tools. Fig.2 shows LEAP architecture.

2.3 Existing Approaches

This subsection presents existing approaches which support healthcare by telemedicine and their problems. Nowadays, due to the growth of techniques of wireless network and sensor, the field was extended to real time field such as emergency healthcare. The MobiHealth Project which is developed by fourteen companies of five European countries supports remote service of chronic patient by BAN which uses sensor and wireless communication [6]. The Politecnico project in Italy is a system that extracts detailed information about serious patients in hospital by physical sensor and transfers extracted signals to the central database of hospital at run time [7]. The project of Tele-medicare in Taiwan is a system that attaches sensor to patient body and transfers extracted signals to Local Patient Computer in patient home by using gateway and transfers live data of patient to remote central hospital server for management [8]. Ubiquitous Healthcare proposed architecture which uses ontology based mobile agent [9]. It also proposed systems that applies context to healthcare [10].

However, since the most of above systems has client-server structure which is simple communication between hospital central database and patients, they cannot avoid the overload of the server. Also since healthcare systems can not support the movement, application scope is limited in specific area. But our system based on JADE supports movements of a patient easily, so a patient checks and monitors his health. Moreover, our system's strong benefit is that it can find and communicate with best doctor for a patient by analyzing basic data such as sensor data, weight, height, age, and sex etc.

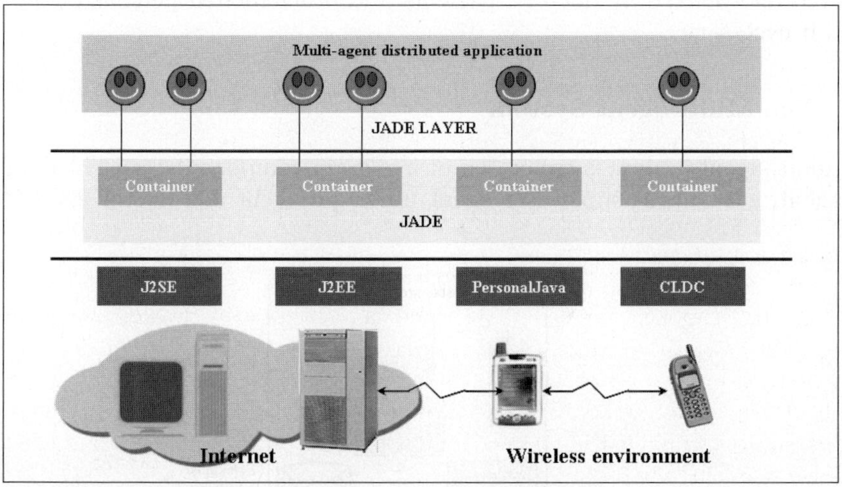

Fig. 2. LEAP Architecture

3 Proposed System Architecture

3.1 System Architecture for Healthcare Services

In this paper, we propose the multi-agent based Real-time Tele Healthcare System that uses JAVA-based application and the JADE, a distributed agent platform. Fig. 3 shows the architecture for a remote diagnosis and prescription system. The system includes three different areas: Body Area Network (BAN) System, Surrogate System, and Hospital System.

The BAN system is the body area to which sensors are attached to capture biosignals, including blood pressure, body temperature, pulse and breathing. These data are transferred to the PDA, a patient's mobile device, through a wireless network. Sensors are connected with wireless interface boards to transfer the captured medical information. Such connection is maintained unless a special event occurs.

The second area is the surrogate system that serves as a hub between a patient and a hospital. An agent determines whether a patient is in a critical condition based on medical data transferred from the BAN System. If it is determined to be an emergency, they are transferred to the hospital system to take an emergency measure, immediately after being stored into the surrogate system. If it is not an emergency, they are just stored into the surrogate system. Of data stored into the surrogate system, necessary data are regularly saved into the central database of the hospital. These real-time data will be deleted after a certain period of time unless it is an emergency. This surrogate system connects the BAN system with the hospital subsystem. Data stored into the surrogate system are available to doctors and other related persons in the hospital.

The third area is a hospital subsystem. Data are registered, retrieved, changed, updated and deleted by doctors, patients and other related persons of the hospital, if necessary.

3.2 The Multi-agent System

The multi-agent system is composed of five main components: patient monitoring agent, gate agent, supervisor agent, manager agent and doctor agent. This

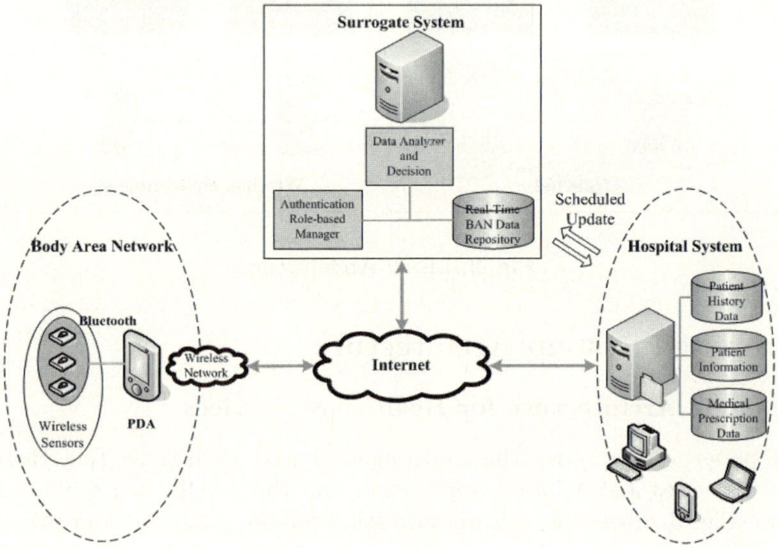

Fig. 3. Healthcare Services Architecture

system has sensors responsible for acquiring information about surrounding devices and services. This information such as blood pressure (systolic/diastolic), body temperature, breathing and pulse, from a patient is provided to a supervisor agent. At this time, the Gate Agent verifies a patient's authentication for his request for services. The Manager Agent stores patient information in the hospital database and searches for the doctor in charge. In addition, this agent transmits messages for request for diagnosis and prescription from the doctor in charge. The Doctor Agent provides the Patient Monitoring Agent with diagnosis and prescription of patient information from charge of data.

Patient Monitoring Agent (PMA). The Patient Monitoring Agent operates on a mobile device with the following functions: First, it uses sensors to detect medical data from a patient, and peripheral data, including temperature and humidity. Second, it transfers the detected data along with sender and device information to a surrogate system via the supervisor agent. Third, it delivers a doctor's observations and diagnosis to his patient via the user interface.

Gate Agent (GA). The Gate Agent verifies a patient's authentication for his request for services. Patients have different rights to have access to Role-Based Access Control (RBAC), in accordance with various privileges given by their roles.

Supervisor Agent (SA). The Supervisor Agent operates between the mobile device and the hospital system, controlling the entire surrogate system. First, it receives real-time medical data including the blood pressure (systolic/diastolic), body temperature, breathing and pulse, from a patient, saves the data into a repository, and then uses a specific pattern recognition module to analyze the data and compare with normal conditions. If the value of data exceeds their normal range (threshold), the agent sends an emergency alert message to a doctor or any other responsible person in the hospital via the manager agent, to take the appropriate emergency measures. If the value falls under the normal range (threshold), services will be discontinued when data are saved into a repository.

Manager Agent (MA). The Manager Agent is operated on a hospital subsystem. If the supervisor agent requests emergency measures, it searches for the doctor in charge, and related hospital workers. This agent sends to message which existed in patient of history data and which requested diagnosis including organism information of patient by the doctor in charge. This agent stores the diagnosis result and opinion with Timestamp and result-id in the medical prescription database. In addition, this Medical History database stores organism information of patient from the Supervisor Agent and manages necessary data for data retrieval, registration and update, and delete of new patient.

Doctor Agent (DA). The doctor diagnosis of a patient is formed through messages from the Manager Agent. As well as this diagnosis, the Manager Agent sends an opinion to the patient. These diagnosis and prescription data is stored in the Medical Prescription database. The stored diagnosis and prescription data are managed and maintained as historic records, for use when required by patients.

4 Implementation

In this paper, we demonstrated the implementation of a multi-agent based Real-time Tele Healthcare System with the suggested architectures, such as the JADE, LEAP, J2SE, and PersonalJava. The Oracle database was used to store patient information, diagnosis and prescription and as a surrogate clinical repository.

We choose the N (N=4, in this case) patients (students), and save patient's one month sensor data to medical history database. It is not easy to recognize special disease by patient's sensor data. To recognize correctly, we must consider all possible cases and use medical diagnosis knowledge provided by professional doctor. The standard of decision in our paper is the difference ($delta_a$) between normal range (table 1) and the present sensing data in each category and the difference ($delta_b$) between the one month average and the present sensing data.

The following is pseudo code for patient disease state decision.

```
/* present disease state of patient is being judged whether it is unusual or not
Input:
  normal range(d1), 1-month average data(d2), present sensing data(d3)
Output:
  state decision (special state(true), normalcy(false))
*/

for each data item ∈ sensing data
{
  delta_a = | d2 - d1 |
  delta_b = | d3 - d1 |
  if  ( delta_a ≥ threshold_a or delta_b ≥ threshold_b )
  then return (true);          // emergency measure
  else return (false);         // normal state
}
```

Fig. 4 shows the sequence diagram of remote medical services based on ACL message transfers among agents. While body temperature, breathing and pulse are detected every one second, the blood pressure every fifteen minutes. All collected medical data are transferred to a mobile device.

Fig. 5 shows the prototype of the implemented healthcare system. Fig. 5 (a) is the user interface diagnosis and prescription of result for a patient via a PDA and Fig. 5 (b) is a Graph that could enable both patient temperature and surrounding temperature to change transition.

Table 1. Data item normal range of patient information

Data Item	Normal Range	Unit
Blood Pressure	140-0/90-60	mmHg
Body Temperature	35-37	℃
Breathing	12-20/1 min	times
Pulse	60-100/1 min	times

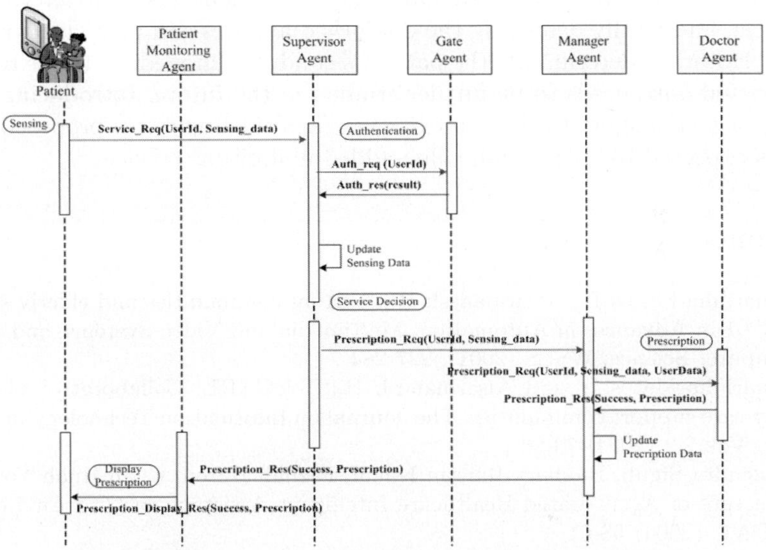

Fig. 4. Sequence Diagram of Multi-Agent based System

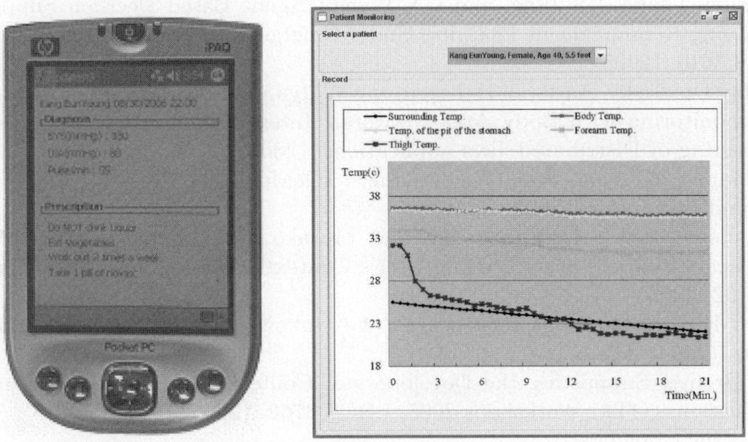

Fig. 5. (a)Patient's diagnosis and prescription (b)Result of Patient monitoring

5 Conclusions

This paper proposed a ubiquitous healthcare system by modeling real-time diagnosis and prescription services provided by a hospital system, based on collected medical and peripheral data. The proposed system provides an interconnection of patients and a hospital in a ubiquitous computing environment. In our system, multi-agents are involved in subsequent operations, such as using sensors to collect medical and peripheral data in real-time, storing the collected data at the surrogate system, determining whether a patient is in a critical condition, transferring to the hospital system data that the patient has been determined to be critical, and finally delivering the doctor's diagnoses and prescriptions to the patient. However, determining the patient's condition in medical terms based on the collected data needs to be further studied in the future. Introducing intelligent agent technologies to the context-aware healthcare system proposed in this paper is expected to create high value-added ubiquitous services.

References

1. Camarinha-Matos L., Afsarmanesh H.,: Virtual communities and elderly support, MIV'01 in Advances in Automation, Multimedia and Video Systems, and Modern Computer Science, WSES (2001), 227-284
2. Camarinha-Matos L. and Afsarmanesh H,: TeleCARE: Collaborative virtual elderly care support communities, The Journal on Information Technology in Healthcare, Vol. 2.2 (2004) 73-86
3. Shailendra Signh, Bukhary Ikhwan Ismail, Fazilah Haron, Chan Huah Yong,: Architecture of Agent-Based Healthcare Intelligent Assistant on Grid Environment, PDCAT, (2004) 58-61
4. H.Castro Oliveira, O. Belo, and Joao Paulo Cunha,: Agents Working on the Integration of Heterogeneous Information Sources in Distributed Healthcare Environments, IBERAMIA-SBIA, 2000, LNAI 1952, (2000) 136-145
5. Devinder Thapa., IS Jung, and GN Wang,: Agent Based Decision Support System Using Reinforcement Learning Under Emergency Circumstances, ICNC 2005, LNCS 3610, (2005) 888-892
6. Nikolay Dokovsky, Aart van Halteren, Ing Widya,: BANip: enabling remote healthcare monitoring with Body Area Networks, International Workshop on scientific engineering of Distributed Java applications, (2003) 27-28
7. Ole Martin Winnem, Stale Walderhaug, "TeleMediCare Project", SINTEF Telecom and Informatics, 2002
8. Kazushuge Ouchi, Takuji Suzuki, and Miwako Doi, "LifeMinder: A Wearable Healthcare Support System Using User's Context", Proceedings of the 22nd ICDCSW'02, 2003
9. Stefan Kim,: Ubiquitous Healthcare: The OnkoNet Mobile Agents Architecture, NODe, (2003) 265-277
10. Tom Brones: Supporting the Developers of Context-Aware Mobile Telemedicine Applications, OTM Workshops 2005, LNCS 3762, (2005).761-770
11. http://www.fipa.org
12. http://leap.crm-paris.com

A Block Based Moving Object Detection Utilizing the Distribution of Noise

M. Ali Akber Dewan, M. Julius Hossain, and Oksam Chae*

Department of Computer Engineering, Kyung Hee University,
1 Seochun-ri, Kiheung-eup, Yongin-si, Kyunggi-do, South Korea, 446-701
dewankhu@gmail.com, mdjulius@yahoo.com, oschae@khu.ac.kr

Abstract. Moving object segmentation in complex scene is the basis for video surveillance, event detection, tracking and development of vision agent in industrial robotics. However, due to presence of camera noise and illumination change, simple background subtraction based techniques are not able to detect moving objects properly. In this paper, we present a novel block based moving object detection method which dynamically quests for both local and global properties of difference image to achieve robustness. Noise model of the difference image is determined analyzing the histogram of difference image and block wise local properties are computed. These local properties are compared with the noise model to extract moving blocks. To remove the stair like artifacts of the segmented result, and to obtain smoothed and accurate boundary, a refinement procedure is employed on the boundary regions of detected moving objects. Experimental results show that the proposed method is robust and achieves better performance in dynamic environment.

1 Introduction

Automatic extraction of moving objects plays an important role in computer vision. Diverse applications of computer vision such as video surveillance, video coding, video indexing, distributed artificial intelligence in video analytics and robotics vision in multi agent environment are getting benefits from this research [1][2][3]. Image differencing approach is one of the most common and simplest approaches for moving object detection. It highlights the moving object regions or content changes while suppressing the static background. However, this approach encounters difficulties to discriminate changes occurred due to presence of moving objects from changes occurred due to presence of noise or illumination variations. A popular method to discriminate these changes is image thresholding. However, a low threshold value tends to create false alarm and a high value tends to swamp significant changes between the frames, where determination of a global threshold value automatically is another difficult task [4]. In general, the optimal threshold value is a time varying and content dependent parameter [5]. Therefore, an autonomous surveillance system based on unsupervised absolute value thresholding may be characterized as inadequate [6].

* Corresponding author.

N.T. Nguyen et al. (Eds.): KES-AMSTA 2007, LNAI 4496, pp. 645–654, 2007.

The efficiency of a moving object detection technique greatly lies in the successful separation of changed regions which are caused by content changes from the changed regions which are caused by noise or illumination changes in the difference image. In compare to plain image differencing approach or absolute difference thresholding technique, block analysis based method provides higher robustness to noise, since it takes into account a pixel neighborhood or localized static information of each blocks when applying the decision rules. Many block-based algorithms and techniques have been developed for moving object segmentation in the last decade. One popular block-based method is proposed in [7]. In this method, the difference image is initially divided into a number of equal sized blocks and after that, blocks are classified into several clusters depending on their local statistical properties. Finally, noise features are computed analyzing the largest cluster and are employed it to detect moving blocks. However, cluster size and number may vary depending on the order of consideration of different blocks. Hence, it cannot ensure extraction of reliable features of noise from the difference image. Moreover, clustering techniques are time consuming and are not feasible for real time processing. In [8], a block matching algorithm for motion estimation is proposed. In this method, the current image frame is first divided into fixed size rectangular blocks, and the motion vector for each block is estimated by finding the closest block of pixels in the previous frame according to a matching criterion. However, searching for the best matching block within a search window requires high computation which limits its application for real time processing. Moreover, smooth region of moving object is another challenge for motion vector generation.

In this paper, we present a novel block comparison based moving object detection method which is reliable in discriminating content changes from the noise level changes. It dynamically computes the threshold value based on both local and global features of difference image. It ensures faster processing to extract moving objects due to simplicity of the algorithm. In our proposed method, difference image is first computed by subtracting the current frame from the reference frame and after that it is divided into $n \times m$ sized blocks, where n and m are integers. Mean value of each block is computed as its local feature and these values are utilized to generate a histogram. Thereafter, a statistical approach is employed on the histogram to compute noise model. This noise model and local features (mean values) of blocks are used in the next processing step with a criterion function to extract moving blocks. Utilization of block means instead of using individual intensity difference values makes the system much robust and faster in noise property computation. However, block based processing produces stair like artifacts on the boundary regions of detected moving objects. So, we employ a further refinement procedure on the boundary blocks to produce smoothed and accurate boundary of moving object.

2 The Proposed Method

In controlled and ideal environment simple image differencing approach is well enough to discriminate the moving object regions from the static background scene. However, practically it is not possible to have such kind of environment due to illumination variation or dynamism in background scene. Moreover, calibration error of capturing device incorporates noise in the image frame [9] [10]. So, difference image is dependent on several factors. Suppose that, I_{diff} denotes the difference image of current frame, I_n and background frame, I_B. Then I_{diff} can be represented with the following equation

$$I_{diff}(x,y)=I_n(x,y)-I_B(x,y)=I_{obj}(x,y)+I_{noise}(x,y) \tag{1}$$

where, I_{obj}, is the intensity changes due to change of contents and I_{noise}, is the intensity changes due to presence of noise. Hence, we need to have some criterion functions that can effectively discriminate I_{obj} from I_{noise} in the difference image.

Mean intensity value of difference image resides outside the moving object region can be considered as noise mean [7]. In our proposed method, we have employed a novel statistical method to compute noise model (μ_{noise}, σ_{noise}) of difference image and after that these noise properties are utilized in further processing steps to decide which regions belong to moving objects and which do not. The overall procedure of the proposed method is described in details in the following subsections.

2.1 Computation of Noise Model Analyzing Histogram

The histogram of signed difference image may contain several modals (concentration of pixel values surrounding a peak and separated from others by valleys), constructed from noise and homogeneous regions of moving objects as well as covered portion of background in current frame. However, largest numbers of pixel values of difference image are concentrated around the peak of the modal which is created from the effect of noise or illumination change. Hence, if we can separate the biggest modal from others, we can easily measure the statistical properties of noise, (μ_{noise}, σ_{noise}) from it.

Fig. 1 depicts histograms of difference image both in presence and absence of moving objects. Histogram shown in Fig. 1(d) is computed from two background frames (Fig. 1(a) and Fig. 1(b)) where content changes are totally absent. In this histogram, concentration of pixel values surrounding a peak occurred due to the presence of noise along with illumination variations. Fig. 1(e) has shown another histogram of difference image computed from Fig. 1(a) and Fig. 1(c). Along with noise and illumination variation, content change is also present in Fig. 1(c). Here, the highest concentration of pixel values occurred surrounding a peak which is also created due to noise and illumination changes. We have utilized this largest concentration of pixel values of difference image for noise model computation.

Considering computational efficiency and robustness against noise, we do not utilize individual pixel value of difference image for histogram computation. Rather, we divide the difference image into a number of fixed sized (*n x m*) image blocks and compute their means. As a result, we get a new mapping of the difference image with these mean values and we termed it as mean preserving difference image (*MPD*). *MPD* is a *2D* array of *rows/m* rows and *cols/n* columns, where *rows* and *cols* are the number of row and column of difference image, respectively. After that, we analyze the histogram of *MPD* with three overlapped sliding windows to find out the largest modal from it. In order to discriminate the positive and negative brightness in the histogram of *MPD*, we allow the gray scale range [-255, 255], which differs from widespread definition of an image histogram, supporting [0, 255].

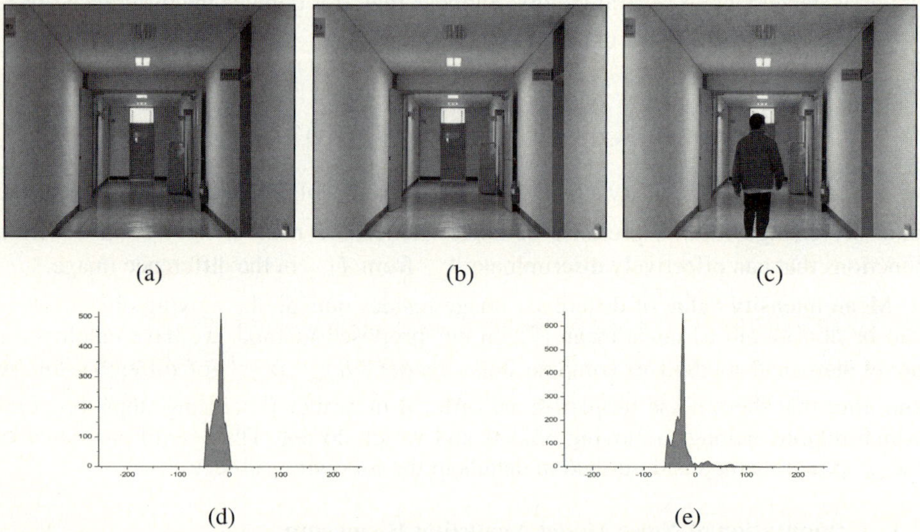

Fig. 1. Difference image histograms in presence or absence of moving objects. (a) Background; (b) Background with different illumination; (c) Moving object in same scenario; (d) Difference image histogram computed from (a) and (b); (e) Difference image histogram computed from (a) and (c).

If we analyze the modals of a histogram carefully, we can see that each modal contains two sharp valleys with a sharp peak inside it. We utilize this characteristic to separate one modal from others in the histogram by detecting two valleys, i.e. starting and ending points of a modal containing one peak inside it. In our method, we have utilized three fixed size overlapping sliding windows for this purpose. Let, *L*, be the length of each sliding window and three sliding windows are placed overlapping each other by *L/2*, shown in Fig. 2(a). In each of the iteration, the mid point of the middle sliding window, x_m is placed on a pixel value of the histogram. The rest of the positions of sliding widows are also placed as well and computes the following measures for each of the sliding windows

$$m(x) = \sum_{i=x-(L/2)}^{x+(L/2)} iP(i) \tag{2}$$

$$\sigma(x) = \sqrt{\sum_{i=x-(L/2)}^{x+(L/2)} i^2 P(i) - \left\{ \sum_{i=x-(L/2)}^{x+(L/2)} iP(i) \right\}^2} \tag{3}$$

where, x represents the center of sliding window. $m(x_l)$, $m(x_m)$, and $m(x_r)$ are means measured by left, middle and right sliding window, respectively. $\sigma(x_l)$, and $\sigma(x_r)$ are standard deviations of left and right sliding window, respectively. *P(i)* is the frequency of *i* in the histogram of *MPD*.

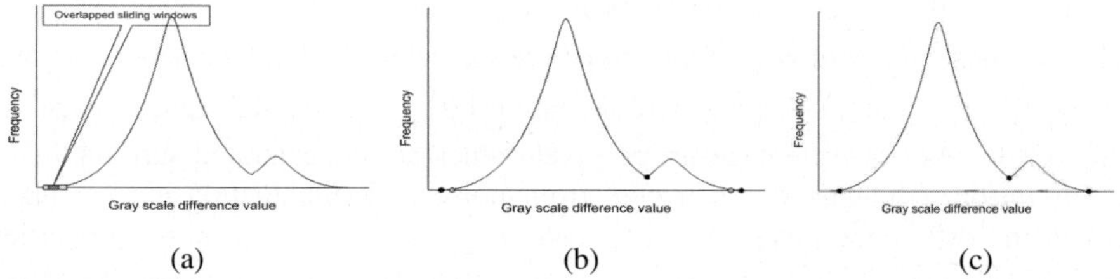

Fig. 2. Modal detection procedure in the difference image histogram. (a) Placement of three sliding windows; (b) Candidate points detected as valleys; (c) Detected valley points.

The overall procedure for noise property calculation is as follows:

1. Three overlapped sliding windows are placed on histogram in such a way that the mid point of the middle sliding window x_m is placed on a pixel position x in the histogram of *MPD*. The rest of the positions of windows are placed to their corresponding positions with respect to x_m (Fig. 2(a)).
2. $m(x_l)$ and $\sigma(x_l)$ for left sliding window, $m(x_m)$ for middle sliding window, $m(x_r)$ and $\sigma(x_r)$ for right sliding window are computed.
3. Three types of valley may appear in histogram (Fig. 2(c)). These three types are discriminated with the following logic functions:
   i. If $m(x_l) < m(x_r)$ and $\sigma(x_l)$ is less than a certain threshold then the position x_m can be considered as a starting valley point of a modal in the histogram.
   ii. If $m(x_l) > m(x_r)$ and $\sigma(x_r)$ is less than a certain threshold then the position x_m can be considered as ending valley point of a modal in the histogram.
   iii. If $m(x_l) > m(x_m) < m(x_r)$, then the position x_m is marked as ending valley point of previous modal and simultaneously, starting valley point of next modal. In Fig. 2(c), the middle one depicts this type of valley point.
4. Each of the point of the histogram is checked iteratively and selects candidate valley points from it. Each of the valley point is marked as either starting (*S*) or ending (*E*) or both (*B*) according to the rules described in step 3.
5. Since, we are using a number of neighboring values of histogram corresponding to the positions covered by sliding windows for valley point detection; it may detect a number of candidate points as starting and ending valley points for each modal (white and dark points shown in Fig. 2(b)). To resolve this problem, we select only those points as starting and ending valley points for a modal which are at maximum distance (dark points shown in Fig. 2(b)). This condition is only applicable for those valley points which are marked as *S* and *E*. For valley points marked as *B*, need to take special care. If a number of consecutive valley points are marked as *B*, then it selects the middle one and uses it as ending valley point for the previous modal and simultaneously, as starting valley point for the following modal (Fig. 2(b)).
6. After detecting the starting and ending valley points of each modal, the size of the modal is measured by counting the number of pixel values inside it.
7. Finally, by comparing the size of modals, the biggest one is selected to determine the noise model of difference image.

After detecting the largest modal, its mean ($\mu_{\max}$) and standard deviation ($\sigma_{\max}$) is computed. These $\mu_{\max}$ and $\sigma_{\max}$ of the modal are very closely approximated to original noise mean (μ_{noise}) and noise standard deviation (σ_{noise}) of the difference image, as the highest number of difference image pixel values are concentrated surrounding a peak (or inside a modal) which occurs from noise and illumination change pixel values in the difference image. So, $\mu_{\max}$ and $\sigma_{\max}$ can be used as noise properties and in rest of the description, these expressions will be termed as μ_{noise} and σ_{noise} for better understanding.

2.2 Detection of Moving Blocks

The noise in the difference image follows Gaussian distribution with μ_{noise} and σ_{noise}, and it occurs due to illumination changes along with the effect incorporated by the capturing device. The noise distribution of the difference image can be modeled mathematically with the following equation:

$$p(I_{diff}(x,y)) = \frac{1}{\sqrt{2\Pi\sigma^2_{noise}}} \exp\left[-(I_{diff}(x,y) - \mu_{noise})^2 / 2\sigma^2_{noise}\right] \tag{4}$$

According to the theory of statistics, there are more than 95% noise pixels whose intensity values in difference image will drop inside the close region $[-2\sigma, +2\sigma]$. Therefore, if a pixel's intensity value in difference image is out of the range $[-2\sigma, +2\sigma]$, the pixel can be classified as a moving pixel. To make the detection procedure more robust and faster, instead of comparing each pixel value of the difference image separately, we have compared block mean with the noise property and detect moving blocks with the following criterion function:

$$\begin{array}{ll} if\,(Mean(Block_i) - \mu_{noise}) > 2 * \sigma_{noise} & Block_i = 1 \\ Otherwise & Block_i = 0 \end{array} \tag{5}$$

where, $Mean(Block_i)$ is the mean value of ith block. Hence, a change mask M_k is generated with fixed size block containing values either 0 or 1. Blocks, containing 1 represent the moving blocks and blocks, containing 0 represent static background blocks in M_k. In this stage, M_k may contain some holes or discrete noisy blocks. Since, we apply block wise comparison for change detection; one or more blocks may responsible to create these holes and discrete noisy regions in M_k. Hence, we have employed a chain-code based approach, which uses block wise morphological operator [11] to remove these noisy regions and loosely connected blocks as well from M_k. The segmentation result using M_k is shown in Fig. 3(a). However, the result shows that further processing steps are required to improve it.

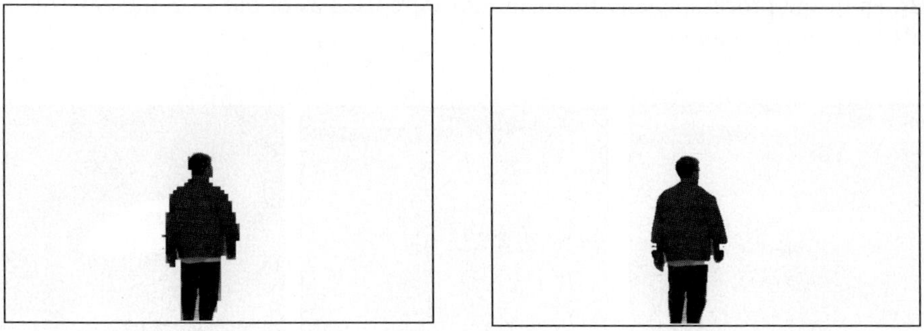

(a) (b)

Fig. 3. Segmented moving object. (a) Moving object before applying boundary refinement procedure; (b) Moving object after applying boundary refinement procedure.

2.3 Refinement of Moving Object Boundary

Since block wise decision is taken in the criterion function for moving region detection, stair like artifacts may appear on the boundary regions of detected moving objects (Fig. 3(a)). Both moving blocks and background blocks on boundary of moving regions are responsible to create these artifacts. Though all pixels inside these blocks are not of fully moving or fully background pixels, they are detected as totally moving or background pixels, due to taking block wise decision in the previous steps. Hence, we select the blocks, neighboring to the boundary regions for reconsideration and employ a pixel wise processing on these blocks using the following functions:

$$I_{mean}(x,y)=\frac{1}{(n\times n)}\sum_{i=x-(n/2)}^{x+(n/2)}\sum_{j=y-(n/2)}^{y+(n/2)} I(i,j) \tag{6}$$

$$\begin{array}{ll} if(I_{mean}(x,y)-\mu_{noise})>2*\sigma_{noise} & I(x,y)=1 \\ Otherwise & I(x,y)=0 \end{array} \tag{7}$$

where $I_{mean}(x,y)$ depicts the mean at position (x, y) in difference image and is computed utilizing its n-neighboring pixels. These mean values are computed for all pixels of the selected boundary blocks and is compared with noise model again and is re-decided whether these pixels should be considered inside moving regions or not. Since this procedure is applied only on the pixels of selected boundary blocks, it performs its operation faster and does not create any problem for real time processing. The result produced after this processing step is shown in Fig. 3(b), where stair like artifacts are removed and produce smoothed boundary of moving object regions.

3 Experimental Results

The proposed method was tested on image sequences acquired in indoor as well as in outdoor environments with various illumination conditions. We applied our proposed method on video sequence of frame size 320 x 240, using a system having Intel Pentium IV 1.6 GHz processor and 512 MB of RAM. Visual C++ 6.0 and MTES

[12], an image processing environment tool were used as of our working environment tools for implementation.

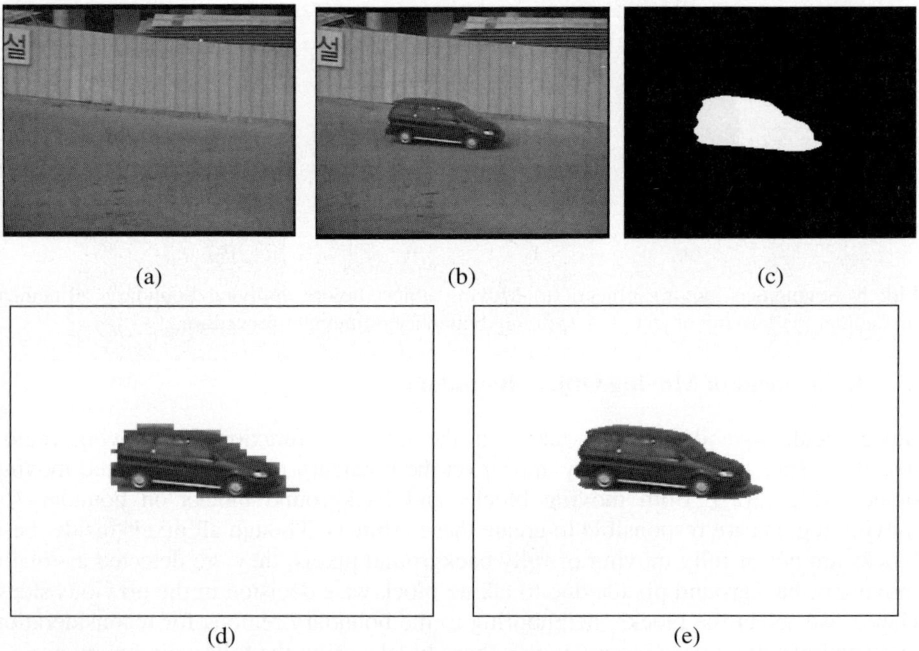

Fig. 4. Moving object segmentation result in outdoor environment of a road scene. (a) Background; (b) Time difference frame in presence of moving object; (c) Change mask after refinement of boundary regions; (d) Segmentation result using moving blocks before refinement of boundary regions; (e) Segmentation result after refinement of boundary regions.

Fig. 4 shows an experimental result of a real road video scene with moving vehicle at outdoor environment. Fig. 4(a) and Fig. 4(b) show the static background and current frame, respectively. The histogram of the difference image of Fig. 4(a) and Fig. 4(b) is employed to compute noise properties and is further used to detect moving blocks from the difference image. The segmented moving object, using moving blocks is shown in Fig. 4(d), where there are stairs like artifacts on boundary regions. Since blocks, located on moving object's boundary and on its neighboring region may contain concurrently both types of (moving and background) pixels, a refinement procedure is applied to these blocks to discriminate the moving pixels from background. The refinement result is shown in Fig. 4(e). Fig. 4(c) shows the refined change mask, which is finally used for moving object segmentation. Fig. 5 shows results of another experiment in an indoor video sequence, having illumination variation to test the accuracy of our proposed method. Final result shows that our method also works well in changing environment.

In order to comprehend the computational efficiency of the algorithm, it should be mentioned that, with the mentioned processing power and the processing steps, execution time for the object detection on grayscale images was approximately 95 ms.

Therefore, the processing speed is 10.52 frames per second. Although this frame rate is about 0.4208 times of the real time frame rate (i.e. 25 frames per second), but this speed is quite reasonable for real time processing. However, using computers with higher CPU speeds which are available this day and in future as well, this frame rate can be improved. Time requires to execute different modules of the whole procedure are shown in Table 1.

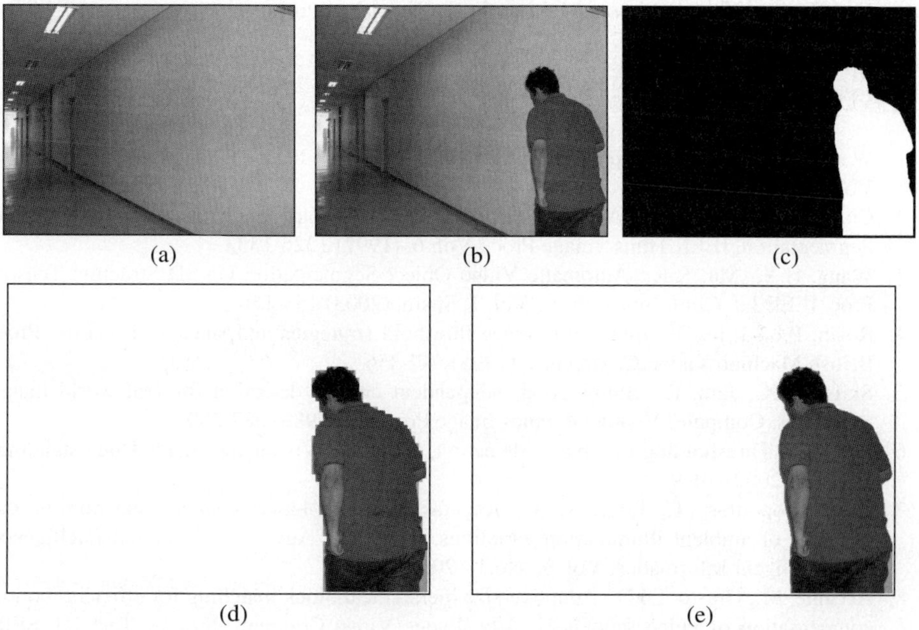

Fig. 5. Moving object segmentation result in indoor environment of building corridor scene. (a) Background; (b) Time difference frame in presence of moving object; (c) Change mask after refinement of boundary regions; (d) Segmentation result using moving blocks before refinement of boundary regions; (e) Segmentation result after refinement of boundary regions.

Table 1. Mean processing time in (MS) for each module

Algorithm	Mean processing time (MS)
Noise mean computation	23
Detection of changed blocks	22
Chain code based post processing for noise reduction	23
Refinement of the boundary regions	27
Total time required	**95**

4 Conclusions

In this paper, we have presented a moving object detection approach for automatic video surveillance system. In this method, noise properties of difference image are computed block wise, by analyzing its histogram. These properties are utilized to detect moving blocks from the difference image. Detection process is dependent on a criterion function, which is dynamic and fully automated. Applied refinement procedure on neighboring blocks of moving object boundary removes the stair like artifact and obtains smoothed and accurate boundary. It improves the accuracy in case of extracting features from segmented moving objects. It reduces higher computational cost by avoiding pixel based processing on whole image that is required for traditional approaches to achieve similar accuracy.

References

1. Wang, J., Adelson, E.: Representing Moving Images with Layer, IEEE Trans. Image Proc., Vol. 3, (1994) 625-638
2. Chang, M.M., Tekalp, A. M., Sezan, M. I.: Simultaneous Motion Estimation and Segmentation, IEEE Trans. Image Proc., Vol. 6, (1997) 1326-1333
3. Wang, H.Y., Ma, K.K.: Automatic Video Object Segmentation via 3D Structure Tensor, Proc. IEEE Int. Conf. Image Proc., Vol. 1, Spain, (2003) 153-156
4. Rosin, P.L., Ellis, T.: Image difference threshold strategies and shadow detection, Proc. British Machine Vision Conference, (1995) 347-356
5. Skifstad, K., Jain, R.: Illumination independent change detection for real world image sequences, Computer Vision Graphics Image Process., (1989) 387-399
6. Rosin, P.: Thresholding for change detection, Computer Vision and Image Understanding, Vol. 86, (2002) 79–95
7. Alexandropoulos, T., Loumos, V., Kayafas,E.: Block-based change detection in the presence of ambient illumination variations, Journal of Adv. Computational Intelligence and Intelligent Informatics, Vol. 9, No.1, (2005) 46-52
8. Accame, M., Giusto, D. D.: Adaptive-size hierarchical block matching for efficient motion compensation of video sequences, Adv. Image Video Commun. Storage Technol. SPIE 2451, (1995) 112-119
9. Radke, R., Andra, S., Al-Kohafi, O., Roysam, B.: Image Change Detection algorithms: A Systematic Survey, IEEE Trans. Image Proc., Vol. 14, Issue 3, (2005) 294-307
10. Foresti,G., Mahoen, P., Regazzoni, C.: Multimedia Video Based Surveillance System, Requirements, Issucs and Solutions, Kluwer Academic Pub., USA (2002)
11. Kim, J. B., Kim, H. J.: Efficient region-based motion segmentation for a video monitoring system, Pattern Recognition Letter, Vol. 24, (2003) 113–128
12. Lee, J., Cho, Y. T., Heo, H., Chae, O.S.: MTES: Visual programming for Teaching and Research in Image Processing, Springer- Verlag Lecture Notes in Computer Science, Vol. 3514, April (2005) 1035-1042

Multi-Agent System for Hierarchical Control with Self-organising Database

Dariusz Choiński, Witold Nocoń, and Mieczysław Metzger

Faculty of Automatic Control, Electronics and Computer Science,
Silesian University of Technology,
ul. Akademicka 16, 44-100 Gliwice, Poland
{dariusz.choinski,witold.nocon,mieczyslaw.metzger}@polsl.pl

Abstract. The paper presents hierarchical and multi-agent control and information system with real-time update of self organising database as well as with negotiation capability for control events and decisions. A practical application is presented, that utilizes the OPC technology in the continuous-time part, and scripts using XML in the discrete-time part of the system for negotiation and cooperation in multi-agent environment. This feature is applied for the improvement of a non-conventional biotechnological process control in the pilot plant.

Keywords: Multi-agent control system; OPC; real time self organising data base, hierarchical control.

1 Introduction

Nowadays automation and information systems designed for industrial plants are complex, large and include a lot of different components such as control instrumentation, control software and communication networks. Integration of the process control system and finally an operation of the process during normal exploitation as well as in emergency situations are difficult tasks. For this reason an advanced control system (apart from standard controllers and computers) should include several additional techniques such as real-time communication with databases or agent-based control algorithms. Fundamentals of agent-based techniques seem to be well-defined in recent publications over the last decade (see for example [1], [2], [3], [4], [5]). Such technology is very convenient for network–based distributed control of manufacturing systems (see for example [6],[7]). Nevertheless, the multi-agent technology can also be helpful for process control as is demonstrated in this paper.

For biotechnological processes, which include very complex biological, chemical and thermodynamical processes, the distributed control system should take into consideration a flexible cooperation with database in different levels of control and remote access for experts, especially in emergency situations. Therefore, biological processes are especially sensible to inadequate decisions. Hence, a complex control systems should bundle different hardware and software technologies, the later being especially important, for accomplishing such tasks. The major problem deals with

N.T. Nguyen et al. (Eds.): KES-AMSTA 2007, LNAI 4496, pp. 655–664, 2007.

appropriate synthesis of flexible network-based computer and control instrumentation system, which should make a flexible access and update of control data, process events, operator and external expert decisions and negotiations possible – all viewed and actualised in real time.

This paper presents a hierarchical and multi-agent control and information system with real-time update of a self organising database, as well as with negotiation capability for control events and decisions. A practical application is presented, that utilizes the OPC technology in the continuous-time part, and scripts using XML in the discrete-time part of the system for negotiation and cooperation in multi-agent environment. The multi-agent feature is applied as an improvement to a nonconventional biotechnological process control in the pilot plant [8], [9]. In the proposed multiagent heterogeneous system the OPC plays an important role as powerful communication technology especially dedicated for real-time distributed control systems (see e.g. [10]). This technology consists of OPC standard and specialized software architectures offered by most of Distributed Control System (DCS) vendors.

2 Architecture of the Multi-Agent System

Davidson and Wernstedt [11] argue for the appropriateness of using software agents for the monitoring and control of bioprocesses. Bioprocesses are getting more and more complex and often involve economical and environmental constraints. This leads to modelling and control problems of increasing complexity when trying to build systems that operate robustly over a wide range of conditions. In order to build control systems for biotechnological process, those are usually decomposed into local problems. This leads to a model or controller that is partitioned into multiple smaller operating entities, each of which is associated with a locally valid model or controller. Advances in software engineering provide a number of solutions, and the concept of software agents is especially suited to facilitate the design and implementation of systems that are partitioned into smaller operating entities. Industrial processes however, are usually equipped with a number of database handlers, control systems or expert systems that are not agent-ready. Hence, wrapper agents are used that provide an approach to 'agentify' those already used systems.

The considered object model is described as a state machine augmented with differential equations [12]. The state can change in two ways:

- instantaneously by discrete transition described by the sequences $\boldsymbol{S}$ of actions from source state to target state
- in a time pass according to a trajectory of the state variables change as a function f of input and output variables

This allows decomposition of complex biotechnological object with control hierarchical system and analysis of this system separately for time-driven and event-driven part. On the other hand, object states are divided into two sets. The first one is a set $\boldsymbol{\Omega}$ of continuous state variables with boundaries depended on physical process parameter constrains, measurement ranges and capabilities of actuators. The second set $\boldsymbol{\Phi}$ contains values describing events for transition conditions. $\boldsymbol{\Phi}$ is divided into two subsets: $\boldsymbol{\Phi}_c$ for controllable events and subset $\boldsymbol{\Phi}_u$ for uncontrollable events. Agents are responsible for transitions, therefore $\boldsymbol{\Phi}$ describes the agent sensors and $\boldsymbol{\Omega}$ describes common agent effectors. Apart from effectors used commonly by all agents, each agent possesses other effectors that are characteristic and dependant on the knowledge source. The type of agent depends on the way the state changes and on the agent's knowledge. The agents are specified on a design level in an iterative process in addition of knowledge needed. The general structure of the presented multi-agent hierarchical control system is shown in Fig. 1. It consists of the following agents:

- Control agent – uses a trajectory for state transition and provides the basic control algorithms for the process. All the closed-loop and open-loop control algorithms are implemented in this agent, hence this part of the control system is time-driven, e.g., the measurements are read and controls are transmitted to the plant in fixed time intervals regardless of the process behaviour. This agent's knowledge is based on implemented control algorithms and realises function $\boldsymbol{f}$ and subset $\boldsymbol{\Phi}_u$ as *Control actions.*
- Supervisory agent – responsible for general supervision of the process performance. It also provides supervisory agent compensation against process fluctuations by executing special sequences of operations, such as biomass dilution or thickening, and by initiation of some sequences of operations in case of well-defied off-nominal situations. In case of some off-nominal situations that can not be dealt with automatically by the supervisory agent, the application sends a report to the remote expert agent. This agent's knowledge depends on phenomenological models of object and operator experience and specific effectors are sequences $\boldsymbol{S}$ of actions and subset $\boldsymbol{\Phi}_c$ as *Supervisory compensator.*
- Expert agent – provides remote expert knowledge in case of some off-nominal situations than can not be dealt with by the supervisory or control agent. The effector for this agent is function $\boldsymbol{f}$ prepared by expert as *Expert compensator.*

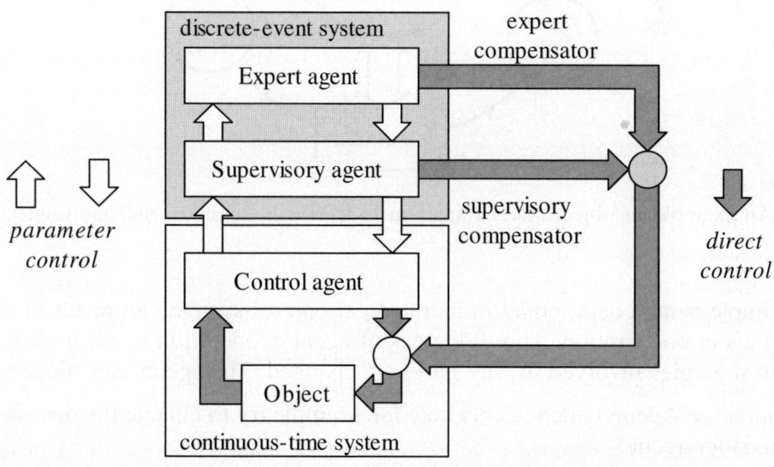

Fig. 1. Architecture of the multi-agent system

Parameters of Control Agent algorithms are set up in order to maintain control theory conditions of local stability. However, object state transition during agents' cooperation can cause the object's instability. Considering the object as hybrid automaton (a state machine augmented with differential equations) increases the range of techniques for stabilizing the multi-agent system [13],[14].

The presented architecture is strictly heterogeneous in two dimensions. Communication is described both as a client-server and producer-consumer protocol. The DCS system is divided into two types of behaviours: time-driven and event-driven. The agents in this multi-agent system not only communicate with the users and with the object, but they also communicate and cooperate with each other, solving problems using direct control interaction and also by improving control system functions. Fig.2 shows an example of object states transitions. Each agent may initiate a state transition, but not every agent may actually enforce this transition without cooperation with other agents. For example, the object state may change from continuous control (Ω_i) to discrete control (Ω_{i+1}) for changing biomass concentration and identification of OTR (oxygen transfer rate) and return to continuous control (Ω_{i+2}) of object with additional function based on calculated OTR coefficient for better dissolved oxygen control.

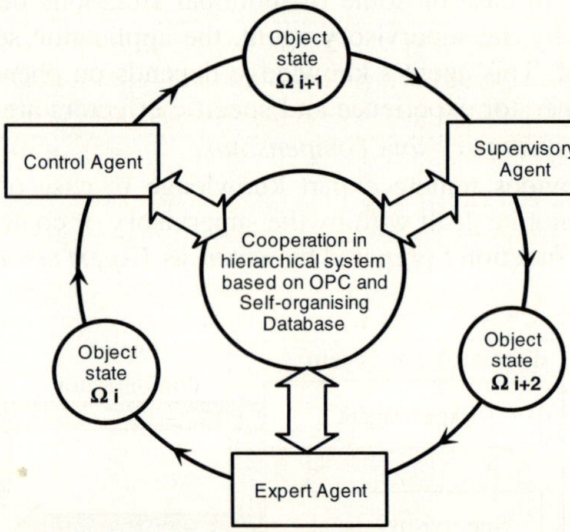

Fig. 2. An example of object states transition in hybrid hierarchical and multi-agent control system

The implemented data model in hierarchical control system can result in three degrees of agent cooperation. Those degrees of agent cooperation result in different interaction strategies involved in way states are changed. The agents can cooperate as:

- Competitive Agents. Such agents may for example try to change the decision of the supervisory agent.
- Collaborative Agents. Such agents share their knowledge in order to maximize benefits of communication in critical system controls. For example, such agents may start negotiating with each other, and try to resolve the parameters of controller using phenomenological model and on-line measurement identification.

- Hostile Agents. Those agents are strongly connected with watchdog systems and are used during object sensors or actuators malfunctions. A hostile agent should interrupt other agents with improper information or agents that are connected to an out-of-order actuator. This is very important for biotechnological systems with live organisms. Most importantly, the control systems should prevent a total loss of the controlled object.

The cooperation is provided in two different levels of abstraction. The real object level of the controlled object is described by OPC configuration as a hierarchical structure. Designer of the DCS for biotechnological object using structure of data model standardised by OPC can decide which part of the system is available for the

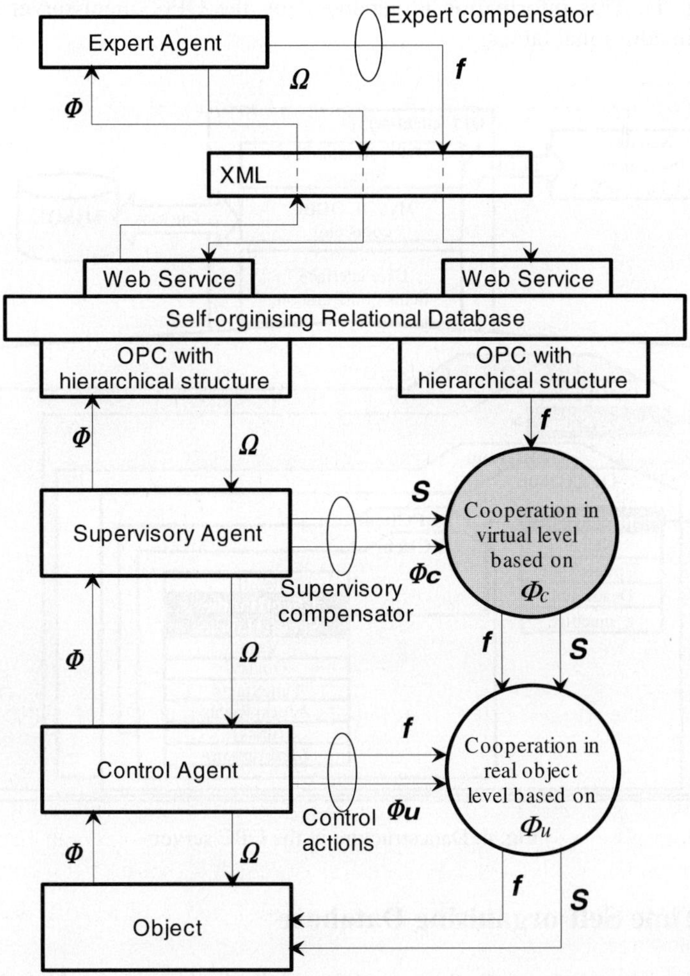

Fig. 3. Cooperation between agents

multi-agent system. This description is converted into the relational database structure. The supervisory agent is equipped with models, methods and algorithms for additional indirect measurements and controls. The off-line measurements, models and knowledge of the expert agent user and its additional algorithms with supervisory agent methodology are a virtual level based on database information (Fig. 3).

3 Data Structure of the OPC Server

Because of the cooperation, in the OPC standard, between connections to the relational database that is surveyed and modified by wrapped agents, it has been necessary to expand the information stored at the OPC server side for the particular OPC clients (Fig. 4). This information is generated by the OPC client-server application and stored in additional tables.

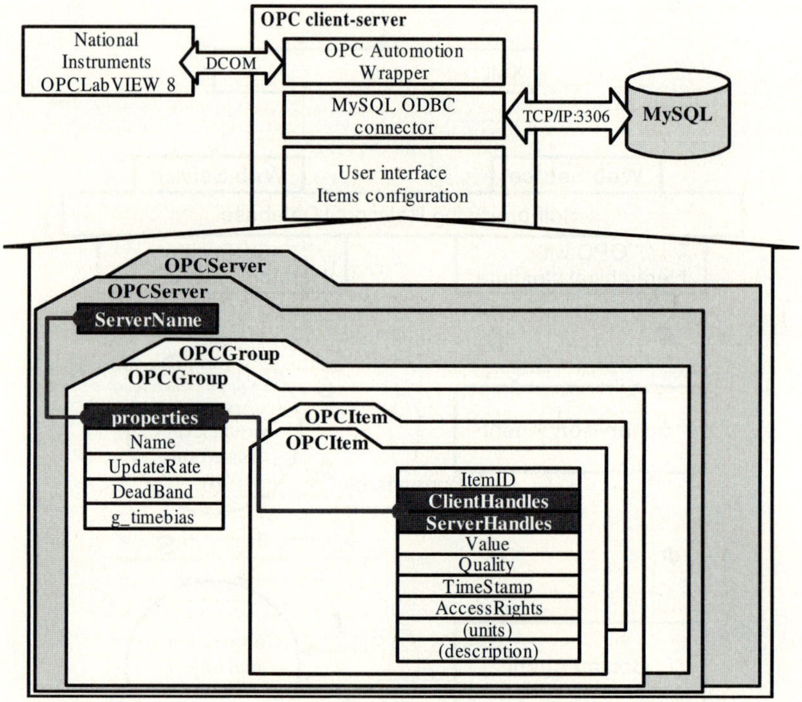

Fig. 4. Data structure in the OPC server

4 Real Time Self-organising Database

Data structure in a relational database is divided into the following tables (Fig. 5).

The `opcservers` table stores unique numeric identifiers (`s_id`) and names (`s_name`) of the OPC servers that are visible locally for computers with OPC client-server applications installed. Those values (`s_id` and `s_name`) are stored because, using *foreign key*, the `groups` table, which possesses a storage engine (of the InnoDB type) is bound with the `opcservers` table. In the current MySQL version only this engine enables a mutual binding of tables using *foreign keys*.

In the `groups` table the following information is stored: unique identifiers (as *primary key*) of the OPC groups defined in the OPC client-server program and basic parameters of $\boldsymbol{\Omega}$. This table is bound by *foreign key* with the `opcservers` table. This enables the hierarchical structure and data coherence to be maintained, which is one of the most important problems for this application. The data structure in the OPC standard is a typical hierarchical structure and stores no information about the mutual data binding. The tree structure of data represented in the OPC server does not correspond to the control system structure. However, it stores the key information about

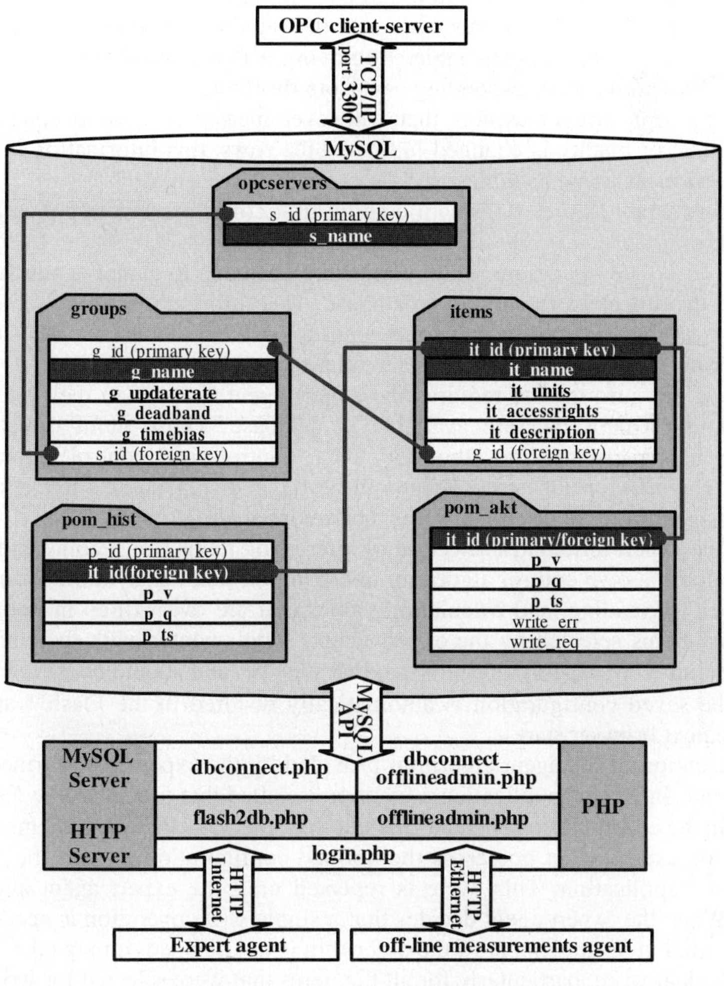

Fig. 5. Data structure in a relational database

the control system state ($\boldsymbol{\Omega}$). Therefore, a mechanism is needed for updating bindings between the relational database and the hierarchical information structure of the control system. In case of removing a row from the `opcservers` table, all groups that were subject to the OPC server will be automatically deleted from the `groups` table.

The `items` table stores information about OPC items defined in the OPC client-server application. In the *primary key* of this table, a unique identifier is present. Additional information about the item is stored and this table is bound using *foreign key*. This enables the hierarchical structure and data coherence to be maintained. In case of removing a row from the table, all groups that were subject to the OPC server will be automatically deleted from the `items` table.

In the `pom_akt` table, the current values of readings bound with OPC items and in addition, write-to-OPC requests sent by the Internet application and the information to be written are stored. The OPC server is able to "answer" such a request by deleting values from the `write_req` column and by writing to the `write_err` column, if necessary. The *primary key* of this table is at the same time bound, using *foreign key*, with the column of the `items` table. Removing a row from the `items` table will automatically remove a corresponding row from this table.

A suitable time trigger assures that whenever measurements are updated, at the moment value or quality is changed in any of the rows, this information is copied to the historical measurements table.

The `users` table stores information about users of the Internet application of expert agent.

All the described operations and bindings are designed to gather as much information about the object as possible in a database. This information, stored in a coherent structure, may than be used by an expert agent for making elementary decisions or for implementing additional algorithms or negotiation.

Usage of this information requires the expert agent to login to the system and to download OPC items that are available in database. Those items directly reflect measurement signals, off-line laboratory data, controls and control algorithms parameters. The structure of items, written in XML, is displayed as a tree, from which the expert agent may choose those items that are interesting.

When the expert terminates selection of interesting items (for reading and writing), it may define it's own control algorithm operating on items that where defined in the OPC server for reading, and calculating values that are overwritten in items defined for writing. Items selected by the expert agent, supplemented with the control algorithm, constitute the expert configuration, that may be, and should be, saved in the database. The saved configuration is automatically restored in the Flash™ application during the next browser start.

The execution of the agent algorithm provided by the expert is performed by calling the script. Inside the application a function is called that processes the XML string included in the configuration file and fills out the `_global` object in memory.

When the expert agent possesses the defined configuration, it runs the read/write mode in the application. This cycle is repeated until the expert agent algorithm is stopped. When the expert agent decides that a single write operation is needed for the selected item, but a supplement of the algorithm is not planned, it may take advantage of an individual write, particularly for all the items that were selected for writing.

This enables the algorithm defined by the expert agent to finally influence the pilot-plant installation and to realize the cooperation between this agent and the control agent. The degree of the negotiation depends on the items definition in the $\boldsymbol{\Phi}$ set at the time of OPC server configuration. Because Internet and asynchronous communication is used for that purpose, one can not consider such a system to be characterized by a constant sample period. Information about the time at which a set of samples has been received is remembered in each cycle. The difference of time, in seconds enables the utilization of algorithms with changing sample period, which are designed for nondeterministic networks.

The appropriate mechanisms in the application code cause the read/algorithm processing/write cycle to have the following form:

1. Download of current values, transfer of "older" values from the `_global.a[]` table into the `_global.aprev[]` table, and storage of current values in the `_global.a[]` table.
2. Execution of the code contained in the algorithm.
3. At this time, the `_global.a[]` table may possess some values changed by the algorithm execution.
4. Copying of values from the `_global.a[]` table into items, that are configured in the user configuration.
5. Sending of values generated by the algorithm to the OPC.

5 User Interface and Implementation

The implemented user interface for the expert agent consists of two windows: **configuration** and **data view** [15]. This is the main application written in Macromedia Flash™ (the source is compatible with the Flash™ MX Professional 2004 version and higher). Reading of the hierarchical structure of items in database, selection of those that are interesting for the Internet user, monitoring of current and historical data and the realization of control algorithms that utilize this data takes place in this application. Mechanisms for establishing connection to the database, sending the defined user configuration and sending values of items with write-to-OPC-server request into the database, are also implemented in this application, together with visualization and coupling of dynamically loaded clips to the measurements.

A tree structure of OPC items and off-line measurements available in the database are displayed in the configuration window. An appropriate option determines the maximum sampling period of data delivery in seconds that is allowed.

6 Concluding Remarks

The presented multi-agent system for hierarchical control with self-organising database enables a safe and effective operation of the pilot wastewater treatment plant installation. This process requires constant control and supervision and because it is not an industrial process, a direct and constant supervision by humans is not possible (no staff available during the night for example). The presented system enables sophisticated and long term experimental studies to be performed – the pilot plant has been in continuous operation for over two years.

Acknowledgement. This work was supported by the Polish Ministry of Scientific Research and Information Technology, using funds for 2006-2008, under grant no. N514 006 31/1739.

References

1. Knapik, M., Johnson, J.B.: Developing intelligent agents for distributed systems: exploring architectures, techniques, and applications. McGraw-Hill Osborne Media (1997)
2. Van Dyke Parunak, H.: A practitioners' review of industrial agent applications. Autonomous Agents and Multi-Agent Systems 3, 4, (2000) 389-407
3. Weiss, G. (ed.).: MultiAgent Systems: A Modern Approach to Distributed Artificial Intelligence, MIT Press (1999)
4. Wooldridge, M.: An Introduction to Multiagent Systems. Wiley (2002).
5. Ferber, J.: Multi-agent systems – an introduction to distributed artifficial intelligence. Addison-Wesley, 1999.
6. Lee, J.: E-manufacturing—fundamental, tools, and transformation. Robotics and Computer-Integrated Manufacturing 19, (2003) 501-507
7. Nahm, Y.-E., Ishikawa, H.: A hybrid multi-agent system architecture for enterprise integration using computer networks. Robotics and Computer-Integrated Manufacturing 21, (2005) 217-234
8. Choiński, D., Nocoń, W., Metzger, M.: Hybrid control system for pilot wastewater treatment plant. Proceedings of the IFAC Workshop on Programmable Devices and Systems – PDeS'06, Brno, (2006) 226-231
9. Nocoń, W., Choiński, D., Metzger, M.: Web-based control and monitoring of the experimental pilot plant installations. Proceedings of the IFAC Workshop on Programmable Devices and Systems – PDS'04, Krakow (2004) 94-99
10. Wang, S., Xu, Z., Cao, J., Zhang, J.: A middleware for web service-enabled integration and interoperation of intelligent building systems. Automation in Constr. Vol. 16. (2007) 112-121
11. Davidsson, P., Wernstedt, F.: Software agents for bioprocess monitoring and control. Journal of Chemical Technology and Biotechnology, Vol. 77. Society of Chemical Industry (2002) 761-766
12. Lynch N., Segala R., Vaandrager F.: Hybrid I/O automata. Information and Computation Vol. 185. (2003) 105-157
13. Decarlo, R. A., Branicky, M. S., Pettersson, S., Lennartson, B.: Perspectives and Results on the Stability and Stabilizability of Hybrid Systems, Proceedings of the IEEE. Vol. 88. (2000) 1069-1082
14. Fregene, K., Kennedy, D. C., Wang, D. W. L.: Toward a Systems- and Control-Oriented Agent Framework. IEEE Transactions on Systems, Man, and Cybernetics—Part B: Cybernetics. Vol. 35. (2005) 999-1012
15. Kraska, J.: An Internet-Based Mobile Control Algorithms"; Master Thesis, Institute of Automatic Control, Silesian University of Technology, 2006 *(in polish)*

Water Floating Self-assembling Agents

Shuhei Miyashita[1], Maik Hadorn[1], and Peter Eggenberger Hotz[1,2]

[1] University of Zurich, Andreasstrasse 15, 8050 Zurich, Switzerland
[2] University of Southern Denmark, Campusvej 55, 5230 Odense M, Denmark

Abstract. In this paper, we present a novel model of autonomous floating agents which can self-assemble on a water surface. We focus especially on the shape of the agent, which is essential as a factor of aggregation but difficult to deal with as finite states. After several experiments, a unique self-assembling behavior was observed by exploiting several physical-level forces; 6 agents gather into a cluster and form a specific pattern, which is unique to the configuration. This is considered crucial to achieve morphogenesis of a multi-agent system, of which we think as a hierarchical aggregation. The results shown here provide a good starting point for speculation about the level of autonomy of a agent in the complex environment when the system creates a self-assembling pattern.

Keywords: self-assembly, autonomous-distributed system.

1 Introduction

It is intuitively obvious that stones on the ground are not alive, whereas plants or animals are. What is the essential difference between physical interaction and life activity? Is there anything behind more than the difference of the characteristics of the components? The definitions of life have been advocated since a long time ago [1] and numerous attempts have been made to explain essential aspects of life, such as self-replicative, dissipative, and autopoietic system [2].

It is no doubt that life sciences have progressed by treating creatures as "materials". They have not only managed to show complex mechanisms of life but also to contribute to the society, such as the medical field, by using it as remedies for diseases. However, although the elucidation of the molecular details of life activities have made a lot of progress in biology, an overall picture is still missing. Overall, most discussions about life converge into the causality of the molecular interaction level and the necessary condition of the subject matter - while it sometimes answers to the sufficient condition - has left the problems on the back burner. It should be taken into account to clarify the life activity that not only temporal parameters but also spatial parameters. But there is an intermediate region between life and non-life where viruses exist. Normally, viruses are considered as non-life, since they cannot reproduce themselves without any help of other species. T4 phage is one of the viruses, which infects E. coli by injecting its DNA [3]. This virus with marked appearance consists of about 70

N.T. Nguyen et al. (Eds.): KES-AMSTA 2007, LNAI 4496, pp. 665–674, 2007.

different kinds of proteins and can use the metabolism of the host cell to replicate itself into about 100 individuals in approximately 40 minutes by taking resources from E. coli. T4 phage has several notable features. For example, this virus is able to self-construct purely by intermolecular forces; no external energy supply is needed during this procedure. In addition, this virus can be synthesized *in vitro* by supplying adequate materials in a well-ordered way, which is remarkable reaction for such a complex mechanism.

With conventional engineering hitting a complexity barrier, it seems very useful to draw inspiration from natural systems. However, no matter how clear the causality of molecular reactions become, the dynamics are still veiled; revealing the morphogenesis of complicated viruses like T4 phage is still one of the biggest challenges of modern science to understand the difference between life and non-life.

Recent advances in robotics reveal the importance of autonomous self construction and embodiment for building intelligent systems. J. von Neumann advocated a distributed system that self-replicates itself [4]. L. S. Penrose proposed a mechanical self-assembling model [5]. The essential issues of how to develop a cellular robotic system (CEBOT) were described by Fukuda *et al.* [6]. S. Murata *et al.* designed a modular robotic system in hardware, which could metamorphose into desired configurations, and showed results of changes in morphology [7, 8, 9, 10]. In the Conro project by A. Castano, A. Behar, and P. M. Will, each module was self-contained (it included its own processor, power supply, communication system, sensors, and actuators) [11]. These modules were designed to work in groups as part of a large configuration. A similar project is presented by M. W. Jorgensen *et al.* with ATRON, which consists of several fully self-contained robot modules [12]. As for the rest, a lot of interesting research about self-assembly, the numerous approaches can be divided into several types: cellular automata [13], cm-scale self-assembly [14, 15, 16, 17, 18, 19, 20, 21, 22, 23, 24, 25]. While currently most robot construction and repair is performed manually, this will be quite difficult when (a) the complexity of the systems exceeds a certain threshold, and (b) if these systems have to be truly adaptive. Taking into account their aspects, there are some work try to deal with this problem by providing a stochastic environment to the system [26, 27], assembly in meso scale [28, 29, 30, 31], field of MEMS [32], nano scale assembly of DNA tile [33]. Many researchers have come up with interesting solutions for some of the problems that future robotics are supposed to deal with, like self-organization and adaptivity to changing environments, fault tolerance and self-repair, self-programming and self-replication, to name but a few.

According to the outcome of these work, we perceive life especially as an autonomous and distributed system, and propose a modular robot called Tribolon, demonstrating one of the activities that we consider important when viruses form their morphology. Our work is based on the notion that the whole discussion about life should be held in the framework of purely physical interactions, by treating life as substances.

2 The Agent Model - Tribolon

2.1 The Architecture

Figure 1 a) shows a prototype of the proposed floating agent called Tribolon. The body is made of foamed rubber. A agent, which can float on the water, weights approximately 3.7 g and covers an area of 12.25 cm^2. It consists of the floating base plate, a vibrator in which a small mass is turning clockwise for the actuator, an antenna which touches an aluminum ceiling, and a rod projecting into the water. This setting causes asymmetry property in the system. A magnet is attached to the bottom of the base plate orthogonally to the symmetry axis so that agents can attract or repel each other. Figure 1 b) shows the experimental setting. By adding a specific amount of electrolyte (salt, 5.56 g/l, 95 mM) to the water, it becomes conductive. And by adding voltage (8.0 V) to the ceiling, current can be supplied constantly to the vibrator via the antenna, getting away into the water through the rod.

Figure 1 c) shows a picture of one vibrating agent on the water. Ripples are created by the vibration of the agent. One of the advantages of this model is that it replaces a mechanical connecting system with magnetic force and the collision of the agents. This enables the system to be smaller and lighter than classical models.

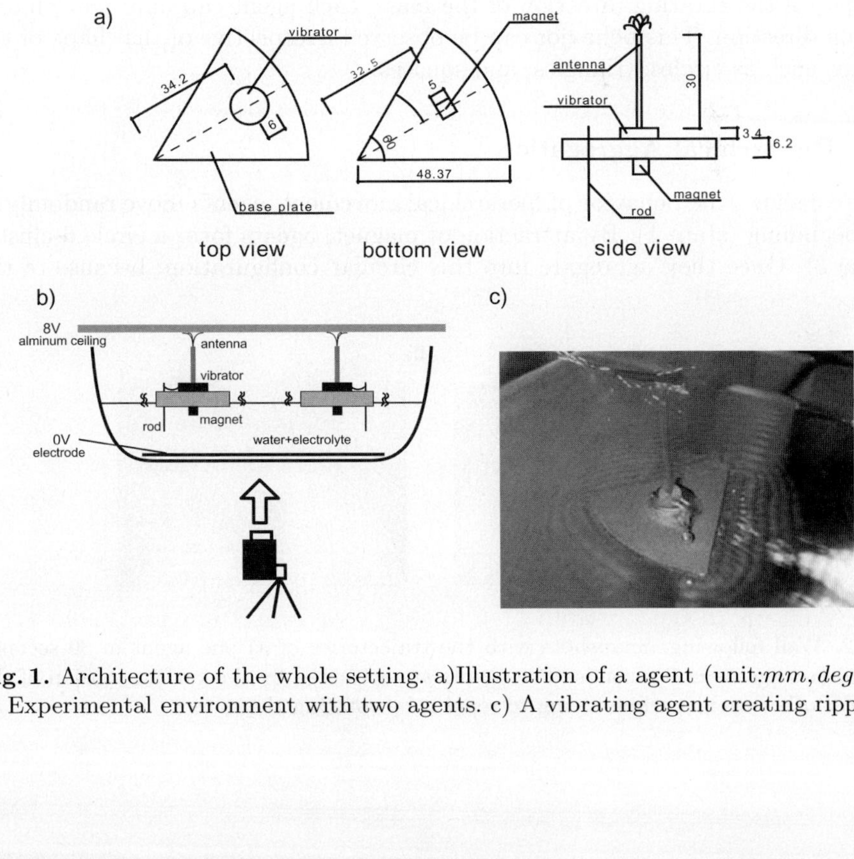

Fig. 1. Architecture of the whole setting. a)Illustration of a agent (unit:mm, $degree$). b) Experimental environment with two agents. c) A vibrating agent creating ripples.

The salt solution can generate the current flow by the following chemical reactions (1). The concentration of salt solution is enough to sustain the current flow during the whole experiment. In order to avoid chemical deposition onto the rods and the electrode, we used platinum for both the rod on the agents and the electrode under the water.

$$2NaCl + H_2O \rightarrow H_2 \uparrow + Cl_2 \uparrow + 2NaOH \quad (1)$$

3 Results and Discussions

As a result of several experiments, we found that the shape of the agent plays an important role in the aggregation behavior of the system.

3.1 Wall Following Behavior

Figure 2 a) and b) show the snapshots with trajectories of one of the characteristic behaviors of one agent and two agents, respectively. It was observed that they kick the wall (or the other agent) to a certain direction and move to one direction by keeping on kicking, which we call the wall following behavior. This usually happens if the vibrating agents are attracted to some objects by specific forces. In this case, the agents are attracted to the wall by hydrophobic interaction, while trying to repel it by rotating the mass. It has to be noted that because of the rotating direction of the mass, each agent can only move into a certain direction. This behavior can be observed irrespective of the shape of the agents, such as circles, triangles, and squares.

3.2 Hierarchical Aggregation

Figure 3 shows the behavior of hierarchical movement. agents move randomly in the beginning (state 1). By attraction of magnet, agents form a circle 6-cluster (state 2). Once they aggregate into this circular configuration, because of the

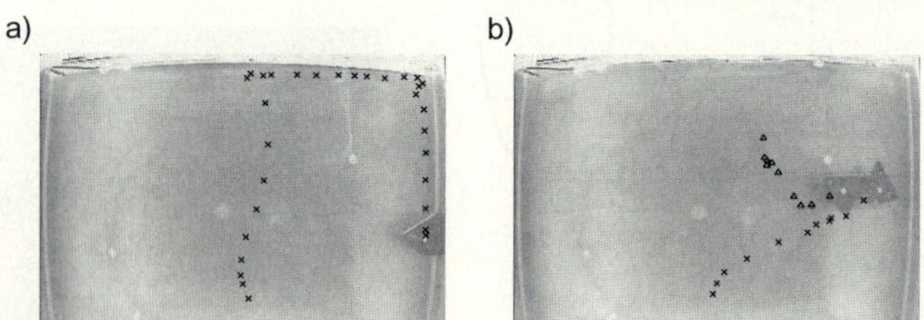

Fig. 2. Wall following. Snapshots with the trajectories of a) one agent in 30 seconds and b) two agents in 9 seconds. The agents are attracted by the surface tension of the wall (or other agent), while trying to repel it by rotating the mass.

repulsion force of the wall following behavior, they form a propeller-like shape and start turning clockwise in constant speed (state 3). This set of sequence represents that the system acquires different level of function by aggregating into one cluster.

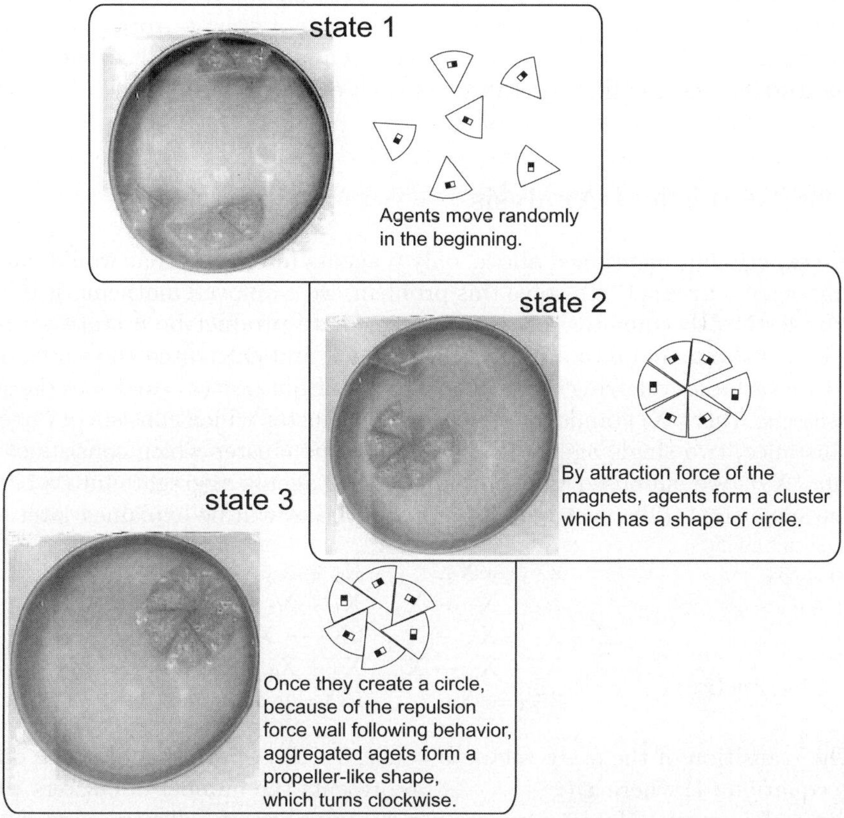

Fig. 3. Hierarchical aggregation. agents move randomly in the beginning (state 1). By magnetic attraction, agents form a circle shape (state 2). Once they create a circle, because of the repulsion force of the wall following behavior, aggregated agents form a propeller-like shape and start turning clockwise (state 3).

Coming back to the initial question, what mechanism enables a virus to self-assemble its form? When the paths of around 70 kinds of protein aggregation are carefully observed, it can be seen that these paths are well ordered [3]. For instance, protein A only can attach to protein B by coupling with protein C (equation (2)). This means that protein A acquires a different level of functionality by coupling with protein C, which then enables protein B to connect to it.

$$\mathrm{A} + \mathrm{B} + \mathrm{C} \rightarrow \mathrm{AC} + \mathrm{B} \rightarrow \mathrm{ABC} \quad (2)$$

We assume that this kind of hierarchical coupling, acquiring different levels of functionality by the aggregation, plays the important role in the process of the morphogenesis such as in a T4 phage; as we see in the aggregation of the head and the tail, only if the DNA(C) is incorporated into the head(A), the tail(B) can attach to the head. In that sense, Figure 3 shows an example of that we consider to be a hierarchical aggregation. It does not start to rotate in constant speed until exactly 6 agents gather together. Only after the 6th agent attaches to the rest of the agents, they can repel each other because of the distribution of agents and form the propeller shape.

4 Mathematical Analysis

In the experiment mentioned above, only 6 agents interact. What would happen if more agents interact? To solve this problem, we employed mathematical analysis by letting the quantity of every intermediate product be a state variable, simplifying the model into a finite states model, and calculated the variation of the state variables referring to the model in [14]. Equation (3) expresses the state transitions, where X_i stands for the states of a cluster which consists of i agents. For instance, two single agents ($2X_1$) become one cluster which consists of two agents (X_2). We supposed that no more than 2 agents aggregate into a cluster at the same time. (The last reaction in parentheses will be explained later.)

$$\begin{array}{ll} 2\mathrm{X}_1 \rightarrow \mathrm{X}_2, & \mathrm{X}_1 + \mathrm{X}_2 \rightarrow \mathrm{X}_3, \\ \mathrm{X}_1 + \mathrm{X}_3 \rightarrow \mathrm{X}_4, & \mathrm{X}_1 + \mathrm{X}_4 \rightarrow \mathrm{X}_5, \\ \mathrm{X}_1 + \mathrm{X}_5 \rightarrow \mathrm{X}_6, & 2\mathrm{X}_2 \rightarrow \mathrm{X}_4, \\ \mathrm{X}_2 + \mathrm{X}_3 \rightarrow \mathrm{X}_5, & \mathrm{X}_2 + \mathrm{X}_4 \rightarrow \mathrm{X}_6, \\ 2\mathrm{X}_3 \rightarrow \mathrm{X}_6, & (2\mathrm{X}_6 \rightarrow 12\mathrm{X}_1), \end{array} \quad (3)$$

The transition of the state vector $\boldsymbol{x} = (x_1, \cdots, x_6)$ obeys the following difference equation (4) where $x_i (i \in 1, \cdots, 6)$ represents the number of clusters which consists of i agents. Here t represents time (number of collision), and F is a transition function.

$$\boldsymbol{x}(t+1) = \boldsymbol{x}(t) + \boldsymbol{F}(\boldsymbol{x}(t)) \quad (4)$$

The transition function F_i is expressed as the following equation (5) where P^b_{ij} represents the bonding probability when two agents i and j collide. The coefficients are decided by referring to chemical reaction equations.

$$\begin{aligned} F_1(\boldsymbol{x}) &= (-2P^b_{11}x_1^2 - 2P^b_{12}x_1x_2 - 2P^b_{13}x_1x_3 - 2P^b_{14}x_1x_4 - 2P^b_{15}x_1x_5)/S^2, \\ F_2(\boldsymbol{x}) &= (P^b_{11}x_1^2 - 2P^b_{12}x_1x_2 - 2P^b_{22}x_2^2 - 2P^b_{23}x_2x_3 - 2P^b_{24}x_2x_4)/S^2, \\ F_3(\boldsymbol{x}) &= (2P^b_{11}x_1x_2 - 2P^b_{13}x_1x_3 - 2P^b_{23}x_2x_3 - 2P^b_{33}x_3^2)/S^2, \\ F_4(\boldsymbol{x}) &= (2P^b_{13}x_1x_3 + P^b_{22}x_2^2 - 2P^b_{14}x_1x_4 - 2P^b_{24}x_2x_4)/S^2, \\ F_5(\boldsymbol{x}) &= (2P^b_{14}x_1x_4 + 2P^b_{23}x_2x_3 - 2P^b_{15}x_1x_5)/S^2, \\ F_6(\boldsymbol{x}) &= (2P^b_{15}x_1x_5 + 2P^b_{24}x_2x_4 + P^b_{33}x_3^2)/S^2 \end{aligned} \quad (5)$$

Table 1 lists the parameters of each collision between agent i and j. (The value in parentheses will be explained later.) These values are calculated depending on the geometrical positions of two agents which collide. We calculated the probability of directing the face of the magnetic North pole side of the agent to the South pole side of the other agent, and visa versa.

Table 1. The bonding probability

i/j	1	2	3	4	5	6
1	0.170	0.146	0.146	0.097	0.049	0
2	0.146	0.125	0.125	0.083	0	0
3	0.146	0.125	0.125	0	0	0
4	0.097	0.083	0	0	0	0
5	0.049	0	0	0	0	0
6	0	0	0	0	0	0(0.017)

Figure 4 a) shows the transition of x_i calculated by a computer. We started with 100 agents as the initial condition. As the figure shows, clusters consisting of 5 agents are produced the most. This phenomenon is referenced as Yield problem in [14].

Taking into account the instability of the 6-cluster of our model (because of the acquisition of the rotational movement in unstable form), we cooperated the disintegration process of the cluster into single agents into the analysis (listed in the parenthesis of Equation (3)). Due to this change, the transition functions F_1 and F_6 are changed into those listed in Equation (6). Since this number depends on the voltage that we supply to the system, we chose the number (0.0170) one tenth of the probability of the corruption of two single agents (0.170). (More concrete value should be measured in the future work.)

$$F_1(\boldsymbol{x}) = (12P^b_{66}x_6^2 - 2P^b_{11}x_1^2 - 2P^b_{12}x_1x_2 - 2P^b_{13}x_1x_3 - 2P^b_{14}x_1x_4 - 2P^b_{15}x_1x_5)/S^2,$$
$$F_6(\boldsymbol{x}) = (2P^b_{15}x_1x_5 + 2P^b_{24}x_2x_4 + P^b_{33}x_3^2 - 2P^b_{66}x_6^2)/S^2 \quad (6)$$

Figure 4 b) shows the result of the calculation with this transition. Surprisingly, even though the 6-cluster is unstable, the system now generates more 6-clusters. This is considered that the efficient supplement of single agents to the 5-clusters from the broken 6-clusters increased the number of 6-clusters in the end.

4.1 Autonomous and Distributed Model

It is not easy to generalize a law of pattern formation of a multi-element system, since it strongly depends on the model used. Focusing on the mechanisms of living things from a viewpoint of autonomous and distributed systems, it is noticed that the components which form the morphology are not always highly autonomous. Generally, living things are built up in a complexity environment with middle range autonomous agents. This implies that it is not the level of autonomy but the scale that is important for the system.

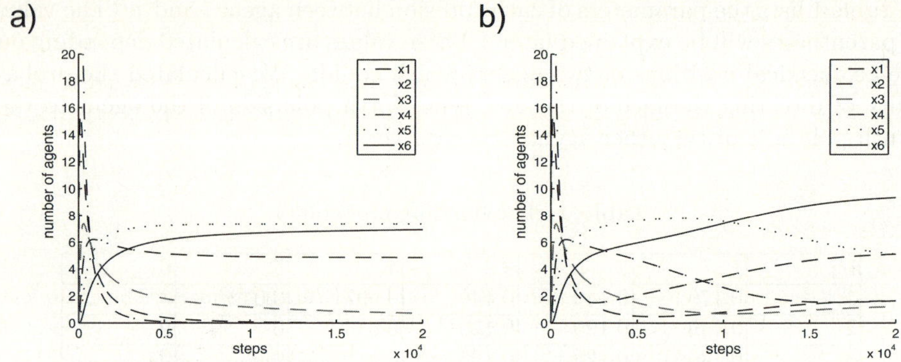

Fig. 4. Comparison of two cases. a) without modeling the corruption of 6-clusters. b) with the corruption.

5 Conclusion

In summary, the following were presented in this work:

1. A novel model of autonomous self-assembling agent which can move on a water
2. Unique self-assembling behavior, especially
   (a) Wall following
   (b) Hierarchical aggregation
3. A solution to the yield problem by un-stabilizing the final state

This work shows an example of hierarchical aggregation, which is considered crucial to achieve morphogenesis of a multi-agent system. The model may also be used to show that an autonomous and distributed system is able to achieve two different hierarchical aggregation behaviors, which we consider it necessary to escape from an specific attractor. We pursued the ability of representation of multi-agent systems focusing on the relation between the autonomy of each agent and the environment, multiple-degree-of-freedom system. Note that the model presented here exploits several physical-level-forces, such as magnetic force and mechanical interactions, although it looks quite simple.

6 Future Work

We are providing a small but significant step towards the clarification of the universal law of self-assembling. There are two major implications of this research. One is that it presents an approach towards tackling the complexity barrier in engineering, and on the other hand it is of theoretical importance because the concept of morphological computation - incorporating morphology and materials - provides a new way of conceptualizing computation. Also towards this end, the size of the individual agents must be reduced significantly (from *cm* to μm).

Acknowledgments

Many thanks to Nathan Labhart for helpful suggestions. This research is supported by the Swiss National Science Foundation, projects #200021-105634/1 and #200021-109864/1.

References

[1] Schrodinger, E.: What is Life?: With Mind and Matter and Autobiographical Sketches. Cambridge University Press (1944)

[2] Maturana, H.R., Varela, F.G.: Autopoiesis and Cognition. Dordrecht, Netherlands: Reidel (1980)

[3] Leiman, P.G., Kanamaru, S., Mesyanzhinov, V.V., Arisaka, F., Rossmann, M.G.: Structure and morphogenesis of bacteriophage t4. Cellular and Molecular Life Sciences **60** (2003) 2356–2370

[4] v. Neumann, J.: Theory of Self-reproducing Automata. Univ. of Illinois Press (1966)

[5] Penrose, L.S.: Self-reproducing. Sci. Amer. **200-6** (1959) 105–114

[6] Fukuda, T., Kawauch, Y.: Cellular robotic system (cebot) as one of the realizations of self-organizing intelligent universal manipulator. In: Proceedings of the 1990 IEEE Conference on Robotics and Automation (ICRA). (1990) 662–667

[7] Murata, S., Kurokawa, H., Kokaji, S.: Self-assembling machine. In: Proceedings of the IEEE Robotics and Automation (ICRA). (1994) 441–448

[8] Murata, S., Kurokawa, H., Tomita, K.: Self-assembly method for mechanical structure. Artificial Life and Robotics **1** (1997) 111–115

[9] Murata, S., Kurokawa, H., Yoshida, E., Tomita, K., Kokaji, S.: A 3-D self-reconfigurable structure. In: Proceedings of the IEEE International Conference on Robotics and Automation (ICRA). (1998) 432–439

[10] Murata, S., Tomita, K., Yoshida, E., Kurokawa, H., Kokaji, S.: Self-reconfigurable robot. In: Proceedings of 6th International Conference on Intelligent Autonomous Systems (IAS-6). (1999) 911–917

[11] Castano, A., Behar, A., Will, P.M.: The conro modules for reconfigurable robots. IEEE/ASME Transactions on Mechatronics **7** (2002) 403–409

[12] Jorgensen, M.W., Ostergaard, E.H., Lund, H.H.: Modular atron: Modules for a self-reconfigurable robot. In: Proceedings of the 2004 IEEE/RSJ International Conference on Intelligent Robots and Systems (IROS). Volume 2. (2004) 2068–2073

[13] Langton, C.G.: Self-reproduction in cellular automata. Physica D: Nonlinear Phenomena **10** (1984) 135–144

[14] Hosokawa, K., Shimoyama, I., Miura, H.: Dynamics of self-assembling systems: Analogy with chemical kinetics. Artificial Life **1** (1994) 413–427

[15] K.Hosokawa, I.Shimoyama, H.Miura: 2-d micro-self-assembly using the surface tension of water. Sensors and Actuators A **57** (1996) 117–125

[16] Yim, M.: New locomotion gaits. In: Proceedings of the IEEE International Conference on Robotics and Automation (ICRA). Volume 3. (1994) 2508–2514

[17] White, P., Kopanski, K., Lipson, H.: Stochastic self-reconfigurable cellular robotics. In: IEEE International Conference on Robotics and Automation (ICRA). Volume 3. (2004) 2888–2893

[18] White, P., Zykov, V., Bongard, J., Lipson, H.: Three dimensional stochastic reconfiguration of modular robots. In: Proceedings of Robotics Science and Systems. (2005) 161–168

[19] Hamlin, G.J., Sanderson, A.C.: Tetrobot modular robotics: Prototype and experiments. In: Proceedings of the 1996 IEEE/RSJ International Conference on Intelligent Robots and Systems (IROS). (1996)

[20] Kotay, K., Rus, D., Vona, M., McGray, C.: The self-reconfiguring robotic molecule. In: Proceedings of the IEEE International Conference on Robotics and Automation (IROS). Volume 1. (1998) 424–431

[21] Rus, D., Vona, M.: Crystalline robots: Self-reconfiguration with compressible unit modules. Autonomous Robots **10** (2001) 107–124

[22] Bojinov, H., Casal, A., Hogg, T.: Multiagent control of self-reconfigurable robots. In: Proceedings of Fourth International Conference on MultiAgent Systems (ICMAS). (2000) 143–150

[23] Detweiler, C., Vona, M., Kotay, K., Rus, D.: Hierarchical control for self-assembling mobile trusses with passive and active links. In: Proceedings of IEEE International Conference on Robotics and Automation (ICRA). (2006) 1483–1490

[24] Zykov, V., Mutilinaios, E., Adams, B., Lipson, H.: Self-reproducing machines. Nature **435** (2005) 163–164

[25] Bhat, P., Kuffner, J., Goldstein, S., Srinivasa, S.: Hierarchical motion planning for self-reconfigurable modular robots. In: Proceedings of IEEE International Conference on Intelligent Robots and Systems (IROS). (2006) 4108–4115

[26] Bishop, J., Burden, S., Klavins, E., Kreisberg, R., Malone, W., Napp, N., Nguyen, T.: Programmable parts: A demonstration of the grammatical approach to self-organization. In: Proceedings of the IEEE/RSJ International Conference on Intelligent Robots and Systems (IROS). (2005) 3684–3691

[27] Shimizu, M., Ishiguro, A.: A modular robot that exploits a spontaneous connectivity control mechanism. In: Proceedings of IEEE International Conference on Robotics and Automation (ICRA). (2005) 2658–2663

[28] Griffith, S., Goldwater, D., Jacobson, J.: Robotics: Self-replication from random parts. Nature **437** (2005) 636

[29] Wolfe, D.B., Snead, A., Mao, C., Bowden, N.B., Whitesides, G.M.: Mesoscale self-assembly: Capillary interactions when positive and negitive menisci have similar amplitudes. Langmuir **19** (2003) 2206–2214

[30] Gracias, D.H., Tien, J., Breen, T.L., Hsu, C., Whitesides, G.M.: Forming electrical networks in three dimensions by self-assembly. Science **289** (2000) 1170–1172

[31] Boncheva, M., Ferrigno, R., Bruzewicz, D.A., Whitesides, G.M.: Plasticity in self-assembly: Templating generates functionally different circuits from a single precursor. Angew. Chem. Int. Ed. **42** (2003) 3368–3371

[32] Bohringer, K.F., Goldberg, K., Cohn, M., Howe, R., Pisano, A.: Parallel microassembly with electrostatic force fields. In: Proceedings of IEEE International Conference on Robotics and Automation. Volume 2. (1998) 1204–1211

[33] Rothemund, P.W.K.: Folding dna to create nanoscale shapes and patterns. Nature **440** (2006) 297–302

Failure Detection Service for Large Scale Systems

Jacek Kobusiński

Institute of Computing Science
Poznań Universitiy of Technology, Poland
jkobusinski@cs.put.poznan.pl

Abstract. This paper addresses the problem of building a failure detection service for large scale distributed systems, as well as multi-agent systems. It describes the failure detector mechanism and defines the roles it plays in the system. Afterwards, the key construction problems that are fundamental in the context of building the failure detection service are presented. Finally, a sketch of general framework for implementing such a service is described. The proposed failure detection service can be used by mobile agents as a crucial component for building fault-tolerant multi-agent systems.

Keywords: failure detector, large scale systems, gossip-style protocol, fault tolerance, multi-agent systems

1 Introduction

As it is known [1], the agent is a running process, having some state and data. It is autonomous, mobile and capable of making its own decisions . Moreover, it can interact with objects, services and finally, communicate with other agents. The communication between agents in the multi-agent systems (MAS) is mainly based on the one-to-one pattern. The agent platform (AP) provides an environment in which agents can execute and perform their tasks.

Nowadays, large scale systems like grids or peer-to-peer networks become more and more interesting environments as agent platforms. They allow to efficiently use and share different resources such as computers, software components and network infrastructure. However, these environments are usually large scale distributed systems that dynamically change their size and topology. These characteristics increase the possibility of the failure occurence (crashes of nodes, links or agents) and leads to new requirements concerning fault tolerance mechanisms. Therefore, fault tolerance is difficult but one of the most important problem in the further development of multi-agent systems. Also, there are some platforms that try to address these problems, they do not propose the perfect solutions [2,3,4,5,6].

In the recent years the failure detectors have gained much attention as a research topic in the context of reliable large scale distributed systems as well as multi-agent systems [5]. They are recognized as an essential component for building fault tolerant multi-agent platforms. The concept of failure detector was

N.T. Nguyen et al. (Eds.): KES-AMSTA 2007, LNAI 4496, pp. 675–684, 2007.

introduced by Chandra and Toueg in [7]. They proposed unreliable failure detector as an abstraction, which encapsulates an additional portion of synchrony to circumvent the widely known FLP impossibility result [8]. Since then, there were many proposals concerning the implementation of failure detectors [7,9,10,11]. In the most of the proposed papers, the local area networks are considered as the base environment for failure detection rather than a wide area distributed systems. Such an assumption has strong consequences in the implementation. Although, the failure detection service plays a very important role in large scale systems, there are only few papers considering building such a service [12,11,13].

In this paper, we propose a general model of failure detection service (FDS) that is flexible, adaptable, and scalable as is required by multi-agent system based on grids and peer-to-peer networks. In general, we take advantage of the latest promising techniques presented in the literature. To meet this objectives, we introduce some modifications and optimizations to the existing solutions. Presented failure detection service can be easily embedded into AP, so agents can benefit from it simillary to [14]. Due to its simplicity and decentralization, the algorithm shown in this paper fits perfectly to the concept of mobile agent autonomy and mobility.

The remaining part of the paper is structured as follows. Related work concerning building FDS is presented in section 2. In section 3, we briefly present unreliable failure detectors and the role they play in a distributed system. Section 4 discusses the implementation aspects of building such a service. In section 5, a base algorithm for the FDS is presented. Finally, in section 6 the concluding remarks and further directions of our work are described.

2 Related Work

Stallings et al. [12] proposed two layers failure detection service for Globus Grid toolkit. The concept of this service is based on *local monitors* that broadcast heartbeats, which are collected by *data collectors*. This hierarchical approach addresses the problem of scalability, but because of partially static architecture it does not adapt well to dynamic environment.

Gossip-style failure detection protocol was proposed by Van Renesse et al. [9]. The authors, inspired by the earlier work of Demers et at. [15] and the way of spreading rumors among people, introduced algorithm that makes a minimal assumption about underlying network and combines efficiency of hierarchical dissemination and robustness of flooding. There is a known modification to this basic scenario that optimizes this protocol for a wide area network. This optimization introduces a hierarchy which allows to perform most of gossiping locally with only a small number of messages exchanged over networks and domains. Gupta et al. [10] implemented the SWIM protocol. To avoid false suspicions, the authors of [10] proposed a redundant mechanism for failure detection. In the case of not receiving positive acknowledgment from the process, the failure detection module asks another host to verify the state of that process. This spreads the responsibility for decision about suspecting process among more then one module.

Adaptive protocols of failure detection are able to adapt to the changing network conditions or application requirements. Chen et al. [16] have proposed several algorithms of failure detectors for both, synchronous and asynchronous systems. In brief, the Chen's failure detector estimates the arrival time of the next heartbeat by obtaining two values, an emission interval and a safety margin, which are computed by a special unit. . Bertier et al. in [17] presented a modification to Chen's work. The main idea of this modification was to use Jacobson RTT estimation algorithm. This failure detector, embedded into DARX multi agent support platform [5], detects failures more aggressively by providing shorter detection time, however, at the same time making more mistakes then Chen's one. The new approach to adaptivity was proposed by Hayashibara at el. [18]. The authors present a new aspect of flexibility, the adaptation to multiple application requirements. In this context, the φ failure detector is flexible, if it can serve applications with various requirements. This is achieved by changing the way a failure detector reports a failure of the process. Traditionally, it simply returns true or false value, but in the solution mentioned above it returns an *accrual value,* which should be interpreted as the degree of confidence that the process failed. Each application can set different threshold for the accrual value received from a failure detector and interprets it in its own way. In general, a smaller threshold means more aggressive detection and, possibly, more false suspicions, while a larger one determines slower but more accurate failure detection. The thresholds can be modified even during runtime without any problems.

3 Unreliable Failure Detectors

The notion of failure detectors was introduced in [7]. Chandra and Toueg defined a failure detector as a distributed oracle that provides information about process failures. Failure detectors are said to be unreliable, because of the mistakes they can do, i.e. not reporting processes that have failed or false suspicions of correct processes. The authors specified two properties of the proposed abstraction: completeness and accuracy. The first one describes the ability of a failure detector to detect a failure of monitored processes. The latter gives the information about a possibility of making mistakes. By analyzing these two properties, one can classify failure detectors into a number of classes. A failure detector consists of a set of modules that are attached in a one-to-one manner to each monitored process. Each module maintains the local lists of suspected processes and, when queried, returns this suspicion list to the processes. The modules exchange information among each other merging the received results with its local ones. This process clearly shows the two roles the failure detector module plays - it detects a failure and propagates information about it to others modules.

3.1 Detection Phase

Detection phase is the most important one, because this is the process of acquiring the new knowledge. The information about the state of monitored objects can be obtained through either a passive or an active scenario [17]. The passive scenario relies on monitoring certain messages that are sent by monitored processes and is often called the heartbeat strategy. The opposite strategy is based on interrogation methodology, where a failure detector asks monitored processes about their state. Both strategies have their advantages and disadvantages. The heartbeat scenario is definitely simpler and faster, i.e. the number of messages it sends is lower and the overall detection time is shorter. On the other hand, it gives no possibility to control, which process will be monitored. The main disadvantage of the latter scenario, is a long time needed to detect a failure because, appropriate messages are send to the monitored process and back. On the other hand, there is a better control of selecting which nodes should be monitored. Recently there is new trend in building failure detectors, which answer in more "fuzzy" ways. Their decision about state of the process is not a binary one, but it is a value that expresses the degree of confidence that the process failed. In this scenario the final decision is moved from the failure detection service layer to the application layer and relys on comparing this probability with a given threshold.

3.2 Propagation Phase

When the failure detector module acquires some information about the state of monitored objects, it should share it with the other modules. This allows to reduce the overhead of failure detection because some failure detector modules do not need to monitor some nodes by themselves, instead they can rely on the shared information. . In LAN, the number of monitored processes is typically not very large and the links between failure detection modules are fast and rather reliable. The situation changes fundamentally in the case of WAN, where a set of monitored objects can be larger and the network connections are much more susceptible to a failure. In a local network, broadcasting is the simplest solution that will not increase the total overhead of failure detection to unacceptable level. The same solution used in WAN will lead to network overloading. In WAN, and similarly, in MAS the spread of the information should be planned more precisely, with utilizing network resources as primary concern [19,20].

4 Implementation Key Points

Some aspects that should be considered during designing FDS for large scale systems, are crucial while others can be considered less important. Taking into account such things like wide geographical area covered by these systems, a number of nodes and network characteristics, we believe that scalability, flexibility and adaptivity are the most important ones. We do not consider security in this paper, because it is not the main field of our research.

4.1 Scalability

The FDS should be very conservative in using network resources. Especially generating new messages, which are bandwidth consuming. In an ideal solution, the failure detection service should be transparent for users and no extra messages should be send. This is, of course, impossible, but the number of messages, generated by FDS may be minimized. In the local area networks the network connections are typically fast and bandwidth, so the amount of messages generated by FDS is not crucial. However, in the context of wide area distributed systems, where some of the network infrastructure can be composed of very slow links, this problem becomes important, because of message storm and message size.

4.2 Flexibility

Many different applications (e.g. agents in MAS) can run simultaneously sharing resources and services. In this context, the output of a failure detector should be flexible, i.e. acceptable and serviceable to all applications that request certain information. This assumption can be difficult to fulfill because the needs of different client applications can be contradictory. The simple example that illustrates this problem, is a situation where one client expects aggressive failure detection, fast reaction to a failure and a possible false suspicion, and the other is rather interested in accurate information and agree to trade-off longer response time for more accurate answers. The conflicting requirements cannot be reconciled by using a traditional approach based on timeouts. In this case, failure detector answer will be hard to express by a simple binary value. One should think about returning the value that can be interpreted by the clients themselves.

4.3 Adaptivity

Typically, large scale systems consist of resources located over a large geographical area, thus they should be treated similarly to WANs. The nodes are connected through unreliable links, the transmission times can vary from short to very long ones. There is no central point of administration and thus, it is difficult to get information about failures in such systems. The connections can periodically diminish their parameters or even stop working. In such a dynamic environment, where a stable static connection path between nodes occurs very seldom, typical implementation of failure detectors proposed for LAN will fail. To successfully implement FDS for large scale environment, one can take into account adaptivity and create service that will adapt to changing network conditions and message loss occurrence.

The comparison of scalability, flexibility and adaptivity of the failure detectors described in Section 2 is presented in Table 1.

5 Architecture of the Failure Detection Service

5.1 System Model

First, we define a model of the system being monitored by our failure detection system. For reasons of complexity and overhead, our system model consists of

Table 1. Comparision of failure detector implementations

	Scalability	Flexibility		Adaptivity
		single	multi	
Globus FD Service	x	-	-	-
Gossip-style FD	x	-	-	x
SWIM FD	x	-	-	x
Chen's FD	-	x	-	-
Bertier's FD	x	x	-	-
φ FD	-	x	x	-

the set of computers and processes residing on them. We do not include network as monitored components, which we treat as an orthogonal problem. To make a distinction between a network failure and a host failure can be hard without an existing alternative communication path. Even if it exists, it will be a challenge to discover such a path because this requires possessing the detailed knowledge of a network topology and providing coordination among distributed processes. For the same reason of simplicity and generality, we do not take into consideration such elements as computer components or low level software abstractions. The primary objective is to enable the construction of a reliable distributed application, not to construct a service that will diagnose the cause the failure.

We assume that every pair of processes is connected by a quasi-reliable communication channel. It means that there is no message corruption, message loss, and finally no creation of spurious messages. With regard to processes, we assume a crash-stop failure model, i.e. failed processes never recover. We consider the system to be asynchronous because there is no bound either on communication delays or on process speed. We also assume that message delays in the communication channels are determined by some random variable whose parameters are unknown, independent of other channel, and whose distribution is positively unbounded. Those parameters can change over time but eventually become stable. These assumptions do not constrain practical implementation and are similar to [17,11].

5.2 Framework

To separate the problem of failure detection and communication among different nodes we assume existence of network overlay layer that is responsible for management of underlying topology of the nodes [21]. By separating this layer we can clearly define the base skeleton for gossip-based failure detector service, which will allow to conduct experiments and make comparison of different strategies of failure detection. At this stage of our research we designed two versions of failure detection service framework described as pull and push model. The first model assumes the activity of the controller node, which assigns monitored nodes by itself, in the latter the supervised node is active and specifies the node that will monitor it.

In general, it is assumed that each module will maintain three subsets of nodes. The first set called *Neighbours* (denoted by $\mathcal{N}$), will consist of all nodes that some node p can currently contact in order to exchange information about the state of other nodes. The next two sets, called *Controllers* ($\mathcal{C}$) and *Supervised* ($\mathcal{S}$) consist of nodes that monitor the node p or are monitored by p, respectively. There is a simple relationship between the nodes from those sets. If node p monitors node q, then p belongs to *Controllers* at q, and q belongs to *Supervised* at p (Fig. 1). Number of items in each of these subsets is limited by some constants due to scalability. Each node can be both the *Controller* as well as the *Supervised* node. This duality can be exploited to propagate results of detection to all nodes in the system.

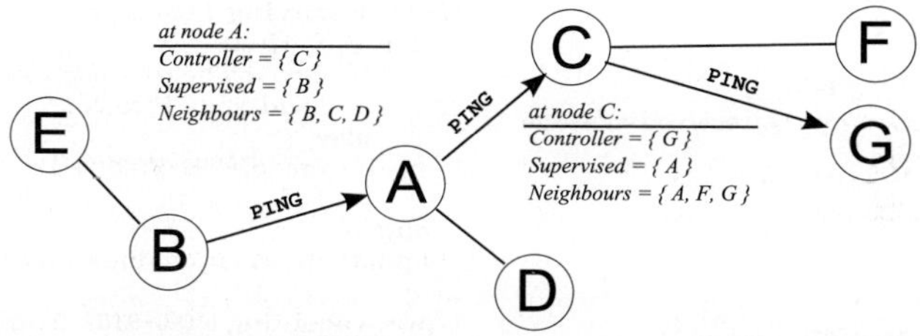

Fig. 1. Relationship beetwen nodes

Every Δt_1 time units each node sends PING message to its own *Controller* nodes to notify them about its correct status. Depending on the model, the node tries to remove nodes from *Supervised*/*Controller* by checking predefined criteria (f.ex. lifetime, interest of the detection results from the application processes). If there is enough space in the reduced set, the node tries to establish a new monitoring relationship with another node by adding it to appropriate set and sending necessary messages. On the other side, the chosen node should agree to play the assigned role.

Every Δt_2 time units, a node can initiate the propagation of information about nodes failures by sending EXCHANGE message. It depends on a current implementation, which nodes will be involved in this process and how much information will be exchanged. It should be emphasized that active participation of both embroiled nodes in such an exchange speeds up dissemination of information. When application process queries failure detector module about some process Q, it will return the most current status information about Q it has.

The proposed frameworks can be easily adapted to existing failure detection protocols. Let us assume, that the node can be monitored only by one node at a time ($|\mathcal{C}| = 1$), the node can remember information about all the other nodes ($|\mathcal{N}| = max$), there is no permanently supervised nodes ($|\mathcal{S}| = 0$) and the exchange of information is done only by PING messages ($\Delta t_1 = 1$, $\Delta t_2 = \infty$). By applying such parameters to the push model presented above the protocol proposed by Van Renesse et al. [9] is obtained.

Pull model:

```
Every time Δt1
  remove nodes from S that fullfill criteria
  send PING-STOP to removed nodes
  if there is space in S then
     select new node and add it to S
     send PING-REQ to added nodes
  end if
  send PING to C

Every time Δt2
  send EXCHANGE to some nodes from N

On querying about process Q
  return status of process Q

Upon receiving EXCHANGE from P
  update N due to received message
  send EXCHANGE to P

Upon receiving PING from P
  update information due to received message

Upon receiving PING-REQ from P
  if agree to being supervised then
     add node P to C
  end if

Upon receiving PING-STOP from P
  remove P from C
```

Push model:

```
Every time Δt1
  remove nodes from C that fullfill criteria
  if there is space in C then
     select new node and add it to C
  end if
  send PING to C

Every time Δt2
  send EXCHANGE to some nodes from N

On querying about process Q
  return status of process Q

Upon receiving EXCHANGE from P
  update N due to received message
  send EXCHANGE to P

Upon receiving PING from P
  if P ∉ S then
     if agree to being controller then
        add node P to S
     else
        send PING-STOP to P
     end if
  end if
  update information due to received message

Upon receiving PING-STOP from P
  remove P from C
```

Algorithm 1: Failure detection service frameworks (pull and push model)

6 Conclusion

One of the most important factors in any distributed application including MAS is the continuous service provision regardless of any failure. Any abnormal behaviour of AP can disturb agents residing on that platform. That is why it is necessary to improve the existing architecture of APs to make them more adaptive and fault tolerant in a large scale distributed environment. This paper presents the concept of the FDS. We believe that embedding such a service in APs will make them more scalable and failure prone.

At this stage of the project, there are any partial experimental results, so they can not be compared with other services, but preliminary tests allow to be optimistic. While designing our FDS, some interesting solutions were incorporated that were published and described in recent years in the context of fault tolerance and distributed systems [18,22,9]. It can be shown that given adequate resources, epidemic strategy can be configured to obtain high reliability, such that a message is delivered to all processes with a high probability. The decentralized nature of this kind of dissemination and failure detection results in protocols that are scalable to a large number of nodes without overloading any single host and adaptable to any application requirements.

The future directions of our research are connected with the construction of the FDS and revealing the most promising combination of parameters of such a service. With this in a view, we plan to perform the extensive set of simulation experiments in order to identify and describe base rules which allow building efficient failure detection service for certain class of applications. In our opinion, MAS platforms are the ones of the possible beneficiaries.

The concept of accrual failure detector [18] seems to be very promising in the context of building FDS. Therefore, we carry on experiments aimed at finding new models of describing the suspicion level of nodes.

Even though we start by assuming existence of the overlay management layer, we plan to conduct some research to estimate the profits from gluing both layers together. In the context of the last research results in the field of peer-to-peer networks [22], this step makes a new possibility to further optimize the communication overhead by using the piggy-back technique.

References

1. Wooldridge, M., Jennings, N.R.: Intelligent agents: Theory and practice. Knowledge Engineering Review **10** (1995) 115–152
2. Alouini, I., Roy, P.V.: Fault-tolerant mobile agents in Mozart. In: 2nd International Symposium on Agent Systems and Applications (ASA2000) and 4th International Symposium on Mobile Agents (MA2000), Zurich, Switzerland (2000)
3. Dellarocas, C., Klein, M.: An experimental evaluation of domain-independent fault handling services in open multi-agent systems. In: Proceedings of the International Conference on Multi-Agent Systems (ICMAS-2000). (2000) 95–102
4. Turner, P.J., Jennings, N.R.: Improving the scalability of multi-agent systems. Lecture Notes in Computer Science **1887** (2000) 246–262
5. Marin, O., Bertier, M., Sens, P.: Darx - a framework for the fault-tolerant support of agent software. In: Proceedings of the 14th IEEE International Symposium on Software Reliability Engineering (ISSRE 2003). (2003) 406–417
6. Ahmad, H.F., Suguri, H., Ali, A., Malik, S., Mugal, M., Shafiq, M.O., Tariq, A., Basharat, A.: Scalable fault tolerant agent grooming environment: Sage. In: AAMAS '05: Proceedings of the 4th International Joint Conference on Autonomous Agents and Multiagent Systems, New York, NY, USA, ACM Press (2005) 125–126
7. Chandra, T.D., Toueg, S.: Unreliable failure detectors for reliable distributed systems. Journal of the ACM **43** (1996) 225–267
8. Fischer, M.J., Lynch, N.A., Paterson, M.S.: Impossibility of distributed consensus with one faulty process. Journal of the ACM **32** (1985) 374–382
9. van Renesse, R., Minsky, Y., Hayden, M.: A gossip-based failure detection service. In: Proc. of the Int. Conf. on Distributed Systems Platforms and Open Distributed Processing (Middleware). (1998) 55–70
10. Gupta, I., Chandra, T.D., Goldszmidt, G.: On scalable and efficient distributed failure detectors. In: Proc. of the 20th Annual Symp. on Principles of Distributed Computing (PODC). (2001) 170–179
11. Hayashibara, N., Cherif, A., Katayama, T.: Failure detectors for large-scale distributed systems. In: Proceeding of the 1st Workshop on Self-Repairing and Self-Configurable Distributed Systems (RCDS), Osaka, Japan (2002) 404–409
12. Stelling, P., DeMatteis, C., Foster, I.T., Kesselman, C., Lee, C.A., von Laszewski, G.: A fault detection service for wide area distributed computations. Cluster Computing **2** (1999) 117–128

13. Horita, Y., Taura, K., Chikayama, T.: A scalable and efficient self-organizing failure detector for grid applications. In: Proceedings of 6th International Workshop on Grid Computing (Grid 2005), Seattle, Washington, USA, IEEE (2005) 202–210
14. Overeinder, B., Brazier, F., Marin, O.: Fault-tolerance in scalable agent support systems: Integrating darx in the agentscape framework. In: Proceedings of the 3rd IEEE/ACM International Symposium on Cluster Computing and the Grid (CCGrid2003). (2003) 688–695
15. Demers, A., Greene, D., Hauser, C., Irish, W., Larson, J., Shenker, S., Sturgis, H., Swinehart, D., Terry, D.: Epidemic algorithms for replicated database maintenance. In: Proceedings of the 6th Annual ACM Symposium on Principles of Distributed Computing, Vancouver, BC, Canada, ACM Press (1987) 1–12
16. Chen, W., Toueg, S., Aguilera, M.K.: On the quality of service of failure detectors. IEEE Trans. Computers **51** (2002) 561–580
17. Bertier, M., Marin, O., Sens, P.: Implementation and performance evaluation of an adaptable failure detector. In: Proceedings of the International Conference on Dependable Systems and Networks (DSN'02), Washington, DC (2002) 354–363
18. Hayashibara, N., Défago, X., Yared, R., Katayama, T.: The φ accrual failure detector. In: SRDS, IEEE Computer Society (2004) 66–78
19. Birman, K.P., Hayden, M., Ozkasap, O., Xiao, Z., Budiu, M., Minsky, Y.: Bimodal multicast. ACM Transactions on Computer Systems **17** (1999) 41–88
20. Eugster, P.T., Guerraoui, R., Handurukande, S.B., Kouznetsov, P., Kermarrec, A.M.: Lightweight probabilistic broadcast. In: Proceedings of the International Conference on Dependable Systems and Networks (DSN 2001), Washington, DC, USA, IEEE Computer Society (2001) 443–452
21. Ganesh, A.J., Kermarrec, A.M., Massoulié, L.: Peer-to-peer membership management for gossip-based protocols. IEEE Trans. Computers **52** (2003) 139–149
22. Jelasity, M., Guerraoui, R., Kermarrec, A.M., van Steen, M.: The peer sampling service: Experimental evaluation of unstructured gossip-based implementations. In: Middleware 2004: ACM/IFIP/USENIX International Middleware Conference, Toronto, Canada, October 18–22, 2004: proceedings. Volume 3231 of Lecture Notes in Computer Science., Springer-Verlag Inc. (2004) 79–98

Container Handling Using Multi-agent Architecture

Meriam Kefi[1], Ouajdi Korbaa[2], Khaled Ghedira[1], and Pascal Yim[2]

[1] Ecole Nationale des Sciences de l'Informatique
Université de Manouba
2010 Manouba, Tunisie
[2] Ecole Centrale de Lille
BP 48, Cité scientifique, F59651Villeneuve d'Ascq, France
Meriem.Elkefi@ec-lille.fr, Khaled.Ghedira@isg.rnu.tn,
{Ouajdi.Korbaa, Pascal.Yim}@ec-lille.fr

Abstract. Container terminals are essential intermodal interfaces in the global transportation network. Efficient container handling at terminals is important in reducing transportation costs and keeping shipping schedules. This paper discusses a multi-agent model to simulate, solve and optimize the amount of storage space for handling container departures within a fluvial or maritime port. The multi-agent model is developed to minimize the expected total number of rehandles while respecting spatio-temporal constraints. Experimental and numerical runs are given and illustrated. These runs carry out the proposed multi-agent model with comparison to the associated centralized version, basing on non-informed search algorithm and informed search algorithm. They show that the multi-agent model reduces rather the number of expected relocation movements especially when basing on informed search algorithm.

Keywords: Container terminal, Space allocation, transportation, Container stacking, Multi-agent systems.

1 Introduction

The improvement of customer service in fluvial or maritime ports constituted of container terminals becomes an important problem facing the big concurrence between ports. One of the measures of performance of this service is the berthing time of ships. This time is composed in major part of container loading and unloading time. To reduce loading duration, it is necessary to choose adequate storage slots either for the arriving containers or for the shifted containers after successive departures of others, in order to be loaded efficiently (minimization of the number of re-handles) within a vessel. This operation allying transfer, stacking and unstacking of containers is claimed to be an important problem among the decision problems met in container terminal [12].

In literature, few studies are dedicated to this problem which is known by "Container Stacking Problem". [6] proposed dynamic programming to attain an ideal configuration while minimizing the number of containers to move and the follow-on travelled distance. The problem is divided into two sub-problems: Bay matching and

N.T. Nguyen et al. (Eds.): KES-AMSTA 2007, LNAI 4496, pp. 685–693, 2007.

move planning problem, and moving tasks sequencing problem. A mathematical model is used for the resolution of each sub-problem.

It is noted worthy that [10] used genetic algorithms in their study to reduce container transfer and handling times and thereafter the berthing time of ship at port quays.

An interesting work, propounded by [7], dealt with the problem of determination of the best storage slot for containers with the aim of minimizing the number of expected relocation movements in loading operation. They deployed the weight of the container as criteria to define certain priority between the different containers to be stacked in the storage yard.

[1] provided an empiric and comparative survey of the different strategies of container storage in a terminal port according to several parameters.

Another work [8] proposed a cost model for determining the optimal storage position and the optimal number of cranes used for container handling.

A recent work [9] has described a model to determine in real time the best slot of an arriving container in the storage yard, while minimizing the number of the expected relocation movements at the phase of loading containers in a ship.

It should be noted that these studies often used the above-mentioned mathematical models to solve the underlined problem. These models were not very efficient and successful seen the complexity and the dynamicity of this problem. Besides, we observe that the multi - agent approach used in [2], [11] and [5] is applied very insufficiently to solve the problem of container stacking.

In this paper, we focus on container handling when container departures occur, while the majority of related works are focused on container arrivals. Furthermore, we opt for the distributed resolution via the multi-agent systems. In fact, these systems prove to lend themselves to model phenomena in which the interactions between various entities are too complex to be apprehended by classic tools of mathematical modelling. Therefore, we set a multi- agent model to solve the problem of container storage. We refer to concepts that are exploited in a multi-agent context to solve combinatorial problems by Ghedira [4], and in eco-solving method applied to solve the Block world Problem [2].

The remainder of this paper is organized as follows. Section 2 presents parameters, hypotheses and objective of the considered problem. Section 3 describes the architecture of the proposed multi-agent model. Section 4 details the dynamic of this model. Section 5 provides corresponding numerical results. Some conclusions and perspectives are drawn in the last section 6.

2 Problem Description

Our essential concern is to arrange the containers in adequate slots with respect to spatio-temporal constraints in order to minimize the relocation movements precisely the unproductive movements at the moment of their departure in order to be loaded on vessels (i.e., when an outside truck or vessel arrives in a random manner and requests a container, one of the difficult problems in the yard operation is that it takes too much handling effort to re-handle containers on the top of the requested container).

It should be mentioned that in port's storage yard, distances are not very big. Therefore, the time that a reach stacker takes to transport a container from one slot to another is very similar whatever the position of the container is. For this reason, we focus on the relocation movements which induce delays.

Thus, our study seeks to answer the following question: after every departure of some containers, how can we rearrange the containers unconcerned by this departure in most adequate manner? The rationale underlying this was to manage the next departures efficiently (i.e., retrieving easily the containers to load on vessel) as well as the next arrivals (i.e., determining easily the appropriate slots for the containers to stack).

To clarify the problem raised, we present next its parameters, its hypotheses, its constraints and its objectives.

2.1 Parameters

The following parameters were taken into account in our study:

- A storage yard composed of *M* Slots,
- *N* ISO Containers stocked in stacks,
- H_{max}: the maximal height of stacks,
- *initConfig*: an initial arrangement of the N containers in the *M* available slots,
- *finConfig*: a final possible arrangement of deranged containers after the departure of some ones. This arrangement must be attained from *initConfig*.

2.2 Hypotheses

We adopt the following hypotheses:

- Container arrival and a departure cannot coincide,
- The departure order of different containers is known, while their arrival order is unknown,
- Only one container arrives at once,
- The containers of different dimensions (1 TEU, 2 TEU etc.) are stacked on two different storage yards,
- A movement consists on removing a container and putting it somewhere,
- Vessels are classified as A, B, C, D, E, and F according to the chronological order of their departure dates which are known in advance,
- Vessels of which the exact departure dates are beyond F are called Y.
- Vessels of which the departure dates is completely unknown are called Z,
- We associate to every container the vessel whereby it is going to leave (A, B, C, D, E, F…Y & Z) which determines implicitly its departure date t. For example, for two different containers C and C', if C is going to leave on vessel A and C' is going to leave on vessel B, C, D,.. or Y then t (C) <t (C '),
- The containers are categorized identically to the associated vessels. For example a container is of category A if it is going to leave on vessel A,
- One container is prioritized to another container if the category of the former is less than the category of the latter (A<B<C<D<E<F).

2.3 Objectives

The objective of the present study is to determine firstly a feasible sequence of successive unstacking–stacking operations, and then approximate one that minimises container rehandles, considering temporal and spatial constraints. To achieve this objective, we have developed a multi-agent model implementing both non informed and informed algorithms. This model will be detailed in the next section.

3 Architecture of Multi-agent Model

3.1 Assumptions and Definitions

- A Container agent corresponds to each container,
- A Container agent cannot be placed on another one unless it is free.
- Container agent *j* blocks another Container agent *i*, if its own category is greater than the category of *i*,
- A Container agent is satisfied, if it is placed in a slot without blocking another one.
- The state of a Container agent is either satisfied or unsatisfied.
- A Container agent is free if he does not have Container agents above.

3.2 Architecture

The agents defined in the proposed multi-agent model are: the Container agent and the Interface agent. The Container agent is defined by its acquaintances formed by the other Container agents and the agent Interface, its static knowledge: identifier, destination port, load, tare and dimension; and its dynamic knowledge: category, the Container agent under, the Container agent on and State. Its behaviour will be described in the following section.

4 Multi-Agent Dynamic

Here, we provide a description of the dynamic of the set up multi-agent model: the main steps of the solving process, and the most important messages exchanged between the different agents and incorporated into their behaviour.

4.1 Solving Process

The solving process associated to the multi-agent model is composed essentially of the following steps:

- Setting the state of the Container agents which block the Container agents A to "Unsatisfied".
- Departure of the Container agents A.
- The Container agents which blocked the Container agents A are deranged, and then they search a new adequate slot by diffusing a message "Search-Place" to their acquaintances to request a new appropriate slot eventually.
- Reception of messages: "I_m_Free" et "I_m_Sorry".

- When receiving messages "I_m_Free", then a Container agent sends the message "Migrate" to the free Container agent.
- Sending "I_am_Leaving": one Container agent leaves its current slot only if it received a message "Confirm" from the Container agent which is accepting it to be on.
- Finishing if all the containers are satisfied.

4.2 Exchanged Messages

The solving process above described is based on a negotiation protocol that involves different types of messages exchanged between two different Container agents *i* and *j*:

- "I_am_Blocked": *i* wants to be free to leave its slot; it is blocked in by *j*.
- "Search_Place»: *i* is looking for an appropriate slot.
- "I_am_Free»: *i* informs j that it is free and then proposes to *j* to take place on it.
- "Apologize": *i* informs j that it can not accept it on because there is an other Container agent which takes the place. This is due to the asynchronous aspect of exchanging messages in a Multi-Agent model generally.
- "Migration": *i* asks *j* if it possible to be above.
- "Confirmation": *i* accepts definitively *j* to be on.
- "Rejection»: *i* refuses the migration request of *j*.
- "I_m_Leaving": *i* informs *j* that it goes away, so *j* becomes free.

These messages are invoked through a global behavior procedure:

```
Procedure Global_Behaviour()
  {
    ACLMessage msg = receive(ACLMessage.INFORM);
    if (msg != null) {
        if ( (msg == ("you_are_Free")) {
          I_m_Free(Sender);
        }
        if ( (msg ==("you_Block_me")) {
          I_Block();
        }


        if ( (msg ==("Search_place")) {
          Search_place_for(Sender));
        }

        if ( (msg ==("on_you")) {
          On_me(Sender);
        }
        if ( (msg ==("I_m_Free")) {
            It_is_Free(Sender);
        }
        if ( (msg ==("appologize")) {
          It_applogizes(Sender);
        }
```

```
        if ((msg == ("CIAO")){
          doDelete();
        }
        if((msg == ("update")){
          update_coordinates(msg);
        }
        if((msg == ("confirm")){
          It_confirms(Sender, SenderState);
        }
        if((msg == ("reject")){
          It_rejects();
        }
      }
    }
```

5 Experimental Results

We used JADE platform: http://www.jade.tilab.com/ to implement the proposed multi-agent model. To evaluate our model, we have first developed the distributed version and the corresponding centralized version, and then we have associated to each one a non informed search algorithm and an informed search algorithm, i.e. in the former, the determination of a new slot for each Container agent is done randomly. By the opposite, in the latter a tested heuristic is applied locally to each Container agent who is searching an appropriate place.

The following graphics show plots comparing these versions. We represent on axis Y the total number of unproductive movements in function of random-generated

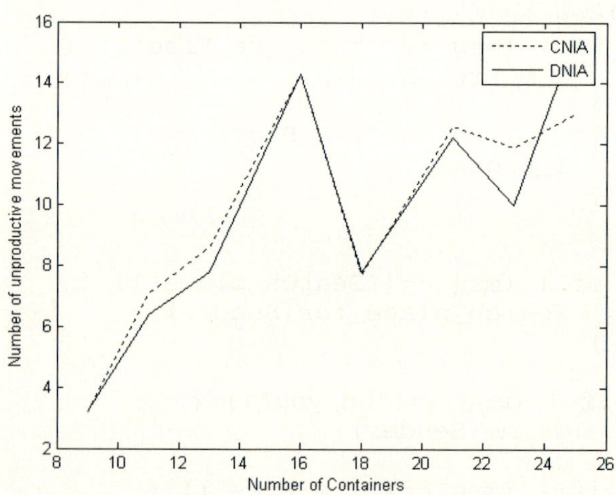

Fig. 1. This figure shows a comparison between the Distributed version DNIA "Distributed Non Informed Algorithm" and the centralized version CNIA "Centralized Non Informed Algorithm", both based on non informed algorithm

instances mentioned on Axis X. Each instance signifies an initial configuration with a certain number of containers. We note that the numerical results correspond to 10-run average for each generated instance.

As shown (Fig. 1), the distributed version outperforms slightly the centralised version given a non informed algorithm to solve the underlying problem. This enhancement is due to cooperation between the agents of our Multi-agent model. In fact, cooperation helps to reach near values in ten runs of the same instance.

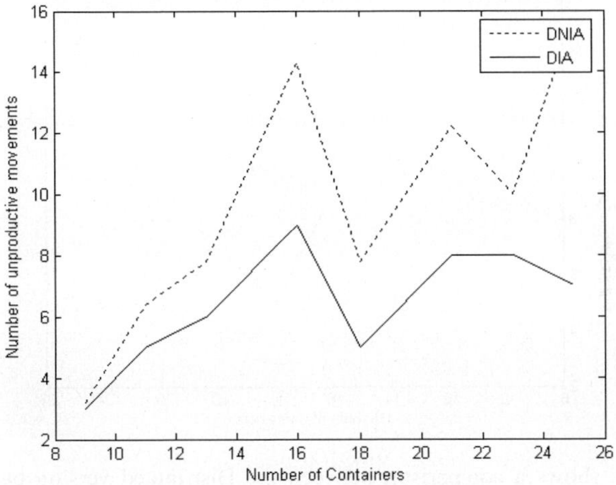

Fig. 2. This figure shows a comparison between two Distributed versions: one based on non informed algorithm DNIA «Distributed Non Informed Algorithm» and the other based on informed algorithm DIA «Distributed Informed Algorithm "

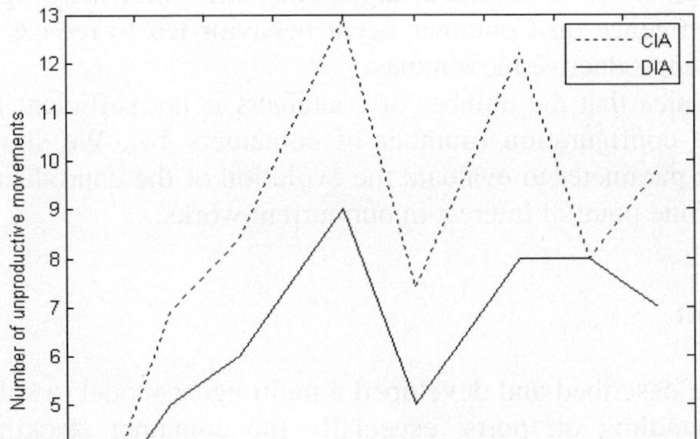

Fig. 3. This figure shows a comparison between the Distributed version DIA "Distributed Informed Algorithm" and the Centralized version CIA "Centralized Informed Algorithm", both based on informed algorithm

To improve the previous results, we have introduced an intelligence degree on the behaviour of the Container agents using an informed algorithm, i.e. when searching a new place; each Container agent does not select randomly a new slot among received proposals, but it applies intelligibly a tested heuristic. Then, as shown (Fig. 2, Fig. 3), a significant improvement is noted particularly when the number of containers is increasing with comparison to the distributed version with non informed algorithm and the centralised version with an informed algorithm.

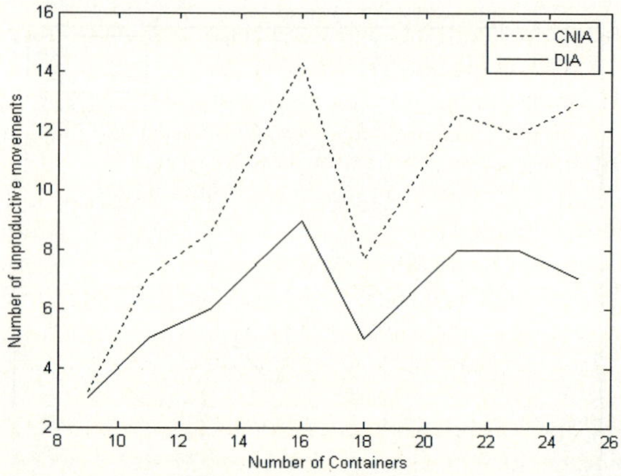

Fig. 4. This figure shows a comparison between the Distributed version based on informed algorithm *DIA* "Distributed Informed Algorithm" and the Centralized version based on non informed algorithm CNIA: Centralized Non Informed Algorithm"

As we clearly observe, the results (Fig. 4) show a larger enhancement which is due to the combination of distributed algorithm and informed algorithm. Indeed, introducing intelligence on Container agent behavior led to reduce significantly the total number of unproductive movements.

We should notice that the number of containers is not sufficient to characterize a generated initial configuration (number of containers 18). We should determine a more significant parameter to evaluate the evolution of the unproductive movements. This constitutes one point of interest in our current works.

6 Conclusion

In this paper, we described and developed a multi-agent model to solve and optimize the container handling on ports, especially the container stacking process after container departures. This model involves uniquely Container agents, which cooperate, coordinate and negotiate within a solving process.

This paper provides empirical and comparative studies showing that multi-agent model where we are introducing intelligence degree within agent behaviour via informed algorithm outperforms considerably the multi-agent model with non informed algorithm which is slightly enhancing the centralised version of the proposed model with non informed algorithm.

Currently, we proceed to develop the proposed multi-agent model, to extend it taking into account both container departures and arrivals.

References

1. Duinkerken, M.B., Evers, J.-J.M., Ottjes, J. A.: A simulation model for integrating quay transport and stacking policies automated containers terminals. Proceedings of the 15th European Simulation Multi-conference, Prague (2001)
2. Ferber, J.: Les systèmes multi-agents: vers une intelligence collective (1995)
3. Gambardella, L.M., Rizzoli, A.E., Zaffalon M.: Simulation and planning of an intermodal container terminal. Simulation, Vol. 71. (1998) 107-116
4. Ghedira, K., Verfaillie, G.: A Multi-Agent Model for the Resource Allocation Problem: A Reactive Approach. Proceedings of the 10th European Conference on Artificial Intelligence, Vienna, Austria (1992) 252-254
5. Henesey, L., Wernstedt, F., Davidsson, P.: Market-Driven Control in Container Terminal Management. Proceedings of the 2nd International Conference on Computer Applications and Information Technology in the Maritime Industries, Hamburg, Germany (2003) 377-386
6. Kim, K.H., Bae, J.W.: Re-marshaling export containers in port container terminals. Computers & Industrial Engineering, Vol. 35. (1998) 655-658
7. Kim, K.H., Park, Y.M., Ryu, K.R.: Deriving decision rules to locate export containers in container yards. European Journal of Operational Research, Vol. 124. (2000) 89-101
8. Kim, K.H., Kim, H.B.: The optimal sizing of the storage space and handling facilities for import containers. Transportation Research, Vol. 36. (2002) 821-835
9. Korbaa, O., Yim, P.: Container Assignment to Stock in a Fluvial Port. International Conference on Systems, Man and Cybernetics, The Netherlands (2004)
10. Kozan, E., Preston, P.: Genetic algorithms to schedule container transfers at multimodal terminals. International Transactions in Operational Research, Vol. 6. (1999) 311-329
11. Rebollo, M., Julian, V., Carrascosa, C., Botti, V.: A Multi-Agent System for the Automation of a Port Container Terminal. Autonomous Agents workshop on Agents in Industry, Barcelona, Spain (2000)
12. Vis, I.F.A., De Koster, R.: Transshipment of containers at a container terminal: an overview. European Journal of Operational Research, Vol. 147. (2003) 1-16

Distributed Code and Data Propagation Algorithm for Longest Common Subsequence Problem Solving

Dariusz Król and Grzegorz Stanisław Kukla

Wrocław University of Technology, Institute of Applied Informatics,
Wybrzeże S. Wyspiańskiego 27, 50-370 Wrocław, Poland
dariusz.krol@pwr.wroc.pl, grzegorz_kukla@o2.pl

Abstract. This paper proposes a distributed code and data propagation algorithm for Longest Common Subsequence problem solving and compares this algorithm performance against the algorithm based on J2EE technology. The new algorithm builds a graph devised to propagate classes and data between different nodes and the client, whereas the J2EE algorithm requires more complex communication and database processing. The proposed algorithm's performance in terms of number of nodes and execution time is better than or comparable to that of the existing algorithms.

1 Introduction

Examining the Longest Common Subsequence (LCS) method in the parallel version in terms of the efficiency is the subject of several papers [1], [3], [11] and [15]. We have chosen two solutions: the first one, based on J2EE and the second, using propagation mechanisms. Using J2EE technology to implement LCS algorithm guarantees the independence of the solution. Additionally, network solutions provided in the J2EE framework are very useful in the case of processing data by the dispersed application [5], [9]. The experiment took advantage of proprietary system using JDK 1.4. The communication between hosts had been established via HTTP protocol.

The distributed code and data propagation (DCDP) is a new version of LCS algorithm worked out by the authors. DCDP is an alternative approach to the classical parallel problem division.

We understand the sequence as the cohesive area of elements of the same type, having a defined length. The LCS algorithm is based on the method of dynamic programming which means that the basic version of the algorithm uses the two-dimensional matrix of numbers. If we assume that the length of sequence X is m and the length of sequence Y is n then the computational complexity will be O(m*n). It is possible to lower the memory complexity to the linear figure if we are caring only about defining the LCS length rather than finding the longest shared sequence.

The paper is organized as follows. The concept details of comparing the sequence in parallel approach are introduced in Section 2. Multi-agent architecture as communication schema is detailed in Section 3. Section 4 describes the main characteristics of DCDP algorithm and empirical evaluation is sketched in Section 5. We conclude with a discussion in Section 6.

N.T. Nguyen et al. (Eds.): KES-AMSTA 2007, LNAI 4496, pp. 694–703, 2007.

2 Concept of Parallel Comparing the Sequence

The algorithm is based on the classic method of comparing the sequences presented in [6]. The main idea of the algorithm consists in using dynamic programming in LCS algorithm implementation. In order to compute the similarity measure of two sequences: X of length m and Y of length n the m x n matrix is created. Rows are indexed by the tokens of the first sequence and columns are indexed by terms of the second one. A value of each matrix element is computed as follows:

$e_{i,j} = e_{i-1,j-1} + 1$ if $X_i = Y_j$
$e_{i,j} = \max(e_{i-1,j}, e_{i,j-1})$ if $X_i != Y_j$

The value $e_{m,n}$ is the searched similarity measure.

In order to take advance of distributed version of this algorithm, one should break the problem into parts. In this case a main part of the algorithm is a division of the matrix of dynamic programming into parts which will be processed by nodes of the system. The matrix appropriately divided is shown in Fig. 1.

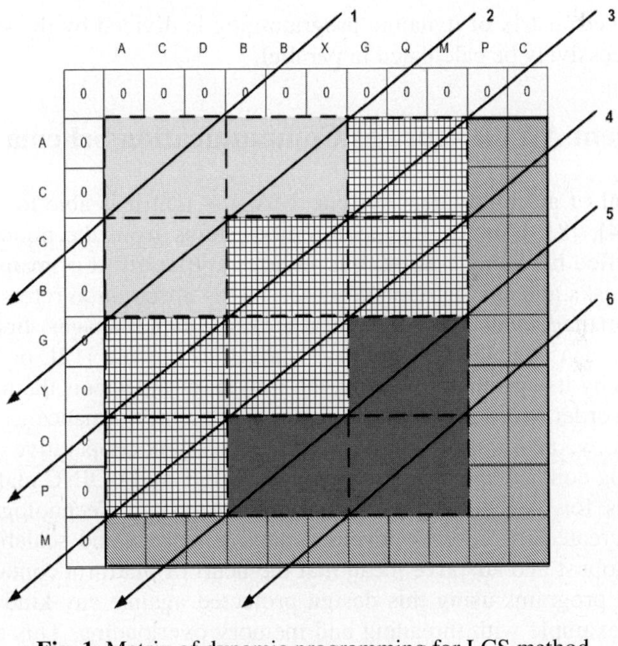

Fig. 1. Matrix of dynamic programming for LCS method

Let us imagine now that the matrix is supposed to be processed by a few processes. Having p processes, each has the same computational possibilities we need to prepare portions for processing appropriately. So that to achieve the state of the matrix for beginning processing, we must make p - 1 steps of preliminary pre-processing. It is possible to divide the matrix into the following logical areas:

1) Area of preliminary processing
   Processing this area is necessary for p processes to do it fully in parallel. The number of stages of processing amounts is p - 1.
2) Area of full processing
   This area is being processed in parallel by p processes. As you can see it is taking place step by step from above down. When processes attain the bottom edge of divided area, then they are going to the next part of the matrix and processing is being continued till the time, when area is too small so that all processes can process it.
3) Area of final processing (divided part)
   Processing this area is taking place when p of processes are not already available to go to the next part of the matrix in order to process it. Similarly, like in the case of preliminary processing, the number of steps and pieces which are being processed are the same.
4) Area of final processing (undivided part)
   After the first three stages one should still process the left over elements of the matrix - both from the ground edge as well as from the right edge.

The introduced matrix of dynamic programming is divided by the system of areas which can successively be calculated in parallel.

3 Multi-agent Architecture as Communication Schema

One of the goal of an multi-agent system (MAS) is that it is able to cooperate with other agents [4], [8]. Usually, this requirement stems from the point that the goal cannot be satisfied by a single agent, but requires a joint effort of many agents, coordinating their tasks and sharing their resources. This observation refers to application such as chain management. For example, a broker (chain manager) first assigns each agent to a part of a complex task via task allocation process [12], or each agent receives requests by its customers. Finally, the broker, or the agents themselves may try to cooperate in order to reduce the total cost or increase performance.

In this paper, we take a special approach to multi-agent cooperative execution. Our research is being conducted within the context of DIET and BOINC platform.

DIET stands for Decentralized Information Ecosystem Technologies [10]. This platform was created in order to develop adaptive, robust and scalable multi-agent applications. Robust and adaptive mean that the heart of platform contains all mechanisms to make programs using this design protected against any kind of exceptions connected for example with threading and memory overloading. This means that the program core can deal with it and change the application behaviour accordingly. Finally it is scalable at global and local level, which means that agents are designed to be light-weight, so this makes a possibility to have big amounts of them running on one machine, and global because there are no size limits to applications based on DIET platform.

The structure of DIET consists of three layers. The most important is the core layer, with minimal software to make multi-agent system running. Second one is the Application Reusable Components with functionality, which is used in computer programs developed on DIET platform. The last one is application layer, which is a set of examples programmed by the authors to show how to use this platform in multi-agent systems.

The most important DIET structure is the element hierarchy: worlds, environments, agents, messages and connections. The heart of this conception are agents. Each of them has to live in an environment, and each environment has to be a part of a world. Agents are able to communicate with each other by establishing a bi-directional connections between a pair of them. Then they can send a message containing a character string and optionally an object. Each agent has an state attribute which mirrors his current action or tells something about his job. As the agents are designed to be lightweight it is possible to set a message buffer capacity, which results in message overload and activity control action done by the DIET kernel, in order not to make ones too active. Messages can be rejected, accepted and of course handled by responding to them and changing state or action. Every agent uses only one thread and none of other agents has access to it Each agent reference is remembered only by the kernel, so that other agents don't have access to it's methods and can not execute it's operational functions. But still an agent is able to create another one and kill it, when that entity isn't needed any more.

When building an DIET application it is necessary to create at least one world and an environment. Usually one world runs on one Java Virtual Machine. It is possible to create worlds on more then one machine. That makes this platform a very good environment to use a multi-agent approach in our project.

On the other hand, the Berkeley Open Infrastructure for Network Computing (BOINC) is the most popular framework for creating distributed computing systems [2]. It is used as a basis for many grid projects such as: SETI@Home, LHC@Home or Rosetta@Home.

Basically the BOINC architecture consists of many independent daemons communicating with themselves via shared memory, disk files or central database. From the deployment point of view the architecture consists of one central server responsible for work generation, distribution and results validation, and many nodes responsible for work execution. This structure is scalable – in some of the projects even hundreds of thousands of nodes are involved. The whole source code of the framework is written in the C++ programming language.

The main disadvantage of this approach is rather complicated project structure. Developer is forced to implement many different applications communicating with themselves in an unnatural way. Applications written in the native language are platform dependent. Thus for a single project there is a need of implementing many client applications, one for each platform (processor and operating system). The second disadvantage is an inability for easy multi-stage processing. Work units generated by one daemon, pre-processed by the nodes don't go back to the creator but are handled by another daemon.

4 DCDP Algorithm

The communication schema in our project is shown in Fig. 2. The DCDP structure consists of three types of elements: client, broker and node. The client, the source of original data does not take part in computation; it takes part in task division, passes classes' code and data to the nodes and receives output data. The broker is a coordination centre for all elements. It is responsible for optimization and task reallocation due to possible error occurrences. The node computes delegated task and passes data to other node or back to client. Nodes can cooperate themselves.

DCDP algorithm was implemented in J2SE platform. Elements of the grid structure communicate with themselves using special socket-based protocol. Data exchange between nodes was inspired by P2P networks.

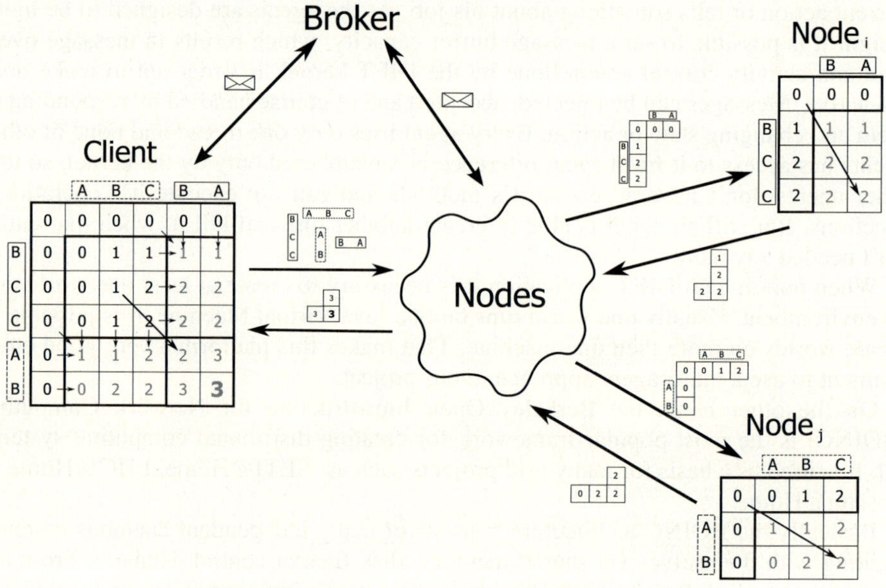

Fig. 2. Communication schema of DCDP algorithm

The execution of the DCDP algorithm is the following:

- The client divides input elements into two entrance sequences for processing into parts about the established length.
- The client records on the broker all parts ensuing this way (the client receives the unique ID for every part and the broker remembers that the part with the given ID is accessible on the client in the given file).
- The client goes through the matrix of calculations one after the other and records tasks on the broker (the recording of the task of calculating the sub matrix consists in sending 2 IDs of parts of tied inscriptions bordering on this sub matrix and 3 IDs of the sub matrix: left, diagonal and upper. The broker sends back the ID to the file in which the result of calculations should be. Thanks to that the client can record all sub matrixes without the need to expect for finishing indirect calculations).
- The broker assigns tasks to the nodes.
- The node receives the task, takes needed data from nodes and the client (asks the broker about locations, about the given ID, gets the list of addresses where from can obtain the needed file).

- The node makes calculations.
- The node records the file containing the result of calculations on the broker and requests the next task. The file of the result receives the ID established earlier from the broker (the same which the client got while recording the task).
- The client waits for calculating the last sub matrix and then it takes it down and reads the result out.

In order to write a working application using a skeleton based in Fig. 2, the programmer has to perform the following steps:

- First, the programmer has to implement tasks classes (extend the task state abstract class to fit his needs, implement a set of code blocks used to carry out a task itself, make a JAR package consisting of the class files, the name of this file will be used as a task name).
- Then, the programmer must extend the abstract client class; this class provides all the functionality needed to communicate with the broker; the task of the programmer is to implement an algorithm carried out by the grid structure.

4.1 Client Code Description

The code listed below is placed on the client of the grid structure. It demonstrates usage of the framework. The `start` must be implemented by the programmer. It contains the main processing algorithm.

The `RemoteTaskInstance` class provides a set of methods to get a result of remote task execution. The `UniqueID` class represents the unique identifier of any of the grid framework elements (including data pieces, task files or computers). All of the features of the grid are represented by proper `UniqueID`.

The `getDataList` is a part of client class. It provides a list of IDs of files placed in a folder containing source data. This folder is specified by the programmer. The system automatically iterates through the folder contents and gains the `UniqueID` for every file it finds. Identifiers are provided by the broker.

The `execTask` is also a part of client class. It registers a task on the broker. The first parameter is the name of the task to be executed. It must be the same as the name of JAR file containing that task. The second argument is an array containing the `UniqueID` of tasks parameters.

```
public void start() {
   ArrayList inputDataList = getDataList(); // (1)
   ArrayList sortedDataList = new ArrayList();
   Iterator inputDataListIterator =
      inputDataList.iterator(); // (2)
   while (inputDataListIterator.hasNext()) {
      UniqueID currentFileID = (UniqueID)
         inputDataListIterator.next();
      RemoteTaskInstance taskInstance =
         execTask("sort", new UniqueID[]
            {currentFileID});
```

```
      sortedDataList.add(taskInstance.getResultID());
   }
   RemoteTaskInstance rti = null; // (3)
   while (sortedDataList.size() > 1) {
      UniqueID part1 = (UniqueID) sortedDataList.get(0);
      UniqueID part2 = (UniqueID) sortedDataList.get(1);
      rti = execTask("merge", new UniqueID[] {part1,
         part2});
      sortedDataList.add(rti.getResultID());
      sortedDataList.remove(part1);
      sortedDataList.remove(part2);
   }// ... (4)
}
/*
(1) Get the list of files to be sorted, create the list of
sorted pieces
(2) For each element of the list execute the remote sorting
task
(3) For every pair of sorted files located on the sorted-
DataList execute the remote merging task
(4) Read the result using rti.getResultStream()*/
```

4.2 Node Code Description

The code listed below is placed in the CodeBlock interface implementation. It is being executed by the node in order to accomplish the task deputed by the broker.

The execute provides the main functionality of the CodeBlock interface. This interface represents an atomic action that can be executed by the node without being interrupted. Between the successive code blocks it is possible for the node to stop the task execution and dump its state to a file. The task can be resumed later using the dumped state.

In order to carry out the task many code blocks must cooperate. Communication between them is provided by the TaskState interface. It controls the task execution process by giving information about code block to be run next. The setNextCodeBlock specifies that code block. Specific implementations of the TaskState interface may also keep data required by code blocks.

```
public void execute(TaskState state) throws TaskExcep-
tion {
 SortTaskState taskState = (SortTaskState) state;//(1)
 ArrayList list = taskState.getList();
 if (list.size() < 2) {
    taskState.setNextCodeBlock("SaveOutputFile");//(2)
    } else {
    ArrayList firstList = (ArrayList) list.get(0);//(3)
    ArrayList secondList = (ArrayList) list.get(1);
    ArrayList resultList = new ArrayList();
    resultList = mergeLists(firstList, secondList);
    list.remove(firstList);
    list.remove(secondList);
    list.add(resultList);
```

```
        taskState.setNextCodeBlock("SortList");
        }
    }
  /*
  (1) Cast the state argument to the proper TaskState implemen-
  tation. Get the list of elements to be sorted
  (2) If the list of elements contains less than two items,
  then set the code block that stops processing and saves the
  result to be executed next
  (3) Else get the two first elements, merge them and put the
  result on the end of the input list */
```

5 Empirical Evaluation

In order to assess project performance, we have done a set of experiments on a Windows XP cluster operated at our department. The cluster used for experiments hosts 16 nodes (1 GHz Pentium). Seven of them were used as nodes, one served as the client and one as the broker in DCDP architecture. J2EE based solution required one host less, because the client and the broker operated on the same computer. The client, the broker and nodes are interconnected by a local Fast Ethernet network (100 mbps).

We have implemented LCS algorithm in two platforms in order to estimate the cost of using: J2EE and DCDP architecture.

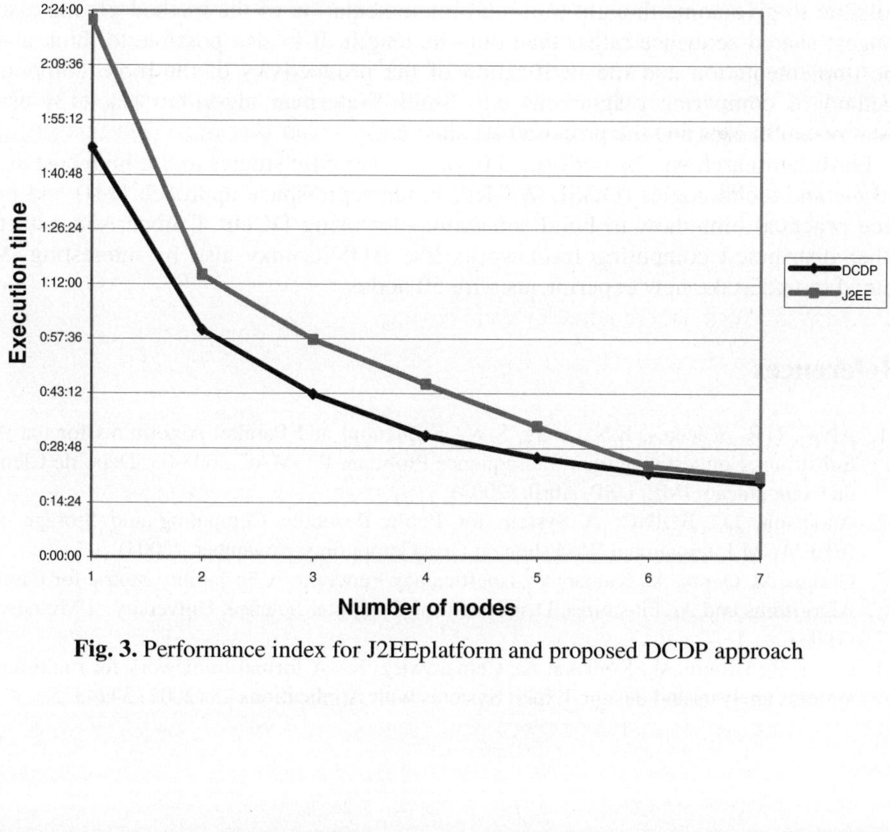

Fig. 3. Performance index for J2EEplatform and proposed DCDP approach

Experiment consisted in finding the longest common subsequence length for two chains for lengths of 500000 elements each. Chains were divided in parts about lengths of 10000 elements. The examination was carried out for the different number of nodes (from 1 to 7).

First of all we measured the application absolute execution time. Performance results are shown in Fig. 3. The speed-ups obtained with the DCDP algorithm show that using this method is a better solution than using J2EE in that case.

6 Conclusions

In this work two architectures were introduced to show the productivity of LCS method implemented in the J2EE and in DCDP architecture. The main conclusion is that on DCDP platform every nodes of the system will be exploited more effectively in order to process data. Another observation is that, that DCDP structure has a dynamic limit to compute algorithms in an efficient way. The more nodes the better efficiency, but costs also grow, in our experiments for more than 5 nodes.

A possibility of introducing improvements to the suggested approach and verifying the productivity exists when clients could take part in preliminary and final processing. It requires the bigger expenditure on the implementation and the alteration of the existing code but it is worth getting more interesting results.

The applied LCS algorithm calculating the length of the longest shared sequence is a simple method of comparing of sequence. On the basis of architecture described it is possible to give some thought to do the implementation of the method getting also a longest shared sequence rather than only its length. It is also possible to think about the implementation and the verification of the productivity of the more compound method of comparing plagiarisms e.g. Smith-Waterman algorithm [7], [13] using local resemblances and the proposed architecture.

Further research will be performed through other case studies to use high cost algorithms and technologies (OGSI, WS-RF, Linda tuple space approach [14]) and provide practical hints how to build supercomputer using DCDP. Comparison with the other distributed computing frameworks like BOINC may also be interesting. We intend to undertake new experiments with 50 nodes.

References

1. Alves, C.R., Cáceres, E.N., Song, S.W.: Sequential and Parallel Algorithms for the All-Substrings Longest Common Subsequence Problem. RT-MAC-2003-03, Dept. de Ciência da Computação, IME, USP, Abril, (2003)
2. Anderson, D.: BOINC: A System for Public-Resource Computing and Storage. 5th IEEE/ACM International Workshop on Grid Computing, November, (2004)
3. Grama, A., Gupta, A., Kumar, V.: Isoefficiency Function: A Scalability Metric for Parallel Algorithms and Architectures, Department of Computer Science, University of Minnesota, (1993)
4. Gruer, P., Hilaire, V., Koukam, A., Cetnarowicz, K.: A formal framework for multi-agent systems analysis and design. Expert Systems with Applications 23 (2002) 349–355

5. Haeuser, J. et al.: A test suite for high-performance parallel Java. Advances in Engineering Software 31 (2000) 687–696
6. Hirschberg, D.S.: Pattern Matching Algorithms: Serial computations of Levenshtein distances, Oxford University Press (1997)
7. Irving, R.W.: Plagiarism and Collusion Detection using Smith-Waterman Algorithm, Tech. Rep. TR-2004-164, University of Glasgow, Computing Science Department Research Report (2004)
8. Kolp, M., Giorgini, P., Mylopoulos, J.: Multi-agent architectures as organizational structures. Auton Agent Multi-Agent Sys 13 (2006) 3–25
9. Laure, E.: OpusJava: A Java framework for distributed high performance computing. Future Generation Computer Systems 18 (2001) 235–251
10. Marrow, P. et. al.: Agents in Decentralized Information Ecosystems: The DIET Approach. Proceedings of the AISB'01 Symposium on Information Agents for Electronic Commerce, AISB'01 Convention, University of York, United Kingdom, March, (2001)
11. Matsuoka, S., Itou, S.: Towards performance evaluation on high-performance computing on multiple Java platforms. Future Generation Computer Systems 18 (2001) 281–291
12. Shehory, O., Kraus, S.: Methods for Task Allocation via Agent Coalition Formation. Artificial Intelligence 101(1-2) (1998) 165-200
13. Smith, T.F., Waterman, M.S.: Identification of Common Molecular Subsequences, J. Mol. Biol. 147 (1981) 195-197
14. Wilkinson, B., Allen, M.: Parallel Programming: Techniques and Applications Using Networked Workstations and Parallel Computers, Prentice Hall (2004)
15. Xu, X., Chen, L., Pan, Y., He, P.: Fast Parallel Algorithms for the Longest Common Subsequence Problem Using an Optical Bus. LNCS 3482 (2005) 338-348

Design and Evaluation of a Converting Patching Agent for VOD Services

Sook-Jeong Ha[1] and Ihn-Han Bae[2]

[1] School of Electrical Engineering and Computer Science, Kyungpook National University, Korea
[2] School of Computer and Information Communication Engineering, Catholic University of Daegu, Korea

Abstract. Patching is a multicast technique that reduces the demand on a server's network-I/O bandwidth by sharing the same video stream with users for true VOD services. This paper proposes a multicast technique called Converting Patching Agent that is based on Greedy Patching and can provide better performance than Grace Patching. Converting Patching Agent changes the latest multicast into a regular multicast for a new special request. It results in the significant reduction of the new multicast data for the new request. Simulation results confirm that the proposed technique is better than grace patching in terms of defection rate and average service latency.

1 Introduction

VOD (Video on Demand) is a critical technology for many important multimedia applications, such as home entertainment, digital video libraries, distance learning, company training, news on demand, and electronic commerce [1, 2]. In most VOD systems, a video server's network-I/O bandwidth is limited. Hence, various multicast techniques have been proposed that share the same video stream with many clients to serve asynchronous video requests. Patching especially reduces the amount of new multicast data by allowing new clients to join an existing multicast, thereby increases the sharing of the video stream.

This paper presents the converting patching agent that is based on greedy patching and converts the latest ongoing patching multicast into a regular multicast for a particular new client. As the result of converting, the new patching stream becomes significantly smaller than that of greedy patching and then the holding time of the new channel for the new client become shorter. The rest of this paper is organized as follows. We describe greedy patching and grace patching in section 2, then introduce the proposed converting patching agent in section 3. We evaluate the performance of the proposed technique in terms of defection rate and average service latency through simulation. Finally, we give some concluding remarks in section 5.

2 Related Work

Various multicast techniques, such as batching, piggybacking and patching, have been proposed for VOD services [2, 3, 4, 5]. These techniques multicast a video stream to

N.T. Nguyen et al. (Eds.): KES-AMSTA 2007, LNAI 4496, pp. 704–710, 2007.

multiple clients on the same channel instead of unicasting the video stream to each client on each channel, and can reduce the demand on the network-I/O bandwidth of the video server. Patching has been proposed to solve the conflicting goals between service latency and sharing of a multicast stream, and have received considerable attention [2, 3, 6]. In patching, a server's communication bandwidth is organized into a set of logical channels, each of which can transmit a video at the playback rate [2]. A channel used to multicast an entire video data (regular stream) is called a regular channel, and a channel used to multicast the beginning portion of the video (patching stream) is called a patching channel. Let us assume that the start time of a regular multicast (R-multicast) for video v is t_r and the current time to schedule a new request is t and the time unit is a minute. The time interval, t-t_r, between the new multicast for the new request and the latest R-multicast, is called skew [2]. If t-t_r is shorter than or equal to a limited time, patching does not schedule an R-multicast, but rather a patching multicast (P-multicast) for the new request. The server services the request with P-multicast, it needs only one new channel for the new P-stream because it is transmitting the R-stream on a regular channel. Then, the new client can exploit an ongoing R-multicast by buffering the future stream from the regular stream (R-stream) while the client downloads and plays a new patching stream (P-stream).

```
t: Current time
t_r: Start time of the latest ongoing R-multicast of video v
PW : Size of patching window, that is, the size of client buffer
D : Portion of video data which should be multicast on a new channel, that is, the
    workload for the new channel
L : Playback duration of video v, that is, video length
v[t_d]: Video data of video v during the period from the beginning to time t_d
GreedyPatching()
    if (t-t_r ≤ PW)        D = v[t-t_r] ;
    else if (t-t_r ≤ L)    D = v[L - Min(PW, L - (t - t_r))] ;
    else                   D = v[L] ;
GracePatching()
    if (t-t_r ≤ PW)        D = v[t-t_r] ;
    else                   D = v[L] ;
```

Fig. 1. Algorithm to determine the video data to deliver on a new channel

In grace patching, the limited time is the size of the regular buffer to buffer an R-stream in a client's station and is called the size of patching window [2]. In contrast, greedy patching schedules a P-multicast as long as the latest R-multicast is ongoing to maximize the sharing of multicast data, therefore the specific period is the playback duration of an entire video [2]. Hereafter, "Request R is within the patching window of the latest R-multicast" means that the skew between a multicast for request R and the latest R-multicast is shorter than or equal to the size of the patching window. Fig. 1 describes greedy patching and grace patching used in server station to determine the video data that should be delivered on a new channel for a new request.

P2Cast is an architecture based on a peer-to-peer approach and use patching to transmit video streams [7]. Unlike patching, P2Cast makes clients buffer the video data to play a server's role, and the server forms an application-level multicast tree over the unicast-only network every session. P2Cast usually outperforms multicast-based patching in terms of defection rate when clients can cache more than 10% of the leading data of a video and the server's workload is high, but every client station in the same session is required to have a failure-tolerant capability.

3 Converting Patching Agent

Let us assume that requests R_1, R_2, R_3, R_4 and R_5 for the same video *v* arrive at a video server at 0, 1, 3, 6, and 7 minutes respectively, the length of the video is 10 minutes, and the size of the patching window, *PW,* is 2 minutes. Figs. 2(a) and 2(b) then show the video streams that the server delivers on channels using greedy patching and grace patching respectively. $skew_{i,j}$ indicates the skew between new request R_j and request R_i that is being served with the latest R-multicast. $Cost\_Greedy_{i,j}$ and $Cost\_Grace_{i,j}$ indicate channel-holding time needed to serve the requests from R_i to R_j using greedy patching and grace patching, respectively.

Since R_1 arrives earliest and there is no ongoing R-multicast, the server initiates an R-multicast. Since $skew_{1,2} \leq PW$, i.e. request R_2 is within the patching window, the server initiates a P-multicast for R_2. In more detail, the client of R_2 buffers the R-stream on channel Ch_1 while it receives and plays the P-stream of *v[1]*. Then, 1 minute later, the client of R_2 plays the buffered stream and continues to download the R-stream, so the client can play the video continuously. In Fig. 2(b), R_3 and R_4 are served with new R-multicasts. However, in Fig. 2(a), both R_3 and R_4 arrive while the latest R-multicast is ongoing. Hence, the server initiates P-steams of *v[L-PW]* for R_3 and R_4.

Since the total cost of serving the 5 requests is $[Cost\_Greedy_{1,5}=10+1+8+8+8] > [Cost\_Grace_{1,5}=10+1+10+10+1]$, grace patching is better. The reason is as follows. Because R_4 is served with an R-multicast in grace patching, $Cost\_Grace_{1,4}$ is larger than $Cost\_Greedy_{1,4}$. However, when request R_5 arrives within the patching window of the latest R-stream on channel Ch_4 in grace patching, the server can transmit a P-stream of *v[1]* as much as the skew that is less than *PW*. Consequently, the total cost is cheaper than greedy patching.

Let us suppose that an R-multicast is ongoing for R_4 instead of a P-multicast when the new R_5 is scheduled in Fig. 2(a). Then, since $skew_{4,5} \leq PW$, the server can transmit a P-stream of *v[1]* like grace patching. $Cost\_Greedy_{1,5}$=10+1+8+10+1=30, hence greedy patching become better than grace patching after all. From the above observation, we can find out the fact that if we can convert the ongoing P-multicast into an ongoing R-multicast when R_5 is scheduled, greedy patching gets better performance. It is worth noting that the server does not finish transmitting the latest P-stream for R_4 when the new request R_5 is scheduled. If the server modifies and extends the end point of the P-stream to the end of the video, the P-stream is converted into an R-stream on the current channel. Hence, this paper proposes an efficient multicast technique called converting patching agent that is based on greedy patching, changes the latest P-multicast into an R-multicast only if the converting is beneficial and can get better performance than those of greedy patching and grace patching.

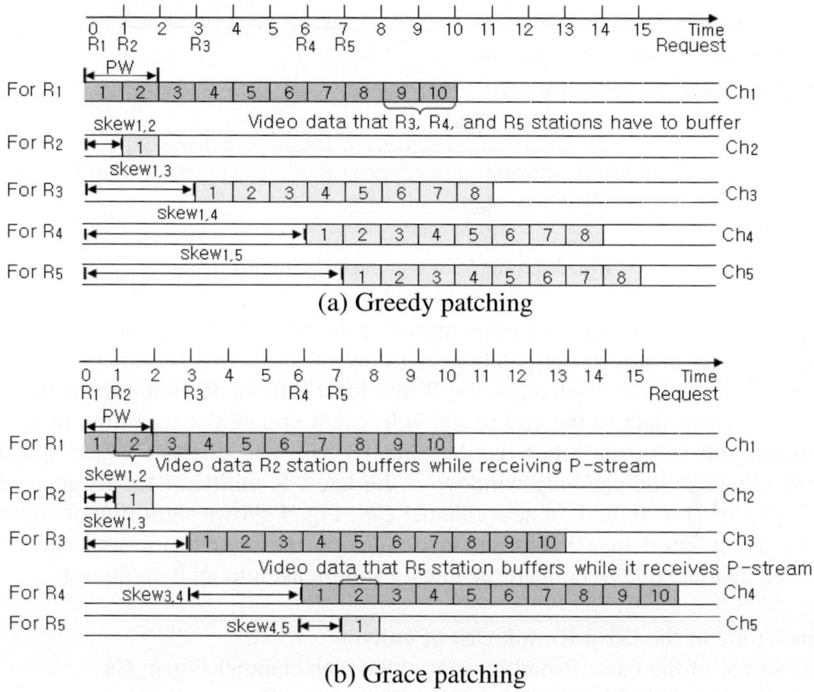

Fig. 2. Examples of greedy patching and grace patching

From Fig. 2, it is clear that converting the latest P-multicast into an R-multicast for a new request is beneficial when the latest P-stream and the new request satisfy two conditions as follows. Suppose that t_r is the start time of the latest R-multicast, t_p is the start time of the latest P-multicast, and t is the current time of scheduling a new request. First, the skew between the latest R-multicast and the latest P-stream is greater than *PW*, that is, the P-multicast is transmitting $v[L\text{-}Min(PW, L\text{-}(t_p\text{-}t_r))]$. If such a P-multicast exits, the additional data that the server should transmit to convert the P-multicast into an R-multicast is only data during the last $Min(PW, L\text{-}(t_p\text{-}t_r))$ minutes. Second, the skew between the new request and the latest P-multicast satisfying the first condition must be shorter than or equal to *PW*. Then, the skew between the new request and the new latest R-multicast that has just been converted from the latest P-multicast become shorter than *PW* and the server can transmit only *v[skew]* for the new request.

In the first condition, the additional transmission cost caused by changing the P-multicast into an R-multicast is $Min(PW, L\text{-}(t_p\text{-}t_r))$, or *PW* minutes at maximum. In the second condition, the reduced transmission cost caused by the new request is $(L\text{-}2{\cdot}PW)$ at minimum. If we consider that the length of a general video is 90 minutes and PW is 5 minutes, then the holding time of the latest patching channel increase by 5 minutes at most and holding time of a new channel decrease by 80 minutes at least. Fig. 3 illustrates converting patching under the same conditions as in Fig. 2(a). The gray-colored area of each channel indicates the video data that has been transmitted when R_5 is scheduled.

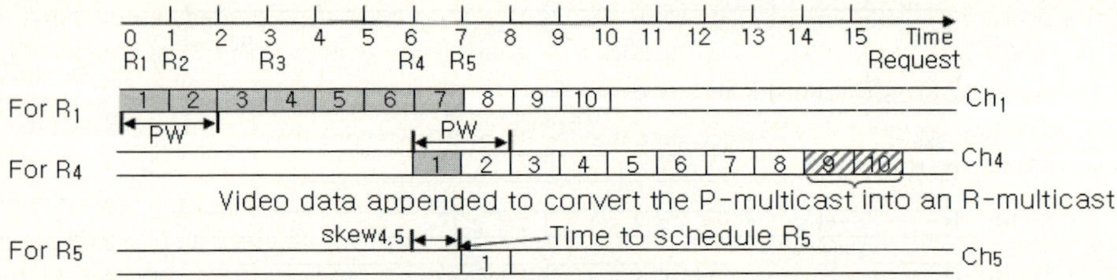

Fig. 3. Example of converting patching agent

Request R_5 is out of the patching window of the latest R-multicast on channel Ch_1. Nevertheless, R_5 is within the patching window of the latest P-multicast on channel Ch_4. Therefore, the server changes the P-multicast into an R-multicast so that it can transmit the video data to the end of the video, and update the start time of the latest R-multicast and id of the latest regular channel to 6 and channel Ch_4, respectively. Now, R_5 is within the patching window of the latest R-multicast for R_4 and only the data of *v[1]* are transmitted on new channel Ch_5. Fig. 4 shows algorithm of converting patching agent, used in a video server, to decide the data to transmit on a newly dispatched channel and to convert the latest P-multicast into an R-multicast.

t_r : Start time of the latest R-multicast of video *v*
t_p: Start time of the latest P-multicast of video *v* on channel *Patch_Ch*
Latest_Regular_Ch: Id of the latest R-multicast of video *v*

ConvertingPatchingAgent()

1. if ($(t-t_r) \leq PW$) $D = v[t-t_r]$;
2. else if ($(t-t_p) \leq PW$ & $(t_p-t_r) > PW$)
   Convert the latest P-multicast into an R-multicast to make sure that channel *Patch_Ch* is used to transmit the video data to the end of the video.
   $t_r = t_p$; $Latest\_Regular\_Ch = Patch\_Ch$; $D = v[t - t_r]$;
3. else $D = v[L - Min(PW, L - (t - t_r))]$

Fig. 4. Converting patching agent for the VOD server

4 Simulation and Performance Evaluation

The performance of converting patching agent is evaluated by simulation and compared with that of grace patching that is better than greedy patching [2]. The

Table 1. Simulation Parameters

Parameter	Values
Number of videos (Length of video)	100 (90 minutes)
Bandwidth of VOD server	1000, 1200, 1400 channels
Size of patching window	5 minutes
Number of requests	100,000
Arrival rate of requests (requests/minute)	30~90 (Poisson distribution)
Defection time	1~5 minutes (Random distribution)

criteria are defection rate and average service latency according to the arrival rates of the video requests, which are generally used to evaluate VOD systems. The parameters used in the simulation are based on those in [2, 3], and are listed in table 1. We assume that popularity of a video follows Zipf-like distribution, and a client watches a video sequentially until it finishes [3, 8, 9].

Fig. 5(a) shows the defection rates of clients with two patching techniques according to the arrival rates of requests. The defection rate is the percentage of clients who canceled requests because the waiting time for services exceeded the client's tolerance [2]. Excluding the cases where the defection rates of both techniques are all zero, converting patching agent improves by 12.6%, 22.6% and 37.5% on average when the server has 1000, 1200, and 1400 channels for video streams, respectively.

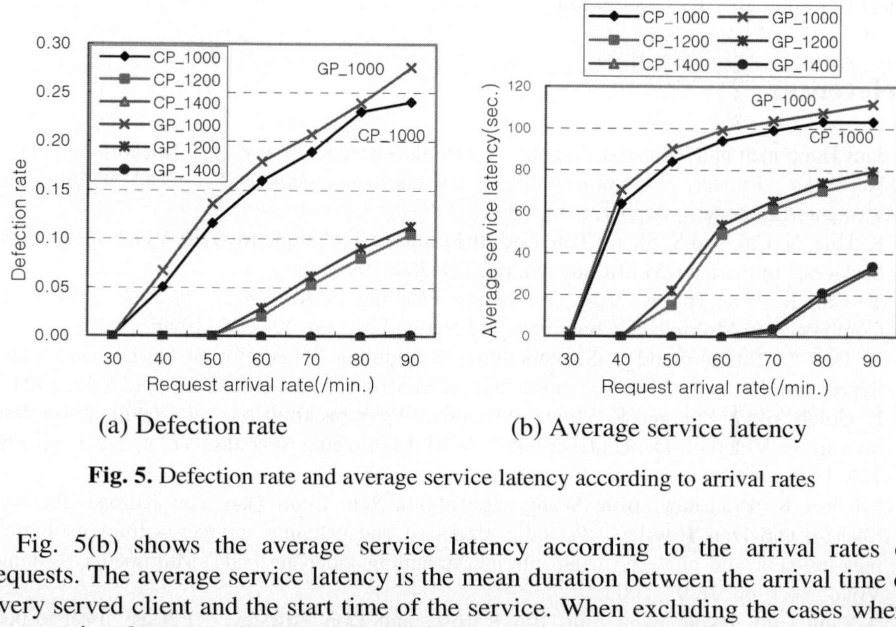

(a) Defection rate (b) Average service latency

Fig. 5. Defection rate and average service latency according to arrival rates

Fig. 5(b) shows the average service latency according to the arrival rates of requests. The average service latency is the mean duration between the arrival time of every served client and the start time of the service. When excluding the cases where the service latency of both techniques are all zero, converting patching agent improves the latency by 18.4%, 11.2%, and 18.5% on average when the server has 1000, 1200, and 1400 channels for video streams, respectively. These improvements result from the following facts. Since converting patching agent is based on greedy patching, it can maximize the sharing of the existing latest R-stream. Moreover, it can detect the case that initiating new R-multicast is more beneficial than persisting in P-multicast to patch the existing R-multicast. Proposed converting patching agent can serve more clients with shorter service latency than grace patching. It means that a VOD server's throughput increases and clients can be served more quickly even the server's workload is high. Hence, we can say that converting patching agent is better than grace patching.

5 Conclusion

Grace patching and greedy patching have been proposed to make efficient use of a video server's limited network-I/O bandwidth by sharing video data stream with multiple clients. In this paper, we proposed converting patching agent that takes advantage of both greedy patching and grace patching. The proposed converting patching agent is based on greedy patching to maximize the sharing of the existing R-multicast. And, it dynamically converts the latest P-multicast into an R-multicast to minimize the cost of a P-multicast for a new request if the new request is within the patching window of the latest P-multicast although it is already out of the patching window of the latest R-multicast. We evaluated the performance of converting patching agent through simulation and compared it with grace patching. The results confirmed that converting patching agent outperformed grace patching in terms of defection rates and service latency.

References

1. Jani Huoponen and Thorsten Wagner, "Video on Demand: A Survey," Telecommunication Networks Project, 1, http://fiddle.visc.vt.edu/courses/ee4984/Pro-jects1996/huoponen _wagner/Euoponen_wagner. html, 1996.
2. K. Hua, Y. Cai, and S. Sheu, "Patching: A Multicast Technique for True Video-on-Demand Services," In Proc. ACM Multimedia, pp. 191-200, 1998.
3. Y. Cai, K. Hua, and K. Vu, "Optimizing Patching Performance," In Proc. SPIE/ACM Conference on Multimedia Computing and Networking, pp. 204-215, 1999.
4. A. Dan, D. Sitaram, and P. Shahabuddin, "Scheduling Polices for an On-Demand Video Server with Batching," In Proc. of the 2nd ACM Multimedia Conference, pp. 25-32, 1994.
5. L. Golubchik, J. Lui, and R. Muntz, "Adaptive Piggy-backing: Arrival Technique for Data Sharing in Video-on-Demand Service," ACM Multimedia Systems, Vol.4, No.3, pp.140-155, 1996.
6. Michael K. Bradshaw, Bing Wang, Subhabrata Sen, Lixin Gao, Jim Kurose, Prashant Shenoy, and Don Towsley, "Periodic broadcast and patching services – implementation, measurement and analysis in an internet streaming video testbed," Multimedia Systems, Vol.4, No.1, pp. 78-93, 2003.
7. A.Yang Guo, Hyoungwon Suh, Jim Kurose, and Don Towsley, "`P2Cast: Peer-to-peer Patching Scheme VoD Service," In Proceedings of the 12th International Conference of World Wide Web, pp. 301-309, 2003.
8. A.Chervenak, D. Patterson, and R. Katz, "Choosing the Best Storage System for Video Service," In Proc. of ACM Multimedia 95, pp. 109-119, Aug. 1995.
9. Christer Samuelesson, "Relating Turing's Formula and Zipf's Law," Proceedings of the Fourth Workshop on Very Large Corpora, 1996.

A Self-organizing Cluster Agent for Routing Algorithm in Wireless Sensor Networks

Sangjoon Jung[1], Younky Chung[1], and Chonggun Kim[2,*]

[1] School of Computer Engineering, Kyungil University
712-701, 33 Buho-ri, Hayang-up, Gyeongsan-si, Gyeongsang buk-do, Korea
{sjjung, ykchung}@kiu.ac.kr
http://computer.kiu.ac.kr

[2] Dept. of Computer Engineering, Yeungnam University
712-749, 214-1, Dae-dong, Gyeongsan-si, Gyeongsangbuk-do, Korea
cgkim@yu.ac.kr
http://nety.yu.ac.kr

Abstract. Developing sensor networks enable us to gather information about specified regions or tasks. Sensing nodes have several characteristics to have constrained energy and limited capacity. So, minimizing energy consumption and maximizing system lifetime have been a major issue for wireless sensor networks. What is unique about our proposed agent approach is that the agent has a learning capability using cluster configuration technique while the fitness criterion is based on energy consumption, election of cluster heads, and the number of clusters. This paper proposes an energy-efficient routing algorithm based on organizing clusters in sensing area by electing agents called cluster heads. This strategy allows an agent as cluster head to keep a routing table about intermediary nodes that can communicate between different clusters. The agent gathers routing information to establish routing paths by intermediary nodes and to decide an appropriate path by the routing table. When a sink node diffuses an interest, an agent finds neighbor clusters to reach to the sink node and transmits a response to intermediary nodes. The intermediary node then transmits the response to the agent. Thus, the agent decides to report to the sink node. This mechanism gives sensor nodes lower consumption of energy and prolonging system lifetime.

1 Introduction

A sensor network is composed of a large number of tiny devices, scattered and deployed in a specified regions. Each sensing device has processing and wireless communication capabilities, which enable it to gather information from the sensing area and to transfer report messages to a base station[1][2]. Each node has several characteristics such as low-power, constrained energy, and limited capacity. To report tasks to the base station, energy-efficient routing algorithm is an important consideration for these networks[3-8].

* Corresponding author.

N.T. Nguyen et al. (Eds.): KES-AMSTA 2007, LNAI 4496, pp. 711–720, 2007.

Current routing algorithms have been proposed to concern both minimizing the total transmit energy, and maximizing the network lifetime. Directed diffusion[3] is a data-centric protocol, i.e. nodes are not identified by IP addresses but by generated data. In data-centric algorithms, the sink sends queries to certain regions and waits for data from the sensors located in the selected regions. Since data is being requested through queries, attribute-based naming which is defined an interest is necessary to specify the properties of data[3]. Network clustering has been proposed to improve the system without additional load and degrading the service. Because, a single-tier network causes a gateway to overload with too much data aggregation[9-15]. In the hierarchical routing protocol, the network is usually divided into clusters where each cluster will have one elected node to be the cluster head, and the remaining nodes will carry out the sensing[9]. LEACH(Low-Energy Adaptive Clustering Hierarchy)[10] proposes a clustering based protocol that utilized randomized rotation of local cluster heads to evenly distribute the energy load among the sensors in the networks.

In this paper, we assume that all sensor nodes have fixed transmission energy and range within a region, allowing them to act routers for other nodes' data in addition to sensing the environment. Our solution is to use an agent-based approach, where the agent has the ability to communicate with all the sensors in the environment, assign the heads of all nodes, learn about the status of all sensors, analyze the decision, and acts autonomously. Some nodes that have opportunity to be elected cluster heads acting as agents broadcast an advertisement message to the rest of the nodes. The non-cluster-head that receives more than two advertisement messages is elected to be an intermediary node. To receive different kinds of messages, the intermediary nodes then report to the two agents that are belong to different clusters. The cluster-head nodes keep information about members that consist of intermediary nodes and cluster-member nodes. In case of reporting tasks to a sink node, the agent that receives data from sensors transmits to the other clusters by way of intermediary nodes. If the neighbor cluster contains the sink node, all sensed data can be reported. Otherwise, the next cluster searches the rest of clusters which can communicate with the sink node. This mechanism gives us to apply to general sensing environments and to outperform classical clustering algorithms by using intermediary nodes. Our mechanism achieves some reduction in the energy dissipation, as all sensors have constrained energy and fixed transmission distance.

2 Related Works for WSNs

In this section, we provide an overview of related routing protocols for sensor networks such as the Directed diffusion, LEACH, PEGASIS and so on.

Routing in sensor networks is based on the data contained in sensor nodes rather than node identification. Directed diffusion suggests the use of attribute-value pairs for the data and queries the sensors in an on demand basis by using those pairs[1,2,3]. In order to demand for sensing tasks, an interest is defined using a list of attribute-value pairs such as name objects, interval, duration, geographical area, etc. The interest is broadcasted by a sink node which is denoted to inject the named task into the network. Each node receiving the interest can do caching to compare the received data with the values in the interests. When a source node has data that match the interest, the data will be reported to the sink node using this interest gradient that was established. A gradient is a reply link to a neighbor from which the interest was received. The sink node will reinforce the shortest path by sending a reinforcement packet with a higher data rate to the neighbor node which forwards it to all the nodes in the shortest path[3]. Since all communication is neighbor-to-neighbor with no need global addressing schemes, each node does not maintain global network topology. However, the applications that require continuous data delivery in continuous monitoring situations to the sink will not work efficiently with query-driven on demand data model. In addition, the naming schemes used in Directed diffusion are application dependent and the matching process for sensing tasks and queries might require some extra overhead at the sensors. Moreover, there is initial and periodic interest and low rate data flooding without geographic routing support.

Low-energy adaptive clustering hierarchy(LEACH)[5] is one of the most popular hierarchical routing algorithms for sensor networks. The idea is to form clusters of the sensor nodes based on the elected cluster heads which are used as routers to the sink. All data processing and aggregation is performed in local cluster heads. Each node transmits directly to the cluster heads and the sink by using single-hop routing. The LEACH includes randomized rotation of the high-energy cluster-head position such that it rotates among the various sensors in order to not drain the battery of a single sensor. The decision to become a cluster head depends on the amount of energy left at the node. This scheme saves energy since the transmissions are done by such cluster heads rather than all nodes. However, it is not applicable to networks deployed in large regions since LEACH uses single-hop routing where each node can transmit directly to the cluster head and the sink. They assume adjustable transmitting power and assume that every cluster head can communicate directly to the sink node. This is different from ours. We assume that each node has fixed transmission power and transmit data by way of cluster heads in a multi-hop network. Furthermore, the idea of dynamic clustering brings extra overhead, for example, head changes, advertisements etc., which may diminish the gain in energy consumption.

PEGASIS(Power-efficient Gathering in Sensor Information Systems)[6] is an improvement of the LEACH protocol by eliminating the overhead of dynamic cluster formation, minimizing the distance non leader-nodes must transmit, limiting the number of transmissions and receives among all nodes, and using only on transmission to the base station. In PEGASIS, a chain among the sensor nodes so that each node receives from and transmits to a close neighbor is formed. Gathered data modes from node to node, aggregated and eventually sent to the sink. Performance of PEGASIS is achieved through the elimination of the overhead caused by dynamic cluster formation as LEACH and through decreasing the number of transmissions and reception by using the chain for aggregating data. However, PEGASIS would require the nodes to consume some extra energy like as excessive delay when delivering data for distant node on the chain. In addition, each node consumes more energy to elects an appropriate neighbor node for construction of the chain, it allows to contain link information about the total topology. The chain in the algorithm to construct routing path is performed in a greedy way, which algorithm needs that all nodes have global knowledge of the network. The applications that require continuous monitoring situations to the sink will not work efficiently because the chain must be configured before delivering data. Another problem is that the single leader can become a bottleneck.

Subramanian and Katz[12] proposed a self-organizing algorithm that develops an addressing, routing and broadcasting infrastructure in the backbone of the network. They described a generic architecture for building a special class of sensor application called self-configurable systems where a large number of sensors coordinate amongst themselves to achieve a large sensing task. The paper listed the general architectural and infra-structural components necessary for building this class of sensor applications. And also, their algorithm for self-organizing a large number of sensor and build a routing infrastructure is described. They mentioned that the size of routing table maintained at every node is reduced. The algorithm mainly targets power constraints and attempts to minimize the power consumed at various stages of the algorithm. The paths and the tree structures are made fault tolerant by constantly making failing node as leaf nodes. However, this paper has several weaknesses. They introduced a large number of router sensors whose only job is to perform data dissemination and interconnect the specialized sensors into a network. Now a day, it is not appropriate classification of sensor nodes, because all nodes have ability to route and to gather data. Moreover, router nodes maintain information about the total number of nodes and some neighboring nodes connected to the router nodes in the network. And the other problem is to introduce extra overhead in the organization phase of algorithm since it is not on-demand.

In our work, we focus on intelligent agent approach with the learning capability using organizing clusters. This provides the agent with the ability to determine the best cluster formation that gives minimum energy consumption during configuring clusters and reporting tasks. The agent achieves its objectives by analyzing the network and applying its learning capability.

3 Routing Algorithm by Self-organizing Clusters

The main goal of using a clustering method by agent is to efficiently control the energy consumption and to configure appropriate clusters for sensor networks by involving them in multi-hop communication between cluster heads. So, we design that agents report to sink node by clustering among neighbors. To form clusters and report data, the agents must maintain the entries in their clusters. In this section, we show how to elect agents, how to configure clusters, and how to maintain members in the clusters.

3.1 Setup Phase

Initially, when clusters are created, each node decides to be an agent as cluster heads. This decision is made by the node n choosing a random number between 0 and 1. With 20%of the nodes to be agents, the agents that are elected randomly broadcast a CHA(Cluster-Head-Advertisement) message to the rest of the nodes. 20%(7/36*100=19.44%) is decided when 7 nodes are selected from 36 nodes under the ideal location. So, agents in our algorithm are elected by 20%.

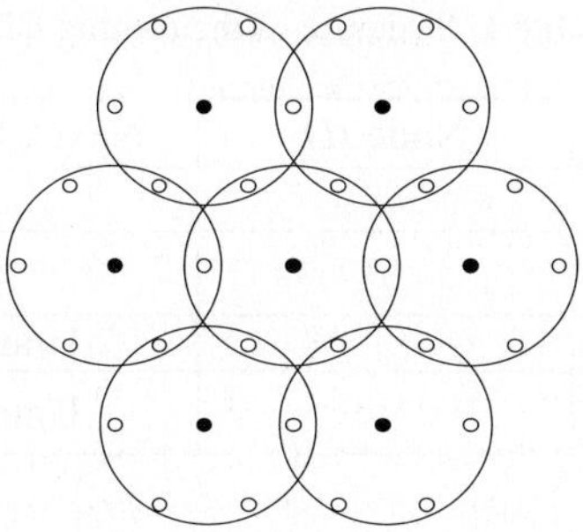

Fig. 1. Ideal location of cluster heads

All agents transmit their CHA(Cluster-Head-Advertisement) messages using the same transmit energy. There are two kinds of nodes received the advertisement messages from the agents. One group is *cluster members*(CM) received only one CHA message from one agent. Another group is *intermediary nodes(IN)* received more than two advertisement messages from agents. All sensor nodes must cache the CHA messages as soon as receiving. Fig.2 shows the procedure of broadcast from agents, and report from cluster members.

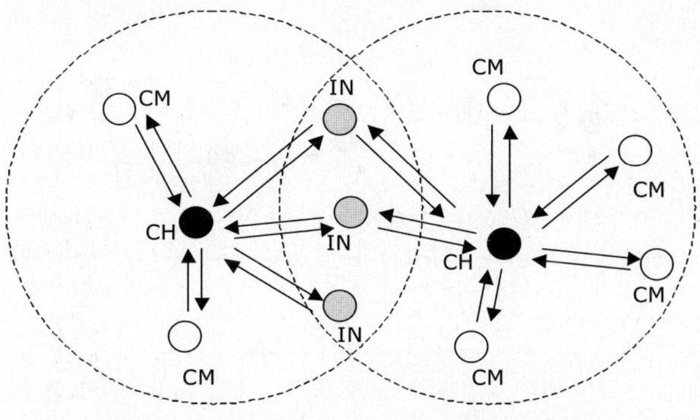

Fig. 2. Procedure of broadcast and report

All intermediary nodes do not respond to the agents because some nodes have a short time expired before reporting. The intermediary nodes that have received two advertisement messages from the different agents report the ACK message with neighbor cluster identifier configured by the cluster head *ID*. Each cluster recognizes which clusters are adjacent. So, each cluster can be connected by intermediary node.

3.2 Organization Phase

Each agent configures a routing table that contains intermediary nodes and cluster members. The node identifies the next cluster since two clusters can communicate between them by way of the intermediary node. Table 1 shows the structure of routing table in the agent.

Table 1. Structure of the routing table

Node type	Node *ID*	Next Cluster Identifier
GW		
⋮	⋮	⋮
CM		Unused (All 0s)
⋮	⋮	Unused (All 0s)

When a sink node diffuses an interest which configures attribute-value pairs for the data and queries, each node saves in the own interest cache. A cluster identifier is contained in the interest, allowing it to be changed in the other clusters. Then, which diffusing an interest, the agent contains the previous cluster identifier. Therefore, each node finds a target cluster since the agent identifies which cluster is adjacent from own cluster. Fig. 3 shows cluster organization configured by routing tables.

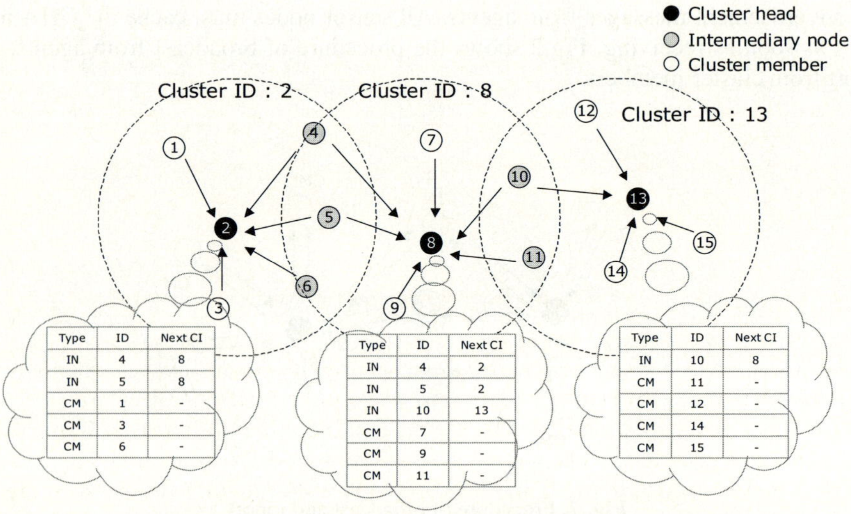

Fig. 3. Cluster organization configured by routing tables

In Fig. 3, node 2 contains the routing table which has two intermediary nodes and three cluster members. If node 2 reports sensing tasks, the node transfers the response to the intermediary node 4 or 5. One of the intermediary nodes among them then transmits the messages to the other agent which is node 8. This procedure is executed repeatedly until the report reaches to the sink node.

3.3 Maintenance Phase

There are two types of maintenance that one can perform in a self-organizing system. They are active and passive monitoring. In the maintenance phase, it would be necessary to keep routing entries in the routing table. In active monitoring, each intermediary node sends an *I am alive* message to the agents. The agents then updates its routing entries. In passive monitoring, an agent checks whether a cluster member is alive only on demand. Passive monitoring is used as a mechanism for saving the energy of a particular node.

3.4 Reorganization Phase

Reorganization happens when a group of nodes are divided from connectives of all groups or when the energy of the agent is reduced to a certain threshold. Each node received the CHA messages saves the cluster *ID*s into own cache, allowing it to know where it is included in groups. If an agent does not have an intermediary node, it should be an isolated group like as left figure in Fig 4.

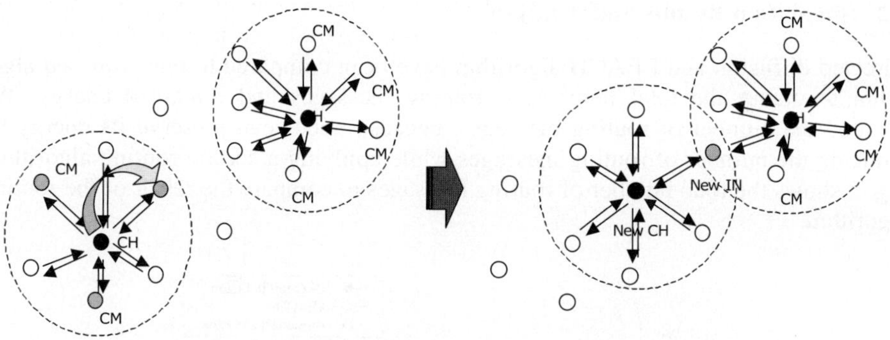

Fig. 4. Procedure of reorganization

The agent sends *reorganization* message to neighbor nodes. The neighbor nodes received the message broadcasts a CHA message to configure another group to connect the other groups. Some node has maintained the other cluster-head information, it can reply to a new agent. Then, the node could be elected to an intermediary node to connect between two groups. If intermediary nodes do not find when executing the first trial, then the procedures are executed until discovering intermediary nodes.

4 Performance Evaluations

To evaluate the performance of the proposed algorithm, we use NS(Network simulator)-2 to compare with Directed diffusion and LEACH algorithm.

4.1 Simulation Environment

We placed from 50 to 400 nodes in 1000m*1000m and executed 10 times respectively, because all node located randomly. To measure the remained energy, a sensor node include the following parameters in the sensor node's node-config routine:

```
-energyModel EnergyModel \
-rxPower 0.175 \
-txPower 0.175 \
-sensePower 0.00000175 \
-idelPower 0.0 \
-initialEnergy 5
```

where

- `rxPower` 0.175 indicates 175*mW* consumed for receiving a packet,
- `txPower` 0.175 indicates 175*mW* consumed for transmitting a packet,
- sensePower 0.00000175 indicates 0.175μW consumed for matching the attributes to the tasks, and
- `initialEnergy` 5 indicates a total energy reserve of 5*J*.

4.2 Simulation Results and Analysis

Directed diffusion and LEACH algorithm have been compared to our proposed algorithm to obtain the total number of routing messages and remained energy. We counted the number of routing messages, because nodes can preserve its energy by reducing the number of routing messages while applying a certain routing algorithm. Fig. 5 shows the total number of routing messages to compare the result of the routing algorithm.

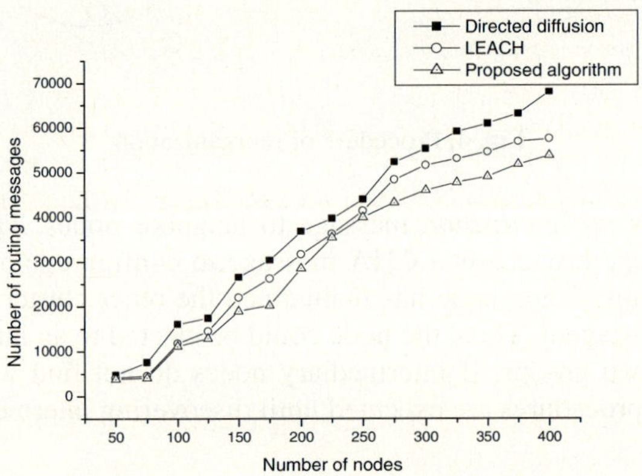

Fig. 5. Total number of routing messages

In Fig. 5, we notice that the number of routing messages of the proposed algorithm is smaller than that of the Directed diffusion and LEACH. Therefore, our algorithm reduces consumption of energy by reducing the routing messages. Fig. 6 shows results about remained energy over three algorithms.

In Fig. 6, all nodes using our algorithm save more energy in compared to the Directed diffusion and LEACH. As all nodes in our simulation located randomly, so differences of three algorithms are not constant. It is found that our algorithm can achieve lower dissipation of energy, better preserving dissipated network energy and effectively postpone network lifetime.

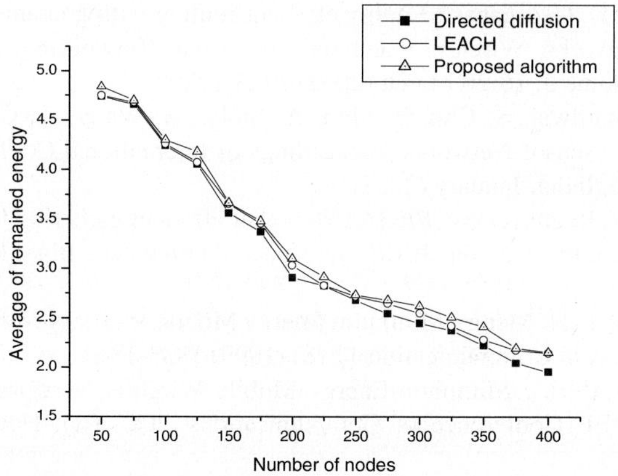

Fig. 6. Average of remained energy

5 Conclusions

In this paper, we have proposed that all sensors have fixed transmission energy and range within a region, allowing them to act routers for other nodes' data. Each node that has been elected to be an agent broadcasts an advertisement message to the rest of the nodes. The non-cluster-heads that receive more than two advertisement messages are elected to be intermediary nodes. The cluster-head nodes keep information about intermediary nodes and cluster-member nodes. In case of reporting tasks to a sink node, the agent that receives data from sensors transmits to the other clusters by way of intermediary nodes. This mechanism gives us to apply to general sensing environments while intermediary nodes contacts with the other clusters. The sensor nodes in the proposed scheme can be applied to various environments where nodes report to the sink node without query, such as parking-lot networks and continuous monitoring. Our mechanism achieves some reduction in the energy dissipation, as all sensors have constrained energy and fixed transmission distance. By using the proposed algorithm, each node has routing information to find a sink node while restricting the number of routing messages and prolonging the network lifetime. Our future plan is to implement the algorithm in a practical sensor node.

References

[1] K. Akkaya and M. Younis : A survey on routing protocols for wireless sensor networks, Ad Hoc Networks, Volume 3, Issue 3, May (2005) 325-349

[2] I.F. Akyildiz, W. Su, Y. Sankarasubramaniam, and E. Cayirci : Wireless Sensor Networks: A Survey. Computer Networks, 38(4), March (2002) 393-422

[3] C. Intanagonwiwat, R. Govindan, and D. Estrin : Directed diffusion: a scalable and robust communication paradigm for sensor networks, IEEE/ACM Mobicom, (2000) 56-67

[4] D. Tian, and N.D. Georganas : Energy efficient routing with guaranteed delivery in wireless sensor networks, Wireless Communications and Networking, 2003. WCNC 2003. 2003 IEEE Volume 3, 16-20 March (2003) 1923-1929

[5] R. Min, M. Bhardwaj, S. Cho, E. Shih, A. Sinha, A. Wang, A. Chandrakasan : Low-Power Wireless Sensor Networks, Proceedings of International Conference on VLSI Design, Bangalore, India, January (2001)

[6] S. Lindsey, C.S. Raghavendra, PEGASIS: power efficient gathering in sensor information systems : Proceedings of the IEEE Aerospace Conference, Big Sky, Montana, March (2002)

[7] V. Rodoplu and T. H. Meng : Minimum Energy Mobile Wireless Networks, IEEE Journal of Selected Areas in Communication 17 (8) (1999) 1333-1344

[8] L. Li and J. Halpern : Minimum-Energy Mobile Wireless Networks Revisited, in Proceedings of IEEE Conference of Communications (ICC '01), Helsinki, Finland, June (2001)

[9] K. Matrouk, B. Landfeldt : Energy-conservation clustering protocol based on heat conductivity for wireless sensor networks, Intelligent Sensors, Sensor Networks and Information Processing Conference 2004, Proceedings of the 2004, 14-17 Dec. (2004) 19 - 24

[10] W. Heinzelman, A. Chandrakasan, H. Balakrishnan : Energy-efficient communication protocol for wireless sensor networks, Proceeding of the Hawaii International Conference System Sciences, Hawaii, January (2000) 3005-3014

[11] Y. Yu, D. Estrin, R. Govindan : Geographical and energy-aware routing: a recursive data dissemination protocol for wireless sensor networks, UCLA Computer Science Department Technical Report, UCLA-CSD TR-01-0023, May (2001)

[12] L. Subramanian, R.H. Katz : An architecture for building self configurable systems, Proceedings of IEEE/ACM Workshop on Mobile Ad Hoc Networking and Computing, Boston, MA, August (2000)

[13] J. Liu and C. Richard Lin : Energy-efficiency clustering protocol in wireless sensor networks, Ad Hoc Networks, Volume 3, Issue 3, May (2005) 371-388

[14] K. Sohrabi, J. Gao, V. Ailawadhi, G. J. Pottie : Protocols for Self-Organization of a Wireless Sensor Network, IEEE Personal Communications 7(5) (2000) 16-27

[15] R. Krishnan and D. Starobinski : Efficient clustering algorithms for self-organizing wireless sensor networks, Ad Hoc Networks, In Press, Corrected Proof, Available online 22 March (2005)

Design of Intelligence Mobile Agent Lifecycle for the Progress Management and Mobile Computing

Haeng-kon Kim[1], Sun-Myoung Hwang[2], and Eun-Ser Lee[3]

[1] Catholic University of Daegu , 330, Hayangup, Kyungsan, Kyungbuk, Korea
hangkon@cu.ac.kr
[2] Daejeon University, 96-3 Yongun-dong, Tong-gu, Taejon 300-716, South Korea
sunhwang@dju.ac.kr
[3] Information & Media Technology Institute, Soongsil University
eslee1@ssu.ac.kr

Abstract. This paper is researched to design agent for the progress management. And this paper is intended to help the design agent and progress management based on the project management. Also, there are many defects that cause the domain problems during the project development. This paper remove the risk of the lifecycle and progress management that check the risk items in the process and also manage the progress and quality problems. Therefore, we are use of the incremental method for the remove of risk items. The analysis is made based on the types required for making use of the defect data. In this case, additional measures must be taken besides merely recording defects. And the theory is apply to the agent web security access control system for the security requirement. This paper provides an design of intelligence agent for the progress management. And the process can also be managed by using lifecycle and progress milestone, which can greatly improve the software process.

1 Introduction

We have attempted to define artificial intelligence through discussion of its major area of research and application. This topic reveals a young and promising field of study whose primary concern is finding an effective way to understand and apply intelligent problem solving, planning, and communication skills to a wide range of practical problems. In spite of the variety of problems addressed in artificial intelligence research, a number of important features emerge that seem common to all divisions of the field. And we are needed to the management for the progress and improvement[15].

When we provide a service or create a product, whether it be developing software, writing a report, or taking a business trip, we always follow a sequence of steps to accomplish a set of task. The tasks are usually performed in the same order each time: for example, you do not usually put up the drywell before the wiring for a house is installed or bake a cake before all the ingredients are mixed together. We can think of a set of ordered tasks as a process: a series of steps involving activities, constraints, and resources that produce an intended output of some kind[8][13].

N.T. Nguyen et al. (Eds.): KES-AMSTA 2007, LNAI 4496, pp. 721–727, 2007.

When the process involves the building of some product, we sometimes refer to the process as a life cycle. Thus, the software development process is sometimes called the software life cycle, because it describes the life of a software product from its conception to its implementation, delivery, use, and maintenance.

This paper provides an design of intelligence agent for the progress management. And the process can also be managed by using lifecycle and progress milestone, which can greatly improve the software process.

2 Related Works

2.1 System Life Span

As software engineers trying to build a maintainable product, the first question we must ask ourselves is whether it possible to build a system right the first time. In other words, if we use highly cohesive components with low coupling, if the documentation is complete and up to date, and if the entire system is cross-referenced, will we need a maintenance phase? Unfortunately, the answer is yes. The reason lie in the nature of the systems themselves. As we have seen, there is no way to guarantee that P-systems and E-systems will not require change. In fact, we must assume that they will change and then build them so that they can be changed easily[16].

2.2 Mobile Agent System

So far, we have approached the problem of building intelligent machines from the view point of mathematics, with the implicit belief of logical reasoning as paradigmatic of intelligence itself, as well as with a commitment of "objective" foundations for logical reasoning[15].

A mobile agent system is a platform that can create, interpret, execute, transfer and terminate mobile agents. A mobile agent system consists of a mobile agent and a

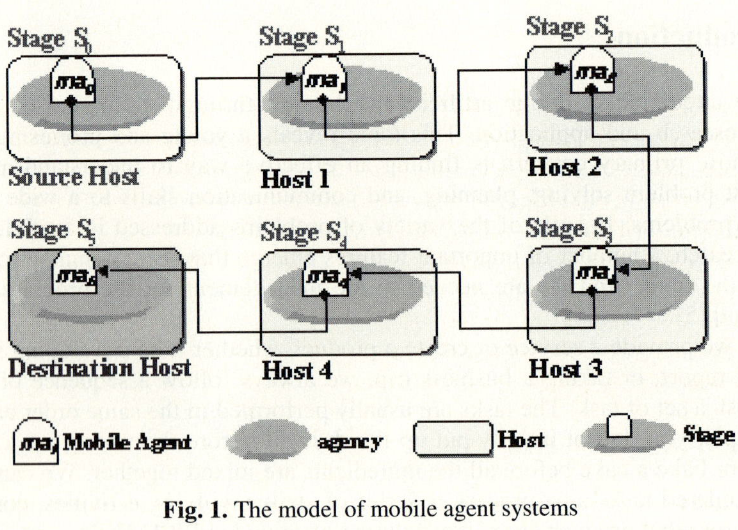

Fig. 1. The model of mobile agent systems

place which is an logical execution environment of mobile agents. A mobile agent is a computer program that acts autonomously on behalf of a user and travels through a network of heterogeneous hosts. When a mobile agent migrates, it chooses next node according to an itinerary which is a predefined travel plan or dynamically according to execution results. A mobile agent executes tasks on a sequence of hosts. That is, a mobile agent logically executes tasks in a sequence of actions. Each action that a mobile agent has to execute on a host (or place) is called a stage. The places where a mobile agent executes in the first step and the last step are called a source and a destination, respectively. A path from a source to a destination is called a migration path.

2.3 MBASE

The difference between failure and success in developing a software-intensive system can often be traced to the presence or absence of clashes among the models used to define the system's product, process, property, and success characteristics [5].

In each case, property models are invoked to help verify that the project's success models, product models, process models, and property levels or models are acceptably consistent. It has been found advisable to do this especially at two particular "anchor point" life cycle process milestones summarized in Table 2 [6][7]. The first milestone is the Life Cycle Objectives (LCO) milestone, at which management verifies the basis for a business commitment to proceed at least through an architecting stage. This involves verifying that there is at least one system architecture and choice of COTS/reuse components which is shown to be feasible to implement within budget and schedule constraints, to satisfy key stakeholder win conditions, and to generate a viable investment business case. The second milestone is the Life Cycle Architecture (LCA) milestone, at which management verifies the basis for a sound commitment to product development (a particular system architecture with specific COTS and reuse commitments which is shown to be feasible with respect to budget, schedule, requirements, operations concept and business case; identification and commitment of all key life-cycle stakeholders; and elimination of all critical risk items) [8].

3 Theory

Mobile agents were evaluated using mobile computing and performance, by way of the web environment. And Evaluation progress was introduced in chapter 2. It should be noted that when evaluation is conducted mobile agent focus on the web. However, the downside of this in terms of process management is that defects with often appear in the next stage. Therefore one should refer to the lifecycle to in order to extract and identify the risk factors of the items. Another problem is that risk factor extraction is a redundant processing job. Furthermore, extraction processing is ambiguous and so firm progress management is required.

This chapter provides identification of risk items and checks the rationale and measurement of total progress management.

3.1 Mobile Agent Lifecycle

The lifecycle is made by protecting the profiles in order to fulfill the mobile agent lifecycle requirement. The next step is the needed items of the mobile agent function.

1. Mobile agent can move through a heterogeneous network
2. Mobile agent can migrate from host to host
3. Mobile agent can interact with other agents
4. Mobile agent can return to its home site when its task is done
5. Mobile agent describe a mobile agent system

Additional we are should check the lifecycle for mobile agent and removal of risk item. Therefore, Lifecycle and identification is as follows:[9][10][11][12]

1. A protection profile developer must be provided to explain the profile of the TOE.
2. Explanation of the TOE must be provided so as to describe its product type and the character of the general IT.
3. The evaluator must have confirmation that the information has satisfied all of the evidence requirements.
4. The evaluator must have confirmation of the explanation of the TOE with regards to quality and consistency.
5. The evaluator must have confirmation of the explanation of the TOE with regards to the relationship of protection profile consistency

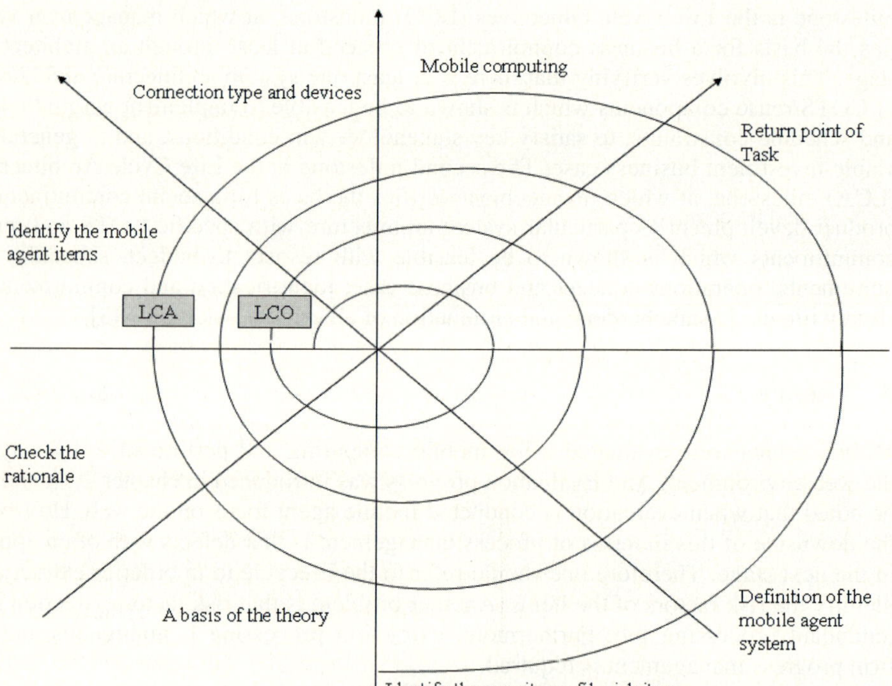

Fig. 2. Mobile agent of lifecycle and milestone

We are describe to the lifecycle of mobile agent view point of progress management and mobile computing. And the next figure is Mobile agent of lifecycle and milestone.

This figure shows the activity of the risk item as it is removed from the security requirement during the extraction process. Each of the factors use the repeatable milestone for progress management. Each stage is as follows:

Table 1. Description of the check point of the mobile agent

Stage	Contents
Identify the mobile agent items	• Identification and analysis the mobile agent items for the extraction of function and risk items
Connection type and devices	• Partially connected devices - mobile computer ▪ laptops & PDA ▪ home computers : occasionally connected to the n/w over SLIP or PPP modem connection • All of these devices - Frequently disconnected from the n/w - Often have low-bandwidth - Unreliable connections into the n/w - Often change their n/w address with each reconnection
Mobile computing	• Agent navigation & adaptation - The world of an agent is dynamic & uncertain ▪ The agent is launched into the world - The sensors ▪ Allow an agent to determine its external state & mechanism • Network sensing - An integral part of our laptop docking system - Network information enables agents to adapt to changing n/w conditions
Return point of Task	• Return to its home site when its task is done
Definition of the mobile agent system	• Coordinates the activities of all local agents • Accepts new agents that are arriving from other machines • All other services are provided by specialized agents • Dock master, traffic monitor, and navigation agents
Identify the security profile risk item	• Identify the security profile risk item at the domain
Basis of the theory	• Build of the repeatable lifecycle for the milestone(LCO, LCA)
Checking of the rationale	• Check the rationale of the repeatable milestone and risk analysis

Therefore, we are gain the check point in this view of mobile agent and computing. And in this milestone explain is the chapter 3.2.

3.2 Milestone for the Progress Management

Use of the milestones (LCO, LCA) is essential for removal of risk and progress management.

Table 2. Description of the milestone element

Milestone Element	Life Cycle Objectives (LCO)	Life Cycle Architecture (LCA)
Definition of Operational Concept	•Top- level system objectives and scope –System boundary –Environment parameters and assumptions –Evolution parameters •Operational concept –Operations and maintenance scenarios and parameters –Organizational life- cycle responsibilities (stakeholders)	•Elaboration of system objectives and scope of increment •Elaboration of operational concept by increment
System Prototype(s)	•Exercise key usage scenarios •Resolve critical risks	•Exercise range of usage scenarios •Resolve major outstanding risks
Definition of System Requirements	•Top- level functions, interfaces, quality attribute levels, including: –Growth vectors and priorities –Prototypes •Stakeholders' concurrence on essentials	•Elaboration of functions, interfaces, quality attributes, and •prototypes by increment –Identification of TBD's((to- be- determined items) •Stakeholders' concurrence on their priority concerns
Definition of System and Software Architecture	•Top- level definition of at least one feasible architecture •Physical and logical elements and relationships •Choices of COTS and reusable software elements •Identification of infeasible architecture options	•Choice of architecture and elaboration by increment •Physical and logical components, connectors, •configurations, constraints •COTS, reuse choices •Domain- architecture and architectural style choices •Architecture evolution parameters
Definition of Life-Cycle Plan	•Identification of life- cycle stakeholders – Users, customers, developers, maintainers, interoperators, general public, others •Identification of life- cycle process model –Top- level stages, increments •Top- level WWWWWHH* by stage	•Elaboration of WWWWWHH* for Initial Operational Capability (IOC) –Partial elaboration, identification of key TBD's for later increments
Feasibility Rationale	•Assurance of consistency among elements above –via analysis, measurement, prototyping, simulation, etc. –Business case analysis for requirements, feasible architectures	•Assurance of consistency among elements above •All major risks resolved or covered by risk management plan

* WWWWWHH: Why, What, When, Who, Where, How, How Much

We are provides that the repeatable cycle based on milestone element. Also, we are checked the progress by the basis of milestone.

Table 3. Setting of the milestone

Cycle 1 (LCO ↓)	Cycle 2	Cycle 3 (LCA ↓)
Determination of top-level concept of operations	Determination of detailed concept of operations	Elaboration of detailed concept of operations by increment, especially IOC
System scope / boundaries /interfaces; top-level requirements	Top-level HW, SW, human requirements	Determination of requirements, growth vector by increment, especially IOC
Small number of feasible candidate architectures (including major COTS, reuse choices)	Provisional choice of top-level information architecture	Choice of life-cycle architecture. Some components of above TBD(low-risk and/or deferrable)
Top-level life cycle responsibilities(stakeholders), process, model, cost/schedule parameters	Make detailed process strategy, responsibilities, cost / schedule allocation	Thorough WWWWWHH plans for IOC; essentials for later increments
Stakeholder concurrence on top-level analysis supporting win-win satisfaction	More detailed analysis supporting win-win satisfaction	Stakeholder concurrence on thorough analysis supporting win-win satisfaction
Top level rationale, including rejected candidate architectures	More detailed rationale underlying system choices	Elaboration of rationale, including risk resolution results

The authors of this paper have provided the repeatable cycle based on the milestone element. Progress was checked using the basis of the milestone.

4 Conclusion

In this paper a new mobile Agent lifecycle was proposed applicable to the extraction of the progress management and mobile computing. And the using milestone LCO and LCA in the MBASE. Therefore, we are using and applicable lifecycle in the similar project. In future studies implementation of network sensing mobile agent applicable to the extract of web information collection will be provided.

References

[1] ISO/IEC 15408-1,2,3 :1999 Information technology - Security techniques - Evaluation criteria for IT security - Part 1: Introduction and general model
[2] The Report of the President's Commission on Critical Infrastructure Protection CCEB (Common Criteria Editorial Board), Common Criteria for Information Technology Security Evaluation, Version 2.0, May 1998.
[3] DOD (U.S. Department of Defense), Trusted Computer System Evaluation Criteria, DOD5200.28-STD, December 1985. 1.0, December 1992.
[4] [ISO96] ISO/IEC Guide 65—General Requirements for Bodies Operating Product Certification Systems, 1996.
[5] Mark Weiser, "The Computer for the Twenty-First Century," Scientific American, pp. 94-10, September 1991
[6] B. Boehm, Software Risk Management, IEEE-CS Press, 1989.
[7] B. Boehm, A. Egyed, J. Kwan, and R. Madachy, "Developing Multimedia Applications with the WinWin Spiral Model," Proceedings, ESEC/ FSE 97, Springer Verlag, 1997.
[8] B. Boehm and P. Bose, "A Collaborative Spiral Process Model Based on Theory W," Proceedings, ICSP3, IEEE, 1994. 17
[9] Eun-Ser Lee and Sun-Myoung Hwang, "Definition of Security Requirement Items and Its Process to Security and Progress Management", LNCIS 344, August 2006.
[10] Eun-Ser Lee and Sun-Myoung Hwang, "Design Implementation of Web Security Access Control System for Semantic Web Ontoloty", LNCS 3481, May 2005.
[11] Eun-Ser Lee and Malrey Lee, "Development System Security Process of ISO/IEC TR 15504 and Security Considerations for Software Process Improvement", LNCIS 344, August 2006.
[12] Eun-Ser Lee and Sang-Ho Lee, "Design progress management for Security Requirements in Ubiqiiuous computing using COQUALMO", LNCS 3984, May 2006.
[13] Roger S. Pressman, "Software Engineering", Mcgraw-hill international edition, 1997.
[14] Nam-deok and Cho,Eun-Ser Lee, "Design and Implementation of Semantic Web Search System using Ontology and Anchor Text", LNCS 3984, May 2006.
[15] George F Luger, "Artificial intelligence", Addison wesley, 2001.
[16] Shari lawrence, "Software engineering", Prentice Hall, 2001.

An Agent Environment for Contextualizing Folksonomies in a Triadic Context

Hong-Gee Kim[1], Suk-Hyung Hwang[2], Yu-Kyung Kang[2], Hak-Lae Kim[3], and Hae-Sool Yang[4]

[1] School of Dentistry, Seoul National University,
28-22 Yeonkun-Dong, Chongno-Ku, Seoul 110-749, Korea
hgkim@snu.ac.kr

[2] Division of Computer and Information Science, SunMoon University,
100 Kal-San-Ri, Tang-Jeong-Myon, A-San, Chung-Nam, 336-840 Korea
{shwang,aquamint99}@sunmoon.ac.kr

[3] Digital Enterprise Research Institute, IDA Business Park, Galway, Ireland
haklae.kim@deri.org

[4] Seoul University of Venture and Information,
37-18 SamSung-Dong, Kang-Nam, Seoul, Korea
hsyang@suv.ac.kr

Abstract. Standardized infrastructure for information or knowledge sharing is required to make autonomous agents interdependent on each other for effective collaboration in a multi-agent system. Folksonomy has become very popular as an enabling technology to provide a common conceptualization of the data that agent systems use. However, there are problems on free-form tagging in folksonomy. Folksonomy is only concerned with a group of instances which are labeled with tags without a formal definition. No available tool provides a way to contextualize folksonomies with respect to users, communities, goals, tasks, and so on. There is no formal approach to classifying and sharing tags that reflect a user's mental model of information resources in terms of folksonomy. We present a novel approach to developing an agent environment for contextualizing folksonomies in a triadic context using Formal Concept Analysis. We conducted an experiment to build concept hieracrhies and contextualize folksonomies from tags of blogosphere.

1 Introduction

A multi-agent system is a network of individual agents that work together to achieve a goal through communication and coordination among each other. Standardized infrastructure for information or knowledge sharing is required to make autonomous agents interdependent on each other for effective collaboration in a multi-agent system. Ontology has become very popular as an enabling technology to provide a common conceptualization of the data that agent systems use. The Semantic Web is the place where software agents perform various intelligent tasks using standard knowledge representational schemes that are named "ontologies."

An ontology is regarded as a semantic tagging system that associates terms (or keywords) with information resources that actually exist in the forms of texts, images,

N.T. Nguyen et al. (Eds.): KES-AMSTA 2007, LNAI 4496, pp. 728–737, 2007.

books, or any artifact. An ontological tagging system is usually heavy-weighted because it expresses information in a formal way, often using a very complicated logical language. However, most digital items should be accessed via WWW in milliseconds, thus a light-weighted way of semantic tagging is rather preferred. Folksonomy is an alternative approach to providing human and software agents with an easy way of semantic tagging that connects remotely existing information resources for common conceptualization. This paper proposes a unique formalism that represents contextualized folksonomies each of which is created by a human or software agent and that are shared by different agents for effective collaboration.

Blogosphere has emerged as a means of decentralized personal publishing and of social sharing in community[1,2]. Tagging and folksonomy are recently popular phenomenon in the blogosphere. Tags are a set of keywords that are attached to resources such as bookmarks, images, and blog entries etc. Such a process is known as tagging. It is not difficult to see attached tags within blog entries on many popular blogs. Folksonomy is a user-generated classification, emerging through bottom-up consensus [3] and provide an approach to address Web-specific classification issues. In general terms, folksonomy is the set of tags with one or more keywords. Users are able to instantly add terms to the folksonomy as they become necessary for a single unit of content. Folksonomy refers to the collaborative but unsophisticated way in which information is being categorized on the web. Instead of using a centralized form of classification, users are encouraged to assign freely chosen tags to pieces of information or data via tagging. The most commonly cited folksonomies in action are web sites such as Flickr, del.icio.us, and Furl. Those sites allow people to store their digital images, bookmarks and share them with friends, or the general public.

However, there are problems using free-form tagging and folksonomy. Folksonomy has only a group of instances which labeled with a tag without a formal definition and relationship among terms. So we cannot identify relationships among the groups. Furthermore, there are critical issues to share folksonomies. All folksonomies exist in vendor specific formats and are stored on vendor sites. There is currently no format for sharing or even expressing one's explicit understanding of the meaning of his/her own folksonomy tags. Those problems happen in blogosphere at a moment, even worse. Although blogs have implicitly or explicitly social relationships between others and they want to share their tags, there is no way to do it.

Our approach in this paper is to build the contextualized folksonomies to provide a shared meaning of tags and to guide methods to use tags. The term contextualized folksonomies refers to a shared collection of tags that is extracted from blogs which has a same group, rather than centralized social systems. It provides a context-centric tag usage to grouped communities. It will recommend members of the group to tag what are the shared terms. If the members can refer the group folksonomies, they can avoid ambiguous word such as synonym, underscore words to describe resources.

Figure 1 shows an example of folksonomy. In order to build contextualized folksonomies from it in blogosphere, we decide to use Formal Concept Analysis(FCA) method. FCA[4] is a conceptual clustering technique which identifies conceptual structures among data sets. These structures are graphically represented as conceptual lattices, allowing the analysis of complex structures and the discovery of dependencies

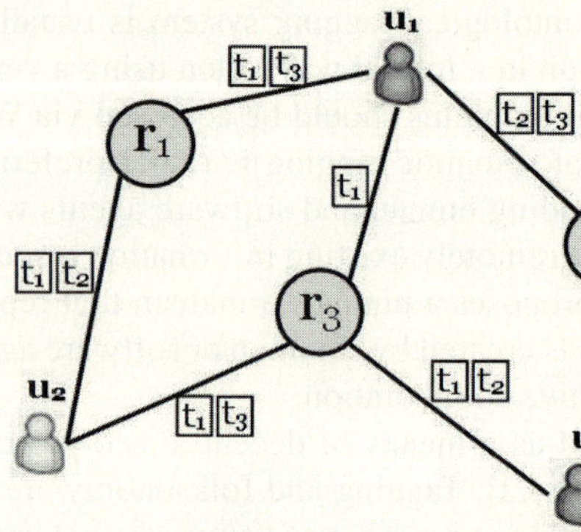

Fig. 1. An example of folksonomy

within the data. The concept lattice of FCA can be provided a convenient hierarchical description with bloggers and their set of tags.

This paper is organized as follows: In Section 2 we introduce briefly the definition of Contextualized Folksonomy based on the FCA and discuss the methods applied in our experiments. In Section 3 we describe the experiments conducted on the collected data, and we report on the results of the experiments. Finally, we sum up our findings and future works in Section 4.

2 Contextualized Folksonomy

Folksonomies allow users(bloggers) to assign tags to resources in order to manage and classify the resources. These resources can be URLs, photos, movies, blog entries or just about anything else on the web. We use this notion of Folksonomy to define "Contextualized Folksonomy" based on the triadic context[4]:

Definition 1. *A Contextualized Folksonomy(CF) is a triadic context* (U, T, R, Y) *that consists of sets* U, T, *and* R *and a ternary relation* Y *between* U, T, *and* R *(i.e.* $Y \subseteq U \times T \times R$*). The elements of* U, T, *and* R *are called users, tags, and resources, respectively, and* $(u, t, r) \in Y$ *is read: the user* u *assigns tag* t *to the resource* r.

Table 1 shows a triadic context representation of the contextualized folksonomy given in figure 1.

Out of combining two of the users, tags and resources in various ways, several dyadic contexts also arise as shown in table 2. A contextualized folksonomy(CF) gives rise to the following dyadic contexts:

$$\mathsf{CF}^U := (U, T \times R, Y^U),$$
$$\mathsf{CF}^T := (T, U \times R, Y^T),$$
$$\mathsf{CF}^R := (R, U \times T, Y^R)$$

where $(u, (t, r)) \in Y^U \Leftrightarrow (t, (u, r)) \in Y^T \Leftrightarrow (r, (u, t)) \in Y^R \Leftrightarrow (u, t, r) \in Y$.

Table 1. Triadic context of figure 1

	t_1	t_2	t_3
u_1	$\{r_1, r_3\}$	$\{r_2\}$	$\{r_1, r_2\}$
u_2	$\{r_1, r_3\}$	$\{r_1\}$	$\{r_3\}$
u_3	$\{r_3\}$	$\{r_3\}$	

Table 2. Dyadic contexts derived from table 1

	t_1			t_2			t_3		
CF^U	r_1	r_2	r_3	r_1	r_2	r_3	r_1	r_2	r_3
u_1	×		×		×		×	×	
u_2	×		×	×					×
u_3			×			×			

	r_1			r_2			r_3		
CF^T	u_1	u_2	u_3	u_1	u_2	u_3	u_1	u_2	u_3
t_1	×	×					×	×	×
t_2		×		×					×
t_3	×			×				×	

	u_1			u_2			u_3		
CF^R	t_1	t_2	t_3	t_1	t_2	t_3	t_1	t_2	t_3
r_1	×		×	×	×				
r_2		×	×						
r_3	×			×		×	×	×	

The dyadic contexts can be used for the formal concept analysis[4] that is a method of exploratory data analysis that aims at the extraction of natural clusters from object-attribute data tables. The clusters, called formal concepts, are naturally interpreted as human-perceived concepts in a traditional sense and can be partially ordered by a subconcept-superconcept hierarchy. The hierarchical structure of formal concepts (so-called concept lattice) represents a structured information obtained automatically from the input data table.

Definition 2. *A* ***dyadic context*** *is a triple $(\mathcal{O}, \mathcal{A}, \mathcal{R})$, where $\mathcal{O}$ is a set of objects and $\mathcal{A}$ is a set of attributes, and $\mathcal{R} \subseteq \mathcal{O} \times \mathcal{A}$ is a binary relation between $\mathcal{O}$ and $\mathcal{A}$. In order to express that an object o is in a relation with an attribute a, we write $(o, a) \in \mathcal{R}$ and read it as "the object o has the attribute a".*

Let $(\mathcal{O}, \mathcal{A}, \mathcal{R})$ be a dyadic context. For $O \subseteq \mathcal{O}$, we define $\text{intent}(O) := \{a \in \mathcal{A} | \forall o \in O : (o, a) \in \mathcal{R}\}$, and, dually for $A \subseteq \mathcal{A}$, we define $\text{extent}(A) := \{o \in \mathcal{O} | \forall a \in A : (o, a) \in \mathcal{R}\}$. The function intent maps a set of objects into the set of attributes common to the objects in O, whereas extent is the dual for attributes sets. These two functions form a *Galois connection* between the objects and attributes of the context.

The central notion of FCA is the *(formal) concept*. Objects from a dyadic context share a set of common attributes and vice versa. Concepts can be imagined as maximal

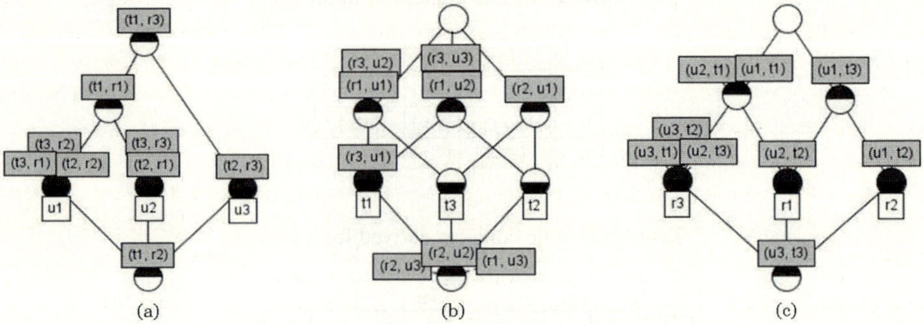

Fig. 2. Concept lattices for the dyadic contexts of Table 1

rectangles in the context table. If we ignore the sequence of the rows and columns we can identify even more concepts. The formal definition of concept is given in the following:

Definition 3. *Let* $\mathcal{C} = (\mathcal{O}, \mathcal{A}, \mathcal{R})$ *be a dyadic context. A* ***formal concept*** *is a pair* (O, A) *with* $O \subseteq \mathcal{O}$ *is called* ***extension****,* $A \subseteq \mathcal{A}$ *is called* ***intension****, and* $(O = \mathsf{extent}(A)) \wedge (A = \mathsf{intent}(O))$*. The set of all concepts of the context* $\mathcal{C}$ *is denoted by* $B(\mathcal{C})$ *i.e.,* $B(\mathcal{C}) = \{(O, A) \in 2^{\mathcal{O}} \times 2^{\mathcal{A}} | O = \mathsf{extent}(A) \wedge A = \mathsf{intent}(O)\}$.

In other words a concept is a pair consisting of a set of objects and a set of attributes which are mapped into each other by the Galois connection.

The set of formal concepts is organized the partial ordering relation $\leq$ -to be read as "is a subconcept of"- as follows: For a dyadic context $\mathcal{C} = (\mathcal{O}, \mathcal{A}, \mathcal{R})$ and two concepts $c_1 = (O_1, A_1)$, $c_2 = (O_2, A_2) \in B(\mathcal{C})$ the *subconcept-superconcept relation* is given by $(O_1, A_1) \leq (O_2, A_2) \Leftrightarrow O_1 \subseteq O_2 (\Leftrightarrow A_1 \supseteq A_2)$. A concept $c_1 = (O_1, A_1)$ is a subconcept of concept $c_2 = (O_2, A_2)$ iff the set of its objects is a subset of the objects of c_2. Or an equivalent expression is iff the set of its attributes is a superset of the attributes of c_2. The set of all formal concepts of a context $\mathcal{C}$ with the *subconcept-superconcept realtion* $\leq$ is always a complete lattice , called the *(formal) concept lattice* of $\mathcal{C}$ and denoted by $\mathcal{L} := (B(\mathcal{C}), \leq)$.

Figure 2 shows the concept lattices for the contexts of Table 2. A concept lattice can be represented graphically using line diagrams. Each node represents a concept with its associated extents and intents. The links connecting nodes represent the subconcept-superconcept relation among them. Attributes propagate along the edges to the bottom of the diagram and dually objects propagate to the top of the diagram.

FCA may be applied to data in which objects are interpreted as having attributes with values. That is, attributes can have values. We call them *many-valued attributes*, in contrast to the *one-valued attributes* considered so far. In this case the basic data is stored in a *many-valued context*.

Definition 4. *A* ***many-valued context*** $(\mathcal{O}, \mathcal{A}, \mathcal{V}, \mathcal{R})$ *is a set of objects* $\mathcal{O}$*, a set of attributes* $\mathcal{A}$*, a set of possible values* $\mathcal{V}$*, and ternarry relation* $\mathcal{R} \subseteq \mathcal{O} \times \mathcal{A} \times \mathcal{V}$ *, where the following holds:* $(o, a, v_1) \in \mathcal{R} \wedge (o, a, v_2) \in \mathcal{R} \Rightarrow (v_1 = v_2)$.

In order to get concepts out of this many-valued context and draw the concept lattice, we have to transform the many-valued context into a one-valued context according to certain rules. The new one-valued context is called the *derived context*. The concepts of derived context are interpreted as the concepts of the many-valued context. The process of transformation is called *conceptual scaling*. And this process is not uniquely determined, but depends on the transformation rules. Formally, a many-valued context is transformed by constructing a *scaling* for each attribute. The scales are used to construct one-valued contexts for each attribute which are then combined or joined to form a one-valued context which represents the original many-valued context.

Definition 5. *A* ***scale*** *for an attribute* $a \in \mathcal{A}$ *from a many-valued context* $(\mathcal{O}, \mathcal{A}, \mathcal{V}, \mathcal{R})$ *is a one-valued context* $S_a = (\mathcal{O}_a, \mathcal{A}_a, \mathcal{R}_a)$ *with* $\mathcal{A}_a \subseteq \mathcal{V}$ *is a set of values of the attribute* $a \in \mathcal{A}$ *and* $\mathcal{O}_a = \{v \in \mathcal{V} | (o, a, v) \in \mathcal{R}\} \subseteq \mathcal{A}_a$.

3 Experiment

In this section, we report upon some experiments using the FCA WIZARD[9][1] for extracting common tags and mining aspectual views of tagging patterns("use of tags") from a given groups of bloggers. The aim of the experiments is to evaluate the feasibility of FCA-based approach when applied to build the group folksonomy for bloggers and to share it between group of bloggers.

3.1 Test Data

The test data used in the following experiments was gathered by hand randomly in blogosphere between August 5th and August 18th, 2006. In order to better coverage about frequency of tags, we collected blogs which provided the tag cloud services or displayed their tag frequency. We collected the following 9 blogs and gathered about 320 tags of them.

- `Channy` : http://www.creation.net/blog
- `mEmOpAd` : http://www.linuxstudy.pe.kr/~kebie/2005/blog/tt/index.php
- `Prak` : http://www.fortytwo.co.kr/tt
- `Weblognara` : http://weblognara.com/
- `Market-trend` : http://www.marketcast.co.kr/blog
- `Oreilly radar` : http://radar.oreilly.com/
- `@hof` : http://www.hof.pe.kr/
- `Channy2` : http://channy.tisory.com
- `Guichan` : http://anihil.cafe24.com/

3.2 Identifying the Common Tags

As a first experiment, to identify common(shared) tags among the given bloggers, we started with one-valued context that captures the tagging information for each blogger.

[1] FCA WIZARD is a formal concept analysis tool developed as part of our project for the data analysis and knowledge discovery in medical domain.

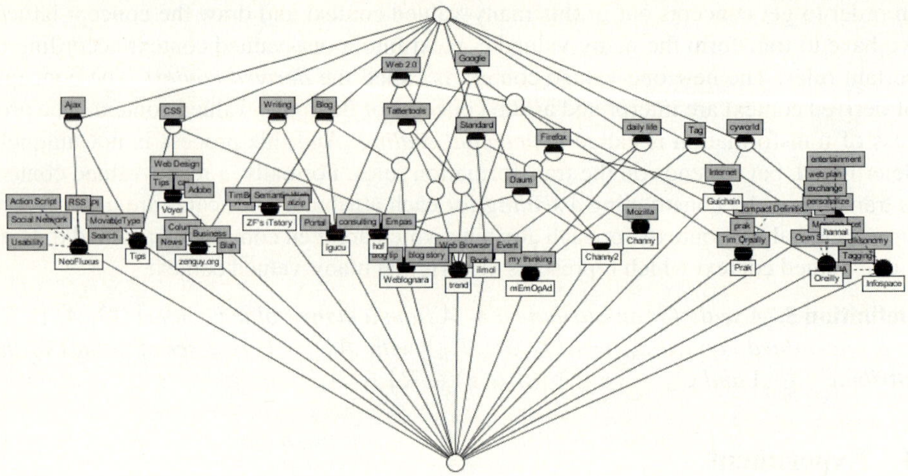

Fig. 3. The result of experiment for identifying the common Tags

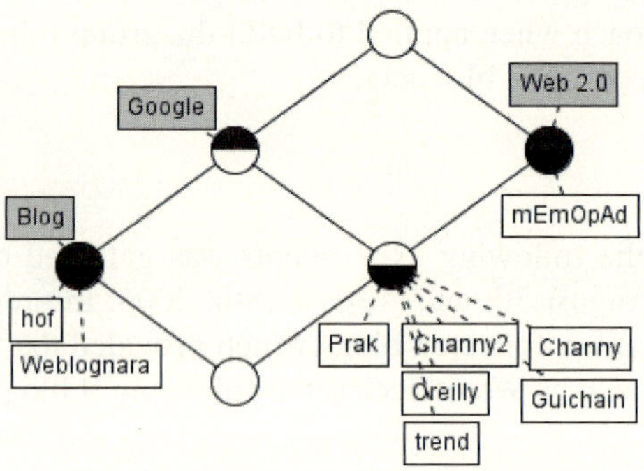

Fig. 4. Simplified concept lattice

That is, the one-valued context is composed of 9 bloggers, 3 tags they used, and binary relations that indicate which tag is used by whom. By applying FCAWIZARD into the one-valued context, we can extract some concepts and display it as concept lattice. From the concept lattice, we can identify some common(shared) tags among the bloggers.

However, the resulting concept lattice is too large to be completely displayed at once(figure 3). In order to analyze it in more detail, we should restrict the set of objects and/or attributes and visualize only the corresponding part of the concept lattice. Figure 4 shows the simplified lattice that we restricted the attribute set into {*Web2.0*, *Google*, *Blog*}. We can identify 4 concepts that group bloggers and tags in ways that are meaningful in identifying common tags. In figure 4, for example, the node labelled by *Blog* with *hof* and *Weblognara* denotes the concept ({*hof, Weglognara*}, {*Blog, Google*}).

Table 3. Many-valued context for the frequency of use of tags

	Google	Web2.0
Channy	5.2	6.6
mEmOpAd	0	6.5
Prak	7.1	44.1
Weblognara	4.2	0
trend	6.4	14.1
Oreilly	9.5	10.9
hof	27	0
Channy2	15.9	11.4
Guichain	8.6	8.6

Table 4. Ordinal scale context for mining the weighted common tags

	VL	L	M	H	VH
$1 \leq f < 5$	×	×	×	×	×
$5 \leq f < 10$		×	×	×	×
$10 \leq f < 15$			×	×	×
$15 \leq f < 20$				×	×
$20 \leq f < 50$					×

It means that bloggers *hof* and *Weglognara* have used common tags *Blog* and *Google*. Of course, this does not mean their actual values for the frequency of use for the tags are the same. Therefore, we need some more consideration for the frequency of use of tags.

3.3 Mining the Weighted Common Tags

The aim of the second experiment is not only to identify the common tags, but also to extract ordinal information from the bloggers' common tags based on the frequency of use of tags. Table 3 shows a many-valued context of the frequency of use of tags for a group of bloggers. The number in each cell of Table 3 denotes the frequency of using of a given tag t, i.e. (Number of times of use of the tag t)/(Number of total posts within a blog).

In order to obtain a more fine-grained view, we apply *conceptual scaling* to the many-valued context. In the conceptual scaling, each attribute("tag") is treated as a separate formal context with the values("frequency of use of tag") as attributes associated with each of the original objects("Bloggers"). We use the ordinal scale context of table 4 in the conceptual scaling. That is, the ordinal scale can be used to interpret each tag whose values("frequency of use of tag") admit a natural ordering such as "Very Low", "Low", "Medium", "High", "Very High". The attribute values then result in a chain of extents, interpreted as a hierarchy.

Figure 5 shows the concept lattices for "Web2.0" and "Google" tags. From the concept lattice, we can identify some groups of bloggers who are using common tags

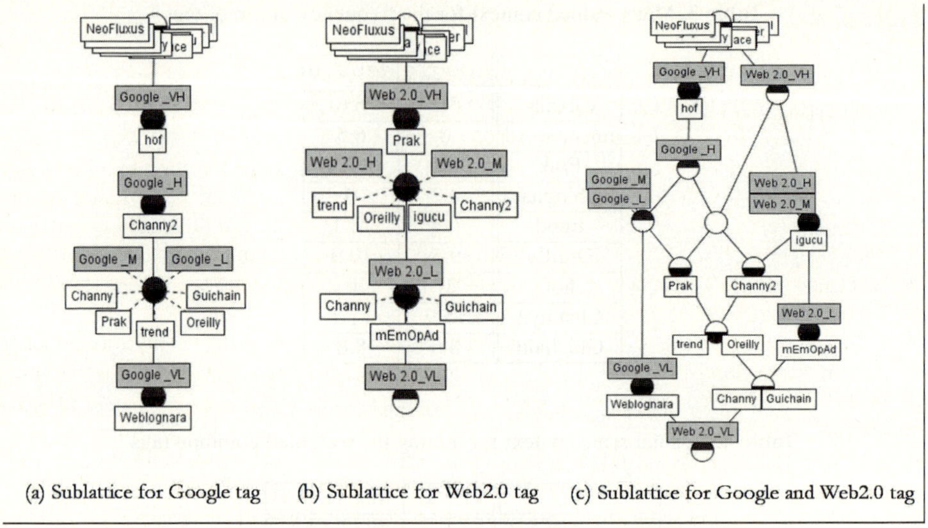

Fig. 5. Concept lattices for "Web2.0" and "Google" tags

as well as classify each bloggers into 5 grades("Very Low", "Low", "Medium" and "High", "Very High"). For example, *Prak* uses `Google` with the frequency of use "Low" as well as `Web2.0` with "Very High". *Channy2* uses `Google` with "High" and `Web2.0` with "Medium".

4 Conclusion

In this paper, we have proposed a formal model of a contextualized folksonomy based on the triadic context. Also we have demonstrated how formal concept analysis can be applied to the contextualized folksonomy. By using the "FCA Wizard", we can have a global view on the conceptual structure of the context with regard to the inherent structures of the value sets and some high potentials of interesting research questions about building and disserminating the contextualized folksonomies both theory and applications. Formal concept analysis provides helpful methods to provide a shared meaning of tags and to guide methods to use tags in the contextualized folksonomies. Therefore, the contextualized folksonomies can provide a context-centric and shared collection of tags to semantically-interlinked online communities.

Acknowledgments

This research was supported by a grant of the Interoperable EHR Research and Development Center(A05-0909-A80405-05N1-00050B), Ministry of Health & Welfare, Republic of Korea.

References

1. Christopher H. Brooks and Nancy Montanez. Improved annotation of the blogosphere via autotagging and hierarchical clustering. In WWW'06: Proceedings of the 15th international conference on World Wide Web, pages 625.632, New York, NY, USA, 2006. ACM Press.
2. Ali-Hasan. Lada A. Adamic. Expressing Social Relationships on the Blog through Links and Comments, http://www-personal.umich.edu/∼ladamic/papers/oc/onlinecommunities.pdf, 2006.
3. Alex Wright. Folksonomy, 2004. http://www.agwright.com/blog/archives/000900.html
4. B. Ganter and R. Wille. Formal Concept Analysis: Mathematical Foundations. Springer-Verlag, 1999.
5. P. du Boucher-Ryan and D. Bridge. Collaborative recommending using formal concept analysis. In F. Coenen M. Bramer and T. Allen, editors, Proceedings of AI-2005, The Twenty-Fifth SGAI International Conference on Artificial Intelligence, pages 205.218. Springer, 2005.
6. James Melzer. A folksonomys types and relationships, 2005. http://www.jamesmelzer.com/bearings/archives2005/02/the debate over.html.
7. David R. Millen and Jonathan Feinberg. Using social tagging to improve social navigation. InWorkshop on the Social Navigation and Community based Adaptation Technologies, 2006.
8. Christoph Schmitz and Andreas Hotho and Robert Jäschke and Gerd Stumme. Mining Association Rules in Folksonomies. Proc. IFCS 2006 Conference, 261-270, Ljubljana, 2006.
9. Y.K.Kang, S.H. Hwang, et al. Development of a FCA Tool for Building Conceptual Hierarchy of Clinical Data. Journal of the Korean Society of Medical Informatics, Vol.11. No.S2, 2005.
10. D. Gruhl, R. Guha, d. Liben-Nowell, and A. Tomkins, "Information diffusion through blogspace", Proceedings of the 13th international conference on the World Wide Web, 2004.
11. David R Millen and Jonathan Feinberg, "Using Social Tagging to Improve Social Navigation", Workshop on the Social Navigation and Community-Based Adaptation Technologies, 2006.
12. Dye, J., "Folksonomy: A game of high-tech (and high-stakes) tag", EContent, 29(3), 38-43, 2006.

Design of Knowledge Discovery Agent for a Bit-Map on Ad Hoc Mobile Networks

HangKon Kim[1], SungSoo Lee[2], and ChongGun Kim[2,*]

[1] Deptartment of Computer Information Communication, Catholic University of Daegu
[2] Department of Computer Engineering, Yeungnam University
hangkon@cu.ac.kr, lssung@chol.com , cgkim@yu.ac.kr

Abstract. A main nature of ad hoc mobile networks is frequent change on their topology that is the source of many problems to be solved. AODV is an on-demand routing protocol for decreasing maintenance overhead on ad hoc networks. But some path breaks can cause significant overhead and transmission delays. If the maintenance overhead of routing table can be reduced, table-driven routing methods could be an efficient substitution. In this paper, we propose a knowledge discovery agent for an effective routing method that is using the simple bit-map topology information. The agent node gathers topology knowledge and creates topology bit-map information. All paths for source to destination can easily be calculated by the bit-map. All the other nodes on the network maintain the bit-map distributed from agent and uses it for routing. Correctness of the proposed agent method is verified by computer simulations.

Keywords: a knowledge discovery agent, Ad hoc networks, AODV, table-driven routing, bit-map table, reducing overhead.

1 Introduction

Since ad hoc network does not rely on existing infrastructure and is self-organized the nodes on such network act as hosts and routers to transmit packets. With its intense change in topology ad hoc network does not relay on pre-established cable network, but it requires special routing algorithm. Routing protocols used in ad hoc network can be divided into two categories: table-driven routing method and on-demand routing method [1].

In table-driven routing method every node maintains information about routing of every node on the network. Nodes perform the routing based on this information. The advantage of this method is that nodes can establish the path without discovering route on need which significantly decreases. But periodical exchange of information between nodes wastes the transmission bandwidth and nodes' energy, and creates another type of overhead. Typical table-driven routing protocols are DSDV (Destination-Sequenced Distance Vector Routing Protocol) [2], OLSR (Optimized Link State Routing Protocol) [3], and TBRPF (Topology Broadcast based on Reverse-Path Forwarding Protocol) [4].

---

[*] Correspoding author.

N.T. Nguyen et al. (Eds.): KES-AMSTA 2007, LNAI 4496, pp. 738–746, 2007.

In on-demand routing method nodes start path discovery process when data is needed to be transmitted. The advantage of this method is that nodes do not periodically exchange information about the routing paths on the network which significantly decreases overhead. But process of finding routes creates transmission delay. Typical protocols which use on-demand routing method are AODV (Ad hoc On-demand Distance Vector) [7], DSR (Dynamic Source Routing) [8], R-AODV (Reverse AODV) [6].

In order to solve problems on on-demand routing method multipath based routing method is proposed. If a substitute path is available, when connection to destination is lost, then nodes energy could be saved, and balanced and long lasted routes could be achieved. Protocols using multipath routing method are node-disjoint, link-disjoint, non-disjoint (partial-disjoint)[12].

An ad hoc network is configured by many moving nodes. If an agent node that represents the network would be applicable for controlling wireless networks, it can be broadly used on organizing tree type network [9].

Because of overhead on table-driven routing method caused by managing routing information, on-demand routing method is studied mostly. If this overhead could be decreased somehow table-driven routing method would be acceptable to apply.

In this paper we propose a knowledge discovery agent for simple link state routing which uses bit-map information. All nodes on the network using identical bit-map information distributed by the agent that represents the network topology can decide routing path from source to destination at any instance. The agent collects network topology information and creates a bit-map information which shows the network topology.

The agent node is can dynamically decided using node connectivity and battery life-time information for network stability.

2 Overview of the Knowledge Discovery Agent and Bit-Map

In proposed method the agent node gathers information for the network topology and delivers knowledge to all nodes that are represented in bit-map table. Using simple bit-map table decreases overhead on table-driven routing. Figure 1 shows the concept of knowledge discovery agent. Using the bit-map table source node can easily check existence of routing path to a destination node. All nodes periodically broadcast hello message in order to find out its neighbor nodes.

The agent node broadcasts query message to all neighboring nodes to collect information about network topology. The node transmitting topology query message becomes parent-node, the receiver becomes a child-node.

The node which received topology query message delivers it to its child-nodes. If it has no child-node, i.e. it is a leaf node, than it sends topology reply message to parent-node which includes node ID, connection and battery life-time information. By repeating this sequence the agent node collects all nodes connection information on the network. After finishing the process the agent node creates knowledge using all this information as a bit-map table.

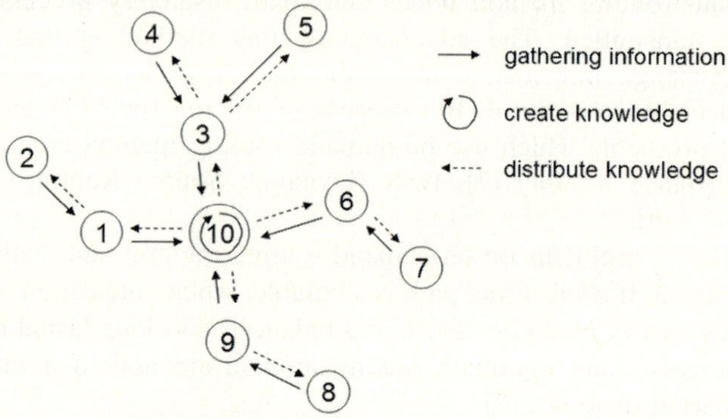

Fig. 1. Knowledge Discovery Agent

On the network shown in Figure 2 agent node is node 1 that collects connection information on the network and creates the topology knowledge on the bit-map table.

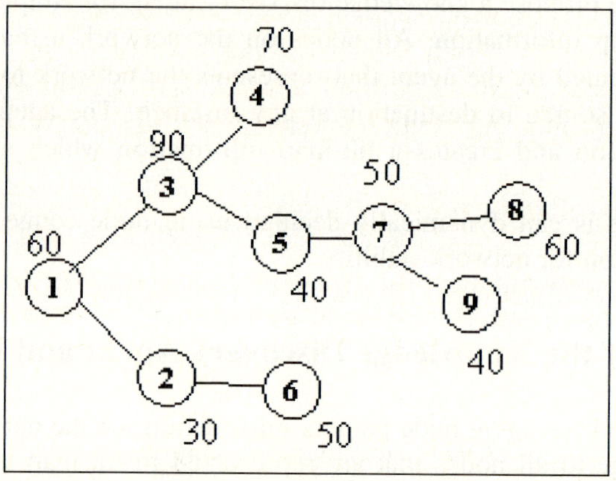

Fig. 2. A network topology

Network topology represented as bit-map table is shown in Table 1 where "1" and "0" stand for "one-hop connection" and "no one-hop connection", respectively.

The agent node elects a substitute agent node using all nodes connection and battery-life time information. In case the agent node disappears from the network, a new agent node has to be selected [11]. Then it sends bit-map and agent node information via topology advertisement message to all nodes. In this way every node on the network will know about all connections on the network. An arbitrary node can be the agent node when the network is initially created. In this initial case selecting process for the proper agent node is main duty.

Table 1. Bit-map table for network topology shown on Figure 1

Node ID	**1**	**2**	**3**	**4**	**5**	**6**	**7**	**8**	**9**
1	0	1	1	0	0	0	0	0	0
2	1	0	0	0	0	0	0	0	0
3	1	0	0	1	1	0	0	0	0
4	0	0	1	0	0	0	0	0	0
5	0	0	1	0	0	0	1	0	0
6	0	1	0	0	0	0	0	0	0
7	0	0	0	0	1	0	0	1	1
8	0	0	0	0	0	0	1	0	0
9	0	0	0	0	0	0	1	0	0

If a new node comes in or leaves the network or any alteration occurs on the network topology, all neighbor nodes that sensed the change send topology reply message to agent node. The agent node has to maintain the bit-map table and transmit the new bit-map table information to all other nodes. Additionally, the agent node periodically sends bit-map information to other nodes on the network.

3 Bit-Map Based Routing Protocol

All nodes on the network periodically send Hello message in order to confirm existence of neighbor nodes. The format of Hello message is shown in Figure 3.

Node's own IP Address
Node's own Sequence Number
Life time

Fig. 3. Hello message format

After receiving hello message from neighbor nodes, a node builds up topology information table. Topology information table consists of neighbor node information and node's own battery life-time, connection and agent node information.

Node ID	Neighbor node	Battery	Number of neighbor nodes	Agent (Yes/No)
3	1,4(0),5(0)	90	3	

Fig. 4. Topology information table of node 3

The Figure 4 shows information collected by an intermediate node 3 based on Figure 2. In the network shown in Figure 2 node 1 is agent and node 3 can have the following information on its table:

a. Receives topology query from node 1.
b. Assigns node 1 as parent-node, and nodes 4 and 5 as child-nodes.
c. Broadcasts topology query message to child-nodes, i.e. to nodes 4 and 5.

For organizing network topology, the agent node broadcasts topology query message (T_query message) to all neighbor nodes. By this way it builds self-centred topology tree. The node which received topology query message assigns the node which has sent the message as parent-node (1), and all other neighbor nodes are assigned as child-node (0). Figure 5 shows topology query message format.

A node after receiving topology query message, if there is no child-node, sends topology reply message to parent-node. The message contains nodes ID, battery life-time, connection and the agent information.

<table>
<tr><td>Type</td><td>T_query ID</td><td>Reserved</td></tr>
<tr><td colspan="3">Agent IP address</td></tr>
<tr><td colspan="3">Agent sequence Number</td></tr>
</table>

Fig. 5. Topology query message format

A parent node after receiving all topology reply messages from its child-nodes records all information on topology information table and transmits the information to its parent-node. Figure 6 shows the format of topology reply message.

A parent-node sends topology reply message to its parent-node only after it receives topology reply message from all its child-nodes. Using this method, the agent node can accumulate topology information for all nodes on the network.

<table>
<tr><td>Type</td><td>T_reply ID</td><td>Reserved</td></tr>
<tr><td colspan="3">Node's own IP address</td></tr>
<tr><td>Battery</td><td>Connectivity</td><td>Agent</td></tr>
<tr><td colspan="3">Neighbor Node's IP address</td></tr>
<tr><td>Battery</td><td>Connectivity</td><td>Agent</td></tr>
<tr><td colspan="3">.
.
.</td></tr>
<tr><td colspan="3">Agent IP address</td></tr>
</table>

Fig. 6. Topology reply message format

Nodes 4, 6, 8 and 9 on Figure 2 have no child-nodes, thus they will promptly send topology reply message (T_reply message) to their parent-nodes. Topology information table of node 5 after receiving topology reply message is shown on Figure 7.

Node ID	Neighbor Node	Battery	Connection information	Agent
5	3(1), 7(0)	40	2	
7	5(1), 8(0), 9(0)	50	3	
8	7(1)	60	1	
9	7(1)	40	1	

Fig. 7. Topology information of node 5

After receiving topology information source node builds up bit-map table. Based on the bit-map table, battery and connection information, agent node selects a agent node. Following rules are applied for selection:

a. Leaf-node can not be a agent node.
b. Agent node should have longest battery life-time.
c. Agent node should have as many as possible neighbor nodes.

By applying above rules node 3 can be selected as a agent node. Nodes 1 and 2 can be candidates.

Arbitrary initiator agent node delivers collected bit-map table which represents network topology to the agent node. The agent node sends the bit-map information to all nodes on the network by topology advertisement message (T_adv message). Figure 8 shows topology advertisement message format.

Nodes that received topology advertisement message checks whether there exists difference between its own previous topology information. If there was any alteration on the topology, the node updates the topology table. The agent node periodically broadcasts topology advertisement message to all other nodes which have to know about network topology and agent node changes.

<table>
<tr><td>Type</td><td>T_adv ID</td><td>Reserved</td></tr>
<tr><td colspan="3">Agent node ID</td></tr>
<tr><td colspan="3">1st candidate-agent node ID</td></tr>
<tr><td colspan="3">2nd candidate-agent node ID</td></tr>
<tr><td colspan="3">Bit-map table</td></tr>
</table>

Fig. 8. Topology advertisement message format

If a node needs to transmit data to another node then it can check whether a routing path exists to destination node by using bit-map table and can quickly establish the connection. The pseudo algorithm of the process is illustrated on Figure 9.

At first source node performs “AND” operation between itself and the destination node. If the source node finds connection between itself and the destination node, then the connection is established immediately.

1. Do "AND" operation between source node and destination node.
   A. If there is a path to the destination node, then finish.
   B. If there is no path, then go to step 2.

2. Find neighbor node.
   A. If there is a neighbor node, do "AND" operation between neighbor node and destination node.
      a. If there is a path to destination node, then finish.
      b. If there is no path, then go to step 2.
   B. If there is no neighbor node, then finish.

Fig. 9. Algorithm of path discovery using "AND" operation

If no connection is found between source and destination node, the source node searches for neighbor node. The neighbor nodes become intermediate nodes as the relay to the destination. Then "AND" operation is repeated between neighbor node and destination node. If no connection is found between neighbor node and destination node, the neighbor node searches for its neighbor nodes. And then "AND" operation is performed to the destination node. This process is continued until a connection is found to the destination.

The method explained above is finding the path with minimum number of hops from source to destination node. Multipaths can also be established between source and destination nodes. Using bit-map table other paths could be assigned as reference paths.

4 Verification of Route Discovery

For verification of the proposed agent method, an example ad hoc network is shown at Figure 10. The created bit-map is shown in Table 2. The agent can be decided by using the algorithm of the Figure 9.

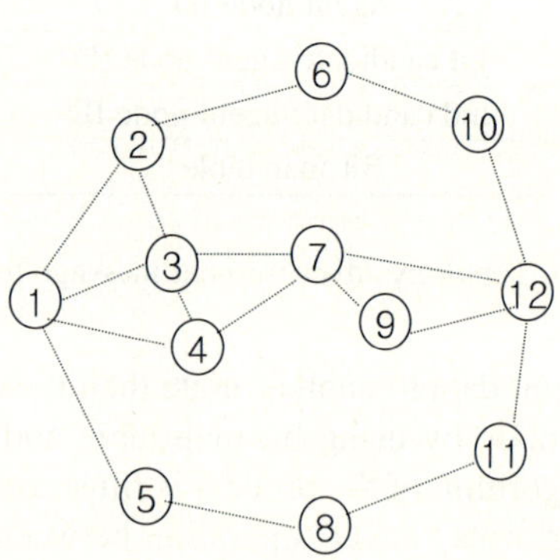

Fig. 10. An experimental network topology

Table 2. Bit-map table for network shown on Figure 11

Node ID	1	2	3	4	5	6	7	8	9	10	11	12
1	0	1	1	1	1	0	0	0	0	0	0	0
2	1	0	1	0	0	1	0	0	0	0	0	0
3	1	1	0	1	0	0	1	0	0	0	0	0
4	1	0	1	0	0	0	1	0	0	0	0	0
5	1	0	0	0	0	0	0	1	0	0	0	0
6	0	1	0	0	0	0	0	0	0	1	0	0
7	0	0	1	1	0	0	0	0	1	0	0	1
8	0	0	0	0	1	0	0	0	0	0	1	0
9	0	0	0	0	0	0	1	0	0	0	0	1
10	0	0	0	0	0	1	0	0	0	0	0	1
11	0	0	0	0	0	0	0	1	0	0	0	1
12	0	0	0	0	0	0	1	0	1	1	1	0

Node 1 is set as a source node. For finding all destinations bit-map operation is performed by the algorithm of Figure 9. The operation results are shown on Figure 11. It shows that paths are established for all destinations. It also shows multiple paths with different number of hops for the same destination.

```
NODE 1
Input SRC & DST: 1 2    1 -> 2

Input SRC & DST: 1 3    1 -> 3

Input SRC & DST: 1 4    1 -> 4

Input SRC & DST: 1 5    1 -> 5

Input SRC & DST: 1 6    1 -> 2 -> 6

Input SRC & DST: 1 7    1 -> 3 -> 7,      1 -> 4 -> 7

Input SRC & DST: 1 8    1 -> 5 -> 8

Input SRC & DST: 1 9    1 -> 3 -> 7 -> 9,      1 -> 4 -> 7 -> 9

Input SRC & DST: 1 10   1 -> 2 -> 6 -> 10

Input SRC & DST: 1 11   1 -> 5 -> 8 -> 11

Input SRC & DST: 1 12   1 -> 3 -> 7 -> 12
```

Fig. 11. Route searching procedure

For data transmission a path with minimum number of hops will be chosen. In practice, when there are multiple paths with same number of hops, one of them will be chosen for the primary path and others will be saved as substitutions. We experimented on various topologies. The probability of finding destinations is 100%.

5 Conclusions

We have proposed a knowledge discovery agent for gathering network information to increase performance on ad hoc networks. The concept of proposed agent and an algorithm for knowledge of network topology are provided. The bit-map knowledge from collected information that includes routing information of the network is created and distributed by the agent using the collected knowledge. Computer simulation on various network topologies shows that suggested knowledge discovery agent method for making bit-map method help sources find destinations with 100% probability. The bit-map table could be also used to search multi-paths.

Further researches should aim to design concrete protocol and compare its performance with the existing ad hoc network routing protocols.

References

1. Yi Lu, Weichao Wang, Yuhui Zhong Bharat Bhargava, "Study of Distance Vector Routing Protocols for Mobile ad hoc networks", Proceedings of the First IEEE International Conference 2003.
2. Charles E. Perkins, Pravin Bhagwat, "Highly Dynamic Destination-Sequenced Distance-Vector Routing(DSDV) for Mobile Computers", Proceedings of ACM SIGMOMM 1994.
3. T. Clausen, P. Jacquet, A, "Optimized link state routing protocol", RFC3626, Experimental, Oct 2003.
4. R.G.Ogier, M.G.Lewis and F.L.Templin, "Topology Broadcast Based on Reverse-Path Forwarding", draft-ietf-manet-tbrpf-08.txt, Apr 2003.
5. Mary Wu, Youngrag Kim, Chonggun Kim "A Path Fault Avoided Routing in Ad Hoc Networks", Journal of KIPS, Dec 2004, pp. 879-888.
6. Chonggun Kim, Elmurod Talipov, and Byoungchul Ahn, "A Reverse AODV Routing Protocol in Ad Hoc Mobile Networks", EUC Workshops 2006, LNCS 4097.
7. C. Perkins, Nokia Research Center, E. Belding-Royer "Ad hoc On-Demand Distance Vector (AODV) Routing", RFC 3561, 2003.
8. D.B. Johnson, D.A. Maltz, and Y-C. Hu, "The Dynamic Source Routing Protocol for Mobile Ad Hoc Networks (DSR)", IETF Mobile Ad Hoc Networks Working Group, Internet Draft, work in progress, 15 Apr. 2003.
9. Sudarshan Vasudevan, Jim Kurose, Don Towsley, "Design and Analysis of a Agent Election Algorithm for Mobile Ad Hoc Networks", UMass Computer Science Technical Report 03-20
10. Chakeres. I. M, Belding-Royer. E. M, "The utility of hello messages for determining link connectivity", The 5th International Symposium on Volume 2, Oct 2002.
11. Vasudevan S., Kurose J., Towsley D., "Design and Analysis of a Agent Election Algorithm for Mobile Ad Hoc Networks", Proceedings of the 12th IEEE International Conference on Network Protocols (ICNP) (2004), pp. 350-360.
12. A. Nasipuri, R.Castaneda, S.R. Dasl, "Performance of Multipath Routing for On-demand Protocols in Mobile Ad hoc Networks", Kluwer Academic Publishers Mobile Networks and Applications, 2001.

Mobile Collaboration Framework for u-Healthcare Agent Services and Its Application Using PDAs

Chang-Won Jeong, Dong-Ho Kim, and Su-Chong Joo

School of Electrical, Electronic and Information Engineering, Wonkwang University, Korea
{mediblue,dhkim1,scjoo}@wonkwang.ac.kr

Abstract. In this paper, we suggested a mobile collaboration framework based on distributed object group framework (DOGF). This paper focuses on the use of this framework to support mobile collaboration. Therefore, we improved the existing work and apply it to the construction of u-healthcare agent services. For supporting mobile collaboration, we divided into two agent types such as the stationary and the moving-typed agents according to the function of mobile devices. The data collected by sensors attached on arbitrary spaces can be shared by 2-typed agents or a home server, and exchanged with each other using the Push and Pull methods. For managing this information, the DOGF provides functions of object group management, storing information and security services to our mobile collaboration framework via defined application interfaces. The agent for executing service functions of mobile devices and an integrated monitoring system on home server are implemented by using TMO scheme. And we used the TMOSM for interactions between distributed components. Finally, we showed via GUI the executablility of healthcare application supporting for medical work in hospitals on our mobile collaboration framework.

1 Introduction

Ubiquitous computing integrates computation into the environment, rather than having computers which are distinct objects. This means that individual physical environment is a share and exchange of information that are collected data from the devices, sensors and desktop computer over wireless network [1, 2]. Recently, research in ubiquitous computing is towards the development of an application environment able to deal with the mobility and interactions of both users and devices [3, 4, 5]. In these researches, collaboration is a very important application of information technology, especially for a u-healthcare environment. With the fast development of mobile computing, wireless mobile networks and mobile devices are becoming widely used in both commercial and academic organizations. Collaboration turns out to be much more useful in a mobile network.

Therefore, this paper presents the mobile collaboration framework based on the distributed object group framework [6]. The DOGF provides functions of object group management, storing information and security services to our mobile collaboration framework via defined application interfaces. With the aim of analysing a complete understanding of the medical work practices, how they manage the

N.T. Nguyen et al. (Eds.): KES-AMSTA 2007, LNAI 4496, pp. 747–756, 2007.

medical information, and how they interact with each other, we considered a workplace study [7]. We decided to scenarios as a way of integrating our researching for u-healthcare framework. These scenarios that require for mobile collaboration include exchange information, sensing data process and group based distributed object collaboration by agent on mobile devices. There are also a lot of applications that can integrate collaboration into healthcare application services.

When considering our environment we found that there were only two mobile devices that would satisfy our demands: Personal Digital Assistants (PDA) and mobile phones. The PDA had a clear advantage over mobile phones when considering sensor data collection and mobility. It was therefore natural to focus on the available PDA in a hospital space and technologies available such as PDAs. According to the function, we divide into two classes; stationary and moving-typed agent. The functions of a stationary-typed agent on PDAs or PC are it collecting related patient health data and environment information about the ward from the healthcare sensors or devices. While the moving-typed agent on PDA can support the nurse's business that provides the interaction with stationary-type agents and home server.

For our implementation environment, we used TMO scheme and TMOSM for interactions between distributed components. Also, we used the Bluetooth technology for interactions between mobile devices. Finally, we showed via GUI the executablility of a healthcare application on our mobile collaboration framework.

The rest of the paper is organized as follows. Section 2 presents related research work. The next section 3 describes the architecture of mobile collaboration framework. Also, we explained the mobile collaboration type which is the type according to the function of mobile devices. Section 4 describes the healthcare application that implements the architecture and demonstrates its features and abilities based on mobile collaboration framework. The last section describes the conclusion and future works.

2 Related Work

Our research enables the design of more flexible and suitable for ubiquitous healthcare environments. Currently, various researchers have been suggested in order to mobile collaboration. This section describes some of the related researches and projects for this paper.

Mobile Collaboration services enable users utilizing mobile device to share information and communicate with each other. Related research has been conducted on various models that support collaboration.

Pervasive Collaborative Computing Environment (PCCE) project[8] that offers an environment for supporting scientific collaborations. This environment houses various tools needed in collaborations such as a synchronous/asynchronous messaging, video conferencing, and file sharing/transfer.

iClouds project is an architecture for supporting spontaneous exchange of information between users that come within each other's digital sphere[9]. The goal of iClouds is to provide a platform for spontaneous collaboration, taking advantage of the observation that people gathered at the same location often have common interests or goals.

JXTA[10] based on J2SE project is java framework for P2Pnetworks that targeted on desktop computers. For solve this platform problem, Sun Microsystems suggests the JXME and JXTA for J2ME for support on small mobile devices.

PROEM[11] is an open computing platform that provides a complete solution for developing and deploying P2P applications for MANETs. The platform is based on experiences from developing a series of mobile applications. It is implemented as a platform independent framework using J2SE.

Such a project constitutes a high level of collaboration between individuals via the use of computers and wireless technologies. But, several approaches have been proposed to provide aspects related to low-level collaboration.

We have developed a distributed object group framework that is component-based framework. Therefore, we improved the existing work and apply it to the construction of u-healthcare agent services for mobile collaboration. For supporting mobile collaboration, we focus on high level collaboration supported by DOGF.

3 Mobile Collaboration Framework

This section presents an overview of the mobile collaboration framework architecture. This framework provides flexible support for defining and configuring the group through the service components of DOGF and behavior of the principal elements of a mobile collaboration environment. Also, we describe the type of mobile collaboration.

3.1 Mobile Collaboration Framework Architecture

Our framework used to component of supporting object group management for domain grouping in distributed object group framework, we consider the interaction of mobile devices and sensors. Also, for information collection and sharing in this environment, we adopted the TMO scheme and TMOSM [12] into the development environment of the healthcare application.

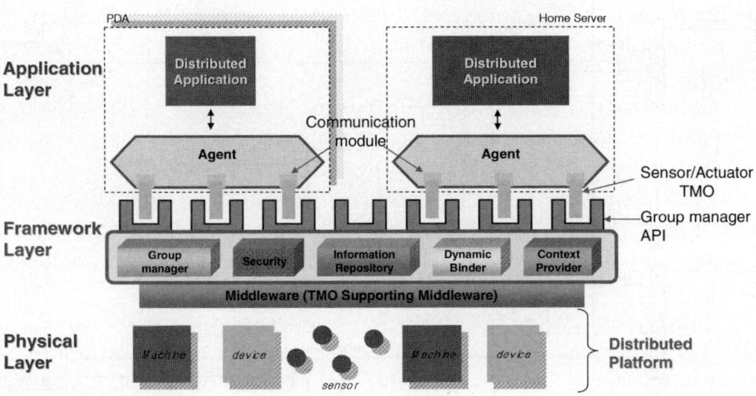

Fig. 1. Architecture of mobile collaboration framework

Figure 1 shows its architecture. It supports a logical single view system environment by grouping them. The group manager API supports the execution of application of appropriate mobile collaboration service on upper layer by using the input information obtained from individual or grouped physical devices on the lower layer as distributed platform. That is, according to the mobile collaboration service or status of the logical domain for collaboration, our framework could configure new groups dynamically by integrating physical devices/sensors or machines on the distributed platform and distributed application and agent on the upper layer. For the middleware of interaction between distributed applications adapted the TMOSMS.

3.2 The Interactions of Component

In the whole mobile collaboration framework, we defined the interaction of components which interacts with the distributed application include the agent, sensors and components of framework. The group manager object provides the interaction of agent in distributed application on PDA or Server and sensor by APIs and service object reference which support collecting real time information from the sensor nodes. Also it supports the security service which is a security object that checks access right for client. When service object is replicated, dynamic binder object provides the reference of service object by binding algorithm. The stationary-typed agent on PDA obtains real time information of sensor node through service object reference. And, the interaction of objects in distributed application returned the result of service by framework components. The suggested framework located at home server supports the security service and manages the devices and related information. Also, home server supports management of collecting information by stationary-typed agent and control of appliance by the context provider object. Figure 2 shows the sequence diagram of relationship with distributed applications, agents and component objects for framework.

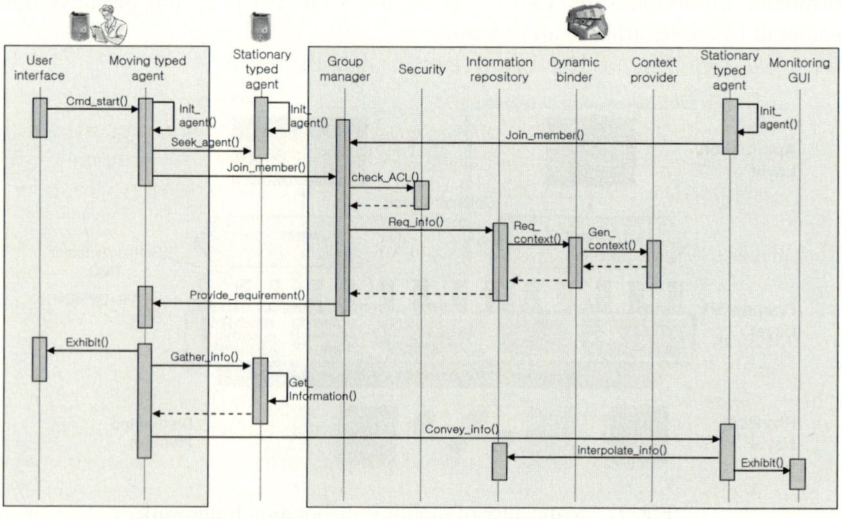

Fig. 2. Sequence diagram for mobile collaboration

3.3 The Agent Type of Mobile Collaboration

This paper suggested mobile collaboration framework, interacts with devices and sensors according to the type of devices. According to the function, we divide it into two classes; stationary and moving-typed agent.

The functions of a stationary-typed agent are collecting related patient health data and environment information about the ward from the healthcare sensors or devices. The moving-typed agent can support the nurse business. Also, its function is similar to the stationary-typed agent. The difference between stationary-typed agent and moving type agent is that the latter only collects the data whereas the moving type agent usually provides the interaction with other agents and home server.

We defined the mobile collaboration type about the interaction of devices. The information collected by sensors can be shared and exchanged by agents or home server in accordance to Push and Pull methods. The interaction between sensors and stationary-typed agent is using push methods that is collecting information using sensor send to moving-typed agent. And, the interaction between stationary-typed agent and moving-typed agent is using push/pull methods. In this case, the push methods arc available to reconfiguration for sensor network and collecting the data from the sensor nodes. Then the pull methods are available to the exchange information between agents.

Also, the mobile device and home server used push/pull methods of agents and communication ways which can use Bluetooth or wireless LAN.

Firstly, Bluetooth way is to exchange data with the home server which is searched Bluetooth dongle. When mobile device has no Bluetooth dongle for the communication way, then mobile device can search the access points by IEEE 802.11b, g. After searching for several devices by AP, it is communicating with searched devices. The above interaction way is according to the system environment. The home server provides information management and service based on context recognition.

4 Implementation of the Healthcare Application

In this section we describe a healthcare application of computer-supported mobile devices cooperation. Also, we implement the healthcare application based on TMO scheme and then show the monitoring of GUI results in collaboration environment.

4.1 Healthcare Application

The healthcare application focuses on the collaboration environment which includes the interaction of management server, patient and nurse.

The hospital has a three wards and each ward located stationary-typed agent on fixed hosts. When nurse's PDA move about the ward, moving-typed agent adds network components and requests the environment information, patient health information to stationary-typed agent in the ward. And then, it provides information which is collected data from sensors, in which case the security information cannot access any information. The PDA of nurse's displays information that is collected data from the stationary type agent. With mobile collaboration application, we used the environment information from sensors such as temperature, humidity and illuminometer. Also, we used the location sensor for location tracking and healthcare sensor such as blood pressure, blood sugar, heart rate and body heat. They can be divided into two types such as security and public. The one is patient health information and the other is environment information in the wards. The moving-typed agent can be obtained public information in the wards, and provide the control of appliances. For example, if using the moving-typed agent, we want to control fan to regulate the temperature, it requires the certification for the control of appliance to home server. Also stationary-typed agent on the ward obtains the public information and security information. And then, it is sent to home server. When information are transmitted a collecting data by agent on PDAs, the home server manages the way data are stored to database. Figure 3 shows the physical environment for mobile collaboration.

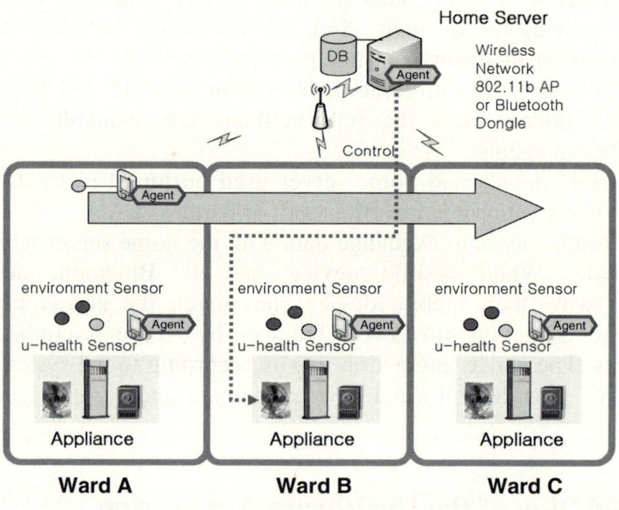

Fig. 3. The system environments of healthcare application for mobile collaboration

4.2 Definition of Healthcare Application Components

The components of the healthcare application based on mobile collaboration framework are defined by the TMO scheme. And we used the TMOSM for interactions between distributed components. Figure 4 describes the interaction of the components for healthcare application.

For the healthcare application, distributed application components of the system(Client, Server) includes the distributed object implemented by TMO scheme and agents that autonomously act according to the perceived context information. In order to facilitate the implementation of autonomous agent for this healthcare Service, we used one more TMOs. Sensor TMO collects environment information, patient health information at the wards and also obtains the location information by agent. Monitoring GUI display the information about the collected data from Sensor TMO. Profile TMO manages user profile, user authority information and Control TMO have responsibility for appliance control. The Context TMO in home server, component for the appliance control service, monitors the action of all information appliances by receiving the information from corresponding appliances. The stationary-typed agents are located on PDAs at wards and home server which are fixed host. The moving-typed agent is located on PDA which is mediate component that interact with PDAs and PCs.

First, for the moving-typed agent on PDA, the Sensor TMO mapping to moving object, called a nurse, in hospital, and sensed by physical sensor (Cricket). It also senses the moving nurse by the periodic time description, stores the location information of home server into information repository. When detecting the moving object, Sensor TMO transfers the location information. And it obtains environment information and patient's health information from sensors or stationary-typed agent at ward. Control TMO controls peripheral appliances by sending those commands in their own proprietary language. In this case, the security object in DOGF needs to grant the access control right this privilege. All of the access control rights specified for the security object need this access privilege.

For stationary-typed agents on PDAs at wards, they also collcct data about the environment information and patient's health information from the environment sensors and health sensors. And then, the information transfers to the Sensor TMO in moving-typed agent.

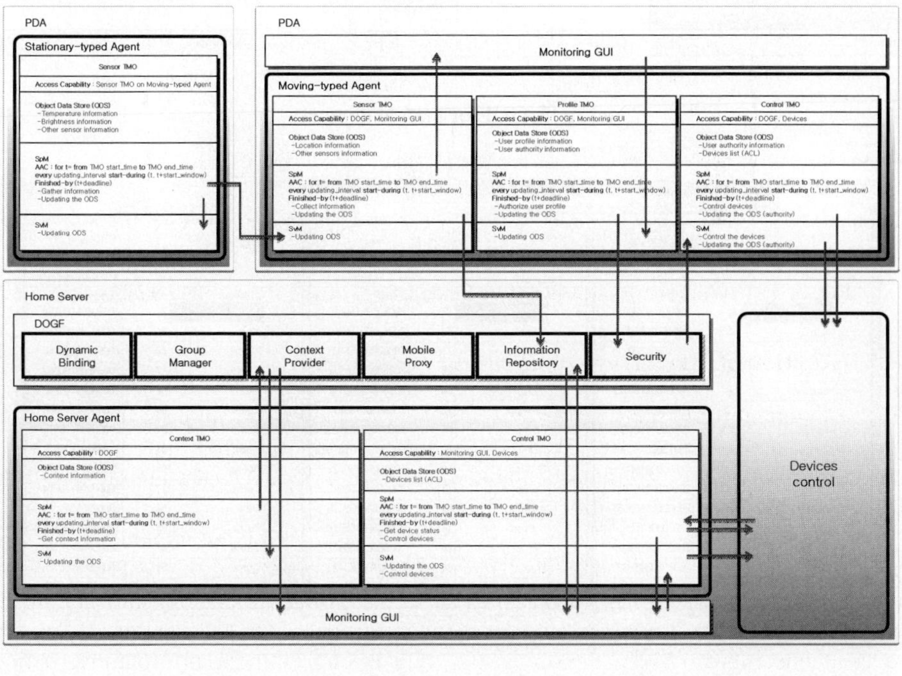

Fig. 4. TMO Class diagram for healthcare application service

4.3 Executing Results of Healthcare Application

We have developed a set of initial healthcare agent services on the mobile collaboration framework to exercise a number of deployment scenarios and collaborative models. Figure 5 presents the healthcare application based on mobile collaboration framework as mentioned above.

First, the GUI of PDA at ward displays collecting data from sensors by stationary type agent and setting environment which add members for collecting data from sensors. The PDA GUI for a nurse shows the information collected data by moving-typed agent for each ward according to location of PDA of nurse. For this, the nurse has to register the collected information to the home server. After then Figure 6 presents the home server GUI displays the collected information for each wards.

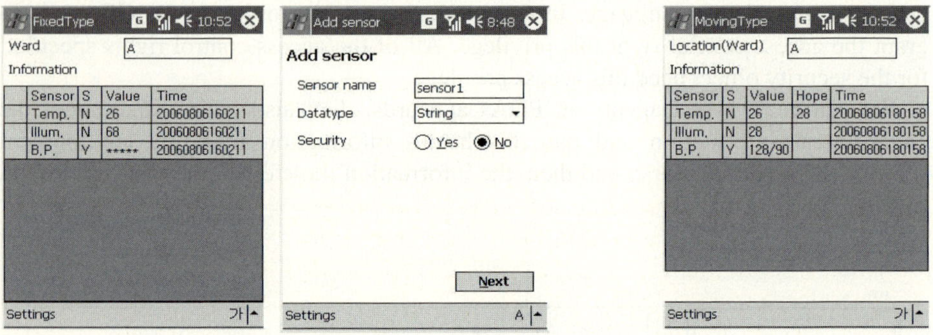

Fig. 5. GUI for mobile devices

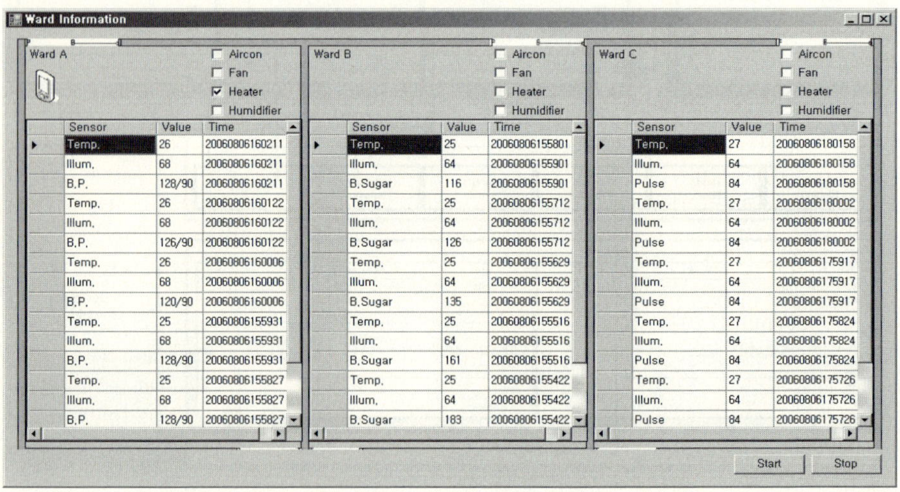

Fig. 6. GUI for integrated monitoring of Home Server

In details, patient health information and the environment information are managed in the home server. It also displays the location for PDA and the status information that is a live home appliance through the hope value on GUI.

5 Conclusions and Future Works

Ubiquitous computing environment includes the various sensors, mobile devices and communication infrastructure. In this environment, research is towards the mobile collaboration that able to deal with the mobility and interactions of both users and devices. Hence, because of limitations on resource and platform, we suggest mobile collaboration framework. It is based on the distributed object group framework that supports the group service and real time service. The mobile collaboration environment includes the sensors, mobile devices and home server. We defined the interaction agent type that interacts with each other. And we applied healthcare service for mobile collaboration such as a hospital ward environment. Each component for executing functions of agent and an integrated monitoring is implemented by TMO scheme. And we used the TMOSM for interactions between distributed components. Finally, we showed via GUI the executablility of healthcare application on our mobile collaboration framework.

Our future work will apply different environments for ubiquitous healthcare service and improving the performance of the framework. We will include studying the mobile agent technologies and then, we will apply to the moving-typed agent.

Acknowledgements. This work was supported by the Regional Research Centers Program of the Korean Ministry of Education & Human Resources Development through the Center for Healthcare Technology Development.

References

[1] Marcela Rodriguez, Jesus Favela, Victor Gonzalez and Miguel Munoz, "Agent Based Mobile Collaboration and Information Access in a Healthcare Environment", Proceedings of Workshop of E-Health: Applications of Computing Science in Medicine and Health Care. ISBN: 970-36-0118-9. Cuernavaca, Mexico, December 2003.

[2] Jukka Riekki, Oleg Davidyuk, Jari Forstadius, Junzhao Sun, Jaakko Sauvola, "Enabling Context-Aware Services for Mobile Users", MCCSIS 2005 April 2005

[3] Y Huang, H Garcia-Molina, "Publish/Subscribe in a Mobile Environment", Wireless Networks, 2004.

[4] TL Pham, G Schneider, S Goose, A Pizano, "Composite Device Computing Environment: A Framework for Situated Interaction using Small Screen Devices", Personal and Ubiquitous Computing, 2001.

[5] L. Kagal, V. Korolev, H. Chen, Anupam Joshi, T. Finin, "Centaurus: A framework for intelligent services in a mobile environment", Distributed Computing Systems Workshop, International Conference, April, 2001, pp.195-201.

[6] Chang-Sun Shin, Chang-Won Jeong, Su-Chong Joo, "Construction of Distributed Object Group Framework and Its Execution Analysis Using Distributed Application Simulation", Embedded and Ubiquitous Computing: International Conference EUC 2004, August, 2004. pp.724-733.

[7] Dong-Suk Kim, Chang-Won Jeong, Su-Chong Joo, "Implementation of Healthcare Application Service in Mobile Collaboration Environment", Proceedings of Korea Computer Congress 2006, Vol. 33, No. 1(D), 2006.6.21-23, pp.88-90.

[8] D.Agarwal, C. McParland, and M. Perry, “Supporting Collaborative Computing and Interaction, “Proceedings of the Grace Hopper Celebration of Women in Computing 2002 Conference, October 9-12, 2002, Vancouver, Canada.

[9] Andreas Heinemann, Jussi Kangasharju, Fernando Lyardet, Max Mühlhäuser, “iClouds - Peer-to-Peer Information Sharing in Mobile Environments”, International Conference on Parallel and Distributed Computing (Euro-Par 2003). Klagenfurt, Austria, 26.-29. August 2003.

[10] Sun Microsystems. jxme : JXTA Java Micro Edition Project. http://jxme.jxta.org/

[11] Gerd Kortuem, jay Schneider, Dustin Preuitt Thaddeus, G. C. Thompson, Stephen Fickas, and Zary Segall, “When Peer-to-Peer comes Face-to-Face: Collaborative Peer-to-peer Computing in Mobile Ad hoc Networks”, In First International Conference on Peer-to-Peer Computing, Linkoping Sweeden, 27-29, August 2001.

[12] Kim, K.H., Ishida, M., and Liu, J.: An Efficient Middleware Architecture Supporting Time-triggered Message-triggered Objects and an NT-based Implementation. In Proceedings of the IEEE CS 2nd International Symposium on Object-oriented Real-time Distributed Computing(ISORC'99) (1999) 54-63.

Developing Load Balancing System for RFID Middlewares Using Mobile Agent Technology

Jian Feng Cui and Heung Seok Chae

Department of Computer Science and Engineering, Pusan National University, 30 Changjeon-dong, Keumjeong-gu, Busan, 609-735, South Korea
{cuijf,hschae}@pusan.ac.kr

Abstract. Mobile agent technology is growing to be an important research topic both in institutions and enterprises. In recent years, RFID middleware technology gains great attentions and load balancing is an important technique to achieve high performance for RFID middlewares. In this paper, we propose an approach to load balancing for RFID middlewares based on Mobile Agent System. Two agents, LGA and RLBA, are designed. LGA is used to gather global workload of RFID middlewares, and RLBA executes load balancing process in case of overloading. This approach fully employed the merits of mobile agent system and provides a scalable and flexible approach to load balancing for RFID middlewares.

Keywords: Load balancing, mobile agent, RFID middleware.

1 Introduction

In recent years, Radio Frequency IDentification (RFID for short) technology has gained great attention due to technology advancements, heightened security concerns and a competitive business environment with emphasis on cost control and affordable RFID tag costs. An RFID system can be used for various applications such as retail, healthcare, logistics, automotive, food industry, etc.

Load balancing is an important technique to enhance the performance of RFID middlewares. In case that some RFID middleware is overloaded, the RFID readers should be reallocated to other under-loaded RFID middlewares to make system resource evenly utilized. As a newly emerging technology, mobile agent systems have been used for various areas [1,2,3,4,7,8,16]. In this paper, we propose a mobile agents based approach to load balancing for RFID middlewares. Two agents, LGA (Load Gathering Agent) and RLBA (RFID Load Balancing Agent), are proposed to perform load balancing. By this approach, system scalability and availability are efficiently improved.

The remainder of the paper is organized as follows. Section 2 is an overview of RFID middlewares. Section 3 discusses the policies of load balancing RFID middlewares. In section 4, the agent based approach to load balancing for RFID middleware is represented. Section 5 gives the related works. Conclusion and future work appear in Section 6.

N.T. Nguyen et al. (Eds.): KES-AMSTA 2007, LNAI 4496, pp. 757–764, 2007.

2 An Overview of RFID Middlewares

RFID middleware is a primary component of EPC Network [14]. The Electronic Product Code (EPC, for short) Network was designed and implemented to enable all objects in the world to be linked via the Internet. Fig. 1 shows the high-level components of the EPC Network Architecture.

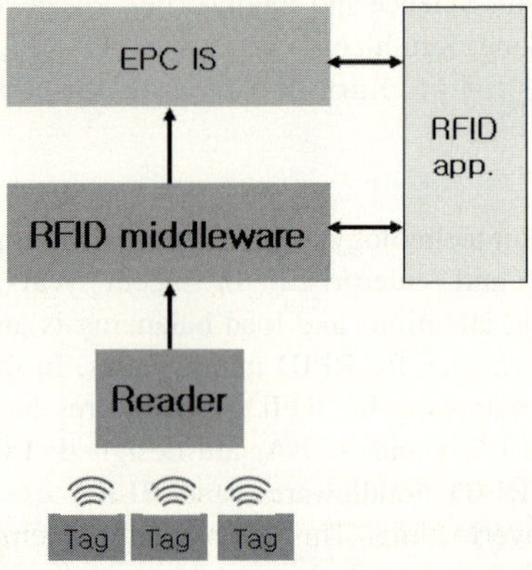

Fig. 1. Overall Architecture of EPC Network

- Tags transmit EPC data using radio frequency. The EPC is a unique number that identifies a specific item and it is stored on a RFID tag. Once the EPC is retrieved from the tag, it can be associated with dynamic data such as its location of origin and the date of its production via EPC IS.
- Readers are devices responsible for detecting when tags enter their read range. They are used to extract the EPCs from the tags. Along with tags, readers enable the automated identification of tagged objects.
- RFID middleware is middleware software designed to process the streams of tag coming from one or more reader devices. RFID middleware performs filtering, aggregation, and counting of tag data, reducing the volume of data prior to sending to RFID Applications. These functions are required in order to handle the extremely large quantities of data that RFID systems can generate through the continuous interrogation of tags.
- The EPC Information Service (EPC IS) makes EPC Network related data available in XML format to requesting services. Data available through the EPC Information Service may include tag read data collected from middleware, instance-level data such as date of manufacture, expiry date, and object class-level data such as product catalog information.
- RFID applications obtain the necessary information from RFID middleware (tag read event) and EPC IS (its associated item data) for business processing.

RFID middleware usually involves a number of readers and tasks. Particularly, readers are arbitrarily positioned in the practical implementation. Tasks are submitted via readers to different RFID middleware in a decentralized fashion. Therefore, a RFID middleware cluster lacks of fixed topology. In addition, because of the distribution feature of RFID middleware, it is hard to globally collect accurate workload information of hosts. Given such a situation, a centralized load balancing scheduler may be not feasible to handle such a complex scheduling problem. Otherwise, the centralized load balancing scheduler will be a bottleneck of the network. Therefore we need to provide a scalable and decentralized scheduling mechanism for RFID middlewares. Mobile agent systems have pretty features to accommodate those situations, such as, agents execute asynchronously and autonomously, mobile agent system is naturally heterogeneous, agents have the ability to survive in network disconnection and they communicate each other in a location-transparent way.

3 Policies of Mobile Agents Based Load Balancing

Load balancing technology is to distribute the amount of work that a host has to do among two or more hosts so that more work gets done in the same amount of time and, in general, all users get served faster.

The design of mobile agents based load balancing platform for RFID middleware system should be compliant to the following policies [3], and the policies must be represented and implemented in appropriate system components.

- Information gathering policy: Maintains information about workload at the hosts in the cluster, including frequency of information exchange and the method for dissemination of the information. There is a tradeoff between having accurate information and minimizing the overhead. It also includes the estimation and specification of workload, e.g. processor load, length of queue, storage utility, etc. In our approach, user can define a proper frequency for information exchange. The number of the readers connected to the RFID middleware and the number of tags read from those readers are candidate of workload in RFID middlewares.
- Initiation policy: Determines who initiates the process of load balancing. The initiator can be the source node, the destination node, or both (symmetric initiations). In our approach, we assume that an overloaded RFID middleware initiates process of load balancing.
- Job transfer policy: Determines when the initiator should reallocate requests to other node. The decision can be made based on only local state or by exchanging global processor load information. The number of tags from RFID readers connected to an RFID middleware can be a good choice as criteria. Once workload exceeds the capability of the RFID middleware, the local RFID middleware starts to distribute workload to remote ones.
- Selection policy: Determines which particular job to reallocate. In our approach, some special agent is designed to obtain the load information of each reader connected to an RFID middleware. Once the workload of an RFID middleware exceeds its allowed capability, we should decide a set of RFID readers so that the overhead of relocation of readers is minimized.

- Location policy: Determines which node the jobs should be reallocated to. The simplest location policy is to choose a node at random. More complicated policies use negotiation, where the initiator negotiates with each member in a subset of node. In our approach, we choose the lightest loaded RFID middleware as the destination. The readers which are selected according to the selection policy are relocated to other RFID middlewares of light workload.

4 The Proposed Approach to Load Balancing for RFID Middlewares

According to our approach, an RFID middleware and an agent container reside on each node above mobile agent platform. RFID readers are dynamically reallocated to RFID middlewares by agents according to local and global workload status.

To support the policies discussed previously, two essential agents, namely Load Gathering Agent (LGA) and Reader Load Balance Agent (RLBA), are designed. RLBA resides on each RFID middleware host, and LGA is a mobile agent traveling within RFID middleware cluster.

4.1 The Underlying Mobile Agent System

In the last few years, mobile agent system technology has become a new exciting field in computer science [6]. There exist a large number of approaches, toolkits, and platforms of different quality and maturity for mobile agent technology, such as Grasshopper [9], IBM Aglet [11] and JADE [13].

JADE is a middleware that facilitates the development of mobile agent system. It could be a proper agent platform for performing load balance for RFID middlewares for the following reasons. JADE runs on Java 2 platform and implements the standard FIPA [12], working with CORBA [10]. For communication mechanism, JADE support ACL [12], inter-platform messaging with plug-in MTPs and XML codec for messages. In addition, JADE Object Manger provides connection authentication, user validation and RPC message encryption. JADE has been studied as a mobile agent platform in many projects.

Each running instance of the JADE runtime environment is called a *container* as it can contain several agents. The set of active containers is called a *platform*. A single special *main container* must always be active in a platform and all other containers register with it as soon as they start. A *main container* differs from normal *containers* as it holds two special agents, *AMS* (Agent Management System) and *DF* (Directory facilitator). *AMS* provides naming service and represents the authority in the platform. DF is used to provide a Yellow Pages service for agents to find each other.

4.2 Load Gathering Agent (LGA)

LGA travels through RFID middlewares in the cluster to gather the workload information from each RFID middleware. As described in Fig. 2, LGA is compliant to information gathering policy. In one RFID middleware cluster, there is only one LGA agent. LGA has two basic functions: first, LGA gets an address list of containers within the RFID middleware cluster and mobile itinerantly among them; and second, when LGA arrives at each node, it could check the load status of current Middleware and tell the latest global load information to it. Each middleware holds a copy of global load information, which will be used to make decision on reader reallocation.

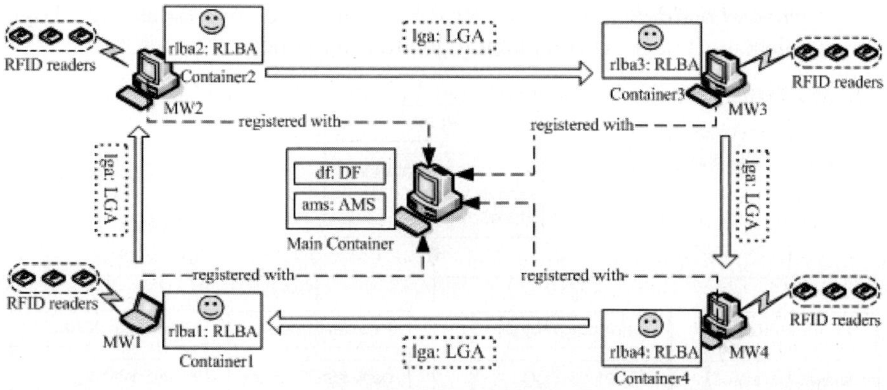

Fig. 2. Itinerary of LGA

The global workload information is organized as Middleware Load Table (for short, MLT). This table gives workload information of each RFID middleware in the cluster. If load information changed, this table will also be updated immediately by LGA. The local copy of MLT at each middleware is used by RLBA to determine the target middleware to which some selected readers should be reallocated.

4.3 RFID Middleware Load Balancing Agent (RLBA)

RLBA is designed to autonomously decide when and where to reallocate overloading readers, and execute reallocation. RLBA resides on each RFID middleware.

Similar with LGA, RLBA also holds a table to track local readers' workload. The load information is organized as Reader Load Table (for short, RLT). Each RFID middleware owns one RLT.

RLBA consists of five components: Load Monitor, Balancing Trigger, Middleware Chooser, Reader Reallocator, and Reader Chooser. MLT and RLT support load balancing decision. Fig. 3 is the class diagram of RLBA. It illustrates how RLBA works with its components to execute load balancing for RFID middleware.

- Load Monitor: Collects local readers' workload information. An operation named *activate(period)* is defined to check the workload of local RFID middleware in a regular period with the parameter *period*, and then update the RLT on local host. Load Monitor is compliant to initiation policy.

  Balancing Trigger: Decides the load balancing criteria and whether to perform load balancing or not. Based on predefined threshold, Balancing Trigger examines the RLT table to determine whether the total workload from its readers is beyond its capacity of processing tag information. Two operations named *activate(period, threshold)* and *ifOverloading()* are defined to perform the above tasks. The parameter *period* defines the cycle period, and *threshold* defines overloading criteria. Once the overall load from readers exceeds the capacity of the middleware, RLBA issues load balancing by signaling Reader Chooser and Middleware Chooser, and the operation *activateLoadBalancing()* of RLBA is invoked by Balancing Trigger. Balancing Trigger is compliant to job transfer policy.

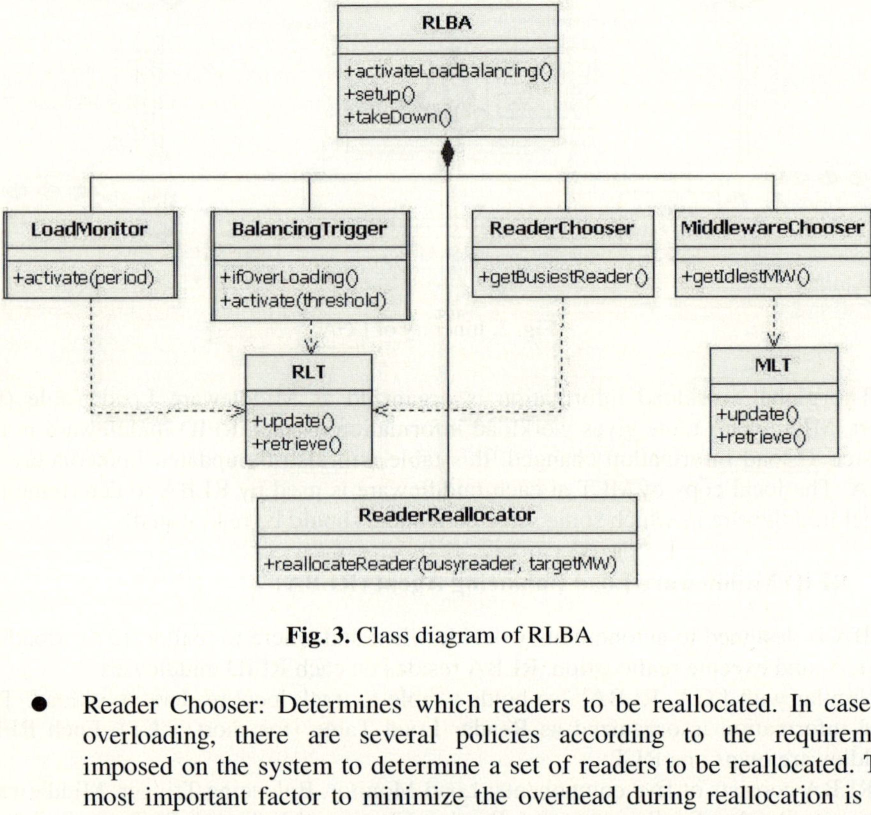

Fig. 3. Class diagram of RLBA

- Reader Chooser: Determines which readers to be reallocated. In case of overloading, there are several policies according to the requirement imposed on the system to determine a set of readers to be reallocated. The most important factor to minimize the overhead during reallocation is the number of reallocated readers, so that we have to determine a set of readers with heavier workload and reallocate them. An operation named *getBusiestReader()* is defined to perform the above tasks. Reader Chooser is compliant to selection policy.
- Middleware Chooser: Determines the destination RFID middleware to which the chosen readers are reallocated. As mentioned earlier, LGA travels through each container, and leaves a copy of MLT at each RFID middleware. In case of overloading, Middleware Chooser accesses MLT to determine the lightest workload RFID middleware for reallocating. An operation named *getIdlestMW()* is defined to perform the above tasks. Middleware Chooser is compliant to location policy.

- Reader Reallocator: Reallocates readers to destination. As soon as the readers to be reallocated and the middleware to be the destination of the reallocation are determined, Reader Reallocator reallocates the selected readers to the destination RFID middleware.

5 Related Works

Mobile agent for network management tends to monitor and control networked devices on site and consequently save the manager capacity and network bandwidth. The MAN architecture [15] is proposed to execute monitoring interface performance and fault diagnosis within the network. There are researches which support load balancing in parallel and distributed system using mobile agent platform. MALD [5] is agent based framework to implement scalable load balancing on distributed web servers. FLASH [7] is a framework for the creation of load-balanced distributed applications in heterogeneous cluster systems. In [16], authors presented a macroscopic characterization of agent-based load balancing in homogeneous minigrid environments. In [8], authors proposed the use of mobile agents as an aid to structure and design distributed and dynamic load-sharing mechanisms for network services.

To improve the scalability and flexibility of RFID middlewares, RFID readers should be dynamically bound to RFID middleware to assure even distribution of input requests. So far to our knowledge, mobile agent technology has not been used to support load balancing for RFID middlewares yet.

6 Conclusion and Future Work

Mobile agent technology is growing to be an important research topic both in institutions and enterprises, and particularly the standardization and formalization of agent communication are hot subjects. In this paper, we choose JADE as underlying mobile agent platform, which is compliant to FIPA standard and based on which we have proposed an agent based approach to load balancing for RFID middleware. Two new agents, LGA (Load Gathering Agent) and RLBA (RFID Load Balancing Agent), are designed to perform load balancing. By this means, readers can be dynamically bound to RFID middlewares depending on each middleware's performance, which greatly improves the scalability and flexibility of RFID middlewares.

Agent oriented programming and agent oriented software engineering are effective technologies to facilitate agent development. In the future work, we'll continue to exploit the features of mobile agents and extend our approach to implement them to the RFID middleware applications. In addition, a more sophisticated load balancing algorithm is expected to exploit rather than simply considering the number of tags on each reader and RFID middlewares.

Acknowledgments

"This work was supported by the Korea Research Foundation Grant funded by the Korean Government(MOEHRD)" (The Regional Research Universities Program/Research Center for Logistics Information Technology).

References

1. Ahmad Naveed, Ali Arshad, Anjum Ashiq, Azim Tahir, Bunn Julian, Hassan Ali, Ikram Ahsan: Distributed Analysis and Load Balancing System for Grid Enabled Analysis on Hand-held devices using Multi-Agents Systems, Proceedings of the 3rd International Conference on Grid and Cooperative Computing, Jun 2004
2. Bin Li, Xiao-Yan Tang, Jian Lv: The Research and Implementation of Services Discovery Agent in Web Services Composition Framework, Machine Learning and Cybernetics, 2005. Proceedings of 2005 International Conference, Aug 2005
3. J.Cao,Y.Sun, X.Wang, K.Das. Sajal: A Framework of Using Cooperating Mobile Agents to Achieve Load Sharing in Distributed Web Server Groups, Proceeding of the fifth inter-
4. Grivas, M., Turner, S.J.: Agent Technology in Load Balancing for Network Applications, International Workshop on Intelligent Agents on the Internet and Web, Mexico, March 1998
5. Jiannong Cao, Yudong Sun, Xianbin Wang, Sajal K. Das: Scalable load balancing on distributed web servers using mobile agents, Journal of Parallel and Distributed Computing, Vol63, Issue 10, Oct 2003
6. Glitho. R.H., Olougouna. E., Pierre, S.: Mobile agents and their use for information retrieval: a brief overview and an elaborate case study, Network IEEE, Volume 16, Issue 1, Jan.-Feb. 2002
7. W. Obeloer, C. Grewe, H. Pals: Load management with mobile agents, in, Proc. 24th EUROMICRO Conference (EUROMICRO 98), Vol.2, Vasteras, Sweden, 1998, pp. 1005-1012
8. Jiannong Cao, Xianbing Wang, Sajal K. Das: A Framework of Using Cooperating Mobile Agents to Achieve Load Sharing in Distributed Web Server Groups", Proc. 5th International Conference on Algorithms and Architectures for Parallel Processing, Oct 2002
9. Grasshopper – the agent platform, http://www.grasshopper.de/
10. CORBA Specification, formal/04-03-12, http://www.omg.org/
11. Aglet Agent Development Toolkit, http://aglets.sourceforge.net/
12. Publicly Available Implementations of FIPA Specifications, http://www.fipa.org/
13. Java Agent Development Framework, http://jade.tilab.com/
14. EPCglobal Inc. (Electronic Product Code), http://www.epcglobalinc.com/
15. Manoj Kumar Kona, Cheng-Zhong Xu: A Framework for Network Management using Mobile Agents, Parallel and Distributed Processing Symposium., Proceedings International, Apr 2002
16. Jiming Liu, Xiaolong Jin, Yuanshi Wang: Agent-based load balancing on homogeneous minigrids: macroscopic modeling and characterization, Parallel and Distributed Systems, IEEE Transactions, Volume16, Issue7, July 2005

Design Agent for the Reliable Web Security Requirement Control

Eun-Ser Lee[1] and Je-Min Bae[2]

[1] Information & Media Technology Institute, Soongsil University
eslee1@ssu.ac.kr
[2] Department of computer Education Kwandong University
gemin@kd.ac.kr

Abstract. This research is intended to design agent for the security requirement to estimate by empirical study. This paper is intended to help the design agent and progress management based on the ISO/IEC 15408. There are many defects that cause the security requirement problems during the web development. This paper remove the risk of the lifecycle and progress management that check the risk items in the web security requirements and also manage the schedule and quality problems. For projects in similar domains, it is possible to remove security risk items and to manage progress by using agent web security lifecycle and progress milestone, which can greatly improve the software process. The analysis is made based on the types required for making use of the defect data. In this case, additional measures must be taken besides merely recording defects. And the theory is apply to the agent web security access control system for the security requirement.

Keywords: Security lifecycle, Agent, Require management.

1 Introduction

Artificial intelligence may be defined as the branch of computer science that is concerned with the automation of intelligent behavior. These principles include the data structures used in knowledge representation, the algorithms needed to apply that knowledge, and languages and programming techniques used in their implementation[17].

However, this definition suffers from the fact that intelligence itself is not very well defined or understood.

The recent advances in information technologies and the proliferation of computing systems and world-wide networks have raised the level of concern about security in the public and private sectors[1][2]. Consumers have access to a growing number of security enhanced IT products with different capabilities and limitations, and should make important decisions about which products provide an appropriate degree of protection of their information.

In order to help consumers select commercial off-the-shelf IT products that meet their security requirements and to help manufacturers of those products gain

N.T. Nguyen et al. (Eds.): KES-AMSTA 2007, LNAI 4496, pp. 765–774, 2007.

acceptance in the global marketplace, the National Institute of Standards and Technology (NIST) and the National Security Agency (NSA) have established a program under the National Information Assurance Partnership (NIAP) to evaluate IT product conformance to international standards. The program, officially known as the NIAP Common Criteria Evaluation and Validation Scheme for IT Security (Common Criteria Scheme in short) is a partnership between the public and private sectors [3][4].

This paper provides an analysis of the efficiency of the removal of agent web security requirement risk. Security risk items can also be removed by using security lifecycle and progress milestone, which can greatly improve the software process.

2 Related Works

2.1 ISO/IEC 15408 (Common Criteria, CC)

The Common Criteria (CC), is meant to be used as the basis for evaluation of security properties of IT products and systems. By establishing such a common criteria base, the results of an IT security evaluation will be meaningful to a wider audience [5][6].

The CC is useful as a guide for the development of products or systems with IT security functions and for the procurement of commercial products and systems with such functions. During evaluation, such an IT product or system is known as a Target of Evaluation (TOE). Such TOEs include, for example, operating systems, computer networks, distributed systems, and applications.

The CC defines three types of requirement constructs: package, PP(Protect Profile) and ST(Security Target). The CC further defines a set of IT security criteria that can address the needs of many communities and thus serve as a major expert input to the production of these constructs. The CC has been developed around the central notion of using, wherever possible, the security requirements components defined in the CC, which represent a well-known and understood domain. Figure 1 shows the relationship between these different constructs [1][2].

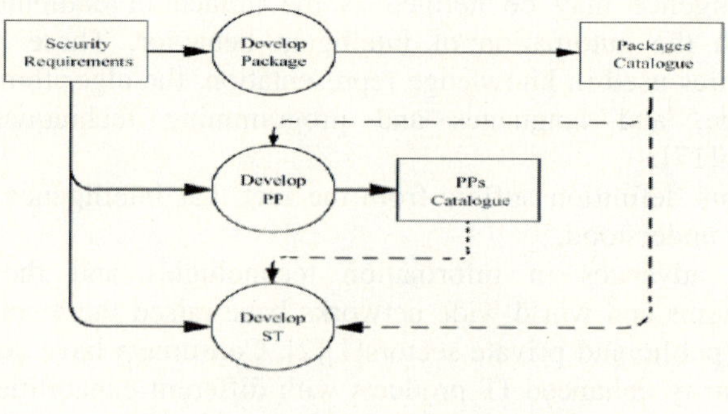

Fig. 1. Use of security requirements

Evaluation criteria are most useful in the context of the engineering processes and regulatory frameworks that are supportive of secure TOE development and evaluation. This sub-clause is provided for illustration and guidance purposes only and is not intended to constrain the analysis processes, development approaches, or evaluation schemes within which the CC might be employed [1][2][3].

The CC requires that certain levels of representation contain a rationale for the representation of the TOE at that level. That is, such a level must contain reasoned and convincing argument that shows that it is in conformance with the higher level, and is itself complete, correct and internally consistent. Rationale directly demonstrating compliance with security objectives supports the case that the TOE is effective in countering the threats and enforcing the organizational security policy.

The CC layers the different levels of representation as described in Figure 2, which illustrates the means by which the security requirements and specifications might be derived when developing a PP or ST. All TOE security requirements ultimately arise from consideration of the purpose and context of the TOE. This chart is not intended to constrain the means by which PPs and STs are developed, but illustrates how the results of some analytic approaches relate to the content of PPs and STs.

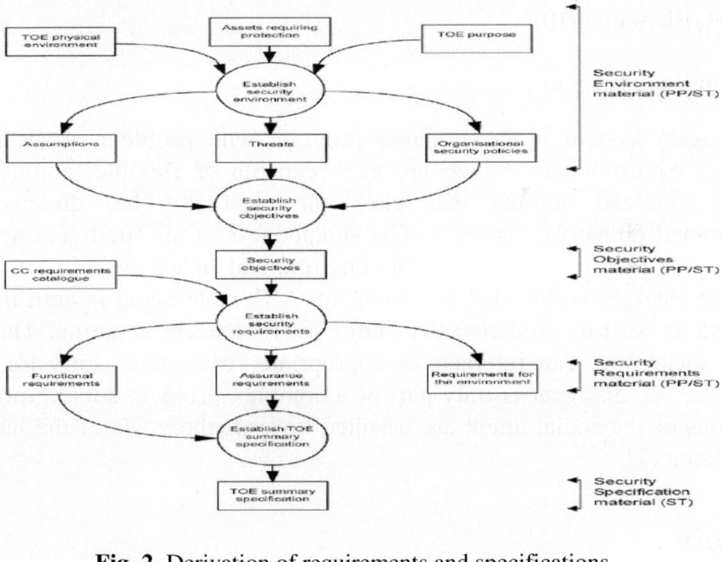

Fig. 2. Derivation of requirements and specifications

An ST shall conform to the content requirements described in this chapter. An ST should be presented as a user-oriented document that minimizes reference to other material that might not be readily available to the ST user [1][2][3].

The rationale may be supplied separately, if that is appropriate. The contents of the ST are shown in Figure 3, which should be used when constructing the structural outline of the ST.

2.2 MBASE

The difference between failure and success in developing a software-intensive system can often be traced to the presence or absence of clashes among the models used to define the system's product, process, property, and success characteristics [7].

In each case, property models are invoked to help verify that the project's success models, product models, process models, and property levels or models are acceptably consistent. It has been found advisable to do this especially at two particular "anchor point" life cycle process milestones summarized in Table 2 [8][9]. The first milestone is the Life Cycle Objectives (LCO) milestone, at which management verifies the basis for a business commitment to proceed at least through an architecting stage. This involves verifying that there is at least one system architecture and choice of COTS/reuse components which is shown to be feasible to implement within budget and schedule constraints, to satisfy key stakeholder win conditions, and to generate a viable investment business case. The second milestone is the Life Cycle Architecture (LCA) milestone, at which management verifies the basis for a sound commitment to product development (a particular system architecture with specific COTS and reuse commitments which is shown to be feasible with respect to budget, schedule, requirements, operations concept and business case; identification and commitment of all key life-cycle stakeholders; and elimination of all critical risk items) [10].

2.3 Definition of Agent

A multi-agent system is a computer program with problem solves situated in interactive environments, which are each capable of flexible, autonomous, yet socially organized actions that can, but need not be, directed towards predetermined objectives or goals. The situatedness of an intelligent agent means that the agent receives input from the environment in which it is active and can also effect changes within that environment. A flexible agent is both intelligently responsive as well as proactive depending on its current situation. Therefore, an agent is social that can interact, as appropriate, with other software or human agent. After all, an agent is only part of a complex problem solving process. The interactions of the social agent are oriented towards the goals of the larger multi-agent system[17].

3 Theory

Security functions were evaluated using security profiling and security targeting, by way of the ubiquitous method. Evaluation progress was introduced in chapter 2. It should be noted that when evaluation is conducted security requirements focus on the functional. However, the downside of this in terms of security is that defects with often appear in the next stage. Therefore one should refer to the lifecycle to in order to extract and identify the risk factors of the security items. Another problem is that risk factor extraction is a redundant processing job. Furthermore, extraction processing is ambiguous and so firm progress management is required. This chapter provides identification of risk items and checks the rationale and measurement of total progress management.

3.1 Security Profile Lifecycle

The lifecycle is made by protecting the profiles in order to fulfill the security requirement. The next step is the removal of the redundant risk items. Protection profile progress and identification is as follows:[11][12][13][14]

1. A protection profile developer must be provided to explain the profile of the TOE.
2. Explanation of the TOE must be provided so as to describe its product type and the character of the general IT.
3. The evaluator must have confirmation that the information has satisfied all of the evidence requirements.
4. The evaluator must have confirmation of the explanation of the TOE with regards to quality and consistency.
5. The evaluator must have confirmation of the explanation of the TOE with regards to the relationship of protection profile consistency

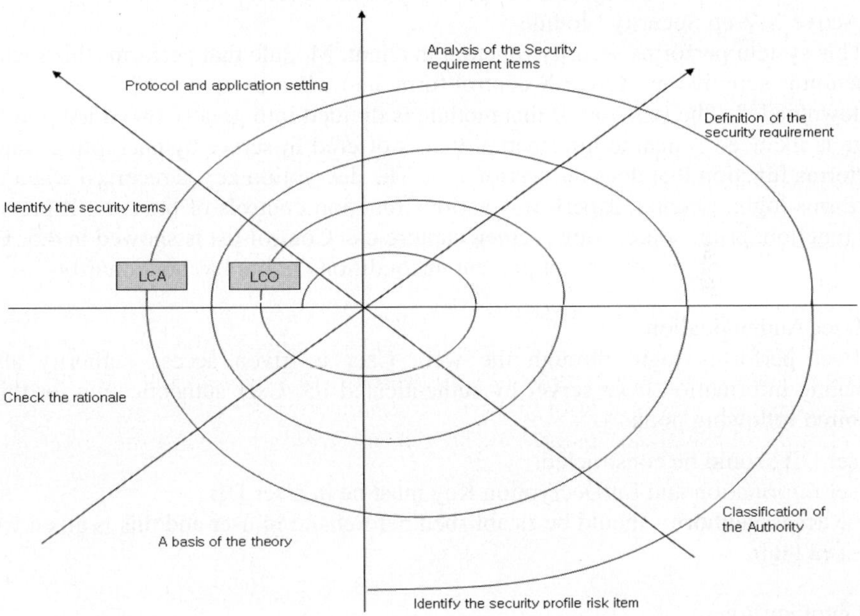

Fig. 3. Web security access control milestone of the security check point

This figure shows the activity of the risk item as it is removed from the security requirement during the extraction process. Each of the factors use the repeatable milestone for progress management. Each stage is as follows:

Table 1. Description of the web security access control milestone protection profile

Stage	Contents
Identify the security items	Identification and analysis the security items for the extraction of the web control
Protocol and application setting	H/W and S/W element setting for the web security control. (e.g protocol and application program)
Analysis of the Security requirement items	Identify and extraction of the considering each of the items relationship
Definition of the security requirement	Definition of the security requirement for system building
Classification of the Authority	Classification of the authority for Access control
Identify the security profile risk item	Identify the security profile risk item at the domain
Basis of the theory	Build of the repeatable lifecycle for the milestone(LCO, LCA)
Checking of the rationale	Check the rationale of the repeatable milestone and risk analysis

Also we will show the system design apply to this check point[16].

1) Active X Web Security Module

This system performs security function in client. Module that performs this security function is supplied by Active X control form, and when user connects to server, user is downloaded. The function of this module is divided into greatly two. First, the web page is included semantic information that is offered in server by encrypting state, it performs function that does decryption this. The decryption key is received when user performs login. Second, it performs various function controls of browser. It prevents the function, print, source view, screen capture etc. Control list is showed in 4.5. User receives limitation in case of not prevent unconditionally but has no authority.

2) User Authentication

User performs login through the web. User is given access authority about semantic information from server by authenticated ID. User authentication for this is required following point.

- User DB should be constructed.
- User information and En/Decryption Key must be in User DB.
- The access authority should be established beforehand in user and this is given when succeed login.

3) Ontology Agent

Ontology Agent accepts information that user wants as keyword and it extracts semantic information that is correct in the keyword in Ontology that already constructed and performs function that makes it by file of html form. Ontology in this paper limits by climate field. And server encrypts this file and sends to User PC. This time, if user has read authority for the semantic information, one can show the information if not, one can't show.

4) En/Dectyption

As preceding section referred, html file that made by Ontology Agent is downloaded being encrypted. User can see this file doing decryption by Active X Web Security Module that is downloaded in client. In this system, en/decryption module can do file whole to doing en/decryption but it can also offer part encryption function to reduce load. This ensures the security of the web page when it is downloaded to the client's PC.

5) Authority Control

Authorized users can get permission or limit the permission assigned during server authentication.

Above control function is performed in client and this system can prevent information leakage by user's mistake as well as information leakage by user's deliberation by performing relevant function. For example, read, print, source view and various function of web browser etc.

Therefore, we are gain the check point in this view of previous items. And in this check point is apply to the efficiency and effectiveness that DRE was calculated at the chapter 4.

3.2 Milestone for the Progress Management

Use of the milestones (LCO, LCA) is essential for removal of risk and progress management.

The authors of this paper have provided the repeatable cycle based on the milestone element. Progress was checked using the basis of the milestone.

Table 2. Description of the milestone element

Milestone Element	Life Cycle Objectives (LCO)	Life Cycle Architecture (LCA)
Definition of Operational Concept	•Top- level system objectives and scope –System boundary –Environment parameters and assumptions –Evolution parameters •Operational concept –Operations and maintenance scenarios and parameters –Organizational life- cycle responsibilities (stakeholders)	•Elaboration of system objectives and scope of increment •Elaboration of operational concept by increment
System Prototype(s)	•Exercise key usage scenarios •Resolve critical risks	•Exercise range of usage scenarios •Resolve major outstanding risks
Definition of System Requirements	•Top- level functions, interfaces, quality attribute levels, including: –Growth vectors and priorities –Prototypes •Stakeholders' concurrence on essentials	•Elaboration of functions, interfaces, quality attributes, and •prototypes by increment –Identification of TBD's((to- be- determined items) •Stakeholders' concurrence on their priority concerns
Definition of System and Software Architecture	•Top- level definition of at least one feasible architecture •Physical and logical elements and relationships •Choices of COTS and reusable software elements •Identification of infeasible architecture options	•Choice of architecture and elaboration by increment •Physical and logical components, connectors, •configurations, constraints •COTS, reuse choices •Domain- architecture and architectural style choices •Architecture evolution parameters
Definition of Life-Cycle Plan	•Identification of life- cycle stakeholders – Users, customers, developers, maintainers, interoperators, general public, others •Identification of life- cycle process model –Top- level stages, increments •Top- level WWWWWHH* by stage	•Elaboration of WWWWWHH* for Initial Operational Capability (IOC) –Partial elaboration, identification of key TBD's for later increments
Feasibility Rationale	•Assurance of consistency among elements above –via analysis, measurement, prototyping, simulation, etc. –Business case analysis for requirements, feasible architectures	•Assurance of consistency among elements above •All major risks resolved or covered by risk management plan

* WWWWWHH: Why, What, When, Who, Where, How, How Much

We are provides that the repeatable cycle based on milestone element. Also, we are checked the progress by the basis of milestone.

Table 3. Setting of the milestone

Cycle 1 (LCO)	Cycle 2	Cycle 3 (LCA)
Determination of top-level concept of operations	Determination of detailed concept of operations	Elaboration of detailed concept of operations by increment, especially IOC
System scope / boundaries /interfaces; top-level requirements	Top-level HW, SW, human requirements	Determination of requirements, growth vector by increment, especially IOC
Small number of feasible candidate architectures (including major COTS, reuse choices)	Provisional choice of top-level information architecture	Choice of life-cycle architecture. Some components of above TBD(low-risk and/or deferrable)
Top-level life cycle responsibilities(stakeholders), process, model, cost/schedule parameters	Make detailed process strategy, responsibilities, cost / schedule allocation	Thorough WWWWWHH plans for IOC; essentials for later increments
Stakeholder concurrence on top-level analysis supporting win-win satisfaction	More detailed analysis supporting win-win satisfaction	Stakeholder concurrence on thorough analysis supporting win-win satisfaction
Top level rationale, including rejected candidate architectures	More detailed rationale underlying system choices	Elaboration of rationale, including risk resolution results

3.3 Configuration of Agent

We are gain the result Fig3 and Table2, 3. And this data is become basic data of the agent configuration. First of all, we are needed the network for the agent navigation and adaptation, network sensing. Also, web agent is running for the web security requirement control. Thus, in this paper propose the next architecture. Agent configuration is like the following.

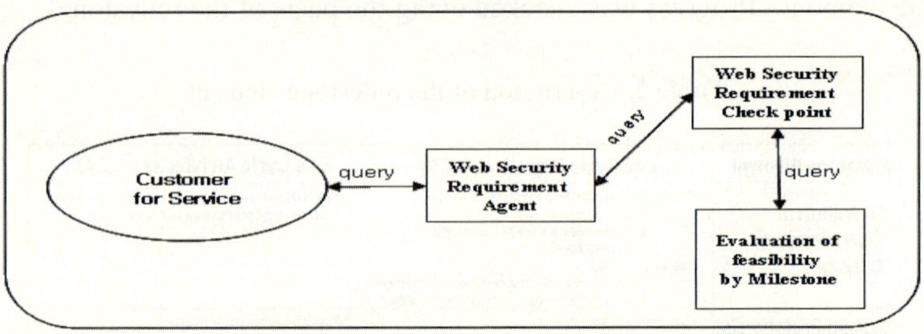

Fig. 4. Configuration of web security requirement control Agent

4 Calculation of Defect Removal Efficiency

Purpose of defect trigger design improves quality of product and heighten productivity. Therefore, when we applied Defect Trigger in actual project, we wish to apply defect removal efficiency to measure ability of defect control activity.

With apply defect trigger, defect removal efficiency analysis [15] investigated defect number found at relevant S/W development step and defect number found at next time step in terms of request analysis, design and coding stage. We show you to compute the defect removal efficiency is as follows

$$DRE = E/(E+D)$$

E= Number of defect found at relevant S/W development step(e.g : Number of defect found at request analysis step)

D= Number of defect found at next S/W development step

(e.g : Defect number that defect found at design step is responsible for defect of request analysis step)

Ideal value of DRE is 1, and this displays that no defect on the project.

Table 4. Tabel of defect removal efficiency

	Number(%) of defect found at relevant S/W development step (E)	Number(%) of defect found at next S/W development step (D)
Inception (Requirements)	30	5
Elaboration (Product Des)	15	3
Construction (Development)	10	2
Transition	3	1
Total	58	10

Table 4 is a table to show up defect number on the each step by inspection with defect trigger application. inspect software development step defect number after Defect Trigger application. We get DRE at each software development step by table 4, it is as following.

0.857 = 30/(30+5) (Inception)
0.833 = 15/(15+3) (Elaboration)
0.833 = 10/(10+2) (Construction)
0.75 = 3/(3+1) (Transition)

Therefore, because DRE is approximated to 1, when we remove defect by Defect Trigger, DRE was analyzed good efficiency.

5 Conclusion

In this paper a new check of Agent lifecycle was proposed applicable to the extraction of the web security access control in the security requirement stage, along with reliable analysis of relevant requirements. The authors propose not only risk item removal, but also progress management. In future studies implementation of agent program applicable to the extract of security requirements will be provided.

References

1. ISO/IEC 15408-1:1999 Information technology - Security techniques - Evaluation criteria for IT security - Part 1: Introduction and general model
2. ISO. ISO/IEC 15408-2:1999 Information technology - Security techniques - Evaluation criteria for IT security - Part 2: Security functional requirements

3. ISO. ISO/IEC 15408-3:1999 Information technology - Security techniques - Evaluation criteria for IT security - Part 3: Security assurance requirements
4. The Report of the President's Commission on Critical Infrastructure Protection CCEB (Common Criteria Editorial Board), Common Criteria for Information Technology Security Evaluation, Version 2.0, May 1998
5. DOD (U.S. Department of Defense), Trusted Computer System Evaluation Criteria, DOD5200.28-STD, December 1985. 1.0, December 1992
6. [ISO96] ISO/IEC Guide 65—General Requirements for Bodies Operating Product Certification Systems, 1996
7. Mark Weiser, "The Computer for the Twenty-First Century," Scientific American, pp. 94-10, September 1991
8. B. Boehm, Software Risk Management, IEEE-CS Press, 1989
9. B. Boehm, A. Egyed, J. Kwan, and R. Madachy, "Developing Multimedia Applications with the WinWin Spiral Model," Proceedings, ESEC/ FSE 97, Springer Verlag, 1997
10. B. Boehm and P. Bose, "A Collaborative Spiral Process Model Based on Theory W," Proceedings, ICSP3, IEEE, 1994. 17
11. Eun-Ser Lee and Sun-Myoung Hwang, "Definition of Security Requirement Items and Its Process to Security and Progress Management", LNCIS 344, August 2006
12. Eun-Ser Lee and Sun-Myoung Hwang, "Design Implementation of Web Security Access Control System for Semantic Web Ontoloty", LNCS 3481, May 2005
13. Eun-Ser Lee and Malrey Lee, "Development System Security Process of ISO/IEC TR 15504 and Security Considerations for Software Process Improvement", LNCIS 344, August 2006
14. Eun-Ser Lee and Sang-Ho Lee, "Design progress management for Security Requirements in Ubiqiiuous computing using COQUALMO", LNCS 3984, May 2006
15. Roger S. Pressman, "Software Engineering", Mcgraw-hill international edition, 1997
16. Nam-deok and Cho,Eun-Ser Lee, "Design and Implementation of Semantic Web Search System using Ontology and Anchor Text", LNCS 3984, May 2006
17. George F Luger, "Artificial intelligence", Addison wesley, 2001

Design and Implementation of an Intelligent Robot Agent System Considering the Server's Workload

Seong-Hoo Kim, Seong-Je Kim, and Kyoo-Seok Park

Div. Of Computer Engineering, KyungNam University
Masan, Korea
{arrayiv,sung0702,kspark}@kyungnam.ac.kr

Abstract. As the Internet sites and users have rapidly been increased, the development for search engines has also been accelerated to satisfy users' expectations. As the result, not only the action of collecting documents through many search engines gave hosts workload, but also regular updating all the information is needed since information is newly added. With any circumstances, the necessity of the technology to collect massive information in hosts has been increased for the speed which is a basic requisite of search systems, and for more accurate collection of documents. Also, the role of search engines grows the bigger for Internet users' various demands and flexible process through World Wide Web. In this paper, we design and implement an intelligent robot agent and a remote control system which doesn't give an excessive workload on a target server and makes the collection of documents done in a short period by considering an average workload rate on the target server and the rate of the workload that a robot experience in collection time, after we compare and analyze the existing Robot Agent Systems and supplement their weak points.

Keywords: Intelligent Robot Agent, Search Engine, Load Balancing.

1 Introduction

Since internet users increasingly want to obtain information of more diverse and in real time, the need of developing a search engine which will promptly provide accurate information classified as a category is arising. Such an engine can reduce network overload.

In this paper, We design and implement an intelligent robot agent scheduler, independent from a platform or controller search engine which can be remote controllable in the internet. The system we propose here is a real time robot control system which can improve the capability of the entire system developed through mixed sorting scheduling methods resulted from monitoring network loads through an intelligent robot agent that can readily collect and renew information scattered in the internet.

Mobile agents support asynchronous and autonomous operations. The nodes can dispatch mobile agents individually that travel independently between the nodes to perform various operations. A mobile agent can encapsulate load balancing policies

N.T. Nguyen et al. (Eds.): KES-AMSTA 2007, LNAI 4496, pp. 775–784, 2007.

and travel to other nodes where it can make decision on load distribution according to the up-to-date states of the nodes. Due to the merits of low network traffic and quick response time, mobile agents can strengthen the scalability of cluster of workstations.

In the following section, relative researches have been examined, then the third section describes designing of the proposed real time search engine, followed by its implementation and evaluation of section 4. Finally, the conclusion is drawn.

2 Related Research

2.1 The Robot Agent

Robot Agent is a program that helps to retrieve documents on the web for desired information and store information, retrieving referable documents recursively, by tracing the structure of hypertext on the web [2][3].

To supply the search services from the internet, at first you have to collect web documents and the collecting method determines the results of the search. You can divide the search method on the basis of robot traversing method as the breath-first traversal and the depth-first traversal. Robots are used for statistical analysis, maintenance, mirroring, finding resources, and various compounding objectives. But there are some possibilities to produce operational errors on robot process, lower the capability when inspecting URL, make problems of controlling indexing results, and the like as it can make overloads to other systems.

To prevent the overloads to the other systems, the robot exclusion rules are required to be observed by inserting a retarding clause between claiming clauses for pages or referring robots.txt files.

2.2 Language Analysis Techniques

As information in the internet consists of multimedia data, most of searching is focused on the information represented by language. Generally, information search is a function to analyze a user's question and then to find the information (web documents) that the user wants[1][2].

In order to find out user's intention of a question, information search system should be able to analyze the composition of the question and the meaning of a key word, and judge what the information the target of search is about and how much the information meets user's intention.

Language processing has its aim to extract as much information as possible from a text given in the form of sentences or clauses. Various processes are applied such as word or clause-based process, substitutive language process, disused language process. Recent web information search engine, however, has a strong tendency of using a dictionary or knowledge information rather than to develop an algorithm to process languages.

2.3 Robot Mixed Sorting Scheduler

The existing search systems are collecting documents by dispatching robots in the order of registration, and it is not using specific robot scheduling methods. This cause the robot to work in overloaded situation as not considering network loading. Apparently the efficiency of robot is decreased.

In contrast, a mixed sorting scheduler creates a priority table by combining lists resulted from detecting optimum load time analyzer and server lists derived from sever manager by alignment scheduler.

2.4 RMI

Using Remote Method Invocation(RMI), the program running at client computers can call methods of objects at remote server computers[13]. RMI consists of three layers of Stub/Skeleton, Remote Reference Layer(RRL), and Transport Layer.

Stub, as an agent of clients which speaks for remote objects, defines all interfaces that remote objects embody and maintains connections to server-side objects being referred as though other local objects within programs. When remote objects are called, Stub has parameter data become serial to deliver them to the remote reference layer as a form of Marshal stream, and after that, when the implementation of remote method is completed, it returns values receiving Marshal stream again from the remote reference layer through reverse serializing.

Marshal stream is used at the communication between server-side RRL and client-side RRL.

Skeleton, as a server-side generator which interfaces with server-side RRL, calls actual objects that are embodied at the side of server after server-side RRL restores parameters forwarded to remote method when requested to call method from client-side RRL. Receiving the return value from remote objects, it delivers again to server-side RRL after marshalling as Marshal stream.

2.5 Load Balancing

Load balancing is an important technique to enhance the performance of multiple nodes. Incoming client requests should be evenly distributed among the nodes to achieve quick response. The load on an overloaded node should be transferred to an under loaded node to enhance the system throughput. Thus, the system resources can get full utilization. Traditional load balancing approaches on cluster of workstation are implemented based on message passing paradigm [7].

A system may be balanced statically at compile/link time or dynamically at runtime. The balancing may be based on processor loading, a tasks data locality and many other factors. The granularity of the tasks also affects balancing. Too much dynamic balancing may actually degrade a systems performance due to the communication, synchronization and complexity of the balancing algorithm used. It is also important to keep in mind that when tasks are migrated from one processor to another to satisfy dynamic balancing requirements, an associated communication and synchronization performance hit is also taken to move the task and its data Figler [8].

3 The Design of System

3.1 Configuration of Robot Agent System

The composition diagram of proposed information search system is as like Fig. 1.

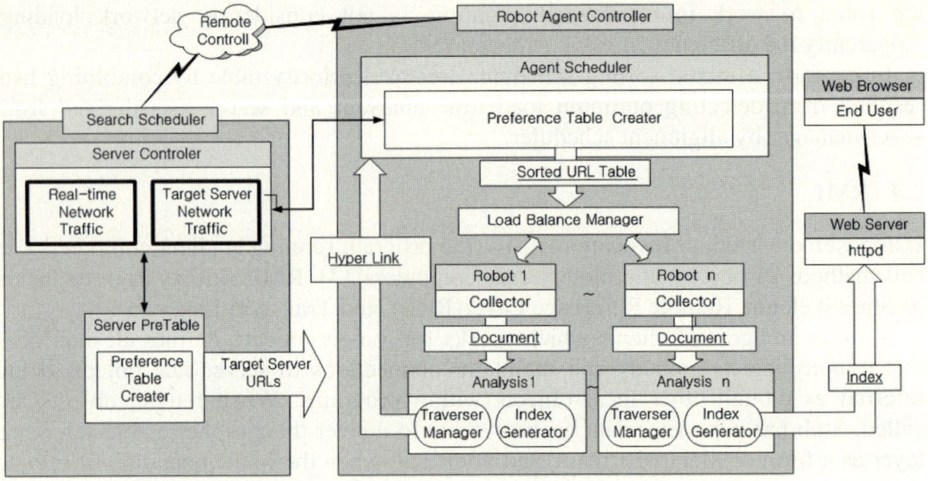

Fig. 1. Composition Diagram of System

Search Scheduler controls the URL of collection target server for information search and makes out schedule so as to perform the collecting work in efficient time zone. Robot Agent Scheduler receives the URL of collection target server and performs the actual work of collecting. Web server transmits the result of collecting to end user.

Search Scheduler consists of Server Controller, which measures the amount of loads of each time zone with the URL of collection target server and Priority Table Creator, which chooses the most efficient targets when performing work with the URL of servers.

Robot Agent Scheduler is composed of Priority Table Creator and Robot Controller, which controls the robots to collect the documents of search target URL reorganized according to the priority, Analyzer, which analyzes the collected documents, and Remote Control System, which uses the RMI. Robot Agent Scheduler is controlled overall by this Remote Control System and is served so as to cope with promptly the circumstance, which changes at all times.

Load Balancing Manager is more time-consuming that involves the interactions between many nodes for gathering load information, negotiating on load reallocation and transporting the workload.

3.2 Server Controller

Fig. 2 shows the composition diagram of Search Scheduler.

Search Scheduler is composed of Server Controller and Priority Table Creator. Server Controller is the module to control the information about many web servers and checks the circumstance of each time zone of web servers and the amount of loads that robot experiences in real time.

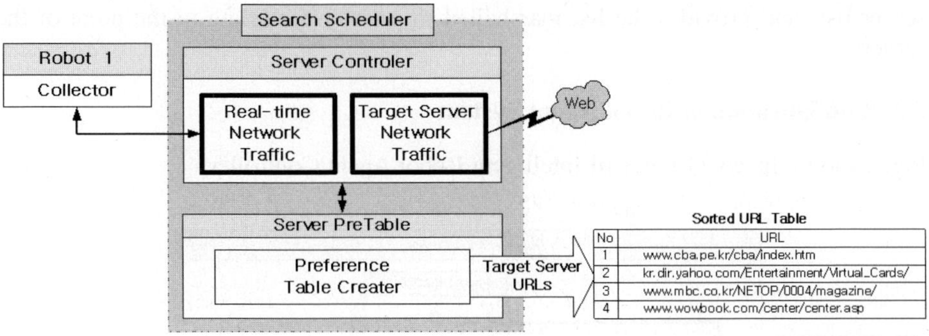

Fig. 2. Composition Diagram of Search Scheduler

Web server's amount of loads are measured per each time zone through Ping each time by Server Controller to save the measured value in average value every time, and gets score together with the value measured during collecting documents. This is the method to secure the averaged base of judging and is used to decide the priority of search target URL.

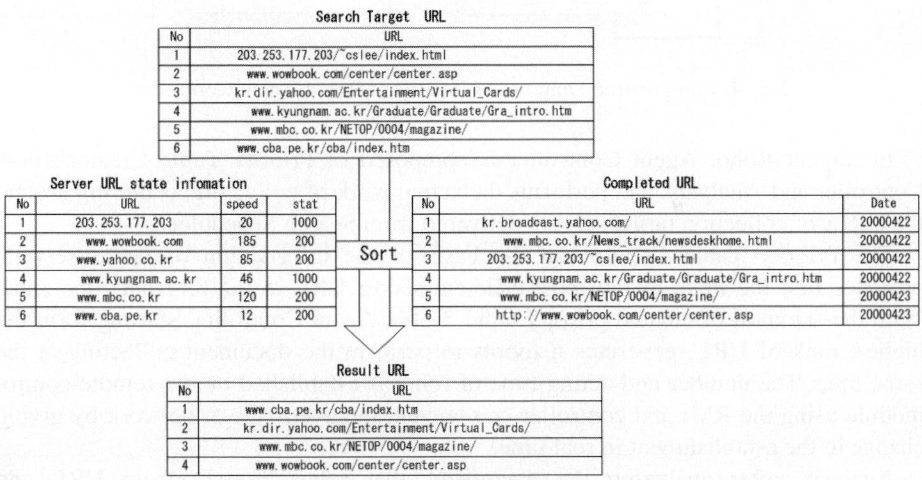

Fig. 3. Priority Deciding Diagram

The method of scoring servers is as follows. Through giving 70% adding point to the amount of loads by each time zone measured by server controller and 30% to the amount of loads which robot experiences in real time and calculating the point by each time zone, it gives high priority to the most efficient time zone. In this method, it establishes each server to have one efficient time zone and provides the list about efficient server of by each time zone.

Priority Table Creator, with these information on server provided by Server Controller, chooses the most efficient server list in the relevant time zone among server lists and provides the list marshalled in priority according to the point of the servers.

3.3 Configuration of Robot Agent System

Fig. 4 shows the consistency of Intelligent Robot Agent Controller.

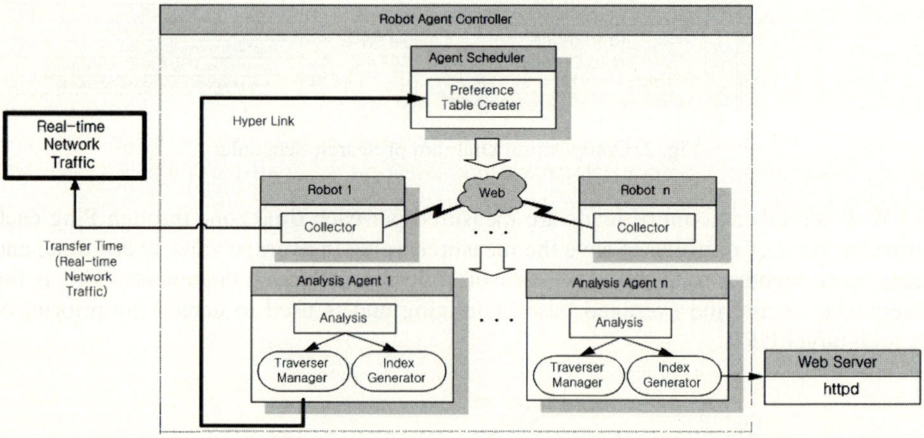

Fig. 4. Composition Diagram of Intelligent Robot Agent Controller

Intelligent Robot Agent Controller is composed of Priority Table Creator/Robot Controller and Analysis, and performs the actual work of collecting, centering around the URLs of collection target servers delivered from Search Scheduler.

The Priority Table Creator, on the basis of the information of target servers, provides first the links of the most efficient server URL in the relevant time zone. With the standard of created priority table, Robot Agent Controller, starting form the highest rank of URL, generates n robots to perform the document collecting at the same time. The number and acting time of robot is established by the remote control module using the RMI and controller can prevent the overloads of network by giving change to the establishment in real time.

Analysis, after analyzing the documents that robot brought from URL and extracting URL linked in other place, sends the URL to Scheduler. It regulates that the infinite recursive loop due to the depth establishment of search target web server is not resulted in. It also, during extracting, delete the URL which goes off the already collected URL and the search target web server. But, this deletion work, by realizing setting changeable by remote control module and making possible the characteristic-added web search, has the type of enterprise search engine which deals with specific network or site only.

After URL extraction and HTML TAG removal, it classifies index word and disused language and creates index DB with vocabulary analyzer.

3.4 Load Balancing Manager

Load Balancing Manager can be realized only when comprehensive and up-to-date load information is available. However, the collection of load information will increase the network traffic and interfere with the services for the clients. The load balancing on wide-area network is more time-consuming that involves the interactions between many nodes for gathering load information, negotiating on load reallocation and transporting the workload. Consequently, the load balancing may not improve but degrade the performance of a cluster of workstation.

Load Balancing Manager can minimize the network traffic and enhance the flexibility.

The system performs admission control for tasks and discovers available resource and places tasks on appropriate machines based on the load balancing scheme. The system is composed of two major components as shown in Fig. 5.

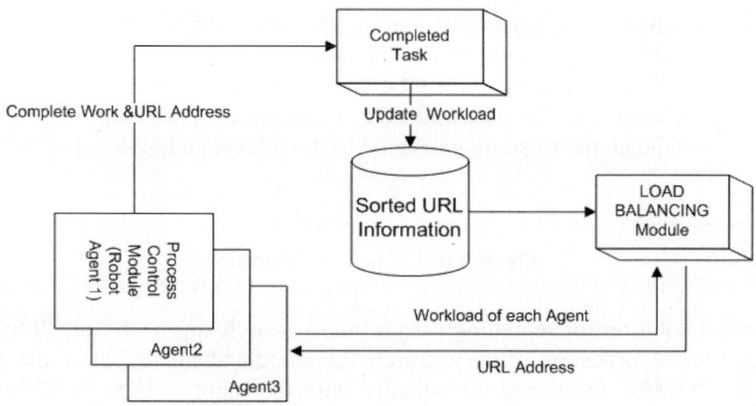

Fig. 5. Load Balancing Manager

(1) The Process Control Module accepts the user's input data and number of agents. Process Control Module compute the workload of each Robot agent and then send amount of workload. The Process Control Module accepts the URL addresses from the Load Balancing Module and starts to create multiple Robot agents and initiate processes to perform the parallel computing at distributed sites and sends them to these URL addresses.
(2) The Load Balancing Module manages the resource information from the database that keeps track of the registry of the master and clients that participate in the computing and decide the Target address for each worker.

3.5 Load Balancing Algorithm of System

A proposed load balancing algorithm using the dynamic policy to decide whether a node is lightly or heavily loaded. The threshold value of a node is the limit value of its workload, and is used to decide whether a node is lightly or heavily loaded. The threshold value of a node may be determined as follows:

Threshold value of a node $(n_i) = \sum_{i=1}^{N} W / N * C_i$

where, W_i = amount of workload of a node

N = number of node in our system

C_i = predefine constant value depends on the processing capability of node n_i

Fig. 6 shows a Load Balancing Algorithm.

```
RequestResources (no-of-resources, amount-of-workload)
{  Rs = Query resources information from the Sorted URL Tables;
   Sort the available resources with amount of workload by ascending order;
   i = 1;
   while ( Rs->next( ) && i <= no-of-resources )
   {
         if ( node.WL + amount-of-workload < threshold_value )
         {  URLLIST[i][0] = node.HOST_NO;
            URLLIST[i][1] = node.HOST_ADDRESS;
            Current_workload = node.WL + amount-of-workload;
            Update the Current_workload to the relevant address;   }
         i++;  }
}
```

Fig. 6. Load Balancing Algorithm

A parallel solution for searching uses a binary search approach. Recall that during each iteration of procedure binary search the middle element *Sm* of the sequence searching is probed and tested for equality with the input x. If *Sm* > x, then all the element larger than *Sm* are discarded; otherwise all the elements smaller than *Sm* are discarded. Thus, the next iteration is applied to a sequence half as long as previously.

The procedure terminates when the probed element equals x or when all elements have been discarded. In the parallel version, there are *N* agents and hence an (*N*+1) array search can be used. At each stage, the sequence is split into *N*+1 subsequences of equal length and the *N* processors simultaneously probe the elements at the boundary between successive subsequences.

4 Implementation and Evaluation

The suggested system consists of a question server which sends search results in HTML file, of a registration server for the input of user's directory, scheduler, Robot controller.

Fig. 7 shows the capacity of the improved mixed scheduling method. The horizontal axis represents time and the vertical axis represents the collecting amount per time. Each method in comparison indicates that the document collecting amount is higher in dawn time zone than in day time zone. The reason is that the network loads is low in dawn time zone and high in day time zone.

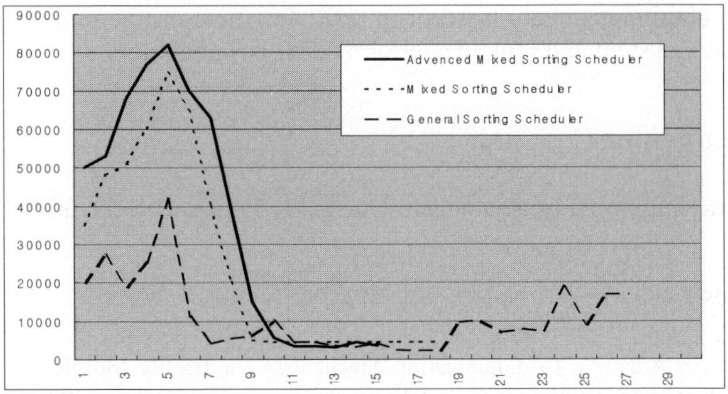

Fig. 7. Comparison of Collecting Amount Between methods

The existing method collected 14213 Kbytes in average per time and it took about 26 hours to collect all the documents. But the improved mixed scheduling method collected 24353 Kbytes and took 18 hours. Accordingly, considering network overloads collected document amount are high.

In addition, the efficiency can be improved even more by increasing the number of robot of the time zone of high efficiency through remote control.

As you can see Fig. 7, in case of Advanced Sorting Scheduling(thin line), it took a time 15hours for collecting total documents, the other way which introduced in this study(thick line) was taken about 14hours.

Therefore, when every Robot collects something it may need, we try to apply the amount of a load for Target Server which already met in real-time mode and meanwhile, try to response to give a priority for more efficiency sites. Finally this causes higher effects as expected.

5 Conclusion

As the existing mixed scheduling method performs the scheduling on the basis of server loads measured during specific time, it has limitation in measuring network loads. The improved mixed scheduling method suggested in this paper to complement that, by deciding specific period and measuring the amount of server loads, and scheduling on the basis of the average value of the amount of loads per every time and the real time information the robot experienced during the target web server connection, is able to measure the network loads much more accurately, so that efficient document collecting to consider the network loads maximally is possible. In addition, it brought great improvement on search system efficiency.

The search system proposed in this study can cope with circumstance actively during specific circumstance outbreak or specific setting alteration because it is possible to control and change the controller in real time through web browser. In addition, this search system has no restriction on platform since it was developed on the basis of JAVA.

Now, the continuous study in security related matters on the remote control work between search system and remote control module is demanding.

Acknowledgement

This work was supported by Kyungnam University Foundation Grant, 2007.

References

1. J. Ahn, “An Adaptive Communication Mechanism for Highly Mobile Agents”, Lecture Notes In Computer Science(ICCS04), Springer Verlag, Vol.3036, pp.192-199, 2004.
2. M. Ranganathan. M. Bednarek and D. Montgomery, “A Reliable Message Delivery Protocol for Mobile Agents”, In Proc. Of the 2nd International Symposium on Agent Systems and Applications and the 4th International Symposium on Mobile Agents, LNCS 1882, pp.206-220, 2000.
3. Young-hun Jung, "A Search Robot Engine using Mixed Sorting Scheduling Method", Dept. of Computer Engineering, The Kyungnam University Master thesis, 1999, 12.
4. A. L. Murphy and G. P. Picco, “Journal of Autonomous Agents and Multi-Agent Systems, Vol5, No.1, pp.81-100, 2002.
5. M. Bui, S. Das, A. Datta, and D. Nguyen. "Randomized Mobile Agent Based Routing in Wireless Networks", International Journal of Foundations of Computer Science, 12 (2001), pp. 365-384.
6. V. Cardellini and M. Colajanni,"Dynamic Load Balancing on Web-server Systems", IEEE Internet Computing, (1999), pp.28-39.
7. J. Figler. "Load Balancing in Concurrent Parallel Applications", May 1999.
8. C. Georgousopoulos and O.F. Rana, "Combining State and Model based Approaches for Mobile Agent Load Balancing", ACM, 2003.
9. K.M.L Tun and T.T. Naing, "Parallel and Distributed Computing Models for Mobile Agent", In Proc. of the 2nd International Conference, 2004.
10. S. Pleisch and A. Schiper, “Falut-Tolerant Mobile Agent Execution” IEEE Trans-actions on Computers, Vol.52, No.2, pp.209-222, 2003.
11. S. Papavassiliou, Jian Ye, O. Tomarchio, A. Puliafito, “Mobile agent-based approach for efficient network management and resource allocation: framework and application”, IEEE Journal on Selected Areas in Communications, Vol.20, 2002, pp.858-872.
12. K. Koukoumpetsos and N. Antonopoulos, “Mobility Patterns: An Alternative Approach to Mobility Management”, Pro. The 6th World Multi-Conference on Systemics, pp.14-18, 2002.
13. PhilipHeller & SimonRoberts, "Inside Secrets Java 2 Developer's Handbook", pp.648-693. 1999.

WANT: A Personal Knowledge Management System on Social Software Agent Technologies

Hak Lae Kim[1], Jae Hwa Choi[2], Hong Gee Kim[3], and Suk Hyung Hwang[4]

[1] Digital Enterprise Research Institute, National University of Ireland, Galway,
IDA Business Park, Upper Newcastle, Galway, Ireland
haklae.kim@deri.org
[2] Dept.of Business Administration, Dankook University
San 29, Anseo-dong, Cheonan-si, Chungnam, Korea
jchoi@dankook.ac.kr
[3] Seoul National University,
28-22 Yeonkun-dong, Jongro-gu, Seoul, Korea
hgkim@snu.ac.kr
[4] Div. of Computer & Information Science, Sun Moon University
100 Kal-San-Ri, Tangjeong-Myeon, Asan-shi, Chungnam, Korea
shwang@sunmoon.ac.kr

Abstract. A multi-agent system is a network individual agent that work together to achieve a goal through communication and collaboration among each other. Standardized infrastructure for information or knowledge sharing is required to make autonomous agents interdependent on each other for effective collaboration in a multi-agent system. In order to enhance productivity of knowledge workers knowledge management tools should support collaborative environments among desktop, web, and even mobile devices. The Semantic Web is the place where software agents perform various intelligent tasks using standard knowledge representational schemes that are named "ontologies." This paper presents a conceptual framework of the social knowledge activities and knowledge processes with regard to the social software agents. Our prototype, called WANT, is a wiki-based semantic tagging system for collaborative and communicative knowledge creation and maintenance by a human or software agent. It can be supported in both desktop and mobile environments.

1 Introduction

The goal of personal knowledge management (PKM) is to enable individual knowledge workers to work better in groups and in organizations [12]. The PKM is focused on individual level of knowledge activities and is to apply collective knowledge of knowledge workers to make optimal decisions in real time. Knowledge workers need tools to capture ideas, thoughts and to manage schedule, address, etc. Most knowledge workers have been using desktop applications. Desktop applications such as email clients, word processors, and web browsers are essential for our daily routine. However, current desktop infrastructures for managing knowledge are ill-suited for personal knowledge-oriented activities and do not support 'anytime' and

N.T. Nguyen et al. (Eds.): KES-AMSTA 2007, LNAI 4496, pp. 785–794, 2007.

'anywhere' access to knowledge. These activities must be supported by KMS, not when they are available but when they want, since knowledge workers do think at any time. New mobile communication technologies and devices allow knowledge workers to capture his or her thoughts or ideas in a relevant, timely manner.

Information and functionality are scattered across applications and websites, making it difficult to aggregate and reuse just the right set of content and operations required for unique user tasks [6]. Therefore, a knowledge management system which combines a desktop part and mobile part could be an effective channel to work better together [18].

PKM has been connected with the social software which is software that supports, extends, or derives value from human social behavior [16]. Social software is not only focused on connecting people, but also on sharing data. Therefore, it plays an important role in building social networking on the web. And the Semantic Desktop and Web2.0 as new computing paradigm provide reliable technologies to enhance functionalities of PKM. The Social Semantic Desktop is a new computing paradigm that provides an advanced way to create, automate and structure information and the technology convergences including the social network, community services, and P2P services [1- 3]. It could provide the transformation of a typical desktop system into a collaborative environment that supports both personal computing and information sharing via social and organizational channels [7].

Web2.0 comprises of technologies and services to enable users to collaborate and share social contents. They include social software, content syndication, messaging protocol such as weblogs, wikis, podcasts, RSS feeds, etc. There exist well-known Web2.0 sites like Flickr[1], del.icio.us[2], Technorati[3]. The majority of such sites are connecting people into communities creating networks of shared experience using folksonomy and RSS [11].

Since social software-based PKM is basically based on open web standards, it not only improves interoperability with different applications, but improves knowledge creation and collaboration as simple conceptual models.

This paper presents a conceptual framework for the social knowledge activities and knowledge processes with regard to the social software and describes the implementation of WANT, Social Software-based Personal Knowledge Management System, to help knowledge activities of knowledge workers

2 Social Knowledge Activities

There are a number of definitions of knowledge and classifications or categorizations of knowledge. According to Spender [14] and Polanyi [15], it can be classified into two types: individual and social knowledge. The knowledge processes transform individual knowledge into social knowledge and the opposite way. Knowledge is created through a social interaction in organizations or communities where knowledge workers are involved.

[1] http://www.flickr.com
[2] http://del.icio.us
[3] http://www.technorati.com

Since knowledge at an individual level has ad-hoc and informal characteristics, it is not easy to manage knowledge by formal and rigid processes, and systems. Knowledge not only exists, but also is continuously created by knowledge workers in response to their adaptive needs. Barth (2004) states that "knowledge management cannot make meaningful outcome, unless every knowledge worker takes *personal responsibility*[14]".

We believe it is necessary to improve an individual knowledge of knowledge workers. Knowledge management happens through both in knowledge process and in social knowledge activities. The knowledge process focuses on procedural orders of knowledge such as creation, dissemination, sharing. The social knowledge activities, on the other hand, can include all kinds of activities to enhance and support socially knowledge processes. Basically these activities should include *collaboration, communication, and connection* based on the knowledge worker's experience.

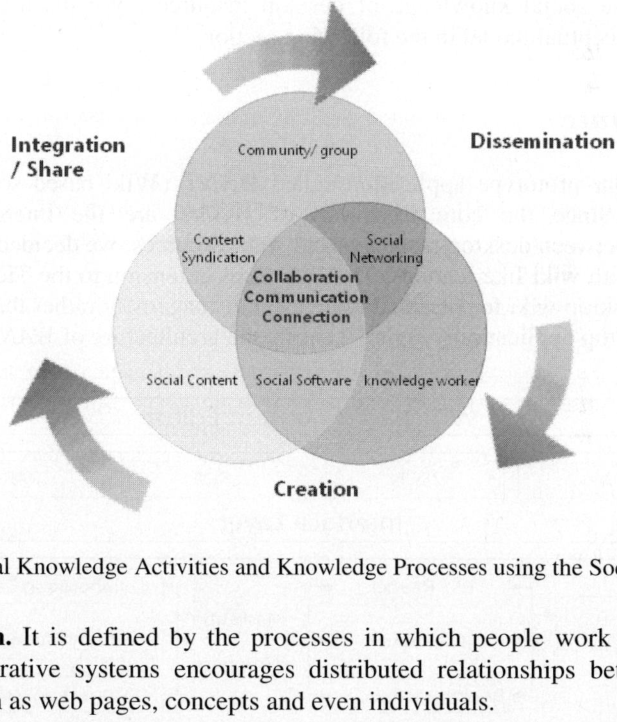

Fig. 1. Social Knowledge Activities and Knowledge Processes using the Social Software

Collaboration. It is defined by the processes in which people work together. Web based collaborative systems encourages distributed relationships between specific resources such as web pages, concepts and even individuals.

Communication. RSS/Atom has become a major part of the communication. Most blogs and even outside the blogosphere are producing RSS/Atom feeds. It can be used to distribute any kind of content- information arriving at irregular intervals - blogs, news, updates to a web site, as well as music, photos and video. A personal RSS/Atom can also be published and shared on the web. RSS/Atom format can be used for interacting desktop contents and web contents.

Connection. This activity is to create or maintain social relationships between knowledge workers. It should focus on uncovering the patterns of people's interconnection and interaction and support knowledge exchange.

A conceptual model for social knowledge activities and knowledge processes for knowledge workers is shown in Figure 1. It explains how desktop content maps to social content. Desktop content can be transformed into social content using social software such as weblogs, wikis and folksonomy based on Web2.0 technologies. These items would be shared across communities via social networking services. Furthermore, knowledge workers are enabled to access this information via syndication technologies.

Most knowledge workers have been spending their time to maintain or create their knowledge in both Desktop and Web environments. It is time-consuming and knowledge workers make their knowledge redundantly in both environments. It is necessary to interconnect or integrate Desktop resources with Web resources and to share and reuse social knowledge of desktop resources. We discuss how we can realize this conceptual model in the following section.

3 Architecture

We describe our prototype application called ***WANT*** (**W**iki based soci**A**l **N**etwork **T**hin client). Since the core principles of ***WANT*** are the interoperation and collaboration between desktop resources and web resources, we decided to implement the interface with wiki-like features. The *WANT*, as extension to the TiddlyWiki[4], is a lightweight desktop wiki for personal knowledge management rather than full-fledged Semantic Desktop applications. Figure 2 shows the architecture of *WANT*.

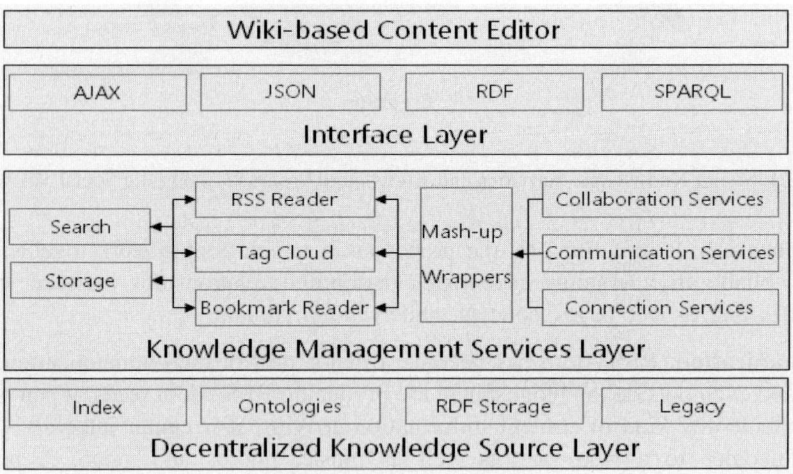

Fig. 2. WANT architecture

[4] http://www.tiddlywiki.com

WANT is implemented in the JavaScript, AJAX, and JSON (interface layer), thus allowing to easily extend with certain functionalities using JavaScript plug-ins and to be easily deployed on web browser environments. Data is stored in HTML itself and RSS formats in a decentralized knowledge source layer. The Knowledge Management Services Layer plays an important role to interact Desktop resources with Web resources using Semantic Web technologies:

Connection to Semantic Desktop. All contents can be saved in HTML files, RSS1.0, RSS2.0, and Atom formats in *WANT*. There are two ways to connect the Semantic Desktop components. Firstly, since *WANT* is a simple and single HTML file, it could be opened by most Semantic Desktop applications. Secondly, all contents of *WANT* would be generated by RSS1.0 format and be saved in RDF storage. When querying with RDF query language such as SPARQL, users would be able to get the data set of RDF storage and to return back as value of the new semantic data.

Connection to Web2.0. Wikis are easy to use as collaboration platform and knowledge management systems. Using Wikis, however, desktop environments will significantly limit the potential for information sharing and collaboration. We overcome some of the weak points of wikis by using Web2.0 technologies. The *WANT* allows knowledge workers to organize their information or knowledge and provides various social content services such as folksonomies, social bookmarking

Fig. 3. User Interface and main functionalities (1: RSS Reader, 2:Social Bookmark Reader, 3:Tagcloud, 4:RSS and Atom, 5: Flickr photo reader, 6: Resource viewer)

and RSS/Atom feeds using HTTP, SOAP, XML RPC, or REST Web Services. For example, when knowledge workers make their own tags in certain content, they can use not only desktop tag which they made before, but folksonomy of weblogs or Technorati.

WANT uses browser-based interfaces (Figure 3). Figure 3 shows a sample content about the picture "simplicity-desktop" from flickr as rendered in *WANT*. It can include pictures, desktop resources, and links. *WANT* includes the RSS aggregator, the Tag reader and the Tag Cloud, and the Social bookmark reader.

4 How It Works

4.1 Enabling Social Collaboration and Authoring

A major aim of WANT is to foster and employ social interactions for knowledge workers through various content services and Web2.0 sites. To leverage the social collaboration, we need to shift from focusing on the individuals to focusing on interactions [17]. Social collaboration within WANT is in particular supported by RSS Reader (number 1 in figure 4), Social Bookmark Reader (number 2 in figure 4), Flickr badge service (number 3 in Figure 4).

The knowledge workers are able to have communities sharing the common interests and to have the links for references. In WANT it enables the users not only to capture web-based information, but also to organize the information together with desktop resources. The collaboratively user-added annotations are to improve social features on desktop resources. For instance, RSS Reader (number 1 in Figure 4) and

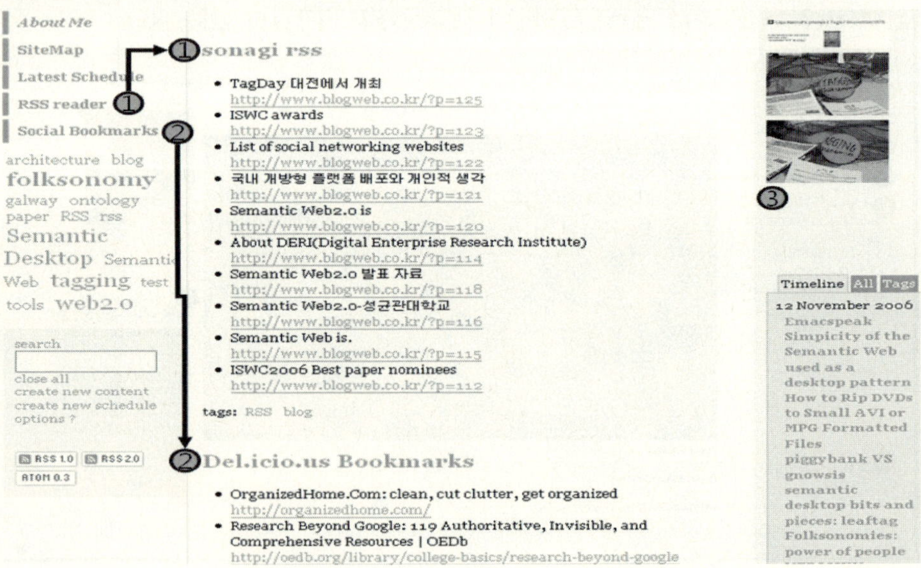

Fig. 4. RSS Reader and Social Bookmark Reader

Social Bookmark Reader (number 2 in Figure 4) can get the data from a user-given URL such as a certain RSS Feed or bookmark URLs. The user can edit this data directly and add user-driven annotations as tags in WANT. These tags can be connected with social communities as specific links. So contents on WANT are easy to reflect social content without user's action and intention.

When users create a new content or double click on certain content, it will create a new link or change to an editing mode. Content in *WANT* contains title, description, created date, and tag items. These items could be enriched with metadata of the content. The description allows knowledge workers to create their own ideas or information about what they want to create. It allows knowledge workers to add their own tags to the tag item with several keywords to help classify and group them. Tags can be assigned to content that can be later used for viewing and finding groups of contents. It allows knowledge workers to add their own tags to the tag item with several keywords to help classify and group them. Tags can be assigned to content that can be later used for viewing and finding groups of contents. If users want to have a tag created from multiple words, each tag can be separated with comma like [*Web2.0, Semantic Desktop*]. The tag Cloud (see Figure 5) allows knowledge workers to help choosing the tags which are good in the content. After creating the entry, they can see the tags at the bottom of the pages like [*tags: Web2.0, Semantic Desktop*] and simultaneously it will be added in the Tag Cloud.

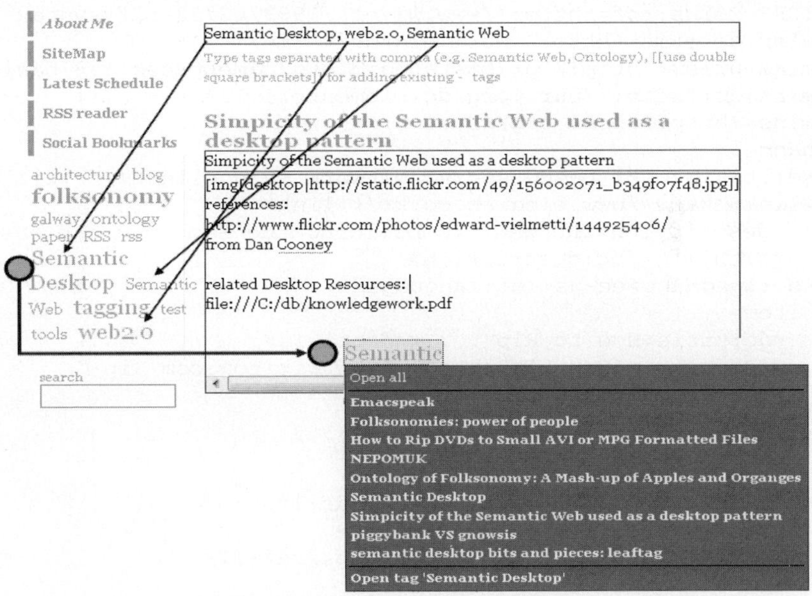

Fig. 5. The Tag cloud and tag based search

4.2 Tag-Based Search

WANT provides two search methods: full-text search and tag based search. The former makes results for matching one or multiple keywords in content. The simple tag based search is to navigate all tag lists in the Tag Cloud with a given tag. The Tag Cloud is a list of the most popular tags used by the knowledge worker. The larger the font size of the tag in the list, the more the tag has been occurred by a greater number of the user.

When users navigate by tag, they are directly connected with other resources using the Tag Cloud, which is clickable. It will improve the findability of content. If users click on certain tag, a list will pop up with links to the other content having the same tag. This is supposed to give an indication of the most popular topics. The availability of new semantic data will allow users to find and make use of relevant data quickly and accurately.

4.3 Enhancing Semantics

What parts of the more valuable information do we want to store, to keep, and to share as a user? If knowledge workers want only to store URLs, one convenient approach is to use the bookmarks functionality in our favorite browser. In *WANT*, we provide various approaches: 1) static HTML files, 2) RSS files, 3) RDF repository, to store information of user-generated contents. Basically the whole content can be stored in HTML files which including URLs, tags, contents, links, date, etc.

```
<?xml version="1.0"?>
<rdf:RDF xmlns:rdf="http://www.w3.org/1999/02/22-rdf-syntax-ns#"
   xmlns:dcterms="http://purl.org/dc/terms/"
   xmlns:prism="http://prismstandard.org/namespaces/1.2/basic/"
   xmlns:dc="http://purl.org/dc/elements/1.1/"
   xmlns="http://purl.org/rss/1.0/">
  <channel>
    <dc:title>Wiki based sociAl Network Thin client</dc:title>
    <link>http://www.blogweb.co.kr/</link>
   <dc:description>The WANT is building the Semantic Web easier
       With....</dc:description>
    <dc:language>en-us</dc:language>
    <item>
      <dc:title>How to Rip ... </dc:title>
      <dc:description>Link:http://www.wikihow.com/Rip-
      </dc:description>
      <dc:subject>
       <rdf:bag>
             <rdf:li>web2.0</rdf:li>
             <rdf:li>Semantic Web</rdf:li>
             <rdf:li>tagging</rdf:li>
             <rdf:li>Semantic Desktop</rdf:li>
       </rdf:bag>
      </dc:subject>

<link>http://www.blogweb.co.kr/#%5B%5BHow%20to%20Rip</link>
      <dc:date>Sun, 12 Nov 2006 23:20:52 GMT</dc:date>
   </item>
  </channel>
```

Fig. 6. Snippet of RSS1.0

It allows users to store as different types of RSS such as RSS1.0, RSS2.0, and ATOM. If users choose RSS1.0 (Figure 6) to store, it can be stored in RDF storage and be reused later. It is a slightly different approach to the other Semantic Wikis. Most of those have a way to generate semantic data in their tools directly. In our approach we create contents and syndication files, then we generate semantic data automatically to reuse and share it. It is a more effective and efficient lightweight mechanism to connect with the Semantic Desktop. In addition, if necessary, it can be integrated with other Semantic Desktop applications because it is a simple HTML.

It enables users to publish and share their contents as an RSS on the Web. Also if they want to collaborate with someone, they exchange the RSS using RSS reader. It is easy to make cooperation in distributed computing environments.

5 Conclusions

In this paper, we propose the conceptual framework of the Social Knowledge Activities and implement the prototype to help knowledge workers for enhancing those activities. Creating knowledge is not the same as sharing knowledge. Sharing knowledge does not automatically lead to collaboration. WANT supports a wide variety of social activities. It can interconnect Desktop and Web. We introduce essential concepts and implement the necessary functionality of collaboration, communication, connection.

The main achievement of this work, distinguishing our approach from existing system, is the establishment of the architecture to interact desktop and web, to create metadata based on RSS feeds, to enrich desktop system with intelligence. Our approach allows knowledge workers to interact and share their resources among their desktop and social software more easily.

However this work has some limitations for collaborate both environments more dynamically. We try to share desktop contents as RSS feeds. But if users want to reflect tags or links on the web coincidently, it is still restricted. Moreover users would be restricted to use open API. Most of them provides limited amount of their folksonomies or restricts access times to get information with it. It is necessary to develop standards or methods to exchange user-oriented folksonomy, because current folksonomies depends on service provider rather than users.

Acknowledgement

The work presented in this paper was partly supported by the Lion project (Science Foundation Ireland, SFI/02/CE1/I131).

References

1. Decker, S., Frank, M.: The networked semantic desktop. In: Workshop on application design, development and implementation issues in the semantic web. 2004
2. KIM. H.L., Kim, H.G., and Decker. S.: How Semantic Desktop and Web2.0 can help Personal Knowledge Management?, In: Workshop on Semantic Web Applications and Tools, 2006

3. Leo Sauermann.: The Gnowsis-Using Semantic Web Technologies to Build a Semantic Desktop. Master's thesis, Technical University of Vienna, 2003
4. Shankaranarayanan, G., Even, A.: The Metadata Enigma. Comm. ACM 49 (2006) 88-94, 2006
5. Ohmukai, I., Hamasaki, M.: A Proposal of Community-based Folksonomy with RDF Metadata. to be published(CEUR Workshop Proceedings, 2005
6. Bakshi, K., Karger, D.: End-User Application Development for the Semantic Web. In: Proc. 3rd ISWC 2005 Workshop on the Semantic Desktop, 2005
7. DesktopLinux.com: Mandriva to help build "Social Semantic Desktop", available at http://desktoplinux.com/news/NS3270716126.html. 2006
8. Quentin Reade, Web users 'only visit six sites'. Available at: http://www. webuser. co.uk/news/81267.html?aff=rss, 2006
9. Alex, B.: How many feeds do you subscribe to?, Available at: http://blogs.msdn.com/ alexbarn/archive/2006/02/01/522004.aspx, 2006
10. Snowtide Informatics: Making Lucene Play Nice with PDF's. available at : http://snowtide.com/home/PDFTextStream/techtips/easy_lucene_integration. 2004
11. Wikipedia: Folksonomy. Available at http://en.wikipedia.org/wiki/Folksonomy, 2006
12. Jason Frand and Carol Hixon, Personal Knowledge Management : Who, What, Why, When, Where, How?, December, 1999. available at : http://www.anderson.ucla.edu/ faculty/jason.frand/researcher/speeches/PKM.htm
13. Steve Barth: "Self-Organization: Taking a Personal Approach to KM" in *Knowledge Management Tools and Techniques: Practitioners and Experts Evaluate KM Solutions* edited by Madanmohan Rao, Butterworth-Heinemann (2004).
14. Spender, J.-C.: Organizational Knowledge, collective Practice and Penrose Rents, In: International Business Review, Vol. 3, No. 4, 1994, 353-367
15. Polanyi, M.: The Tacit Dimension, London, 1966
16. Tom, C.: My working definition of social software, available at: http://www. plasticbag.org/archives/2003/05/my_working_definition_of_social_software/, 2003
17. Peter M. Senge, The Fifth Discipline: The Art & Practice of The Learning Organization, 2006
18. Niklas E.G.D, Using Mobile Knowledge in a Knowledge Management System, available at: http://www.ituniv.se/program/mob/magisteruppsatser/2005/danell.pdf

Mobile Agents Using Data Mining for Diagnosis Support in Ubiquitous Healthcare*

Romeo Mark A. Mateo, Louie F. Cervantes, Hae-Kwon Yang, and Jaewan Lee

School of Electronic and Information Engineering, Kunsan National University
68 Miryong-dong, Kunsan, Chonbuk 573-701, South Korea
{rmmateo,lfcervantes,hkyang,jwlee}@kunsan.ac.kr

Abstract. Recent research topics in healthcare including intelligent decision support services, expert medical services and autonomous management are based on multi-agent systems. The cooperation of these software agents provides efficient monitoring, analyzing, and managing the data of patient where abnormal patterns are detected to have an advance treatment and prevent loss of life. In this paper, a framework for ubiquitous healthcare based on multi-agent is presented. This paper proposes a mobile agent for diagnosis support in ubiquitous healthcare. The expert mobile agent (EMA) classifies the data of patient by using neuro-fuzzy algorithm for consultation report. A pre-processing method based on the profile of an expert is used to filter the data from the history of patient. Result of neuro-fuzzy from cross-validation test shows a high accurate classification in data compared to other highly accurate classifiers.

1 Introduction

Agent-based healthcare system addresses the importance of intelligent programs to substitute the real person's functions in healthcare services and management. This technique benefits most of individuals through their decision making and automation of their tasks. Most of these implementations are used for decision support system [1]. The use of agent-based intelligent support systems is important in medical industries because it allows doctors and nurses gather quick information. This information is processed in various ways to assist with making diagnosis and treatment decision. Also, because software agents deals with distributed systems, it assist in diversity of storing and retrieving medical records, analysis of real-time data gathered and other necessary information retrieval in distributed environment.

Techniques and algorithms are integrated to the agent-based healthcare system for medical diagnosis. Neural network is a common technique for medical diagnosis [2, 3]. Successful application examples show that neural diagnostic systems are better than human diagnostic capabilities. Moreover, neural network are used to analyze medical images [4, 5]. These research articles survey various approaches and techniques to improved diagnosis in medical images, including mammography, ultrasound and

* This work is supported by the Korea Research Foundation Grant funded by the Korean Government (MOEHRD) (KRF-2006-521-D00372).

N.T. Nguyen et al. (Eds.): KES-AMSTA 2007, LNAI 4496, pp. 795–804, 2007.

magnetic resonance imaging. Neural network-based agents are used for discovering rules in medical database [6]. Medical databases, which consist of patient histories, specialist's conclusions, laboratory results, etc., are typically distributed set of semi-structured data, and because agent technology is well-suited approach to develop the medicine decision supporting systems, the integration of neural network in the agent is necessary. Medical databases are dynamically changed because the structures of features characterizing the diseases are continuously updated. The features of diseases depend on tools and technologies those doctors and specialists currently use to diagnose and treat the patients. Even though the integration of neural networks within agents is well-researched, it still needs intensive research on using hybrid systems [20] because of the vast changes of information and the classical methods may not solve the problem of classification.

In this paper, we propose an expert mobile agent using data mining to support the diagnosis of the patient in ubiquitous healthcare. Moreover, a framework of ubiquitous healthcare based on multi-agent is presented. The framework supports the mobility of the mobile agent which executes classification algorithm to the data of patient. The paper investigated efficient classifiers on data mining to integrate with the proposed expert mobile agent. The proposed expert mobile agent (EMA) uses neuro-fuzzy classification for consultation of patient. On first phase, the fuzzy system of EMA is trained from the previous data of other patients. A pre-processing method based on the profile of an expert is used to filter the relevant data from its expertise. After the training, the EMA are deployed to execute classification of data. Result from cross-validation test shows that the neuro-fuzzy classification provides a high accuracy in classifying the data compared to other highly accurate classifiers.

2 Related Works

Intelligent agent for healthcare plays a crucial role on giving correct information for diagnosis and providing immediate medical services. Home healthcare services provide information to a consumer of the necessary diagnosis and continuous monitoring of patient to acquire immediate response and save lives in case of abnormal indications. Agent-based intelligent decision support is proposed for the home healthcare environment [7]. The multi-agent platform is combined with artificial neural network for the intelligent decision support system in a group of medical specialists collaborating in the pervasive management of care for a patient. Mobile agents are used to serve the collaboration of services for mobile users [8]. An agent is an autonomous, social, reactive and proactive entity, sometimes also mobile. Since telemedicine is grounded on communication and sharing of resources, agents are suitable for its analysis and implementation, and these are adopted for developing a prototype telemedical agent.

Data mining aims to extract interesting information from large databases is used for decision support in the field of medicine. In order to have mobility, data mining framework for mobile environment are proposed by researchers [9, 10, 11]. A context-awareness on data mining is used to maximize the adaptive capacity of data mining [9]. The use of decision support PDA supported by data mining facility can be a great asset to the medical professionals while working on an emergency or while rushing to attend an emergency. Data mining in mobile environment using mobile agents is found in the work of Lee, et. al. [10]. This is done by sending a mobile agent to the LBS and then it performs the classification mining in the database. In the HCARD model of Gerardo, et. al. [11], proposed an Integrator agent to perform knowledge discovery in the heterogeneous server in the distributed environment. Data mining are essential in extracting rules from databases and provide decision support knowledge in healthcare environment.

2.1 Neuro-fuzzy Classification

Fuzzy systems are used to handle uncertainty from the data that cannot be handled by classical methods. It uses the fuzzy set to represent a suitable mathematical tool for modeling of imprecision and vagueness [12]. The pattern classification of fuzzy classifiers provides a means to extract fuzzy rules for information mining that leads to comprehensible method for knowledge extraction from various information sources. The fuzzy algorithm is also a popular tool for information retrieval. Fuzzy *c*-means classifier (FCM) uses an iterative procedure that starts with an initial random allocation of the objects to be classified to *c* clusters. Neuro-fuzzy systems are the hybrid of artificial neural networks and fuzzy systems. The algorithm borrows the learning ability of neural networks to determine the membership values. It is among the most popular data mining techniques used in recent research [13, 14]. There are many types of neuro-fuzzy rule generation algorithm [15]. FuNE-I is a neuro-fuzzy model that is based on the architecture of feed-forward neural network with five layers which uses only rules with one or two variables in antecedents [16]. A Sugeno-Type neuro-fuzzy system is used for a scheme to construct an *n*-link robot manipulator to achieve high-precision position tracking [17]. A neuro-fuzzy classification (NEFCLASS) is a fuzzy classifier that creates fuzzy rule from data by a single run through the data set [14].

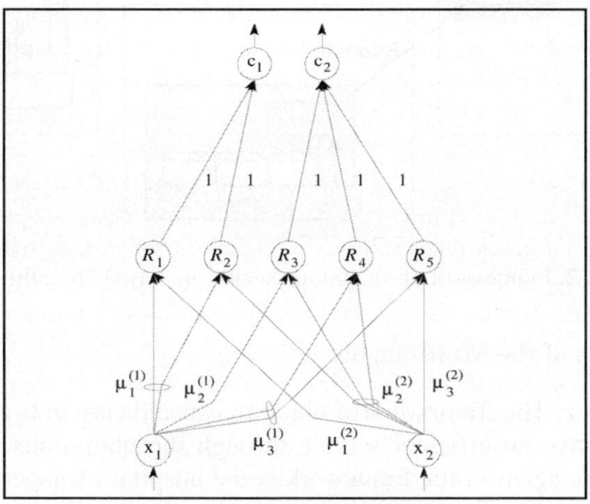

Fig. 1. A NEFCLASS system with two inputs, five rules and two output classes

3 Framework of Ubiquitous Healthcare Based on Multi-agents

In this study, we propose a framework for the ubiquitous healthcare. The proposed framework consists of multi-agents managing the hospital shown in Figure 2. A ubiquitous healthcare in [18] proposes a method of accessing healthcare services by individual consumers applying to mobile computing device. The OnkoNet Mobile Agents Architecture was developed and consists of cooperation protocols, inference model and health ontology to provide efficient ubiquitous healthcare environment. In our framework, mobility support for expert mobile agent (EMA) and data mining support for the ubiquitous healthcare are considered. Figure 2 illustrates the proposed architecture based on multi-agent system. Each doctors and specialist have their own EMA. In Figure 2, there are three different rooms consists of monitor agent (MA) to monitor the readings of the sensors in the patient and triggers the room manager (RM) to communicate with the hospital manager (HM) for necessary diagnosis or actions to be taken by the doctors. The movement of EMA from room 1 to room 2 requires communication with RM to initialize the interaction to the agents inside the room. This procedure considers verification of accessing data for security.

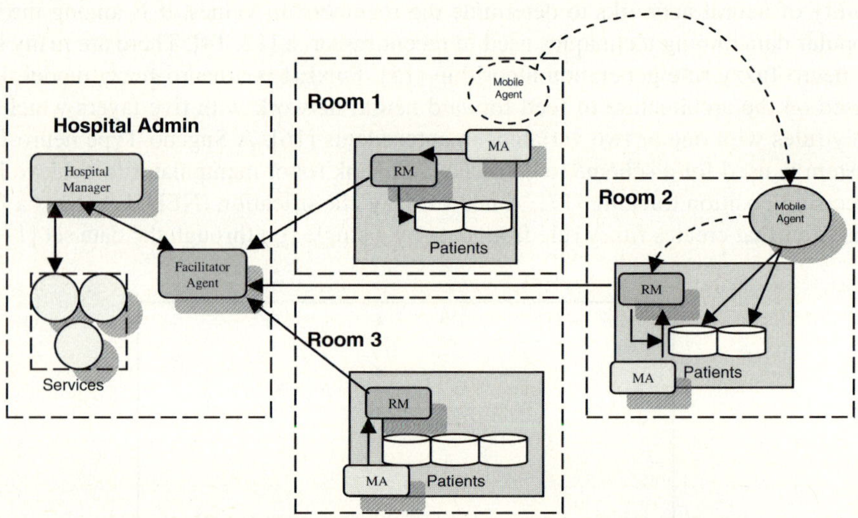

Fig. 2. Framework of ubiquitous healthcare based on multi-agent

3.1 Components of the Multi-agents

Hospital Manager. The framework of ubiquitous healthcare in Figure 2 is consists of multi-agents to have an efficient service through the ubiquitous healthcare system. The main software agent in the framework is the hospital manager (HM). It concerns in managing the services in the hospital supporting the decision making and management. Moreover, it communicates to other agent component through facilitator agent. The deployments of the services in rooms are done by the hospital manager.

Facilitator Agent. The main function of the facilitator agent (FA) is a broker between the HM and room manager (RM). All negotiations of requesting services from the room manager are done with the FA before the HM deploys its service in the room. The confirmation of deploying the services are received by FA and the RM is informed if the request of service is possible or not. Also, FA receives the alert message from the room manager if there are needs of attention with the patient.

Room Manager. The framework of the ubiquitous healthcare is consisted of physical room for the patients shown in Figure 2. A room manager (RM) coordinates the task of agents within the room. The software agents communicate to RM for every event that needs attention of the individuals inside the hospital. RM also request for services needed by the patients to hospital manager via FA. After the negotiation of FA, the service is deployed and adds to the RM.

Monitor Agent. Healthcare sensors and other sensors used for monitoring the patient are handled by monitor agents. These are programmed to detect abnormal patterns from readings of the patient. This study assumes that these sensors are used for monitoring of patient and send the signal for analysis.

Service Modules. In our proposed system, the services modules are used to support the diagnosis of patient, decision making and management of the hospital. These are managed by the HM. FA negotiates the request from the RM before HM deploys the service in the RM.

Expert Mobile Agent. The expert mobile agent (EMA) uses the proposed framework. Doctors and specialist uses their PDA as the host of EMA. The main function of the EMA is to help on the diagnosis of a patient by checking the current data and processed it with the data mining tool. EMA moves to all allowed patient for the service. Before deploying, the EMA request verification to RM for security reason so that the data will not be altered by malicious attack.

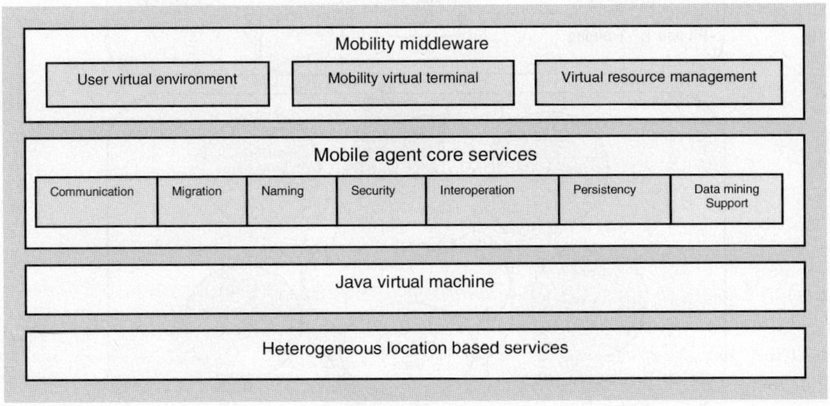

Fig. 3. Proposed mobile agent middleware

3.2 Mobile Agent Middleware

Mobile agent-based middleware is one of the issues of research for providing an advanced infrastructure that integrates protocols, mechanism, and tools to permit communication to mobile agents. SOMA in [19] discusses more issues of the mobile agent middleware. In our research, the design of mobile agent middleware is a Java-based platform. The infrastructure is divided in layers of service for designing, implementing, and deploying mobile agent-based applications. As shown in Figure 3, our proposed middleware consists of four layers. We focused more on the last component which is the data mining support. This additional service is provided to operate the data mining of the mobile agent on the data of patient.

4 Data Mining Model for Diagnosis Support

The expert mobile agent or EMA performs the consultation to the patients for advance diagnosis which is based on the proposed data mining model. Our data mining model has two phases shown in Figure 4. The first phase includes the training of EMA's fuzzy system based on data pre-processing by selecting relevant information from the profile of an expert. Also, the training phase provides EMA to have an accurate classification based on the expert profile of the doctor or specialist. The second phase processes the data of the patient to neuro-fuzzy classification. The procedure is done by deploying the EMA from the PDA of the doctor in the room and classifies the data of patient. The security configuration of deployment is also considered in this process. After the process, the results are returned display the result of consultation.

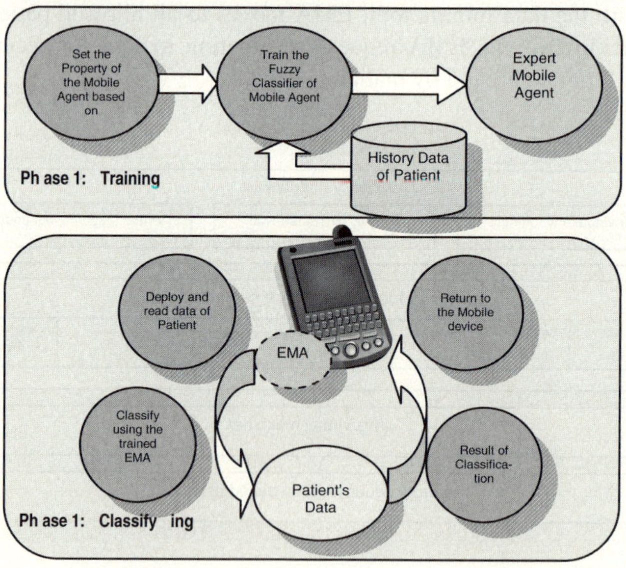

Fig. 4. Data mining model using neuro-fuzzy for support of diagnosing the patient

The proposed data mining approach considers the profile of an expert in the mobile device as the basis of extracting the relevant information from Phase 1. Now let us consider the set of profiles that will be used in the preprocessing data mining: $P = \{p_1, p_2 \ldots p_x\}$. After collecting the profiles, the mobile agent uses these features to select the relevant attributes of C where it is the raw data from the patient history database. Let D as the set of the selected tuples from C. Equation 1 represent the pre-processing algorithm. The following are the phases of our proposed algorithm.

$$D = \sum_{i=1}^{n} C\{c_1, c_2, \ldots, c_n\} \tag{1}$$
$$\text{where } c_n attribute\,(value) = p_x\,(value)$$

Let us say the EMA is a cardiologist then the cases related to heart disease are gathered by D. D is used to train the fuzzy system of EMA. The structure of the neuro-fuzzy system consists of three layered perceptron. The 1st layer is for inputs ($U_1 = \{x_1, \ldots, x_n\}$), 2nd layer is for generating rules ($U_2 = \{R_1, \ldots, R_k\}$), and 3rd layer is an output layer ($U_3 = \{c_1, \ldots, c_m\}$). The system also contains weights from the input layer (U_1) to rule layer (U_2) and from rule layer (U_2) to the output layer (U_3). Each connection between units $x_i \in U_1$ and $R_k \in U_2$ is labeled with a linguistic term A $A_{jr}^{(i)}$ ($j_r \in \{1, \ldots, qi\}$). The values from the input layer are mapped through the fuzzy sets of the weights. $W(R, c) \in \{0, 1\}$ holds for all, $R \in U_2$, $c \in U_3$. The values from the input and rule layer are evaluated in the connection of the hidden and output layer. For all output units, $c \in U_3$ the net input net_c is calculated Equation 2.

$$net_c = \frac{\sum_{R \in U_2} W(R,c) \cdot o_R}{\sum_{R \in U_2} W(R,c)} \tag{2}$$

To train the fuzzy sets from the input, Equation 3 is used. After the training, the EMA is ready to classify the data from the patient. A Java codes is shown in Figure 5.

$$\delta_R = o_R(1 - o_R) \sum_{c \in U_3} W(R,c)\delta c \tag{3}$$

```
INPUT:   profile, preprocessdata
OUTPUT:  NeurofuzzyClassification(preprocessdata)
public class ExpertMobileAgent extends Aglets {
   public void Preprocess(String[] profile, String[] val) {
         while(rsData.next())
         {
         if rsData.getObject(profile)=val; AddInfo(rowset)
         }
         NeurofuzzyClassification(preprocessdata);
   }
   public void ClassifyData(int in1, double[] pattern) {
   }
```

Fig. 5. Neuro-fuzzy classification algorithm integrated in EMA

5 Simulation Result

The proposed framework of multi-agents was simulated using the JADE platform. Neuro-fuzzy algorithm was coded in Java and embedded it to the expert mobile agents. The environment OS platform used here are Windows OS, Red Hat Linux and Sun Solaris 8 to simulate the heterogeneity of system. To test the performance of algorithms, we used data mining tools which are the NEFCLASS and Weka data mining. We chose the data of heart disease from UCI machine learning repository used by the machine learning community for the empirical analysis of machine learning algorithms.

5.1 Classification Accuracy

Precision and recall are two typical measures for evaluating the performance of information retrieval systems. Given a discovered cluster γ and the associated reference cluster Γ, precision (PγΓ) and recall (RγΓ) applied to evaluate the performance of clustering algorithms. In classifier algorithm, recall and precision is performed by cross-validation test of the classified instances. To evaluate the performance of the algorithms, these measurements were used. This is done by calculating the average precisions in Equation 4 where *AvgP* is the summation of precision (P_n) of classes divided by the number of classes. Average of recall is computed in Equation 5 where *AvgR* is the summation of recall (R_n) of classes divided by the number of classes. The number of correctly classified instances was used to determine accuracy. The processing time of modeling of the algorithm and cross-validation of the classifier were observed to determine the time constraint and classification accuracy respectively. The classical methods used for comparison are simple logistic (SL), multi-layered perceptron (MLP) and classifier decision tree (J48) [9, 10] which are highly accurate classification methods.

$$AvgP = \frac{\sum_{i=1}^{n} P_n}{n} \tag{4}$$

$$AvgR = \frac{\sum_{i=1}^{n} R_n}{n} \tag{5}$$

5.2 Result

Comparison of classical methods for performance is shown in Figure 4. The bar graphs present the comparison of processing time and accuracy of neuro-fuzzy classification and other classical methods. In Figure 4a, the processing time of neuro-fuzzy is much faster than the MLP while SL and J48 classifier has less processing time. In accuracy, we can justify the performance of neuro-fuzzy is better than the other classical methods in the sense that even though it has a high processing time than the SL and J48, it is more accurate of classifying patterns shown in Figure 6b.

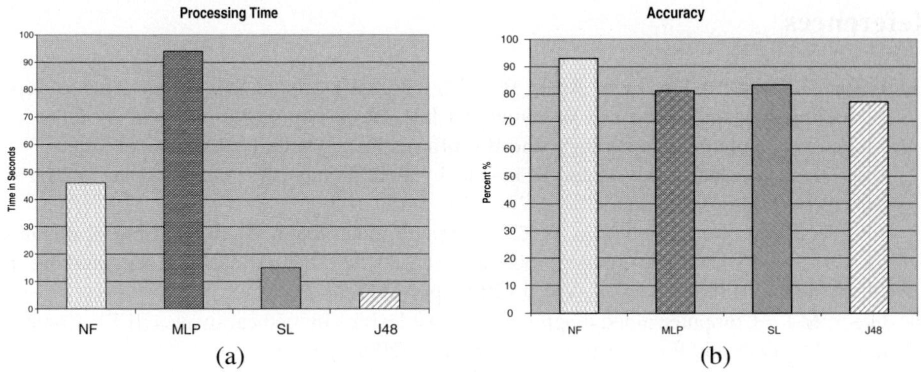

Fig. 6. Bar graphs showing the processing time and accuracy of each algorithm

The result of precision and recall are presented in Table 1 and 2, respectively. It is important that the classifier has a high accurate in classification to be used in consultation procedure of EMA. Neuro-fuzzy has the highest precision which has an average of 0.91 and recall which has an average of 0.91 compared to MLP (0.81, 0.83), and SL (0.38, 0.35), and J48 (0.77, 0.77). Most of these classical methods were able to predict testing data with the number of misclassified patterns between 51 to 63 while neuro-fuzzy has only 25 misclassified patterns out of 270 tuples.

Table 1. Precision

Classes	NF	MLP	SL	J48
present	0.89	0.828	0.843	0.793
absent	0.92	0.79	0.821	0.742
Average	0.91	0.809	0.832	0.768

Table 2. Recall

Classes	NF	MLP	SL	J48
present	0.90	0.833	0.86	0.793
absent	0.91	0.783	0.8	0.742
Average	0.91	0.808	0.83	0.768

6 Conclusion

Ubiquitous healthcare shows more researchable topics and it includes the integration of multi-agent systems. In this paper, we present the framework of ubiquitous healthcare based on multi-agent which supports the mobility and data mining of the mobile agent. We propose the expert mobile agent (EMA) that performs data mining to support the diagnosis of a patient. The EMA uses the neuro-fuzzy to process the consultation function. Also, a pre-processing of the relevant data based on the expert profile is shown to train the fuzzy system more efficiently. Result from simulations shows that neuro-fuzzy outperformed other high accurate classifiers. Future work will be more on the functionality of the proposed multi-agent framework in ubiquitous healthcare.

References

1. Foster D., et al.: A Survey of Agent-Based Intelligent Decision Support Systems to Support Clinical Management and Research. 1st Intl. Workshop on Multi-Agent Systems for Medicine, Computational Biology, and Bioinformatics, Utrecht, Netherlands (2005)
2. Brause, R.: Medical Analysis and Diagnosis by Neural Networks. Medical Data Analysis. Springer-Verlag, Berlin Heidelberg (2001) pp. 1-13
3. Joo, S., Moon, W. K., and Kim, H. C.: Computer-aided diagnosis of solid breast nodules on ultrasound with digital image processing and artificial neural network. Engineering in Medicine and Biology Society, Vol. 1 (2004) pp. 1397-1400
4. Giger, M.L.: Computer-aided diagnosis of breast lesions in medical images. IEEE Computational Science and Engineering, Vol. 2, Issue 5 (2000) pp. 39-45
5. Verma, B.; Zakos, J.: A computer-aided diagnosis system for digital mammograms based on fuzzy-neural and feature extraction techniques. IEEE Transactions on Information Technology in Biomedicine, Vol. 5, Issue 1, (2001) pp. 46-54
6. Schetinin, V.: Neural Network based Agent for Discovering Rules in Medical Databases: available at http://citeseer.ist.psu.edu/update/605607
7. Cervantes, L., Lee, Y. S., Yang, H, Ko, S. H., and Lee, J.: Agent-based Intelligent Decision Support for the Home Healthcare Environment. ICHIT 2006, Jeju, South Korea (2006)
8. Della Mea, V.: Agents Acting and Moving in Healthcare scenario: A Paradigm of Telemedical Collaboration. IEEE Transaction on Information technology in Biomedicine, Vol. 5, Issue 1 (2001) pp. 10-15
9. Vajirkar, P., Singh, S., and Lee, Y.: Context-Aware Data Mining Framework for Wireless Medical Application. Database and Expert Systems Applications, LNCS 2736 Springer-Verlag (2003) pp. 381-391
10. Lee, J., Mateo, R. M., Gerardo, B. D., and Go, S. H.: Location-Aware Agent Using Data Mining for the Distributed Location-Based Services. ICCSA 2006, Part V, LNCS 3984, Springer-Verlag (2006) pp. 867-876
11. Gerardo, B. D., Lee, J. W. and Joo, S.: The HCARD Model using an Agent for Knowledge Discovery. International Journal of KAIS, Vol. 14, Special Issue (2005) pp. 53-58
12. Zadeh, L. A.: Fuzzy Sets. Information and Control (1965) pp. 338-353
13. Klose, A., Nürnberger, A., Nauck , D., and Kruse R.: Data Mining with Neuro-Fuzzy Models. Data Mining and Computational Intelligence, Springer-Verlag (2001) pp. 1-36
14. Nauck, D., and Kruse, R.: NEFCLASS - A Neuro-Fuzzy Approach for the Classification of Data. In Proceedings of ACM Symposium on Applied Computing, Nashville (1995)
15. Mitra, S., and Hayashi, Y.: Neuro-Fuzzy Rule Generation: Survey in Soft Computing Framework. IEEE Trans. Neural Networks, Vol. 11 (2000) pp. 748-768
16. Halgamuge, S. K., and Glesner, M.: Neural Networks in Designing Fuzzy Systems for Real World Applications. Fuzzy Sets and Systems, Vol. 65, No. 1 (1994) pp. 1-12
17. Wai, R. J., and Chen, P. C.: Intelligent Tracking Control for Robot Manipulator Including Actuator Dynamics via TSK-type Fuzzy Neural Network. IEEE Trans. Fuzzy Systems Vol. 12 (2004) pp. 552-560
18. Kirn, S.: Ubiquitous Healthcare: The OnkoNet Mobile Agents Architecture. NetObjectDays (2002) pp. 105-118
19. Bellavista, P., Corradi, A., and Stenfalli, S.: Mobile Agent Middleware for Mobile Computing. Computer Journal. (March 2001) pp. 73-81
20. Hardala, F., Kiri, A., Özdemir, H., Yilmaz, T., and Güler, I.: A Neurofuzzy Classification System for the Effects of Diabetes Mellitus on Ophthalmic Artery. Journal of Medical Systems, Vol. 28, Issue 2 (2004) pp. 167-176

Agent-Based Approach to Distributed Ensemble Learning of Fuzzy ARTMAP Classifiers*

Louie Cervantes, Jung-sik Lee, and Jaewan Lee

School of Electronic and Information Engineering, Kunsan National University, 68 Miryong-dong, Kunsan, Chonbuk, 573-701, South Korea
{lfcervantes,leejs,jwlee}@kunsan.ac.kr

Abstract. This paper presents a parallel and distributed approach to ensemble learning of Fuzzy ARTMAP classifiers based on the multi-agent platform. Neural networks have been used successfully in a broad range of non-linear problems that are difficult to solve using traditional techniques. Training a neural network for practical applications is often time consuming thus extensive research work is being carried out to accelerate this process. Fuzzy ARTMAP (FAM) is one of the fastest neural network architectures given its ability to produce neurons on demand to represent new classification categories. FAM can adapt to the input data without having to specify an arbitrary structure. However, FAM is vulnerable to noisy data which can rapidly degrade network performance. Due to its fast learning features, FAM is sensitive to the sequence of input sample presentations. In this paper we propose a parallel and distributed approach to ensemble learning for FAM networks as a means to improve the over-all performance of the classifier and increase its resilience to noisy data. We use the multi-agent platform to distribute the computational load of the ensemble to several hosts. The multi-agent platform is a robust environment that can support large-scale neural network ensembles. Our approach also demonstrates the feasibility of large-scale ensembles. The experimental results show that ensemble learning substantially improved the performance of fuzzy ARTMAP classifiers.

1 Introduction

Neural networks have been used successfully in a broad range of non-linear problems that are difficult to solve using traditional techniques. Neural networks have been successfully used in pattern recognition, classification, prediction, control and optimization problems. Fuzzy ARTMAP (FAM) is a neural network architecture that performs incremental supervised learning of recognition categories and multi-dimensional maps for both binary and analog input patterns.

* This research was supported by the Program for the Training of Graduate Students in Regional Innovation which was conducted by the Ministry of Commerce Industry and Energy of the Korean Government.

N.T. Nguyen et al. (Eds.): KES-AMSTA 2007, LNAI 4496, pp. 805–814, 2007.

FAM is affected by two network parameters - the choice parameter and the baseline vigilance parameter. Higher values in these parameters tend create more nodes in the category representation layer of the FAM. Fewer nodes are desired in order to keep a minimum number of clusters in which the data are categorized. For this reason, a single pass over the training set is often preferred to avoid creating unnecessary classes. Unfortunately, this approach results in the increased sensitivity of the network to the order of presentation of the training set. The fast learning feature of FAM is also the reason for its sensitivity to input ordering.

Ensembles of learning machines have emerged as an important area of machine learning research and have shown encouraging and meaningful results in a number of applications[1]. We shall use the term ensemble to encompass many similar terms in the literature that refer to various methods of combining a set of learning machines that work together to solve a machine learning problem.

A neural network ensemble is a paradigm where a collection of a finite number of neural networks is trained for the same tasks [2]. In [3], the generalization ability of neural network system was substantially improved by training multiple neural networks and combining their results. The central concern of this paper is to develop a high performance machine learning application that leverages the capabilities of FAM while overcoming its weaknesses by means of the ensemble method. Agent technology is the suitable framework for this application because the knowledge is distributed and the cooperation among several entities is needed.

The following section will introduce the fuzzy ARTMAP neural network, the multi-agent platform and the ensemble learning method. Section three will describe the architecture of the agent-based ensemble application and the agents' descriptions and interactions. Section four will present the experimental evaluation. In the final section we conclude with a brief discussion of the future directions of this research work.

2 Related Works

2.1 Fuzzy ARTMAP

The FAM algorithm corresponds to a family of neural network architectures introduced by Carpenter et al. [4] and has proven to be among the superior neural network architectures for machine learning problems.

FAM can produce new neurons on demand to represent classification categories. This property allows the FAM to automatically adapt to the database without having to arbitrarily and a priori specify its network structure, but it also has the undesirable side effect that for a large database, it produces a large network size that can slow down the algorithm's training time. It would be desirable to have a method capable of keeping the network convergence time manageable, without affecting the generalization performance of the network or its result size when the training completes. The important advantages of FAM over other neural classifiers includes the fact that are that it learns the required task fast, it has the capability to do on-line learning, and its learning structure allows the explanation of the answers that the neural network produces.

Castro et al. [5] proposed two partitioning approach for the FAM algorithm: the network partitioning approach and the data partitioning approach. The data partitioning approach for FAM reduced the training time without a significant effect on the classification performance of the network.

2.2 Fuzzy ARTMAP Architecture

The Fuzzy ARTMAP architecture consists of four layers of field nodes as shown in Figure 1. The layers that are worth describing are the input layer F_1^a, the category representation layer F_2^a, and the output layer F_2^b. The input layer of Fuzzy ARTMAP is the layer where an input vector of dimensionality of the following form is applied:

$$I = (a, a^c) = (a_1, a_2, ..., a_{M_a}, a_1^c, a_2^c, ...a_{m_a}^c) \quad (1)$$

$$a_1^c = 1 - a_i, \quad \forall \in (1, 2, ...M_a) \quad (2)$$

The assumption is that the input vector a has each of its components in the interval [0, 1]. The layer F_2^a of Fuzzy ARTMAP is referred to as the category representation layer, this is where the categories (or groups) of input patterns are formed. Finally, the output layer is the layer that produces the outputs of the network. An output of the network represents the output to which the input applied at the input layer of FAM is supposed to be mapped to.

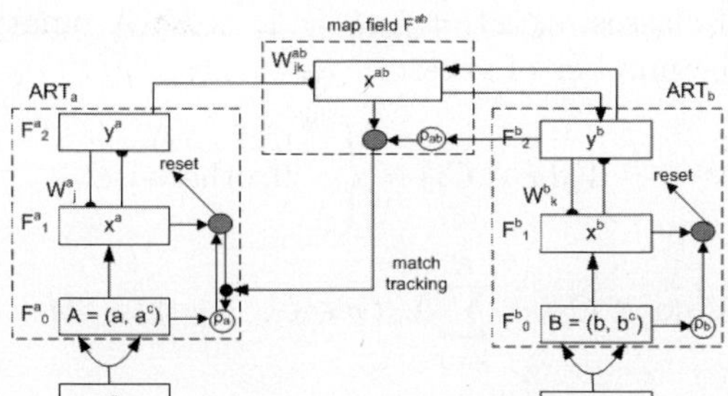

Fig. 1. Fuzzy ARTMAP Architecture

There are two sets of weights worth mentioning in FAM. The first set of weights are weights from F_2^a to F_1^a designated as $W_{ij}^a, 1 \leq j \leq N_a, 1 \leq i \leq M_a$, referred to as top-down weights. The output pattern that this node j is mapped to corresponds to the the vector of the weights $W_j^a = (W_{j1}^a, W_{j2}^a, ...W_{j,2Ma}^a)$. The FAM is trained to map every input pattern of the training list to its corresponding output pattern. The task is considered accomplished or the learning is complete when the weights do not change during a list presentation. This training scenario is known as offline learning.

Approaches to solving the category proliferation problem in ART neural systems can be divided into two groups: those that seek an off-line solution and those that preserve the original online characteristic of ART systems. The former solutions make use of post-processing methods and the latter solutions introduce some changes in the original Fuzzy ARTMAP architecture. The work described in this paper uses the unmodified FAM.

2.3 Ensemble Learning

In general, a neural network ensemble is constructed in two steps: training a number of component neural networks and then combining the component predictions [2]. Ensemble methods combine the output of several neural networks. The output of an ensemble is the weighted average of the outputs of each network. The resulting ensemble often outperforms the constituent networks.

Despite their obvious advantages, these methods have at least three weaknesses: 1) increased storage, 2) increased computation and 3) decreased comprehensibility. There are several methods for combining outputs from multiple neural networks such as for example bagging and boosting [12]. The ensembing method used in the current work is based on the simple voting strategy [13]. A common and simple way for resolving conflicts in opinions in human social life is by voting. The same principle can be applied to achieve a combined outcome for predictions from multiple classifiers. The most conservative method is the unanimous voting strategy where x is classified to C_i if and only if all classifiers predict C_i, otherwise x is rejected.

Given an input sample, x, if there are K classifiers denoted by $e_1, ...e_K$, from all the K predicted classes, $e_k(x) = j_k, k = 1, ..., K$. A binary function can be used to represent the number of votes:

$$V_k(x \in C_i) = \begin{cases} 1, \text{ if } e_k(x)=i, i \in \Lambda \\ 0, \text{otherwise} \end{cases} \tag{3}$$

$$V_E(x \in C_i) = \sum_{k=1}^{K} V_k(x \in C_i \quad i = 1, ..., M \tag{4}$$

Our procedure for the voting method is summarized below:

1. A number of networks are training on different orderings of the training data.
2. During testing, each of the individual network makes its prediction for a test item in the normal way.
3. The number of predictions made for each category is counted and the one with the highest score or the most number of votes is the final predicted category outcome.
4. The voting strategy provides an indication of the confidence of a particular prediction, since the larger the voting majority, the more certain is the prediction.
5. The voting strategy however is not appropriate for data that have temporal sequence unless appropriately method to batch the data is used.

2.4 Java Agent Development Framework

The agent-based platform [9] provides the environment to build distributed systems with intelligent local components that are designed to both cooperate and coordinate their activities. In this paper the inter-agent messages are structured data units that are modeled through the use of a high level declarative the agent communication language (ACL).

Java Agent Development Framework (JADE) is a FIPA-compliant, agent-oriented tool implemented in JAVA. FIPA is a standards organization that promotes agent-based technology and the interoperability of its standards with other technologies. JADE is composed of an agent platform (execution environment) and a set of packages which provide the basic support for multi-agent construction. Jade can be distributed among several computers and is customizable by means of a remote graphical interface. In addition it provides the mandatory components (FIPA) for agent management:

- Agent Management System (AMS) : responsible for providing the white page, agent lifecycle and agent directory services
- Directory Facilitator (DF) : provides the yellow pages service
- Agent Communication Channel (ACC) : controls the message passing mechanism

The development environment incorporates a set of graphical support tools which facilitate the platform management, providing support for agent debugging and execution. These tools are: RMA (Remote Monitoring Agent), Dummy Agent, Sniffer Agent, Introspector Agent, GUI-DF. In order to create a JADE agent, it is necessary to instantiate a class that extends the Agent class. Each agent is implemented in a thread following the lifecycle proposed by FIFA. Codifying an agent means defining the tasks it must accomplish during its execution, each task is an instance of the *Behaviours* class. In addition, each Agent class has methods to add or delete behaviours; this action can be done anytime in its lifecycle.

Communication architecture offers a flexible and efficient message-passing mechanism; for each agent, the framework creates and manages a queue for private messages. The Communication paradigm used is based on the asynchronous message, which conforms with the FIPA Agent Communication Language (ACL) standard. Each message is an object of the ACLMessage class, and has methods for managing the message parameters, for sending and receiving messages, for message filtering, etc.

To facilitate the development of agent conversations, JADE offers a library of interaction protocols defined by FIPA (Query, Request, Contract-Net, etc). Interaction protocols define the order and type of message involved in a conversation. For each task, it is necessary to add a behavior in the agent which initiates the conversation and in the other agent which will respond. In an open and distributed agent-based ensemble of classifiers, standard specifications are vital for ensuring interoperability of the autonomous agents.

3 Distributed Ensemble Learning of Fuzzy ARTMAP Classifiers

3.1 Agent Platform Architecture

The agent platform of the distributed ensemble FAM classifiers include the FIPA specified agents (AAC, AMS and DF), provided by Jade. Figure 2 show the architecture of the multi-agent FAM ensemble. The three agent classes are shown with the important internal modules that support their functions with the Jade environment. All agent communication is performed through message transfer. Message representation is based on the Agent Communication Language (ACL) formulated by FIPA. The ACL contains two distinct parts - the communicative act, and the content of the message. Communication acts have precise declarative

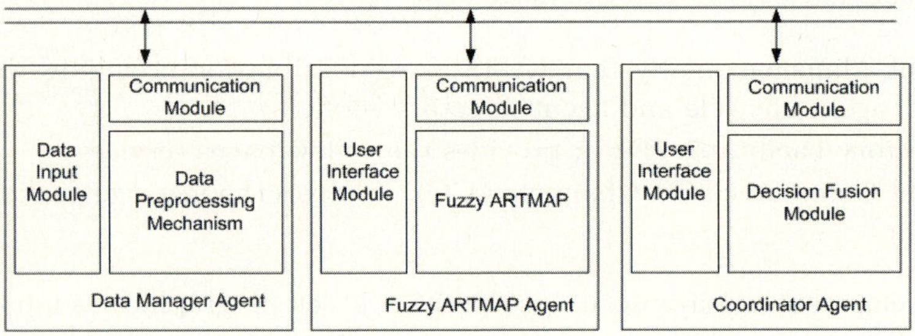

Fig. 2. Multi-agent FAM Ensemble Architecture

meaning independent of the content of a message and extends the intrinsic meaning of the message content. The software architecture is based on Java Virtual Machines that run simultaneously on several hosts communicating with each other through Java Remote Method Invocation. An agent container provides a complete runtime environment for agent execution and allows agents to concurrently execute on the same host. Each agent container is a multithreaded execution environment composed of one thread per agent. Each agent is an active object having more than one behavior and can engage in multiple, simultaneous activities. The architectural structure of the system is based hierarchies of agent containers distributed across the network. The agent classes are:

Coordinator Agent - The CA handles the management and over-all coordination of the system. It derives from the GuiAgent class which provides this agent with a graphical interface. It contains other useful modules such as for example the preprocessing to normalize the data. The FAM network expects all inputs to be normalized in the range [0, 1]. The interface provides a means to specify the source files for the training and test datasets. It communicates with the DF in Jade to obtain the name and location (AID) of the other agents in the platform. The main role of this agent is to initiate the classification task by sending the appropriate message to the data manager agent.

Data Manager Agent - The Data Manager Agent (DMA) uses a TickerBehaviour to periodically check for incoming classification tasks from the CA. It reads the training and testing datasets from the file system. To start the classification task, the DMA creates the ACL message that holds a Java Serializable object containing the arrays of training and testing data values. It also communicates with the DF to obtain a list of available FAM classifier agents just before sending the data to these agents. DMA uses the ParallelBehaviour to manage separate listeners on individual threads to handle incoming message containing the classification results from the classifiers agents. The ensemble process to combine the independent results is invoked by DMA. It wraps the data in a serializable object before sending in an ACL message to the CA thus completing the classification job.

FAM Classifier Agent - The FAM Classifier Agent (FCA) contains the code to carry out the classification task using the fuzzy ARTMAP neural network. Each FCA runs in a separate thread and can be distributed across several hosts in the agent platform. There are two implementation of the FCA used in the experiment. The first used a graphical use interface which is used by the agent to display its local classification result. The other implementation does not use any interfacc to increase its responsiveness particularly when working with large training or testing datasets. The GuiAgent class of Jade provides a convenient way for agent and non-agent elements within the FAM classifier to interact effectively. Only one instance each for the coordinator agent and the data manager agent is needed for the whole platform. The FCA can have many instances distributed across several agent-containers in the participating host machines. FCAs can be added and removed at run time. This robustness of the multi-agent platform enables the DMA to still perform the ensemble method even if some FCAs fail to submit a result within a specified timeout such as for example when an agent host goes down.

3.2 System Implementation

In the distributed ensemble platform, the roles of every agent type in the classification task were used as basis for defining the details of each conversation. The proposed neural network ensemble is shown in figure 3. The ensemble consists of N independent fuzzy ARTMAP networks inside the individual FAM agents. All of the FCA receive randomly sorted input values from the DMA and independently process them to produce outputs. The output of the respective network is sent to the CA in a serializable Java object contained in an ACL Message. The relevant sequence of ACL Messages is outlined below:

1. The JADE Main Platform is initialized and the CA, DMA and required number of FCA agents are loaded in their respective agent containers distributed to several machine hosts in the network.
2. The CA prompts for the filename of the training and test data. The ACL REQUEST is sent to the DMA to read the training and test data.
3. The DMA reads the data and creates a serializable object which is attached as the content object to an INFORM message sent to the FCA.

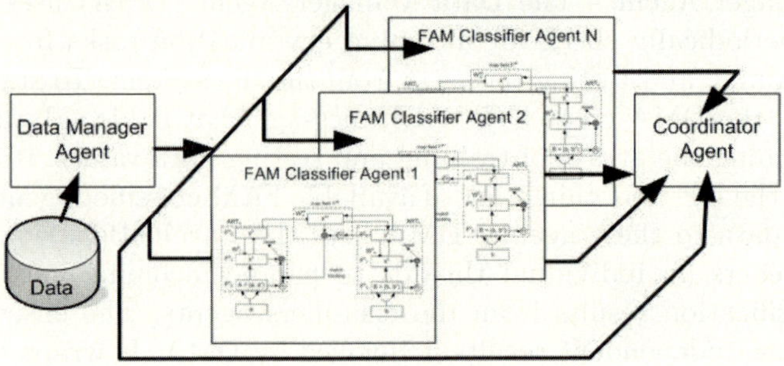

Fig. 3. FAM Ensemble Framework

4. Each FCA goes through the training and testing phase and outputs the classification results. Graphical type FCAs display their individual output in their interface.
5. Upon completing the classification task, each FCA sends an REPLY message to the DMA. The result data is obtained from the de-serialized object and the DMA proceeds to carry out the ensemble method to combine all the results.
6. The final classification output of the system is sent to the CA for display in the user interface.

4 Experimental Results

The practicability and performance of the proposed FAM ensemble was evaluated using the Iris data set obtained from the UCI Machine Learning Repository [13]. There were 150 samples with three output classes in the iris dataset. Each class contained 50 samples and each sample comprised of four input features. For all the simulation runs, the vigilance parameter is set to 0 and the single epoch run was used. From each class, 30 samples were randomly selected to be included in the training dataset for a total of 90 training samples after the sample from the 3 classes were combined. The remaining 60 samples comprised the testing dataset.

We performed simulation runs using the ensemble size of 3, 5, 7, and 9 FAM networks. The respective graphs of the classification rates of the various ensemble sizes are shown in figure 4. For each run, the training sample is presented at random order to the FAM. For each of the simulation runs, the number of training samples correctly classified by the each of the FAM networks ranged from 45 to 50 out of the 60 samples. As expected when the same FAM network was introduced with a new random sample, its classification rate tends to be different from previous values. This illustrates the sensitivity of the individual FAM network to the ordering of training samples. The ensemble yielded better classification results than any of its member FAM network. This indicates

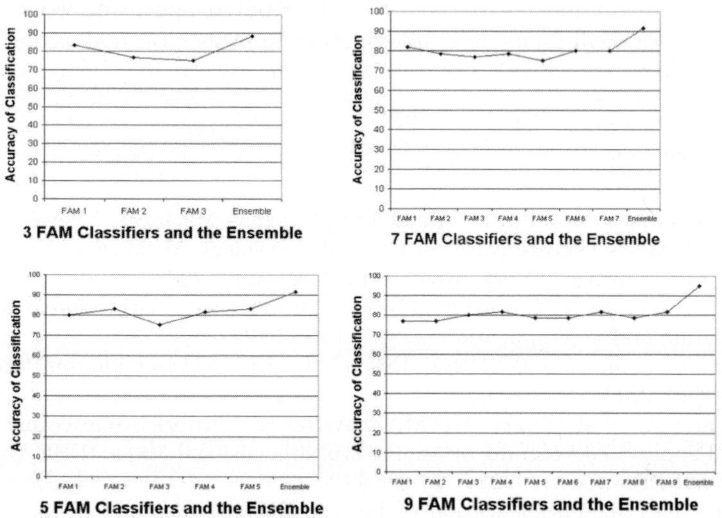

Fig. 4. Classification Performance of Individual FAM Classifier

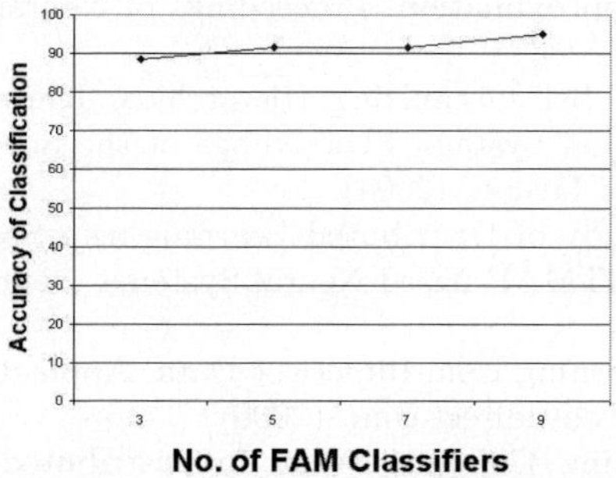

Fig. 5. Classification performance and Ensemble Size

that our ensemble approach that combined the classification of individual FAM networks could indeed produce better classification performance.

In figure 5, the classification performance is compared across various ensemble sizes. Our results indicate that the ensemble generally improved its performance for each increment of the ensemble size.

5 Conclusions

This work demonstrates that better classification performance can be achieved using the ensemble approach that combines the result of several neural networks using a voting strategy. We also introduced a multi-agent implementation for the distributed neural network ensemble. The multi-agent platform provided a robust environment for the parallel and distributed neural network ensemble. Our future works will focus on employing our approach on real world classification problems.

References

1. Valentini G. and Masulli F.: Ensembles of Learning Machines. Neural Nets WIRN Vietri-02, Series Lecture Notes in Computer Sciences, M. Marinaro and R. Tagliaferri, Eds.: Springer-Verlag, Heidelberg (Germany), 2002.
2. Maqsood, I. et.al: An Ensemble of Neural Networks for weather forcasting. Neural Computing and its Applications (2004) 13 112-122
3. Hansen, L., et al.: Neural Network Ensembles. IEEE Transactions on Pattern Analysis (1990) 12 (10) 993-1001
4. Carpenter A. et al.: Fuzzy ARTMAP: A Neural Network Architecture for Incremental Supervised Learning of Analog Multidimensional Maps. IEEE Transactions on Neural Networks. (1992) 3 (5) 698-713
5. Castro, J. et al.: Parallelization of Fuzzy ARTMAP to improve its convergence speed: The network partitioning approach and the data partitioning approach. Non-Linear Analysis. Available online at www.sciencedirect.com. (2005)
6. Andonie, A. et al.: A Modified Fuzzy ARTMAP Architecture for Incremental Learning Function Approximation. Proceedings of Neural Networks and Computational Intelligence. (2003)
7. Carpenter, A. et al.: Self-Organizing Hierarchical Knowledge Discovery by an ARTMAP Image Fusion System. Proceedings of the Seventh International Conference on Information Fusion. (2004)
8. Hernandez, et al.: Study of Distributed Learning as a Solution to Category Proliferation in Fuzzy ARTMAP based Neural Systems. Neural Networks (2003) (16) 1039-1057
9. Hartono, P. et al.: Learning from Imperfect Data. Applied Soft Computing. Available online at www.sciencedirect.com. (2006)
10. Calisti, M.: Abstracting Communication in Distributed Agent-Based Systems. ECOOP2002. Workshop on Concrete Communication Abstractions of The Next 701 Distributed Object Systems (2002)
11. Lim, C. et al.: Applications of Fuzzy ARTMAP and Fuzzy C-Means Clustering to Pattern Classification with Incomplete Data. Neural Computing and Applications (2005) 14: 104-113
12. Zeng X, et al.: Using Neural Network to Approximate an Ensemble of Classifiers. Neural Processing Letters (2000) 12: 225-237
13. Downs, J. et al.: Application of the Fuzzy ARTMAP Neural Network Model to Medical Pattern Classification Tasks. Artificial Intelligence in Medicine (1996) 8: 403-428
14. Blake, C. et al.: UCI Repository of machine learning databases. Available at http://www.ics.uci.edu/ mlearn/MLRepository.html, University of California, Department of Information and Computer Science, Irvine, California

Framework for Data Aggregation and Insurance Service in Vehicle Telematics Using Multi-agents*

Bobby D. Gerardo[1], Jaewan Lee[1], and Malrey Lee[2]

[1] School of Electronic and Information Engineering, Kunsan National University
68 Miryong-dong, Kunsan, Chonbuk 573-701, Korea
{bgerardo, jwlee}@kunsan.ac.kr

[2] School of Electronic and Information Engineering, Chonbuk National University
664-14 Deokjin-dong, Jeonju, Chonbuk 561-756, Korea
mrlee@mail.chonbuk.ac.kr

Abstract. The emerging technology in vehicle telematics encourage stakeholders to consider services that can be beneficial to both clients and the telematics service providers. This paper proposes a novel framework for insurance telematics in Korea using a mobile aggregation agent (AA) and data mining agent (DA). To our knowledge, this model is recent of its kind in this country and the baseline information from driver's characteristics serves as reference for the flexible insurance policies. We are able to present a use-case scenario and examples to demonstrate our model. With this flexible insurance framework, customers can manage their own insurance premiums and lower the cost of motoring. This framework is perceived to have promising applications ranging from academic to industrial uses.

1 Introduction

Emerging technologies in telematics provide opportunities for data mining, cross-selling and even new products and services based on better perception of customer behavior. Several companies with promising visions and innovative ideas are beginning to see this wirelessly enabled and computer-enhanced set of solutions as an essential tool for doing business.

The current trends in vehicle telematics drives several stakeholders in this field to consider services that could be beneficial for both clients and the telematics service providers. Vehicle data recorders and telematics are not a future possibility, but currently available technologies [1], [3], [5], [7]. As costs of telematics devices decrease, many industries, including insurance, are taking a second look at utilizing these devices to improve services, and develop innovative solutions. On the other hand, recording devices such as black boxes [1], [7], [12] can range from passive to real-time where it can record vehicle conditions like speed, acceleration, and position. Also it can record crash data that includes information prior to the accident.

* This research was supported by grant R01-2006-000-10147-0 from the Basic Research Program of the Korea Science and Engineering Foundation.

N.T. Nguyen et al. (Eds.): KES-AMSTA 2007, LNAI 4496, pp. 815–824, 2007.

Multi-agents are intended for communication and cooperation in which they have the ability to behave socially, to interact and communicate with other agents like exchange information, receive instructions and give responses and cooperate when it helps them fulfill their own goals [13]. The role of agents for distributed information management, which include resource discovery, information integrity and navigation assistance is perceived to be important. Multi-agents can then be employed in ubiquitous environment such as those intended for telematics application.

This paper proposes a novel framework for data aggregation and insurance service in vehicle telematics for Korean motorists using multi-agents. To our knowledge, this framework is new of its kind in this country and the baseline information from driver's characteristics and patterns serves as reference for the development of flexible insurance policies which in return could benefit vehicles owners for a reduced premiums. The proposed aggregation agent resides at the vehicle side and we assume that the vehicle is mobile and could have the ability to transmit data in both real and non-real-time conditions. We will use some transformation and visualization of data to discover driving patterns and in this case propose flexible insurance policies.

2 Related Studies

Vehicle telematics allows the deployment of several new services and applications integrating wireless communication technology into an automobile. As a result, the vehicle acquires new capabilities and offers more services to its users. Both [5] and [17] presented examples of telematics services such as: (1) Navigation and traffic information system, (2) Voice recognition and wireless Internet service, (3) Safety systems, (4) Security systems, and (5) Diagnostics and maintenance services.

One drawback of implementing insurance telematics is the cost of recording device. Earlier implementations of the systems did not progress well due to high cost of devices which eventually forced some companies like Progressive to stop the project [6]. Insurance companies must be aware of what is possible these days and in the future. Black boxes used for vehicle recording come into varieties such as in the form of trip-logging, passive GPS tracking, and crash data recording devices. Nowadays, these devices are already affordable and widely available.

2.1 Motivations

A classic method for data retrieval from vehicle is done through plugging-in of telematics device to a PC to obtain the data stored in it. There are modern approach can be used by means of emerging technology for telecommunications. Insurance telematics is not yet popular in Korea, so it is likely that flexible insurance or pay as you drive scheme will be widely acceptable. We can develop proprietary scheme for insurance telematics based on drivers' behavior. This framework can encourage TSP and telecommunications collaboration which will yield rapid deployment of the system. Aggregation of information can yield comprehensive data warehouse that can be utilized for driving pattern discovery, traffic prediction and forecasting. In addition, this information can help road safety and reduce car accidents.

2.2 Insurance Telematics

Auto insurer initiatives can be traced back as early as 1998 where Progressive [6] launched the pay as you drive service using GPS tracking device. In the same year Norwich Union [6], [11], [15] in UK piloted the same pay as you drive based on Progressive methodology. Norwich initiative is using IBM Black Boxes for flexible insurance. This method of paying motor insurance will not only appeal to drivers who recorded low mileage but it could also attract high mileage drivers if they avoid busy city centers or drive at off-peak hours. GMAC [14] in the USA also piloted an insurance discount based on mileage. Other initiatives were piloted by AXA of Ireland which gives discounts to young drivers if they can maintain a designated speed limit. Japan's AIOI insurance had also initiated the pay as you drive service.

2.3 Multi-agents in Vehicle Telematics

Mobile agents [2] are used to tamper resistant hardware to protect privacy. Parties requesting access to private information receive mobile agents that encapsulate private data and access control policies. The agents are executed in a protected environment to access data. A paper on control system tasks using multi-agents [13] demonstrates that cooperative behavior can be achieved without extensive inter-agent communications. This means that multi-agent design for mobile object where resource is scarce is promising because overhead cost for communication can be reduced.

3 System Architecture

The proposed global architecture for telematics system is shown in Figure 1. There are ranges of wireless technologies for vehicle telematics that are described by [5] where its viability will depend on the different target applications that they were optimized for. Some examples are Bluetooth, ZigBee, UWB, and Wi-Fi. On the other hand, InternetCar [1] proposes architecture to connect car to the Internet.

Some other workable communication medium to transmit data from a mobile vehicle is described in the primer on real-time traffic system [12], which includes cellular technologies such as GSM, CDMA, GPRS, EDGE, WCDMA, TDMA, iDEN, and WiDEN. These technologies are more suitable for real-time data delivery due to its availability and the types of network it can support.

In this framework, we chose CDMA as a medium for data delivery of real-time vehicle information. Collection methods can be done through on-board sensors, acoustic, embedded, radar and video sensors. Moreover, Floating Car Data (FDC) [1], [12] can be used for more accurate recording of vehicle data. FDC refers to a set of protocols, services and data formats by which cars transmit information to a server [1]. The use of FDC can be traced back from the mid-80's using programs like Ali and Euroscout from Siemens and Socrates from Philips. In this program, vehicles arc equipped with GPS and cellular modems to transmit speed and position data to remote data centers.

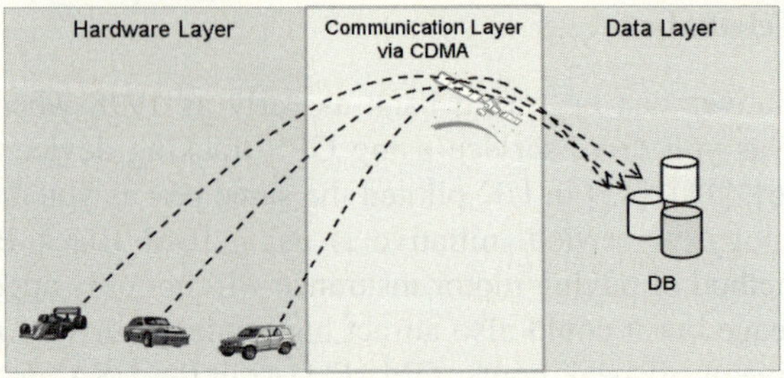

Fig. 1. Global Architecture for Data Aggregation in a Telematics System

The aggregation function (AA) shown in Figure 2 will act as coordinator to organize, extract and filter the sensor data. The design for AA is influenced and anchored on the FDC framework. The major task of AA is to prepare the data prior to data delivery. Three specific tasks undertaken by AA are extraction, filtering and communications. The extraction function of AA uses feature extraction and data fusion algorithm to collect data from sensors while the filtering algorithm removes noise, handle outlier values, and perform data transformation. The communication task of AA is to prepare the data in appropriate communication format so that in can be transmitted to the data center through CDMA channel.

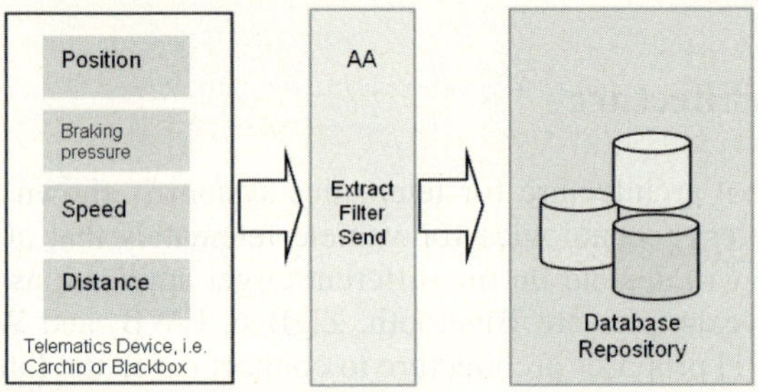

Fig. 2. Architecture of the Aggregation Agent (AA)

Figure 3 shows the process diagram of AA and DA. It presents the AA components, storage of data, initiating the DA and generating the driving patterns. Finally, the generated patterns will be correlated to corresponding insurance policies.

The components of data mining agent (DA) shown in Figure 4 are summarization and rules-generation functions. Summarization is extracting relevant data of selected or chosen cases from the management table and tabulating all the events associated to it. Once the tabulation is done, the rule function will be initiated to calculate for the driving patterns based on the predefined rules presented in Table 3. The parameters used for the rule functions are values obtained from the driving attributes.

As soon as the DA generated the driving pattern of a particular driver it will be correlated to pre-defined insurance scheme. In this scenario, the concerned driver can benefit lower premiums based on his driving characteristics. Our framework also

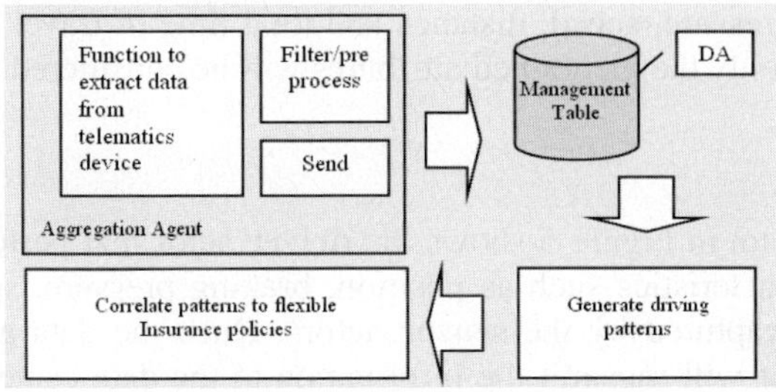

Fig. 3. Process diagram for the aggregation and data mining agent

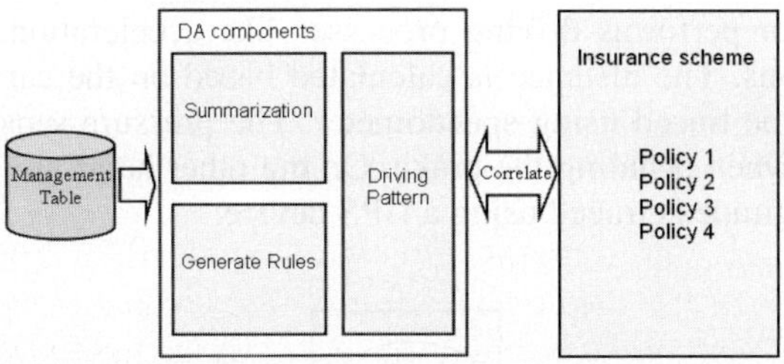

Fig. 4. Components of Data Mining Agent (DA)

considers important aspects of the system like data type and its availability whether public or private. Other aspects include method of storage, access controls and data security, transmission between users or system, and regulatory requirements for data.

4 Data Collection Using Vehicle Telematics Devices

The data is collected using a trip logging [6] device, which will show the date and time that the car is started, stopping date, and total time of driving. In the parameter, distance traveled throttle and speed are indicated. The device can also record car mileage and as well as the average monthly mileage. The passive GPS tracking mentioned by [6] can collect driving date and time, distance traveled and location arrived.

Also, it can derive the departure and arrival time and total driving time. There are devices that can also monitor the real-time condition of a car. The information from this device includes dates, latitude and longitude, car location, and status of the car which is either running or stationary. Some papers like in [1], [2], [6], [17] proposed other data types to be collected by the vehicle recording devices, which can be used for analysis of vehicle crashes. This includes acceleration, speed, engine RPM and throttle, ignition cycle, air bag and safety belts.

5 Use Case-Scenario and Evaluations

To illustrate the work around of our framework, we considered the data presented in Tables 1 to 4 for the data processing and analysis. This is limited to information about position, braking pressure, speed, distance and total time of travel. Due to the limitations of our study, only the mentioned attributes will be considered and tested.

5.1 Use Case

The use case diagram in Figure 5 shows the **driver** actor that performs driving interaction and its characteristics such as position, braking pressure, speed, distance and time duration are captured by the **sensor** actors. Once the data are aggregated, the **mobile phone** actor will transmit the information to the data center and the **receiver** actor will acquire and then store it in the database. The AA performs the data accumulation and it extracts data from the different sensors and prepares it for delivery to a data center.

The **driver** actor performs driving processes like acceleration, braking, stopping and road navigations. The distance is calculated based on the car mileage while the vehicle speed can be traced using speedometer. The pressure sensor will capture the amount of forced when pedaling the brake. On the other hand, the position expressed in latitude and longitude is traced using a GPS device.

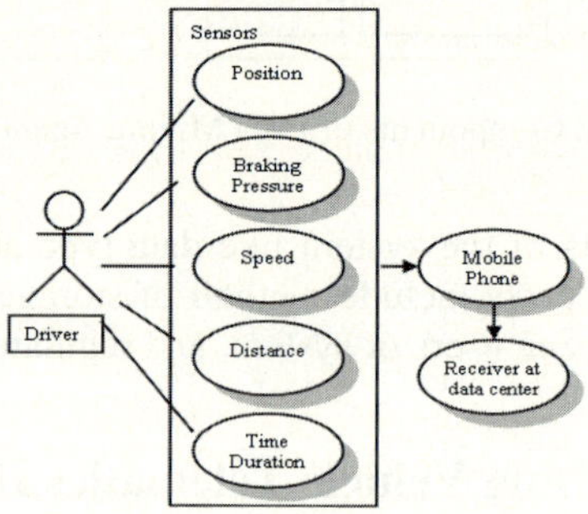

Fig. 5. Use Case diagram for data Aggregation Agent

5.2 Obtaining Base-Line Data Prior to Actual Deployment

The most appropriate method to get base-line data prior to actual implementations of the telematics system is through personal interview or survey. Some questions to answers through this stages are: driving characteristics, willingness to use vehicle telematics, acceptability to share the information for research and commercial purposes. The data delivery that comprises the communication component of the infrastructure can be coordinated with the telecommunications company that offers cellular service in the region. This can be included in actual deployment of the system.

A randomized survey is appropriate of at least 5,000 samples and the information to be learned are: (1) driving safety such as accidents, breakdown, and emergency services, (2) driving responsiveness, (3) positions such as where they usually go, (4) acceptability of the flexible premium, and (5) how much is paid and possible discount that can be availed for variable insurance scheme.

This research can be coordinated to the telematics service providers (TSP) or third party institutions that maybe interested for the telematics data. A reward system can be given to the driver participants that are joining the study. For example, full disclosure of data for use by the third parties will give him 20% discounts in one year insurance premium. Selected disclosure of data will correspond to lesser premium discounts.

5.3 Evaluations and Testing

Driving characteristics of Driver 1 can be collected from the start and stop points of driving events. Other information can refer to the date and time, distance traveled, and vehicle position. The management table can be populated with several records based on events accumulated on a time domain. Such table will also include the records of other drivers. The computed average for sensor data of 5 drivers is shown in Table 1.

Latitude and longitude values are translated to actual location showing addresses relative to the previous positions. The final position refers to the last position that the vehicle rested after the navigations. The values in the tables are averages for all the events in a certain time range. Heterogeneous devices would provide unformatted data prior to uploading to a data center, hence, AA will be used to pre-process it.

Table 1. Illustration of cumulative data derived from car sensors

Driver	Position (final pos)	Braking Pres., lbs.	Speed Km/hr	Distance Km.	Total Time, Hrs.
1	37.57/126.97	Hard	87	451.11	91.21
2	37.48/126.89	Harder	85	358.23	85.70
3	37.56/126.99	Regular	58	147.68	70.32
4	37.54/126.95	Regular	44	512.12	95.65
5	37.53/126.92	Regular	76	342.50	72.33

Table 2 shows the transformed equivalent of the values in Table 1. The attribute classes are provided by the domain experts. We can compute for the ranges of the given attribute using our function in (1):

$$\delta = \frac{Max(I) - Min(I)}{\varepsilon} \tag{1}$$

Where δ is the discrete cut points, ***Max (I)*** is the highest score in the given attribute, ***Min (I)*** is the lowest score in the given attribute, ε is the expert opinion on particular attribute. This will follow that the classes of the given attribute can be determined using a function. The attribute that belongs to the range defined by the cut point, C_1 is calculated as ***Min (I)*** + ε, C_2 is calculated as $C_1 + \varepsilon$, C_3 is calculated as $C_2 + \varepsilon$, and C_n is calculated as $C_{n-1} + \varepsilon$ and until ***Max (I)*** $\in C_n$. For example, if the $\varepsilon = 3$, then the δ for attribute Speed is 14. In this case, the three classes formed are: Regular (44 to 58), Fast (59 to 73) and very fast (74 to 87). Other approaches of the choice of cut-points for the classes are trends in driving, expert opinion, averaging, or using a function. Here, we considered the domain expert opinion for the ranges.

Table 2. Information stored in Management Table

Driver	Position	Braking Pressure	Speed	Distance	Total Time
1	Near	Hard	Very Fast	Longer	Long
2	Far	Harder	Very Fast	Longer	Long
3	Far	Regular	Regular	Short	Regular
4	Near	Regular	Regular	Longer	Long
5	Near	Regular	Fast	Longer	Regular

Flexible or discounted insurance premiums can range from Policy 1 which is associated to regular premium to Policy 3 which is a highly discounted rate. A "SAFE DRIVING" pattern can receive full benefits by getting highest premium discounts. On the other hand, RISKY DRIVING pattern have no discount at all. The insurance policies are given as: (1) Policy 1 [RISKY DRIVING] for regular premium; (2) Policy 2 [NORMAL] for insurance premium discounted at 10%; and (3) Policy 3 [SAFE DRIVING] for insurance premium discounted at 20%.

One concern of insurance companies is safety. This implies that the safer the driver is the lesser is the liability of the company for insurance claims. In addition, this will insure road safety and reduce road accidents. According to Finnegan et al [6], Berlin taxi accident fell by 66% after installing telematics tracking while European Union telematics found 28% reduction in accident and 40% in cost. Table 3 shows the driving patterns provided by the domain experts per driving category. Experts can base the patterns on the norms of data aggregated from the motorists.

Table 3. Driving patterns and descriptions

No.	Driving Patterns-(P, BP, S, D, TD)	Descriptions
A	Near, Regular, Regular, Short, Short	SAFER DRIVING
B	Far, Hard, Fast, Long, Long, Regular	NORMAL
C	Farther, Harder, Very Fast, Longer, Long	RISKY DRIVING

As soon as a driving pattern is determined for a particular driver, a correlation will be computed and the computation will be based from the records in Table 2 and 3. While Table 4 shows the cumulative data for Driver 1. The events refer to driving actions which constitute the start and stop events of driving. The final position is computed relative to the starting position. Table entries correspond to the descriptive equivalent of the continuous data generated per event.

The Chi-square test is used to compute the correlation and determine if the generated pattern do not differ significantly to the expected driving patterns. Where O is the observed frequency, E is the expected frequency, and k is the number of cases. In our example, we are able to determine the driving pattern of Driver 1 described as "RISKY DRIVING" using equation 2. Hence, the scheme assigned to this pattern is "POLICY 1". This implies that the driving pattern of Driver 1 is likely prone to accident. In this scenario, the insurance company charges regular premium to Driver 1.

$$\chi^2 = \sum_{i=1}^{k} \frac{(O-E)^2}{E} \tag{2}$$

Granting that the driving pattern is "NORMAL", and then the insurance company can give a discount of 10% on insurance premium. On the other hand, 20% discount is given to driver with "SAFE DRIVING" pattern.

Table 4. Data for a Driver 1

Events	Position	Braking Pressure	Speed	Distance	Total Time	Driving Pattern
1	Near	Regular	Regular	Short	Long	RISKY DRIVING
2	Near	Hard	Very Fast	Short	Long	
3	Near	Regular	Fast	Short	Regular	
4	Far	Regular	Very Fast	Longer	Long	
Ave.	Near	Hard	Very Fast	Longer	Long	

5.4 Applications

This framework has a promising applications ranging from academic to industrial uses. It is probable to collaborate with insurance companies in Korea for the disposition of information obtained from the study. There are possibilities of commencing an initiative to encourage flexible insurance premium based on how safe the driver is. Insurance companies can support the initiative of developing an algorithm for flexible policy scheme in Korea based on the data to be collected from actual deployment of the system.

Also, we can improve this by continual extraction of data from samples for a certain period until the norm is achieved. In exchange, the participants can be granted benefits like knowing their driving patterns or providing them reduced insurance premiums. It is estimated that parties involve can be benefited at several aspects. The clients can have choices of flexible or reduced insurance based on driving pattern. Companies can win the trust of clients by providing them up-to-date data, diagnostics and road-side assistance and reduced insurance rates. In addition, they can gather data for future traffic prediction, customer trends and for decision support purposes.

6 Conclusion and Recommendations

In this paper we are able to show the framework for data aggregation and insurance service using multi-agents and perceived it to have promising applications ranging from academic to industrial uses. We are able to present a use-case scenario and illustrative examples to demonstrate our model.

With flexible insurance framework, customers can manage their own insurance premiums and as well as the cost of motoring. Safer driving will imply discounted insurance and as well as road safety. However, there are some constraints that have to be addressed by this study like the cost of the study including methodology, sampling and analysis. Another is the willingness of the participant to cooperate in the study, which eventually affects their privacy and security. Others refer to investment such as cost of telematics devices and the availability of GPS on board a vehicle.

References

1. Chen, et al., Scaling Real-Time Telematics Applications using Programmable Middle boxes: A Case Study in Traffic Prediction, (2003)
2. S. Duri, et al., Data Protection and Data Sharing in Telematics, MONET-ACM Mobile Networks and Applications Journal 9(6): 693-701, (2004)
3. J. Munson, et al., A Rule-based System for Sense-and-Respond Telematics Services", International Workshop on Mobile Commerce, Proceedings of the 2nd international workshop on Mobile commerce, pp. 40-44, (2005)
4. T. Bauer, et al., A Flexible Integration Strategy for In-Car Telematics Systems, Proc. of 2nd Int. Workshop on Software Eng. for automotive systems, Vol. 30 No. 4, (2005)
5. T. Nolte, H. Hansson, and L. Bello, Wireless Automotive Communications, available at www.mrtc.mdh.se/publications/0903.pdf (2004)
6. D. Finnegan and C. Sirota, "Is Vehicle Data Recording Auto Insurance's Future?" Available at http://www.qualityplanning.com/ qpc_resources_public/ reports/ ISO.QPC.Vehicle Data Recording. v5links.pdf (2004)
7. Bisdikian, et al., Intelligent Pervasive Middleware for Context-Based and Localized Telematics Services, Proc. of 2nd int'l workshop on Mobile commerce, ACM Press, 2002
8. P. Kumar and S. Swarup, Business Intelligence and Insurance, available at http://datawarehouse.ittoolbox.com/pub/ND102201.pdf (2001)
9. Egil Juliussen, The Future of Automotive Telematics, available at www.touchbriefings.com/pdf/11/ auto031_r_juliussen.pdf (2003)
10. J. Garabedian, et al., Winning with the Wireless: A challenge for Auto Insurers, available at http://www.financetech.com/ showArticle.jhtml? articleID=14705766 (2002)
11. IBM United Kingdom Limited, Norwich Union's Pay As You Drive insurance initiative using IBM Black Box, available at http://uk.builder.com/whitepapers/ (2004)
12. ABI Research, A primer on real-time traffic system, available at www.abiresearch.com/whitepaper
13. M. Helm et al., Simulation of Cooperative Control System Tasks using Hedonistic Multi-agents, available at: www.cs.ttu.edu/ ~mhelm/ file. pdf (2006)
14. GMAC Insurance and OnStar Create Innovative Insurance Products, available at http://www.telematicsupdate.com
15. Norwich Union gets high-tech help for telematics-based car insurance, available at http://networks.silicon.com , (2003)
16. Norwich Union Launches, Pay-As-You-Drive Insurance., available at http://www.telematicsjournal.com, (2004)
17. P. Moskowitz, et al., Framework for Security and Privacy in Automotive Telematics WMC'02, ACM Press, (2002)

Business Model and Comparasion of S/W Source Code vs. Digital License for IPRs

ByungRae Cha

Dept. of Computer Eng., Honam Univ., Korea
chabr@honam.ac.kr

Abstract. The IPR(Intellectual Property Rights) system plays a core role in the development of information society in the 21st century as it did in the birth and growth of previous industrial society. The IPR system and technology for the extended software source code from digital contents are very significant to enhance international competitiveness. A problem happens if the original software source code is made public at dispute on its ownership. This study has suggested the Business model and comparasion of S/W source code vs. Digital License to protect the copyright infringement and technique leakage caused by its opening. We can prevent IPRs violation if index and search technology of digital license attached to mobile agents.

1 Introduction

Recently the academic circle's interests in economic effects of copyright and the policy unit's interests in the quantitative measures of national-economic role of copyright are heightened globally. The system of Intellectual Property Rights (IPR) has played a crucial role for the progress of information society in the 21st century as it did for the birth and development of previous industrial society. In link to the source code of extended software in digital contents, the well-organized IPR system and description have very important meaning in Korea as well with a view to enhancing international competitiveness. The management technology of copyright of software source code stays at beginning stage, compared with the studies on the copyright protection of digital contents. The problem comes from opening the software source code to prove the ownership if a dispute for software source code happens. This study is purposed to suggest digital license prototypes and business models for software source code in order to prevent the copyright infringement and the leakage of technique after publicizing the original software source code. The suggested digital license for software source code includes the information on the developer and structural information of source code. Accordingly the information on the developer's ownership and the structural information of source code enable to judge the originality of software on dispute without providing the source code firsthand. And we can do to attach index and search function in mobile agents. Mobile agents index and search SW source code as move the Internet and create/store digital license equivalent to SW source code.

N.T. Nguyen et al. (Eds.): KES-AMSTA 2007, LNAI 4496, pp. 825–832, 2007.

2 Related Works

DRM technology is a management system to protect the proper copyright of digital contents produced on various purposes during the production, the distribution and the utilization. DRM technology can be categorized into two section largely. One is involved in the copyright management, and the other is in the copyright protection. The technique of the copyright protection takes up the majority of commercial DRM technology at the moment. It is understood as the technique to force a series of principles and scenarios defined in the technique of copyright management. The technique of the copyright management is to establish a globally unified management system for a set of digital works. Many organizations drive to standardize the description language for contents identifier and contents meta data. The standardization of DRM technology is proceeded in the following orientation.

- Specification of DRM platform: MPEG-21, TV Anytime, AAP, OeBF, OMA
- Technology of identification system: DOI[1], URI, MPEG-21, DII
- Right presentation technology : XrML[2], ODRL[3, 4] XMCL, MPEG-21, RDD/REL
- Technology of meta data management of IPR : INDECS[5], ONIX[6], DC[7]

And Study of pattern matching of Software source code was studied by Rieger[8] and Yang[9]. Rieger achieved study that software source code detects visualizing part that is duplicated or is reproduced, Two different programs study that find equal code as syntactic achieve by Yang. We have gone SIM[10], Dup[11], Plague, YAP[12], YAP3, MOSS[13] in abroad, and clonechecker of KAIST, LOFC of Busan Univ., exEyesLight[14] in domestic for program plagiarism detection S/W.

This study is carried out to support to DRM for software source code. The DRM description for software source code is not available. The encryption technology of IPR protection technology takes an optional effect on 1:1 trade, which stays at the initial stage. The software source code has a meaning far above a plant's design sheet, and it is an important resource of digital contents as much as production line. But the DRM technology to support this resource falls far behind.

3 Necessity of Source Code Protection Technology and Business Model

It is assumed that a suit for illegal leakage of software source code is instituted among software developers A, B and C at court D (Fig. 1). The dispute occurs between A, owner of original source code, and B, illegal leaker (Fig. 1, (1)). The court orders developer A and B to submit their source code for the settlement(Fig. 1, (2)). The software source code needs to be inspected by an expert, and the court D inquires the differentiation between original and leaked code to special software developer C(Fig. 1, (3)). Special software developer C has to

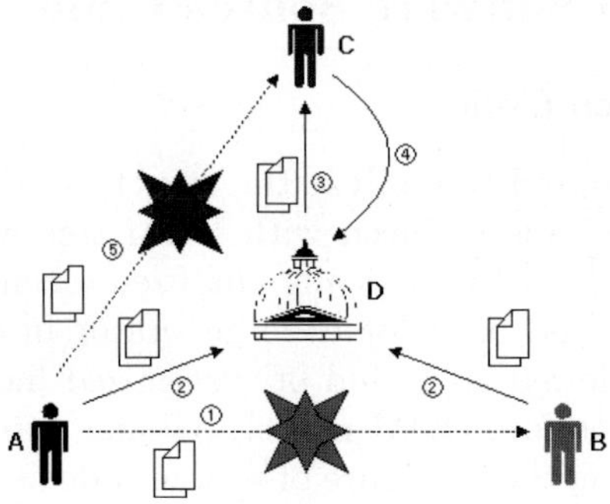

Fig. 1. The problem in the secondary IPR infringement

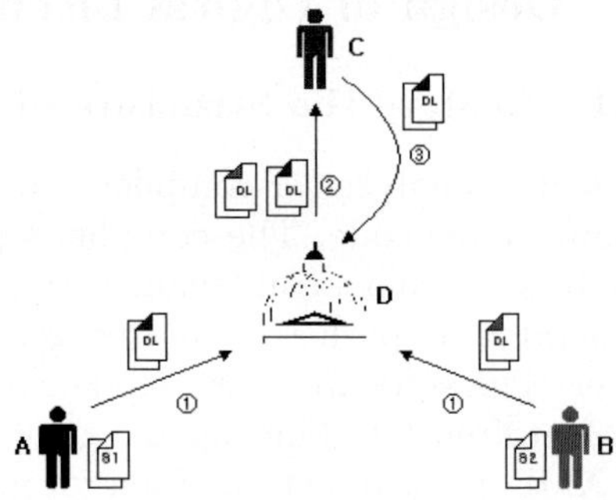

Fig. 2. IPR infringement Protection by Digital License

analyze the source code to compare the original and leaked code. The differentiation between original and leaked code depends on the C's opinions completely, so the objectivity is deficient(Fig. 1, (4)). Further, it is secondary source code leakage(Fig. 1, (5)) and unintentional infringement of the IPR for source code. That is, the source code has to be exposed automatically in the middle of clarifying the differentiation between original and leaked one, which may bring about secondary disputes.

The problem in the IPR infringement(Fig. 1) can be ruled out through the application of digital license (Fig. 2). The source code is not informed to the court at the dispute between A and B, but it is generated in digital license and furnished to the court instead(Fig. 2, (1)). The court can obtain objective report on the matching degree via pattern matching program and get additional advice from the developer C for the decision(Fig. 2, (2)). Using digital license may help removing the apple of the secondary dispute. The developer C shall verify the digital license instead of source code and give advice how much the leaked software source code is matched with the original one(Fig. 2, (3)).

The business model using digital license will protect the A's IPR, facilitate the identification of the source code leaker B and prevent a secondary potential dispute on unintentional leakage. Besides, the decision can base on the objective and obvious report from pattern matching program, not on C's subjective views. It is not mainly affected by the C's advice after long-lasting analysis, and so the settlement of the dispute shall be shortened thanks to quick analysis of the pattern matching program.

We can make extend this business model to computing environment of mobile agents. Mobile agent has some features of autonomy, proactivity, learning, cooperation and mobility. Specially, Mobility and cooperation are important features in this business model environment. Because mobile agent integrates two facet(secondary objects) of index and search functions dynamically, Mobile agent can accomplish duty(search and index of SW source code, create and store digital license) using ability of mobility and cooperation.

4 Design of Digital License of Software Source Code

4.1 Analyze the Structure of Source Code

A lexical analyzer of compiler can be adopted to analyze the structure of program source code. The compiler's path tree is replaced with DOM tree which enables to use neutral language in this study, while lexically analyzed tokens are generated as path tree in tree generator and used for code generator in complier. The software source code can be viewed as a kind of systemized file like XML(eXtensible Markup Language). Therefore DOM model, parsing outcome of XML file, is able to be used for modelling the structure of source code as well. This study is also to propose a building technique of reserved words, library and function as DOM tree via parsing of software source code (Fig. 3). 32 reserved words are defined in ANSI standard C programming language. Many C compilers add other reserved words to support the efficiency of compiler environment, joint programming with other language, interrupt and memory maintenance. The programs in C programming language are composed of main() function and subfunctions. In consequence, main() function becomes the root node of software architecture DOM tree of digital license, and subfunctions shall consist of child node of main() function. Each node has the data of library and parameters and pattern information of operators.

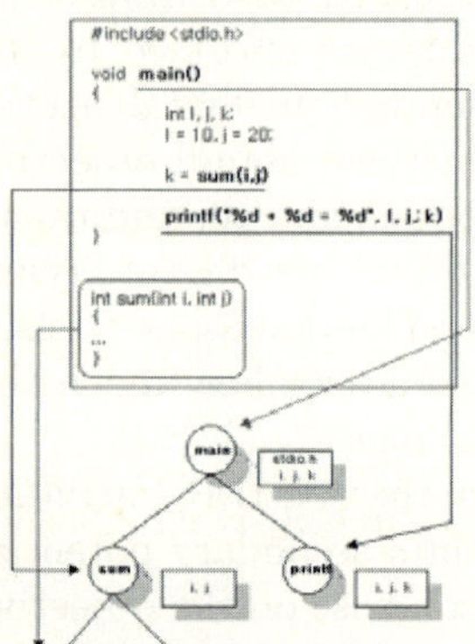

Fig. 3. Transformation Process to Software Architecture DOM Tree from Source Code

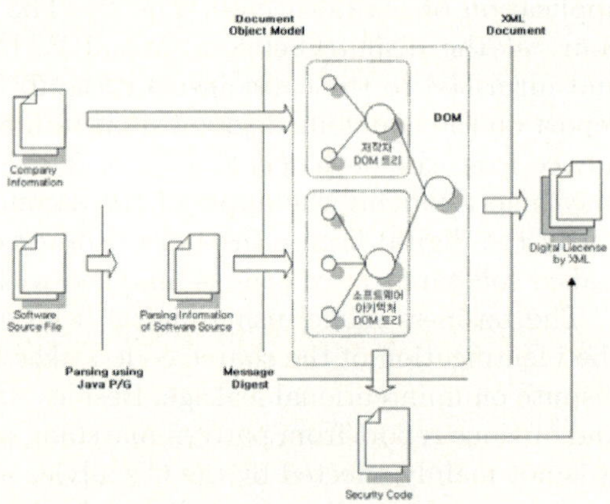

Fig. 4. Digital License Generation Process of Software Source Code

4.2 Generation Process of Source Code Digital License

The digital license of software source code has the root node called software digital license(SWDL) and two child nodes under it. The child node is constituted of author DOM tree for the information on the author of software source code and software architecture DOM tree indicating the architecture of software source code as shown in the fig. 4. Message Digest(MD)[15] is added to author DOM tree to insure the integrity of contents and information of general software registration. The core of digital license of software source code is the software architecture DOM tree. This child node is a tree- formed architecture of reserved token and software structure generated through parsing of software source code as shown in the fig. 3.

4.3 DOM Child Node Generation

The root of software architecture node undertakes the main() function. Several child nodes are allocated under the root node of main() function. The child node consists of the block of software source code and function call:

- Child Node Generation by block marked({, })
- Child Node Generation by Control Statement
- Child Node Generation by Function Call

The software source code comprises the block marked in brace ({, }). The block is generated to child node as code package to be executed. The block may comprise the subfunction, the command sentence to be executed subsequently and the block. The block is created as child node. The control statement of programming language is classified into conditional statement, iterative statement and branch statement. In the control statement, the programming jobs' order or procedure is altered under given conditions. The implementation of program shall differ from given conditions even for the execution of the same program. In this case, the control statement is generated as child node. C programming language is composed of more than 1 function. It is largely divided into main() function and subfunction and generates the subfunction as child node. The subfunction can call the other subfunctions and itself to create them as child node.

4.4 Pattern Generation of Operator and Input/Output Data Type in Child Node

The child node is not made in the absence of operator of function code. The function with operator is able to generate new output corresponding to the input. In other words, the function with operator is thought as a processing, while the code without operator is merely a display on the screen, or it is considered as blank code to disguise the original code in ill-intended case. Therefore, the tree digest is required to cope with this code. Each node-specialized pattern is to be produced. The pattern shown in the node is to create the node pattern in internal format, using the input data type and number, operator and output data type and number. If the pattern data of node is applied, the difference in the functions of program modules can be identified via the node's pattern data, even though the software architectures are same.

5 Simulation of Digital License and Complemented

5.1 Digital License Indexing and Searching

The digital license of software source code gives the information on software writer, programming language, registration date, registration number, version and DOM information of software source code in regards to XML-based software source code. Parts of XML-based digital license code needs to be encrypted for the integrity and confidentiality of digital license of software source code.[16] In this study, the message digest is implemented via JCE[15] package as shown in the figure 5. And, Figure 6 shows search procedure of Digital License. Search procedure was consisted of 3 steps : Index Pattern, Architecture Pattern and Node Pattern.

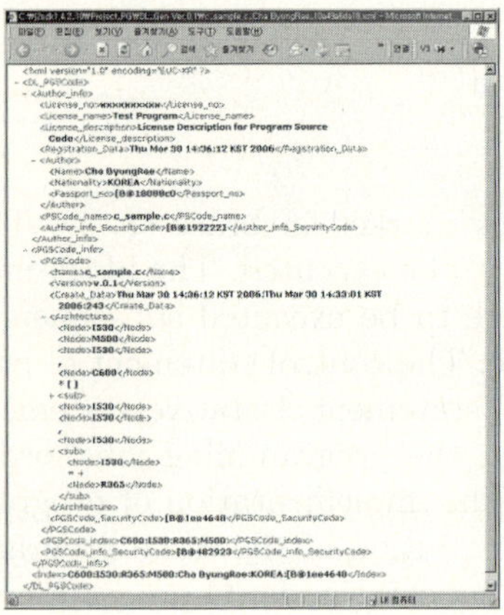

Fig. 5. Digital License Prototyping of Software Source Code

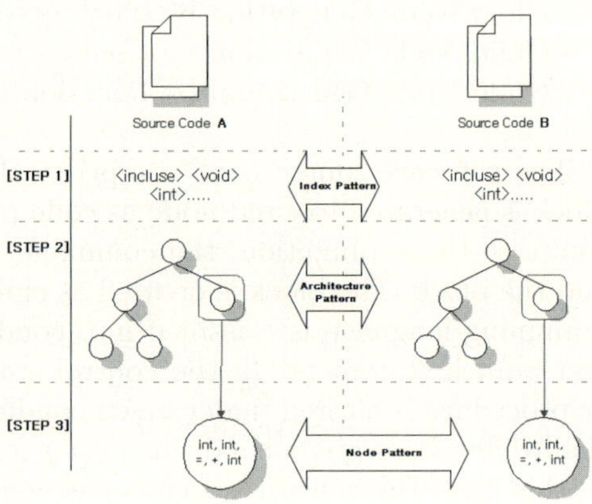

Fig. 6. Searching 3 Steps of Digital License

First, we search digital license instead of software source code in file system and create index of invert file. Figure 7 shows result of Digital License Indexing. And then searcher retrieves information of digital license and position using created invert file. Figure 8 shows result of Digital License Searching. We can attach index and search function of SW source code in mobile agents and mobile agents create/store digital license equivalent to SW source code. Therefore, we can cope IPRs problem of SW source code using mobile agents spontaneously.

5.2 Digital License vs. S/W Source Code

The result of file size comparisons for 160 source code samples are shown in Figure 9. According as size of S/W source code increases, size of digital license changed monotone increasing with noise. We inferred noise from Digital License vs. S/W source code size in Figure 9. Noise is produced by remark(comment) in Source code and coding style of S/W developers.

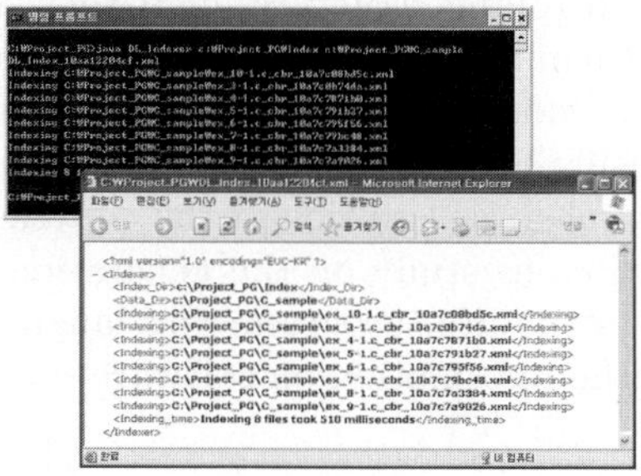

Fig. 7. Result of Digital License Indexing

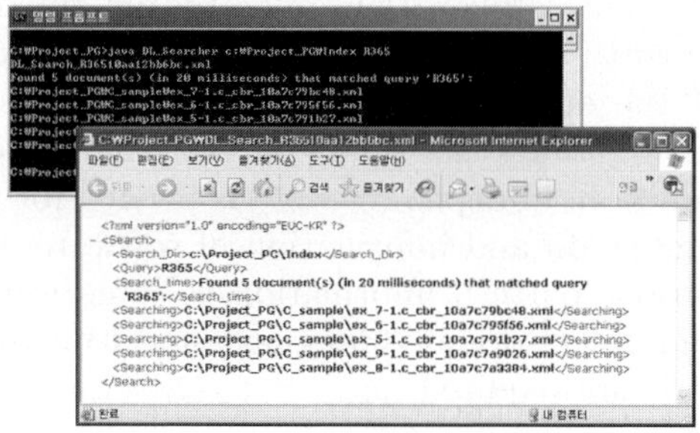

Fig. 8. Result of Digital License Searching

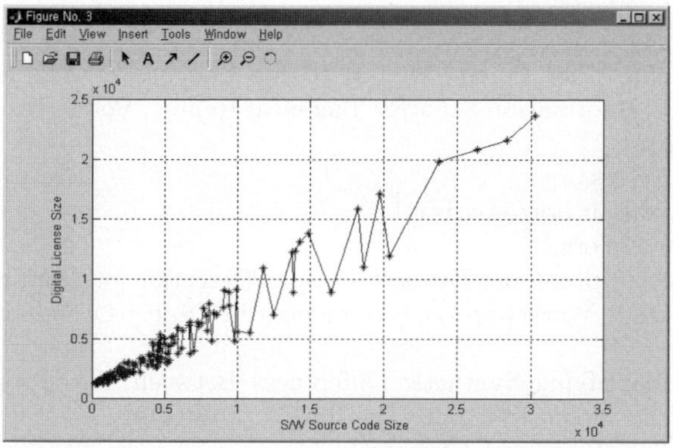

Fig. 9. Size of Digital License vs. S/W source code

6 Conclusion

The IPR system plays a core role in the development of information society in the 21st century as it did in the birth and growth of previous industrial society. The IPR system and description for the extended software source code from digital contents are very significant to enhance international competitiveness. Currently other methods than the encryption do not exist in connection with the software source code, a resource factor of digital works, and the research for this is at early phase. The encrypted means are applied only to the first trade between the software source code developer and buyer, and they do not help to protect the copyright for the following trades. A problem happens if the original software source code is made public at dispute on its ownership. This study has suggested the design and business model of digital license of software source code to protect the copyright infringement and technique leakage caused by its opening. The software source code is significantly long and complicated. It is supposed to be hard to find the uniqueness in the middle of inspecting the ownership if different softwares are operated by similar controlling structure, or if they are constituted of similar algorithms, or if one of them is a partially modified version of the other. There might be other cases to judge the uniqueness evidently. Therefore further study on the pattern to ensure the uniqueness is required. In addition, in-depth research is needed for more clear description of the copyright protection and information of software. Continuous study on the generation of various pattern information to integrate the structural programming language and object-oriented language and to include the information on software source code are required.

References

1. http://www.doi.org
2. http://www.xrml.org
3. http://odrl.net/
4. Renato Iannella, "The Opend Digital Rights Language: XML for Digital Rights Management", Information Security Technical Report. Vol. 9, No. 3, Elsevier 1363-4127, April 2004.
5. http://www.indecs.org/
6. http://www.editeur.org/onix.html
7. http://dublincore.org/
8. Matthias Rieger, Stephane Ducasse, Visual Detection of Duplicated Code, Proceedings ECOOP Workshop on Experiences in Object-Oriented Re-Engineering, 1988.
9. Wuu Yang, Identifying Syntactic Differences Between Two Programs, Software-Practice and Experience, Vol. 21(7), 739-755. July 1991.
10. http://www.few.vu.nl/ dick/sim.html
11. http://glimpse.arizona.edu/javadup.html
12. http://www.ccsr.cam.ac.uk/ mw263/YAP.html
13. http://www.cs.berkeley.edu/ aikem/moss.html
14. Cho dong-uk, "S/W ProgramPlagiarism Inspection Techniques and Analysis of S/W Tools for Protection of Digital Properties", Proceeding of the Korea Contents Association, Vol.1, No.1, 2003.
15. Jess Garms and Daniel Somerfield, "Professional Java Security", Wrox, 2001.
16. Berlin Lautenbach, "Introduction to XML Encryption and XML Signature", Information Security Technical Report. Vol. 9, No. 3, Elsevier 1363-4127, April 2004.

PKG-MIB: Private-Mib for Package-Based Linux Systems in a Large Scale Management Domain

Jong-Hyouk Lee[1,⋆], Young-Ju Han[1], Jong-Hyun Kim[2], Jung-Chan Na[2], and Tai-Myoung Chung[1]

[1] Internet Management Technology Laboratory,
Dept. of Computer Engineering, Sungkyunkwan University,
300 Cheoncheon-dong, Jangan-gu,
Suwon-si, Gyeonggi-do, 440-746, Korea
{jhlee, yjhan, tmchung}@imtl.skku.ac.kr

[2] Electronics and Telecommunications Research Institute (ETRI),
161 Gajeong-dong, Yeseong-gu, Daejeon-si, 305-700, Korea
{jhk, njc}@etri.re.kr

Abstract. Most of current Linux systems have package management tools which provide the easy way - installing, updating, and removing - without compiling source code within package-based Linux systems. However, any of package management tools does not provide a network-based management mechanism because the tools are only designed for the sole Linux system, not for multiple Linux systems managed in a large management domain. Thus, in this paper, we propose a private-mib to manage package-based Linux systems, called PKG-MIB. With PKG-MIB, an administrator would efficiently manage package-based Linux systems in a large scale management domain by Simple Network Management Protocol (SNMP). According to the result of performance evaluation of implementation, a Linux system with PKG-MIB reduces the cost for package management up to 8 times compared to a normal Linux system.

1 Introduction

Over the past several years there has been increasing the number of Linux systems. Due to some of reasons such as keeping up-to-date technologies and the activity of hundreds of Linux developers around the world, Linux systems are used in a broad range of communities, universities, laboratories, and industries. Technically speaking, Linux is a true Unix kernel, although it is not a full Unix operating system (OS). Moreover, Linux systems are compliant with the IEEE POSIX standard since the 2.6 version of the Linux kernel and includes all the features of a modern Unix OS such as virtual memory, a virtual filesystem,

⋆ Jong-Hyouk Lee is the corresponding author and he is also a student member of ACM and IEEE.

N.T. Nguyen et al. (Eds.): KES-AMSTA 2007, LNAI 4496, pp. 833–840, 2007.

lightweight processes, Unix signals, SVR interprocess communications, support for Symmetric Multiprocessor (SMP) systems, and so on [1]. Thus, Linux systems are expected that they will play major roles in the both personal and business computing areas.

From the software management point of view, the early Linux systems used a source-based software, called tarball[1]. Making use of source-based softwares has been requiring that users have to be aware of which libraries will be need to build softwares. It also has an inconvenience such as unpacking tarball and compiling the source-code extracted from tarball. However, the software management in Linux system has been changed after appearing package-based Linux systems. A package-based Linux system provides package management tools such as APT (Advances Packaging Tool) or YUM (Yellow dog Updater) [2,3]. The tools are used to install, update, remove, and manage packages including software and information about software.

Although a package-based Linux system brought up the efficient software management, there is still lack of convenience. For instance, within the package-based Linux system, an administrator cannot manage packages of the Linux systems by remotely. This problem is a critical in case of a large scale management domain. If an administrator need to manage a large scale Linux-based system domain, the administrator has to connect each Linux system and then execute separately APT in order to manage packages of the Linux systems. This operation would be impracticable when the managed domain has lots of Linux systems. To solve this problem, we propose PKG-MIB which can be used to manage package-based Linux systems in a large scale management domain by SNMP [4]. Due to the operation of PKG-MIB is achieved by SNMP, PKG-MIB is easily included into SNMP based network management systems as a module [5].

The rest of the paper is organized as follows. In section 2, we discuss the current problem of package management in Linux systems and then our proposed PKG-MIB to provide an efficient remote package management mechanism is introduced in section 3. In section 4, we evaluate the performance of PKG-MIB. The final section gives our conclusions.

2 Problem Statements

Advances Packaging Tool (APT) is one of package management tools in Linux environments that provides easy installing, updating, and removing of packages in the local system via a remote package repository [2,6]. APT mechanism checks dependency of a package, and solves the dependency problem. APT also provides semi-automatic package updating which means APT does not start to update local packages, but an administrator starts updating, then APT updates selectively local packages with new packages via the remote package

[1] An archive, created with the Unix tar utility, containing myriad related files. “Here, I'll just ftp you a tarball of the whole project.” Tarballs have been the standard way to ship around source-code distributions since the mid-1980s; in retrospect it seems odd that this term did not enter common usage until the late 1990s.

repository. This package management tool helps personal users or administrators of Linux systems who think source-based Linux system is inconvenience themselves and too much like hard work - installing or updating new packages, removing the packages have vulnerabilities, and checking dependency problems.

Even though APT is a useful tool for package management, the tool is only used to be locally. The question we have to ask here is what APT could be operated by remotely. In other words, APT does not provide any remote package management mechanisms. If there are one hundred Linux systems, the administrator have to connect each a Linux system and then execute a package management tool to install a package. It would take too much time and it is impossible to finish if the administrator has to update some packages including vulnerabilities in a short time. Moreover, APT does not provide any remote monitoring mechanisms by SNMP since as we mentioned, APT is designed for its local Linux system. There has been no study that tried to develop any remote package management mechanisms using SNMP. So, in this paper, we introduce a private-mib and its agent module to provide the remote package management mechanism.

3 PKG-MIB: The Private-Mib for Package Management

We introduce the private-mib and its agent module in this section. In the beginning, we present an overview of PKG-MIB and then the defined structure information, table information, and entry information are introduced. Finally, we describe the agent module which needs to support the enterprise subtree in case of PKG-MIB in SNMP agent.

3.1 PKG-MIB and Its Structure, Table, and Entry Information

The structure of management information version 2 (SMIv2), which is specified in [7], defines the general framework within which a MIB can be defined and constructed. In the SMIv2, the private node used to identify objects defined unilaterally. The private node's subtree has only one child node defined, the enterprise node. This portion of the subtree is used to allow vendors to enhabce the management of their devices and to share this information with other users and vendors who might need to inter-operate with their systems [8]. A branch within the enterprise subtree is allocated to each vendor that registers for an enterprise object identifier. So, we applied the private enterprise number (PEN) to manage package-based linux systems. Internet Assigned Numbers Authority (IANA) has assigned the following PEN[2] to us: 27315.

In order to fill the information related packages in PKG-MIB, we used the information listed in Table 1. Based on the information listed in Table 1, we

---

[2] The assignment information would be confirmed in the following registry link: http://www.iana.org/assignments/enterprise-numbers

Table 1. The information related packages

Field name	Description
package	the name of package (e.g., telnet, xserver-xorg-driver-chips)
status	the status of package (e.g., installed, not-installed, unpacked)
priority	the priority of package (e.g., important, required, optional, extra)
section	the section of package (e.g., base, util, libs, x11, misc)
installed-sized	the file size of package
maintainer	the maintainer or author of package
architecture	the architecture of package (e.g., all, i386, i686, ia64, powerpc, mips)
source	its necessary packages to install or update
version	the version of package
depends	its required packages to install or update
description	extra information or description of package

wrote PKG-MIB. PKG-MIB defines objects that describe the behavior of the information related packages. PKG-MIB only has *packageStats* group, which has its own object identifier (OID): *1.3.6.1.4.1.27315.1.0*. Table 2 shows *packageStats* group. The objects of *packageStats* group relate to packages and are used by a SNMPv2 entity acting in an agent role to describe those object resources. The agent controls the objects of *packageStats* group for dynamic configuration by a manager.

Table 2 lists the objects contained in *packageStats* group. There are two tables - *packageStatsTable (1.3.6.1.4.1.27315.1.0.1)* and *packageInfoTable (1.3.6.1.4.1.27315.1.0.2)*. Those tables are read-only tables consisting of one entry for each object resource that can be dynamically configured when the information of packages is changed by APT. For instance, if packages are updated by APT, related objects are updated. Another example is that if an administrator is interested in the status information of installed package, then the OID is *packageStats.1.1.5* or *1.3.6.1.4.1.27315.1.0.1.1.5*.

3.2 The Agent Module for PKG-MIB

In order to support the enterprise subtree in case of PKG-MIB, its agent module has to be implemented. So, we have built an agent module of PKG-MIB within net-snmp v5.2 on a package-based Linux system (Ubuntu Linux - code name is dapper) which has APT as a package management tool. As indicated earlier, PKG-MIB contains the packages information listed in Table 1. Thus, the agent module collects the packages information from a package status file[3].

The agent module is consisted by functions which handle the package status file and maintain the objects contained in *packageStats* group. The functions are summarized in Table 3.

[3] This file contains information about whether a package is marked for removing or not, whether it is installed or not, etc. This file has all information listed in Table 1 and exists /var/lib/dpkg/status on Ubuntu Linux.

Table 2. Information of packageStats group

Object name	Entry name	Sub-object name	Syntax
packageStatsTable (1)	-	-	Sequence of
	packageStatsEntry (1)	-	Sequence
		packageIndexStats (1)	CounterIndex
		packageName (2)	DisplayString
		packageVersion (3)	DisplayString
		packageDesc (4)	DisplayString
		packageStatus (5)	DisplayString
packageInfoTable (2)	-	-	Sequence of
	packageInfoEntry (1)	-	Sequence
		packageIndexInfo (1)	CounterIndex
		packagePriority (2)	DisplayString
		packageSection (3)	DisplayString
		packageMaintainer (4)	DisplayString
		packageSource (5)	DisplayString
		packageDepends (6)	DisplayString
		packageRecommends (7)	DisplayString
		packageSuggests (8)	DisplayString
		packageSize (9)	DisplayString
		packageMD5Sum (10)	DisplayString

Table 3. Functions for handling the package status file and maintaining the objects

Function name	Description
main function	the main function of agent module
retrieve functions	the retrieve function to get some information
get_packages	used to get all listed in Table 1 from the status file
get_config	used to get the configuration
record functions	the record functions to write the retrieved information
record_package_stats_index	used for packageStatsEntry's packageIndexStats object
record_package_stats_data	used for packageStatsEntry's the rest of objects
record_package_info_index	used for packageInfoEntry's packageIndexInfo object
record_package_info_data	used for packageInfoEntry's the rest of objects
misc functions	miscellaneous functions
set_command	used for SNMP SetRequest from a manager
open_log	log function
debugging	debugging function

4 Performance Evaluation

In this section, we verify the operation of PKG-MIB and its agent module and then we show the result of performance evaluation. To do the performance evaluation, we used a MG-SOFT mib-browser on Windows system as a manager system (IP address is 10.22.11.168) and net-snmp v5.2 included PKG-MIB and its agent module on 10 of Ubuntu Linux systems as agent systems (IP address is 10.22.11.171-180).

4.1 Verification of the Operation

Fig. 1 shows the result of snmpwalk operation which retrieves a subtree of management values using SNMP GetNextRequests from agent including PKG-MIB and its agent module. As we can see in Fig. 1, all information needed for managing packages is implemented as the objects of *packageStats* group.

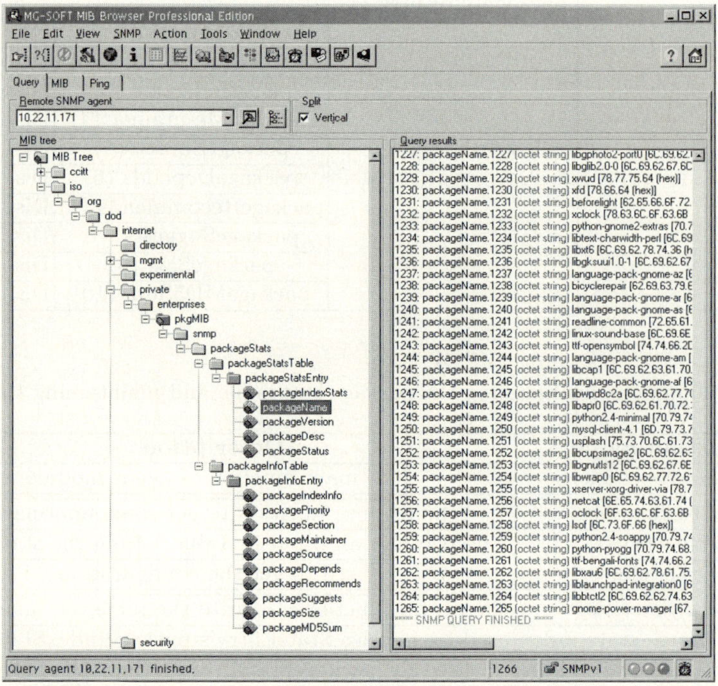

Fig. 1. The screen capture of SNMP operation

4.2 Measurement for Package Updating Time

Each package updating time - manual case and using PKG-MIB case - was measured. The procedures of measurement was as follows:

- *Manual case*: An administrator connects from the manager to an agent and execute APT to update 5 packages. This procedure have to repeat for 10 agents.
- *Using PKG-MIB case*: An administrator opens MG-SOFT mib-browser and execute SNMP SetRequest to update 5 packages for 10 agents. This procedure is done by 10 times click on the MG-SOFT mib-browser.

Table 4. The measured packages updating time

Manual case	Using PKG-MIB case
200 seconds	25 seconds

The measured packages updating time is shown in Table 4. The measured packages updating time should be subjective. The updating time may vary according to another. We believe, however, that it is one of the valid method for measuring.

Based on the time in Table 4, it is reasonable to suppose that using PKG-MIB is more efficient. Also, we can represent the result of measurement for package updating time in the line graph as shown Fig. 2. As we can see, the cumulative packages updating time for the manual case is rapidly increased by increasing the number of nodes in the management domain. On the contrary, the result of using PKG-MIB case shows that the cumulative packages updating time is slowly increased due to taking advantage of SNMP.

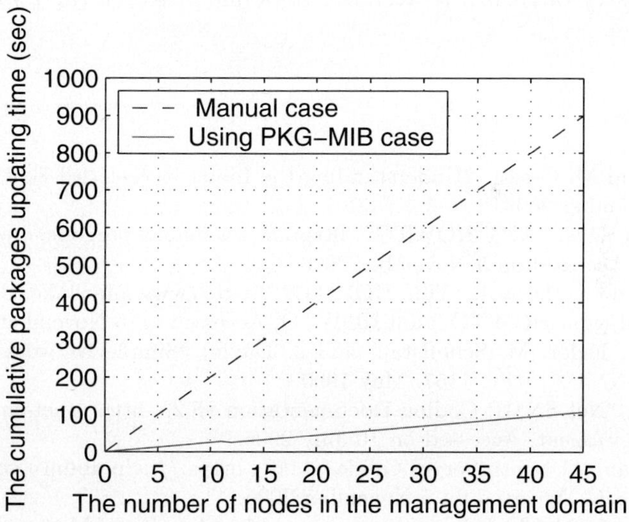

Fig. 2. Comparison between manual case and using PKG-MIB case in the respect of cumulative packages updating time

5 Conclusions

As we have seen, within the narrow limits of the software management, the current Linux systems have a critical problem. In a large scale management domain, the package management tools cannot lift a burden from managing packages since any of the tools does not have the faculty for remote managing mechanism. In order to extend the faculty of package management tools, in this paper, we proposed PKG-MIB. The defined structure, table, and entry information for PKG-MIB were introduced and the agent module for PKG-MIB was implemented. Also, the performance evaluation was performed. The result of performance evaluation clearly shows that the Linux system with PKG-MIB reduces the cost for package management up to 8 times compared to the normal Linux system if there are 40 Linux systems having needed to update packages. The result is sufficient to show that PKG-MIB should be used to manage package-based Linux systems in a large scale management domain. Especially, PKG-MIB would be easily included into SNMP based network management systems as a module.

Acknowledgment

We would gratefully like to thank to IANA for their efforts for the PEN assignment for PKG-MIB and the management of Internet domain names and addresses. This research has been supported by the Network Security Technology Research Project, ETRI, Korea and a grant of the Korea Health 21 R&D Project, Ministry of Health & Welfare, Republic of Korea (02-PJ3-PG6-EV08-0001).

References

1. D. Bovet and M. Cesati, "Understanding the Linux Kernel, 3rd Edition", O'Reilly Media, November 2005.
2. G. Noronha Silva, "APT HOWTO", http://www.debian.org/doc/manuals/ apt-howto, Accessed on 5 November 2006.
3. G. Brown and J. Pickard, "Yum HOWTO", http://www.phy.duke.edu rgb/General/yum_HOWTO/yum_HOWTO, Accessed on 5 November 2006.
4. J. Case, M. Fedor, M. Schoffstall, and J. Davin, "Simple Network Management Protocol (SNMP)", RFC 1157, May 1990.
5. Net-SNMP, "Net-SNMP Coding Documentation v5.2", http://net-snmp.source forge.net/dev/agent, Accessed on 10 July 2006.
6. B. Arumugam, "Ubuntu Server Guide v6.06", https://help.ubuntu.com/ubuntu /serverguide/C, Accessed on 5 November 2006.
7. M. Rose and K. McCloghrie. "Structure and Identification of Management Information for TCP/IP-based Internets", RFC 1155, May 1990.
8. W. Stallings, "SNMP, SNMPv2, SNMPv3, and RMON 1 and 2, 3rd Edition", Addison-Wesley, December 1998.

Perormance of TPR*-Trees for Predicting Future Positions of Moving Objects in U-Cities

Min-Hee Jang[1], Sang-Wook Kim[1], and Miyoung Shin[2]

[1] Division of Information and Communications, Hanyang University, 17 Haengdang-dong, Sungdong-goo, Seoul 133-791, Korea
zzmini@hanmail.net, wook@hanyang.ac.kr
[2] School of Electrical Engineering and Computer Science, Kyungpook National University, 1370 Sankyuk-dong, Buk-goo, Daegu 702-701, Korea
shinmy@mail.knu.ac.kr

Abstract. The TPR*-tree is the most widely-used index structure for effectively predicting the future positions of moving objects. The TPR*-tree, however, has the problem that both of the *dead space* in a bounding region and the overlap among bounding regions become larger as the prediction time point in the future gets farther. This makes more nodes within the TPR*-tree accessed in query processing time, which incurs serious performance degradation. In this paper, we examine the performance problem quantitatively via a series of experiments. First, we show how much the performance deteriorates as a prediction time point gets farther from the present, and also show how the frequent updates of positions of moving objects alleviate this problem. Our contribution would help provide important clues to devise strategies improving the performance of TPR*-trees further.

1 Introduction

The recent advances of technologies in mobile communications and global positioning systems have increased people's attentions to an effective use of information on the objects that move in two-dimensional space. Those moving objects send their current positions to a server periodically. This position information has the property of spatio-temporal data where spatial locations of objects change with time[3]. The database that stores the information of a lot of objects' locations changing with time is called a *moving object database*[13]. Moving object management for telematics in U-cities is a typical application of a moving object databases.

Users' queries issued on moving object databases can be categorized into two types: *past-time queries* and *future-time queries*[9]. The past-time query is to retrieve the history of dynamic objects' movements in the past[11,14], while the future-time query is to predict movements of dynamic objects in the future[9]. An example of the future-time query is "Retrieve all the vehicles passing over the Golden Gate Bridge at 1 pm". In this paper, we discuss processing of future-time queries in an effective way.

N.T. Nguyen et al. (Eds.): KES-AMSTA 2007, LNAI 4496, pp. 841–850, 2007.

To support fast processing of users' queries in a large volume of moving object databases, an effective index structure is needed[5]. For future-time queries, index structures such as the VCI-tree[6], the TPR-tree[8], and the TPR*-tree[10] have been proposed. Among these, the TPR*-tree overcomes the limitation of the TPR-tree, and thus is the most widely-used for predicting the future locations of moving objects. The TPR*-tree stores the maximum and minimum speeds at each axis in the space where a set of objects move and uses them to predict the future locations of moving objects. However, the TPR*-tree still suffers from the problem that both of the *dead space* [1] in a bounding region and the overlap among bounding regions get larger as the future prediction time point gets farther. This makes more nodes within the TPR*-tree accessed from disk in query processing, which incurs the serious performance degradation[10].

In this paper, we investigate the effects of various factors on the query performance of the TPR*-tree empirically. First, we quantitatively show how much the performance of query processing with the TPR*-tree deteriorates as a prediction time point gets farther from the present, and also show how much the frequent updates of the TPR*-tree alleviate this performance deterioration. Based on the results, we examine the relationship between the updates of the TPR*-tree and the performance of query processing. To the best of our knowledge, the quantitative study on the update effects in query processing with the TPR*-tree have never been addressed previously. We believe that this result would provide important clues to devise novel strategies that improve the performance of the TPR*-trees significantly.

This paper is organized as follows. As related work, Section 2 briefly reviews the existing index structures for processing of the future-time queries, and discusses their pros and cons. Section 3 points out the primary cause of the performance degradation of the TPR*-tree, which becomes the motivation of this work. Section 4 performs a quantitative study via a variety of experiments for examining the relationship between the updates and the query performance of the TPR*-tree. Finally, Section 5 summarizes and concludes this paper.

2 Related Work

In this section, we introduce prior index structures for future prediction, and discuss their characteristics.

2.1 TPR-Tree

The TPR-tree[8], which has been devised based on a *multidimensional file structure*[2,15] called the R*-tree[1], predicts the future locations of moving objects by storing the location and the speed of each object at a given time point[8]. The locations of moving objects are indexed using CBRs(conservative bounding rectangles) instead of MBRs(minimum bounding rectangles), the concept employed in the R*-tree. A CBR is composed of an MBR, representing the region that encloses a set of moving objects at a specific time point, and the maximum

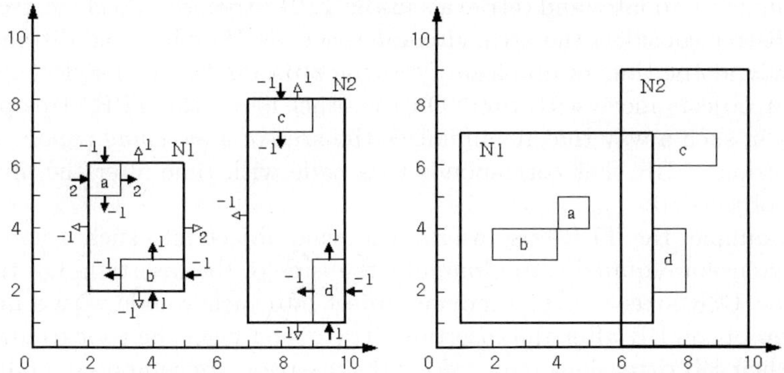

(a) time 0. (b) time 1.

Fig. 1. An extension of a CBR in the TPR-tree for future prediction

and minimum moving speeds of the objects within an MBR at each axis. The location of a moving object at any future time point can be easily predicted with the location and moving speed stored in a CBR. The predicted region of a node computed by using a current location of an MBR and its maximum and minimum speeds at each axis is defined as a BR(bounding rectangle).

Figure 1 shows an extension of a CBR in the TPR-tree for future prediction. In Figure 1(a), *a*, *b*, *c*, and *d* denote moving objects, and N1, N2 denote MBRs corresponding to the nodes storing moving objects, respectively. Here, the maximum speed of every moving object is assumed to be the same as the minimum speed of them. The black arrows with numerals denote the directions and speeds of moving objects at each axis, respectively. The white arrows with numerals denote either the maximum of the maximum speed or the minimum of the minimum speed of the moving objects located within the corresponding CBR. A CBR consists of an MBR and the white arrows. Figure 1(a) shows two CBRs stored in a node at time point 0. Figure 1(b) shows their BRs at time point 1, which are extended by considering the movements of moving objects.

A CBR of the TPR-tree maintains only the maximum and the minimum speed, regardless of the moving objects' actual locations in its corresponding MBR. As shown in Figure 1, therefore, the BR induced from a CBR has the property that it monotonically grows with time. To prevent an infinite growth of the BR, the TPR-tree employs a strategy that an MBR is recomputed by considering the actual locations of objects when position updates of moving objects are requested.

2.2 TPR*-Tree

The TPR*-tree[10] basically uses the same structure as the TPR-tree. During the updates, however, the TPR-tree employs the insertion and the deletion algorithms of the R*-tree as they are, while the TPR*-tree employs their modified versions that reflect objects' movability. This makes it possible to improve the performance of updates and retrievals in the TPR*-tree over the TPR-tree. Since the TPR-tree considers the area, circumference, overlapping, and distance of an MBR only at the time of updates of moving objects, it cannot reflect the property that objects move with time. On the other hand, the TPR*-tree performs updates in such a way that it minimizes the area of a *sweeping region*, which is an extension of BR that corresponds to a node with time after the updates of moving objects[10].

For example, the TPR-tree inserts a moving object into such a node whose MBR extension required is minimum at the time of the insertion. On the other hand, the TPR*-tree inserts a moving object into such a node with a minimum extension of the BR after the insertion. Traversing from the root to lower-level nodes, their BR extensions required for the insertion are computed, and also are stored into a priority queue. Among these, the optimal node for the insertion is the one having the smallest value.

With this strategy, the TPR*-tree requires a cost higher than the TPR-tree for updates. However, owing to its compactness of BRs, it greatly improves the overall query performance. According to the experimental results in [10], the TPR*-tree shows 1.5 times to 5 times better performance than the TPR-tree.

3 Motivation

In this section, we discuss why the performance of query processing with the TPR*-tree degrades as a future prediction time point gets farther. Also, we analyze the reason why the updates of objects' locations affect the query performance.

3.1 Performance Problem

As mentioned earlier, the BR of a node continuously grows with time since the CBR of the TPR*-tree stores only the maximum and the minimum speeds of the objects at each axis The reason why the CBR stores only the maximum and minimum speeds at each axis as a representative of objects' movements is to avoid the excessive storage overhead caused by keeping the positions and speeds of all the objects within the descendent nodes of the CBR. This makes the overlap among BRs become larger, leading to a huge dead space. Accordingly, the number of disk accesses gets higher for node findings, which eventually deteriorates the performance of query processing.

Figure 2 shows that the CBR for a node at time point 0 is extended into the BR at time point 1. The locations of objects a and b in Figure 2(a) were moved as in Figure 2(b), according to their directions and speeds. On the other hand, since the CBR of the TPR*-tree has only the maximum and minimum speeds at each axis for all the moving objects, MBR N is extended into N' at time point 1 as in Figure 2(b), while the actual locations of the two moving objects at time point 1 is moved into N''. Such an unnecessary enlargement incurs a large dead space and an overlap with other BRs. These two factors are all undesirable in that

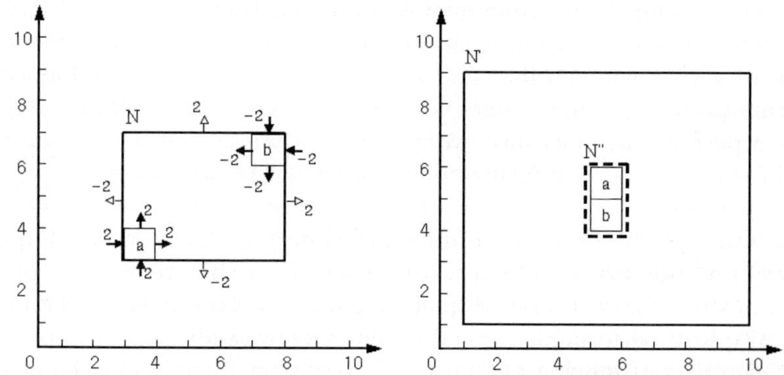

(a) time 0. (b) time 1.

Fig. 2. An extension of a BR

they cause performance deterioration in query processing. Section 4 examines this problem quantitatively via a series of experiments.

3.2 Update Effect

To prevent an excessive extension of a CBR, the TPR*-tree employs such a strategy that the MBR contained in the CBR is recomputed by reflecting their actual positions whenever the one of moving objects within the CBR is updated. In Figure 2(b), The BR N', including the objects a and b at time point 1, incurs a large dead space. If the positions of some objects are updated at time point 1, however, the TPR*-tree can reduce the BR N' to N'' by using the revised algorithm. Both the dead space and the overlap of BRs get decreased, and so overall query performance is improved. In Section 4, we examine the relationship between the updates and the query performance of the TPR*-tree via a series of experiments.

4 Experiments

In this section, we investigate both the performance degradation of the TPR*-tree for queries with a farther prediction time point in the future via a variety of experiments and also show the effect of the frequent updates in the TPR*-tree on the query performance. First, we describe the environments for performance study, and then analyze the experimental results.

The dataset used in our experiments was generated with GSTD[12]. GSTD is a data generator to produce various types of moving object data with different characteristics, and thus has been used in several previous works on moving objects[7,4]. We used 100,000 points generated in such a way that each point object has a random speed within the range of [0, 50] and is located in a normalized two-dimensional space of 10,000 x 10,000, where the locations of these objects have a Gaussian distribution in each axis.

Query regions for future-time queries were generated to have 0.01%, 0.16%, 0.64%, and 2.56% area of entire space, respectively, and their locations were set to have a uniform distribution. The generated queries were built to have their future prediction time points within the range of [10, 100] at the querying time. As a performance measure, we used the *averaged number of node accesses* required for processing 100 future-time queries of the same type.

In this paper, we performed three types of experiments to study the effect of different future prediction time points on the query performance. In Experiment 1, we measured the changes of query performances for different future prediction time points with different sizes of query regions. In Experiment 2, we measured changes of query performances for different future prediction time points with different numbers of moving objects. In Experiment 3, we measured changes of performances for different future prediction time points with different frequencies of updates on the TPR*-tree, which is targeted for dynamic environments.

Experiment 1 is to examine how different sizes of query regions affect the query performance for different future prediction time points. The sizes of query regions were set to 0.01%, 0.16%, 0.64%, and 2.56% area of entire space, respectively. All future-time queries were generated at time point 0, which is the time point right after the TPR*-tree generation. The future prediction time points were set to 20, 40, 60, 80, and 100, respectively. Figure 3 shows the results. The horizontal axis denotes a future prediction time point, and the vertical axis does the average number of node accesses for performing 100 future-time queries.

The results show that the number of node accesses get largely increased as future prediction time point becomes farther from the querying time. When a query region is 2.56%, 24%, 50%, 86%, and 126%, more node accesses are required for future-time queries of future prediction time points 40, 60, 80, and 100, respectively, in comparison with that of the future prediction time point 20.

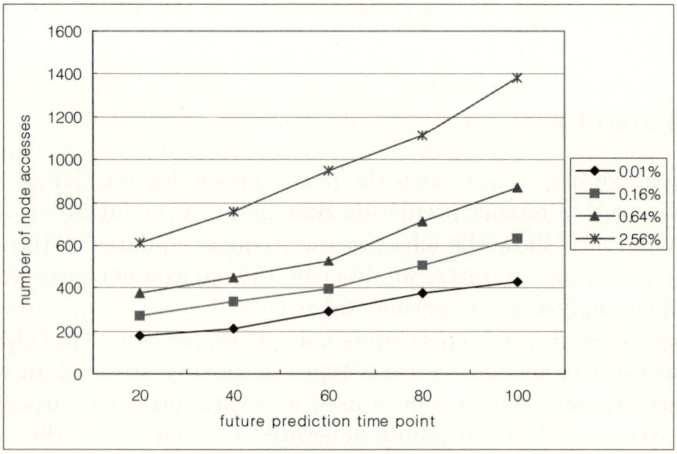

Fig. 3. Performance comparisons with different sizes of query regions

Even when query regions are taken to be 0.01%, 0.16%, and 0.64%, respectively, their increasing tendency of the number of node accesses is quite similar.

Experiment 2 is to examine how different numbers of moving objects affect the query performance for different future prediction time points. The numbers of moving objects were chosen to be 20,000, 50,000, and 100,000, respectively. As in Experiment 1, the future prediction time points were set to 20, 40, 60, 80, and 100, respectively. The size of query regions was set to 0.16%. Figure 4 shows the results. The horizontal axis denotes a future prediction time point and the vertical axis does the average number of node accesses for performing 100 future-time queries.

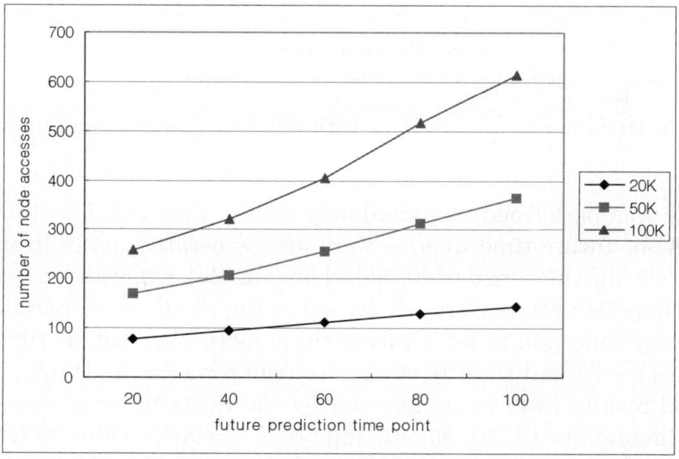

Fig. 4. Performance comparisons with different numbers of moving objects

The results show that, in all cases, the number of node accesses increases considerably as the future prediction time point gets farther. When the number of moving objects is 20,000, 10%, 27%, 50%, and 83%, additional numbers of node accesses occur at future prediction time points 40, 60, 80, and 100, respectively, in comparison with the future prediction time point of 20. When the number of moving objects are 50,000 and 100,000, respectively, the increasing rates of the number of node accesses are even worse. When the number of moving objects is 50,000, the numbers of additional node accesses for future prediction time points 40, 60, 80, and 100 are 16%, 42%, 75%, and 114% higher compared with the future prediction time point of 20. When the number of moving objects is 100,000, the numbers become 18%, 51%, 90%, and 137%. This means that as a moving object database increases in size, the query performance degradation gets much higher for farther future prediction time points.

Experiment 3 is to investigate how different frequencies of updates affect the query performance for different future prediction time points. Updates frequencies for every moving object were chosen to be 0, 5, 10, 25, and 50, respectively.

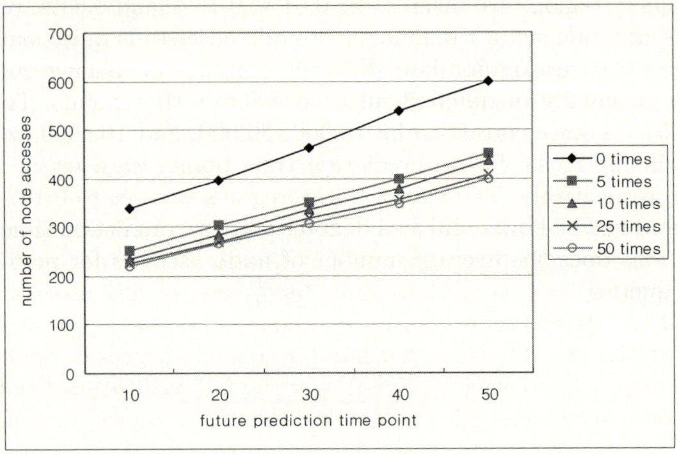

Fig. 5. Performance comparisons with different frequencies of updates

Each update was performed at a randomly chosen time point within the range of [0, 50]. Also, future-time queries were also generated at arbitrarily chosen time points within the range of [0, 50]. Thus, in this experiment, updates and future-time queries were performed, accessing the database in parallel. The future prediction time points for a future-time query were set to 10, 20, 30, 40, and 50, respectively, and the size of a query region was set to 0.16%. To prevent experimental results from being affected by the querying time, we set each of the future time points 10, 20, 30, 40, and 50 as a relative value to the querying time point, and computed the average time for processing future-time queries with the same future prediction time point.

Figure 5 shows the results of Experiment 3. The horizontal axis denotes a future prediction time point and the vertical axis does the average number of node accesses for performing 100 future-time queries. When a future prediction time point is 50, the number of node accesses decreases significantly by 75%, 73%, 69%, and 66% for 5, 10, 25, and 50 updates cases in comparison with the case of no updates. This implies that frequent updates have a good impact on the performance of future-time queries. The reason is explained as follows. Updates tend to reduce the BR size at a future prediction time point, thereby making both the overlap and the dead space among BRs smaller. This causes the query performance to improve greatly.

5 Conclusions

In this paper, we have examined the performance of future-time queries processed with the TPR*-tree via a series of experiments, which is to analyze what kinds of factors are significant for improving the query performance.

The critical problem of the TPR*-tree is that both of the dead space in a BR and the overlap among BRs become greatly larger as the future prediction time point gets farther. To have a deep understanding of this problem, we have showed how much the query performance of the TPR*-tree deteriorates as a prediction time point in the future gets farther. Also, we have showed how the frequent updates of the CBR affect the query performance with the TPR*-tree positively.

The experimental results show that the number of node accesses for query processing with the TPR*-tree increases seriously as a future prediction time point gets farther. Also, they show that, for larger moving objects databases, the performance degradation gets much higher as a prediction time point in the future gets farther. On the other hand, frequent updates does alleviate this problem. That is because such updates make the BR at a future prediction time point decreased, leading to the reduction of both the overlap among BRs and the dead space. Based on the important clues obtained from these results, we are conducting a research on devising a novel approach to alleviate the problem of the query performance degradation with the TPR*-tree more significantly.

Acknowledgement

This research was partially supported by the MIC(Ministry of Information and Communications) of Korea under the ITRC(Information Technology Research Center) support program supervised by the IITA(IITA-2005-C1090-0502-0009).

References

1. N. Beckmann, H. Kriegel, R. Schneider, and B. Seeger: The R-tree: An Efficient and Robust Access Method for Points and Rectangles. In Proc. the ACM Int'l. Conf. on Management of Data(ACM SIGMOD). (1990) 322–331
2. Sang-Wook Kim, Sanghyun Park, and Wesley W. Chu: An Index-Based Approach for Similarity Search Supporting Time Warping in Large Sequence Databases. In Proc. IEEE Int'l. Conf. on Data Engineering (IEEE ICDE 2001) 607–614
3. D. L. Lee, J. Xu, B. Zheng, and W. C. Lee: Data Management in Location-Dependent Information Services. In Proc. IEEE Pervasive Computing. Vol. 1, No. 3. (2002) 65–72
4. B. Lin and J. Su: On Bulk Loading TPR-Tree. In Proc. of IEEE Int'l. Conf. on Mobile Data Management. (2004) 395–406
5. M. F. Mokbel, T. M. Ghanem, and W. G. Aref: Spatio-Temporal Access Methods. In Proc. Bulletin of the IEEE Computer Society Technical Committee on Data Engineering. Vol. 26, No. 2. (2003) 40–49
6. S. Prabhakar, Y. Xia, D. V. Kalashnikov, W. G. Aref, and S. E. Hambrusch: Query Indexing and Velocity Constrained Indexing: Scalable Techniques for Continuous Queries on Moving Objects. In Proc. IEEE Trans. on Computers. Vol. 51, No. 10. (2002) 1124–1140
7. D. Pfoser, C. S. Jensen, and Y. Theodoridis: Novel Approaches in Query Processing for Moving Objects. In Proc. the Int'l. Conf. on Very Large Data Bases(VLDB). (2000) 395–406

8. S. Saltenis, C. S. Jensen, S. T. Leutenegger, and M. A. Lopez: Indexinng the Positions of Continuously Moving Objects. In Proc. the ACM Int'l. Conf. on Management of Data(ACM SIGMOD). (2000) 331–342
9. A. P. Sistla, O. Wolfson, S. Chamberlain, and S. Dao: Modeling and Querying Moving Objects," In Proc. IEEE Conf. on Data Engineering(ICDE). (1997) 422–432
10. Y. Tao, D. Papadias, and J. Sun: The TPR*-Tree: An Optimized Spatio-Temporal Access Method for Predictive Queries. In Proc. the Int'l. Conf. on Very Large Data Bases(VLDB). (2003) 790–801
11. Y. Theodoridis, M. Vazirgiannis, and T. Sellis: Spatio-Temporal Indexing for Large Multimedia Applications. In Proc. the IEEE Conf. on Multimedia Computing and Systems(ICMCS). (1996) 441–448
12. Y. Theodoridis, R. Silva, and M. Nascimento: On the Generation of Spatiotemporal Datasets," In Proc. the Int'l. Symp. on Spatial Databases. (1999) 147–164
13. O. Wolfson, B. Xu, S. Chamberlain, and L. Jiang: Moving Objects Databases: Issues and Solutions. In Proc. the Int'l. Conf. on Scientific and Statistical Database Management(SSDBM). (1998) 111–122
14. X. Xu, J. Han, and W. Lu: RT-Tree: An Improved R-Tree Indexing Structure for Temporal Spatial Databases. In Proc. the Int'l. Symp. on Spatial Data Handling(SDH). (1990). 1040–1049
15. K. Y. Whang, S. W. Kim, and G. Wiederhold: Dynamic Maintenance of Data Distribution for Selectivity Estimation. The VLDB Journal. Vol. 3, No. 1. (1994) 29–51

Strategy of Positioning for LBS on U-Campus

Jaegeol Yim, Ilseok Ko, and Jaesu Do

Dept. of Computer and Multimedia Dongguk University at GyeongJoo, GyeongBuk, 780-714
Republic of Korea
{yim, isko, dojesu}@dongguk.ac.kr

Abstract. Location-based service is one of the most popular buzzwords in the field of U-city. Positioning a user is an essential ingredient of a location-based system on a U-city. For the outdoor positioning, GPS based practical solutions have been introduced. However, GPS measurement error is too big to be used for U-campus services because the size of a campus is relatively smaller than a city. We propose Relative-Interpolation Method in order to improve the correctness of outdoor positioning. Besides, indoor positioning is necessary for U-campus while GPS signal is not available inside buildings. For the indoor positioning, Cricket, Active Badge, and so on have been introduced. These methods require special equipments dedicated for positioning. Our method does not require such equipments because it determines the user's position based on the receiver signal strength indicators (RSSI) from access points (AP) which are already installed for WLAN. The algorithm we are using for indoor positioning is a kind of finger prints method. However our algorithm builds a *decision tree* instead of a look-up table in the *off-line* phase. Therefore, our method is faster than existing indoor positioning methods in the *real-time* phase. We have integrated our indoor and outdoor positioning methods and implemented a prototype of our indoor-outdoor positioning method on a laptop. Our experimental results are discussed in this paper.

Keywords: Ubiquitous-Campus, LBS, Positioning, Fingerprint method, Decision Tree.

1 Introduction

A location based service (LBS) provides useful information to the users based on the geographic location the user designates or the user is currently located. Examples of LBS include directory service, gateway service, location utility service, presentation service, route service, and so on [1-8].

Positioning users is an essential technique for developing location-based services. Positioning techniques can be classified into outdoor and indoor. Thanks to Global Positioning System (GPS), outdoor positioning is in the level of practical use. GPS's disadvantages include its significant position measurement inaccuracy [9] and the limitation of its availability [10]. Outdoor positioning methods using cellular communication infrastructures [11] and wireless communication infrastructures [4] have also been introduced.

N.T. Nguyen et al. (Eds.): KES-AMSTA 2007, LNAI 4496, pp. 851–858, 2007.

Many researches regarding indoor positioning have been performed. Among them, Cricket [12] and Active Badge [13] are pioneers in this field. They are prominently accurate. However, they require special equipments dedicated for positioning. RADAR [14] is a radio-frequency based system for indoor positioning. RADAR requires a set of base stations.

In this paper, we are introducing a both indoor and outdoor positioning system. A prototype of this system is currently implemented on a laptop computer. Our outdoor positioning algorithm is called Relative-Interpolation Method. We use two reference points as other interpolation methods do. However, the position of our reference point is x-y coordinates of the window in which the campus map is being rendered instead of absolute latitude and longitude. Our indoor positioning method is essentially similar to RADAR's method. However, our method is significantly different from RADAR's in the following two aspects: 1) we use RSSI from APs, 2) we build a decision tree in the *off-line* phase.

2 Related Works

This paper introduces a both outdoor and indoor positioning method. GPS is successfully used for outdoor positioning in practice. One of the disadvantages of GPS is its inaccuracy. Many efforts have been made to improve the accuracy of GPS. Differential GPS (DGPS) is one of the efforts [9]. Another effort was made in [15] where they use a mobile phone to assist GPS. Many researchers have applied Kalman filters on GPS positioning and made improvement [16, 17].

Cricket and Active Badge are most famous indoor positioning for LBS. There are so many other indoor positioning methods published. Friis-based positioning method has been introduced in [18-20]. Our work was directly inspired by RADAR [14]. They placed three base stations (desktop pc), BS_1, BS_2 and BS_3 on the floor. Their mobile host was a laptop computer. If the mobile host broadcasts a UDP packet then the base stations can receive the packet and read the signal strength accompanying the packet. The positioning system uses the signal strength to construct a *look-up* table during off-line analysis as well as to infer the location of a user in *real time*.

Our indoor method is similar to RADAR's. The major difference is that we are using RSSIs from APs instead of base stations. Since APs are already installed for wireless LAN, we do not require any extra equipment dedicated for the purpose of positioning. Another difference is that we are building a *decision tree* instead of a *look-up table*. Constructing a decision tree is more time consuming. However making a decision on a decision tree is much faster. Therefore, our method is faster in the *real-time* phase.

3 Design of an Integrated Positioning System

This paper is proposing an integrated positioning system which can be used outdoor as well as indoor.

3.1 System Structure

The hardware structure of our system is shown in Figure 1. It is basically a PDA equipped with a WLAN card and a GPS receiver.

The proposed indoor-outdoor positioning system is running on a PDA. It reads RSSIs when turned on and determines if the current position is inside of a building referring to the RSSIs. There always is an RSSI greater than -50dBm when it is inside of a building. Therefore, if all the RSSIs are less than the threshold (-60dBm) then the system decides that it is outdoor, otherwise it decides that it is indoor. This implies that our strategy sometimes misjudges that it is indoor when it is actually outdoor. In this case we determine user's position with indoor positioning method. Even when it is outdoor, GPS data is invalid if the line of sight to the satellites is interfered. In this case, we are applying indoor positioning method. Our positioning strategy is summarized in Figure 2.

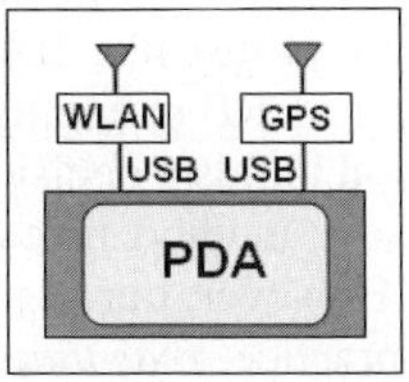

Fig. 1. The hardware structure of our indoor-outdoor positioning system

```
Indoor-Outdoor Positioning Algorithm
  1)  Reads RSSIs
  2)  if all RSSIs < Threshold {
        A.  Read GPS;
        B.  If GPS is valid {
              i.   Outdoor Positioning
        C.  }
        D.  else {
              i.   Apply indoor positioning
        E.  }
  3)  else {
        A.  Indoor positioning
4)  }
```

Fig. 2. Indoor-Outdoor Positioning Algorithm

3.2 Outdoor Positioning

We are using GPS receiver for outdoor positioning. At reference positions A and B, we measure GPS coordinates, longitude and latitude as well as X, Y coordinates on the window. Let X_x, X_y, X_{lon} and X_{lat} be X's x(y)-coordinates, longitude and latitude. Let C be the unknown user's current position and C_{lon} and C_{lat} be the GPS data measured at the current position. Then we estimate user's x, y coordinates on the map with the following expressions.

$$C_x = \left(\frac{C_{lon} - A_{lon}}{B_{lon} - A_{lon}} \right)(B_x - A_x) + A_x \quad (1) \qquad C_y = \left(\frac{C_{lat} - A_{lat}}{B_{lat} - A_{lat}} \right)(B_y - A_y) + A_y \quad (2)$$

We call our method Relative-Interpolation because the coordinates of the reference point is not absolute longitude and latitude but relative x-y coordinates to the (0, 0) point of the window in which the map is rendered. One thing we have to note is that north to south line of the map must coincide with y axis of the window.

3.3 Indoor Positioning

Our indoor positioning method is basically a finger print method just like RADAR. However our finger print consists of RSSIs from APs instead of RF strengths from base stations. The more significant difference is that our method builds a decision tree instead of a look-up table in the off-line phase. Therefore, our method takes more time than existing finger print methods in the off-line phase. However, our method is much faster than existing methods in the *real-time* phase. In practice, *real-time* phase is time critical whereas *off-line* phase is not. Therefore, positioning service must be fast in the *real-time* phase.

In the off-line phase of our method, we build a decision tree with training data. An entry T of the training data set is a 6-tuple, T=(cp, I1, I2, I3, I4, I5), where cp is an integer identifying a candidate point and Ii, for 1=<i=<5, is discretized value of RSSI of i-th AP. An example of discretizing policy can be $I1 = \{x \mid x > -30\}$, $I2 = \{x \mid -40 < x \leq -30\}$, and so on.

Given a set of training data, we build a decision tree with the algorithm Construct_DT shown in Fig. 3. Step (4) of the algorithm computes *I* where *I* is the expected information needed to classify a given sample and is given by

$$I(s_1, s_2, \ldots, s_m) = -\sum_{i=1}^{m} p_i \log_2(p_i) \tag{3}$$

where, m is the number of candidate points, S is the number of tuples in the training data set (rows of Table in the algorithm), s_i is the number of rows of S in class CP_i, $p_i = {}^{s_i}\!/_{s}$.

Step (5) of the algorithm computes the entropy, or expected information based on the partitioning into subsets by CP_i. Let CP_i have v distinct values, $\{a_1, a_2, \ldots, a_v\}$. CP_i can be used to partition S into v subsets, $\{S_1, S_2, \ldots, S_v\}$, where S_j contains those samples in S that have value a_j of CP_i. Let S_{ij} be the number of samples of class CP_i in a subset S_j. The entropy E(CP_i) is given by

$$E(CP_i) = \sum_{j=1}^{v} \frac{s_{ij} + \ldots + s_{mj}}{s} I(s_{ij}, \ldots, s_{mj}). \tag{4}$$

At step (6) the algorithm computes information gain G(CP_i) by the following expression, $Gain(CP_i) = I(s_1, s_2, \ldots, s_m) - E(CP_i)$.

```
Algorithm Construct_DT(int[][] Table, ListType MacList, int Index)
    (1)     If (Number of rows in Table != 1)
            A. If all the CPs of Table are the same, CPi, then Tree[Index] = CPi and return
            B. else if (Number of rows in Table == 0) then Tree[Index] = Empty and return
            C. else if (Number of columns in Table == 1) then Tree[Index] = for each CPi in
            Table, probability of CPi and return
    (2)     else Tree[Index] = CP and return
    (3)     end if
  // Number of rows != 1 and !A and !B and !C
    (4)     Compute I
    (5)     Compute Entropies for each AP
    (6)     Tree[Index] = MacAddress of the AP with maximum Gain and array P. P[i]
    is the probability of CPi in Table
    (7)     Construct subMacList
    (8)     loop(i=1; i<=number of Intervals; i++)
            A. generate subTable
            B. subIndex = Index*number of Interval + i
            C. Construct_DT (subTable, subMacList, subIndex)
    (9)     end loop
end Construct_DT
```

Fig. 3. Algorithm to construct a decision tree

4 Experiments

We have implemented a prototype of indoor-outdoor positioning system for U-campus. Our program is written in C# and running on a portable laptop computer. The operating system of the computer is Microsoft Windows XP. The laptop is equipped with Intel(R) PRO/Wireless 2200BG Network Connection and Model X-150 GPS receiver from Jacom Inc. We have performed experiments for outdoor positioning and indoor positioning.

4.1 Experiments for Outdoor Positioning

Our outdoor positioning experiments were performed outside of the Natural Science Building. The area is shown in Fig. 4 zoomed. The measured x, y coordinate, latitude and longitude of reference points are shown in Table 1. We have performed experiments of comparing the x-y coordinates of current position obtained by clicking mouse on the window and obtained from our outdoor positioning program 200 times and the results are summarized in Table 2. The table says that among the 200 experiments, less than 1 m error occurred 11 times, and so on. The average error was 4.875m.

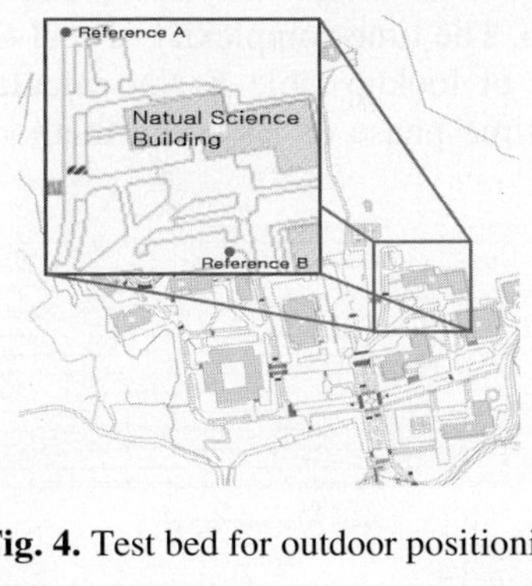

Fig. 4. Test bed for outdoor positioning

Table 1. x,y coordinates, latitude and longitude of reference points

	Coordinates		GPS data	
	X	Y	latitude	longitude
A	1842	1140	N 35°51′48.7″	E 29°11′44.72″
B	2112	1566	N 5°51′45.01″	E 129°11′47.5″

Table 2. Summary of our outdoor positioning experiments

error (m)	0 ~ 1	1 ~ 2	2 ~ 4	4 ~ 6	6 ~ 8
Occurrence	11	17	61	51	33
Probability	5.5%	8.5%	30.5%	25.5%	16.5%
Average Error = 4.875 m					

4.2 Experiments of Indoor Positioning

We have implemented K-NN, Bayesian and our decision tree methods in order to compare the performances of them. The test bed for our indoor positioning experiments is the Micro LAB on the 4-th floor of Natural Science Building. We performed experiments of running 1-NN, Bayesian and decision tree methods on training data of N=5, I=6, and M=96, where N is the number of APs, I is the number of intervals, and M is the number of candidate points, in order to compare their accuracies. The test results are shown in Fig. 5. In the figure, 'number of samples' is the same as the number of measurements we have performed at a candidate point to obtain training data. An entry of lookup table of 1-NN is the average of the measurements. When the number of samples is 10, 1-NN method is much more accurate than others. However, the difference gets decreases as the number of samples increases and when the number of samples is 50 the accuracies of the three methods are almost same.

The advantage of *our decision tree method* is the fast *real-time phase* process time. We have measured real-time phase execution times of 1-NN, Bayesian, and our decision tree methods. The result is shown in Fig. 6. The time complexity of real time phase of K-NN is *O(N*M)* because for each row of lookup table K-NN calculates Euclidean distance. The time complexity of real time phase of Bayesian method is

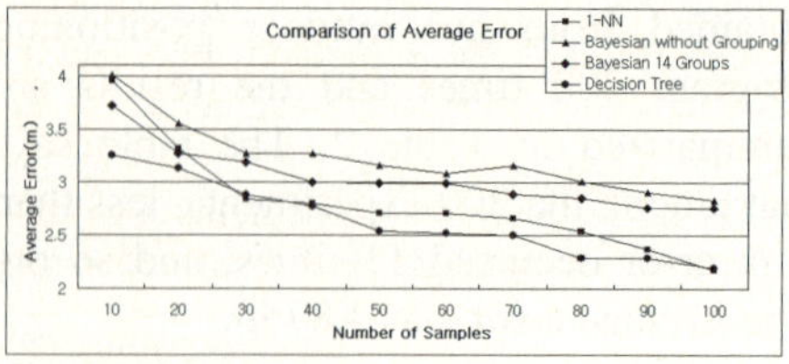

Fig. 5. Comparison of accuracy

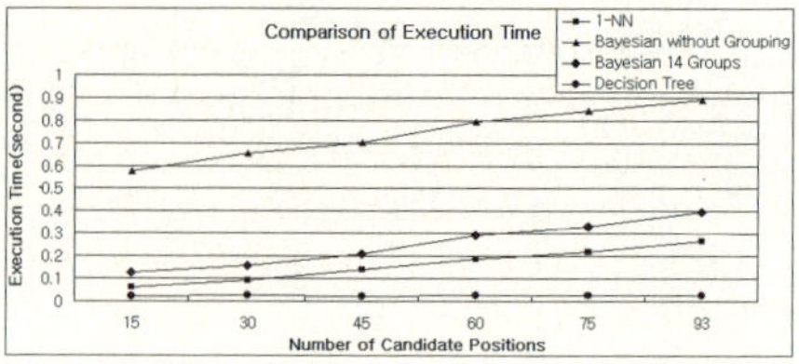

Fig. 6. Comparison of execution times

*O(I*N*M)* because it counts the number of samples belong to the same interval for each AP. In the case of decision tree, the execution time of real time phase is not affected by the number of candidate points and it is *O(I*N)*.

5 Conclusion

Positioning is an essential technique for Location Based Services (LBS). We have proposed an indoor-outdoor positioning system for U-campus. For outdoor positioning, we have proposed Relative-Interpolation Method. Our experiments showed that the average error of the proposed method is 4.875m.

For the indoor positioning, we have proposed decision tree method. Our algorithm builds a *decision tree* instead of a look-up table in the *off-line* phase. Therefore, our method is faster than existing indoor positioning methods in the *real time* phase. Our experimental results showed that our method is much faster than existing methods yet its accuracy is not inferior to the other's accuracies

References

1. Marwa Mabrouk, OpenGIS Location Services(OpenLS):Core Services OGC 03-006r http://www.opengis.org/
2. Krishnamurthy, N.,"Using SMS to Deliver Location-based Services," Proceedings of 2002 IEEE International Conference on Personal Wireless Communications (Proceedings of ICPWC'2002), Dec. 15-17 2002, pp. 177 - 181.
3. Virrantaus, K., J. Veijalainen, and J. Markkula, "Developing GIS-Supported Location-Based Services", Proceedings of the Second International Conference on Web Information Systems Engineering, 2001, Vol. 2, Dec. 3-6 2001, pp. 66-75.
4. Koo, S. and C. Rosenberg, "Location-based E-campus Web Services: From Design to Development", Proceedings of the First IEEE International Conference on Pervasive Computing and Communications, 2003 (PerCom 2003), March 23-26 2003, pp. 207-215.
5. Bravo, A.M. Moreno, J.I. Soto, I., "Advanced positioning and location based services in 4G mobile-IP radio access networks," 15th IEEE International Symposium on Personal, Indoor and Mobile Radio Communications, 2004. PIMRC 2004, 5-8 Sept. 2004, vol. 2, pp. 1085 – 1089.
6. Jose, R., Moreira, A., Meneses, F and Coulson, G., "An open architecture for developing mobile location-based applications over the Internet," Proceedings of the Sixth IEEE Symposium on Computers and Communications, 3-5 July 2001, pp. 500 – 505.
7. Ahmad, U., Nasir, U., Iqbal, M., Lee, Y., Lee, S., and Hwang, I., "On building a reflective middleware service for location-awareness," Proceedings of the 11th IEEE International Conference on Embedded and Real-Time Computing Systems and Applications, 2005, 17-19 Aug. 2005, pp. 439 – 442.
8. Wu, X. and Schulzrinne, H., "Location-based services in Internet telephony," Second IEEE Consumer Communications and Networking Conference, 2005, 3-6 Jan. 2005, pp. 331 – 336.
9. Wilson, T, Barth, J., Pierce, S., Kosro, P. and Waldorf, B., "A Lagrangian drifter with inexpensive wide area differential GPS positioning," Proceedings of the MTS/IEEE. Prospects for the 21st Century, 23-26 Sept. 1996, Vol. 2, pp. 851 – 856.

10. Zheng, J., Wang, Y. and Nihan, N.,_"Quantitative evaluation of GPS performance under forest canopies," Proceedings of the IEEE Networking, Sensing and Control 2005, 19-22 March 2005, pp. 777 – 782.
11. Ygnace, J., Drane, C., "Cellular telecommunication and transportation convergence: a case study of a research conducted in California and in France on cellular positioning techniques and transportation issues," Proceedings of the IEEE Intelligent Transportation Systems, 2001, 25-29 Aug. 2001, pp.16 – 22.
12. Priyanthat, N., Chakraborty, A. and Balakrishnan, H, "The Cricket Location-Support System," Proc. of 6th ACM International Conference on Mobile Computing and Networking, Boston, MA, Aug. 2000.
13. Want, R., Hopper, A., Falco, V. and Gibbons, J., "The Active Badge Location System", ACM Transactions on Infomation Systems 10, Jan. 1992. pp. 91-102.
14. Bahl, P. and Padmanabhan, V., "RADAR:An in-building RF-based user location and tracking system", INFOCOM 2000, Mar. 2000, pp. 775-784.
15. Feng, S. and Law, C._"Assisted GPS and its impact on navigation in intelligent transportation systems," Proceedings of the IEEE 5th International Conference on Intelligent Transportation Systems, 2002, pp. 926 – 931.
16. Mao, X, Wada, M. and Hashimoto, H., "Nonlinear iterative algorithm for GPS positioning with bias model," Proceedings of the 7th International IEEE Conference on Intelligent Transportation Systems, 2004, 3-6 Oct. 2004, pp. 684 – 689.
17. Chen, G. and Harigae, M., "Using IMM adaptive estimator in GPS positioning," Proceedings of the 40th SICE Annual Conference, 25-27 July 2001, pp. 78 – 83.
18. Lassabe, F.; Canalda, P.; Chatonnay, P.; Spies, F.;, "A Friis-based calibrated model for WiFi terminals positioning," Sixth IEEE International Symposium on a World of Wireless Mobile and Multimedia Networks, 2005. WoWMoM 2005. 13-16 June 2005, pp. 382 – 387.
19. Chunhan Lee; Yushin Chang; Gunhong Park; Jaeheon Ryu; Seung-Gweon Jeong; Seokhyun Park; Jae Whe Park; Hee Chang Lee; Keum-shik Hong; Man Hyung Lee;, "Indoor positioning system based on incident angles of infrared emitters," 30th Annual Conference of IEEE Industrial Electronics Society, 2004. IECON 2004. Vol. 3, 2-6 Nov. 2004, pp. 2218 – 2222.

Indirect DiffServ QoS for SIP in Broadband Access Networks

Seungchul Park

School of Internet Media Engineering, Korea University of Technology and Education
Byeongcheon-myun, Cheonan, Chungnam, Republic of Korea
scpark@kut.ac.kr

Abstract. From both technical and economic viewpoints, DiffServ IP QoS architecture is accepted as a more practical solution in Internet world because the other IntServ architecture is more complex and has scalability problem. This paper presents a dynamic DiffServ Model called Indirect DiffServ to support dynamic QoS requirements of SIP-based ASP multimedia applications in broadband access networks. An operational model of QoS-enabled broadband access networks to support dynamic QoS for SIP is firstly proposed. Then the procedures to establish QoS-enabled SIP sessions of ASP applications are presented. The signaling interfaces for dynamic provisioning of DiffServ QoS are designed to be based on COPS and COPS-PR protocols. The inter-layer mapping of QoS attributes between SIP and underlying layers is also discussed in this paper. The underling broadband access networks are assumed to support largely-deployed Metro Ethernet 802.1 D/Q QoS.

1 Introduction

Largely emerging SIP(Session Initiation Protocol)-based multimedia applications such as Internet telephony, multimedia conference, IP TV, and so on may consist of multiple media streams of different QoS(Quality of Service) requirements. The SIP allows QoS requirement of each constituent media stream of a SIP session to be dynamically negotiated through the execution of SDP(Session Description Protocol) Offer/Answer model during the session establishment[1,2]. There are two different types of SIP applications, P2P(Peer to Peer) type and ASP(Application Service Provider) type respectively. In the case of P2P(Peer-to-Peer) applications, end-systems responsible for being charged will become QoS clients, and the QoS will be directly requested by the application entities of end systems according to the dynamically negotiated SIP QoS. But it is different in the case of ASP applications where there are QoS proxy servers responsible for being charged. SIP Default proxy server, H.323 default gatekeeper, and RTSP streaming server may become QoS proxy servers, and QoS will be indirectly requested by a QoS proxy server on behalf of the application entities of end systems.

This paper presents a dynamic DiffServ model called Indirect DiffServ for SIP-based ASP multimedia applications in broadband access networks. Since most broadband access networks support non-trustful subscribers of being charged, differently from enterprise networks, we need to more carefully consider user authentication,

N.T. Nguyen et al. (Eds.): KES-AMSTA 2007, LNAI 4496, pp. 859–866, 2007.

QoS admission control, and QoS policing issues when supporting QoS in broadband access networks. In this paper, a topology of QoS-enabled broadband access networks reflecting those consideration factors and its operational model to support dynamic QoS requirements of SIP-based ASP applications are firstly suggested. Then the procedures to support DiffServ QoS for both assured-QoS type and enabled-QoS type SIP sessions are presented on the basis of the proposed operational model. The signaling interfaces for dynamic provisioning of DiffServ in broadband access networks are designed to be based on COPS(Common Open Policy Service)[3] and COPS-PR(COPS PRovisioning)[4] protocols, and corresponding signaling interactions are appropriately integrated into the procedures to setup QoS-enabled SIP sessions. Recently most broadband access networks are developed to be based on Metro Ethernet because of its inexpensiveness, easiness in maintenance, and high bandwidth. Metro Ethernet-based broadband access networks allow IEEE 802.1D/Q VLAN(Virtual Local Area Network)-based traffic filtering and priority(traffic type)-based differentiated traffic processing[5,6,7]. This paper assumes Metro Ehernet-based broadband access networks, and the inter-layer mapping of QoS attributes between SIP and underlying layers will be also discussed in this paper.

2 Related Works

There were some trials to propose RSVP-based access networks under the assumption that RSVP would be widely deployed and most multimedia applications would be developed assuming RSVP as the resource reservation protocol[8,9]. But they are all for P2P applications where end systems are responsible for supporting QoS, and it is still believed that using RSVP even in the access network will introduce unneeded complexity. DSL Forum specified architectural requirements for DiffServ-enabled DSL networks and proposed multi-service delivery framework for home networks in QoS-enabled BAN environments, in TR-059[10] and TR-094[11] respectively. The QoS architecture suggested by DSL Forum is based on the DiffServ of static provisioning method because of its simplicity and compatibility problem with legacy broadband access networks, but the dynamic DiffServ architecture which is more effective to support dynamic SIP QoS is remained as a future second phase QoS solution. Though ENRICO model of [12] proposes a dynamic QoS solution for both ATM-aggregated and Ethernet-aggregated DSL networks, it is not enough to support SIP QoS because it does not provide solutions for the signaling binding between SIP QoS and IP QoS(dynamic DiffServ) and the mapping between IP DiffServ and the QoS of underlying broadband access networks. [13] proposes a COPS protocol-based signaling mechanism to support dynamic DiffServ for SIP QoS, but its coverage is limited to only edge-to-edge backbone network, but does not include broadband access networks. It neither cover the mapping issue between IP DiffServ QoS and the QoS of underlying layer of broadband access networks(i.e., 802.1D/Q QoS).

3 Indirect DiffServ Model for SIP in Broadband Access Networks

Fig. 1 shows the proposed Indirect DiffServ model to support dynamic QoS requirements of SIP ASP applications in broadband access networks. In order to support dynamic QoS requirements of SIP sessions in an effective way, the QoS-enabled broadband access networks need to provide signaling interfaces between QoS client and QoS server for IP DiffServ request and admission.

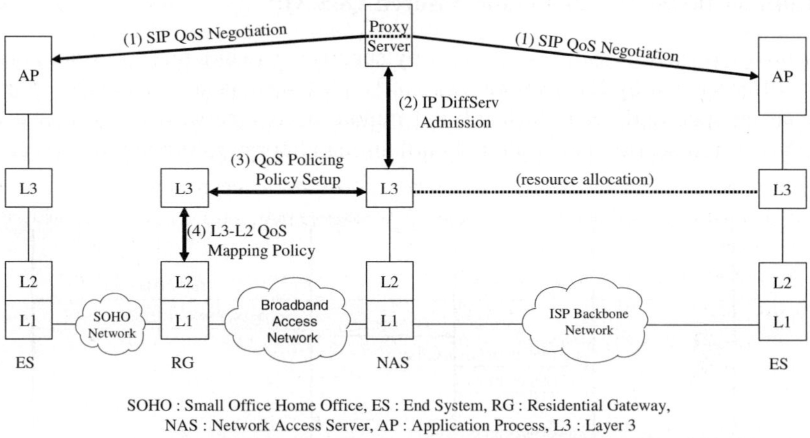

Fig. 1. Indirect DiffServ Model for SIP in Broadband Access Networks

NAS(Network Access Server) which maintains subscriber profile information and QoS policy information in a broadband access network will become the QoS server and provide admission control service of DiffServ QoS as well as DiffServ classification, traffic shaping and policing, DSCP(Differentiated Service Code Point) marking and remarking, class-based queuing and scheduling, and so on. In the case of ASP(Application Service Provider) applications where there are QoS proxy servers(i.e., default SIP proxy servers) of being charged, the QoS proxy servers will become the QoS clients. Thus, the corresponding IP DiffServ QoS will be indirectly requested by the QoS proxy server on behalf of the application process of end system to the NAS. In the QoS-enabled broadband access network, the CPE modem is operating as a L3(Layer 4) RG(Routing Gateway) which is capable of handling IP DiffServ traffic. For the purpose of supporting reliable QoS in the broadband access network, the RG provides the policing function of the IP DiffServ admitted by NAS and the mapping function between the IP DiffServ and the QoS of underlying layer of broadband access network.

4 Indirect DiffServ Support for QoS-Enabled SIP

SIP supports precondition mechanism for coordinating the session setup and QoS support. There may be two QoS types for the interaction between session setup and QoS support: assured-QoS type which is implemented through mandatory QoS preconditions and enabled-QoS type implemented through optional ones[8,14]. In case of assured-QoS type session, the session will not be established unless the QoS requirements of mandatory preconditions are successfully met. But in enabled-QoS type session, session establishment and QoS support are decoupled and may proceed concurrently. Even if the QoS requirements of optional preconditions are not met, a session of best-effort QoS will be established.

4.1 Indirect DiffServ Support for Assured-QoS SIP

An assured-QoS SIP session is initiated by specifying mandatory QoS preconditions in SDP offer carried by INVITE message of UAC(User Agent Client). QoS preconditions can be specified for both local and remote access networks, and callee of the assured-QoS SIP session is not alerted until all mandatory QoS preconditions are met.

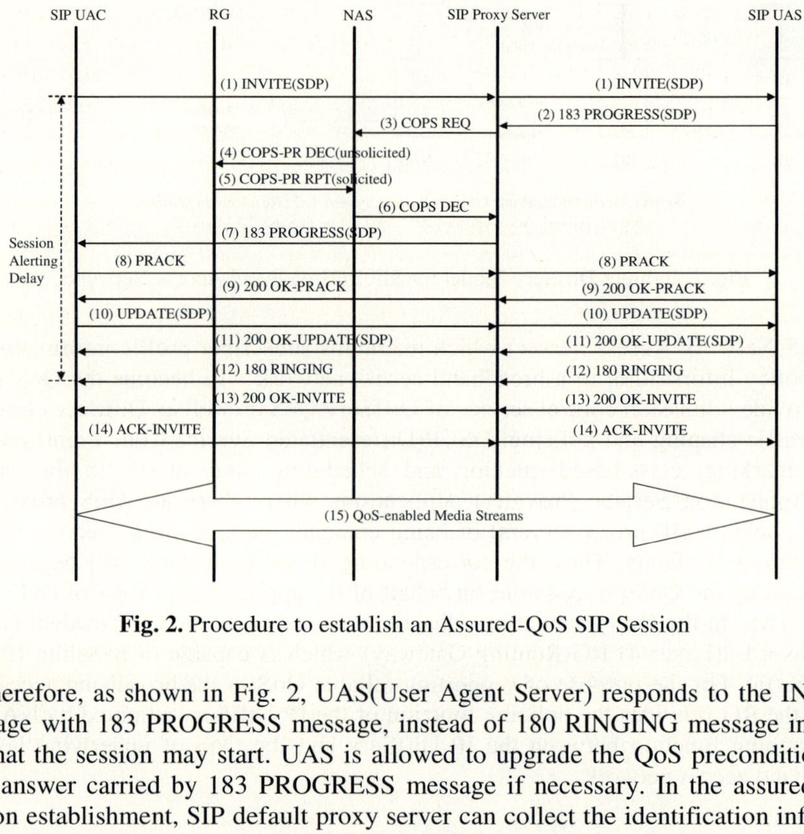

Fig. 2. Procedure to establish an Assured-QoS SIP Session

Therefore, as shown in Fig. 2, UAS(User Agent Server) responds to the INVITE message with 183 PROGRESS message, instead of 180 RINGING message indicating that the session may start. UAS is allowed to upgrade the QoS preconditions of SDP answer carried by 183 PROGRESS message if necessary. In the assured-QoS session establishment, SIP default proxy server can collect the identification information and QoS attributes such as media type, codec type, transmission type, bandwidth, and so on by capturing the INVITE and 183 PROGRESS messages of SIP. Based on the collected QoS requirements of a SIP session, the SIP proxy server starts the QoS signaling interactions to support IP DiffServ QoS. In the procedure to support IP DiffServ QoS, the SIP proxy server is performing the role of QoS PEP(Policy Enforcement Point) which outsources the decision for supporting IP Diffserv QoS from NAS which is accordingly performing the role of QoS PDP(Policy Decision Point). Thus, the signaling interface between SIP proxy server and NAS for IP DiffServ QoS can be realized by using standard PEP-PED COPS protocol. SIP proxy server can request NAS to admit IP DiffServ QoS by sending COPS REQ message which contains identification information, QoS attributes, and corresponding DiffServ class for each media stream of a session((3) of Fig. 2). NAS makes a decision on whether the IP DiffSrev QoS will be admitted or not after checking the predefined QoS policy, profile information for the requestor, status of network resources, and so on. And then it downloads the DiffServ QoS information to RG and enables the RG to setup the policing policies for the IP DiffServ traffic and map the IP DiffServ QoS into the QoS of underlying layer of broadband access network.

The COPS protocol is inappropriate for the signaling interface between NAS and RG because, in that case, each RG should be operating as a server and NAS have to keep status information about every RG. Instead, we propose COPS-PR protocol, a provisioning version of COPS protocol, for this signaling interface. After initialization of COPR-PR protocol by RG, NAS downloads the DiffServ QoS information to RG via an unsolicited COPS-PR DEC message, and RG reports the execution result of the downloaded DiffServ QoS policy via a solicited COPS-PR RPT message((4) & (5) of Fig. 2). NAS, after receiving the COPS-PR RPT message, finally admits the IP DiffServ QoS by sending COPS DEC message to the SIP proxy server((6) of Fig. 2). At this time, the SIP proxy server forwards 183 PROGRESS message to UAC. Since, in the assured-QoS session establishment, UAC responds with PRACK message. The same admission procedure may be concurrently performed at UAS side.

UAS is informed of the status of QoS support in UAC side via an UPDATE message that has a SDP body indicating the status of each precondition as "success" or "failure". The SIP proxy server is responsible for specifying the status of QoS preconditions of SDP body of the UPDATE message. When receiving the UPDATE message from UAC and upon the success of its preconditions similarly indicated, the UAS determines all preconditions have been met, alerts the callee, and sends 180 RINGING message to the UAC, indicating the session may start. When the callee accepts the session, UAS sends 200 OK message for INVITE message and the procedure to establish an assured-QoS SIP session is finally completed.

4.2 Indirect DiffServ Support for Enabled-QoS SIP

Since callee is not alerted about the session establishment until all mandatory QoS preconditions are met, the session alerting delay of an assured-QoS session, interval between the departure time of INVITE message and the arrival time of 180 RINGING message, is somewhat long. The session alerting delay can be shown as follows.

$$\begin{aligned} \textit{Session Alerting Delay} \quad & \geq\ 3 \times RTT^{ee} + RTT^{en} + RTT^{rn}\ , \\ RTT^{ee} &= \textit{End to End Round Trip Time}, \\ RTT^{sn} &= \textit{SIP Proxy Server to NAS Round Trip Time}, \\ RTT^{rn} &= \textit{RG to NAS Round Trip Time} \end{aligned}$$

The long session alerting delay can influence more negatively than non-support of QoS on some applications which are sensitive to initial session setup delay. In case of an enabled-QoS session, as shown in Fig. 3, UAS sends the 180 RINGING message to the UAC, indicating the session may start, immediately after receiving INVITE message containing a SDP offer. Though the optional preconditions are not met during the IP DiffServ admission procedure, the session establishment proceeds so as to support best-effort QoS. SIP proxy server executes the admission procedure for IP DiffServ QoS immediately after receiving 180 RINGING message containing a SDP answer, but it does not inform UAS of the status of QoS preconditions. The procedure to support IP DiffServ QoS in an enabled-QoS SIP session is same as the case of assured-QoS one.

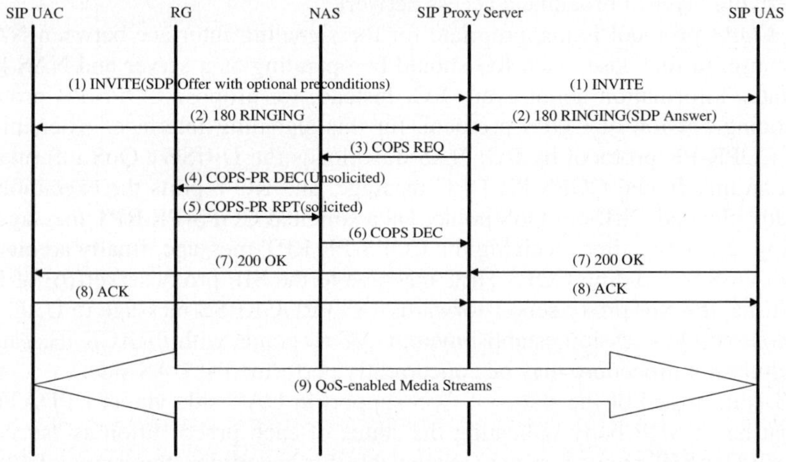

Fig. 3. Procedure to establish an Enabled-QoS SIP Session

4.3 Inter-layer Mapping of QoS Attributes

The attributes to specify QoS requirements of SIP are different from the QoS parameters of underlying layers, IP DiffServ layer and 802.1 D/Q layer of broadband access networks. In order to support SIP QoS in broadband access networkss, the SIP QoS requirements specified by the attributes such as media type, transmission type, and protocol type needs to be appropriately mapped into IP DiffServ classes first, and then into 802.1 D/Q traffic priority types. The inter-layer QoS mapping can be done in a flat way or in a structured way. Whereas, in Flat QoS mapping, the SIP QoS requirement of a media stream is independently mapped into the QoS of underlying layers from other media streams of same SIP session, Structured QoS mapping allows the QoS relationship among some or all media streams of a SIP session to be kept in the QoS of lower layers. Table 1 shows an example of inter-layer Flat QoS mapping. In the Flat QoS mapping, a DiffServ class can be independently allocated to the SIP QoS of each media stream of a session, and a VLAN traffic type specified in the 3 bit priority field of 820.1 D/Q can be accordingly matched with the DiffServ class.

A multimedia conference application which requires synchronization among its constituent audio, video, and shared application media streams can be a good example of Structured QoS mapping. In this application, as shown in Table 2, AF3 DiffServ class may be commonly allocated to the SIP QoSs of audio, video, and shared application media streams, which enables those media streams to be processed through a common queue of a network node. The processing priority of each media stream within a common queue in case of congestion can be specified in DP(Drop Precedence) such as low(1), medium(2), and high(3).

Table 1. Inter-layer Flat QoS Mapping

SIP QoS (Media Attributes)	IP DiffServ (DiffServ Classes)	802.1 D/Q BAN QoS (Traffic Type)
Media type = Control Transport = TCP Transmission = sendrecv	EF (101110)	NC (111)
Media type = Audio Transport = RTP/AVP Transmission = sendrecv	AF31 (011010))	VO (110)
Media type = Video Transport = RTP/AVP Transmission = sendrecv	AF21 (010010))	V1 (101)
Media type = Application Transport = TCP Transmission = sendrecv	AF11 (001010))	EE (011)
Media type = Data Transport = TCP Transmission = sendonly	BE (000000))	BE (000)

Table 2. Inter-layer Structured QoS Mapping

SIP QoS (Media Attributes)	IP DiffServ (DiffServ Classes)	802.1 D/Q BAN QoS (Traffic Type)
Media type = Control Transport = TCP Transmission = sendrecv	EF (101110)	VID {EF class VID} Priority = NC
Media type = Audio Transport = RTP/AVP Transmission = sendrecv	AF31 (011010))	VID {AF3 class VID} Priority = VO
Media type = Video Transport = RTP/AVP Transmission = sendrecv	AF32 (011100))	VID {AF3 class VID} Priority = VI
Media type = Application Transport = TCP Transmission = sendrecv	AF33 (011110))	VID {AF3 class VID} Priority = EE
Media type = Data Transport = TCP Transmission = sendonly	BE (000000))	VID {BE class VID} Priority = BE

IP DiffServ - EF : Expedited Forwarding, AF : Assured Forwarding, BE : Best Effort
802.1 D/Q - NC : Network Contol, VO : Voice, VI : Video, EE : Excellent Effort, BE : Best Effort

Problem occurs when mapping the common DiffServ class into the VLAN traffic type of underlying broadband access network because 802.1 D/Q Metro Ethernet does not support the DP. In this paper, assignment of VID(VLAN ID) in accordance with corresponding DiffServ class is suggested so as to support Structured QoS mapping. That is, VIDs are classified into several groups, each of which represents a VLAN class correspondent with a DiffServ class, and a VID is selected from a group correspondent with DiffServ class. Then VLAN traffic type of the 3 bit priority field is used to specify the DP. Metro Ethernet switches supporting this Structured QoS mapping can allocate queues according to VIDs representing VLAN classes, and determine the drop precedence of each frame based on its VLAN traffic type. This Structured QoS mapping may result in increase in number of VIDs, but it can be solved through the VLAN stacking technology[15].

5 Conclusion

Since SIP-based multimedia applications such as audiovisual telephony and IP TV are likely to be rapidly spread in Internet world, it is important to smoothly support dynamic QoS requirements of SIP session so as to keep backward compatibility with legacy end systems. In this paper, we proposed a dynamic DiffServ model call Indirect DiffServ for SIP-based ASP multimedia applications in broadband access networks. The model requires NAS(Network Access Server)s and RG(Residential Gateway)s of broadband access networks to be appropriately upgraded for the purpose of supporting QoS signaling, but legacy end systems do not have to be upgraded. The procedures to establish QoS-enabled SIP sessions and inter-layer mapping of QoS attributes presented in this paper will enable the backward compatible Indirect

DiffServ model to be promptly deployed in broadband access networks to effectively support QoS-sensitive SIP-based ASP multimedia applications.

References

1. Rosenberg, J. et al. : SIP : Session Initiation Protocol. IETF RFC 3261(June 2002)
2. Rosenberg, J., Schulzrinne, H. : An Offer/Answer Model with Session Description Protocol(SDP). IETF RFC 3264(June 2002)
3. Durham, D. et al. : The COPS(Common Open Policy Service) Protocol. RFC 2748(Jan. 2000)
4. Chan, K., et al. : COPS Usage for Policy Provisioning(COPS-PR). RFC 3084(March 2001)
5. Ooghe, S. et. al. : Impact of the Evolution of the Metropolitan Network on the DSL Access Architecture. IEEE Communications Magazine(Feb. 2003)
6. Zier, L., Fisher, W., and Brockners, F. : Ethernet-Based Public Communication Services: Challenge and Opportunity. IEEE Communications Magazine(March 2004)
7. Fineberg, V. : A Practical Architecture for Implementing End-to-End QoS in an Ip Network. IEEE Communications Magazine(Jan. 2002)
8. Sargento, S. et al. : IP-Based Access Networks for Broadband Multimedia Services. IEEE Communications Magazine(Feb. 2003)
9. Bernet, Y. : The Complementary Roles of RSVP and Differentiated Services in the Full-Service QoS Network. IEEE Communications Magazine(Feb. 2000)
10. DSL-Forum TR-059 : DSL Evolution - Architecture Requirements for the Support of QoS-Enabled IP Services. (Sept. 2003)
11. DSL-Forum TR-094 : Multi-Service Delivery Framework for Home Networks(Aug. 2004)
12. Bouchat, C., Bosch, S., and Pollet, T. : QoS in DSL Access," IEEE Communications Magazine(Sept. 2003)
13. Salsano, S., Veltri, L. : QoS Control by Means of COPS to Support SIP-Based Applications. IEEE Network(March/April 2002)
14. Gamarillo, G., Marshall, W., and Rosenberg, J. : Integration of Resource Management and SIP. IETF RFC 3312(Oct. 2002)
15. Ali, M., Chiruvolu, G., and Ge, A. : Traffic Engineering in Metro Ethernet. IEEE Network(March/April 2005)

An Intelligent Positioning Scheme for Mobile Agents in Ubiquitous Networks for U-City*

Pora Kim and Sekchin Chang

Dept. of Electrical and Computer Engineering, University of Seoul, Seoul, Korea

Abstract. Various advanced schemes have been proposed for the establishment of u-city. Especially, most of the schemes are based on ubiquitous networks. In this paper, it is assumed that the ubiquitous networks include some mobile agents. Therefore, an intelligent positioning scheme is presented for the efficient positioning of the mobile agents in the ubiquitous networks. The approach consists of location detection and location tracking. Simulation results indicate that using the algorithm, the excellent detection and tracking performance can be achieved in ubiquitous networks for u-city.

1 Introduction

A lot of attention has recently been paid to the establishment of the services such as intelligent disaster prevention, intelligent building management, intelligent health care, intelligent traffic control, and so on in metropolitan cities [1]. The services can be converged more efficiently and can be accessed more easily via ubiquitous networks [2]. The city which can offer the services based on ubiquitous networks is generally called u-city. The wireless sensor nodes are usually utilized for the realization of ubiquitous networks [3]. In the design of the wireless sensor nodes, the most important design metrics are cost and power [3, 4]. In other words, low power and low cost are the inevitable requirements for the commercial deployment of the wireless sensor nodes. However, the requirements often prevent the delicate algorithms from being adopted in the implementation of the sensor node, which may lead to a considerable performance degradation. To overcome the design limitation of the sensor nodes, mobile agents were introduced in wireless sensor networks [5]. In this paper, a new structure is presented for realistic implementation of the mobile agents. The structure assumes the wireless sensor node to be implemented in a cellular phone. Since the sensor node can exploit the resources of the cellular phone in such case, the node can achieve the desirable performance without high power and high cost overhead. In addition, the user who holds the cellular phone with the sensor node can access the u-city services if the phone holder is in a relevant ubiquitous network area. By employing the structure, the wireless sensor node can act as a realistic mobile agent in ubiquitous networks. However, the accurate positioning of the mobile

* This work was supported by Smart (Ubiquitous) City Consortium under Seoul R&BD Program.

N.T. Nguyen et al. (Eds.): KES-AMSTA 2007, LNAI 4496, pp. 867–874, 2007.

agent is required for the efficient usage of the agent in ubiquitous networks. Therefore, in this paper an intelligent positioning scheme is proposed for the accurate location detection and tracking of mobile agents in ubiquitous networks for u-city. The intelligent positioning scheme consists of location detection which is further classified into coarse detection and fine detection, and location tracking. Simulation results indicate that the proposed scheme can achieve excellent performance in location detection and location tracking of mobile agents.

2 The Mobile Agents in Ubiquitous Networks

For the realistic implementation of the mobile agent, the sensor node module can be included in a cellular modem as shown in Fig. 1. Since most of the recent cellular modems are based on the system-on-chip (SoC) technology in the design, the cellular modem mainly consists of embedded processor, embedded DSP, and hardware accelerators as illustrated in Fig. 1. The embedded processor, the embedded DSP, and the hardware accelerators usually include the control and simple computations, the highly intensive computations, and the high-speed and parallel computations of the cellular modem algorithms, respectively. In addition, graphics accelerator and image accelerator can be included in the cellular modem for multimedia services as indicated in Fig. 1. In the view of such SoC structure, the hardware accelerator can be considered a kind of peripheral module. This indicates that the SoC structure relatively easily allows the addition of the hardware accelerators with low complexity. Therefore, the sensor node algorithm can be added in the cellular modem as a simple hardware accelerator as shown in Fig. 1 since the algorithm exhibits relatively low complexity [6].

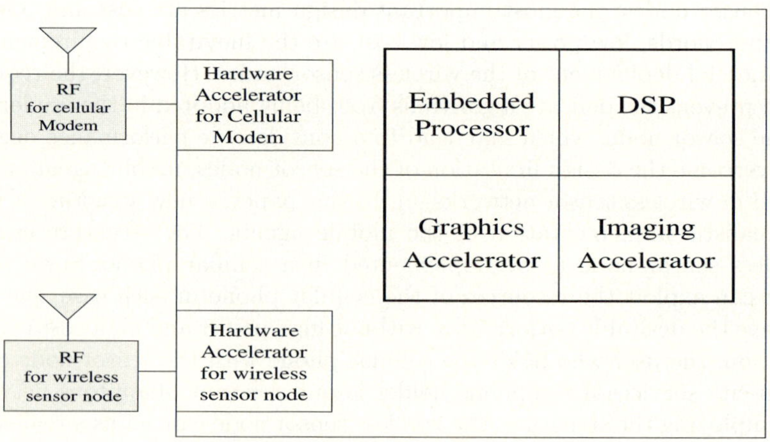

Fig. 1. The modem structure for mobile agents

Once the wireless sensor node is implemented in the cellular modem, the u-city services can easily be offered to the cellular phone holder in ubiquitous networks.

3 The Intelligent Positioning Scheme for Mobile Agents

The mobile agent can afford various u-city services to the user in ubiquitous networks. However, the location of the mobile agent should accurately be detected and tracked to fully utilize the functionality of the agent because the agent can move in ubiquitous networks. Fig. 2 depicts the intelligent positioning scheme for mobile agents. As illustrated in the figure, the intelligent positioning can interactively be made by the reference sensor nodes and the target mobile agent. As an example for a mobile agent in a ubiquitous network which consists of wireless sensor nodes, Fig. 3 illustrates the moving of the mobile agent from region 1 to region 2. In Fig. 3, the R1 through R6 indicate the reference sensor nodes for the location detection and tracking of the mobile agent. In the figure, the region 1 and the region 2 are covered by the reference group I of R1, R2, R3, and R4, and the reference group II of R3, R4, R5, and R6, respectively. In the region 1 of the figure, the location detection and tracking are interactively made by the mobile agent and the reference group I. In the shared region, the tracking of the mobile agent is interactively made by the agent itself, and the reference group I and II. In addition, the mobile agent and the reference group II interactively track the agent itself in the region 2.

3.1 The Location Detection

The location detection consists of coarse detection and fine detection. As indicated in Fig. 2, the reference sensor nodes roughly detect the location of the

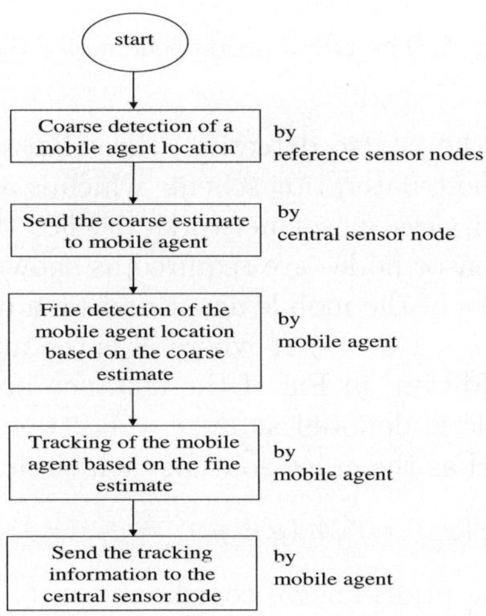

Fig. 2. The intelligent positioning scheme for mobile agents

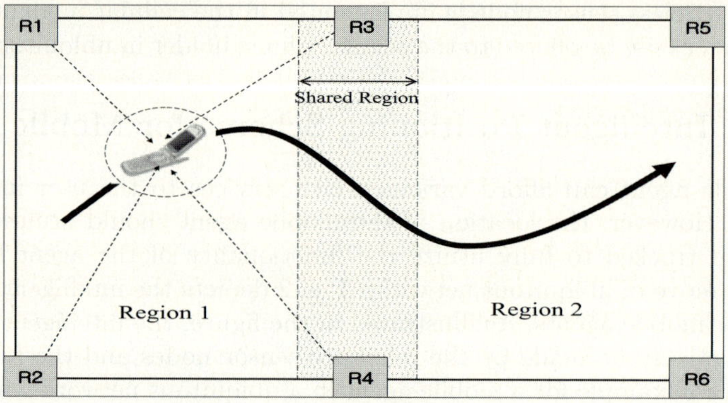

Fig. 3. The mobile agent in ubiquitous networks

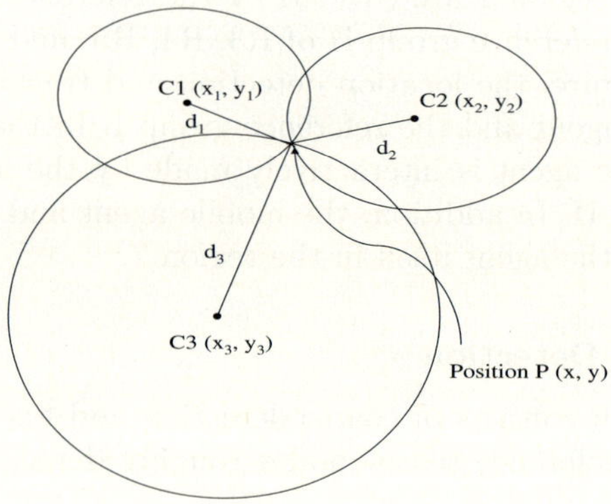

Fig. 4. The trilateration scheme for coarse detection

mobile agent. As the coarse detection, the trilateration scheme [7] is utilized. Fig. 4 illustrates the trilateration scheme which is a conventional location detection method for wireless sensor networks [7]. For the trilateration approach, at least 3 reference sensor nodes are required as shown in the figure. In Fig. 4, the location coordinates of the mobile agent and each reference node are denoted as (x, y) and (x_i, y_i), $i = 1, 2, \cdots, N$ where N is the number of the reference nodes, respectively. In addition, in Fig. 4 the distance between the mobile agent and each reference node is denoted as d_i, $i = 1, 2, \cdots, N$. As shown in Fig. 4, the (x, y) is determined as the cross point of each circle whose expression is given as

$$(x - x_i)^2 + (y - y_i)^2 = d_i^2, i = 1, 2, \cdots, N \tag{1}$$

where (x_i, y_i) is the priori known coordinate. In (1), d_i can be calculated in the ith reference node using the time-of-arrival (TOA) [8, 9] as follows:

$$d_i = TOA_i \times c, i = 1, 2, \cdots, N \tag{2}$$

where TOA_i and c indicate the TOA for the ith reference node and the speed of the light, respectively. For the measurement of TOA_i, a pilot signal is transmitted to the ith reference node by the mobile agent. Since the measured TOA includes some error due to the channel fading [7], the trilateration scheme just produces a coarse estimate of the location coordinate. Therefore, for more accurate estimation a fine detection scheme is required.

Since the fine detection usually requires high computation, it is assumed in this paper that the detection is performed by the mobile agent which can exploit the resources of the cellular phone. In addition, the proposed fine detection scheme is based on the coarse estimate. For the fine detection, one central node which is selected from the reference nodes transmits the coarse estimate to the mobile agent. After receiving the estimate, the mobile agent can increase the accuracy of the estimate using the steepest descent algorithm [10]. Usually, the algorithm just requires relatively simple computation, and exhibits fairly good performance when the initial value is close to the optimal value. Since the coarse estimate can be used as the initial value, the algorithm can generate a fairly accurate estimate. In the fine estimation approach, the steepest descent algorithm is utilized as follows:

$$\mathbf{P}^{(k+1)} = \mathbf{P}^{(k)} - \frac{\lambda}{2}\nabla f(\mathbf{P}^{(k)}) \tag{3}$$

where k denotes the iteration time, and λ indicates the step size. In (3), $\mathbf{P}^{(k)}$ is the location estimate at the kth iteration and is given as

$$\mathbf{P}^{(k)} = \begin{bmatrix} x^{(k)} \\ y^{(k)} \end{bmatrix} \tag{4}$$

In addition, in (3) $f(\mathbf{P}^{(k)})$ is the cost function defined as

$$f(\mathbf{P}^{(k)}) = \sum_{i=1}^{N}[d_i - \sqrt{(x^{(k)} - x_i)^2 + (y^{(k)} - y_i)}]^2 \tag{5}$$

In (5), d_i is calculated in the mobile agent using (2). The TOA_i of (2) is also measured using a pilot signal. In the case of fine detection, the pilot signal is transmitted to the mobile agent by the ith reference node, which is a contrast to the case of coarse detection. In (3), $\nabla f(\mathbf{P}^{(k)})$ is the derivative of (5), and is defined as

$$\nabla f(\mathbf{P}^{(k)}) = \begin{bmatrix} \frac{\partial f(\mathbf{P}^{(k)})}{\partial x^{(k)}} \\ \frac{\partial f(\mathbf{P}^{(k)})}{\partial y^{(k)}} \end{bmatrix} \tag{6}$$

In (3) to (6), the coordinate of the coarse estimate is used as $\mathbf{P}^{(0)}$. The steepest descent algorithm is continued until $|x^{(k)} - x^{(k-1)}| + |y^{(k)} - y^{(k-1)}| < \epsilon$ where ϵ is a predefined threshold. As soon as the mobile agent achieves the fine estimate, it sends the estimate back to the central node.

3.2 The Location Tracking

Based on the fine estimate, the mobile agent can track itself in motion. For the tracking, a cost function is defined as follows:

$$J(l,m) = \sum_{i=1}^{N}[d_i - \sqrt{(x^{(n)} + l \cdot \triangle - x_i)^2 + (y^{(n)} + m \cdot \triangle - y_i)}]^2 \quad (7)$$

where n and $\triangle$ denote the nth time index and the incremental of the coordinate for the tracking, respectively. In (7), l and m are the integer indices ranging from -3 to 3. The d_i of (7) is also calculated in the mobile agent using (2). Like the case of fine detection, TOA_i of (2) is also measured using a pilot signal which is transmitted to the mobile agent by the ith reference node. For the tracking, the integer indices which minimize the cost function of (7) are selected as the optimal indices as shown below:

$$(\hat{l}, \hat{m}) = \arg\min_{l,m} J(l,m), l, m = -3, -2, \cdots, 2, 3 \quad (8)$$

where $\hat{l}$ and $\hat{m}$ indicate the optimal indices. Then, using the optimal indices, the coordinate of the mobile agent is updated as follows:

$$x^{(n+1)} = x^{(n)} + \hat{l} \cdot \triangle, y^{(n+1)} = y^{(n)} + \hat{m} \cdot \triangle \quad (9)$$

In (9), $x^{(0)}$ and $y^{(0)}$ are set to the coordinate values of the fine estimate. Whenever the update of (9) occurs, the mobile agent sends the updated coordinate back to the central node.

4 Simulation Result

Simulation results exhibit the effectiveness of the proposed intelligent positioning scheme. For the simulation, the environment of Fig. 3 is considered, which was already described in section 3. In addition, 2.4 GHz fading and additive white Gaussian noise (AWGN) under which wireless sensor nodes are usually deployed are assumed as the channel environments for the simulation.

For the performance evaluation of the proposed location detection scheme, the mean squared error (MSE) values between exact and estimated location position are given in Fig. 5. Under the environment of Fig. 3, N is set to 4 in (1), (2) and (5), As shown in Fig. 5, the fine detection achieves the improvement of about 1.5 dB over the coarse detection. Since wireless sensor networks generally rely on only the coarse detection method to locate a target sensor node, the detection performance can significantly be enhanced using the suggested location detection approach.

For the performance evaluation of the proposed location tracking scheme, Fig. 6 exhibits the comparison of the exact and estimated coordinates of the mobile agent in motion under the environment of Fig. 3. In the case, N of (7) is set to 4 and 6 in the region 1 and 2, and in the shared region, respectively. As illustrated

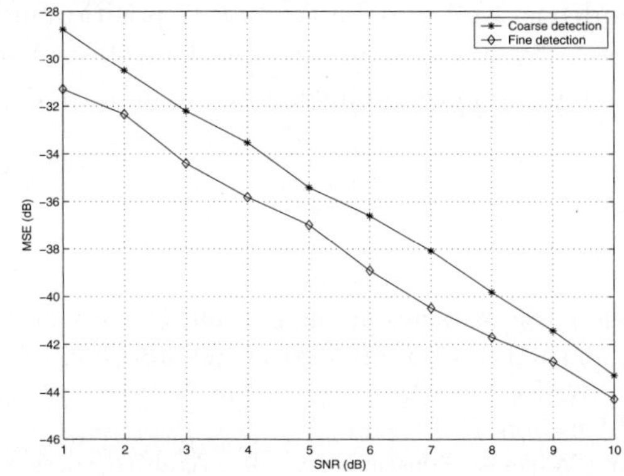

Fig. 5. The MSE performances for coarse and fine location detection

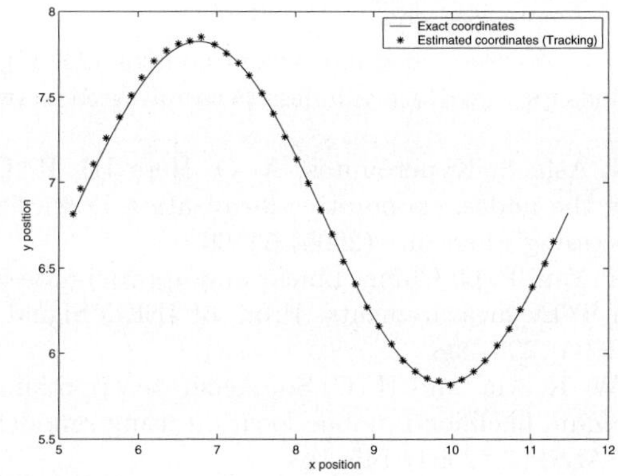

Fig. 6. The estimated and exact coordinates of the mobile agent in motion

in Fig. 6, the estimated coordinates are very close to the exact coordinates, which indicates that the proposed tracking scheme performs well in tracking the mobile agent in ubiquitous networks.

5 Conclusion

In this paper, an intelligent positioning scheme is proposed to achieve the excellent performances in locating and tracking the mobile agent in ubiquitous networks. The SoC-based structure is also presented for the realization of the mobile agent in this paper. The intelligent positioning technique consists of location detection and location tracking. The location detection is also further classified into coarse detection and fine detection. The coarse location detection is made by the reference sensor nodes. On the other hand, the fine detection and the tracking are performed by the mobile agent, which considerably alleviates the load of the sensor nodes as well as achieves better positioning performance. The simulation results also reveal that the proposed intelligent scheme can significantly improve the detection performance, and can achieve the excellent tracking performance.

References

1. I. Akyildiz, W. Su, Y. Sankarasubramaniam, and E. Cayirci, Wireless Sensor Networks: A Survey. Journal of Computer Networks **38** (2002) 393–422
2. N. S. Correal, and N. Patwari, Wireless Sensor Networks: Challenges and Opportunities. Proc. of Virginia Tech Symp. Wireless Personal Comm. (2001) 1–9
3. E. H. Callaway Jr., Wireless Sensor Networks: Architectures and Protocols. Auerbach. (2003)
4. C. M. Cordeiro and D. P. Agrawal, Ad Hoc & Sensor Networks: Theory and Applications. World Scientific. (2006)
5. L. Tong, Q. Zhao, and S. Adireddy, Sensor networks with mobile agents. Proc. of IEEE MILCOM **1** (2003) 688–693
6. IEEE Std 802.15.4: Wireless medium access control (MAC) and physical layer (PHY) specifications for Low-Rate Wireless Personal Area Networks (LR-WPANs). (2003)
7. N. Patwari, J. N. Ash, S. Kyperountas, A. O. Hero III, R. L. Moses, and N. S. Correal, Locating the nodes: cooperative localization in wireless sensor networks. IEEE Signal Processing Magazine (2005) 54–69
8. Y. T. Chan, C. H. Yau, P. C. Ching, Linear and approximate maximum likelihood localization from TOA measurements. Proc. of IEEE Signal Processing and Its Applications **2** (2003) 295–298
9. K. W. Cheung, W. K. Ma, and H. C. So, Accurate approximation algorithm for TOA-based maximum likelihood mobile location using semidefinite programming. Proc. of IEEE ICASSP **2** (2004) 145–148
10. S. Haykin, Adaptive Filter Theory, 4th Edition. Prentice-Hall. (2001)

A User Interface for Controlling Information Appliances in Smart Homes

Seong Joon Lee[1], Ilseok Ko[2,*], and Min Wook Kil[3]

[1] Department of Computer Engineering, Kyungpook National University, 1370 Sankyuk-dong, Buk-gu Daegu, South Korea
imggaibi@hotmail.com

[2] School of Computer and Multimedia, Dongguk University, 707 Seokjang-dong, Kyungju, Kyungsangbukdo, South Korea
isko@dongguk.edu

[3] Department of Medical Information, Mun Kyung College, HoGyeMyun, Mun Kyung, Kyungsangbukdo, South Korea
isko@dongguk.edu

Abstract. The web user interface offers many advantages; however, a fixed IP address is required to manage each server, it does not have a push function for completion or error messages, it suffers from problems associated with firewalls, and it is slow. To address these issues, we propose an efficient method for controlling information appliances in smart homes that uses a natural language processing technique and an instant messaging system (IMS). The proposed agent-based management system for smart homes provides a user-friendly interface, and can make use of add-on functionalities of the IMS such as voice chatting, SMS, and multimedia. Moreover, the proposed method can be applied using middleware technologies such as Jini [1] UPnP [2], OSGi [4], and HAVi [5].

1 Introduction

A smart home (SH) system consists of a home network, a home server, and a re-mote controller. The home network is an internal network for sharing resources and enabling communication between the various appliances and services in the SH [10]. Such networking can be achieved with appliances, known as information appliances (IAs) or network appliances, which have the computational power to perform predefined functions and the ability to share information. When the home server is connected to the Internet, such networking enables the remote and indirect control of the IAs [10]. Control includes both IA operation and monitoring.

Advances in Internet technologies have prompted the development of various Internet-based SH remote controls, most of which adopt a web interface between people and machines. Although the use of this type of user interface (UI) has many advantages, it also has the following problems: using the physical IP address, lack of push technology, non-real-time processing, and significant transmission delay. Push

* Correspondent author.

N.T. Nguyen et al. (Eds.): KES-AMSTA 2007, LNAI 4496, pp. 875–883, 2007.

technology enables authorized users who are subscribers to a server to automatically receive up-to-date information via the home server. In a web-based SH, each client must first download all state information on the smart devices to be controlled, which consists of CGI scripts, Java applets, and HTML documents. This downloaded information may include information for a variety of devices, even though the user may be interested in only one particular device. These methods are slow and tedious because when the IA has many functions users have to recursively choose between many items displayed on the screen, which is difficult for beginners and individuals not accustomed to operating such systems. Further, users have usually already decided what to do before they interact with the system.

In our approach we rewrite user commands, which in a web interface require hierarchically clicked menus, as a single sentence. We propose a cost-effective system that uses natural language processing (NLP) to improve the efficiency and accuracy of the user interface.

2 Related Work

2.1 The Problem of Previous System

One of the main goals in SH system development is that the system provides the user with an interface that is independent of their location. To this end, most UIs for SHs are phone- or web-based. The UI of phone-based system make use of dual tone multiple frequency (DTMF) [6] or the wireless application protocol (WAP)[8]. DTMF produces poor UIs and operation difficulties, particularly because it requires a dial-up connection. The WAP does enable push functionality, but it does not support real Internet integration, has complex layer architecture, and uses a web-based UI.

The advantages of the Internet-based Client/Server model are its simple design and the availability of a multitude of development tools. However, its disadvantages are that beginners must be trained in its use, and it has less scalability and flexibility. In this model, the server is responsible for managing resources, so the installation of various programs for the control of each device is required. An additional problem is that not all client devices, such as cellular phones, PCs, and PDAs, are fitted with the same type of UI.

In web-based systems, web servers use the HTTP protocol to send information. Although the use of a web-based interface has many advantages, such as scalability, visibility and a uniform UI, it has four main problems. First, it cannot provide push functionality for supplying the homeowner with status information, because the HTTP protocol is defined for one-way continuous transmission. A second problem is that, because of the TCP delay, it cannot be used to send data in real-time, such as audio or time-dependent information. The third problem is that a home server using a web server must be implemented with an additional authentication system, and the fourth problem is that it must maintain a security system for protection from various types of hacking attacks.

In the paper [10], HASMuIM have proposed a novel method using an IM system that is lightweight and supports real-time processing. This method can receive modified state information to occur at an IA without reconnecting to Home Server. One possible disadvantage of this approach is the production of many data packets that transfer changes in the state information around the home network. Then when home-owners reconnect to the server, they may receive many such packets. However, the total number of packets can be reduced by instructing the server to send the metadata list of updated IAs, i.e., the modified IA state information, only when the homeowner requires it. When home-owners interact with their home network, most are only interested in controlling one device.

2.2 Jini Network Technology

Jini consists of a set of APIs and high level network protocols for distributed systems. It provides a simple way to perform tasks in home networks such as discovering, registering, and removing devices and services. Jini creates software infrastructure, called a federation of services, which shares access to the services and engages in interactions without the prior knowledge of the other systems or any need for human intervention [1]. In the next section, we propose a method for controlling information appliances in a home network that uses an IM system.

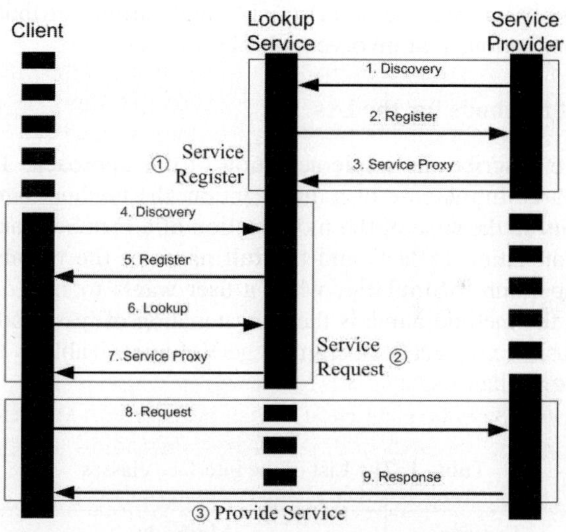

Fig. 1. Jini Technology Process

The system for enabling a Jini service consists of three protocols—discovery, join, and lookup—which provide a dynamic configuration, and requires a pre-established network to function. When a service is plugged into the Jini network, it uses a multicast request to find the local lookup service through the discovery protocol, and then registers the proxy service object in the lookup table via the join protocol (step 1 in Figure 2). Clients can also use the discovery protocol to find the lookup service. Subsequently, when a client requests a search for a service, the lookup service returns matching proxy service objects to the client (step 2). Finally, the object downloaded from the lookup server communicates directly with the service provider (step 3) [1].

3 Proposed Technique

In this paper, we present an efficient real-time method, named SHMS (Smart Home Management System), for home automation systems, which uses Jini network technology and IM system technology to provide a uniform graphic UI (GUI) for end users. In this approach, state information is immediately sent from the home network to the homeowner via the Internet.

There are three main agents in this implementation: the Mobile Messenger Agent (MMA) for tools installed in the portable devices, the Home Messenger Agent (HMA) that provides an interface between the MMA and the Information Appliance Manager Agent (IAMA), which manages the IAs. These agents have been designed independently in terms of programming environment as well as platform.

3.1 Assumptions

We assume that the home network uses a virtual private network (VPN), and that the residential gateway uses ADSL with dynamic IP addressing. Further, each sentence is assumed to have only one service object (or IA) and various attribute values, i.e., we do not consider instructions that involve multiple services.

3.2 Definition of Methods for the IAs

In this section, we describe the implementation of our approach. The method is divided into three parts. In the case of a direct action, the method name is the verb describing that action. In the case of the modification of a member variable, the method name is the concatenation of 'set' and the full name of the member variable (e.g., setAngle, setTemperature). Similarly, when a user wants to find out the value of a member variable, the method name is the concatenation of 'get' and the full name of the member variable (e.g., getTemperature, getVolume). Table 1 lists the interface classes of example appliances.

Table 1. The List of the interface classes

Service	Method
Lamp	turnOn()
	turnoff()
Boiler	switchOn()
	switchOff()
	setTime(Calendar dateTime)
	Calendar getTime()
	setTemperature(float degrees)
	float getTemperature

3.3 Mobile Messenger Agent

The MMA is the agent that is executed on a personal computer or the homeowner's portable devices, such as a PDA, mobile phone, or smart phone, and is used to indirectly control home appliances connected to the home network. Because the proposed method is based on sentences, the MMA can use a GUI of a general IM system However, in legacy systems, there is no cryptographic algorithm for protecting text messages. To solve this problem, the proposed system is designed to encode messages with the 3DES method. This key is delivered from the home server once the client's messenger is started.

3.4 Home Messenger Agent

The HMA is the agent that acts as the interface between the MMA and the IAMA, and that manages the homeowners including any additions to the homeowner group. When there is a user request to add a principal to the forward list of the HMA, the HMA allow the candidate for acquiring user's nickname, immediately block off the candidate, and sends a message with the nickname and e-mail address to the MMA. The indoor administrator can then look through the homeowner management list.

In order to deal with user requests, this agent creates the Order Document (OD) XML code after the message received from the MMA has been analyzed, and transfers the OD to the IAMA. The OD is the context that means certain behavior information of user. This context consists of 5 Ws (who, what, where, when, and why) and 1 H (how) [13]. However, not all context factors are used but only those relevant to each purpose. The proposed method does not include 'Why', because our method does not consider the relationships between different objects, and the user is expected to directly instruct the object to do something. There is then no reason to consider the 'Why' content. Table 2 summarizes the concepts and factors in the use of 4W1H to create an OD.

Table 2. Configuration of Order Document

Behavior context	Concept	Information factor
Who	User ID	Login ID of the homeowner that transmits the OD
What	Object or device	Name of object or device to be controlled by 'who'
Where	Room	Room where object or device is located
When	Time to execute	Either scheduled time or immediate execution
How	Operation	Methods described in the event which will occur in 'what'

The first stage of the creation of the OD document by the HMA is divided into five subphases: *decodeMessage*, *parseMessage*, *selectObject*, *decideFunction*, and *generateODD*. If the message transferred from the MMA is a sentence encoded with 3DES, decodeMessage deciphers the received message, and returns a plaintext. *parseMessage* uses the lexicalized probabilistic parser [14] to create the parse tree from this text, extracts verbs and nouns from the parse tree, and sends them back. Using the noun set, selectObject decides which service the user wants to control. By using the remaining nouns, verb set, and service define document predefined by the IAMA, *decideFunction* attributes domains and methods by using a knowledge-based AI system, and re-turns the result vector. *generateODD* then generates the OD XML code.

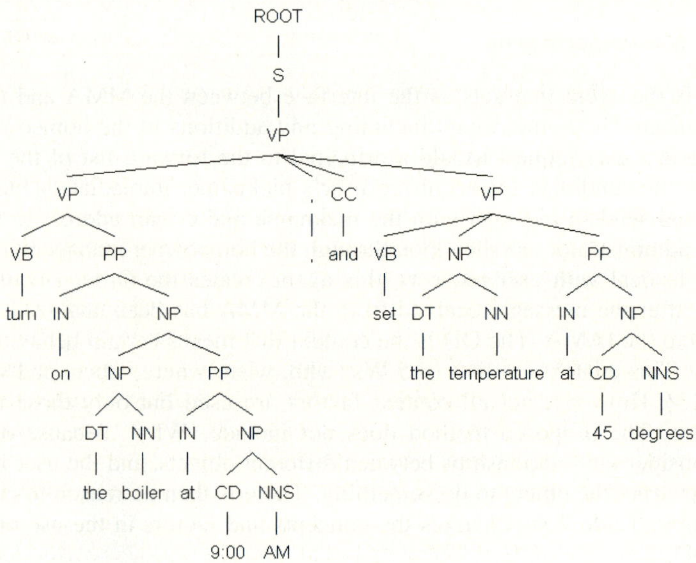

Fig. 2. Parse tree for the plaintext

For example, if the plaintext is "turn on the boiler at 9:00 AM, and set the temperature at 45 degrees," then *parseMessage* returns the following results:

selectObject searches in the NN attributes of the plaintext to determine the required service, and records the selected word, "boiler," in the 'what' item of the OD document and deletes this sub-tree. *decideFunction* searches the method and synonym tables for the word in the each NN category. In the method and synonym tables, if the word is not detected, this plaintext is in formal error. If detected, this noun is added to the 'how' list and this sub-tree is deleted. Using pattern matching, this function also decides the domain of each NP sub-tree, and clusters the NP sub-trees according to their domains. The domain decided by this stage use the value for a method. If the method to use the domain exist, it record as the parameter of the method. If it does not exist, this domain is appended to the 'how' list and the value is filled in.

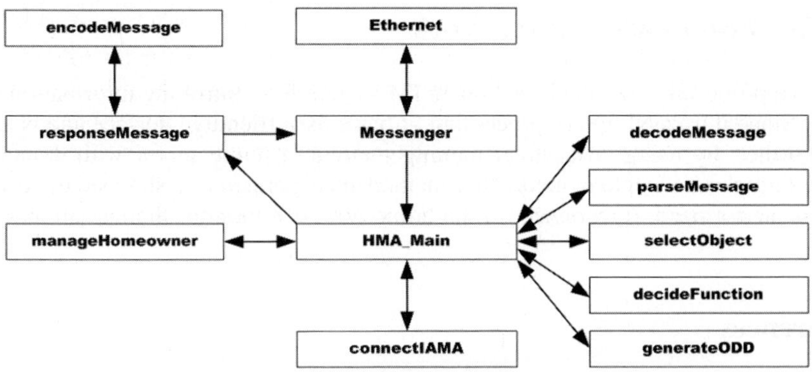

Fig. 3. The Architecture of Home Messenger Agent

3.5 Information Appliance Manager Agent

The IAMA is to create the new object necessary for connecting with middleware based on OD document transmitted from the HMA, and to monitor the appliance. Because Jini is centralized System, to download a certain service in the different device must be connected to Lookup server. To complete the proposed system, the IAMA is designed so as to include the Lookup server of Jini to directly manage the Lookup table. So, all IAs, when firstly connecting to Lookup server, is made into OD documents by the *createODD* method called by the modified register function, the *registerAppliance*, in the IAMA, and store local storage device. For checking a message, these documents should also send to HMA.

If a problem is encountered at the object created by the *manageControl* method in IAMA, it informs the homeowner of an urgent error message. Otherwise, if the task is completed, it must send a result message to the homeowner. These processes must be performed as quickly as possible. The Fig. 6 shows the architecture of IAMA.

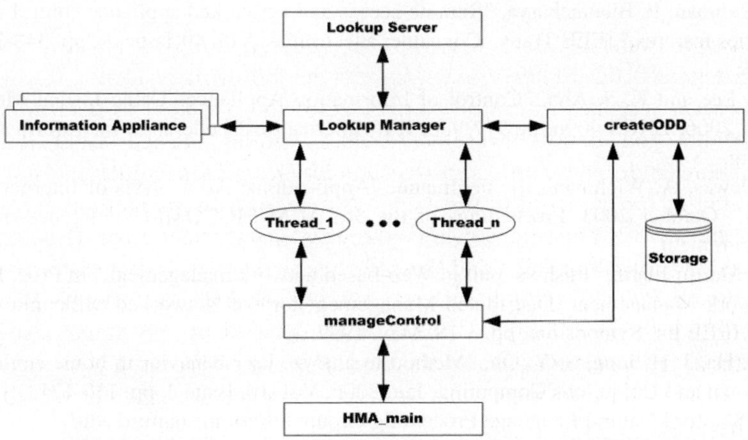

Fig. 4. The Architecture of Information Appliance Manager Agent

4 Conclusions and Future Work

We have proposed a method based on an IM system for controlling information appliances. Since it uses NL query processing, it has a user-friendly interface and is easy to use. Further, by using an abstract naming method, a home server with dynamic IP addressing can be implemented. This method incorporates a push system, so homeowners do not need to reconnect to the home server to monitor the operation of their IAs.

References

1. S. I. Kumaran, I. Kumaran, Jini technology: an overview, PHPTR, 2002.
2. "Universal Plug and Play Device Architecture Reference Specification Version 1.0," Jun. 2000. [Online]. Available : http://upnp.org/download/UPnPDA10_20000613.htm
3. B.A. Miller, B.A., T. Nixon, C. T., Tai, C., M.D. Wood, M.D., "Home networking with Universal Plug and Play," *Communications Magazine*, IEEE, Vol. 39, Issue 12, Dec. 2001 Page(s): 104 – 109.
4. "About the OSGi Service Platform technical Whitepaper Revision 4.1," Nov. 2005. [Online] Available : http://www.osgi.org/documents/collateral/TechnicalWhitePaper2005 osgi-sp-overview.pdf
5. "The HAVi Specification Version 1.1," May. 2001. [Online]. Available : http://www.havi.org
6. I. Coskun, H. Ardam, "A Remote Controller for Home and Office Appliances by Telephone," *Consumer Electronics*, IEEE Transactions on, Nov. 1998. Vol. 44, No. 4, pp. 1291-1297
7. R.S. Ivanov, "Controller for mobile control and monitoring via short message services," *Telecommunications in Modern Satellite, Cable and Broadcasting Service*, 6th International Conference on, Oct. 2003,Vol. 1, pp.108 – 111.
8. M. Nikolova, F. Meijs, P. Voorwinden, "Remote mobile control of home appliances," *Consumer Electronics*, IEEE Transactions on, Feb. 2003. Vol. 49, Issue 1, pp. 123 – 127
9. M. Rahman, P. Bhattacharya, "Remote access and networked appliance control using biometrics features," IEEE Trans. Consumer Electronics, Vol. 49(Issue 2), pp. 348-353, May. 2003.
10. S. J. Lee and K. S. Ahn, "Control of Information Appliances Using Instant Messaging," EUC 2006, LNCS 4096, pp.977-986, IFIP International Federation for Information Processing.
11. C. Dewes, A. Wichmann, A. Feldmann, "Applications: An analysis of Internet chat systems," October 2003. Proceedings of the 3rd ACM SIGCOMM conference on Internet measurement.
12. J. P. Martin-Flatin, "Push vs. pull in Web-based network management," in Proc. Integrated Network Management, Distributed Management for the Networked Millennium, the 6th IFIP/IEEE Int. Symposium, pp. 3-18, May. 1999.
13. T. S. Ha, J. H. Jung, S. Y. Oh, "Method to analyze user behavior in home environment," Personal and Ubiquitous Computing, Jan. 2006, Vol. 10, Issue 2, pp. 110-121.
14. The Stanford Natural Language Processing Group, http://nlp.stanford.edu/
15. N. S. Liang, L. C. Fu, C. L. Wu, "An integrated, flexible, and Internet-based control architecture for home automation system in the Internet era," *Robotics and Automation*, IEEE International Conference on, May. 2002, Vol. 2, pp. 1101 – 1106.

16. B.A. Miller, T. Nixon, C. Tai, M.D. Wood, "Home networking with Universal Plug and Play," *Communications Magazine*, IEEE, Vol. 39, Issue 12, Dec. 2001 pp. 104 – 109.
17. R. Lea, S. Gibbs, A. Dara-Abrams, E. Eytchison, "Networking home entertainment devices with HAVi," *Computer*, Vol. 33, Issue 9, Sep. 2000 pp. 35 - 43.
18. P. Dobrev, D. Famolari, C. Kurzke, B.A. Miller, "Device and service discovery in home networks with OSGi, " *Communications Magazine*, IEEE, Vol. 40, Issue 8, Aug. 2002, pp. 86 – 92.
19. M. Yasumura, R. Yoshida, M. Yoshida, "Prototyping and Evaluation of New Remote Controls for People with Visual Impairment," ICCHP 2006, LNCS 4061, pp.461-468
20. MSN Messenger Protocol Documentation. Downloaded from http://www.hypothetic.org/docs/msn

A Design of the Simulator for Web-Based Load Balancing

Yun Ji Na[1] and Il Seok Ko[2,*]

[1] Convergence Information Technology Research Center, 707 Seokjang-dong, Gyeongju-si, Gyeongsangbuk-do, 780-714 Korea
yjna@korea.com

[2] Division of Computer and Multimedia, Dongguk University, 707 Seokjang-dong, Gyeongju-si, Gyeongsangbuk-do, 780-714 Korea
isko@dongguk.edu

Abstract. Web services are based on a web-based system in which the performance of this system becomes a major factor in determining the quality of services.An improvement in the system performance can be achieved using a typeof load uniformity (hereinafter called load balancing) by producing a cluster system using several server systems from a web service provider's point of view. In addition, certain frequently used web objects or objects that will be frequently applied during web caching from a web service user's point of view are to be closely related to a nearby web server in clients. This study investigates a method that improves the acceleration performance in a web-based service. A method that distributes loads applied to a web system itself using a dynamic load balancing method, which uses various types of information between nearby servers, was investigated from a web service provider's point of view.

Keywords: cluster system, dynamic load balancing.

1 Introduction

Web services are based on a web-based system in which the performance of this system becomes a major factor in determining the quality of services. An improvement in the system performance can be achieved using a typeof load uniformity (hereinafter called load balancing) by producing a cluster system using several server systems from a web service provider's point of view. In addition, certain frequently used web objects or objects that will be frequently applied during web caching from a web service user's point of view are to be closely related to a nearby web server in clients. Studies from both web service provider and web service using systems' point of view are required to improve the performance of a web system.

* Corresponding author.

N.T. Nguyen et al. (Eds.): KES-AMSTA 2007, LNAI 4496, pp. 884–891, 2007.

An improvement method using a load distribution method presents an effective way to improve the performance rather than an improvement in the performance of an individual system itself from a web service provider system's point of view. This study configures a group using nearby servers that are geographically close to clients using a dynamic load balancing method and proposes a load balancing method that adaptively distributes users' loads to this grouped server system. The proposed technique reflects certain geographical problems and various costs that occur in several servers in order to distribute loads, which are applied to a web server. In addition, this technique can reduce costs by sharing a server and improve the processing capability of a server and response time for the overload applied in a server.

This study investigates a method that improves the acceleration performance in a web-based service. A method that distributes loads applied to a web system itself using a dynamic load balancing method, which uses various types of information between nearby servers, was investigated from a web service provider's point of view.

2 Methods of a Load Balancing Test

Round-robin scheduling, weighted round-robin scheduling, and other various algorithms have been applied as a load balancing algorithm [1, 2, 3, 4]. However, there are some difficulties in the use of these algorithms in a web system because it doesn't reflect geographical information between servers.

Round-Robin Scheduling method uniformly distributes the user request to servers one after another in which a load balancer uniformly connects the new request of clients to servers. This method has the merit that it presents a fast processing in general configurations by managing all servers in a group as a same manner regardless of the sever connection acceptance response time compared to other algorithms. Although the Round-robin DNS analyzes a single domain as a different IP, the basis of scheduling is based on a host. In addition, it is difficult to effectively use the algorithm due to the caching. Thus, a serious dynamic load imbalance between real servers may occur.

Four different tests were used to evaluate the performance as follows.

1) Deterministic Modeling

This method applies an analytic evaluation method using work loads. In this method, factors that affect the performance, such as average waiting times, can be evaluated. This method presents an easy and fast evaluation but has the demerit that the precise estimation of work loads is difficult.

2) Queueing Models

A queueing analysis method can be used as an analytical method to analyze the performance. This method uses a type of probability distribution function, such as uniform, Gaussian, Poisson, exponential, and Rayleigh functions. Although this method is an adequate means to compare certain algorithms, it has the demerit that the process is very complicate due to their mathematical characteristics and presents a non-realistic result in arrival and processing distributions.

3) Simulations

Simulations are the most general method to evaluate the performance. A random number generator may be used to generate tasks. However, a simulation can be performed using a certain scenario, which is similar to a real operation environment, as a test vector.

4) Implementations

This method was implemented using a coding process and obtained certain datas during that process. Although this method presents a reliable result, it has a disadvantage of high costs.

Thus, it is necessary to properly mix these methods in order to evaluate the performance in an actual test. This study uses both simulation and implementation methods in a test. Figure 1 presents methods used in a test. The test applied to evaluate the performance of this study designs a structure of test vectors and implements it using a coding process. Then, a test module is designed and implemented to verify these test vectors. The test module outputs the result of the test that is applied to a load balancing test based on LSG_i and LS_i, respectively, for the input test vector and response times.

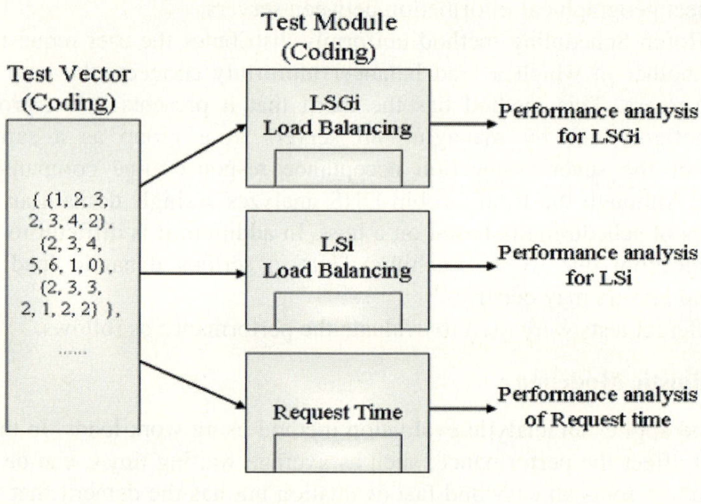

Fig. 1. Test methods

3 Data Expression and Calculation

This study proposes a structure expression in a load balancing and test vector expression method for test data. The proposed method has the merit that it can modify and extend for the change in network topologies.

1) Expression of LS_i

The ith processor can be expressed as Pi if the number of total processors are n where the range of i is $1 \leqq i \leqq n$. Each processor in this distribution system has their own group, and each group consists of several streams. Coding methods used in a simulation to express the stream are binary coding, tree coding, and program coding.

This paper presents the load information of n LS_i as a binary-coded vector in order to execute a simulation. Table 1 presents the structure of a binary-coded vector used in this simulation.

Table 1. Structure of binary vector

LS1	LS2	LS3	LS4	LS5	LS6	LS7

The load state L_i of LSi expresses $V(L_i)$ which is 2-dimensional vector-form. Being the number of LS is 7, the simulation example have $7 \leqq n$ matrix-form as belows.

$$V(L_i) = \{ \{v_{11}, v_{12}, \ldots, v_{1n-1}, v_1 n\},$$
$$\{v_{21}, v_{22}, \ldots, v_{2n-1}, v_2 n\},$$
$$\ldots\ldots$$
$$\{v_{m1}, v_{m2}, \ldots, v_{mn-1}, v_{mn}\},$$
$$\}$$

The notation n and m can express as follows.

$1 \leqq n \leqq s$, s expresses the number of local servers.

$1 \leqq m \leqq t$, t expresses the number of test vectors.

So $V(L_i)$ in the simulation example have 7 X 1 matrix as follow form.
$V(L_i)= \{ \{3,3,2,1,4,2,2\} \}$

2) Expression of $C_{distance}$ and $C_{traffic}$, δ

$C_{distance}$ and $C_{traffic}$ can express $V(C_{distance})$ and $V(C_{traffic})$ those are test vector expression of themselves.

$$V(C_{distance}) = \{ \{v_{11}, v_{12}, \ldots, v_{1n-1}, v_{1n}\},$$
$$\{v_{21}, v_{22}, \ldots, v_{2n-1}, v_{2n}\},$$
$$\ldots\ldots$$
$$\{v_{m1}, v_{m2}, \ldots, v_{mn-1}, v_{mn}\},$$
$$\}$$

$V(C_{traffic}) = \{ \{v_{11}, v_{12}, \dots, v_{1n-1}, v_{1n}\},$
$\{v_{21}, v_{22}, \dots, v_{2n-1}, v_{2n}\},$
......
$\{v_{o1}, v_{o2}, \dots, v_{on-1}, v_{on}\},$
$\}$

And data structure of the performance index δ of each server have n≦1 matrix structure. n is the number of local server.

$V(\delta) = \{ \{v_{11}, v_{12}, \dots, v_{1n-1}, v_{1n}\} \}$

n and m, o are as follows.
1≦n≦s, s is number of local server
1≦m≦t, t is number of test vector of $C_{distance}$
1≦o≦r, r is number of test vector of $C_{traffic}$

So the data structures of simulation example are as follows.

$V(C_{distance}) = \{ \{0, 1, 2, 3, 4, 5, 6\} \}$

$V(C_{traffic}) = \{ \{0, 3, 4, 1, 1, 3, 2\} \}$

$V(\delta) = \{ \{1.6, 1.2, 1.4, 1.5, 1, 1.1, 1.5\} \}$

3) Calculation of data

As previously mentioned, four information tables, $V(L_i)$, $V(C_{distance})$, $V(C_{traffic})$, and $V(\delta)$, for the state of a server can be used to calculate a table type of vector values, $V(C_{local})$, for the transmission cost, C_{local}, according to the geographical location as follows.

$V((C_{local}) = V(C_{distance}) + V(C_{traffic})$

In addition, the vector value, $V(C_{load})$, of the load for each local vector, C_{load}, can be calculated using the following equation.

$V(C_{load}) = V(L_i) \times V(\delta)$
$V(C_{total}) = V(C_{local}) + V(C_{load})$

We can present the algorithm for this operation.

Algorithm3)
input : $V(L_i)$, $V(C_{distance})$, $V(C_{traffic})$, $V(\delta)$
n = number of local server
m = number of test vector
output : $V(C_{local})$, $V(C_{load})$, $V(C_{total})$

```
procedure run_make_V(C_local)()
{
  for k = 1 to n
  {
    for l = to m
      V(C_local)(k,l) = V(C_distance)(k,l) + V(C_traffic)(k,l)
  }
}

procedure run_make_V(C_load)()
{
  for k = 1 to n
  {
    for l = to m
      V(C_load)(k,l) = V(L_i)(k,l) * V(δ))(k)
  }
}

procedure run_make_V(C_total)()
{
  for k = 1 to n
  {
    for l = to m
      V(C_total)(k,l) = V(C_local)(k,l) + V(C_load)(k,l)
  }
}
```

4 Configuration of a Test Module

Figure 2 presents the configuration of a module used in this test. Each local server (LS) generates Requests (run_request()) and calculates basic loads (Execute load_value()) using their own load information (float load_state_value). In addition, the return time information can be managed to calculate the return time of Requests (Execute turn_arround_time()).

① Request Manager

A Request Manager is a module that processes users' Requests in which it receives users' Requests (Accept Requests()) and selects a server to process the request (Choose Server()) based on the results of the process, and then it finally processes the Request (run request_processing()) or load distribution (run redirect()).

② Load Manager

A Load Manager is a module that manages loads of a server using the information of a state table and provides information to determine a server to process Requests, which are to be processed by a Request Manager.

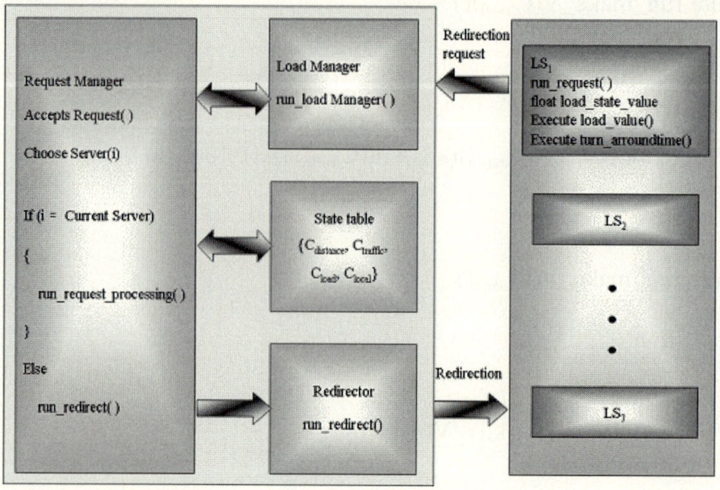

Fig. 2. Configuration of a test module

Table 2. Present the functions of these modules

Table 2 Functions of the module	
Module	**Function**
Request Manager	Processing users' Requests
Load Manager	Managing loads of a server using the information of a State Table
Redirector	Load distribution
state Table	Managing various values in a table

③ Redirector

A Redirector is a module that distributes loads if it is necessary.

④ State Table

A State Table is a module that manages various values in a table to distribute loads.

5 Conclusion

This study investigates a method that improves the acceleration performance in a web-based service. A method that distributes loads applied to a web system itself using a dynamic load balancing method, which uses various types of information between nearby servers, was investigated from a web service provider's point of view.

References

[1] Il Seok Ko, Choon Seong Leem, "An Improvement of Response Time for Electronic Commerce System," Information Systems Frontiers, vol.6, no.4, pp.313-323, 2004.

[2] V. Cardellini, M. Colajanni, P. S. Yu "Dynamic Load Balancing on Web-Server Systems," IEEE Internet Computing, pp.28-39, May, 1999.

[3] Cardellini V., Colajanni M., Yu P. S., "Redirection Algorithms for Load Sharing in Distributed Web-server Systems," Distributed Computing Systems, Proceedings, 19th IEEE International Conference, pp.528-523, 1999.

[4] Jian Liu, Longlu Xu, Baogen Gu, Jing Zhang, "A Scalable High Performance Internet Cluster Server," High Performance Computing in the Asia-Pacific Region 2000. Proceedings, The Fourth International Conference/Exhibition, vol.2, pp.941-944, 2000.

[5] Kangasharju J., Ross K. W., "A Replicated Architecture for the Domain Name System," INFOCOM 2000, Nineteenth Annual Joint Conference of the IEEE Computer and Communications Societies, vol.2, pp.660-669, 2000.

[6] Il Seok Ko, Yun Ji Na, Choon Seong Leem, "ACASH: An Adaptive Web Caching Methid with Heterogenity od Web Object and Reference Characteristics," Journal of KISS: Information networking, vol.31, no.3, pp.305-313, 2004.

Remote Control of Smart Homes Using Korean Natural Language Processing

Seong Joon Lee, Seung Woon Kim, and Kwang Seon Ahn

Department of Computer Engineering, Kyungpook National University
1370 Sankyuk-dong, Buk-gu Daegu 702-701, Korea
{imggaibi,zeroth79,gsahn}@knu.ac.kr

Abstract. Recently, with the rapid growth of digital and Internet technologies, the demand for using the Internet in the home have increased. Also, the development of various Internet-based remote controller for smart homes have been prompted, most of which adopt a web interface between people and machines. In this paper, we proposed new user interface (UI) for smart homes with natural language processing (NLP) agent. The aim in the proposed method is to provide the homeowner with the naturalness of the communication for the end-user. In the proposed method, the home server shows to be able to remotely control and monitor appliances, after analyze the plaintext, by using the natural language. Therefore, using the proposed method, almost all appliances of smart homes can be effectively managed through the integrated UI.

1 Introduction

A smart home (SH) system consists of a home network, a home server, and a remote controller. The home network is an internal network for sharing resources and enabling communication between the various appliances and services in the SH [10]. Communications within a home network rely on middleware to mediate between separate units; implementations of home networks have typically utilized well-known middleware technologies such as Jini [1], uPnP [2], OSGi [4], and HAVi [5]. Such networking can be achieved with appliances, known as information appliances (IAs) or network appliances, which have the computational power to perform predefined functions and the ability to share information. The home server acts as a gateway connecting the home network to the Internet and as a server to manage the appliances within the home. When the home server is connected to the Internet, the IAs within the network can be remotely controlled [10]. The devices used to control and monitor home appliances from distant locations, known as remote controllers, can be used either when the user is at home or away from home. Remote controllers can be used both to monitor the applications on the network, and to change their status.

Advances in Internet and digital technologies have prompted the development of various Internet-based remote controllers for smart homes, most of which adopt a Web-based interface between people and machines. Web used Client-Server model carry out the important role that can overcome the difference of communication between operating systems or program languages. Also, the user interface provides the

N.T. Nguyen et al. (Eds.): KES-AMSTA 2007, LNAI 4496, pp. 892–900, 2007.

graphic user interface (GUI) that is easy and convenient to the user. However, due to slow and inconsistent transmission rates, which result in systems instability, the Internet has not been adopted in real time environment. In the various environments, the more the efficient user interfaces is required, the more the programming code is complicated. Generally, because there are not user-friendly interfaces, the difficulties are particularly acute for the elderly. Also, the users already decide what to do before control the device. This determination can rewrite the order, which hierarchically clicked menus, to single sentence.

As using the system with Natural Language Processing, we propose a cost-effective system that uses natural language processing (NLP) to improve the efficiency and accuracy of the user interface. The proposed method applies NLP module to legacy Web-based System.

2 Related Work

2.1 The Problem of Previous System

The user interface (UI) of a phone-based system makes use of dual tone multiple frequency (DTMF) [6], the wireless application protocol (WAP) [8], or the session initiation protocol (SIP) [9]. Because these systems have the push function, the homeowner does not need to reconnect to the home server to see the state information of the IA that was directed to perform some action. However, DTMF-based UIs are poor and suffer from operational difficulties, particularly because they require a dial-up connection. In addition, the WAP does not support real Internet integration or real-time operation, and can only be used over the phone. Moreover, when homeowners use multimedia in conjunction with the SIP, the diversity problem may arise, whereby the homeowners have diverse terminals according to environment and convenience. In this situation, media negotiation (MN) is required because each device has a different communication and processing power. Although the SIP already has MN defined by IETF (Internet Engineering Task Force), it expends considerable time on the round trip between clients. The WAP or SIP can, however, be used as the protocol for instant messaging systems (IMSs).

In the paper [10], HASMuIM have proposed a novel method using an IM system that is lightweight and supports real-time processing. This method can receive modified state information to occur at an appliance without reconnecting to Home Server. One possible disadvantage of this approach is the production of many data packets that transfer changes in the state information around the home network. Then when homeowners reconnect to the server, they may receive many such packets. However, the total number of packets can be reduced by instructing the server to send the metadata list of updated appliances, i.e., the set of name of the modified appliance, only when the homeowner requires it. When homeowners interact with their home network, most are only interested in controlling one device.

2.2 Question Answering Systems

In NLP, the computer processes human language and returns a response that people expect [11]. NLP methods typically accept a sentence, generate appropriate parsing trees, and translate it into defined structures. When other technologies are combined with NLP, the user interface can be simple, and provide powerful and uniform tools for carrying out the user's desired operations.

Our proposed system is similar to that of other QA systems. A user's command or query written in NL arrives at the home server. The home server analyzes the delivered message with shallow NLP technology that does not use semantic analysis, and then creates a new object using the products (or keywords). This object is then supplied to a Jini network. Through the creation of this object, users can control all the Jini-enabled devices.

2.3 Jini Network Technology

Jini consists of a set of APIs and high level network protocols for distributed systems. It provides a simple way to perform tasks in home networks such as discovering, registering, and removing devices and services. Jini creates software infrastructure, called a federation of services, which shares access to the services and engages in interactions without the prior knowledge of the other systems or any need for human intervention [1]. In the next section, we propose a method for controlling information appliances in a home network that uses an IM system.

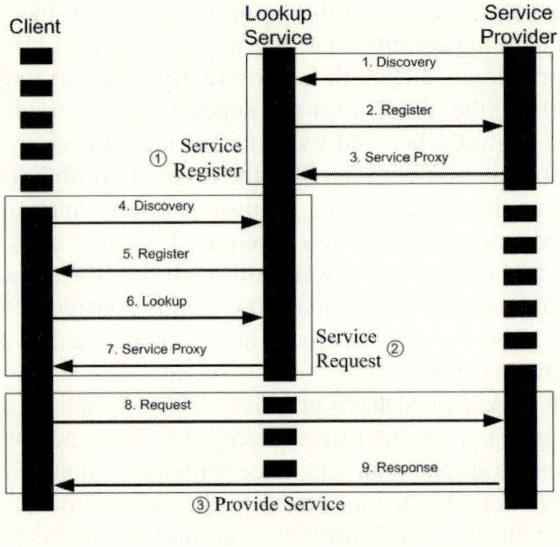

Fig. 1. Jini Technology Process

The system for enabling a Jini service consists of three protocols—discovery, join, and lookup—which provide a dynamic configuration, and requires a pre-established network to function. When a service is plugged into the Jini network, it uses a multicast request to find the local lookup service through the discovery protocol, and then registers the proxy service object in the lookup table via the join protocol (step 1 in Fig. 1). Clients can also use the discovery protocol to find the lookup service. Subsequently, when a client requests a search for a service, the lookup service returns matching proxy service objects to the client (step 2). Finally, the object downloaded from the lookup server communicates directly with the service provider (step 3) [1].

3 Smart Home Management System

In this paper, we present an efficient method, named Smart Home Management System (SHMS), for home automation systems, which uses Jini network technology and NLP method. We assume that each sentence is assumed to have only one service object (or IA) and various attribute values. Although the proposed system is not provided with a very impressive performance, it provides usability and user-friendly interface.

3.1 Define of Appliances Method

In this section, we describe the implementation of our approach. The method is divided into three parts. In the case of a direct action, the method name is the verb describing that action. In the case of the modification of a member variable, the method name is the concatenation of 'set' and the full name of the member variable (e.g., setAngle, setTemperature). Similarly, when a user wants to find out the value of a member variable, the method name is the concatenation of 'get' and the full name of the member variable (e.g., getTemperature, getVolume). Table 1 lists the interface classes of example appliances.

Table 1. Interface classes of example appliances

Service	Method
Lamp	turnOn()
	turnoff()
Boiler	switchOn()
	switchOff()
	setTime(Calendar dateTime)
	Calendar getTime()
	setTemperature(float degrees)
	float getTemperature

3.2 The Client Browser

The follow figure 2 is the client browser for homeowner. The homeowner is two ways to control and manage appliances networked a smart home. One is the way to click the icon of device to control, and fill in the control box (Fig. 2). Another is the way to directly write the plaintext made by Korean language in the textbox in the bottom of the browser (Fig. 3).

Fig. 2. The Legacy Method

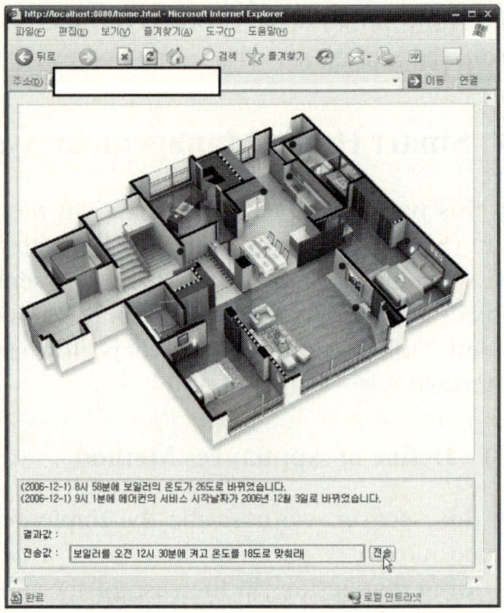

Fig. 3. The Expanded Method

3.3 The Home Server

The Home Server is the agent that acts as the interface between the browser and the IAMA. The functionality of this agent is to:

1. If the data transformed from the client is the plaintext, analysis the plaintext.
2. Create the Order Document for IAMA and send OD to IAMA

In order to deal with user requests, this agent creates the Order Document (OD) XML code after the message received from the MMA has been analyzed, and transfers the OD to the IAMA. The OD is the context that means certain behavior information of user. This context consists of 5 Ws (who, what, where, when, and why) and 1

Table 2. Configuration of Order Document

Behavior context	Concept	Information factor
Who	User ID	Login ID of the homeowner that transmits the OD
What	Object or device	Name of object or device to be controlled by 'who'
Where	Room	Room where object or device is located
When	Time to execute	Either scheduled time or immediate execution
How	Operation	Methods described in the event which will occur in 'what'

H (how) [13]. However, not all context factors are used but only those relevant to each purpose. The proposed method does not include 'Why', because our method does not consider the relationships between different objects, and the user is expected to directly instruct the object to do something. There is then no reason to consider the 'Why' content. Table 2 summarizes the concepts and factors in the use of 4W1H to create an OD.

The first stage for the creation of the OD document call generateODDDoc agent. This agent is divided into five subphases: *parseMessage*, *selectObject*, *decideFunction*, and *generateODD*. *parseMessage* uses the KLT(Korean Language Technology) library, Korean lexical analyzer, [15] to create the parse tree from this text, extracts verbs and nouns from the parse tree, and sends them back. Using the noun set, *selectObject* decides which service the user wants to control. By using the remaining nouns, verb set, and service define document predefined by the IAMA, *decideFunction* attributes domains and methods by using a knowledge-based AI system, and re-turns the result vector. *generateODD* then generates the OD XML code.

For example, if the Korean plaintext is "BoilerReul Ojun 12si 30boonE KyeGo OndoReul 18doRo MajchuRa," then *parseMessage* returns the following results:

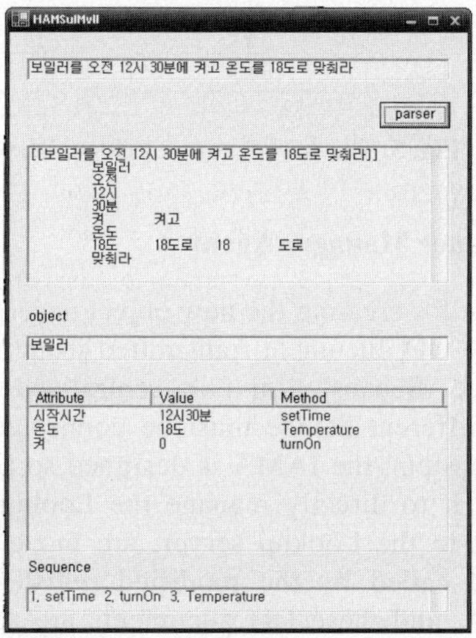

Fig. 4. Parser Tree of Korea Plaintext

selectObject searches in the noun attributes of the plaintext to determine the required service, and records the selected word, "boiler," in the 'what' item of the OD document and deletes this sub-tree. *decideFunction* searches the method and synonym tables for the word in the each noun category. In the method and synonym tables, if the word is not detected, this plaintext is in formal error. If detected, this noun is added to the 'how' list and this sub-tree is deleted. Using pattern matching, this function also decides the domain of each noun attribute, and clusters the noun attribute according to their domains. The domain decided by this stage use the value for a method. If the method to use the domain exist, it record as the parameter of the method. If it does not exist, this domain is appended to the 'how' list and the value is filled in.

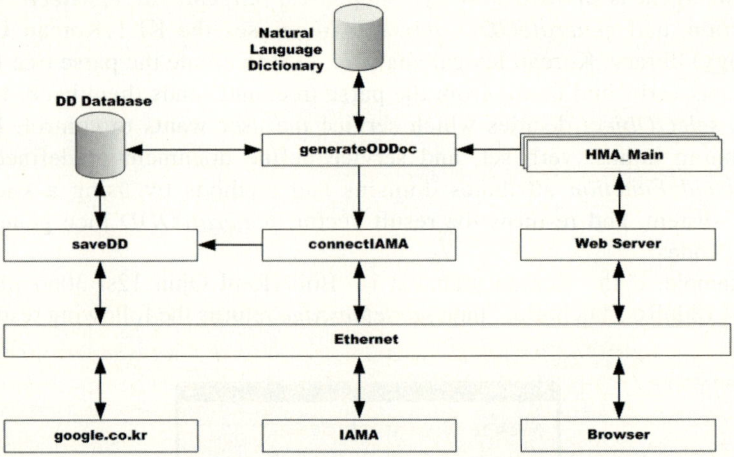

Fig. 5. The Architecture of Home Server

3.4 Information Appliance Manager Agent

The IAMA is responsible for creating the new object necessary for connecting with middleware based on the OD document transmitted from the home server, and for monitoring the appliance. Because Jini is a centralized system, to download a certain service on the different device must be connected to Lookup server. To complete the proposed system, the IAMA is designed so as to include the Lookup server of Jini, allowing it to directly manage the Lookup table. Hence, all IAs, when they first connect to the Lookup server, are made into OD documents by the *createODD* method called by the modified register function, the *register Appliance*, in the IAMA, and these OD documents are stored on a local storage device. For checking a message, these documents should also be sent to the home server.

If any problem is encountered at the object created by the *manageControl* method in the IAMA, the agent sends the homeowner an urgent error message. Otherwise, if the task is completed, a result message is sent to the homeowner. These processes must be performed as quickly as possible. Fig. 6 shows the architecture of the IAMA.

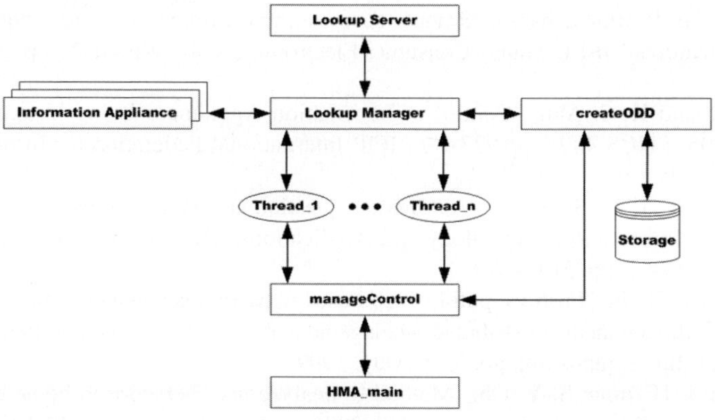

Fig. 6. The Architecture of Information Appliance Manager Agent

4 Conclusions and Future Work

We have proposed a method based on Web technology and natural language processing (NLP) for controlling information appliances. Since it includes natural language query processing, it has a user-friendly interface and is easy to use. Directions for future research include adding expansion functions for the voice remote control (RC), and verifying the universality of our RC by adapting it to the different middlewares. Although the proposed system is not provided with a very impressive performance, it provides usability and user-friendly interface.

References

1. S. I. Kumaran, I. Kumaran, Jini technology: an overview, PHPTR, 2002.
2. "Universal Plug and Play Device Architecture Reference Specification Version 1.0," Jun. 2000. [Online]. Available : http://upnp.org/download/UPnPDA10_20000613.htm
3. B.A. Miller, B.A., T. Nixon, C. T., Tai, C., M.D. Wood, M.D., "Home networking with Universal Plug and Play," *Communications Magazine*, IEEE, Vol. 39, Issue 12, Dec. 2001 Page(s): 104 – 109.
4. "About the OSGi Service Platform technical Whitepaper Revision 4.1," Nov. 2005. [Online] Available : http://www. osgi.org/documents/collateral/ TechnicalWhitePaper2005 osgi-sp-overview.pdf
5. "The HAVi Specification Version 1.1," May. 2001. [Online]. Available : http://www. havi.org
6. I. Coskun, H. Ardam, "A Remote Controller for Home and Office Appliances by Telephone," *Consumer Electronics*, IEEE Transactions on, Nov. 1998. Vol. 44, No. 4, pp. 1291-1297
7. R.S. Ivanov, "Controller for mobile control and monitoring via short message services," *Telecommunications in Modern Satellite, Cable and Broadcasting Service*, 6th International Conference on, Oct. 2003,Vol. 1, pp.108 – 111.
8. M. Nikolova, F. Meijs, P. Voorwinden, "Remote mobile control of home appliances," *Consumer Electronics*, IEEE Transactions on, Feb. 2003. Vol. 49, Issue 1, pp. 123 – 127

9. M. Rahman, P. Bhattacharya, "Remote access and networked appliance control using biometrics features," IEEE Trans. Consumer Electronics, Vol. 49(Issue 2), pp. 348-353, May. 2003.
10. S. J. Lee and K. S. Ahn, "Control of Information Appliances Using Instant Messaging," EUC 2006, LNCS 4096, pp.977-986, IFIP International Federation for Information Processing.
11. A. Andrenucci, E. Sneiders, "Automated question answering: review of the main approaches," Information Technology and Applications, Third International Conference on, July 2005, Vol 1, pp. 514 – 519.
12. J. P. Martin-Flatin, "Push vs. pull in Web-based network management," in Proc. Integrated Network Management, Distributed Management for the Networked Millennium, the 6th IFIP/IEEE Int. Symposium, pp. 3-18, May. 1999.
13. T. S. Ha, J. H. Jung, S. Y. Oh, "Method to analyze user behavior in home environment," Personal and Ubiquitous Computing, Jan. 2006, Vol. 10, Issue 2, pp. 110-121.
14. N. S. Liang, L. C. Fu, C. L. Wu, "An integrated, flexible, and Internet-based control architecture for home automation system in the Internet era," *Robotics and Automation*, IEEE International Conference on, May. 2002, Vol. 2, pp. 1101 – 1106.
15. Korean morphological analyzer, Natural Language Processing & Information Retrieval Lab, http://nlp.kookmin.ac.kr/

A VDS Based Traffic Accident Prediction Analysis and Future Application

Chungwon Lee[1], Bong Gyou Lee[2], Kisu Kim[3], and Hye Sun Lee[2]

[1] University of Seoul, Dept of Transportation Engineering
90 Jeonnong-Dong Dongdaemun-Gu, Seoul 130-743 Korea
[2] Graduate School of Information, Yonsei University
134 Shinchon-Dong, Seodaemun-Gu, Seoul, 120-749, Korea
[3] Korean Institute of Construction Technology, Highway Research Department
2311 Daewha-Dong Ilsan-Gu Goyang-Si Gyeonggi-Do 411-712 Korea
chungwon@uos.ac.kr, bglee@yonsei.ac.kr, kisu9525@kict.re.kr, emailme@yonsei.ac.kr

Abstract. The purpose of this study is to present the prediction analysis of traffic accident and future applications by calculating accident risks based on ITS (Intelligent Transport Systems) data via VDS (Vehicle Detecting System). Many researches have described various traffic accident predictions and reductions. However, until recently, empirical studies considering real traffic environments are still insufficient. Site selections in this study were made from accident related data and collected data from VDS to calculate traffic accident rates. Results from analysis showed occurrence of traffic accident by time and explore possibilities for its application in future traffic accident management.

Keywords: traffic accident, prediction analysis, accident risk rate, VDS (Vehicle Detecting System), ITS (Intelligent Transport System).

1 Introduction

Traffic accidents along with the automobile growth have been one of the serious social issues in Korea. The traffic accident rates have reached into 220,755 cases within a year, year 2004 caused millions of injuries and billions of social cost. With these serious circumstances, Korean government has begun to build ITS (Intelligent Transport Systems) since 1997 for reducing traffic congestions and accidents.

Decentralized traffic volume via ITS have reduced number of automobiles within managing sectors which resulted to degrade chances of accidents. However, according to the RQCM (Rate Quality Control Method), it is very difficult to find out the relationship between traffic volume and accident occurrences. Also, chances of multiple accident' occurrences at same time are almost zero except for post-traffic accidents from the first accident. Thus, traffic accident prevention plan via ITS have resulted to operate for only traffic management function through traffic controls and processes which responds after the accidents. Therefore, it is feeble to perform as traffic accident management function for a correspondence before the accidents occur.

N.T. Nguyen et al. (Eds.): KES-AMSTA 2007, LNAI 4496, pp. 901–909, 2007.

This paper presents the prediction analysis of traffic accident and future applications by calculating accident risks based on VDS data. It consists of five parts. After Introduction, section two presents the hypothesis and method for analyzing traffic accident risk rates. Section three describes how to collect and analyze VDS data. Section four explains future applications and final section draws conclusions.

2 Hypothesis and Analysis Method

Traffic components largely consist into automobiles, drivers, and roads. Throughout various accident results, previous studies claim that the cause of the accidents refers to geometric structure of the roads, traffic environments, and drivers' conditions. Therefore, accident occurrence causes can be found in traffic components which can be replace into traffic conditions of automobile malfunctions, drivers' conditions, and traffic environments as Fig.1.

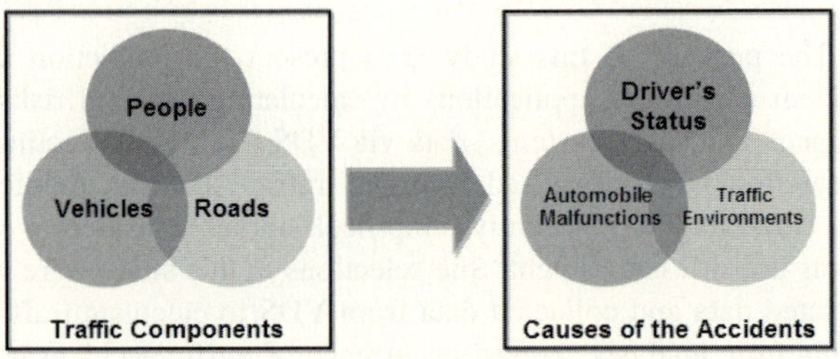

Fig. 1. Cause of the traffic accidents

Accidents due to the drivers' conditions or automobile malfunction are absolute for their solutions. However, indefinite traffic circumstances are hard to confront preliminary accidents. Therefore, it is necessary to analyze accident risk rates base on indefinite circumstances like driving behaviors in order to reduce the risks.

In traffic condition research, physical element, such as the relationship between geometric structure of road and accident has been made clear; under same circumstance accident dangerousness is related to traffic characteristics (driving behavior). Also, in geometric structure related research the method used to evaluate accident dangerousness is done through use of accident data and treats dangerousness and number of cases as equal which bring limitation for comparison, most of all utilization of data from post accident is not possible for considering measurements at accident prevention level. Of course there has been researches where physical elements such as vehicle characteristics and geometric structure reflected as design elements before the accident occur and latent accident dangerousness prediction have been made, but this also require hardware management of reflections made at the time of design and results can unfold differently in actual situations.

Therefore, this study with use of data previous to actual accident occurrence that considers the driver's driving behavioral characteristics and accident dangerousness warning for caution and alert as means for accident prevention could be made through data analysis. The distance between driving vehicles are located within the stop distance and for computation purposes of accident dangerousness following assumptions are made. First, during driving unexpected situations can occur; vehicle slows speed down and then stop. Second, in cases of unexpected situations the front vehicle and rear vehicle's speed decrease is consistent with driving speed. Third, the distance between front vehicles on the road driving within accident occurrence distance it further increases accident dangerousness.

In general accidents occur in forms of collision, rear-end collision, overturn etc. and dangerousness of these accidents occur when the vehicle looses its steering capability and run off the road or spin. This is contributed by mechanical, physical characteristics or cases where outside force is transmitted and when such situations occur actual accident takes place.

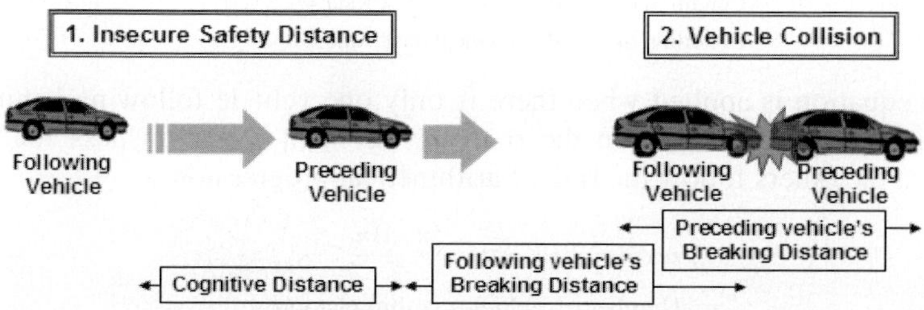

Fig. 2. Process of the Vehicle Collision

In this study accident dangerousness is determined base on unsecured safety distance, the safety distance mentioned here is not the legal standard but as shown in the drawing is a collision probability from comparison of between distance of the front vehicle and rear vehicle and stop distance.

Different from established loop detector the domestically introduced VDS that was installed with ITS has special feature called image recognition which can collect traffic length or characteristics aside from detecting traffic quantity and speed data. Therefore, with DVS you can calculate the exact between distances of vehicles in sequence as they pass, from it safety distance is calculated and accident dangerousness is assessed. With data collected from VDS the between distance of front vehicle and the rear vehicle is calculated by following equation.

$$d_0 = v_1 \times t_{(2-1)} - (l_1 + l_2) \tag{1}$$

d_0: Distance Between Vehicles (m) v_1 : Speed of Preceing Vehicle (m/s)
l_1 : Length of Preceding Vehicle (m) l_2 : Length of Following Vehicle (m)
$t_{(2-1)}$: Difference of Passing time for preciding and Following Vehicles (s)

For determining accident dangerousness the safety distance was compared with calculated between distances of vehicle from below equation.

$$d_s = v_2 \times t_{ir} + b_{d1-d2} \tag{2}$$

d_s: Safety Distance (m) v_2 : Speed of Following Vehicle (m/s)
t_{ir} : Cognitive Response Time (s)
b_{d1-d2} : Difference of Breaking Distance for preciding and Following Vehicles (m)

Applied in the above equation, result from previous study done by Fergenson (1971) of 0.47 sec. which was the minimum time for cognitive response time, based on driving speed and breaking distance from each of the front vehicle and the rear vehicle a numerical formula is computed.

$$b_{d1-d2} = \frac{v_1^2 - v_2^2}{2 \times a} \tag{3}$$

b_{d1-d2} : Distance between Vehicles

v_1 : Speed of Preceding Vehicle v_2 : Speed of Following Vehicle

a : Reduction Rate of Preceding and Following Vehicles

Accident dangerousness presented in this research is calculated from comparison of previously calculated between distances of vehicle and safety distance.

$$\begin{aligned} &\text{Condition of } d_v \geq d_s \text{, Accident Risk Rate} = 1 \\ &\text{Condition of } d_v \leq d_s \text{, Accident Risk Rate} = 0 \end{aligned} \tag{4}$$

Above equation is applied when there is only one vehicle following behind, when there is a platoon of vehicle on the road the accident dangerousness for the front vehicle and the others follow the rule of arithmetical progression.

$$accident\ \ risk\ \ rate = \frac{(n-1)n}{2} \tag{5}$$

n: Number of Vehicles within platoon

When collision occur between the front vehicle and one following vehicle accident dangerousness is one, when there are two vehicles following behind and collision occur between the front vehicle and one rear vehicle, collision between rear vehicle one and rear vehicle two, and between front vehicle and first rear vehicle, collision occur in succession and then the accident dangerousness is three. Of course, statistical method in determining probability of traffic accident occurrence and assessment of dangerousness used in this study is not desirable. The social cost of traffic accident is too large and because it involves human lives. But it would be inefficient to send out alert or warning for accident dangerousness when there is only one vehicle in a hundred that is not keeping safety distance.

The purpose of this study is to explore application of ITS for prevention of traffic accident through VDS collected data processing to determine accident dangerousness and provide alert information to the driver on road electronic board.

First, accident dangerousness level is based on detected vehicle's risk rate in percentages. In other words, if 60 cars have passed by in five minutes the computation value is 6 from accident dangerousness equation and the accident dangerousness rate for the zone and time is 10% which can be used as traffic safety management index.

3 Data Collection and Analysis

The purpose of this study was to evaluate traffic accident dangerousness through VDS, and develop measures to prevent traffic accidents by calculating relevant data. Therefore, selected site is equipped with installed ITS, site was selected where there was frequent and serious traffic accident.

First, with roads divided into express national road, general national road, local read, special/metropolitan road and city/district road other than local road all other roads have ITS installed and can become selected site. Next, for site selection based on seriousness of the accident Figure 3 is reviewed to see traffic accident by road type.

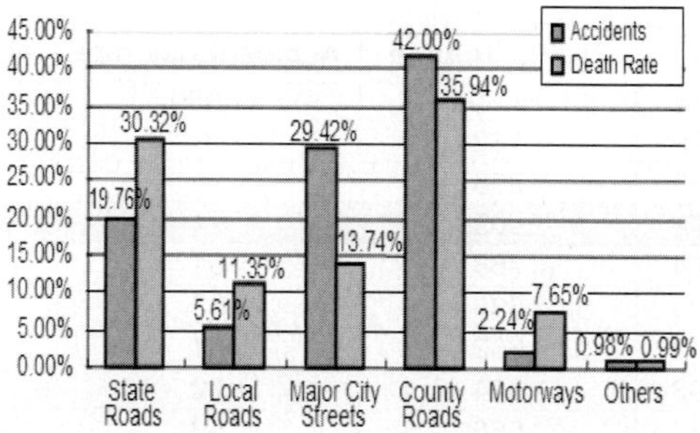

Fig. 3. Accident comparison by Road Type

Also, the examination of trends show seriousness of traffic accident with accident death rate which show express national road with highest date rate, then local and general national road showed higher numbers, followed by city/district road, special city and metropolitan city road. Considering high death toll at the time of accident on express national road, the severity of the accident is higher on local and national road. VDS data collection was conducted Figure 4 on national road 3 (Keongiam) with installment of site equipped test bed.

Fig. 4. Block diagram of data collection site

Generally one VDS installed on site has the capability of detecting all four lanes but with use of site test bed only single lane was detected. During the preparation stage, before research data collection was made VDS detection zone and processor correction was conducted to verify 95% accuracy for optimal data collection. Applicable data for research was collected after preparations were all completed at the selected site (national road 3) from Wednesday May 24, 2006 at 05:00 hours to 21:00 hours. The data collected from the selected site measured individual vehicle speed and traffic length to calculate the between distance of the front vehicle and the rear vehicle safety distance and comparison is made to calculate accident dangerousness and accident dangerousness and accident dangerousness rate is broken down by hour in Table 1.

Table 1. Accident dangerousness by hours

Time	Speed	Traffic Volume	Accident risk rate	Ratio of the Risk Rate
Total	64.94	10,694	1,050	9.82%
05am~06am	70.70	132	1	0.76%
06am~07am	73.93	338	26	7.69%
07am~08am	67.72	603	74	12.27%
08am~09am	64.24	768	91	11.85%
09am~10am	67.20	635	53	8.35%
10am~11am	66.42	662	61	9.21%
11am~12pm	64.93	624	72	11.54%
12pm~01pm	65.78	556	58	10.43%
01pm~02pm	62.21	725	70	9.66%
02pm ~03pm	63.18	821	74	9.01%
03pm~04pm	62.09	825	87	10.55%
04pm~05pm	62.05	888	106	11.94%
05pm~06pm	65.22	860	81	9.42%
06pm~07pm	60.26	905	84	9.28%
07pm~08pm	59.61	785	69	8.79%
08pm~09pm	63.42	567	43	7.58%

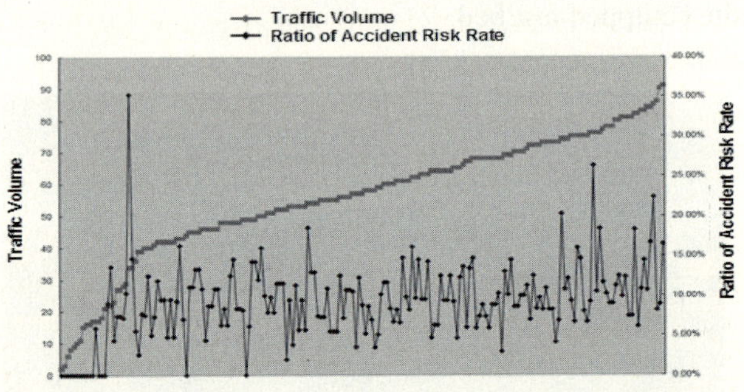

Fig. 5. Traffic volume – Relations of risk rate ratio in 5 minuets of sequences

Analysis of hourly breakdown, first noticeable fact shows that accident dangerousness is not proportional to traffic quantity. The peak traffic quantity of 905 cars was reached during evening traffic hours (18:00~19:00 hours) but the ratio between traffic quantity and accident dangerousness ratio is highest during morning traffic hours (07:00~08:00 hours). When each hourly data is converted to unite of 5 minutes and enumerated by traffic quantity the comparison of results as seen on Figure 5 are the same. Results from Table 1 and Figure 6 on vehicle speed also show that there is no correlation between traffic quantity and accident dangerousness.

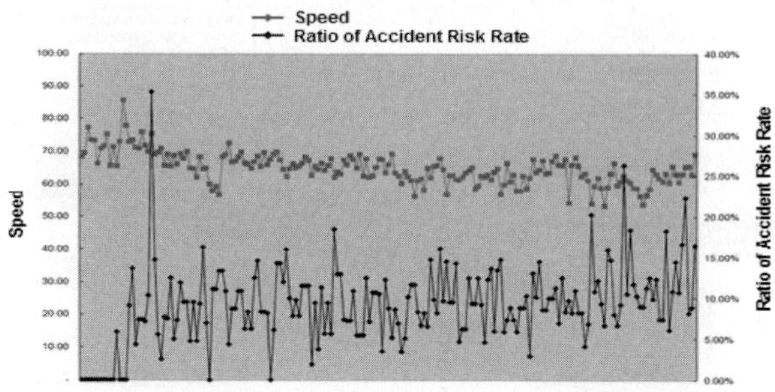

Fig. 6. Speed-accident dangerousness ratio (5min. unit)

On the other hand, as it gets closer to evening hours(17:00~21:00 hours) accident dangerousness rate decreases, it is determines that the drivers maintain safety distance as it gets closer to sunset the intensity of illumination goes down and visibility becomes limited.

From Table 1 accident dangerousness ratio that exceeds 10%, it is during these hours (07:00~09:00 hours, 11:00~13:00 hours, 15:00~17:00 hours) that we can explore safety management schemes with ITS. First, by utilizing data through out the year driving characteristics during weekdays, weekends, particular day of the week, a particular month and etc. from shown patterns, safety management is possible by providing hourly safety information on the electronic road board to drivers.

Also, consider application in providing caution notice information to the drivers for zones that have no particular patterns for weekdays, weekend, day of the week, monthly characteristics and etc. can make real time accident dangerousness assessment by evaluating the accident dangerousness with data for 5 minutes at each hour.

4 Future Application

In traffic accidents are more preventable by humans than unexpected natural disasters, such are factors that make counter measurements possible, it is the unexpected situations that actually inhibit traffic flow and contributes to traffic accident. Therefore, by analyzing road accident dangerousness and provide caution or notice information to the users, not only user safety but also can increase traffic convenience and efficiency.

This can be possible by applying it to system that will automatically keep distance with front car with IHCC (Intelligent Highway Cruise Control System), collision prevention system to minimize collision damage called CMS (Collision Mitigation break System), ACC (Automatic Cruise Control System) and etc. that support safety and auto drive.

ACC continues to develop for the sake of traffic convenience and safety, the United States and Japan are in the process of developing radar or laser applied ACC. It is thought that domestically, installed VDS for the purpose of traffic information

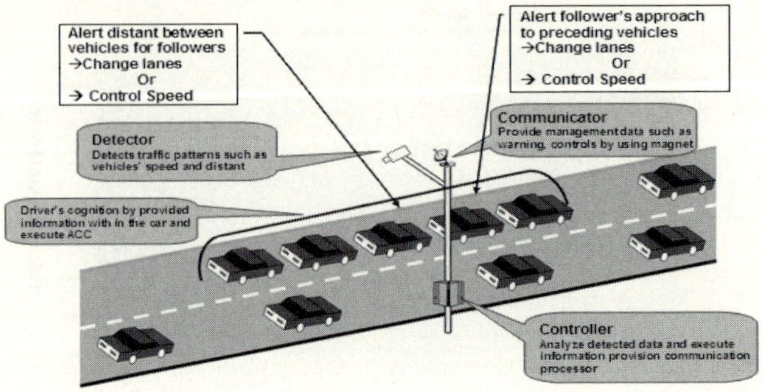

Fig. 7. Real time traffic management concept

collection site can provide real time information. Of course the site will require communication technology for traffic management to caution or alert the drivers and this can be obtained with no major investment in costs when done through utilization of adjacent communication network of grafting and establish a communication network MANET (Mobile Ad-hoc Network).

Figure 7 is ACC grafting concept to VDS utilized information with communication technology MANET. As shown in the drawing information compiled through VDS is analyzed in real time with site-equipped controller and the information is passed on to communication network through MANET to the vehicle terminal. With the provided data the driver is given caution, warning, etc. and made aware, with use of ACC control data safety management and traffic flow control is possible. ACC that graft together VDS and MANET doesn't need extra mounting of laser or radar on the car, more than anything else through overall management of traffic flow has the advantage in preventing traffic accident and increased traffic efficiency possible.

5 Conclusion

Of the causes for traffic accident, this study purpose was to come up with scheme to prevent traffic accident, it aimed traffic condition as a major cause and to access data for analysis information gathered from VDS was utilized to analysis accident dangerousness and with results from it applications for transportation operation safety management were explored. Of course, accident dangerousness rate calculated in this study is significantly higher in number than the actual rate of traffic accident rate, but suggested accident dangerousness is for accident prevention index and should not be a problem when used in actual cases since it is relative numbers and are not from actual accident data. Therefore, accident dangerousness index in this study reflects traffic condition including driving characteristics and can be used at preventative level as a basis for transportation operations safety management.

Presented method in this study in accident dangerousness index can be provided in real time through mobile ad-hoc network to build advanced road system through automatic cruise gear.

Acknowledgements

This work was supported by the Seoul Industry-University-Research Cooperation and the Brain Korea 21 Project of Ministry of Education & Human Resources Development in 2006.

References

1. D. Sun.R. F. Benekohal(2003), "Analysis of Car Following Characteristics for Estimating Work Zone Safety", Annul Meeting of the Transportation Research Board.
2. G. J. Andersen(2003), "Visual Information for Car Following by Drivers: The role of Scene Information", Annul Meeting of the Transportation Research Board.
3. Ministry of Construction and Transportation, "Development of Traffic Safety Evaluation Model for Advanced Traffic Management", (2003)
4. Paul G.(2000), "Headway on urban streets : observational data and an intervention to decrease tailgating", Transportation Research Part F.
5. Road Traffic Authority, "Statistic of Traffic in 2005", (2005)
6. Road Traffic Authority, "Comparisons of Accidents for OECD Member Countries", (2005)
7. Road Traffic Authority, "Analysis of Traffic Accidents Statistics", (2005)
8. Ruediger L.(1999), "Highway Design and Traffic Safety Engineering Handbook", McGraw-Hill.
9. Sang Woo, Nam, "Trend of Mobile Ad-Hoc Network and Wireless Access Technologies", Vol.108. Ministry of Information and Communication Radio Research Libratory, (2002)
10. Seung Gul, Pek, "Accident Prediction in Motorways Considering Traffic Volumes and Length", Vol.23. Korean Society of Transportation, (2005)
11. Seung Lim, Kang, "A GIS-based Traffic Accident Analysis on Highways using Alignment Related Risk Indices", Vol. 21. Korean Society of Transportation, (2003)

An Intelligent Diversity Scheme for Accurate Positioning of Mobile Agents for U-City*

Junseok Lee and Sekchin Chang

Dept. of Electrical and Computer Engineering, University of Seoul, Seoul, Korea

Abstract. Based on ubiquitous networks, many schemes have been proposed for realization of u-city. For abundance of u-city services to be available, mobile agents can be utilized in ubiquitous networks. However, highly accurate positioning is required for the efficient use of the mobile agents in ubiquitous networks. In this paper, an intelligent diversity scheme is proposed to enhance the positioning performance. Simulation results indicate that using the suggested diversity scheme, the positioning performance can significantly be improved in mobile agents for u-city.

1 Introduction

A lot of metropolitan cities have recently tried to offer the intelligent services such as as intelligent disaster prevention, intelligent building management, intelligent health care, and intelligent traffic control [1]. When the services are available on ubiquitous networks [2], the city which offers such services is generally called u-city. Usually, the wireless sensor node is considered a realistic basis for implementation of ubiquitous networks [3]. However, the sensor node mainly suffers from some design limitation such as low cost and low power [3, 4]. Mobile agents were introduced in wireless sensor networks to overcome the design limitation [5]. For realization of the mobile agent, the sensor node algorithm can be implemented in a cellular modem [6]. Using the implementation method, the design limitation of the sensor node can also be overcome since the node is able to utilize the resources of the cellular phone. However, highly accurate positioning of the mobile agent is required for the efficient use of the agent in ubiquitous networks. In this paper, an intelligent diversity scheme is proposed to increase the positioning accuracy. The intelligent diversity can be achieved using multiple antennas. In the intelligent scheme, transmitter diversity or receiver diversity is utilized according to the positioning procedure, which results in the improvement of the positioning performance. Simulation results indicate that the proposed diversity scheme can considerably enhance the positioning performance for mobile agents.

* This work was supported by Smart (Ubiquitous) City Consortium under Seoul R&BD Program.

N.T. Nguyen et al. (Eds.): KES-AMSTA 2007, LNAI 4496, pp. 910–917, 2007.

2 The Mobile Agents with Diversity

For the realistic implementation of the mobile agent, the sensor node module can be included in a cellular modem [6]. To achieve a diversity for the mobile agent, the sensor node can employ multiple antennas. Fig. 1 shows the cellular modem structure for the mobile agent with diversity. In addition to the structure shown in [6], multiple antennas are utilized for the sensor node as illustrated in Fig. 1, which produces a diversity gain for the mobile agent. Unlike the general structures for diversity [7], the proposed structure utilizes only one RF module for two antennas in Fig. 1, which indicates that little hardware overhead is required for the diversity gain. Therefore, for the optimal selection of one antenna, the intelligent diversity scheme is based on an antenna switching approach which will be explained in more detail in section 3.

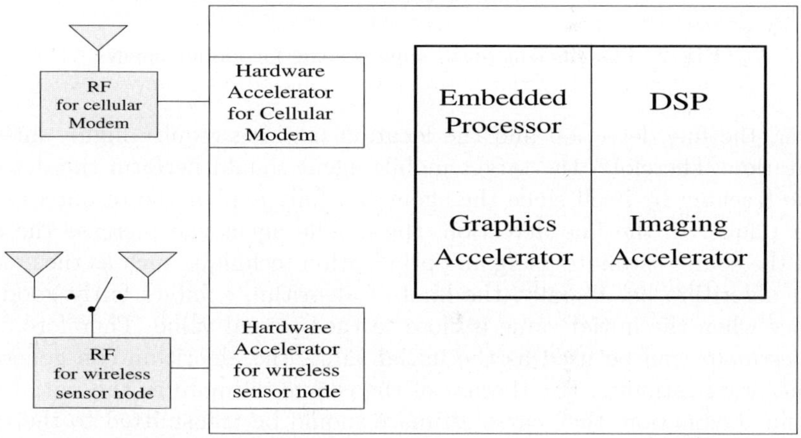

Fig. 1. The modem structure for the mobile agent with diversity

3 The Intelligent Diversity Scheme for Accurate Positioning of Mobile Agents

The mobile agent can afford various u-city services to the user in ubiquitous networks. However, the location of the mobile agent should accurately be detected and tracked to fully utilize the functionality of the agent because the agent can move in ubiquitous networks. Fig. 2 depicts an efficient positioning scheme for mobile agents [6]. The positioning scheme consists of coarse and fine location detection, and location tracking as described in [6]. As indicated in Fig. 2, the coarse detection is usually utilized by the reference sensor nodes which are selected from general wireless sensor nodes [8]. Since the coarse detection usually exhibits simple computation, the general sensor nodes can perform the detection.

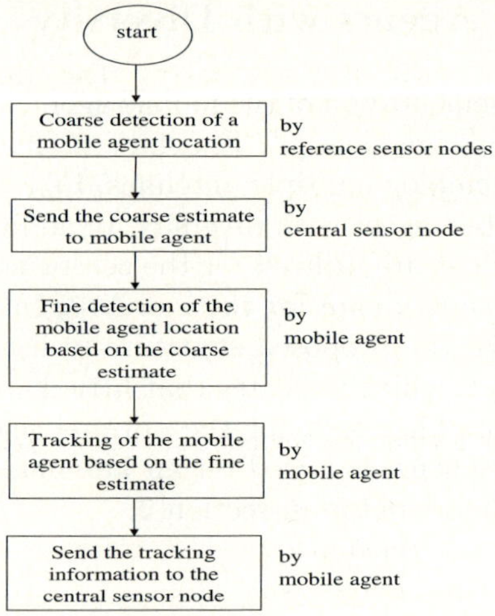

Fig. 2. The efficient positioning scheme for mobile agents

However, the fine detection and the location tracking require highly intensive computation. Therefore, the target mobile agent should perform the detection and the tracking by itself since the agent can fully exploit the resources of the cellular phone. In the fine detection, the mobile agent can increase the accuracy of the coarse estimate using an optimization technique such as the steepest descent algorithm [9]. Usually, the kind of algorithm exhibits fairly good performance when the initial value is close to the optimal value. Therefore, if the coarse estimate can be used as the initial value, the algorithm can generate a fairly accurate estimate. For the use of the coarse estimate as the initial value in the fine estimation, the coarse estimate should be transmitted to the target mobile agent by a central sensor node which is selected from the reference nodes. However, the received estimate usually includes the error due to the channel fading, which degrades the performance of the fine location detection. To overcome the performance degradation, the intelligent diversity can be utilized in the mobile agent. For the efficient utilization of the diversity when the mobile agent employs two antennas as in Fig. 1, the central node send the coarse estimate using the message format as shown in Fig. 3.

As illustrated in the figure, the message format includes location information field where the same two coarse estimation values are consecutively placed. The first coarse estimate and the second estimate are located at the discrete time-index of n_1 and n_2, respectively. As depicted in Fig. 3, the time length for each coarse estimate is N. When the central sensor node transmits the coarse estimates in the message format, the received coarse estimate signal at the kth antenna $r_k(n)$ is expressed as

$$r_k(n) = \alpha_k \cdot s(n) + \eta_k(n) \tag{1}$$

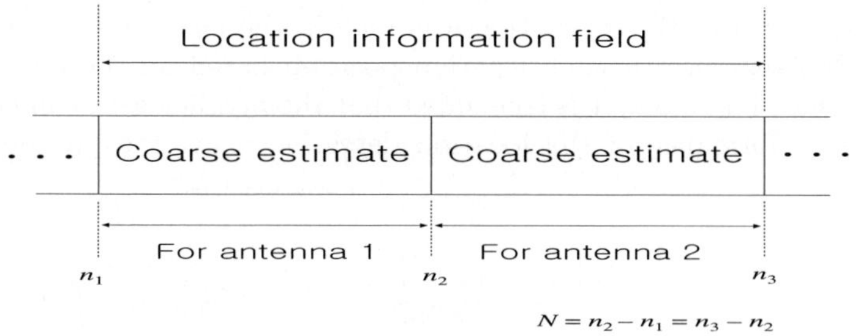

Fig. 3. The message format for the efficient utilization of the intelligent diversity

where $s(n)$ denotes the transmitted coarse estimation signal, and α_k and $\eta_k(n)$ indicate the channel parameter and additive white Gaussian noise (AWGN), respectively at the kth antenna. In (1), the flat fading is considered the channel effect. In addition, in (1) $k = 1$ when $n_1 \leq n < n_2$, and $k = 2$ when $n_2 \leq n < n_3$. In other words, the mobile agent receives the first coarse estimate at the 1st antenna, and then switches to the 2nd antenna and receives the second coarse estimate at the 2nd antenna. At the kth antenna, the power of the received estimate signal pwr_k is calculated as

$$pwr_k = E[|r_k(n)|^2] = G_k \cdot \sigma_s^2 + \sigma_n^2 \tag{2}$$

where $E[\cdot]$, σ_s^2, and σ_n^2 denote the expected value operator, the power of $s(n)$, and the power of $\eta_k(n)$, respectively. Since the power of AWGN is same for all antenna branches [7], the antenna index k can be omitted in σ_n^2 of (2). In (2), G_k indicates the channel gain at the kth antenna, and is expressed as

$$G_k = E[|\alpha_k|^2] \tag{3}$$

From (2), the signal-to-noise ratio (SNR) at the kth antenna SNR_k can be derived as follows:

$$SNR_k = \frac{pwr_k - \sigma_n^2}{\sigma_n^2} \tag{4}$$

The expression of (4) implies that the received signal (at the antenna) with higher power leads to that with higher SNR. Therefore, to achieve the coarse estimate signal with higher SNR, the mobile agent selects the estimate signal from the two received estimate signals as follows:

if $pwr_1 > pwr_2$, *then select* `the first coarse estimate signal;`
otherwise, select `the second coarse estimate signal`

Since the signal with higher SNR usually causes better receiver performance, the mobile agent can achieve higher probability for decoding the received coarse estimate signal correctly by choosing the coarse estimate signal with higher SNR. This surely increases the estimation accuracy in the location detection because the detection searches the fine location position based on the decoded coarse estimate value. Therefore, it is concluded that the mobile agent can enhance the estimation performance of the location detection using the proposed receiver diversity.

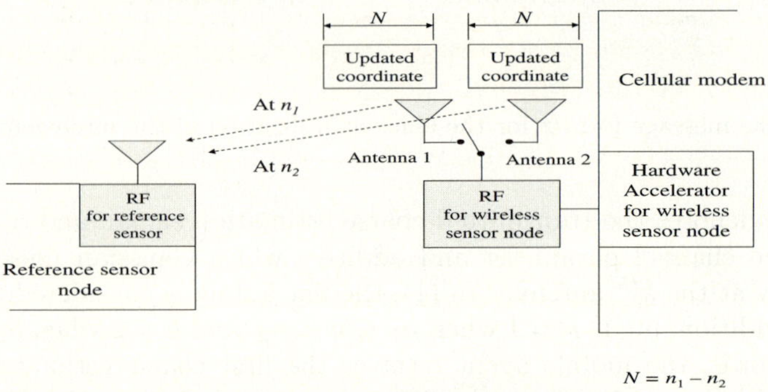

Fig. 4. The transmission of the updated estimate using transmitter diversity in mobile agents

If the mobile agent is in motion, the fine estimate should continuously be updated in the location tracking as follows [6]:

$$x^{(n+1)} = x^{(n)} + \hat{l} \cdot \triangle, y^{(n+1)} = y^{(n)} + \hat{m} \cdot \triangle \tag{5}$$

where $x^{(n)}$ and $y^{(n)}$ are the updated coordinates of the mobile agent at time index of n. In (5), $\hat{l}$ and $\hat{m}$ are the integer indices to indicate the motion direction and amount of the mobile agent. In addition, $\triangle$ denotes the coordinate incremental for tracking. Since high computation is usually required to determine $\hat{l}$ and $\hat{m}$, the mobile agent also performs the location tracking by itself. Therefore, the mobile agent should send the updated coordinate back to the central node whenever the update of (5) occurs. For the central node to decode the received coordinate more correctly, the transmitter diversity is utilized in the mobile agent as shown in Fig. 4. In other words, the mobile agent transmits the updated coordinate through the antenna 1 at the time index of n_1, and then switches to antenna 2 and transmits the same coordinate through antenna 2 at the time index of n_2. In Fig. 4, N denotes the time length of the updated coordinate data, and is defined as $n_2 - n_1$. If the central node selects the coordinate signal with higher SNR (or power) from the received 2 coordinates, the probability for correctly decoding the updated coordinate signal can be increased as in the case of the fine detection.

4 Simulation Result

Simulation results exhibit the effectiveness of the proposed intelligent diversity scheme. For the simulation, the environment of Fig. 5 [6] is considered. This figure illustrates the moving of the mobile agent from region 1 to region 2. In addition, 2.4 GHz flat fading and additive white Gaussian noise (AWGN) are assumed as the channel environments for the simulation.

For the performance evaluation of the location detection based on the diversity scheme, the mean squared error (MSE) values between exact and estimated location position are given in Fig. 6. As shown in Fig. 6, the location detection with the diversity achieves the improvement of about 1 dB over the location detection without the diversity. For investigating the effects of the proposed transmitter

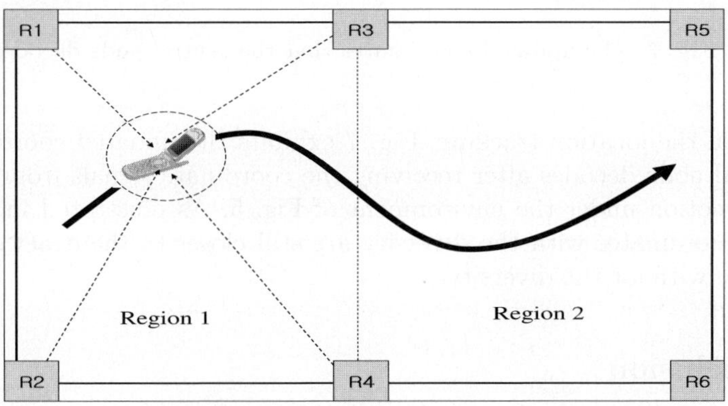

Fig. 5. The mobile agent in ubiquitous networks

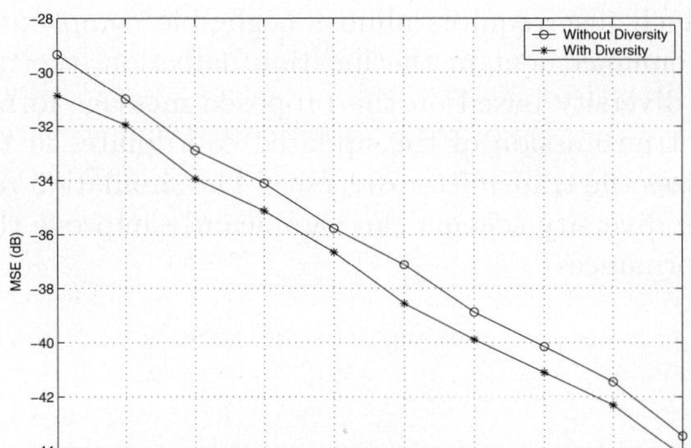

Fig. 6. The MSE performances for the location detection

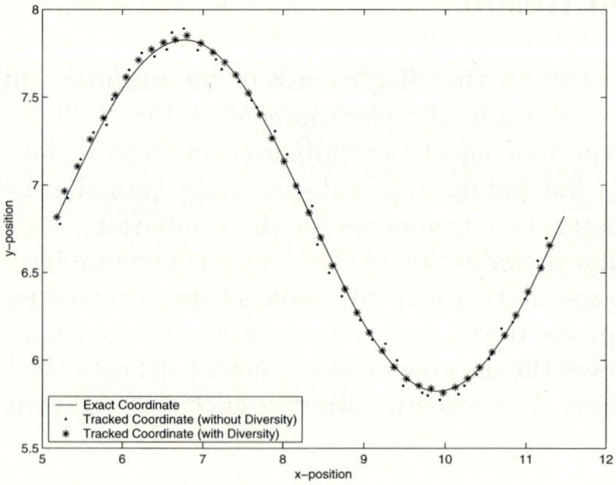

Fig. 7. The updated coordinates that the central node decodes

diversity in the location tracking, Fig. 7 exhibits the updated coordinates that the central node decodes after receiving the coordinate signals from the mobile agent in motion under the environment of Fig. 5. As illustrated in Fig. 7, the decoded coordinates with the diversity are still closer to the exact coordinates than those without the diversity.

5 Conclusion

In this paper, an intelligent diversity scheme is proposed to improve the performances in locating and tracking the mobile agent in ubiquitous networks. To acquire the diversity, the mobile agent performs the antenna switching between two antennas, which just requires almost negligible complexity overhead. For the performance enhancement in the location detection, the mobile agent utilizes the receiver diversity based on the proposed message format. In addition, for more reliable transmission of the updated coordinates in the tracking, the mobile agent utilizes the transmitter diversity. The simulation results also reveal that the proposed diversity scheme can significantly improve the detection and the tracking performances.

References

1. I. Akyildiz, W. Su, Y. Sankarasubramaniam, and E. Cayirci, Wireless Sensor Networks: A Survey. Journal of Computer Networks **38** (2002) 393–422
2. N. S. Correal, and N. Patwari, Wireless Sensor Networks: Challenges and Opportunities. Proc. of Virginia Tech Symp. Wireless Personal Comm. (2001) 1–9
3. E. H. Callaway Jr., Wireless Sensor Networks: Architectures and Protocols. Auerbach. (2003)

4. C. M. Cordeiro and D. P. Agrawal, Ad Hoc & Sensor Networks: Theory and Applications. World Scientific. (2006)
5. P. Venkitasubramaniam, S. Adireddy, and L. Tong, Sensor networks with mobile agents: optimal random access and coding. IEEE Journal on Sel. Areas in Comm. **22** (2004) 1058–1068
6. P. Kim and S. Chang, An intelligent positioning scheme for mobile agents in ubiquitous networks for u-city. *submitted to* KES AMSTA 2007
7. G. L. Stüber, Principles of Mobile Communication, Second Edition. Kluwer Academic Publishers. (2001)
8. N. Patwari, J. N. Ash, S. Kyperountas, A. O. Hero III, R. L. Moses, and N. S. Correal, Locating the nodes: cooperative localization in wireless sensor networks. IEEE Signal Processing Magazine (2005) 54–69
9. S. Haykin, Adaptive Filter Theory, 4th Edition. Prentice-Hall. (2001)

A Relay-Transmission of the RFID Tag ID over the Wireless and TCP/IP with a Security Agent

Heung-Kuk Jo and Hoon-Jae Lee

Division of Computer & Information Eng.,
DongSeo University, Sa-Sang Gu, Ju-Rae Dong, San 69-1,
Busan, Korea
{Heung-Kuk Jo,hkjo@gdsu,dongseo.ac.kr}

Abstract. Radio Frequency Identification (RFID) systems can be applied to various areas. A RFID system is composed of tag, reader and host computer. The reader has an antenna for transferring the energy and sending/receiving of data. Cables are used for communication between reader of RFID system and host computer. If a RFID system is implemented in the existing building or gate, the data communication line should be installed in the wall of building or inside of the wall, and this causes some problems. We suggest a solution to this problem, as ID data from tag is transferred to host computer by wireless communication in the reader. In addition, TCP/IP communication can be used for recognizing ID data in long distance.

In this paper, we show a device that sends ID Data from reader to the receiver over wireless network using CC1020 chip and a device that sends ID data from the receiver to the host computer over internet network using W3100A chip. From this, the paper explains the system transferring ID data to the host computer over wireless and Internet network. There are main circuits for each device and block diagram for programming shown in the picture. The system contains a security agent for security managements and services. In addition, the actual data waveform and experimental device are shown in the picture as well as the result of identification in the host computer.

1 Introduction

Most of the existing RFID systems are used for recognizing objects. These systems use wired data communication between reader and host computers. This kind of data communication using the wire has problems in the connection between devices and it gives low efficiency to the communication. To solve such problem, the communication between two devices should use wireless networks. Wireless communication can give more mobility between devices and its multi-communication per each point has more advantages compared to the wired networks. In addition, remote data-monitoring devices can be a part of a home network system. In order to identify a person entering to the house in other area and to open and close the door after the identification, TCP/IP communication should be utilized.

N.T. Nguyen et al. (Eds.): KES-AMSTA 2007, LNAI 4496, pp. 918–927, 2007.

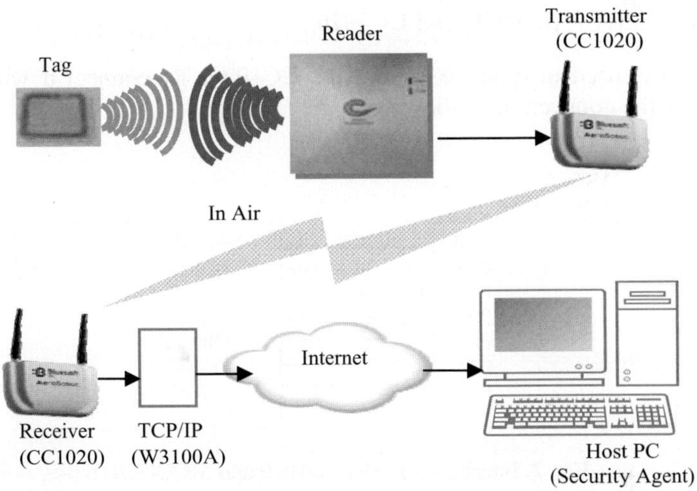

Fig. 1. System Concept

This paper explains an architecture, which is to transfer tag ID and data to the host computer over the wireless network and the internet as shown in Figure 1.

CC1020 chip is used in the wireless communication and on the other hand, W3100A chip is used in the TCP/IP communication. This paper briefly describes two chips, main circuit and the method of setting up the register for the program. As a result, the paper also shows the actual waveform of data pin in the wireless chip, the result of monitoring the host computer and the manufactured device.

In this paper, we explains the system transferring ID data to the host computer over wireless and Internet network. The system contains a security agent for security managements and services (confidential service, authentication service, and access-control services, etc.) [15-19]. In addition, the actual data waveform and experimental device are shown in the picture as well as the result of identification in the host computer.

2 Wireless System Using CC1020

2.1 Introduction of CC1020

CC1020 chip of Chipcon Company is a single UHF Transceiver chip and the possible bandwidths are 400MHz and 915MHz. The multi-channel chip can be used in each different frequency through the program. With the program, channel separation also can be set up in narrow-band 12.5KHz or 25KHz. As for the strong point of this chip, it can set up all frequencies, RF Power, etc. by inputting the value in the register within CC1020. Atmega128 microcontroller is used to input data to the register [1, 2].

2.2 Interfaces Between MCU and CC1020

Atmega128 was used in order to control the CC1020. To connect it with MCU, it should follow the connection as shown in Figure 2 [3].

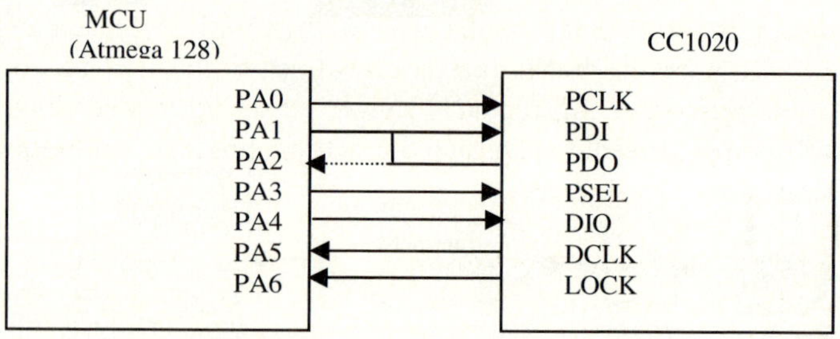

Fig. 2. Interface between CC1020 and MCU[3, 4]

2.3 Programming for Operation of CC1020

2.3.1 Time Chart for Write Mode into MCU for Programming

Fig. 3 (a) and (b) show the Time Chart to input the data in CC1020 at MCU. As shown in Figure 3 (a) and (b), PSEL can write or read in "Low" condition. Also, the data of PCLK should be made in MCU and sent to CC1020.

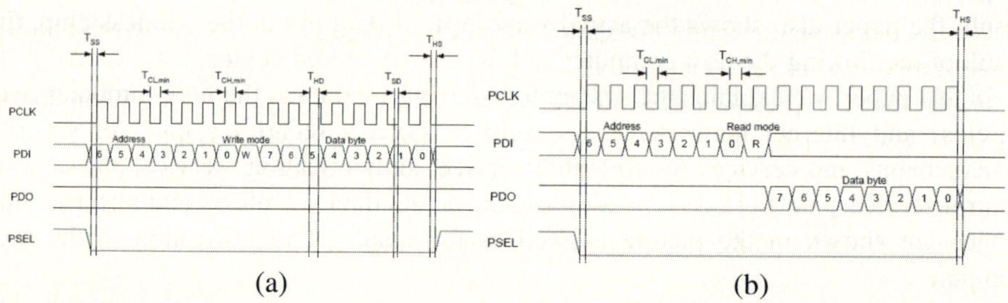

Fig. 3. Time Chart (a) For Data Writing CC1020 (b) For Data Reading CC1020[3,4]

The 7th Bit of PDI decides Write or Read condition. The data will be written when MCU in CC1020 is "High" in condition and will be read when MCU is setting to "Low" condition. Data Read is required to confirm whether CC1020 is in the sending or receiving state and also it is required to confirm whether CC1020 is in "LOCK" state. CC1020 chip has registers ranging from 00h to 45h. Each register decides the necessary state in order to drive the chip.

2.3.2 Process of Register Setting

CC1020 chip has registers, addressed by ranging from 00h to 45h [3]. Each register decides the necessary state in order to drive the chip. MSB in Main register should be "0" to put CC1020 chip in receiving state, and "1" to put it in sending state. The next

6 addresses are used for the frequency selection register. CC1020 can select two frequencies A and B and this register should be set up if it selects the frequency A to 900MHz and the frequency B to 400MHz. Two bits of address 4 and 5 are used for setting the power-down mode. Address 1 in Register 3 is for power-down in each part and last LSB is the bit for initializing the register. When Main register is set up in 0xC1, the register is 1100 0001 so that CC1020 should be in sending mode and selects frequency B, meaning that it does not reset. With this method, all registers are set up in the following order. All registers should be set up into sending or receiving state as the following procedure. During the setting process, calibration should be operated more than or at least once.

Procedure of TX_Status or RX_Status :

```
Begin
  Setup All register
  TX_Setup
  Calibration
  RX_Setup
  Calibration
  Power_UP
  Calibration
  Mode Select of TX or RX
End
```

For the next transmission, TX-Status should be called for sending and RX_Status should be called for receiving. Thus, the data is transmitted through this process. In the experiment, the data is transmitted to DIO Pin (NRZ type) by DCLK signal as shown in Figure 4 (a). If the data is transmitted with UART communication, it is not necessary to follow this signal and DCLK pin receives the data in the transmission. Figure 4 (b) shows the wave form sending the register set value to CC1020. This waveform sends the data by coding in Manchester type.

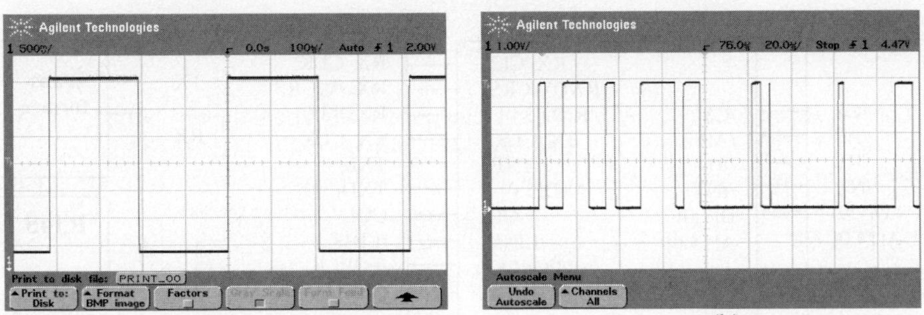

(a) (b)

Fig. 4. Waveform (a) DCLK Waveform, (b) PDI waveform

3 TCP/IP Communication Using W3100A

3.1 Introduction of W3100A

W3100A has total of 64 PINs including Address PIN of A [14:0], Data PIN of D [7:0] and MODE PIN, etc. And it has interfaces with Atmega128 as MCU and RTL8201[9]. W3100A, the LAN LSI of WIZNET company, has TCP/UDP, ICMP, IPv4, ARP, and Driver Program implemented in OS Kernel (Software) and also has a Chip (Hardware logic) implemented to MAC/PHY part in Hardware [9].

3.2 Interfaces Between MCU, W3100A and RTL8201

At first, Figure 5 shows the interface between MCU (Atmega128), W3100A and RTL8021. W3100A provides three types of MCU Interface and the left side of Figures 5 is the Direct Interface between MCU and W3100A with Memory through Address Bus among them. For each connected PIN, /CS is chip select, /WR is Write, /RD is Read enable and /INT is Interrupt PIN, and all of them are in Active Low. In addition, the rest of them are Data and Address PINs. MCU has a total of 64Pins, in which port A and port C are used as Address and Data Pin. The circuit is made by the following structures: /RD, and /WR of Atmega128 are connected to /RD, and /WR of W3100A, /CS is in low and /INT is connected with PE4 Pin of Atmega128. In here, Clocked Mode is used for analyzing MCU bus signal by using Clock. RTL8201 is Receiver in low power and low voltage supporting all Ethernet Physical-layers of 10/100M, which interface with W3100A as shown in Right Side of the figure. In the Figure, RTL8201 is connected to each of TX_CLK(transferring clock), RX_CLK(receiving clock), TXD[3:0] (transferring data), RX[3:0](receiving data), TXE (transferring enable), RXDV/CRS (carrier sensing) and COL(error sensing) Pins in W3100A. The speed of TX_CLK is 2.5MHz in 10BASE-T and 25MHz in 100BASE-T. And TXD[3:0] sends the data when TX_CLK is at rising edge and senses the carrier with RXDV/CRS re-synchronizing RX_CLK. RX[3:0] receives the data at falling edge of RX_CLK and COL senses the error.

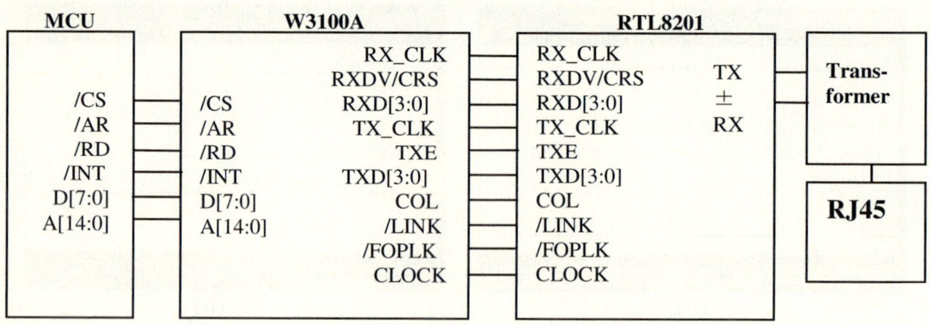

Fig. 5. Interface between MCU, W3100A and RTL8201

TPRX- and TPRX+, the 30th and 31st Pin of Figure 5, and TPTX-, TPTX+, the 33rd and 34th Pin, are connected with RJ45 by passing a TRANSFORMER.

3.3 Programming for MCU

3.3.1 Memory Map and Register Map of W3100A

Memory of W3100A consists of Control registers from 0x0000 up to 0x0200, Tx_data buffer(8Kbyte) from 0x4000 up to 0x6000, and Rx_data buffer(8Kbyte) from 0x6000 up to 0x8000 [9]. Tx_data buffer only performs the write function as the memory during sending/receiving. Rx_data buffer only performs the read function as a buffer during receiving. Address ranging from 0x8000 to 8FFF in MCU is used for W3100A Control register, address ranging from 0xC000 to 0xDFFF is used for W3100A Tx_data buffer, and address ranging from 0xE000 to 0xFFFF is used for Rx_data buffer. The buffer size of the entire Channel in W3100A is 8Kbye and the buffer value of each Channel can be set up with TMSR (Tx_data Memory Size Register) and RMSR (Rx_data Memory Size Register)[9].

Register map in Control Register of W3100A for MCU Program is explained below. Channel Control Register, the same as Cn_CR (n is 0~3), is used for transferring each Channel, initializing receiving data, accessing and terminating. Meanwhile Sys_Init command of C0_CR is used for setting gateway, subnet mask, source IP and source H/W Address. Interrupt State Register, the same as Cn_ISR (n is 0~3), informs the result of Channel 0 Socket command, and Init_OK informs the completion of Sys_Init command. IR (interrupt register) is used for generating interrupt and classifying channel. Each bit of C0, C1, C2, or C3 informs which interrupt occurs in a channel 0, 1, 2, or 3. MCU can verify the occurrence of interruption by testing the register in the channel interrupt status of response channel. Each bit of C0R, C1R, C2R, or C3R informs the occurrence of data transmission in a channel 0, 1, 2, or 3. IMR (interrupt mask register) is the register, used as the mask of interruption from the interrupt register in response register. And the others of IDM_OR, IDM_ARQ, IDM_AR1 and IDM_DR are used for setting IDM (Indirect mode) and the rest of registers, not shown in the table, are used for storing Gateway, Subnet Mask, Source Hardware Address, Source IP address, etc. and Pointer of each Channel.

3.3.2 MCU Program with TCP Client

Figure 6 is depicted by a TCP Client. First of all, TCP Client is in INIT(initialization) status by Socket_init command register in CLOSED state. At this time, it sets up MAC, IP, Gateway and Subnet mask and requests ARP for accessing to the server with destination IP. If there is a response for the request, it becomes in SYNSENT state after sending SYN, otherwise, it returns to CLOSED state. In SYNSENT state, if it receives the response of SYN(SYN, ACK), it accesses to the server and goes to ESTABLISHED state (connection mode) after sending ACK.

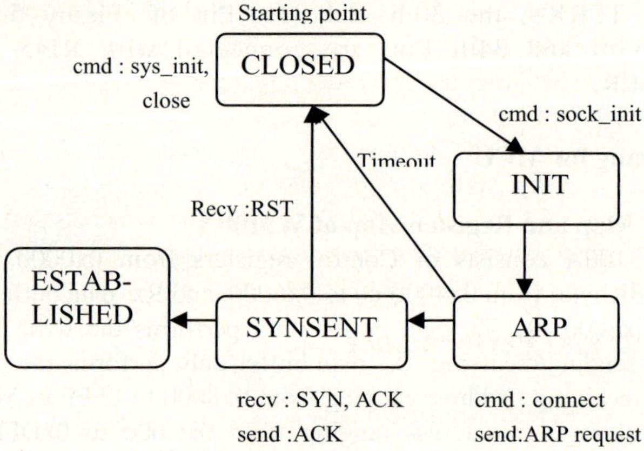

Fig. 6. Construction of TCP Client

```
#include <mega128.h>
#include<stdio.h>
#include"main.h"
#include"socket.c"
int main()

{ void Mega128_Init(void); // Initialize MCU
...
void Socket_Init(unsigned char Value);
{...
//-----W3100A Function-------//
extern void W3100A_Init(void);
extern void sysinit();
extern voidsetsubmask();
extern void setgateway();
extern void setMACAddr();
extern void setIP();
extern void setIPprotocol();
...
//--------Socket API--------//
extern char socket();
extern char connect();
......
extern void wait_10ms();
}
void LanServer_Init(void);
void LanServer_Wait(void);
void LanServer_Run(void);
}
```

Fig. 7. TCP/IP Program

Figure 7 shows TCP Client MCU Programming. As TCP/IP communication is implemented as Interrupt type, it is composed of ISR(Interrupt Service Routine) and Main Routine. The Programs of Figure 7 are programs according to the TCP Client of Figure 6, which initializes W3100A after initializing Atmega128, sets up IP, GW and SM and then generates the next socket and accesses to the desired Server IP. It sends the data at every 10ms using Timer/Count Interrupt. And ISR stores the Interrupt status value of Channel in IR and returns the value so that it can classify each Channel and check whether data should be transmitted.

4 Experiment and Performance

Figure 8 shows a wireless Sender/Receiver module by using CC1020, a TCP/IP communication module by using W3100A, and a Atmega 128 MCU device. The Sender/Receiver frequency of CC1020 was calibrated to the antenna bandwidth between 433.92MHz and 400MHz. In operation process, the system reads tag's ID in reader at first and sends the tag's ID from the actualized wireless Sender/Receiver to the actualized receiver with CC1020 chip. MCU receives Tag's ID in serial by W3100A TCP Client programming. In next step, the received ID will be sent to the Host computer through the Internet network by accessing Server program. In final step, Server Program outputs the personal information to the screen by inter-working with DB of the host computer

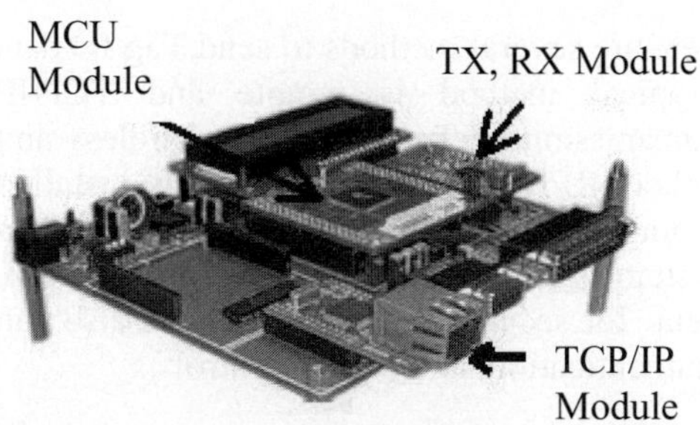

Fig. 8. Device for Experiment

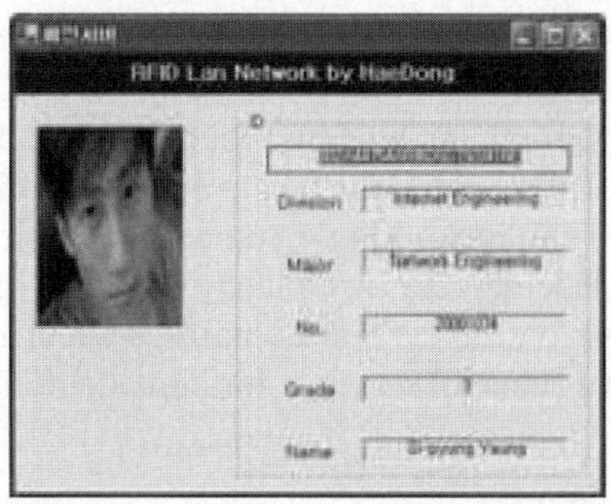

Fig. 9. Monitoring Result in Host PC

Figure 9 is the execution screen of Server program. Server program is implemented by using .NET programming language. When received Tag data, Host computer compares Tag ID with personal information of DB and outputs the personal information identified with ID. With experiments, the proposed system can be installed in various places (applications) for monitoring visitors or employees without changing the existing building. This means that it can be applied to RFID systems and other various systems.

Table 1. The Baud Rate of each Section (Wireless, TCP/IP)

<table>
<tr><td rowspan="2">Item</td><td colspan="3">Performance</td></tr>
<tr><td>Part</td><td>Baud Rate</td><td>Buffer Size</td></tr>
<tr><td rowspan="3">Data Transmission Section</td><td>RFID → Transmitter</td><td>9600bps</td><td>30Byte</td></tr>
<tr><td>Transmitter → Receiver</td><td>4.8k bps</td><td>30Byte</td></tr>
<tr><td>Receiver → Host PC</td><td>50M bps</td><td>-</td></tr>
<tr><td>Security Agent</td><td colspan="3">1) Security management for RFID database
2) Security services for confidential, authentication, and access control
- data confidentiality for communication links
- authentication service for legitimate devices
- access control for legitimate users (password)</td></tr>
</table>

Table 1 shows the Baud Rate in each part. Baud rate is higher in Internet network and lower from RFID reader to Transmitter. In this Process, buffer is needed in each part, because the preamble is attached or removed by the wireless communication. The system contains a security agent for security managements and also can give security services for confidentiality, authentication, and access control [15-19].

5 Conclusion

RFID system is applied in many areas. There are several methods to send Tag ID data to host computer. Among them, a typical method is remote and TCP/IP communication. This paper shows the transmission of Tag's ID over wireless and TCP/IP communication. By using this method, RFID system can be simply installed and can make long distance monitoring. From the results of experiment, Tag's ID data can be sent to any place whenever the system is connected to the Internet network. Finally, the system contains a security agent for security managements and also can give security services for confidentiality, authentication, and access control.

References

1. Chipcon Tech. Support, "User Manual Rev. 2.0", Chip AS,(2004), www.chipcon.com.
2. Chipcon Tech. Support, "User Manual Rev. 2.3", Chip AS, 2003
3. Chipcon, Tech. Support, "CC1020 Datasheet (rev. 1.5)", Chipcon AS
4. Chipcon Tech. Support "Application Note AN025", Chipcon AS, 2004
5. Hwang-Hae Kyun, Sung-Jun Bae, "I love ATMEGA128", Bok-Du Verlag, (2005)
6. Michael J. Donahoo, Kenneth L. Calvert, "TCP/IP Socket Programing version C", (2001)
7. Sung-Woo Yun, "TCP/IP Socket Programming",(2003)
8. Duck-Yong Yun , "AVR ATmega128 Master", (2004)
9. Technical Datasheet V1.32, www.WIZnet.co.kr
10. Klaus Finkenzeller, " RFID Handbook", John Wiley & Son, LTD, (1999)

11. Jonathan Collins, “FTC Asks RFID Users to Self-Regulate”, RFID Journal, Mar. 10, (2005)
12. E. Kaasinen, “User Acceptance of Mobile Services-Value. Ease of Use, Trust and Ease of Adoption,” doctoral dissertation, VTT 566, VTT Publication,(2005)
13. Vinay Deolalikar, Malena Mesarina, John Recker, Salil Pradon,”Perturbative time and frequency allocations for RFID reader networks” (2006) USN
14. Kyung-Hyup Lee, Yoon-Young An, Hee-Dong Park, You-Ze Cho,” A Data-centric Self-organization Schema for Energy-Efficient Wireless Sensor Networks” (2006) USN (USN 2006 in Seoul).
15. Soo-Cheol Kim, Sang-Soo Yeo, Sung Kwon Kim, "**MARP: Mobile Agent for RFID Privacy Protection**", 7th Smart Card Research and Advanced Application IFIP Conference (CARDIS '06), *Lecture Notes in Computer Science*, vol.3928, pp.300-312, April 2006.
16. "FIPA Agent Management Specification," FIPA 2004 Specification, Foundation for Intelligent Physical Agents, March 2004. <URL: http://www.fipa.org/specs/fipa00023/SC00023k.pdf>
17. W. Jansen, T. Karygiannis, “NIST Special Publication 800-19 – Mobile Agent Security, “ <URL: http://csrc.**nist**.gov/**publication**s/**nist**pubs/**800-19**/sp**800-19**.pdf>
18. W. Stalling, *Cryptography and Network Security (4th edition)*, Prentice-Hall, 2006.
19. W. Trappe & L. Washington, *Introduction to Cryptography with coding theory (2nd edition)*, Prentice-Hall, 2006.

An Agent-Based Web Service Composition Using Semantic Information and QoS

Eunjoo Lee[1] and Byungjeong Lee[2,*]

[1] Department of Computer Engineering, Kyungpook National University, Korea
ejlee@knu.ac.kr
[2] School of Computer Science, University of Seoul, Korea
bjlee@uos.ac.kr

Abstract. Applications based on Web service technology have grown rapidly and it has become more and more necessary to select appropriate services. Semantic information and QoS (Quality of Service) are important to choose appropriate services. Since it is hard to match a user's needs with only a single Web service, the composition of Web services based on workflows is required. To achieve this composition, we utilized our prior research, which suggested that a framework can support a single-service selection based on QoS properties. In this paper, we propose a framework that supports a Web service composition based on semantic information and QoS properties. Three additional elements are extended and added to the prior one: A Broker Agent, a Composition Server and a Composition Agent. This framework is independent of specific composition algorithm. Furthermore, dynamic aspects are considered and supported by a Composition Agent in this framework, which enables a requester to work smoothly with a few loads.

Keywords: Web service, QoS, agent, composition framework.

1 Introduction

Currently, applications based on Web service technology have grown rapidly. This is due to their ability to reuse autonomous services over the Web and to enable application-to-application communication [1]. Due to the abundance of Web service providers, it is important to choose the proper service provider and semantics and QoS of Web service should be considered in such circumstances.

Recently, Semantic Web service uses ontology to describe services and supports automation of service-related processing. For automatic discovery and composition of Web services, matchmaking input/output type matching, and precondition/effect analysis using ontology are performed [2]. Some works have studied semantic discovery and composition, which check the interoperability and compatibility of services according to both syntactic and semantic descriptions [3] [4]. The input and output of given two service type are used to check compatibility. However, since it

* Corresponding author.

N.T. Nguyen et al. (Eds.): KES-AMSTA 2007, LNAI 4496, pp. 928–937, 2007.

does not support semantic matching level of ontology concepts or weights of functional properties, it tends to limit elaborate service discovery and composition.

Several studies have also been conducted on a QoS-aware framework for Web services [5] [6], however, they do not consider the end users' condition and most of them require loads on users' side. Also, it is hard to determine the validity of providers [5]. A framework that complements those limitations is suggested [7]. In Lee et al.'s framework [7], a few elements are added to the original Web service framework: A QoS Agent and a QoS Server. The history data of each Web service are utilized for more flexible queries, which are collected by a QoS Agent. Users can query based on their specific conditions with a few loads, and the QoS properties can be gained in a Provider-independent fashion. Lee et al.'s framework, however, supports only single Web service selection using QoS and does not consider Web service composition using semantic information. Some studies have focused on Web service composition based on QoS properties, such as response time, reliability, execution time, reputation, and rank [8] [9] [10] [11]. Most studies adopt a workflow-based approach, where some QoS-suitable services are selected for each task on a predefined workflow, according to several optimization algorithms, such as genetic algorithm [11], a simulated annealing algorithm [10], Integer programming [8], and fuzzy logic [9]. In those studies, it is argued that the adopted algorithm has advantages over others.

In this paper, we propose a framework that supports Web service composition based on semantic information and QoS properties. Since there are tradeoffs in adapting optimization algorithms, we do not select a specific algorithm and define a simple generic model that can be utilized with optimization algorithms. We focus on the architecture of the framework using both semantic information and QoS properties. We extend Lee et al.'s framework by extending and adding three additional elements: A Broker Agent, a Composition Server and a Composition Agent. A Broker Agent performs semantic matching using ontology. A Composition Agent monitors the execution state of the Web service application and it reacts under some conditions specified in the statechart model. A Composition Server performs a composition of Web services using a certain algorithm and a predefined workflow, according to the Composition Agent's messages.

The remaining parts of this paper are organized as follows. Section 2 presents some models that provide the basis of the framework. Section 3 describes our proposed framework and a working process. Finally, Sect. 4 outlines a summary, contribution, and limitations of our study.

2 Web Service Composition Model

This section defines terms and models that can be used in this composition framework. They are as follows.

Definition 1. Task Model (TM)
A task model consists of elements to specify functionality of a task. The elements include input, output, precondition and effect of a task.

TM = <input, output, precondition, effect>

Definition 2. Abstract Workflow Model
An abstract workflow model of workflow *m*, $AWF_{m,}$ is composed of individual 'abstract work'. It means that a workflow template is defined but any concrete services are not allocated to the workflow.

$$AWF_m = <W_1, W_2, \ldots, W_n>$$

where
W_k is an abstract work, that is, it only specifies a particular task in TM.

We assume that each numbered work W_i is mapped onto a concrete work in a specific workflow model. We describe it using an activity diagram in UML (Fig. 1). In Fig. 1, W3 and W4 are executed concurrently. W5 and W6 are optional work under some conditions. All other works are sequentially executed. Figure 1.(a) shows an example of an abstract workflow model *wf*. An abstract workflow model AWF_{wf} is $<W_1, W_2, W_3, W_4, W_5, W_6>$.

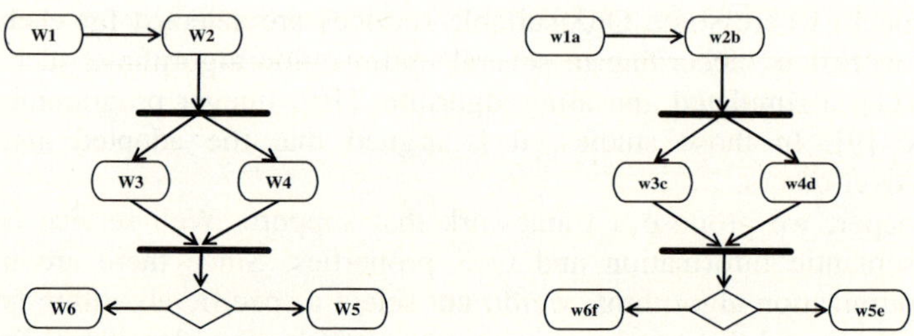

(a) An abstract workflow (b) A concrete workflow

Fig. 1. An example of a workflow

Definition 3. Web Service (WS)
A Web service WS consists of its functionality, QoS properties and its locality information.

$$WS=< IOPE, QoS, Loc>$$

where
IOPE = <input, output, precondition, effect>
IOPE is composed of input, output, precondition, and effects of a Web service.
$QoS=<q_1, q_2, \ldots, q_m>$
q_k is a QoS properties, such as reliability, rank, or response time.
Loc is locality information.

The functionality of a Web service is specified by input, output, precondition, and effect of the service [12]. IOPE is an abstract characterization of what a service can do. It relates and builds upon the type of content in UDDI, describing properties of a service necessary for automatic discovery and composition [13]. IOPE is specified by using ontology written in OWL-S to understand semantics of a Web service.

Definition 4. Concrete Workflow Model
A concrete workflow CWF_m for workflow *m* is composed of individual Web services.

$$CWF_m = <w_{1a}, w_{2b}, \ldots, w_{nc}>$$

where
$w_{ki} \in candW_k$ ($candW_k$ is a candidate set composed of Web services that match the task of W_k.)

w_k indicates a Web service that matches the functional task of W_k. Usually, several candidate Web services exists which are functionally suitable to W_k, and then, w_k is selected by a specific strategy. Since the state of a Web service varies under different circumstances, we assume that the candidate Web services can be changed.

The inputs, outputs, preconditions, and effects are used for matching between tasks in Abstract Workflow Model and Web services in Concrete Workflow Model. We modify the matching levels proposed in [3]. The matching levels are Exact, Subsume, Relaxed, and Fail.

- Exact: Exact matching between IOPE types of Web services and TM types of tasks.
- Subsume: TM types of tasks are subclasses of IOPE types of Web services.
- Relaxed: TM types of tasks are superclasses of IOPE types of Web services.
- Fail: Not matching between IOPE types of Web services and TM types of tasks.

The matching level orders are the followings:

$$\text{Exact} > \text{Subsume} > \text{Relaxed} > \text{Fail}$$

The inputs, outputs, preconditions, and effects between Web services must be compatible for composition. For instance, in Fig.1.(b) the output and effect of w_{1a} must have one of Exact, Subsume, or Relaxed relationships with the input and precondition of w_{2b}, respectively.

Users can choose the QoS properties in which they are interested. All QoS properties can be scaled from zero to one for further processing [8]. Since Web service applications may be used in dynamic environments like a mobile platform and that the set of available services can change [14], we added location information for future usage. We can obtain each QoS property value of the whole concrete workflow by using the values of individual services and their workflow style. For example, CWF_{wf}'s *execution time* in a concrete workflow (Fig.1.(b)), one of QoS properties, can be obtained by the following equation.

$$T_{exe}(CWF_{wf})=t_{exe}(w_{1a}) + t_{exe}(w_{2b}) + \max(w_{3c}, w_{4d}) + \max(w_{5e}, w_{6f})$$

where
$T_{exe}(M)$ is the execution time of a concrete workflow model M
$t_{exe}(w)$ is the execution time of service w.

In this example, the execution time of conditional branch connected with w_{5e}, w_{6f}, may be executed in other ways, such as using probability or averaging two values. Such methods are dependent on a coordinator's principle. It is not hard to acquire other QoS properties. For further details, see [8].

3 Proposed Framework

In this section, we outline the extended framework (Fig. 2). Our reasons for using this composition framework include the following:

1) Algorithm independence - There are various optimization algorithms which can be used to compose Web services and they have advantage and disadvantage in several aspects. Therefore, we can not fix a specific algorithm in instantiating an abstract workflow.
2) Frequent reconfiguration - Web services and their compositions have dynamic features. They indicate that there may require the frequent instantiation of an abstract workflow using an optimization algorithm, which can lead to problems in a real application.
3) Unexpected results - Expected result and the actual results may be different, especially due to some unavailable services in that time. We reflect on the actual results in the composition.
4) Mobile environment – The reliable execution of Web services in a mobile environment requires active deployment, and dynamic reconfiguration [15].

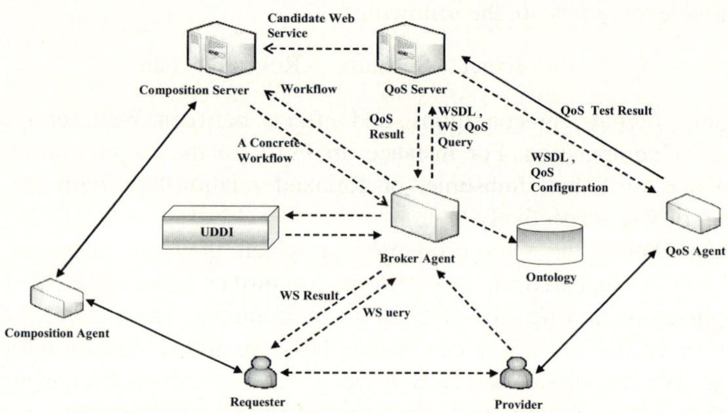

Fig. 2. A Web service composition framework

We extend the work of our prior research [7] whereby requesters can get QoS-suitable Web services which match their specific conditions. A general Web service framework consists of a Provider, Requester, and UDDI. In [7], the elements including QoS Server and QoS Agent are added to the original one. The retrieval strategy of QoS information, for each Web service, is based on the prior framework. Herein, we briefly explain the role of the elements (Fig. 2).

- **A QoS Server** calculates the metric values of QoS properties for each Web service by using the history data that a QoS Agent has retrieved. It selects QoS-suitable Web services for its evaluating scheme, among its registered Web services and it forwards selected Web services to Broker Agent.

- **A QoS Agent** regularly tests Web services registered in a QoS Server and it stores the test results for each Web service in a QoS Server. The necessary information for testing is given by a QoS Server.

We can say the main advantage of Lee et al.'s framework has two factors. At first, requesters can obtain suitable Web services that match their specific usage patterns. This is done by using history data collected by a QoS Agent. A usage pattern reflects a requster's specific conditions. For example, a requester uses a Web service during night time and that it does not care about the performance of the service. In that case, the data tested at night are more meaningful than those during day times. Second, a requester has comparatively fewer loads by virtue of the elements. For further information, please refer to [7].

3.1 Extended Framework

The Broker Agent classifies user's queries into functional and non-functional (QoS) parts. It searches UDDI and obtains a list of functionally-suitable Web services matching to a task. Then, it passes the functionally-suitable ones and the non-functional parts to a QoS Server for further processing. The Broker Agent also accesses an ontology repository containing domain ontology written in OWL and service ontology written in OWL-S. Figure 3 shows a description of a 'BookSearch'

```
<profile:Profile rdf:ID="request">
<profile:serviceProduct rdf:ID="example.owl\bookSearch"> </profile:serviceProduct>
<profile:hasInput rdf:ID="example.owl#bookISBN"> </profile:hasInput>
<profile:hasOutput rdf:ID="example.owl#bookName"> </profile:hasOutput>
<profile:hasPrecondition rdf:ID="example.owl#validBookISBN">
</profile:hasPrecondition>
<profile:hasResult rdf:ID="example.owl#knownBookISBNValidity"> </profile:hasResult>
<profile:textDescription> (QoS {time=1; priority=high}) (Loc {pos=korea})
</profile:textDescription> </profile:Profile>
```

Fig. 3. A Web service in OWL-S

```
iopeMatch(tm, iope) {
   score =  w_i*degreeMatch(tm.input, iope.input);
   score += w_o*degreeMatch(tm.output, iope.output);
   score += w_p*degreeMatch(tm.precondition, iope.precondition);
   score += w_e*degreeMatch(tm.effect, iope.effect);
   return score;
}
degreeMatch(tm_type, iope_type) {
   if tm_type equivalence iope_type then degree += Exact;
   if tm_type subClassOf iope_type then degree += Subsume;
   if tm_type superClassOf iope_type then degree += Relaxed;
   return degree;
}
```

Fig. 4. A matching algorithm

Web service written in OWL-S. The input and output of the BookSearch Web service are a concept bookISBN and bookName in example.owl file, respectively. The precondition indicates that bookISBN should be valid. Figure 3 shows that the execution time and priority of the service is 1 and high, respectively.

The Broker Agent evaluates the matching degrees between IOPE and TM and makes a list of candidate Web services. We extend the matching algorithm proposed in [3]. Figure 4 shows a matching algorithm, where matching scores are computed for each of input, output, precondition and effect. Web services are sorted according to the scores and a list of candidate Web services is obtained. In Fig. 4, w_i, w_o, w_p, ard w_e are weights of input, output, precondition, and effect, respectively.

The Composition Server's main role is to instantiate a workflow which is returned from Broker Agent by using a selected algorithm. It interacts with a Composition Agent; when it gets a 'fully configuring' message from the Composition Agent, it resets the Web service composition model which conforms to that time. When the Composition Server receives a 'partly configuring' message with a service identifier to be removed, it replaces the specific service with the most appropriate service at that time. The replaced service is the principal one that degrades the whole quality of the instantiated workflow. As was previously asserted, a Composition Server has the up-to-date candidates for each abstract work. This can be achieved with the interaction between a Broker Agent, a QoS Server, and UDDI. The QoS Server registers the abstract workflow and it sends a query to the Broker Agent to get functionally-suitable Web services for each abstract work. Next, the QoS Server pushes the changed candidate sets to the Composition Server.

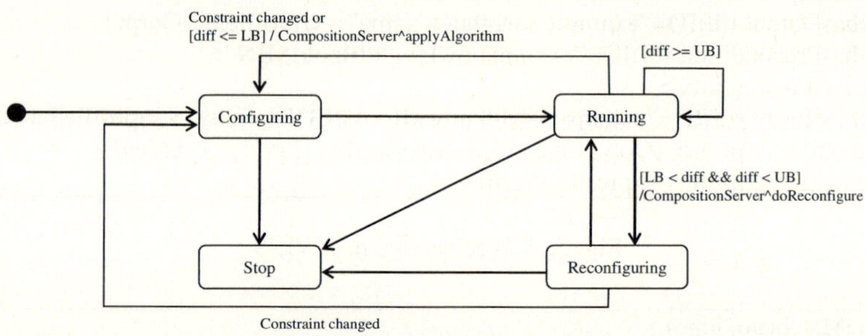

Fig. 5. A statechart model of a Composition Agent

A Composition Agent resides in a requester's part and it monitors the execution state. It triggers the action of the Component Server by passing messages according to the monitored results. Figure 5 shows the changing state of the application via the Statechart diagram in UML, however, the transition action that notifies a Composition Server is performed by a Composition Agent, not by the application. An expected QoS values are calculated by the process described in the previous section 2. An actual QoS vlaue is handled by a QoS Agent. A QoS Agent regards the whole concrete workflow as a single service, and it tests the workflow as in the same way of the QoS Agent. One difference is that the Composition Server notifies the target QoS property by referring to the weight that the Requester sets. For example, if a

Requester selects the execution time as the most important property, a Composition Agent tests the application to get the execution time. A QoS configuration [4], containing test data, is delivered from a Composition Server to a Composition Agent. We assume that the most weighted property is selected, however, K-properties can be considered of course.

Definition 5. QoS Gap

$$diff = 1 - C_{act}/C_{exp}$$

where
UB is upper bound for '*diff*'.
LB is lower bound for '*diff*'.
LB and UB are given in advance by a Composition Server (default values) or by a requester.
C_{exp}, C_{act} are, respectively, an expected QoS value and an actual QoS value for a specific QoS property in an instantiated workflow.

The '*diff*' value, which is scaled between zero and one, indicates the gap between the expected and actual results. There are three cases as follows:

Case 1. $diff \geq UB$
In this case, we assume that the situation operates smoothly and that a Composition Server does not take or engage in any action.

Case 2. $diff \leq LB$
This case indicates that there are some serious problems in executing the composed Web services. And so the Composition Agent passes a message to a Composition Server to construct the whole concrete workflow. This step is almost the same as the first instantiation of an abstract workflow.

Case 3. $LB < diff < UB$
In this case, we consider the execution result of an application to be somewhat degraded. This is because some Web services become unavailable at that time, either temporarily or permanently. In particular, the set of localized services changes as the users move [9]. From this, we can infer that a permanently-unavailable service implies that it is a localized service and that the application is executed under a mobile environment. To handle this situation, a QoS Agent directs that a Composition Server finds a deteriorating service and replaces it with a better one. After that, a Composition Server consults with a QoS Server about the list of Web services to be composed. A QoS Server can determine which service is the degraded one by using part of the latest test data among the history data. A QoS Server maintains an up-to-date execution profile for each service. This is because a QoS Agent regularly tests Web services. Finally, a Composition Server determines the most appropriate one in the candidate Web services of a specific abstract work W_i in which the degrading one belongs. This is done in ways similar to [7]. Asides from those cases, the concrete workflow is newly constructed when a requester changes the initial QoS constraints.

3.2 Process

This subsection presents the working process of a requester by using a Web services composition to meet both its functional and non-functional requirements in the application of this framework. We assume that QoS configurations and a query form are used based on the prior research [7]. Steps 1 to 5 are the initial setup process.

1. A requester sends a query to the Broker Agent. A query contains an abstract workflow, keywords for each abstract works that shows the functional requirements, each weight value of the QoS properties and IOPE, and several bits of information reflecting a user's condition. The latter two elements are in common with those of a query [7].
2. A Broker Agent extracts a functional task for each abstract work and searches UDDI and returns a list of functionally-suitable Web services matching to the task.
3. A QoS Server registers returned Web services if necessary, and sends them to a Composition Server.
4. A Composition Server applies a predefined optimization algorithm by using the concrete workflow model and it returns the resulting concrete workflow model and C_{exp} to the Broker Agent.
5. The Broker Agent passes them to a requester.
6. A Composition Agent monitors the execution state and triggers a Composition Server according to the defined state chart model (Fig. 3), as in the previous subsection.

4 Conclusion

We have proposed a Web service composition framework based on semantic information and QoS properties. A Broker Agent, A Composition Agent and a Composition Server have been extended and added to our prior work [7]. The main contribution of our framework can be summarized as follows:

- We did not limit the framework to a specific composition algorithm. Hence, various optimization algorithms are utilized in our framework. To do this, we have defined a simple notation that can be used in several algorithms.
- We consider the dynamic aspects of Web services. A composition algorithm [8] considers some dynamic situation, such as changing constraints, however, it does not consider changing candidate sets under their execution environments. The latter is often the case in a mobile environment. Our framework contains those concepts.
- It may be costly to apply an optimization algorithm frequently. We limit this condition to two cases: 1) a QoS of an application which degrades significantly; and 2) a user changes the constraints of QoS properties during execution.
- The gap is reduced between the expected and actual QoS properties using a Composition Agent. The statechart model describes the state of the executing application.

Our framework does have some limitations.

- A QoS Agent operates according to the statechart model on a discrete range with UB, LB, and *diff*, however, the statechart model may be elaborated so as to reflect various aspects of an application in a dynamic environment.
- Security aspects should be considered, because only authorized agents can access Requesters and Providers in the extended framework.

In future research, we will refine the workflow model. Then, we will show that the refined model can be used in any algorithms which can be applied to Web service composition.

Acknowledgments

This research was supported by Kyungpook National University Research Fund, 2006.

References

1. Patel, C., Supekar, K., and Lee, Y.: A QoS Oriented Framework for Adaptive Management of Web Service based Workflows. Lecture Notes in Computer Science, Vol. 2736. Springer-Verlag (2003) 826–835
2. Chaiyakul, S., Limapichat, K., Dixit, A. and Nantajeewarawat, E.: A Framework for Semantic Web Service Discovery and Planning. In Proc. of IEEE Conference on Cybernetics and Intelligent Systems. (2006) 1-5
3. Gue, R., Le, J. and Xia, X..: Capability Matching of Web Services Based on OWL-S. In Proc. of International Workshop on Database and Expert Systems Applications. (2005) 653-657
4. Karakoc, E., Kardas, K. and Senkul, P.: A Workflow-Based Web Service Composition System. In Proc. of International Conference on Web Intelligence and International Agent Technology Workshops. (2006) 113-116
5. Kalepu, S., Krishnaswamy, S. and Loke, S. W.: Verity: A QoS Metric for Selecting Web services and Providers. In Proc. of the International Conference on Web Information Systems Engineering Workshops (2003) 131–139
6. Yu, T. and Lin, K.: The Design of QoS Broker Algorithms for QoS-Capable Web services. In Proc. of IEEE Conference on e-Technology, e-Commerce and e-Service (2004) 17–24
7. Lee, E. J., Jung, W. S., Lee, W. J., Park, Y. J., Lee, B. J., Kim, H. C. and Wu, C. S.: A Framework to Support QoS-Aware Usage of Web services. Lecture Notes in Computer Science, Vol. 3579. Springer-Verlag, (2005) 318-327
8. Zeng, L., Benatallah, B., Ngu, A. H. H. , Dumas, M, Kalagnanam, J., and Chang, H.: QoS-Aware Middleware for Web services Composition. IEEE Transactions on Software Engineering, Vol. 30, No. 5. (2004) 311–327
9. Lin, M., Xie, J., Guo, H. and Wang, H.: Solving QoS-driven Web service Dynamic Composition as Fuzzy Constraint Satisfaction. In Proc. of the IEEE International Conference on e-Technology, e-Commerce and e-Service (2005) 9–14
10. Chen, H., Jin, H., Ning, X. and Lü, Z.: Q-SAC: Toward QoS Optimized Service Automatic Composition. In Proc. of IEEE International Symposium on Cluster Computing and the Grid Cardiff (2005) 623–630
11. Canfora, G., Di Penta, M., Esposito, R., and Villani, M. L.: An Approach for QoS-aware Service Composition based on Genetic Algorithms. In Proc. of the Genetic and Computation Conference (2005) 1069–1075
12. DAML: Web Ontology Languages for Services, http://www.daml.org/services/owl-s/. (2006)
13. Singh, M., and Huhns, M.: Service-Oriented Computing. John Wiley & Sons. (2005)
14. Loke, S. W.: Proactive and Reactive Discovery, Composition, and Activation of Localized Services Accessed from Mobile Devices. In Proc. of the AI Moves to IA: Workshop on Artificial Intelligence, Information Access, and Mobile Computing (2003)
15. Chuang, S., Chan, A.T.S., Cao, J. and Cheung, R.: Dynamic Service Reconfiguration for Wireless Web Access. In Proc. of WWW2003 (2003) 58–67

Traffic Signal Planning Using a Smart Agent System

You-Sik Hong[1], Geuk Lee[2], Cheonshik Kim[3], and Jong Won Kim[4]

[1] Dept. of Computer Science, Sangji Univ.,Korea
yshong@sangji.ac.kr
[2] Dept. of Computer Engineering Hannam Univ., Korea
leegeuk@hannam.ac.kr
[3] Major in Digital Media Engineering Anyang Univ.,Korea
kimcsik@emapl.com
[4] Dept of Electronics Eng., Daegu Univ.,Korea
jwkim@daegu.ac.kr

Abstract. The multi-agent systems approach can provide a new desirable solution to the problems of traffic congestion and traffic accidents. Currently, a traffic simulator is needed to understand and explore the difficulties in agent-oriented traffic control. Therefore, in this paper, we propose an electro-sensitive traffic light using a smart agent algorithm to reduce traffic congestion and traffic accidents. Specifically, we designed and implemented a system to create optimum traffic signals in congested conditions using smart agent algorithms. In order to solve these problems, our approach antecedently creates an optimal traffic cycle of passenger car units at the bottom traffic intersection. Mistakes are possible due to different car lengths, car speeds, and the length of the intersection. Therefore, our approach consequently reduces car waiting time and start-up delay time using fuzzy control of feedback data. Urban traffic situations are extremely complex and highly interactive. The proposed method adaptively controls the cycle of traffic signals even though the traffic volume varies. The effectiveness of this method was shown through the simulation of multiple intersections.

Keywords: Traffic Signal, Smart Agent, Car, Accident.

1 Introduction

Most urban areas currently experience severe traffic jams on street networks. As traffic congestion spreads, there is a need to apply intelligent algorithms to diminish the waste of time, air pollution, and so on. Agent-oriented fuzzy traffic control allows inexact traffic data to be manipulated as a useful tool in designing traffic signal timing plans adaptively. With the development of computation technologies, such as distributed artificial intelligence (DAI), the so called multi-agent systems (MAS) approach, and cooperative problem solving approaches, it may be possible to handle this problem. They offer certain advantages for problem solving: faster response, increased flexibility, robustness, resource sharing,

N.T. Nguyen et al. (Eds.): KES-AMSTA 2007, LNAI 4496, pp. 938–947, 2007.

and better adaptability [1-2]. Congestion wastes fuel and increases air pollution due to increased idling, acceleration, and braking. Since drive time is a non-productive activity, congestion also impacts the regional economy by increasing drive times. Traffic incidents are events that disrupt the normal flow of traffic, usually by physical impediments in travel lanes. Events such as vehicular crashes, breakdowns, and debris in travel lanes are the most common form of incidents. Fuzzy logic is one of these new methods. Before the 1970s, applications of this method were few; after the 1990s, this method became widely used all over the world. Traffic signal control is also one of these applications (Pappis and Mamdani 1997, Naktsuyama et. All 1984; Tzes, Mcshane and Kim 995, Niittymaki 1998) [3,4]. Many other papers (Favilla et al., 1993; Hoyer and Jumar, 1994; Kelsey et al., 1993; Skowronski and Shaw, 1993) [5,6,7,8] present fuzzy systems for a multi-way single intersection. The control problem for a network of intersections, however, is still an important issue in the field of traffic engineering. Longley (1968) [9] identified two types of congestion and proposed a signal control and queue management procedure that aims at reducing secondary congestion. The FLC(Fuzzy Logic Controller) uses three linguistic input variables and one linguistic output. The fuzzy input variables are the elapsed time of the current interval, the number of vehicles crossing an intersection during the green phase, and the length of queuing from the red direction. The output is the extension time, which is calculated using 27 fuzzy rules. This FLC was simulated at less critical intersections. Gomide et al. proposed an FLC with adaptive strategies for fuzzy urban traffic systems. The FLC adjusts its membership functions according to the traffic conditions to optimize the controller's performance. However, all membership functions are adapted concurrently. As a result, the relation from one membership function to the others remains the same. Jamshidi et al. developed a simulator for fuzzy control of traffic systems [10]. In all the work described above, the control and queue handling strategy is static. Traffic signals are not a cure-all for every problem intersection. A signal in the wrong location can contribute to rear-end collisions, excessive delays, unnecessary travel on alternate routes and a more congested traffic flow. We can see a lot of traffic congestion in downtown intersections. In Korea, traffic congestion is so severe that incorrect traffic control signals are actually one of the causes of congestion. Therefore, in this paper, we will analyze traffic circumstances in real-time and alleviate the problem of traffic congestion. With the increasing number of vehicles on restricted roads, there is an increased amount of wasted time as well as a decreased average car speed. This paper proposes a new concept for coordinating green time, which controls 10 traffic intersection systems. For instance, if we have a baseball game at 8 pm today, traffic volume in the direction of the baseball game will be increased 1 hour or 1 hour and 30 minutes before the baseball game. At that time we cannot predict optimal green time even with a smart electro-sensitive traffic light system. Therefore, in this paper, to improve average vehicle speed and reduce average vehicle waiting time, we created optimal green time using fuzzy rules and a neural network. The computer simulation results showed reduced average vehicle wait time, suggesting that it coordinated green time better than the electro-sensitive traffic light system, which does not consider coordinating green time. This paper is organized as follows: section 2 describes the motive for this paper . Section 3 presents optimal green time for traffic intersections. Section 4 explains the determination of an optimal traffic cycle using a smart agent algorithm. Finally, section 5 will give conclusions.

2 Motivations

Traffic engineers use signals to avoid traffic congestion and improve safety for both motorists and pedestrians. Traffic signals are not a cure-all for every problem intersection. A signal in the wrong location can contribute to rear-end collisions, excessive delays, unnecessary travel on alternate routes and a more congested traffic flow. We can see a lot of traffic congestion happening at downtown intersections. In Korea, traffic congestion is so bad that incorrectly used traffic control signal is one of its causes. Therefore, in this paper, we will analyze traffic circumstances in real-time and solve the problem of traffic congestion. system in place. Therefore, in this paper, to improve average vehicle speed and reduce average vehicle waiting time, we created an optimal green time using a smart agent algorithm. The computer simulation results showed reduced average vehicle waiting time, which suggested that it coordinated green time better than the electro-sensitive traffic light system, which does not consider coordinating green time. In this paper, we propose an optimal green time to analyze traffic network flow in cases of saturated flow. Another important feature of saturated networks is the effect of the queue spillback on travel times in up stream links. The concept of agents has been the outgrowth of research during the past half-century on AI. The idea of a software entity that could perform tasks on behalf of a user was well established by the mid 1970s. Using this background, it is possible to apply concepts of reasoning, knowledge representation, machine learning, and especially planning to traffic control. The practical applications of agents have more pragmatic origins. Currently, many problems are very complex, both within the area of traffic control and beyond. A multi-agent system (MAS), which refers to all types of systems composed of multiple autonomous components, is utilized for many applications. There have been several studies on agent-oriented traffic control systems. Traffic accidents on icy roads are reduced by half, on average, with the introduction of this smart agent road system. A sudden change in the weather, such as sudden heavy fog and rain, is another cause of accidents which endanger drivers' safety. The driver's risk can be reduced significantly if it is possible to predict the road conditions in advance, such as knowing that the road is wet, icy or foggy. The most important individual smart agent road systems are meant to prevent traffic accidents.

3 Optimal Green Times for Traffic Intersection

In this section, we define the basis of the fuzzy rule for traffic congestion circumstances and describe our calculation method. The traffic volume balance is held at each signalized intersection of the traffic network for a certain sampling period. It can be described by the following equation.

$$C_e(green) = G_{rte}(car) + G_{rti}(car) - G_{rto}(car) \tag{1}$$

where :
$G_{rte}(car)$: Excess incoming traffic cars
$G_{rti}(car)$: Incoming traffic cars
$G_{rto}(car)$: Outgoing traffic cars

If there are so many vehicles in the line, we cannot know how many vehicles are going to go straight or turn right. Therefore, to determine optimal green time, the algorithm must predict the number of cars going straight, rather than turning right

$$G_e(green){=}C_{xe}(in){*}R_{tn}(exp\ in){+}S_{tr}(exp\ in), G_{tr}(green){=}N_1{*}W_L{*}C_{xt}(in, out) \tag{2}$$

where:
$C_{xe}(in)$: Excess incoming traffic cars
$R_{tn}(exp\ in)$:expected cars for right turn
$S_{tr}(exp\ in)$:expected cars for straight

Figure 1 explains how to create optimal green time, offset, red interval and waiting queue depending on the different lengths of the lower and upper traffic intersections.

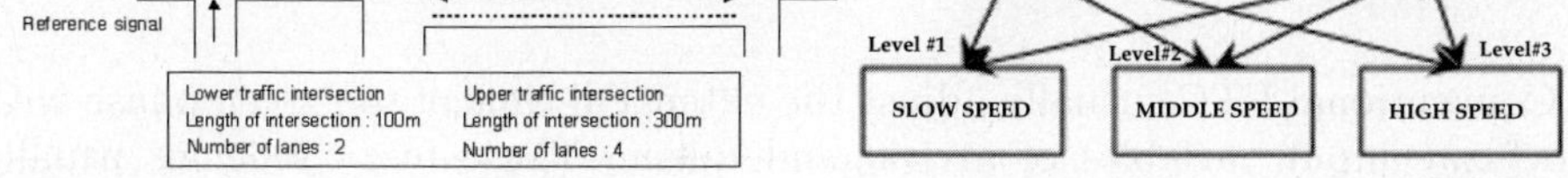

Fig. 1. Architectural view of proposed agent-oriented traffic simulation system

The fuzzy traffic control simulator was developed by an agent-oriented paradigm. We implemented the simulator suitable for traffic junction networks in Visual C++ to analyze traffic conditions and to calculate the optimized traffic signals. Agent-oriented programming allows the simulation to be built in a modular fashion, making it easily expandable and maintainable. We designed an agent-oriented fuzzy traffic control system in the form of Figure 1. As you can see in Figure 1, there are three important modules, specifically the user interface

(UI), fuzzy traffic control (FTC), and coordination modules. The user interface module presents the user with a graphical display representing the simulated traffic control environment. The fuzzy traffic control module adaptively controls the cycle of traffic signals, regardless of traffic volume sets. The coordination module cooperates and coordinates traffic signals between neighboring intersections. There are Car, Lane, Detector, Signal, and Crossroad agents in the FTC module of Figure 1. Figure 2 shows the overall architecture of each single intersection, which is a member consisting of target 4 * 4 crossroads. The information about cars detected between the front and rear detector for each lane is used as an input parameter for the signal control algorithm module in the crossroad agent. The detector agent is responsible for perceiving cars. The signal control algorithm module of the crossroad agent uses a predefined signal control algorithm with the detected information about cars to calculate the extension time of the current controller. The extension time adapts the signal state of the Signal agent. There are many types of control for which the Crossroad agent is responsible. We can add several control algorithms to the Crossroad agent easily by changing only the signal control algorithm module of this agent.

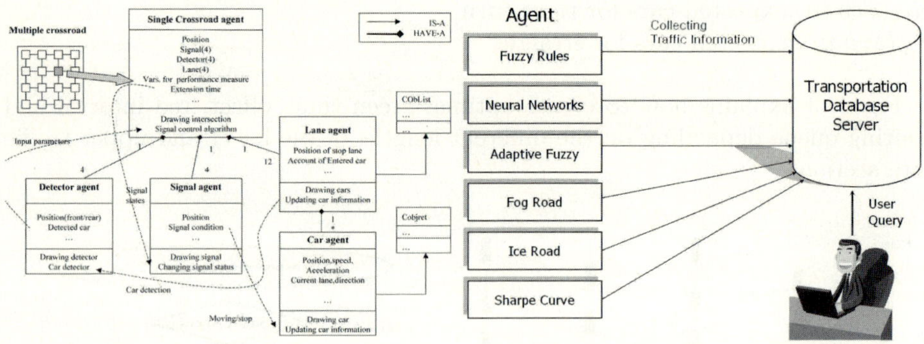

Fig. 2. The overall architecture of each single intersection consisting of 3 * 3 crossroads using smart agent

Conventional FTCs usually adjust the extension time of the green phase with the fuzzy input variables of arrival and queue. These fuzzy schemes usually utilize arrival and queue values while the controllers are in operation. Thus, these methods are inadequate for an intersection where traffic volume varies. Some traffic elements can be taken into consideration using the fuzzy control rules to diminish congestion. For an FLC to be practical, it must be designed to diminish the traffic congestion of intersections with variable traffic volumes. Thus, in this paper, we present an FLC using different control rules and different maximum extension times according to traffic volume; because of this variability, vehicles can flow smoothly at intersections.

4 Proposed Algorithms for Traffic Congestion

In this section, we described algorithms and calculation methods to supply optimal green time signals. In figure 3, a, b, c, and d represent different road conditions. The numbers 0.8 -1.0 implicate a heavily rainy road or frozen road, a regular rainy road can be expressed by 05-0.7. Hereby, as explained in Figure 7 P1, P2, P3 show the road's incline considering whether conditions. The connection lines represent the vehicle type. Therefore, with heavily rainy road conditions, the additional factors of road incline and vehicle type, in case of a large vehicle (0.8), or a regular car (0.6) will also be included in the calculation.

In this paper, the neural network consists of one input layer, one hidden layer, and one output layer. We use a supervised learning process, which adjusts weights to reduce the error between the desired output and real output for green time. This is depicted as follows. (1) Initialize weights and offset (2) Establish training pattern (3) Compute the error between the target pattern output layer neural cell (t_j) and the output layer neural cell (a_j)

$$e_j = t_j - a_j \tag{3}$$

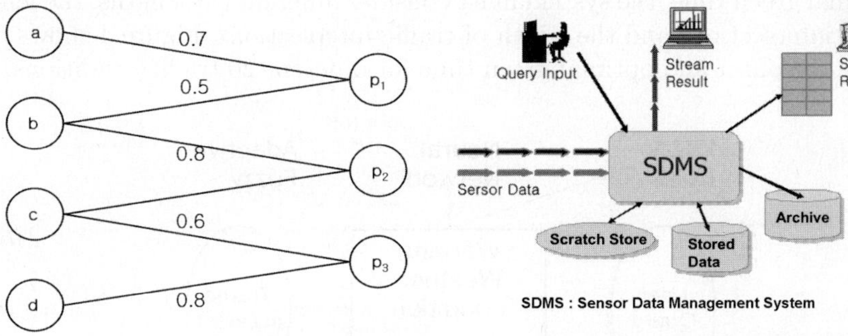

Fig. 3. Fuzzy conversion factor considering whether conditions

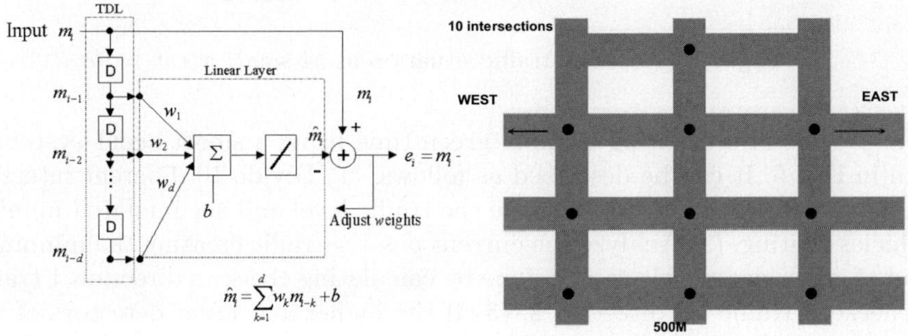

Fig. 4. Optimal traffic cycle of 10 Traffic intersections

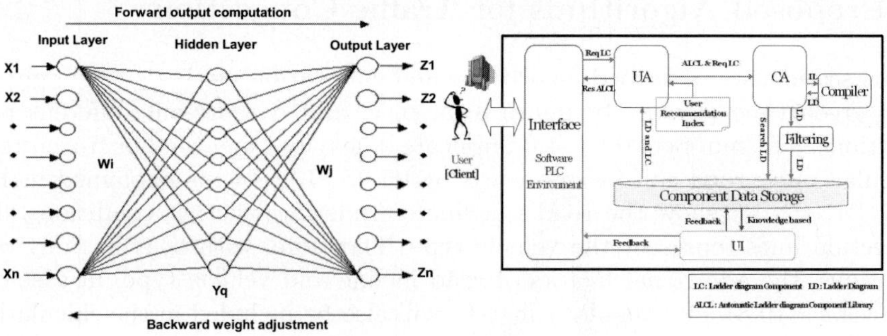

Fig. 5. Simulation of smart agent system using fuzzy-neural

(4) Calculate weights between the input neural cell(i, j) by the following equation

$$W(new)_{ij} = W(old)_{ij} + ae_{iaj} \tag{4}$$

(5) Repeat the process from number (2) above.

The process is repeated until optimal green time is calculated. In order to create optimal green time, the system must consider different car lengths, the length of traffic intersections and the width of traffic intersections. Figure 4 shows how to create the offset and optimal green time for different 20 traffic conditions.

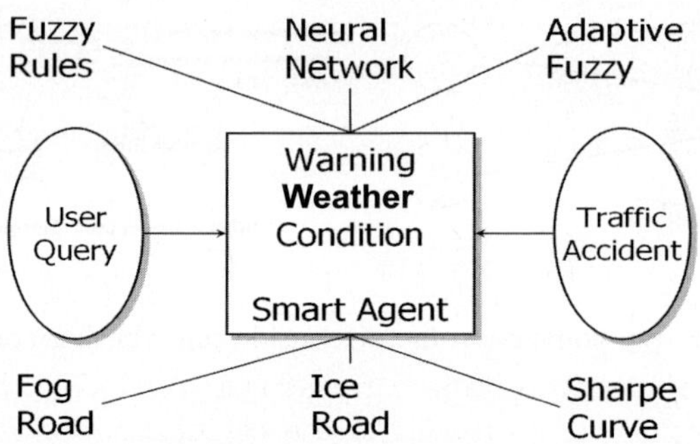

Fig. 6. Forecasting traffic situation using smart agent

The process of estimating optimal green time using a smart agent system is shown in Fig. 5. It can be described as follows: (1) Divide 10 different intersections into 3 by 3 amounts and analyze the traffic level and accumulated number of vehicles waiting. (2) Analyze the current passing traffic creating the minimum period of green signal; this can be done by calculating check-in direction 1 traffic and checkout traffic for direction 2. (3) If the higher and lower detectors of the intersection are both reading "on," and the accumulated number of vehicles is on "high," the intersection is overloaded with cars. (4) Calculate the minimum period of green.

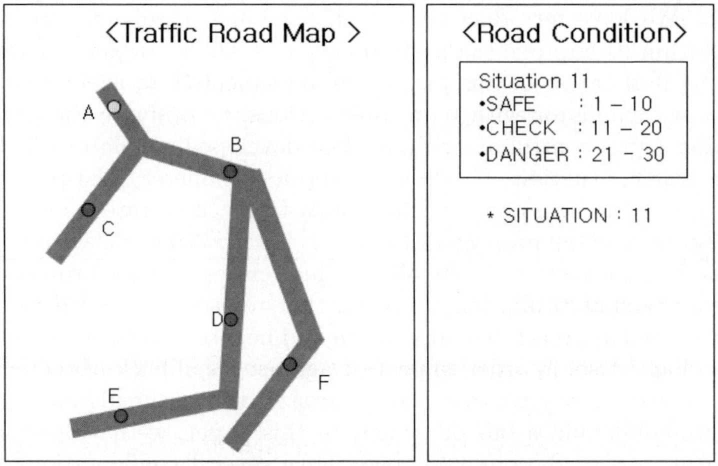

Fig. 7. Simulation result using smart agent

Table 1. Simulation result using smart agent

Traffic Road	Weather conditions				Speed conditions			Smart agent	
Road I.D.	Rain	sun	Snow	Fog	Lv1	Lv2	Lv3	Danger	Normal
ABC	44	100	34	70	83	17	21	O	X
CDE	83	88	66	50	16	24	81	O	X
ABC	64	100	23	46	91	30	10	O	X
EFG	86	91	56	23	12	88	14	X	O
ABC	94	100	34	78	21	31	87	X	O
CDE	82	97	45	56	13	92	17	O	X
ABC	74	88	58	55	18	24	91	X	O

Figure 6 shows the composition of the sensor database. This system collects traffic circumstance at real-time. The collected information is stored in a database. In addition, this stored information is used to forecast future traffic conditions. Figure 7 shows that while a vehicle is being driven, the driver will be apprised of accidents occurs or obstacles ahead by the smart agent

In this paper, the system tested can automatically calculate an optimal green time and speed level when it identifies an obstacle, traffic accident or poor weather conditions using smart agent algorithm. Table 1 shows simulation result using the smart agent.

5 Conclusion

A new agent-oriented FLC has been proposed. Conventional fuzzy controllers are designed to improve the performance of controllers in the case of uneven traffic flow. We have introduced the traffic volume parameter into the smart agent algorithm to improve the performance of FTC for urban crossroads with time-varying flow rates. In this paper, we implemented an agent-oriented multiple crossroad simulator with n n intersections, to evaluate the performance of the traffic signal control algorithms. The developed simulator can be tested in specific traffic situations by changing input parameter. The proposed FLC has been applied to crossroads with various traffic flow rates. The simulation results indicate that the proposed FLC yields lower average delays and lower average costs than conventional controllers. This verifies that the proposed method fits the real needs at traffic junctions. Further research is needed to develop a more coordinated approach to cooperation and negotiation between neighboring traffic junctions. Also, in order to protect against a spillback phenomenon from multiple crossroads, we have to consider car size such as the distinction between a small automobile and a bus or trailer. In this paper, we have performed our simulation at crossroads, but real roads have several configurations including crossroads, 3-branch streets, 5-branch streets, and so on. Thus, experiments are needed for various practical traffic conditions. This paper proposed that we can establish a safe priority order using fuzzy rules. As a result, we can create optimal green time using the smart agent algorithm. This paper shows that the smart agent algorithm will be able to forecast optimal traffic information, estimation of destination arrival time, roads under construction and dangerous roads. With a computer simulation, we showed that the traffic congestion phenomenon generated under highly saturated traffic conditions is improved using the smart agent algorithm.

References

1. N. Findler, G. Elder, "Multiagent Coordination and Cooperation in a Distributed Dynamic Environment with Limited Resources," Artificial Intelligence in Engineering, Vol.9, pp.229-238, 1995.
2. Rodriquez, J., Noriega, P., Sierra, C., Padget, J., FM96.5 A Java-based Electronic Auction House, Proc. of the Second International Conference on the Practical Application of Intelligent Agents and Multi-Agent Technology, 1997
3. Nakatsuyama, M.,Nagahashi, H., and Nishizuka, N., (1984). Fuzzy Logic Phase Controller for Traffic Junctions in the One-way Arterial Road, IFAC-World Congress, preprints, Budapest 1984, pp13-18.
4. Tzes, A., McShane, W.R., and Kim, S., (1995). Expert Fuzzy Logic Traffic Signal Control for Transportation Networks, Institute of Transportation Engineer 65th Annual Meeting, Denver USA. pp154-158.
5. Niittymaki, Jarkko, P., (1997). Isolated Traffic Signals-Vehicle Dynamics and Fuzzy Control, Ph.D. Thesis, Helsinki University of Technology, Civil and Environmental Engineering.
6. Favilla, J., Machion, A. and Gomide, F. (1993). Fuzzy Traffic control: Adaptive strategies, Second IEEE International Conference on Fuzzy Systems, San Francisco, CA, pp506-511.
7. Hoyer, R. and Jumar, U. (1994). Fuzzy control of traffic lights, Third IEEE International Conference on Fuzzy Systems, Orlando, FL, pp1526-1531.

8. Kelsey, R., Bisset, K. and Jamshidi, M. (1993). A simulation environment of fuzzy control of traffic systems, XII IFAC-World Congress, Sydney, Australia, pp553-556.
9. Skowronski, W. and Shaw, L. (1993). Self-learning fuzzy traffic controller for a traffic junction, I. European Congress on Intelligent Techniques and Soft Computing-EUFIT 93, Aachen, Germany, pp751-761.
10. Longely, D., 1968. A control strategy for a congested computer-controlled traffic network. Transportation Research 2, pp391-408. M.
11. Jamshidi, R. Kelsey, and K. Bisset, Traffic fuzzy control: software and hardware implementations, Proc. of fifth IFSA World Congress, pp.907-910, 1993

Entity Based Message Authentication System

Young-Soo Kim, Sung-Ja Choi, Gang-Soo Lee, and Geuk Lee

Department of Computer Engineering, Han Nam University, Tae-jeon, 300-791, Korea
{experkim,sjchoi,gslee,leegeuk}@hannam.ac.kr

Abstract. A critical promotion factor of e-Commerce is message authentication, the procedure that allows communicating parties to verify received messages as authentic. The client-server system has been traditionally used for authentication service. As distributed computing environment has been complicated more and more, Agent system is developed and used extensively for security service. We have designed and successfully tested an agent model for security service which we have named NMAP(New Message Authentication Protocol) to deal with message authentication in the public key cryptosystem. The NMAP's performance has been shown by various trials of message authentication based on file size and key length. The NMAP's overall performance on the processing speed is shown to be superior to that of client-server system named PGP. This suggested system will provide an active e-Commerce and non-interactive authentication service.

Keywords: authentication, security, agent, public key cryptosystem, entity based cryptosystem..

1 Introduction

The most important variables in active e-Commerce are message unforgery, message non-repudiation, message unalteration and message authentication. We analyzed and made a comparison of client-server system named PGP and agent system named NMAP(New Message Authentication Protocol) for security service[1],[2],[3]. Therefore we designed the message authentication protocol for the sole proprietorship or small-medium sized company and named it NMAP as security agent system.

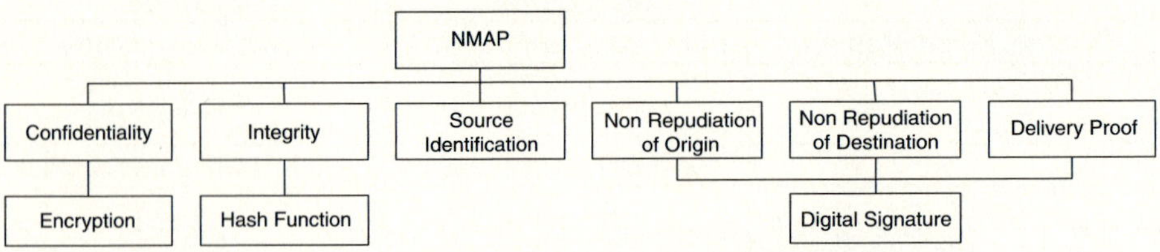

Fig. 1. Basic Structure of Security Agent System named NMAP

As seen in Fig. 1, security agent system named NMAP provides the confidentiality, integrity, source identification, non-repudiation of origin, non-repudiation of destination and delivery proof required by the OSI security architecture and

N.T. Nguyen et al. (Eds.): KES-AMSTA 2007, LNAI 4496, pp. 948–955, 2007.

X.411[4],[5]. The design of the authentication header was mainly studied to add security to the message of the entity based cryptosystem protocol that is capable of configuring and safely transferring the encryption message to the unspecified persons by using only public information[6].

2 Problems and Improvement Schemes of Client-Server System with Security Model

X.400 recommends two methods[7]: the signature method after encryption and the transmission method after saving at the time of message token configuration. Both have problems in being exposed to the risk of forgery of the encrypted message and signature value, and the system knowing the receiver's encryption key[8]. As well, the authentication based public key cryptosystem requires much more processing time and storage for the certificate and it has limited transfer ability not to transfer the message to any user who could not receive the issued public key certificate from the certificate agency. As PGP may have multiple numbers of ID in which the same user is related to the multiple public key, the public key ring was established to efficiently manage the public key. In addition, with multiple public keys distributed, so the management of public key is highly complicated[9]. The comparison of client-server system named PGP and security agent system named NMAP is shown in Table 1.

In some cases, it is the Contact Volume Editor that checks all the pdfs. In such cases, the authors are not involved in the checking phase.

Table 1. Comparison of Message Security System(Client-Server System and Agent System)

Classification	PGP(Client-Server System)	NMAP(Agent System)
Message Processing time	a little slower	a little faster
Transmission	Public key owner	Unspecified users
User Identification	public key	Identifier
Key Generation	public key created by private key	public key created by private key
Encryption Method	session key used	Secret key used
Token Configuration	signature method after encryption	Encryption method after signature
Transmission Method	transmitted after saving	direct transmission
Deliverycertifying Calculation	Plain text	cipher text
Key Ring	establishment needed	establishment not needed
Certificate	Necessary	not necessary
Content Confidentiality	IDEA, RSA	DES, RSA
Content Integrity	MD5, RSA	MD5, RSA
Origin Authentication	encrypt-then-sign	Sign and encrypt
Non-Repudiation of Origin	encrypt-then-sign	Sign and encrypt
Non-Repudiation of Delivery	encrypt-then-sign	Sign and encrypt

3 Problems and Improvement Schemes of Client-Server System with Security Model

The authentication is classified into the entity authentication and message authentication. The entity authentication during the connection of communication has been actively studied, while the non-interactive message authentication has not been sufficiently studied. The developed NMAP deals with the message authentication, but it does not include a reliable third party. A message authentication is the process in which the received message's authenticity is identified by the parties exchanging the messages. The message authenticity is composed of non-forgery, unalteration and non-repudiation and it is performed through the public key[10].

The message authentication, as seen in Fig. 2, implements the authentication by placing the authentication-related parameters on the message header part and by defining the digital authenticated data structure by placing the various security-related parameters within this structure[11].

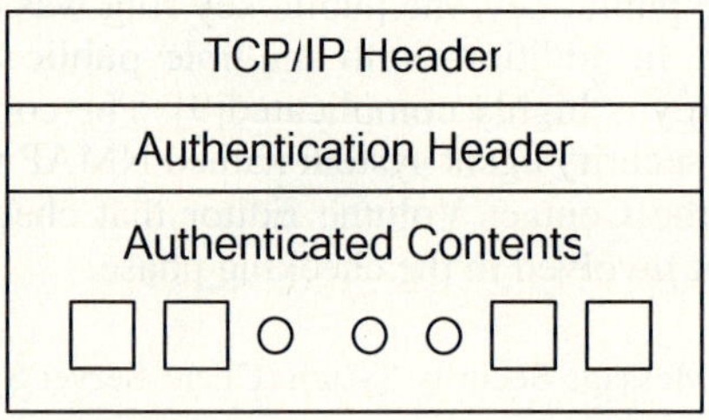

Fig. 2. Capsule Structure of Authentication Parameter

The methods of protocol implementation are as follows: the combined authentication header and message to send, the message sending after first sending the authentication header and then waiting for the authentication and processing by sending the parameter comprising the header to receive the positive response[12]. The NMAP applied the method to transfer the combined authentication header and message to minimize overhead caused by the hand shaking, in the transmission environment of a sole proprietorship and small-medium sized company.

There are Encrypt-and-Sign, Encrypt-then-Sign, and Sign-then-Encrypt, In the digital authentication method which are performed by combining the public key cryptosystem and electronic signature[13].

The Encrypt-and-Sign method requires the overhead for the additional encryption to provide the electronic signature with the confidentiality. The Encrypt-then-Sign is open to dispute as to the validity of the electronic signature created by the encryption text, forged through the electronic signature and tested by the encryption text. Sign-then-Encryption has a weak point in which the electronic signature cannot be notarized by the third party because only the decryption key owner can test the signature. NMAP applied the Sign-then-Encryption method which is designed to provide the public service of the electronic signature based on the future certificate.

4 Agent Design and Performance Analysis of Security Agent System Named NMAP

4.1 Protocol Design of Security Agent System Named NMAP

In NMAP, the sender transmits the encrypted message using the identifier in the form of character text, and the receiver receives the issued private key, calculated as its ID and KDC secret information from KDC, to recover the message as shown in Fig. 3.

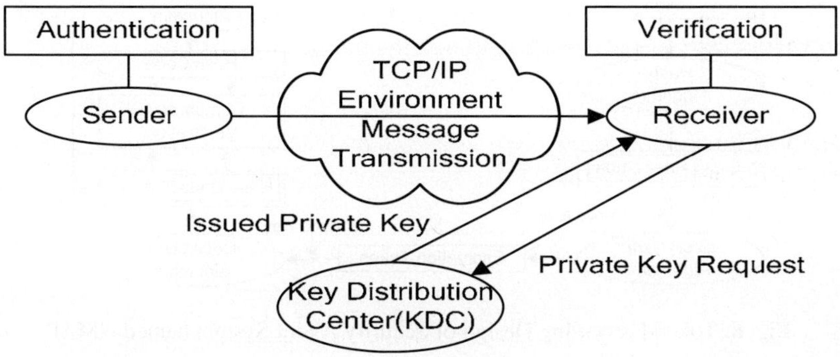

Fig. 3. Concept Map of Security Agent System named NMAP

NMAP maintains the message security through a series of continuous digital signatures for the calculated hash as shown in Fig. 4.

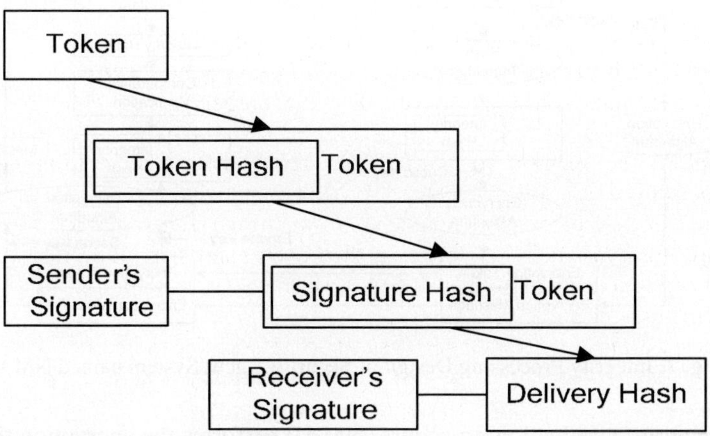

Fig. 4. Hash Sequence of Security Agent System named NMAP

NMAP public key creation method creates the private key from the public key, unlike the certificate based public key encryption as shown in Fig. 5.

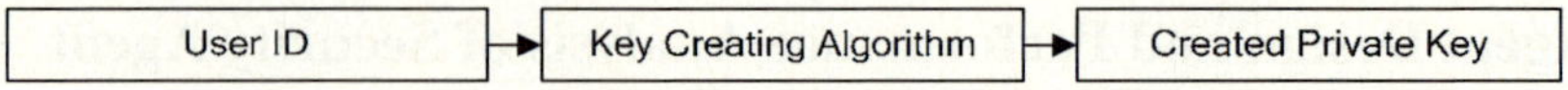

Fig. 5. Key Creation Method of Security Agent System named NMAP

In addition, it defines the digital signed data structure, within the message token, and places all the various authentication-related parameters in this structure, to deliver them to the receiver and to perform the authentication as shown in Fig. 6.

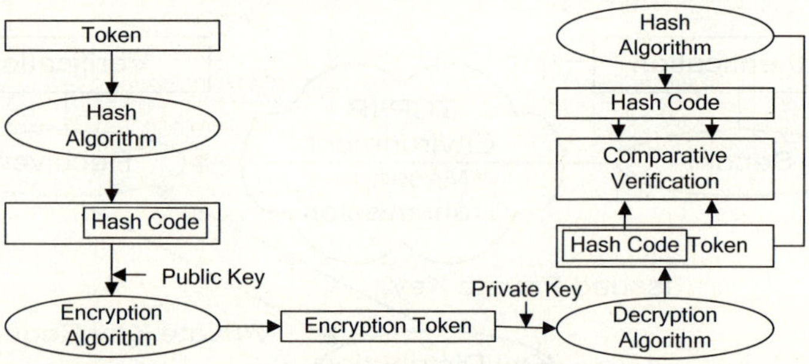

Fig. 6. Token Processing Design of Security Agent System named NMAP

The NMAP integrity processing verifies that the message content is not changed during the transmission, calculates the fixed size of hash code as the message digested, adds it to the message and then sends it after the encryption as shown in Fig. 7.

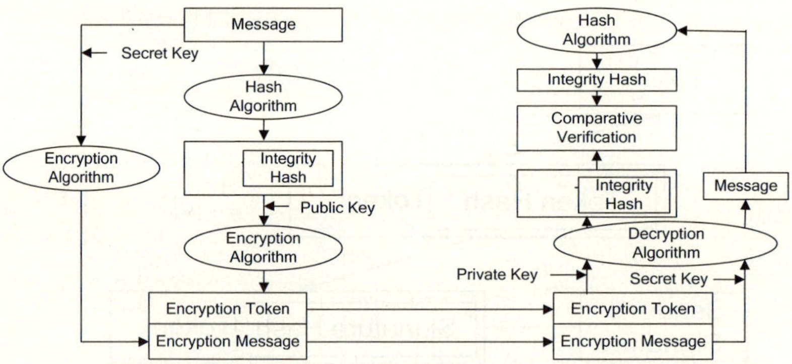

Fig. 7. Integrity Processing Design of Security Agent System named NMAP

In the integrity verification procedure, NMAP performs the encryption through the electronic signature for the hash code by using the private key and public key, and performs the encryption into the secret key again, together with the message to send. The receiver verifies the integrity in comparison with the hash code after identifying the electronic signature as shown in Fig. 8.

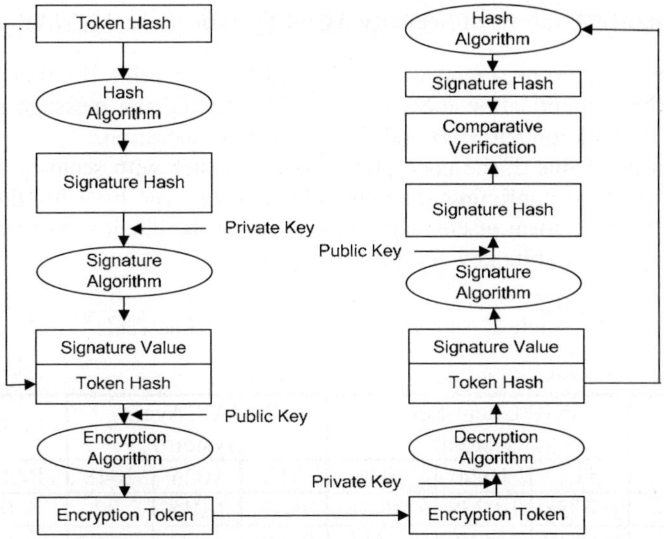

Fig. 8. Electronic Signature Processing Design of Security Agent System named NMAP

For proof of delivery not to deny what the receiver has actually received, the sender is required to check the sent receipt by sending the receipt after calculating the delivery proof value when the receiver reads the received message and after requesting the delivery proof service when delivering the message as seen in Fig. 9.

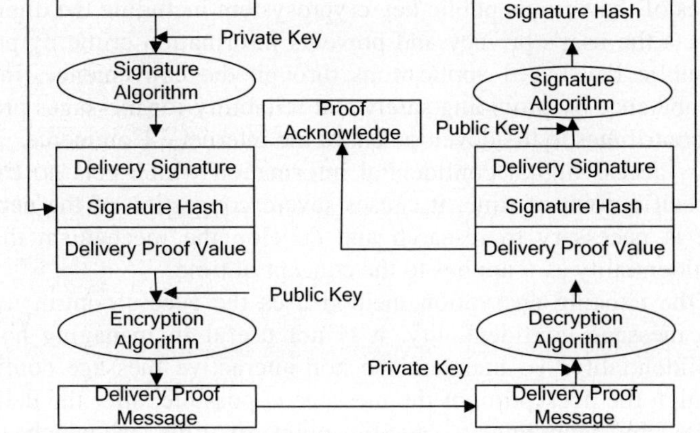

Fig. 9. Delivery Proof Processing Design of Security Agent System named NMAP

4.2 Performance Analysis of Security Agent System Named NMAP

Using a mock simulation, a trial was performed by measuring the time required for configuring the authentication message by differentiating the message size and key length under the Celeron 850MHz and Windows 2000 operation.

As seen in the Table 2, the encryption speed is faster with security agent system named NMAP which configures the authentication message by using the public key and secret key in the form of character string than in the client-server system named PGP. which uses the public key cryptosystem method.

Table 2. Authentication Submission Time Table

(Unit : second)

Classification	PGP(Client-Server System)			NMAP(Agent System)			PGP-NMAP
	512	1024	2048	512	1024	2048	PGP-NMAP
100K	0.4358	1.0338	2.8418	0.42	1.018	2.826	0.0158
200K	0.8718	2.0678	5.6858	0.838	2.034	5.652	0.0338
300K	1.3078	3.1018	8.5298	1.258	3.052	8.48	0.0498
400K	1.7436	4.1356	11.3736	1.676	4.068	11.306	0.0676
500K	2.1796	5.1696	14.2176	2.096	5.088	14.134	0.0836

5 Conclusion

The NMAP named security agent system's protocol uses only public information to configure the encryption message and to safely send the message to unspecified targets. It reduces the complexity in the key management by creating a private key at the time of decryption, which basically uses the user identifier in the form of character string and delivers the various security service as an user-friendly convenience with the advantages of the various public key cryptosystem including the digital signature. NMAP protects the user's privacy and prevents information crime by promoting the use of the public key based applications through the convenience. In addition, it enhances competence by providing safety and reliability for messages processed over the internet, contributes to the development of the internet e-Commerce.

Many users access digital confidential information which gets no confidentiality value at a specific point of time. It causes severe congestion of the network traffic. Therefore, it is necessary to research and develop the mechanism that maintains message confidentiality as it applies to the concept of time.

Although the existing encryption method uses the access control mechanism to maintain the message confidentiality, it is not useful in managing non-interactive message confidentiality. To maintain the non-interactive message confidentiality, a method in which the decryption of the message is possible, after the defined specific point of time, by including a specific point of time in which the message confidentiality value does not exist any more at the time of encryption, is required.

Acknowledgments. The Author (Young-Soo Kim, Sung Ja Choi, Gang-Soo Lee and Geuk Lee) supported by fond of the second stage BK21 of korea ministry of education and human resource and this work was supported by a grant from Security Engineering Research Center of Ministry of Commerce, Industry and Emergy.

References

1. Puliafto, A., S. Riccobene, and M. Scarpa, “An analytical comparison of the client-server, remote evaluation and mobile agents paradigms,” Proceedings of First International Symposium on Agents Systems and Applications(1999)248-292
2. Denning, D., "Timestamps in Key Distribution Protocols," Communications of the ACM(1981)
3. Kaufman, C., P. Radia and S. Mike, Network Security : Private Communication in a Public World, Prentice Hall(1995)
4. Mitchell, C., M. Wallker, and D. Rush, "CCITT/ISO Standards for Security Message Handling," IEEE J.Sel. Areas in Comm.(1989)51-524
5. Stallings, W., Network and Internetwork Security: Principles and Practice, Prentice Hall(1995)
6. Shamir, A., "Identity-based cryptosystems and signature schemes," Advances in Cryptology: Crypto(1985)47-53
7. King, J., "X.400 Security," Computers & Security(1992)707-710
8. Manros, C., The X.400 Blue Book Companion. Twickenham, England: Technology Appraisals(1981)
9. Schneider, B., E-Mail Security : How to Keep Your Electronic Messages Private, John Wiley & Sons, Inc(1995)
10. Study on the Authentication Implementation and Access Control, Korea Electronic & Communication Research Center(1996)
11. Burrows, M, M. Abadi, and R. Needham, "A logic of authentication," ACM Trans. on Computer Systems(1990)18-36
12. Kille, S., "Implementing X.400 and X500 : The PP and QUIPU Systems," Artoch House inc(1991)
13. Bellare, M., and C. Namprempre, "Authenticated encryption," Springer-Verlag, Berlin Germany(2000)531-545

The Design and Testing of Automated Signature Generation Engine for Worms Detection

Sijung Kim[1], Geuk Lee[2], and Bonghan Kim[3]

[1] Dept. of Computer Science, ChungJu National University, Chungju , Chungbuk, Korea
sjkim6183@hanmir.com
[2] Dept. of Computer Engineering, Hannam University, Taejeon, Korea
leegeuk@hannam.ac.kr
[3] Dept. of Computer & Information Engineering, Chongju University, Chongju, Chungbuk, Korea
bhkim@chongju.ac.kr

Abstract. We have proposed automated signature generation engine for unknown attack detection. For this proposal, we have studied signature engine divided into header field and payload field. Especially, in payload field, we proposed signature generation agent which can be presented by using Suffix tree, and Longest Common Subsequence(LCSeq) among them is used to generate new signature automatically. Through the test, Snort signature and generated signature by using Longest Common Subsequence(LCSeq) are compared and evaluated.

1 Introduction

Intrusion detection system is the system that detects and manages intrusions in advance by observing suspicious intrusions as an intrusion attempts. Current intrusion detection system causes a lot of problems such as traffic heavy network, switched network, asymmetry network, response impossible after detection, long time required to response, too much analyzed data accumulation, judgment error and so on[1].

Active Response System, an intrusion system, has been developing in order to solve these problems. Established information protection methods are not reflected in the step of system design, so they have original limits to detect various attacks that are possible to be happened after being served and to present an effective response to them. Therefore, active response intrusion detection system in the terms of network infra is able to perform functions such as real-time intrusion detections, reverse tracking, restoring and so on.

Now, international research on active response intrusion detection system is the early stage. There has been researching on both an intrusion detection system as a model and an active response policy linked to. And it has been mentioned shortly on the technology guide.

Therefore, this paper has studied a agent that can generate signatures automatically for active response approach which is needed in intrusion detection tools. Through

N.T. Nguyen et al. (Eds.): KES-AMSTA 2007, LNAI 4496, pp. 956–963, 2007.

this foundation research, a signature generation engine that automatically generates signature headers and signature payloads by detecting unknown attack has been proposed and advantages and disadvantages has been presented by testing.

2 Related Works

Various international researches have been making progress for active signature generation. Kai Hwang has developed CAIDS(Cooperative Anomaly and Intrusion Detection System)[2]. CAIDS builds anomaly detection system(ADS) and network-based intrusion detection that is operated in a way of dialog via signature generator. Experimental discovery of CAIDS clearly shows an advantage that unites NIDS and ADS. Jinqiao. Yu has developed TRINETR. TRINETR is a cooperative architecture and consisted of three main elements which are alert aggregation, knowledge-based alert evaluation, and alert correlation[3]. Zian Zhang has developed RAIRS(Rollbackable Automated Intrusion Response System)[4]. RAIRS is a rollback mechanism that decides automatically whether the response will be roll-backed or not. RAIRS is able to manage false alarm originated in False Positive and it roll-backs any unnecessary response. This makes to overcome disadvantages of conservative response system. Hyang-Ah Kim has developed Autograph. Autograph suggests a way to detect worm signature automatically through the parts of suspicious flow selection and signature generation[5]. Ke Wang has proposed anomalous payload-based worm detection and signature generation through PAYL[6].

3 Automated Signature Generation Engine

Automated signature generation engine covers a function that generates an attack packet which is received from traffic anomaly detection engine and protocol anomaly detection engine into a signature rule. A signature rule is consisted of a field area defined IDS, a packet, a header and payload area.

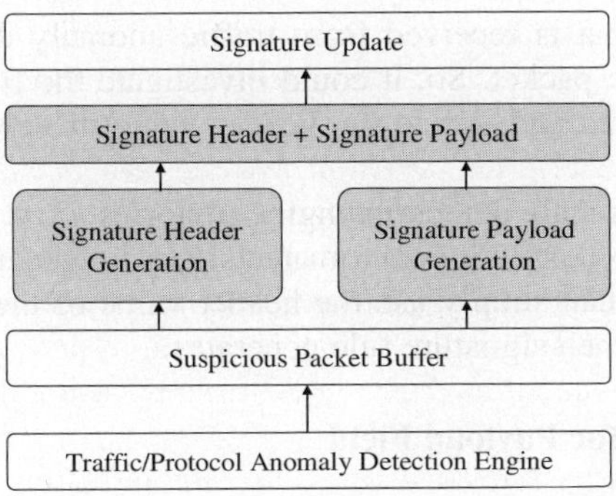

Fig. 1. The Procedure of Signature Rule generation

As shown in Fig.1, it should be divided into rule generation for header area and rule generation for payload in order to generate unknown packet which is detected from traffic anomaly detection engine and protocol anomaly engine into new signature rule. ID, Level, Name will be defined upon system policy if the rule is generated by defined field of the system.

3.1 Rule Generation for Header Field

Among the headers of unknown packet which are received when rules for header area are generated, the fields that are referred can be defined as in Fig.2.

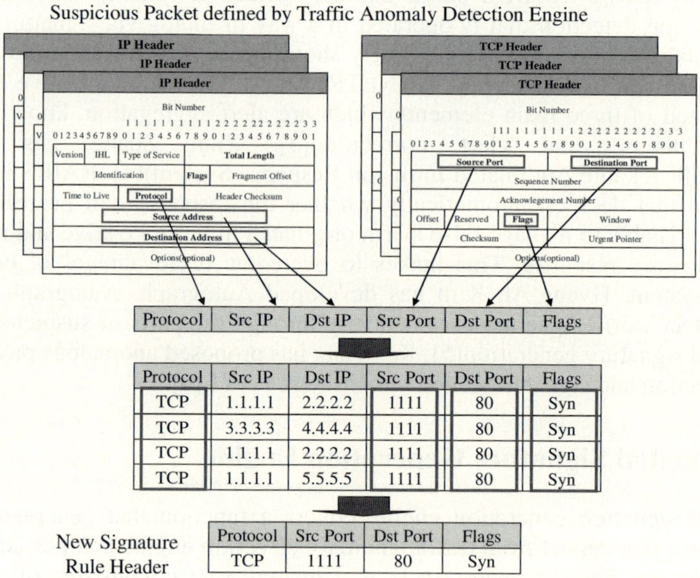

Fig. 2. Traffic Anomaly Packet Header Rule Generation

Suspicious packet that is received from traffic anomaly detection engine is consisted of more than one packet. So, it could investigate the field which has the same value of rule header area and define the header value of new signature value. Fig.2 shows that process.

Because protocol anomaly detection engine performs a test as Packet-By-Packet, it is a united packet that is passed into automated signature generation. Therefore, signature generation engine can simply use the header value of the packet which is applicable to the rule field when signature rule generates.

3.2 Rule Generation for Payload Field

The united string should be extracted among packets in order to store specific string from payload for a few packets received from traffic anomaly detection engine into signature field of signature rule. To extract identical strings, all strings should be dealt. Having done that, Suffix Tree will be used among index data structure that was built the results as a data structure.

The approach method of extracting identical string between two strings by using the results of whole procedure on strings using Suffix Tree can be defined as three kinds. There are some tries to go around pattern through these methods, that is, identical signature are been able to generated for Fragmentation and Mutated packet. And also the approach method that is to minimize false positives has been considered[7].

String Equality(SE): SE has the most intuitive approach. It extracts continued identical strings among suspicious packets. This method is able to extract highly accurate identical strings and it is very useful in the point of reducing false positives

However, if packet has been changed in detail, identical string could not be found. When a worm passes, if it changes its payload or cause to occur fragmentation in detail, there is no way to extract signature.

Longest Common Substring(LCS): LCS has in common with SE to extract identical string accurately. SE only arranges identical string when there are a few identical strings. But, LCS is different because of that extracts the longest string among identical strings. Once the longest string has been extracted, it is possible to extract signature even if there are fragmentation and mutation of some packets in detail.

Longest Common Subsequence(LCSeq): LCSeq is similar to LCS. LCS is an Order-insensitive method that looks for CS without regard the order of identical strings, but LCSeq is an Order-insensitive method that considers the order of identical strings. And also, LCS looks for continued identical string, but LCSeq looks for discontinued identical string.

Therefore, it is good to look for signature of polymorphic worm that has a transformed form when it duplicates by itself. However, there is a disadvantage to happen higher false positive than LCS. Next is an instance of the process that looks for continued identical string by LCS. The identical string is extracted from two given strings, S1= 'ABCDEFCBGH' and S2= 'FECBGAGFHE'. If it extracts signature by using LCSeq, the identical string ‘FCBGH’ is extracted like Fig.3.

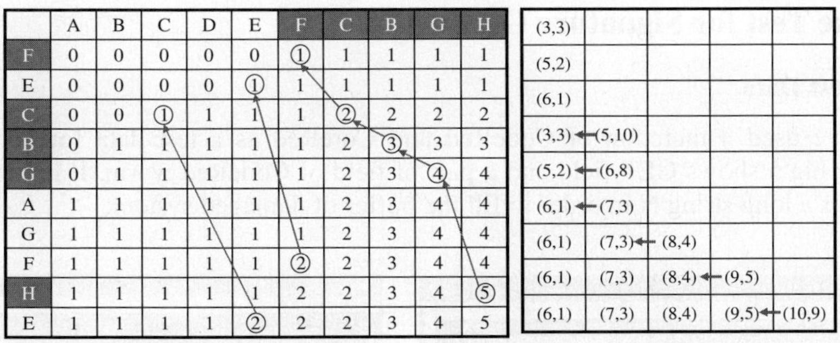

Fig. 3. Signature Extraction by Using LCSeq

In the chapter, we have studied on signature generation with applying three of string algorithms that have input suspicious packets. When applied these algorithms, normal string used in common is also included in signature. For example, in the case of HTTP, there are some fields ‘GET’, ‘HTTP/1.1’, ‘HOST’, and so on. If these strings would be generated as signatures, there would be quite a lot of false positives to happen.

Therefore, in order to decrease those, normal packet and suspicious packet should be applied at the same time. And then if except signatures that are generated from normal packet among signatures from anomalous packet, it is possible to decrease false positive pretty much.

In the example of Fig.4, generated signatures from the result of applying suspicious packet first in order 'HTTP/1.1', 'GET', '\xff\bf', and '\xde\xad'. Also, generated signature from normal packets are 'HTTP/1.1', 'GET', etc. therefore, when it deceases candidates generated from normal packet among signatures which are consisted of secondary candidates, '\xff\bf' will be generated.

To do so, it is able to decrease much false positives compared to signature generated applying just suspicious packet, if except signature generated through normal packet.

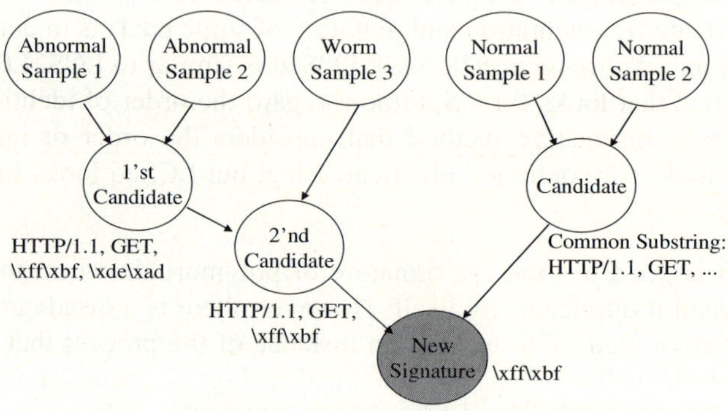

Fig. 4. New Signature Generation

4 The Test for Signature Generation

4.1 Test Data

We have used a mutation of CodeRed and CoreRed as a test data for anomalous packet. Fig.5 shows GET field and a part of field of CodeRed worm. It has a feature that uses a long string N in order to fill the buffer of damaged system.

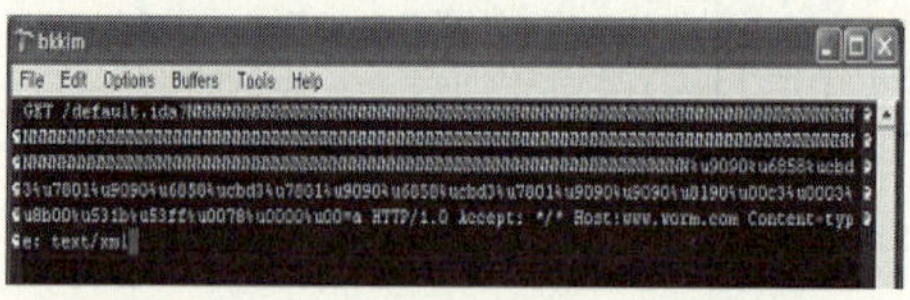

Fig. 5. Payload of CodeRed Worm

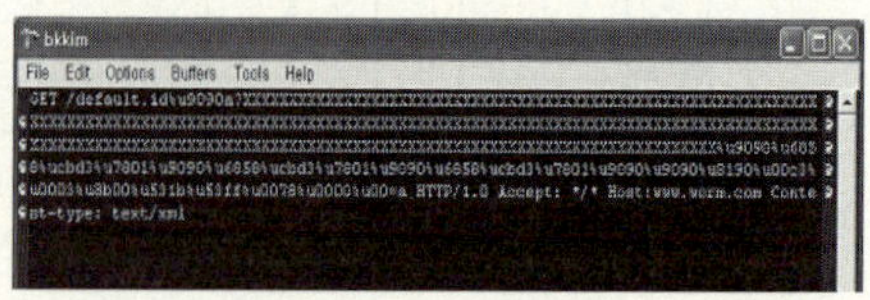

Fig. 6. Payload of CodeRed Worm Mutation

Fig.6 is shown the difference that it inputs any value into ida values and uses X string rather than N string. The reason to use any value into ida values is to test whether it generates signature for the worm which inserts NOP(Non Operation) code into strings and is duplicated to go around established signature.

4.2 New Signature Generation Test

The test order is as followed.

1. Extraction of Common Substring from Normal packet
2. Signature Candidate Extraction by applying LCSeq to Mutation of CodeRed and CoreRed.
3. Common String Extraction between Common Substring and the first candidate
4. Generation of new signature list from the first candidates after eliminating Common String

Outputs data part of normal packet with priority. Then, it outputs the length of the identical string between two packets and the identical string. If it is not identical, it displays '*'. The identical string is consisted as Table 1. It is except from identical strings under 3byte.

Table 1. The identical string that was extracted

1: GET
2: /default.html
3: HTTP/1.0
4: Accept:
5: Referer:
6: http://www.
7: Accept-Language: ko
8: Accept-Encoding:gzip, deflat
9: User-Agent: Mozilla/4.0(compatible; MSIE 6.0; Windows NT 5.1;SV1)
10: Host:
11: .com
12: Content-type:text/html
13: Connection: Keep-Alive
14: Cookie: EmpasPCID=112373632918726888; RMID=81fef54742f;

Table 2 shows the identical string which confirms Common Substring of the first signature candidates and as input.

Table 2. Identical String between Signature Candidates and Common Substring

1: GET /default.
2: HTTP/1.0
3: Accept:
4: host: www.
5: .com
6: Content-type: text/

Therefore, when it extracts apart from strings above, the identical string generates four of new signatures like Table 3.

Table 3. New Signature Generation

1: ida*a? 2: %u9090%u6858%ucbd3%u7801%u9090%u6858%ucbd3%u7801%u9090%u6858 %ucbd3%u7801%u9090%u9090%u8190%u00c3%u8b00%u531b%u53ff%u0078% u0000%u00=a 3: worm 4: xml

4.3 Comparison to Snort Signature

Table 4 below shows a comparison signature generated through Snort signature and the test.

Table 4. Comparison between Snort Signature and New Signature

Snort Signature	New Signature
1: .ida 2: ida?	1: ida*a? 2: %u9090%u6858%ucbd3... 3: worm 4: xml

For the detection of CodeRed worm, Uricontent of Snort uses '.ida' or 'ida' and so on, and new signature shows four of signature generations. Snort signature is a rule made by manual work by specialist after coded worm happened, however, new signature is a signature made by using LCSeq, so it is short of elaboration. But, it has an advantage that generates new signature automatically. Also, for the worm type with a feature which has a transformed form itself while being duplicated, it is possible to detect polymorphic worms by marking as '*' for the distance between 'ida' and 'a' like number1 item of new signature. However, as has considered the result, there is an expectation to happen false positive because multiple signatures happen. So, in order to apply this try to the system, how much false positive on signature generated automatically would be decreased will be the main issue.

5 Conclusion

This paper has researched on the automated signature generation of intrusion detection tool. In order to generate auto response policy against attacks, it has presented the agent that generates signature organically on the base of united model of NIDS and ADS.

To generate signature automatically through unknown Attack detection, it has proposed an automated signature generation. For this proposal, signature generation engine has been divided into header area and payload field to study. Especially, in the payload field, signature generation methods have been proposed which are presented by Suffix Tree, and has generated new signature automatically by using Longest Common Subsequence(LCSeq) among them. Through the test, snort signature and the signature generated by using Longest Common Subsequence has been compared and evaluated.

References

1. Carl Endorf, Eugene Schultz, Jim Mellander, "*Intrusion Detection & Prevention*", McGrawHill, 2004.
2. Kai Hwang, Ying Chen, Hua Liu, "*Defending Distributed Systems Against Malicious Intrusions and Network Anomalies"*, IPDPS'05, pp.286a, 2005.
3. Jinqiao Yu, Y. V. Ramana Reddy, Sentil Selliah, Srinivas Kankanahalli, Sumitra Reddy, Vijayanand Bharadwaj, "*TRINETR: An Intrusion Detection Alert Management System*", 13th IEEE (WETICE'04), pp.235-240, 2004.
4. Jian Zhang, Jian Gong, Yong Ding, "*Research on automated rollbackability of intrusion response*", Journal of Computer Security, Vol.12, No.5, pp.737-751, 2004.
5. Hyang-Ah Kim, Bread Karp, "*Autograph: Toward Automated, Distributed Worm Signature Detection*", 13th Usenix Security Symposium (Security 2004), August, 2004.
6. Ke Wang, Gabriela Cretu, Salvatore J. Stolfo, "*Anomalous Payload-based Worm Detection and Signature Generation*", RAID 2005, September, 2005.
7. Newsome. J, Karp. B, Song. D, "*Polygraph: automatically generating signatures for polymorphic worm*", Security and Privacy 2005 IEEE Symposium, pp.226- 241, May 2005.

Data Conversion Agents Between Heterogeneous Using IC Information

Gui-Jung Kim[1], Jung-Min Oh[2], and Jung-Soo Han[3]

[1] Department of Bio Medical Engineering, Konyang University
26, Nae-Dong, Nonsan, Chungnam, Republic of Korea
gjkim@konyang.ac.kr

[2] Department of Computer Engineering, Hannam University
133, Ojeong-Dong, Daedeuk-Gu, Daejeon, Republic of Korea
mimdein@hanmail.net

[3] Division of Information & Communication, Baekseok University
115, Anseo-Dong, Cheonan, Chungnam, Republic of Korea
jshan@bu.ac.kr

Abstract. This paper proposes the solution of compatibility problem between heterogeneous by converting automatically CAD data into proper data for the PCB production process by using IC information. It also proposes a method for data conversion agent by analyzing data of each IC item designed by CAD from the assembling process. This paper has a purpose of maximizing work efficiency with NC Programming automation and contributing to maximize productivity by NC program conversion of different Mounters and Mounter Data & Line Balance optimization.

1 Introduction

This paper proposes the solution of compatibility problem between heterogeneous by automating manual labor for converting design data into process data in the PCB(Printed Circuit Board) production. PCB production process is produced by facilities like Loader, Surface Mount Clean Machine, Screen Printer, Chip Mounter, and so on. However, these facilities do not recognize data designed by CAD. A lot of minor enterprises operate data conversion by hand and there are errors to happen and the precision to fall, so faulty goods are often produced and it costs a lot. This paper proposes the method for data conversion agent by analyzing each IC's information designed by CAD in the PCB assembly process. That is, to maximize work efficiency with NC(Numerical Control) programming automation is the purpose of this paper. It develops error extraction algorithm to prevent data error. It makes item, locations, angle data add and correct them as suitable data for PCB production process in order to prevent CAD designer's human error. It takes location information from IC information as a base key IC and converts data by join computing, angle modification, coordinates modification, layer separation. As the result, NC program conversion is done between different Mounters and the productivity will be the maximum due to the optimization of Mounter Data & Line Balance. It makes PCB mass production possible and also helps get more accurate PCB production and high reliability.

N.T. Nguyen et al. (Eds.): KES-AMSTA 2007, LNAI 4496, pp. 964–971, 2007.

2 Related Work

The purpose of NC software is to maximize work efficiency by NC programming automation with CAD data. It makes NC programming conversion possible for heterogeneous facilities and maximizes the productivity by removing human error. There are two big kinds of NC software developed theses days. First, the software is that can do data conversion with linking up with CAD. If someone knows technical knowledge of graphic design and CAD, she or he can use the software easily. Also, it has an advantage to check PCB product designed now on the graphic[1]. However, if the user does not have any technical knowledge of graphic design, it is not easy to use. The matter of fact, it is difficult for minor enterprises that produce PCB production to secure experts. Second, it is the software that aims at instructions[2]. It has been developed as simple NC software, so it is strong to convert data in a short time. But, users have to input instructions into software by themselves. That means they should know instructions very well and operate. Even though NC software developed now is a very expensive facility, it is not easy for users to approach. The software has been designed for certain people, so it does not really satisfy minor enterprises. Therefore, it is quite common for general minor enterprises to prefer to operate work by hand. NC software from now on should minimize the operating time and be easy approachable from users. It also should remove ambiguity and contradiction by reducing human error. And it should offer interface that makes users confirm the process of data conversion and minimize the data conversion time, and let users trust the result of conversion 100%[3].

3 Data Conversion Agents

In this paper, we compare the data about the IC with CAD data and propose data conversion agent to transfer the data. The data conversion agent converts CAD data with PCB data. The steps was as follows.

Step 1 - Data error detection agent
Step 2 - Data extraction agent
Step 3 - Join operation and Angle modification agent
Step 4 - Coordinates modification agent

3.1 Item Repository

IC item information repository is consisted of information such as item location, angles (X,Y values), standard, item name, composition amount and so on. This information is the extract of necessary information out of CAD data from the conversion process. The repository has two Tables. Because the CAD designer makes CAD data as a form of two files, those two files will be stored as themselves at the Tables. Each Table will be united through error detect algorithm, data addition and modification after. This united Table will be output two files according to Layer's value. These two files will be applied to different PCB production line [4][5][6]. Data structure that is converted with the information of the IC item repository to data that can be recognized by PCB facility is the same as Table 1. 'Location No' of 'Table1' and 'RefDes' of 'Table2' have location information each other [7]. 'Q ty' of 'Table1' means the number of location information of 'Location No', which was used for data extraction earlier. The other fields of 'Table1' are Vendor information. 'ParType'' field and 'PartDecal' field of 'Table2' are specific information of item. The field 'Pins' means the number of item pins. The field 'Layer' is the information that the item should be connected to front or back of the PCB board and it will be used as data to choose PCB production line. 'Orient' is an angle of item. 'X' and 'Y' fields mean coordinates. The other fields include specific information of item.

Table 1. Data Structure of PCB Item Repository

Table	Field Name	Data Type	Explanation
1	Specification	String	resistance
	Q_ty	Long	Amount
	Location_No	String	location
	Approved Vendor 1_st	String	Vendor information
	Approved Vendor Parts No_1	String	Vendor information
	Approved Vendor 2_nd	String	Vendor information
	Approved Vendor Parts No_2	String	Vendor information
2	PartType	String	Item information
	RefDes	String	location
	PartDecal	Long	Item information
	Pins	Long	Pin numbers
	Layer	Bool	Front, back
	Orient	Long	Angle
	X	Long	coordinates
	Y	Long	coordinates
	SMD	Bool	Item information
	Glued	Bool	Item information

3.2 Data Conversion Agents Process

Step 1 - Data error detection agent: There is a possibility for CAD data to include human error because it is made by CAD designer. In order to convert CAD data into PCB data, it should first extract and modify error of CAD data itself. Error detection agent grasps error type of CAD data and modifies it. It inspects if item name, location, angle's value are agreed to the information stored in the PCB production process. If they do not, it regards it to human error. When item information does not agree with stored database of material management system among CAD data, agent outputs error message and gets a new item name at the same time. The origin of CAD data should be taken in the middle of PCB board. That's why there is no more than 70 of coordinate value of (X, Y) which is centering of the origin. Therefore, if it happens to be more than 70, it will be regarded as an error and print out a message. Angles exist only 0°, 90°, 180°, and 270°. However, if there is other value or no value found, it will be regarded as an error and print out a message. Fig. 1 shows error detection agent. It is agent that seeks for errors by item name, coordinates, and angle among CAD data stored two Tables. It seeks errors on the each field of one table and then it repeats until the last Record.

```
Specification, Location_No, 1_st, Parts_No_01, 2_nd,
Parts_No_02 save in Database Table1.
PartType, RefDes, PartDecal, Pins, Layer, Orient, X, Y save
in Database Table2.
FOR(Last Recode of Table 1)
     IF(Item Name of Material Management DB != Item name
      of Table1)
          Agent event (output message)
     ENDIF
Save Item Name in Database Table1
FOR(Last Recode of Table 2)
     IF(Orient of Table2!=0 || Orient of Table2!=90 ||
        Orient of Table2!=180 || Orient of Table2!=270)
          Agent event(output message)
     EN IF
     IF(Coordinate of Table2>70||Coordinate of Table==NULL)
          Agent event(output message)
     ENDIF
Save Orient, Coordinate in Database Table2
```

Fig. 1. Error Detection Agent

Step 2 - Data extraction agent: Because Cad data are designed by designer's convenience, it has unnecessary data for the PCB production process. This unnecessary data will be picked out through the PCB production process analysis. According to the analysis result, the data that is needed indeed among CAD data will be extracted. Sometimes it happens that the data deleted by user's convenience could be useful information for the PCB production process. So, it extracts data by deleting data unneeded and adding data needed. Fig. 2 shows the agent process of storing CAD data after extracting at repository. After passing theses steps, data with no error will be stored at repository.

Step 3 - Join operation and Angle modification agent: Using agent, Item information will be converted by comparing IC item data and CAD data. It operates Join operation by seeking the same data from comparing 'Location_No' field of 'Table1' and 'RefDes' field of 'Table2'. The values of 'Location_No' and 'RefDes' have information of that it assembles IC item to which location of PCB board. This data is taken as the primary key. If there is the same data after comparing two Tables, it stores 'Specification' of 'Table1', 'Q_ty', 'Orient' of 'Table2', 'X', 'Y', and 'Layer' at another

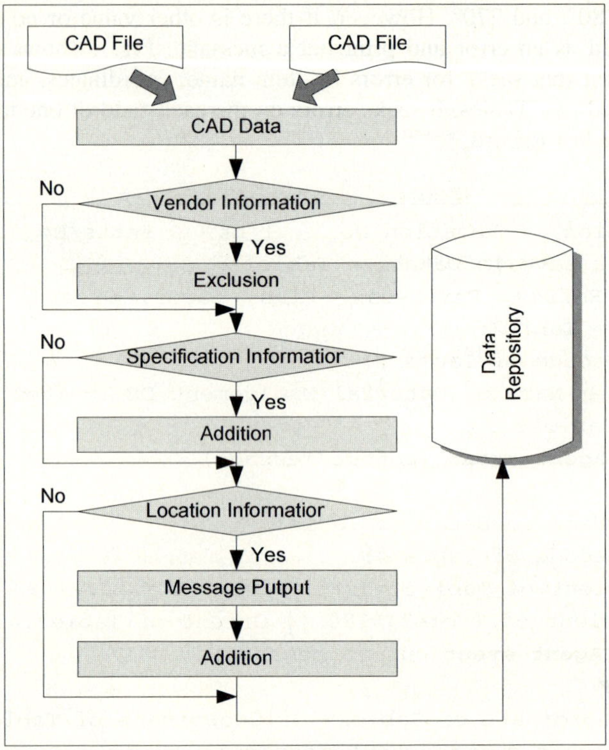

Fig. 2. CAD Data Extraction Agent

Table inside the database. Surely the values of 'Location_No' and 'RefDes' are stored at the field of Join. These values are the minimum information that tells which item goes which location on the PCB board. In the case of ‘Orient’, data item angle, CAD designer might give any angle value. That is why the designer applies the standard angle of IC item irrespective of PCB board. Therefore, he/she converts the standard angle by PCB board and stores it at database. The standard angles of IC item are 0°, 90°, 180°, and 270°. Contrary to designer’s expectation, IC item might be connected to the front or the of PCB board. Then this layer happens to be changed. Let’s take a look at Fig. 3. If the designer who would expect to connect IC item to the front decides to do this to the back, + pole and – pole will be exchanged. Therefore, when it happens to modify layer necessarily, the angles of IC item are changed 0° to 180° and 90° to 270°. Angles are stored at the ‘Orient’ field of Table generated after computing Join of database.

Step 4 - Coordinates modification agent : Data conversion agent processes data error detection, data extraction, join- operation, angle amendment, coordinate amendment, and layer separation. And it converts CAD data into PCB data. Each IC item data has a coordinate value (X,Y) centering around any origin of designer[8]. The coordinate values used by CAD data and PCB production process have different system. Therefore, it should confirm that coordinate value of CAD data has been

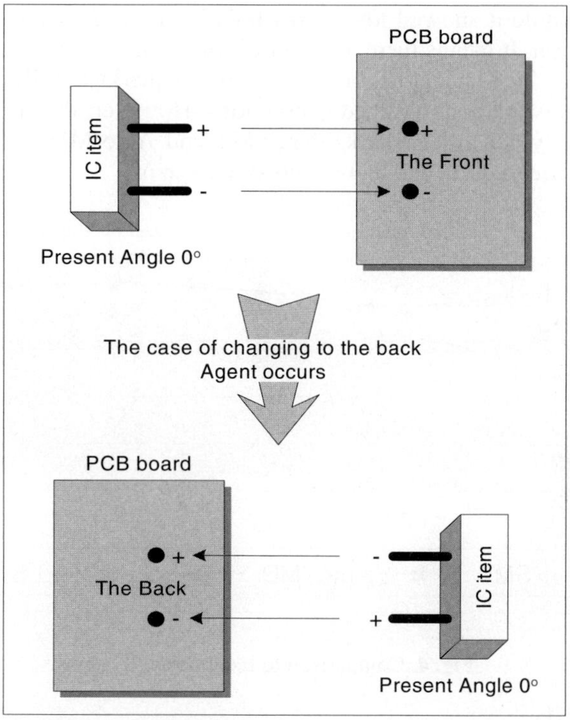

Fig. 3. The Case of changing connection layer of IC item

designed centering on which location and modifies it as suitable coordinate value for the PCB production process. The designer decides a coordinate value somewhere in the middle of the board, while a corner on the board is used as an origin in the PCB production process. So you can find where the designer decided the origin by the origin algorithm. And then search for the closest corner from this origin among four corners of PCB board and change this closest corner to origin. After deciding new origin, the coordinate value of IC item will be replaced by new origin.

4 Evaluation of Performance

The method proposed in this paper has been compared to the established data conversion system. The comparison criterion was the conversion process that converts CAD data that has more than 500 item data to suitable data for the PCB production process. The conversion process has been done total 35 times. The comparison criterion evaluates the quality of software by according to how much system helps operate and it is easy and convenient. Evaluation list includes efficiency increase (UI), productivity improvement (PI), training decrease (TD), and user expropriation increase (EI)[9][10][11]. Proposed item takes 100 for satisfaction and has been asked to 8 users. Fig. 4 shows comparison of the established system to the proposed method by this paper. First I chose SMT system that is possible to link to CAD and Auto-MD system that aims instructions among established systems. SMT showed high marks for UI and ED, while it showed low marks for PI, ED, and TD. SMT especially got very low estimation. It means there are a lot of users who do not use CAD. Auto-MD got high marks for PI, TD, and EI, but it showed low marks for UI, and ED. It is estimated as a complex method of aimed instructions. However, the proposed method by this paper generally got higher marks than SMT and Auto-MD. It says that it offers convenient interface to users and there is no system error.

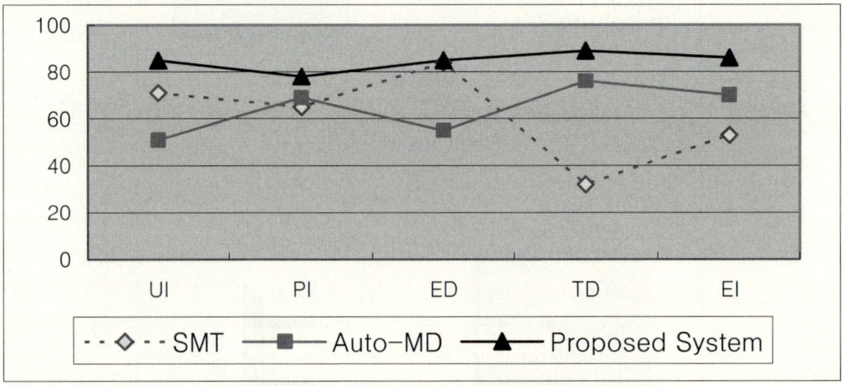

Fig. 4. Comparison to Established Systems

5 Conclusions

This paper aims to convert CAD data to suitable data for the PCB production process by using IC item information. It modifies the data automatically by extracting data needed in the PCB production process from CAD data and confirming location information and omitted information. It converts CAD data after extracting and removing errors to the data which is used in the PCB production process. Also, the proposed system has been verified that it is how easy and convenient from the viewpoint of established systems and users through comparing to established systems. According to that, this paper could effectively solve the problem of data compatibility between heterogeneous with IC item information. Therefore, there is a NC program conversion between different Mounters and Human errors can be removed in advance. It is also possible to maximize the productivity by Mounter Data & line Balance optimum. This makes massive production of PCB possible and there will be more accurate PCB production and high reliability achieved.

References

1. http://www.smtkorea.co.kr
2. http://www.changil21c.co.kr/
3. Kim Kyung-Soo, "PCB Circuit Design", HongReung Science Publish, 2004.
4. E. Berkean and E. Kinnen, "IC layout planning and placement by dimensional relaxation", IEEE Int Conference on Computer Design, pp.449-451, 1995.

5. Watanable, Hiroyuki, "IC Layout Generation and Compaction using Mathematical Optimization", The University of Rochester, Ph.D. 1994.
6. Crama Y, Klundert J, Spieksma F, "Production planning models for printed circuit board assembly", 1999.
7. Craig S. Mullins, "Database Administration: The Complete Guide to Practices and Procedures", Addison-Wesley Professional, 2002.
8. Park S.S and Sohn J.H, "Efficient operation of a surface mounting machine with a multi thread turret", International Journal of Production Research", pp.1131-1143, 1996.
9. M.lea, "Evaluation User Interface Design, user Interface Design for Computer System", Halstead Press, 1988.
10. T.Mandel, "The Elements of User Interface Design", Wiley, 1997.
11. Song Young-Jea, "Object-Oriented Modeling and Software Engineering of CBD focus", E-Han Publish, 2004.

User Adaptive Game Characters Using Decision Trees and FSMs

Tae Bok Yoon[1], Kyo Hyeon Park[2], Jee Hyong Lee[3,*], and Keon Myung Lee[4]

[1,2,3] School of Information & Communication Engineering
Sungkyunkwan University, Suwon, Korea
[4] School of Electrical and Computer Engineering,
Chungbuk National University, Korea
{[1]tbyoon,[2]megagame}@skku.edu, [3]jhlee@ece.skku.ac.kr,
[4]kmlee@cbnu.ac.kr

Abstract. Recently, various ways are being explored for enhancing the fun of computer games and lengthening the life cycle of them. Some games, add realistic graphic effect and excellent acoustic effect, and make the tendencies of game players reflected. This paper suggests the method to collect and analyze the action patterns of game players. The game players' patterns are modeled using FSM (Finite State Machine). The result obtained by analyzing the data on game players is used for creating game-agents which show new action patterns by altering the FSM defined previously. This characters are adaptable game-agent which is learnable the action patterns of game players. The proposal method can be applied to create characters which play the role of partners with game players or the role of enemies against game players.

1 Introduction

User Modeling is the user cognitive process in which the data on users occurred under a specific system environment are collected and analyzed to achieve the objective of the system more effectively [1]. User modeling technique is being exploited diversely to analyze the profile information of users in the many fields such as games, educations, ubiquitous technologies and web mining [2].

Particularly in the game environment, the services based on analyzing gaming data of players, collected in the process of game activity, and utilizing the user modeling technique is being researched actively. In the game 'Black & White by Peter Molyneux', a partner game-agent named Creature shows the adapted appearance in accordance with the game operation of the game player, and in the game Sims, an game-agent also shows relatively different appearance according to the player's game operation [3]. However, until now, in the field of games, user modeling techniques which have been used in console games and package games, rely on specific features of individual games.

This paper suggests D-FSM (Dynamic Finite State Machine) method which can create new game-agents behavior patterns to which the game players` tendencies are

---

[*] Corresponding author.

N.T. Nguyen et al. (Eds.): KES-AMSTA 2007, LNAI 4496, pp. 972–981, 2007.

reflected. This method alters the FSMs for game-agents which were set with static action patterns in the initial phase of games, by adding information on the initial FSMs based on the data on game players which are collected while of gaming. In this paper, for analysis of the collected data, decision tree method which is a kind of machine learning methods is used; the initial FSMs are altered with the analysis result. Along with this, script languages which set action patterns are defined so that D-FSM can be used flexibly for games in various categories D-FSM suggested in this paper can be used for following cases:

First, the suggested D-FSM can be used as a tool for designing game-agents in the initial phase of game production. Ordinarily game developers spend a lot of time and efforts to create game-agents in the initial phase of game production, in creating game-agents' behavior patterns. In designing computer games, if real game players` information is used, it will be much easier to create game-agents and thus the developing time and effort will be reduced. Moreover, it is possible to create more realistic game-agents of which behaviors are similar to human players.

Second, the suggested D-FSM can be used for creating partner game-agents in games. There are cases where game-agents share the same objectives with game players and cooperate with them. Since the D-FSM can build game-agents from human players' play record, game-agents which has abilities similar to game players and play games can easily be created.

Third, the suggested D-FSM can be used for creating hostile game-agents in games. Since the action patterns of monster game-agents in most games are also static, human players easily penetrate the patterns. This fact can be a factor which dwindles the life cycle of the games. The model dynamically generated from human players' play record can be used for creating hostile game-agents. Then, the game-agents will have a similar the abilities and action patterns adaptive to game players.

In Section 2, the related research is introduced and in Section 3, the player modeling using D-FSM and its application to game-agents is described. The experiment result is examined in Section 4. Finally, Section 5 concludes this paper and suggests future works.

2 Dynamic-FSM Method Using Player's Gaming Data

2.1 Decision Tree in Games

Decision tree method, which is a technique widely used for data classification in data mining illustrates the patterns in the data in a tree structure by analyzing the given data. The reason why it is frequently used is that its' output is in a simple formal and can be easily understood by users. It has been applied to 'Black & White' and has a big favor from users. In the game, the player can give positive or negative feedbacks by caressing or beating the creature. A decision tree is generated from the feedbacks. Sometimes the creature checks if given objects will release his hunger using the decision tree and sets the degree of aggressiveness. For example, if the player treats a creature with violence rather than praise more often, the creature gets violent more and more.2.1 FSMs for game-agents.

2.2 FSM (Finite State Machine)

The state machine or finite state machine, which is one of widely used software designing patterns, is a machine which has finite numbered states [4]. An FSM is defined with the finite collection of states and transitions, between states. In applications, each state of an FSM may be attached with a specific action. Game-agents conduct the actions attached to the present event state as their tasks. When a certain occurs, a state transits into another state will follow. FSMs are most widely used technique in games as it has a simple structure. FSMs are sometimes used mixed with other artificial intelligence techniques. Figure 1 illustrates an FSM.

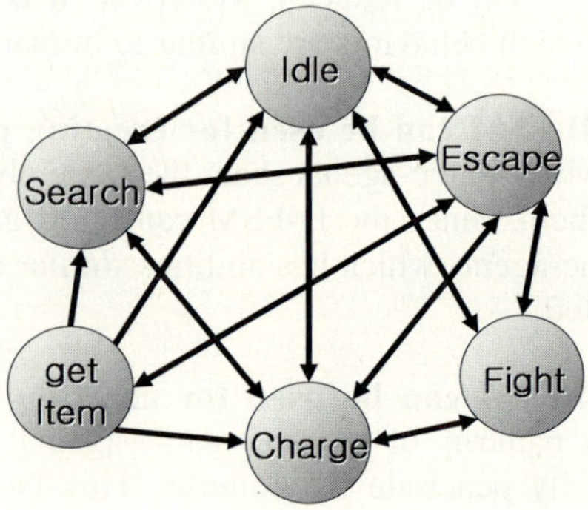

Fig. 1. 6 states and transition relation

Here it has 6 states and each state can be transited to the other states. For example, if the game-agent in a state of "Idle" is attacked by the player, its' state will change into a state of "Fight" and if the game-agent is injured, its' state will change into a state of "Escape".

2.3 Elementary Game-Agents and Improved Game-Agents

Game-agents are the characters which response to game players among various elements in games. They can be merchants in stores, guides showing ways, teammates with game players, monsters, environmental elements such as the climate and geographic varieties and the tool items which game players use. The behavior patterns of these game-agents are directly set in source code or script, so never change during game playing. Game-agents created by an improved method, usually show various responses according to the users` tendencies.

Table 1 and 2 illustrate the functions of game-agents in usual games and the examples of games to which elementary and improved game-agents are applied.

Table 1. Elementary Game-agents

Description	· Game players\` abilities are not reflected to by game-agents. · Game-agents show the same response to the same situation. · Game-agents \` game patterns can be grasped. · Game-agents show the same static action patterns according to situations.
Applied Examples	Space Invaders, Pac-Man , Donkey Kong and the like

Table 2. Improved Game-agents

	Ability-Adaptable Game-agents	Behavior-Adaptable Game-agents
Description	· Game-agents \` abilities var y according to game player s' abilities. (hitting accurac y rate, speed, energy and th e like) · Game-agents control the le vel of game difficulty dyna mically.	· Game-agents \` action patterns var y according to the tendencies and abilities of game players. · Game-agents are dynamically de veloped in games. · Game-agents contribute to enhan cing fun by raising the realistic fe eling of games.
Applied Ex amples	FIFA, Call of Duty, Meda l of Honor and the like	Black & White, The Sims and t he like

For elementary game-agents, the action patterns are defined in source codes directly in most cases. This causes the disadvantage that improving game-agents is difficult and that game-agents have static action patterns. Adaptable game-agents adjust their ability and play with the new ability in the next games. Ability-adaptable game-agents change only numeric parameters, such as the level of difficult, hit ratio, power, etc. and do not change the behavior patterns. On the contrary, behavior-adaptable game-agents show dynamic response to the game players by analyzing the game player's gaming data. Behavior-adaptable game-agents change behavior patterns or even take the creative and more developed actions which were not designed in the initial phase of game production. The D-FSM suggested in this paper can be utilized as a fundamental technique for providing behavior- adaptable game-agents through adjusting the initially defined FSM of game-agents using the gaming data of game players.

3 Dynamic-FSM Method Using Player's Gaming Data

This paper proposes the D-FSM which is the method to alter transition rules of FSMs defined in the initial phase using the gaming data of players collected in games. To create game-agents which can adapt to game players, the gaming data of players should be collected first. The modeling on game players are conducted by analyzing the collected data. The model of game players will be created in a form of FSMs. The

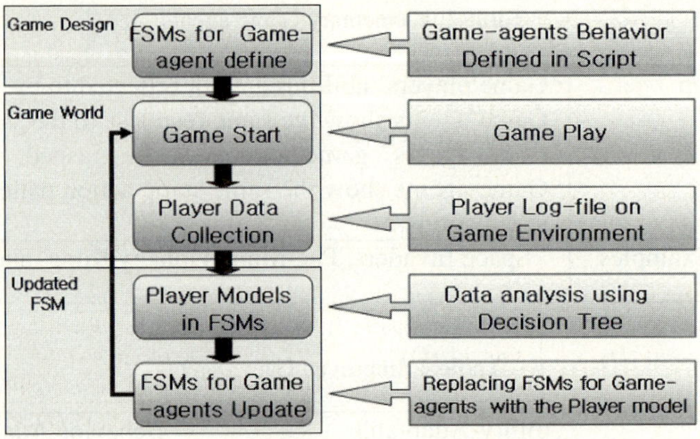

Fig. 2. Work-flow for D-FSM

FSMs will be applied to game-agents to change the behavior patterns of game-agents. The diagram in Figure 2 illustrates the D-FSM process.

3.1 FSMs for Game-Agents

Figure 1 shows the state transition of game-agents which is generally used in FPS (First Person Shooter) or MMORPG (Massively Multi-player Online Role Playing Game).

Table 3. State transition rules

S_{from}	PH	NH	HD	D	SD	PD	S_{to}
Search	H	H	0	O	1	1	Search
Search	M	H	0	I	-1	0	Charge
Search	M	M	1	I	1	-1	Idle
Search	H	L	1	I	0	0	Escape
Charge	H	H	1	I	0	-1	Charge
Charge	H	M	1	I	1	-1	Idle
Charge	L	L	1	I	1	-1	Fight
Charge	L	M	0	O	1	0	Search
…	…	…	…	…	…	…	…
getItem	M	M	0	O	0	0	Idle
getItem	M	H	0	O	1	1	Search

Game-agents take the actions defined in its' current state. The transitions between states are usually defined with situation variables.

The representative situation variables are the body strength of PC (Player Character) and game-agent, and the difference of body strength, attack power, moving speed and distance between PC and game-agent. As additional variables, the items which

Table 4. Variables used for defining transitions and their description

Variable	Attribute values	Description on Attribute values
PC Health Point (**PH**)	Low(L), Mid(M), High(H)	0~40=L, 41~70=M, 71~100=H
Game-agent Health Point (**NH**)	Low(L), Mid(M), High(H)	0~40=L, 41~70=M, 71~100=H
Health Difference of PC and Game-agent (**HD**)	0 ,1	PH < NH = 0, PH > NH = 1,
Distance of PC and Game-agent (**D**)	Near(N), Inside(I), Outside(O)	N : attack possible distance. I : inside of view sight. O : outside of view sight.
Speed Difference of PC and Game-agent (**SD**)	-1,0,1	If PC is faster than game-agent, 1. If game-agent is faster than game-agent, -1. If the speed is the same, 0.
Power Difference of PC and Game-agent (**PD**)	-1,0,1	If PC is stronger than game-agent, 1. If game-agent is stronger than game-agent, -1. If the power is the same, 0.

characters own can be considered. Various other abilities of characters and the existence of comrade characters in the vicinity also may be counted.

After the states for game-agents are defined and the variables which are used for state transitions are identified, the state transition rules as shown in Table 4 should be generated.

The transitions in Table 3 are defined using the values of 6 variables shown in Table 4. For example if the game-agent is currently in the state of "Search", the health of PC and game-agent are 80 and 39 respective, PC is in the view sight and the moving speed and the attack power of PC and game-agent are the same, the next state of game-agent will be "Escape", which is what the fourth rule says. Defining state transitions by scripts is one of most frequently used techniques because it is easy in the maintenance and expansion of game-agents.

3.2 Modeling Players with FSMs

This section will describe how to model players' playing patterns with FSMs which will be applied to game-agents. Players' playing data can be modeled in various ways. The reason for modeling with FSMs is to make the model easily adapted for updating game-agents' FSMs. If players' models are applied to game-agents, an game-agent will play intelligently or in a way similar to how players' play in the game.

In order to model players' playing patterns with FSM, it is assumed that players also have a hidden FSM and choose actions based on the FSM. However, it is not easy to infer what states the player has it is also assume that the player's FSM has the same states as game-agents'. The current state of the player is inferred from observations and some heuristics.

Game-agents in FPS games usually have six states shown as in Figure 1. The description of states and heuristics for inferring the player's current state are as follows:

Idle – In this state is that the game player takes no action. He may wait for the comrade or the enemy, or takes a rest on a specific location. If situation variables do not change for a certain time interval (5~10 sec.), the player is regarded as in "Idle" state.

Charge - This is the state to approach toward the enemies. If the player is in enemy's view and approaching to the enemy, the player is regarded as in "Charge" state.

Fight - This is the state to combat with enemies which are in attackable distance. If an enemy is within attackable distance and attack keys are inputted, the player is regarded as in "Fight" state.

Escape - This is the state to retreat from enemies when game players feel disadvantageous in combat or when game players were in low body strength condition. If the player is in enemy's view and the distance between them is increasing, the player is regarded as in "Escape" state.

getItem - This is the state where game players search tool items near from him to recover the body strength. If the player gets near to items and the player's current state is "Escape". The player is regarded as in "getItem" state.

Search - This is the state to move around to search enemies. So, if the player is out of enemy's sight and moves around, the player is regarded as in "Search" state.

When a game player's action and the values of situation variables are consistent with the heuristic condition of a state, the game player is regarded as in the state.

For example, if the enemy is out of the player's sight and the player is moving around, it is concluded that the player is currently "Search" state. If the enemy gets into the player's sight and the player gets closer to the enemy, the player is regarded charging the enemy, i.e., in "Charge" state.

The next step is acquiring the player's transition rules between states. For acquiring transition rules, the values of situation variables are recorded whenever the player's state changes. That is, the values is collected in a form of (S_{from}, PH, NH, HD, D, DS, PD, S_{to}). The collected data is pre-processed to remove duplications and decision tree method is applied to find out transition rules in the data.

The result of decision tree learning is expressed in a tree structure which can be described in a table format like tables. The rules newly created will replace the game-agent's transition rules. That is, it is used to make game-agents intelligent or human-like. The newly created game-agent rules are not static rules. These rules can be created from the game player's gaming data whenever new rules need. .

4 Experiment

A Half-Life by Valve Software[7] game and Jeffrey's HPB Bot[8] where a PC and an game-agent combat is used for the experiment of the method suggested in this paper. This game initially sets the states and transition rules of the game-agent by reading the scripts and collects the player's gaming data while game is going on. The player can move forward, backward, left and right, and attacks the game-agent.

The PC and the game-agent initially have the same moving speed and power. The power given to the PC and the game-agent in the initial phase is 100. If the values of situation variables do not change for more than 5 seconds, it is defined that the player is in "Idle".

Fig. 3. DeathMatch Class Game in MOD of Half-life and Game Screenshot

During game play information on the body strength of the PC and the game-agent, the difference of body strength, the distance between the PC and the game-agent, and the difference of moving speed and attack power are collected, whenever the state of the game player.

The game recognizes the state of the PCs on the basis of the heuristics previously described. After one round of the game, the game-agent learns the collected data using decision tree method and updates its' transition rules. Figure 4 illustrates the initial transition rule in Table 3. Those are represented in a tree format. Figure 5 represents the newly learned transition rules from the player's gaming data. It is noted that the newly learned transition rules are quite different from the initial one. It has more complex rule for "Charge" than the original but simpler one for "Search". The game-agent takes actions more intelligently than with the initial one.

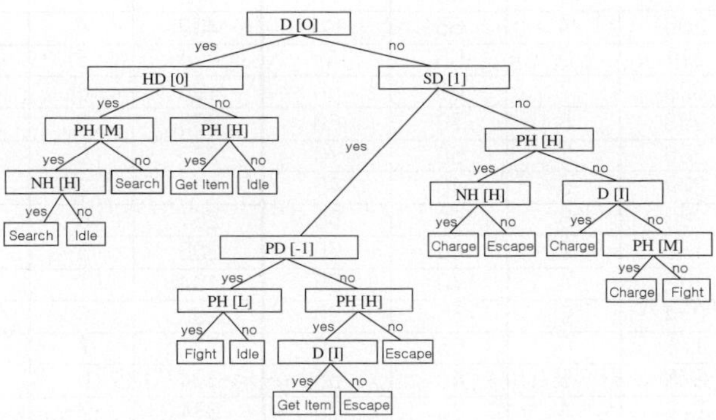

* Meaning of Node : Variable [Attribute value]
* Learning method : Decision Tree (ID3).
* Training data : Initial transition rule in Table 4.

Fig. 4. Initial transition rules in a tree format

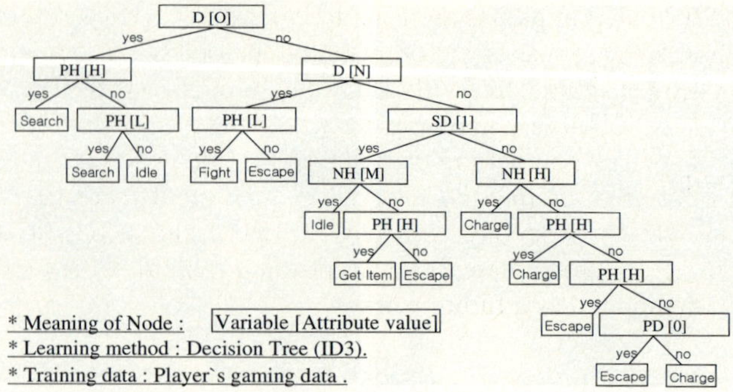

Fig. 5. A newly learned transition rules from the player's data

5 Conclusion and Future Works

In experiment, players answered that more difficult degree was felt constraint as 23 than existent method. Also, answered that interesting degree improved as 30, and extended game play time averagely[Table 5].

Table 5. Experiment result using D-FSM

Players	Game-agent of initial transition rules				Game-agent of learned transition rules			
	Game Data		Satisfaction (0~100)		Game Data		Satisfaction (0~100)	
	Game progress time(sec.)	Death number of times	Degree of difficulty	Degree of enjoyment	Game progress time(sec.)	Death number of times	Degree of difficulty	Degree of enjoyment
Player 1	184	5	55	65	272	7	80	85
Player 2	130	5	60	70	172	5	90	90
Player 3	205	4	65	60	402	4	100	100
Player 4	271	9	55	45	221	5	70	98
Player 5	328	7	60	50	196	6	60	95
Player 6	253	8	70	60	265	9	80	95
Player 7	266	7	50	70	287	6	90	85
Player 8	204	6	40	45	131	2	70	90
Player 9	200	5	50	55	261	6	40	80
Player 10	133	2	60	50	205	3	80	85
Player 11	182	4	70	65	184	7	90	90
Player 12	234	5	55	50	290	5	70	70
Player 13	178	6	65	30	221	7	80	60
Player 14	278	5	40	50	324	6	80	80
Player 15	213	4	30	70	334	7	90	85
Avg.	217.3	5.5	55.0	55.7	251.0	5.7	78.0	85.9

* Mission of this game is killing game-agents 10 times.

* Game progress time: It is cost time to kill game-agents 10 times.

* Death number of times: It is number of times that player dies during mission achievement.

* Degree of difficulty: If it is near to 0, easy. If it is near to 100, difficulty.

* Degree of enjoyment: If it is near to 0, boring. If it is near to 100, interesting

Game-agents provided in games are usually have static action patterns. These static actions of game-agents are one of factors which degrades the fun of games and makes the life-cycle of games short. This study is on an approach which can provide creative and diversified game-agents by updating FSMs through collecting and analyzing the gaming data of players. This method can be used in other the process of game development, such as designing initial FSMs for game-agents, creating partner game-agents of players which cooperate with players and creating hostile game-agents in games. To generate a more compact rule set pruning method which reduces the number of rules will be applied as a future work.

References

[1] G. Brajnik, G. Guida, C. Tasso, "User modeling in expert man-machine interfaces: a case study in intelligent information retrieval", IEEE Trans. SMC, Vol. 20, No. 1, 1990, pp.166–185.
[2] H. J. Cha, Y. S. Kim, J. H. Lee, T. B. Yoon, "An Adaptive Learning System with Learning Style Diagnosis based on Interface Behaviors", International Conference on E-learning and Games, Edutainment 2006.
[3] S. Rabin, AI Game Programming Wisdom 2, Charles River Media, 2002.
[4] M. D. Loura, Game Programming Gems, Charles River Media, 2000.
[5] J. E. Laird, "Using a Computer Game to Develop Advanced AI", Computer IEEE Journal, Vol. 34, No. 7, 2001, pp.70-75.
[6] P. Spronck, I. S. Kuyper, E. Postma, "Online Adaptation of Game Opponent AI in Simulation and in Practice", Proceedings of the 4th international Conference on Intelligent Games and Simulation, GAME-ON 2003, pp.93 – 100.
[7] http://www.half-life.com : Half-Life Game site.
[8] http://hpb-bot.bots-united.com : PHB Bot site for Half-Life game.

A Location-Aware Error Control Scheme of Route Multicast for Moving Agents*

Mikyung Kang[1], Sang-Wook Kim[2], Gyung-Leen Park[1], Min-Jae Kang[3], Ho-Young Kwak[4], Hoon Kwon[4], and Junghoon Lee[1,**]

[1] Dept. of Computer Science and Statistics, Cheju National University
[2] College of Information and Communications, Hanyang University
[3] Dept. of Electronic Engineering, Cheju National University
[4] Dept. of Computer Engineering, Cheju National University
{mkkang, glpark, minjk, kwak, dreamerz, jhlee}@cheju.ac.kr,
wook@hanyang.ac.kr

Abstract. This paper proposes a route multicast scheme for the moving agent such as a robot or telematics built on top of a wireless network which runs CFP (Contention Free Period) and CP (Contention Period) alternately. With the mappings of route multicast to CFP, error report to CP, and retransmission to redundant bandwidth of CFP, the proposed scheme can guarantee timely delivery of route information under the strict traffic control of the access point. Combined with beacon-aided timely error report mechanism, the location-aware error control scheme is able to reduce the network load by making an agent perform the error control only for the route information it may use in the near future. The simulation results demonstrate that the proposed scheme enhances the effective success ratio of the route multicast by 8.1 % for the given experiment parameters, compared with the legacy multicast protocols.

1 Introduction

As an example of the moving agent system, a networked robot is a group of robotic devices connected to a communication network such as Internet or LAN[1]. The mobile team navigates to solve a multi-objective task, having many useful applications such as search and rescue, exploration and hazard detection, and so on[2]. To achieve a given goal, mobile robots essentially cooperate through wireless networks under the well-defined coordination. Moreover, as fast navigation and dynamic environment require that control inputs are acquired in a timely manner, each agent operates under strict real-time constraints, making their communications also have real-time constraints. In the mean time, the IEEE 802.11 WLAN (Wireless LAN) does not only have capabilities to provide a real-time service via CFP (Contention Free Period) but also supports a probabilistic access to the medium via CP (Contention Period)[3]. Thus, many robot systems exploit WLAN as their inter-robot communication infrastructure.

* This research was supported by the MIC, Korea, under the ITRC support program supervised by the IITA (IITA-2006-C1090-0603-0040).

** Corresponding author.

N.T. Nguyen et al. (Eds.): KES-AMSTA 2007, LNAI 4496, pp. 982–989, 2007.

Each application of a networked robot has its own communication characteristics. One of the most common applications is the garbage collection: a robot finds and collects garbages while it establishes a trajectory without colliding with obstacles, and returns to the base before it runs out of energy[4]. In this system, each robot interacts with one another by a multicast primitive which can send efficiently the same data to multiple recipients with a minimal overhead. In particular, some robots that find a good route to a target or the base need to multicast route information to the interested collaborators. This route information is cached at receivers to be used in determining their routes.

While multicast is the essential communication primitive, the wireless channels are subject to unpredictable location-dependent and time-varying errors[5]. In case of multiple clients, each client will have different channel conditions, processing powers, and only limited feedback channel capabilities. The error control should take into account these characteristics and should not affect the transmission of other guaranteed traffic such as real-time messages or newly generated route information. However, the error control essentially accompanies additional messages including the report of damaged frames and corresponding retransmissions. Therefore, error control should be tightly controlled by a coordinator such as an AP (Access Point). In addition, in case of route information multicast, every robot does not want to retransmit damaged frames as some route is far away from its current location. Hence, the error control scheme for route multicast should try to recover just the necessary messages, not all lost ones.

Based on such a requirement, this paper proposes an error control scheme for multicast route information on IEEE 802.11 WLAN and analyzes its performance. The main idea of this paper is as follows: First, to regulate the error control, retransmission requests should be delivered via CP while the retransmission is mapped to the redundant bandwidth of CFP. The load control combined with efficient allocation of network bandwidth is critical to the performance of the communication system. Second, to avoid the per-frame ACK, errors are reported via an error list. Though a variable message size makes it hard to decide when to report the error list, receivers can determine the completion of each message transmission by counting *Beacon* frames that are periodically generated by AP to announce the beginning of new CFP. Finally, a location-aware error control makes a robot participate in the error control only when its current location is within a given bound.

The rest of this paper is organized as follows: After Section 2 introduces some related works and backgrounds, Section 3 explains basic assumptions on a network model, an error model, and a robot operation. Section 4 describes the proposed route multicast scheme in detail, and Section 5 shows the result of performance measurement. Finally, Section 6 concludes this paper.

2 Related Works

This section introduces previous works regarding robot control and a multicast protocol on wireless networks. First, as for the work of robot control system,

Vaughan et al. presented a method to evaluate control and coordination strategies for a group of wireless networked robots[6]. An integrated robot and network simulation tool was developed to investigate the behavior of robot controllers. Based on this framework, the authors demonstrated that trail-laying behaviors can be usually implemented in shared localization space, rather than directly in the real world. As an example of more sophisticated robot control, Michelan et al. investigated an autonomous control system of mobile robots based on the *artificial immune network* theory[4]. To perform a garbage collection task, the information captured by sensors enters the robot as an antigen. Possible antigens in a system are related to the current direction of a robot, direction of obstacles, distance, direction to bases, and a level of internal energy.

Second, as an example of MAC layer error control for multicast, SPEED system proposed a real-time area multicast protocol that directly considers geographic information in designing a multicast protocol in ad-hoc networks[7]. End-to-end real-time communication guarantees are achieved by using a novel combination of feedback control and non-deterministic QoS-aware geographic forwarding with a bounded hop count. In addition, Lu et al. proposed a timestamp-based content-aware adaptive retry mechanism[8]. MAC dynamically determines whether to send or discard a packet by its retransmission deadline, which is assigned to each packet according to its temporal relationship and error propagation characteristics with respect to other packets within the same multicast group. However, their scheme is too complex to be exploited in the WLAN standard as it crosses the protocol layer boundaries. These schemes are mainly built on top of CP, and just adjusting the transmission rate without an explicit feedback control. In addition, Lee et al. proposed an error control scheme for video streaming in WLAN based on the bandwidth allocation method[9]. In their scheme, under the complete control of AP, receiver nodes send an error report in a best-effort manner via DCF while AP makes a packet retransmitted on an overallocated slot.

3 Background and Basic Assumptions

WLAN divides its time axis into CFP and CP, which are mapped into PCF (Point Coordination Function) and DCF (Distributed Coordination Function), respectively, as shown in Fig. 1. To provide the deterministic access to each node during CFP, AP polls each node according to the predefined order, and only the polled node can transmit its frame. In the DCF interval, every node including AP contends the medium via the CSMA/CA (Carrier Sense Multiple Access with Collision Avoidance) protocol. The AP periodically initiates CFP by broadcasting a *Beacon* frame that has the precedence in transmission via shorter IFS (InterFrame Space).

This paper assumes that the main control policy of a garbage collecting task follows the work of an Ant-style routing scheme[6]. Robots start from a base position and must search the goal. Each robot records its movement trail as a sequence of landmarks in localization space and shares the path with its teammates after each successful traverse. Instead of directly modifying their physical environment like real-world ants, robots communicate localization space landmarks over a wireless network. Whenever a robot reaches a goal location or successfully returns to a base, it broadcasts its list of landmarks over the network so that other robots can follow. Each landmark records the robot's position, direction, and the current time. The current position can be obtained with the support of GPS (Global Positioning System) receiver, location sensor, and so on, being expressed according to the appropriate coordination system[6].

Robots will follow the heading hint given by averaging the indications of any nearby landmarks. Common paths are created and reinforced as they are repeatedly followed. In this way, robots tend to converge to a reasonably efficient forward and backward paths from a source to a goal. Each robot keeps route information only for the specified time interval, and then destroys, making it up to date. If there are no nearby landmarks, a robot just drives forwards. A simple steering strategy can be defeated by tricky arrangements of obstacles, so occasionally a robot will stop when it faces a wall or another robot. In this case, the robot selects the path randomly.

A robot node is associated with a channel which has either of two states, namely, *error state* and *error-free state* at any time instant. A channel is defined between each mobile and AP, and can be modeled as a Gilbert channel[5]. We can denote the transition probability from state *good* to state *bad* by p while the one from *bad* to *good* by q, as shown in Fig. 2. The pair of p and q representing a range of channel conditions has been obtained by using the trace-based channel estimation. The average error probability and the average length of a burst of errors are derived to $\frac{p}{p+q}$ and $\frac{1}{q}$, respectively. A packet is received correctly if the corresponding channel remains in state *good* for the whole duration of packet transmission. Otherwise, it is received in error.

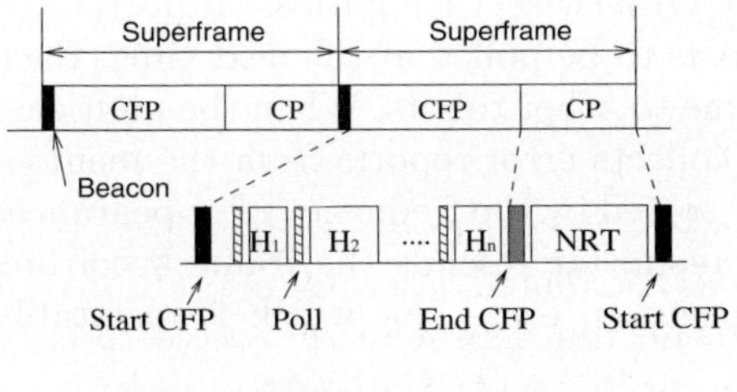

Fig. 1. Time axis of wireless LAN

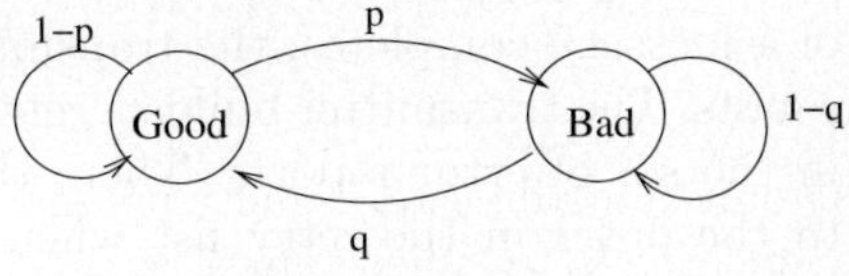

Fig. 2. Error model

4 Proposed Error Control Scheme

4.1 Communication Strategy

The landmark list gets longer if the route contains many turns, random choices, and wrong decisions during navigation. A couple of landmark broadcast strategies would permit more robust system performance. One extreme approach is to pack the entire sequence of landmarks into a single long packet and to send it only once. However, if the packet is lost, some or all of the robots will not receive anything at all. The reasonable alternative is to send as a data stream. By a data stream, we mean that a packetized message periodically arrives at the network interface during the life time of the stream. Even if some packets are lost, others could still get through. Thus, some robots may receive at least partial information about what trail to follow. The information is valid only when all frames belonging to a packet are correctly received.

The cause of frame loss is collisions or overflows on the interface queue. To deal with this situation, for each data frame, the MAC layer transmits several control frames for the purpose of collision avoidance. Furthermore, if these control frames are lost due to a collision or bad propagation, the MAC layer will try to retransmit them again. However, according to the standard, the automatic MAC layer ACK from the receiver is not mandatory in multicast. To enhance the ratio of successful data multicast, some kind of error control function is required on top of underlying MAC. Additionally, a partitioned transmission makes it possible for a retransmission to be performed for the information part that is actually damaged.

4.2 Traffic Mapping

As a landmark list is the most critical information for robots to fulfill a given task, it seems reasonable to send this information via safe and deterministic channel, namely, CFP. In addition, as will be shown shortly, multicast via CFP can make it easy for a receiver to build an error report. If a robot wants to start multicasting its route information, it first gets the admission from AP by sending a polling request via CP. AP polls the robot from the next CFP only if it has correctly received the request and the total amount of allocation does not exceed a system-defined CFP interval. Otherwise, the multicast is postponed and the request enters a waiting queue so as to be polled at the next time. Each time a robot is polled, it broadcasts a frame to other robots. When the multicast of a message completes, the transmitter collects error reports from the member robots. The transmitter builds a retry list sorted by the frequency of appearances in the set of error reports. Then, the transmitter resends the frame according to the order in the retry list when it meets an extra slot which is inevitably generated by the bandwidth reservation.

4.3 Error Report

Per-frame response to the receiver increases the total number of control frames, leading to a substantial increase in the network load. However, if the receiver reports the list of damaged frames after estimating the end of delivery of one message, this waste can be dramatically reduced. It is desirable to send an error report via the contention period so as not to interfere the transmission of other streams. We make all frames include a field specifying the number of frames in their belonging message so that the receiver can decide when to report an error frame list as long as it receives at least one frame.

To construct the error report, the receiver initializes an error frame list when it receives a frame of a message for the first time. If the sequence number of this frame is not 1 but k, the receiver appends the numbers of 1 through $(k-1)$ to the error list. From then, the receiver appends each number of the erroneously received or omitted frames. As the receiver hears *Beacon* and a stream sends a frame per a polling round, the receiver can recognize the frame loss. Fig. 3 shows an example, where a message is transmitted with u frames and each of them has information that specifies u as well as its sequence number and message identifier. When the counter reaches u, the receiver sends an error report back to the sender via a contention period.

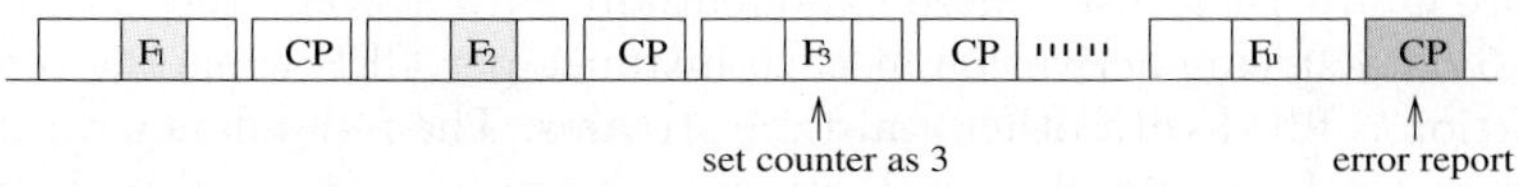

Fig. 3. An example of error reporting

The valid scope area refers to the geometric area of a data value and somehow reflects the access probabilities that a robot needs the route information. Only those receivers within this area request the retransmission to obviate unnecessary traffic load fluctuation. When a robot receives a frame belonging to a specific data stream for the first time, it also extracts the location information recorded in the frame. This information is used as the criteria by which the robot decides whether to participate in a subsequent error control. This paper calculates the valid scope area as shown in Fig. 4. Robot A is located at L_A while robot B is located at L_B, when they receive the first frame belonging to the stream multicasted by robot C. Robot C has moved from L_C to $L_C^{'}$, the base location, so it broadcasts route information. Though Fig. 4 plots the movement of C as the horizontal dotted line, the route may be complex, containing many turns. Robot A and B calculate the area of $\triangle L_A L_C L_C^{'}$ and $\triangle L_B L_C L_C^{'}$, respectively. Intuitively, the larger the area is, the less useful the information is for the robot. However, such area criteria are incomplete because the closer a robot to the base, the smaller the area, whichever point is selected for the third point of the triangle. Alternatively, we choose the distance from the receiver (i.e., L_A or L_B) to the line between the sender (i.e., L_C) and the base, say *distance criteria*. After all, only robot A sends an error report to robot C, if it encounters network errors.

5 Performance Measurement

This section measures the performance of the proposed error control scheme for route multicast via simulation using ns-2 event simulator[10]. In the simulation, we assumed that 10 robots traverse a square area consisting of 50 by 50 grids at

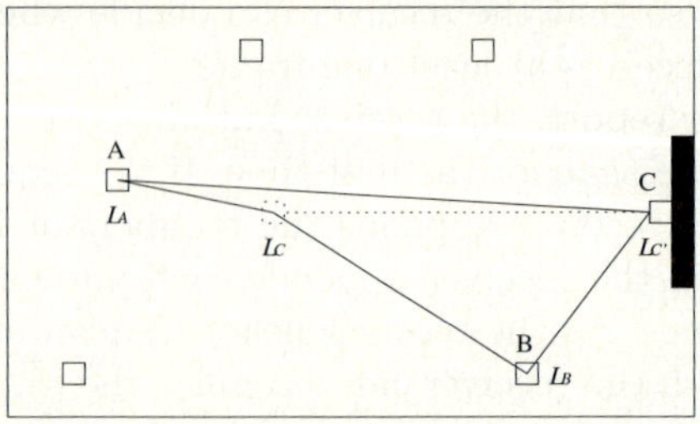

Fig. 4. Area and distance criteria

equal speed, simplifying the experiment. In addition, the number of frames in each route multicast is distributed exponentially with average 100. There can be at most 5 data stream activation at each instance per CFP while the remaining CFP portion is filled with other real-time streams. The redundancy ratio means the average time amount that each stream may meet additional polls. Finally, the bound of distance criteria is set to 20 % of the edge length.

Fig. 5 shows the success ratio for the frame error rate of 0.0 through 0.5. In this experiment, the redundancy ratio is set to 0.2. The *BasicCast* means the legacy multi/broadcast mechanism without error recovery while the *RandomRecover* makes each receiver report all the errors and randomly selects the frames to be retransmitted when it meets additional polls. The curve marked as *AreaCast* corresponds to the proposed scheme. The performance gaps between the proposed scheme and other schemes get larger as the error rate increases. Fig. 5 shows that the proposed scheme enhances the successful delivery of route multicast by up to 8.1 % when the frame error rate (not bit error rate) is around 0.3. Fig. 6 plots the success ratio measured by changing redundancy ratio from 0 to 0.3 for the respective frame error rate, 0.0, 0.1, ..., 0.5. As expected, the performance improvement increases when the redundancy ratio gets higher.

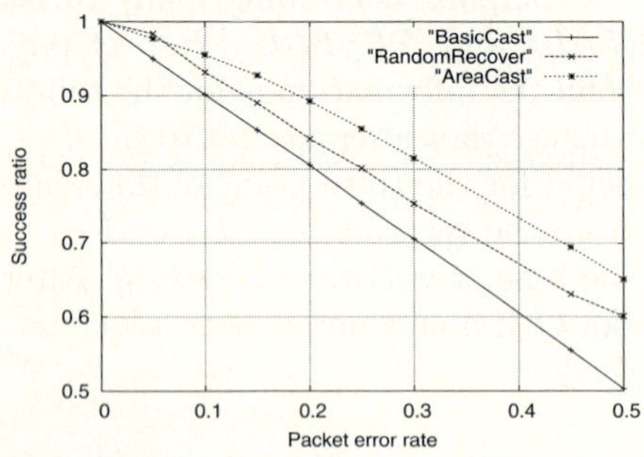

Fig. 5. Effect of frame error rate

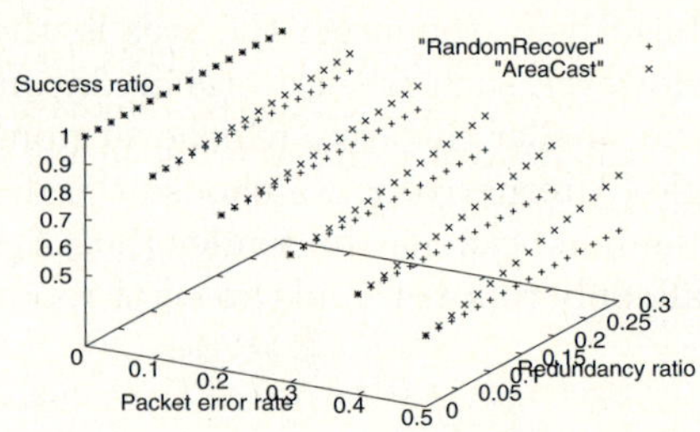

Fig. 6. Effect of redundancy ratio

6 Conclusion

In this paper, we have proposed a route multicast scheme for the networked robot system built on top of IEEE 802.11 WLAN and have also evaluated its performance. With the mappings of route multicast to CFP, error report to CP, and retransmission to overallocated bandwidth, this scheme is able to eliminate the interference to the guaranteed stream transmission under the complete control of AP. In addition, the message size field contained in each frame enables the timely report of an error list as long as at least a frame arrives at the receiver for each message. Based on this framework, a location-aware error control scheme has been proposed. According to its current position along with the start point of route multicast stream, each robot decides whether to participate in the error control of a specific route multicast stream. The proposed scheme enhances the effective success ratio of the route multicast by 8.1 % compared with legacy multicast protocols.

As a future work, we are first planning to develop a route multicast protocol for a robot system that exploits dual channel WLANs. Dual channel WLAN has many benefits in scheduling time-sensitive messages in addition to the basic bandwidth expansion. The sophisticated mapping of data streams will enhance the multicast performance if a location based information is combined.

References

1. Das, S.M., Hu, Y.C., Lee, C.S.G., Lu, Y.: An efficient group communication protocol for mobile robots. Proceedings of the 2005 IEEE International Conference on Robotics and Automation (2005)
2. Kannan, B., Parker, L. E.: Adaptive causal models for fault diagnosis and recovery in multi-robot teams. Proceedings of IEEE International Conference on Intelligent Robots and Systems (2006)
3. IEEE 802.11-1999: Part 11 - Wireless LAN Medium Access Control (MAC) and Physical Layer (PHY) Specifications (1999) also available at *http://standards.ieee.org/getieee802.*
4. Michelan, R. and Zuben, F.J.V.: Decentralized control system for autonomous navigation based on an evolved artificial immune network. The 2002 IEEE World Congress on Computational Intelligence - WCCI'2002 (2002) 1021–1026
5. Shah, S., Chen, K., Nahrstedt, K.: Dynamic bandwidth management for single-hop ad hoc wireless networks. ACM/Kluwer Mobile Networks and Applications (MONET) Journal. 10 (2005) 199-217
6. Vaughan, R.T., Stoy, K., Sukhatme, G.S., Mataric, M.J.: Whistling in the dark: cooperative trail following in uncertain localization space. Proc. Int. Conf. Autonomous Agents (2000) 187–194
7. He, T., Stankovic, J., Lu, C., Abdelzaher, T.: SPEED: A real-time routing protocol for sensor networks. University of Virginia Tech. Report CS-2002-09 (2002)
8. Lu, A., Chen, T., Steenkiste, P.: Video streaming over 802.11 WLAN with context-aware adaptive retry. IEEE International Conference on Multimedia and Expo (2005)
9. Lee, J., Kang, M., Park, G., Kim, H., Kim, C., Kim, S.: An error control scheme for multicast video streaming on the last hop wireless LANs. Lecture Notes in Computer Science, Vol. 3994. Springer-Verlag, Berlin Heidelberg New York (2006) 962–969
10. Fall, K., Varadhan, K.: Ns notes and documentation. Technical Report, VINT project, UC-Berkeley and LBNL (1997)

Blur Detection of Digital Forgery Using Mathematical Morphology

Linna Zhou[1,2], Dongming Wang[3], Yunbiao Guo[2], and Jingfei Zhang[2]

[1] Beijing University of Posts and Telecommunication, Beijing 100876, China;
[2] Beijing Application Institute of Electronic Technology, Beijing 100091, China;
[3] China Securities Regulatory Commission Information Center, Beijing 100032, China
linnazhou1972@hotmail.com

Abstract. Recent studies on agent techniques for digital content management have been focused on the guarantee of safety and intelligent management. The digital forensic to detect content tampering is a typical application of safety digital content management. Based on the edge processing and analysis using edge preserving smoothing filtering and mathematical morphology, a new blur edge detection scheme is proposed in this paper, which can indicate possible tampering and locate tampered regions without any embedded information such as watermarking technique. Experimental results demonstrate the effectiveness of the proposed scheme.

Keywords: image forensics; security agent; safety content management application; mathematical morphology; blur.

1 Introduction

Web based large amount digital content such as image, video and audio have been the most popular service in the Internet. However, illegally duplication and tampering of the distributed content in the Internet may cause some troubles or even great economic loss to the digital contents providers. Due to these problems, studies on security techniques have been increasingly stressed. Recent studies on agent techniques for digital content management have been focused on the guarantee of safety and intelligent management. The digital forensic is a typical application of safety digital content management. Nowadays, the advent of low-cost and high-resolution digital cameras, and sophisticated photo-editing software make it remarkably easy to manipulate and alter digital images. The saying "seeing is believing" is no longer true in this digital world, and one would naturally ask whether the photo he receives is a real one[1]. This also becomes a serious issue when it comes to photographic evidence presented in the court or for insurance claims. Therefore, we need a reliable way to examine the authenticity of images, even in a situation where the images look real and unsuspicious to human. The detection of digital tampering has become a crucial requirement.

N.T. Nguyen et al. (Eds.): KES-AMSTA 2007, LNAI 4496, pp. 990–998, 2007.

2 Related Work

In the past, several techniques[2-4] based on data hiding in images have been designed as means for detecting tampering. However, in practice, very few images are created with watermarks. Under most circumstances active approaches fail because there is no watermark to detect. This gives rise to research activities in passive blind image authentication that handle images with no previously added hidden information.

A common manipulation when altering an image is to copy and paste portions of the image to conceal a person or an object in the scene. There are many previous works to detect and locate forged regions in an image; for example, Chang et al.[5-8] introduce a passive-blind approach, by bicoherence, without prior knowledge as embedded information to detect spliced images. A method that could be able to detect duplicate regions, which are produced by copy and paste operations, is proposed by Fridrich[9]. Farid[10] has proposed several useful techniques to expose digital image forgeries such as the features of higher-order correlations in the frequency domain for natural signals, and the trace of image re-sampling. There are more works[11-13] in this domain, such as the study about detecting digital tampering by an EM-based color filter interpolation approach. These approaches could be able to detect simple duplicate regions in fake image. However, in order to make a seamless and plausible fake image, most of digital forgeries applying blurring to remove unwanted boundaries. In contrast to these approaches, we propose a new blur detection scheme using edge preserving smoothing filtering and mathematical morphology in this paper. It can not only judge whether or not a given image is blurred, but also determine to what extent the given image is blurred. The proposed scheme takes advantage of the ability of edge preserving smoothing filtering in both sharpening the blurred edge and eliminating noise. It is very essential for discriminating defocus blur from manual blur, which can indicate possible tampering and locate tampered regions in the absence of any digital watermark or signature. Experimental results demonstrate the effectiveness of the proposed scheme.

The rest of the paper is organized as follows: we analysis the blur operation and edge characteristic in section 3 and present our scheme in section 4; experimental results are given in section 5; finally, we conclude the paper in section 6.

3 Blurring Process and Image Edge Characteristic

To eliminate statistical or vision distortion on spliced edge during the making of digital image frauds, forgers usually apply retouching methods such as blurring, desalting, shading to remove unwanted traces caused by copy and paste operations on the fake images. Of the retouching skills, blurring is a very common process in digital image manipulation. Its fundamental is to acquire smooth effectiveness by neighborhood averaging grayscale value of partial pixels. It is used for reducing the degree of discontinuity or even concealing spliced boundaries.

3.1 Mathematical Model for Manual Blur

Since blurring process is necessary, if we could trace such a process, the existence of digital image tampering could be exposed. In order to trace blurring process, we should know what a blur look like and how it occurs in an image. When pictures are blurred with image processing software, for example, Photoshop, users may be asked to specify the blur mode. Since some blur modes support user-defined filters, people can manipulate images with diverse blur radiuses and intensity. From its algorithm realized by Photoshop, blurring is motion averaging filtering of chosen image regions to smooth the image. Different blur modes involve different motion averaging filtering functions. The width of filter window decides blur radius, and filter parameters reflect blur extent. In other words, blurring is the process of replacing the original pixel point value $f(i, j)$ with the neighborhood average value $g(i, j)$ in the pixel's partial region, which is spatial filtering all pixels with weighed matrix multiplied by $1/n^2$. When the chosen region is square, blurring can be formulated as:

$$g(i,j)=\frac{1}{n^2}\sum_{k=-[n/2]}^{[n/2]}\sum_{l=-[n/2]}^{[n/2]} f(i+k,j+l) \tag{1}$$

Where *n*, a non-negative integer, is the width of filter window, whose value indicates the size of the blur region. The larger neighborhood region *n* for grayscale average, the greater blur intensity of the chosen image region.

3.2 Image Defocus Blur and Manual Blur

During the detection, we can find defocus blur blocks in some natural images on account of wide screen. When it comes to digital tampering, it often refers to manual blur. Therefore, it is important to draw the line between common image operations (e.g. defocus blur) and malicious attacks (e.g. manual blur) for image forensics accurately.

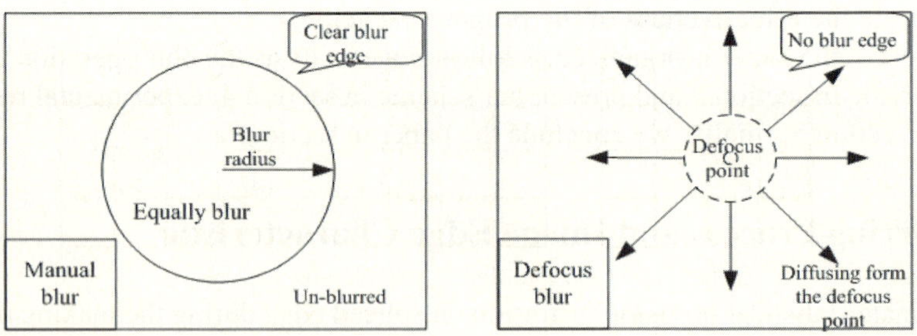

Fig. 1.

In Equation (1), we can see that manual blur is the result of windowed filtering pixels in the region whose blur radius is decided manually. So the outputs of blur filtering are equivalent within the blur radius, correspondingly pixels outside the blur radius are non-processed at all. Camera defocus blur, which doesn't have blur radius, results in radiate decrease of pixels' grayscale value relative to defocus point gradually. The effects of the two blurring processes are illustrated in Fig.1.

With the two diagrams, the most difference between manual blur and camera defocus blur is that, manual blur has obvious blur edge, while the pixels outside and inside the blur region have clear distinctness, opposite to camera defocus blur without blur edge and more smoothing changes.

Hence if proposed image process can be applied to counteract accurately the manual blur effect of neighborhood average grayscale filtering, and expose the blur boundaries as well as the distinction between the outside and inside pixels, people can discriminate manual blur from defocus blur. With enough analysis and experiments, edge preserving smoothing filtering[14] is proved to be a suitable method.

3.3 Edge Preserving Smoothing Filtering Reveals Manual Blur Boundary

The reason for edge blur caused by motion averaging during manually blurring, is that calculating grayscale average regardless of the existence of local regions' edges. Thus, to diminish the blur, we can choose local regions without edges in the surround of each pixel, and treat grayscale average as the output of specified pixel. And edge preserving smoothing filtering is the effective method. The detailed procedure is calculating grayscale covariance σ_l of point $f(i, j)$'s nine neighborhood regions which includes four pentagons, four hexagons, and one square, just like Figure 2 shows, then treating the average grayscale in the region with least covariance as the point's output. The process can eliminate noise caused by manually blurring, and sharpen the blur boundary to expose blurred regions.

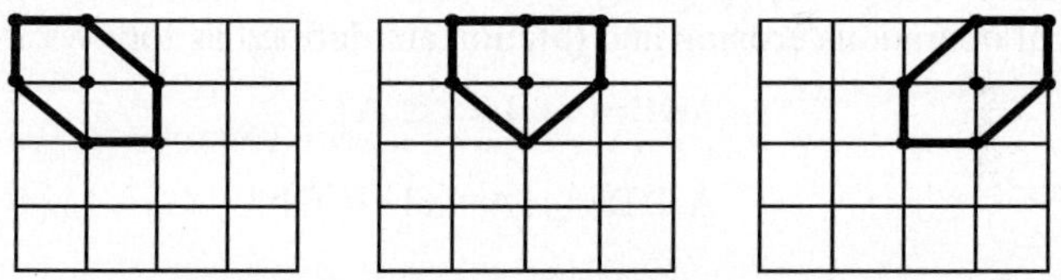

(a) top left (b) topside (c)top right

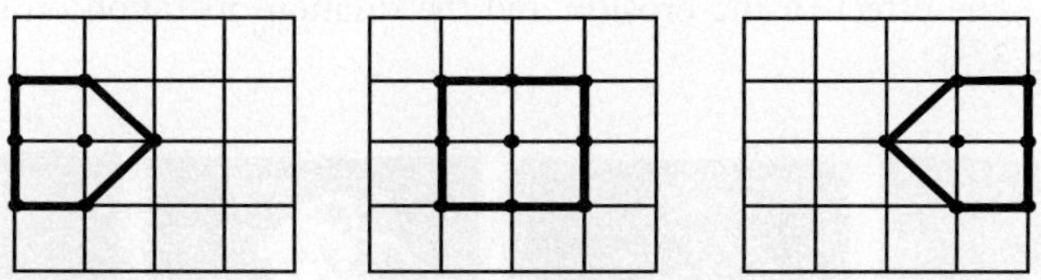

(d) left (e) middle (f) right

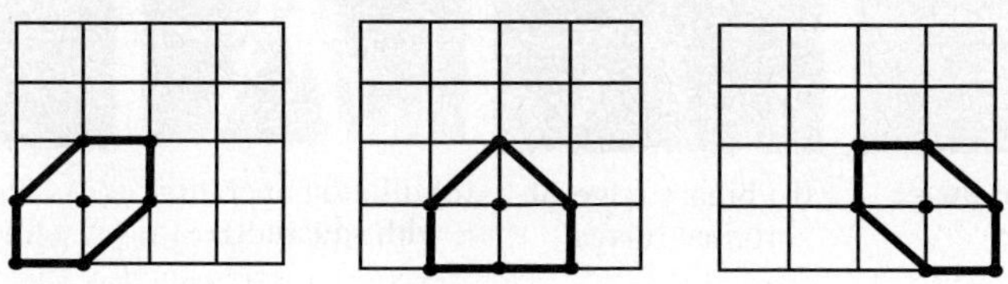

(g) bottom left (h) bottom (i) bottom right

Fig. 2.

Given that grayscale covariance σ_l of point $f(i,j)$ in the above nine local regions can defined as follows:

$$\sigma_l = \frac{1}{n_l}\sum_{k=0}^{n_l}\left| f(i+k, j+k) - \overline{f(i,j)} \right|^2 \quad (l = a, b, c \cdots i) \tag{2}$$

Then preserving edge smooth filtering can be formulated like this:

$$g(i,j) = \min(\sigma_a, \sigma_b, \sigma_c, \cdots, \sigma_i) \tag{3}$$

After preserving edge smooth filtering, the spliced edge can be shown clearer. At the moment, detectors can select appropriate structuring element (SE), apply erosion operation in mathematical morphology to exclude natural edges and discern the spliced fraud regions in images.

4 Blur Detection Using Mathematical Morphology

From the proposed detection scheme in the previous section, we can discriminate effectively manual blur from defocus blur. They are different in binary image mathematical morphology edge. We can use mathematical morphology operation to detect the manual blurred by eliminating regular edge.

An image is represented by a set of pixels. The morphological operations can be visualized as working with two images. The processed image is referred as the active image and the other kernek image is referred as the structuring element (SE). We can filter the active image by probing it with various SEs. The two major basic morphological operations erosion and dilation are defined as follows:

$$A \Theta B = \{x \mid B + x \subseteq A\} \tag{4}$$

$$A \oplus B = \bigcup \{A + x \mid x \in B\} \tag{5}$$

Where Θ and $\oplus$ denote the erosion and the dilation operation. A is the active image and B is the SE. The effect of the erosion and the dilation operation on binary image is shown in Fig.3.

(a) original image (b) binary edge of original image (c) dilation operation with square SE (d) erosion operation with square SE

Fig. 3.

Manually blurring is the process of filtering pixels within the set blur radius with windowed filter function, whose foundational is neighborhood grayscale averaging of pixels within the blur radius, which is equal to structuring element dilation operation on local edges in mathematical morphology. Shown in Figure 4 are an image's mathematical morphology filtering edges before and after manual blur. Thus, erosion operation on binary edge images processed with edge preserving smoothing filter can weaken image natural region's edge, and preserve enhanced edge of tampered region, thereby the fraudulent region can be detected easily.

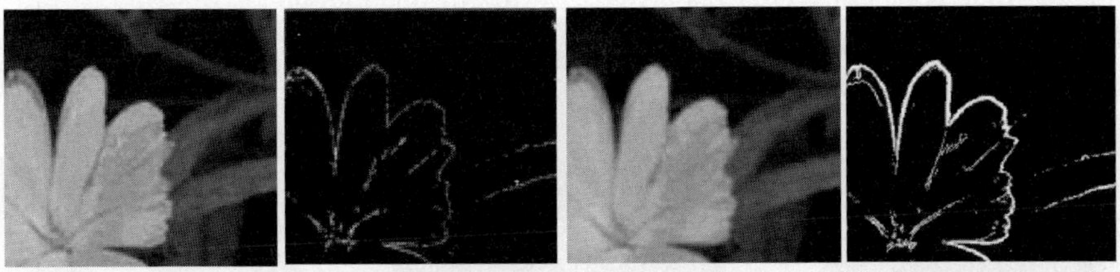

(a)original image (b) binary edge of original image (c)Gaussian blur image (d) binary edge of blur image

Fig. 4.

Now, mathematical morphology filter can be used to detect the manual blurred by eliminating regular edge and preserving the blurred edge that has been sharpened by edge preserving smoothing filtering. When we process binary edge image using operations erosion and dilation, it is very important to construct SE. The form and size of SE is the key to take out the useful information. We have done some experiments with the several SEs, and the best result is obtained by the 3 × 3 square SE.

$$SE = \begin{bmatrix} 1 & 1 & 1 \\ 1 & 1 & 1 \\ 1 & 1 & 1 \end{bmatrix} \tag{6}$$

Once mathematic morphology filtering eliminates regular edge and preserves the blurred edge, we can set appropriate threshold and determine whether an image is blurred or not and locate tampered regions. The detection scheme can be directly applied to single image as follow steps:

Filter the original image by using edge preserving smoothing filtering, which can discriminate effectively regular edge from manual blur edge.

Convert the original image to binary image and get the edge binary image.

Process the edge binary image by using erosion operation with SE.

Set appropriate threshold, locate the possible tampered regions and determine whether an image is blurred or not.

5 Experimental Results

The performance of the blur detection scheme has been tested by simulation experimental using Matlab6.5[15]. Different from other prior studies, some of our test images are collected from the Internet (download from *Worth1000.com*), and some test pictures were taken by digital camera. We believed that if we want to achieve a more practical method for digital image forgery detection, we have to use real-life images to test. The experimental results are listed in Table1. An example demonstrating the process of our scheme for determining blur detection is shown in Fig.5.

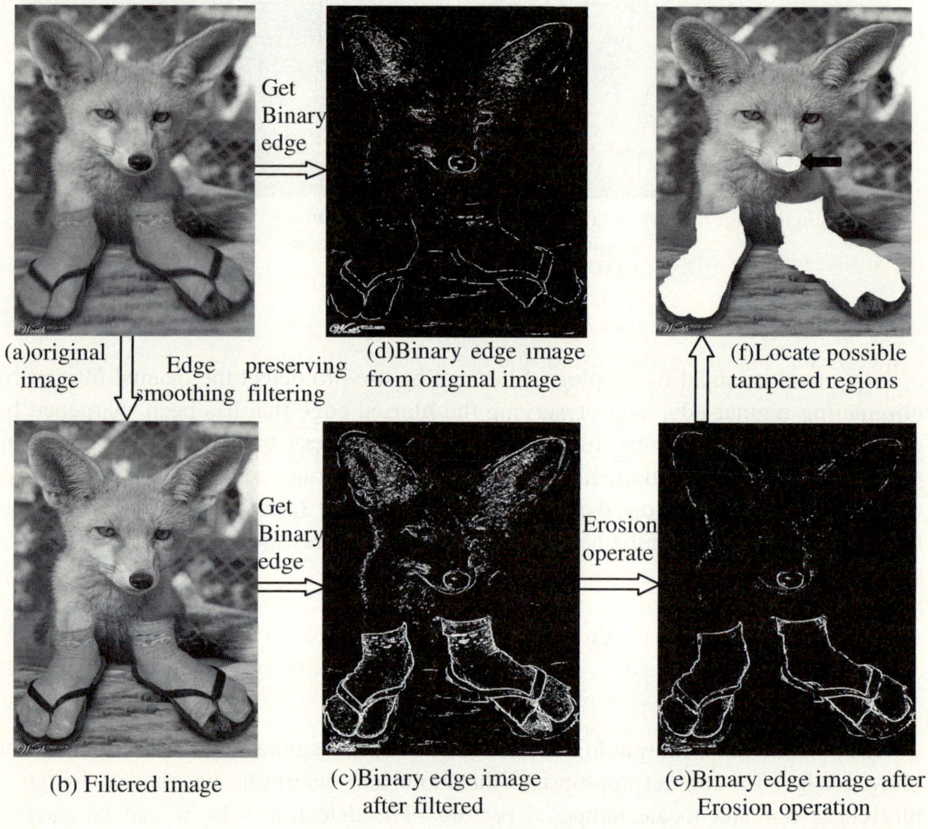

Fig. 5.

In Figure 5(f), the area covered with white color is the blurred local region detected with the above method. But the small area pointed out with black arrow is mistakenly detected blurred. Based on the results, the detection of manual blur edge using mathematical morphology after edge preserving smoothing filtering can exactly identify digital forensics, and locate fraudulent regions precisely.

Table 1.

Image source for detecting	Image amount	Right numbers	Error numbers	Detect rate(%)
Original image download from Worth1000	20	19	1	95
Compounded image download from Worth1000	55	46	9	83.6
Sony N1 Camera Original image	118	108	10	91.5
Kodak DC290 Original image	125	112	13	89.6
Sony N1 Camera Compounded image	28	25	3	89.3
Nikon E5700 Camera Compounded image	15	13	2	86.6

6 Conclusion and Future Direction

In this paper, we have proposed a new blur detection scheme for images taken by digital cameras. The scheme makes use of the ability of edge preserving smoothing filtering and mathematical morphology in edge processing and fake analysis. The scheme judges can indicate possible tampering and locate tampered regions without any embedded information such as watermarking technique. Experimental results demonstrate the effectiveness of the proposed scheme.

Digital tampering may affect image characteristics in many aspects, and the work based on a single characteristic described in this paper reveals just a small fraction of the image forensics detection. With the development of digital forgery technology, digital detection keeping pace with digital tampering is difficulty only depending on single digital forensic tool. Though still having a lot of room for improvement, the scheme we propose can serve as a well-posed starting point for future exploration in this direction. The future digital forensic direction would be multiplex forensic tools in conjunction with awareness and sensible policy and law to create convincing digital forgeries.

Acknowledgments. This work was supported by the National Science Foundation of China (Grant Number: 60372083).

References

1. K. Hafner: The Camera Never Lies, but the Software Can[N], New York Times, 2004.
2. I. J. Cox, M. L. Miller, and J. A. Bloom: Digital Watermarking[M], The Morgan Kaufmann Series in Multimedia Information and Systems, Morgan Kaufmann, 2002.

3. E. T. Lin, C. I. Podilchuk, and E. J. Delp: Detection of image alterations using semi-fragile watermarks[A], SPIE International Conference on Security and Watermarking of Multimedia Contents II, vol.3971, San Jose, CA, 2000.
4. J. Fridrich, M. Goljan, and A. C. Baldoza: New fragile authentication watermark for images[A], IEEE International Conference on Image Processing[C], Vancouver, Canada, 2000.
5. Ng T. T., Chang S. F.: A model for image splicing, IEEE International Conference on Image Processing, 2004, vol. 2, pp. 1169 – 1172.
6. Ng T. T., Chang S. F., and Sun Q.: Blind detection of photomontage using higher order statistics[A], IEEE Proceedings of International Symposium on Circuits and Systems[C], Vancouver, Canada, May 23-26, 2004, Vol.5, pp. 688-691.
7. Ng T. T., Chang S. F.: A Data Set of Authentic and Spliced Image Blocks[R], ADVENT Technical Report, No.203-2004-3, Electrical Engineering Department, Columbia University, New York, June 8, 2004.
8. Ng T. T., Chang S. F.: Blind Detection of Digital Photomontage using Higher Order Statistics[R], ADVENT Technical Report, No. 203-2004-3, Electrical Engineering Department, Columbia University, New York, 2004.
9. Fridrich J, Soukal D, and Lukáš J: Detection of copy-move forgery in digital images[DB/OL], Proceedings of DFRWS, 2003. http://www.ws.binghamton.edu/fridrich/publications.html
10. A. C. Popescu, H. Farid: Exposing digital forgeries by detecting traces of resampling[J], IEEE Transactions on Signal Processing, 2005, 53(2):758–767.
11. J. Lukáš, J. Fridrich, and M. Goljan: Detecting digital image forgeries using sensor pattern noise[A], Proceedings of the SPIE[C], volume 6072, 2006.
12. Hany Farid: Creating and Detecting Doctored and Virtual Images[J], Implications to The Child Pornography Prevention Act, 2004.
13. Popescu A.C, Farid H: Exposing Digital Forensics by Detecting Duplicated Image Regions[R], Technical Report, TR2004-515, Dartmouth College, Computer Science 2004.
14. Gong ShengRong, Liu ChunPing, and Wang Qiang: Digital Image Processing and Analysis[M], Beijing, Tsinghua University Press, 2006.
15. Sun ZhaoLin: MATLAB 6.X Image Processing[M], Beijing, Tsinghua University Press, 2006.

An Efficient Authentication Protocol for Low-Cost RFID Systems*

Zhen-Ai Jin, Hyun-Ju Cho, and Kee-Young Yoo**

Department of Computer Engineering Graduate School,
Kyungpook National University Daegu, Korea
{sarah, hannah}@infosec.knu.ac.kr,
yook@knu.ac.kr

Abstract. Recently, RFID (Radio Frequency IDentification) has been receiving a great deal of attention. RFID is a very simple technology that involves communication between a microchip called a RFID tag and an electronic reader. RFID systems can be used in a variety of applications. There are some security properties that RFID systems require, such as location privacy and replay-attack resistance. In order to satisfy these security properties, many authentication protocols have been proposed. In these security protocols, most of them are easily attacked by attackers because tags cannot perform high-cost computations due to their limited resources. In this paper, we propose an efficient authentication protocol called SKU (Symmetric Key Update), which can provide replay-attack resistance and location privacy in low-cost RFID systems.

Keywords: Authentication, RFID, symmetric encryption, security, location privacy.

1 Introduction

Recently, RFID has been receiving a lot of attention. RFID is a very simple technology that involves radio frequency (RF) communication between a microchip called a RFID tag and an electronic reader. RFID technology is non-touch identification that use radio frequency signals (RF) to identify objects without the need of physical contact [1]. RFID systems can be used in variety of applications due to their low cost and compact size. RFID systems consist of three components: a back-end database, RFID readers and RFID tags. The Back-end database, which stores information associated with objects and readers, is a computer such as a server. In this paper, Back-end database is notated as a server. Readers are devices, which can read from or write to RFID tags through RF communication. RFID tags are made of silicon chips that carry object identification data and communicate with readers. RFID tags can be classified by service level and their power sources. MIT Auto-ID center has classified RFID tags

* This work was supported by grant No. R01-2006-000-10614-0 from the Basic Research Program of the Korea Science & Engineering Foundation.

** Corresponding author.

N.T. Nguyen et al. (Eds.): KES-AMSTA 2007, LNAI 4496, pp. 999–1006, 2007.

into five classes by their service level [2]. RFID tags are also classified into three types of RFID, passive tags, semi-passive tags and active tags by their power sources. Passive tags have no on-board power source, whereas semi-passive tags and active tags have batteries. Class 0 to class 2 RFID tags belong to passive tags. Class 3 RFID tags belong to semi-passive tags. Class 4 RFID tags belong to active tags. Active tags can initiate communication.

There are also two types of fields in communication area between RFID tags and readers. One is near field and the other is far field. Near field is the range in which readers can receive tag responses. Far field is the range which the reader can receive signals of the tag located in the near field. Signals from readers are much stronger than those from tags due to the owned power resources.

In RFID systems, there are some serious threats originated from RFID systems characters. For example, anybody with a reader can read the information from RFID tags and trace tags. The information stored in RFID tag must be protected from loss, unauthorized access, destruction, modification or disclosure. In order to satisfy these security properties, it is important to prohibit unauthorized readers from reading the information of tags and tracing tags in RFID systems [3]. There have been several authentication protocols such as that provide customer privacy: re-encryption based schemes [4,5], hash-based scheme [6,7,8] and XOR based scheme [9]. However most of these approaches are susceptible to attack. Because tags have limited resources due to their implementation cost. In related-work section, we will provide an overview on these schemes and a security analysis.

In this paper, an efficient authentication protocol using SKU is proposed, which can provide replay-attack resistance and location privacy. A symmetric key between a reader and a tag is one time key that is generated and updated by the server. The server generates a new symmetric key and saves it into tags for the next session.

This paper is organized as follows: currently existing protocols are introduced in section 2. The proposed secure protocol is described in section 3. In section 4, we analyze the security of our protocol with regard to several types of attacks. Concluding remarks are presented in section 5.

2 Related Approaches

Since the security of RFID systems was proposed, there have been several authentication protocols that can satisfy security properties. In this section, we describe and analyze these protocols: Re-encryption based schemes [4,5], hash based schemes [6,7,8], and XOR based scheme [9].

2.1 Re-encryption Based Schemes

This scheme uses ELGamal which is an asymmetric key encryption technology. Re-encryption based scheme checks the cipher text which is the encryption of unique ID of a tag, and re-encrypts it; this is done by using one-time random numbers. First, the RFID tag issues an encrypted ID-cipher text. After a reader reads a RFID tag, the reader re-encrypt the cipher text with a new randomly generated number. The reader uses a random number as an access key in order to protect the information of RFID tags. In order to reduce encryption cost, the scheme uses symmetric encryption to encrypt the cipher text with a random number as a key [10].

Even though the cipher text and the random number are updated, the cipher text can be decrypted all at once due to property of ELGamal. The re-encryption based schemes can not provide location privacy if the access key is disclosed by attackers. Attackers can use the access key to trace tags [11].

2.2 Hash-Based Schemes

In these schemes, each RFID Tag generates a random number and the secret ID is hashed. These schemes are cheaper than the re-encryption scheme. Readers send request to tags. Upon receiving the hash numbers and random numbers, the reader sends it to the server. The server generates a hash number from the secret ID in the server and random numbers which are received from the reader, then the server verifies tags by comparing the generated hash number and the one which is received from readers.

Each RFID tag encrypts its ID by using the one-way hash function to prevent the exposure of its ID. It, however, can not reduce the searching time from the server [12]. When attackers receive hash numbers or random numbers that are used to calculate the hash numbers, location privacy can not provide in this schemes because attackers can use hashed IDs to trace tags [12].

2.3 XOR-Based Scheme

In [9], each tag generates some random numbers and performs XOR-operation of them with the random number from the reader. Each tag generates some random numbers and makes secret number sets. A secret set consists of three random numbers (a, b, c). Tags and the server hold same secret sets. The reader selects a random number a, and sends it to the tags. The tags find the secret set which contains a. The tags do XOR operation of a and b in the same entry, and response the result to the reader. The reader does XOR operation of b and c, and sends the result to the tags. The reader and tag verify the result each other by checking the modified numbers [13]. However, this scheme can not provide customer privacy. Because attacker can eavesdrop and save the secret sets under the insecure network. We suppose that tag and server have m secret sets of random number. If attackers eavesdrop $2m$-random numbers and save them, then they can know the period of secret sets and all the values. Therefore the scheme can not provide location privacy. And also, if an attacker sends number zero to the tag, attackers can intercept the correct random number from a reader [14].

3 Efficient Authentication Protocol

An efficient authentication protocol is proposed that can provide customer privacy in semi-passive RFID systems. In the proposed protocol, we assume that attackers are located in a far field. Attackers can receive messages from the reader, but they can not receive messages from tags. In this section, we describe the efficient authentication protocol in detail. The server has two symmetric keys, which one is for tags and the other is for the reader, and the server has two tables to store these keys.

The notations that will be used throughout this paper are the following;

- T : tag.
- S : server.
- R : reader.
- E : attacker.
- ID : tag's id.
- R_{ID} : reader's id.
- N_R : random number which is generated by a reader.
- N_T : random number which is generated by a tag.
- K_{ST} : the symmetric key for each tag.
- K_{SR} : long term key for each reader.
- K'_{ST} : new symmetric key.
- K_S : secrete share between server and tag.
- $\{M\}K$: Encryption of message M with symmetric key K.

A server and a reader share a long term key (denoted by K_{SR}), which is stored in the database. Each tag has a symmetric key with server, denoted by K_{ST}. When tags are issued, a server generates K_{SR} and K_{ST}. As shown in Table 1 and Table 2, K_{SR} and K_{ST} are stored in the server. The key K'_{ST} is a new symmetric key and will be used in the next session. A server has two tables; one for tags and the other for the reader, and the contents of two the tables follow Table 1 and Table 2.

Table 1. Tag data Table which is stored in the server

ID	H(ID)	Key(K_{ST})	K_{RT}	S_K	…
11001	101	Key	K_{RT}	S_K	
01101	010	Key			

Table 2. Reader data Table which is stored in the server

ID	Key(K_{SR})	…
11001	Key	
01101	Key	

The contents of RFID tags presented in Fig 1. As shown in Fig 1, key K_{ST} is written in RFID tags.

ID	H(ID)	K_{ST}	S_K	…

Fig. 1. Information which is stored in a RFID tag

In our protocol, we use Tiny Encryption Algorithm (TEA) [15] which provides good security for mobile systems such as RFID in order to encrypt messages between a reader and tags, because tags have limited source of low-cost RFID tags. In the communication between server and reader, an authenticated encryption algorithm such as PKI or AES is used for providing authentication and data confidentiality. The Fig .2 shows an efficient authentication protocol using TEA.

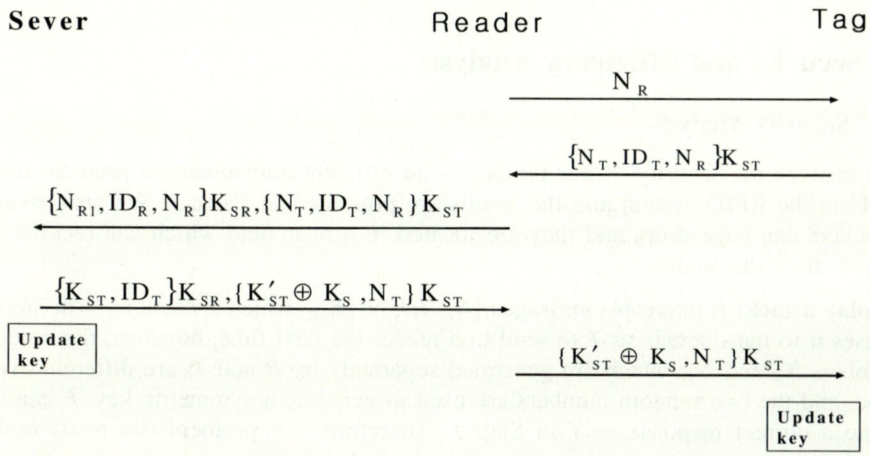

Fig. 2. An Efficient authentication protocol using SKU for low-cost RFID systems

In the description of protocols, the notation "=>" means the telecommunication direction. For example, "*R* =>" means that a reader sends a message to tags.

Step 1. *R* =>*T*: Reader send a random numbers N_R to tags. Upon receiving the random number *N*r, a tag generates a random number N_T and encrypts random numbers N_R, N_T and its ID, with key K_{ST}.

Step 2. *T* => *R*: Tag sends the encrypted message $\{N_R, ID, N_T\}K_{ST}$ to the reader. The reader encrypts its ID, random number N_R and new random number N_R,with long-term key K_{SR}.

Step 3. *R* => *S*: The reader sends its ID and encrypted message $\{N_R, ID_R, N_R\}K_{SR}$ and message which had been sent from the tag to server. R_{ID} is to reduce the searching time. In using the reader ID as a lookup key, the server can find the key K_{SR} and the message can be decrypted. If the sever can decrypt the message, then the server can verify that it is valid. The server decrypts message which is from the tag, then the server obtains tag's ID. The server verifies the reader and tag by comparing random number N_R, which is from the tag, with random number N_R, which is from reader. If the tag and the reader are valid, then the server encrypts key K_{RT} and tag's ID with K_{SR} for the reader. The server generates update key K'_{ST}, then N_T and ($K'_{ST} \oplus K_S$) are encrypted with key K_{ST} for the tag.

Step 4. S => R: Server sends encryption messages $\{K_{RT}, ID\}K_{SR}$ and $\{K'_{ST} \oplus K_S, N_T\}K_{ST}$ to the reader. Upon receiving the encryption message, the reader decrypts it and can obtain a symmetric key K_{RT} and the tag's ID.

Step 5. R => T: Reader send encryption message $\{K'_{ST} \oplus K_S, N_T\}K_{ST}$ to the tags. Then the tag decrypts and check the N_T, if the random number is correct, the tag updates the symmetric key K'_{ST} for the next session.

4 Security and Efficiency Analysis

4.1 Security Analysis

We analyze the security of our protocol - an efficient authentication protocol using SKU in the RFID system and the results are presented in Table 3. We also assume attackers can eavesdrop, and they are located in a near field which can receive the signals from the reader.

Replay attack: E intercepts messages $\{N_R, ID, N_T\}K_{SR}$ which are sent by T in Step 2. E uses it to masquerade as T to send to a reader the next time, however, the random numbers N_R and N_T, which are generated separately by R and T, are different every time, and the two random numbers are used to generate a symmetric key, E can not make a correct response to T in Step 5. Therefore, our protocol can resist replay attack.

Location privacy: In Step 2, a tag sends the encryption message $\{N_R, ID, N_T\}K_{SR}$ which is included in the tags' ID, thus the ID was not exposed to the attacker. The attacker can not trace the tags.

Mutual authentication: Reader decrypt the encryption message $\{N_R, ID, N_T\}K_{SR}$ from the tag, and check the random number N_R. If it is the correctly random number, then the reader authenticates the tag. The tag decrypt the encryption message $\{K'_{ST}, N_T\}K_{ST}$, and verify the random number that tags was sent to the reader N_T. If it is the correct random number, then the tag authenticates the reader. Through this process, tag and the reader authenticate each other.

Forward security: An attacker can not obtain the symmetric key even thought he has the long-term key K_{ST}. An attacker can get the secrete share S_K, which is set between the server and the tag, thus he can not computed the update key K'_{ST}.

Table 3. A comparison of protocols security measures

	Location privacy	Replay attack	Authentication		Forward security
			Reader	Tag	
Re-encryption based	×	O	×	O	×
Hash-based	×	O	O	O	×
XOR-based	×	O	O	O	×
Our protocol	O	O	O	O	O

4.2 Efficiency

The level of efficiency was determined by count the number of operations performed by the protocols.

The results are shown in Table 4. In comparison with other protocols, our protocol requires the same or less number of operations, but it provides more security.

Table 4. A comparison of protocols operation at tags

	Re-encryption based	Hash-based	XOR- based	Our protocol
The number of symmetric encryptions/decryptions	2	0	0	2
The number of hash operations	1	2	0	0
The number of exchange information with tags	4	3	3	3
The number of XOR operations	0	2	2	1

5 Conclusion

In this paper, an efficient authentication protocol using SKU for low-cost RFID systems has been proposed. Due to the limitation of the resources in tags, Server computes symmetric keys for the next session and sends it to the tags. Tags just receive a new session key from the reader and update the next session key. Our efficient authentication protocol can provide location privacy and mutual authentication, as well as replay-attack resistance and forward security can be provided in low-cost RFID systems. Therefore, we expect that our protocol can be used in low-cost RFID systems.

References

1. K. Finkenzeller.: RFID HandBook; Fundamentals and applications in Contactless smart Cards and Identification. Second Edition. John Wiley & Sons Ltd (2003)
2. Auto-ID Center. : Draft Protocol Specification for a Class 0 Radio Frequency Identification tag (2003)
3. RFID Technology, Privacy Issues. (2004) http://itpro.nikkeibp.co.jp/free/NBY/RFID/20040823/1/
4. Junichiro Saito., Jae-Cheol Ryou., and Kouichi Sakurai. : Enhancing Privacy of Universal Re-encryption Scheme for RFID tags. L.T. Yang et al. (eds.): EUC 2004 LNCS 3207 (2004) 879-890
5. Ari Juels and Ravikanth Pappu. :Squealing Euros. : Privacy Protection in RFID-Enabled Banknotes. Rebecca N. Wright, editor, Financial cryptography-FC'03, 2742 (2003) 1 03-121

6. Stephen A, Weis, Sanjay E. Sanma, Ronald L. Rivest, and Daniel W. Engels, "Security and privacy Aspects of Low-Cost Radio Frequency Identification Systems," first international conference on security in pervasive computing (2003) http://theory.lcs.mit.edu/sweis/spc-rfid.pdf
7. M. Ohkubo, K. Suxuki and S. Kinoshita. : Efficient hash- chain based RFID Privacy Protection Scheme. Ubcomp2004 workshop (2004)
8. D. Henrici and Palu Muller. : Hash-based Enhancement of Location Privacy for Radio-Frequency Identification Deveices using Varying Identifiers. PerSec'04 at IEEE PerCom. (2004) 149-153
9. A. Juels. : Minimalist Cryptography for Low-Cost RFID Tags. the Fourth International conference on Security in Communication Networks-SCN 2004, 3352 LNCS. (2004) 149-164
10. Ari Juels. : RFID Security and Privacy: A Research Survey. IEEE Journal, Vol. 24. issue 2. (2006) 381- 394
11. G. Avoine. : Privacy Issues in RFID Banknote Protection Scheme. J.-J.Quisquater, P. Paradinas, Y. Deswarte, and A. Abou E1 Kadam, editors, Sixth International Conference on Smart Card Research and Advanced Applications – CARDIS. Kluwer Academic Publishers. (2004) 33-48
12. Jianhua Ma, Akito Nakamura, Runhe Huang. : A random id update scheme to protect location privacy in RFID-based student administration systems. DEXA Workshops. (2005) 67-71
13. G. Avoine. : adversarial Model for Radio Frequency Identification. 2005 Cryptology ePrint Archive, Report 2005/049. Referenced (2005)
14. A. Juels. : Authentication Pervasive Devices with Human Protocols. Advances in Cryptologh-CRYPTO 2005, Vol. 3621. Springer-Verlag, Lecture Notes in Computer Science (2005) 293-308
15. P. Israsena. : Securing Ubiquitous and Low-Cost RFID Using Tiny Encryption Algorithm. Wireless Pervasive Computing, 2006 1st International Symposium, (2006)1-4

A Countermeasure of Fake Root Key Installation Using One-Time Hash Chain

Younggyo Lee[1], Jeonghee Ahn[2], Seungjoo Kim[1], and Dongho Won[1,⋆,⋆⋆]

[1] Information Security Group, Department of Computer Engineering, Sungkyunkwan University, 300 Chunchun-dong, Jangan-gu, Suwon, Kyunggi-do, 440-746, Korea
{yglee, skim, dhwon}@security.re.kr
[2]Department of Computer Science, Doowon Technical College, 678 Jangwon-ri, Juksan-myoun, Anseong, Kyunggi-do, 456-718, Korea
jhpro@doowon.ac.kr

Abstract. Adil Alsaid and Chris J. Mitchell proposed the possibility of fake root public key installation by an attacker in user's PC and showed its countermeasures in 2005. The root public keys are used to verify the certificates for applet providers. Therefore the insertion of false public keys allows arbitrary numbers of rogue application to be executed on a user's PC. We propose a protection method for installing fake root keys in a user's PC. The method uses the one-time hash chain based on NOVOMODO. In the proposal, as the computation costs and the storage amounts for hash chain are very small, the proposed method will not be a big load to the publisher of the browser, the applet provision agent and the user. The implementation of the method is simple and it offers convenient authentication of the root to the user. Therefore, the method can provide a intelligent and secure agent technique for the digital content distribution.

Keywords: root public key, hash chain, publisher of the browser, web browser, applet provider, applet distribution agent.

1 Introduction

Adil Alsaid and Chris J. Mitchell proposed the possibility of fake root key installation by an attacker in user's PC and shown its countermeasures in 2005 [1]. The famous web browsers such as Microsoft Internet Explorer or Netscape have a repository of root public keys in user's PC. These public keys are used to verify the certificates for applet providers (or digital content distribution agents). The applet is signed digitally by the applet provision agents and is downloaded to a user's PC. The signature in the applet is verified using corresponding public key certificate in a user' PC. As the certificate is signed using a root private key, it must be verified using a stored root public key (certificate) in a user's PC.

⋆ This work was supported by the University IT Research Center Project funded by the Korean Ministry of Information and Communication.

⋆⋆ Corresponding author: Dongho Won, Presenting author: Younggyo Lee.

N.T. Nguyen et al. (Eds.): KES-AMSTA 2007, LNAI 4496, pp. 1007–1016, 2007.

As the insertion of false public keys is simpler than the insertion of a rogue application directly, it allows arbitrary number of rogue applications to be executed on a user's PC. A stored false public key can not be detected by current attack detection software. As the availability of certificate creation tool such as 'makecert.exe' is free, an attacker can easily install a false public key in a user's PC.

Adil Alsaid and Chris J. Mitchell presented a practical method of fake root key installation in a user's PC by window hooking of 'security warning' message box on the Microsoft Windows 2000 operating system and the Microsoft Internet Explorer web browser. And they discussed its countermeasures in [1]. However, the countermeasures are inefficient and unpractical. In this paper, we propose protection method for installing fake root keys in user's PC. The method uses the one-time hash chain based on NOVOMODO [8]. As the computation costs and the storage amounts for hash chain are very small, the method will not be a big load to the publisher of the browser, the applet provision agent and the user. The rest of this paper is organized as follows. In section 2, we analyze the countermeasures proposed by Adil Alsaid and Chris J. Mitchell in detail. In section 3, we propose the protection method for installing fake root keys using one-time hash chain. In section 4, we analyze the characteristics of our proposal and compare it to other countermeasures. Finally, in section 5, we conclude our paper.

2 Countermeasures and Their Analyses

In this section, we show the countermeasures to the threat of installation of a fake root certificate in a user's PC proposed by Adil Alsaid and Chris J. Mitchell and analyze them in more detail.

2.1 Re-authentication (Prevention)

"When carrying out such a security sensitive task, users should always be re-authenticated."[1] Even if the prevention is an ideal solution, this prevention requires a long time because of the modification to the Windows environment (mostly software). It may be an approach based on a simple idea but is an inefficient solution. Is re-authentication an ideal solution for preventing every similar security sensitive attack? Is the re-authentication of the user by the same method an ideal solution nevertheless a user is authenticated at log-in time? What do you say to take re-authentication by the different method?

2.2 Restricting Access to the List of Root Public Keys (Prevention or Detection)

"The attack could also be prevented by restricting access to the list of root public keys to special privileged users or processes."[1] When the new root public key is installed, the solution can prevent a user's PC from installing a false root certificate using the list of root public keys (prevention). Small software can detect

the installed malicious third party certificates by periodical scanning using the list of root public keys (detection). If an attacker could acquire the access to the list of root public keys by hacking, he/she can install the fake root certificates successfully by its modification (adding or changing). Therefore the list of root public keys must be protected from the attacker. One such approach is for the list to get the integrity when it is offered with the browser. However, the solution is difficult to implement because it requires adding new items to the list later.

2.3 Verifying the Status of Certificate Using OCSP (Detection)

The method is to verify the status of a root certificate before it is used by OCSP (Online Certificate Status Protocol). Therefore, the approach is not for prevention but detection. The method is very strong and one of the best solutions. It is free for adding new root public keys later. And the possibility of faking status response of a certificate asserted by Adil Alsaid and Chris J. Mitchell is not nearly to happened because the OCSP request and response are signed and validated and CA(Certification Authority) is TTP(Trusted Third Party) [1]. However, the solution requires the resources of a CA, the communication with the CA, the storage amounts for CA's certificate, and the computation costs of signature and signature validation in user and CA. And the root certificate is not easily revoked like user's certificate by the loss or compromise of the associated private key. If a root certificate would be trusted at first, its trust can nearly not be changed until it is expired. Therefore, the solution verifying root certificate whenever using it may be inefficient.

2.4 Two Lists of Root Keys (Detection)

The method requires the browser to maintain two lists of root keys. *"One list for genuine root keys that were verified by the publisher of the browser, i.e. shipped with the browser. A second list will contain root public keys that were added by the user and that were not shipped with the browser."*[1] Although this method is similar to the second countermeasure above, it seems that this countermeasure is more efficient than previous countermeasure. One list is verified by the publisher of the browser and offered with the browser. The other is made by adding the root keys to user's PC. For the time of transaction engagement that one of the root keys is used in the second lists, the browser will check if the root key being used is included in the first list or not. The approach is not for prevention but for detection. When a new root key is installed, checking if the root key is included in first list may be more efficient. The solution is also difficult for adding of new items to the first list later. And the first list must be stored securely (e.g. integrity, encryption) in order to prevent from adding of a false root certificate item by an attacker.

3 A Countermeasure of Fake Root Key Installation Using One-Time Hash Chain

Earlier, we have discussed about the countermeasures of fake root key installation in a user's PC. We see that each countermeasure has some advantages and some disadvantages. The hash chain of NOVOMODO can give the message authentication and integrity using small computation costs (at least 10,000 times faster in computation than a digital signature operation) and the small resource of 20 bytes [7]. Therefore, in this section, we propose a new countermeasure of fake root key installation using the hash chain. The countermeasure requires small modifications to the Windows software and do not need the CA or lists of root keys. The method uses the one-time hash chain based on NOVOMODO [8].

3.1 Proposal (1)

In NOVOMODO by Micali [8], a user's certificate $Cert_{user}$ includes two 20-byte (160-bit) hash values in addition. The one (X_{365}) is used as "validity target" and the other (Y_1) is used as "revocation target." These values are produced by applying one-way hash function to two different 20-byte values randomly selected in CA. When the time interval is one day, the value X_{365} is computed by 365 hash operations from $X_0 : X_1 = h(X_0), X_2 = h(X_1), \cdots, X_{365} = h(X_{364})$; and Y_1 by one hash operation from $Y_0 : Y_1 = h(Y_0)$. The CA keeps secretly the initial values X_0, Y_0 and all the intermediate values X_i. The CA releases the corresponding intermediate hash value to each user as a certificate's "validity proof" (X_i) or "revocation proof" (Y_0) at initial time of each interval. In E-commerce or E-business based on PKI, the client getting the user's certificate and corresponding hash value compares the two values using the hash function. If the result of comparison is the same, the client gets the message authentication and integrity concerning the received hash value (X_i or Y_0) and confirms the user's certificate status by the hash value. The hash value is small. However, since the inversing of the hash function is practically impossible, someone (e.g., attacker) cannot forge it.

As Younggyo Lee *et al* use the one-time hash chain and the counters in [9,10,11], an attacker acquiring user's (session or partial) private cannot reuse a hash value in the same interval. We use the one-time hash chain based on NOVOMODO. Suppose that the number of root is 1,000 sufficiently.

[Procedure]

1. At first, the publisher of the browser computes R_0 by 1,000 hash operations from random input value R_{1000} using hash function h as follows.

$$R_{1000} \xrightarrow{h} R_{999} \xrightarrow{h} \cdots R_i \cdots R_j \cdots \xrightarrow{h} R_1 \xrightarrow{h} R_0$$

2. He delivers the final hash value (R_0) with the browser to each user.
3. And he distributes the input value and all intermediate values ($R_{1000} \cdots R_1$) in step 1 securely to roots one by one.

4. Later, each root tries the root key installation using a hash value to a user's PC. The certificate management program then confirms the authentication to the root using hash operation instead of displaying a set of dialog boxes to allow the user to manage the root key installing process.
5. The certificate management program repeatedly computes the hash operation until the hash operation result of R_i is equal to the hash values R_0 contained in the provided browser as follows.

$$R_0 \stackrel{?}{=} h^i(R_i)$$

If this holds, the certificate management program can confirm the authentication of the root key. It continues to install the root key in own PC.

The countermeasure uses only small storage amounts (20 bytes) and small computation costs (hash operation) and offers the secure root key installation without interaction with user (or attacker). Therefore the method is considered to be an efficient solution.

However, the countermeasure has some problems. Suppose that the root having R_i tries to install root key before the root having $R_j (i > j)$. An attacker can steal the hash value R_i during the communication or in a user's PC. As he/she easily compute the last hash chain $(R_{i-1}, \cdots, R_j, \cdots, R_1)$, he/she can install fake root keys in a user's PC successfully. Therefore, each root has to take turns in this method. And when the root tries to re-install root key due to the some reasons (the loss or compromise of the associated private key, etc.), the re-installation is impossible because the hash value is valid only one-time.

3.2 Proposal (2)

In NOVOMODO, if a certificate includes 10 validity targets $(A_{365}, B_{365}, C_{365}, \cdots, J_{365})$ and one revocation target (Y_1) and CA releases a different validity proof periodically, a user can use the certificate as 10 certificates with different levels [7]. The implementation has the same effect that the user has 10 certificates. In the proposal (2), we use the implementation of NOVOMODO. Suppose that the number of root is 1,000 sufficiently.

[Procedure]

1. The publisher of the browser computes $({}_1R_0, \cdots, {}_{1000}R_0)$ by 5 hash operations from 1,000 random input values $({}_1R_5, \cdots, {}_{1000}R_5)$ using hash function h as follows.

$$
\begin{gathered}
{}_1R_5 \xrightarrow{h}{}_1 R_4 \cdots {}_1R_1 \xrightarrow{h}{}_1 R_0 \\
\vdots \\
{}_kR_5 \xrightarrow{h}{}_k R_4 \cdots {}_kR_1 \xrightarrow{h}{}_k R_0 \\
\vdots \\
{}_{1000}R_5 \xrightarrow{h}{}_{1000} R_4 \cdots {}_{1000}R_1 \xrightarrow{h}{}_{1000} R_0
\end{gathered}
$$

2. He delivers the final hash values $({}_1R_0, \cdots, {}_{1000}R_0)$ with the browser to each user.
3. And he distributes the input values and all intermediate values in step 1 securely to roots one by one as follows.

$$({}_1R_5, \cdots, {}_1R_1) \rightarrow 1^{st}\ root$$
$$\vdots$$
$$({}_kR_5, \cdots, {}_kR_1) \rightarrow k^{th}\ root$$
$$\vdots$$
$$({}_{1000}R_5, \cdots, {}_{1000}R_1) \rightarrow 1000^{th}\ root$$

4. When *k-th* root tries the root key installation using hash value ${}_kR_1$ to a user's PC, the certificate management program repeatedly computes the hash operation (1 hash operation at the first installation) until the hash operation result of ${}_kR_1$ is equal to the hash value ${}_kR_0$ contained in the provided browser as follows.

$$ {}_kR_0 \stackrel{?}{=} h({}_kR_1)$$

   If this holds, the certificate management program can confirm the authentication about *k-th* root. It continue to install the root key in own PC.
5. And when the root tries to re-install root key, he tries the root key installation using hash value ${}_kR_2$. The certificate management program repeatedly computes the hash operation (2 hash operation at the second installation) until the hash operation result of ${}_kR_2$ is equal to the hash value ${}_kR_0$ as follows.

$$ {}_kR_0 \stackrel{?}{=} h^2({}_kR_2)$$

   This approach needs smaller computation costs (1∼5 hash operations) than the proposal (1). But it requires more much storage amounts (20×1000 bytes) in a user's PC than the proposal (1). Contrary to the proposal (1), each root does not need the installation order in this proposal. And each root can re-install the root key until 5 times.

4 Characteristics and Comparisons

In this section, we explain the characteristics of the proposals and compare them to other solutions. The detailed characteristics are as follows:

[The authentication to the root using hash value]
When the security sensitive task such as the installation of root key is carried

out, user always must be re-authenticated [1]. Therefore, the authentication to the root is required in the installation of root key, as stated in section 2. The application of different authentication methods is more efficient than the application of same authentication method. Our proposals authenticate the root using hash value unlike the general authentication methods (e.g. at log-in). The hash function operation is at least 10,000 times faster in computation than a digital signature operation generally [7]. And it produces always 20-byte output [8].

[Giving facility for the user]
Even if the malicious installation of a root key is not established silently, the certificate management program does not look good for the user. 'Adding a root certificate' message box asks the user for confirmation that the user wishes to add the new certificate to the root store. The message box shows the issuer name and thumbprint (20-byte hash value by sha-1) for the certificate. The user can obtain the corresponding thumbprint (from a file in CD or a list in browser manual) by the certificate issuer and compare this with that. As the user must compare 40 hexadecimal characters with own eye, this confirmation method is primitive and can be inconvenience to the user. Moreover, the user can make a mistake in the confirmation process. However, the proposals would not be inconvenient for the user. The thumbprint confirmation is established automatically in the proposals. Therefore the proposals afford facility for the user.

[The storage amounts for a user]
Table 1 shows the storage amounts for hash value in a user by the number of root. We see from Table 1 that the required storage amounts are very small and uniform in the proposal (1) but they are in proportion to the number of root in the proposal (2). For the case of 1,000 roots, the proposal (2) requires 1,000

Table 1. The storage amounts for a user

The number of root	100	1,000	5,000	10,000	50,000
Proposal (1)	20 bytes	20 bytes	20 bytes	20 bytes	20 bytes
Proposal (2)	20 × 100 bytes	20 × 1000 bytes	20 × 5000 bytes	20 × 10000 bytes	20 × 50000 bytes

times more much in storage amounts than the proposal (1). However, as 20×1000 bytes $\simeq$ 20 K bytes, this amount is small one which will not be a big load to the user' PC. For the case of 10,000 roots, the storage amounts are at most about 200 K bytes.

[The computation costs in a publisher of the browser]
Table 2 shows the computation costs for the generation of hash chain in a publisher of the browser by the number of root. We see from Table 2 that the

Table 2. The computation costs in a publisher of the browser

The number of root	100	1,000	5,000	10,000	50,000
Proposal (1)	100	1,000	5,000	10,000	50,000
Proposal (2)	5 × 100	5 × 1,000	5 × 5,000	5 × 10,000	5 × 50,000

Unit : hash operations.

computation costs are in proportion to the number of root in the proposal (1) and (2) and the proposal (2) requires 5 times of computations compared to the proposal (1). As the hash operation speed for our proposal is at least 10,000 times faster in computation than a digital signature operation and the hash chain generation is established in off-line before the distribution of the browser, the computation costs will not be a big load to the publisher of the browser.

[The computation costs for a user]
Table 3 shows the computation costs required for the authentication in a user. We see from Table 3 that the computation costs are proportional to the number of root in the proposal (1). In case 5,000 roots in the proposal (1), the user computes 1 hash operation in the installation of first root and computes 5,000 hash operations in the installation of 5,000th root. Therefore, the user computes average 2,500 hash operations for one installation. As the hash operation speed is at least 10,000 times faster in computation than a digital signature operation, this computation costs will not be a big load to a user. But in the proposal (2), the computation costs are constant (1∼5 hash operations). The user computes 1 hash operation in the first installation and computes 5 hash operations in the 5th re-installation. This computation costs are very small.

Table 3. The computation costs for a user

The number of root	100	1,000	5,000	10,000	50,000
Proposal (1)	ave. 50	ave. 500	ave. 2,500	ave. 5,000	ave. 25,000
Proposal (2)	1~5	1~5	1~5	1~5	1~5

Unit : hash operations.

We compare the proposals to the other countermeasures such as in Table 4. We see that the proposal (1) and (2) require the small computation costs and the storage amounts because they use the hash chain. As they require small modifications to the Windows environments (the certificate management program in detail), these implementation can be achieved in a short-term. Especially,

Table 4. The comparison of the proposals with the other countermeasures

	Proposal (1)	Proposal (2)	re-authentication	restricting access	OCSP	two lists
Application type	prevention	prevention	prevention	prevention or detection	detection	detection
Implementation method	authentication using hash chain	authentication using hash chain	re-authentication to security sensitive task	restricting access to the list of root public keys	verifying the status of certificate using OCSP	2 lists of applet providers
Need of other party	-	-	-	-	CA	-
Costs	hash operation	hash operation	authentication operation	restricting access	communication, signature and validation	compare
Addition or modification of software	small	small	big	big	big	medium
At install (publisher of the browser)	1000 hash operations	5 × 1000 hash operations	-	-	issue certificate (CA)	make list
At install (user)	ave. 500 hash operations/1 install	1~5 hash operations/1 install	1 authentication/1 install	1 restricting access/1 install	1 sign and 1 validation/1 install	scan, update list and compare periodically
Storage amounts (user)	20 bytes	20 × 1000 bytes	-	-	CA's certificate	2 lists
Scalability	medium	medium	medium	medium	big	small
Reinstall	difficult	easy	medium	medium	easy	difficult
Authentication	support (different)	support (different)	support (same)	-	not	not

the proposal (2) is useful for re-installation of the root key. And they have an advantage of authenticating the root by the different method with login.

5 Conclusions

Adil Alsaid and Chris J. Mitchell presented the practical method of fake root key installation in a user's PC by window hooking of 'security warning' message box on the Microsoft Windows 2000 operating system and the Microsoft Internet Explorer web browser. And they showed its countermeasures in [1]. However, the countermeasures are inefficient and unpractical. In this paper, we discuss the countermeasures in more detail. And we propose protection methods of fake root key installation using one-time hash chain.

In the proposals, as the computation costs and the storage amounts for hash chain are very small, the proposals will not be a big load to the publisher of the browser, the applet provision agent and the user. Their implementation are simple and they offer the required authentications. Consequently, we see that the proposals not only protect the fake root key installation but also afford facility for the user to compare the thumbprint. Therefore, the method can provide a intelligent and secure agent technique for the digital content distribution based on Web.

References

1. Adil Alsaid and Chris J. Mitchell .: Installing Fake Root Keys in a PC, Euro PKI 2005, LNCS 3545, pp. 227-239, 2005.
2. Jianying Zhou, Feng Bao and Robert Deng.: An Efficient Public-Key Framework, ICICS 2003, LNCS 2836, pp.88-99, 2003.
3. Jong-Phil Yang, Chul Sur, Hwa-Sik Jang and Kyung-Hyune Rhee.: Practial Modification of An Efficient Public-Key Framework, 2004 IEEE International Conference on e-Technology, e-Commerce and e-Service, March 2004
4. Leo Reyzin.:General Time/Storage Tradeoffs for Hash-Chain Re-comoutation, unpublished manuscript.
5. M. Myers, R. Ankney, A. Mappani, S. Galperin, C. Adams.: X.509 Internet Public Key Infrastructure Online Certificate Status Protocol - OCSP, IETF RFC 2560, June, 1999.
6. R. Housley, W. Ford, W. Polk and D. Solo.: Internet X.509 Public Key Infrastructure Certificate and CRL Profile, IETF RFC 3280, April, 2002.
7. Satoshi Koga, Kouichi Sakurai.:A Distributed Online Certificate Status Protocol with a Single Public Key, Public Key Cryptography 2004, LNCS 2947, pp.389-401, 2004.
8. Silvio Micali.:NOVOMODO ; Scable Certificate Validation And Simplified PKI Management, 1st Annual PKI Research Workshop Preproceedings, pp.15-25, 2002.
9. Younggyo Lee, Injung Kim, Seungjoo Kim, and Dongho Won.: A Method for Detecting the Exposure of an OCSP Responder's Session Private Key in D-OCSP-KIS, Euro PKI 2005, LNCS 3545, pp. 215-226, 2005.
10. Younggyo Lee, Jeonghee Ahn, Seungjoo Kim, and Dongho Won.: A Method for Detecting the Exposure of an OCSP Responder's Private Key using One-Time Hash Value, IJCSNS International Journal of Computer Science and Network Security, VOL. 5 No.8, pp. 179-186, August 2005.
11. Younggyo Lee, Jeonghee Ahn, Seungjoo Kim, and Dongho Won.:A PKI System for Detecting the Exposure of a User's Secret Key, Euro PKI 2006, LNCS 4043, pp. 248-250, 2006.

Proposal for a Ubiquitous Virtual Enterprise Reference Model on Next-Generation Convergence Network

Yonghee Shin[1], Hyori Jeon[2], and Munkee Choi[2]

[1] IT Mangement Research Group, Electronics and Telecommunications Research Institute, 161 Gajeong-dong, Yuseong-gu, Daejeon, 305-350, Korea
[2] School of IT Business, Information and Communication University, 119 Mungi-ro, Yuseong-gu, Daejeon, 305-714, Korea
syong@etri.re.kr, hrjeon@icu.ac.kr, mkchoi@icu.ac.kr

Abstract. Nowadays, market environments can be expressed in terms of increased competition, the diverse demands of customers, and the rapid innovation of IT technologies and so forth. These environmental changes have left companies facing a highly competitive market, and company revenues are gradually decreasing. Virtual enterprise is considered as the one of the alternatives available to facilitate adjustment to the new digital economy and save companies from falling into a red ocean. The contents of this research include the definition of virtual enterprise, and the characteristics and technologies for embodying it in a company. The paper suggests a new flow model for a decision support system in a virtual enterprise set-up, assesses critical success factors and weights as per the factors from case studies. From the previous analysis, paper defines the needs of collaborative activities in virtual enterprise and suggests the ubiquitous virtual enterprise reference model.

Keywords: virtual enterprise, digital economy, flow model, decision making, DSS model, critical success factor, firm relation type.

1 Introduction

A fiercely competitive market creates the strongest firms according to market economy theory. For this reason, firms which want to maintain their existence concentrate their activities upon cost saving, quality improvement, R&D activity and so forth. In order to cope with the rapid environmental changes, they need a flexible and collaborative relationship. The way to overcome reduced profits and ensure a firm's survival in the center of a radically changing market is by cooperating with others, even one's competitors. In this respect, competitors will no longer be enemies. Famous examples of this relationship between rivals include M&A (through the buying of a rival's stocks) and strategic alliance (through contracting). Such relationships embrace outsourcing, joint ventures, strategic alliance, M&A, and virtual enterprise according to the need to procure external resources. Herein lies the paper's objective: to find a way to construct a virtual enterprise as an effective method of pursuing cooperation between one's own firm and others in the same

N.T. Nguyen et al. (Eds.): KES-AMSTA 2007, LNAI 4496, pp. 1017–1026, 2007.

market. The paper's contents are organized into three parts. First, it introduces the literature survey results with regard to the virtual enterprise concept and defines the concept used in the study. Second, it suggests a checklist to be used when a virtual enterprise is formed. The checklist is composed of the critical success factors extracted from virtual enterprise case studies and literature surveys. Thus, a new decision support flow model based upon this checklist is proposed. Third, it analyzes the checklist factor weights by the AHP method. Finally, a new virtual enterprise reference model is suggested, which aims for, expanding existing enterprise system and meeting the various needs. The ultimate object it to make it easier to construct the collaborative relationships in virtual enterprise.

2 The Concepts and Technologies of Virtual Enterprise

Increased competition among systems and the development of IT technologies has brought a cooperative system structure to many firms. These changes request the firms' structures and strategies reformed. The newly shaped firms are referred to as 'virtual enterprises'. In this respect, the causation leading to change is composed of two aspects: one concerns management environment features, the other concerns the management paradigm. Firms that want to survive the changes have to meet customer needs and provide competitive products of high quality at a low price, compared to their competitors. For this purpose, firms promptly use all the resources available without considering their physical locations. A virtual enterprise with a virtual structure is the new paradigm that displays adaptation to this situation. Since the Iacoca Research Center at Rehei University suggested the theory of virtual enterprise in 1991, that theory has been accepted by many managers and is now suggested as an essential concept. The various concepts of virtual enterprise are very ambiguous; a list of these concepts as defined by scholars from around the world follows below.

Table 1. Various Concepts of Virtual Enterprise

	Definition
Wexler, Joanie (1993)[4]	An association of employers and employees who are not united at all times by a physical work environment.
Byme, J. A.(1993) [5]	A temporary network formed by independent firms has the purpose of accessing other markets and reducing costs.
Anonymous(1994) [13]	A team is composed of personnel who work as if they were a part of one company to achieve common goal.
Flieding (1998)[10]	VE is an organization unconstrained by geographic location, and a membership intersection multiple traditional organization.
Pappaionnou(1998) [7]	VE is made up of a number of cooperating companies who are over a wide area but work together to meet some market demand.
unikarisruhe web (2000) [14]	A cooperation firm network built on an understanding of the common shared goal.
Korea National Computerization Agency (2002) [15]	Firms act like a firm using extra- and intra-network and IT technology; coordinating firms must have core competences and help each other as partners.

In summarizing the above definitions, the research suggests a new virtual enterprise definition. "For the purpose of achieving common goals, firms and persons in possession of core competences create instant value chains by contract and then focus on their core competence efficiency and effects." A virtual enterprise uses market power to develop, manufacture, exchange, distribute and maintain products and services by a method that is not mimicked by traditional firms; the method involves the use of external technologies and resources. A virtual enterprise is able to access required resources based on flexible organizational structures.

IT support is used to develop and integrate in order to achieve virtual characteristics. To developing and integrating software is most important in the realization of a virtual enterprise. Basically, the realization of a virtual enterprise requires next technologies, i.e. telecommunication and protocol technologies to transfer messages, standard object-oriented technologies to support application system interaction among firms, technologies of information modeling and exchange, and management technologies to control the integrated virtual enterprise workflow and manage knowledge. The software functions required to embody a virtual enterprise include a database, knowledgebase, telecommunications software between the client and server, middleware, workflow management, and agent. The requirements for these factors follow. [9][10][15][16]

- Persistency: created objects can be reused at the time, if required.
- Concurrency control, naming, query processing, recovery: they must support the distribution system.
- Security: the information created by an virtual enterprise is protected and properly dispersed when the virtual enterprise achieves its goals.
- Rule processing and trigger management
- Active objects: The virtual enterprise resources are executed automatically, and are programmed as active objects

Table 2. The Fundamental Technologies Required to Set Up a Virtual Enterprise

Technology	Explain
Telecommunications	Firms in VE are dispersed at many other physical locations Telecommunications technologies such as Internet, Wireless, All-IP and so forth are fundamental.
Security	3 needs of virtual enterprise to need security - information sent by each partner is protected - the information created by a virtual enterprise needs security. - the information assets are dispersed by a secret method.
Objects	Standard objective technologies are essential to secure the application between a virtual enterprise system and application system. Therefore, a framework that is able to interact and reuse based on the standard objective technologies must be proposed.
MIS	Technologies of job cooperation, and knowledge control are important - such as strategic planning, scheduling, projector controls, workflow controls, products design, and manufacturing process control. The success of VE depends on easily accessing resources such as data, human, systems, tools, and software through a heterogeneous network.
Agent	Functions of the virtual enterprise agent is the act instead of others' entities and help manage operations such as user agent, service agent, resource agent, information agent, group agent and so on.

3 Empirical Study to Suggest Virtual Enterprise Decision Flow Model with Decision Factors Weights

3.1 Suggestion of Virtual Enterprise DSF Model

For the success of the virtual enterprise, we must check up the whole process step by step. In other words, before constructing a virtual enterprise, firms that want to create a virtual enterprise should consider their own core competences and problems, and then decide on partners who can cover problems and accelerate core competences based on analyzing their own firm's situation. Thereafter, they must measure their work performances by criteria for measuring success. The paper suggests a decision flow model that is composed of critical success factors from virtual enterprise case studies and literature surveys. The original decision flow model (Fig.1) is suggested in this paper.

In the first stage, a firm reevaluates its resources and examines closely its core competencies and problems. It must choose a partner that is actually able to make maximum use of one's own core competences and cover one's problems to form a

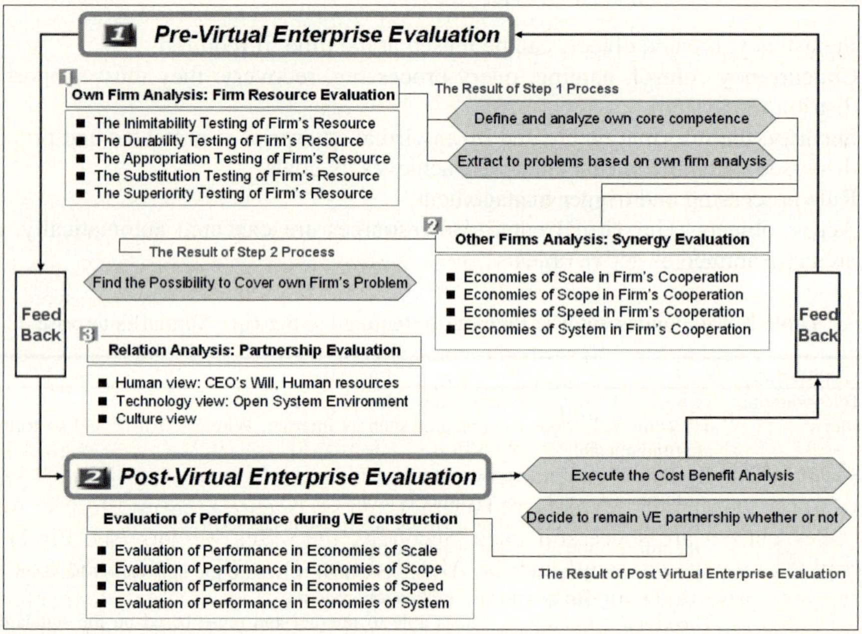

Fig. 1. To form a virtual enterprise, a firm has to decide on a partner. This shows the process of the decision flow model that explains the process of choosing a partnership. Two stages are involved: pre-virtual enterprise evaluation and post-virtual enterprise. The first stage involves choosing a partnership by evaluating one's own and the other firm's core competences. The second stage involves evaluating the results of the virtual enterprise operation and deciding whether to maintain the relationship.

virtual enterprise. A firm examines its core competences by inimitability testing, durability testing, appropriation testing, substitution testing and superiority testing. It is considered to connect between other firms' analysis and performance. In other words, if 4 factors (economies of scope, scale, speed, and system) are increased as a result of forming a virtual enterprise, then we can say that the virtual enterprise's construction has been a success. So, an analysis is executed to select firms that have the excellent qualifications required to contribute to performance in terms of those 4 factors. The consideration of maximizing one's own core competences and selecting other firms that cover one's problems are completed by examining the analysis of both firms. Now, in this section, we want to deal with determining the optimal substitute (the firms as proper partners to make a virtual enterprise) based on the above two considerations. So, forming a virtual enterprise is based not on the individual effort of one's own and another firm, but on the relationship between the firms.

If one's own firm selects a partner according to the conditions and orders outlined in the previous section, namely the CSFs(Critical Success Factors) in the Pre-Virtual Enterprise, then they will work together as a virtual enterprise. At this point, the examination of performance is very important. If there is no such examination, they will not have any adapted feedback about the virtual enterprise's activities. So, in the long-term, we cannot assure success in forming and operating a virtual enterprise. In this paper, we set the criteria for examination as economies of scale, scope, speed, and system.

3.2 Empirical Study to Extract Weights Per CSFs

In this section, it is assumed that the weights of CSFs are different because a virtual enterprise is created through the coordinative network among firms. If the coordination between the firms is strong, then certain factors will exercise greater influence than others. A case study and the AHP method are used to prove the assumption. Firstly, we study the successful virtual enterprise. Secondly, we carry out an expert survey on the CSFs weights using the AHP method. Finally, we suggest a checklist to be used when forming a virtual enterprise from the decision flow model and the result of empirical study.

Before a expert survey, firstly, we choose 18 such cases: Burpee, Amex, Airotech, Nokia, EGL(Eagle Global Logistic), Electropia, Ameritech, Softbank, Mazda, Toyota, Ford, Benz, Cisco, Dell, Menichetti, Federal Express, Boeing, General Electric. All of the cases chosen are successful ones. Secondly, the CSFs weights (see Fig.1) are determined by expert survey1 and the AHP method. The CSFs are defined from the virtual enterprise decision flow model proposed in the paper. The AHP method is used to determine the weights of the criteria. First of all, we determine the hierarchy of the CSFs (see Fig.2). An expert survey is conducted to evaluate the relative significant of the CSFs using a dual comparison method. After completing the whole process, the CSFs weights are finally determined.[12]

---

[1] We conducted an expert survey in March 2006. The survey of expertise was conducted by interview, and yielded 26 respondents who employed as venture employees, researchers, and professors in the IT industry.

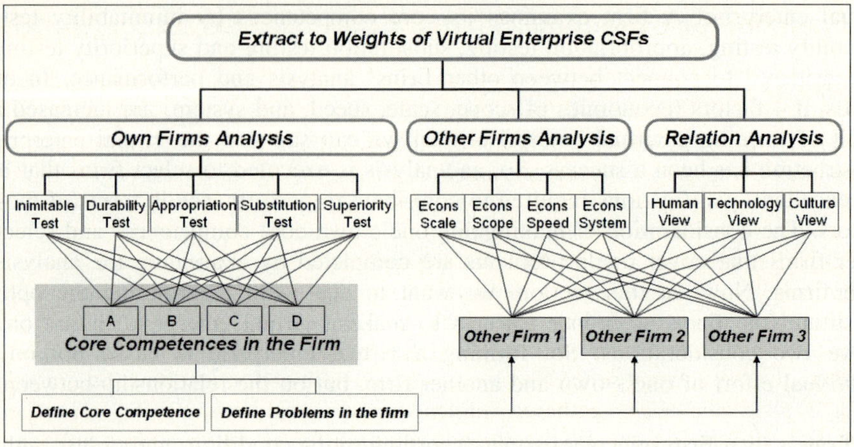

Fig. 2. In the AHP method, factors in the 2nd step are compared as pairs and their rank is determined through the survey of 26 experts, and the weights are calculated by these ranks. The 9 degree score is used for the expert survey. Level 1 means two factors are similar. Level 9 means one factor is superior to the other.

The paper suggests the weights of the 1st and 2nd levels' factors without considering the 3rd factors[2]. The result of the CSFs weights is shown in Table 4.

Table 3. The CSFs Weights to Construct Virtual Enterprise

1st Level	Weight	2nd Level Factors	Weight
Inter-firm Analysis	0.379	**Inimitability Testing**	**0.183**
		Durability Testing	0.038
		Appropriation Testing	0.030
		Substitution Testing	0.013
		Superiority Testing	**0.116**
Extra-firms Analysis	0.113	Economies of Scale	0.008
		Economies of Scope	0.010
		Economies of Speed	0.028
		Economies of System	0.067
Relationship Analysis	0.508	**Human View**	**0.107**
		Technology View	**0.122**
		Culture View	**0.279**

After analyzing, we study the results and get some implications. Virtual enterpise must focus on the relationship among firms. The most important factor is culture. However, the performance of relation is the most important factor in high rank relation type. Specially, the first one is culture view and the second is technology view in relation analysis. Other factors such as inimitability and superiority are also important when making the collaborative relationship with other firms.

2 The 3rd level factors weights remain open to further study. In this paper, we consider the 3rd level factors to be the same as 2nd level factors.

4 Proposal for Ubiquitous Virtual Enterprise Reference Model

Through the analysis of virtual enterprise decision flow model and weights of CSFs, relationship analysis factor is most important one. To establish effectively virtual enterprise, numerous network related technologies, theories are examined, and USN, Ubiquitous Computing, Web service, and IMS are derived as effective tools for improving the relationship among virtual enterprise entities. In this context, through combining above technologies, Ubiquitous Virtual Enterprise Reference model is suggested.

Recently, one can predict that the ubiquitous environment of the future will comprise publicly and privately deployed sensor network; USN (Ubiquitous Sensor Network)[24]. Thus, USN based services such as enabling sensor-based services in mobile network are being studied[19]. USN based services have the property of context awareness because USN collects status, and location of object by embedding RFID sensor, and those gathered information are analyzed to offer user specific responses. In this context, Web service is required for effective service provisioning because to manage all those collected and analyzed information, which is from

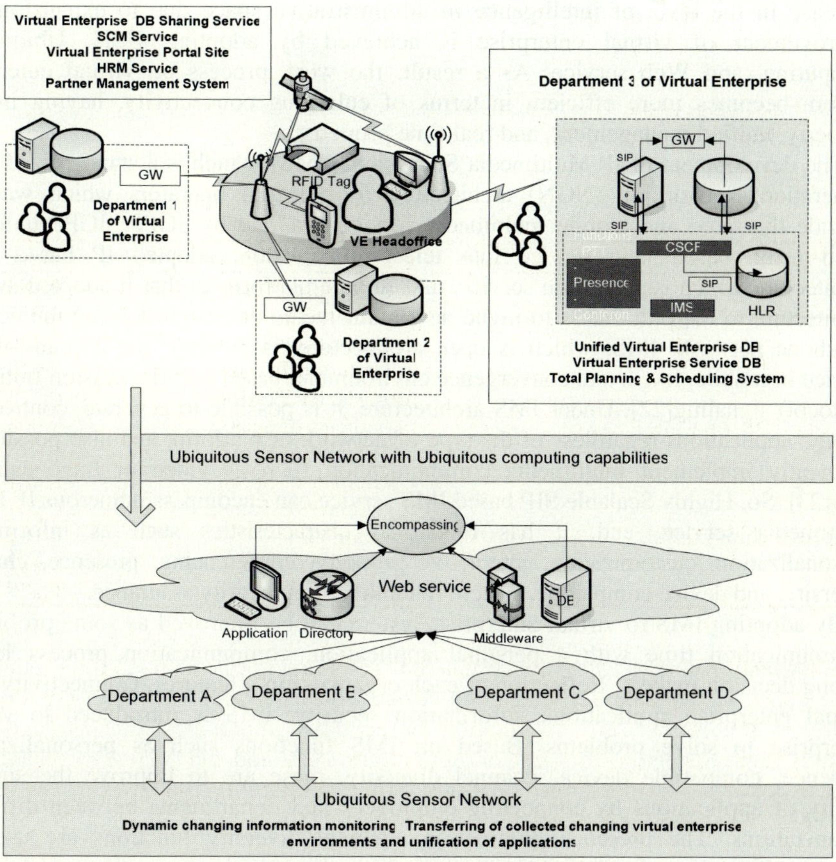

Fig. 3. Ubiquitous Virtual Enterprise Reference Model

distributed source and heterogeneous systems, Integration is required. Web service enables unified management through supporting collaborative operations among heterogeneous service, and applications in different system domains[20]. Therefore, Web service is recognized as strong tool for USN based service. Furthermore, Ubiquitous computing under USN is getting important. Ubiquitous computing can be considered as the core of USN because it can offer real time intelligent service[21].

By adopting above technologies to virtual enterprise system can be improved as some problems; instant respond of changing affairs, connectivity with virtual enterprise service, efficient handling of integrated affairs, supporting existing main computing operations, access to other departments' data. USN with ubiquitous computing is introduced to virtual enterprise to solve these problems. Based on U-context awareness, offering changing information of work domain in real time is possible and offering Optimized action such as Work assignment, scheduling is possible. Based on real time information sharing, interface unification, and elimination of system confliction, Effective integrated work execution is possible. Based on web service, Transformation of existing main computerized operation to newly adopted system and access to the Heterogeneous DB are possible. The increase in the level of intelligence in administrative space due to expansion and improvement of virtual enterprise is achieved by adopting USN, Ubiquitous computing, and Web service. As a result, the work process on virtual enterprise system becomes more efficient in terms of enlarging connectivity, having instant property, unified management, and real time property.

The development of IP Multimedia Subsystem (IMS), which is a standardized Next Generation Networking (NGN) architecture for telecom operators which want to provide the fixed and mobile multimedia service, was led by 3GPP. 3GPP unfolded multi-media based new service into telecommunication, adopted IP based open architecture which supports the service, and service platform so that it adopted system architecture to the pith that is to invite new multi-media domain to telecommunication backbone network. IMS, which is open architecture, is possible for IP multi-media service in mobile, fixed, and convergence environment, based on SIP (Session Initiation Protocol) signaling[22]. Under IMS architecture, it is possible to generate, control and change applications regardless of the type of network or platform, and also possible to efficiently implement multimedia communication such as video or huge capacity data[23]. So, Highly Scalable SIP based IMS service can encompass numerous IP based multimedia service, and, it has functional characteristics such as information personalization, customization, interactive property, conferencing, presence, channel diversity, and device compatible, system reliability, and security assurance.

By adopting IMS to virtual enterprise system can be improved as some problems; communication time with a personal application, communication process length among decision makers, Reflection of each organization's features, Connectivity with virtual enterprise applications, information security. IMS is introduced to virtual enterprise to solve problems. Based on IMS functions such as personalization, presence, compatible device, channel diversity, these are to improve the support ability of applications by connecting employees and departments between different organizations. The presence, interaction, channel diversity functions are used to remove the delay of administration by supporting quick connection of decision makers in moving or absence situations. The personalization and customization functions are used to support of efficient execution by taking into account each department, organization tasks and needs. The presence, interaction, channel diversity, and conferencing functions are used to achieve efficiency by removing the delay in connection by offering various channels. The QoS and security assurance functions is used to strengthen the security related to USN and the usage of Web service and remove privacy weakness of previous system through all IP based executions. Having those numerous functional advantages, IMS plays a role as enabling leverage or gateway of ubiquitous virtual enterprise reference model. Specifically, collected heterogeneous type of information is unified as homogeneous type, and those are classified, or assigned to proper system, service, and DB in IMS. Then functional characteristics of IMS are combined with information. Through above process, final service process in USN is supported.

5 Conclusions

Nowadays, firms must survive and grow fast amidst keen competition, verified customer preferences, and a changeable market situation. As such, the present market requires firms that have the ability to cope with the situation in a short time. To meet this requirement, competitive power through firms' relations is necessary. This paper introduces the virtual enterprise as a good firm-type capable of responding to market needs and changes. The concept of the virtual enterprise is ambiguous and has not been clearly defined compared with other related concepts such as M&A, strategic alliance, and so forth. This paper's contents are to introduce the virtual enterprise as an effective solution, to determine the weight of the critical success factors based on virtual enterprise decision flow model when firms choose their partners and to suggest the ubiquitous virtual enterprise reference model. The fact that by using the reference model which can be suggested by combining USN, Ubiquitous computing, Web service, and IMS, can provide the basis for meeting numerous needs found on virtual enterprise case-study, is much meaningful. It is expected that based on the combining ability which can strengthen the virtual enterprise collaborative system provided by various next generation technologies. The technologies offer quick and seamless communications among intra and extra organization, access to the information of other department. In the aspect that the innovation of process inside the virtual enterprise has direct relationship with the improvement of virtual enterprise service to other entities, the importance of developing efficient virtual enterprise relation system rises and the ubiquitous virtual enterprise reference model has its implication by suggesting the possibility for it.

Acknowledgments. This research work has been supported by the MIC (Ministry of Information and Communication), Korea, under the ITRC (Information Technology Research Center) program supervised by the IITA (Institute of Information Technology Assessment).

References

1. Gichan Kim.: The Change of 4's Economics Concepts and Relation Competitiveness, Center World. (1997)
2. Bleecker., Samuel E.: The Virtual Organization. Futurist. Vol.28, Iss.2. p.10 (1994)

3. Barner R.: The New Millennium Workplace; Seven Changes That Will Challenge Managers and Workers. Futurist, Vol.30, Iss.2, pp. 14-16 (1996)
4. Wexler, Joanie M.: Ties That Bind. Computerworld, Vol.27, Iss26, p97 (1993)
5. J.Byrne: The Virtual Corporation. Business Week, pp.36-41 (1993)
6. R.T. Fielding, E.J. Whitehead, K.M. Anderson, G.A. Bolcher, et al.: Web Based Development of Complex Information Products. Association for Computing Machinery, CACM, NewYork (1998)
7. T. Pappaioannou, J. Edwards.: Mobile Agent Technology Enabling the Virtual Enterprise,. Pattern for Database Query (1998)
8. Zhou Q., Besant CB.: Information Management in Production Planning for a Virtual Enterprise. International Journal of Production Research, Vol.37, No.1, pp.207-218 (1999)
9. R. Heckman.: Planning to Solve the Skills Problem in the Virtual Information Management Organization. International Journal of Information and Management, Vol.18, No.1, pp. 3-16 (1998)
10. Michael Weyrich., Paul Drews.: An Interactive Environment for Virtual Manufacturing. Computer in Industry, Vol. 38, pp.149-158 (1999)
11. Olga Volkoff., Yolande E. Chan., E.F. Peter Newson.: Leading the Development and Implementation of Collaborative Interorganizational system. Information and Management, Vol.35, pp. 63-75 (1999)
12. Satty,L.T.: A Scaling Method for Priorities in hierarchical Strcuture. Jounal of Mathematical psychology, Vol.15, pp.234-281 (1977)
13. Anonymous: What is a Virtual Cooperation? (http://www.intellaction.com)
14. The Unikarisruhe Web (2000)
15. Report: The Model of Virtual Enterprise in CALS/EC. National Computerization Agency, 2002
16. The Concept of Virtual Enterprise: www.kcals.or.kr
17. . NIIP Report: www.nca.or.kr
18. Christopher K. Hess, Roy H. Campbell.: A Context-Aware Data Management System for Ubiquitous Computing Applications. Proceedings of the 23rd International Conference on Distributd Computing Systems. IEEE Press (2003)
19. Tsetsos. V, et al.: Commercial wireless sensor networks: technical and business issues. Proceedings of the Second Annual Conference on WONS, IEEE Press (2005)
20. Hoang PHAN Huy, Takahiro KAWAMURA, Tetsuo HASEGAWA.: Web service Gateway - a step forward to e-business. Proceedings of the IEEE International Conference on Web Services. IEEE Press (2004)
21. .Hyunjung Park, Jeehyong Lee.: A Framework of context awareness for ubiquitous computing middlewares. Proceedings of the 4th annual ACIS International Conference on Computer and Information science. IEEE Press (2005)
22. Khin Phyo Thant, Thinn Thu Naing.: A Migration Framework for Ubiquitous Computing Applied in Mobile Applications. Proceedings of 6th Asia-Pacific Symposium on Information and Telecommunication Technologies. IEEE Press (2005)
23. Niklas Blum, Thomas Magedanz.: PTT + IMS = PTM Towards Community/Presence based IMS Multimedia services. Proceedings of the seventh IEEE international Symposium on Multimedia. IEEE Press (2005)
24. Young yong Kim.: Research Directions in the Area of USN (Ubiquitous Sensor Network) Towards Practical UE (Ubiquitous Environments). Lecture Notes in Computer Science, Vol. 4097, Springer-Verlag, Berlin Heidelberg (2006)

The Hiding of Secret Data Using the Run Length Matching Method

Ki-Hyun Jung[1], Jae-Gil Yu[2], Se-Min Kim[2], Ki-Jong Kim[1], Jin-Yong Byun[1], and Kee-Young Yoo[*]

[1] Division of Computer Information, Yeungjin College, 218 Bokhyun-Dong, Buk-Gu, Daegu 702-721, Republic of Korea
[2] Department of Computer Engineering, Kyungpook National University, 1370 Sankyuk-Dong, Buk-Gu, Daegu 702-701, Republic of Korea
{kingjung, kjkim, jybyun}@yjc.ac.kr,
{eobiya, resemin}@infosec.knu.ac.kr,
yook@knu.ac.kr

Abstract. This study proposes a data hiding method based on run length encoding. This proposed method uses the location of accumulated run length values, where the cover data run length are compared with the secret data run length. The run length matching (RLM) method uses the run length table which is constructed from the cover and secret data. The experimental results demonstrated that the RLM has advantages with respect to different types of data and run length encoding value match.

Keywords: steganography, data hiding, run length encoding, embedded data.

1 Introduction

Steganography conceals the existence of a secret message while cryptography protects the content of messages. Steganography derived from Greek, literally means "*covered writing*", is the art of hiding information in ways that prevent the detection of hidden messages. It includes a vast array of secret communication methods that can conceal a message's very existence [1, 7, 8].

Embedding data, which need to be hidden, into an image requires two sources. The first is innocent-looking data that will hold hidden information, called cover data. The second is the message, which is the information that needs to be hidden. A message may come in the form of plain text, ciphertext, image, video, or in any other way that can be embedded in a bit stream. When combined, the cover data and the embedded message make stego data. A stego key, like as password, may also be used to hide. Information can be hidden in many different ways. In order to hide information, straight message insertion may encode every bit of information in the cover data or it may selectively embed the message in noisy areas that draw less attention. A message may also be scattered randomly throughout the cover data. There are a number of

[*] Corresponding author.

N.T. Nguyen et al. (Eds.): KES-AMSTA 2007, LNAI 4496, pp. 1027–1034, 2007.

ways to hide information; the most common methods are least significant bit (LSB) insertion, masking and filtering, and algorithms and transformations [1].

The definition of breaking a steganographic system is different from that of a cryptographic system. In cryptography, the system is broken when the attacker can read a secret message, whereas there are two stages in breaking a steganographic system; the attacker can detect that steganography has been used and in addition, it does this mean the attacker is able to read the embedded message [2].

This paper proposes a data hiding method based on run length encoding. The proposed method uses the length of the location, which compares the cover data run length and secret data run length.

This paper is organized as follows. Section 2 reviews four recently reported methods relating to the proposed method. In Section 3, the detail of our newly-proposed scheme Run Length Matching (RLM) is described. In Section 4, experimental results are presented and discussed. Finally, concluding remarks are presented in Section 5.

2 Related Work

Wu and Tsai proposed a pixel-value differencing method [3], whereby a cover image is partitioned into non-overlapping blocks of two consecutive pixels. A difference value is calculated from the values of the two pixels in each block. Secret data are embedded into a cover image by replacing the difference values of the two-pixel blocks of the cover image with similar ones, in which bits of the embedded data are included.

Chang and Tseng's method [4] employs two-sided, three-sided and four-sided side match schemes. The two-sided side match method uses the side information of the upper and left neighboring pixels in order to make estimates. The three-sided side match scheme utilizes not only the upper and left pixels, but also one of the other neighboring pixels, right or bottom. In order to make more precise estimates, the four-sided side match method uses all neighboring sides; upper, left, right and bottom of a given pixel, instead of only two of them in the two-sided side match scheme.

Chang, Lin and Wang proposed the BRL and GRL data hiding methods [5] which incorporate both run length encoding and modular arithmetic. The BRL which hides bitmap files by run length embeds simple data with long streams of repeating bits. The GRL method, which hides general files by run length, embeds complicated data with short streams of repeating bits. In the BRL method, the least significant bit of the first pixel in each block holds a bit value 0 or 1 of the secret data, and the second pixel indicates the number of embedded secret bits.

The run length encoding technique was initiated in the 1950s and has become the compression standard in fax transmissions and bitmap file coding [6].

3 Proposed Method

Two secret data hiding schemes are presented in the following subsections. The proposed method uses the location of the accumulated run length value, which compares the cover data run length encoding values of, say Table *C*, with those of the secret data run length encoding of values of, say Table *S*. If a matching value pair exists, say (v_i, c_i), where v_i is the run count value and c_i is the run value in the cover data's run length encoding table and (v_i', c_i') value pair, where v_i' is the run count and c_i' is the run value in the secret data's run length encoding table, then the total length, say TL_i, is stored in the index of Table *I*.

The Run Length Matching (RLM) method uses the run length encoding values of cover data and secret data. Since the run length encoding technique is utilized, cover and secret data are supported by various types of formats, for example, binary, gray and color image.

Fig. 1 shows a flowchart of the proposed RLM method. After calculating the run length encoding values from cover and secret data, the first value of the *S* table is selected. This value is compared with all values in Table *C*. Then, two cases, i.e. $(v_i, c_i) = (v_i', c_i')$ or $(v_i, c_i) \neq (v_i', c_i')$, occur. This process will be described, in detail, in the following section. The process is completed after all secret data are matched and the index table is composed.

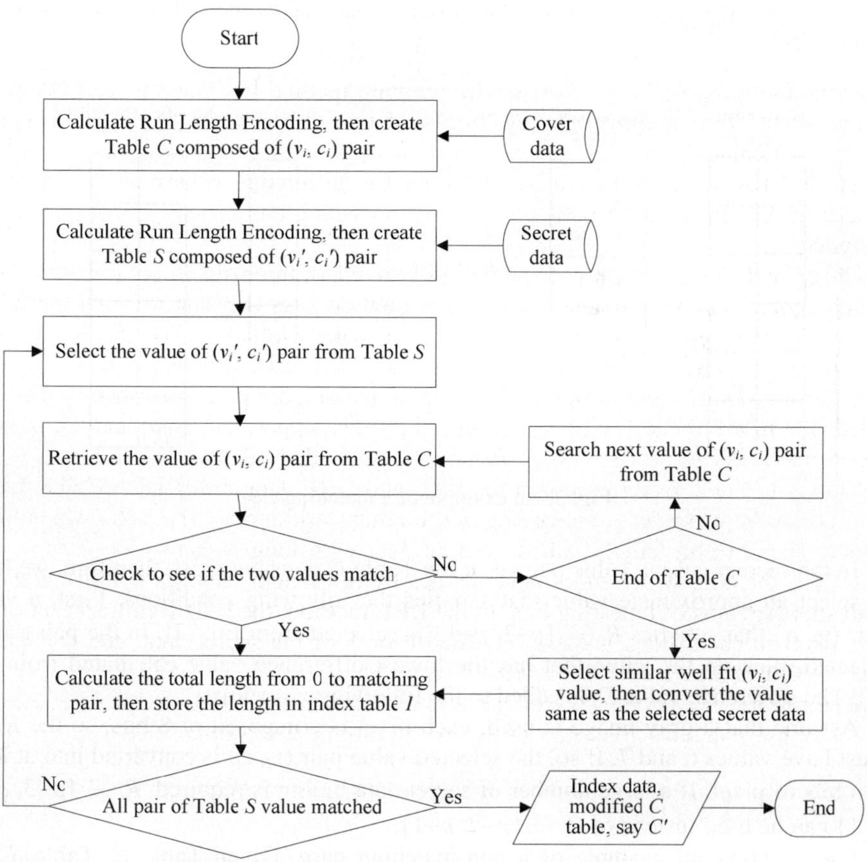

Fig. 1. The data embedding process of the RLM scheme

Given a run length value v_i, the total length, with index i, is calculated by Eq. (1).

$$TL_i = \sum_{k=0}^{i-1} v_k \ . \tag{1}$$

If p is the bit length of a pixel, the range element set R_i, is defined to be

$$R_i = \{\ X \mid X \in (TL_i \bmod p) \ \sim \ X \in (TL_{i+1} \bmod p)\ \}. \tag{2}$$

Finally, the difference value, D_i is declared by the following,

$$D_i = \mid v_i - v_i' \mid. \tag{3}$$

For example, consider two cases separately. In the first case, suppose that a value pair (v_i, c_i) exists in Table C, which matches with the selected value (v_i', c_i').

Fig. 2 gives an example of a matching case. Consider a value $(v_i', c_i') = (4, 1)$. In the cover data of Table C, the matching value is present, so the $TL_2 = 8$ is stored in index I table. If the values match, the process that makes the index table I is simple and direct.

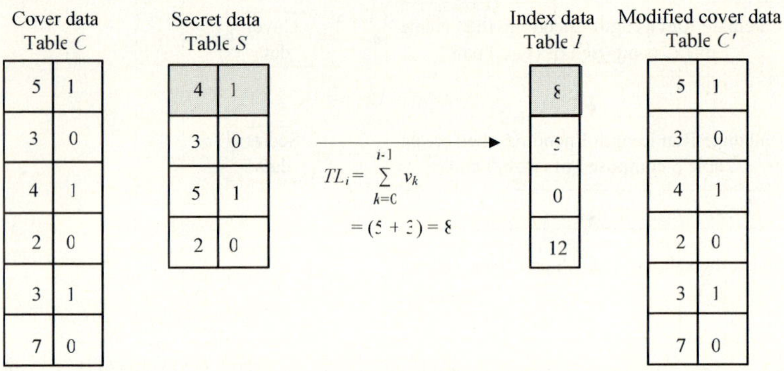

Fig. 2. An example of a matching case

In the second case, a value pair (v_i, c_i) in Table C does not exist, therefore, we have to select an approximate value that satisfies the following conditions: First, a value pair (v_i, c_i) that satisfies $R_i \supseteq \{p-2, p-1\}$ is selected from Eq. (2). In the pair values selected, pick up the value that has the lowest difference value calculated from Eq. (3). The difference value D_i is added to the following v_{i+1} value.

Assume that a gray image is used, each pixel is composed of 8 bits, so the R_i set must have values 6 and 7. If so, the selected value pair (v_i, c_i) is converted into at least two bits of pixel. If a large number of secret data hiding is required, $R_i \supseteq \{p-3, p-2, p-1\}$ can be used instead of $R_i \supseteq \{p-2, p-1\}$.

Fig. 3 shows an example of a non-matching case. Given Table C, Table S and value $(v_0', c_0') = (4, 1)$, on the c_0' value is 1, find R_i set which satisfies $R_i \supseteq \{6, 7\}$.

The two values are selected, in that (10, 1) has {6, 7, 0} and (6, 1) has {2, 3, 4, 5, 6, 7, 0}. Finally, we selected (6, 1) because the D_i of (6, 1) is $| v_i - v_i' | = | 6 - 4 |$, which is less than that of (10, 1). The TL_i value of (6, 1) is 34, which is the value stored in the index Table *I*.

After finishing the value of (v_i, c_i), Table *C* is modified to match the value (v_i', c_i') in table *S*, so the (6, 1) value is converted to (4, 1), then the difference value 2 is added to (8, 0), which consequently results in (10, 0).

In the next calculation, we select (1, 0) for the Table *S*, where the c_i' value is 0 and (2, 0) and (10, 0) value's set are {5, 6, 7}, {6, 7, 0}, respectively, which satisfies the conditions $R_i \supseteq \{6, 7\}$. The final selected value is (2, 0) because the $| 2 - 1 |$ value is less than the $| 10 - 1 |$ value. Value 5 is stored in the index Table *I*. The values (2, 0) and (3, 1) are replaced with (1, 0) and (4, 1), respectively.

In the worst case, although all values in table *C* are consumed, but there exists what is hidden, the proposed method utilizes the existing long run length encoding values, which have not been used to hide the remaining secret data.

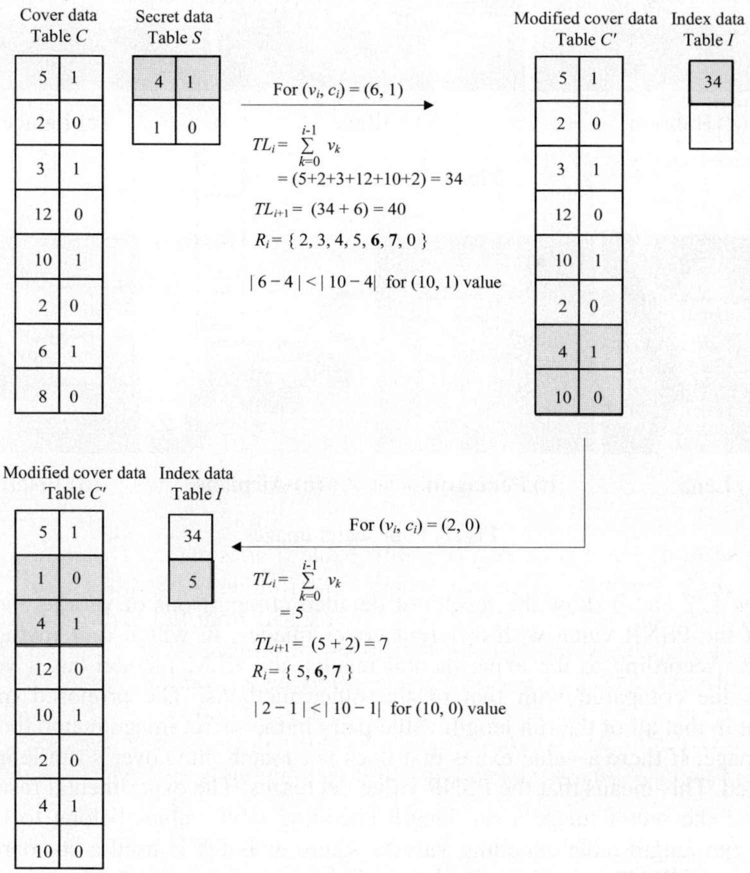

Fig. 3. An example of a non-matching case

4 Experimental Results

In our experiments, the three 512x512 gray images shown in Fig. 4 were used as cover images. The four 160x160 secret images shown in Fig. 5 were used. The reason that we used the 160x160 secret image size is that the maximum embedding data are restricted in the BRL method. If the BRL method is not included in the experiments, the size of the secret image can be enlarged, as shown in Table 4. We employed the peak signal-to-noise ratio (PSNR) as a measure of stego image quality.

(a) Baboon (b) Village (c) Palace

Fig. 4. Three cover images

(a) Lena (b) Pentagon (c) Airplane (d) Fishingboat

Fig. 5. Four secret images

Tables 1, 2 and 3 show the results of detailed comparisons of various methods in terms of the PSNR value with different cover images, in which four methods were involved. According to the experimental results, the RLM method has a very good PSNR value compared with that of the other methods. The proposed method is excellent in that all of the run length value pairs in the secret image match those of the cover image. If there a value exists that does not match, the cover's run length value is changed. This means that the PSNR value decreases. The experimental results show that all of the secret image's run length encoding table values belong to the cover image's run length table encoding values, where or not it is used a standard image. Therefore, the PSNR value is maintained as it is

Table 1. Comparisons of PSNR values using Baboon as the cover image

Method	Lena	Pentagon	Airplane	Fishing boat
LSB 2bits-LSB	48.41	48.61	48.70	48.63
LSB 3bits-LSB	43.95	43.82	44.06	44.00
BRL k=8	43.41	43.61	43.30	43.36
Two-sided side match	36.90	36.70	36.41	36.95
RLM	∞	∞	∞	∞

Table 2. Comparisons of PSNR values using Village as the cover image

Method	Lena	Pentagon	Airplane	Fishing boat
LSB 2bits-LSB	48.41	48.59	48.08	48.59
LSB 3bits-LSB	43.88	43.83	43.99	43.93
BRL k=8	43.38	43.58	43.27	43.32
Two-sided side match	39.05	38.84	38.71	39.17
RLM	∞	∞	∞	∞

Table 3. Comparisons of PSNR values using Palace as the cover image

Method	Lena	Pentagon	Airplane	Fishing boat
LSB 2bits-LSB	48.38	48.58	48.09	48.57
LSB 3bits-LSB	43.92	43.83	44.04	43.97
BRL k=8	43.47	43.66	43.37	43.45
Two-sided side match	40.17	40.05	39.85	40.38
RLM	∞	∞	∞	∞

Table 4 shows that the size of the secret image is enlarged from 160x160 to 256x256. Although the PSNR grows from 2 to 5 in the other schemes, the value of the proposed RLM method is still the same. This means that other methods are affected by the size of the secret image. The RLM method is not affected. The only factor that is affected is whether or not a matching value exists.

Table 4. A Comparison of PSNR values using Baboon as the cover image and the 256x256 image size for four secret images

Method	Lena	Pentagon	Airplane	Fishing boat
LSB 2bits-LSB	44.35	44.51	44.03	44.53
LSB 3bits-LSB	39.87	39.75	39.95	38.89
Two-sided side match	35.34	35.27	34.84	35.35
RLM	∞	∞	∞	∞

5 Conclusion

We have proposed a secret data hiding method based on run length encoding. The proposed method uses the length of the location, whereby the cover data run length encoding value is compared with the secret data run length encoding value.

Our experimental results have shown that the proposed method provides a better way to hide data, as compared with the four other methods. The more the two selected images are similar, the better the PSNR value becomes. The experimental results show that all secret image's run length encoding table values belong to the cover image's run length table encoding value, whether or not it is used as a standard image in the data hiding field. There is no need to refer to the original image during the extraction process. Although the PSNR values were excellent, the RLM method has to send additional information. What is useful to apply that the secret data is smaller.

Acknowledgements

This research was supported by the MIC of Korea, under the ITRC support program supervised by the IITA (IITA-2006-C1090-0603-0026).

References

1. N. F. Johnson, S. Jajodia, Exploring Steganography : Seeing the Unseen, Computer Practices (1998) 26-34.
2. J. Zollner, H. Federrath, H. Klimant, A. Pfitzmann, R. Piotraschke, A. Westfeld, G. Wicke, G. Wolf, Modeling the security of steganographic systems, 2nd Workshop on Information Hiding (1998) 345-355.
3. D. C. Wu, W. H. Tsai, A steganographic method for images by pixel-differencing, Pattern Recognition Letters 24 (9-10) (2003) 1613–1626.
4. C. C. Chang, H. W. Tseng, A steganographic method for digital images using side-match, Pattern Recognition Letters 25 (12) (2004) 1431-1437.
5. C. C. Chang, C. Y. Lin, Y. Z. Wang, New image steganographic methods using run-length approach, Information Sciences (2006).
6. R. C. Gonzalez, R. E. Woods, Digital Image Processing, Prentice Hall, Upper Saddle River, NJ, 2002.
7. R. J. Anderson, F.A.P. Petitcolas, On the limits of steganography, IEEE Journal on Selected Areas in Communications 16 (1998) 474-481.
8. F. A. P. Petitcolas, R. J. Anderson, M. G. Kuhn, Information hiding – a survey, Proceedings of the IEEE, special issue on protection of multimedia content, 87(7) (1999) 1062-1078.

Agent-Based Connection Control for Digital Content Service

Dragan Jevtic[1], Marijan Kunstic[1], and Kresimir Cunko[2]

[1] University of Zagreb, Faculty of Engineering and Computing, Unska 3,
HR-10000 Zagreb, Croatia
dragan.jevtic@fer.hr, marijan.kunstic@fer.hr
[2] Envox lab d.o.o. Zagreb, Slavonska avenija 6/7,
HR-10000 Zagreb, Croatia
Kresimir.Cunko@envox-lab.hr

Abstract. This paper explores improvements that can be achieved by applying intelligent agent techniques to solve the problem of self-adaptive routing. A potential redundancy has been recognized in telecommunication network configuration, hidden in the routing method between the user access points and service providers. The main idea presented here is perpetual transfer adaptation for all requests that are sent from a user to a service location over all the network elements. Self-adaptation is based on the continuous monitoring of the available communication channel capacity between the user and the service. The actions are based on perpetually seeking the optimal throughput via the nodes that maximize exploitation of the communication channel. From the user's point of view, accumulation and exploration of knowledge concerning throughput properties in the network can optimally utilize redundant capacities thus providing service more rapidly.

Keywords: adaptive routing, Q-learning, reinforcement.

1 Introduction

Information routing is crucial in telecommunication networks, particularly when user requests need to complete a collection of tasks. Due to the ever-increasing amount of data carried in such networks, perpetual discovery of superior connection paths helps improve the response performance of user requests. In contrast, current methods for information routing are based on deterministic and preset rules of choice. This implies a hierarchical collection of predetermined values for all traffic situations, given in so-called routing tables, which comprise alternatives for undesired events.

Optimal choices in such systems depend on current node and link capacities which are more frequently related to current traffic distributions between network nodes than to physical characteristics of connections. To increase service resource availability, however, it is of crucial importance that each requested task is dealt with as fast as possible. We feel that this existing inflexible way of information forwarding through the network is not satisfactory for an agent environment and for efficient digital content management.

N.T. Nguyen et al. (Eds.): KES-AMSTA 2007, LNAI 4496, pp. 1035–1042, 2007.

The model presented in this paper is based in the notion of discovering and exploiting additional capacities hidden in the redundancy of network nodes and links, which show variable properties over time. This is attributed not only to different properties of local and wide area traffic, but also to nodes operating in different time zones. The task at hand was to transfer packets to the desired destinations while minimizing their stay-time in the network. Reducing their stay-time improves connection performance by providing faster service. Specific digital content characteristics were excluded from consideration. In this paper, we deem the term 'client' to be a user device used to access the network and the term 'server' to be a device that stores digital content.

The novelty and premise in our approach is in its freedom to choose various transfer nodes for a requested client - server connection [3, 5]. Furthermore, this choice is confided to an intelligent agent, placed at the client side, i.e. located outside of the network nodes. Learning was on-line, concomitantly interacting with the environment and continuously reducing overall network load and bandwidth requirements.

This does not have repercussions on the structural arrangement of the network and shows that self-trained methods can utilize network capacities better than traditional methods. This approach is better suited for the current Internet world with the potential to better exploit the inherent redundancy of network capacity. The basic elements of this model include the client, the server, routing points (*RP*) and the link capacities between them. The original intention was to integrate all nodes specialized in information routing and to check if self-training was the best method for managing traffic between multiple nodes [3, 4].

This paper is organized in the following way: Reinforcement learning and a general description of the *Q*-algorithm, along with an agent learning procedure, are given in Section 2. In Section 3, we elaborate upon the characteristics of the network elements and traffic properties of the model. Section 4 describes the learning and network parameters, followed by simulation results and concluding remarks in Section 5.

2 Review of Reinforcement Learning

Reinforcement learning (RL) overcomes three important drawbacks inherent in traditional programming approaches. Firstly, RL requires very little programming effort since an automatic training process, rather than programming, does the most important work. Secondly, as the environment changes, training can be re-run without any additional programming. Thirdly, under certain assumptions, it is mathematically guaranteed that RL will converge to an optimal policy. Automatic training in a changing environment with mathematically guaranteed convergence to a desired policy is fundamental property of RL, i.e. self-adaptation.

Reinforcement learning is in order: consider a discrete-time system ($t = 0, 1, 2, \ldots$) in which the state transitions depend on the actions performed by an agent. A new state s_{t+1} (resulting from a state transition from the previous state s_t) after action a_t is determined by Markovian process with probability p:

$$p(s_{t+1}|s_t,a_t,s_{t-1},a_{t-1}, \ldots) = p(s_{t+1}|s_t,a_t) \tag{1}$$

In this process, costs/rewards are accumulated with a discount factor and can occur in certain states,

$$Q(s_o,a_o) = \lim_{N\to\infty} E[\sum_{t=0}^{N-1} \gamma^t g(s_t,a_t,s_{t+1}) \mid s_o,a_o] \tag{2}$$

where Q denotes the cost/reward estimate for a starting pair state/action, s_o/a_o, E denotes expectation, subsequent actions a_t's are determined by an action policy, such as $a_t = argmax_a Q(s_t,a)$, g is cost/reward, and γ is a discount factor.
To make a cost/reward estimate that results from a particular policy of action, there is a whole number of algorithms. In a Q-learning algorithm (Watkins 1989) updating is on-line, without explicitly using probability estimates [1, 2]. The updating is based on actual state transitions and is incremental,

$$\begin{aligned} Q_{t+1}(s_t,a_t) &= (1-\alpha)Q_t(s_t,a_t) + \\ &\alpha[REWARD\ \ (s_t,a_t) + \gamma \max_{a_{t+1}} Q_t(s_{t+1},a_{t+1})], \end{aligned} \tag{3}$$

where $\alpha \in (0,1)$ is the learning rate.
Agent actions were designed as show the following steps:

```
 1. Set initial state Q(s,a) to 0 for all s and a
    2. Get one packet from agent queue
    3. Establish current state
       (scan level of available link capacity)
       (scan current time)
       (read destination server identity)
    4. Select routing point - RP considering greedy value
        (exploration or not)
    5. Execute action (send packet toward selected RP)
    6. Calculate change in link capacity
    7. Calculate Reward signal r
    8. Calculate new Q value using expression (3)
    9. Return to point 2
```

3 Traffic, Routing and Model Configuration

The simulation model and its features are given in Figure 1. A client is connected to the network by a channel with capacity C. Clients' activities generate packets that need to be handled by servers. These packets must be transported from the client to the server through a fixed RP or one of several available RPs. A single server can receive only one packet at a time.

The main conflict which arises in the network is that of clients' requirements to increase their individual capacities in a state of increasingly reduced network node and link capacity. This is usually a consequence of the overall network traffic characteristics, i.e. the packets generated by other clients, computer applications, etc.

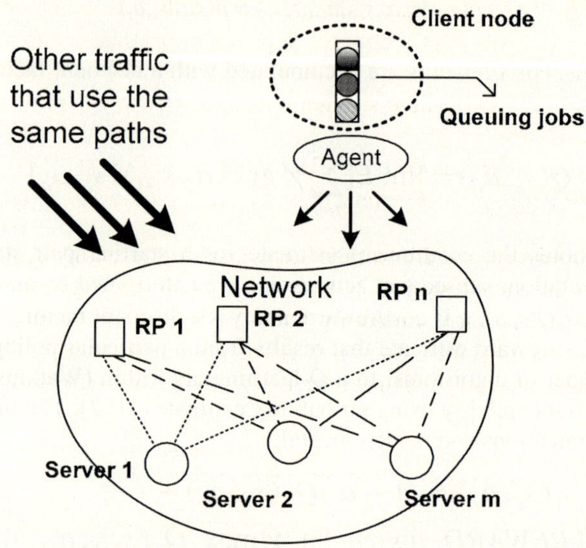

Fig. 1. The routing process controlled by an agent

Sufficient network resources may be available to support current demands, but the challenge is to ensure continuous adaptation to changing demands.

The client node in our model was created to have an unlimited queue and the following features. If a packet is generated while the corresponding channel is unavailable, it will be placed in the queue (FIFO). Once the channel becomes available, the packet is removed from the queue and activated, thus assuming part of the available channel capacity. Stay-time in the queue is not limited. Once activated, the packet occupies a certain amount of capacity. The time necessary for information transfer depends on packet characteristics and the transfer speed.

Furthermore, an increase in network traffic can reduce channel capacity and the current transfer speed for any active packet. Consequently, the speed can drop to 0% of C. For such a case, the packet is placed into an active queue and is time controlled. If a predefined maximal waiting time is exceeded, the packets in the active queue are completely lost (cancelled).

Since the speed of an end-to-end connection is limited by the channel capacity C of the client-network interface, other capacities in the network are expressed as n% of C, where n ranges from 0 to 100. In reality, from the client's point of view, nodes and links that offer higher speeds than the peak capacity of the client interface do not contribute to service quality. Increased traffic can significantly reduce the free capacities of network nodes and links. Choosing the optimal *RP* for a given packet is fundamental, along with defining a possible path from the client to the specified *RP*. This action is important, particularly if a choice exists beyond the current time zone. In our model, the reference point is the client node. Thus, to simulate, self-trained selection was assumed:

- The client connection represents the referent capacity of the model,
- the capacity of the client's channel depends on local network traffic and changes during the simulation,
- *RP*s are placed in the client's domain (without time shift) and the remote domain (with time shift),
- a channel enabled by an *RP* has capacity *n*% *C*, and
- connections between an *RP* and a server have capacity *n*% *C*, were *n* is set to 100.

The simulation was divided into time steps, each representing one second. The entire simulation covered the period of 24 hours, representing the duration of one day.

Packets were generated randomly with the time distribution shown in Figure 2. Each packet was given an identifier for a particular server. The identifiers were generated randomly using a uniform distribution over *RPs*, together with packet generation.

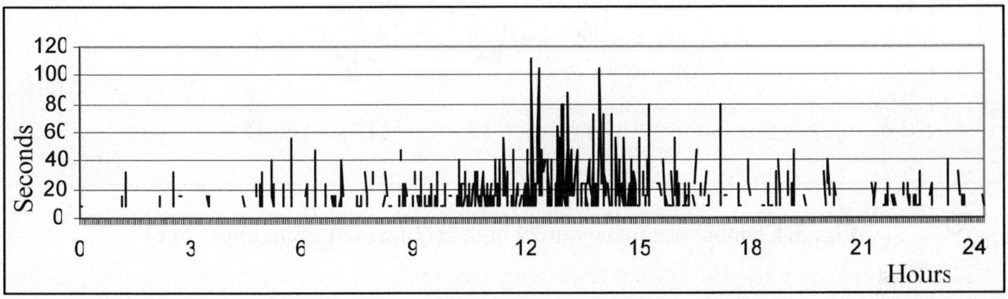

Fig. 2. Distribution of packets duration generated over 24 hours in one minute step

The channel capacity changed gradually over 24 hours, in 2 hour intervals (Table 1). In the case of low traffic, 100% of the channel capacity was assumed to be available. The available capacity gradually decreased, reaching a low of 14% in subinterval 12 - 14 h. From 14 to 24 h, the available capacity increased stepwise to nearly full capacity as a consequence of decreased network traffic (Table 1).

Table 1. Assumed reduction of the channel capacity over a time period of 24 h

Hours	*C*	Hours	*C*	Hours	*C*
0-2	100%	8-10	39%	16-18	42%
2-4	82%	10-12	25%	18-20	57%
4-6	69%	12-14	14%	20-22	71%
6-8	54%	14-16	28%	22-24	85%

The assumed (link volume, connection volume) channel availability distribution was built into our simulation model and corresponds well to reality, simplified by setting invariable capacities within 2-hour subintervals to ease simulation.

4 Results

Simulation experiments were performed on the above described system. There was one client, two *RPs* and three *SP* nodes. An agent equipped with a routing method based on *Q*-learning and a queuing feature was placed at the client node.

The time distribution of available capacities at *RP*1 and *RP*2 followed that given in Table 1, but *RP*1 had 80% and *RP*2 60% of each subinterval capacity. It was also assumed that *RP*2 was geographically remote and belonged to a time zone which was 6 hours ahead of the client. For example, during the 00-02h subinterval at the client node, *RP*1's capacity was 80% of the available capacity *C* in the 00-02h subinterval, while *RP*2's capacity was 60% of the available capacity in the 06-08h subinterval.

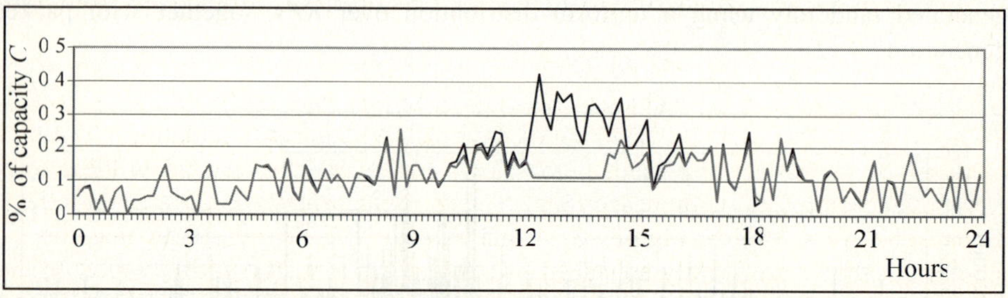

Fig. 3. Channel occupancy in 24 hours (*Q*-learned agent upper line)

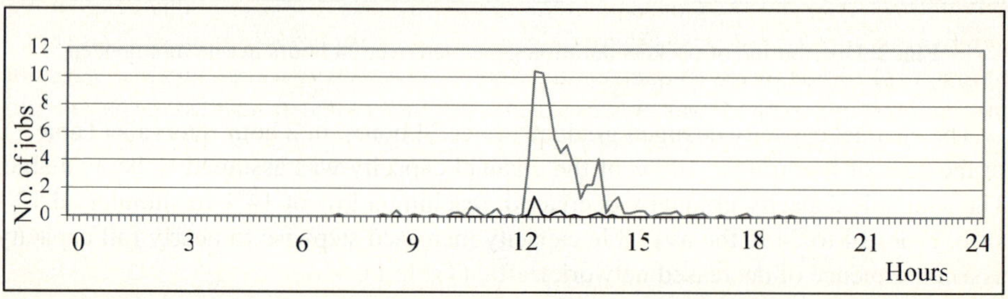

Fig. 4. Number of packets in the passive queue (*Q*-learned agent lower line)

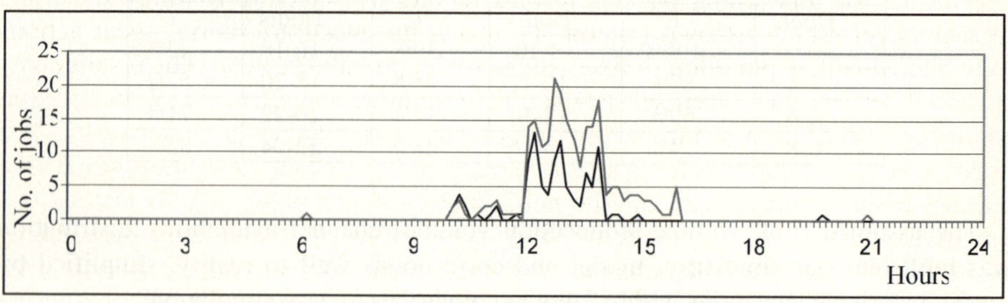

Fig. 5. Packets lost during simulation (*Q*-learned agent lower line)

Obviously, the minimal capacities of *RP*1 and *RP*2 in subintervals 12-16h and 18-20h were 8.4%*C* and 11.2%, respectively.

The self-trained routing method was responsible for selecting the appropriate *RP* for a requested packet, taking into account the current load and time shift between *RP*s with respect to the client node. As already mentioned, the model uses a passive unlimited queue and an active queue with a time limit, preset to 40 seconds in our simulations.

The reward signal r used by the Q-learning algorithm is computed in the following way:

$$r = 10 \text{ x } a\% \quad (4)$$

where a represents an increase in link occupancy Thus, for an increase in link occupancy of 5% (e.g. from 10% to 15%),_the reward is 0.5. An increase from 50% to 90% yields a reward of 4.

For the case where there is no increase, i.e. the increase is 0%, the reward is –5. Generally, we found these values to be optimal considering the situations in which irregular states can occur, i.e. breaking of the links. We ran two simulations. The first one utilized the Q-learning method to select paths while the second utilized only distribution point *RP*1 (without learning), placed in the same time zone as the client. The results are shown in Figures 3, 4 and 5. Figure 3 clearly shows that the Q-learning method significantly enhanced packet transfer regulation with respect to the method without learning capabilities. We can see that channel capacity was utilized better using Q-learning during subintervals with high traffic which also reflects a significant decrease in the number of packets in the passive queue for these subintervals (Figure 4). Consequently, there was also a notable decrease in the number of canceled packets (Figure 5). For the 24-hour period simulated, an average of 0.505311 packets were in queue and a total of 232 packets were cancelled when the on-line learning method was not used. For the case when Q-learning was applied, there was on average 0.022883 packets in queue and a total of 103 packets were cancelled.

5 Conclusion

A software simulation of the self-adaptive selection of *RP*s was used to measure the method proposed in this paper. An intelligent agent located at the user device was responsible for forwarding packets toward servers by selecting routing points and frequently refreshing its own criteria. To renew its selection criteria, agent actions were altered with exploration in fixed amount with probability 10%. The results show upgraded performance when self-adaptation by reinforcement learning is included in continuous and non-hierarchically organized routing in the network. The improvement is evident with respect to parameters such as the utilization of existing channel capacity, queue size and the number of cancelled packets. This method can economize service management by selecting optimal *RP*s and thus maximizing utilization of the available channel capacity. This is a promising property for managing a multi-node network, although the gain in service speed depends on the traffic load and distribution and the relation factor between agent exploration and knowledge utilization. For real traffic loads, this factor depends on the expected fluctuation of traffic. The proposed model shows that the method can be efficiently implemented in a client node and that it can work without interfering with other nodes.

References

1. C.J.C.H. Watkins, P. Dayan, , "Q-learning", Machine Learning, Vol 8, pp. 55-68, Kluwer Academic Publishers, 1992.
2. Long-Ji Lin, "Self-Improving Reactive Agents Based On Reinforcement Learning, Planning and Teaching", Machine Learning, Vol 8, pp. 69-97, Kluwer Academic Publishers, 1992.
3. M. Kunštić, D. Jevtić, D. Sablić, "Self-trained agents optimize communication service by intelligent selection", Proc. KES2000, Vol 2, pp. 687-690, Brigthon, England, 2000.
4. D. Jevtić, D. Sablić , "Intelligent call transfer based on reinforcement learning", Proc. IJCNN2000, pp. 120-123, Como, Italy
5. D. Jevtić, , M. Kunštić, N. Jerković, "The Intelligent Agent-Based Control of Service Processing Capacity", Knowledge-Based Intelligent Information and Engineering Systems, Part 2, pp. 668-674, ISSN 0302-9743, Springer-Verlag 2003.

Author Index

Printing: Mercedes-Druck, Berlin
Binding: Stein + Lehmann, Berlin

Lecture Notes in Artificial Intelligence (LNAI)

Vol. 4509: Z. Kobti, D. Wu (Eds.), Advances in Artificial Intelligence. XII, 552 pages. 2007.

Vol. 4496: N.T. Nguyen, A. Grzech, R.J. Howlett, L.C. Jain (Eds.), Agent and Multi-Agent Systems: Technologies and Applications. XXI, 1046 pages. 2007.

Vol. 4483: C. Baral, G. Brewka, J. Schlipf (Eds.), Logic Programming and Nonmonotonic Reasoning. IX, 327 pages. 2007.

Vol. 4482: A. An, J. Stefanowski, S. Ramanna, C.J. Butz, W. Pedrycz, G. Wang (Eds.), Rough Sets, Fuzzy Sets, Data Mining and Granular Computing. XIV, 585 pages. 2007.

Vol. 4481: J. Yao, P. Lingras, W.-Z. Wu, M. Szczuka, N.J. Cercone, D. Ślęzak (Eds.), Rough Sets and Knowledge Technology. XIV, 576 pages. 2007.

Vol. 4452: M. Fasli, O. Shehory (Eds.), Agent-Mediated Electronic Commerce. VIII, 249 pages. 2007.

Vol. 4451: T.S. Huang, A. Nijholt, M. Pantic, A. Pentland (Eds.), Artifical Intelligence for Human Computing. XVI, 359 pages. 2007.

Vol. 4438: L. Maicher, A. Sigel, L.M. Garshol (Eds.), Leveraging the Semantics of Topic Maps. X, 257 pages. 2007.

Vol. 4429: R. Lu, J.H. Siekmann, C. Ullrich (Eds.), Cognitive Systems. X, 161 pages. 2007.

Vol. 4426: Z.-H. Zhou, H. Li, Q. Yang (Eds.), Advances in Knowledge Discovery and Data Mining. XXV, 1161 pages. 2007.

Vol. 4411: R.H. Bordini, M. Dastani, J. Dix, A.E.F. Seghrouchni (Eds.), Programming Multi-Agent Systems. XIV, 249 pages. 2007.

Vol. 4410: A. Branco (Ed.), Anaphora: Analysis, Algorithms and Applications. X, 191 pages. 2007.

Vol. 4399: T. Kovacs, X. Llorà, K. Takadama, P.L. Lanzi, W. Stolzmann, S.W. Wilson (Eds.), Learning Classifier Systems. XII, 345 pages. 2007.

Vol. 4390: S.O. Kuznetsov, S. Schmidt (Eds.), Formal Concept Analysis. X, 329 pages. 2007.

Vol. 4389: D. Weyns, H.V.D. Parunak, F. Michel (Eds.), Environments for Multi-Agent Systems III. X, 273 pages. 2007.

Vol. 4384: T. Washio, K. Satoh, H. Takeda, A. Inokuchi (Eds.), New Frontiers in Artificial Intelligence. IX, 401 pages. 2007.

Vol. 4371: K. Inoue, K. Satoh, F. Toni (Eds.), Computational Logic in Multi-Agent Systems. X, 315 pages. 2007.

Vol. 4369: M. Umeda, A. Wolf, O. Bartenstein, U. Geske, D. Seipel, O. Takata (Eds.), Declarative Programming for Knowledge Management. X, 229 pages. 2006.

Vol. 4342: H. de Swart, E. Orłowska, G. Schmidt, M. Roubens (Eds.), Theory and Applications of Relational Structures as Knowledge Instruments II. X, 373 pages. 2006.

Vol. 4335: S.A. Brueckner, S. Hassas, M. Jelasity, D. Yamins (Eds.), Engineering Self-Organising Systems. XII, 212 pages. 2007.

Vol. 4334: B. Beckert, R. Hähnle, P.H. Schmitt (Eds.), Verification of Object-Oriented Software. XXIX, 658 pages. 2007.

Vol. 4333: U. Reimer, D. Karagiannis (Eds.), Practical Aspects of Knowledge Management. XII, 338 pages. 2006.

Vol. 4327: M. Baldoni, U. Endriss (Eds.), Declarative Agent Languages and Technologies IV. VIII, 257 pages. 2006.

Vol. 4314: C. Freksa, M. Kohlhase, K. Schill (Eds.), KI 2006: Advances in Artificial Intelligence. XII, 458 pages. 2007.

Vol. 4304: A. Sattar, B.-H. Kang (Eds.), AI 2006: Advances in Artificial Intelligence. XXVII, 1303 pages. 2006.

Vol. 4303: A. Hoffmann, B.-H. Kang, D. Richards, S. Tsumoto (Eds.), Advances in Knowledge Acquisition and Management. XI, 259 pages. 2006.

Vol. 4293: A. Gelbukh, C.A. Reyes-Garcia (Eds.), MICAI 2006: Advances in Artificial Intelligence. XXVIII, 1232 pages. 2006.

Vol. 4289: M. Ackermann, B. Berendt, M. Grobelnik, A. Hotho, D. Mladenič, G. Semeraro, M. Spiliopoulou, G. Stumme, V. Svátek, M. van Someren (Eds.), Semantics, Web and Mining. X, 197 pages. 2006.

Vol. 4285: Y. Matsumoto, R.W. Sproat, K.-F. Wong, M. Zhang (Eds.), Computer Processing of Oriental Languages. XVII, 544 pages. 2006.

Vol. 4274: Q. Huo, B. Ma, E.-S. Chng, H. Li (Eds.), Chinese Spoken Language Processing. XXIV, 805 pages. 2006.

Vol. 4265: L. Todorovski, N. Lavrač, K.P. Jantke (Eds.), Discovery Science. XIV, 384 pages. 2006.

Vol. 4264: J.L. Balcázar, P.M. Long, F. Stephan (Eds.), Algorithmic Learning Theory. XIII, 393 pages. 2006.

Vol. 4259: S. Greco, Y. Hata, S. Hirano, M. Inuiguchi, S. Miyamoto, H.S. Nguyen, R. Słowiński (Eds.), Rough Sets and Current Trends in Computing. XXII, 951 pages. 2006.

Vol. 4253: B. Gabrys, R.J. Howlett, L.C. Jain (Eds.), Knowledge-Based Intelligent Information and Engineering Systems, Part III. XXXII, 1301 pages. 2006.

Vol. 4252: B. Gabrys, R.J. Howlett, L.C. Jain (Eds.), Knowledge-Based Intelligent Information and Engineering Systems, Part II. XXXIII, 1335 pages. 2006.

Vol. 4251: B. Gabrys, R.J. Howlett, L.C. Jain (Eds.), Knowledge-Based Intelligent Information and Engineering Systems, Part I. LXVI, 1297 pages. 2006.

Vol. 4248: S. Staab, V. Svátek (Eds.), Managing Knowledge in a World of Networks. XIV, 400 pages. 2006.

Vol. 4246: M. Hermann, A. Voronkov (Eds.), Logic for Programming, Artificial Intelligence, and Reasoning. XIII, 588 pages. 2006.

Vol. 4223: L. Wang, L. Jiao, G. Shi, X. Li, J. Liu (Eds.), Fuzzy Systems and Knowledge Discovery. XXVIII, 1335 pages. 2006.

Vol. 4213: J. Fürnkranz, T. Scheffer, M. Spiliopoulou (Eds.), Knowledge Discovery in Databases: PKDD 2006. XXII, 660 pages. 2006.

Vol. 4212: J. Fürnkranz, T. Scheffer, M. Spiliopoulou (Eds.), Machine Learning: ECML 2006. XXIII, 851 pages. 2006.

Vol. 4211: P. Vogt, Y. Sugita, E. Tuci, C.L. Nehaniv (Eds.), Symbol Grounding and Beyond. VIII, 237 pages. 2006.

Vol. 4203: F. Esposito, Z.W. Raś, D. Malerba, G. Semeraro (Eds.), Foundations of Intelligent Systems. XVIII, 767 pages. 2006.

Vol. 4201: Y. Sakakibara, S. Kobayashi, K. Sato, T. Nishino, E. Tomita (Eds.), Grammatical Inference: Algorithms and Applications. XII, 359 pages. 2006.

Vol. 4200: I.F.C. Smith (Ed.), Intelligent Computing in Engineering and Architecture. XIII, 692 pages. 2006.

Vol. 4198: O. Nasraoui, O. Zaïane, M. Spiliopoulou, B. Mobasher, B. Masand, P.S. Yu (Eds.), Advances in Web Mining and Web Usage Analysis. IX, 177 pages. 2006.

Vol. 4196: K. Fischer, I.J. Timm, E. André, N. Zhong (Eds.), Multiagent System Technologies. X, 185 pages. 2006.

Vol. 4188: P. Sojka, I. Kopeček, K. Pala (Eds.), Text, Speech and Dialogue. XV, 721 pages. 2006.

Vol. 4183: J. Euzenat, J. Domingue (Eds.), Artificial Intelligence: Methodology, Systems, and Applications. XIII, 291 pages. 2006.

Vol. 4180: M. Kohlhase, OMDoc – An Open Markup Format for Mathematical Documents [version 1.2]. XIX, 428 pages. 2006.

Vol. 4177: R. Marín, E. Onaindía, A. Bugarín, J. Santos (Eds.), Current Topics in Artificial Intelligence. XV, 482 pages. 2006.

Vol. 4160: M. Fisher, W. van der Hoek, B. Konev, A. Lisitsa (Eds.), Logics in Artificial Intelligence. XII, 516 pages. 2006.

Vol. 4155: O. Stock, M. Schaerf (Eds.), Reasoning, Action and Interaction in AI Theories and Systems. XVIII, 343 pages. 2006.

Vol. 4149: M. Klusch, M. Rovatsos, T.R. Payne (Eds.), Cooperative Information Agents X. XII, 477 pages. 2006.

Vol. 4140: J.S. Sichman, H. Coelho, S.O. Rezende (Eds.), Advances in Artificial Intelligence - IBERAMIA-SBIA 2006. XXIII, 635 pages. 2006.

Vol. 4139: T. Salakoski, F. Ginter, S. Pyysalo, T. Pahikkala (Eds.), Advances in Natural Language Processing. XVI, 771 pages. 2006.

Vol. 4133: J. Gratch, M. Young, R. Aylett, D. Ballin, P. Olivier (Eds.), Intelligent Virtual Agents. XIV, 472 pages. 2006.

Vol. 4130: U. Furbach, N. Shankar (Eds.), Automated Reasoning. XV, 680 pages. 2006.

Vol. 4120: J. Calmet, T. Ida, D. Wang (Eds.), Artificial Intelligence and Symbolic Computation. XIII, 269 pages. 2006.

Vol. 4118: Z. Despotovic, S. Joseph, C. Sartori (Eds.), Agents and Peer-to-Peer Computing. XIV, 173 pages. 2006.

Vol. 4114: D.-S. Huang, K. Li, G.W. Irwin (Eds.), Computational Intelligence, Part II. XXVII, 1337 pages. 2006.

Vol. 4108: J.M. Borwein, W.M. Farmer (Eds.), Mathematical Knowledge Management. VIII, 295 pages. 2006.

Vol. 4106: T.R. Roth-Berghofer, M.H. Göker, H.A. Güvenir (Eds.), Advances in Case-Based Reasoning. XIV, 566 pages. 2006.

Vol. 4099: Q. Yang, G. Webb (Eds.), PRICAI 2006: Trends in Artificial Intelligence. XXVIII, 1263 pages. 2006.

Vol. 4095: S. Nolfi, G. Baldassarre, R. Calabretta, J.C.T. Hallam, D. Marocco, J.-A. Meyer, O. Miglino, D. Parisi (Eds.), From Animals to Animats 9. XV, 869 pages. 2006.

Vol. 4093: X. Li, O.R. Zaïane, Z. Li (Eds.), Advanced Data Mining and Applications. XXI, 1110 pages. 2006.

Vol. 4092: J. Lang, F. Lin, J. Wang (Eds.), Knowledge Science, Engineering and Management. XV, 664 pages. 2006.

Vol. 4088: Z.-Z. Shi, R. Sadananda (Eds.), Agent Computing and Multi-Agent Systems. XVII, 827 pages. 2006.

Vol. 4087: F. Schwenker, S. Marinai (Eds.), Artificial Neural Networks in Pattern Recognition. IX, 299 pages. 2006.

Vol. 4068: H. Schärfe, P. Hitzler, P. Øhrstrøm (Eds.), Conceptual Structures: Inspiration and Application. XI, 455 pages. 2006.

Vol. 4065: P. Perner (Ed.), Advances in Data Mining. XI, 592 pages. 2006.

Vol. 4062: G.-Y. Wang, J.F. Peters, A. Skowron, Y. Yao (Eds.), Rough Sets and Knowledge Technology. XX, 810 pages. 2006.

Vol. 4049: S. Parsons, N. Maudet, P. Moraitis, I. Rahwan (Eds.), Argumentation in Multi-Agent Systems. XIV, 313 pages. 2006.

Vol. 4048: L. Goble, J.-J.C.. Meyer (Eds.), Deontic Logic and Artificial Normative Systems. X, 273 pages. 2006.